MULTIVARIABLE CALCULUS, LINEAR ALGEBRA, AND DIFFERENTIAL EQUATIONS

SECOND EDITION

MULTIVARIABLE CALCULUS, LINEAR ALGEBRA, AND DIFFERENTIAL EQUATIONS

SECOND EDITION

STANLEY I. GROSSMAN
University of Montana

Harcourt Brace Jovanovich, Publishers
and its subsidiary, Academic Press
San Diego New York Chicago Austin
London Sydney Tokyo Toronto

TO AARON, ERIK, AND KERSTIN

ISBN: 0-15-564751-2
Library of Congress Catalog Card Number: 85-70253

Printed in the United States of America

Contents

4 DIFFERENTIATION OF FUNCTIONS OF TWO OR MORE VARIABLES 153

5 MULTIPLE INTEGRATION 264

6 INTRODUCTION TO VECTOR ANALYSIS 322

7 MATRICES AND LINEAR SYSTEMS OF EQUATIONS 398

8 DETERMINANTS 448

9 VECTOR SPACES AND LINEAR TRANSFORMATIONS 478

10 CALCULUS IN $\mathbb{R}^n$ 576

11 ORDINARY DIFFERENTIAL EQUATIONS 619

12 MATRICES AND SYSTEMS OF DIFFERENTIAL EQUATIONS 719

13 TAYLOR POLYNOMIALS, SEQUENCES, AND SERIES 784

Preface

In 1977 the first edition of my book *Calculus* was published. It, like the second and third editions that followed, contained a comprehensive introduction to the calculus of one and several variables. Many instructors suggested that since a large number of students stopped after studying one variable calculus, a shorter version of my text should be available.

To meet that need, Academic Press published *Calculus, Part I* in 1981, with a second edition in 1986. This one-variable calculus text contains the first fourteen chapters of the main book. The original plan was to publish, simultaneously, a second "short book" containing the last seven chapters of the main text.

Most first-year calculus courses cover similar material. However, I soon found that this was not the case for the second-year course. Some schools cover only multivariable calculus in the second year. Others include some linear algebra, some differential equations, or both. Moreover, some universities include more advanced calculus material in the second year: topics such as Taylor's theorem in n variables and mappings from $\mathbb{R}^n$ to $\mathbb{R}^m$.

Thus, we made the decision to write a book that would be usable in a wide variety of courses. My goal has been to retain the flavor of the original calculus book while making a large number of traditionally "post-calculus" topics accessible to sophomores.

To accomplish this, I have done a number of things. Most important, I have continued to include large numbers of examples and exercises. Most mathematicians agree that the only way to learn calculus well is by solving problems, and this book is designed to encourage students to learn by this method. *Multivariable Calculus, Linear Algebra, and Differential Equations* contains over 730 examples—many more than are commonly found in texts at this level. Each example includes all the algebraic steps needed to complete the solution. As a student, I was infuriated by statements like "it now easily follows that . . ." when it was not at all easy for me. Students have a right to see the "whole hand," so to speak, so that they always know how to get from "a" to "b." In many instances, explanations are highlighted in color to make a step easier to follow.

The text includes approximately 4300 exercises—including both drill and applied-type problems. More difficult problems are marked with an asterisk (*) and a few especially difficult ones are marked with a double asterisk (**). The exercises provide the most important learning tool in any undergraduate mathematics textbook. I stress to my students that no matter how well they think they understand my lectures or the textbook, they do not really know the material until they have worked problems. A vast difference exists between understanding someone else's solution and solving a new problem by yourself. Learning mathematics without doing problems is about as easy as learning to ski without going to the slopes.

I have also tried to introduce difficult concepts by first discussing simple cases and applications. As an example, I begin the section on linear transformations (Section 9.6) with two very simple applications. There are plenty of difficult theorems proved in this book, but I have tried to put them off until the concepts are sufficiently exemplified.

Mathematics becomes more interesting if one knows something about the historical development of the subject. I try to convince my students that, contrary to what they may believe, many great mathematicians lived interesting and often controversial lives. Thus, to make the subject more interesting and, perhaps, more fun, I have included a number of full-page biographical sketches of mathematicians who helped develop the calculus. In these sketches students will learn, for example, about the dispute between Newton and Leibniz, the gift for languages of Hamilton, and the love life of Lagrange. It is my hope that these notes will bring the subject to life.

The answers to most odd-numbered exercises appear at the back of the book. In addition, a student's manual containing detailed solutions to all odd-numbered problems and an instructor's manual containing detailed solutions to all even-numbered problems are available. These were prepared by Professor Leon Gerber at St. John's University in New York City.

This text covers a wide range of topics, but the only prerequisite is a course in one-variable calculus. Principally, I expect that a student using this book will know the following:

- ☐ how to compute limits
- ☐ how to differentiate any elementary function
- ☐ the basic techniques of integration including, especially, integration by parts and integration by a variety of substitutions
- ☐ the basic applications of differentiation and integration including curve sketching, computing areas, and computing volumes

The student who comes to the course with these skills will do well.

The book is divided, roughly, into five parts. The first part consists of Chapters 1–6. These are similar to Chapters 15–20 in *Calculus, Third Edition,* and include basic multivariable calculus material. The basic difference is that I introduce the space $\mathbb{R}^n$ in Section 3.7 and then, in later sections, I generalize to $\mathbb{R}^n$ basic topics in $\mathbb{R}^2$ and $\mathbb{R}^3$. However, the approach is gradual. Vectors in the plane are discussed in Chapter 1, with vector functions in the plane in Chapter 2. Chapter 3 contains an extension of the material in Chapters 1 and 2 to three and more dimensions. Computer drawn graphs of certain quadric surfaces are given in Section 3.6.

Chapter 4 contains an introduction to the calculus of two and more variables. The gradient is introduced in Section 4.5 as the natural extension of the ordinary derivative. Chapter 5 provides an introduction to multiple integration with an emphasis on applications. Chapter 6 contains a detailed introduction to vector analysis including a discussion, with proofs and applications, of Green's, Stokes's and the divergence theorems.

The second part of the book is an introduction to linear algebra in Chapters 7, 8, and 9. This material requires no multivariable calculus except a familiarity with vectors in $\mathbb{R}^2$, $\mathbb{R}^3$ and $\mathbb{R}^n$. These chapters can be covered any time after Chapter 1, Sections 3.1–3.5, and Section 3.7.

Chapters 7 and 8 contain introductions to matrices, determinants, and the Gauss–Jordan technique for solving systems of equations. Chapter 9 includes more advanced material on vector spaces and linear transformations.

The third part of the book consists of the single Chapter 10. The chapter combines techniques from calculus and linear algebra and contains discussions of some of the most elegant results in the calculus including Taylor's theorem in n variables, the multivariable mean value theorem, and the implicit function theorem. None of the results here are found in standard calculus texts.

Chapters 11 and 12 comprise the fourth part of the book and provide a one-quarter or semester introduction to ordinary differential equations. Chapter 11 is independent of Chapters 1–10 and can be covered at any time. It contains detailed discussions of first-order and linear second-order equations. Also included are optional discussions of electric circuits and vibratory motion.

Chapter 12, on systems of differential equations, begins with three sections that require no matrix theory. The remainder of the chapter combines matrix theory and linear systems. The diagonalization technique is used in Section 12.6 to compute e^{At}, the principal matrix solution of a linear homogeneous system of differential equations.

Some one-variable calculus courses cover infinite series and others do not. For that reason I have included, in Chapter 13, a discussion of Taylor's theorem, sequences, and series. This material, except for Section 13.14 on power series solutions to differential equations, can be covered at any time. This is the fifth part of the text.

In discussing eigenvalues and eigenvectors, it is necessary to know something about complex numbers. For that reason I have provided a discussion of complex numbers in Appendix 3. Also included, in Appendix 5, is a proof of the Picard theorem regarding the existence and uniqueness of a wide variety of first-order differential equations.

Numbering in the book is fairly standard. Within each section examples, problems, theorems, and equations are numbered consecutively, starting with 1. Reference to an example, problem, theorem, or equation outside the section in which it appears is referenced by chapter, section, and number. Thus, for example, Example

4 in Section 2.3 is called, simply, Example 4 in that section, but outside the section is referred to as Example 2.3.4. The more difficult problems are marked (*) or occasionally (**), and the problems where the use of a calculator is advisable are marked . Sections which are more difficult and can be omitted without loss of continuity are labeled "optional." Finally, the ends of proofs of theorems are marked with a ■.

Acknowledgments

I am grateful to many individuals who helped in the preparation of this text. Many of the reviewers of *Calculus, Third Edition,* provided useful criticism that improved the material in Chapters 1–6. Five reviewers painstakingly worked their way through the entire text of the first edition of this book and provided hundreds of detailed, insightful suggestions. I am particularly grateful to Professor George Cain of the Georgia Institute of Technology, Professor Art Copeland at the University of New Hampshire, Professor Carl Cowen at Purdue University, Professor Charles Denlinger at Millersville State College, and Professor Keith Yale at the University of Montana.

In preparing this second edition, I was fortunate to obtain many suggestions both for improving the logical order of material and for correcting the kinds of pedagogical errors that creep into any first edition. One major change suggested by a number of reviewers was to include $\mathbb{R}^n$ material in Chapters 3 and 4, rather than postpone it until Chapter 10.

I wish to thank the following individuals for their invaluable help in preparing the second edition: Alfred D. Andrew, Georgia Institute of Technology; James M. Edmondson, Santa Barbara Community College; Nathaniel Grossman, UCLA; Daniel S. Kahn, Northwestern University; T. J. Ransford, University of Leeds; John Venables, Lakewood Community College, White Bear Lake; and Paul Yearout, Brigham Young University.

Professor Leon Gerber, who prepared the Student's and Instructor's manuals, made many useful suggestions for the improvement of the problem sets. This book is a better teaching tool because of him.

I am also very grateful to the Wadsworth Publishing Company, Inc. for permission to use material from my book *Elementary Linear Algebra, Second Edition* (1984), in Chapters 7, 8, and 9, to the Addison-Wesley Publishing Company, Inc. for permission to use material from my book (with William R. Derrick), *Elementary Differential Equations with Applications, Second Edition* (1981), in Chapters 11, 12, and Appendix 5, and to the Saunders Publishing Company, Inc. for permission to use material from their book, *An Introduction to the History of Mathematics, Fifth Edition* (1983), by Howard Eves, in the biographical sketch of Bernoulli on page 649.

Finally, I owe a considerable debt to the editorial and production staffs of Academic Press, who provided help in a great number of ways in the writing and production of this book.

Stanley I. Grossman
January 1986

1 Vectors in the Plane

With this chapter we begin a new subject in the study of calculus—the study of vectors and vector functions. This will lead us, in Chapter 4, to the study of functions of two and more variables, a sharp departure from your study of one-variable calculus.

The modern study of vectors began essentially with the work of the great Irish mathematician Sir William Rowan Hamilton (1805–1865) who worked with what he called quaternions.[†] After Hamilton's death, his work on quaternions was supplanted by the more adaptable work on vector analysis by the American mathematician and physicist Josiah Willard Gibbs (1839–1903) and the general treatment of ordered n-tuples by the German mathematician Herman Grassman (1809–1877).

Throughout Hamilton's life and for the remainder of the nineteenth century, there was considerable debate over the usefulness of quaternions and vectors. At the end of the century, the great British physicist Lord Kelvin wrote that quaternions ". . . although beautifully ingenious, have been an unmixed evil to those who have touched them in any way . . . vectors . . . have never been of the slightest use to any creature."

But Kelvin was wrong. Today nearly all branches of classical and modern physics are represented using the language of vectors. Vectors are also used with increasing frequency in the social and biological sciences. Quaternions, too, have recently been used in physics—in particle theory and other areas.

[†]See the accompanying biographical sketch.

1805–
1865

Sir William Rowan Hamilton

Sir William Rowan Hamilton
The Granger Collection

Born in 1805 in Dublin, where he spent most of his life, William Rowan Hamilton was without question Ireland's greatest mathematician. Hamilton's father (an attorney) and mother died when he was a small boy. His uncle, a linguist, took over the boy's education. By his fifth birthday, Hamilton could read English, Hebrew, Latin, and Greek. By his 13th birthday he had mastered not only the languages of continental Europe, but also Sanscrit, Chinese, Persian, Arabic, Malay, Hindi, Bengali, and several others as well. Hamilton liked to write poetry, both as a child and as an adult, and his friends included the great English poets Samuel Taylor Coleridge and William Wordsworth. Hamilton's poetry was considered so bad, however, that it is fortunate that he developed other interests—especially in mathematics.

Although he enjoyed mathematics as a young boy, Hamilton's interest was greatly enhanced by a chance meeting at the age of 15 with Zerah Colburn, the American lightning calculator. Shortly afterwards, Hamilton began to read important mathematical books of the time. In 1823, at the age of 18, he discovered an error in Simon Laplace's *Mécanique céleste* and wrote an impressive paper on the subject. A year later he entered Trinity College in Dublin.

Hamilton's university career was astonishing. At the age of 21, while still an undergraduate, he had so impressed the faculty that he was appointed Royal Astronomer of Ireland and Professor of Astronomy at the University. Shortly thereafter, he wrote what is now considered a classical work on optics. Using only mathematical theory, he predicted conical refraction in certain types of crystals. Later this theory was confirmed by physicists. Largely because of this work, Hamilton was knighted in 1835.

Hamilton's first great purely mathematical paper appeared in 1833. In this work he described an algebraic way to manipulate pairs of real numbers. This work gives rules that are used today to add, subtract, multiply, and divide complex numbers. At first, however, Hamilton was unable to devise a multiplication for triples or n-tuples of numbers for $n > 2$. For 10 years he pondered this problem, and it is said that he solved it in an inspiration while walking on the Brougham Bridge in Dublin in 1843. The key was to discard the familiar commutative property of multiplication. The new objects he created were called *quaternions,* which were the precursors of what we now call vectors.

For the rest of his life, Hamilton spent most of his time developing the algebra of quaternions. He felt that they would have revolutionary significance in mathematical physics. His monumental work on this subject, *Treatise on Quaternions,* was published in 1853. Thereafter, he worked on an enlarged work, *Elements of Quaternions.* Although Hamilton died in 1865 before his *Elements* was completed, the work was published by his son in 1866.

Students of mathematics and physics know Hamilton in a variety of other contexts. In mathematical physics, for example, one encounters the Hamiltonian function, which often represents the total energy in a system and the Hamilton-Jacobi differential equations of dynamics. In matrix theory, the Cayley-Hamilton theorem states that every matrix satisfies its own characteristic equation.

Despite the great work he was doing, Hamilton's final years were a torment to him. His wife was a semi-invalid and he was plagued by alcoholism. It is therefore gratifying to point out that during these last years, the newly formed American National Academy of Sciences elected Sir William Rowan Hamilton to be its first foreign associate.

In this chapter we will explore properties of vectors in the plane. When going through this material, the reader should keep in mind that, like most important discoveries, vectors have been a source of great controversy—a controversy that was not resolved until well into the twentieth century.[†]

1.1 VECTORS AND VECTOR OPERATIONS

In many applications of mathematics to the physical and biological sciences and engineering, scientists are concerned with entities that have both magnitude (length) and direction. Examples include the notions of force, velocity, acceleration, and momentum. It is frequently useful to express these quantities geometrically.

Let P and Q be two different points in the plane. Then the **directed line segment** from P to Q, denoted $\overrightarrow{PQ}$, is the straight-line segment that extends from P to Q (see Figure 1a). Note that the directed line segments $\overrightarrow{PQ}$ and $\overrightarrow{QP}$ are different since they point in opposite directions (Figure 1b).

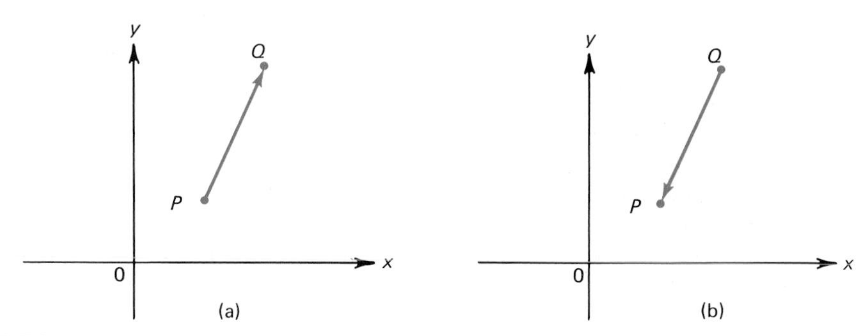

FIGURE 1

The point P in the directed line segment $\overrightarrow{PQ}$ is called the **initial point** of the segment and the point Q is called the **terminal point.** The two important properties of a directed line segment are its magnitude (length) and its direction. If two directed line segments $\overrightarrow{PQ}$ and $\overrightarrow{RS}$ have the same magnitude and direction, we say that they are **equivalent** no matter where they are located with respect to the origin. The directed line segments in Figure 2 are all equivalent.

Definition 1 GEOMETRIC DEFINITION OF A VECTOR The set of all directed line segments equivalent to a given directed line segment is called a **vector.** Any directed line segment in that set is called a **representation** of the vector.

REMARK. The directed line segments in Figure 2 are all representations of the same vector.

[†]For interesting discussions of the development of modern vector analysis, consult the book by M. J. Crowe, *A History of Vector Analysis* (Univ. of Notre Dame Press, Notre Dame, 1967), or Morris Kline's excellent book *Mathematical Thought from Ancient to Modern Times* (Oxford Univ. Press, New York, 1972), Chapter 32.

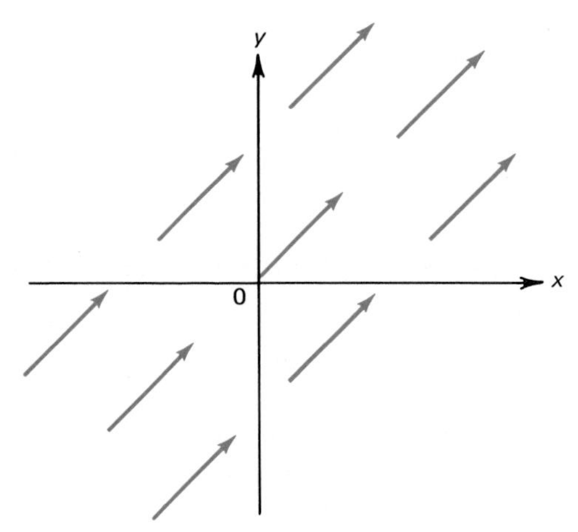

FIGURE 2

NOTATION. We will denote vectors by lowercase boldface letters such as **v**, **w**, **a**, **b**.

From Definition 1 we see that a given vector **v** can be represented in many different ways. In fact, let $\overrightarrow{PQ}$ be a representation of **v**. Then without changing magnitude or direction, we can move $\overrightarrow{PQ}$ in a parallel way so that its initial point is shifted to the origin. We then obtain the directed line segment $\overrightarrow{0R}$, which is another representation of the vector **v** (see Figure 3). Now suppose that R has the Cartesian coordinates (a, b). Then we can describe the directed line segment $\overrightarrow{0R}$ by the coordinates (a, b). That is, $\overrightarrow{0R}$ is the directed line segment with initial point (0, 0) and terminal point (a, b). Since one representation of a vector is as good as another, we can write the vector **v** as (a, b). In sum, we see that a vector can be thought of as a point in the xy-plane.

Definition 2 ALGEBRAIC DEFINITION OF A VECTOR A **vector v** in the xy-plane is an ordered pair of real numbers (a, b). The numbers a and b are called the **components** of the vector **v**. The **zero vector** is the vector (0, 0) and is denoted **0**.

Definition 3 SCALAR Since we will often have to distinguish between real numbers and vectors (which are pairs of real numbers), we will use the term **scalar**[†] to denote a real number.

Definition 4 MAGNITUDE OF A VECTOR Since a vector is really a set of equivalent directed line segments, we define the **magnitude** or **length** of a vector as the length of any one of its representations.

[†]The term "scalar" originated with Hamilton. His definition of the quaternion included what he called a *real part* and an *imaginary part.* In his paper, "On Quaternions, or on a New System of Imaginaries in Algebra," in *Philosophical Magazine,* 3rd Ser., **25,** 26–27 (1844), he wrote

> The algebraically *real* part may receive . . . all values contained on the one *scale* of progression of numbers from negative to positive infinity; we shall call it therefore the *scalar part,* or simply the *scalar* of the quaternion. . . .

Moreover, in the same paper Hamilton went on to define the imaginary part of his quaternion as the *vector* part. Although this was not the first usage of the word *vector,* it was the first time it was used in the context of Definitions 1 and 2. In fact, it is fair to say that the paper from which the above quotation was taken marks the beginning of modern vector analysis.

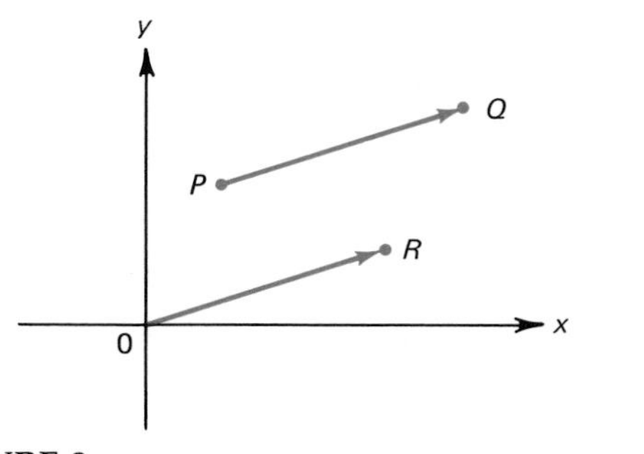

FIGURE 3

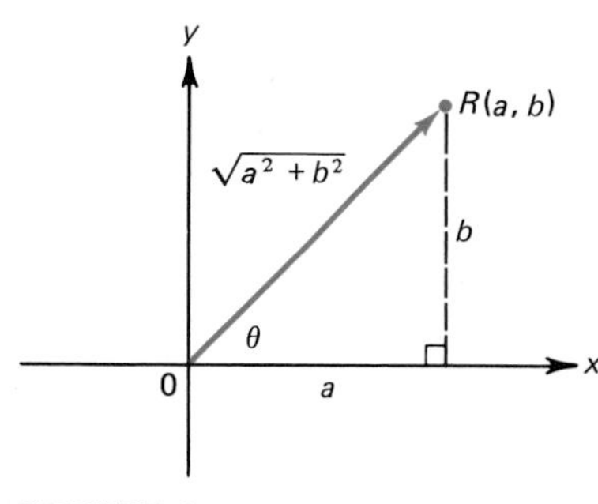

FIGURE 4

Using the representation $\overrightarrow{0R}$, and writing the vector $\mathbf{v} = (a, b)$, we find that

$$|\mathbf{v}| = \text{magnitude of } \mathbf{v} = \sqrt{a^2 + b^2}. \tag{1}$$

This follows from the Pythagorean theorem (see Figure 4). We have used the notation $|\mathbf{v}|$ to denote the magnitude of $\mathbf{v}$. Note that $|\mathbf{v}|$ *is a scalar.*

EXAMPLE 1 Calculate the magnitudes of the vectors (a) $(2, 2)$; (b) $(2, 2\sqrt{3})$; (c) $(-2\sqrt{3}, 2)$; (d) $(-3, -3)$; (e) $(6, -6)$.

Solution.

(a) $|\mathbf{v}| = \sqrt{2^2 + 2^2} = \sqrt{8} = 2\sqrt{2}$

(b) $|\mathbf{v}| = \sqrt{2^2 + (2\sqrt{3})^2} = 4$

(c) $|\mathbf{v}| = \sqrt{(-2\sqrt{3})^2 + 2^2} = 4$

(d) $|\mathbf{v}| = \sqrt{(-3)^2 + (-3)^2} = \sqrt{18} = 3\sqrt{2}$

(e) $|\mathbf{v}| = \sqrt{6^2 + (-6)^2} = \sqrt{72} = 6\sqrt{2}$ ■

Definition 5 DIRECTION OF A VECTOR We now define the **direction** of the vector $\mathbf{v} = (a, b)$ to be the angle θ, measured in radians, that the vector makes with the positive x-axis. By convention, we choose θ such that $0 \le \theta < 2\pi$.

It follows from Figure 4 that if $a \neq 0$, then

$$\tan \theta = \frac{b}{a}. \tag{2}$$

REMARK 1. The zero vector has a magnitude of 0. Since the initial and terminal points coincide, we say that *the zero vector has no direction.*

REMARK 2. If the initial point of $\mathbf{v}$ is not at the origin, then the direction of $\mathbf{v}$ is defined as the direction of the vector equivalent to $\mathbf{v}$ whose initial point *is* at the origin.

REMARK 3. It follows from Remark 2 that *parallel vectors have the same direction.*

EXAMPLE 2 Calculate the directions of the vectors in Example 1.

Solution. We depict these five vectors in Figure 5.

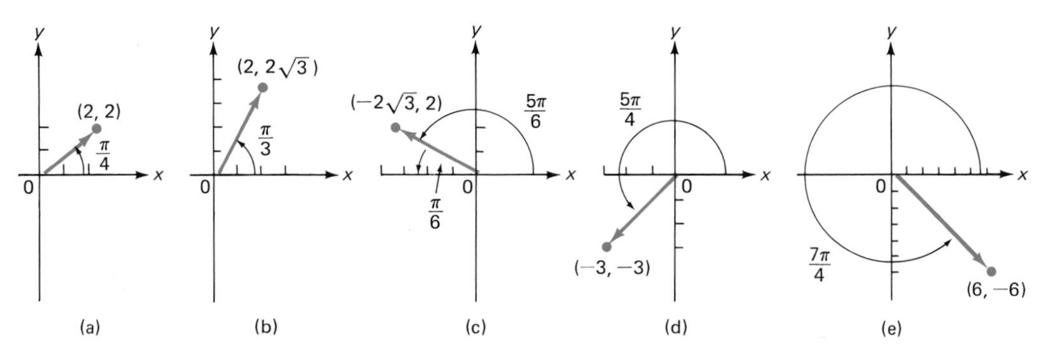

FIGURE 5

(a) Here **v** is in the first quadrant, and since $\tan\theta = 2/2 = 1$, $\theta = \pi/4$.

(b) Here $\theta = \tan^{-1} 2\sqrt{3}/2 = \tan^{-1}\sqrt{3} = \pi/3$.

(c) We see that **v** is in the second quadrant, and since $\tan^{-1} 2/(2\sqrt{3}) = \tan^{-1} 1/\sqrt{3} = \pi/6$, we see from the figure that $\theta = \pi - (\pi/6) = 5\pi/6$.

(d) Here **v** is in the third quadrant, and since $\tan^{-1} 1 = \pi/4$, $\theta = \pi + (\pi/4) = 5\pi/4$.

(e) Since **v** is in the fourth quadrant, and since $\tan^{-1}(-1) = -\pi/4$, $\theta = 2\pi - (\pi/4) = 7\pi/4$. ■

Let $\mathbf{v} = (a, b)$. Then as we have seen, **v** can be represented in many different ways. For example, the representation of **v** with the initial point (c, d) has the terminal point $(c + a, d + b)$. This is depicted in Figure 6. It is easy to show that the directed line segment $\overrightarrow{PQ}$ in Figure 6 has the same magnitude and direction as the vector **v** (see Problems 48 and 49).

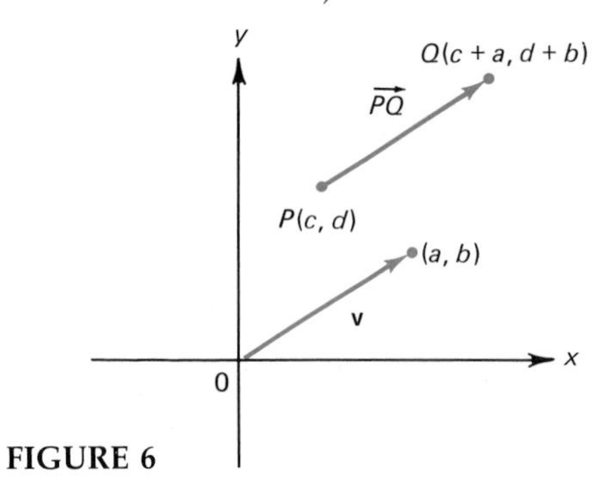

FIGURE 6

EXAMPLE 3 Find a representation of the vector $(2, -1)$ whose initial point is the point $P = (5, -4)$.

Solution. Let $Q = (5 + 2, -4 - 1) = (7, -5)$. Then $\overrightarrow{PQ}$ is a representation of the vector $(2, -1)$. This is illustrated in Figure 7. ■

We now turn to the question of adding vectors and multiplying them by scalars.

Definition 6 ADDITION AND SCALAR MULTIPLICATION OF A VECTOR Let $\mathbf{u} = (a_1, b_1)$ and $\mathbf{v} = (a_2, b_2)$ be two vectors in the plane and let α be a scalar. Then we define

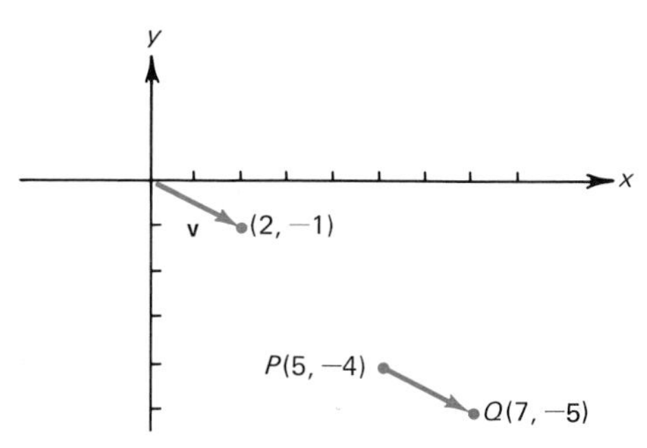

FIGURE 7

(i) $\mathbf{u} + \mathbf{v} = (a_1 + a_2, b_1 + b_2)$
(ii) $\alpha\mathbf{u} = (\alpha a_1, \alpha b_1)$
(iii) $-\mathbf{v} = (-1)\mathbf{v} = (-a_2, -b_2)$
(iv) $\mathbf{u} - \mathbf{v} = \mathbf{u} + (-\mathbf{v}) = (a_1 - a_2, b_1 - b_2)$

We have

To add two vectors, we add their corresponding components, and to multiply a vector by a scalar, we multiply each of its components by that scalar.

EXAMPLE 4 Let $\mathbf{u} = (1, 3)$ and $\mathbf{v} = (-2, 4)$. Calculate (a) $\mathbf{u} + \mathbf{v}$; (b) $3\mathbf{u}$; (c) $-\mathbf{v}$; (d) $\mathbf{u} - \mathbf{v}$; and (e) $-3\mathbf{u} + 5\mathbf{v}$.

Solution.

(a) $\mathbf{u} + \mathbf{v} = (1 + (-2), 3 + 4) = (-1, 7)$
(b) $3\mathbf{u} = 3(1, 3) = (3, 9)$
(c) $-\mathbf{v} = (-1)(-2, 4) = (2, -4)$
(d) $\mathbf{u} - \mathbf{v} = \mathbf{u} + (-\mathbf{v}) = (1 + 2, 3 - 4) = (3, -1)$
(e) $-3\mathbf{u} + 5\mathbf{v} = (-3, -9) + (-10, 20) = (-13, 11)$ ■

There are interesting geometric interpretations of vector addition and scalar multiplication. First, let $\mathbf{v} = (a, b)$ and let α be any scalar. Then

$$|\alpha\mathbf{v}| = |(\alpha a, \alpha b)| = \sqrt{\alpha^2 a^2 + \alpha^2 b^2} = |\alpha|\sqrt{a^2 + b^2} = |\alpha||\mathbf{v}|.$$

That is, multiplying a vector by a scalar has the effect of multiplying the length of the vector by the absolute value of that scalar.

Moreover, if $\alpha > 0$, then $\alpha\mathbf{v}$ is in the same quadrant as $\mathbf{v}$ and, since $\tan^{-1}(\alpha b/\alpha a) = \tan^{-1}(b/a)$, the direction of $\alpha\mathbf{v}$ is the same as the direction of $\mathbf{v}$. If $\alpha < 0$, then the direction of $\alpha\mathbf{v}$ is equal to the direction of $\mathbf{v}$ plus π (which is the direction of $-\mathbf{v}$).

EXAMPLE 5 Let $\mathbf{v} = (1, 1)$. Then $|\mathbf{v}| = \sqrt{1 + 1} = \sqrt{2}$ and $|2\mathbf{v}| = |(2, 2)| = \sqrt{2^2 + 2^2} = \sqrt{8} = 2\sqrt{2} = 2|\mathbf{v}|$. Also, $|-2\mathbf{v}| = \sqrt{(-2)^2 + (-2)^2} = 2\sqrt{2} = 2|\mathbf{v}|$. Moreover, the direction of $2\mathbf{v}$ is $\pi/4$, while the direction of $-2\mathbf{v}$ is $5\pi/4$. This is illustrated in Figure 8. ■

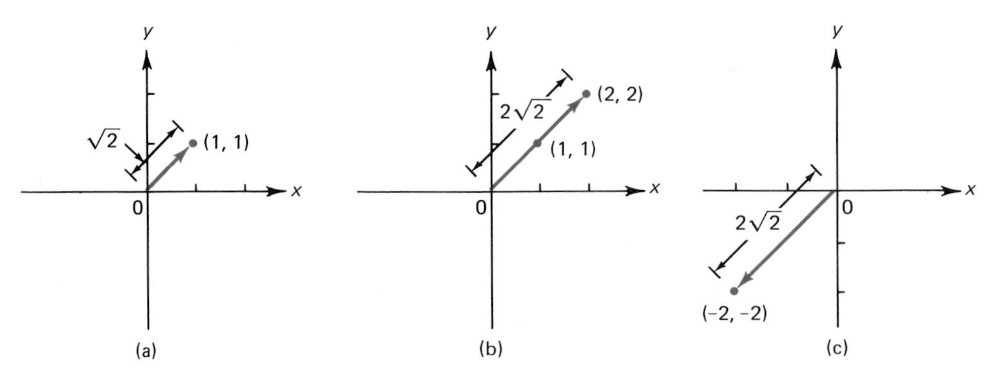

FIGURE 8

Now suppose we add the vectors $\mathbf{u} = (a_1, b_1)$ and $\mathbf{v} = (a_2, b_2)$, as in Figure 9. From the figure we see that the vector $\mathbf{u} + \mathbf{v} = (a_1 + a_2, b_1 + b_2)$ can be obtained by shifting the representation of the vector $\mathbf{v}$ so that its initial point coincides with the terminal point (a_1, b_1) of the vector $\mathbf{v}$. We can therefore obtain the vector $\mathbf{u} + \mathbf{v}$ by drawing a parallelogram with one vertex at the origin and sides $\mathbf{u}$ and $\mathbf{v}$. Then $\mathbf{u} + \mathbf{v}$ is the vector that points from the origin along the diagonal of the parallelogram.

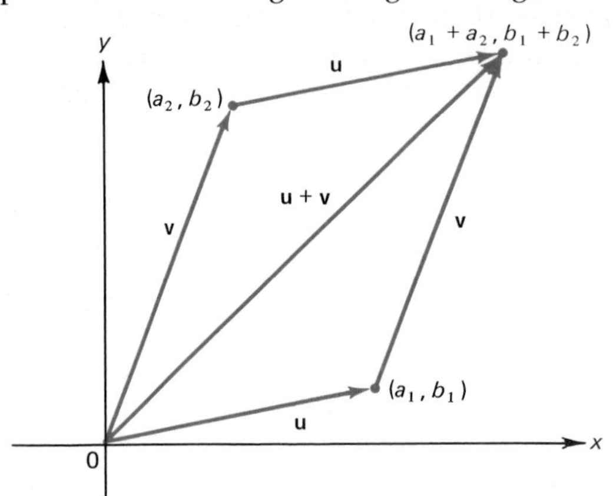

FIGURE 9

NOTE. Since a straight line is the shortest distance between two points, it immediately follows from Figure 9 that

$$|\mathbf{u} + \mathbf{v}| \leq |\mathbf{u}| + |\mathbf{v}|.$$

For obvious reasons this inequality is called the **triangle inequality.**

We can also obtain a geometric representation of the vector $\mathbf{u} - \mathbf{v}$. Since $\mathbf{u} = \mathbf{u} - \mathbf{v} + \mathbf{v}$, the vector $\mathbf{u} - \mathbf{v}$ is the vector that must be added to $\mathbf{v}$ to obtain $\mathbf{u}$. This is illustrated in Figure 10a.

The following theorem lists several properties that hold for any vectors, $\mathbf{u}$, $\mathbf{v}$, and $\mathbf{w}$ and any scalars α and β. Since the proof is easy, we leave it as an exercise (see Problem 58). Some parts of this theorem have already been proven.

FIGURE 10

Theorem 1 Let **u**, **v**, and **w** be any three vectors in the plane, let α and β be scalars, and let **0** denote the zero vector.

(i) $\mathbf{u} + \mathbf{v} = \mathbf{v} + \mathbf{u}$

(ii) $\mathbf{u} + (\mathbf{v} + \mathbf{w}) = (\mathbf{u} + \mathbf{v}) + \mathbf{w}$

(iii) $\mathbf{v} + \mathbf{0} = \mathbf{v}$

(iv) $0\mathbf{v} = \mathbf{0}$ (here the 0 on the left is the scalar zero)

(v) $\alpha\mathbf{0} = \mathbf{0}$

(vi) $(\alpha\beta)\mathbf{v} = \alpha(\beta\mathbf{v})$

(vii) $\mathbf{v} + (-\mathbf{v}) = \mathbf{0}$

(viii) $1\mathbf{v} = \mathbf{v}$

(ix) $(\alpha + \beta)\mathbf{v} = \alpha\mathbf{v} + \beta\mathbf{v}$

(x) $\alpha(\mathbf{u} + \mathbf{v}) = \alpha\mathbf{u} + \alpha\mathbf{v}$

(xi) $|\alpha\mathbf{v}| = |\alpha||\mathbf{v}|$

(xii) $|\mathbf{u} + \mathbf{v}| \le |\mathbf{u}| + |\mathbf{v}|$

Many of the properties above can be illustrated geometrically. For example, rule (i), which is called the **commutative law for vector addition,** is illustrated in Figure 11. Similarly, rule (ii), which is called the **associative law for vector addition,** is illustrated in Figure 12.

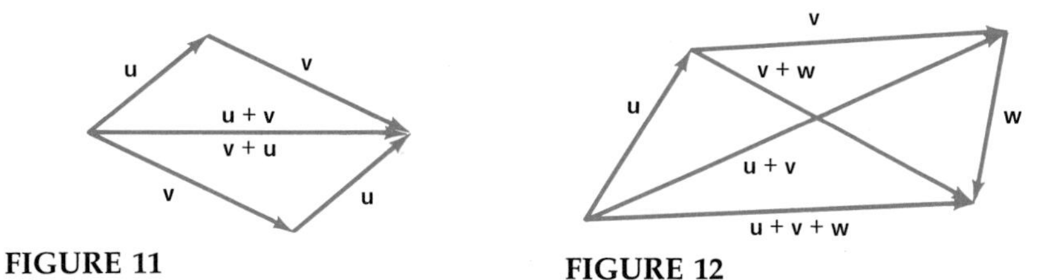

FIGURE 11 **FIGURE 12**

When a set of vectors together with a set of scalars and the operations of addition and scalar multiplication have the properties given in Theorem 1(i)–(x), we say that the vectors form a **vector space.** The set of vectors of the form (a, b), where a and b are real numbers, is denoted $\mathbb{R}^2$. We will not discuss properties of abstract vector spaces here, except to say that all abstract vector spaces have properties very similar to the properties of the vector space $\mathbb{R}^2$. We will discuss abstract vector spaces in Chapter 9.

There are two special vectors in $\mathbb{R}^2$ that allow us to represent other vectors in $\mathbb{R}^2$ in a convenient way. We will denote the vector (1, 0) by the vector symbol **i** and the vector (0, 1) by the vector symbol **j** (see Figure 13).[†] If (a, b) denotes any other vector in $\mathbb{R}^2$, then since $(a, b) = a(1, 0) + b(0, 1)$, we may write

$$\mathbf{v} = (a, b) = a\mathbf{i} + b\mathbf{j}. \qquad \textbf{(3)}$$

[†]The symbols **i** and **j** were first used by Hamilton. He defined his quaternion as a quantity of the form $a + b\mathbf{i} + c\mathbf{j} + d\mathbf{k}$, where a was the "scalar part" and $b\mathbf{i} + c\mathbf{j} + d\mathbf{k}$ the "vector part." In Section 3.2 we will write vectors in space in the form $b\mathbf{i} + c\mathbf{j} + d\mathbf{k}$.

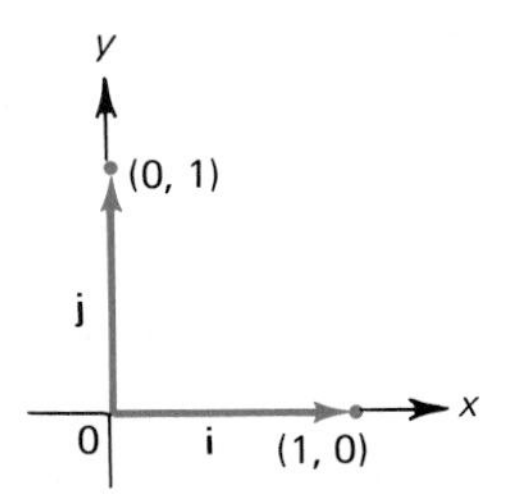

FIGURE 13

Moreover, any vector in $\mathbb{R}^2$ can be represented in a unique way in the form $a\mathbf{i} + b\mathbf{j}$ since the representation of (a, b) as a point in the plane is unique. (Put another way, a point in the xy-plane has one and only one x-coordinate and one and only one y-coordinate.) Thus Theorem 1 holds with this new representation as well.

When the vector $\mathbf{v}$ is written in the form $\mathbf{v} = a\mathbf{i} + b\mathbf{j}$, we say that $\mathbf{v}$ *is resolved into its horizontal and vertical components,* since a is the horizontal component of $\mathbf{v}$ while b is its vertical component. The vectors $\mathbf{i}$ and $\mathbf{j}$ are called **basis vectors** for the vector space $\mathbb{R}^2$.

Now suppose that a vector $\mathbf{v}$ can be represented by the directed line segment $\overrightarrow{PQ}$, where $P = (a_1, b_1)$ and $Q = (a_2, b_2)$. (See Figure 14.) If we label the point (a_2, b_1) as R, then we immediately see that

$$\mathbf{v} = \overrightarrow{PQ} = \overrightarrow{PR} + \overrightarrow{RQ}. \tag{4}$$

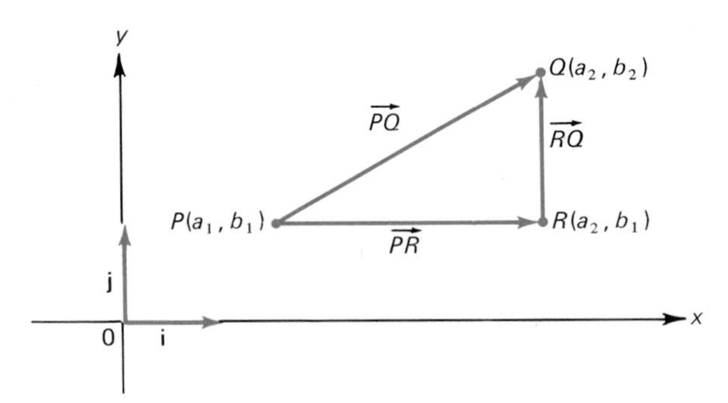

FIGURE 14

If $a_2 \geq a_1$, then the length of $\overrightarrow{PR}$ is $a_2 - a_1$, and since $\overrightarrow{PR}$ has the same direction as $\mathbf{i}$ (since they are parallel), we can write

$$\overrightarrow{PR} = (a_2 - a_1)\mathbf{i}. \tag{5}$$

If $a_2 < a_1$, then the length of $\overrightarrow{PR}$ is $a_1 - a_2$, but then $\overrightarrow{PR}$ has the same direction as $-\mathbf{i}$ so $\overrightarrow{PR} = (a_1 - a_2)(-\mathbf{i}) = (a_2 - a_1)\mathbf{i}$ again. Similarly,

$$\overrightarrow{RQ} = (b_2 - b_1)\mathbf{j}, \tag{6}$$

and we may write [using (4), (5), and (6)]

$$\mathbf{v} = (a_2 - a_1)\mathbf{i} + (b_2 - b_1)\mathbf{j}. \tag{7}$$

EXAMPLE 6 Resolve the vector represented by the directed line segment from $(-2, 3)$ to $(1, 5)$ into its vertical and horizontal components.

Solution. Using (7), we have

$$\mathbf{v} = (a_2 - a_1)\mathbf{i} + (b_2 - b_1)\mathbf{j} = [1 - (-2)]\mathbf{i} + (5 - 3)\mathbf{j} = 3\mathbf{i} + 2\mathbf{j}. \blacksquare$$

We conclude this section by defining a kind of vector that is very useful in certain types of applications.

Definition 7 UNIT VECTOR A **unit vector u** is a vector that has length 1.

EXAMPLE 7 The vector $\mathbf{u} = (1/2)\mathbf{i} + (\sqrt{3}/2)\mathbf{j}$ is a unit vector since

$$|\mathbf{u}| = \sqrt{\left(\frac{1}{2}\right)^2 + \left(\frac{\sqrt{3}}{2}\right)^2} = \sqrt{\frac{1}{4} + \frac{3}{4}} = 1. \blacksquare$$

Let $\mathbf{u} = a\mathbf{i} + b\mathbf{j}$ be a unit vector. Then $|\mathbf{u}| = \sqrt{a^2 + b^2} = 1$, so $a^2 + b^2 = 1$ and $\mathbf{u}$ is a point on the unit circle (see Figure 15). If θ is the direction of $\mathbf{u}$, then we immediately see that $a = \cos\theta$ and $b = \sin\theta$. Thus any unit vector $\mathbf{u}$ can be written in the form

$$\mathbf{u} = (\cos\theta)\mathbf{i} + (\sin\theta)\mathbf{j} \tag{8}$$

where θ is the direction of $\mathbf{u}$.

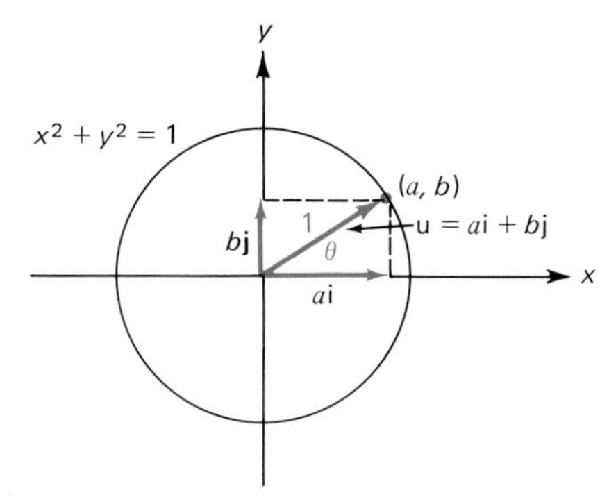

FIGURE 15

EXAMPLE 8 The unit vector $\mathbf{u} = (1/2)\mathbf{i} + (\sqrt{3}/2)\mathbf{j}$ of Example 7 can be written in the form (8) with $\theta = \cos^{-1}(1/2) = \pi/3$. Note that since $\cos\theta = 1/2$ and $\sin\theta = \sqrt{3}/2$, θ is in the first quadrant. We need this fact to conclude that $\theta = \pi/3$. It is also true that $\cos 5\pi/3 = 1/2$, but $5\pi/3$ is in the fourth quadrant. ■

Finally:

Let $\mathbf{v}$ be any nonzero vector. Then $\mathbf{u} = \mathbf{v}/|\mathbf{v}|$ is the unit vector having the same direction as $\mathbf{v}$.

(See Problem 31.)

EXAMPLE 9 Find the unit vector having the same direction as $\mathbf{v} = 2\mathbf{i} - 3\mathbf{j}$.

Solution. Here $|\mathbf{v}| = \sqrt{4 + 9} = \sqrt{13}$, so $\mathbf{u} = \mathbf{v}/|\mathbf{v}| = (2/\sqrt{13})\mathbf{i} - (3/\sqrt{13})\mathbf{j}$ is the required unit vector. ■

EXAMPLE 10 Find the vector $\mathbf{v}$ whose direction is $5\pi/4$ and whose magnitude is 7.

Solution. A unit vector $\mathbf{u}$ with direction $5\pi/4$ is given by

$$\mathbf{u} = \left(\cos \frac{5\pi}{4}\right)\mathbf{i} + \left(\sin \frac{5\pi}{4}\right)\mathbf{j} = -\frac{1}{\sqrt{2}}\mathbf{i} - \frac{1}{\sqrt{2}}\mathbf{j}.$$

Then $\mathbf{v} = 7\mathbf{u} = -(7/\sqrt{2})\mathbf{i} - (7/\sqrt{2})\mathbf{j}$. This vector is sketched in Figure 16a. In Figure 16b we have translated $\mathbf{v}$ so that it points toward the origin. This representation of $\mathbf{v}$ will be useful in Section 1.3. ■

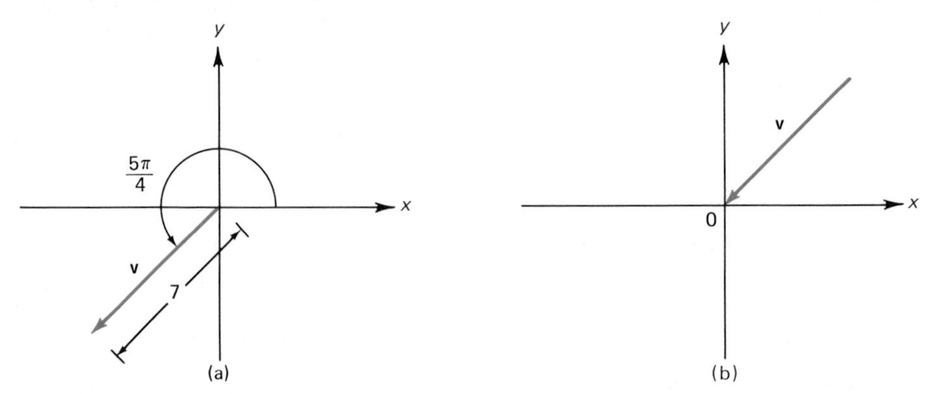

FIGURE 16

We conclude this section with a summary of properties of vectors, given in Table 1.

TABLE 1

Object	Intuitive definition	Expression in terms of components if $\mathbf{u} = u_1\mathbf{i} + u_2\mathbf{j} = (u_1, u_2)$ and $\mathbf{v} = v_1\mathbf{i} + v_2\mathbf{j} = (v_1, v_2)$
Vector $\mathbf{v}$	Magnitude and direction ↗ $\mathbf{v}$	$v_1\mathbf{i} + v_2\mathbf{j}$ or (v_1, v_2)
$\lvert\mathbf{v}\rvert$	Magnitude or length of $\mathbf{v}$	$\sqrt{v_1^2 + v_2^2}$
$\alpha\mathbf{v}$	↗$\mathbf{v}$ ↗ $\alpha\mathbf{v}$ (Here $\alpha = 2$)	$\alpha v_1\mathbf{i} + \alpha v_2\mathbf{j}$ or $(\alpha v_1, \alpha v_2)$
$-\mathbf{v}$	↗ $\mathbf{v}$ ↙ $-\mathbf{v}$	$-v_1\mathbf{i} - v_2\mathbf{j}$ or $(-v_1, -v_2)$ or $-(v_1, v_2)$
$\mathbf{u} + \mathbf{v}$	$\mathbf{u} + \mathbf{v}$, $\mathbf{v}$, $\mathbf{u}$	$(u_1 + v_1)\mathbf{i} + (u_2 + v_2)\mathbf{j}$ or $(u_1 + v_1, u_2 + v_2)$
$\mathbf{u} - \mathbf{v}$	$\mathbf{v}$, $\mathbf{u} - \mathbf{v}$, $\mathbf{u}$	$(u_1 - v_1)\mathbf{i} + (u_2 - v_2)\mathbf{j}$ or $(u_1 - v_1, u_2 - v_2)$

PROBLEMS 1.1

In Problems 1–6 a vector **v** and a point P are given. Find a point Q such that the directed line segment $\overrightarrow{PQ}$ is a representation of **v**. Sketch **v** and $\overrightarrow{PQ}$.

1. $\mathbf{v} = (2, 5)$; $P = (1, -2)$
2. $\mathbf{v} = (5, 8)$; $P = (3, 8)$
3. $\mathbf{v} = (-3, 7)$; $P = (7, -3)$
4. $\mathbf{v} = -\mathbf{i} - 7\mathbf{j}$; $P = (0, 1)$
5. $\mathbf{v} = 5\mathbf{i} - 3\mathbf{j}$; $P = (-7, -2)$
6. $\mathbf{v} = e\mathbf{i} + \pi\mathbf{j}$; $P = (\pi, \sqrt{2})$

In Problems 7–18, find the magnitude and direction of the given vector.

7. $\mathbf{v} = (4, 4)$
8. $\mathbf{v} = (-4, 4)$
9. $\mathbf{v} = (4, -4)$
10. $\mathbf{v} = (-4, -4)$
11. $\mathbf{v} = (\sqrt{3}, 1)$
12. $\mathbf{v} = (1, \sqrt{3})$
13. $\mathbf{v} = (-1, \sqrt{3})$
14. $\mathbf{v} = (1, -\sqrt{3})$
15. $\mathbf{v} = (-1, -\sqrt{3})$
16. $\mathbf{v} = (1, 2)$
17. $\mathbf{v} = (-5, 8)$
18. $\mathbf{v} = (11, -14)$

In Problems 19–26, write in the form $a\mathbf{i} + b\mathbf{j}$ the vector **v** that is represented by $\overrightarrow{PQ}$. Sketch $\overrightarrow{PQ}$ and **v**.

19. $P = (1, 2)$; $Q = (1, 3)$
20. $P = (2, 4)$; $Q = (-7, 4)$
21. $P = (5, 2)$; $Q = (-1, 3)$
22. $P = (8, -2)$; $Q = (-3, -3)$
23. $P = (7, -1)$; $Q = (-2, 4)$
24. $P = (3, -6)$; $Q = (8, 0)$
25. $P = (-3, -8)$; $Q = (-8, -3)$
26. $P = (2, 4)$; $Q = (-4, -2)$

27. Let $\mathbf{u} = (2, 3)$ and $\mathbf{v} = (-5, 4)$. Find the following:
   (a) $3\mathbf{u}$ **(b)** $\mathbf{u} + \mathbf{v}$
   (c) $\mathbf{v} - \mathbf{u}$ **(d)** $2\mathbf{u} - 7\mathbf{v}$
   Sketch these vectors.
28. Let $\mathbf{u} = 2\mathbf{i} - 3\mathbf{j}$ and $\mathbf{v} = -4\mathbf{i} + 6\mathbf{j}$. Find the following:
   (a) $\mathbf{u} + \mathbf{v}$ **(b)** $\mathbf{u} - \mathbf{v}$
   (c) $3\mathbf{u}$ **(d)** $-7\mathbf{v}$
   (e) $8\mathbf{u} - 3\mathbf{v}$ **(f)** $4\mathbf{v} - 6\mathbf{u}$
   Sketch these vectors.
29. Show that the vectors **i** and **j** are unit vectors.
30. Show that the vector $(1/\sqrt{2})\mathbf{i} + (1/\sqrt{2})\mathbf{j}$ is a unit vector.
31. Show that if $\mathbf{v} = a\mathbf{i} + b\mathbf{j} \neq \mathbf{0}$, then $\mathbf{u} = (a/\sqrt{a^2 + b^2})\mathbf{i} + (b/\sqrt{a^2 + b^2})\mathbf{j}$ is a unit vector having the same direction as **v**.

In Problems 32–37, find a unit vector having the same direction as the given vector.

32. $\mathbf{v} = 2\mathbf{i} + 3\mathbf{j}$
33. $\mathbf{v} = \mathbf{i} - \mathbf{j}$
34. $\mathbf{v} = (3, 4)$
35. $\mathbf{v} = (3, -4)$
36. $\mathbf{v} = -3\mathbf{i} + 4\mathbf{j}$
37. $\mathbf{v} = (a, a)$, $a \neq 0$

38. If $\mathbf{v} = a\mathbf{i} + b\mathbf{j} \neq \mathbf{0}$, show that $a/\sqrt{a^2 + b^2} = \cos\theta$ and $b/\sqrt{a^2 + b^2} = \sin\theta$, where θ is the direction of **v**.
39. For $\mathbf{v} = 2\mathbf{i} - 3\mathbf{j}$, find $\sin\theta$ and $\cos\theta$.
40. For $\mathbf{v} = -3\mathbf{i} + 8\mathbf{j}$, find $\sin\theta$ and $\cos\theta$.

A vector **v** has a direction opposite to that of a vector **u** if $|\text{direction } \mathbf{v} - \text{direction } \mathbf{u}| = \pi$. In Problems 41–46, find a unit vector **u** that has a direction opposite the direction of the given vector **v**.

41. $\mathbf{v} = \mathbf{i} + \mathbf{j}$
42. $\mathbf{v} = 2\mathbf{i} - 3\mathbf{j}$
43. $\mathbf{v} = (-3, 4)$
44. $\mathbf{v} = (-2, 3)$
45. $\mathbf{v} = -3\mathbf{i} - 4\mathbf{j}$
46. $\mathbf{v} = (8, -3)$

47. Let $\mathbf{u} = 2\mathbf{i} - 3\mathbf{j}$ and $\mathbf{v} = -\mathbf{i} + 2\mathbf{j}$. Find a unit vector having the same direction as the following:
   (a) $\mathbf{u} + \mathbf{v}$ **(b)** $2\mathbf{u} - 3\mathbf{v}$
   (c) $3\mathbf{u} + 8\mathbf{v}$
48. Let $P = (c, d)$ and $Q = (c + a, d + b)$. Show that the magnitude of $\overrightarrow{PQ}$ is $\sqrt{a^2 + b^2}$.
49. Show that the direction of $\overrightarrow{PQ}$ in Problem 48 is the same as the direction of the vector (a, b). [*Hint:* If $R = (a, b)$, show that the line passing through the points P and Q is parallel to the line passing through the points 0 and R.]

In Problems 50–57, find a vector **v** having the given magnitude and direction. [*Hint:* See Example 10.]

50. $|\mathbf{v}| = 3$; $\theta = \pi/6$
51. $|\mathbf{v}| = 8$; $\theta = \pi/3$
52. $|\mathbf{v}| = 7$; $\theta = \pi$
53. $|\mathbf{v}| = 4$; $\theta = \pi/2$
54. $|\mathbf{v}| = 1$; $\theta = \pi/4$
55. $|\mathbf{v}| = 6$; $\theta = 2\pi/3$
56. $|\mathbf{v}| = 8$; $\theta = 3\pi/2$
57. $|\mathbf{v}| = 6$; $\theta = 11\pi/6$

58. Prove Theorem 1. [*Hint:* Use the definitions of addition and scalar multiplication of vectors.]
59. Show algebraically (i.e., strictly from the definitions of vector addition and magnitude) that for any two vectors **u** and **v**, $|\mathbf{u} + \mathbf{v}| \leq |\mathbf{u}| + |\mathbf{v}|$.
60. Show that if neither **u** nor **v** is the zero vector, then $|\mathbf{u} + \mathbf{v}| = |\mathbf{u}| + |\mathbf{v}|$ if and only if **u** is a positive scalar multiple of **v**.

1.2 THE DOT PRODUCT

In Section 1.1 we showed how a vector could be multiplied by a scalar but not how two vectors could be multiplied. Actually, there are several ways to define the product of two vectors, and in this section we will discuss one of them. We will discuss a second product operation in Section 3.4.

Definition 1 DOT PRODUCT Let $\mathbf{u} = (a_1, b_1) = a_1\mathbf{i} + b_1\mathbf{j}$ and $\mathbf{v} = (a_2, b_2) = a_2\mathbf{i} + b_2\mathbf{j}$. Then the **dot product** of **u** and **v**, denoted $\mathbf{u} \cdot \mathbf{v}$, is defined by

$$\mathbf{u} \cdot \mathbf{v} = a_1a_2 + b_1b_2. \tag{1}$$

REMARK. The dot product of two vectors is a *scalar*. For this reason the dot product is often called the **scalar product.** It is also called the **inner product.**

EXAMPLE 1 If $\mathbf{u} = (1, 3)$ and $\mathbf{v} = (4, -7)$, then

$$\mathbf{u} \cdot \mathbf{v} = 1(4) + 3(-7) = 4 - 21 = -17. \blacksquare$$

Theorem 1 For any vectors **u**, **v**, **w**, and scalar α,

(i) $\mathbf{u} \cdot \mathbf{v} = \mathbf{v} \cdot \mathbf{u}$ **(ii)** $(\mathbf{u} + \mathbf{v}) \cdot \mathbf{w} = \mathbf{u} \cdot \mathbf{w} + \mathbf{v} \cdot \mathbf{w}$
(iii) $(\alpha\mathbf{u}) \cdot \mathbf{v} = \alpha(\mathbf{u} \cdot \mathbf{v})$ **(iv)** $\mathbf{u} \cdot \mathbf{u} \geq 0$; and $\mathbf{u} \cdot \mathbf{u} = 0$ if and only if $\mathbf{u} = \mathbf{0}$
(v) $|\mathbf{u}| = \sqrt{\mathbf{u} \cdot \mathbf{u}}$

Proof. Let $\mathbf{u} = (u_1, u_2)$, $\mathbf{v} = (v_1, v_2)$, and $\mathbf{w} = (w_1, w_2)$.

(i) $\mathbf{u} \cdot \mathbf{v} = u_1v_1 + u_2v_2 = v_1u_1 + v_2u_2 = \mathbf{v} \cdot \mathbf{u}$
(ii) $(\mathbf{u} + \mathbf{v}) \cdot \mathbf{w} = (u_1 + v_1, u_2 + v_2) \cdot (w_1, w_2) = (u_1 + v_1)w_1 + (u_2 + v_2)w_2$
$= u_1w_1 + u_2w_2 + v_1w_1 + v_2w_2 = \mathbf{u} \cdot \mathbf{w} + \mathbf{v} \cdot \mathbf{w}$
(iii) $(\alpha\mathbf{u}) \cdot \mathbf{v} = (\alpha u_1, \alpha u_2) \cdot (v_1, v_2) = \alpha u_1v_1 + \alpha u_2v_2 = \alpha(u_1v_1 + u_2v_2) = \alpha(\mathbf{u} \cdot \mathbf{v})$
(iv) $\mathbf{u} \cdot \mathbf{u} = u_1^2 + u_2^2 \geq 0$; and $\mathbf{u} \cdot \mathbf{u} = \mathbf{0}$ if and only if $u_1 = u_2 = 0$.
(v) $\sqrt{\mathbf{u} \cdot \mathbf{u}} = \sqrt{(u_1, u_2) \cdot (u_1, u_2)} = \sqrt{u_1^2 + u_2^2} = |\mathbf{u}| \blacksquare$

The dot product is useful in a wide variety of applications. An interesting one follows.

Definition 2 ANGLE BETWEEN TWO VECTORS Let **u** and **v** be two nonzero vectors. Then the **angle** φ between **u** and **v** is defined to be the smallest angle† between the representations of **u** and **v** that have the origin as their initial points. If $\mathbf{u} = \alpha\mathbf{v}$ for some scalar α, then we define $\varphi = 0$ if $\alpha > 0$ and $\varphi = \pi$ if $\alpha < 0$.

Theorem 2 Let **u** and **v** be two nonzero vectors. Then if φ is the angle between them,

†The smallest angle will be in the interval $[0, \pi]$.

$$\cos\varphi = \frac{\mathbf{u}\cdot\mathbf{v}}{|\mathbf{u}||\mathbf{v}|}. \tag{2}$$

Proof. The law of cosines states that in the triangle of Figure 1,

$$c^2 = a^2 + b^2 - 2ab\cos C. \tag{3}$$

We now place the representations of **u** and **v** with initial points at the origin so that $\mathbf{u} = (a_1, b_1)$ and $\mathbf{v} = (a_2, b_2)$ (see Figure 2). Then from the law of cosines,

$$|\mathbf{v} - \mathbf{u}|^2 = |\mathbf{v}|^2 + |\mathbf{u}|^2 - 2|\mathbf{u}||\mathbf{v}|\cos\varphi.$$

But using Theorem 1 several times, we have

$$|\mathbf{v} - \mathbf{u}|^2 = (\mathbf{v} - \mathbf{u})\cdot(\mathbf{v} - \mathbf{u}) = \mathbf{v}\cdot\mathbf{v} - 2\mathbf{u}\cdot\mathbf{v} + \mathbf{u}\cdot\mathbf{u} = |\mathbf{v}|^2 - 2\mathbf{u}\cdot\mathbf{v} + |\mathbf{u}|^2.$$

Thus, after simplification, we obtain

$$-2\mathbf{u}\cdot\mathbf{v} = -2|\mathbf{u}||\mathbf{v}|\cos\varphi,$$

from which the theorem follows. ■

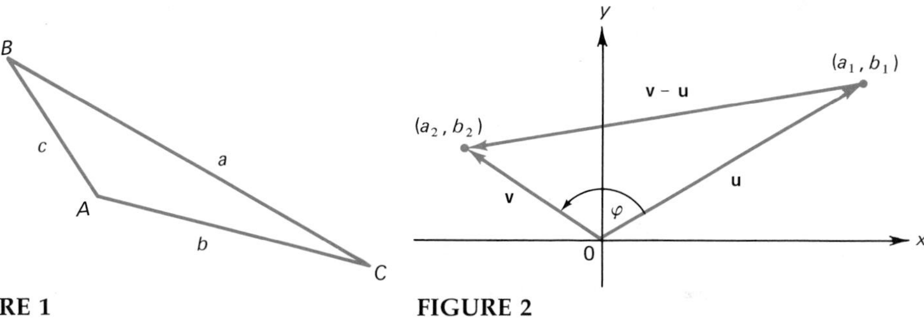

FIGURE 1

FIGURE 2

REMARK. Using Theorem 2, we could define the dot product $\mathbf{u}\cdot\mathbf{v}$ by

$$\mathbf{u}\cdot\mathbf{v} = |\mathbf{u}||\mathbf{v}|\cos\varphi. \tag{4}$$

EXAMPLE 2 Find the cosine of the angle between the vectors $\mathbf{u} = 2\mathbf{i} + 3\mathbf{j}$ and $\mathbf{v} = -7\mathbf{i} + \mathbf{j}$.

Solution. $\mathbf{u}\cdot\mathbf{v} = -14 + 3 = -11$, $|\mathbf{u}| = \sqrt{2^2 + 3^2} = \sqrt{13}$, and $|\mathbf{v}| = \sqrt{(-7)^2 + 1^2} = \sqrt{50}$; so

$$\cos\varphi = \frac{\mathbf{u}\cdot\mathbf{v}}{|\mathbf{u}||\mathbf{v}|} = \frac{-11}{\sqrt{13}\sqrt{50}} = \frac{-11}{\sqrt{650}} \approx -0.4315. \ ■$$

Definition 3 PARALLEL VECTORS Two nonzero vectors **u** and **v** are **parallel** if the angle between them is 0 or π.

EXAMPLE 3 Show that the vectors $\mathbf{u} = (2, -3)$ and $\mathbf{v} = (-4, 6)$ are parallel.

Solution.

$$\cos \varphi = \frac{\mathbf{u} \cdot \mathbf{v}}{|\mathbf{u}||\mathbf{v}|} = \frac{-8 - 18}{\sqrt{13}\sqrt{52}} = \frac{-26}{\sqrt{13}(2\sqrt{13})} = \frac{-26}{2(13)} = -1,$$

so $\varphi = \pi$. ■

Theorem 3 If $\mathbf{u} \neq \mathbf{0}$, then $\mathbf{v} = \alpha\mathbf{u}$ for some nonzero constant α if and only if $\mathbf{u}$ and $\mathbf{v}$ are parallel.

Proof. This follows from the last part of Definition 2 (see also Problem 43). ■

Definition 4 ORTHOGONAL VECTORS The nonzero vectors $\mathbf{u}$ and $\mathbf{v}$ are called **orthogonal** (or **perpendicular**) if the angle between them is $\pi/2$.

EXAMPLE 4 Show that the vectors $\mathbf{u} = 3\mathbf{i} - 4\mathbf{j}$ and $\mathbf{v} = 4\mathbf{i} + 3\mathbf{j}$ are orthogonal.

Solution. $\mathbf{u} \cdot \mathbf{v} = 3 \cdot 4 - 4 \cdot 3 = 0$. This implies that $\cos \varphi = (\mathbf{u} \cdot \mathbf{v})/(|\mathbf{u}||\mathbf{v}|) = 0$. Since φ is in the interval $[0, \pi]$, $\varphi = \pi/2$.

Theorem 4 The nonzero vectors $\mathbf{u}$ and $\mathbf{v}$ are orthogonal if and only if $\mathbf{u} \cdot \mathbf{v} = 0$.

Proof. This proof is also easy and is left as an exercise (see Problem 44). ■

REMARK. The condition $\mathbf{u} \cdot \mathbf{v} = 0$ is often given as the *definition* of orthogonal.

A number of interesting problems involve the notion of the *projection* of one vector onto another. Before defining this term, we prove the following theorem.

Theorem 5 Let $\mathbf{v}$ be a nonzero vector. Then for any other vector $\mathbf{u}$, the vector

$$\mathbf{w} = \mathbf{u} - [(\mathbf{u} \cdot \mathbf{v})/|\mathbf{v}|^2]\mathbf{v} \text{ is orthogonal to } \mathbf{v}.$$

Proof.

$$\mathbf{w} \cdot \mathbf{v} = \left(\mathbf{u} - \frac{(\mathbf{u} \cdot \mathbf{v})\mathbf{v}}{|\mathbf{v}|^2}\right) \cdot \mathbf{v} = \mathbf{u} \cdot \mathbf{v} - \frac{(\mathbf{u} \cdot \mathbf{v})(\mathbf{v} \cdot \mathbf{v})}{|\mathbf{v}|^2}$$

$$= \mathbf{u} \cdot \mathbf{v} - \frac{(\mathbf{u} \cdot \mathbf{v})|\mathbf{v}|^2}{|\mathbf{v}|^2} = \mathbf{u} \cdot \mathbf{v} - \mathbf{u} \cdot \mathbf{v} = 0$$

The vectors $\mathbf{u}$, $\mathbf{v}$, and $\mathbf{w}$ are illustrated in Figure 3. ■

Definition 5 PROJECTION Let $\mathbf{u}$ and $\mathbf{v}$ be nonzero vectors. Then the **projection of u onto v** is a vector, denoted $\text{Proj}_{\mathbf{v}}\,\mathbf{u}$, which is defined by

$$\text{Proj}_{\mathbf{v}}\,\mathbf{u} = \frac{\mathbf{u} \cdot \mathbf{v}}{\mathbf{v} \cdot \mathbf{v}}\mathbf{v} = \frac{\mathbf{u} \cdot \mathbf{v}}{|\mathbf{v}|^2}\mathbf{v} = \left(\frac{\mathbf{u} \cdot \mathbf{v}}{|\mathbf{v}|}\right)\frac{\mathbf{v}}{|\mathbf{v}|}. \quad (5)$$

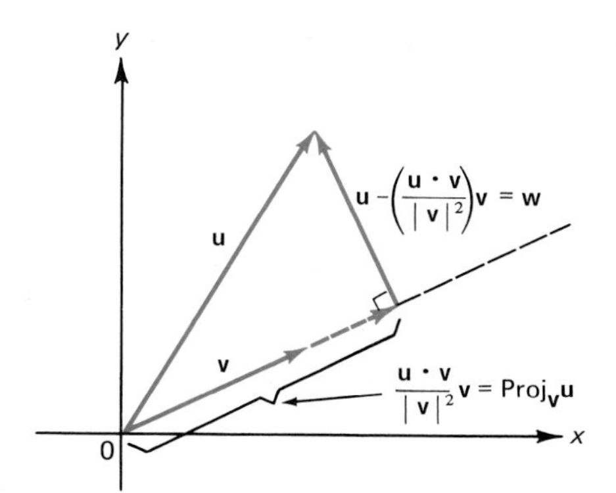

FIGURE 3

> *The* **component** *of* **u** *in the direction* **v** is **u** · **v**/|**v**|. **(6)**

Note that $\mathbf{v}/|\mathbf{v}|$ is a unit vector in the direction of **v**.

If we use the alternative definition of the dot product given in equation (4), we see that

$$\text{component of } \mathbf{u} \text{ in the direction } \mathbf{v} = \frac{\mathbf{u} \cdot \mathbf{v}}{|\mathbf{v}|} = |\mathbf{u}| \cos \varphi, \tag{7}$$

where φ is the angle between **u** and **v**. We immediately see, from (7), that the component of **u** in the direction **v** is greatest in absolute value when **u** and **v** lie on the same straight line (so that $\varphi = 0$ or π and $\cos \varphi = 1$ or -1) and is smallest (zero) when **u** and **v** are perpendicular. In Figure 4 we illustrate the component of a unit vector **u** on a unit vector **v** when φ varies between 0 and $\pi/2$.

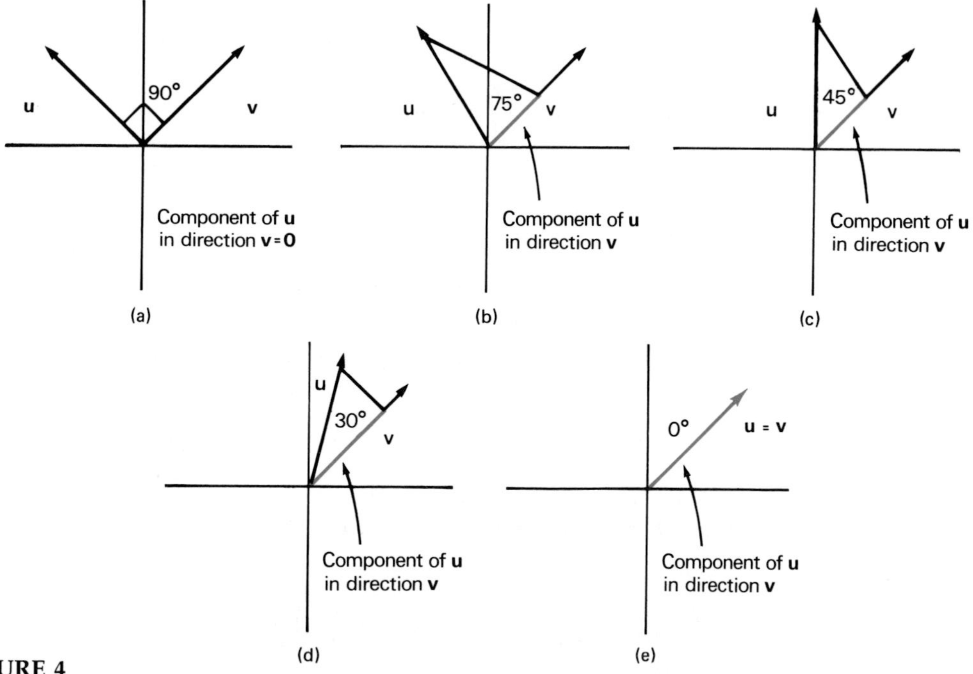

FIGURE 4

We summarize some properties of projection:

1. From Figure 3 and the fact that $\cos \varphi = (\mathbf{u} \cdot \mathbf{v})/(|\mathbf{u}||\mathbf{v}|)$, we find that

> $\mathbf{v}$ and $\text{Proj}_{\mathbf{v}}\, \mathbf{u}$ have
>
> **(i)** the same direction if $\mathbf{u} \cdot \mathbf{v} > 0$ and
> **(ii)** opposite directions if $\mathbf{u} \cdot \mathbf{v} < 0$.

2. $\text{Proj}_{\mathbf{v}}\, \mathbf{u}$ can be thought of as the "**v**-component" of the vector **u**. We will see an illustration of this in our discussion of force in the next section.

3. If **u** and **v** are orthogonal, then $\mathbf{u} \cdot \mathbf{v} = 0$, so that $\text{Proj}_{\mathbf{v}}\, \mathbf{u} = \mathbf{0}$.

4. An alternative definition of projection is: If **u** and **v** are nonzero vectors, then $\text{Proj}_{\mathbf{v}}\, \mathbf{u}$ is the unique vector having the properties

> **(i)** $\text{Proj}_{\mathbf{v}}\, \mathbf{u}$ is parallel to **v** and **(ii)** $\mathbf{u} - \text{Proj}_{\mathbf{v}}\, \mathbf{u}$ is orthogonal to **v**.

EXAMPLE 5 Let $\mathbf{u} = 2\mathbf{i} + 3\mathbf{j}$ and $\mathbf{v} = \mathbf{i} + \mathbf{j}$. Calculate $\text{Proj}_{\mathbf{v}}\, \mathbf{u}$.

Solution. $\text{Proj}_{\mathbf{v}}\, \mathbf{u} = (\mathbf{u} \cdot \mathbf{v})\mathbf{v}/|\mathbf{v}|^2 = [5/(\sqrt{2})^2]\mathbf{v} = (5/2)\mathbf{i} + (5/2)\mathbf{j}$. This is illustrated in Figure 5. ■

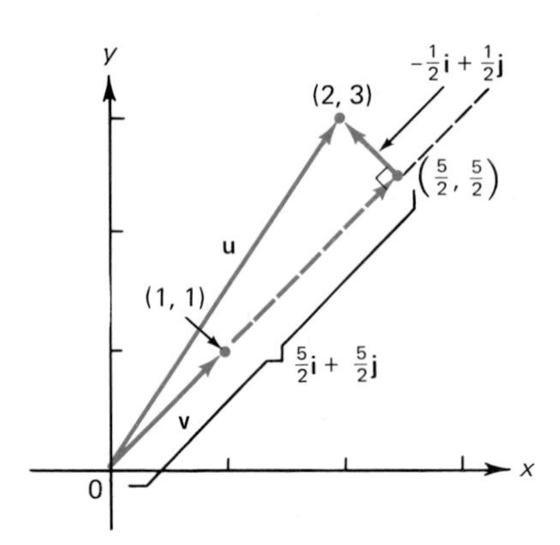

FIGURE 5

EXAMPLE 6 Let $\mathbf{u} = 2\mathbf{i} - 3\mathbf{j}$ and $\mathbf{v} = \mathbf{i} + \mathbf{j}$. Find $\text{Proj}_{\mathbf{v}}\, \mathbf{u}$.

Solution. Here $(\mathbf{u} \cdot \mathbf{v})/|\mathbf{v}|^2 = -\frac{1}{2}$, so that $\text{Proj}_{\mathbf{v}}\, \mathbf{u} = -\frac{1}{2}\mathbf{i} - \frac{1}{2}\mathbf{j}$. This is illustrated in Figure 6. ■

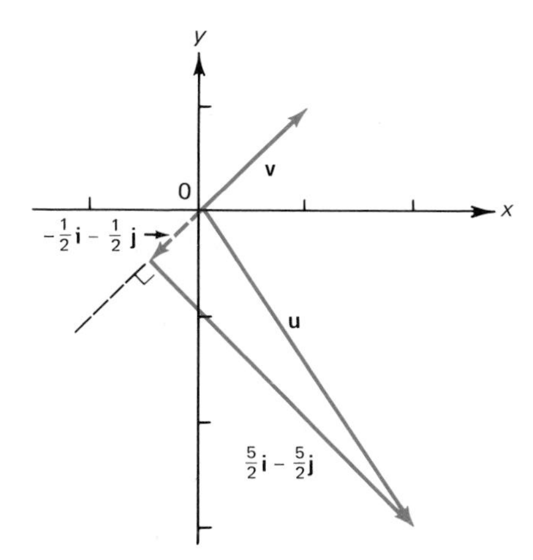

FIGURE 6

PROBLEMS 1.2

In Problems 1–10, calculate both the dot product of the two vectors and the cosine of the angle between them.

1. $\mathbf{u} = \mathbf{i} + \mathbf{j}$; $\mathbf{v} = \mathbf{i} - \mathbf{j}$
2. $\mathbf{u} = 3\mathbf{i}$; $\mathbf{v} = -7\mathbf{j}$
3. $\mathbf{u} = -5\mathbf{i}$; $\mathbf{v} = 18\mathbf{j}$
4. $\mathbf{u} = \alpha\mathbf{i}$; $\mathbf{v} = \beta\mathbf{j}$; α, β real and nonzero
5. $\mathbf{u} = 2\mathbf{i} + 5\mathbf{j}$; $\mathbf{v} = 5\mathbf{i} + 2\mathbf{j}$
6. $\mathbf{u} = 2\mathbf{i} + 5\mathbf{j}$; $\mathbf{v} = 5\mathbf{i} - 3\mathbf{j}$
7. $\mathbf{u} = -3\mathbf{i} + 4\mathbf{j}$; $\mathbf{v} = -2\mathbf{i} - 7\mathbf{j}$
8. $\mathbf{u} = 4\mathbf{i} + 5\mathbf{j}$; $\mathbf{v} = 7\mathbf{i} - 4\mathbf{j}$
9. $\mathbf{u} = 11\mathbf{i} - 8\mathbf{j}$; $\mathbf{v} = 4\mathbf{i} - 7\mathbf{j}$
10. $\mathbf{u} = -13\mathbf{i} + 8\mathbf{j}$; $\mathbf{v} = 2\mathbf{i} + 11\mathbf{j}$

11. Show that for any nonzero real numbers α and β, the vectors $\mathbf{u} = \alpha\mathbf{i} + \beta\mathbf{j}$ and $\mathbf{v} = \beta\mathbf{i} - \alpha\mathbf{j}$ are orthogonal.
12. Let $\mathbf{u}$, $\mathbf{v}$, and $\mathbf{w}$ denote three arbitrary vectors. Explain why the product $\mathbf{u} \cdot \mathbf{v} \cdot \mathbf{w}$ is *not defined*.

In Problems 13–20, determine whether the given vectors are orthogonal, parallel, or neither. Then sketch each pair.

13. $\mathbf{u} = 3\mathbf{i} + 5\mathbf{j}$; $\mathbf{v} = -6\mathbf{i} - 10\mathbf{j}$
14. $\mathbf{u} = 2\mathbf{i} + 3\mathbf{j}$; $\mathbf{v} = 6\mathbf{i} - 4\mathbf{j}$
15. $\mathbf{u} = 2\mathbf{i} + 3\mathbf{j}$; $\mathbf{v} = 6\mathbf{i} + 4\mathbf{j}$
16. $\mathbf{u} = 2\mathbf{i} + 3\mathbf{j}$; $\mathbf{v} = -6\mathbf{i} + 4\mathbf{j}$
17. $\mathbf{u} = 7\mathbf{i}$; $\mathbf{v} = -23\mathbf{j}$
18. $\mathbf{u} = 2\mathbf{i} - 6\mathbf{j}$; $\mathbf{v} = -\mathbf{i} + 3\mathbf{j}$
19. $\mathbf{u} = \mathbf{i} + \mathbf{j}$; $\mathbf{v} = \alpha\mathbf{i} + \alpha\mathbf{j}$; α real
20. $\mathbf{u} = -2\mathbf{i} + 3\mathbf{j}$; $\mathbf{v} = -\mathbf{i} + 2\mathbf{j}$

21. Let $\mathbf{u} = 3\mathbf{i} + 4\mathbf{j}$ and $\mathbf{v} = \mathbf{i} + \alpha\mathbf{j}$.
    (a) Determine α such that $\mathbf{u}$ and $\mathbf{v}$ are orthogonal.
    (b) Determine α such that $\mathbf{u}$ and $\mathbf{v}$ are parallel.
    (c) Determine α such that the angle between $\mathbf{u}$ and $\mathbf{v}$ is $\pi/4$.
    (d) Determine α such that the angle between $\mathbf{u}$ and $\mathbf{v}$ is $\pi/3$.
22. Let $\mathbf{u} = -2\mathbf{i} + 5\mathbf{j}$ and $\mathbf{v} = \alpha\mathbf{i} - 2\mathbf{j}$.
    (a) Determine α such that $\mathbf{u}$ and $\mathbf{v}$ are orthogonal.
    (b) Determine α such that $\mathbf{u}$ and $\mathbf{v}$ are parallel.
    (c) Determine α such that the angle between $\mathbf{u}$ and $\mathbf{v}$ is $2\pi/3$.
    (d) Determine α such that the angle between $\mathbf{u}$ and $\mathbf{v}$ is $\pi/3$.
23. In Problem 21 show that there is no value of α for which $\mathbf{u}$ and $\mathbf{v}$ have opposite directions.
24. In Problem 22 show that there is no value of α for which $\mathbf{u}$ and $\mathbf{v}$ have the same direction.

In Problems 25–38, calculate $\text{Proj}_{\mathbf{v}}\,\mathbf{u}$.

25. $\mathbf{u} = 3\mathbf{i}$; $\mathbf{v} = \mathbf{i} + \mathbf{j}$
26. $\mathbf{u} = -5\mathbf{j}$; $\mathbf{v} = \mathbf{i} + \mathbf{j}$
27. $\mathbf{u} = 2\mathbf{i} + \mathbf{j}$; $\mathbf{v} = \mathbf{i} - 2\mathbf{j}$
28. $\mathbf{u} = 2\mathbf{i} + 3\mathbf{j}$; $\mathbf{v} = 4\mathbf{i} + \mathbf{j}$
29. $\mathbf{u} = \mathbf{i} + \mathbf{j}$; $\mathbf{v} = 2\mathbf{i} - 3\mathbf{j}$
30. $\mathbf{u} = \mathbf{i} + \mathbf{j}$; $\mathbf{v} = 2\mathbf{i} + 3\mathbf{j}$
31. $\mathbf{u} = 4\mathbf{i} + 5\mathbf{j}$; $\mathbf{v} = 2\mathbf{i} + 4\mathbf{j}$
32. $\mathbf{u} = 4\mathbf{i} + 5\mathbf{j}$; $\mathbf{v} = 2\mathbf{i} - 4\mathbf{j}$
33. $\mathbf{u} = -4\mathbf{i} + 5\mathbf{j}$; $\mathbf{v} = 2\mathbf{i} - 4\mathbf{j}$
34. $\mathbf{u} = -4\mathbf{i} - 5\mathbf{j}$; $\mathbf{v} = -2\mathbf{i} - 4\mathbf{j}$

35. $\mathbf{u} = \alpha\mathbf{i} + \beta\mathbf{j}$; $\mathbf{v} = \mathbf{i} + \mathbf{j}$

36. $\mathbf{u} = \mathbf{i} + \mathbf{j}$; $\mathbf{v} = \alpha\mathbf{i} + \beta\mathbf{j}$

37. $\mathbf{u} = \alpha\mathbf{i} - \beta\mathbf{j}$; $\mathbf{v} = \mathbf{i} + \mathbf{j}$

38. $\mathbf{u} = \alpha\mathbf{i} - \beta\mathbf{j}$; $\mathbf{v} = -\mathbf{i} + \mathbf{j}$

39. Let $\mathbf{u} = a_1\mathbf{i} + b_1\mathbf{j}$ and $\mathbf{v} = a_2\mathbf{i} + b_2\mathbf{j}$. Give a condition on a_1, b_1, a_2, and b_2 that will ensure that $\mathbf{v}$ and $\text{Proj}_{\mathbf{v}}\, \mathbf{u}$ have the same direction.

40. In Problem 39, give a condition that will ensure that $\mathbf{v}$ and $\text{Proj}_{\mathbf{v}}\, \mathbf{u}$ have opposite directions.

41. Let $P = (2, 3)$, $Q = (5, 7)$, $R = (2, -3)$, and $S = (1, 2)$. Calculate $\text{Proj}_{\overrightarrow{PQ}}\, \overrightarrow{RS}$ and $\text{Proj}_{\overrightarrow{RS}}\, \overrightarrow{PQ}$.

42. Let $P = (-1, 3)$, $Q = (2, 4)$, $R = (-6, -2)$, and $S = (3, 0)$. Calculate $\text{Proj}_{\overrightarrow{PQ}}\, \overrightarrow{RS}$ and $\text{Proj}_{\overrightarrow{RS}}\, \overrightarrow{PQ}$.

43. Prove that the nonzero vectors $\mathbf{u}$ and $\mathbf{v}$ are parallel if and only if $\mathbf{v} = \alpha\mathbf{u}$ for some nonzero constant α. [*Hint:* Show that $\cos\varphi = \pm 1$ if and only if $\mathbf{v} = \alpha\mathbf{u}$.]

44. Prove that the nonzero vectors $\mathbf{u}$ and $\mathbf{v}$ are orthogonal if and only if $\mathbf{u} \cdot \mathbf{v} = 0$.

45. Show that the vector $\mathbf{v} = a\mathbf{i} + b\mathbf{j}$ is orthogonal to the line $ax + by + c = 0$.

46. Show that the vector $\mathbf{u} = b\mathbf{i} - a\mathbf{j}$ is parallel to the line $ax + by + c = 0$.

47. A triangle has vertices $(1, 3)$, $(4, -2)$, and $(-3, 6)$. Find the cosine of each of its angles.

48. A triangle has vertices (a_1, b_1), (a_2, b_2), and (a_3, b_3). Find a formula for the cosines of each of its angles.

***49.** The **Cauchy-Schwarz inequality** states that for any real numbers a_1, a_2, b_1, and b_2,

$$\left|\sum_{k=1}^{2} a_k b_k\right| \le \left(\sum_{k=1}^{2} a_k^2\right)^{1/2}\left(\sum_{k=1}^{2} b_k^2\right)^{1/2}.$$

Use the dot product to prove this formula. Under what circumstances can the inequality be replaced by an equality? [*Hint:* Show that $|\mathbf{a} \cdot \mathbf{b}| \le |\mathbf{a}|\,|\mathbf{b}|$.]

50. Prove that the shortest distance between a point and a line is measured along a line through the point and perpendicular to the line.

51. Find the distance between $P = (2, 3)$ and the line through the points $Q = (-1, 7)$ and $R = (3, 5)$. [*Hint:* Draw a picture and use the Pythagorean theorem.]

52. Find the distance between $(3, 7)$ and the line along the vector $\mathbf{v} = 2\mathbf{i} - 3\mathbf{j}$ that passes through the origin.

53. Use the dot product and the Cauchy-Schwarz inequality (Problem 49) to prove the triangle inequality: $|\mathbf{u} + \mathbf{v}| \le |\mathbf{u}| + |\mathbf{v}|$.

1.3 SOME APPLICATIONS OF VECTORS (OPTIONAL)

In this section we discuss some elementary applications of vectors. Instead of attempting to give a set of rules for solving problems involving vectors, we will give a variety of examples. Further examples of the use of vectors will be given in the next chapter, after we discuss the differentiation of vector functions.

EXAMPLE 1 Show that the midpoints of the sides of a quadrilateral are the vertices of a parallelogram.

Solution. The situation is sketched in Figure 1. Using the fact that B, D, F, and H are midpoints, we have

$$\overrightarrow{BD} = \overrightarrow{BC} + \overrightarrow{CD} = \tfrac{1}{2}\overrightarrow{AC} + \tfrac{1}{2}\overrightarrow{CE} = \tfrac{1}{2}(\overrightarrow{AC} + \overrightarrow{CE}) = \tfrac{1}{2}\overrightarrow{AE}$$

and

$$\overrightarrow{HF} = \overrightarrow{HG} + \overrightarrow{GF} = \tfrac{1}{2}\overrightarrow{AG} + \tfrac{1}{2}\overrightarrow{GE} = \tfrac{1}{2}(\overrightarrow{AG} + \overrightarrow{GE}) = \tfrac{1}{2}\overrightarrow{AE}.$$

Thus $\overrightarrow{BD} = \overrightarrow{HF}$. Similarly, $\overrightarrow{HB} = \overrightarrow{FD}$, so $BDFH$ is a parallelogram. ■

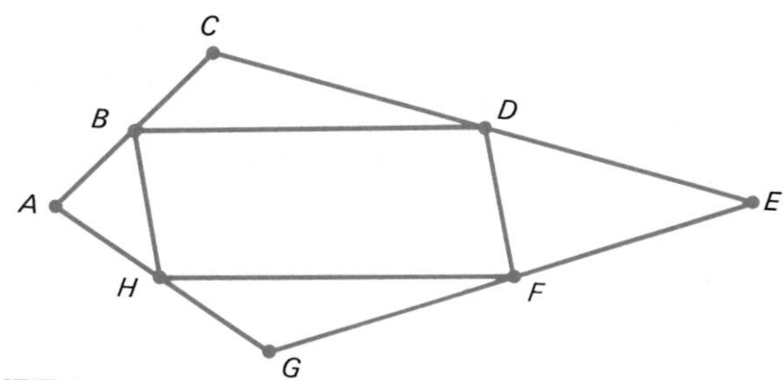

FIGURE 1

EXAMPLE 2 Show that the line segment joining the midpoints of two sides of a triangle has the length of half the third side and is parallel to it.

Solution. We refer to Figure 2. From the figure we see that

$$\overrightarrow{ST} = \overrightarrow{PT} - \overrightarrow{PS}.$$

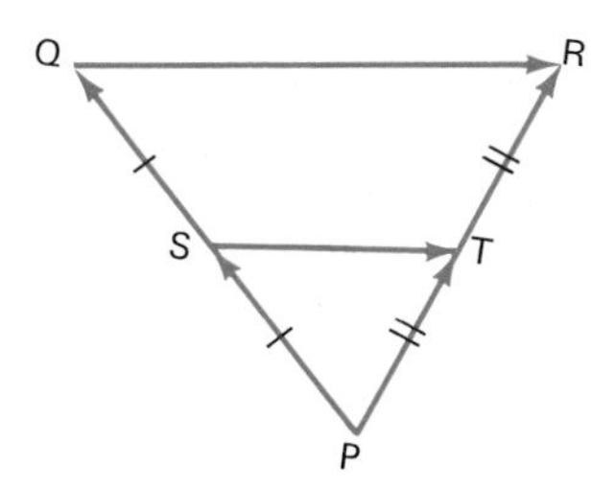

FIGURE 2

But $\overrightarrow{PT} = \frac{1}{2}\overrightarrow{PR}$ and $\overrightarrow{PS} = \frac{1}{2}\overrightarrow{PQ}$. Thus

$$\overrightarrow{ST} = \tfrac{1}{2}\overrightarrow{PR} - \tfrac{1}{2}\overrightarrow{PQ} = \tfrac{1}{2}(\overrightarrow{PR} - \overrightarrow{PQ}) = \tfrac{1}{2}\overrightarrow{QR}.$$

Hence from Theorem 1.2.3, $\overrightarrow{ST}$ is parallel to $\overrightarrow{QR}$. Moreover $|\overrightarrow{ST}| = \frac{1}{2}|\overrightarrow{QR}|$. This is what we wanted to show. ■

In studying calculus of one variable, you probably solved problems involving the notion of force. We usually start such problems with a statement like "a force of x newtons is applied to. . . ." Implicit in such a problem is the idea that a force with a certain magnitude (measured in pounds or newtons) is exerted in a certain direction. In this context force can (and should) be thought of as a vector.

If more than one force is applied to an object, then we define the **resultant** of the forces applied to the object to be the *vector sum* of these forces. We can think of the resultant as the *net* applied force.

EXAMPLE 3 A force of 3 N[†] is applied to the left side of an object, one of 4 N is applied from the bottom, and a force of 7 N is applied from an angle of $\pi/4$ to the horizontal. What is the resultant of forces applied to the object?

[†]1 newton (N) is the force that will accelerate a 1-kg mass at the rate of 1 m/sec^2; 1 N = 0.2248 lb.

Solution. The forces are indicated in Figure 3. We write each force as a magnitude times a unit vector in the indicated direction. For convenience we can think of the center of the object as being at the origin. Then $\mathbf{F}_1 = 3\mathbf{i}$; $\mathbf{F}_2 = 4\mathbf{j}$; and $\mathbf{F}_3 = -(7/\sqrt{2})(\mathbf{i} + \mathbf{j})$. This last vector follows from the fact that the vector $-(1/\sqrt{2})(\mathbf{i} + \mathbf{j})$ is a unit vector pointing toward the origin making an angle of $\pi/4$ with the x-axis (see Example 1.1.10). Then the resultant is given by

$$\mathbf{F} = \mathbf{F}_1 + \mathbf{F}_2 + \mathbf{F}_3 = \left(3 - \frac{7}{\sqrt{2}}\right)\mathbf{i} + \left(4 - \frac{7}{\sqrt{2}}\right)\mathbf{j}.$$

The magnitude of $\mathbf{F}$ is

$$|\mathbf{F}| = \sqrt{\left(3 - \frac{7}{\sqrt{2}}\right)^2 + \left(4 - \frac{7}{\sqrt{2}}\right)^2} = \sqrt{74 - \frac{98}{\sqrt{2}}} \approx 2.17 \text{ N}.$$

The direction θ can be calculated by first finding the unit vector in the direction of $\mathbf{F}$:

$$\begin{aligned}\frac{\mathbf{F}}{|\mathbf{F}|} &= \frac{3 - (7/\sqrt{2})}{\sqrt{74 - (98/\sqrt{2})}}\mathbf{i} + \frac{4 - (7/\sqrt{2})}{\sqrt{74 - (98/\sqrt{2})}}\mathbf{j} \\ &= (\cos\theta)\mathbf{i} + (\sin\theta)\mathbf{j}.\end{aligned}$$

Then

$$\cos\theta = \frac{3 - (7/\sqrt{2})}{\sqrt{74 - (98/\sqrt{2})}} \approx -0.8990 \quad \text{and} \quad \sin\theta = \frac{4 - (7/\sqrt{2})}{\sqrt{74 - (98/\sqrt{2})}} \approx -0.4379.$$

This means that θ is in the third quadrant, and $\theta \approx 3.5949 \approx 206°$ (or $-154°$). This is illustrated in Figure 4. ■

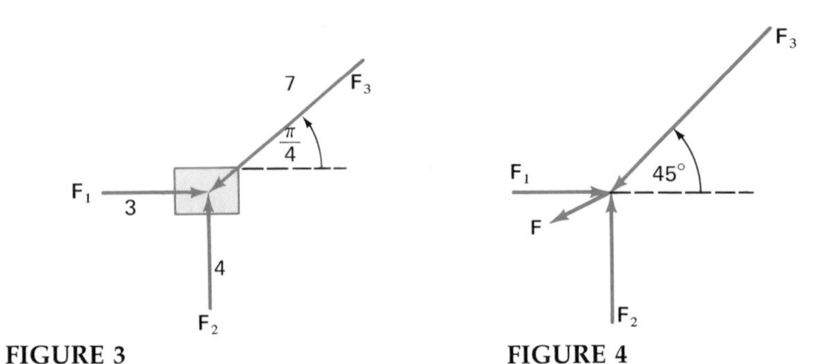

FIGURE 3

FIGURE 4

The work done by a force F in moving an object a distance d is given by

$$W = Fd \tag{1}$$

where units of work are newton-meters or foot-pounds. In formula (1) it is assumed that the force is applied in the same direction as the direction of motion. However, this is not always the case. In general, we may define

$$W = (\text{component of } \mathbf{F} \text{ in the direction of motion}) \times (\text{distance moved}). \tag{2}$$

If the object moves from P to Q, then the distance moved is $|\overrightarrow{PQ}|$. The vector $\mathbf{d}$, one of whose representations is $\overrightarrow{PQ}$, is called a **displacement vector.** Then from equation (6) on page 17,

$$\text{component of } \mathbf{F} \text{ in direction of motion} = \frac{\mathbf{F} \cdot \mathbf{d}}{|\mathbf{d}|}. \tag{3}$$

Finally, combining (2) and (3), we obtain

$$W = \frac{\mathbf{F} \cdot \mathbf{d}}{|\mathbf{d}|}|\mathbf{d}| = \mathbf{F} \cdot \mathbf{d}. \tag{4}$$

That is, *the work done is the dot product of the force* $\mathbf{F}$ *and the displacement vector* $\mathbf{d}$. Note that if $\mathbf{F}$ acts in the direction $\mathbf{d}$ and if φ denotes the angle (which is zero) between $\mathbf{F}$ and $\mathbf{d}$, then $\mathbf{F} \cdot \mathbf{d} = |\mathbf{F}|\,|\mathbf{d}| \cos \varphi = |\mathbf{F}|\,|\mathbf{d}| \cos 0 = |\mathbf{F}|\,|\mathbf{d}|$, which is formula (1).

EXAMPLE 4 A force of 4 N has the direction $\pi/3$. What is the work done in moving an object from the point (1, 2) to the point (5, 4), where distances are measured in meters?

Solution. A unit vector with direction $\pi/3$ is given by $\mathbf{u} = (\cos \pi/3)\mathbf{i} + (\sin \pi/3)\mathbf{j} = (1/2)\mathbf{i} + (\sqrt{3}/2)\mathbf{j}$. Thus $\mathbf{F} = 4\mathbf{u} = 2\mathbf{i} + 2\sqrt{3}\mathbf{j}$. The displacement vector $\mathbf{d}$ is given by $(5 - 1)\mathbf{i} + (4 - 2)\mathbf{j} = 4\mathbf{i} + 2\mathbf{j}$. Thus

$$W = \mathbf{F} \cdot \mathbf{d} = (2\mathbf{i} + 2\sqrt{3}\mathbf{j}) \cdot (4\mathbf{i} + 2\mathbf{j}) = (8 + 4\sqrt{3}) \approx 14.93 \text{ N} \cdot \text{m}.$$

The component of $\mathbf{F}$ in the direction of motion is sketched in Figure 5. ■

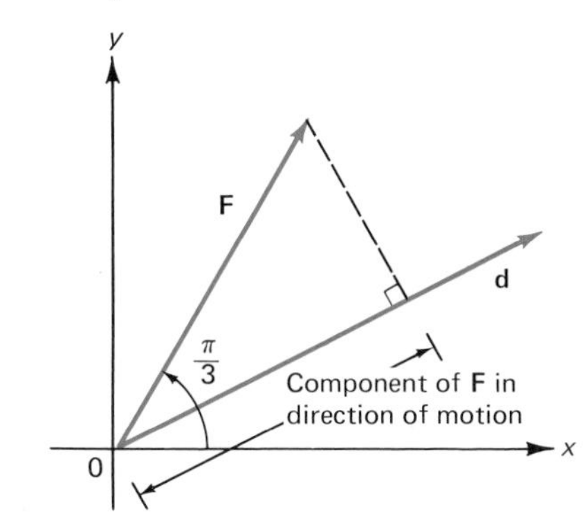

FIGURE 5

PROBLEMS 1.3

1. Show that the diagonals of a parallelogram bisect each other.
2. Show that the diagonals of a rhombus are orthogonal.
3. Show that in a trapezoid the line segment that joins the midpoints of the two sides that are not parallel is parallel to the parallel sides and has a length equal to the average of the lengths of the parallel sides.
4. Use vector methods to show that the angles opposite the equal sides in an isosceles triangle are equal.

*5. Prove that the medians of a triangle intersect at a fixed point that is two-thirds the distance from each vertex to the opposite side.

6. Use the result of Problem 5 to find the point of intersection of the medians of the triangle with vertices at (3, 4), (2, −1), and (−3, 2).

In Problems 7 – 15, find the resultant of the forces acting on an object. Then find the force vector that must be applied so that the object will remain at rest.

7. 2 N (from right), 5 N (from above)
8. 2 N (from left), 5 N (from below)
9. 3 N (from left), 5 N (from right), 3 N (from above)
10. 10 lb (from right), 8 lb (from below)
11. 5 lb (from above), 4 lb (from direction $\pi/6$)
12. 6 N (from left), 4 N (from direction $\pi/4$), 2 N (from direction $\pi/3$)
13. 2 N (from above), 3 N (from direction $3\pi/4$)
14. 5 N (from direction $\pi/3$), 5 N (from direction $2\pi/3$)
15. 7 N (from direction $\pi/6$), 7 N (from direction $\pi/3$), 14 N (from direction $5\pi/4$)

In Problems 16–24, find the work done when the force with given magnitude and direction moves an object from P to Q. All distances are measured in meters. (Note that work can be negative.)

16. $|\mathbf{F}| = 3$ N; $\theta = 0$; $P = (2, 3)$; $Q = (1, 7)$
17. $|\mathbf{F}| = 2$ N; $\theta = \pi/2$; $P = (5, 7)$; $Q = (1, 1)$
18. $|\mathbf{F}| = 6$ N; $\theta = \pi/4$; $P = (2, 3)$; $Q = (-1, 4)$
19. $|\mathbf{F}| = 4$ N; $\theta = \pi/6$; $P = (-1, 2)$; $Q = (3, 4)$
20. $|\mathbf{F}| = 7$ N; $\theta = 2\pi/3$; $P = (4, -3)$; $Q = (1, 0)$
21. $|\mathbf{F}| = 3$ N; $\theta = 3\pi/4$; $P = (2, 1)$; $Q = (1, 2)$
22. $|\mathbf{F}| = 6$ N; $\theta = \pi$; $P = (3, -8)$; $Q = (5, 10)$
23. $|\mathbf{F}| = 4$ N; θ is the direction of $2\mathbf{i} + 3\mathbf{j}$; $P = (2, 0)$; $Q = (-1, 3)$
24. $|\mathbf{F}| = 5$ N; θ is the direction of $-3\mathbf{i} + 2\mathbf{j}$; $P = (1, 3)$; $Q = (4, -6)$
25. Two tugboats are towing a barge (see Figure 6). Tugboat 1 pulls with a force of 500 N at an angle of 20° with the horizontal. Tugboat 2 pulls with a force of x newtons at an angle of 30°. The barge moves horizontally (i.e., $\theta = 0$). Find x.

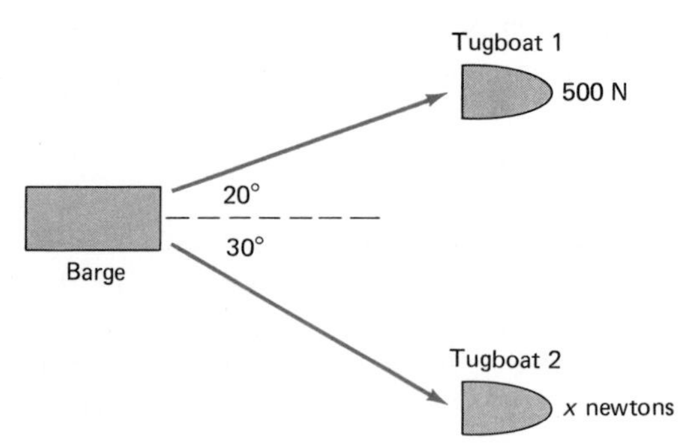

FIGURE 6

26. Answer the question of Problem 25 if the angles are 50° and 75°, respectively, and all other data remain the same.
27. In Problem 25 how much work is done by each tugboat in moving the barge a distance of 750 m?
28. In Problem 26 how much work is done by each tugboat in moving the barge a distance of 2 km?

REVIEW EXERCISES FOR CHAPTER ONE

In Exercises 1–6, find the magnitude and direction of the given vector.

1. $\mathbf{v} = (3, 3)$
2. $\mathbf{v} = -3\mathbf{i} + 3\mathbf{j}$
3. $\mathbf{v} = (2, -2\sqrt{3})$
4. $\mathbf{v} = (\sqrt{3}, 1)$
5. $\mathbf{v} = -12\mathbf{i} - 12\mathbf{j}$
6. $\mathbf{v} = \mathbf{i} + 4\mathbf{j}$

In Exercises 7–10, write the vector $\mathbf{v}$ that is represented by $\overrightarrow{PQ}$ in the form $a\mathbf{i} + b\mathbf{j}$.

7. $P = (2, 3)$; $Q = (4, 5)$
8. $P = (1, -2)$; $Q = (7, 12)$
9. $P = (-1, -6)$; $Q = (3, -4)$
10. $P = (-1, 3)$; $Q = (3, -1)$

11. Let $\mathbf{u} = (2, 1)$ and $\mathbf{v} = (-3, 4)$. Find the following:
 (a) $5\mathbf{u}$ **(b)** $\mathbf{u} - \mathbf{v}$
 (c) $-8\mathbf{u} + 5\mathbf{v}$
12. Let $\mathbf{u} = -4\mathbf{i} + \mathbf{j}$ and $\mathbf{v} = -3\mathbf{i} - 4\mathbf{j}$. Find the following:
 (a) $-3\mathbf{v}$ **(b)** $\mathbf{u} + \mathbf{v}$
 (c) $3\mathbf{u} - 6\mathbf{v}$

In Exercises 13–19, find a unit vector having the same direction as the given vector.

13. $\mathbf{v} = \mathbf{i} + \mathbf{j}$
14. $\mathbf{v} = -\mathbf{i} + \mathbf{j}$
15. $\mathbf{v} = 2\mathbf{i} + 5\mathbf{j}$
16. $\mathbf{v} = -7\mathbf{i} + 3\mathbf{j}$
17. $\mathbf{v} = 3\mathbf{i} + 4\mathbf{j}$
18. $\mathbf{v} = -2\mathbf{i} - 2\mathbf{j}$
19. $\mathbf{v} = a\mathbf{i} - a\mathbf{j}$

20. For $\mathbf{v} = 4\mathbf{i} - 7\mathbf{j}$, find $\sin\theta$ and $\cos\theta$, where θ is the direction of $\mathbf{v}$.
21. Find a unit vector with direction opposite to that of $\mathbf{v} = 5\mathbf{i} + 2\mathbf{j}$.
22. Find two unit vectors orthogonal to $\mathbf{v} = \mathbf{i} - \mathbf{j}$.
23. Find a unit vector with direction opposite to that of $\mathbf{v} = 10\mathbf{i} - 7\mathbf{j}$.

In Exercises 24–27, find a vector $\mathbf{v}$ having the given magnitude and direction.

24. $|\mathbf{v}| = 2$; $\theta = \pi/3$
25. $|\mathbf{v}| = 1$; $\theta = \pi/2$
26. $|\mathbf{v}| = 4$; $\theta = \pi$
27. $|\mathbf{v}| = 7$; $\theta = 5\pi/6$

In Exercises 28–31, calculate the dot product of the two vectors and the cosine of the angle between them.

28. $\mathbf{u} = \mathbf{i} - \mathbf{j}$; $\mathbf{v} = \mathbf{i} + 2\mathbf{j}$
29. $\mathbf{u} = -4\mathbf{i}$; $\mathbf{v} = 11\mathbf{j}$
30. $\mathbf{u} = 4\mathbf{i} - 7\mathbf{j}$; $\mathbf{v} = 5\mathbf{i} + 6\mathbf{j}$
31. $\mathbf{u} = -\mathbf{i} - 2\mathbf{j}$; $\mathbf{v} = 4\mathbf{i} + 5\mathbf{j}$

In Exercises 32–37, determine whether the given vectors are orthogonal, parallel, or neither. Then sketch each pair.

32. $\mathbf{u} = 2\mathbf{i} - 6\mathbf{j}$; $\mathbf{v} = -\mathbf{i} + 3\mathbf{j}$
33. $\mathbf{u} = 4\mathbf{i} - 5\mathbf{j}$; $\mathbf{v} = 5\mathbf{i} - 4\mathbf{j}$
34. $\mathbf{u} = 4\mathbf{i} - 5\mathbf{j}$; $\mathbf{v} = -5\mathbf{i} + 4\mathbf{j}$
35. $\mathbf{u} = -7\mathbf{i} - 7\mathbf{j}$; $\mathbf{v} = \mathbf{i} + \mathbf{j}$
36. $\mathbf{u} = -7\mathbf{i} - 7\mathbf{j}$; $\mathbf{v} = -\mathbf{i} + \mathbf{j}$
37. $\mathbf{u} = -7\mathbf{i} - 7\mathbf{j}$; $\mathbf{v} = -\mathbf{i} - \mathbf{j}$

38. Let $\mathbf{u} = 2\mathbf{i} + 3\mathbf{j}$ and $\mathbf{v} = 4\mathbf{i} + \alpha\mathbf{j}$.
(a) Determine α such that $\mathbf{u}$ and $\mathbf{v}$ are orthogonal.
(b) Determine α such that $\mathbf{u}$ and $\mathbf{v}$ are parallel.
(c) Determine α such that the angle between $\mathbf{u}$ and $\mathbf{v}$ is $\pi/4$.
(d) Determine α such that the angle between $\mathbf{u}$ and $\mathbf{v}$ is $\pi/6$.

In Exercises 39–44, calculate $\text{Proj}_{\mathbf{v}}\, \mathbf{u}$.

39. $\mathbf{u} = 14\mathbf{i}$; $\mathbf{v} = \mathbf{i} + \mathbf{j}$
40. $\mathbf{u} = 14\mathbf{i}$; $\mathbf{v} = \mathbf{i} - \mathbf{j}$
41. $\mathbf{u} = 3\mathbf{i} - 2\mathbf{j}$; $\mathbf{v} = 3\mathbf{i} + 2\mathbf{j}$
42. $\mathbf{u} = 3\mathbf{i} + 2\mathbf{j}$; $\mathbf{v} = \mathbf{i} - 3\mathbf{j}$
43. $\mathbf{u} = 2\mathbf{i} - 5\mathbf{j}$; $\mathbf{v} = -3\mathbf{i} - 7\mathbf{j}$
44. $\mathbf{u} = 4\mathbf{i} - 5\mathbf{j}$; $\mathbf{v} = -3\mathbf{i} - \mathbf{j}$

45. Let $P = (3, -2)$, $Q = (4, 7)$, $R = (-1, 3)$, and $S = (2, -1)$. Calculate $\text{Proj}_{\overrightarrow{PQ}}\, \overrightarrow{RS}$ and $\text{Proj}_{\overrightarrow{RS}}\, \overrightarrow{PQ}$.

In Exercises 46–48, calculate the resultant of the forces acting on an object.

46. 3 N (from left), 2 N (from below)
47. 5 N (from right), 2 N (from left), 3 N (from above)
48. 2 N (from direction $\pi/4$), 5 N (from left), 4 N (from direction $2\pi/3$), 6 N (from direction $3\pi/4$)

In Exercises 49–52, find the work done when the force with given magnitude and direction moves an object from P to Q. All distances are measured in meters.

49. $|\mathbf{F}| = 2$ N; $\theta = \pi/4$; $P = (1, 6)$; $Q = (2, 4)$
50. $|\mathbf{F}| = 3$ N; $\theta = \pi/2$; $P = (3, -5)$; $Q = (2, 7)$
51. $|\mathbf{F}| = 11$ N; $\theta = \pi/6$; $P = (-1, -2)$; $Q = (-7, -4)$
52. $|\mathbf{F}| = 8$ N; $\theta = 2\pi/3$; $P = (-1, 4)$; $Q = (5, -6)$

2 Vector Functions, Vector Differentiation, and Parametric Equations in $\mathbb{R}^2$

2.1 VECTOR FUNCTIONS AND PARAMETRIC EQUATIONS

In Chapter 1 we considered vectors (in the plane) that could be written as

$$\mathbf{v} = (a, b) = a\mathbf{i} + b\mathbf{j}. \tag{1}$$

In this chapter we see what happens when the numbers a and b in (1) are replaced by functions $f_1(t)$ and $f_2(t)$.

Definition 1 VECTOR FUNCTION Let f_1 and f_2 be functions of the real variable t. Then for all values of t for which $f_1(t)$ and $f_2(t)$ are defined, we define the **vector-valued function f** by

$$\mathbf{f}(t) = (f_1(t), f_2(t)) = f_1(t)\mathbf{i} + f_2(t)\mathbf{j}. \tag{2}$$

The **domain** of **f** is the intersection of the domains of f_1 and f_2.

REMARK. For simplicity we will refer to vector-valued functions as **vector functions.**

EXAMPLE 1 Let $\mathbf{f}(t) = f_1(t)\mathbf{i} + f_2(t)\mathbf{j} = (1/t)\mathbf{i} + \sqrt{t+1}\mathbf{j}$. Find the domain of **f**.

Solution. The domain of **f** is the set of all t for which f_1 and f_2 are defined. Since $f_1(t)$ is defined for $t \neq 0$ and $f_2(t)$ is defined for $t \geq -1$, we see that the domain of **f** is the set $\{t: t \geq -1 \text{ and } t \neq 0\}$. ■

Let **f** be a vector function. Then for each t in the domain of **f**, the endpoint of the vector $f_1(t)\mathbf{i} + f_2(t)\mathbf{j}$ is a point (x, y) in the xy-plane, where

$$x = f_1(t) \quad \text{and} \quad y = f_2(t). \tag{3}$$

Definition 2 PLANE CURVES AND PARAMETRIC EQUATIONS Suppose that the interval $[a, b]$ is in the domain of the function **f** and that both f_1 and f_2 are continuous in $[a, b]$. Then the set of points $(f_1(t), f_2(t))$ for $a \leq t \leq b$ is called a **plane curve** C. Equation (2) is called the **vector equation** of C, while equations (3) are called the **parametric equations** or **parametric representation** of C. In this context the variable t is called a **parameter.**

REMARK. We will usually refer to a plane curve simply as a **curve.**

EXAMPLE 2 Describe the curve given by the vector equation

$$\mathbf{f}(t) = (\cos t)\mathbf{i} + (\sin t)\mathbf{j}, \qquad 0 \leq t \leq 2\pi. \tag{4}$$

Solution. We first see that for every t, $|\mathbf{f}(t)| = 1$ since $|\mathbf{f}(t)| = \sqrt{\cos^2 t + \sin^2 t} = 1$. Moreover, if we write the curve in its parametric representation, we find that

$$x = \cos t, \qquad y = \sin t, \tag{5}$$

and since $\cos^2 t + \sin^2 t = 1$, we have

$$x^2 + y^2 = 1,$$

which is, of course, the equation of the unit circle. This curve is sketched in Figure 1. Note that in the sketch the parameter t represents both the length of the arc from $(1, 0)$ to the endpoint of the vector and the angle (measured in radians) the vector makes with the positive x-axis. The representation $x^2 + y^2 = 1$ is called the *Cartesian equation* of the curve given by (5).

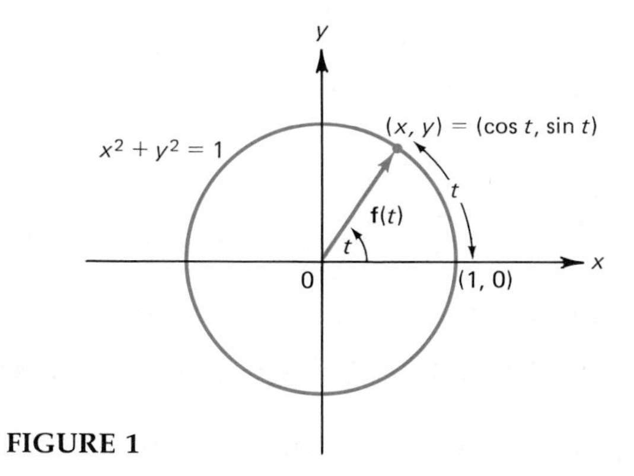

FIGURE 1

REMARK 1. The curve drawn in Figure 1 is *not* the graph of the function $\mathbf{f}(t) = (\cos t)\,\mathbf{i} + (\sin t)\mathbf{j}$. To draw this graph would require *three* dimensions, one each for the values taken by t, x, and y. Rather, the curve drawn is a graph of the (vector) values the function $\mathbf{f}$ takes; that is, the *range* of the function. We will discuss three dimensional graphs in Chapters 3 and 4.

REMARK 2. As t increases from 0 to 2π, we move around the unit circle in the counterclockwise direction. If t were instead restricted to the range $0 \leq t \leq \pi$, then we would not get the entire circle. Rather, we would stop at the point $(\cos \pi, \sin \pi) = (-1, 0)$, which would give us the upper semicircle only. ■

Definition 3 CARTESIAN EQUATION OF A PLANE CURVE A **Cartesian**† **equation** of the curve $f(t) = x(t)\mathbf{i} + y(t)\mathbf{j}$ is an equation relating the variables x and y only.

EXAMPLE 3 Describe and sketch the curve given parametrically by $x = t + 3$, $y = t^2 - t + 2$.

Solution. With problems of this type the easiest thing to do is to write t as a function of x or y, if possible. Since $x = t + 3$, we immediately see that $t = x - 3$ and $y = t^2 - t + 2 = (x - 3)^2 - (x - 3) + 2 = x^2 - 7x + 14$. This is the Cartesian equation of the curve and is the equation of a parabola. It is sketched in Figure 2. Note that in the Cartesian equation of the parabola, the parameter t does not appear. ■

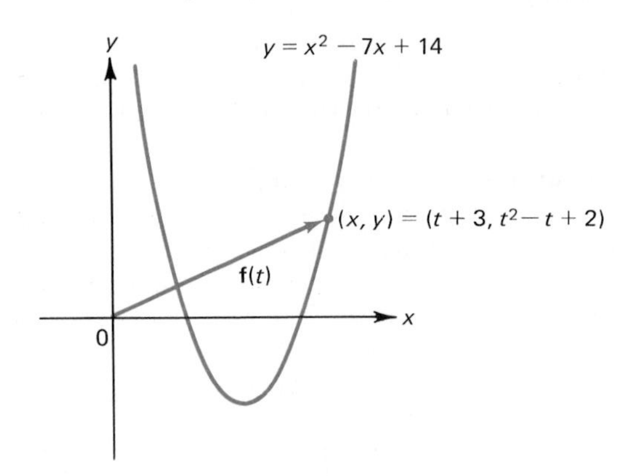

FIGURE 2

REMARK. As in Example 2, the curve in Figure 2 is not a graph of the function $\mathbf{f}(t) = x(t)\mathbf{i} + y(t)\mathbf{j}$. It is a graph of the range of the function.

EXAMPLE 4 Describe and sketch the curve given by the vector equation

$$\mathbf{f}(r) = (1 - r^4)\mathbf{i} + r^2\mathbf{j}. \tag{6}$$

Solution. First, we note that the parameter in this problem is r instead of t. This makes absolutely no difference since, as in the case of the variable of integration, the parameter is a "dummy" variable. Now to get a feeling for the shape of this curve, we display in Table 1, values of x and y for various values of r. Plotting some of these

†Named after the French philosopher and mathematician René Descartes (1596–1650). See the accompanying biographical sketch.

1596–
1650

René Descartes

René Descartes
The Granger Collection

The Cartesian plane is named after the great French mathematician and philosopher René Descartes. Born near the city of Tours in 1596, Descartes received his education first at the Jesuit school at La Flèche and later at Poitier, where he studied law. He had delicate health and, while still in school, developed the habit of spending the greater part of each morning in bed. Later, he considered these morning hours the most productive period of the day.

At the age of 16, Descartes left school and moved to Paris, where he began his study of mathematics. One year later, in 1617, he joined the army of Maurice, Prince of Nassau. He also served with Duke Maximillian I of Bavaria and with the French army at the siege of La Rochelle.

Descartes was not a professional soldier, however, and his periods of military service were broken by periods of travel and study in various European cities. After leaving the army for good, he resettled in Paris to continue his mathematical studies and then moved to Holland where he lived for 20 years.

Much stimulated by the scientists and philosophers he met in France, Holland, and elsewhere, Descartes later became known as the "father of modern philosophy." His statement "Cogito ergo sum" ("I think, therefore I am") played a central role in his philosophical writings.

Descartes's program for philosophical research was enunciated in his famous *Discours de la méthode pour bien conduire sa raison et chercher la vérité dans les sciences* (A Discourse on the Method of Rightly Conducting the Reason and Seeking Truth in the Sciences) published in 1637. This work was accompanied by three appendixes: *La dioptrique* (in which the law of refraction—discovered by Snell—was first published), *Les météores* (which contained the first accurate explanation of the rainbow), and *La géométrie. La géométrie,* the third and most famous appendix, took up about a hundred pages of the *Discours.* One of the major achievements of *La géométrie* was that it connected figures of geometry with the equations of algebra. The work established Descartes as the founder of analytic geometry.

In 1649 Descartes was invited to Sweden by Queen Christina. He agreed, reluctantly, but was unable to survive the harsh, Scandinavian winter. He died in Stockholm in early 1650.

points leads to the sketch in Figure 3. To write the Cartesian equation for this curve, we square both sides of the equation $y = r^2$ to obtain $y^2 = r^4$ and $x = 1 - r^4 = 1 - y^2$, which is the equation of the parabola sketched in Figure 4. Note that this curve is *not* the same as the curve sketched in Figure 3 since the parametric representation $y = r^2$ requires that y be nonnegative. Thus the curve described by (6) is only the *upper half* of the parabola described by the equation $x = 1 - y^2$.

TABLE 1

r	0	$\pm\frac{1}{2}$	± 1	$\pm\frac{3}{2}$	± 2
$x = 1 - r^4$	1	$\frac{15}{16}$	0	$-\frac{65}{16}$	-15
$y = r^2$	0	$\frac{1}{4}$	1	$\frac{9}{4}$	4

■

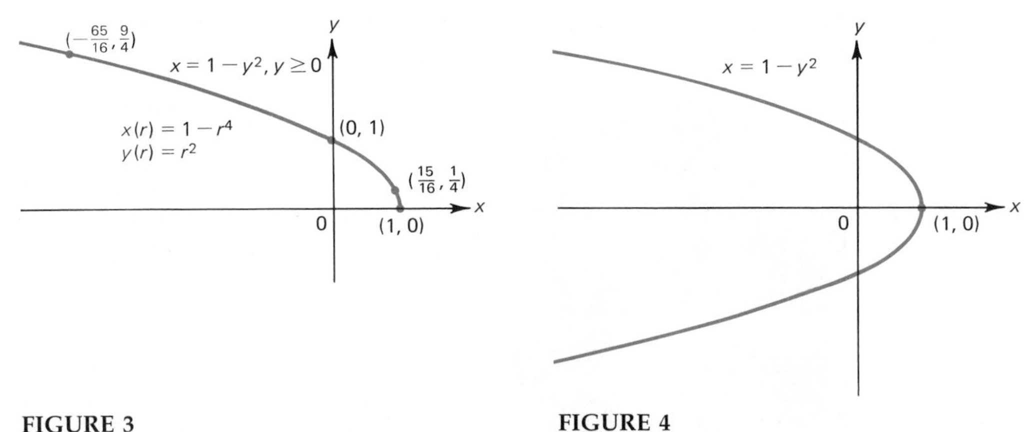

FIGURE 3 **FIGURE 4**

We now obtain a parametric representation of a line L. Let $P = (x_1, y_1)$ and $Q = (x_2, y_2)$ be two points on the line and let R be a third point on the line (see Figure 5). Since the vectors $\overrightarrow{PQ}$ and $\overrightarrow{PR}$ have the same or opposite directions, they are parallel and by Theorem 1.2.3, we have

$$\overrightarrow{PR} = t\overrightarrow{PQ} \tag{7}$$

for some real number t (t is positive if $\overrightarrow{PQ}$ and $\overrightarrow{PR}$ have the same direction and t is negative if $\overrightarrow{PQ}$ and $\overrightarrow{PR}$ have opposite directions). Then from Figure 5 we see that

$$\overrightarrow{0R} = \overrightarrow{0P} + \overrightarrow{PR} = \overrightarrow{0P} + t\overrightarrow{PQ}. \tag{8}$$

But $\overrightarrow{PQ} = (x_2 - x_1)\mathbf{i} + (y_2 - y_1)\mathbf{j}$ and $\overrightarrow{0P} = x_1\mathbf{i} + y_1\mathbf{j}$, so from (8), if $R = (x, y)$, we obtain

$$x\mathbf{i} + y\mathbf{j} = x_1\mathbf{i} + y_1\mathbf{j} + t(x_2 - x_1)\mathbf{i} + t(y_2 - y_1)\mathbf{j}$$

or

$$x\mathbf{i} + y\mathbf{j} = [x_1 + t(x_2 - x_1)]\mathbf{i} + [y_1 + t(y_2 - y_1)]\mathbf{j}.$$

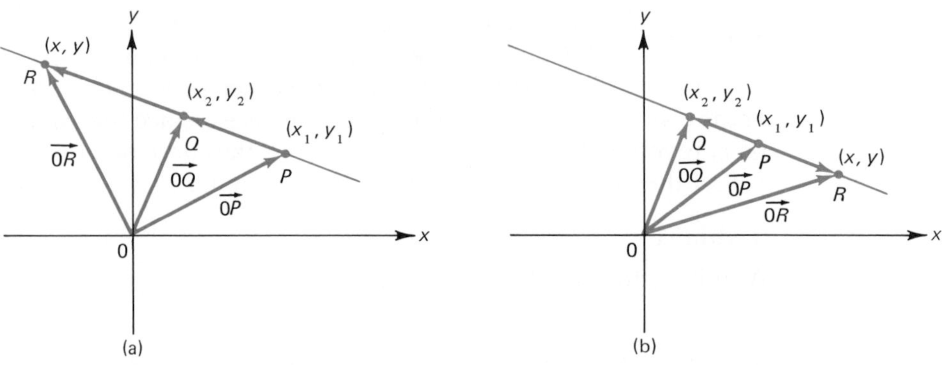

FIGURE 5

Thus a parametric equation of the line passing through the points (x_1, y_1) and (x_2, y_2) is

$$x = x_1 + t(x_2 - x_1) \quad \text{and} \quad y = y_1 + t(y_2 - y_1). \tag{9}$$

EXAMPLE 5 Find a parametric representation of the line passing through the points $P = (2, 5)$ and $Q = (-3, 9)$.

Solution. Using (9), we immediately find that

$$x = 2 - 5t \quad \text{and} \quad y = 5 + 4t. \tag{10}$$

Note that the problem asked for *a* parametric representation of the line, since there are several such representations. For example, reversing the roles of P and Q, we obtain the representation [from the new $P = (-3, 9)$ and $Q = (2, 5)$]

$$x = -3 + 5s \quad \text{and} \quad y = 9 - 4s. \tag{11}$$

Although these representations look different, they do represent the same line, as is seen by inserting a few sample values for t and s. For example, if $t = 1$, from (10) we obtain the point $(-3, 9)$. This point is obtained from (11) by setting $s = 0$.

To obtain another parametric representation, we first observe that the slope-intercept form of the line passing through the points (2, 5) and $(-3, 9)$ is given by

$$\frac{y - 5}{x - 2} = \frac{9 - 5}{-3 - 2} = -\frac{4}{5}, \quad \text{or} \quad y = -\frac{4}{5}x + \frac{33}{5}.$$

Then setting $t = x$, we obtain the "slope-intercept" parametrization

$$x = t, \quad y = -\tfrac{4}{5}t + \tfrac{33}{5}. \blacksquare$$

The following example shows how vector analysis can be useful in a simple physical problem.

EXAMPLE 6 A cannonball shot from a cannon has an initial velocity of 600 m/sec. The muzzle of the cannon is inclined at an angle of 30°. Ignoring air resistance, determine the path of the cannonball.

Solution. We place the x- and y-axes so that the mouth of the cannon is at the origin (see Figure 6). The velocity vector $\mathbf{v}$ can be resolved into its vertical and horizontal components:

$$\mathbf{v} = v_x\mathbf{i} + v_y\mathbf{j}.$$

A unit vector in the direction of $\mathbf{v}$ is

$$\mathbf{u} = (\cos 30°)\mathbf{i} + (\sin 30°)\mathbf{j} = \frac{\sqrt{3}}{2}\mathbf{i} + \frac{1}{2}\mathbf{j}$$

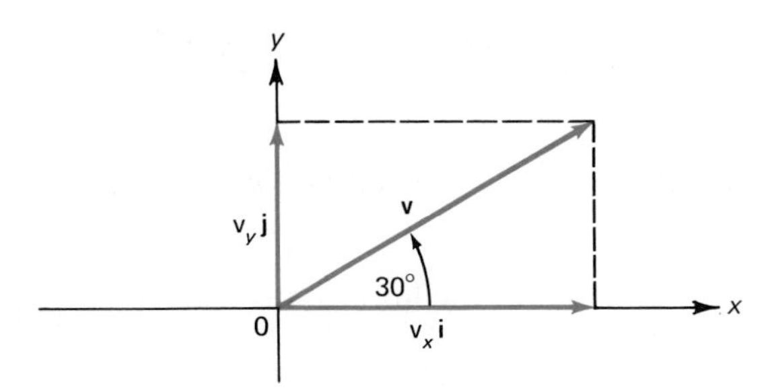

FIGURE 6

So initially,

$$\mathbf{v} = |\mathbf{v}|\mathbf{u} = 600\mathbf{u} = 300\sqrt{3}\mathbf{i} + 300\mathbf{j}.$$

The scalar $|\mathbf{v}|$ is called the **speed** of the cannonball. Thus initially, $v_x = 300\sqrt{3}$ m/sec (the initial speed in the horizontal direction) and $v_y = 300$ m/sec (the initial speed in the vertical direction). Now the vertical acceleration (due to gravity) is

$$a_y = -9.81 \text{ m/sec}^2 \quad \text{and} \quad v_y = \int a_y\, dt = -9.81t + C.$$

Since, initially, $v_y(0) = 300$, we have $C = 300$ and

$$v_y = -9.81t + 300.$$

Then

$$y(t) = \int v_y\, dt = -\frac{9.81t^2}{2} + 300t + C_1$$

and since $y(0) = 0$ (we start at the origin), we find that

$$y(t) = -\frac{9.81t^2}{2} + 300t.$$

To calculate the x-component of the position vector, we note that, ignoring air resistance, the velocity in the horizontal direction is constant;† that is,

$$v_x = 300\sqrt{3} \text{ m/sec}.$$

Then $x(t) = \int v_x\, dt = 300\sqrt{3}t + C_2$, and since $x(0) = 0$, we obtain

$$x = 300\sqrt{3}t.$$

Thus the position vector describing the location of the cannonball is

†There are no forces acting on the ball in the horizontal direction since the force of gravity acts only in the vertical direction.

$$\mathbf{s}(t) = x(t)\mathbf{i} + y(t)\mathbf{j} = 300\sqrt{3}t\mathbf{i} + \left(300t - \frac{9.81t^2}{2}\right)\mathbf{j}.$$

To obtain the Cartesian equation of this curve, we start with

$$x = 300\sqrt{3}t$$

so that

$$t = \frac{x}{300\sqrt{3}}$$

and

$$y = 300\left(\frac{x}{300\sqrt{3}}\right) - \frac{9.81}{2}\frac{x^2}{(300\sqrt{3})^2} = \frac{x}{\sqrt{3}} - \frac{9.81}{540{,}000}x^2.$$

This parabola is sketched in Figure 7. ■

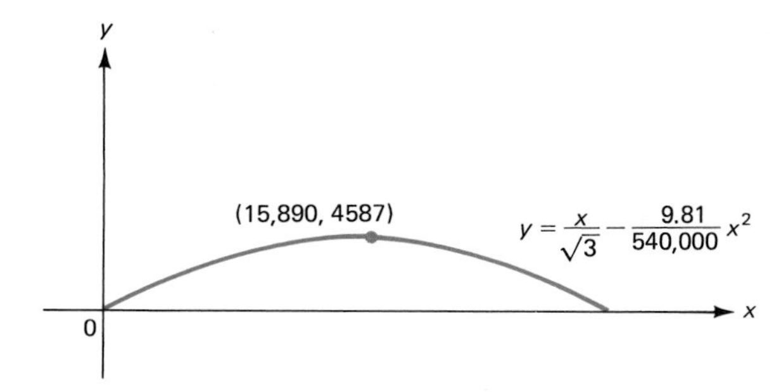

FIGURE 7

We close this section by describing, parametrically, a curve that historically has been of great interest. Suppose that a wheel of radius r is rolling in a straight line without slipping. Let P be a fixed point on the wheel a distance s from the center. As the wheel moves, the curve traced by the point P is called a **trochoid**† and if $s = r$ (i.e., if P is on the circumference of the wheel), then the curve traced is called a **cycloid.**‡

To simplify our computations, we place the wheel so that it rolls in a clockwise fashion on the x-axis (see Figure 8a). Suppose that P moves through an angle of α radians to reach its position in Figure 8b. Then the new position vector $\overrightarrow{0P}$ is given by

$$\overrightarrow{0P} = \overrightarrow{0R} + \overrightarrow{RC} + \overrightarrow{CP}. \tag{12}$$

†From the Greek word "trochos," meaning "wheel."

‡From the Greek word "kyklos," meaning "circle." The cycloid was a source of great controversy in the seventeenth century. Many of its properties were discovered by the French mathematician Gilles Personne de Roberval (1602–1675), although the curve was first discussed by Galileo. Unfortunately, Roberval, for unknown reasons, did not publish his discoveries concerning the cycloid, which meant that he lost credit for most of them. The ensuing arguments over who discovered what were so bitter that the cycloid became known as the "Helen of geometers" (after Helen of Troy—the source of the intense jealousy that led to the Trojan war).

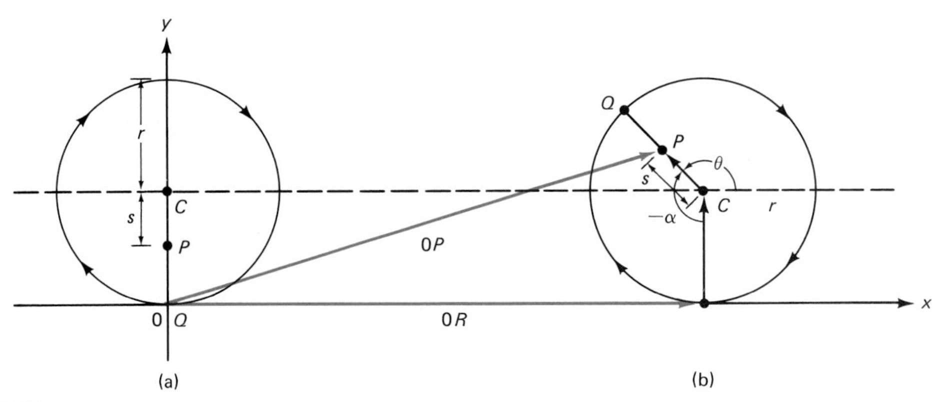

FIGURE 8

The length of $\overrightarrow{0R}$ is the length of the arc of the circle that is generated by moving through an angle of α radians and is equal to α times the radius of the circle $= \alpha r$. Thus

$$\overrightarrow{0R} = \alpha r\mathbf{i}.$$

Clearly, $\overrightarrow{RC} = r\mathbf{j}$, and if θ denotes the angle the vector $\overrightarrow{CP}$ makes with the positive x-axis, then

$$\overrightarrow{CP} = s[(\cos\theta)\mathbf{i} + (\sin\theta)\mathbf{j}].$$

Hence using (12), we have

$$\begin{aligned}\overrightarrow{0P} &= \alpha r\mathbf{i} + r\mathbf{j} + s(\cos\theta)\mathbf{i} + s(\sin\theta)\mathbf{j}\\ &= [\alpha r + s(\cos\theta)]\mathbf{i} + [r + s(\sin\theta)]\mathbf{j}.\end{aligned} \tag{13}$$

But since $\alpha + \theta = 3\pi/2$, we have $\theta = 3\pi/2 - \alpha$, so

$$\cos\theta = \cos\left(\frac{3\pi}{2} - \alpha\right) = -\sin\alpha, \qquad \sin\theta = \sin\left(\frac{3\pi}{2} - \alpha\right) = -\cos\alpha,$$

and the equation of the trochoid becomes

$$x\mathbf{i} + y\mathbf{j} = \overrightarrow{0P} = (r\alpha - s\sin\alpha)\mathbf{i} + (r - s\cos\alpha)\mathbf{j} \tag{14}$$

or

$$x = r\alpha - s\sin\alpha, \qquad y = r - s\cos\alpha. \tag{15}$$

For example, if $r = 3$ and $s = 2$, (15) becomes

$$x = 3\alpha - 2\sin\alpha, \qquad y = 3 - 2\cos\alpha.$$

A computer-drawn sketch of this curve is given in Figure 9.

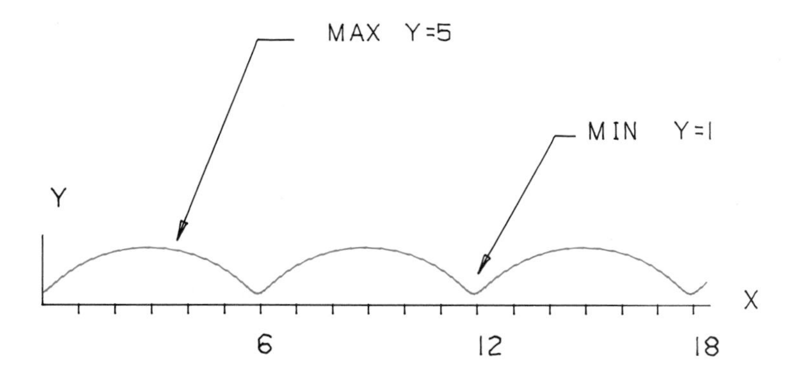

FIGURE 9

When $r = s$, P is on the circumference of the wheel, and we obtain the cycloid given parametrically by

$$x = r(\alpha - \sin \alpha), \qquad y = r(1 - \cos \alpha). \tag{16}$$

Note that $y \geq 0$ in (16) and $y = 0$ when $\cos \alpha = 1$, which occurs when α is a multiple of 2π. For these values of α, $x = r\alpha = 2\pi nr$. Moreover, y takes on its maximum value $2r$ when $\alpha = \pi, 3\pi, 5\pi, \ldots$. With this information, we can sketch the cycloid, as in Figure 10.

REMARK. Here is one instance in which it is much easier to represent a curve parametrically. If you doubt it, try to write the Cartesian equation of the trochoid [equation (14)] or the simpler cycloid (see Problem 49). You will quickly learn to appreciate the advantages of parametric representation. In fact, while it is possible to find the Cartesian equation of the trochoid, the parametric representation is much easier to use. ■

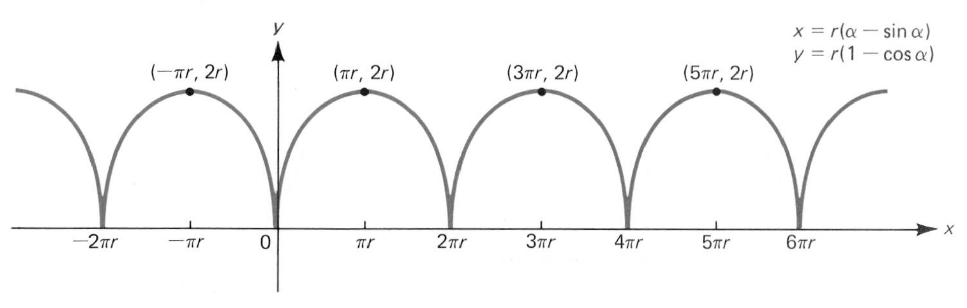

FIGURE 10

PROBLEMS 2.1

In Problems 1–8, find the domain of each vector-valued function.

1. $\mathbf{f}(t) = \dfrac{1}{t}\mathbf{i} + \dfrac{1}{t-1}\mathbf{j}$
2. $\mathbf{f}(t) = \sqrt{t}\,\mathbf{i} + \dfrac{1}{t}\mathbf{j}$
3. $\mathbf{f}(s) = \dfrac{1}{s^2-1}\mathbf{i} + (s^2-1)\mathbf{j}$
4. $\mathbf{f}(u) = e^{1/u}\mathbf{i} + e^{-1/(u+1)}\mathbf{j}$
5. $\mathbf{f}(s) = (\ln s)\mathbf{i} + \ln(1-s)\mathbf{j}$
6. $\mathbf{f}(r) = (\sin r)\mathbf{i} + r\mathbf{j}$
7. $\mathbf{f}(t) = (\sec t)\mathbf{i} + (\csc t)\mathbf{j}$
8. $\mathbf{f}(w) = (\tan w)\mathbf{i} + (\cot w)\mathbf{j}$

In Problems 9–24, find the Cartesian equation of each curve, and then sketch the curve in the xy-plane.

9. $\mathbf{f}(t) = t^2\mathbf{i} + 2t\mathbf{j}$
10. $\mathbf{f}(t) = (2t-3)\mathbf{i} + t^2\mathbf{j}$
11. $\mathbf{f}(t) = t^2\mathbf{i} + t^3\mathbf{j}$
12. $\mathbf{f}(t) = 3(\sin t)\mathbf{i} + 3(\cos t)\mathbf{j}$
13. $\mathbf{f}(t) = (2t-1)\mathbf{i} + (4t+3)\mathbf{j}$
14. $\mathbf{f}(t) = 2(\cosh t)\mathbf{i} + 2(\sinh t)\mathbf{j}$
15. $\mathbf{f}(t) = (t^4+t^2+1)\mathbf{i} + t^2\mathbf{j}$
16. $\mathbf{f}(t) = t^2\mathbf{i} + t^8\mathbf{j}$
17. $\mathbf{f}(t) = t^3\mathbf{i} + (t^9-1)\mathbf{j}$
18. $\mathbf{f}(t) = t\mathbf{i} + e^t\mathbf{j}$
19. $\mathbf{f}(t) = e^t\mathbf{i} + t^2\mathbf{j}$
20. $\mathbf{f}(t) = (t^2+t-3)\mathbf{i} + \sqrt{t}\,\mathbf{j}$

*21. $\mathbf{f}(t) = e^t(\sin t)\mathbf{i} + e^t(\cos t)\mathbf{j}$ [*Hint:* Show that $|\mathbf{f}(t)| = e^t$.]

22. $\mathbf{f}(t) = \mathbf{i} + (\tan t)\mathbf{j}$
23. $\mathbf{f}(t) = e^t\mathbf{i} + e^{2t}\mathbf{j}$

*24. $\mathbf{f}(t) = \dfrac{6t}{1+t^3}\mathbf{i} + \dfrac{6t^2}{1+t^3}\mathbf{j}$

In Problems 25–30, find two parametric representations of the straight line that passes through the given points.

25. (2, 4); (1, 6)
26. (−3, 2); (0, 4)
27. (3, 5); (−1, −7)
28. (4, 6); (7, 9)
29. (−2, 3); (4, 7)
30. (−4, 0); (3, −2)
31. The equation $(x^2/a^2) + (y^2/b^2) = 1$ is the equation of an ellipse. Show that the curve given by the vector equation $\mathbf{f}(t) = a(\cos t)\mathbf{i} + b(\sin t)\mathbf{j}$ is an ellipse.
32. A cannonball is shot upward from ground level at an angle of 45° with an initial speed of 1300 ft/sec. Find a parametric representation of the path of the cannonball. Then find the Cartesian equation of this path.
33. How many feet (horizontally) does the cannonball in Problem 32 travel before it hits the ground?
34. How many meters (horizontally) does the cannonball in Example 6 travel before it hits the ground?
35. An object is thrown down from the top of a 150-m building at an angle of 30° (below the horizontal) with an initial velocity of 100 m/sec. Determine a parametric representation of the path of the object. [*Hint:* Draw a picture.]
36. When the object in Problem 35 hits the ground, how far is it from the base of the building?
37. A point is located 25 cm from the center of a wheel 1 m in diameter. Find the parametric representation of the curve traced by that point as the wheel rolls.
38. Answer the question in Problem 37 if the point is located on the circumference of the wheel.

*39. A **hypocycloid**† is a curve generated by the motion of a point P on the circumference of a circle that rolls internally, without slipping, on a larger circle (see Figure 11). Assume that the radius of the large circle is a, while that of the smaller circle is b. If θ is the angle in Figure 11, show that a parametric representation of the hypocycloid is

$$x = (a-b)\cos\theta + b\cos\left(\frac{a-b}{b}\right)\theta \quad \text{and}$$

$$y = (a-b)\sin\theta - b\sin\left(\frac{a-b}{b}\right)\theta.$$

[*Hint:* First show that $a\theta = b\alpha$.]

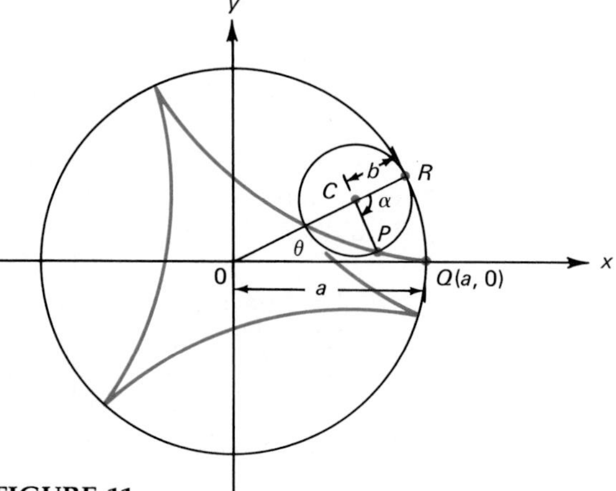

FIGURE 11

†The "hypo" comes from the Greek "hupo," meaning "under."

40. If $a = 4b$ in Problem 39, then the curve generated is called a **hypocycloid of four cusps.** Sketch this curve.

***41.** Show that the Cartesian equation of the hypocycloid of four cusps is

$$x^{2/3} + y^{2/3} = a^{2/3}.$$

[*Hint:* Use the identities $\cos 3\theta = 4\cos^3\theta - 3\cos\theta$ and $\sin 3\theta = 3\sin\theta - 4\sin^3\theta$ and the fact that $a - b = 3b$ and $(a - b)/b = 3$.]

42. If $a = nb$ in Problem 39, then the curve generated is called a **hypocycloid of *n* cusps.** Sketch the hypocycloid of seven cusps.

43. Show that the hypocycloid of two cusps is a straight-line segment.

44. Calculate the area bounded by the hypocycloid of four cusps. [*Hint:* Use the result of Problem 41.]

***45.** An **epicycloid**† is a curve generated by the motion of a point on the circumference of a circle that rolls externally, without slipping, on a fixed circle. Show that the parametric representation of an epicycloid is given by

$$x = (a + b)\cos\theta - b\cos\left(\frac{a+b}{b}\right)\theta \quad \text{and}$$

$$y = (a + b)\sin\theta - b\sin\left(\frac{a+b}{b}\right)\theta,$$

where a and b are as in Problem 39.

46. If $a = 4b$, show that the epicycloid generated has the sketch given in Figure 12.

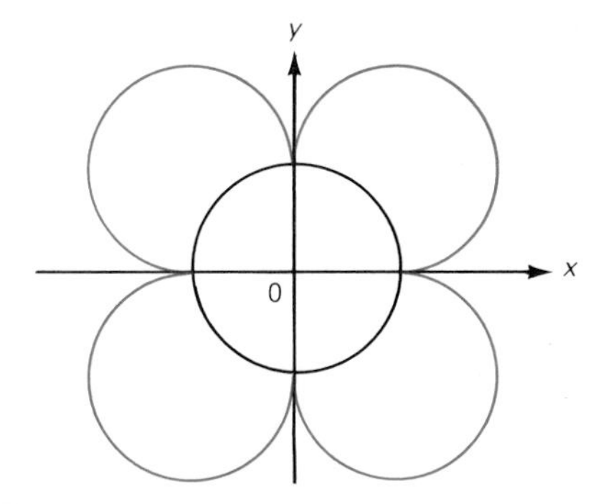

FIGURE 12

47. Sketch the epicycloid in the case $a = 5b$.

48. Sketch the epicycloid in the case $a = 8b$.

***49.** Show that the Cartesian equation of the cycloid is

$$x = r\cos^{-1}\left(\frac{r-y}{y}\right) - \sqrt{2ry - y^2}.$$

50. Show that a parametric representation of the hyperbola $(x^2/a^2) - (y^2/b^2) = 1$ is given by

$$x = a\sec\theta, \qquad y = b\tan\theta.$$

51. Use the substitution $\tan\theta = \sin t$ to show that the lemniscate $r^2 = \cos 2\theta$ has the Cartesian parametric representation $x = \cos t/(1 + \sin^2 t)$, $y = \sin t\cos t/(1 + \sin^2 t)$.

2.2 THE EQUATION OF THE TANGENT LINE TO A PARAMETRIC CURVE

Suppose that $x = f_1(t)$ and $y = f_2(t)$ is the parametric representation of a curve C. We would like to be able to calculate the equation of the line tangent to the curve without having to determine the Cartesian equation of the curve. However, there are complications that might occur since the curve could intersect itself. This happens if there are two numbers $t_1 \neq t_2$ such that $f_1(t_1) = f_1(t_2)$ and $f_2(t_1) = f_2(t_2)$. Thus there are three possibilities. At a given point the curve could have the following:

(i) a unique tangent
(ii) no tangent
(iii) two or more tangents

This is illustrated in Figure 1. A condition that ensures that there is at least one tangent

†From the Greek "epi," meaning "upon" or "over."

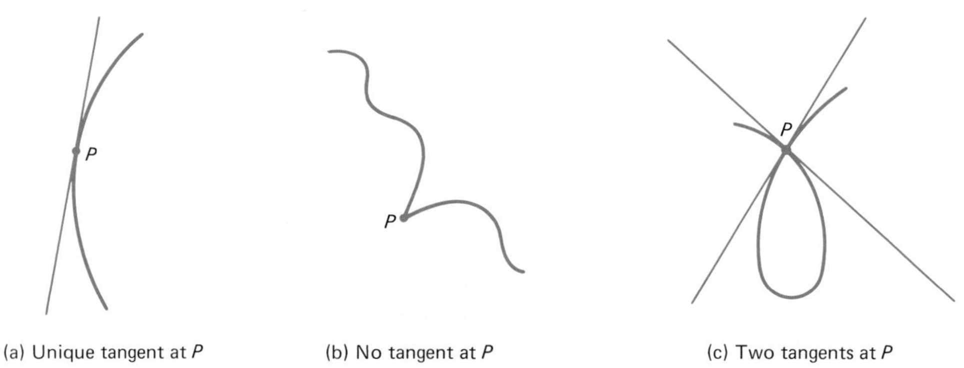

FIGURE 1

line at each point is that

$$f_1' \text{ and } f_2' \text{ exist and } [f_1'(t)]^2 + [f_2'(t)]^2 \neq 0. \tag{1}$$

REMARK. Condition (1) simply states that the derivatives of f_1 and f_2 are not zero at the same value of t.

Theorem 1 Let (x_0, y_0) be on the curve C given by $x = f_1(t)$ and $y = f_2(t)$. If the curve passes through (x_0, y_0) when $t = t_0$,† then the slope m of the line tangent to C at (x_0, y_0) is given by

$$m = \lim_{t \to t_0} \frac{f_2'(t)}{f_1'(t)} \tag{2}$$

provided that this limit exists.

Proof. We refer to Figure 2. From the figure we see that

$$\frac{dy}{dx} = \lim_{\Delta t \to 0} \frac{f_2(t_0 + \Delta t) - f_2(t_0)}{f_1(t_0 + \Delta t) - f_1(t_0)} \overset{\text{L'Hôpital's rule}}{=} \lim_{\Delta t \to 0} \frac{f_2'(t_0 + \Delta t)}{f_1'(t_0 + \Delta t)} \overset{\text{let } t = t_0 + \Delta t}{=} \lim_{t \to t_0} \frac{f_2'(t)}{f_1'(t)} \quad \blacksquare$$

Corollary. If $f_2'(t_0)$ exists and $f_1'(t_0) \neq 0$, then

$$m = \frac{f_2'(t_0)}{f_1'(t_0)} \tag{3}$$

Proof.

$$\frac{dy}{dx} = \lim_{\Delta t \to 0} \frac{f_2(t_0 + \Delta t) - f_2(t_0)}{f_1(t_0 + \Delta t) - f_1(t_0)} = \lim_{\Delta t \to 0} \frac{[f_2(t_0 + \Delta t) - f_2(t_0)]/\Delta t}{[f_1(t_0 + \Delta t) - f_1(t_0)]/\Delta t} \overset{\text{limit theorem for quotients}}{=} \frac{f_2'(t_0)}{f_1'(t_0)}. \quad \blacksquare$$

†The curve may pass through the point (x_0, y_0) for other values of t as well, and therefore it may have other tangent lines at that point, as in Figure 1c.

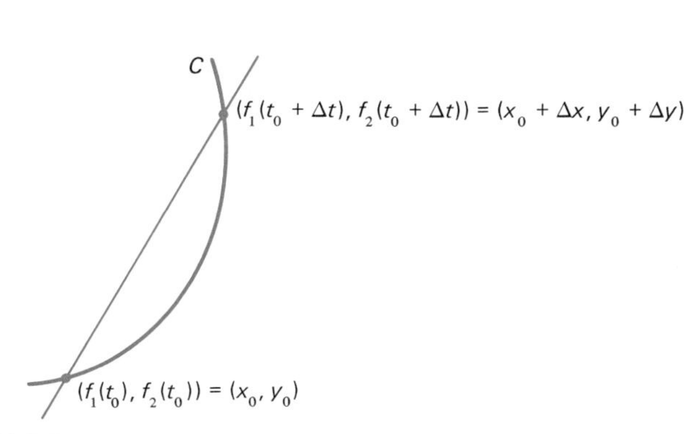

FIGURE 2

Using the corollary, we find that the tangent line to C at t_0 is given by

$$\frac{y - y_0}{x - x_0} = \frac{f_2'(t_0)}{f_1'(t_0)},$$

or after simplification,

$$f_2'(t_0)(x - x_0) - f_1'(t_0)(y - y_0) = 0. \tag{4}$$

Note that if $f_1'(t_0) = 0$, then equation (4) is still valid, and we obtain

$$f_2'(t_0)(x - x_0) = 0. \tag{5}$$

Since by assumption $[f_1'(t_0)]^2 + [f_2'(t_0)]^2 \neq 0$, the assumption $f_1'(t_0) = 0$ implies that $f_2'(t_0) \neq 0$. Thus we can divide by $f_2'(t_0)$ in (5) to obtain the equation

$$x - x_0 = 0, \qquad \text{or} \qquad x = x_0.$$

In this case the tangent line is vertical. If $f_2'(t_0) = 0$, then C has a horizontal tangent at t_0.

In sum, if condition (1) holds, then

(i) C has a vertical tangent line at t_0 if

$$\left.\frac{dx}{dt}\right|_{t=t_0} = f_1'(t_0) = 0 \qquad \text{and} \qquad f_2'(t_0) \neq 0.$$

(ii) C has a horizontal tangent line at t_0 if

$$\left.\frac{dy}{dt}\right|_{t=t_0} = f_2'(t_0) = 0 \qquad \text{and} \qquad f_1'(t_0) \neq 0. \tag{6}$$

EXAMPLE 1 Find the equation of a line (or lines) tangent to the curve $x = e^t$, $y = e^{-t}$ at the point (1, 1).

Solution. The point (1, 1) is reached only when $t = 0$. Then

$$\frac{dy}{dx} = \frac{d(e^{-t})/dt}{d(e^t)/dt} = -\frac{e^{-t}}{e^t}\bigg|_{t=0} = -1,$$

and the equation of the tangent line is

$$\frac{y-1}{x-1} = -1, \qquad \text{or} \qquad y = -x + 2. \ \blacksquare$$

EXAMPLE 2 Find the equation of the line tangent to the curve

$$x = 2t^3 - 15t^2 + 24t + 7, \qquad y = t^2 + t + 1 \qquad \text{at} \quad t = 2.$$

Solution. Here $f_1'(2) = (6t^2 - 30t + 24)|_{t=2} = -12$ and $f_2'(2) = (2t + 1)|_{t=2} = 5$. When $t = 2$, $x = 11$ and $y = 7$. Then using (4), we obtain

$$5(x - 11) + 12(y - 7) = 0, \qquad \text{or} \qquad 5x + 12y = 139. \ \blacksquare$$

EXAMPLE 3 Find all horizontal and vertical tangents to the curve of Example 2.

Solution. There are vertical tangents when

$$f_1'(t) = 6t^2 - 30t + 24 = 0 = 6(t^2 - 5t + 4) = 6(t - 4)(t - 1),$$

or when $t = 1$ and $t = 4$, since $f_2'(t)$ is nonzero at these points. When $t = 1$, $x = 18$; and when $t = 4$, $x = -9$. Thus the vertical tangents are the lines $x = 18$ and $x = -9$. There is a horizontal tangent when $2t + 1 = 0$, or $t = -\frac{1}{2}$. When $t = -\frac{1}{2}$, $y = \frac{3}{4}$; and the line $y = \frac{3}{4}$ is a horizontal tangent. $\blacksquare$

EXAMPLE 4 Find all points at which the curve $x = 2 + 7\cos\theta$, $y = 8 + 3\sin\theta$ has a vertical or horizontal tangent.

Solution. We find that $f_1'(\theta) = x'(\theta) = -7\sin\theta$, which is 0 when $\theta = n\pi$ for some integer n. If $\theta = n\pi$, then

$$x = \begin{cases} 9, & n \text{ even} \\ -5, & n \text{ odd} \end{cases} \qquad \text{and} \qquad y = 8,$$

so there are vertical tangents at (9, 8) and (−5, 8). Similarly, $f_2'(\theta) = y'(\theta) = 3\cos\theta$, which is 0 when $\theta = (n + \frac{1}{2})\pi$ for some integer n. If $\theta = (n + \frac{1}{2})\pi$, then

$$x = 2 \qquad \text{and} \qquad y = \begin{cases} 11, & n \text{ even} \\ 5, & n \text{ odd,} \end{cases}$$

so the curve has horizontal tangents at (2, 11) and (2, 5). You should verify that when $f_1'(\theta) = 0$, $f_2'(\theta) \neq 0$, and vice versa.

REMARK. It is not difficult to verify that x and y satisfy the equation

$$\frac{(x-2)^2}{49} + \frac{(y-8)^2}{9} = 1.$$

This is the equation of an ellipse centered at (2, 8). ■

PROBLEMS 2.2

In Problems 1–16, find the slope of the line tangent to the given curve for the given value of the parameter.

1. $x = t^2 - 2;\ y = 4t;\ t = 3$
2. $x = t^3;\ y = t^4 - 5;\ t = -1$
3. $x = t^2;\ y = \sqrt{1-t};\ t = \frac{1}{2}$
4. $x = t + 4;\ y = t^3 - t + 4;\ t = 2$
5. $x = e^{2t};\ y = e^{-2t};\ t = 1$
6. $x = \cos\theta;\ y = \sin\theta;\ \theta = \pi/4$
7. $x = \cos 2\theta;\ y = \sin 2\theta;\ \theta = \pi/4$
8. $x = \tan\theta;\ y = \sec\theta;\ \theta = \pi/4$
9. $x = \sec\theta;\ y = \tan\theta;\ \theta = \pi/4$
10. $x = \cosh t;\ y = \sinh t;\ t = 0$
11. $x = \sqrt{1 - \sin\theta};\ y = \sqrt{1 + \cos\theta};\ \theta = 0$
12. $x = 8\cos\theta;\ y = -3\sin\theta;\ \theta = 2\pi/3$
13. $x = \cos^3\theta;\ y = \sin^3\theta;\ \theta = \pi/6$
14. $x = \cos^3\theta;\ y = \sin^3\theta;\ \theta = \pi/2$
15. $x = \theta;\ y = \dfrac{1}{\theta};\ \theta = \dfrac{\pi}{4}$
16. $x = t\cosh t;\ y = \dfrac{\tanh t}{1+t};\ t = 0$

In Problems 17–22, find the equation of the tangent line.

17. the curve of Problem 1
18. the curve of Problem 4
19. the curve of Problem 5
20. the curve of Problem 7
21. the curve of Problem 11
22. the curve of Problem 13

In Problems 23–32, find the points [in the form (x, y)] at which the given curves have vertical and horizontal tangents.

23. $x = t^2 - 1;\ y = t^2 - 4$
24. $x = 2\cos\theta;\ y = 3\sin\theta$
25. $x = \sin 3\theta;\ y = \cos 5\theta$
26. $x = \sin\theta + \cos\theta;\ y = \sin\theta - \cos\theta$
27. $x = e^{3t};\ y = e^{-5t}$
28. $x = \sinh t;\ y = \cosh t$
29. $x = \cosh t;\ y = \sinh t$
***30.** $x = \theta\sin\theta;\ y = \theta\cos\theta$
31. $x = \dfrac{1}{\sqrt{1-t^2}};\ y = \sqrt{1-t^2}$
32. $x = \ln(1 + t^2);\ y = \ln(1 + t^3)$

33. Suppose a curve is given in polar coordinates by the equation $r = f(\theta)$. Show that the curve can then be given parametrically by

$$x = f(\theta)\cos\theta \quad \text{and} \quad y = f(\theta)\sin\theta.$$

In Problems 34–41, use the result of Problem 33 to determine the slope of the tangent line to the given curve for the given value of θ.

34. $r = 5\sin\theta;\ \theta = \pi/6$
35. $r = 5\cos\theta + 5\sin\theta;\ \theta = \pi/4$
36. $r = -4 + 2\cos\theta;\ \theta = \pi/3$
37. $r = 4 + 3\sin\theta;\ \theta = 2\pi/3$
38. $r = 3\sin 2\theta;\ \theta = \pi/6$
39. $r = 5\sin 3\theta;\ \theta = \pi/4$
40. $r = e^{\theta/2};\ \theta = 0$
41. $r^2 = \cos 2\theta;\ \theta = \pi/6$

42. Find the equation of the two tangents to the curve

$$x = t^3 - 2t^2 - 3t + 11, \qquad y = t^2 - 2t - 5$$

at the point $(11, -2)$.

***43.** Calculate the equations of the three lines that are tangent to the curve $x = t^3 - 2t^2 - t + 3$, $y = 2t - t^2 - (2/t)$ at the point $(1, -1)$.

44. Calculate the slope of the tangent to the cycloid given in equations (2.1.16) and show that the tangent is vertical when α is a multiple of 2π.

45. For what values of α does the trochoid given by equations (2.1.15) have a vertical tangent?

2.3 THE DIFFERENTIATION AND INTEGRATION OF A VECTOR FUNCTION

In Section 2.2 we showed how the derivative dy/dx could be calculated when x and y were given parametrically in terms of t. In this section we will show how to calculate the derivative of a vector function.

Since a derivative is a special kind of limit, we first need to define the limit of a vector function. This definition is what you might expect.

Definition 1 LIMIT OF A VECTOR FUNCTION Let $\mathbf{f}(t) = f_1(t)\mathbf{i} + f_2(t)\mathbf{j}$. Let t_0 be any real number, $+\infty$, or $-\infty$. If $\lim_{t\to t_0} f_1(t)$ and $\lim_{t\to t_0} f_2(t)$ both exist, then we define

$$\lim_{t\to t_0} \mathbf{f}(t) = \left[\lim_{t\to t_0} f_1(t)\right]\mathbf{i} + \left[\lim_{t\to t_0} f_2(t)\right]\mathbf{j}. \tag{1}$$

That is, *the limit of a vector function is determined by the limits of its component functions.* Thus in order to calculate the limit of a vector function, it is only necessary to calculate two ordinary limits.

EXAMPLE 1 Let $\mathbf{f}(t) = [(\sin t)/t]\mathbf{i} + [\ln(3 + t)]\mathbf{j}$. Calculate $\lim_{t\to 0} \mathbf{f}(t)$.

Solution.

$$\lim_{t\to 0} \mathbf{f}(t) = \left[\lim_{t\to 0} \frac{\sin t}{t}\right]\mathbf{i} + \left[\lim_{t\to 0} \ln(3 + t)\right]\mathbf{j} = \mathbf{i} + (\ln 3)\mathbf{j} \quad ■$$

Definition 2 CONTINUITY OF A VECTOR FUNCTION **f** is **continuous** at t_0 if the component functions f_1 and f_2 are continuous at t_0. Thus, **f** is continuous at t_0 if

(i) **f** is defined at t_0.
(ii) $\lim_{t\to t_0} \mathbf{f}(t)$ exists.
(iii) $\lim_{t\to t_0} \mathbf{f}(t) = \mathbf{f}(t_0)$.

Definition 3 DERIVATIVE OF A VECTOR FUNCTION Let f be defined at t. Then **f** is **differentiable** at t if

$$\lim_{\Delta t\to 0} \frac{\mathbf{f}(t + \Delta t) - \mathbf{f}(t)}{\Delta t} \tag{2}$$

exists and is finite. The vector function $\mathbf{f}'$ defined by

$$\mathbf{f}'(t) = \frac{d\mathbf{f}}{dt} = \lim_{\Delta t\to 0} \frac{\mathbf{f}(t + \Delta t) - \mathbf{f}(t)}{\Delta t} \tag{3}$$

is called the **derivative** of **f**, and the domain of $\mathbf{f}'$ is the set of all t such that the limit in (2) exists.

Definition 4 DIFFERENTIABILITY IN AN OPEN INTERVAL The vector function $\mathbf{f}$ is **differentiable** on the open interval I if $\mathbf{f}'(t)$ exists for every t in I.

Before giving examples of the calculation of derivatives, we prove a theorem that makes this calculation no more difficult than the calculation of "ordinary" derivatives.

Theorem 1 If $\mathbf{f}(t) = f_1(t)\mathbf{i} + f_2(t)\mathbf{j}$, then at any value t for which $f_1'(t)$ and $f_2'(t)$ exist,

$$\mathbf{f}'(t) = f_1'(t)\mathbf{i} + f_2'(t)\mathbf{j}. \quad \textbf{(4)}$$

That is, the derivative of a vector function is determined by the derivatives of its component functions.

Proof.

$$\begin{aligned}
\mathbf{f}'(t) &= \lim_{\Delta t \to 0} \frac{\mathbf{f}(t + \Delta t) - \mathbf{f}(t)}{\Delta t} \\
&= \lim_{\Delta t \to 0} \frac{[f_1(t + \Delta t)\mathbf{i} + f_2(t + \Delta t)\mathbf{j}] - [f_1(t)\mathbf{i} + f_2(t)\mathbf{j}]}{\Delta t} \\
&= \lim_{\Delta t \to 0} \frac{[f_1(t + \Delta t) - f_1(t)]\mathbf{i} + [f_2(t + \Delta t) - f_2(t)]\mathbf{j}}{\Delta t} \\
&= \lim_{\Delta t \to 0} \left[\frac{f_1(t + \Delta t) - f_1(t)}{\Delta t}\right]\mathbf{i} + \lim_{\Delta t \to 0} \left[\frac{f_2(t + \Delta t) - f_2(t)}{\Delta t}\right]\mathbf{j} \\
&= f_1'(t)\mathbf{i} + f_2'(t)\mathbf{j} \quad \blacksquare
\end{aligned}$$

EXAMPLE 2 Let $\mathbf{f}(t) = (\cos t)\mathbf{i} + e^{2t}\mathbf{j}$. Calculate $\mathbf{f}'(t)$.

Solution.

$$\mathbf{f}'(t) = \frac{d}{dt}(\cos t)\mathbf{i} + \frac{d}{dt}e^{2t}\mathbf{j} = -(\sin t)\mathbf{i} + 2e^{2t}\mathbf{j} \quad \blacksquare$$

Once we know how to calculate the first derivative of $\mathbf{f}$, we can calculate higher derivatives as well.

Definition 5 SECOND DERIVATIVE If the function $\mathbf{f}'$ is differentiable at t, we define the **second derivative** of $\mathbf{f}$ to be the derivative of $\mathbf{f}'$. That is,

$$\mathbf{f}'' = (\mathbf{f}')'. \quad \textbf{(5)}$$

EXAMPLE 3 Let $\mathbf{f}(t) = (\ln t)\mathbf{i} + (1/t)\mathbf{j}$. Calculate $\mathbf{f}''(t)$.

Solution. $\mathbf{f}'(t) = (1/t)\mathbf{i} - (1/t^2)\mathbf{j}$, so

$$\mathbf{f}''(t) = -\frac{1}{t^2}\mathbf{i} + \frac{2}{t^3}\mathbf{j}.$$

Note that $\mathbf{f}'$ and $\mathbf{f}''$ are defined for all $t > 0$. ($\mathbf{f}'$ and $\mathbf{f}''$ are not defined for $t \le 0$ because $\ln t$ is not defined for $t \le 0$.) ■

GEOMETRIC INTERPRETATION OF f′ We now seek a geometric interpretation for $\mathbf{f}'$. As we have seen, the set of vectors $\mathbf{f}(t) = f_1(t)\mathbf{i} + f_2(t)\mathbf{j}$ for t in the domain of $\mathbf{f}$ form a curve C in the plane (see Figure 1a). For fixed t, $\mathbf{f}(t)$ is the vector drawn in this figure. If φ denotes the direction of $\mathbf{f}'$, then [see equation (1.1.2)]

$$\tan \varphi = \frac{f_2'(t)}{f_1'(t)} = \frac{dy/dt}{dx/dt} = \frac{dy}{dx}. \tag{6}$$

The last step is justified by the chain rule. This result implies that $\mathbf{f}'(t)$ is tangent to the curve $\mathbf{f}$ at the point $(f_1(t), f_2(t))$ (see Figure 1b). This is a natural extension of the usual notion of a tangent line to the case of a vector function.

To see this in another way, we see in Figure 2 that the vector $\mathbf{f}(t + \Delta t) - \mathbf{f}(t)$ is a secant vector whose direction approaches that of the tangent vector as $\Delta t \to 0$.

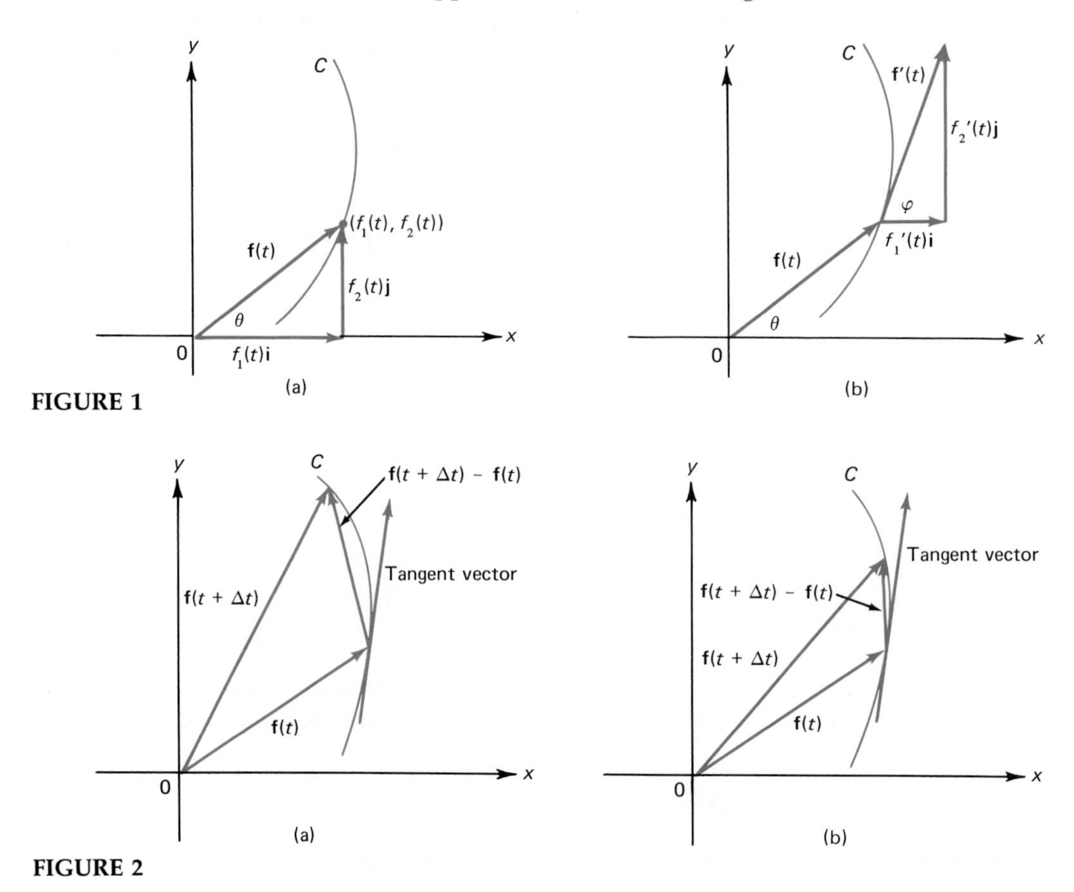

FIGURE 1

FIGURE 2

Definition 6 UNIT TANGENT VECTOR It is sometimes useful to calculate a **unit tangent vector** to a curve C. This vector is a tangent vector with a magnitude of 1. The unit tangent vector is usually denoted $\mathbf{T}$ and can be calculated by the formula

$$\mathbf{T}(t) = \frac{\mathbf{f}'(t)}{|\mathbf{f}'(t)|} \tag{7}$$

for any number t so long as $\mathbf{f}'(t) \neq 0$. This follows since $\mathbf{f}'(t)$ is a tangent vector and $\mathbf{f}'/|\mathbf{f}'|$ is a unit vector.

EXAMPLE 4 Find the unit tangent vector to the curve $\mathbf{f} = (\ln t)\mathbf{i} + (1/t)\mathbf{j}$ at $t = 1$.

Solution. $\mathbf{f}'(t) = (1/t)\mathbf{i} - (1/t^2)\mathbf{j} = \mathbf{i} - \mathbf{j}$ when $t = 1$. Since $|\mathbf{f}'| = |\mathbf{i} - \mathbf{j}| = \sqrt{2}$, we find that

$$\mathbf{T} = \frac{1}{\sqrt{2}}\mathbf{i} - \frac{1}{\sqrt{2}}\mathbf{j}. \quad \blacksquare$$

We now turn to the integration of vector functions. We mention these here for the sake of completeness. However, we will not need to integrate vector functions until Chapter 12.

Definition 7 INTEGRAL OF A VECTOR FUNCTION

(i) Let $\mathbf{f}(t) = f_1(t)\mathbf{i} + f_2(t)\mathbf{j}$ and suppose that the component functions f_1 and f_2 have antiderivatives. Then we define the **antiderivative,** or **indefinite integral,** of **f** by

$$\int \mathbf{f}(t)\,dt = \left(\int f_1(t)\,dt\right)\mathbf{i} + \left(\int f_2(t)\,dt\right)\mathbf{j} + \mathbf{C}, \tag{8}$$

where **C** is a constant vector of integration.

(ii) If f_1 and f_2 are integrable over the interval $[a, b]$, then we define the **definite integral** of **f** by

$$\int_a^b \mathbf{f}(t)\,dt = \left(\int_a^b f_1(t)\,dt\right)\mathbf{i} + \left(\int_a^b f_2(t)\,dt\right)\mathbf{j}. \tag{9}$$

REMARK 1. The antiderivative (or indefinite integral) of **f** is a new vector function and is *not unique* since $\int f_1(t)\,dt$ and $\int f_2(t)\,dt$ are not unique.

REMARK 2. The definite integral of **f** is a constant vector since $\int_a^b f_1(t)\,dt$ and $\int_a^b f_2(t)\,dt$ are constants.

EXAMPLE 5 Let $\mathbf{f}(t) = (\cos t)\mathbf{i} + (\sin t)\mathbf{j}$. Calculate the following:

(a) $\int \mathbf{f}(t)\,dt$ **(b)** $\int_0^{\pi/2} \mathbf{f}(t)\,dt$

Solution.

$$\textbf{(a)} \int \mathbf{f}(t)\,dt = \left(\int \cos t\,dt\right)\mathbf{i} + \left(\int \sin t\,dt\right)\mathbf{j} = (\sin t + C_1)\mathbf{i} + (-\cos t + C_2)\mathbf{j}$$

$$= (\sin t)\mathbf{i} - (\cos t)\mathbf{j} + \mathbf{C},$$

where $\mathbf{C} = C_1\mathbf{i} + C_2\mathbf{j}$ is a constant vector.

$$\textbf{(b)} \int_0^{\pi/2} \mathbf{f}(t)\,dt = \left(\sin t\Big|_0^{\pi/2}\right)\mathbf{i} + \left(-\cos t\Big|_0^{\pi/2}\right)\mathbf{j} = \mathbf{i} + \mathbf{j} \quad \blacksquare$$

EXAMPLE 6 For $\mathbf{f}'(t) = t\mathbf{i} + t^3\mathbf{j}$ and $\mathbf{f}(0) = \mathbf{i} - 2\mathbf{j}$, calculate $\mathbf{f}(t)$.

Solution. We first find that

$$\mathbf{f}(t) = \int \mathbf{f}'(t)\,dt = \left(\frac{t^2}{2} + C_1\right)\mathbf{i} + \left(\frac{t^4}{4} + C_2\right)\mathbf{j} = \frac{t^2}{2}\mathbf{i} + \frac{t^4}{4}\mathbf{j} + \mathbf{C}. \tag{10}$$

To evaluate $\mathbf{C}$, we substitute the value $t = 0$ into (10) to find that $\mathbf{f}(0) = \mathbf{C} = \mathbf{i} - 2\mathbf{j}$, so that

$$\mathbf{f}(t) = \frac{t^2}{2}\mathbf{i} + \frac{t^4}{4}\mathbf{j} + \mathbf{i} - 2\mathbf{j} = \left(\frac{t^2}{2} + 1\right)\mathbf{i} + \left(\frac{t^4}{4} - 2\right)\mathbf{j}. \quad \blacksquare$$

PROBLEMS 2.3

In Problems 1–10, calculate the first and second derivatives of the given vector function.

1. $\mathbf{f}(t) = t\mathbf{i} - t^5\mathbf{j}$
2. $\mathbf{f}(t) = (1 + t^2)\mathbf{i} + \dfrac{2}{t}\mathbf{j}$
3. $\mathbf{f}(t) = (\sin 2t)\mathbf{i} + (\cos 3t)\mathbf{j}$
4. $\mathbf{f}(t) = \dfrac{t}{1+t}\mathbf{i} - \dfrac{1}{\sqrt{t}}\mathbf{j}$
5. $\mathbf{f}(t) = (\ln t)\mathbf{i} + e^{3t}\mathbf{j}$
6. $\mathbf{f}(t) = e^t(\sin t)\mathbf{i} + e^t(\cos t)\mathbf{j}$
7. $\mathbf{f}(t) = (\tan t)\mathbf{i} + (\sec t)\mathbf{j}$
8. $\mathbf{f}(t) = (\tan^{-1} t)\mathbf{i} + (\sin^{-1} t)\mathbf{j}$
9. $\mathbf{f}(t) = (\ln \cos t)\mathbf{i} + (\ln \sin t)\mathbf{j}$
10. $\mathbf{f}(t) = (\cosh t)\mathbf{i} + (\sinh t)\mathbf{j}$

In Problems 11–20, find the unit tangent vector to the given curve for the given value of t.

11. $\mathbf{f}(t) = t^2\mathbf{i} + t^3\mathbf{j};\ t = 1$
12. $\mathbf{f}(t) = t\mathbf{i} + \dfrac{1}{t}\mathbf{j};\ t = 1$
13. $\mathbf{f}(t) = (\cos t)\mathbf{i} + (\sin t)\mathbf{j};\ t = 0$
14. $\mathbf{f}(t) = (\cos t)\mathbf{i} + (\sin t)\mathbf{j};\ t = \pi/2$
15. $\mathbf{f}(t) = (\cos t)\mathbf{i} + (\sin t)\mathbf{j};\ t = \pi/4$
16. $\mathbf{f}(t) = (\cos t)\mathbf{i} + (\sin t)\mathbf{j};\ t = 3\pi/4$
17. $\mathbf{f}(t) = (\tan t)\mathbf{i} + (\sec t)\mathbf{j};\ t = 0$
18. $\mathbf{f}(t) = (\ln t)\mathbf{i} + e^{2t}\mathbf{j};\ t = 1$
19. $\mathbf{f}(t) = \dfrac{t}{t+1}\mathbf{i} + \dfrac{t+1}{t}\mathbf{j};\ t = 2$
20. $\mathbf{f}(t) = \dfrac{t+1}{t}\mathbf{i} + \dfrac{t}{t+1}\mathbf{j};\ t = 2$

In Problems 21–28, calculate the indicated integral.

21. $\displaystyle\int_0^2 (t^2\mathbf{i} + t^4\mathbf{j})\,dt$
22. $\displaystyle\int [(\sin 2t)\mathbf{i} + e^t\mathbf{j}]\,dt$
23. $\displaystyle\int (t^{-1/2}\mathbf{i} + t^{1/2}\mathbf{j})\,dt$
24. $\displaystyle\int_0^{\pi/4} [(\cos 2t)\mathbf{i} - (\sin 2t)\mathbf{j}]\,dt$

25. $\int_0^1 [(\sinh t)\mathbf{i} - (\cosh t)\mathbf{j}]\, dt$

26. $\int_1^e \left[\left(\frac{1}{t}\right)\mathbf{i} - \left(\frac{3}{t}\right)\mathbf{j}\right] dt$

27. $\int [(\ln t)\mathbf{i} + te^t\mathbf{j}]\, dt$

28. $\int [(\tan t)\mathbf{i} + (\sec t)\mathbf{j}]\, dt$

29. Let $\mathbf{f}'(t) = t^3\mathbf{i} - t^5\mathbf{j}$ and let $\mathbf{f}(0) = 2\mathbf{i} + 5\mathbf{j}$. Find $\mathbf{f}(t)$.

30. Let $\mathbf{f}'(t) = (1/\sqrt{t})\mathbf{i} + \sqrt{t}\mathbf{j}$ and let $\mathbf{f}(1) = -2\mathbf{i} + \mathbf{j}$. Find $\mathbf{f}(t)$.

31. Let $\mathbf{f}'(t) = (\cos t)\mathbf{i} + (\sin t)\mathbf{j}$ and let $\mathbf{f}(\pi/2) = \mathbf{i}$. Find $\mathbf{f}(t)$.

32. Find $\mathbf{f}(t)$ in Problem 31 for $\mathbf{f}(\pi/2) = \mathbf{j}$.

33. The ellipse $(x^2/a^2) + (y^2/b^2) = 1$ can be written parametrically as $x = a \cos \theta$, $y = b \sin \theta$. Find a unit tangent vector to the ellipse at $\theta = \pi/4$.

34. Find a unit tangent vector to the cycloid $\mathbf{f}(\alpha) = r(\alpha - \sin \alpha)\mathbf{i} + r(1 - \cos \alpha)\mathbf{j}$ for $\alpha = 0$.

35. Find a unit tangent vector to the cycloid of Problem 34 for $\alpha = \pi/2$.

36. Find a unit tangent vector to the cycloid of Problem 34 for $\alpha = \pi/3$.

37. Find, for $\theta = \pi/6$, a unit tangent vector to the hypocycloid of Problem 2.1.39, assuming that $a = 5$ and $b = 2$.

2.4 SOME DIFFERENTIATION FORMULAS

Many of the rules used in the differentiation of scalar functions carry over to vector functions.

Theorem 1 Let $\mathbf{f}$ and $\mathbf{g}$ be vector functions that are differentiable in an interval I. Let the scalar function h be differentiable in I. Finally, let α be a scalar and let $\mathbf{v}$ be a constant vector. Then, on I, we have

(i) $\mathbf{f} + \mathbf{g}$ is differentiable and

$$\frac{d}{dt}(\mathbf{f} + \mathbf{g}) = \frac{d\mathbf{f}}{dt} + \frac{d\mathbf{g}}{dt} = \mathbf{f}' + \mathbf{g}'. \tag{1}$$

(ii) $\alpha\mathbf{f}$ is differentiable and

$$\frac{d}{dt}\alpha\mathbf{f} = \alpha\frac{d\mathbf{f}}{dt} = \alpha\mathbf{f}'. \tag{2}$$

(iii) $\mathbf{v} \cdot \mathbf{f}$ is differentiable and

$$\frac{d}{dt}\mathbf{v} \cdot \mathbf{f} = \mathbf{v} \cdot \frac{d\mathbf{f}}{dt} = \mathbf{v} \cdot \mathbf{f}' \tag{3}$$

(iv) $h\mathbf{f}$ is differentiable and

$$\frac{d}{dt}h\mathbf{f} = h\frac{d\mathbf{f}}{dt} + \frac{dh}{dt}\mathbf{f} = h\mathbf{f}' + h'\mathbf{f}. \tag{4}$$

(v) $\mathbf{f} \cdot \mathbf{g}$ is differentiable and

$$\frac{d}{dt}\mathbf{f} \cdot \mathbf{g} = \mathbf{f} \cdot \frac{d\mathbf{g}}{dt} + \frac{d\mathbf{f}}{dt} \cdot \mathbf{g} = \mathbf{f} \cdot \mathbf{g}' + \mathbf{f}' \cdot \mathbf{g}. \tag{5}$$

Proof.

(i) $$\frac{d}{dt}(\mathbf{f} + \mathbf{g}) = \frac{d}{dt}(f_1\mathbf{i} + f_2\mathbf{j} + g_1\mathbf{i} + g_2\mathbf{j})$$
$$= \frac{d}{dt}[(f_1 + g_1)\mathbf{i} + (f_2 + g_2)\mathbf{j}] = (f_1 + g_1)'\mathbf{i} + (f_2 + g_2)'\mathbf{j}$$
$$= (f_1' + g_1')\mathbf{i} + (f_2' + g_2')\mathbf{j}$$
$$= (f_1'\mathbf{i} + f_2'\mathbf{j}) + (g_1'\mathbf{i} + g_2'\mathbf{j}) = \mathbf{f}' + \mathbf{g}'$$

(ii) $$\frac{d}{dt}(\alpha\mathbf{f}) = \frac{d}{dt}(\alpha f_1\mathbf{i} + \alpha f_2\mathbf{j}) = \alpha f_1'\mathbf{i} + \alpha f_2'\mathbf{j} = \alpha(f_1'\mathbf{i} + f_2'\mathbf{j}) = \alpha\mathbf{f}'$$

(iii) Let $\mathbf{v} = v_1\mathbf{i} + v_2\mathbf{j}$, where v_1 and v_2 are constants. Then

$$\mathbf{v} \cdot \mathbf{f} = v_1f_1 + v_2f_2 \qquad \text{and} \qquad (\mathbf{v} \cdot \mathbf{f})' = v_1f_1' + v_2f_2' = \mathbf{v} \cdot \mathbf{f}'.$$

(iv) $$\frac{d}{dt}h\mathbf{f} = \frac{d}{dt}(hf_1\mathbf{i} + hf_2\mathbf{j}) = (hf_1)'\mathbf{i} + (hf_2)'\mathbf{j} = (hf_1' + h'f_1)\mathbf{i} + (hf_2' + h'f_2)\mathbf{j}$$
$$= h(f_1'\mathbf{i} + f_2'\mathbf{j}) + h'(f_1\mathbf{i} + f_2\mathbf{j}) = h\mathbf{f}' + h'\mathbf{f}$$

(v) $\mathbf{f} \cdot \mathbf{g} = (f_1\mathbf{i} + f_2\mathbf{j}) \cdot (g_1\mathbf{i} + g_2\mathbf{j}) = f_1g_1 + f_2g_2$, so that

$$\frac{d}{dt}(\mathbf{f} \cdot \mathbf{g}) = f_1g_1' + f_1'g_1 + f_2g_2' + f_2'g_2 = f_1g_1' + f_2g_2' + f_1'g_1 + f_2'g_2$$
$$= (f_1\mathbf{i} + f_2\mathbf{j}) \cdot (g_1'\mathbf{i} + g_2'\mathbf{j}) + (f_1'\mathbf{i} + f_2'\mathbf{j}) \cdot (g_1\mathbf{i} + g_2\mathbf{j})$$
$$= \mathbf{f} \cdot \mathbf{g}' + \mathbf{f}' \cdot \mathbf{g}. \quad ■$$

EXAMPLE 1 Let $\mathbf{f}(t) = t\mathbf{i} + t^3\mathbf{j}$, $\mathbf{g}(t) = (\cos t)\mathbf{i} + (\sin t)\mathbf{j}$ and $\mathbf{v} = 2\mathbf{i} - 3\mathbf{j}$. Calculate (a) $(\mathbf{f} + \mathbf{g})'$, (b) $(\mathbf{v} \cdot \mathbf{f})'$, and (c) $(\mathbf{f} \cdot \mathbf{g})'$.

Solution.

(a) $$(\mathbf{f} + \mathbf{g})' = \mathbf{f}' + \mathbf{g}' = (\mathbf{i} + 3t^2\mathbf{j}) + [-(\sin t)\mathbf{i} + (\cos t)\mathbf{j}]$$
$$= (1 - \sin t)\mathbf{i} + (3t^2 + \cos t)\mathbf{j}$$

(b) $$(\mathbf{v} \cdot \mathbf{f})' = \mathbf{v} \cdot \mathbf{f}' = (2\mathbf{i} - 3\mathbf{j}) \cdot (\mathbf{i} + 3t^2\mathbf{j}) = 2 - 9t^2$$

(c) $$\begin{aligned}(\mathbf{f}\cdot\mathbf{g})' &= \mathbf{f}\cdot\mathbf{g}' + \mathbf{f}'\cdot\mathbf{g}\\ &= (t\mathbf{i} + t^3\mathbf{j})\cdot[-(\sin t)\mathbf{i} + (\cos t)\mathbf{j}] + (\mathbf{i} + 3t^2\mathbf{j})\cdot[(\cos t)\mathbf{i} + (\sin t)\mathbf{j}]\\ &= -t\sin t + t^3\cos t + \cos t + 3t^2\sin t\\ &= (\cos t)(t^3 + 1) + (\sin t)(3t^2 - t) \quad \blacksquare\end{aligned}$$

EXAMPLE 2 Let $\mathbf{f}(t) = (\cos t)\mathbf{i} + (\sin t)\mathbf{j}$. Calculate $\mathbf{f}\cdot\mathbf{f}'$.

Solution.

$$\begin{aligned}\mathbf{f}\cdot\mathbf{f}' &= [(\cos t)\mathbf{i} + (\sin t)\mathbf{j}]\cdot[-(\sin t)\mathbf{i} + (\cos t)\mathbf{j}]\\ &= -\cos t\sin t + \sin t\cos t = 0 \quad \blacksquare\end{aligned}$$

Example 2 can be generalized to the following interesting result.

Theorem 2 Let $\mathbf{f}(t) = f_1\mathbf{i} + f_2\mathbf{j}$ be a differentiable vector function such that $|\mathbf{f}(t)| = \sqrt{f_1^2(t) + f_2^2(t)}$ is constant. Then

$$\mathbf{f}\cdot\mathbf{f}' = 0. \tag{6}$$

Proof. Suppose $|\mathbf{f}(t)| = C$, a constant. Then

$$\mathbf{f}\cdot\mathbf{f} = f_1^2 + f_2^2 = |\mathbf{f}|^2 = C^2,$$

so

$$\frac{d}{dt}(\mathbf{f}\cdot\mathbf{f}) = \frac{d}{dt}C^2 = 0.$$

But

$$\frac{d}{dt}(\mathbf{f}\cdot\mathbf{f}) = \mathbf{f}\cdot\mathbf{f}' + \mathbf{f}'\cdot\mathbf{f} = 2\mathbf{f}\cdot\mathbf{f}' = 0,$$

so

$$\mathbf{f}\cdot\mathbf{f}' = 0. \quad \blacksquare$$

NOTE. In Example 2, $|\mathbf{f}(t)| = \sqrt{\cos^2 t + \sin^2 t} = 1$ for every t.

UNIT NORMAL VECTOR There is an interesting geometric application of Theorem 2. Let $\mathbf{f}(t)$ be a differentiable vector function. For all t for which $\mathbf{f}'(t) \neq 0$, we let $\mathbf{T}(t)$ denote the unit tangent vector to the curve $\mathbf{f}(t)$. Then since $|\mathbf{T}(t)| = 1$ by the definition of a *unit* tangent vector, we have, from Theorem 2 [assuming that $\mathbf{T}'(t)$ exists],

$$\mathbf{T}(t)\cdot\mathbf{T}'(t) = 0. \tag{7}$$

That is, $\mathbf{T}'(t)$ is *orthogonal* to $\mathbf{T}(t)$. Recall† that a line perpendicular to a tangent line is called a **normal line.** We call the vector $\mathbf{T}'$ a **normal vector** to the curve $\mathbf{f}$. Finally, whenever $\mathbf{T}'(t) \neq 0$, we can define the **unit normal vector** to the curve $\mathbf{f}$ at t as

$$\mathbf{n}(t) = \frac{\mathbf{T}'(t)}{|\mathbf{T}'(t)|}. \tag{8}$$

EXAMPLE 3 Calculate a unit normal vector to the curve $\mathbf{f}(t) = (\cos t)\mathbf{i} + (\sin t)\mathbf{j}$ at $t = \pi/4$.

Solution. First, we calculate $\mathbf{f}'(t) = -(\sin t)\mathbf{i} + (\cos t)\mathbf{j}$, and since $|\mathbf{f}'(t)| = 1$, we find that

$$\mathbf{T}(t) = \frac{\mathbf{f}'(t)}{|\mathbf{f}'(t)|} = -(\sin t)\mathbf{i} + (\cos t)\mathbf{j}.$$

Then

$$\mathbf{T}'(t) = -(\cos t)\mathbf{i} - (\sin t)\mathbf{j} = \mathbf{n}(t)$$

since $|\mathbf{T}'(t)| = 1$, so that at $t = \pi/4$,

$$\mathbf{n} = -\frac{1}{\sqrt{2}}\mathbf{i} - \frac{1}{\sqrt{2}}\mathbf{j}.$$

This is sketched in Figure 1 [remember that $\mathbf{f}(t) = (\cos t)\mathbf{i} + (\sin t)\mathbf{j}$ is the parametric equation of the unit circle]. The reason the vector $\mathbf{n}$ points inward is that it is the negative of the position vector at $t = \pi/4$. ■

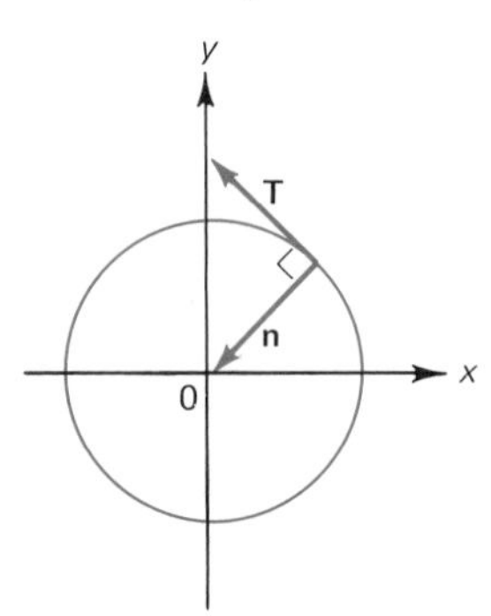

FIGURE 1

EXAMPLE 4 Calculate a unit normal vector to the curve $\mathbf{f}(t) = [(t^3/3) - t]\mathbf{i} + t^2\mathbf{j}$ at $t = 3$.

Solution. Here $\mathbf{f}'(t) = (t^2 - 1)\mathbf{i} + 2t\mathbf{j}$ and

$$|\mathbf{f}'(t)| = \sqrt{(t^2-1)^2 + 4t^2} = \sqrt{t^4 - 2t^2 + 1 + 4t^2} = \sqrt{t^4 + 2t^2 + 1} = t^2 + 1.$$

†See Stanley Grossman, *Calculus,* Third Edition (Orlando: Academic Press, 1984), or *Calculus of One Variable,* Second Edition (Orlando: Academic Press, 1986), page 141.

Thus

$$\mathbf{T}(t) = \frac{\mathbf{f}'(t)}{|\mathbf{f}'(t)|} = \frac{t^2 - 1}{t^2 + 1}\mathbf{i} + \frac{2t}{t^2 + 1}\mathbf{j}.$$

Then

$$\mathbf{T}'(t) = \frac{d}{dt}\left(\frac{t^2 - 1}{t^2 + 1}\right)\mathbf{i} + \frac{d}{dt}\left(\frac{2t}{t^2 + 1}\right)\mathbf{j} = \frac{4t}{(t^2 + 1)^2}\mathbf{i} + \frac{2 - 2t^2}{(t^2 + 1)^2}\mathbf{j}.$$

Finally,

$$\begin{aligned} |\mathbf{T}'(t)| &= \left\{\left[\frac{4t}{(t^2 + 1)^2}\right]^2 + \left[\frac{2 - 2t^2}{(t^2 + 1)^2}\right]^2\right\}^{1/2} = \frac{1}{(t^2 + 1)^2}(16t^2 + 4 - 8t^2 + 4t^4)^{1/2} \\ &= \frac{1}{(t^2 + 1)^2}\sqrt{4t^4 + 8t^2 + 4} = \frac{2}{(t^2 + 1)^2}\sqrt{t^4 + 2t^2 + 1} \\ &= \frac{2(t^2 + 1)}{(t^2 + 1)^2} = \frac{2}{t^2 + 1}, \end{aligned}$$

so that

$$\mathbf{n}(t) = \frac{\mathbf{T}'(t)}{|\mathbf{T}'(t)|} = \frac{t^2 + 1}{2}\left[\frac{4t}{(t^2 + 1)^2}\mathbf{i} + \frac{2 - 2t^2}{(t^2 + 1)^2}\mathbf{j}\right] = \frac{2t}{t^2 + 1}\mathbf{i} + \frac{1 - t^2}{t^2 + 1}\mathbf{j}.$$

At $t = 3$, $\mathbf{T}(t) = \frac{4}{5}\mathbf{i} + \frac{3}{5}\mathbf{j}$ and $\mathbf{n} = \frac{3}{5}\mathbf{i} - \frac{4}{5}\mathbf{j}$. Note that $\mathbf{T}(3) \cdot \mathbf{n}(3) = 0$. This is sketched in Figure 2. ■

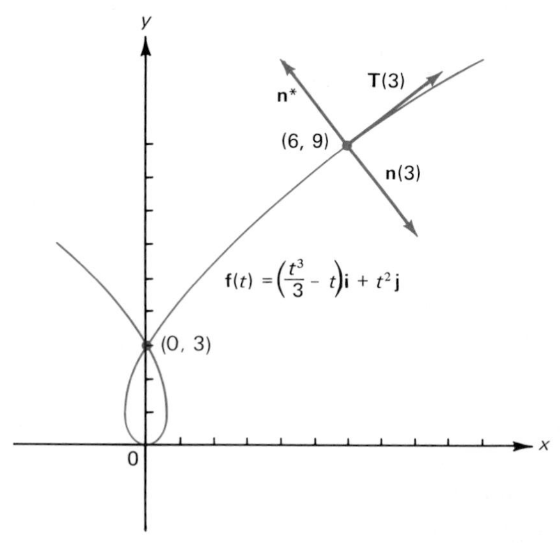

FIGURE 2

In Figure 2 we see that there are two vectors perpendicular to $\mathbf{T}$ at the point P: one pointing "inward" ($\mathbf{n}$) and one pointing "outward" ($\mathbf{n}^*$). In general, the situation is as shown in Figure 3. How do we know which of the two vectors is $\mathbf{n}$? If φ is the direction of $\mathbf{T}$, then since $\mathbf{T}$ is a unit vector,

$$\mathbf{T} = (\cos \varphi)\mathbf{i} + (\sin \varphi)\mathbf{j},$$

so that

$$\frac{d\mathbf{T}}{dt} = \frac{d\mathbf{T}}{d\varphi}\frac{d\varphi}{dt}$$

and

$$\left|\frac{d\mathbf{T}}{dt}\right| = \left|\frac{d\mathbf{T}}{d\varphi}\right|\left|\frac{d\varphi}{dt}\right| = \left|\frac{d\varphi}{dt}\right| \qquad \left(\text{since } \left|\frac{d\mathbf{T}}{d\varphi}\right| = 1\right).$$

Then

$$\mathbf{n}(t) = \frac{d\mathbf{T}/dt}{|d\mathbf{T}/dt|} = \frac{d\mathbf{T}}{d\varphi}\frac{d\varphi/dt}{|d\varphi/dt|} = \frac{d\varphi/dt}{|d\varphi/dt|}[-(\sin \varphi)\mathbf{i} + (\cos \varphi)\mathbf{j}].$$

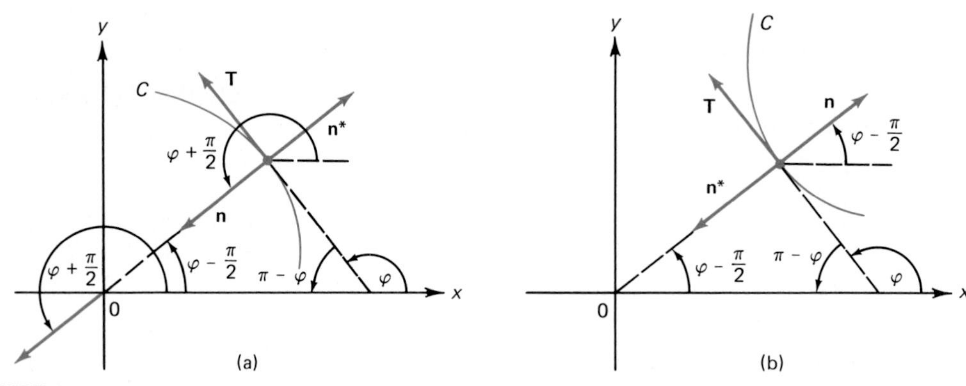

FIGURE 3

Since $d\varphi/dt$ is a scalar, $(d\varphi/dt)/|d\varphi/dt| = \pm 1$. It is $+1$ in Figure 3a since φ increases as t increases so that $d\varphi/dt > 0$, while it is -1 in Figure 3b since φ decreases as t increases. In Figure 3a, then,

$$\mathbf{n}(t) = -(\sin \varphi)\mathbf{i} + (\cos \varphi)\mathbf{j} = \cos[\varphi + (\pi/2)]\mathbf{i} + \sin[\varphi + (\pi/2)]\mathbf{j}.$$

That is, the direction of $\mathbf{n}$ is $\varphi + (\pi/2)$ and $\mathbf{n}$ must point inward (on the concave side of C) as in Figure 3a. In the other case, $(d\varphi/dt)/|d\varphi/dt| = -1$, so

$$\mathbf{n}(t) = (\sin \varphi)\mathbf{i} - (\cos \varphi)\mathbf{j} = \cos[\varphi - (\pi/2)]\mathbf{i} + \sin[\varphi - (\pi/2)]\mathbf{j}.$$

That is, the direction of **n** is $\varphi - (\pi/2)$ and so **n** points as in Figure 3b. In either case:

The unit normal vector **n** *always points in the direction of the concave side of the curve.*

PROBLEMS 2.4

In Problems 1–15, calculate the indicated derivative.

1. $\dfrac{d}{dt}\{[2t\mathbf{i} + (\cos t)\mathbf{j}] + [(\tan t)\mathbf{i} - (\sec t)\mathbf{j}]\}$
2. $\dfrac{d}{dt}\{e^t[(\cos t)\mathbf{i} + (\sin t)\mathbf{j}]\}$
3. $\dfrac{d}{dt}\left[(3\mathbf{i} - 4\mathbf{j}) \cdot \left(\sqrt{t}\mathbf{i} - \dfrac{1}{\sqrt{t}}\mathbf{j}\right)\right]$
4. $\dfrac{d}{dt}\left[(t^3\mathbf{i} - t^5\mathbf{j}) \cdot \left(-\dfrac{1}{t^3}\mathbf{i} + \dfrac{1}{t^5}\mathbf{j}\right)\right]$
5. $\dfrac{d}{dt}[(t^3\mathbf{i} - t^5\mathbf{j}) + (-t^3\mathbf{i} + t^5\mathbf{j})]$
6. $\dfrac{d}{dt}\{[(\sinh t)\mathbf{i} + (\cosh t)\mathbf{j}] \cdot (-2\mathbf{i} + 7\mathbf{j})\}$
7. $\dfrac{d}{dt}\{3 \sin t[(\sinh t)\mathbf{i} + (\cosh t)\mathbf{j}]\}$
8. $\dfrac{d}{dt}\{[(\sinh t)\mathbf{i} + (\cosh t)\mathbf{j}] \cdot [(\sin t)\mathbf{i} + (\cos t)\mathbf{j}]\}$
9. $\dfrac{d}{dt}\left[\left(\dfrac{t+1}{t}\mathbf{i} + \dfrac{t}{t+1}\mathbf{j}\right) \cdot \left(t^{10}\mathbf{i} - \dfrac{1}{t^{10}}\mathbf{j}\right)\right]$
10. $\dfrac{d}{dt}\{[e^{2t}(\cos t)\mathbf{i} + e^{2t}(\sin t)\mathbf{j}] \cdot (t\mathbf{i} + t\mathbf{j})\}$
11. $\dfrac{d}{dt}\{[(\ln t)\mathbf{i} + (\ln t^3)\mathbf{j}] \cdot [(\tanh t)\mathbf{i} + (\operatorname{sech} t)\mathbf{j}]\}$
12. $\dfrac{d}{dt}\{[(\sin^{-1} t)\mathbf{i} + (\cos^{-1} t)\mathbf{j}] \cdot (2\mathbf{i} - 10\mathbf{j})\}$
13. $\dfrac{d}{dt}\{[\sin^{-1} t)\mathbf{i} + (\cos^{-1} t)\mathbf{j}] + [(\cos t)\mathbf{i} + (\sin t)\mathbf{j}]\}$
14. $\dfrac{d}{dt}\{(\cos t)[(\sin^{-1} t)\mathbf{i} + (\cos^{-1} t)\mathbf{j}]\}$
15. $\dfrac{d}{dt}\{[(\sin^{-1} t)\mathbf{i} + (\cos^{-1} t)\mathbf{j}] \cdot [(\tan^{-1} t)\mathbf{i} + (\cos t)\mathbf{j}]\}$

In Problems 16–29, find $\mathbf{T}(t)$, $\mathbf{n}(t)$, and the particular vectors **T** and **n** when $t = t_0$. Then sketch the curve near $t = t_0$ and include the vectors **T** and **n** in your sketch.

16. $\mathbf{f} = (\cos 3t)\mathbf{i} + (\sin 3t)\mathbf{j}$; $t = 0$
17. $\mathbf{f} = (\cos 5t)\mathbf{i} + (\sin 5t)\mathbf{j}$; $t = \pi/2$
18. $\mathbf{f} = 2(\cos 4t)\mathbf{i} + 2(\sin 4t)\mathbf{j}$; $t = \pi/4$
19. $\mathbf{f} = -3(\cos 10t)\mathbf{i} - 3(\sin 10t)\mathbf{j}$; $t = \pi$
20. $\mathbf{f} = 8(\cos t)\mathbf{i} + 8(\sin t)\mathbf{j}$; $t = \pi/4$
21. $\mathbf{f} = 4t\mathbf{i} + 2t^2\mathbf{j}$; $t = 1$
22. $\mathbf{f} = (2 + 3t)\mathbf{i} + (8 - 5t)\mathbf{j}$; $t = 3$
23. $\mathbf{f} = (4 - 7t)\mathbf{i} + (-3 + 5t)\mathbf{j}$; $t = -5$
24. $\mathbf{f} = (a + bt)\mathbf{i} + (c + dt)\mathbf{j}$, a, b, c, d real; $t = t_0$
25. $\mathbf{f} = 2t\mathbf{i} + (e^{-t} + e^t)\mathbf{j}$; $t = 0$

*26. $\mathbf{f} = (t - \cos t)\mathbf{i} + (1 - \sin t)\mathbf{j}$; $t = \pi/2$

*27. $\mathbf{f} = (t - \cos t)\mathbf{i} + (1 - \sin t)\mathbf{j}$; $t = \pi$

*28. $\mathbf{f} = (t - \cos t)\mathbf{i} + (1 - \sin t)\mathbf{j}$; $t = \pi/4$

29. $\mathbf{f} = (\ln \sin t)\mathbf{i} + (\ln \cos t)\mathbf{j}$; $t = \pi/6$

2.5 ARC LENGTH REVISITED

In a one-variable calculus course the *arc length* of the curve $y = f(x)$ between $x = a$ and $x = b$ is defined by[†]

$$\text{arc length} = s = \int_a^b \sqrt{1 + [f'(x)]^2}\, dx$$

where it is assumed that f is differentiable in $[a, b]$. We now derive a formula for arc

[†]See *Calculus* or *Calculus of One Variable,* page 543.

length in a more general setting. Let the curve C be given parametrically by

$$x = f_1(t), \qquad y = f_2(t). \tag{1}$$

We will assume in this section that f_1' and f_2' exist. Let t_0 be a fixed number, which fixes a point $P_0 = (x_0, y_0) = (f_1(t_0), f_2(t_0))$ on the curve (see Figure 1). The arrows in Figure 1 indicate the direction in which a point moves along the curve as t increases. We define the function $s(t)$ by

$$s(t) = \text{length along the curve } C \text{ from } (f_1(t_0), f_2(t_0)) \text{ to } (f_1(t), f_2(t)).$$

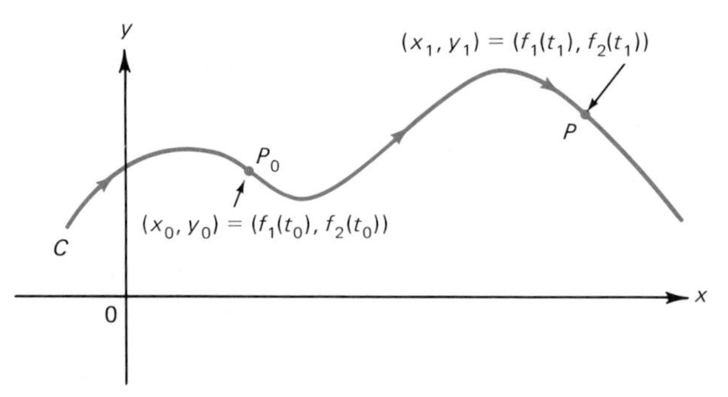

FIGURE 1

The following theorem allows us to calculate the length of a curve given parametrically.

Theorem 1 Suppose that f_1' and f_2' are continuous in the interval $[t_0, t_1]$. Then $s(t)$ is a differentiable function of t for $t \in [t_0, t_1]$ and

$$\frac{ds}{dt} = \sqrt{[f_1'(t)]^2 + [f_2'(t)]^2} = \sqrt{\left(\frac{dx}{dt}\right)^2 + \left(\frac{dy}{dt}\right)^2}. \tag{2}$$

Sketch of Proof. The proof of this theorem is quite difficult. However, it is possible to give an intuitive idea of what is going on. Consider an arc of the curve between t and $t + \Delta t$ (see Figure 2). First, we note that as $\Delta t \to 0$, the ratio of the length of the secant line L to the length of the arc Δs between the points $(x(t_0), y(t_0))$ and $(x(t_0 + \Delta t), y(t_0 + \Delta t))$ approaches 1. That is,

$$\lim_{\Delta t \to 0} \frac{\Delta s}{L} = 1. \tag{3}$$

Then since $L = \sqrt{\Delta x^2 + \Delta y^2}$, we have

$$\frac{\Delta s}{\Delta t} = \left(\frac{\Delta s}{L}\right)\left(\frac{L}{\Delta t}\right) = \left(\frac{\Delta s}{L}\right)\left(\frac{\sqrt{\Delta x^2 + \Delta y^2}}{\Delta t}\right) = \left(\frac{\Delta s}{L}\right)\sqrt{\left(\frac{\Delta x}{\Delta t}\right)^2 + \left(\frac{\Delta y}{\Delta t}\right)^2}$$

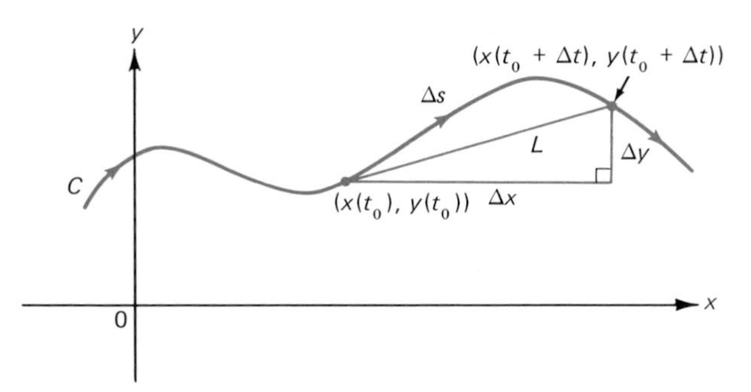

FIGURE 2

and

$$\frac{ds}{dt} = \lim_{\Delta t \to 0} \frac{\Delta s}{\Delta t} = \lim_{\Delta t \to 0} \frac{\Delta s}{L} \cdot \lim_{\Delta t \to 0} \sqrt{\left(\frac{\Delta x}{\Delta t}\right)^2 + \left(\frac{\Delta y}{\Delta t}\right)^2} = \sqrt{\left(\frac{dx}{dt}\right)^2 + \left(\frac{dy}{dt}\right)^2}.$$

The last step can be justified under the assumption that dx/dt and dy/dt are continuous, but its proof is beyond the scope of this book.† ■

We use equation (2) to define the length of the arc from t_0 to t_1.

Definition 1 ARC LENGTH Suppose that f_1' and f_2' are continuous in the interval $[t_0, t_1]$. Then the **arc length** of the curve $(f_1(t), f_2(t))$ is given by

$$\begin{aligned} s(t_1) &= \text{length of arc from } t_0 \text{ to } t_1 \\ &= \int_{t_0}^{t_1} \left(\frac{ds}{dt}\right) dt = \int_{t_0}^{t_1} \sqrt{\left(\frac{dx}{dt}\right)^2 + \left(\frac{dy}{dt}\right)^2}\, dt. \end{aligned} \tag{4}$$

REMARK 1. If we can write y as a differentiable function of x, then (4) follows from the definition of arc length given at the beginning of this section. To see this, we start with

$$s = \int_a^b \sqrt{1 + \left(\frac{dy}{dx}\right)^2}\, dx,$$

where $x(t_0) = a$ and $x(t_1) = b$. Then by the chain rule,

$$\frac{dy}{dx} = \frac{dy/dt}{dx/dt} \qquad \text{and} \qquad dx = \frac{dx}{dt}\, dt.$$

†See, for example, R. C. Buck, *Advanced Calculus* (McGraw-Hill, New York, 1965), page 321.

Thus

$$\begin{aligned} s &= \int_a^b \sqrt{1 + \left(\frac{dy}{dx}\right)^2}\, dx = \int_{t_0}^{t_1} \sqrt{1 + \left(\frac{dy/dt}{dx/dt}\right)^2}\, \frac{dx}{dt}\, dt \\ &= \int_{t_0}^{t_1} \sqrt{\frac{(dx/dt)^2 + (dy/dt)^2}{(dx/dt)^2}}\, \frac{dx}{dt}\, dt = \int_{t_0}^{t_1} \left(\frac{\sqrt{(dx/dt)^2 + (dy/dt)^2}}{dx/dt}\right) \frac{dx}{dt}\, dt \quad \left(\text{if } \frac{dx}{dt} > 0\right) \\ &= \int_{t_0}^{t_1} \sqrt{\left(\frac{dx}{dt}\right)^2 + \left(\frac{dy}{dt}\right)^2}\, dt. \end{aligned}$$

We emphasize that although this last result may be interesting, it does not prove the theorem because we cannot always write y as a function of x. For example, if $x = \cos t$ and $y = \sin t$, then $x^2 + y^2 = 1$ and $y = \pm\sqrt{1 - x^2}$, which is not a function of x since for every x in $(-1, 1)$ there are two values of y.

Equation (4) can be rewritten in a slightly different form. Using the chain rule (applied to differentials), we have

$$\frac{ds}{dt}\, dt = ds, \tag{5}$$

and so (4) becomes

$$s(t_1) = \int_{t_0}^{t_1} ds, \tag{6}$$

where

$$ds = \sqrt{\left(\frac{dx}{dt}\right)^2 + \left(\frac{dy}{dt}\right)^2}\, dt. \tag{7}$$

In this context the variable s is called the **parameter of arc length.** We will discuss this further in the next section. Note, however, that while x measures distance along the horizontal axis and y measures distance along the vertical axis, s measures distance *along the curve* given parametrically by equations (1).

REMARK 2. The function $\frac{ds}{dt}$ given by (2) represents **speed.** Theorem 1 says that *arc length is the integral of speed.* You should not confuse speed with velocity. The velocity function is a vector-valued function (defined in Example 2.1.6 on p. 31) while the speed function is a scalar function.

EXAMPLE 1 Calculate the length of the curve $x = \cos t$, $y = \sin t$ in the interval $[0, 2\pi]$.

Solution.

$$ds = \sqrt{\left(\frac{dx}{dt}\right)^2 + \left(\frac{dy}{dt}\right)^2}\, dt = \sqrt{(-\sin t)^2 + (\cos t)^2}\, dt = dt,$$

so that

$$s = \int_0^{2\pi} ds = \int_0^{2\pi} dt = 2\pi.$$

NOTE. Since $x = \cos t$, $y = \sin t$ for t in $[0, 2\pi]$ is a parametric representation of the unit circle, we have verified the accuracy of formula (4) in this instance, since the circumference of the unit circle is 2π. ■

EXAMPLE 2 Let the curve C be given by $x = t^2$ and $y = t^3$. Calculate the length of the arc from $t = 0$ to $t = 3$.

Solution. Here

$$\sqrt{\left(\frac{dx}{dt}\right)^2 + \left(\frac{dy}{dt}\right)^2} = \sqrt{(2t)^2 + (3t^2)^2} = \sqrt{4t^2 + 9t^4} = t\sqrt{4 + 9t^2},$$

so that

$$s = \int_0^3 t\sqrt{4 + 9t^2}\, dt = \frac{1}{27}(4 + 9t^2)^{3/2}\Big|_0^3 = \frac{85^{3/2} - 8}{27}. \quad ■$$

Suppose now that a curve is given in polar coordinates by $r = f(\theta)$. Since $x = r\cos\theta$ and $y = r\sin\theta$, we have

$$x = f(\theta)\cos\theta \qquad \text{and} \qquad y = f(\theta)\sin\theta,$$

so that

$$\frac{dx}{d\theta} = f'(\theta)\cos\theta - f(\theta)\sin\theta \qquad \text{and} \qquad \frac{dy}{d\theta} = f'(\theta)\sin\theta + f(\theta)\cos\theta.$$

Therefore

$$\begin{aligned}\left(\frac{dx}{d\theta}\right)^2 + \left(\frac{dy}{d\theta}\right)^2 &= [f'(\theta)]^2\cos^2\theta - 2f'(\theta)f(\theta)\sin\theta\cos\theta + [f(\theta)]^2\sin^2\theta \\ &\quad + [f'(\theta)]^2\sin^2\theta + 2f'(\theta)f(\theta)\sin\theta\cos\theta + [f(\theta)]^2\cos^2\theta \\ &= [f(\theta)]^2 + [f'(\theta)]^2 = r^2 + \left(\frac{dr}{d\theta}\right)^2.\end{aligned}$$

Thus the formula for arc length in polar coordinates becomes

$$s = \int_{\theta=\theta_0}^{\theta=\theta_1} \sqrt{r^2 + \left(\frac{dr}{d\theta}\right)^2}\, d\theta. \tag{8}$$

EXAMPLE 3 Calculate the arc length of the cardioid $r = a(1 + \cos\theta)$, $a > 0$.

Solution. The curve is sketched in Figure 3. Since the curve is symmetric about the polar axis, we calculate the arc length between $\theta = 0$ and $\theta = \pi$ and then multiply the result by 2. We have

$$\begin{aligned}\frac{s}{2} &= \int_0^\pi \sqrt{r^2 + \left(\frac{dr}{d\theta}\right)^2}\, d\theta = \int_0^\pi a\sqrt{(1+\cos\theta)^2 + \sin^2 d\theta} \\ &= a\int_0^\pi \sqrt{1 + 2\cos\theta + \cos^2\theta + \sin^2\theta\, d\theta} \\ &= a\int_0^\pi \sqrt{(2 + 2\cos\theta)}\, d\theta = a\int_0^\pi \sqrt{\frac{4 + 4\cos\theta}{2}}\, d\theta \\ &= 2a\int_0^\pi \sqrt{\frac{1+\cos\theta}{2}}\, d\theta = 2a\int_0^\pi \cos\frac{\theta}{2}\, d\theta = 4a\sin\frac{\theta}{2}\Big|_0^\pi = 4a.\end{aligned}$$

The total arc length is therefore $8a$.

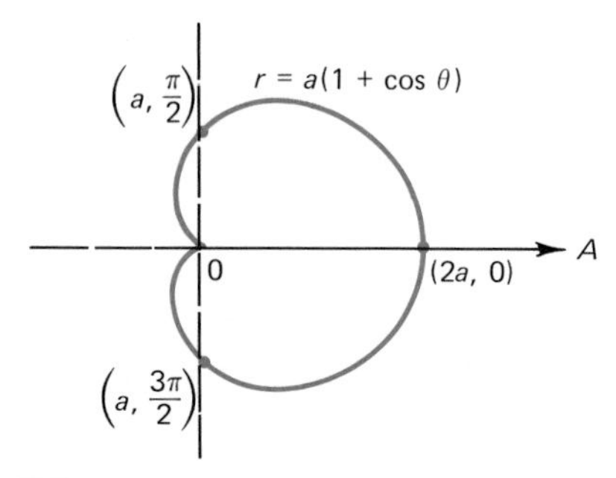

FIGURE 3

REMARK. If we ignored symmetry and tried to integrate from 0 to 2π, we would obtain

$$s = 2a\int_0^{2\pi} \sqrt{\frac{1+\cos\theta}{2}}\, d\theta = 2a\int_0^{2\pi} \cos\frac{\theta}{2}\, d\theta = 4a\sin\frac{\theta}{2}\Big|_0^{2\pi} = 0.$$

The problem is that $\cos\theta/2 \geq 0$ for $0 \leq \theta \leq \pi$ and $\cos\theta/2 \leq 0$ for $\pi \leq \theta \leq 2\pi$. We see that $\sqrt{(1 + \cos\theta/2} = \cos(\theta/2)$ for $0 \leq \theta \leq \pi$ and $\sqrt{(1 + \cos\theta)/2} = -\cos(\theta/2)$ for $\pi \leq \theta \leq 2\pi$. Thus it would be necessary to write

$$\begin{aligned}s &= 2a\int_0^{2\pi} \left|\cos\frac{\theta}{2}\right| d\theta = 2a\left(\int_0^\pi \cos\frac{\theta}{2}\, d\theta - \int_\pi^{2\pi} \cos\frac{\theta}{2}\, d\theta\right) \\ &= 4a\left(\sin\frac{\theta}{2}\Big|_0^\pi - \sin\frac{\theta}{2}\Big|_\pi^{2\pi}\right) = 4a[1 - (-1)] = 8a,\end{aligned}$$

as before. In other problems the reader is cautioned to pay attention to signs when taking square roots. ■

There is a more concise way to write our formula for arc length by using vector notation. Let the curve C be given by

$$\mathbf{f}(t) = f_1(t)\mathbf{i} + f_2(t)\mathbf{j}. \tag{9}$$

Then

$$\mathbf{f}'(t) = f_1'(t)\mathbf{i} + f_2'(t)\mathbf{j}$$

and

$$|\mathbf{f}'(t)| = \sqrt{[f_1'(t)]^2 + [f_2'(t)]^2} = \frac{ds}{dt}, \tag{10}$$

so that the length of the arc between t_0 and t_1 is given by

$$s = \int_{t_0}^{t_1} |\mathbf{f}'(t)|\, dt. \tag{11}$$

EXAMPLE 4 Calculate the length of the arc of the curve $\mathbf{f}(t) = (2t - t^2)\mathbf{i} + \frac{8}{3}t^{3/2}\mathbf{j}$ between $t = 1$ and $t = 3$.

Solution. $f'(t) = (2 - 2t)\mathbf{i} + 4\sqrt{t}\,\mathbf{j}$ and

$$|\mathbf{f}'(t)| = \sqrt{4 - 8t + 4t^2 + 16t} = 2\sqrt{t^2 + 2t + 1} = 2(t + 1),$$

so

$$s = \int_1^3 2(t + 1)\, dt = (t^2 + 2t)\Big|_1^3 = 12. \blacksquare$$

PROBLEMS 2.5

In Problems 1–20, find the length of the arc over the given interval or the length of the closed curve.

1. $x = t^3;\ y = t^2;\ 1 \le t \le 4$
2. $x = \cos 2\theta;\ y = \sin 2\theta;\ 0 \le \theta \le \pi/2$
3. $x = t^3 + 1;\ y = 3t^2 + 2;\ 0 \le t \le 2$
4. $x = 1 + t;\ y = (1 + t)^{3/2};\ 0 \le t \le 1$
5. $x = \dfrac{1}{\sqrt{t + 1}};\ y = \dfrac{t}{2(t + 1)};\ 0 \le t \le 4$
6. $x = e^t \cos t;\ y = e^t \sin t;\ 0 \le t \le \pi/2$
7. $x = \sin^2 t;\ y = \cos^2 t;\ 0 \le t \le \pi/2$
8. The hypocycloid of four cusps $x = a\cos^3\theta$, $y = a\sin^3\theta$, $a > 0$. [*Hint:* Calculate the length in the first quadrant and multiply by 4.]

*9. The cardioid $r = a(1 + \sin\theta)$. [*Hint:*

$$\int \sqrt{1 + \sin\theta}\, d\theta = \int \sqrt{1 + \sin\theta} \cdot \frac{\sqrt{1 - \sin\theta}}{\sqrt{1 - \sin\theta}}\, d\theta.$$

Pay attention to signs.]

10. One arc of the cycloid $x = a(\theta - \sin\theta)$, $y = a(1 - \cos\theta)$, $a > 0$.

11. $x = t^3$; $y = t^2$; $-1 \le t \le 1$ [*Hint:* $\sqrt{t^2} = -t$ for $t < 0$.]
12. $r = a \sin\theta$; $0 \le \theta \le \pi/2$, $a > 0$
13. $r = a \cos\theta$; $0 \le \theta \le \pi$, $a > 0$
***14.** $r = a\theta$; $0 \le \theta \le 2\pi$, $a > 0$
15. $r = e^{\theta}$; $0 \le \theta \le 3$
16. $r = \theta^2$; $0 \le \theta \le \pi$
17. $r = 6\cos^2(\theta/2)$; $0 \le \theta \le \pi/2$
18. $r = \sin^3(\theta/3)$; $0 \le \theta \le \pi/2$ [*Hint:* $\sin^2(\theta/3) = \frac{1}{2}(1 - \cos(2\theta/3))$.]
19. $\mathbf{f}(t) = e^t(\sin t)\mathbf{i} + e^t(\cos t)\mathbf{j}$; $0 \le t \le \pi/2$
20. $\mathbf{f}(t) = 3(\cos\theta)\mathbf{i} + 3(\sin\theta)\mathbf{j}$; $0 \le \theta \le 2\pi$
21. The parametric representation of the ellipse $(x^2/a^2) + (y^2/b^2) = 1$ is given by $x = a\cos\theta$, $y = b\sin\theta$. Find an integral that represents the length of the circumference of an ellipse but do not try to evaluate it. The integral you obtain is called an **elliptic integral,** and it arises in a variety of physical applications. It cannot be integrated (except numerically) unless $a = b$.
***22.** A tack is stuck in the front tire of a bicycle wheel with a diameter of 1 m. What is the total distance traveled by the tack if the bicycle moves a total of 30π m?

2.6 ARC LENGTH AS A PARAMETER

In many problems it is convenient to use the arc length s as a parameter. We can think of the vector function $\mathbf{f}$ as the *position vector* of a particle moving in the xy-plane. Then if $P_0 = (x_0, y_0)$ is a fixed point on the curve C described by the vector function $\mathbf{f}$, we may write

$$\mathbf{f}(s) = x(s)\mathbf{i} + y(s)\mathbf{j}, \tag{1}$$

where s is the distance along the curve measured from P_0 in the direction of increasing s. In this way we can determine the x- and y-components of the position vector as we move s units *along the curve.*

EXAMPLE 1 Write the vector $\mathbf{f}(t) = (\cos t)\mathbf{i} + (\sin t)\mathbf{j}$ (which describes the unit circle) with arc length as a parameter. Take $P_0 = (1, 0)$.

Solution. We have $ds/dt = 1$ (from Example 2.5.1), so that since $(1, 0)$ is reached when $t = 0$, we find that

$$s = \int_0^t ds = \int_0^t \frac{ds}{du}\,du = \int_0^t du = t.$$

Thus we may write

$$\mathbf{f}(s) = (\cos s)\mathbf{i} + (\sin s)\mathbf{j}.$$

For example, if we begin at the point $(1, 0)$ and move π units along the unit circle (which is half the unit circle), then we move to the point $(\cos\pi, \sin\pi) = (-1, 0)$. This is what we would expect. See Figure 1. ■

EXAMPLE 2 Let $\mathbf{f}(t) = (2t - t^2)\mathbf{i} + \frac{8}{3}t^{3/2}\mathbf{j}$, $t \ge 0$. Write this curve with arc length as a parameter.

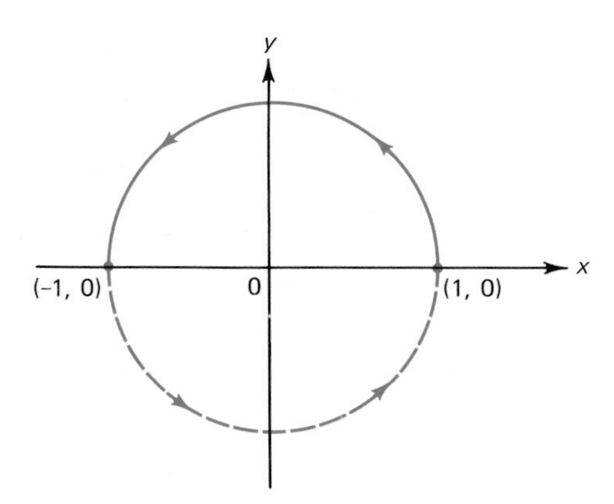

FIGURE 1

Solution. Suppose that the fixed point is $P_0 = (0, 0)$ when $t = 0$. Then from Example 2.5.4 we have

$$\frac{ds}{dt} = 2(t + 1),$$

so that

$$s = \int_0^t \frac{ds}{du}\, du = \int_0^t 2(u + 1)\, du = t^2 + 2t.$$

This leads to the equations

$$t^2 + 2t - s = 0 \qquad \text{and} \qquad t = \frac{-2 + \sqrt{4 + 4s}}{2} = \sqrt{1 + s} - 1.$$

We took the positive square root here since it is assumed that t starts at 0 and increases. Then

$$x = 2t - t^2 = 4\sqrt{1 + s} - 4 - s$$
$$y = \tfrac{8}{3}t^{3/2} = \tfrac{8}{3}(\sqrt{1 + s} - 1)^{3/2},$$

and we obtain

$$\mathbf{f}(s) = (4\sqrt{1 + s} - 4 - s)\mathbf{i} + \tfrac{8}{3}(\sqrt{1 + s} - 1)^{3/2}\mathbf{j}. \quad ■$$

As Example 2 illustrates, writing **f** explicitly with arc length as a parameter can be tedious (or, more often, impossible).

There is an interesting and important relationship between position vectors, tangent vectors, and normal vectors that becomes apparent when we use s as a parameter.

Theorem 1 If the curve C is parametrized by $\mathbf{f}(s) = x(s)\mathbf{i} + y(s)\mathbf{j}$, where s is arc length and x and y have continuous derivatives, then the unit tangent vector $\mathbf{T}$ is given by

$$\mathbf{T}(s) = \frac{d\mathbf{f}}{ds}. \tag{2}$$

Proof. With *any* parametrization of C, the unit tangent vector is given by [see equation (2.3.7) on page 45]

$$\mathbf{T}(t) = \frac{\mathbf{f}'(t)}{|\mathbf{f}'(t)|}.$$

So choosing $t = s$ yields

$$\mathbf{T}(s) = \frac{d\mathbf{f}/ds}{|d\mathbf{f}/ds|}.$$

But from equation (2.5.10), $|\mathbf{f}'(t)| = ds/dt$, so

$$\left|\frac{d\mathbf{f}}{ds}\right| = \left|\frac{d\mathbf{f}/dt}{ds/dt}\right| = \left|\frac{\mathbf{f}'(t)}{ds/dt}\right| = 1,$$

and the proof is complete. ■

Theorem 1 is quite useful in that it provides a check of our calculation of the parametrization in terms of arc length. For if

$$\mathbf{f}(s) = x(s)\mathbf{i} + y(s)\mathbf{j},$$

then

$$\mathbf{T} = \frac{d\mathbf{f}}{ds} = \frac{dx}{ds}\mathbf{i} + \frac{dy}{ds}\mathbf{j}.$$

But $|\mathbf{T}| = 1$, so that $|\mathbf{T}|^2 = 1$, which implies that

$$\left(\frac{dx}{ds}\right)^2 + \left(\frac{dy}{ds}\right)^2 = 1. \tag{3}$$

We can apply this result in Example 2. We have

$$x(s) = 4\sqrt{1+s} - 4 - s \quad \text{and} \quad y(s) = \tfrac{8}{3}(\sqrt{1+s} - 1)^{3/2},$$

so

$$\frac{dx}{ds} = \frac{2}{\sqrt{1+s}} - 1, \qquad \frac{dy}{ds} = \frac{2}{\sqrt{1+s}}(\sqrt{1+s} - 1)^{1/2},$$

and

$$\left(\frac{dx}{ds}\right)^2 + \left(\frac{dy}{ds}\right)^2 = \frac{4}{1+s} - \frac{4}{\sqrt{1+s}} + 1 + \frac{4}{1+s}(\sqrt{1+s} - 1) = 1,$$

as expected.

PROBLEMS 2.6

In the following problems, find parametric equations in terms of the arc length s measured from the point reached when $t = 0$. Verify your solution by using formula (3).

1. $\mathbf{f} = 3t^2\mathbf{i} + 2t^3\mathbf{j}$
2. $\mathbf{f} = t^3\mathbf{i} + t^2\mathbf{j}$
3. $\mathbf{f} = (t^3 + 1)\mathbf{i} + (t^2 - 1)\mathbf{j}$
4. $\mathbf{f} = (3t^2 + a)\mathbf{i} + (2t^3 + b)\mathbf{j}$
5. $\mathbf{f} = 3(\cos\theta)\mathbf{i} + 3(\sin\theta)\mathbf{j}$
6. $\mathbf{f} = a(\sin\theta)\mathbf{i} + a(\cos\theta)\mathbf{j}$
7. $\mathbf{f} = a(\cos\theta)\mathbf{i} + a(\sin\theta)\mathbf{j}$
8. $\mathbf{f} = 3(\cos t + t\sin t)\mathbf{i} + 3(\sin t - t\cos t)\mathbf{j}$
9. $\mathbf{f} = (a + b\cos\theta)\mathbf{i} + (c + b\sin\theta)\mathbf{j}$
10. $\mathbf{f} = ae^t(\cos t)\mathbf{i} + ae^t(\sin t)\mathbf{j}$
11. One cusp of the hypocycloid of four cusps

$$\mathbf{f} = a(\cos^3\theta)\mathbf{i} + a(\sin^3\theta)\mathbf{j}, \qquad 0 \le \theta \le \frac{\pi}{2}, \quad a > 0.$$

*12. The cycloid $x = a(\theta - \sin\theta)$, $y = a(1 - \cos\theta)$, $a > 0$.

2.7 VELOCITY, ACCELERATION, FORCE, AND MOMENTUM

Suppose that an object is moving in the plane. Then we can describe its motion parametrically by the vector function

$$\mathbf{f}(t) = f_1(t)\mathbf{i} + f_2(t)\mathbf{j}. \tag{1}$$

In this context **f** is called the **position vector** of the object, and the curve described by **f** is called the **trajectory** of the object. We then have the following definition.

Definition 1 VELOCITY AND ACCELERATION VECTOR If $\mathbf{f}'$ and $\mathbf{f}''$ exist, then

$$\text{(i)}\quad \mathbf{v}(t) = \mathbf{f}'(t) = f_1'(t)\mathbf{i} + f_2'(t)\mathbf{j} \tag{2}$$

is called the **velocity vector** of the moving object at time t.

$$\text{(ii)}\quad \mathbf{a}(t) = \frac{d\mathbf{v}}{dt} = \mathbf{f}''(t) = f_1''(t)\mathbf{i} + f_2''(t)\mathbf{j} \tag{3}$$

is called the **acceleration vector** of the object.

This definition is, of course, not surprising. It simply extends to the vector case our notion of velocity as the derivative of position and acceleration as the derivative of velocity.

Definition 2 SPEED AND ACCELERATION SCALAR

(i) The **speed** $v(t)$ of a moving object is the magnitude of the velocity vector.

(ii) The **acceleration scalar** $a(t)$ is the magnitude of the acceleration vector.

REMARK 1. Since we have already shown that $|\mathbf{f}'(t)| = ds/dt$ [equation (2.5.10)], we have, since $v(t) = |\mathbf{v}(t)| = |\mathbf{f}'(t)|$,

$$v(t) = \frac{ds}{dt}. \tag{4}$$

REMARK 2. Although $\mathbf{a}(t)$ is the derivative of $\mathbf{v}(t)$, it is *not true* in general that $a(t)$ is the derivative of the speed $v(t)$. For example, consider the motion along the unit circle given by

$$\mathbf{f}(t) = (\cos t)\mathbf{i} + (\sin t)\mathbf{j}.$$

Then

$$\mathbf{v}(t) = -(\sin t)\mathbf{i} + (\cos t)\mathbf{j} \quad \text{and} \quad \mathbf{a}(t) = -(\cos t)\mathbf{i} - (\sin t)\mathbf{j}.$$

But $v(t) = |\mathbf{v}(t)| = 1$, so that $dv/dt = 0$. But $a(t) = |\mathbf{a}(t)| = 1$, which is, evidently, not equal to dv/dt.

REMARK 3. It follows from Theorem 2.4.2 that *if speed is constant, then the velocity and acceleration vectors are orthogonal.*

EXAMPLE 1 A particle is moving along the circle with the position vector $\mathbf{f} = 3(\cos 2t)\mathbf{i} + 3(\sin 2t)\mathbf{j}$. Calculate $\mathbf{v}(t)$, $\mathbf{a}(t)$, $v(t)$, and $a(t)$, and find the velocity and acceleration vectors when $t = \pi/6$. Assume that distance is measured in meters.

Solution. Here $\mathbf{v}(t) = \mathbf{f}'(t) = -6(\sin 2t)\mathbf{i} + 6(\cos 2t)\mathbf{j}$ and $\mathbf{a}(t) = -12(\cos 2t)\mathbf{i} - 12(\sin 2t)\mathbf{j}$. Then $v(t) = |\mathbf{v}(t)| = \sqrt{36 \sin^2 2t + 36 \cos^2 2t} = 6$ m/sec and $a(t) = |\mathbf{a}(t)| = 12$ m/sec^2. Finally, $\mathbf{v}(\pi/6) = -3\sqrt{3}\mathbf{i} + 3\mathbf{j}$ and $\mathbf{a}(\pi/6) = -6\mathbf{i} - 6\sqrt{3}\mathbf{j}$. We sketch these vectors in Figure 1. Note that $\mathbf{a}(t) \cdot \mathbf{v}(t) = 0$. This follows from Theorem 2.4.2 and the fact that $|\mathbf{v}(t)|$ is constant. ■

We now calculate the vectors that describe the motion of an object in a vertical plane. First, we make the simplifying assumption that the only force acting on the object is the force of gravity. We ignore, for example, the frictional force due to air resistance. The force of gravity is directed vertically downward. There is no force acting in the horizontal direction. Thus the acceleration vector is given by

$$\mathbf{a} = -g\mathbf{j}, \tag{5}$$

where the constant $g = 9.81$ m/sec^2 $= 32.2$ ft/sec^2. In addition,

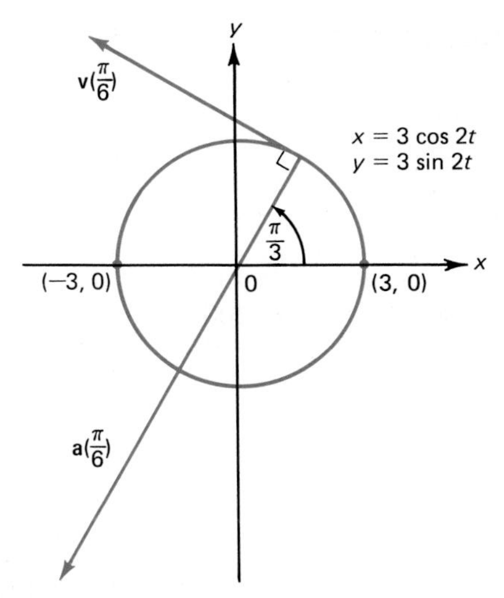

FIGURE 1

$$\mathbf{F} = m\mathbf{a} = -mg\mathbf{j},$$

where m is the mass of the object. To obtain the velocity vector $\mathbf{v}(t)$, we simply integrate (5) to obtain

$$\mathbf{v}(t) = -gt\mathbf{j} + \mathbf{C}, \tag{6}$$

where $\mathbf{C}$ is a constant vector. But from (6) we find that

$$\mathbf{v}(0) = \mathbf{C},$$

so that $\mathbf{C}$ is the initial velocity vector $\mathbf{v}_0$ (the velocity at $t = 0$) and (6) becomes

$$\mathbf{v}(t) = -gt\mathbf{j} + \mathbf{v}_0. \tag{7}$$

Integrating (7), we obtain the position vector

$$\mathbf{f}(t) = -\tfrac{1}{2}gt^2\mathbf{j} + \mathbf{v}_0t + \mathbf{D}, \tag{8}$$

where $\mathbf{D}$ is another constant vector. Evaluating (8) at $t = 0$ yields $\mathbf{D} = \mathbf{f}(0) =$ the initial position vector $\mathbf{f}_0$, so (8) becomes

$$\mathbf{f}(t) = -\tfrac{1}{2}gt^2\mathbf{j} + \mathbf{v}_0t + \mathbf{f}_0. \tag{9}$$

EXAMPLE 2 A cannon whose muzzle is tilted upward at an angle of 30° shoots a ball at an initial velocity of 600 m/sec.

(a) Find the position vector at all times $t \geq 0$.
(b) How much time does the ball spend in the air?

(c) How far does the cannonball travel?
(d) How high does the ball get?
(e) How far from the cannon does the ball land?
(f) What is the speed of the ball at the time of impact with the earth? Assume that the mouth of the cannon is at ground level and ignore air resistance.

Solution.

(a) From Example 2.1.6 we find, resolving the initial velocity vector into its vertical and horizontal components, that $\mathbf{v}_0 = 300\sqrt{3}\mathbf{i} + 300\mathbf{j}$. Moreover, if we place the origin so that it coincides with the mouth of the cannon, then the initial position vector $\mathbf{f}_0$ is $(0, 0) = 0\mathbf{i} + 0\mathbf{j}$. From (9)

$$\begin{aligned} \mathbf{f}(t) &= -\tfrac{1}{2}gt^2\mathbf{j} + 300\sqrt{3}t\mathbf{i} + 300t\mathbf{j} + \mathbf{0} \\ &= 300\sqrt{3}t\mathbf{i} + (300t - \tfrac{1}{2}gt^2)\mathbf{j}. \end{aligned}$$

(b) The ball hits the ground when the vertical component of the position vector is zero, that is, when $300t - \frac{1}{2}gt^2 = 0$, which occurs at $t = 600/g$ seconds.

(c) The total distance traveled is

$$s = \int_0^{600/g} \left(\frac{ds}{dt}\right) dt.$$

But

$$\begin{aligned} \frac{ds}{dt} &= v(t) = |\mathbf{v}(t)| = |\mathbf{f}'(t)| = |300\sqrt{3}\mathbf{i} + (300 - gt)\mathbf{j}| \\ &= \sqrt{270{,}000 + (300 - gt)^2} \end{aligned}$$

Thus using formula (2.5.4), we have

$$s = \int_0^{600/g} \left(\frac{ds}{dt}\right) dt = \int_0^{600/g} \sqrt{270{,}000 + (300 - gt)^2}\, dt \text{ meters}$$

Let $300 - gt = \sqrt{270{,}000} \tan\theta = 300\sqrt{3}\tan\theta$. Then

$$-g\, dt = 300\sqrt{3}\sec^2\theta\, d\theta$$

and

$$dt = \left(\frac{-300\sqrt{3}}{g}\right)\sec^2\theta\, d\theta.$$

Also,

$$\sqrt{270{,}000 + (300 - gt)^2} = \sqrt{270{,}000(1 + \tan^2\theta)} = 300\sqrt{3}\sec\theta.$$

When $t = 0$, $300 = 300\sqrt{3}\tan\theta$, $\tan\theta = 1/\sqrt{3}$, and $\theta = \pi/6$. When $t = 600/g$,

$\tan\theta = -1/\sqrt{3}$ and $\theta = -\pi/6$. Thus

$$s = -\frac{1}{g}\int_{\pi/6}^{-\pi/6} (300\sqrt{3})(300\sqrt{3})\sec^3\theta\,d\theta$$

$$= \frac{270{,}000}{g}\int_{-\pi/6}^{\pi/6} \sec^3\theta\,d\theta = \frac{540{,}000}{g}\int_{0}^{\pi/6} \sec^3\theta\,d\theta$$

Entry 147 in the Table of Integrals at the back of the book

$$= \frac{270{,}000}{g}\left(\ln|\sec\theta + \tan\theta| + \sec\theta\tan\theta\right)\Big|_0^{\pi/6}$$

$$= \frac{270{,}000}{g}\left(\ln\left|\frac{2}{\sqrt{3}} + \frac{1}{\sqrt{3}}\right| + \frac{2}{\sqrt{3}}\frac{1}{\sqrt{3}}\right) = \frac{270{,}000}{g}\left(\ln\sqrt{3} + \frac{2}{3}\right) \approx 33.467 \text{ km.}$$

(d) The maximum height is achieved when $dy/dt = 0$. That is, when $300 - gt = 0$ or when $t = 300/g$. For that value of t, $y_{\max} = 45{,}000/g \approx 4587.2$ m $= 4.5872$ km.

(e) In $600/g$ seconds, the x-component of **f** increases from 0 to $(300\sqrt{3})\cdot(600/g) \approx 31{,}780.7$ m ≈ 31.78 km.

(f) Since speed $= v(t) = \sqrt{270{,}000 + (300 - gt)^2}$ meters per second, we find that upon impact $t = 600/g$ seconds, so that

$$v\left(\frac{600}{g}\right) = \sqrt{270{,}000 + \left(300 - g\cdot\frac{600}{g}\right)^2} = \sqrt{270{,}000 + 90{,}000}$$

$$= \sqrt{360{,}000} = 600 \text{ m/sec.}$$

These results are illustrated in Figure 2. ■

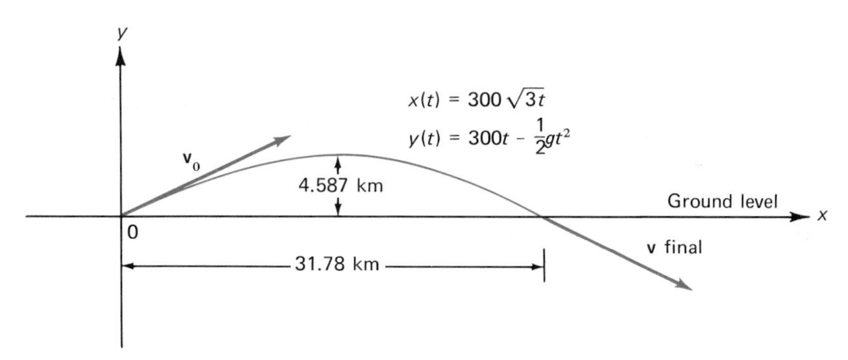

FIGURE 2

There is another convenient application of vectors applied to the motion of an object.

Definition 3 MOMENTUM The **momentum P** of a particle at any time t is a vector defined as the product of the mass m of the particle and its velocity **v**. That is,

$$\mathbf{P} = m\mathbf{v}. \tag{10}$$

NEWTON'S[†] SECOND LAW OF MOTION Newton's second law of motion states that the *rate of change of momentum of a moving object is proportional to the resultant force and is in the direction of that force*. That is,

$$\mathbf{F} = \frac{d\mathbf{P}}{dt}. \tag{11}$$

If the mass of the object is constant, then using (10) and (11), we have the familiar law

$$\mathbf{F} = \frac{d}{dt}m\mathbf{v} = m\frac{d\mathbf{v}}{dt} = m\mathbf{a}. \tag{12}$$

If mass is not constant, then (12) becomes

$$\mathbf{F} = \frac{d}{dt}m\mathbf{v} = m\frac{d\mathbf{v}}{dt} + \mathbf{v}\frac{dm}{dt}. \tag{13}$$

EXAMPLE 3 The cannonball of Example 2 has a mass of 8 kg. Find the force acting on the ball at any time t.

Solution. Here m is constant, so that $\mathbf{F} = m\mathbf{a} = (8 \text{ kg})(-g\mathbf{j}) = -8g\mathbf{j}$, which is a force with a magnitude of $8g$ newtons acting vertically downward. There is no force in the horizontal direction. ■

EXAMPLE 4 An object of mass m moves in the elliptical orbit given by $\mathbf{f}(t) = a(\cos \alpha t)\mathbf{i} + b(\sin \alpha t)\mathbf{j}$. Find the force acting on the object at any time t.

Solution. We easily find that $\mathbf{a}(t) = \mathbf{f}''(t) = -\alpha^2 a(\cos \alpha t)\mathbf{i} - \alpha^2 b(\sin \alpha t)\mathbf{j} = -\alpha^2\mathbf{f}(t)$. Since m is constant, $\mathbf{F} = m\mathbf{a} = -m\alpha^2\mathbf{f}(t)$. Thus the force always acts in the direction opposite to the direction of the position vector (thereby pointing *toward* the origin) and has a magnitude proportional to the distance of the object from the origin. Such a force is called a **central force** (see Figure 3). ■

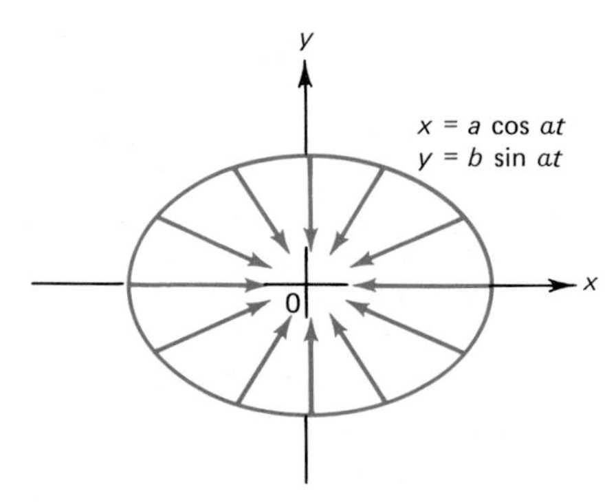

FIGURE 3

[†]See the accompanying biographical sketch.

1642–
1727

Sir Isaac Newton

Isaac Newton
The Granger Collection

Isaac Newton was born in the small English town of Woolsthorpe on Christmas Day 1642, the year of Galileo's death. His father, a farmer, had died before Isaac was born. His mother remarried when he was three and, thereafter, Isaac was raised by his grandmother. As a boy, Newton showed great cleverness and inventiveness—designing a water clock and a toy gristmill, among other things. One of his uncles, a Cambridge graduate, took an interest in the boy's education, and as a result, Newton entered Trinity College, Cambridge in 1661. His primary interest at that time was chemistry.

Newton's interest in mathematics began with his discovery of two of the great mathematics books of his day: Euclid's *Elements* and Descartes's *La géométrie.* He also became aware of the work of the great scientists who preceded him, including Galileo and Fermat.

By the end of 1664, Newton seems to have mastered all the mathematical knowledge of the time and had begun adding substantially to it. In 1665 he began his study of the rates of change, or *fluxions,* of quantities, such as distances or temperatures that varied continuously. The result of this study was what today we call *differential calculus.*

Newton disliked controvery so much that he delayed the publication of many of his findings for years. An unfortunate result of one of these delays was a conflict with Leibniz over who first discovered calculus. Leibniz made similar discoveries at about the same time as Newton, and to this day there is no universal agreement as to who discovered what first. The conflict stirred up so much ill will that English mathematicians (supporters of Newton) and continental mathematicians (supporters of Leibniz) had virtually no communication for more than a hundred years. English mathematics suffered greatly as a result.

Newton made many of the discoveries that governed physics until the discoveries of Einstein early in this century. In 1679 he used a new measurement of the radius of the earth, together with an analysis of the earth's motion, to formulate his universal law of gravitational attraction. Although he made many other discoveries at that time, he communicated them to no one for five years. In 1684 Edmund Halley (after whom Halley's comet is named) visited Cambridge to discuss his theories of planetary motion with Newton. The conversations with Halley stimulated Newton's interest in celestial mechanics and led him to work out many of the laws that govern the motion of bodies subject to the forces of gravitation. The result of this work was the 1687 publication of Newton's masterpiece, *Philosophiae naturalis principia mathematica* (known as the *Principia*). It was received with great acclaim throughout Europe.

Newton is considered by many the greatest mathematician the world has ever produced. He was the greatest "applied" mathematician, determined by his ability to discover a physical property and analyze it in mathematical terms. Leibniz once said, "Taking mathematics from the beginning of the world to the time when Newton lived, what he did was much the better half." The great English poet Alexander Pope wrote,

> Nature and Nature's laws lay hid in night;
> God said, 'Let Newton be,' and all was light.

Newton, by contrast, was modest about his accomplishments. Late in life he wrote, "If I have seen farther than Descartes, it is because I have stood on the shoulders of giants." All who study mathematics today are standing on Isaac Newton's shoulders.

If the vector sum of forces acting on a system is zero, then, since $\mathbf{0} = \mathbf{F} = \mathbf{P}'(t)$, we find that the *total momentum of the system is constant*. This fact is called the **principle of the conservation of linear momentum.** We now see how that principle can be applied to the motion of a rocket.

Let the mass of a rocket be denoted by m, and suppose that the rocket is traveling with velocity $\mathbf{v}(t)$. The rocket is propelled by gas emissions. We assume that the exhaust velocity of the gases ejected from the rocket, relative to the rocket, is $\mathbf{u}(t)$. Then the total velocity of the gas (relative to the earth, say) is $\mathbf{u} + \mathbf{v}$. The rate of change of momentum on the rocket is

$$\mathbf{P}'(t) = \frac{d}{dt}(m\mathbf{v}) = m\frac{d\mathbf{v}}{dt} + \mathbf{v}\frac{dm}{dt}.$$

As the rocket loses mass, the exhaust gains it, so the rate of change of momentum of the gases is given by

$$-(\mathbf{u} + \mathbf{v})\frac{dm}{dt}.$$

Here dm/dt is the rate at which exhaust gas accumulates. The total force acting on the rocket-gas system is the sum of the rates of change of momentum for each of the two components of the system; that is,

$$\mathbf{F} = m\frac{d\mathbf{v}}{dt} + \mathbf{v}\frac{dm}{dt} - \mathbf{u}\frac{dm}{dt} - \mathbf{v}\frac{dm}{dt} = m\frac{d\mathbf{v}}{dt} - \mathbf{u}\frac{dm}{dt}$$

or

$$\mathbf{F} + \mathbf{u}\frac{dm}{dt} = m\frac{d\mathbf{v}}{dt}. \tag{14}$$

The term $\mathbf{u}\ dm/dt$ is called the **thrust** of the rocket. Note that $dm/dt < 0$ (since the rocket is losing mass), so that the thrust has the direction *opposite* to that of the velocity $\mathbf{u}$ of the exhaust gases.

Now if the rocket is free of any gravitational field, then there are no external forces acting on the system, so $\mathbf{F} = \mathbf{0}$ and (14) becomes

$$\mathbf{u}\frac{dm}{dt} = m\frac{d\mathbf{v}}{dt}. \tag{15}$$

If $\mathbf{u}$ is constant, then (15) becomes

$$\frac{d\mathbf{v}}{dt} = \frac{\mathbf{u}}{m}\frac{dm}{dt}$$

and integration yields

$$\mathbf{v}(t) - \mathbf{v}_0 = \mathbf{u} \int_{m_0}^{m} \frac{dm}{m} = \mathbf{u} \ln m \Big|_{m_0}^{m}$$

or

$$\mathbf{v} - \mathbf{v}_0 = -\mathbf{u} \ln\left(\frac{m_0}{m}\right). \tag{16}$$

Here $\mathbf{v}_0$ is the initial velocity vector and m_0 is the initial mass of the rocket and the fuel combined. Thus once the rocket has escaped all gravitational fields, we can calculate its velocity at any time if we know the velocity of the escaping gases and the proportion of gases that have been expelled.

EXAMPLE 5 A 1000-kg rocket carrying 2000 kg of fuel is motionless in space. Its engine starts and its exhaust velocity is 0.7 km/sec. What is its speed when all its fuel is consumed?

Solution. Here $\mathbf{v}_0 = 0$. Then

$$|\mathbf{v}| = |\mathbf{u}| \ln \frac{m_0}{m} = 0.7 \ln \frac{3000}{1000} = 0.7 \ln 3 \approx 0.769 \text{ km/sec}$$

$$\approx 2769 \text{ km/hr } (\approx 1720 \text{ mi/hr}). \blacksquare$$

EXAMPLE 6 The calculation above was made in the absence of gravitational forces. Let us now assume that the rocket is fired upward from the earth and assume that gas is ejected at the constant rate of 100 kg/sec. Then the force propelling the rocket upward is equal to the thrust and is given by

$$F_{\text{thrust}} = \left| \mathbf{u} \frac{dm}{dt} \right| = |(700 \text{ m/sec})(-100 \text{ kg/sec})| = 70{,}000 \text{ N}.$$

The thrust is opposed by the force of gravity:

$$F_{\text{grav}} = -mg = (-3000 \text{ kg})(9.81 \text{ m/sec}^2) = -29{,}430 \text{ N}.$$

(Note that we have treated **u** and **F** as scalars since we are assuming motion in only one direction: straight up.) We see that the initial net upward force is given by

$$F_{\text{thrust}} + F_{\text{grav}} = 70{,}000 - 29{,}430 = 40{,}570 \text{ N}.$$

Finally, just before all the fuel is used up, the force due to gravity is

$$F_{\text{grav}} = (-1000 \text{ kg})(9.81 \text{ m/sec}^2) = -9810 \text{ N},$$

so the net force is $70{,}000 - 9810 = 60{,}190$ N. Of course, when the fuel burns out, the rocket will fall back to earth.

REMARK. In this example we assumed, as usual, that the earth's gravitational attraction could be represented by the constant $-mg$; that is, the rocket did not go high enough to escape or partially escape the effect of the earth's gravitational field. ■

PROBLEMS 2.7

In Problems 1–14, the position vector of a moving particle is given. For the indicated value of t, calculate the velocity vector, the acceleration vector, the speed, and the acceleration scalar. Then sketch the portion of the trajectory showing the velocity and acceleration vectors.

1. $\mathbf{f} = (\cos 3t)\mathbf{i} + (\sin 3t)\mathbf{j};\ t = 0$
2. $\mathbf{f} = (\cos 5t)\mathbf{i} + (\sin 5t)\mathbf{j};\ t = \pi/2$
3. $\mathbf{f} = 2(\cos 4t)\mathbf{i} + 2(\sin 4t)\mathbf{j};\ t = \pi/6$
4. $\mathbf{f} = -3(\cos 10t)\mathbf{i} - 3(\sin 10t)\mathbf{j};\ t = \pi$
5. $\mathbf{f} = 4t\mathbf{i} + 2t^2\mathbf{j};\ t = 1$
6. $\mathbf{f} = (2 + 3t)\mathbf{i} + (8 - 5t)\mathbf{j};\ t = 3$
7. $\mathbf{f} = (4 - 7t)\mathbf{i} + (-3 + 5t)\mathbf{j};\ t = -5$
8. $\mathbf{f} = (a + bt)\mathbf{i} + (c + dt)\mathbf{j};\ t = t_0,\ a, b, c, d$ real
9. $\mathbf{f} = 2t\mathbf{i} + (e^{-t} + e^t)\mathbf{j};\ t = 0$
10. $\mathbf{f} = \cosh t\mathbf{i} + 2t\mathbf{j};\ t = 2$
11. $\mathbf{f} = (t - \cos t)\mathbf{i} + (1 - \sin t)\mathbf{j};\ t = \pi/2$
12. $\mathbf{f} = (t - \cos t)\mathbf{i} + (1 - \sin t)\mathbf{j};\ t = \pi$
13. $\mathbf{f} = (t - \cos t)\mathbf{i} + (1 - \sin t)\mathbf{j};\ t = \pi/4$
14. $\mathbf{f} = (\ln \sin t)\mathbf{i} + (\ln \cos t)\mathbf{j};\ t = \pi/6$

*15. A bullet is shot from a gun with an initial velocity of 1200 m/sec. The gun is inclined at an angle of 45°. Find (a) the total distance traveled by the bullet, (b) the horizontal distance traveled by the bullet, (c) the maximum height it reaches, and (d) the speed of the bullet at impact. Assume that the gun is held at ground level.

*16. Answer the questions of Problem 15 if the angle of inclination of the gun is 60° and the initial velocity is 3000 ft/sec.

17. A man is standing at the top of a 200-m building. He throws a ball horizontally with an initial speed of 20 m/sec.
    - **(a)** How far does the ball travel in its path?
    - **(b)** How far from the base of the building does the ball hit the ground?
    - **(c)** At what angle with the horizontal does the ball hit the ground?
    - **(d)** With what speed does the ball hit the ground?

18. An airplane, flying horizontally at a height of 1500 m and with a speed of 450 km/hr, releases a bomb.
    - **(a)** How long does it take the bomb to hit the ground?
    - **(b)** How far does it travel in the horizontal direction?
    - **(c)** What is its speed of impact?

19. A girl is standing 20 ft from a tall building. She throws a ball at an angle of 60° toward the building with an initial speed of 60 ft/sec. Her hand is 4 ft above the ground when she releases the ball. How high up the building does the ball hit?

*20. A man can throw a football a maximum distance of 60 yd. What is the maximum speed, in feet per second, that he can throw the football?

21. A 2000-kg rocket carrying 3000 kg of fuel is initially motionless in space. Its exhaust velocity is 1 km/sec. What is its velocity when all its fuel is consumed?

22. The rocket in Problem 21 is launched from the earth, and gas is emitted at a constant rate of 75 kg/sec.
    - **(a)** Find the net initial force acting on the rocket.
    - **(b)** Find the net force just before all the fuel is expended.

23. In Problem 22, what is the minimum thrust needed to get the rocket off the ground?

2.8 CURVATURE AND THE ACCELERATION VECTOR (OPTIONAL)

The derivative dy/dx of a curve $y = f(x)$ measures the rate of change of the vertical component of the curve with respect to the horizontal component. As we have seen, the derivative ds/dt represents the change in the length of the arc traced out by the vector $\mathbf{f} = x(t)\mathbf{i} + y(t)\mathbf{j}$ as t increases. Another quantity of interest is the rate of change of the direction of the curve with respect to the length of the curve. That is, how much does the direction change for every one-unit change in the arc length? We are thus led to the following definitions.

Definition 1 CURVATURE

(i) Let the curve C be given by the differentiable vector function $\mathbf{f}(t) = f_1(t)\mathbf{i} + f_2(t)\mathbf{j}$. Let $\varphi(t)$ denote the direction of $\mathbf{f}'(t)$. Then the **curvature** of C, denoted $\kappa(t)$, is the absolute value of the rate of change of direction with respect to arc length; that is,

$$\kappa(t) = \left|\frac{d\varphi}{ds}\right|. \tag{1}$$

Note that $\kappa(t) \geq 0$.

(ii) The **radius of curvature** $\rho(t)$ is defined by

$$\rho(t) = \frac{1}{\kappa(t)} \quad \text{if} \quad \kappa(t) > 0. \tag{2}$$

REMARK 1. The curvature is a measure of *how fast* the curve turns as we move along it.

REMARK 2. If $\kappa(t) = 0$, we say that the radius of curvature is *infinite.* To understand this idea, note that if $\kappa(t) = 0$, then the "curve" does not bend and so is a straight line (see Example 2). A straight line can be thought of as an arc of a circle with *infinite* radius.

In Chapter 3 we will need a definition of curvature that does not depend on the angle φ.

Theorem 1 If $\mathbf{T}(t)$ denotes the unit tangent vector to $\mathbf{f}$, then

$$\kappa(t) = \left|\frac{d\mathbf{T}}{ds}\right|. \tag{3}$$

Proof. By the chain rule (which applies just as well to vector-valued functions),

$$\frac{d\mathbf{T}}{ds} = \frac{d\mathbf{T}}{d\varphi}\frac{d\varphi}{ds}.$$

But since φ is the direction of $\mathbf{f}'$, and therefore also the direction of $\mathbf{T}$, we have

$$\mathbf{T} = (\cos\varphi)\mathbf{i} + (\sin\varphi)\mathbf{j},$$

so that

$$\frac{d\mathbf{T}}{d\varphi} = -(\sin\varphi)\mathbf{i} + (\cos\varphi)\mathbf{j} \quad \text{and} \quad \left|\frac{d\mathbf{T}}{d\varphi}\right| = \sqrt{\sin^2\varphi + \cos^2\varphi} = 1.$$

Thus

$$\left|\frac{d\mathbf{T}}{ds}\right| = \left|\frac{d\mathbf{T}}{d\varphi}\right|\left|\frac{d\varphi}{ds}\right| = 1\left|\frac{d\varphi}{ds}\right| = \kappa(t),$$

and the theorem is proved. ■

We now derive an easier way to calculate $\kappa(t)$.

Theorem 2 With the curve C given as in Definition 1, the curvature of C is given by the formula

$$\kappa(t) = \frac{|(dx/dt)(d^2y/dt^2) - (dy/dt)(d^2x/dt^2)|}{[(dx/dt)^2 + (dy/dt)^2]^{3/2}} = \frac{|x'y'' - y'x''|}{[(x')^2 + (y')^2]^{3/2}} \tag{4}$$

where $x(t) = f_1(t)$ and $y(t) = f_2(t)$.

Proof. By the chain rule,

$$\frac{d\varphi}{ds} = \frac{d\varphi}{dt}\frac{dt}{ds} = \frac{d\varphi/dt}{ds/dt}. \tag{5}$$

From equation (2.5.10), on page 59,

$$\frac{ds}{dt} = \sqrt{\left(\frac{dx}{dt}\right)^2 + \left(\frac{dy}{dt}\right)^2}, \tag{6}$$

and from equation (2.3.6) on page 44, we obtain

$$\tan\varphi = \frac{dy/dt}{dx/dt} \tag{7}$$

or

$$\varphi = \tan^{-1}\frac{dy/dt}{dx/dt}. \tag{8}$$

Differentiating both sides of (8) with respect to t, we obtain

$$\frac{d\varphi}{dt} = \frac{1}{1 + \left(\frac{dy/dt}{dx/dt}\right)^2}\frac{d}{dt}\left(\frac{dy/dt}{dx/dt}\right) = \frac{\left(\frac{dx}{dt}\right)^2}{\left(\frac{dx}{dt}\right)^2 + \left(\frac{dy}{dt}\right)^2}\cdot\frac{\frac{dx}{dt}\frac{d^2y}{dt^2} - \frac{dy}{dt}\frac{d^2x}{dt^2}}{\left(\frac{dx}{dt}\right)^2} \tag{9}$$

Substitution of (6) and (9) into (5) completes the proof of the theorem. ■

EXAMPLE 1 We certainly expect that the curvature of a circle is constant and that its radius of curvature is its radius. Show that this is true.

Solution. The circle of radius r centered at the origin is given parametrically by

$$x = r\cos t, \qquad y = r\sin t.$$

Then $dx/dt = -r\sin t$, $d^2x/dt^2 = -r\cos t$, $dy/dt = r\cos t$, and $d^2y/dt^2 = -r\sin t$, so from (4),

$$\kappa(t) = \frac{|r^2\sin^2 t + r^2\cos^2 t|}{[r^2\sin^2 t + r^2\cos^2 t]^{3/2}} = \frac{r^2}{r^3} = \frac{1}{r}$$

and

$$\rho(t) = \frac{1}{\kappa(t)} = r,$$

as expected. ■

EXAMPLE 2 Show that for a straight line $\kappa(t) = 0$.

Solution. A line can be represented parametrically [see equation (2.1.9)] by $x = x_1 + t(x_2 - x_1)$ and $y = y_1 + t(y_2 - y_1)$. Then $dx/dt = x_2 - x_1$, $d^2x/dt^2 = 0$, $dy/dt = y_2 - y_1$, and $d^2y/dt^2 = 0$. Substitution of these values into (4) immediately yields $\kappa(t) = 0$. ■

EXAMPLE 3 Find the curvature and radius of curvature of the curve given parametrically by $x = (t^3/3) - t$ and $y = t^2$ (see Example 2.4.4) at $t = 2\sqrt{2}$.

Solution. We have $dx/dt = t^2 - 1$, $d^2x/dt^2 = 2t$, $dy/dt = 2t$, and $d^2y/dt^2 = 2$, so

$$\kappa(t) = \frac{|2(t^2-1) - 2t(2t)|}{[(t^2-1)^2 + (2t)^2]^{3/2}} = \frac{2(t^2+1)}{(t^4+2t^2+1)^{3/2}} = \frac{2(t^2+1)}{(t^2+1)^3} = \frac{2}{(t^2+1)^2}.$$

Then $\kappa(2\sqrt{2}) = 2/81$, and the radius of curvature $\rho(2\sqrt{2}) = 1/\kappa(2\sqrt{2}) = 81/2$. A portion of this curve near $t = 2\sqrt{2}$ is sketched in Figure 1. For reference purposes the unit tangent vector **T** at $t = 2\sqrt{2}$ is included. ■

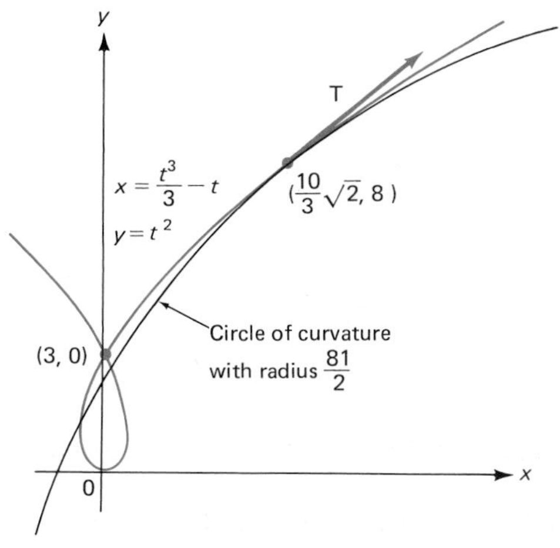

FIGURE 1

If we are given the Cartesian equation of a curve, $y = f(x)$, then the formula for the curvature, now denoted by $\kappa(x)$, is simpler. Proceeding as before, we have

$$\frac{d\varphi}{dx} = \frac{d\varphi}{ds}\frac{ds}{dx} \qquad \text{or} \qquad \frac{d\varphi}{ds} = \frac{d\varphi/dx}{ds/dx}.$$

From the formula for arc length†

$$\frac{ds}{dx} = \sqrt{1 + \left(\frac{dy}{dx}\right)^2}.$$

Also, since $\tan \varphi = dy/dx$, we have

$$\varphi = \tan^{-1}\frac{dy}{dx} \qquad \text{and} \qquad \frac{d\varphi}{dx} = \frac{1}{1 + (dy/dx)^2} \cdot \frac{d^2y}{dx^2}$$

so that

$$\kappa(x) = \left|\frac{d\varphi}{dx}\right| = \frac{|d^2y/dx^2|}{[1 + (dy/dx)^2]^{3/2}}. \tag{10}$$

EXAMPLE 4 Calculate the curvature and radius of curvature of the curve $y = 3/x$ at the point (1, 3).

Solution. Here $dy/dx = -3/x^2$, $d^2y/dx^2 = 6/x^3$, and

$$\kappa(x) = \frac{|6/x^3|}{[1 + (9/x^4)]^{3/2}} = \frac{|6/x^3|}{[(9 + x^4)/x^4]^{3/2}} = \frac{|6x^3|}{(9 + x^4)^{3/2}}.$$

When $x = 1$, $\kappa(1) = 6/10\sqrt{10} = 3\sqrt{10}/50 \approx 0.19$. Then $\rho(1) = 50/3\sqrt{10} \approx 5.27$. The curve is sketched in Figure 2. ■

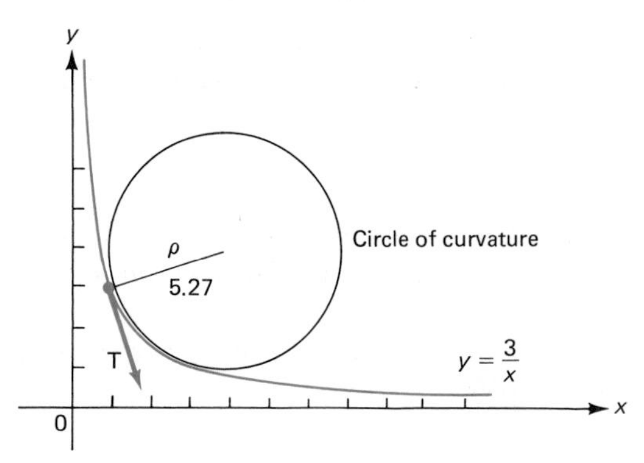

FIGURE 2

†See the introduction to Section 2.5.

NOTE 1. In Figures 1 and 2 the circle with radius of curvature $\rho(t)$ that lies on the concave side of C is called the **circle of curvature,** or **osculating circle.**

NOTE 2. The curve $y = 3/x$ is concave up for $x > 0$. Graphically, it seems as if the curve is "flattening out" as x gets larger. This is easy to prove since $\lim_{x\to\infty} \kappa(x) = 0$. In fact, it can be shown that if any curve is concave up (or down) for x sufficiently large, then $\lim_{x\to\infty} \kappa(x) = 0$.

TANGENTIAL AND NORMAL COMPONENTS OF ACCELERATION

There is an interesting relationship between curvature and acceleration vectors. If a particle is moving along the curve C with position vector

$$\mathbf{f}(t) = x(t)\mathbf{i} + y(t)\mathbf{j},$$

then from the last section the acceleration vector is given by

$$\mathbf{a}(t) = \frac{d^2x}{dt^2}\mathbf{i} + \frac{d^2y}{dt^2}\mathbf{j}. \tag{11}$$

The representation (11) resolves **a** into its horizontal and vertical components. However, there is another representation that is often more useful. Imagine yourself driving on the highway. If the car in which you are riding accelerates forward, you are pressed to the back of your seat. If it turns sharply to one side, you are thrown to the other. Both motions are due to acceleration. The second force is related to the rate at which the car turns, which is, of course, related to the curvature of the road. Thus we would like to express the acceleration vector as a component in the direction of motion and a component somehow related to the curvature of the path. How do we do so? A glance at either Figure 1 or Figure 2 reveals the answer. The radial line of the circle of curvature is perpendicular to the unit tangent vector **T** since **T** is tangent to the circle of curvature, and in a circle tangent and radial lines at a point are orthogonal. (See Example 2.4.3.) But a vector that is perpendicular to **T** has the direction of the unit normal vector **n**. Thus the component of acceleration in the direction **n** will be a measure of the acceleration due to turning. We would like to write

$$\mathbf{a} = a_T\mathbf{T} + a_n\mathbf{n}, \tag{12}$$

where a_T and a_n are, respectively, the components of **a** in the tangential and normal directions. Before calculating a_T and a_n, we need the following result.

Theorem 3 For a curve C, let **T** and **n** be the unit tangent and unit normal vectors, respectively, and let φ be as before. Then

$$\text{(i)} \quad \frac{d\mathbf{T}}{d\varphi} = \pm\mathbf{n} \tag{13}$$

and

(ii) $$\frac{d\mathbf{T}}{ds} = \kappa\mathbf{n}, \tag{14}$$

where κ is the curvature of C. In (13) the plus sign is taken if φ increases as s increases and the minus sign is taken if φ decreases as s increases.

Proof.

(i) By the definition of $\mathbf{T}$,

$$\mathbf{T} = (\cos\varphi)\mathbf{i} + (\sin\varphi)\mathbf{j}.$$

Then

$$\frac{d\mathbf{T}}{d\varphi} = -(\sin\varphi)\mathbf{i} + (\cos\varphi)\mathbf{j} = \pm\mathbf{n}^{\dagger}$$

since $\mathbf{T}\cdot(d\mathbf{T}/d\varphi) = 0$ and $|-(\sin\varphi)\mathbf{i} + (\cos\varphi)\mathbf{j}| = 1$.

(ii) If $d\varphi/ds > 0$, then

$$\frac{d\mathbf{T}}{ds} = \frac{d\mathbf{T}}{d\varphi}\frac{d\varphi}{ds} = \mathbf{n}\kappa$$

from the definition of κ. If $d\varphi/ds < 0$, then

$$\frac{d\mathbf{T}}{ds} = \frac{d\mathbf{T}}{d\varphi}\frac{d\varphi}{ds} = (-\mathbf{n})(-\kappa) = \kappa\mathbf{n}. \blacksquare$$

We can now show how to write $\mathbf{a}(t)$ in its tangential and normal components.

Theorem 4 Let the curve C be given by the vector function $\mathbf{f}(t)$. Then the acceleration vector $\mathbf{a}(t) = \mathbf{f}''(t)$ can be written as

$$\mathbf{a} = \frac{d^2s}{dt^2}\mathbf{T} + \left(\frac{ds}{dt}\right)^2\kappa\mathbf{n} \tag{15}$$

where $\mathbf{T}$ and $\mathbf{n}$ are the unit tangent and unit normal vectors to the curve, s is the arc length measured from the point reached when $t = 0$, and κ is the curvature.

NOTE. Since $v = ds/dt$ is the speed, we can write (15) as

† To see why $d\mathbf{T}/d\varphi = -\mathbf{n}$ if $d\varphi/ds < 0$, look again at the discussion on page 52. There we showed that if $d\varphi/ds < 0$ (Figure 2.4.3b), then

$$\mathbf{n} = \sin\varphi\mathbf{i} - \cos\varphi\mathbf{j} = -[-\sin\varphi\mathbf{i} + \cos\varphi\mathbf{j}] = -\frac{d\mathbf{T}}{d\varphi}.$$

$$\mathbf{a} = \frac{dv}{dt}\mathbf{T} + v^2\kappa\mathbf{n} = \frac{dv}{dt}\mathbf{T} + \frac{v^2}{\rho}\mathbf{n}. \tag{16}$$

Thus the tangential component of acceleration is $a_T = dv/dt$ and the normal component of acceleration is $a_n = v^2/\rho$.

Proof of Theorem 4. We have

$$\mathbf{v} = \frac{d\mathbf{f}}{dt} = \frac{d\mathbf{f}}{ds}\frac{ds}{dt} = \mathbf{T}\frac{ds}{dt}$$

(from Theorem 2.6.1, $\mathbf{T} = d\mathbf{f}/ds$). Then

$$\mathbf{a} = \frac{d\mathbf{v}}{dt} = \frac{d}{dt}\left(\mathbf{T}\frac{ds}{dt}\right) = \frac{d^2s}{dt^2}\mathbf{T} + \frac{ds}{dt}\frac{d\mathbf{T}}{dt}.$$

But

$$\frac{ds}{dt}\frac{d\mathbf{T}}{dt} = \frac{ds}{dt}\left(\frac{d\mathbf{T}}{ds}\frac{ds}{dt}\right) = \left(\frac{ds}{dt}\right)^2\frac{d\mathbf{T}}{ds} = \left(\frac{ds}{dt}\right)^2\kappa\mathbf{n}.$$

The last step follows from Theorem 3, and the theorem is proved. ■

The result given by (15) or (16) is very important in physics. If an object of constant mass m is traveling along a trajectory, then the force acting to keep the object on that trajectory is given by

$$\mathbf{F} = m\mathbf{a} = m\frac{dv}{dt}\mathbf{T} + mv^2\kappa\mathbf{n}. \tag{17}$$

The term $mv^2\kappa$ is the magnitude of the force necessary to keep the object from "moving off" the trajectory because of the force exerted on the object caused by turning.

EXAMPLE 5 A 1500-kg race car is driven at a speed of 150 km/hr on a circular race track of radius 50 m. What frictional force must be exerted by the tires on the road surface to keep the car from skidding?

Solution. The frictional force exerted by the tires must be equal to the component of the force (due to acceleration) normal to the circular race track. That is,

$$F = mv^2\kappa = \frac{mv^2}{\rho} = (1500\text{ kg})\frac{(150{,}000\text{ m})^2}{(3600\text{ sec})^2}\cdot\frac{1}{50\text{ m}}$$

$$= 52{,}083\tfrac{1}{3}(\text{kg})(\text{m})/\text{sec}^2 = 52{,}083\tfrac{1}{3}\text{ N}. \ ■$$

EXAMPLE 6 Let the car of Example 5 have the **coefficient of friction** μ. That is, the maximum frictional force that can be exerted by the car on the road surface is μmg, where mg is the **normal force** of the car on the road (the force of the car on the road due to gravity). What is the minimum value μ can take in order that the car not slide off the road?

Solution. We must have $\mu mg \geq 52{,}083\frac{1}{3}$ N. But $\mu mg = \mu(9.81)(1500)$, so we obtain

$$\mu \geq \frac{52{,}083\frac{1}{3}}{(9.81)(1500)} \approx 3.54. \blacksquare$$

PROBLEMS 2.8

In Problems 1–20, find the curvature and radius of curvature for each curve. Sketch the unit tangent vector and the circle of curvature.

1. $\mathbf{f} = 2\cos t\mathbf{i} + 2\sin t\mathbf{j};\ t = \pi/4$
2. $\mathbf{f} = 2\cos t\mathbf{i} + 2\sin t\mathbf{j};\ t = \pi/2$
3. $\mathbf{f} = t\mathbf{i} + t^2\mathbf{j};\ t = 1$
4. $\mathbf{f} = 3\sin t\mathbf{i} + 4\cos t\mathbf{j};\ t = 0$
5. $\mathbf{f} = 3\sin t\mathbf{i} + 4\cos t\mathbf{j};\ t = \pi/2$
6. $\mathbf{f} = 3\sin t\mathbf{i} + 4\cos t\mathbf{j};\ t = \pi/4$
7. $\mathbf{f} = (\cos t + t\sin t)\mathbf{i} + (\sin t - t\cos t)\mathbf{j};\ t = \pi/6$
8. $y = x^2;\ (0, 0)$
9. $y = x^2;\ (1, 1)$
10. $xy = 1;\ (1, 1)$
11. $y = e^x;\ (0, 1)$
12. $y = e^x;\ (1, e)$
13. $y = \ln x;\ (1, 0)$
14. $y = \cos x;\ (\pi/3, \frac{1}{2})$
15. $y = ax^2 + bx + c;\ (0, c),\ a \neq 0$
16. $y = \ln\cos x;\ (\pi/4, \ln(1/\sqrt{2}))$
17. $y = \sqrt{1 - x^2};\ (0, 1)$

*18. $y = \sin^{-1} x;\ (1, \pi/2)$

19. $x = \cos y;\ (0, \pi/2)$
20. $x = y^3;\ (1, 1)$

21. At what point on the parabola $y = ax^2$ is the curvature a maximum?
22. At what point on the curve $y = \ln x$ is the curvature a maximum?
23. For what value of t in the interval $[0, \pi/2]$ is the curvature of the curve $\mathbf{f}(t) = a(\cos^3 t)\mathbf{i} + a(\sin^3 t)\mathbf{j}$ a minimum? For what value is it a maximum?

*24. Let $r = f(\theta)$ be the equation of a curve in polar coordinates. Show that

$$\kappa(\theta) = \frac{|r^2 + 2(dr/d\theta)^2 - r\,d^2r/d\theta^2|}{[r^2 + (dr/d\theta)^2]^{3/2}}.$$

In Problems 25–30, use the result of Problem 24 to calculate the curvature. In all cases assume that $a > 0$.

25. $r = a\sin 2\theta;\ \theta = \pi/8$
26. $r = a\theta;\ \theta = 1$
27. $r = 2a\cos\theta;\ \theta = \pi/3$
28. $r = a(1 + \sin\theta);\ \theta = \pi/2$
29. $r = a(1 - \cos\theta);\ \theta = \pi$
30. $r = e^{a\theta};\ \theta = 1$

In Problems 31–38, find the tangential and normal components of acceleration for each of the given position vectors.

31. $\mathbf{f} = (\cos 2t)\mathbf{i} + (\sin 2t)\mathbf{j}$
32. $\mathbf{f} = 2(\cos t)\mathbf{i} + 3(\sin t)\mathbf{j}$
33. $\mathbf{f} = t\mathbf{i} + t^2\mathbf{j}$
34. $\mathbf{f} = t\mathbf{i} + (\cos t)\mathbf{j}$
35. $\mathbf{f} = (t^3 - 3t)\mathbf{i} + (t^2 - 1)\mathbf{j}$
36. $\mathbf{f} = e^{-t}\mathbf{i} + e^t\mathbf{j}$
37. $\mathbf{f} = t^2\mathbf{i} + t^3\mathbf{j}$
38. $\mathbf{f} = (\sin t^2)\mathbf{i} + (\cos t^2)\mathbf{j}$

39. Show that if a particle is moving at a constant speed, then the tangential component of acceleration is zero.
40. Suppose that the driver of the car of Examples 5 and 6 reduces his speed by a factor of M. Show that the frictional force needed to keep the car from skidding is reduced by a factor of M^2.
41. A truck traveling at 80 km/hr and weighing 10,000 kg is moving on a curved stretch of track. The equation of the curved section is the parabola $y = x^2 - x$ meters. What is the frictional force exerted by the wheels of the truck on the track at the "point" (0, 0)?
42. If the coefficient of friction for the truck in Problem 41 is 2.5, what is the maximum speed it can achieve at the point (0, 0) without going off the road?

43. If the race car of Example 5 is placed on a track with half the radius of the original one, how much slower would it have to be driven so as not to increase the normal component of acceleration?

*44. A woman swings a rope attached to a bucket containing 3 kg of water. The pail rotates in the vertical plane in a circular path with a radius of 1 m. What is the smallest number of revolutions that must be made every minute in order that the water stay in the pail? [*Hint:* Calculate the pressure of the water on the bottom of the pail. This can be determined by first calculating the normal force of the motion. Then the water will stay in the bucket if this normal force exceeds the force due to gravity.]

REVIEW EXERCISES FOR CHAPTER TWO

In Exercises 1–8, find the Cartesian equation of each curve, and then sketch the curve in the xy-plane.

1. $\mathbf{f}(t) = t\mathbf{i} + 2t\mathbf{j}$
2. $\mathbf{f}(t) = (2t - 6)\mathbf{i} + t^2\mathbf{j}$
3. $\mathbf{f}(t) = t^2\mathbf{i} + (2t - 6)\mathbf{j}$
4. $\mathbf{f}(t) = t^2\mathbf{i} + t^4\mathbf{j}$
5. $\mathbf{f}(t) = (\cos 4t)\mathbf{i} + (\sin 4t)\mathbf{j}$
6. $\mathbf{f}(t) = 4(\sin t)\mathbf{i} + 9(\sin t)\mathbf{j}$
7. $\mathbf{f}(t) = t^6\mathbf{i} + t^2\mathbf{j}$
8. $\mathbf{f}(t) = e^t(\cos t)\mathbf{i} + e^t(\sin t)\mathbf{j}$

In Exercises 9–16, find the slope of the line tangent to the given curve for the given value of the parameter, and then find all points at which the curve has vertical and horizontal tangents.

9. $x = t^3;\ y = 6t;\ t = 1$
10. $x = t^7;\ y = t^8 - 5;\ t = 2$
11. $x = \sin 5\theta;\ y = \cos 5\theta;\ \theta = \pi/3$
12. $x = \cos^2 \theta;\ y = -3\theta;\ \theta = \pi/4$
13. $x = \cosh t;\ y = \sinh t;\ t = 0$
14. $x = \dfrac{2}{\theta};\ y = -3\theta;\ \theta = 10\pi$
15. $x = 3 \cos \theta;\ y = 4 \sin \theta;\ \theta = \pi/3$
16. $x = 3 \cos \theta;\ y = -4 \sin \theta;\ \theta = 2\pi/3$

17. Find the slope of the line tangent to the polar curve $r = -3 \cos \theta$ for $\theta = \pi/3$.
18. Find the slope of the line tangent to the polar curve $r = \sin 2\theta$ for $\theta = \pi/8$.
19. Find the slope of the line tangent to the spiral of Archimedes $r = \theta$ for $\theta = \pi$.
20. Calculate the equation of the line that is tangent to the curve $x = t^2 - t - 2$, $y = t^2 - t + 1$ at the point $(4, 7)$.

In Exercises 21–24, find the first and second derivatives of the given vector functions.

21. $\mathbf{f}(t) = 2t\mathbf{i} - t^3\mathbf{j}$
22. $\mathbf{f}(t) = \left(\dfrac{1}{t^2}\right)\mathbf{i} + \sqrt[3]{t}\mathbf{j}$
23. $\mathbf{f}(t) = (\cos 5t)\mathbf{i} + 2(\sin 4t)\mathbf{j}$
24. $\mathbf{f}(t) = (\tan t)\mathbf{i} + (\cot t)\mathbf{j}$

In Exercises 25–30, find the unit tangent and unit normal vectors to the given curve for the given value of t.

25. $\mathbf{f}(t) = t^4\mathbf{i} + t^5\mathbf{j};\ t = 1$
26. $\mathbf{f}(t) = (\cos 2t)\mathbf{i} + (\sin 2t)\mathbf{j};\ t = \pi/6$
27. $\mathbf{f}(t) = (\sin 5t)\mathbf{i} + (\cos 5t)\mathbf{j};\ t = \pi/30$
28. $\mathbf{f}(t) = (\cosh 2t)\mathbf{i} + (\sinh 2t)\mathbf{j};\ t = 0$
29. $\mathbf{f}(t) = (\ln t)\mathbf{i} + \sqrt{t}\mathbf{j};\ t = 1$
30. $\mathbf{f}(t) = (\tan t)\mathbf{i} + (\cot t)\mathbf{j};\ t = \pi/3$

In Exercises 31–34, calculate the integral.

31. $\displaystyle\int_0^3 (t^3\mathbf{i} + t^5\mathbf{j})\,dt$
32. $\displaystyle\int [(\cos 3t)\mathbf{i} + (\sin 3t)\mathbf{j}]\,dt$
33. $\displaystyle\int (\sqrt{t}\mathbf{i} + \sqrt[3]{t}\mathbf{j})\,dt$
34. $\displaystyle\int_0^{\pi/3} [(\cos t)\mathbf{i} + (\tan t)\mathbf{j}]\,dt$

35. For $\mathbf{f}'(t) = t^7\mathbf{i} - t^6\mathbf{j}$ and for $\mathbf{f}(0) = -\mathbf{i} + 3\mathbf{j}$, find $\mathbf{f}(t)$.

In Exercises 36–40, calculate the derivative.

36. $\dfrac{d}{dt}[(2t\mathbf{i} + \sqrt{t}\mathbf{j}) \cdot (4\mathbf{i} - 3\mathbf{j})]$
37. $\dfrac{d}{dt}\{[(\cos t)\mathbf{i} + (\sin t)\mathbf{j}] \cdot [(\cosh t)\mathbf{i} - (\sinh t)\mathbf{j}]\}$
38. $\dfrac{d}{dt}[(e^t)(t^2\mathbf{i} - t^{3/2}\mathbf{j})]$

39. $\dfrac{d}{dt}\left[\left(\dfrac{t}{1+t}\mathbf{i} - \dfrac{t+1}{t}\mathbf{j}\right) \cdot (e^t\mathbf{i} + e^{-t}\mathbf{j})\right]$

40. $\dfrac{d}{dt}\{(2\mathbf{i} - 11\mathbf{j}) \cdot [-(\tan t)\mathbf{i} + (\sec t)\mathbf{j}]\}$

In Exercises 41–46, find the length of the arc over the given interval or the length of the closed curve.

41. $x = \cos 4\theta$; $y = \sin 4\theta$; $0 \le \theta \le \pi/12$
42. $x = e^t \sin t$; $y = e^t \cos t$; $0 \le t \le \pi/2$
43. $r = 2(1 + \cos\theta)$
44. $r = 5 \sin\theta$
45. $r = \theta^2$; $1 \le \theta \le 5$
46. $r = 2\theta$; $0 \le \theta \le 20\pi$

In Exercises 47–50, find parametric equations in terms of the arc length s measured from the point reached when $t = 0$. Verify your answer by using equation (2.6.3).

47. $\mathbf{f} = 3t\mathbf{i} + 4t^{3/2}\mathbf{j}$
48. $\mathbf{f} = \frac{2}{9}t^{9/2}\mathbf{i} + \frac{1}{3}t^3\mathbf{j}$
49. $\mathbf{f} = 2(\cos 3t)\mathbf{i} + 2(\sin 3t)\mathbf{j}$
50. $\mathbf{f} = e^t(\sin t)\mathbf{i} + e^t(\cos t)\mathbf{j}$

In Exercises 51–56 the position vector of a moving particle is given. For the indicated value of t, calculate the velocity vector, the acceleration vector, the speed, and the acceleration scalar. Then sketch a portion of the trajectory showing the velocity and acceleration vectors.

51. $\mathbf{f} = (\cos 2t)\mathbf{i} + (\sin 2t)\mathbf{j}$; $t = \pi/6$
52. $\mathbf{f} = 6t\mathbf{i} + 2t^3\mathbf{j}$; $t = 1$
53. $\mathbf{f} = (2^t + e^{-t})\mathbf{i} + 2t\mathbf{j}$; $t = 0$
54. $\mathbf{f} = (1 - \cos t)\mathbf{i} + (t - \sin t)\mathbf{j}$; $t = \pi/2$
55. $\mathbf{f} = 2(\sinh t)\mathbf{i} + 4t\mathbf{j}$; $t = 1$
56. $\mathbf{f} = (3 + 5t)\mathbf{i} + (2 + 8t)\mathbf{j}$; $t = 6$

57. A boy throws a ball into the air with an initial speed of 30 m/sec. The ball is thrown from shoulder level (1.5 m above the ground) at an angle of 30° with the horizontal. Ignoring air resistance, find the following:
   (a) The total distance traveled by the ball.
   (b) The horizontal distance traveled by the ball.
   (c) The maximum height the ball reaches.
   (d) The speed of the ball at impact with the ground.
   (e) The angle at which the path of the ball meets the ground at the point of impact.
58. A 1500-kg rocket carrying 1000 kg of fuel starts from a position motionless in space. Its exhaust velocity is 1.5 km/sec. What is its velocity when all its fuel is consumed?
59. The rocket in Exercise 58 is launched from the earth, and gas is emitted at a constant rate of 40 kg/sec.
   (a) Find the net initial force acting on the rocket.
   (b) Find the net force just before all the fuel is expended.
60. In Exercise 59, what is the minimum thrust needed to get the rocket off the ground?

In Exercises 61–70, find the curvature and radius of curvature for each curve. Sketch the unit tangent vector and the circle of curvature.

61. $\mathbf{f} = (\cos 2t)\mathbf{i} + (\sin 2t)\mathbf{j}$; $t = \pi/3$
62. $\mathbf{f} = t^2\mathbf{i} + 2t\mathbf{j}$; $t = 2$
63. $\mathbf{f} = 4(\cos t)\mathbf{i} + 9(\sin t)\mathbf{j}$; $t = \pi/4$
64. $y = 2x^2$; $(0, 0)$
65. $xy = 1$; $(2, \frac{1}{2})$
66. $y = e^{-x}$; $(1, 1/e)$
67. $y = \sqrt{x}$; $(4, 2)$
68. $r = 3 \cos 2\theta$; $\theta = \pi/6$
69. $r = 1 + \sin\theta$; $\theta = \pi/2$
70. $r = 3\theta$; $\theta = \pi$

In Exercises 71–74, find the tangential and normal components of acceleration for each of the given position vectors.

71. $\mathbf{f} = 2(\sin t)\mathbf{i} + 2(\cos t)\mathbf{j}$
72. $\mathbf{f} = 4(\cos t)\mathbf{i} + 9(\sin t)\mathbf{j}$
73. $\mathbf{f} = 3t^2\mathbf{i} + 2t^3\mathbf{j}$
74. $\mathbf{f} = (\cos t^2)\mathbf{i} + (\sin t^2)\mathbf{j}$

75. A 1300-kg race car is driven at a speed of 175 km/hr on a circular race track of radius 65 m. What frictional force must be exerted by the tires on the road surface to keep the car from skidding?

3 Vectors in Space

In Chapters 1 and 2 we developed a theory of vectors and vector-valued functions in the plane. In this chapter we will extend that development to vectors in space (three dimensions). We will also show how these ideas carry over to vector functions with four or more components. Throughout this chapter we will stress the many similarities between vectors with two components (vectors in the plane) and those with three or more components (vectors in space).

3.1 THE RECTANGULAR COORDINATE SYSTEM IN SPACE

Any point in a plane can be represented as an ordered pair of real numbers. It is not surprising, then, that any point in space can be represented by an **ordered triple** of real numbers

$$(a, b, c), \tag{1}$$

where a, b, and c are real numbers.

Definition 1 THREE-DIMENSIONAL SPACE $\mathbb{R}^3$ The set of ordered triples of the form (1) is called **real three-dimensional space** and is denoted $\mathbb{R}^3$.

There are many ways to represent a point in $\mathbb{R}^3$. Two other methods will be given in Section 3.9. However, the most common representation (given in Definition 1) is very similar to the representation of a point in the plane by its x- and y-coordinates.

We begin, as before, by choosing a point in $\mathbb{R}^3$ and calling it the **origin,** denoted by 0. Then we draw three mutually perpendicular axes, called the **coordinate axes,** which we label the **x-axis,** the **y-axis,** and the **z-axis.** These axes can be selected in a variety of ways, but the most common selection has the x- and y-axes drawn horizontally with the z-axis vertical. On each axis we choose a positive direction and measure distance along each axis as the number of units in this positive direction measured from the origin.

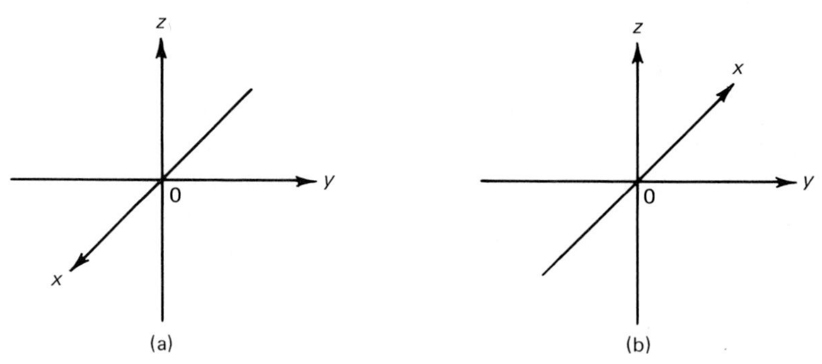

FIGURE 1

The two basic systems of drawing these axes are depicted in Figure 1. If the axes are placed as in Figure 1a, then the system is called a **right-handed system;** if they are placed as in Figure 1b, the system is a **left-handed system.** In the figures the arrows indicate the positive directions on the axes. The reason for this choice of terms is as follows: In a right-handed system, if you place your right hand so that your index finger points in the positive direction of the x-axis while your middle finger points in the positive direction of the y-axis, then your thumb will point in the positive direction of the z-axis. This is illustrated in Figure 2. For a left-handed system the same rule will work for your left hand. For the remainder of this text we will follow common practice and depict the coordinate axes using a right-handed system.

If you have trouble visualizing the placement of these axes, do the following. Face any uncluttered corner (on the floor) of the room in which you are sitting. Call the corner the origin. Then the x-axis lies along the floor, along the wall, and to your left; the y-axis lies along the floor, along the wall, and to your right; and the z-axis lies along the vertical intersection of the two perpendicular walls. This is illustrated in Figure 3.

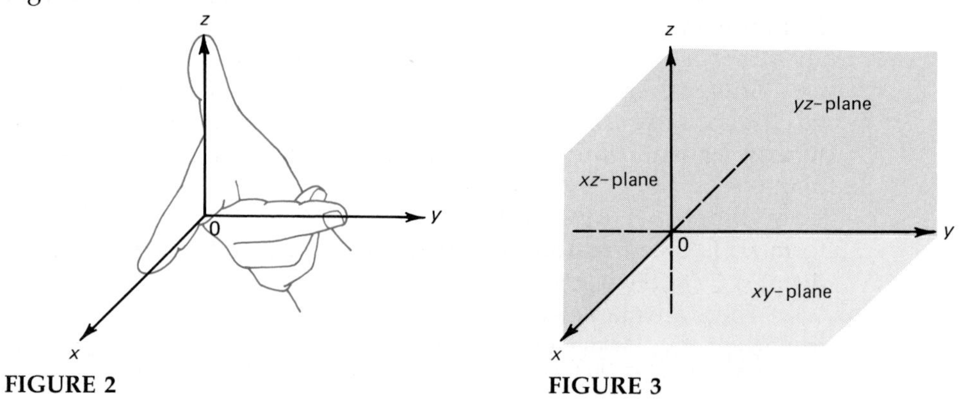

FIGURE 2 **FIGURE 3**

The three axes in our system determine three **coordinate planes** that we will call the ***xy*-plane,** the ***xz*-plane,** and the ***yz*-plane.** The xy-plane contains the x- and y-axes and is simply the plane with which we have been dealing in one-variable calculus. The xz- and yz-planes can be thought of in a similar way.

Having built our structure of coordinate axes and planes, we can describe any point P in space in a unique way:

$$P = (x, y, z), \tag{2}$$

where the first coordinate x is the distance from the yz-plane to P (measured in the positive direction of the x-axis), the second coordinate y is the distance from the xz-plane to P (measured in the positive direction of the y-axis), and the third coordinate z is the distance from the xy-plane to P (measured in the positive direction of the z-axis). Thus, for example, any point in the xy-plane has z-coordinate 0; any point in the xz-plane has y-coordinate 0; and any point in the yz-plane has x-coordinate 0. Some representative points are sketched in Figure 4.

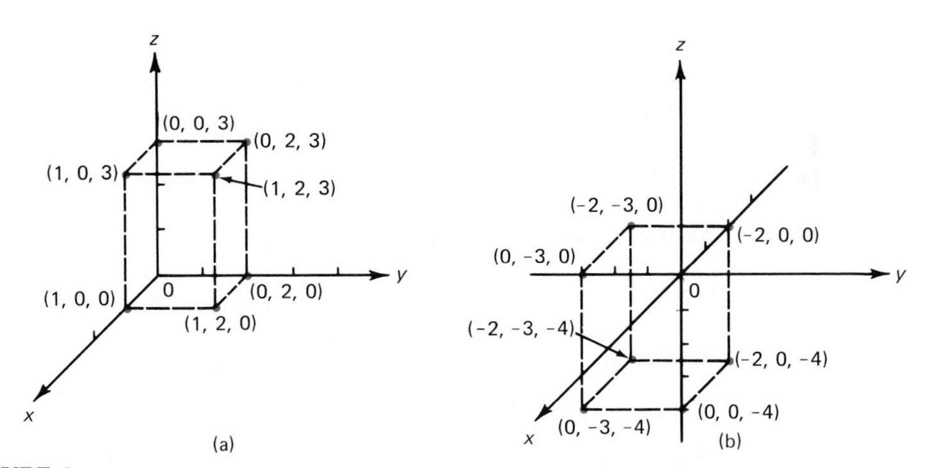

FIGURE 4

In this system the three coordinate planes divide $\mathbb{R}^3$ into eight **octants,** just as in $\mathbb{R}^2$ the two coordinate axes divide the plane into four quadrants. The first octant is always chosen to be the one in which the three coordinates are positive.

The coordinate system we have just established is often referred to as the **rectangular coordinate system,** or the **Cartesian coordinate system.** Once we are comfortable with the notion of depicting a point in this system, then we can generalize many of our ideas from the plane.

Theorem 1 Let $P = (x_1, y_1, z_1)$ and $Q = (x_2, y_2, z_2)$ be two points in space. Then the distance $\overline{PQ}$ between P and Q is given by

$$\overline{PQ} = \sqrt{(x_1 - x_2)^2 + (y_1 - y_2)^2 + (z_1 - z_2)^2}. \tag{3}$$

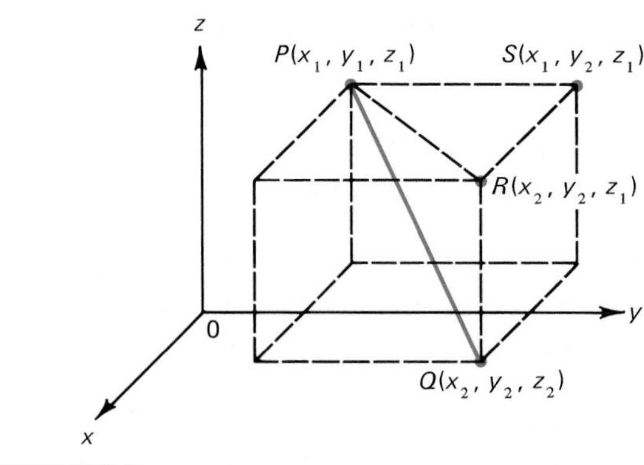

FIGURE 5

Proof. The two points are sketched in Figure 5. From the Pythagorean theorem, since the line segments PR and RQ are perpendicular, the triangle PQR is a right triangle and

$$\overline{PQ}^2 = \overline{PR}^2 + \overline{RQ}^2. \tag{4}$$

But using the Pythagorean theorem again, we have

$$\overline{PR}^2 = \overline{PS}^2 + \overline{SR}^2, \tag{5}$$

so combining (4) and (5) yields

$$\overline{PQ}^2 = \overline{PS}^2 + \overline{SR}^2 + \overline{RQ}^2. \tag{6}$$

Since the x- and z-coordinates of P and S are equal,

$$\overline{PS}^2 = (y_2 - y_1)^2. \tag{7}$$

Similarly,

$$\overline{RS}^2 = (x_2 - x_1)^2 \tag{8}$$

and

$$\overline{RQ}^2 = (z_2 - z_1)^2. \tag{9}$$

Thus using (7), (8), and (9) in (6) yields

$$\overline{PQ}^2 = (x_2 - x_1)^2 + (y_2 - y_1)^2 + (z_2 - z_1)^2,$$

and the proof is complete. ■

EXAMPLE 1 Calculate the distance between the points $(3, -1, 6)$ and $(-2, 3, 5)$.

Solution. $\overline{PQ} = \sqrt{[3 - (-2)]^2 + [-1 - 3]^2 + (6 - 5)^2} = \sqrt{42}$. ■

Definition 2 GRAPH IN $\mathbb{R}^3$ The **graph** of an equation in $\mathbb{R}^3$ is the set of all points in $\mathbb{R}^3$ whose coordinates satisfy the equation.

One of our first examples of a graph in $\mathbb{R}^2$ was the graph of the unit circle $x^2 + y^2 = 1$. This example can easily be generalized.

Definition 3 SPHERE A **sphere** is the set of points in space at a given distance from a given point. The given point is called the **center** of the sphere, and the given distance is called the **radius** of the sphere.

EXAMPLE 2 Suppose that the center of a sphere is the origin (0, 0, 0) and the radius of the sphere is 1. Let (x, y, z) be a point on the sphere. Then from (3)

$$1 = \sqrt{(x - 0)^2 + (y - 0)^2 + (z - 0)^2}.$$

Simplifying and squaring, we obtain

$$x^2 + y^2 + z^2 = 1, \tag{10}$$

which is the equation of the **unit sphere.** ■

In general, if the center of a sphere is (a, b, c), the radius is r, and if (x, y, z) is a point on the sphere, we obtain

$$r = \sqrt{(x - a)^2 + (y - b)^2 + (z - c)^2},$$

or

$$(x - a)^2 + (y - b)^2 + (z - c)^2 = r^2. \tag{11}$$

This is sketched in Figure 6.

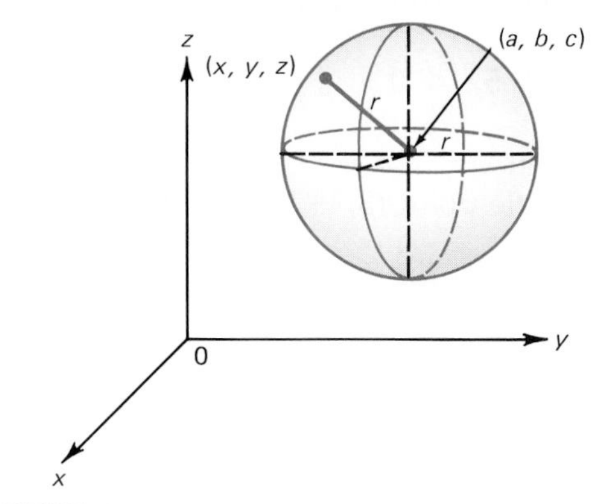

FIGURE 6

EXAMPLE 3 Find the equation of the sphere with center at $(1, -3, 2)$ and radius 5.

Solution. From (11) we obtain

$$(x - 1)^2 + (y + 3)^2 + (z - 2)^2 = 25. \quad ■$$

EXAMPLE 4 Show that

$$x^2 - 6x + y^2 + 2y + z^2 + 10z + 5 = 0 \tag{12}$$

is the equation of a sphere, and find its center and radius.

Solution. Completing the squares, we obtain

$$x^2 - 6x = (x - 3)^2 - 9$$
$$y^2 + 2y = (y + 1)^2 - 1$$
$$z^2 + 10z = (z + 5)^2 - 25,$$

so that

$$0 = x^2 - 6x + y^2 + 2y + z^2 + 10z + 5$$
$$= (x - 3)^2 - 9 + (y + 1)^2 - 1 + (z + 5)^2 - 25 + 5$$

and

$$(x - 3)^2 + (y + 1)^2 + (z + 5)^2 = 30,$$

which is the equation of a sphere with center $(3, -1, -5)$ and radius $\sqrt{30}$. ■

■ WARNING: Not every second-degree equation in a form similar to (12) is the equation of a sphere. For example, if the number 5 in (12) is replaced by 40, we obtain

$$(x - 3)^2 + (y + 1)^2 + (z + 5)^2 = -5.$$

Clearly, the sum of squares cannot be negative, so there are *no* points in $\mathbb{R}^3$ that satisfy this equation. On the other hand, if we replaced the 5 by 35, we would obtain

$$(x - 3)^2 + (y + 1)^2 + (z + 5)^2 = 0.$$

This equation can hold only when $x = 3$, $y = -1$, and $z = -5$. In this case the graph of the equation contains the single point $(3, -1, -5)$.

PROBLEMS 3.1

In Problems 1–15, sketch the given point in $\mathbb{R}^3$.

1. $(1, 4, 2)$
2. $(3, -2, 1)$
3. $(-1, 5, 7)$
4. $(8, -2, 3)$
5. $(-2, 1, -2)$
6. $(1, -2, 1)$
7. $(3, 2, -5)$
8. $(-2, -3, -8)$
9. $(2, 0, 4)$
10. $(-3, -8, 0)$
11. $(0, 4, 7)$
12. $(1, 3, 0)$
13. $(3, 0, 0)$
14. $(0, 8, 0)$
15. $(0, 0, -7)$

In Problems 16–25, find the distance between the two points.

16. $(8, 1, 6)$; $(8, 1, 4)$
17. $(3, -4, 3)$; $(3, 2, 5)$
18. $(3, -4, 7)$; $(3, -4, 9)$
19. $(-2, 1, 3)$; $(4, 1, 3)$
20. $(2, -7, 5)$; $(8, -7, -1)$
21. $(1, 3, -2)$; $(4, 7, -2)$
22. $(3, 1, 2)$; $(1, 2, 3)$

23. $(5, -6, 4)$; $(3, 11, -2)$
24. $(-1, -7, -2)$; $(-4, 3, -5)$
25. $(8, -2, -3)$; $(-7, -5, 1)$

26. Find the equation of the sphere with center at $(2, -1, 4)$ and radius 2.
27. Find the equation of the sphere with center at $(-1, 8, -3)$ and radius $\sqrt{5}$.
28. Show that the equation $x^2 + y^2 + z^2 - 4x - 4y + 8z + 8 = 0$ is the equation of a sphere, and find its center and radius.
29. Do the same for the equation $x^2 + y^2 + z^2 + 3x - y + 2z - 1 = 0$.
30. Find a number α such that the equation $x^2 + y^2 + z^2 - 2x + 8y - 5z + \alpha = 0$ has exactly one solution.
31. In Problem 30 give a condition on α in order that the equation have no solution.
32. The equation $x^2 + y^2 + z^2 + ax + by + cz + d = 0$ is a second-degree equation in three variables. Let $e = d - (a/2)^2 - (b/2)^2 - (c/2)^2$. Show that the equation **(a)** is the equation of a sphere if $e < 0$, **(b)** has exactly one solution if $e = 0$, and **(c)** has no solutions if $e > 0$.
33. Three points P, Q, and R are **collinear** if they lie on the same straight line. Show that, in $\mathbb{R}^2$, P, Q, and R are collinear if $\overline{PR} = \overline{PQ} + \overline{QR}$ or $\overline{PQ} = \overline{PR} + \overline{RQ}$ or $\overline{QR} = \overline{QP} + \overline{PR}$. Use this last fact in $\mathbb{R}^3$ to show that the points $(-1, -1, -1)$, $(5, 8, 2)$, and $(-3, -4, -2)$ are collinear.
34. Show that the points $(3, 0, 1)$, $(0, -4, 0)$, and $(6, 4, 2)$ are collinear.
***35.** Let $P = (x_1, y_1, z_1)$ and $Q = (x_2, y_2, z_2)$. Show that the midpoint of PQ is the point $R = ((x_1 + x_2)/2, (y_1 + y_2)/2, (z_1 + z_2)/2)$. [*Hint:* Show that P, Q, and R are collinear and that $\overline{PR} = \overline{RQ}$.]
36. Find the midpoint of the line joining the points $(2, -1, 4)$ and $(5, 7, -3)$.
37. Find the equation of the sphere that has a diameter with endpoints $(3, 1, -2)$ and $(4, 1, 6)$. [*Hint:* Find the center and radius of the sphere by using the result of Problem 35.]
38. One sphere is said to be **inscribed** in a second sphere if it has a smaller radius and the same center as the second sphere. Find the equation of a sphere of radius 1 inscribed in the sphere given by $x^2 + y^2 + z^2 - 2x - 4y + z - 2 = 0$.
39. Find the volumes of the spheres of Problems 37 and 38.
40. Find the surface areas of the spheres of Problems 37 and 38.

3.2 VECTORS IN $\mathbb{R}^3$

In Chapter 1 we developed properties of vectors in the plane $\mathbb{R}^2$. Given the similarity between the coordinate systems in $\mathbb{R}^2$ and $\mathbb{R}^3$, it should come as no surprise to learn that vectors in $\mathbb{R}^2$ and $\mathbb{R}^3$ have very similar structures. In this section we will develop the notion of a vector in space. The development will closely follow the development in Sections 1.1 and 1.2 and, therefore, some of the details will be omitted.

Let P and Q be two distinct points in $\mathbb{R}^3$. Then the **directed line segment** $\overrightarrow{PQ}$ is the straight line segment that extends from P to Q. Two directed line segments are **equivalent** if they have the same magnitude and direction. A **vector** in $\mathbb{R}^3$ is the set of all directed line segments equivalent to a given directed line segment, and any directed line segment $\overrightarrow{PQ}$ in that set is called a **representation** of the vector.

So far, our definitions are identical. For convenience we will choose P to be the origin and label the endpoint of the vector R, so that the vector $\mathbf{v} = \overrightarrow{0R}$ can be described by the coordinates (x, y, z) of the point R. Then the **magnitude** of $\mathbf{v} = |\mathbf{v}| = \sqrt{x^2 + y^2 + z^2}$ (from Theorem 3.1.1).

EXAMPLE 1 Let $\mathbf{v} = (1, 3, -2)$. Find $|\mathbf{v}|$.

Solution. $|\mathbf{v}| = \sqrt{1^2 + 3^2 + (-2)^2} = \sqrt{14}$. ■

Let $\mathbf{u} = (x_1, y_1, z_1)$ and $\mathbf{v} = (x_2, y_2, z_2)$ be two vectors and let α be a real number (scalar). Then we define

$$\mathbf{u} + \mathbf{v} = (x_1 + x_2, y_1 + y_2, z_1 + z_2)$$

and

$$\alpha\mathbf{u} = (\alpha x_1, \alpha y_1, \alpha z_1).$$

This is the same definition of vector addition and scalar multiplication we had before and is illustrated in Figure 1.

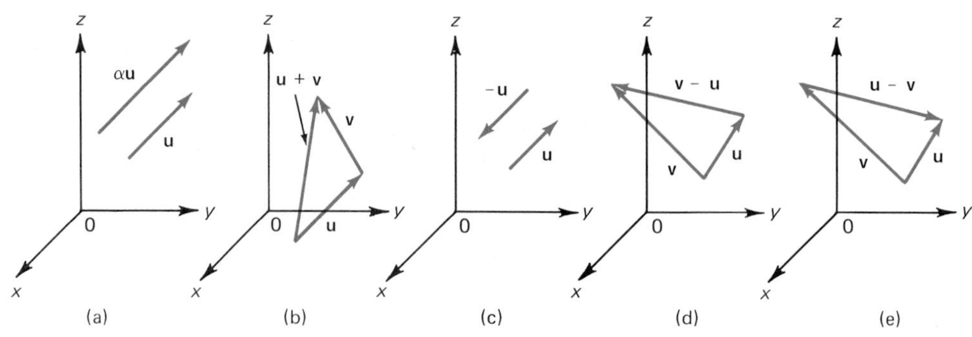

FIGURE 1

EXAMPLE 2 Let $\mathbf{u} = (2, 3, -1)$ and $\mathbf{v} = (-6, 2, 4)$. Find (a) $\mathbf{u} + \mathbf{v}$, (b) $3\mathbf{v}$, (c) $-\mathbf{u}$, and (d) $4\mathbf{u} - 3\mathbf{v}$.

Solution.

(a) $\mathbf{u} + \mathbf{v} = (2 - 6, 3 + 2, -1 + 4) = (-4, 5, 3)$
(b) $3\mathbf{v} = (-18, 6, 12)$
(c) $-\mathbf{u} = (-2, -3, 1)$
(d) $4\mathbf{u} - 3\mathbf{v} = (8, 12, -4) - (-18, 6, 12) = (26, 6, -16)$ ■

The following theorem extends to three dimensions the results of Theorem 1.1.1. Its proof is easy and is left as an exercise (Problem 54).

Theorem 1 Let **u**, **v**, and **w** be any three vectors in space, let α and β be scalars, and let **0** denote the zero vector (0, 0, 0).

(i) $\mathbf{u} + \mathbf{v} = \mathbf{v} + \mathbf{u}$ **(ii)** $\mathbf{u} + (\mathbf{v} + \mathbf{w}) = (\mathbf{u} + \mathbf{v}) + \mathbf{w}$
(iii) $\mathbf{v} + \mathbf{0} = \mathbf{v}$ **(iv)** $0\mathbf{v} = \mathbf{0}$
(v) $\alpha\mathbf{0} = \mathbf{0}$ **(vi)** $(\alpha\beta)\mathbf{v} = \alpha(\beta\mathbf{v})$
(vii) $\mathbf{v} + (-\mathbf{v}) = \mathbf{0}$ **(viii)** $(1)\mathbf{v} = \mathbf{v}$
(ix) $(\alpha + \beta)\mathbf{v} = \alpha\mathbf{v} + \beta\mathbf{v}$ **(x)** $\alpha(\mathbf{u} + \mathbf{v}) = \alpha\mathbf{u} + \alpha\mathbf{v}$
(xi) $|\alpha\mathbf{v}| = |\alpha|\,|\mathbf{v}|$ **(xii)** $|\mathbf{u} + \mathbf{v}| \le |\mathbf{u}| + |\mathbf{v}|$

A **unit vector u** is a vector with magnitude 1. If **v** is any nonzero vector, then $\mathbf{u} = \mathbf{v}/|\mathbf{v}|$ is a unit vector having the same direction as **v**.

EXAMPLE 3 Find a unit vector having the same direction as $\mathbf{v} = (2, 4, -3)$.

Solution. Since $|\mathbf{v}| = \sqrt{2^2 + 4^2 + (-3)^2} = \sqrt{29}$, we have

$$\mathbf{u} = \left(\frac{2}{\sqrt{29}}, \frac{4}{\sqrt{29}}, \frac{-3}{\sqrt{29}}\right). \quad \blacksquare$$

We can now formally define the direction of a vector in $\mathbb{R}^3$. We cannot define it to be the angle θ the vector makes with the positive x-axis, since, for example, if $0 < \theta < \pi/2$, then there are an *infinite number* of unit vectors making the angle θ with the positive x-axis, and these together form a cone (see Figure 2).

Definition 1 DIRECTION OF A VECTOR The **direction** of a nonzero vector $\mathbf{v}$ in $\mathbb{R}^3$ is defined to be the unit vector $\mathbf{u} = \mathbf{v}/|\mathbf{v}|$.

REMARK. We could have defined the direction of a vector $\mathbf{v}$ in $\mathbb{R}^2$ in this way. For if $\mathbf{u} = \mathbf{v}/|\mathbf{v}|$, then $\mathbf{u} = (\cos\theta, \sin\theta)$, where θ is the direction of $\mathbf{v}$ (according to the $\mathbb{R}^2$ definition).

DIRECTION COSINES It would still be useful to define the direction of a vector in terms of some angles. Let $\mathbf{v}$ be the vector $\overrightarrow{0P}$ depicted in Figure 3. We define α to be the angle between $\mathbf{v}$ and the positive x-axis, β the angle between $\mathbf{v}$ and the positive y-axis, and γ the angle between $\mathbf{v}$ and the positive z-axis. The angles α, β, and γ are called the **direction angles** of the vector $\mathbf{v}$. Then from Figure 3

$$\cos\alpha = \frac{x_0}{|\mathbf{v}|}, \qquad \cos\beta = \frac{y_0}{|\mathbf{v}|}, \qquad \cos\gamma = \frac{z_0}{|\mathbf{v}|}. \tag{1}$$

If $\mathbf{v}$ is a unit vector, then $|\mathbf{v}| = 1$ and

$$\cos\alpha = x_0, \qquad \cos\beta = y_0, \qquad \cos\gamma = z_0. \tag{2}$$

By definition, each of these three angles lies between 0 and π. The cosines of these angles are called the **direction cosines** of the vector $\mathbf{v}$.

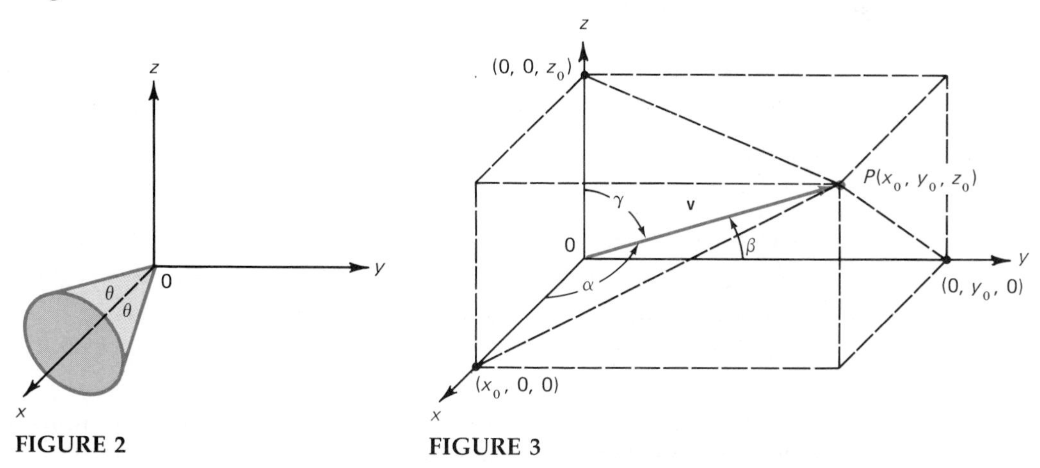

FIGURE 2 FIGURE 3

Note from (1) that

$$\cos^2 \alpha + \cos^2 \beta + \cos^2 \gamma = \frac{x_0^2 + y_0^2 + z_0^2}{|\mathbf{v}|^2} = \frac{x_0^2 + y_0^2 + z_0^2}{x_0^2 + y_0^2 + z_0^2} = 1. \tag{3}$$

If α, β, and γ are any three numbers between 0 and π such that condition (3) is satisfied, then they uniquely determine a unit vector given by $\mathbf{u} = (\cos \alpha, \cos \beta, \cos \gamma)$.

REMARK. If $\mathbf{v} = (a, b, c)$ and $|\mathbf{v}| \neq 0$, then the numbers a, b, and c are called **direction numbers** of the vector $\mathbf{v}$.

EXAMPLE 4 Find the direction cosines of the vector $\mathbf{v} = (4, -1, 6)$.

Solution. The direction of $\mathbf{v}$ is $\mathbf{v}/|\mathbf{v}| = \mathbf{v}/\sqrt{53} = (4/\sqrt{53}, -1/\sqrt{53}, 6/\sqrt{53})$. Then $\cos \alpha = 4/\sqrt{53} \approx 0.5494$, $\cos \beta = -1/\sqrt{53} \approx -0.1374$, and $\cos \gamma = 6/\sqrt{53} \approx 0.8242$. From these we use a calculator to obtain $\alpha \approx 56.7° \approx 0.989$ radian, $\beta \approx 97.9° \approx 1.71$ radians, and $\gamma = 34.5° \approx 0.602$ radian. The vector, along with the angles α, β, and γ, is sketched in Figure 4. ■

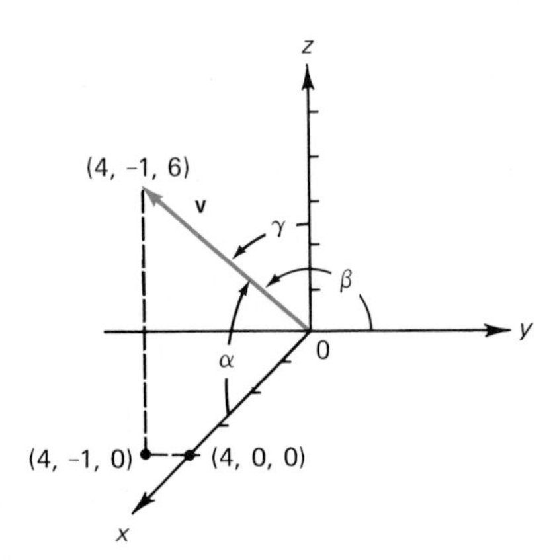

FIGURE 4

EXAMPLE 5 Find a vector $\mathbf{v}$ of magnitude 7 whose direction cosines are $1/\sqrt{6}$, $1/\sqrt{3}$, and $1/\sqrt{2}$.

NOTE. We can solve this problem because $(1/\sqrt{6})^2 + (1/\sqrt{3})^2 + (1/\sqrt{2})^2 = 1$.

Solution. Let $\mathbf{u} = (1/\sqrt{6}, 1/\sqrt{3}, 1/\sqrt{2})$. Then $\mathbf{u}$ is a unit vector since $|\mathbf{u}| = 1$. Thus the direction of $\mathbf{v}$ is given by $\mathbf{u}$, and so

$$\mathbf{v} = |\mathbf{v}|\mathbf{u} = 7\mathbf{u} = \left(\frac{7}{\sqrt{6}}, \frac{7}{\sqrt{3}}, \frac{7}{\sqrt{2}}\right). \quad ■$$

It is interesting to note that if $\mathbf{v}$ in $\mathbb{R}^2$ is written

$$\mathbf{v} = (\cos \theta)\mathbf{i} + (\sin \theta)\mathbf{j},$$

where θ is the direction of $\mathbf{v}$, then $\cos \theta$ and $\sin \theta$ are the direction cosines of $\mathbf{v}$. Here

$\alpha = \theta$ and we define β to be the angle that **v** makes with the y-axis (see Figure 5). Then $\beta = (\pi/2) - \alpha$, so that

$$\cos\beta = \cos\left(\frac{\pi}{2} - \alpha\right) = \sin\alpha,$$

and **v** can be written in the "direction cosine" form

$$\mathbf{v} = (\cos\alpha)\mathbf{i} + (\cos\beta)\mathbf{j}.$$

In Chapter 1 we showed how any vector in the plane can be written in terms of the basis vectors **i** and **j**. To extend this idea to $\mathbb{R}^3$, we define

$$\mathbf{i} = (1, 0, 0), \qquad \mathbf{j} = (0, 1, 0), \qquad \mathbf{k} = (0, 0, 1) \tag{4}$$

Here **i**, **j**, and **k** are unit vectors. The vector **i** lies along the x-axis, **j** along the y-axis, and **k** along the z-axis. These vectors are sketched in Figure 6. If $\mathbf{v} = (x, y, z)$ is any vector in $\mathbb{R}^3$, then

$$\mathbf{v} = (x, y, z) = (x, 0, 0) + (0, y, 0) + (0, 0, z) = x\mathbf{i} + y\mathbf{j} + z\mathbf{k}. \tag{5}$$

That is, *any vector* **v** *in* $\mathbb{R}^3$ *can be written in a unique way in terms of the vectors* **i**, **j**, *and* **k**.

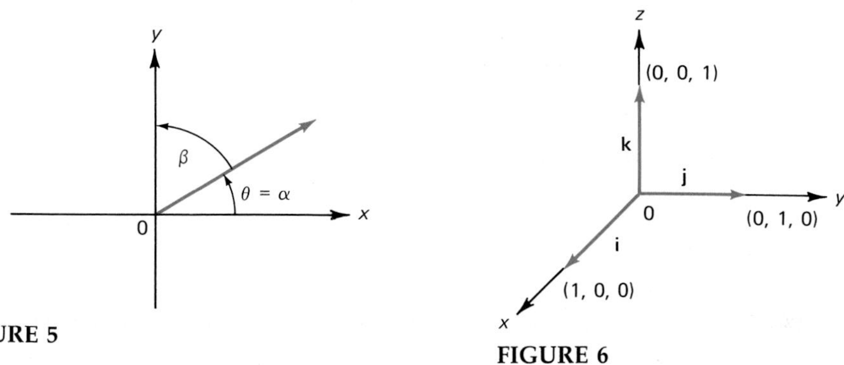

FIGURE 5

FIGURE 6

Let $P = (a_1, b_1, c_1)$ and $Q = (a_2, b_2, c_2)$. Then as in Section 1.1, the vector $\mathbf{v} = \overrightarrow{PQ}$ can be written

$$\mathbf{v} = (a_2 - a_1)\mathbf{i} + (b_2 - b_1)\mathbf{j} + (c_2 - c_1)\mathbf{k} \tag{6}$$

(see Problem 55).

EXAMPLE 6 Find a vector in space that can be represented by the directed line segment from $(2, -1, 4)$ to $(5, 1, -3)$.

Solution. $\mathbf{v} = (5 - 2)\mathbf{i} + [1 - (-1)]\mathbf{j} + (-3 - 4)\mathbf{k} = 3\mathbf{i} + 2\mathbf{j} - 7\mathbf{k}$. ■

We now turn to the notion of dot product (or scalar product) in $\mathbb{R}^3$.

Definition 2 DOT PRODUCT If $\mathbf{u} = x_1\mathbf{i} + y_1\mathbf{j} + z_1\mathbf{k}$ and $\mathbf{v} = x_2\mathbf{i} + y_2\mathbf{j} + z_2\mathbf{k}$, then we define the **dot product** (or **scalar product** or **inner product**) by

$$\mathbf{u} \cdot \mathbf{v} = x_1x_2 + y_1y_2 + z_1z_2. \qquad (7)$$

As before, the dot product of two vectors is a *scalar*. Note that $\mathbf{i} \cdot \mathbf{i} = 1$, $\mathbf{j} \cdot \mathbf{j} = 1$, $\mathbf{k} \cdot \mathbf{k} = 1$, $\mathbf{i} \cdot \mathbf{j} = 0$, $\mathbf{j} \cdot \mathbf{k} = 0$, and $\mathbf{i} \cdot \mathbf{k} = 0$.

EXAMPLE 7 For $\mathbf{u} = 2\mathbf{i} - 3\mathbf{j} - 4\mathbf{k}$ and $\mathbf{v} = -3\mathbf{i} + \mathbf{j} - 2\mathbf{k}$, calculate $\mathbf{u} \cdot \mathbf{v}$.

Solution. $\mathbf{u} \cdot \mathbf{v} = 2(-3) + (-3)(1) + (-4)(-2) = -1$. ■

Theorem 2 For any vectors $\mathbf{u}$, $\mathbf{v}$, and $\mathbf{w}$ in space, and for any scalar α, we have

(i) $\mathbf{u} \cdot \mathbf{v} = \mathbf{v} \cdot \mathbf{u}$
(ii) $(\mathbf{u} + \mathbf{v}) \cdot \mathbf{w} = \mathbf{u} \cdot \mathbf{w} + \mathbf{v} \cdot \mathbf{w}$
(iii) $(\alpha\mathbf{u}) \cdot \mathbf{v} = \alpha(\mathbf{u} \cdot \mathbf{v})$
(iv) $\mathbf{u} \cdot \mathbf{u} \geq 0$, and $\mathbf{u} \cdot \mathbf{u} = 0$ if and only if $\mathbf{u} = \mathbf{0}$
(v) $|\mathbf{u}| = \sqrt{\mathbf{u} \cdot \mathbf{u}}$

Proof. The proof is almost identical to the proof of Theorem 1.2.1 and is left as an exercise (see Problem 56). ■

Theorem 3 If φ denotes the angle between two nonzero vectors $\mathbf{u}$ and $\mathbf{v}$, we have

$$\cos \varphi = \frac{\mathbf{u} \cdot \mathbf{v}}{|\mathbf{u}|\,|\mathbf{v}|}. \qquad (8)$$

Proof. The proof is almost identical to the proof of Theorem 1.2.2 and is left as an exercise (see Problem 57). For an interesting corollary to this theorem, see Problem 61. ■

EXAMPLE 8 Calculate the cosine of the angle between $\mathbf{u} = 3\mathbf{i} - \mathbf{j} + 2\mathbf{k}$ and $\mathbf{v} = 4\mathbf{i} + 3\mathbf{j} - \mathbf{k}$.

Solution. $\mathbf{u} \cdot \mathbf{v} = 7$, $|\mathbf{u}| = \sqrt{14}$, and $|\mathbf{v}| = \sqrt{26}$, so that $\cos \varphi = 7/\sqrt{(14)(26)} = 7/\sqrt{364} \approx 0.3669$ and $\varphi = 68.5° \approx 1.2$ radians. ■

Definition 3 PARALLEL AND ORTHOGONAL VECTORS

(i) Two nonzero vectors $\mathbf{u}$ and $\mathbf{v}$ are **parallel** if the angle between them is 0 or π.
(ii) Two nonzero vectors $\mathbf{u}$ and $\mathbf{v}$ are **orthogonal** (or **perpendicular**) if the angle between them is $\pi/2$.

Theorem 4

(i) If $\mathbf{u} \neq 0$, then $\mathbf{u}$ and $\mathbf{v}$ are parallel if and only if $\mathbf{v} = \alpha\mathbf{u}$ for some constant α.

(ii) If $\mathbf{u}$ and $\mathbf{v}$ are nonzero, then $\mathbf{u}$ and $\mathbf{v}$ are orthogonal if and only if $\mathbf{u} \cdot \mathbf{v} = 0$.

Proof. Again the proof is easy and is left as an exercise (see Problem 58). ■

EXAMPLE 9 Show that the vectors $\mathbf{u} = \mathbf{i} + 3\mathbf{j} - 4\mathbf{k}$ and $\mathbf{v} = -2\mathbf{i} - 6\mathbf{j} + 8\mathbf{k}$ are parallel.

Solution. Here $(\mathbf{u} \cdot \mathbf{v})/(|\mathbf{u}||\mathbf{v}|) = -52/(\sqrt{26}\sqrt{104}) = -52/(\sqrt{26} \cdot 2\sqrt{26}) = -1$, so $\cos\theta = -1$, $\theta = \pi$, and $\mathbf{u}$ and $\mathbf{v}$ are parallel (but have opposite directions). An easier way to see this is to note that $\mathbf{v} = -2\mathbf{u}$ so that by Theorem 4, $\mathbf{u}$ and $\mathbf{v}$ are parallel. The vectors are sketched in Figure 7. ■

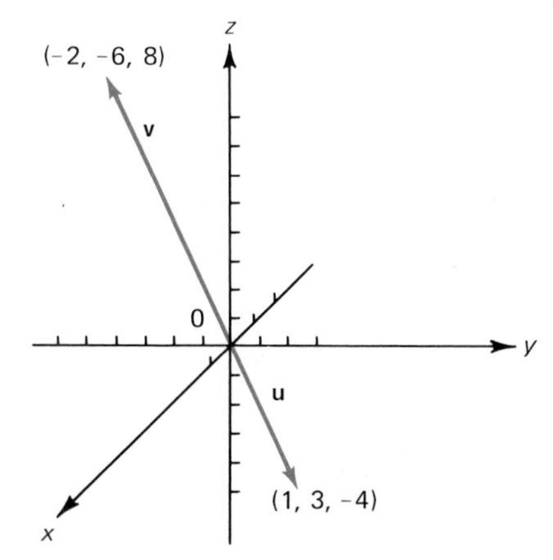

FIGURE 7

We now turn to the definition of the projection of one vector on another. First, we state the theorem that is the analog of Theorem 1.2.5 (and has an identical proof).

Theorem 5 Let $\mathbf{v}$ be a nonzero vector. Then for any other vector $\mathbf{u}$,

$$\mathbf{w} = \mathbf{u} - \frac{\mathbf{u} \cdot \mathbf{v}}{|\mathbf{v}|^2}\mathbf{v}$$

is orthogonal to $\mathbf{v}$.

Definition 4 PROJECTION Let $\mathbf{u}$ and $\mathbf{v}$ be nonzero vectors. Then the **projection**† **of u onto v**, denoted $\text{Proj}_{\mathbf{v}}\,\mathbf{u}$, is defined by

$$\text{Proj}_{\mathbf{v}}\,\mathbf{u} = \frac{\mathbf{u} \cdot \mathbf{v}}{\mathbf{v} \cdot \mathbf{v}}\mathbf{v} = \frac{\mathbf{u} \cdot \mathbf{v}}{|\mathbf{v}|^2}\mathbf{v} = \left(\frac{\mathbf{u} \cdot \mathbf{v}}{|\mathbf{v}|}\right)\frac{\mathbf{v}}{|\mathbf{v}|}. \quad (9)$$

†The projection vector in $\mathbb{R}^2$ is sketched in Figure 1.2.3. The derivation and a sketch of a projection vector in $\mathbb{R}^3$ are virtually the same.

The **component** of **u** in the direction **v** is given by $(\mathbf{u} \cdot \mathbf{v})/|\mathbf{v}|$.

EXAMPLE 10 Let $\mathbf{u} = 2\mathbf{i} + 3\mathbf{j} + \mathbf{k}$ and $\mathbf{v} = \mathbf{i} + 2\mathbf{j} - 6\mathbf{k}$. Find $\text{Proj}_{\mathbf{v}}\, \mathbf{u}$.

Solution. Here $(\mathbf{u} \cdot \mathbf{v})/|\mathbf{v}|^2 = \frac{2}{41}$, so

$$\text{Proj}_{\mathbf{v}}\, \mathbf{u} = \frac{2}{41}\mathbf{i} + \frac{4}{41}\mathbf{j} - \frac{12}{41}\mathbf{k}.$$

The component of **u** in the direction **v** is $(\mathbf{u} \cdot \mathbf{v})/|\mathbf{v}| = 2/\sqrt{41}$. ■

Note that, as in the planar case, $\text{Proj}_{\mathbf{v}}\, \mathbf{u}$ is a vector that has the same direction as **v** if $\mathbf{u} \cdot \mathbf{v} > 0$ and the direction opposite to that of **v** if $\mathbf{u} \cdot \mathbf{v} < 0$.

PROBLEMS 3.2

In Problems 1–20, find the magnitude and the direction cosines of the given vector.

1. $\mathbf{v} = 3\mathbf{j}$
2. $\mathbf{v} = -3\mathbf{i}$
3. $\mathbf{v} = 14\mathbf{k}$
4. $\mathbf{v} = -8\mathbf{j}$
5. $\mathbf{v} = 4\mathbf{i} - \mathbf{j}$
6. $\mathbf{v} = \mathbf{i} + 2\mathbf{k}$
7. $\mathbf{v} = -2\mathbf{i} + 3\mathbf{j}$
8. $\mathbf{v} = \mathbf{i} + \mathbf{j} + \mathbf{k}$
9. $\mathbf{v} = \mathbf{i} - \mathbf{j} + \mathbf{k}$
10. $\mathbf{v} = \mathbf{i} + \mathbf{j} - \mathbf{k}$
11. $\mathbf{v} = -\mathbf{i} + \mathbf{j} + \mathbf{k}$
12. $\mathbf{v} = \mathbf{i} - \mathbf{j} - \mathbf{k}$
13. $\mathbf{v} = -\mathbf{i} + \mathbf{j} - \mathbf{k}$
14. $\mathbf{v} = -\mathbf{i} - \mathbf{j} + \mathbf{k}$
15. $\mathbf{v} = -\mathbf{i} - \mathbf{j} - \mathbf{k}$
16. $\mathbf{v} = 2\mathbf{i} + 5\mathbf{j} - 7\mathbf{k}$
17. $\mathbf{v} = -7\mathbf{i} + 2\mathbf{j} - 13\mathbf{k}$
18. $\mathbf{v} = \mathbf{i} + 7\mathbf{j} - 7\mathbf{k}$
19. $\mathbf{v} = -3\mathbf{i} - 3\mathbf{j} + 8\mathbf{k}$
20. $\mathbf{v} = -2\mathbf{i} - 3\mathbf{j} - 4\mathbf{k}$

21. The three direction angles of a certain unit vector are the same and are between 0 and $\pi/2$. What is the vector?
22. Find a vector of magnitude 12 that has the same direction as the vector of Problem 21.
23. Show that there is no unit vector whose direction angles are $\pi/6$, $\pi/3$, and $\pi/4$.
24. Let $P = (2, 1, 4)$ and $Q = (3, -2, 8)$. Find a unit vector in the direction of $\overrightarrow{PQ}$.
25. Let $P = (-3, 1, 7)$ and $Q = (8, 1, 7)$. Find a unit vector whose direction is opposite that of $\overrightarrow{PQ}$.
26. In Problem 25 find all points R that satisfy $\overrightarrow{PR} \perp \overrightarrow{PQ}$.

*27. Show that the set of points R that satisfy the condition of Problem 26 and the condition $|\overrightarrow{PR}| = 1$ form a circle.

In Problems 28–46 let $\mathbf{u} = 2\mathbf{i} - 3\mathbf{j} + 4\mathbf{k}$, $\mathbf{v} = -2\mathbf{i} - 3\mathbf{j} + 5\mathbf{k}$, $\mathbf{w} = \mathbf{i} - 7\mathbf{j} + 3\mathbf{k}$, and $\mathbf{t} = 3\mathbf{i} + 4\mathbf{j} + 5\mathbf{k}$.

28. Calculate $\mathbf{u} + \mathbf{v}$.
29. Calculate $2\mathbf{u} - 3\mathbf{v}$.
30. Calculate $-18\mathbf{u}$.
31. Calculate $\mathbf{w} - \mathbf{u} - \mathbf{v}$.
32. Calculate $\mathbf{t} + 3\mathbf{w} - \mathbf{v}$.
33. Calculate $2\mathbf{u} - 7\mathbf{w} + 5\mathbf{v}$.
34. Calculate $2\mathbf{v} + 7\mathbf{t} - \mathbf{w}$.
35. Calculate $\mathbf{u} \cdot \mathbf{v}$.
36. Calculate $|\mathbf{w}|$.
37. Calculate $\mathbf{u} \cdot \mathbf{w} - \mathbf{w} \cdot \mathbf{t}$.
38. Calculate the angle between **u** and **w**.
39. Calculate the angle between **t** and **w**.
40. Calculate the angle between **v** and **t**.
41. Calculate $\text{Proj}_{\mathbf{v}}\, \mathbf{u}$.
42. Calculate $\text{Proj}_{\mathbf{u}}\, \mathbf{v}$.
43. Calculate $\text{Proj}_{\mathbf{t}}\, \mathbf{w}$.
44. Calculate $\text{Proj}_{\mathbf{w}}\, \mathbf{t}$.
45. Calculate $\text{Proj}_{\mathbf{w}}\, \mathbf{u}$.
46. Calculate $\text{Proj}_{\mathbf{t}}\, \mathbf{v}$.

47. Find the distance between the point $P = (2, 1, 3)$ and the line passing through the points $Q = (-1, 1, 2)$ and $R = (6, 0, 1)$. [*Hint:* See Problem 1.2.50.]
48. Find the distance from the point $P = (1, 0, 1)$ to the line passing through the points $Q = (2, 3, -1)$ and $R = (6, 1, -3)$.
49. Show that the points $P = (3, 5, 6)$, $Q = (1, 2, 7)$, and $R = (6, 1, 0)$ are the vertices of a right triangle.
50. Show that the points $P = (3, 2, -1)$, $Q = (4, 1, 6)$, $R = (7, -2, 3)$, and $S = (8, -3, 10)$ are the vertices of a parallelogram.

*51. A polyhedron in space with exactly four vertices is called a **tetrahedron** (see Figure 8). Let **P** represent the vector $\overrightarrow{0P}$, **Q** the vector $\overrightarrow{0Q}$, and so on. A line is drawn from each vertex to the centroid of the opposite side. Show that these four lines meet at the endpoint of the vector

$$\mathbf{v} = \frac{\mathbf{P} + \mathbf{Q} + \mathbf{R} + \mathbf{S}}{4}.$$

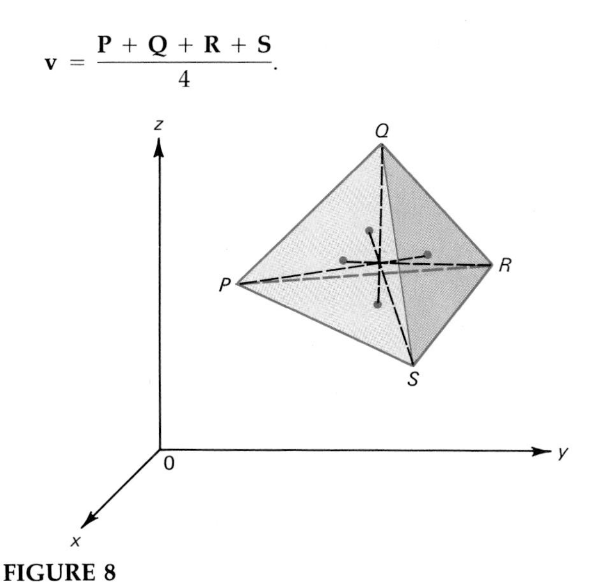

FIGURE 8

52. A force of 3 N acts in the direction of the vector with direction cosines $(1/\sqrt{6}, 1/\sqrt{3}, 1/\sqrt{2})$. Find the work done in moving the object from the point (1, 2, 3) to the point (2, 8, 11), where distance is measured in meters. [*Hint:* See Example 1.3.4.]

53. Find the work done when a force of 3 N acting in the direction of the vector $\mathbf{v} = \mathbf{i} + \mathbf{j} - \mathbf{k}$ moves an object from $(-1, 3, 4)$ to $(3, 7, -2)$. Again distance is measured in meters.

54. Prove Theorem 1.

55. Prove that formula (6) is correct. [*Hint:* Follow the steps leading to formula (1.1.7).]

56. Prove Theorem 2.

57. Prove Theorem 3.

58. Prove Theorem 4.

***59.** Let PQR be a triangle in $\mathbb{R}^3$. Show that if a force of N newtons of constant direction moves an object around the triangle, then the total work done by that force is zero.

***60.** Find the angle between the diagonal of a cube and the diagonal of one of its faces.

61. (a) Use Theorem 3 to prove the **Cauchy-Schwarz inequality:**

$$\left(\sum_{i=1}^{3} u_i v_i\right)^2 \le \sum_{i=1}^{3} u_i^2 \sum_{i=1}^{3} v_i^2, \tag{10}$$

where u_i and v_i are real numbers.

(b) Show that equality holds in (10) if and only if at least one of the vectors (u_1, u_2, u_3) and (v_1, v_2, v_3) is a multiple of the other.

62. Use the dot product to find two unit vectors perpendicular to the vectors (1, 2, 3) and $(-4, 1, 5)$.

63. Find two unit vectors perpendicular to $(-2, 0, 4)$ and $(3, -2, -1)$.

3.3 LINES IN $\mathbb{R}^3$

Recall that the equation of a line can be determined if we know either (i) two points on the line, or (ii) one point on the line and the direction (slope) of the line. In Section 2.1 we showed how the equation of a line could be written parametrically. In $\mathbb{R}^3$ our intuition tells us that the basic ideas are the same. Since two points determine a line, we should be able to calculate the equation of a line in space if we know two points on it. Alternatively, if we know one point and the direction of a line, we should also be able to find its equation.

We begin with two points $P = (x_1, y_1, z_1)$ and $Q = (x_2, y_2, z_2)$ on a line L. A vector parallel to L is a vector with representation $\mathbf{v} = \overrightarrow{PQ}$, or [from formula (3.2.6)]

$$\mathbf{v} = (x_2 - x_1)\mathbf{i} + (y_2 - y_1)\mathbf{j} + (z_2 - z_1)\mathbf{k}. \tag{1}$$

Now let $R = (x, y, z)$ be another point on the line. Then $\overrightarrow{PR}$ is parallel to $\overrightarrow{PQ}$, which is parallel to $\mathbf{v}$, so that by Theorem 3.2.4(i),

$$\overrightarrow{PR} = t\mathbf{v} \tag{2}$$

for some real number t. Now look at Figure 1. From the figure we have (in each of the

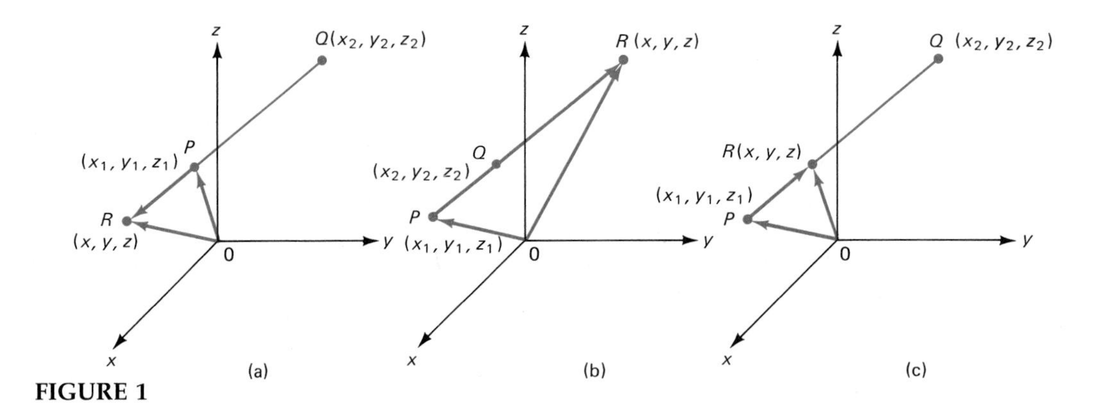

FIGURE 1

three possible cases)

$$\overrightarrow{0R} = \overrightarrow{0P} + \overrightarrow{PR}, \tag{3}$$

and combining (2) and (3)

$$\overrightarrow{PR} = \overrightarrow{0R} - \overrightarrow{0P} = t\mathbf{v}.$$

Thus

$$\overrightarrow{0R} = \overrightarrow{0P} + t\mathbf{v}, \tag{4}$$

or

$$(x, y, z) = (x_1, y_1, z_1) + t(x_2 - x_1, y_2 - y_1, z_2 - z_1). \tag{4a}$$

Equation (4) or (4a) is called the **vector equation** of the line L. For if R is on L, then (4) is satisfied for some real number t. Conversely, if (4) is satisfied, then reversing our steps, we see that $\overrightarrow{PR}$ is parallel to $\mathbf{v}$, which means that R is on L.

NOTE. Since $\mathbf{v} = \overrightarrow{PQ} = \overrightarrow{0Q} - \overrightarrow{0P}$, (4) can be rewritten as

$$\overrightarrow{0R} = \overrightarrow{0P} + t(\overrightarrow{0Q} - \overrightarrow{0P}),$$

or

$$\overrightarrow{0R} = (1 - t)\overrightarrow{0P} + t\overrightarrow{0Q}. \tag{5}$$

The vector equation (5) is sometimes very useful.

If we write out the components of equation (4), we obtain

$$x\mathbf{i} + y\mathbf{j} + z\mathbf{k} = x_1\mathbf{i} + y_1\mathbf{j} + z_1\mathbf{k} + t(x_2 - x_1)\mathbf{i} + t(y_2 - y_1)\mathbf{j} + t(z_2 - z_1)\mathbf{k},$$

or

$$x = x_1 + t(x_2 - x_1),$$
$$y = y_1 + t(y_2 - y_1), \qquad (6)$$
$$z = z_1 + t(z_2 - z_1).$$

The equations (6) are called the **parametric equations** of a line.

Finally, solving for t in (6), and defining $x_2 - x_1 = a$, $y_2 - y_1 = b$, and $z_2 - z_1 = c$, we find that

$$\frac{x - x_1}{a} = \frac{y - y_1}{b} = \frac{z - z_1}{c}. \qquad (7)$$

The equations (7) are called the **symmetric equations** of the line.

Here a, b, and c are direction numbers of the vector $\mathbf{v}$. Of course, equations (7) are valid only if a, b, and c are nonzero.

EXAMPLE 1 Find a vector equation, parametric equations, and symmetric equations of the line L passing through the points $P = (2, -1, 6)$ and $Q = (3, 1, -2)$.

Solution. First, we calculate

$$\mathbf{v} = (3 - 2)\mathbf{i} + [1 - (-1)]\mathbf{j} + (-2 - 6)\mathbf{k} = \mathbf{i} + 2\mathbf{j} - 8\mathbf{k}.$$

Then from (4), if $R = (x, y, z)$ is on the line,

$$\overrightarrow{0R} = x\mathbf{i} + y\mathbf{j} + z\mathbf{k} = \overrightarrow{0P} + t\mathbf{v} = 2\mathbf{i} - \mathbf{j} + 6\mathbf{k} + t(\mathbf{i} + 2\mathbf{j} - 8\mathbf{k}),$$

or

$$x = 2 + t, \qquad y = -1 + 2t, \qquad z = 6 - 8t.$$

Finally, since $a = 1$, $b = 2$, and $c = -8$, we find the symmetric equations

$$\frac{x - 2}{1} = \frac{y + 1}{2} = \frac{z - 6}{-8}.$$

To check this, we verify that $(2, -1, 6)$ and $(3, 1, -2)$ are indeed on the line. We have

$$\frac{2 - 2}{1} = \frac{-1 + 1}{2} = \frac{6 - 6}{-8} = 0$$

and

$$\frac{3 - 2}{1} = \frac{1 + 1}{2} = \frac{-2 - 6}{-8} = 1.$$

Other points on the line can be found. For example, if $t = 3$, we obtain

$$3 = \frac{x-2}{1} = \frac{y+1}{2} = \frac{z-6}{-8},$$

which yields the point $(5, 5, -18)$. ■

EXAMPLE 2 Find symmetric equations of the line passing through the point $(1, -2, 4)$ and parallel to the vector $\mathbf{v} = \mathbf{i} + \mathbf{j} - \mathbf{k}$.

Solution. We simply choose $\overrightarrow{0P} = \mathbf{i} - 2\mathbf{j} + 4\mathbf{k}$ and $\mathbf{v}$ as above. Then $a = 1$, $b = 1$, $c = -1$, and we obtain

$$\frac{x-1}{1} = \frac{y+2}{1} = \frac{z-4}{-1}. \quad ■$$

What happens if one of the direction numbers a, b, or c is zero?

EXAMPLE 3 Find symmetric equations of the line containing the points $P = (3, 4, -1)$ and $Q = (-2, 4, 6)$.

Solution. Here $\mathbf{v} = -5\mathbf{i} + 7\mathbf{k}$, and $a = -5$, $b = 0$, $c = 7$. Then a parametric representation of the line is

$$x = 3 - 5t, \qquad y = 4, \qquad z = -1 + 7t.$$

Solving for t, we find that

$$\frac{x-3}{-5} = \frac{z+1}{7} \qquad \text{and} \qquad y = 4.$$

The equation $y = 4$ is the equation of a plane parallel to the xz-plane, so we have obtained an equation of a line in that plane. ■

EXAMPLE 4 Find symmetric equations of the line in the xy-plane that passes through the points $(x_1, y_1, 0)$ and $(x_2, y_2, 0)$.

Solution. Here $\mathbf{v} = (x_2 - x_1)\mathbf{i} + (y_2 - y_1)\mathbf{j}$, and we obtain

$$\frac{x - x_1}{x_2 - x_1} = \frac{y - y_1}{y_2 - y_1} \qquad \text{and} \qquad z = 0.$$

We can rewrite this as

$$y - y_1 = \left(\frac{y_2 - y_1}{x_2 - x_1}\right)(x - x_1).$$

Here $(y_2 - y_1)/(x_2 - x_1) = m$, the slope of the line, and when $x = 0$, $y = y_1 - [(y_2 - y_1)/(x_2 - x_1)]x_1 = b$, the y-intercept of the line. That is, $y = mx + b$, which is the slope-intercept form of a line in the xy-plane. Thus we see that the symmetric equations of a line in space are really a generalization of the equation of a line in the plane. ■

Now, what happens if two of the direction numbers are zero?

EXAMPLE 5 Find the symmetric equations of the line passing through the points $P = (2, 3, -2)$ and $Q = (2, -1, -2)$.

Solution. Here $\mathbf{v} = -4\mathbf{j}$, so that $a = 0$, $b = -4$, and $c = 0$. A parametric representation of the line is, by equations (6), given by

$$x = 2, \qquad y = 3 - 4t, \qquad z = -2.$$

Now $x = 2$ is the equation of a plane parallel to the yz-plane, while $z = -2$ is the equation of a plane parallel to the xy-plane. Their intersection is the line $x = 2$, $z = -2$, which is parallel to the y-axis. In fact, the equation $y = 3 - 4t$, says, essentially, that y can take on any value (while x and z remain fixed). ■

The results of Examples 1, 3, and 5 are summarized in Theorem 1.

Theorem 1 Let L be a line passing through the point (x_1, y_1, z_1) and parallel to the vector $\mathbf{v} = a\mathbf{i} + b\mathbf{j} + c\mathbf{k}$. Then symmetric equations of the line are as follows:

(i) $$\frac{x - x_1}{a} = \frac{y - y_1}{b} = \frac{z - z_1}{c}$$

if a, b, and c are all nonzero.

(ii) $$x = x_1, \qquad \frac{y - y_1}{b} = \frac{z - z_1}{c}$$

if $a = 0$. Then the line is parallel to the yz-plane. If either b or $c = 0$, but $a \neq 0$, similar results hold.

(iii) $$x = x_1, \qquad y = y_1, \qquad z = z_1 + ct$$

if a and b are 0. Then the line is parallel to the z-axis. If a and c or b and c are 0, similar results hold.

■ WARNING: The parametric or symmetric equations of a line are *not* unique. To see this, simply choose two other points on the line.

EXAMPLE 6 In Example 1, the line contains the point $(5, 5, -18)$. Choose $P = (5, 5, -18)$ and $Q = (3, 1, -2)$. We find that $\mathbf{v} = -2\mathbf{i} - 4\mathbf{j} + 16\mathbf{k}$, so that

$$x = 5 - 2t, \qquad y = 5 - 4t, \qquad z = -18 + 16t.$$

[Note that if $t = \frac{3}{2}$, we obtain $(x, y, z) = (2, -1, 6)$.] The symmetric equations are now

$$\frac{x - 5}{-2} = \frac{y - 5}{-4} = \frac{z + 18}{16}. \quad ■$$

PROBLEMS 3.3

In Problems 1–22, find a vector equation, parametric equations, and symmetric equations of the indicated line.

1. Containing (2, 1, 3) and (1, 2, −1).
2. Containing (1, −1, 1) and (−1, 1, −1).
3. Containing (1, 3, 2) and (2, 4, −2).
4. Containing (−2, 4, 5) and (3, 7, 2).
5. Containing (−4, 1, 3) and (−4, 0, 1).
6. Containing (2, 3, −4) and (2, 0, −4).
7. Containing (1, 2, 3) and (3, 2, 1).
8. Containing (7, 1, 3) and (−1, −2, 3).
9. Containing (1, 2, 4) and (1, 2, 7).
10. Containing (−3, −1, −6) and (−3, 1, 6).
11. Containing (2, 2, 1) and parallel to $2\mathbf{i} - \mathbf{j} - \mathbf{k}$.
12. Containing (−1, −6, 2) and parallel to $4\mathbf{i} + \mathbf{j} - 3\mathbf{k}$.
13. Containing (1, 0, 3) and parallel to $\mathbf{i} - \mathbf{j}$.
14. Containing (2, 1, −4) and parallel to $\mathbf{i} + 4\mathbf{k}$.
15. Containing (−1, −2, 5) and parallel to $-3\mathbf{j} + 7\mathbf{k}$.
16. Containing (−2, 3, −2) and parallel to $4\mathbf{k}$.
17. Containing (−1, −3, 1) and parallel to $-7\mathbf{j}$.
18. Containing (2, 1, 5) and parallel to $3\mathbf{i}$.
19. Containing (a, b, c) and parallel to $d\mathbf{i} + e\mathbf{j}$, $d, e \neq 0$.
20. Containing (a, b, c) and parallel to $d\mathbf{k}$, $d \neq 0$.
21. Containing (4, 1, −6) and parallel to $(x - 2)/3 = (y + 1)/6 = (z - 5)/2$.
22. Containing (3, 1, −2) and parallel to $(x + 1)/3 = (y + 3)/2 = (z - 2)/(-4)$.

23. Let L_1 be given by
$$\frac{x - x_1}{a_1} = \frac{y - y_1}{b_1} = \frac{z - z_1}{c_1}$$
and L_2 be given by
$$\frac{x - x_2}{a_2} = \frac{y - y_2}{b_2} = \frac{z - z_2}{c_2}.$$
Show that a direction vector of L_1 is orthogonal to a direction vector of L_2 if and only if $a_1a_2 + b_1b_2 + c_1c_2 = 0$.

24. Show that direction vectors of the lines
$$L_1: \quad \frac{x - 3}{2} = \frac{y + 1}{4} = \frac{z - 2}{-1}$$
and
$$L_2: \quad \frac{x - 3}{5} = \frac{y + 1}{-2} = \frac{z - 3}{2}$$
are orthogonal.

25. Show that the lines
$$L_1: \quad \frac{x - 1}{1} = \frac{y + 3}{2} = \frac{z + 3}{3}$$
and
$$L_2: \quad \frac{x - 3}{3} = \frac{y - 1}{6} = \frac{z - 3}{9}$$
are equations of the same straight line.

In the plane, two lines that are not parallel have exactly one point of intersection. In $\mathbb{R}^3$, this is not the case. For example, the lines L_1: $x = 2$, $y = 3$ (parallel to the z-axis) and L_2: $x = 1$, $z = 3$ (parallel to the y-axis) are not parallel and have no points in common. It takes a bit of work to determine whether two lines in $\mathbb{R}^3$ do have a point in common (they usually do not).

26. Determine whether the lines
$$L_1: \quad x = 1 + t, \qquad y = -3 + 2t, \qquad z = -2 - t$$
and
$$L_2: \quad x = 17 + 3s, \qquad y = 4 + s, \qquad z = -8 - s$$
have a point of intersection. [*Hint:* If (x, y, z) is a point common to both lines, then $x = 1 + t = 17 + 3s$, $y = -3 + 2t = 4 + s$, $z = -2 - t = -8 - s$. Find (if possible) numbers s and t that satisfy all three of these equations.

27. Determine whether the lines
$$x = 2 - t, \qquad y = 1 + t, \qquad z = -2 - t$$
and
$$x = 1 + s, \qquad y = -2s, \qquad z = 3 + 2s$$
have a point in common.

In Problems 28–33, determine whether the given pair of lines has a point of intersection. If so, find it.

28. L_1: $x = 2 + t$, $y = -1 + 2t$, $z = 3 + 4t$; L_2: $x = 9 + s$, $y = -2 - s$, $z = 1 - 2s$
29. L_1: $x = 3 + 2t$, $y = 2 - t$, $z = 1 + t$; L_2: $x = 4 - s$, $y = -2 + 3s$, $z = 2 + 2s$
30. L_1: $\dfrac{x - 4}{-3} = \dfrac{y - 1}{7} = \dfrac{z + 2}{-8}$;
$$L_2: \quad \frac{x - 5}{1} = \frac{y - 3}{-1} = \frac{z - 1}{2}$$

31. L_1: $\frac{x-2}{-5} = \frac{y-1}{1} = \frac{z-3}{4}$;

L_2: $\frac{x+3}{4} = \frac{y-2}{-1} = \frac{z-7}{6}$

32. L_1: $x = 4 - t, y = 7 + 5t, z = 2 - 3t$; L_2: $x = 1 + 2s, y = 6 - 2s, z = 10 + 3s$

33. L_1: $x = 1 + t, y = 2 - t, z = 3t$; L_2: $x = 3s, y = 2 - s, z = 2 + s$

34. Let L be given in its vector form $\overrightarrow{0R} = \overrightarrow{0P} + t\mathbf{v}$. Find a number t such that $\overrightarrow{0R}$ is perpendicular to $\mathbf{v}$.

35. Use the result of Problem 34 to find the distance between the line L (containing P and parallel to $\mathbf{v}$) and the origin.

(a) $P = (2, 1, -4)$; $\mathbf{v} = \mathbf{i} + \mathbf{j} + \mathbf{k}$
(b) $P = (1, 2, -3)$; $\mathbf{v} = 3\mathbf{i} - \mathbf{j} - \mathbf{k}$
(c) $P = (-1, -4, 2)$; $\mathbf{v} = -\mathbf{i} + \mathbf{j} + 2\mathbf{k}$

***36.** Show that the lines L_1: $x = x_1 + a_1t, y = y_1 + b_1t, z = z_1 + c_1t$ and L_2: $x = x_2 + a_2s, y = y_2 + b_2s, z = z_2 + c_2s$ have a point in common or are parallel if and only if the determinant

$$\begin{vmatrix} a_1 & a_2 & x_1 - x_2 \\ b_1 & b_2 & y_1 - y_2 \\ c_1 & c_2 & z_1 - z_2 \end{vmatrix} = 0.$$

37. Apply the result of Problem 36 to the lines in Problem 28.

38. Do the same for the lines in Problem 29.

3.4 THE CROSS PRODUCT OF TWO VECTORS

To this point the only product of vectors we have considered has been the dot or scalar product. We now define a new product, called the *cross product*[†] (or *vector product*), which is defined only in $\mathbb{R}^3$.

Definition 1 CROSS PRODUCT Let $\mathbf{u} = a_1\mathbf{i} + b_1\mathbf{j} + c_1\mathbf{k}$ and $\mathbf{v} = a_2\mathbf{i} + b_2\mathbf{j} + c_2\mathbf{k}$. Then the **cross product (vector product)** of $\mathbf{u}$ and $\mathbf{v}$, denoted $\mathbf{u} \times \mathbf{v}$, is a new vector defined by

$$\mathbf{u} \times \mathbf{v} = (b_1c_2 - c_1b_2)\mathbf{i} + (c_1a_2 - a_1c_2)\mathbf{j} + (a_1b_2 - b_1a_2)\mathbf{k}. \quad (1)$$

Note that *the result of the cross product is a vector, while the result of the dot product is a scalar.*

Here the cross product seems to have been defined somewhat arbitrarily. There are obviously many ways to define a vector product. Why was this definition chosen? We will answer that question in this section by demonstrating some of the properties of the cross product and illustrating some of its uses.

EXAMPLE 1 Let $\mathbf{u} = \mathbf{i} - \mathbf{j} + 2\mathbf{k}$ and $\mathbf{v} = 2\mathbf{i} + 3\mathbf{j} - 4\mathbf{k}$. Calculate $\mathbf{w} = \mathbf{u} \times \mathbf{v}$.

Solution. Using formula (1), we have

$$\begin{aligned}\mathbf{w} &= [(-1)(-4) - (2)(3)]\mathbf{i} + [(2)(2) - (1)(-4)]\mathbf{j} + [(1)(3) - (-1)(2)]\mathbf{k} \\ &= -2\mathbf{i} + 8\mathbf{j} + 5\mathbf{k}. \ \blacksquare\end{aligned}$$

[†]The cross product was defined by Hamilton in one of a series of papers discussing his quaternions, which were published in *Philosophical Magazine* between the years 1844 and 1850.

NOTE. In this example $\mathbf{u} \cdot \mathbf{w} = \mathbf{v} \cdot \mathbf{w} = 0$. That is, $\mathbf{u} \times \mathbf{v}$ is orthogonal to both **u** and **v**. As we will shortly see, the cross product of **u** and **v** is always orthogonal to both **u** and **v**.

Before continuing our discussion of the uses of the cross product, we remark that there is an easy way to remember how to calculate $\mathbf{u} \times \mathbf{v}$ if you are familiar with the elementary properties of 3×3 determinants. If you are not, we suggest that you turn to Sections 8.1 and 8.2 where these properties are discussed.

Theorem 1

$$\mathbf{u} \times \mathbf{v} = \begin{vmatrix} \mathbf{i} & \mathbf{j} & \mathbf{k} \\ a_1 & b_1 & c_1 \\ a_2 & b_2 & c_2 \end{vmatrix}^{\dagger}$$

Proof.

$$\begin{vmatrix} \mathbf{i} & \mathbf{j} & \mathbf{k} \\ a_1 & b_1 & c_1 \\ a_2 & b_2 & c_2 \end{vmatrix} = \mathbf{i}\begin{vmatrix} b_1 & c_1 \\ b_2 & c_2 \end{vmatrix} - \mathbf{j}\begin{vmatrix} a_1 & c_1 \\ a_2 & c_2 \end{vmatrix} + \mathbf{k}\begin{vmatrix} a_1 & b_1 \\ a_2 & b_2 \end{vmatrix} \tag{2}$$

$$= (b_1c_2 - c_1b_2)\mathbf{i} + (a_2c_1 - a_1c_2)\mathbf{j} + (a_1b_2 - b_1a_2)\mathbf{k},$$

which is equal to $\mathbf{u} \times \mathbf{v}$ according to Definition 1. ■

EXAMPLE 2 Calculate $\mathbf{u} \times \mathbf{v}$, where $\mathbf{u} = 2\mathbf{i} + 4\mathbf{j} - 5\mathbf{k}$ and $\mathbf{v} = -3\mathbf{i} - 2\mathbf{j} + \mathbf{k}$.

Solution.

$$\mathbf{u} \times \mathbf{v} = \begin{vmatrix} \mathbf{i} & \mathbf{j} & \mathbf{k} \\ 2 & 4 & -5 \\ -3 & -2 & 1 \end{vmatrix} = (4 - 10)\mathbf{i} - (2 - 15)\mathbf{j} + (-4 + 12)\mathbf{k}$$

$$= -6\mathbf{i} + 13\mathbf{j} + 8\mathbf{k}. \quad ■$$

The following theorem summarizes some properties of the cross product.

Theorem 2 Let **u**, **v**, and **w** be vectors in $\mathbb{R}^3$, and let α be a scalar.

(i) $\mathbf{u} \times \mathbf{0} = \mathbf{0} = \mathbf{0} \times \mathbf{u}$.
(ii) $\mathbf{u} \times \mathbf{v} = -(\mathbf{v} \times \mathbf{u})$.
(iii) $(\alpha\mathbf{u} \times \mathbf{v}) = \alpha(\mathbf{u} \times \mathbf{v})$.
(iv) $\mathbf{u} \times (\mathbf{v} + \mathbf{w}) = (\mathbf{u} \times \mathbf{v}) + (\mathbf{u} \times \mathbf{w})$.
(v) $(\mathbf{u} \times \mathbf{v}) \cdot \mathbf{w} = \mathbf{u} \cdot (\mathbf{v} \times \mathbf{w})$. (This product is called the **scalar triple product** of **u**, **v**, and **w**.)
(vi) $\mathbf{u} \cdot (\mathbf{u} \times \mathbf{v}) = \mathbf{v} \cdot (\mathbf{u} \times \mathbf{v}) = 0$. (That is, $\mathbf{u} \times \mathbf{v}$ is orthogonal to both **u** and **v**.)
(vii) If **u** and **v** are parallel, then $\mathbf{u} \times \mathbf{v} = \mathbf{0}$.

---

† The determinant is defined as a real number, not a vector. This use of the determinant notation is simply a convenient way to denote the cross product.

Proof.

(i) Let $\mathbf{u} = a_1\mathbf{i} + b_1\mathbf{j} + c_1\mathbf{k}$. Then

$$\mathbf{u} \times \mathbf{0} = \begin{vmatrix} \mathbf{i} & \mathbf{j} & \mathbf{k} \\ a_1 & b_1 & c_1 \\ 0 & 0 & 0 \end{vmatrix} = 0\mathbf{i} + 0\mathbf{j} + 0\mathbf{k} = \mathbf{0}.$$

Similarly, $\mathbf{0} \times \mathbf{u} = \mathbf{0}$.

(ii) Let $\mathbf{v} = a_2\mathbf{i} + b_2\mathbf{j} + c_2\mathbf{k}$. Then

$$\mathbf{u} \times \mathbf{v} = \begin{vmatrix} \mathbf{i} & \mathbf{j} & \mathbf{k} \\ a_1 & b_1 & c_1 \\ a_2 & b_2 & c_2 \end{vmatrix} = -\begin{vmatrix} \mathbf{i} & \mathbf{j} & \mathbf{k} \\ a_2 & b_2 & c_2 \\ a_1 & b_1 & c_1 \end{vmatrix} = -(\mathbf{v} \times \mathbf{u}),$$

since interchanging the rows of a determinant has the effect of multiplying the determinant by -1 [see Property 4 in Section 8.2].

(iii)

$$(\alpha\mathbf{u}) \times \mathbf{v} = \begin{vmatrix} \mathbf{i} & \mathbf{j} & \mathbf{k} \\ \alpha a_1 & \alpha b_1 & \alpha c_1 \\ a_2 & b_2 & c_2 \end{vmatrix} = \alpha\begin{vmatrix} \mathbf{i} & \mathbf{j} & \mathbf{k} \\ a_1 & b_1 & c_1 \\ a_2 & b_2 & c_2 \end{vmatrix} = \alpha(\mathbf{u} \times \mathbf{v})$$

The second equality follows from Property 2 in Section 8.2.

(iv) Let $\mathbf{w} = a_3\mathbf{i} + b_3\mathbf{j} + c_3\mathbf{k}$. Then

$$\mathbf{u} \times (\mathbf{v} + \mathbf{w}) = \begin{vmatrix} \mathbf{i} & \mathbf{j} & \mathbf{k} \\ a_1 & b_1 & c_1 \\ a_2 + a_3 & b_2 + b_3 & c_2 + c_3 \end{vmatrix} = \begin{vmatrix} \mathbf{i} & \mathbf{j} & \mathbf{k} \\ a_1 & b_1 & c_1 \\ a_2 & b_2 & c_2 \end{vmatrix} + \begin{vmatrix} \mathbf{i} & \mathbf{j} & \mathbf{k} \\ a_1 & b_1 & c_1 \\ a_3 & b_3 & c_3 \end{vmatrix}$$

$$= (\mathbf{u} \times \mathbf{v}) + (\mathbf{u} \times \mathbf{w}).$$

The second equality is easily verified by direct calculation.

(v)
$$\begin{aligned}(\mathbf{u} \times \mathbf{v}) \cdot \mathbf{w} &= [(b_1c_2 - c_1b_2)\mathbf{i} + (c_1a_2 - a_1c_2)\mathbf{j} + (a_1b_2 - b_1a_2)\mathbf{k}] \\ &\quad \cdot (a_3\mathbf{i} + b_3\mathbf{j} + c_3\mathbf{k}) \\ &= b_1c_2a_3 - c_1b_2a_3 + c_1a_2b_3 - a_1c_2b_3 + a_1b_2c_3 - b_1a_2c_3.\end{aligned}$$

We can easily show that $\mathbf{u} \cdot (\mathbf{v} \times \mathbf{w})$ is equal to the same expression (see Problem 35).

NOTE. For an interesting geometric intrepretation of the scalar triple product, see Problem 41.

(vi) We know that $\mathbf{u} \cdot (\mathbf{u} \times \mathbf{v}) = (\mathbf{u} \times \mathbf{v}) \cdot \mathbf{u}$ [since the dot product is commutative—see Theorem 3.2.2(i)]. But from parts (ii) and (v) of this theorem,

$$(\mathbf{u} \times \mathbf{v}) \cdot \mathbf{u} = \mathbf{u} \cdot (\mathbf{v} \times \mathbf{u}) = \mathbf{u} \cdot (-\mathbf{u} \times \mathbf{v}) = -\mathbf{u} \cdot (\mathbf{u} \times \mathbf{v}).$$

Thus $\mathbf{u} \cdot (\mathbf{u} \times \mathbf{v}) = -\mathbf{u} \cdot (\mathbf{u} \times \mathbf{v})$, which can only occur if $\mathbf{u} \cdot (\mathbf{u} \times \mathbf{v}) = 0$. A similar computation shows that $\mathbf{v} \cdot (\mathbf{u} \times \mathbf{v}) = 0$.

(vii) If $\mathbf{u}$ and $\mathbf{v}$ are parallel, then $\mathbf{v} = \alpha\mathbf{u}$ for some scalar α [from Theorem 3.2.4(i)], so that

$$\mathbf{u} \times \mathbf{v} = \begin{vmatrix} \mathbf{i} & \mathbf{j} & \mathbf{k} \\ a_1 & b_1 & c_1 \\ \alpha a_1 & \alpha b_1 & \alpha c_1 \end{vmatrix} = \mathbf{0}$$

since the third row is a multiple of the second row [see Property 6 in Section 8.2]. ■

NOTE. We could have proved this theorem without using determinants, but the proof would have involved many more computations.

What happens when we take cross products of the basis vectors $\mathbf{i}$, $\mathbf{j}$, $\mathbf{k}$? It is easy to verify the following:

$$\begin{aligned} &\mathbf{i} \times \mathbf{i} = \mathbf{j} \times \mathbf{j} = \mathbf{k} \times \mathbf{k} = \mathbf{0}, \\ &\mathbf{i} \times \mathbf{j} = \mathbf{k}, \quad \mathbf{k} \times \mathbf{i} = \mathbf{j}, \quad \mathbf{j} \times \mathbf{k} = \mathbf{i}, \\ &\mathbf{j} \times \mathbf{i} = -\mathbf{k}, \quad \mathbf{i} \times \mathbf{k} = -\mathbf{j}, \quad \mathbf{k} \times \mathbf{j} = -\mathbf{i}. \end{aligned} \tag{3}$$

To remember these results, consider the circle in Figure 1. The cross product of two consecutive vectors in the clockwise direction is positive, while the cross product of two consecutive vectors in the counterclockwise direction is negative. Note that the formulas above show that the cross product is *not* associative, since, for example, $\mathbf{i} \times (\mathbf{i} \times \mathbf{j}) = \mathbf{i} \times \mathbf{k} = -\mathbf{j}$ while $(\mathbf{i} \times \mathbf{i}) \times \mathbf{j} = \mathbf{0} \times \mathbf{j} = \mathbf{0}$, so that

$$\mathbf{i} \times (\mathbf{i} \times \mathbf{j}) \neq (\mathbf{i} \times \mathbf{i}) \times \mathbf{j}.$$

In general,

$$\mathbf{u} \times (\mathbf{v} \times \mathbf{w}) \neq (\mathbf{u} \times \mathbf{v}) \times \mathbf{w}.$$

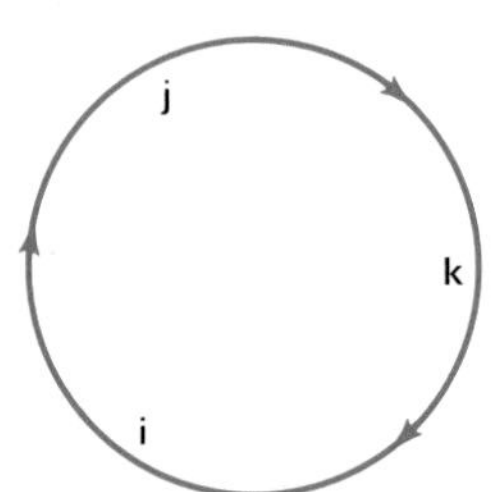

FIGURE 1

EXAMPLE 3 Calculate $(3\mathbf{i} + 4\mathbf{k}) \times (2\mathbf{i} - 3\mathbf{j})$.

Solution. This is a good example of the usefulness of Theorem 2 and the formulas (3). We have

$$\begin{aligned} (3\mathbf{i} + 4\mathbf{k}) \times (2\mathbf{i} - 3\mathbf{j}) &= (3 \cdot 2)(\mathbf{i} \times \mathbf{i}) + (4 \cdot 2)(\mathbf{k} \times \mathbf{i}) - (3 \cdot 3)(\mathbf{i} \times \mathbf{j}) \\ &\quad + 4(-3)(\mathbf{k} \times \mathbf{j}) \\ &= \mathbf{0} + 8\mathbf{j} - 9\mathbf{k} + 12\mathbf{i} = 12\mathbf{i} + 8\mathbf{j} - 9\mathbf{k}. \end{aligned}$$ ■

EXAMPLE 4 Find a line whose direction vector is orthogonal to the direction vectors of the lines $(x - 1)/3 = (y + 6)/4 = (z - 2)/-2$ and $(x + 2)/-3 = (y - 3)/4 = (z + 1)/1$ and that passes through the point $(2, -1, 1)$.

Solution. The directions of these lines are

$$\mathbf{v}_1 = 3\mathbf{i} + 4\mathbf{j} - 2\mathbf{k} \qquad \text{and} \qquad \mathbf{v}_2 = -3\mathbf{i} + 4\mathbf{j} + \mathbf{k}.$$

A vector orthogonal to these vectors is

$$\mathbf{w} = \mathbf{v}_1 \times \mathbf{v}_2 = \begin{vmatrix} \mathbf{i} & \mathbf{j} & \mathbf{k} \\ 3 & 4 & -2 \\ -3 & 4 & 1 \end{vmatrix} = 12\mathbf{i} + 3\mathbf{j} + 24\mathbf{k}.$$

Then symmetric equations of a line satisfying the requested conditions are given by

$$L_1: \quad \frac{x - 2}{12} = \frac{y + 1}{3} = \frac{z - 1}{24}.$$

NOTE. $\mathbf{w}_1 = \mathbf{v}_2 \times \mathbf{v}_1 = -(\mathbf{v}_1 \times \mathbf{v}_2)$ is also orthogonal to $\mathbf{v}_1$ and $\mathbf{v}_2$, so symmetric equations of another line are given by

$$L_2: \quad \frac{x - 2}{-12} = \frac{y + 1}{-3} = \frac{z - 1}{-24}.$$

However, L_1 and L_2 are really the same line. (Explain why.) ■

The preceding example leads to a basic question. We know that $\mathbf{u} \times \mathbf{v}$ is a vector orthogonal to $\mathbf{u}$ and $\mathbf{v}$. But there are always *two* unit vectors orthogonal to $\mathbf{u}$ and $\mathbf{v}$ (see Figure 2). The vectors $\mathbf{n}$ and $-\mathbf{n}$ ($\mathbf{n}$ stands for normal, of course) are both orthogonal to $\mathbf{u}$ and $\mathbf{v}$. Which one is in the direction of $\mathbf{u} \times \mathbf{v}$? The answer is given by the **right-hand rule.** If the right hand is placed so that the index finger points in the direction of $\mathbf{u}$ while the middle finger points in the direction of $\mathbf{v}$, then the thumb points in the direction of $\mathbf{u} \times \mathbf{v}$ (see Figure 3).

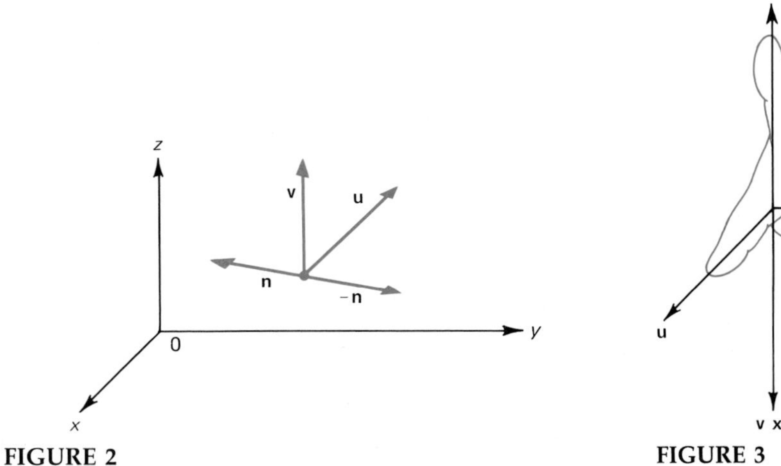

FIGURE 2

FIGURE 3

Having discussed the direction of the vector $\mathbf{u} \times \mathbf{v}$, we now turn to a discussion of its magnitude.

Theorem 3 If φ is the angle between $\mathbf{u}$ and $\mathbf{v}$, then

$$|\mathbf{u} \times \mathbf{v}| = |\mathbf{u}||\mathbf{v}| \sin \varphi. \tag{4}$$

Proof. It is easy to show (by comparing components) that

$$|\mathbf{u} \times \mathbf{v}|^2 = |\mathbf{u}|^2|\mathbf{v}|^2 - (\mathbf{u} \cdot \mathbf{v})^2 \tag{5}$$

(see Problem 36). Then since $(\mathbf{u} \cdot \mathbf{v})^2 = |\mathbf{u}|^2|\mathbf{v}|^2 \cos^2 \varphi$ (from Theorem 3.2.3),

$$\begin{aligned} |\mathbf{u} \times \mathbf{v}|^2 &= |\mathbf{u}|^2|\mathbf{v}|^2 - |\mathbf{u}|^2|\mathbf{v}|^2 \cos^2 \varphi = |\mathbf{u}|^2|\mathbf{v}|^2(1 - \cos^2 \varphi) \\ &= |\mathbf{u}|^2|\mathbf{v}|^2 \sin^2 \varphi, \end{aligned}$$

and the theorem follows after taking square roots of both sides. ■

There is an interesting geometric interpretation of Theorem 3. The vectors $\mathbf{u}$ and $\mathbf{v}$ are sketched in Figure 4 and can be thought of as two adjacent sides of a parallelogram. Then from elementary geometry we see that

$$\text{area of the parallelogram} = |\mathbf{u}||\mathbf{v}| \sin \varphi = |\mathbf{u} \times \mathbf{v}|. \tag{6}$$

EXAMPLE 5 Find the area of a parallelogram with consecutive vertices at $P = (1, 3, -2)$, $Q = (2, 1, 4)$, and $R = (-3, 1, 6)$.

Solution. One such parallelogram is sketched in Figure 5 (there are two others). We have

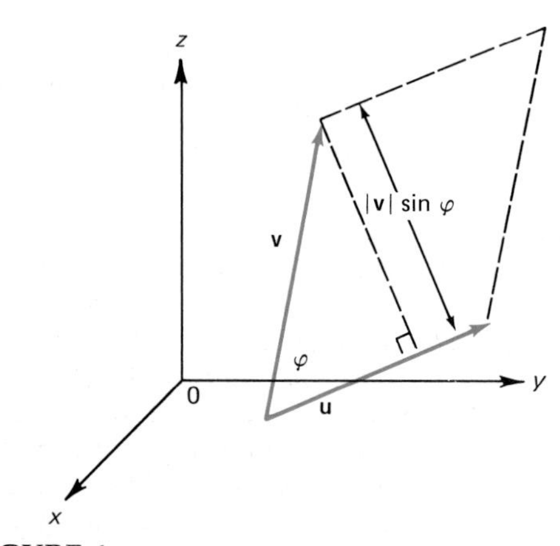

FIGURE 4

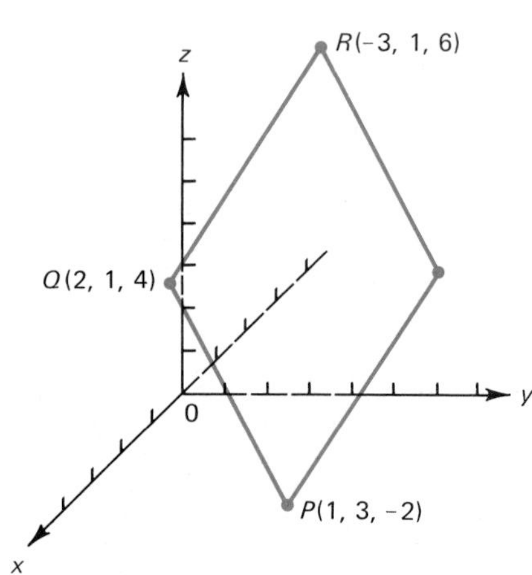

FIGURE 5

$$\text{area} = |\overrightarrow{PQ} \times \overrightarrow{QR}| = |(\mathbf{i} - 2\mathbf{j} + 6\mathbf{k}) \times (-5\mathbf{i} + 2\mathbf{k})|$$

$$= \left\| \begin{matrix} \mathbf{i} & \mathbf{j} & \mathbf{k} \\ 1 & -2 & 6 \\ -5 & 0 & 2 \end{matrix} \right\| = |-4\mathbf{i} - 32\mathbf{j} - 10\mathbf{k}| = \sqrt{1140} \text{ square units.} \blacksquare$$

PROBLEMS 3.4

In Problems 1–20, find the cross product $\mathbf{u} \times \mathbf{v}$.

1. $\mathbf{u} = \mathbf{i} - 2\mathbf{j}$; $\mathbf{v} = 3\mathbf{k}$
2. $\mathbf{u} = 3\mathbf{i} - 7\mathbf{j}$; $\mathbf{v} = \mathbf{i} + \mathbf{k}$
3. $\mathbf{u} = \mathbf{i} - \mathbf{j}$; $\mathbf{v} = \mathbf{j} + \mathbf{k}$
4. $\mathbf{u} = -7\mathbf{k}$; $\mathbf{v} = \mathbf{j} + 2\mathbf{k}$
5. $\mathbf{u} = -2\mathbf{i} + 3\mathbf{j}$; $\mathbf{v} = 7\mathbf{i} + 4\mathbf{k}$
6. $\mathbf{u} = a\mathbf{i} + b\mathbf{j}$; $\mathbf{v} = c\mathbf{i} + d\mathbf{j}$
7. $\mathbf{u} = a\mathbf{i} + b\mathbf{k}$; $\mathbf{v} = c\mathbf{i} + d\mathbf{k}$
8. $\mathbf{u} = a\mathbf{j} + b\mathbf{k}$; $\mathbf{v} = c\mathbf{i} + d\mathbf{k}$
9. $\mathbf{u} = 2\mathbf{i} - 3\mathbf{j} + \mathbf{k}$; $\mathbf{v} = \mathbf{i} + 2\mathbf{j} + \mathbf{k}$
10. $\mathbf{u} = 3\mathbf{i} - 4\mathbf{j} + 2\mathbf{k}$; $\mathbf{v} = 6\mathbf{i} - 3\mathbf{j} + 5\mathbf{k}$
11. $\mathbf{u} = -3\mathbf{i} - 2\mathbf{j} + \mathbf{k}$; $\mathbf{v} = 6\mathbf{i} + 4\mathbf{j} - 2\mathbf{k}$
12. $\mathbf{u} = \mathbf{i} + 7\mathbf{j} - 3\mathbf{k}$; $\mathbf{v} = -\mathbf{i} - 7\mathbf{j} + 3\mathbf{k}$
13. $\mathbf{u} = \mathbf{i} - 7\mathbf{j} - 3\mathbf{k}$; $\mathbf{v} = -\mathbf{i} + 7\mathbf{j} - 3\mathbf{k}$
14. $\mathbf{u} = 2\mathbf{i} - 3\mathbf{j} + 5\mathbf{k}$; $\mathbf{v} = 3\mathbf{i} - \mathbf{j} - \mathbf{k}$
15. $\mathbf{u} = 10\mathbf{i} + 7\mathbf{j} - 3\mathbf{k}$; $\mathbf{v} = -3\mathbf{i} + 4\mathbf{j} - 3\mathbf{k}$
16. $\mathbf{u} = 2\mathbf{i} + 4\mathbf{j} - 6\mathbf{k}$; $\mathbf{v} = -\mathbf{i} - \mathbf{j} + 3\mathbf{k}$
17. $\mathbf{u} = 2\mathbf{i} - \mathbf{j} + \mathbf{k}$; $\mathbf{v} = 4\mathbf{i} + 2\mathbf{j} + 2\mathbf{k}$
18. $\mathbf{u} = 3\mathbf{i} - \mathbf{j} + 8\mathbf{k}$; $\mathbf{v} = \mathbf{i} + \mathbf{j} - 4\mathbf{k}$
19. $\mathbf{u} = a\mathbf{i} + a\mathbf{j} + a\mathbf{k}$; $\mathbf{v} = b\mathbf{i} + b\mathbf{j} + b\mathbf{k}$
20. $\mathbf{u} = a\mathbf{i} + b\mathbf{j} + c\mathbf{k}$; $\mathbf{v} = a\mathbf{i} + b\mathbf{j} - c\mathbf{k}$

21. Find two unit vectors orthogonal to both $\mathbf{u} = 2\mathbf{i} - 3\mathbf{j}$ and $\mathbf{v} = 4\mathbf{j} + 3\mathbf{k}$.
22. Find two unit vectors orthogonal to both $\mathbf{u} = \mathbf{i} + \mathbf{j} + \mathbf{k}$ and $\mathbf{v} = \mathbf{i} - \mathbf{j} - \mathbf{k}$.
23. Use the cross product to find the sine of the angle φ between the vectors $\mathbf{u} = 2\mathbf{i} + \mathbf{j} - \mathbf{k}$ and $\mathbf{v} = -3\mathbf{i} - 2\mathbf{j} + 4\mathbf{k}$.
24. Use the dot product to calculate the cosine of the angle between the vectors of Problem 23. Then show that for the values you have calculated, $\sin^2 \varphi + \cos^2 \varphi = 1$.

In Problems 25–28, find a line that has direction vector orthogonal to the direction vectors of the two given lines and that passes through the given point.

25. $\dfrac{x+2}{-3} = \dfrac{y-1}{4} = \dfrac{z}{-5}$; $\dfrac{x-3}{7} = \dfrac{y+2}{-2} = \dfrac{z-8}{3}$; $(1, -3, 2)$

26. $\dfrac{x-2}{-4} = \dfrac{y+3}{-7} = \dfrac{z+1}{3}$; $\dfrac{x+2}{3} = \dfrac{y-5}{-4} = \dfrac{z+3}{-2}$; $(-4, 7, 3)$

27. $x = 3 - 2t$, $y = 4 + 3t$, $z = -7 + 5t$; $x = -2 + 4s$, $y = 3 - 2s$, $z = 3 + s$; $(-2, 3, 4)$
28. $x = 4 + 10t$, $y = -4 - 8t$, $z = 3 + 7t$; $x = -2t$, $y = 1 + 4t$, $z = -7 - 3t$; $(4, 6, 0)$

In Problems 29–34, find the area of a parallelogram with the given adjacent vertices.

29. $(1, -2, 3)$; $(2, 0, 1)$; $(0, 4, 0)$
30. $(-2, 1, 1)$; $(2, 2, 3)$; $(-1, -2, 4)$
31. $(-2, 1, 0)$; $(1, 4, 2)$; $(-3, 1, 5)$
32. $(7, -2, -3)$; $(-4, 1, 6)$; $(5, -2, 3)$
33. $(a, 0, 0)$; $(0, b, 0)$; $(0, 0, c)$
34. $(a, b, 0)$; $(a, 0, b)$; $(0, a, b)$

35. Show that if $\mathbf{u} = (a_1, b_1, c_1)$, $\mathbf{v} = (a_2, b_2, c_2)$, and $\mathbf{w} = (a_3, b_3, c_3)$, then

$$\mathbf{u} \cdot (\mathbf{v} \times \mathbf{w}) = \begin{vmatrix} a_1 & b_1 & c_1 \\ a_2 & b_2 & c_2 \\ a_3 & b_3 & c_3 \end{vmatrix}.$$

36. Show that $|\mathbf{u} \times \mathbf{v}|^2 = |\mathbf{u}|^2|\mathbf{v}|^2 - (\mathbf{u} \cdot \mathbf{v})^2$. [*Hint:* Write out in terms of components.]
37. Show that the area of the triangle PQR is given by $A = \frac{1}{2}|\overrightarrow{PQ} \times \overrightarrow{QR}|$.
38. Use the result of Problem 37 to calculate the area of the triangle with vertices at $(2, 1, -4)$, $(1, 7, 2)$, and $(3, -2, 3)$.
39. Calculate the area of the triangle with vertices at $(3, 1, 7)$, $(2, -3, 4)$, and $(7, -2, 4)$.
40. Calculate the area of the triangle with vertices at $(1, 0, 0)$, $(0, 1, 0)$, and $(0, 0, 1)$. Sketch this triangle.
***41.** Let $\mathbf{u}$, $\mathbf{v}$, and $\mathbf{w}$ be three vectors that are not in the same plane. Then they form the sides of a *parallelepiped* in space (see Figure 6). Prove that the volume of the parallelepiped is given by $V = |(\mathbf{u} \times \mathbf{v}) \cdot \mathbf{w}|$. [*Hint:* The area of the base is $|\mathbf{u} \times \mathbf{v}|$.]

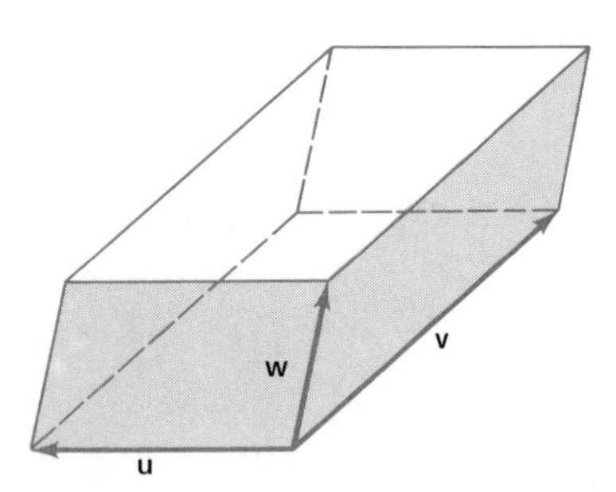

FIGURE 6

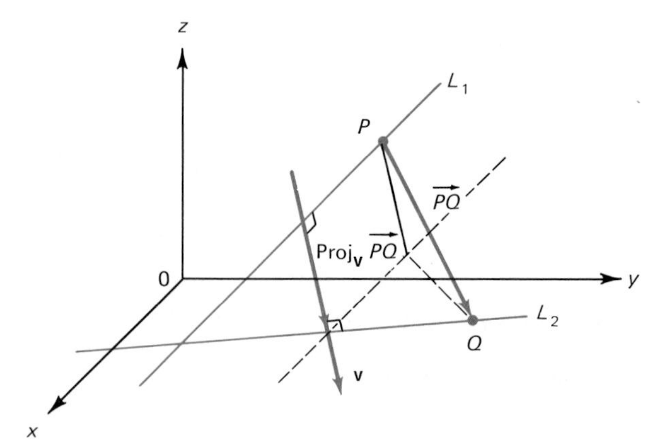

FIGURE 7

42. Calculate the volume of the parallelepiped determined by the vectors $\mathbf{u} = 2\mathbf{i} - \mathbf{j} + \mathbf{k}$, $\mathbf{v} = 3\mathbf{i} + 2\mathbf{j} - 2\mathbf{k}$, $\mathbf{w} = 3\mathbf{i} + 2\mathbf{j}$.

43. Calculate the volume of the parallelepiped determined by the vectors $\mathbf{i} - \mathbf{j}$, $3\mathbf{i} + 2\mathbf{k}$, $-7\mathbf{j} + 3\mathbf{k}$.

44. Calculate the volume of the parallelepiped determined by the vectors $\overrightarrow{PQ}$, $\overrightarrow{PR}$, and $\overrightarrow{PS}$, where $P = (2, 1, -1)$, $Q = (-3, 1, 4)$, $R = (-1, 0, 2)$, and $S = (-3, -1, 5)$.

***45.** Calculate the distance between the lines

$$L_1: \frac{x-2}{3} = \frac{y-5}{2} = \frac{z-1}{-1} \quad \text{and}$$

$$L_2: \frac{x-4}{-4} = \frac{y-5}{4} = \frac{z+2}{1}.$$

[*Hint:* The distance is measured along a vector **v** that is perpendicular to both L_1 and L_2. Let P be a point on L_1 and Q a point on L_2. Then the length of the projection of $\overrightarrow{PQ}$ on **v** is the distance between the lines, measured along a vector that is perpendicular to them both. See Figure 7.]

***46.** Find the distance between the lines

$$L_1: \frac{x+2}{3} = \frac{y-7}{-4} = \frac{z-2}{2} \quad \text{and}$$

$$L_2: \frac{x-1}{-3} = \frac{y+2}{4} = \frac{z+1}{1}.$$

***47.** Find the distance between the lines $x = 2 - 3t$, $y = 1 + 2t$, $z = -2 - t$, and $x = 1 + 4s$, $y = -2 - s$, $z = 3 + s$.

***48.** Find the distance between the lines $x = -2 + 5t$, $y = -3 - 2t$, $z = 1 + 4t$, and $x = 2 + 3s$, $y = -1 + s$, $z = 3s$.

3.5 PLANES

In Section 3.3 we derived the equation of a line in space by specifying a point on the line and a vector *parallel* to this line. We can derive the equation of a plane in space by specifying a point in the plane and a vector orthogonal to every vector in the plane. This orthogonal vector is called a **normal vector** and is denoted by **N**. (See Figure 1.)

Definition 1 PLANE Let P be a point in space and let **N** be a given nonzero vector. Then the set of all points Q for which $\overrightarrow{PQ}$ and **N** are orthogonal constitutes a **plane** in $\mathbb{R}^3$.

NOTATION: We will usually denote a plane by the symbol Π.

Let $P = (x_0, y_0, z_0)$ and $\mathbf{N} = a\mathbf{i} + b\mathbf{j} + c\mathbf{k}$. Then if $Q = (x, y, z)$,

$$\overrightarrow{PQ} = (x - x_0)\mathbf{i} + (y - y_0)\mathbf{j} + (z - z_0)\mathbf{k}.$$

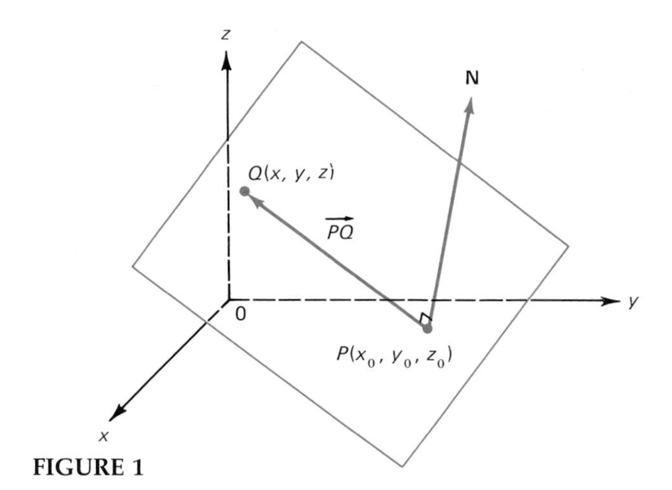

FIGURE 1

If $\overrightarrow{PQ} \perp \mathbf{N}$, then $\overrightarrow{PQ} \cdot \mathbf{N} = 0$. But this implies that

$$a(x - x_0) + b(y - y_0) + c(z - z_0) = 0. \tag{1}$$

A more common way to write the equation of a plane is easily derived from (1):

$$ax + by + cz = d,$$

where

$$d = ax_0 + by_0 + cz_0 = \overrightarrow{OP} \cdot \mathbf{N}. \tag{2}$$

EXAMPLE 1 Find an equation of the plane Π passing through the point (2, 5, 1) and normal to the vector $\mathbf{N} = \mathbf{i} - 2\mathbf{j} + 3\mathbf{k}$.

Solution. From (1) we immediately obtain

$$(x - 2) - 2(y - 5) + 3(z - 1) = 0,$$

or

$$x - 2y + 3z = -5. \tag{3}$$

This plane is sketched in Figure 2. ■

REMARK. The plane can be sketched by setting $x = y = 0$ in (3) to obtain $(0, 0, -\frac{5}{3})$, $x = z = 0$ to obtain $(0, \frac{5}{2}, 0)$, and $y = z = 0$ to obtain $(-5, 0, 0)$. These three points all lie on the plane.

The three coordinate planes are easily represented as follows:

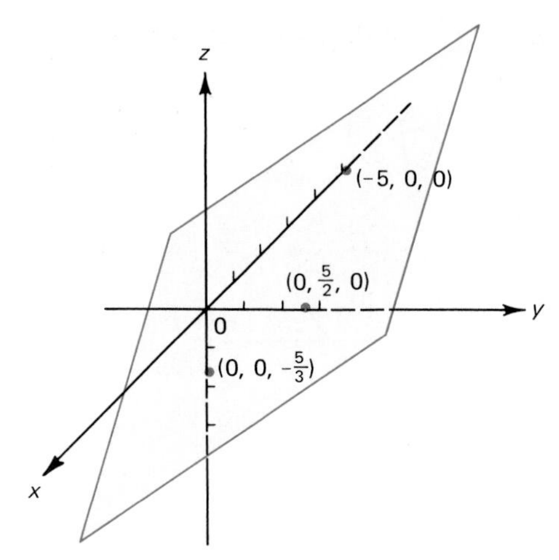

FIGURE 2

(i) The xy-plane. This plane passes through the origin (0, 0, 0) and any vector lying along the z-axis is normal to it. The simplest such vector is **k**. Thus from (1) we obtain

$$0(x - 0) + 0(y - 0) + 1(z - 0) = 0,$$

which yields

$$z = 0 \tag{4}$$

as the equation of the xy-plane. (This result should not be very surprising.)

REMARK. The equation $z = 0$ is really a shorthand notation for the equation of the xy-plane. The full notation is

$$\text{the } xy\text{-plane} = \{(x, y, z) : z = 0\}.$$

The shorthand notation is fine as long as we don't lose sight of the fact that we are in $\mathbb{R}^3$.

(ii) The xz-plane has the equation

$$y = 0. \tag{5}$$

(iii) The yz-plane has the equation

$$x = 0. \tag{6}$$

Three points that are not collinear determine a plane since they determine two nonparallel vectors that intersect at a point.

EXAMPLE 2 Find an equation of the plane Π passing through the points $P = (1, 2, 1)$, $Q = (-2, 3, -1)$, and $R = (1, 0, 4)$.

Solution. The vectors $\overrightarrow{PQ} = -3\mathbf{i} + \mathbf{j} - 2\mathbf{k}$ and $\overrightarrow{QR} = 3\mathbf{i} - 3\mathbf{j} + 5\mathbf{k}$ lie on the plane and are therefore orthogonal to the normal vector. Thus

$$\mathbf{N} = \overrightarrow{PQ} \times \overrightarrow{QR} = \begin{vmatrix} \mathbf{i} & \mathbf{j} & \mathbf{k} \\ -3 & 1 & -2 \\ 3 & -3 & 5 \end{vmatrix} = -\mathbf{i} + 9\mathbf{j} + 6\mathbf{k},$$

and we obtain

$$\Pi: \quad -(x - 1) + 9(y - 2) + 6(z - 1) = 0,$$

or

$$-x + 9y + 6z = 23.$$

Note that if we choose another point, say Q, we get the equation

$$-(x + 2) + 9(y - 3) + 6(z + 1) = 0,$$

which reduces to

$$-x + 9y + 6z = 23.$$

The plane is sketched in Figure 3. ■

Definition 2 PARALLEL PLANES Two planes are **parallel** if their normal vectors are parallel.

Two parallel planes are drawn in Figure 4.

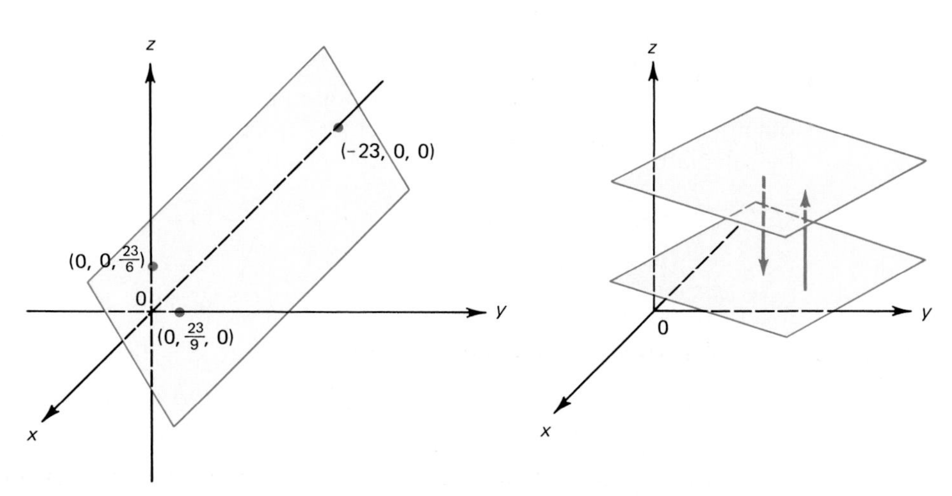

FIGURE 3 **FIGURE 4**

EXAMPLE 3 The planes Π_1: $2x + 3y - z = 3$ and Π_2: $-4x - 6y + 2z = 8$ are parallel since $\mathbf{N}_1 = 2\mathbf{i} + 3\mathbf{j} - \mathbf{k}$, $\mathbf{N}_2 = -4\mathbf{i} - 6\mathbf{j} + 2\mathbf{k}$, and $\mathbf{N}_2 = -2\mathbf{N}_1$ (and $\mathbf{N}_1 \times \mathbf{N}_2 = \mathbf{0}$). ■

If two distinct planes are not parallel, then they intersect in a straight line.

EXAMPLE 4 Find all points of intersection of the planes $2x - y - z = 3$ and $x + 2y + 3z = 7$.

Solution. When the planes intersect, we have

$$2x - y - z = 3$$

and

$$x + 2y + 3z = 7.$$

Multiplying the first equation by 2 and adding it to the second, we obtain

$$\begin{array}{r} 4x - 2y - 2z = \ \ 6 \\ \underline{x + 2y + 3z = \ \ 7} \\ 5x + \qquad\quad z = 13 \end{array}$$

or $z = -5x + 13$. Then from the first equation

$$y = 2x - z - 3 = 2x - (-5x + 13) - 3 = 7x - 16.$$

Then setting $x = t$, we obtain the parametric representation of the line of intersection:

$$x = t, \qquad y = -16 + 7t, \qquad z = 13 - 5t.$$

This line is sketched in Figure 5. Note that this line is orthogonal to both normal vectors. ■

We conclude this section by indicating how the distance from a plane to a point can be calculated. Look at Figure 6. If Q is the point, then the required distance is the distance measured along a line orthogonal to Π. That is, the shortest distance is obtained by "dropping a perpendicular" from the point to the plane, which is done by calculating (for any point P on the plane)

$$D = |\text{Proj}_{\mathbf{N}}\, \overrightarrow{PQ}| = \frac{|\overrightarrow{PQ} \cdot \mathbf{N}|}{|\mathbf{N}|}. \tag{7}$$

EXAMPLE 5 Find the distance D between the plane $2x - y + 3z = 6$ and the point $Q = (3, 5, -7)$.

Solution. One point on the plane is $P = (3, 0, 0)$ and $\mathbf{N} = 2\mathbf{i} - \mathbf{j} + 3\mathbf{k}$. Then $\overrightarrow{PQ} = 5\mathbf{j} - 7\mathbf{k}$, $|\overrightarrow{PQ} \cdot \mathbf{N}| = 26$ and $|\mathbf{N}| = \sqrt{14}$, so that

$$D = \frac{26}{\sqrt{14}}. \ ■$$

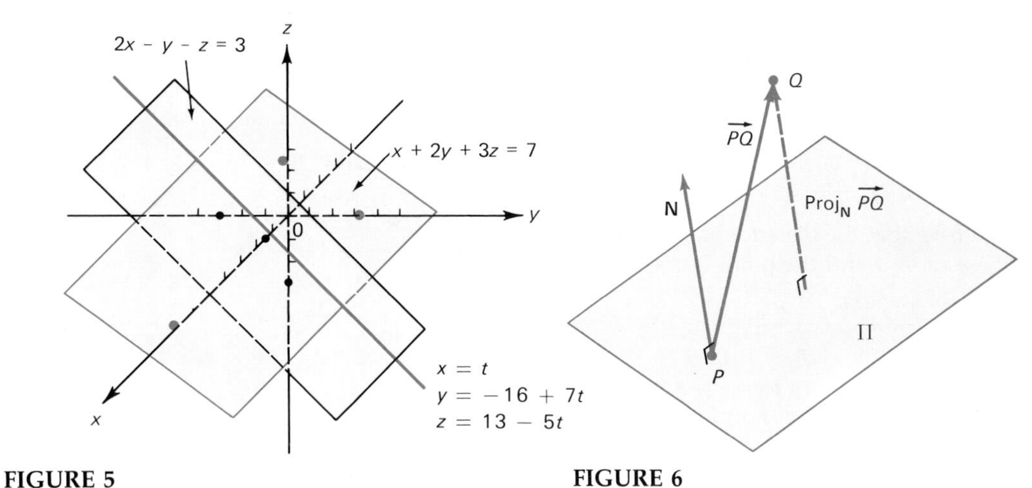

FIGURE 5 **FIGURE 6**

EXAMPLE 6 Find the distance from the plane $ax + by + cz = d$ to the origin.

Solution. If $a \neq 0$, one point on the plane is $P = (d/a, 0, 0)$. [If $a = 0$ but $b \neq 0$, a point on the plane is $(0, d/b, 0)$, leading to the same result.] Then $\mathbf{N} = a\mathbf{i} + b\mathbf{j} + c\mathbf{k}$ and $\overrightarrow{P0} = (d/a)\mathbf{i}$, so that $|\overrightarrow{0P} \cdot \mathbf{N}| = |d|$, $\mathbf{N} = \sqrt{a^2 + b^2 + c^2}$, and

$$D = \frac{|d|}{\sqrt{a^2 + b^2 + c^2}}. \blacksquare$$

PROBLEMS 3.5

In Problems 1–16, find the equation of the plane and sketch it.

1. $P = (0, 0, 0)$; $\mathbf{N} = \mathbf{i}$
2. $P = (0, 0, 0)$; $\mathbf{N} = \mathbf{j}$
3. $P = (0, 0, 0)$; $\mathbf{N} = \mathbf{k}$
4. $P = (1, 2, 3)$; $\mathbf{N} = \mathbf{i} + \mathbf{j}$
5. $P = (1, 2, 3)$; $\mathbf{N} = \mathbf{i} + \mathbf{k}$
6. $P = (1, 2, 3)$; $\mathbf{N} = \mathbf{j} + \mathbf{k}$
7. $P = (2, -1, 6)$; $\mathbf{N} = 3\mathbf{i} - \mathbf{j} + 2\mathbf{k}$
8. $P = (-4, -7, 5)$; $\mathbf{N} = -3\mathbf{i} - 4\mathbf{j} + \mathbf{k}$
9. $P = (-3, 11, 2)$; $\mathbf{N} = 4\mathbf{i} + \mathbf{j} - 7\mathbf{k}$
10. $P = (3, -2, 5)$; $\mathbf{N} = 2\mathbf{i} - 7\mathbf{j} - 8\mathbf{k}$
11. $P = (4, -7, -3)$; $\mathbf{N} = -\mathbf{i} - \mathbf{j} - \mathbf{k}$
12. $P = (8, 1, 0)$; $\mathbf{N} = -7\mathbf{i} + \mathbf{j} + 2\mathbf{k}$
13. containing $(1, 2, -4)$, $(2, 3, 7)$, and $(4, -1, 3)$
14. containing $(-7, 1, 0)$, $(2, -1, 3)$, and $(4, 1, 6)$
15. containing $(1, 0, 0)$, $(0, 1, 0)$, and $(0, 0, 1)$
16. containing $(2, 3, -2)$, $(4, -1, -1)$, and $(3, 1, 2)$

Two planes are **orthogonal** if their normal vectors are orthogonal. In Problems 17–23, determine whether the given planes are parallel, orthogonal, coincident (i.e., the same), or none of these.

17. Π_1: $x + y + z = 2$; Π_2: $2x + 2y + 2z = 4$
18. Π_1: $x - y + z = 3$; Π_2: $-3x + 3y - 3z = -9$
19. Π_1: $2x - y + z = 3$; Π_2: $x + y - z = 7$
20. Π_1: $2x - y + z = 3$; Π_2: $x + y + z = 3$
21. Π_1: $3x - 2y + 7z = 4$; Π_2: $-2x + 4y + 2z = 16$
22. Π_1: $-4x + 4y - 6z = 7$; Π_2: $2x - 2y + 3z = -3$
23. Π_1: $-4x + 4y - 6z = 6$; Π_2: $2x - 2y + 3z = -3$

In Problems 24–26, find the equation of the set of all points of intersection of the two planes.

24. Π_1: $x - y + z = 2$; Π_2: $2x - 3y + 4z = 7$
25. Π_1: $3x - y + 4z = 3$; Π_2: $-4x - 2y + 7z = 8$
26. Π_1: $-2x - y + 17z = 4$; Π_2: $2x - y - z = -7$

In Problems 27–30, find the distance from the given point to the given plane.

27. $(2, -1, 4)$; $3x - y + 7z = 2$
28. $(4, 0, 1)$; $2x - y + 8z = 3$
29. $(-7, -2, -1)$; $-2x + 8z = -5$
30. $(-3, 0, 2)$; $-3x + y + 5z = 0$

31. Prove that the distance between the plane $ax + by + cz = d$ and the point (x_0, y_0, z_0) is given by

$$D = \frac{|ax_0 + by_0 + cz_0 - d|}{\sqrt{a^2 + b^2 + c^2}}.$$

The *angle between two planes* is defined to be the acute angle between their normal vectors. In Problems 32–34, find the angle between the two planes described in the indicated problem.

32. Problem 24 **33.** Problem 25
34. Problem 26

***35.** Let **u** and **v** be two nonparallel vectors in a plane Π. Show that if **w** is any other vector in Π, then there exist scalars α and β such that

$$\mathbf{w} = \alpha\mathbf{u} + \beta\mathbf{v}.$$

This expression is called the **parametric representation** of the plane Π. [*Hint:* Draw a parallelogram in which $\alpha\mathbf{u}$ and $\beta\mathbf{v}$ form adjacent sides and the diagonal vector is **w**.]

36. Three vectors **u**, **v**, and **w** are called **coplanar** if they all lie in the same plane Π. Show that if **u**, **v**, and **w** all pass through the origin, then they are coplanar if and only if the scalar triple product equals zero:

$$\mathbf{u} \cdot (\mathbf{v} \times \mathbf{w}) = 0.$$

In Problems 37–41, determine whether the three given position vectors (i.e., one endpoint at the origin) are coplanar. If they are coplanar, find the equation of the plane containing them.

37. $\mathbf{u} = 2\mathbf{i} - 3\mathbf{j} + 4\mathbf{k}$; $\mathbf{v} = 7\mathbf{i} - 2\mathbf{j} + 3\mathbf{k}$; $\mathbf{w} = 9\mathbf{i} - 5\mathbf{j} + 7\mathbf{k}$
38. $\mathbf{u} = -3\mathbf{i} + \mathbf{j} + 8\mathbf{k}$; $\mathbf{v} = -2\mathbf{i} - 3\mathbf{j} + 5\mathbf{k}$; $\mathbf{w} = 2\mathbf{i} + 14\mathbf{j} - 4\mathbf{k}$
39. $\mathbf{u} = 2\mathbf{i} + \mathbf{j} - 2\mathbf{k}$; $\mathbf{v} = 2\mathbf{i} - \mathbf{j} - 2\mathbf{k}$; $\mathbf{w} = 2\mathbf{i} - \mathbf{j} + 2\mathbf{k}$
40. $\mathbf{u} = 3\mathbf{i} - 2\mathbf{j} + \mathbf{k}$; $\mathbf{v} = \mathbf{i} + \mathbf{j} - 5\mathbf{k}$; $\mathbf{w} = -\mathbf{i} + 5\mathbf{j} - 16\mathbf{k}$
41. $\mathbf{u} = 2\mathbf{i} - \mathbf{j} - \mathbf{k}$; $\mathbf{v} = 4\mathbf{i} + 3\mathbf{j} + 2\mathbf{k}$; $\mathbf{w} = 6\mathbf{i} + 7\mathbf{j} + 5\mathbf{k}$

***42.** Let $P(x_1, y_1, z_1)$, $Q(x_2, y_2, z_2)$, and $R(x_3, y_3, z_3)$ be three points in $\mathbb{R}^3$ that are not collinear. Show that an equation of the plane passing through the three points is

$$\begin{vmatrix} x & y & z & 1 \\ x_1 & y_1 & z_1 & 1 \\ x_2 & y_2 & z_2 & 1 \\ x_3 & y_3 & z_3 & 1 \end{vmatrix} = 0.$$

43. Let $a_1x + b_1y + c_1z = d_1$, $a_2x + b_2y + c_2z = d_2$, and $a_3x + b_3y + c_3z = d_3$ be equations of three planes. Show that the planes have a unique point in common if

$$\begin{vmatrix} a_1 & b_1 & c_1 \\ a_2 & b_2 & c_2 \\ a_3 & b_3 & c_3 \end{vmatrix} \neq 0.$$

3.6 QUADRIC SURFACES

A **surface** in space is defined as the set of points in $\mathbb{R}^3$ satisfying the equation $F(x, y, z) = 0$. For example, the equation

$$F(x, y, z) = x^2 + y^2 + z^2 - 1 = 0 \qquad \textbf{(1)}$$

is the equation of the unit sphere, as we saw in Section 3.1. In this section we will take a brief look at some of the most commonly encountered surfaces in $\mathbb{R}^3$. We will take a more detailed look at general surfaces in $\mathbb{R}^3$ in Chapter 4.

Having already discussed the sphere, we turn our attention to the cylinder.

Definition 1 CYLINDER Let a line L and a plane curve C be given. A **cylinder** is the surface generated when a line parallel to L moves around C, remaining parallel to L. The line L is called the **generatrix** of the cylinder, and the curve C is called its **directrix**.

EXAMPLE 1 Let L be the z-axis and C the circle $x^2 + y^2 = a^2$ in the xy-plane. Sketch the cylinder.

Solution. As we move a line along the circle $x^2 + y^2 = a^2$ and parallel to the z-axis, we obtain the **right circular cylinder** $x^2 + y^2 = a^2$ sketched in Figure 1a. A computer-drawn sketch of the right circular cylinder $x^2 + y^2 = 4$ is given in Figure 1b. ■

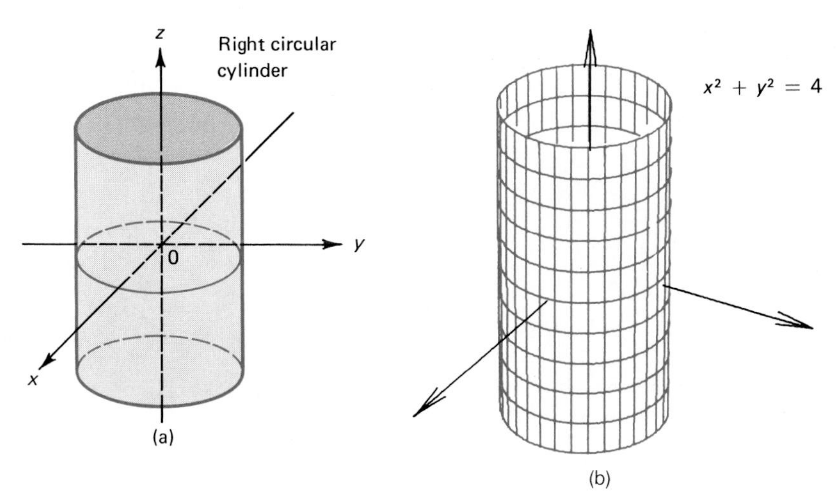

FIGURE 1

REMARK. We can write the equation $x^2 + y^2 = 4$ as

$$x^2 + y^2 + 0 \cdot z^2 = 4.$$

This illustrates that we are talking about a cylinder in $\mathbb{R}^3$ rather than the circle $x^2 + y^2 = 4$ in $\mathbb{R}^2$.

EXAMPLE 2 Suppose L is the x-axis and C is given by $y = z^2$. Sketch the resulting cylinder.

Solution. The curve $y = z^2$ is a parabola in the yz-plane. As we move along it, parallel to the x-axis, we obtain the **parabolic cylinder** sketched in Figure 2. ■

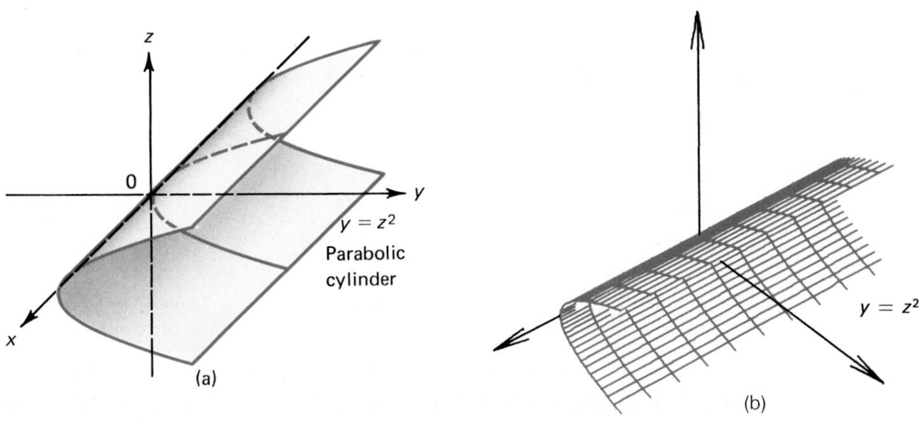

FIGURE 2

EXAMPLE 3 Suppose L is the x-axis and C is given by $y^2 + (z^2/4) = 1$. Sketch the resulting cylinder.

Solution. $y^2 + (z^2/4) = 1$ is an ellipse in the yz-plane. As we move along it parallel to the x-axis, we obtain the **elliptic cylinder** sketched in Figure 3. ■

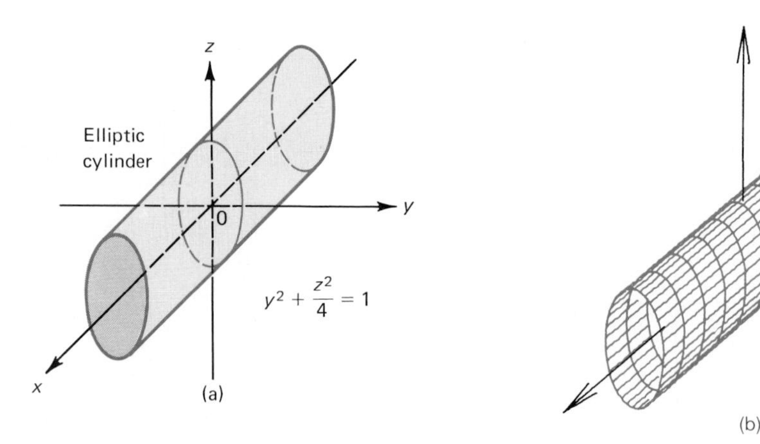

FIGURE 3

As you know from your study of the conic sections,[†] a second-degree equation in the variables x and y forms a circle, parabola, ellipse, or hyperbola (or a degenerate form of one of these such as a single point or a straight line or a pair of straight lines) in the plane. In $\mathbb{R}^3$ we have the following definition.

Definition 2 QUADRIC SURFACE A **quadric surface** in $\mathbb{R}^3$ is the graph of a second-degree equation in the variables x, y, and z. Such an equation takes the form

$$Ax^2 + By^2 + Cz^2 + Dxy + Exz + Fyz + Gx + Hy + Jz + K = 0. \tag{2}$$

We have already seen sketches of several quadric surfaces. We list below the standard forms[‡] of eleven types of nondegenerate[§] quadric surfaces. Although we will not prove this result here, any quadric surface can be written in one of these forms by a translation or rotation of the coordinate axes. Here is the list:

1. sphere
2. right circular cylinder
3. parabolic cylinder
4. elliptic cylinder
5. hyperbolic cylinder
6. ellipsoid
7. hyperboloid of one sheet

[†]See *Calculus* or *Calculus of One Variable,* Chapter 10.

[‡]A standard form is one in which the surface has its center or vertex at the origin and its axes parallel to the coordinate axes.

[§]By degenerate we again mean a surface whose graph contains a finite number of points (or none) or consists of a pair of planes. We saw examples of degenerate spheres (zero or one point) in Section 3.1.

8. hyperboloid of two sheets
9. elliptic paraboloid
10. hyperbolic paraboloid
11. elliptic cone

We have already seen sketches of surfaces 1, 2, 3, and 4.

5. **The hyperbolic cylinder:** $(y^2/b^2) - (x^2/a^2) = 1$. This is the equation of a hyperbola in the xy-plane. See Figure 4a. A computer-drawn sketch of $(y^2/4) - (x^2/9) = 1$ is given in Figure 4b.

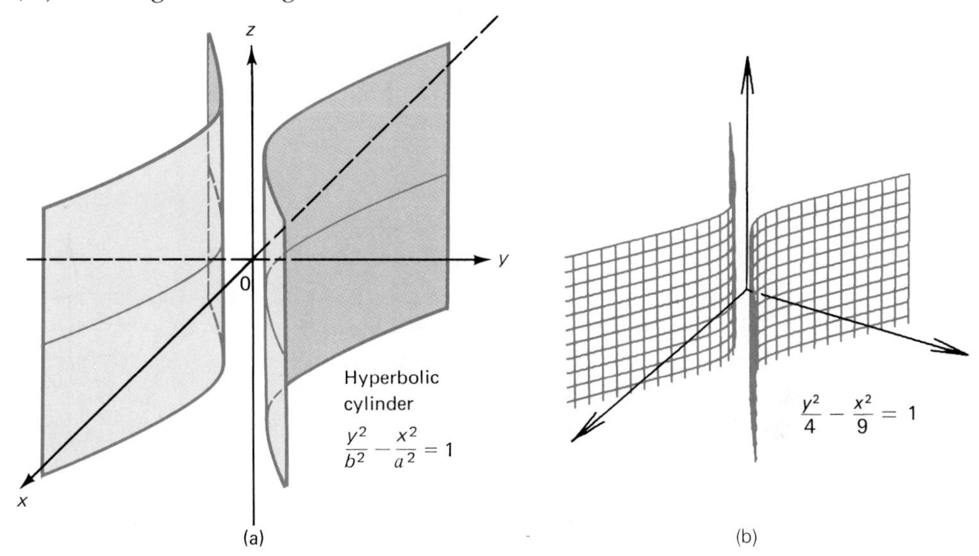

FIGURE 4

We can best describe the remaining six surfaces by looking at **cross sections** parallel to a given coordinate plane. For example, in the unit sphere $x^2 + y^2 + z^2 = 1$, cross sections parallel to the xy-plane are circles. To see this, let $z = c$, where $-1 < c < 1$ (this is a plane parallel to the xy-plane). Then

$$x^2 + y^2 = 1 - c^2,$$

which is the equation of a circle in the xy-plane.

6. **The ellipsoid:** $(x^2/a^2) + (y^2/b^2) + (z^2/c^2) = 1$. This surface is sketched in Figure 5a. It is a closed "watermelon-shaped" surface. Cross sections parallel to the xy-plane, the xz-plane, and the yz-plane are all ellipses. A computer-drawn sketch of the ellipsoid $(x^2/4) + (y^2/9) + (z^2/16) = 1$ is given in Figure 5b.

NOTE. If the ellipse $(x^2/a^2) + (y^2/b^2) = 1$ is revolved about the x-axis, the resulting surface is the ellipsoid $(x^2/a^2) + (y^2/b^2) + (z^2/c^2) = 1$. What do you get if the ellipse is rotated around the y-axis? (See Problem 35.)

7. **The hyperboloid of one sheet:** $(x^2/a^2) + (y^2/b^2) - (z^2/c^2) = 1$. This surface is sketched in Figure 6a. Cross sections parallel to the xy-plane are ellipses. Cross

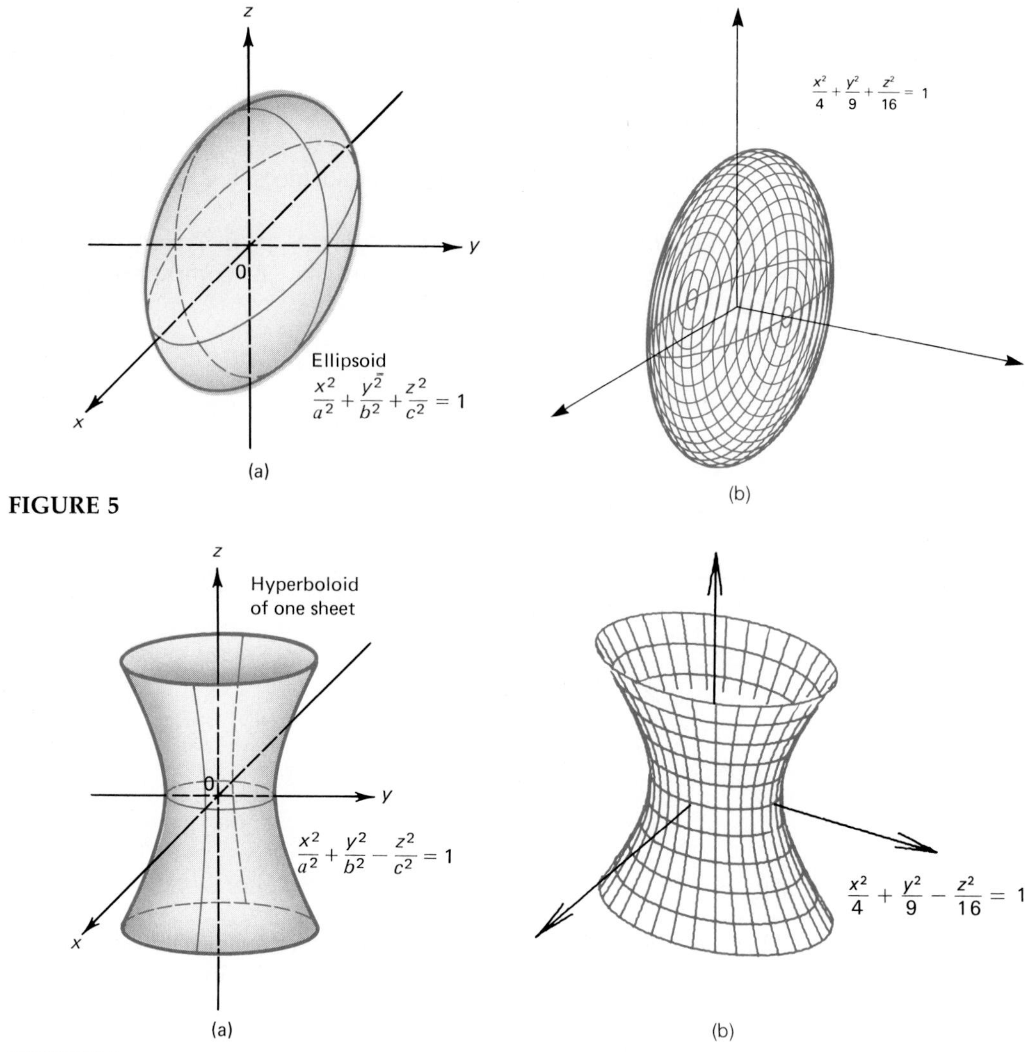

FIGURE 5

FIGURE 6

sections parallel to the xz-plane and the yz-plane are hyperbolas. A computer-drawn sketch of $(x^2/4) + (y^2/9) - (z^2/16) = 1$ is given in Figure 6b.

8. **The hyperboloid of two sheets:** $(z^2/c^2) - (x^2/a^2) - (y^2/b^2) = 1$. This surface is sketched in Figure 7a. Cross sections are the same as those for the hyperboloid of one sheet. Note that the equation implies that $|z| \geq |c|$ (explain why). A computer-drawn sketch of $(z^2/4) - (x^2/4) - (y^2/9) = 1$ is given in Figure 7b.

9. **The elliptic paraboloid:** $z = (x^2/a^2) + (y^2/b^2)$. This surface is sketched in Figure 8a. For each positive fixed z, $(x^2/a^2) + (y^2/b^2) = z$ is the equation of an ellipse. Hence cross sections parallel to the xy-plane are ellipses. If x or y is fixed, then we obtain parabolas. Hence cross sections parallel to the xz- or yz-planes are parabolas. A computer-drawn sketch of $z = (x^2/4) + (y^2/9)$ is given in Figure 8b.

10. **The hyperbolic paraboloid:** $z = (x^2/a^2) - (y^2/b^2)$. This surface is sketched in Figure 9a. For each fixed z we obtain a hyperbola parallel to the xy-plane. Hence

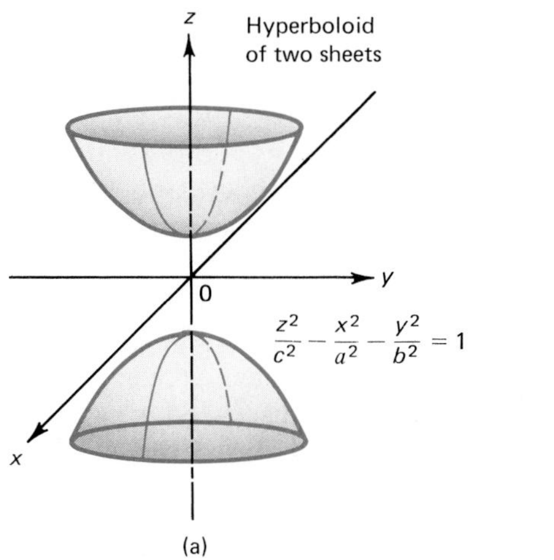

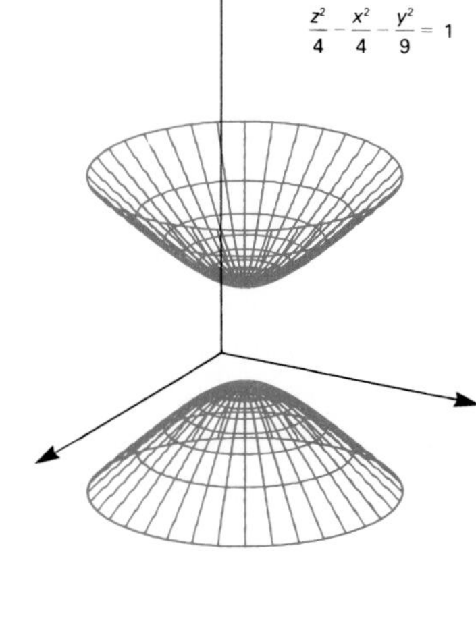

FIGURE 7

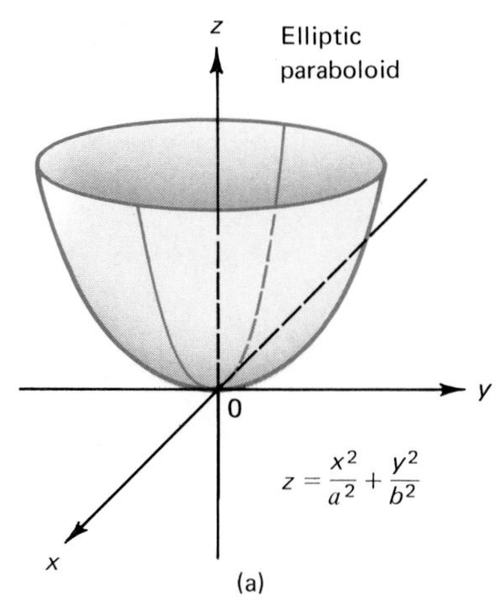

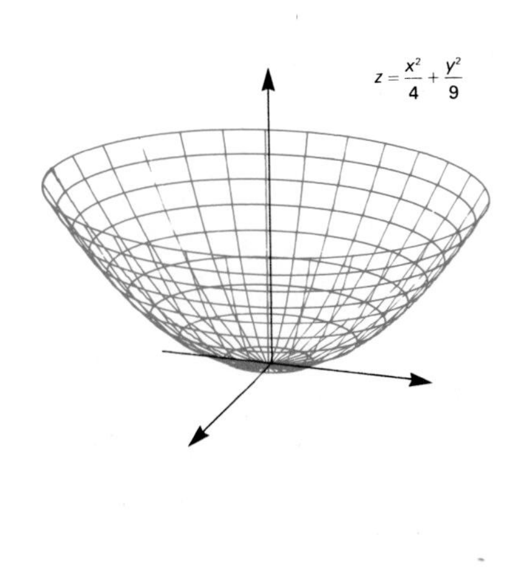

FIGURE 8

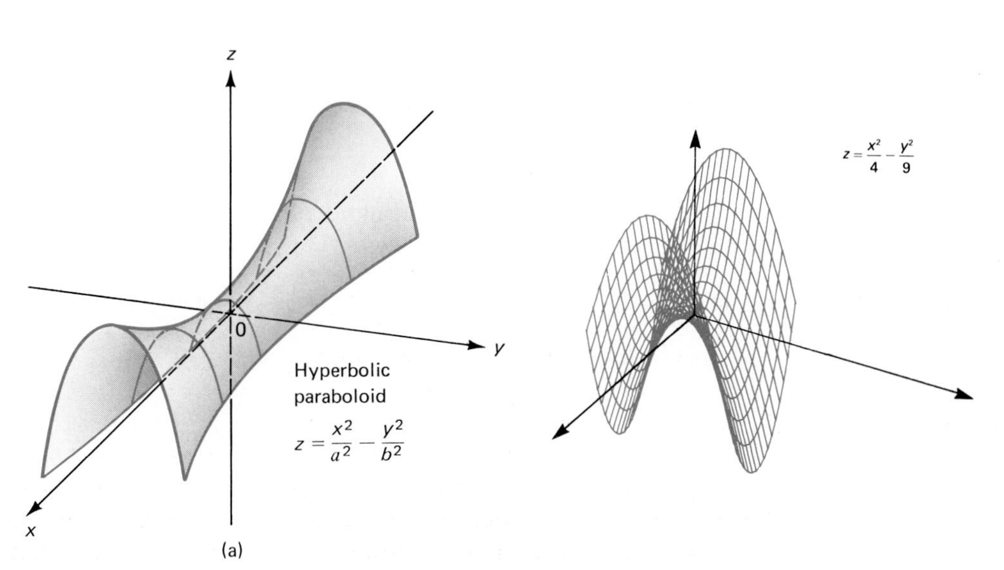

FIGURE 9

cross sections parallel to the xy-plane are hyperbolas. If x or y is fixed, we obtain parabolas. Thus cross sections parallel to the xz- and yz-planes are parabolas. The shape of the graph suggests why the hyperbolic paraboloid is often called a **saddle surface.** A computer-drawn sketch of $z = (x^2/4) - (y^2/9)$ is given in Figure 9b.

11. **The elliptic cone:** $(x^2/a^2) + (y^2/b^2) = z^2$. This surface is sketched in Figure 10a. We get one cone for $z > 0$ and another for $z < 0$. Cross sections cut by planes not passing through the origin are either parabolas, circles, ellipses, or hyperbolas. That is, the cross sections are the **conic sections** discussed in one-variable calculus courses. If $a = b$, we obtain the equation of a **circular cone.** A computer-drawn sketch of $(x^2/4) + (y^2/9) = z^2$ is given in Figure 10b.

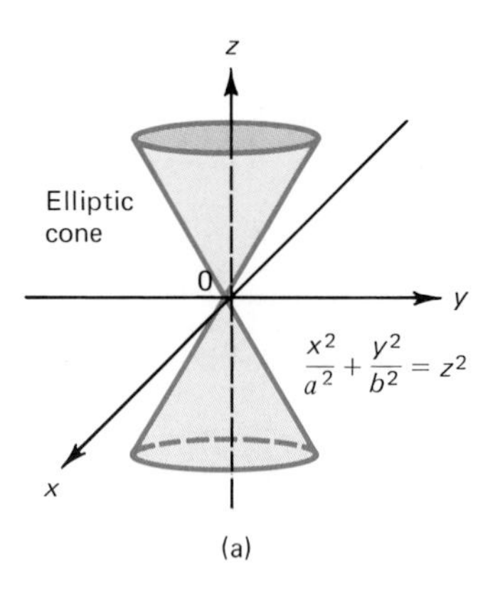

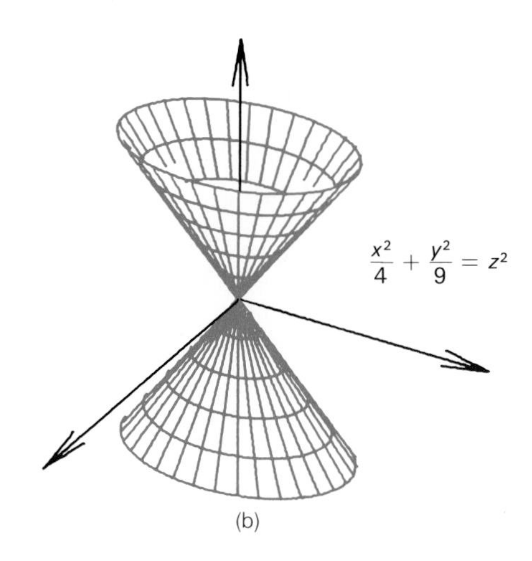

FIGURE 10

EXAMPLE 4 Describe the surface given by $x^2 - 4y^2 - 9z^2 = 25$.

Solution. Dividing by 25, we obtain

$$\frac{x^2}{25} - \frac{4y^2}{25} - \frac{9z^2}{25} = 1, \quad \text{or} \quad \frac{x^2}{25} - \frac{y^2}{(5/2)^2} - \frac{z^2}{(5/3)^2} = 1.$$

This is the equation of a hyperboloid of two sheets. Note that here cross sections parallel to the yz-plane are ellipses. ■

EXAMPLE 5 Describe the surface given by $4x - 4y^2 - z^2 = 0$.

Solution. Dividing by 4, we obtain

$$x = y^2 + \frac{z^2}{4}.$$

This is the equation of an elliptic paraboloid. The only difference between this surface and the ones described under surface 9 is that now cross sections parallel to the yz-plane are ellipses while cross sections parallel to the other coordinate planes are parabolas. This surface is sketched in Figure 11. ■

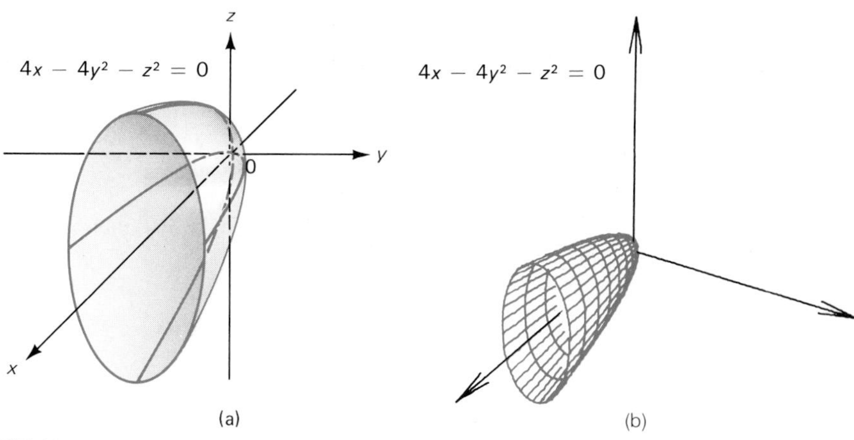

FIGURE 11

EXAMPLE 6 Describe the surface $x^2 + 8x - 2y^2 + 8y + z^2 = 0$.

Solution. Completing the squares, we obtain

$$(x + 4)^2 - 2(y - 2)^2 + z^2 = 8,$$

or dividing by 8,

$$\frac{(x + 4)^2}{8} - \frac{(y - 2)^2}{4} + \frac{z^2}{8} = 1.$$

This surface is a hyperboloid of one sheet, which, however, is centered at $(-4, 2, 0)$ instead of the origin. Moreover, cross sections parallel to the xz-plane are ellipses, while those parallel to the other coordinate planes are hyperbolas. ■

As a rule, to identify a quadric surface with x, y, and z terms, completing the square will generally resolve the problem. If there are terms of the form cxy, dyz, or exz present, then it is necessary to rotate the axes, a technique that we will not discuss here except to note that it is similar to the $\mathbb{R}^2$ technique discussed in your one-variable calculus text.

■ WARNING: Watch out for degenerate surfaces. For example, there are clearly no points that satisfy

$$x^2 + \frac{y^2}{4} + \frac{z^2}{9} + 1 = 0.$$

PROBLEMS 3.6

In Problems 1–7, draw a sketch of the given cylinder. Here the directrix C is given. The generatrix L is the axis of the variable missing in the equation.

1. $y = \sin x$
2. $z = \sin y$
3. $y = \cos z$
4. $y = \cosh x$
5. $z = x^3$
6. $z = |y|$
7. $x^2 + y^2 + 2y = 0$ [*Hint:* Complete the square.]

In Problems 8–34, identify the quadric surface and sketch it.

8. $x^2 + y^2 = 4$
9. $y^2 + z^2 = 4$
10. $x^2 + z^2 = 4$
11. $\frac{x^2}{4} - \frac{y^2}{9} = 1$
12. $x^2 + 4z^2 = 1$
13. $y^2 - 2y + 4z^2 = 1$

14. $x^2 - z^2 = 1$

15. $3x^2 - 4y^2 = 4$

16. $x^2 + y^2 + z^2 = 1$

17. $x^2 + y^2 - z^2 = 1$

18. $y^2 + 2y - z^2 + x^2 = 1$

19. $x^2 + 2y^2 + 3z^2 = 4$

20. $x + 2y^2 + 3z^2 = 4$

21. $x^2 - 2y - 3z^2 = 4$

22. $x^2 + 4x + y^2 + 6y - z^2 - 8z = 2$

23. $4x - x^2 + y^2 + z^2 = 0$

24. $5x^2 + 7y^2 + 8z^2 = 8z$

25. $x^2 - y^2 - z^2 = 1$

26. $x^2 - 2y^2 - 3z^2 = 4$

27. $z^2 - x^2 - y^2 = 2$

28. $y^2 - 3x^2 - 3z^2 = 27$

29. $x^2 + y^2 + z^2 = 2(x + y + z)$

30. $4x^2 + 4y^2 + 16z^2 = 16$

31. $4y^2 - 4x^2 + 8z^2 = 16$

32. $z + x^2 - y^2 = 0$

33. $-x^2 + 2x + y^2 - 6y = z$

34. $x + y + z = x^2 + z^2$

35. Identify and sketch the surface generated when the ellipse $(x^2/a^2) + (y^2/b^2) = 1$ is revolved about the y-axis.

3.7 THE SPACE $\mathbb{R}^n$ AND THE SCALAR PRODUCT

In Chapter 1 and in this chapter we have discussed vectors in $\mathbb{R}^2$ and $\mathbb{R}^3$. Recall that a vector in $\mathbb{R}^2$ is an ordered pair that we can write as $\mathbf{v} = (x, y)$ where x and y are real numbers. Similarly, a vector in $\mathbb{R}^3$ is an ordered triple $\mathbf{v} = (x, y, z)$ where x, y, and z are real numbers.

There are many practical situations where we would not like to be limited to vectors in $\mathbb{R}^2$ or $\mathbb{R}^3$. For example, the buyer for a large department store might have to place orders for varying quantities of 13 different items. He can represent his purchase order by a vector having 13 components. Such a vector might be given as

$$\mathbf{p} = (130, 40, 12, 46, 120, 5, 8, 250, 80, 60, 75, 310, 50).$$

Thus, the buyer would order 130 units of the first item, 40 units of the second item, 12 units of the third item, and so on. Note that the order in which the numbers in $\mathbf{p}$ are written is important. If the first two numbers were interchanged, for example, the buyer would then be ordering 40 units of the first item and 130 units of the second item—quite a change from the original order.

One important thing to note about the example above is that the vector given cannot be represented geometrically—at least not in terms of anything we can draw on a piece of paper. Nevertheless, such a vector has great practical significance.

Definition 1 n-VECTOR **(i)** An **n-component row vector** is an *ordered* set of n numbers written as

$$(x_1, x_2, \ldots, x_n) \tag{1}$$

(ii) An **n-component column vector** is an ordered set of n numbers written as

$$\begin{pmatrix} x_1 \\ x_2 \\ \vdots \\ x_n \end{pmatrix}. \tag{2}$$

In (1) or (2), x_1 is called the **first component** of the vector, x_2 is the **second component,** and so on. In general, x_k is called the ***k*th component** of the vector.

For simplicity, we shall often refer to an n-component row vector as a **row vector** or an ***n*-vector.** Similarly, we shall use the term **column vector** (or ***n*-vector**) to denote an n-component column vector. Any vector whose entries are all zero is called a **zero vector.**

EXAMPLE 1 The following are examples of vectors:

(i) (3, 6) is a row vector (or a 2-vector).

(ii) $\begin{pmatrix} 2 \\ -1 \\ 5 \end{pmatrix}$ is a column vector (or a 3-vector).

(iii) (2, -1, 0, 4) is a row vector (or a 4-vector).

(iv) $\begin{pmatrix} 0 \\ 0 \\ 0 \\ 0 \\ 0 \end{pmatrix}$ is a column vector and a zero vector. ■

WARNING: The word "ordered" in the definition of a vector is essential. Two vectors with the same components written in different orders are *not* the same. Thus, for example, the row vectors (1, 2) and (2, 1) are not equal.

As before, we shall denote vectors with boldface lowercase letters such as **u**, **v**, **a**, **b**, and **c**. A zero vector is denoted **0**.

It is time to describe some properties of vectors. Since it would be repetitive to do so first for row vectors and then for column vectors, we shall give all definitions in terms of column vectors. Similar definitions hold for row vectors.

The components of all the vectors in this text are either real or complex numbers.[†]

$\mathbb{R}^n$ We use the symbol $\mathbb{R}^n$ to denote the set of all n-vectors $\begin{pmatrix} a_1 \\ a_2 \\ \vdots \\ a_n \end{pmatrix}$, where each a_i is a real number. Similarly, we use the symbol $\mathbb{C}^n$ to denote the set of all n-vectors $\begin{pmatrix} c_1 \\ c_2 \\ \vdots \\ c_n \end{pmatrix}$, where each c_i is a complex number. In Chapters 1 and 3 we discussed the sets $\mathbb{R}^2$ and $\mathbb{R}^3$. In Chapter 9 we shall examine arbitrary sets of vectors.

[†]A complex number is a number of the form $a + ib$, where a and b are real numbers and $i = \sqrt{-1}$. A description of complex numbers is given in Appendix 3. We shall not encounter complex vectors again until Chapter 11. Therefore, unless otherwise stated, we assume, for the time being, that all vectors have real components.

Definition 2 EQUAL VECTORS Two column (or row) vectors **a** and **b** are **equal** if and only if they have the same number of components and their corresponding components are equal. In symbols, the vectors $\mathbf{a} = \begin{pmatrix} a_1 \\ a_2 \\ \vdots \\ a_n \end{pmatrix}$ and $\mathbf{b} = \begin{pmatrix} b_1 \\ b_2 \\ \vdots \\ b_n \end{pmatrix}$ are equal if and only if $a_1 = b_1, a_2 = b_2, \ldots, a_n = b_n$.

Definition 3 ADDITION OF VECTORS Let $\mathbf{a} = \begin{pmatrix} a_1 \\ a_2 \\ \vdots \\ a_n \end{pmatrix}$ and $\mathbf{b} = \begin{pmatrix} b_1 \\ b_2 \\ \vdots \\ b_n \end{pmatrix}$ be n-vectors. Then the sum of **a** and **b** is defined by

$$\mathbf{a} + \mathbf{b} = \begin{pmatrix} a_1 + b_1 \\ a_2 + b_2 \\ \vdots \\ a_n + b_n \end{pmatrix}. \tag{3}$$

EXAMPLE 2 $\begin{pmatrix} 1 \\ 2 \\ 4 \end{pmatrix} + \begin{pmatrix} -6 \\ 7 \\ 5 \end{pmatrix} = \begin{pmatrix} -5 \\ 9 \\ 9 \end{pmatrix}$. ■

EXAMPLE 3 $\begin{pmatrix} 2 \\ -1 \\ 3 \\ -4 \end{pmatrix} + \begin{pmatrix} -2 \\ 1 \\ -3 \\ 4 \end{pmatrix} = \begin{pmatrix} 0 \\ 0 \\ 0 \\ 0 \end{pmatrix} = \mathbf{0}$. ■

■ WARNING: It is essential that **a** and **b** have the same number of components. For example, the sum $\begin{pmatrix} 2 \\ 3 \end{pmatrix} + \begin{pmatrix} 1 \\ 2 \\ 3 \end{pmatrix}$ is not defined, since 2-vectors and 3-vectors are different kinds of objects and cannot be added together. Moreover, it is not possible to add a row and a column vector together. For example, the sum $\begin{pmatrix} 1 \\ 2 \end{pmatrix} + (3, 5)$ is *not* defined.

When dealing with vectors, we shall refer to numbers as **scalars** (which may be real or complex, depending on whether the vectors in question are real or complex).

Definition 4 SCALAR MULTIPLICATION OF VECTORS Let $\mathbf{a} = \begin{pmatrix} a_1 \\ a_2 \\ \vdots \\ a_n \end{pmatrix}$ be a vector and α a scalar. Then the product $\alpha\mathbf{a}$ is given by

$$\alpha \mathbf{a} = \begin{pmatrix} \alpha a_1 \\ \alpha a_2 \\ \vdots \\ \alpha a_n \end{pmatrix}. \tag{4}$$

That is, to multiply a vector by a scalar, we simply multiply each component of the vector by the scalar.

EXAMPLE 4 $3\begin{pmatrix} 2 \\ -1 \\ 4 \end{pmatrix} = \begin{pmatrix} 6 \\ -3 \\ 12 \end{pmatrix}$. ■

NOTE. Putting Definition 3 and Definition 4 together, we can define the difference of two vectors by

$$\mathbf{a} - \mathbf{b} = \mathbf{a} + (-1)\mathbf{b}. \tag{5}$$

This means that if $\mathbf{a} = \begin{pmatrix} a_1 \\ a_2 \\ \vdots \\ a_n \end{pmatrix}$ and $\mathbf{b} = \begin{pmatrix} b_1 \\ b_2 \\ \vdots \\ b_n \end{pmatrix}$, then $\mathbf{a} - \mathbf{b} = \begin{pmatrix} a_1 - b_1 \\ a_2 - b_2 \\ \vdots \\ a_n - b_n \end{pmatrix}$.

EXAMPLE 5 Let $\mathbf{a} = \begin{pmatrix} 4 \\ 6 \\ 1 \\ 3 \end{pmatrix}$ and $\mathbf{b} = \begin{pmatrix} -2 \\ 4 \\ -3 \\ 0 \end{pmatrix}$. Calculate $2\mathbf{a} - 3\mathbf{b}$. ■

Solution. $2\mathbf{a} - 3\mathbf{b} = 2\begin{pmatrix} 4 \\ 6 \\ 1 \\ 3 \end{pmatrix} + (-3)\begin{pmatrix} -2 \\ 4 \\ -3 \\ 0 \end{pmatrix} = \begin{pmatrix} 8 \\ 12 \\ 2 \\ 6 \end{pmatrix} + \begin{pmatrix} 6 \\ -12 \\ 9 \\ 0 \end{pmatrix} = \begin{pmatrix} 14 \\ 0 \\ 11 \\ 6 \end{pmatrix}$. ■

Once we know how to add vectors and multiply them by scalars, we can prove a number of facts relating these operations. Several of these facts are given in Theorem 1. We prove parts (ii) and (iii) and leave the remaining parts as exercises (see Problems 21–23).

Theorem 1 Let $\mathbf{a}$, $\mathbf{b}$, and $\mathbf{c}$ be n-vectors and let α and β be scalars. Then:

(i) $\mathbf{a} + \mathbf{0} = \mathbf{a}$.
(ii) $0\mathbf{a} = \mathbf{0}$.
(iii) $\mathbf{a} + \mathbf{b} = \mathbf{b} + \mathbf{a}$ (commutative law).
(iv) $(\mathbf{a} + \mathbf{b}) + \mathbf{c} = \mathbf{a} + (\mathbf{b} + \mathbf{c})$ (associative law).
(v) $\alpha(\mathbf{a} + \mathbf{b}) = \alpha\mathbf{a} + \alpha\mathbf{b}$ (distributive law for scalar multiplication).
(vi) $(\alpha + \beta)\mathbf{a} = \alpha\mathbf{a} + \beta\mathbf{a}$.
(vii) $(\alpha\beta)\mathbf{a} = \alpha(\beta\mathbf{a})$.

Proof of **(ii)** *and* **(iii)**.

(ii) If $\mathbf{a} = \begin{pmatrix} a_1 \\ a_2 \\ \vdots \\ a_n \end{pmatrix}$, then $0\mathbf{a} = 0\begin{pmatrix} a_1 \\ a_2 \\ \vdots \\ a_n \end{pmatrix} = \begin{pmatrix} 0 \cdot a_1 \\ 0 \cdot a_2 \\ \vdots \\ 0 \cdot a_n \end{pmatrix} = \begin{pmatrix} 0 \\ 0 \\ \vdots \\ 0 \end{pmatrix} = \mathbf{0}$.

(iii) Let $\mathbf{b} = \begin{pmatrix} b_1 \\ b_2 \\ \vdots \\ b_n \end{pmatrix}$. Then $\mathbf{a} + \mathbf{b} = \begin{pmatrix} a_1 + b_1 \\ a_2 + b_2 \\ \vdots \\ a_n + b_n \end{pmatrix} = \begin{pmatrix} b_1 + a_1 \\ b_2 + a_2 \\ \vdots \\ b_n + a_n \end{pmatrix} = \mathbf{b} + \mathbf{a}$.

Here we used the fact that for any two numbers x and y, $x + y = y + x$ and $0 \cdot x = 0$.

NOTE. In (ii), the zero on the left is the scalar zero (i.e., the real number 0) and the zero on the right is the zero vector. These two things are different. ■

EXAMPLE 6 To illustrate the associative law, we note that

$$\left[\begin{pmatrix} 3 \\ 1 \\ 2 \end{pmatrix} + \begin{pmatrix} -2 \\ 4 \\ -1 \end{pmatrix}\right] + \begin{pmatrix} 6 \\ -3 \\ 5 \end{pmatrix} = \begin{pmatrix} 1 \\ 5 \\ 1 \end{pmatrix} + \begin{pmatrix} 6 \\ -3 \\ 5 \end{pmatrix} = \begin{pmatrix} 7 \\ 2 \\ 6 \end{pmatrix}$$

while

$$\begin{pmatrix} 3 \\ 1 \\ 2 \end{pmatrix} + \left[\begin{pmatrix} -2 \\ 4 \\ -1 \end{pmatrix} + \begin{pmatrix} 6 \\ -3 \\ 5 \end{pmatrix}\right] = \begin{pmatrix} 3 \\ 1 \\ 2 \end{pmatrix} + \begin{pmatrix} 4 \\ 1 \\ 4 \end{pmatrix} = \begin{pmatrix} 7 \\ 2 \\ 6 \end{pmatrix}. \quad ■$$

Example 6 illustrates the importance of the associative law of vector addition, since if we wish to add three or more vectors, we can do so only by adding them two at a time. The associative law tells us that we can do this in two different ways and still come up with the same answer. If this were not the case, the sum of three or more vectors would be more difficult to define since we would have to specify whether we wanted $(\mathbf{a} + \mathbf{b}) + \mathbf{c}$ or $\mathbf{a} + (\mathbf{b} + \mathbf{c})$ to be the definition of the sum $\mathbf{a} + \mathbf{b} + \mathbf{c}$.

In Sections 1.2 and 3.2 we discussed the dot or scalar product of vectors in $\mathbb{R}^2$ and $\mathbb{R}^3$. We now generalize this notion to vectors in $\mathbb{R}^n$. We shall use the term *scalar product* when referring to this operation.

Definition 5 SCALAR PRODUCT Let $\mathbf{a} = \begin{pmatrix} a_1 \\ a_2 \\ \vdots \\ a_n \end{pmatrix}$ and $\mathbf{b} = \begin{pmatrix} b_1 \\ b_2 \\ \vdots \\ b_n \end{pmatrix}$ be two n-vectors.

Then the **scalar product** of $\mathbf{a}$ and $\mathbf{b}$, denoted $\mathbf{a} \cdot \mathbf{b}$, is given by

$$\mathbf{a} \cdot \mathbf{b} = a_1b_1 + a_2b_2 + \cdots + a_nb_n. \quad \textbf{(6)}$$

■ WARNING: When taking the scalar product of **a** and **b**, it is necessary that **a** and **b** have the same number of components.

We shall often be taking the scalar product of a row vector and column vector. In this case we have

$$(a_1, a_2, \ldots, a_n) \cdot \begin{pmatrix} b_1 \\ b_2 \\ \vdots \\ b_n \end{pmatrix} = a_1b_1 + a_2b_2 + \cdots + a_nb_n. \tag{7}$$

EXAMPLE 7 Let $\mathbf{a} = \begin{pmatrix} 1 \\ -2 \\ 3 \end{pmatrix}$ and $\mathbf{b} = \begin{pmatrix} 3 \\ -2 \\ 4 \end{pmatrix}$. Calculate $\mathbf{a} \cdot \mathbf{b}$.

Solution. $\mathbf{a} \cdot \mathbf{b} = (1)(3) + (-2)(-2) + (3)(4) = 3 + 4 + 12 = 19.$ ■

EXAMPLE 8 Let $\mathbf{a} = (2, -3, 4, -6)$ and $\mathbf{b} = \begin{pmatrix} 1 \\ 2 \\ 0 \\ 3 \end{pmatrix}$. Compute $\mathbf{a} \cdot \mathbf{b}$.

Solution. Here $\mathbf{a} \cdot \mathbf{b} = (2)(1) + (-3)(2) + (4)(0) + (-6)(3) = 2 - 6 + 0 - 18 = -22.$ ■

EXAMPLE 9 Suppose that a manufacturer produces four items. The demand for the items is given by the demand vector $\mathbf{d} = (30, 20, 40, 10)$. The price per unit that he receives for the items is given by the price vector $\mathbf{p} = (\$20, \$15, \$18, \$40)$. If he meets his demand, how much money will he receive?

Solution. The demand for the first item is 30 and the manufacturer receives \$20 for each of the first item sold. He therefore receives (30)(20) = \$600 from the sale of the first item. By continuing this reasoning we see that the total cash received is given by $\mathbf{d} \cdot \mathbf{p}$. Thus income received $= \mathbf{d} \cdot \mathbf{p} = (30)(20) + (20)(15) + (40)(18) + (10)(40) = 600 + 300 + 720 + 400 = \$2020.$ ■

The next result follows directly from the definition of the scalar product (see Problem 48).

Theorem 2 Let **a**, **b**, and **c** be n-vectors and let α and β be scalars. Then:

(i) $\mathbf{a} \cdot \mathbf{0} = 0$.
(ii) $\mathbf{a} \cdot \mathbf{b} = \mathbf{b} \cdot \mathbf{a}$ (commutative law for scalar product).
(iii) $\mathbf{a} \cdot (\mathbf{b} + \mathbf{c}) = \mathbf{a} \cdot \mathbf{b} + \mathbf{a} \cdot \mathbf{c}$ (distributive law for scalar product).
(iv) $(\alpha\mathbf{a}) \cdot \mathbf{b} = \alpha(\mathbf{a} \cdot \mathbf{b})$.

Note that there is *no* associative law for the scalar product. The expression $(\mathbf{a} \cdot \mathbf{b}) \cdot \mathbf{c} = \mathbf{a} \cdot (\mathbf{b} \cdot \mathbf{c})$ does not make sense because neither side of the equation is defined. For the left side, this follows from the fact that $\mathbf{a} \cdot \mathbf{b}$ is a scalar and the scalar product of the scalar $\mathbf{a} \cdot \mathbf{b}$ and the vector $\mathbf{c}$ is not defined.

Recall that in $\mathbb{R}^2$ the vector $\mathbf{v} = (x, y)$ has length $|\mathbf{v}| = \sqrt{x^2 + y^2}$. In $\mathbb{R}^3$ the vector $\mathbf{v} = (x, y, z)$ has length $|\mathbf{v}| = \sqrt{x^2 + y^2 + z^2}$. Another term for length is *norm*. We now extend this concept to $\mathbb{R}^n$.

Definition 6 NORM OF A VECTOR The **norm** of a vector $\mathbf{v}$ in $\mathbb{R}^n$, denoted $|\mathbf{v}|$, is given by

$$|\mathbf{v}| = \sqrt{\mathbf{v} \cdot \mathbf{v}}. \tag{8}$$

NOTE. If $\mathbf{v} = (a_1, a_2, \ldots, a_n)$, then $\mathbf{v} \cdot \mathbf{v} = a_1^2 + a_2^2 + \cdots + a_n^2$, so that

$$|\mathbf{v}| = \sqrt{a_1^2 + a_2^2 + \cdots + a_n^2}. \tag{9}$$

Previously, the norm of a vector in $\mathbb{R}^2$ or $\mathbb{R}^3$ was called the **magnitude** of that vector.

EXAMPLE 10 Let $\mathbf{v} = (7, -1, 2, 4, 5)$. Compute $|\mathbf{v}|$.

Solution. $|\mathbf{v}| = \sqrt{7^2 + (-1)^2 + 2^2 + 4^2 + 5^2} = \sqrt{95}$. ■

PROBLEMS 3.7

In Problems 1–10 perform the indicated computation with $\mathbf{a} = \begin{pmatrix} -3 \\ 1 \\ 4 \end{pmatrix}$, $\mathbf{b} = \begin{pmatrix} 5 \\ -4 \\ 7 \end{pmatrix}$, and $\mathbf{c} = \begin{pmatrix} 2 \\ 0 \\ -2 \end{pmatrix}$.

1. $\mathbf{a} + \mathbf{b}$
2. $3\mathbf{b}$
3. $-2\mathbf{c}$
4. $\mathbf{b} + 3\mathbf{c}$
5. $2\mathbf{a} - 5\mathbf{b}$
6. $-3\mathbf{b} + 2\mathbf{c}$
7. $0\mathbf{c}$
8. $\mathbf{a} + \mathbf{b} + \mathbf{c}$
9. $3\mathbf{a} - 2\mathbf{b} + 4\mathbf{c}$
10. $3\mathbf{b} - 7\mathbf{c} + 2\mathbf{a}$

In Problems 11–20 perform the indicated computation with $\mathbf{a} = (3, -1, 4, 2)$, $\mathbf{b} = (6, 0, -1, 4)$, and $\mathbf{c} = (-2, 3, 1, 5)$. Of course, it is first necessary to extend the definitions in this section to row vectors.

11. $\mathbf{a} + \mathbf{c}$
12. $\mathbf{b} - \mathbf{a}$
13. $4\mathbf{c}$
14. $-2\mathbf{b}$
15. $2\mathbf{a} - \mathbf{c}$
16. $4\mathbf{b} - 7\mathbf{a}$
17. $\mathbf{a} + \mathbf{b} + \mathbf{c}$
18. $\mathbf{c} - \mathbf{b} + 2\mathbf{a}$
19. $3\mathbf{a} - 2\mathbf{b} + 4\mathbf{c}$
20. $\alpha\mathbf{a} + \beta\mathbf{b} + \gamma\mathbf{c}$
21. Let $\mathbf{a} = \begin{pmatrix} a_1 \\ a_2 \\ \vdots \\ a_n \end{pmatrix}$ and let $\mathbf{0}$ denote the n-component zero column vector. Use Definitions 3 and 4 to show that $\mathbf{a} + \mathbf{0} = \mathbf{a}$ and $0\mathbf{a} = \mathbf{0}$.
22. Let $\mathbf{a} = \begin{pmatrix} a_1 \\ a_2 \\ \vdots \\ a_n \end{pmatrix}$, $\mathbf{b} = \begin{pmatrix} b_1 \\ b_2 \\ \vdots \\ b_n \end{pmatrix}$, and $\mathbf{c} = \begin{pmatrix} c_1 \\ c_2 \\ \vdots \\ c_n \end{pmatrix}$. Compute $(\mathbf{a} + \mathbf{b}) + \mathbf{c}$ and $\mathbf{a} + (\mathbf{b} + \mathbf{c})$ and show that they are equal.
23. Let $\mathbf{a}$ and $\mathbf{b}$ be as in Problem 22 and let α and β be scalars. Compute $\alpha(\mathbf{a} + \mathbf{b})$ and $\alpha\mathbf{a} + \alpha\mathbf{b}$ and show that they are equal. Similarly, compute $(\alpha + \beta)\mathbf{a}$ and $\alpha\mathbf{a} + \beta\mathbf{a}$ and show that they are equal. Finally, show that $(\alpha\beta)\mathbf{a} = \alpha(\beta\mathbf{a})$.
24. Find numbers α, β, and γ such that $(2, -1, 4) + (\alpha, \beta, \gamma) = \mathbf{0}$.

25. In the manufacture of a certain product, four raw materials are needed. The vector $\mathbf{d} = \begin{pmatrix} d_1 \\ d_2 \\ d_3 \\ d_4 \end{pmatrix}$ represents a given factory's demand for each of the four raw materials to produce one unit of its product. If $\mathbf{d}_1$ is the demand vector for factory 1 and $\mathbf{d}_2$ is the demand vector for factory 2, what is represented by the vectors $\mathbf{d}_1 + \mathbf{d}_2$ and $2\mathbf{d}_1$?

26. Let $\mathbf{a} = \begin{pmatrix} 1 \\ 3 \\ 2 \end{pmatrix}$, $\mathbf{b} = \begin{pmatrix} -2 \\ 4 \\ 1 \end{pmatrix}$, and $\mathbf{c} = \begin{pmatrix} 0 \\ 1 \\ 4 \end{pmatrix}$. Find a vector $\mathbf{v}$ such that $2\mathbf{a} - \mathbf{b} + 3\mathbf{v} = 4\mathbf{c}$.

In Problems 27–33 calculate the scalar product of the two vectors.

27. $\begin{pmatrix} 2 \\ 3 \\ -5 \end{pmatrix}; \begin{pmatrix} 3 \\ 0 \\ 4 \end{pmatrix}$

28. $(1, 2, -1, 0); (3, -7, 4, -2)$

29. $\begin{pmatrix} 5 \\ 7 \end{pmatrix}; \begin{pmatrix} 3 \\ -2 \end{pmatrix}$

30. $(8, 3, 1); (7, -4, 3)$

31. $(a, b); (c, d)$

32. $\begin{pmatrix} x \\ y \\ z \end{pmatrix}; \begin{pmatrix} y \\ z \\ x \end{pmatrix}$

33. $(-1, -3, 4, 5); (-1, -3, 4, 5)$

34. Let $\mathbf{a}$ be an n-vector. Show that $\mathbf{a} \cdot \mathbf{a} \geq 0$.

35. Find conditions on a vector $\mathbf{a}$ such that $\mathbf{a} \cdot \mathbf{a} = 0$.

In Problems 36–40 perform the indicated computation with $\mathbf{a} = \begin{pmatrix} 1 \\ -2 \\ 4 \end{pmatrix}$, $\mathbf{b} = \begin{pmatrix} 0 \\ -3 \\ -7 \end{pmatrix}$, and $\mathbf{c} = \begin{pmatrix} 4 \\ -1 \\ 5 \end{pmatrix}$.

36. $(2\mathbf{a}) \cdot (3\mathbf{b})$

37. $\mathbf{a} \cdot (\mathbf{b} + \mathbf{c})$

38. $\mathbf{c} \cdot (\mathbf{a} - \mathbf{b})$

39. $(2\mathbf{b}) \cdot (3\mathbf{c} - 5\mathbf{a})$

40. $(\mathbf{a} - \mathbf{c}) \cdot (3\mathbf{b} - 4\mathbf{a})$

Two vectors $\mathbf{a}$ and $\mathbf{b}$ are said to be **orthogonal** if $\mathbf{a} \cdot \mathbf{b} = 0$. In Problems 41–45 determine which pairs of vectors are orthogonal.

41. $\begin{pmatrix} 2 \\ -3 \end{pmatrix}; \begin{pmatrix} 3 \\ 2 \end{pmatrix}$

42. $\begin{pmatrix} 1 \\ 4 \\ -7 \end{pmatrix}; \begin{pmatrix} 2 \\ 3 \\ 2 \end{pmatrix}$

43. $\begin{pmatrix} 5 \\ 4 \\ 6 \\ -1 \end{pmatrix}; \begin{pmatrix} 3 \\ -4 \\ 1 \\ 5 \end{pmatrix}$

44. $(1, 0, 1, 0); (0, 1, 0, 1)$

45. $\begin{pmatrix} a \\ 0 \\ b \\ 0 \\ c \end{pmatrix}; \begin{pmatrix} 0 \\ d \\ 0 \\ e \\ 0 \end{pmatrix}$

46. Determine a number α such that $(1, -2, 3, 5)$ is orthogonal to $(-4, \alpha, 6, -1)$.

47. Determine all numbers α and β such that the vectors $\begin{pmatrix} 1 \\ -\alpha \\ 2 \\ 3 \end{pmatrix}$ and $\begin{pmatrix} 4 \\ 5 \\ -2\beta \\ 7 \end{pmatrix}$ are orthogonal.

48. Using the definition of the scalar product, prove Theorem 2.

In Problems 49–53 compute the norm of the given vector.

49. $\begin{pmatrix} 1 \\ 2 \\ 3 \\ 4 \end{pmatrix}$

50. $(0, 0, 1, 0, 0, 0)$

51. (a, b, c, d, e)

52. $\begin{pmatrix} 1 \\ \frac{1}{2} \\ \frac{1}{3} \\ \frac{1}{4} \end{pmatrix}$

53. $(1, 1, 1, 1, 1, 1, 1, 1, 1, 1)$

54. Let α be a scalar and let $\mathbf{v}$ be a vector in $\mathbb{R}^n$. Show that $|\alpha\mathbf{v}| = |\alpha||\mathbf{v}|$.

3.8 VECTOR FUNCTIONS AND PARAMETRIC EQUATIONS IN $\mathbb{R}^3$

Having seen how vectors in $\mathbb{R}^2$ generalize to vectors in $\mathbb{R}^3$, we can imagine properties of vector functions in $\mathbb{R}^3$. We will describe these briefly, as almost all their properties have already been proved in Chapter 2. We will then extend these results to $\mathbb{R}^n$.

Definition 1 VECTOR FUNCTIONS IN $\mathbb{R}^3$ Let f_1, f_2, and f_3 be functions of the real variable t. Then for all values of t for which $f_1(t)$, $f_2(t)$, and $f_3(t)$ are defined, we define the **vector-valued function f of a real variable** t by

$$\mathbf{f}(t) = (f_1(t), f_2(t), f_3(t)) = f_1(t)\mathbf{i} + f_2(t)\mathbf{j} + f_3(t)\mathbf{k}. \tag{1}$$

If f_1, f_2, and f_3 are continuous over an interval I, then as t varies over I, the set of points traced out by the end of the vector $\mathbf{f}$ is called a **curve** in space.

EXAMPLE 1 Sketch the curve $\mathbf{f}(t) = (\cos t)\mathbf{i} + 4(\sin t)\mathbf{j} + t\mathbf{k}$.

Solution. Here $x = \cos t$ and $y = 4 \sin t$, so eliminating t, we obtain $x^2 + (y^2/16) = 1$, which is the equation of an ellipse in the xy-plane. Since $z = t$ increases as t increases, the curve is a spiral that climbs up the side of an elliptical cylinder, as sketched in Figure 1. This curve is called an **elliptical helix.** ■

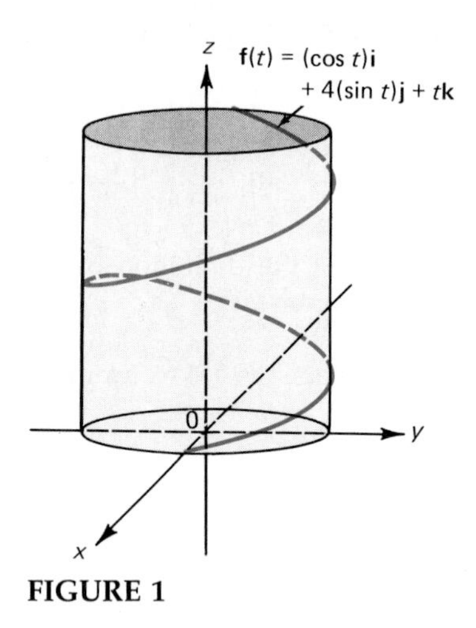

FIGURE 1

Definition 2 LIMIT If $\mathbf{f}(t) = f_1(t)\mathbf{i} + f_2(t)\mathbf{j} + f_3(t)\mathbf{k}$, and if $\lim_{t \to t_0} f_1(t)$, $\lim_{t \to t_0} f_2(t)$, and $\lim_{t \to t_0} f_3(t)$ exist, then we define

$$\lim_{t \to t_0} \mathbf{f}(t) = \left[\lim_{t \to t_0} f_1(t)\right]\mathbf{i} + \left[\lim_{t \to t_0} f_2(t)\right]\mathbf{j} + \left[\lim_{t \to t_0} f_3(t)\right]\mathbf{k}. \tag{2}$$

Definition 3 DERIVATIVE

$$\frac{d\mathbf{f}(t)}{dt} = \mathbf{f}'(t) = \lim_{\Delta t \to 0} \frac{\mathbf{f}(t + \Delta t) - \mathbf{f}(t)}{\Delta t}, \tag{3}$$

provided that this limit exists.

Theorem 1 $\mathbf{f}$ is differentiable if and only if its component functions f_1, f_2, and f_3 are differentiable, and in that case

$$\mathbf{f}'(t) = f_1'(t)\mathbf{i} + f_2'(t)\mathbf{j} + f_3'(t)\mathbf{k} = (f_1'(t), f_2'(t), f_3'(t)) \tag{4}$$

EXAMPLE 2 Find the derivative of $\mathbf{f}(t) = (\cos t)\mathbf{i} + (\sin t)\mathbf{j} + t\mathbf{k}$.

Solution. $\mathbf{f}'(t) = -(\sin t)\mathbf{i} + (\cos t)\mathbf{j} + \mathbf{k}$. ■

The following theorem is new.

Theorem 2 If $\mathbf{f}$ and $\mathbf{g}$ are differentiable, then $\mathbf{f} \times \mathbf{g}$ is differentiable and

$$(\mathbf{f} \times \mathbf{g})' = (\mathbf{f}' \times \mathbf{g}) + (\mathbf{f} \times \mathbf{g}'). \tag{5}$$

Proof. The proof is an easy application of the product rule of differentiation and is left as an exercise (see Problem 25). ■

NOTE. Since $\mathbf{g} \times \mathbf{f} = -(\mathbf{f} \times \mathbf{g})$, it follows that $(\mathbf{f} \times \mathbf{g})' = -(\mathbf{g} \times \mathbf{f})'$. Thus the order in which things are written is important in (5).

EXAMPLE 3 Let $\mathbf{f}(t) = (\cos t)\mathbf{i} + (\sin t)\mathbf{j} + t\mathbf{k}$ and $\mathbf{g}(t) = t^2\mathbf{i} + t^3\mathbf{j} + t^4\mathbf{k}$. Calculate

$$\frac{d}{dt}(\mathbf{f} \times \mathbf{g}).$$

Solution.

$$\begin{aligned}
(\mathbf{f} \times \mathbf{g})' &= (\mathbf{f}' \times \mathbf{g}) + (\mathbf{f} \times \mathbf{g}') \\
&= [-(\sin t)\mathbf{i} + (\cos t)\mathbf{j} + \mathbf{k}] \times (t^2\mathbf{i} + t^3\mathbf{j} + t^4\mathbf{k}) \\
&\quad + [(\cos t)\mathbf{i} + (\sin t)\mathbf{j} + t\mathbf{k}] \times (2t\mathbf{i} + 3t^2\mathbf{j} + 4t^3\mathbf{k}) \\
&= \begin{vmatrix} \mathbf{i} & \mathbf{j} & \mathbf{k} \\ -\sin t & \cos t & 1 \\ t^2 & t^3 & t^4 \end{vmatrix} + \begin{vmatrix} \mathbf{i} & \mathbf{j} & \mathbf{k} \\ \cos t & \sin t & t \\ 2t & 3t^2 & 4t^3 \end{vmatrix} \\
&= (t^4 \cos t - t^3)\mathbf{i} + (t^2 + t^4 \sin t)\mathbf{j} - (t^3 \sin t + t^2 \cos t)\mathbf{k} \\
&\quad + (4t^3 \sin t - 3t^3)\mathbf{i} + (2t^2 - 4t^3 \cos t)\mathbf{j} + (3t^2 \cos t - 2t \sin t)\mathbf{k} \\
&= (4t^3 \sin t + t^4 \cos t - 4t^3)\mathbf{i} + (t^4 \sin t - 4t^3 \cos t + 3t^2)\mathbf{j} \\
&\quad + [2t^2 \cos t - (t^3 + 2t)\sin t]\mathbf{k}. \quad ■
\end{aligned}$$

Facts about arc length and tangent vectors hold exactly as in $\mathbb{R}^2$. Proofs of the parts of the following theorem are identical to the proofs given in Chapter 2.

Theorem 3 Let $\mathbf{f}$ have a continuous derivative and let $\mathbf{T}(t)$ and $s(t)$ denote the unit tangent vector and arc length, respectively.

(i) If $\mathbf{f}'(t) \neq \mathbf{0}$, $\mathbf{T}(t) = \mathbf{f}'(t)/|\mathbf{f}'(t)|$ [see equation (2.3.7)].
(ii) If $|\mathbf{f}(t)|$ is constant over an interval I, then $\mathbf{f} \cdot \mathbf{f}' = 0$ on I (see Theorem 2.4.2).
(iii) $ds/dt = |\mathbf{f}'(t)|$ [see Theorem 2.5.1 and equation (2.5.10)].
(iv) $s(t_1) = \int_{t_0}^{t_1} |\mathbf{f}'(t)|\, dt$ [see equation (2.5.4)]; here $s(t)$ measures arc length from a reference point $P_0 = (x(t_0), y(t_0), z(t_0))$.
(v) $\mathbf{T} = d\mathbf{f}/ds$ (see Theorem 2.6.1).

REMARK. Formula (iv) can be derived as in Section 2.5 if we start with an intuitive notion of arc length. However, formula (iv) can be used as the *definition* of arc length. The precise definition is given in Definition 4.

Definition 4 ARC LENGTH Suppose f has a continuous derivative in the interval $[t_0, b]$ and suppose that for every t_1 in $[t_0, b]$, $\int_{t_0}^{t_1} |\mathbf{f}'(t)|\, dt$ exists. Then $\mathbf{f}$ is said to be **rectifiable** in the interval $[t_0, b]$, and the **arc length** of the curve $\mathbf{f}(t)$ in the interval $[t_0, t_1]$ is given by

$$s(t_1) = \int_{t_0}^{t_1} |\mathbf{f}'(t)|\, dt,$$

where $t_0 \leq t_1 \leq b$.

EXAMPLE 4 Let $\mathbf{f}(t) = (\cos t)\mathbf{i} + (\sin t)\mathbf{j} + t\mathbf{k}$. (This surface is called a **circular helix.**)

(a) Calculate $\mathbf{T}(t)$ at $t = \pi/3$.
(b) Find the length of the arc from $t = 0$ to $t = 4$.

Solution. **(a)** $\mathbf{f}'(t) = -(\sin t)\mathbf{i} + (\cos t)\mathbf{j} + \mathbf{k}$ and

$$|\mathbf{f}'(t)| = \sqrt{\sin^2 t + \cos^2 t + 1} = \sqrt{2},$$

so that

$$\mathbf{T}(t) = -\frac{\sin t}{\sqrt{2}}\mathbf{i} + \frac{\cos t}{\sqrt{2}}\mathbf{j} + \frac{1}{\sqrt{2}}\mathbf{k}.$$

At $t = \pi/3$,

$$\mathbf{T}\left(\frac{\pi}{3}\right) = -\frac{\sqrt{3}}{2\sqrt{2}}\mathbf{i} + \frac{1}{2\sqrt{2}}\mathbf{j} + \frac{1}{\sqrt{2}}\mathbf{k}.$$

(b) Since $ds/dt = |\mathbf{f}'(t)| = \sqrt{2}$,

$$s(4) = \int_0^4 \sqrt{2}\, dt = 4\sqrt{2}. \quad \blacksquare$$

The curvature $\kappa(t)$ and the unit normal vector $\mathbf{n}(t)$ are defined differently in $\mathbb{R}^3$. They would have to be. For one thing, there is a whole "plane-full" of unit vectors orthogonal to $\mathbf{T}$. Also, curvature cannot be defined as $|d\varphi/ds|$ since there are now three angles that define the direction of C. Therefore, we use an alternative definition of curvature (see Theorem 2.8.1).

Definition 5 CURVATURE If $\mathbf{f}$ has a continuous derivative, then the **curvature** of $\mathbf{f}$ is given by

$$\kappa(t) = \left|\frac{d\mathbf{T}}{ds}\right|. \tag{6}$$

REMARK. By Theorem 2.8.1, equation (6) is equivalent in $\mathbb{R}^2$ to the definition $\kappa(t) = |d\varphi/ds|$, which was our $\mathbb{R}^2$ definition.

Definition 6 PRINCIPAL UNIT NORMAL VECTOR For any value of t for which $\kappa(t) \neq 0$, the **principal unit normal vector n** is defined by

$$\mathbf{n}(t) = \frac{1}{\kappa(t)}\frac{d\mathbf{T}}{ds}. \tag{7}$$

REMARK. Having defined a unit normal vector, we must show that it is orthogonal to the unit tangent vector in order to justify its name.

Theorem 4 $\mathbf{n}(t) \perp \mathbf{T}(t)$.

Proof. Since $1 = \mathbf{T}\cdot\mathbf{T}$, we differentiate both sides with respect to s to obtain

$$0 = \mathbf{T}\cdot\frac{d\mathbf{T}}{ds} + \frac{d\mathbf{T}}{ds}\cdot\mathbf{T} = 2\mathbf{T}\cdot\frac{d\mathbf{T}}{ds} = (2\kappa)\mathbf{T}\cdot\mathbf{n}. \quad \blacksquare$$

EXAMPLE 5 Find the curvature and principal unit normal vector, $\kappa(t)$ and $\mathbf{n}(t)$, for the curve of Example 4 at $t = \pi/3$.

Solution. Since $ds/dt = \sqrt{2}$, $s = \sqrt{2}t$ and $t = s/\sqrt{2}$, so that

$$\mathbf{T}(s) = -\frac{\sin(s/\sqrt{2})}{\sqrt{2}}\mathbf{i} + \frac{\cos(s/\sqrt{2})}{\sqrt{2}}\mathbf{j} + \frac{1}{\sqrt{2}}\mathbf{k}$$

and

$$\frac{d\mathbf{T}}{ds} = -\frac{\cos(s/\sqrt{2})}{2}\mathbf{i} - \frac{\sin(s/\sqrt{2})}{2}\mathbf{j}.$$

Hence

$$\kappa(t) = \sqrt{\frac{\cos^2(s/\sqrt{2})}{4} + \frac{\sin^2(s/\sqrt{2})}{4}} = \frac{1}{2}$$

and

$$\mathbf{n}(t) = \frac{1}{\kappa(t)}\frac{d\mathbf{T}}{ds} = -(\cos t)\mathbf{i} - (\sin t)\mathbf{j}.$$

At $t = \pi/3$,

$$\mathbf{n}\left(\frac{\pi}{3}\right) = -\frac{1}{2}\mathbf{i} - \frac{\sqrt{3}}{2}\mathbf{j}. \quad \blacksquare$$

REMARK. When it is inconvenient to solve for t in terms of s, we can calculate $d\mathbf{T}/ds$ directly by the relation

$$\frac{d\mathbf{T}}{ds} = \frac{d\mathbf{T}/dt}{ds/dt}. \tag{8}$$

There is a third vector that is often useful in applications.

Definition 7 BINORMAL VECTOR The **binormal vector B** to the curve **f** is defined by

$$\mathbf{B} = \mathbf{T} \times \mathbf{n}. \tag{9}$$

From this definition we see that **B** is orthogonal to both **T** and **n**. Moreover, since $\mathbf{T} \perp \mathbf{n}$, the angle θ between **T** and **n** is $\pi/2$, and

$$|\mathbf{B}| = |\mathbf{T} \times \mathbf{n}| = |\mathbf{T}|\,|\mathbf{n}| \sin\theta = 1,$$

so that **B** is a unit vector. Also, **T**, **n**, and **B** form a right-handed system.

EXAMPLE 6 Find the binormal vector to the curve of Example 5 at $t = \pi/3$.

Solution.

$$\mathbf{B} = \mathbf{T} \times \mathbf{n} = \begin{vmatrix} \mathbf{i} & \mathbf{j} & \mathbf{k} \\ -\dfrac{\sin t}{\sqrt{2}} & \dfrac{\cos t}{\sqrt{2}} & \dfrac{1}{\sqrt{2}} \\ -\cos t & -\sin t & 0 \end{vmatrix} = \frac{\sin t}{\sqrt{2}}\mathbf{i} - \frac{\cos t}{\sqrt{2}}\mathbf{j} + \frac{1}{\sqrt{2}}\mathbf{k}$$

When $t = \pi/3$, $\mathbf{B} = (\sqrt{3}/2\sqrt{2})\mathbf{i} - (1/2\sqrt{2})\mathbf{j} + (1/\sqrt{2})\mathbf{k}$. The vectors **T**, **n**, and **B** are sketched in Figure 2. $\blacksquare$

Our definitions of velocity and acceleration vectors remain unchanged in $\mathbb{R}^3$. Suppose that an object is moving in space. Then we can describe its motion parametrically by the vector function

$$\mathbf{f}(t) = f_1(t)\mathbf{i} + f_2(t)\mathbf{j} + f_3(t)\mathbf{k}. \tag{10}$$

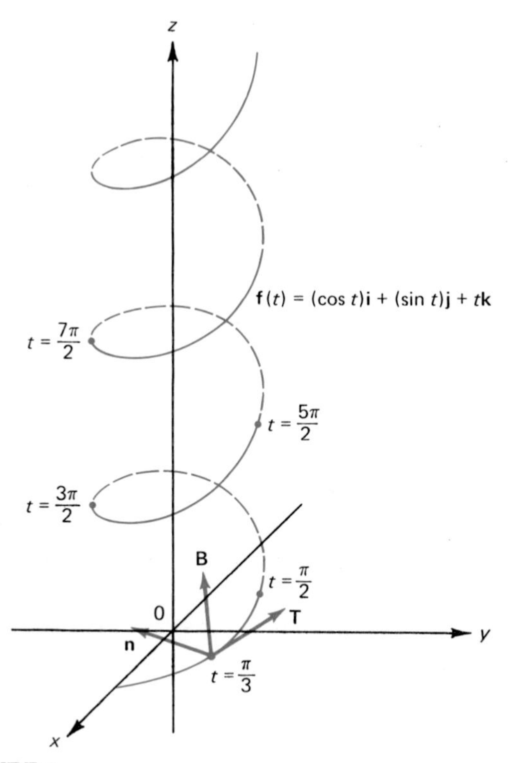

FIGURE 2

In this context $\mathbf{f}$ is again called the **position vector** of the object, and the curve described by $\mathbf{f}$ is called the **trajectory** of the point.

Definition 8 VELOCITY AND ACCELERATION VECTORS If $\mathbf{f}'$ and $\mathbf{f}''$ exist, then

$$\text{(i)}\quad \mathbf{v}(t) = \mathbf{f}'(t) = f_1'(t)\mathbf{i} + f_2'(t)\mathbf{j} + f_3'(t)\mathbf{k} \tag{11}$$

is called the **velocity vector.**

$$\text{(ii)}\quad \mathbf{a}(t) = \frac{d\mathbf{v}}{dt} = \mathbf{f}''(t) = f_1''(t)\mathbf{i} + f_2''(t)\mathbf{j} + f_3''(t)\mathbf{k} \tag{12}$$

is called the **acceleration vector.**

(iii) The **speed** of the object $v(t)$ is the magnitude of the velocity vector:

$$v(t) = |\mathbf{v}(t)|. \tag{13}$$

(iv) The **acceleration scalar** $a(t)$ is the magnitude of the acceleration vector:

$$a(t) = |\mathbf{a}(t)|. \tag{14}$$

EXAMPLE 7 Find the velocity, acceleration, speed, and acceleration scalar of an object whose position vector is given by $\mathbf{f}(t) = (\cos t)\mathbf{i} + (\sin t)\mathbf{j} + t^3\mathbf{k}$.

Solution.

$$\begin{aligned}\mathbf{v}(t) &= -(\sin t)\mathbf{i} + (\cos t)\mathbf{j} + 3t^2\mathbf{k}\\ \mathbf{a}(t) &= -(\cos t)\mathbf{i} - (\sin t)\mathbf{j} + 6t\mathbf{k}\\ v(t) &= \sqrt{\sin^2 t + \cos^2 t + (3t^2)^2} = \sqrt{1 + 9t^4}\\ a(t) &= \sqrt{\cos^2 t + \sin^2 t + (6t)^2} = \sqrt{1 + 36t^2}\end{aligned}$$

For example, at $t = 1$, $v(1) = \sqrt{10}$ and $a(1) = \sqrt{37}$. ■

As in the $\mathbb{R}^2$ case, the acceleration vector $\mathbf{a}(t)$ can be resolved into its tangential and normal components.

Theorem 5

$$\mathbf{a}(t) = \frac{d^2s}{dt^2}\mathbf{T} + \left(\frac{ds}{dt}\right)^2 \kappa\mathbf{n}.$$

Proof. $\mathbf{v}(t) = (d\mathbf{f}/dt) = (d\mathbf{f}/ds)(ds/dt) = \mathbf{T}(ds/dt)$ [from Theorem 3(v)]. Then

$$\begin{aligned}\mathbf{a}(t) &= \frac{d\mathbf{v}}{dt} = \mathbf{T}\frac{d^2s}{dt^2} + \frac{d\mathbf{T}}{dt}\frac{ds}{dt} = \mathbf{T}\frac{d^2s}{dt^2} + \left(\frac{d\mathbf{T}}{ds}\frac{ds}{dt}\right)\left(\frac{ds}{dt}\right)\\ &= \frac{d^2s}{dt^2}\mathbf{T} + \kappa\mathbf{n}\left(\frac{ds}{dt}\right)^2\end{aligned}$$

and the theorem is proved. ■

Using this result, we can prove the next theorem.

Theorem 6

$$\kappa = \frac{|\mathbf{f}' \times \mathbf{f}''|}{|\mathbf{f}'|^3}.$$

Proof. $\mathbf{f}' \times \mathbf{f}'' = (\mathbf{T}\, ds/dt) \times [(d^2s/dt^2)\mathbf{T} + \kappa\mathbf{n}(ds/dt)^2]$. Now $\mathbf{T} \times \mathbf{T} = \mathbf{0}$ and $\mathbf{T} \times \mathbf{n} = \mathbf{B}$. Thus using Theorem 3.4.2(iv), we have

$$\mathbf{f}' \times \mathbf{f}'' = \left(\frac{ds}{dt}\right)^3 \kappa\mathbf{B}.$$

Then taking absolute values and using the fact that $|ds/dt| = |\mathbf{f}'|$ yields

$$|\mathbf{f}' \times \mathbf{f}''| = \kappa|\mathbf{f}'|^3,$$

from which the result follows. ■

Theorem 6 is useful for calculating curvature when it is not easy to write $\mathbf{f}$ in terms of the arc length parameter s.

EXAMPLE 8 Calculate the curvature of $\mathbf{f} = t^2\mathbf{i} + t^3\mathbf{j} + t^4\mathbf{k}$ at $t = 1$.

Solution. Here

$$\mathbf{f}' = 2t\mathbf{i} + 3t^2\mathbf{j} + 4t^3\mathbf{k}$$

$$|\mathbf{f}'| = \sqrt{4t^2 + 9t^4 + 16t^6} = t\sqrt{4 + 9t^2 + 16t^4}$$

$$\mathbf{f}'' = 2\mathbf{i} + 6t\mathbf{j} + 12t^2\mathbf{k}$$

$$\mathbf{f}' \times \mathbf{f}'' = \begin{vmatrix} \mathbf{i} & \mathbf{j} & \mathbf{k} \\ 2t & 3t^2 & 4t^3 \\ 2 & 6t & 12t^2 \end{vmatrix} = 12t^4\mathbf{i} - 16t^3\mathbf{j} + 6t^2\mathbf{k}$$

$$|\mathbf{f}' \times \mathbf{f}''| = \sqrt{144t^8 + 256t^6 + 36t^4} = t^2\sqrt{144t^4 + 256t^2 + 36}.$$

Thus

$$\kappa = \frac{\sqrt{144t^4 + 256t^2 + 36}}{t(4 + 9t^2 + 16t^4)^{3/2}}.$$

At $t = 1$, $\kappa = \sqrt{436}/29^{3/2} \approx 0.1337$. ■

It is easy to extend the results of this section to a function whose range is in $\mathbb{R}^n$.

VECTOR-VALUED FUNCTION

Let $f_1, f_2, \ldots, f_n$ be functions of the real variable t. Then for all values of t for which $f_1(t), f_2(t), \ldots, f_n(t)$ are defined, the **vector-valued function f** is given by

$$\mathbf{f}(t) = (f_1(t), f_2(t), \ldots, f_n(t)). \qquad \textbf{(15)}$$

The functions $f_1, f_2, \ldots, f_n$ are called the **component functions** of **f**.

NOTATION. We will denote a vector-valued function by

$$\mathbf{f}\colon \mathbb{R} \to \mathbb{R}^n.$$

CURVE IN $\mathbb{R}^n$

Suppose that the interval $[a, b]$ is in the domain of each of the functions $f_1, f_2, \ldots, f_n$. If $\mathbf{f}: \mathbb{R} \to \mathbb{R}^n$ is given by (15), then the set of vectors

$$C = \{f_i(t): a \leq t \leq b\} \tag{16}$$

is called a **curve in $\mathbb{R}^n$.**

DIFFERENTIABILITY

Let $\mathbf{f}(t) = (f_1(t), f_2(t), \ldots, f_n(t))$ where each $f_i(t)$ is differentiable at t_0.

(i) Then $\mathbf{f}$ is **differentiable** at t_0 and

$$\mathbf{f}'(t_0) = (f_1'(t_0), f_2'(t_0), \ldots, f_n'(t_0)). \tag{17}$$

(ii) If f_i' is continuous at t_0 for $i = 1, 2, \ldots, n$, then f is said to be **continuously differentiable** at t_0.

EXAMPLE 9 If $\mathbf{f}(t) = (t, t^2, \ldots, t^n)$, compute $\mathbf{f}'(t)$.

Solution. $\mathbf{f}'(t) = (1, 2t, 3t^2, \ldots, nt^{n-1})$. ■

EXAMPLE 10 If $\mathbf{f}(t) = (c_1, c_2, \ldots, c_n)$, a constant vector, then $\mathbf{f}'(t) = (0, 0, \ldots, 0) = \mathbf{0}$. ■

EXAMPLE 11 If $\mathbf{f}(t) = f(t)\mathbf{x}$ where $\mathbf{x} = (x_1, x_2, \ldots, x_n)$ is a constant vector and f is a scalar function, then

$$\mathbf{f}'(t) = \frac{d}{dt}(f(t)x_1, f(t)x_2, \ldots, f(t)x_n) = (f'(t)x_1, f'(t)x_2, \ldots, f'(t)x_n) = f'(t)\mathbf{x}.$$

In particular, if $f(t) = t$, then

$$\frac{d}{dt}(t\mathbf{x}) = \mathbf{x}.$$ ■

The proof of the following theorem is virtually identical to the proof of Theorem 2.4.1 on page 47 and is omitted.

Theorem 7 Let $\mathbf{f}: \mathbb{R} \to \mathbb{R}^n$ and $\mathbf{g}: \mathbb{R} \to \mathbb{R}^n$ be differentiable on an open interval (a, b). Let the scalar function h be differentiable on (a, b). Let α be a scalar and let $\mathbf{v}$ be a constant vector in $\mathbb{R}^n$. Then

(i) $\mathbf{f} + \mathbf{g}$ is differentiable and

$$\frac{d}{dt}(\mathbf{f} + \mathbf{g}) = \frac{d\mathbf{f}}{dt} + \frac{d\mathbf{g}}{dt} = \mathbf{f}' + \mathbf{g}'. \tag{18}$$

(ii) $\alpha\mathbf{f}$ is differentiable and

$$\frac{d}{dt}\alpha\mathbf{f} = \alpha\frac{d\mathbf{f}}{dt} = \alpha\mathbf{f}'. \tag{19}$$

(iii) $\mathbf{v} \cdot \mathbf{f}$ is differentiable and

$$\frac{d}{dt}\mathbf{v} \cdot \mathbf{f} = \mathbf{v} \cdot \frac{d\mathbf{f}}{dt} = \mathbf{v} \cdot \mathbf{f}'. \tag{20}$$

(iv) $h\mathbf{f}$ is differentiable and

$$\frac{d}{dt}h\mathbf{f} = h\frac{d\mathbf{f}}{dt} + \frac{dh}{dt}\mathbf{f} = h\mathbf{f}' + h'\mathbf{f}. \tag{21}$$

(v) $\mathbf{f} \cdot \mathbf{g}$ is differentiable and

$$\frac{d}{dt}\mathbf{f} \cdot \mathbf{g} = \mathbf{f} \cdot \frac{d\mathbf{g}}{dt} + \frac{d\mathbf{f}}{dt} \cdot \mathbf{g} = \mathbf{f} \cdot \mathbf{g}' + \mathbf{f}' \cdot \mathbf{g}. \tag{22}$$

Theorem 8 Let $\mathbf{f}: \mathbb{R} \to \mathbb{R}^n$ be differentiable in $\mathbb{R}$ and suppose that $|\mathbf{f}(t)|$ is constant. Then $\mathbf{f}(t) \cdot \mathbf{f}'(t) = 0$ for every t in $\mathbb{R}$.

Proof. We have $\mathbf{f} \cdot \mathbf{f} = |\mathbf{f}|^2 = C$, a constant, so that

$$0 = \frac{d}{dt}(\mathbf{f} \cdot \mathbf{f}) = \underbrace{\mathbf{f} \cdot \mathbf{f}' + \mathbf{f}' \cdot \mathbf{f}}_{\text{from (22)}} = 2(\mathbf{f} \cdot \mathbf{f}'),$$

so that $\mathbf{f} \cdot \mathbf{f}' = 0$. ■

We can use Theorem 8 to generalize some interesting ideas in $\mathbb{R}^2$ and $\mathbb{R}^3$. Let $\mathbf{f}(t)$ be a differentiable curve in $\mathbb{R}^n$ for $t \in [a, b]$. Then at any point $t \in (a, b)$ with $\mathbf{f}'(t) \neq \mathbf{0}$, we define the **unit tangent vector** $\mathbf{T}(t)$ to the curve by

$$\mathbf{T}(t) = \frac{\mathbf{f}'(t)}{|\mathbf{f}'(t)|}. \tag{23}$$

Since $|\mathbf{T}(t)| = 1$, $|\mathbf{T}(t)|$ is constant and, by Theorem 8,

$$\mathbf{T}(t) \cdot \mathbf{T}'(t) = 0. \tag{24}$$

Finally, if $\mathbf{T}'(t) \neq \mathbf{0}$, we define the **unit normal vector** $\mathbf{n}(t)$ by

$$\mathbf{n}(t) = \frac{\mathbf{T}'(t)}{|\mathbf{T}'(t)|}. \tag{25}$$

From (24) it is clear that $\mathbf{T}(t) \cdot \mathbf{n}(t) = 0$.

Thus the notions of unit tangent and normal vectors can be extended to $\mathbb{R}^n$. The only real difference is that for $n > 3$, we cannot draw them.

PROBLEMS 3.8

In Problems 1–6, find the unit tangent vector $\mathbf{T}$ for the given value of t.

1. $\mathbf{f}(t) = t\mathbf{i} + t^2\mathbf{j} + t^3\mathbf{k}$; $t = 1$
2. $\mathbf{f}(t) = t^3\mathbf{i} + t^5\mathbf{j} + t^7\mathbf{k}$; $t = 1$
3. $\mathbf{f}(t) = t\mathbf{i} + e^t\mathbf{j} + e^{-t}\mathbf{k}$; $t = 0$
4. $\mathbf{f}(t) = t^2\mathbf{i} + t^2\mathbf{j} + t^{5/2}\mathbf{k}$; $t = 4$
5. $\mathbf{f}(t) = 4(\cos 2t)\mathbf{i} + 9(\sin 2t)\mathbf{j} + t\mathbf{k}$; $t = \pi/4$
6. $\mathbf{f}(t) = (\cosh t)\mathbf{i} + (\sinh t)\mathbf{j} + t^2\mathbf{k}$; $t = 0$

7. Find the arc length of the curve $\mathbf{f} = 2(\cos 3t)\mathbf{i} + 2(\sin 3t)\mathbf{j} + t^2\mathbf{k}$ between $t = 0$ and $t = 10$.
8. Find the arc length of the curve $\mathbf{f} = e^t(\cos 2t)\mathbf{i} + e^t(\sin 2t)\mathbf{j} + e^t\mathbf{k}$ between $t = 1$ and $t = 4$.
9. Find the arc length of the curve $\mathbf{f} = \frac{2}{3}t^3\mathbf{i} + (1 + t^{9/2})\mathbf{j} + (1 - t^{9/2})\mathbf{k}$ between $t = 0$ and $t = 2$.

In Problems 10–15, find the velocity vector, the speed, the acceleration vector, and the acceleration scalar for the given value of t.

10. $\mathbf{f} = t^2\mathbf{i} + t^3\mathbf{j} + t^4\mathbf{k}$; $t = 2$
11. $\mathbf{f} = (\cos t)\mathbf{i} + (\sin t)\mathbf{j} + t^4\mathbf{k}$; $t = 1$
12. $\mathbf{f} = (\ln t)\mathbf{i} + \dfrac{1}{t}\mathbf{j} + \dfrac{1}{t^2}\mathbf{k}$; $t = 1$
13. $\mathbf{f} = (\cosh t)\mathbf{i} + (\sinh t)\mathbf{j} + t\mathbf{k}$; $t = 0$
14. $\mathbf{f} = e^t\mathbf{i} + e^{-t}\mathbf{j} + \sqrt{t}\mathbf{k}$; $t = 4$
15. $\mathbf{f} = \sqrt{t}\mathbf{i} + \sqrt[3]{t}\mathbf{j} + \sqrt[4]{t}\mathbf{k}$; $t = 1$

In Problems 16–23, find the unit tangent vector $\mathbf{T}$, the unit normal vector $\mathbf{n}$, the binormal vector $\mathbf{B}$, and the curvature κ at the given value of t. Check that $\mathbf{n} \cdot \mathbf{T} = 0$.

16. $\mathbf{f} = a(\sin t)\mathbf{i} + a(\cos t)\mathbf{j} + t\mathbf{k}$; $t = \pi/4$; $a > 0$
17. $\mathbf{f} = a(\sin t)\mathbf{i} + a(\cos t)\mathbf{j} + t\mathbf{k}$; $t = \pi/6$; $a > 0$
18. $\mathbf{f} = a(\cos t)\mathbf{i} + b(\sin t)\mathbf{j} + t\mathbf{k}$; $t = 0$; $a > 0$, $b > 0$, $a \neq b$
19. $\mathbf{f} = a(\cos t)\mathbf{i} + b(\sin t)\mathbf{j} + t\mathbf{k}$; $t = \pi/2$; $a > 0$, $b > 0$, $a \neq b$
20. $\mathbf{f} = t\mathbf{i} + t^2\mathbf{j} + t^3\mathbf{k}$; $t = 0$
21. $\mathbf{f} = e^t(\cos 2t)\mathbf{i} + e^t(\sin 2t)\mathbf{j} + e^t\mathbf{k}$; $t = 0$
22. $\mathbf{f} = e^t(\cos 2t)\mathbf{i} + e^t(\sin 2t)\mathbf{j} + e^t\mathbf{k}$; $t = \pi/4$
23. $\mathbf{f} = \mathbf{i} + t\mathbf{j} + t^2\mathbf{k}$; $t = 1$

24. Let $\mathbf{f} = a(\cos t)\mathbf{i} + a(\sin t)\mathbf{j} + t\mathbf{k}$ (a circular helix). Show that the angle between the unit tangent vector $\mathbf{T}$ and the z-axis is constant.
25. Show, by writing out the component functions, that $(\mathbf{f} \times \mathbf{g})' = (\mathbf{f}' \times \mathbf{g}) + (\mathbf{f} \times \mathbf{g}')$.
26. Show that the curvature of a straight line in space is zero.

*27. Let $\mathbf{B}$ be the binormal vector to a curve $\mathbf{f}$, and let s denote arc length. Show that if $d\mathbf{B}/ds \neq \mathbf{0}$, then $d\mathbf{B}/ds \perp \mathbf{B}$ and $d\mathbf{B}/ds \perp \mathbf{T}$. [*Hint:* Differentiate $\mathbf{B} \cdot \mathbf{T} = 0$ and $\mathbf{B} \cdot \mathbf{B} = 1$.]

*28. Since, from Problem 27, $d\mathbf{B}/ds$ is orthogonal to $\mathbf{B}$ and $\mathbf{T}$, it must be parallel to $\mathbf{n}$. Hence

$$\frac{d\mathbf{B}}{ds} = -\tau\mathbf{n}$$

for some number τ. This number is called the **torsion** of the curve C and is a measure of how much the curve twists. Show that

$$\frac{d\mathbf{n}}{ds} = -\kappa\mathbf{T} + \tau\mathbf{B}.$$

**29. Show that

$$\mathbf{f}'''(t) = \left[\frac{d^3s}{dt^3} - \left(\frac{ds}{dt}\right)^3\kappa^2\right]\mathbf{T} + \left[3\left(\frac{d^2s}{dt^2}\right)\frac{ds}{dt}\kappa + \left(\frac{ds}{dt}\right)^2\frac{d\kappa}{dt}\right]\mathbf{n} + \left(\frac{ds}{dt}\right)^3\kappa\tau\mathbf{B}.$$

**30. Use the result of Problem 29 to show that

$$\tau = \frac{\mathbf{f}''' \cdot (\mathbf{f}' \times \mathbf{f}'')}{\kappa^2 (ds/dt)^6} = \frac{\mathbf{f}''' \cdot (\mathbf{f}' \times \mathbf{f}'')}{|\mathbf{f}' \times \mathbf{f}''|^2}.$$

*31. Calculate the torsion of $\mathbf{f} = (\cos t)\mathbf{i} + (\sin t)\mathbf{j} + t\mathbf{k}$ at $t = \pi/6$.

*32. Calculate the torsion of $\mathbf{f} = e^t(\cos t)\mathbf{i} + e^t(\sin t)\mathbf{j} + e^t\mathbf{k}$ at $t = 0$.

In Problems 33–43, compute the derivative of the given function.

33. $\mathbf{f}(t) = (t, \sin t, \cos t, t^3)$
34. $\mathbf{f}(t) = (\tan t, \ln t, e^t, t^5, \sqrt{t})$
35. $\mathbf{f}(t) = (e^t, e^{t^2}, \ldots, e^{t^n})$
36. $\mathbf{g}(t) = \left(\frac{1}{t}, \frac{1}{t^2}, \frac{1}{t^3}, \ldots, \frac{1}{t^n}\right)$
37. $\mathbf{f}(t) = (\ln t, \ln 2t, \ldots, \ln nt)$
38. $\mathbf{f}(t) = (1, 2, 3, 4, \ldots, n)$
39. $\mathbf{f}(t) = t^2\mathbf{x}$ where $\mathbf{x}$ is a constant vector
40. $h(t) = \mathbf{f}(t) + \mathbf{g}(t)$ where $\mathbf{f}$ and $\mathbf{g}$ are as in Problems 35 and 36
41. $\mathbf{h}(t) = 3\mathbf{f}(t)$ where $\mathbf{f}$ is as in Problem 35.
42. $h(t) = \mathbf{g}(t) \cdot (n, n-1, \ldots, 3, 2, 1)$ where $\mathbf{g}$ is as in Problem 36.
43. $h(t) = \mathbf{f}(t) \cdot \mathbf{g}(t)$ where $\mathbf{f}$ and $\mathbf{g}$ are as in Problems 35 and 36.

In Problems 44–47, compute the unit tangent vector and the unit normal vector to the given curve.

44. $\mathbf{f}(t) = (1, t, t^2, t^3)$
45. $\mathbf{f}(t) = (t, t^2, \ldots, t^n)$
46. $\mathbf{f}(t) = (\sin t, \cos t, \sin t, \cos t, \sin t, \cos t)$
47. $\mathbf{f}(t) = (e^t, e^{2t}, \ldots, e^{nt})$

We define the **arc length** $s(t)$ of the curve $\mathbf{f}(t)$ for $t \in [a, b]$ by

$$s(t) = \int_a^t |\mathbf{f}'(u)| du.$$

Note that, as in $\mathbb{R}^2$ and $\mathbb{R}^3$, $s(t)$ exists whenever f is continuously differentiable. In Problems 48–50 compute the arc length of the given curve.

48. $\mathbf{f}(t)$ is the curve in Problem 46 $t \in [0, 2\pi]$
49. $\mathbf{f}(t) = (1, t, 2t, 3t, \ldots, nt)$; $t \in [1, 5]$
50. $\mathbf{f}(t) = (t, t^2, t, t^2, t)$; $t \in [0, 1]$

3.9 CYLINDRICAL AND SPHERICAL COORDINATES

In this chapter, so far, we have represented points using rectangular (Cartesian) coordinates. However, there are many ways to represent points in space, some of which are quite bizarre. In this section we will briefly introduce two common ways to represent points in space. The first is the generalization of the polar coordinate system in the plane.

CYLINDRICAL COORDINATES

In the **cylindrical coordinate system** a point P is given by

$$P = (r, \theta, z), \quad \textbf{(1)}$$

where $r \geq 0$, $0 \leq \theta < 2\pi$, r and θ are polar coordinates of the projection of P onto the xy-plane, called the **polar plane,** and z is the distance (measured in the positive direction) of this plane from P (see Figure 1). In this figure $\overrightarrow{0Q}$ is the projection of $\overrightarrow{0P}$ on the xy-plane.

EXAMPLE 1 Discuss the graphs of the equations in cylindrical coordinates:
(a) $r = c, c > 0$ **(b)** $\theta = c$ **(c)** $z = c$
(d) $r_1 \leq r \leq r_2$, $\theta_1 \leq \theta \leq \theta_2$, $z_1 \leq z \leq z_2$

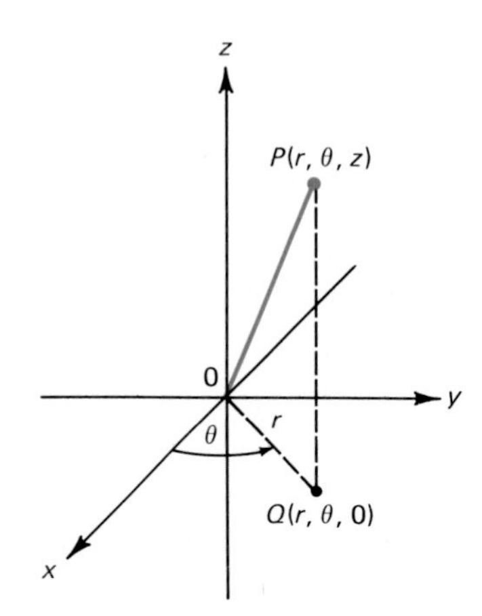

FIGURE 1

Solution. **(a)** If $r = c$, a constant, then θ and z can vary freely, and we obtain a right circular cylinder with radius c. (See Figure 2a.) This is the analog of the circle whose equation in polar coordinates is $r = c$ and is the reason the system is called the cylindrical coordinate system.

(b) If $\theta = c$, we obtain a half plane through the z-axis (see Figure 2b).

(c) If $z = c$, we obtain a plane parallel to the polar plane. This plane is sketched in Figure 2c.

(d) $r_1 \le r \le r_2$ gives the region between the cylinders $r = r_1$ and $r = r_2$. $\theta_1 \le \theta \le \theta_2$ is the wedge-shaped region between the half planes $\theta = \theta_1$ and $\theta = \theta_2$. Finally, $z_1 \le z \le z_2$ is the "slice" of space between the planes $z = z_1$ and $z = z_2$. Putting these results together, we get the rectangularly shaped solid in Figure 3. ■

To translate from cylindrical to rectangular coordinates and back again, we have the formulas

$$x = r\cos\theta, \qquad y = r\sin\theta, \qquad z = z \tag{2}$$

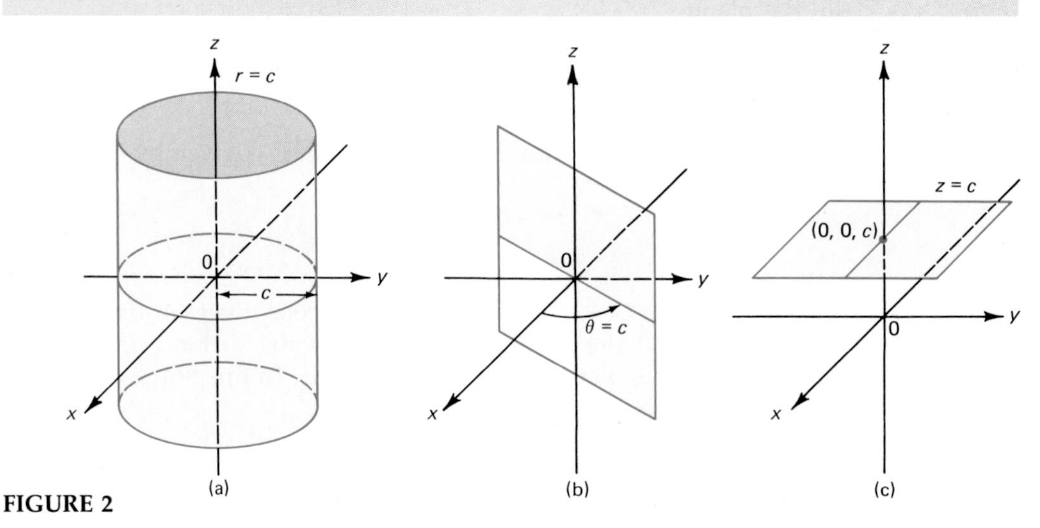

FIGURE 2

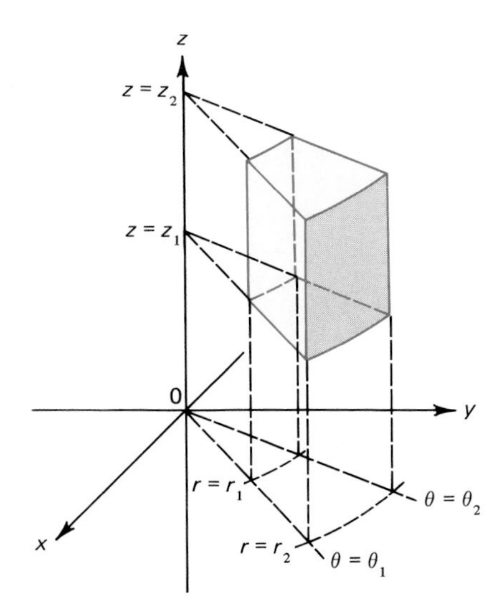

FIGURE 3

and

$$r = \sqrt{x^2 + y^2}, \qquad \tan\theta = \frac{y}{x}, \qquad z = z. \tag{3}$$

These need no proof as they follow from the formulas for polar coordinates.

EXAMPLE 2 Convert $P = (2, \pi/3, -5)$ from cylindrical to rectangular coordinates.

Solution. $x = r\cos\theta = 2\cos(\pi/3) = 1$, $y = r\sin\theta = 2\sin(\pi/3) = \sqrt{3}$, and $z = -5$. Thus in rectangular coordinates $P = (1, \sqrt{3}, -5)$. ■

EXAMPLE 3 Convert $(-3, 3, 7)$ from rectangular to cylindrical coordinates.

Solution. $r = \sqrt{3^2 + 3^2} = 3\sqrt{2}$, $\theta = \tan^{-1}(3/-3) = \tan^{-1}(-1) = 3\pi/4$ [since $(-3, 3)$ is in the second quadrant], and $z = 7$, so $(-3, 3, 7) = (3\sqrt{2}, 3\pi/4, 7)$ in cylindrical coordinates. ■

SPHERICAL COORDINATES

The second new coordinate system in space is the **spherical coordinate system.** This system is not a generalization of any system in the plane. A typical point P in space is represented as

$$P = (\rho, \theta, \varphi) \tag{4}$$

where $\rho \geq 0$, $0 \leq \theta < 2\pi$, $0 \leq \varphi \leq \pi$, and

ρ is the (positive) distance between the point and the origin.

θ is the same as in cylindrical coordinates.

φ is the angle between $\overrightarrow{0P}$ and the positive z-axis.

This system is illustrated in Figure 4.

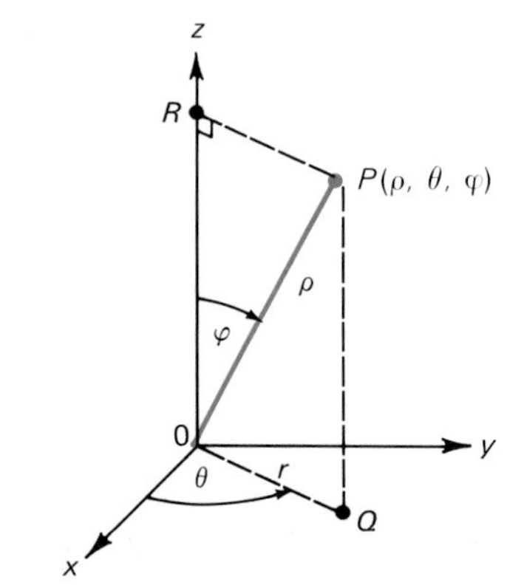

FIGURE 4

EXAMPLE 4 Discuss the graphs of the equations in spherical coordinates:
(a) $\rho = c, c > 0$ **(b)** $\theta = c$ **(c)** $\varphi = c$
(d) $\rho_1 \leq \rho \leq \rho_2,\ \theta_1 \leq \theta \leq \theta_2,\ \varphi_1 \leq \varphi \leq \varphi_2$

Solution. **(a)** The set of points of which $\rho = c$ is the set of points c units from the origin. This set constitutes a sphere centered at the origin with radius c and is what gives the coordinate system its name. This surface is sketched in Figure 5a.

(b) As with cylindrical coordinates, the graph is a half plane containing the z-axis. See Figure 5b.

(c) If $0 < c < \pi/2$, then the graph of $\varphi = c$ is obtained by rotating the vector $\overrightarrow{0P}$ around the z-axis. This yields the circular cone sketched in Figure 6a. If $\pi/2 < c < \pi$, then we obtain the cone of Figure 6b. Finally, if $c = 0$, we obtain the positive z-axis;

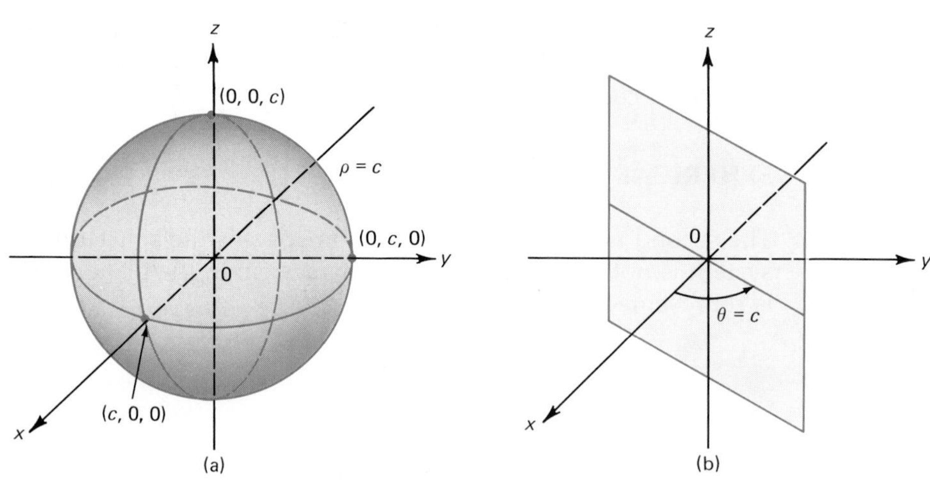

FIGURE 5

if $c = \pi$, we obtain the negative z-axis; and if $c = \pi/2$, we obtain the xy-plane.

(d) $\rho_1 \leq \rho \leq \rho_2$ gives the region between the sphere $\rho = \rho_1$ and the sphere $\rho = \rho_2$. $\theta_1 \leq \theta \leq \theta_2$ consists, as before, of the wedge-shaped region between the half planes $\theta = \theta_1$ and $\theta = \theta_2$. Finally, $\varphi_1 \leq \varphi \leq \varphi_2$ yields the region between the cones $\varphi = \varphi_1$ and $\varphi = \varphi_2$. Putting these results together, we obtain the solid (called a **spherical wedge**) sketched in Figure 7. ■

To convert from spherical to rectangular coordinates, we have, from Figure 4,

$$x = \overline{0Q} \cos \theta \qquad \text{and} \qquad y = \overline{0Q} \sin \theta.$$

But from Figure 4

$$\overline{0Q} = \overline{PR} = \rho \sin \varphi, \tag{5}$$

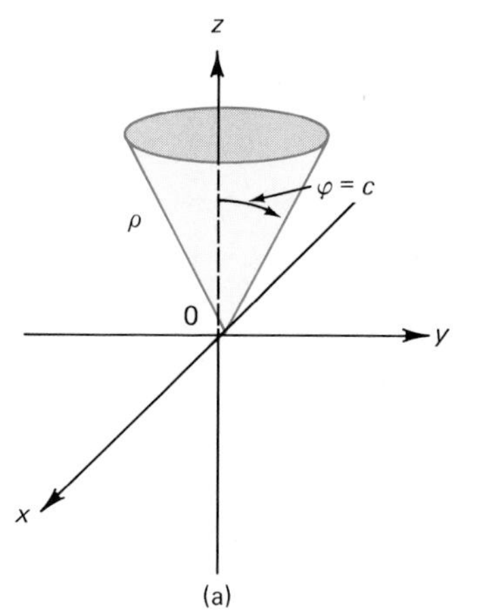

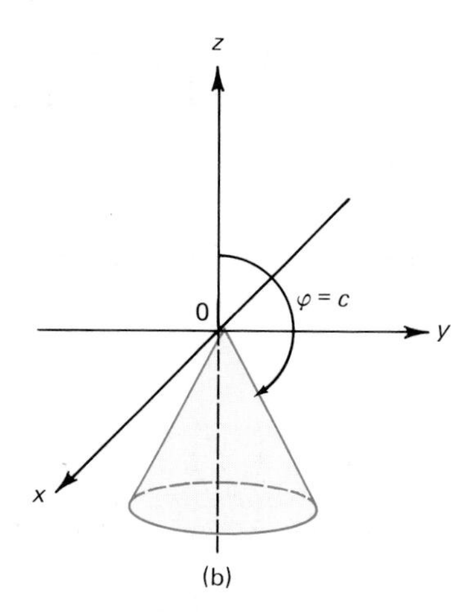

FIGURE 6

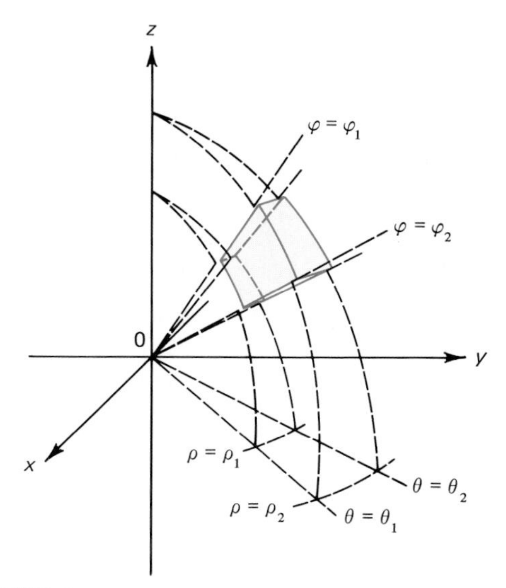

FIGURE 7

so that

$$x = \rho \sin \varphi \cos \theta \tag{6}$$

$$y = \rho \sin \varphi \sin \theta. \tag{7}$$

Finally, from triangle $0PR$ we have

$$z = \rho \cos \varphi. \tag{8}$$

To convert to spherical coordinates from rectangular coordinates, we have

$$\overline{0P} = \sqrt{x^2 + y^2 + z^2} = \rho, \tag{9}$$

so that from (8)

$$\cos \varphi = \frac{z}{\rho} = \frac{z}{\sqrt{x^2 + y^2 + z^2}}.$$

Knowing ρ and φ, we can then calculate θ from equation (6) and (7).

EXAMPLE 5 Convert $(4, \pi/6, \pi/4)$ from spherical to rectangular coordinates.

Solution.

$$x = \rho \sin \varphi \cos \theta = 4 \sin\frac{\pi}{4}\cos\frac{\pi}{6} = 4\frac{\sqrt{2}}{2}\frac{\sqrt{3}}{2} = \sqrt{6},$$

$$y = \rho \sin \varphi \sin \theta = 4 \sin\frac{\pi}{4}\sin\frac{\pi}{6} = 4\frac{\sqrt{2}}{2}\frac{1}{2} = \sqrt{2},$$

and

$$z = \rho \cos \varphi = 4 \cos\frac{\pi}{4} = 2\sqrt{2}.$$

Therefore, in rectangular coordinates $(4, \pi/6, \pi/4)$ is $(\sqrt{6}, \sqrt{2}, 2\sqrt{2})$. ■

EXAMPLE 6 Convert $(1, \sqrt{3}, -2)$ from rectangular to spherical coordinates.

Solution. $\rho = \sqrt{x^2 + y^2 + z^2} = \sqrt{1 + 3 + 4} = \sqrt{8} = 2\sqrt{2}$. Then from (8)

$$\cos \varphi = \frac{z}{\rho} = \frac{-2}{2\sqrt{2}} = -\frac{1}{\sqrt{2}},$$

so that $\varphi = 3\pi/4$. Finally, from (6)

$$\cos\theta = \frac{x}{\rho\sin\varphi} = \frac{1}{(2\sqrt{2})(1/\sqrt{2})} = \frac{1}{2};$$

so that $\theta = \pi/3$, and we find that $(1, \sqrt{3}, -2)$ is $(2\sqrt{2}, \pi/3, 3\pi/4)$ in spherical coordinates. ■

We will not discuss the graphs of different kinds of surfaces given in cylindrical and spherical coordinates as that would take us too far afield. Both coordinate systems are useful in a wide variety of physical applications. For example, points on the earth and its interior are much more easily described by using a spherical coordinate system than a rectangular system.

We will return to cylindrical and spherical coordinates in Chapter 5 in our discussion of multiple integration. For convenience, we reproduce below the formulas for changing from one coordinate system to another that we derived in this section.

CONVERSION FORMULAS

(i) From cylindrical to rectangular coordinates:

$$x = r\cos\theta$$
$$y = r\sin\theta$$
$$z = z.$$

(ii) From rectangular to cylindrical coordinates:

$$r = \sqrt{x^2 + y^2}$$
$$\tan\theta = \frac{y}{x}, \qquad 0 \le \theta < 2\pi$$
$$z = z.$$

(iii) From spherical to rectangular coordinates:

$$x = \rho\sin\varphi\cos\theta$$
$$y = \rho\sin\varphi\sin\theta$$
$$z = \rho\cos\varphi.$$

(iv) From rectangular to spherical coordinates:

$$\rho = \sqrt{x^2 + y^2 + z^2}$$
$$\varphi = \cos^{-1}\frac{z}{\rho}, \qquad 0 \le \varphi \le \pi$$
$$\cos\theta = \frac{x}{\rho\sin\varphi} \quad \text{or} \quad \sin\theta = \frac{y}{\rho\sin\varphi}, \qquad 0 \le \theta < 2\pi.$$

PROBLEMS 3.9

In Problems 1–9, convert from cylindrical to rectangular coordinates.

1. $(2, \pi/3, 5)$
2. $(1, 0, -3)$
3. $(8, 2\pi/3, 1)$
4. $(4, \pi/2, -7)$
5. $(3, 3\pi/4, 2)$
6. $(10, 5\pi/3, -3)$
7. $(10, \pi, -3)$
8. $(13, 3\pi/2, 4)$
9. $(7, 5\pi/4, 2)$

In Problems 10–18, convert from rectangular to cylindrical coordinates.

10. $(1, 0, 0)$
11. $(0, 1, 0)$
12. $(0, 0, 1)$
13. $(1, 1, 2)$
14. $(-1, 1, 4)$
15. $(2, 2\sqrt{3}, -5)$
16. $(2\sqrt{3}, -2, 8)$
17. $(2, -2\sqrt{3}, 4)$
18. $(-2\sqrt{3}, -2, 1)$

In Problems 19–27, convert from spherical to rectangular coordinates.

19. $(2, 0, \pi/3)$
20. $(4, \pi/4, \pi/4)$
21. $(6, \pi/2, \pi/3)$
22. $(3, \pi/6, 5\pi/6)$
23. $(7, 7\pi/4, 3\pi/4)$
24. $(3, \pi/2, \pi/2)$
25. $(4, \pi/3, 2\pi/3)$
26. $(4, 2\pi/3, \pi/3)$
27. $(5, 11\pi/6, 5\pi/6)$

In Problems 28–36, convert from rectangular to spherical coordinates.

28. $(1, 1, 0)$
29. $(1, 1, \sqrt{2})$
30. $(1, -1, \sqrt{2})$
31. $(-1, -1, \sqrt{2})$
32. $(1, -\sqrt{3}, 2)$
33. $(-\sqrt{3}, -1, 2)$
34. $(2, \sqrt{3}, 4)$
35. $(-2, \sqrt{3}, -4)$
36. $(-\sqrt{3}, -2, -4)$

37. Write the equation $x^2 + y^2 + z^2 = 25$ in cylindrical and spherical coordinates.
38. Write the plane $ax + by + cz = d$ in cylindrical and spherical coordinates.
39. Write the equation $r = 9 \sin \theta$ in rectangular coordinates.
40. Write the equation $r^2 \sin 2\theta = z^3$ in rectangular coordinates.
41. Write the surface $x^2 + y^2 - z^2 = 1$ in cylindrical and spherical coordinates.
42. Write the surface $x^2 + 4y^2 + 4z^2 = 1$ in cylindrical and spherical coordinates.
43. Write the equation $z = r^2$ in rectangular coordinates.
44. Write the equation $\rho \cos \varphi = 1$ in rectangular coordinates.
45. Write the equation $\rho^2 \sin \varphi \cos \varphi = 1$ in rectangular coordinates.
46. Write the equation $\rho = \sin \theta \cos \varphi$ in rectangular coordinates.
47. Write the equation $z = r^2 \sin 2\theta$ in rectangular coordinates.
48. Write the equation $\rho = 2 \cot \theta$ in rectangular coordinates.
49. Write the equation $x^2 + (y - 3)^2 + z^2 = 9$ in spherical coordinates.
50. Using the dot product, show that $\cos \varphi = z/\rho$.

REVIEW EXERCISES FOR CHAPTER THREE

In Exercises 1–4, sketch the two given points and then find the distance between them.

1. $(3, 1, 2); (-1, -3, -4)$
2. $(4, -1, 7); (-5, 1, 3)$
3. $(-2, 4, -8); (0, 0, 6)$
4. $(2, -7, 0); (0, 5, -8)$

5. Find the equation of the sphere centered at $(-1, 4, 2)$ with radius 3.
6. Show that $x^2 + y^2 + z^2 - 4x + 8y + 10z = 12$ is the equation of a sphere, and find its center and radius.
7. Show that the points $(1, 3, 0)$, $(3, -1, -2)$, and $(-1, 7, 2)$ are collinear.

In Exercises 8–11, find the magnitude and direction cosines of the given vector.

8. $\mathbf{v} = 2\mathbf{i} - \mathbf{k}$
9. $\mathbf{v} = 3\mathbf{j} + 11\mathbf{k}$
10. $\mathbf{v} = \mathbf{i} - 2\mathbf{j} - 3\mathbf{k}$
11. $\mathbf{v} = -4\mathbf{i} + \mathbf{j} + 6\mathbf{k}$

12. Find a unit vector in the direction of $\overrightarrow{PQ}$, where $P = (3, -1, 2)$ and $Q = (-4, 1, 7)$.
13. Find a unit vector whose direction is opposite to that of $\overrightarrow{PQ}$, where $P = (1, -3, 0)$ and $Q = (-7, 1, -4)$.

In Exercises 14–19, let $\mathbf{u} = \mathbf{i} - 2\mathbf{j} + 3\mathbf{k}$, $\mathbf{v} = -3\mathbf{i} + 2\mathbf{j} + 5\mathbf{k}$, and $\mathbf{w} = 2\mathbf{i} - 4\mathbf{j} + \mathbf{k}$. Calculate the following:

14. $3\mathbf{v} + 5\mathbf{w}$
15. $\text{Proj}_{\mathbf{v}}\, \mathbf{w}$
16. $\text{Proj}_{\mathbf{w}}\, \mathbf{u}$
17. $2\mathbf{u} - 4\mathbf{v} + 7\mathbf{w}$
18. $\mathbf{u} \cdot \mathbf{w} - \mathbf{w} \cdot \mathbf{v}$
19. The angle between $\mathbf{u}$ and $\mathbf{v}$

20. Compute the scalar product of $(1, 3, -1, 2, 4)$ and $(5, 0, 1, 2, -3)$.

21. Find a number α such that $\begin{pmatrix} 2 \\ -1 \\ 4 \\ 6 \end{pmatrix}$ and $\begin{pmatrix} 1 \\ 5 \\ \alpha \\ 4 \end{pmatrix}$ are orthogonal.

22. Find the distance from the point $P = (3, -1, 2)$ to the line passing through the points $Q = (-2, -1, 6)$ and $R = (0, 1, -8)$.

23. Find the work done when a force of 4 N acting in the direction of the vector $\mathbf{v} = -\mathbf{i} + \mathbf{j} + \mathbf{k}$ moves an object from $(2, 1, -6)$ to $(3, 5, 8)$ (distance in meters).

In Exercises 24–27, find a vector equation, parametric equations, and symmetric equations of the given line.

24. Containing $(3, -1, 4)$ and $(-1, 6, 2)$.

25. Containing $(-4, 1, 0)$ and $(3, 0, 7)$.

26. Containing $(3, 1, 2)$ and parallel to $3\mathbf{i} - \mathbf{j} - \mathbf{k}$.

27. Containing $(1, -2, -3)$ and parallel to the line $(x + 1)/5 = (y - 2)/-3 = (z - 4)/2$.

28. Show that the lines L_1: $x = 3 - 2t$, $y = 4 + t$, $z = -2 + 7t$ and L_2: $x = -3 + s$, $y = 2 - 4s$, $z = 1 + 6s$ have no points of intersection.

29. Find the distance from the origin to the line passing through the point $(3, 1, 5)$ and having the direction $\mathbf{v} = 2\mathbf{i} - \mathbf{j} + \mathbf{k}$.

In Exercises 30–33, find the cross product $\mathbf{u} \times \mathbf{v}$.

30. $\mathbf{u} = 3\mathbf{i} - \mathbf{j}$; $\mathbf{v} = 2\mathbf{i} + 4\mathbf{k}$

31. $\mathbf{u} = 7\mathbf{j}$; $\mathbf{v} = \mathbf{i} - \mathbf{k}$

32. $\mathbf{u} = 4\mathbf{i} - \mathbf{j} + 7\mathbf{k}$; $\mathbf{v} = -7\mathbf{i} + \mathbf{j} - 2\mathbf{k}$

33. $\mathbf{u} = -2\mathbf{i} + 3\mathbf{j} - 4\mathbf{k}$; $\mathbf{v} = -3\mathbf{i} + \mathbf{j} - 10\mathbf{k}$

34. Find two unit vectors orthogonal to both $\mathbf{u} = \mathbf{i} - \mathbf{j} + 3\mathbf{k}$ and $\mathbf{v} = -2\mathbf{i} - 3\mathbf{j} + 4\mathbf{k}$.

35. Find the equation of the line passing through $(-1, 2, 4)$ and orthogonal to L_1: $(x - 1)/4 = (y + 6)/3 = z/-2$ and L_2: $(x + 3)/5 = (y - 1)/1 = (z + 3)/4$.

36. Calculate the area of a parallelogram with the adjacent vertices $(1, 4, -2)$, $(-3, 1, 6)$, and $(1, -2, 3)$.

37. Calculate the area of a triangle with vertices at $(2, 1, 3)$, $(-4, 1, 7)$, and $(-1, -1, 3)$.

38. Calculate the volume of the parallelepiped determined by the vectors $\mathbf{i} + \mathbf{j}$, $2\mathbf{i} - 3\mathbf{k}$, and $2\mathbf{j} + 7\mathbf{k}$.

In Exercises 39–41, find the equation of the plane containing the given point and orthogonal to the given normal vector.

39. $P = (1, 3, -2)$; $\mathbf{N} = \mathbf{i} + \mathbf{k}$

40. $P = (1, -4, 6)$; $\mathbf{N} = 2\mathbf{j} - 3\mathbf{k}$

41. $P = (-4, 1, 6)$; $\mathbf{N} = 2\mathbf{i} - 3\mathbf{j} + 5\mathbf{k}$

42. Find the equation of the plane containing the points $(-2, 4, 1)$, $(3, -7, 5)$, and $(-1, -2, -1)$.

43. Find all points of intersection of the planes Π_1: $-x + y + z = 3$ and Π_2: $-4x + 2y - 7z = 5$.

44. Find all points of intersection of the planes Π_1: $-4x + 6y + 8z = 12$ and Π_2: $2x - 3y - 4z = 5$.

45. Find all points of intersection of the planes Π_1: $3x - y + 4z = 8$ and Π_2: $-3x - y - 11z = 0$.

46. Find the distance from $(1, -2, 3)$ to the plane $2x - y - z = 6$.

47. Find the angle between the planes of Exercise 43.

48. Show that the position vectors $\mathbf{u} = \mathbf{i} - 2\mathbf{j} + \mathbf{k}$, $\mathbf{v} = 3\mathbf{i} + 2\mathbf{j} - 3\mathbf{k}$, and $\mathbf{w} = 9\mathbf{i} - 2\mathbf{j} - 3\mathbf{k}$ are coplanar, and find the equation of the plane containing them.

In Exercises 49–51, draw a sketch of the given cylinder. The directrix C is given. The generatrix L is the axis of the variable missing in the equation.

49. $x = \cos y$

50. $y = z^2$

51. $z = \sqrt[3]{x}$

In Exercises 52–58, identify the quadric surface and sketch it.

52. $x^2 + y^2 = 9$

53. $x^2 - \dfrac{y^2}{4} + \dfrac{z^2}{9} = 1$

54. $x^2 - \dfrac{y^2}{4} - \dfrac{z^2}{9} = 1$

55. $-9x^2 + 16y^2 - 9z^2 = 25$

56. $4x^2 + y^2 + 4z^2 - 8y = 0$

57. $x = \dfrac{y^2}{4} - \dfrac{z^2}{9}$

58. $y^2 + z^2 = x^2$

In Exercises 59–61, find the unit tangent vector $\mathbf{T}$ for the given value for t.

59. $\mathbf{f}(t) = t^3\mathbf{i} - \sqrt{t}\mathbf{j} + 2t\mathbf{k}$; $t = 1$

60. $\mathbf{f}(t) = e^t\mathbf{i} + e^{-t}\mathbf{j} + t^3\mathbf{k}$; $t = 0$

61. $\mathbf{f}(t) = (\cos 3t)\mathbf{i} - (\sin 3t)\mathbf{j} + t\mathbf{k}$; $t = \pi/9$

62. Find the arc length of the curve $\mathbf{f} = 3(\sin 2t)\mathbf{i} + 3(\cos 2t)\mathbf{j} + t^2\mathbf{k}$ between $t = 0$ and $t = 5$.

63. The position vector of a moving particle is given by $\mathbf{f} = t^3\mathbf{i} + (\cos t)\mathbf{j} - (\sin t)\mathbf{k}$. Find the velocity, the speed, the acceleration vector, and the acceleration scalar when $t = \pi/4$.

64. Answer the questions in Exercise 63 for $\mathbf{f} = e^t\mathbf{i} + (\ln t)\mathbf{j} - t^{3/2}\mathbf{k}$ at $t = 1$.

In Exercises 65–67, find the unit tangent vector $\mathbf{T}$, the unit normal vector $\mathbf{n}$, the binormal vector $\mathbf{B}$, and the curvature κ for the given value of t.

65. $\mathbf{f} = 2(\cos t)\mathbf{i} + 2(\sin t)\mathbf{j} + t\mathbf{k}$; $t = \pi/6$
66. $\mathbf{f} = t^2\mathbf{i} + t^3\mathbf{j} - t\mathbf{k}$; $t = 0$
67. $\mathbf{f} = e^t\mathbf{i} + e^t(\cos t)\mathbf{j} + e^t(\sin t)\mathbf{k}$; $t = 0$

68. Find the torsion of the function in Exercise 66 at $t = 0$ (see Problem 28 on page 142).
69. Compute the derivative of $\mathbf{f}(t) = (t^3, \cos t, \ln t, e^{2t})$.
70. Compute the derivative of $\mathbf{f}(t) = (\sqrt{t}, t^{5/3}, (1/t), t^5 - 2t + 2)$.
71. Compute the unit tangent vector and the unit normal vector to the curve $\mathbf{f}(t) = (t^2, t^4, t^6, t^8)$ at $t = 1$.
72. Compute the length of the curve $\mathbf{f}(t) = (t^2, t^2, t, t)$ for $t \in [0, 1]$.

In Exercises 73–80, convert from one coordinate system to another as indicated.

73. $(3, \pi/6, -1)$, cylindrical to rectangular
74. $(2, 2, -4)$, rectangular to cylindrical
75. $(2, 2\pi/3, 4)$, cylindrical to rectangular
76. $(-2, 2\sqrt{3}, -4)$ rectangular to cylindrical
77. $(3, \pi/3, \pi/4)$, spherical to rectangular
78. $(2, 7\pi/3, 2\pi/3)$, spherical to rectangular
79. $(-1, 1, -\sqrt{2})$, rectangular to spherical
80. $(2, -\sqrt{3}, 4)$, rectangular to spherical

81. Write the equation $x^2 + y^2 + z^2 = 25$ in cylindrical and spherical coordinates.
82. Write the equation $r^2 \cos 2\theta = z^3$ in rectangular coordinates.
83. Write the equation $x^2 - y^2 + z^2 = 1$ in cylindrical and spherical coordinates.
84. Write the equation $z = r^3$ in rectangular coordinates.
85. Write the equation $\rho \sin \varphi = 1$ in rectangular coordinates.

4 Differentiation of Functions of Two or More Variables

For most of the functions we have so far encountered in this book, we have been able to write $y = f(x)$. This means that we could write the variable y explicitly in terms of the single variable x. However, in a great variety of applications it is necessary to write the quantity of interest in terms of two or more variables. We have already encountered this situation. For example, the volume of a right circular cylinder is given by

$$V = \pi r^2 h,$$

where r is the radius of the cylinder and h is its height. That is, V is a function of the *two* variables r and h.

As a second example, the **ideal gas law,** which relates pressure, volume, and temperature for an ideal gas is given by

$$PV = nRT,$$

where P is the pressure of the gas, V is the volume, T is the absolute temperature (i.e., in degrees kelvin), n is the number of moles of the gas, and R is a constant. Solving for P, we find that

$$P = \frac{nRT}{V}.$$

That is, we can write P as a function of the *three* variables n, T, and V.

As a third example, according to **Poiseuille's law,** the resistance R of a blood vessel of length l and radius r is given by

$$R = \frac{\alpha l}{r^4}$$

where α is a constant of proportionality. If l and r are allowed to vary, then R is written as a function of the *two* variables l and r.

As a final example, let the vector $\mathbf{v}$ be given by

$$\mathbf{v} = x\mathbf{i} + y\mathbf{j} + z\mathbf{k}.$$

Then the magnitude of $\mathbf{v}$, $|\mathbf{v}|$, is given by

$$|\mathbf{v}| = \sqrt{x^2 + y^2 + z^2}.$$

That is, the magnitude of $\mathbf{v}$ is written as a function of the *three* variables, x, y, and z.

There are many examples like the four cited above. It is probably fair to say that very few physical, biological, or economic quantities can be properly expressed in terms of one variable alone. Often we write these quantities in terms of one variable simply because functions of only one variable are the easiest functions to handle.

In this chapter we will see how many of the operations we have studied in our discussion of the "one-variable" calculus can be extended to functions of two or more variables. We will begin by discussing the basic notions of limits and continuity and will go on to differentiation and applications of differentiation. In Chapter 5 we will discuss the integration of functions of two and three variables.

4.1 FUNCTIONS OF TWO OR MORE VARIABLES

In Chapters 2 and 3 we discussed the notion of vector-valued functions. For example, if $\mathbf{f}(t) = f_1(t)\mathbf{i} + f_2(t)\mathbf{j}$, then for every t in the domain of $\mathbf{f}$, we obtain a vector $\mathbf{f}(t)$. In defining functions of two or more variables, we obtain a somewhat reversed situation. For example, in the formula for the volume of a right circular cylinder, we have

$$V = \pi r^2 h. \tag{1}$$

Here the volume V is written as a function of the two variables r and h. Put another way, for every ordered pair of positive real numbers (r, h), there is a unique positive number V such that $V = \pi r^2 h$. To indicate the dependence of V on the variables r and h, we write $V(r, h)$. That is, V can be thought of as a function that assigns a positive real number to every ordered pair of real numbers. Thus we see what we meant when we said that the situation is "reversed." Instead of having a vector function of one variable (a scalar), we have a scalar function of an ordered pair of variables (which is a vector in $\mathbb{R}^2$). We now give a definition of a function of two variables.

Definition 1 FUNCTION OF TWO VARIABLES Let D be a subset of $\mathbb{R}^2$. Then a **real-valued function of two variables** f is a rule that assigns to each point (x, y) in D a unique real number that we denote $f(x, y)$. The set D is called the **domain** of f. The set $\{f(x, y): (x, y) \in D\}$, which is the set of values the function f takes on, is called the **range** of f.

EXAMPLE 1 Let $D = \mathbb{R}^2$. For each point $(x, y) \in \mathbb{R}^2$, we assign the number $f(x, y) = x^2 + y^4$. Since $x^2 + y^4 \geq 0$ for every pair of real numbers (x, y), and since every positive number is the square of some positive number, we see that the range of f is $\mathbb{R}^+$, the set of nonnegative real numbers. ■

When the domain D is not given, we will take the domain of f to be the largest subset of $\mathbb{R}^2$ for which the expression $f(x, y)$ makes sense.

EXAMPLE 2 Find the domain and range of the function f given by $f(x, y) = \sqrt{4 - x^2 - y^2}$, and find $f(0, 1)$ and $f(-1, 1)$.

Solution. Clearly, f is defined when the expression under the square root sign is nonnegative. Thus $D = \{(x, y): x^2 + y^2 \leq 4\}$. This is the disk[†] centered at the origin with radius 2. It is sketched in Figure 1. Since x^2 and y^2 are nonnegative, $4 - x^2 - y^2$ is largest when $x = y = 0$. Thus the largest value of $\sqrt{4 - x^2 - y^2} = \sqrt{4} = 2$. Since $x^2 + y^2 \leq 4$, the smallest value of $4 - x^2 - y^2$ is 0, taken when $x^2 + y^2 = 4$ (at all points on the circle $x^2 + y^2 = 4$). Thus the range of f is the closed interval $[0, 2]$. Finally,

$$f(0, 1) = \sqrt{4 - 0^2 - 1^2} = \sqrt{3} \quad \text{and} \quad f(-1, 1) = \sqrt{4 - (-1)^2 - 1^2} = \sqrt{2}.$$

■

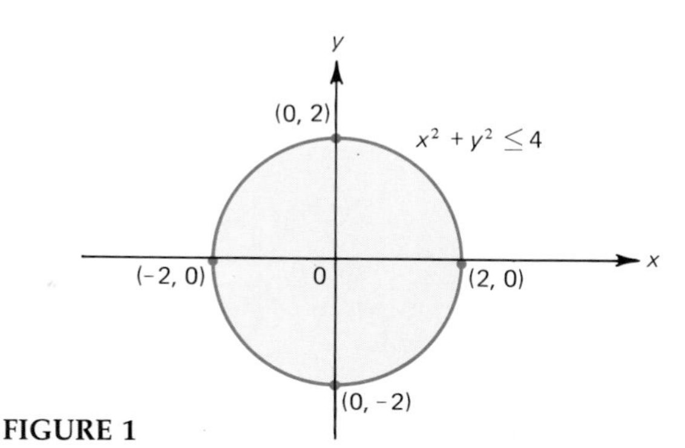

FIGURE 1

REMARK. We emphasize that the domain of f is a subset of $\mathbb{R}^2$, while the range is a subset of $\mathbb{R}$, the real numbers.

CAUTION. In Figure 1 we sketched the *domain* of the function. We did *not* sketch its graph. The graph of a function of two or more variables is more complicated and will be discussed shortly.

In one-variable calculus we write $y = f(x)$. That is, we use the letter y to denote the value of a function of one variable. Here we can use the letter z (or any other letter

[†]The *disk* of radius r is the circle of radius r together with all points interior to that circle.

for that matter) to denote the value taken by f, which is now a function of two variables. We then write

$$z = f(x, y). \tag{2}$$

EXAMPLE 3 Find the domain and range of the function

$$z = f(x, y) = \frac{\sqrt{x^2 + y^2 - 9}}{x - y}.$$

Solution. $\sqrt{x^2 + y^2 - 9}$ is defined only when $x^2 + y^2 \geq 9$. This is the circle $x^2 + y^2 = 9$ with radius 3 and its exterior. However, $f(x, y)$ is also not defined when $x - y = 0$ or when $y = x$. Thus all points on the line $y = x$ are excluded from the domain of f. We have the domain of $f = \{(x, y): x^2 + y^2 \geq 9 \text{ and } x \neq y\}$. This domain is sketched in Figure 2. The points on the dotted line are excluded.

We now calculate the range of f. We first note that $\sqrt{x^2 + y^2 - 9}$ can take any nonnegative value. Since $x - y$ can be positive or negative, we see that $\sqrt{x^2 + y^2 - 9}/(x - y)$ can take on any real value, so the range of f is $\mathbb{R}$. Note that $f(x, y) = 0$ at any point on the circle except at the points $(3/\sqrt{2}, 3/\sqrt{2})$ and $(-3/\sqrt{2}, -3/\sqrt{2})$. The function is not defined at these last two points. ■

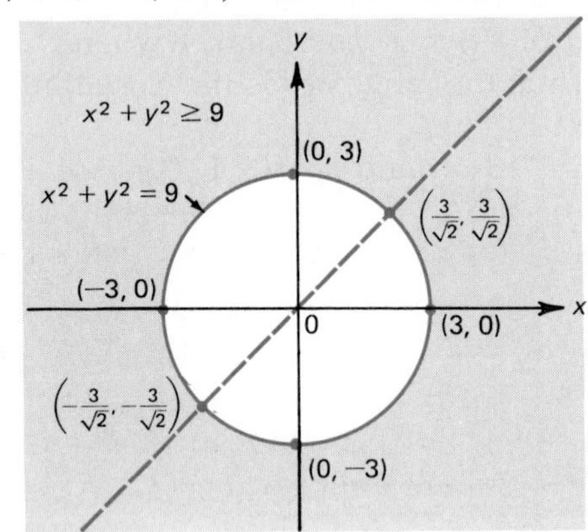

FIGURE 2

EXAMPLE 4 Let $z = f(x, y) = 3/\sqrt{x^2 - y}$. Find the domain and range of f.

Solution. $f(x, y)$ is defined if $x^2 - y > 0$ or $y < x^2$. This is the region in the plane "outside" the parabola $y = x^2$ and is sketched in Figure 3. Note that the parabola $y = x^2$ is *not* in the domain of f. The range of f is the set of positive real numbers since when $x^2 - y > 0$, $3/\sqrt{x^2 - y}$ can take on any positive real number. ■

EXAMPLE 5 Find the domain and range of $f(x, y) = \tan^{-1}(y/x)$.

Solution. $\tan^{-1}(y/x)$ is defined so long as $x \neq 0$. Hence the domain of $f = \{(x, y): x \neq 0\}$. From the definition of $\tan^{-1} x$, we find that the range of f is the open interval $(-\pi/2, \pi/2)$.† ■

---

†Recall that in order to define $\tan^{-1}x$, we must restrict $\tan x$ to an interval over which it is one-to-one. The usual such interval is $\left(-\frac{\pi}{2}, \frac{\pi}{2}\right)$.

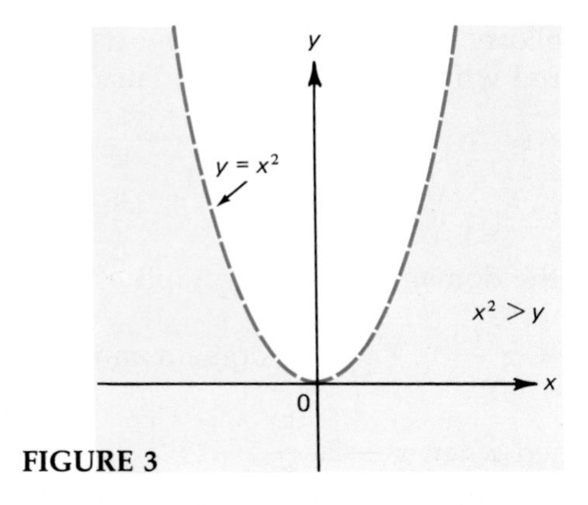

FIGURE 3

We now turn to the definition of a function of three variables.

Definition 2 FUNCTION OF THREE VARIABLES Let D be a subset of $\mathbb{R}^3$. Then a **real-valued function of three variables** f is a rule that assigns to each point (x, y, z) in D a unique real number that we denote $f(x, y, z)$. The set D is called the **domain** of f, and the set $\{f(x, y, z): (x, y, z) \in D\}$, which is the set of values the function f takes on, is called the **range** of f.

NOTE. We will often use the letter w to denote the values that a function of three variables takes. We then have

$$w = f(x, y, z). \tag{3}$$

EXAMPLE 6 Let $w = f(x, y, z) = \sqrt{1 - x^2 - (y^2/4) - (z^2/9)}$. Find the domain and range of f. Calculate $f(0, 1, 1)$.

Solution. $f(x, y, z)$ is defined if $1 - x^2 - (y^2/4) - (z^2/9) \geq 0$, which occurs if $x^2 + (y^2/4) + (z^2/9) \leq 1$. From Section 3.6 we see that the equation $x^2 + (y^2/4) + (z^2/9) = 1$ is the equation of an ellipsoid. Thus the domain of f is the set of points (x, y, z) in $\mathbb{R}^3$ that are on and interior to this ellipsoid (sketched in Figure 4). The range of f is

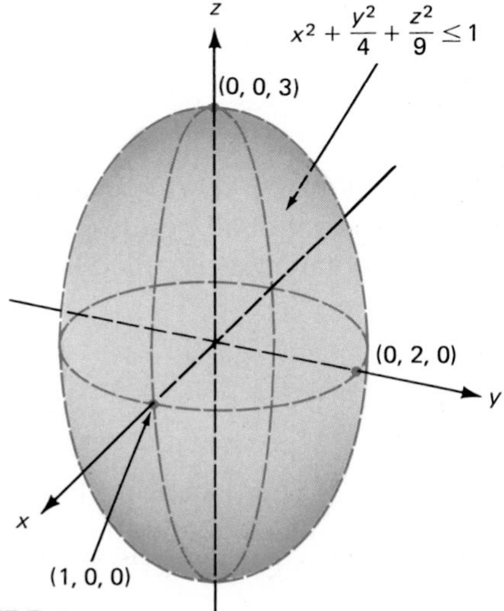

FIGURE 4

the closed interval [0, 1]. This follows because $x^2 + (y^2/4) + (z^2/9) \geq 0$, so that $1 - x^2 - (y^2/4) - (z^2/9)$ is maximized when $x = y = z = 0$. Finally,

$$f(0, 1, 1) = \sqrt{1 - 0^2 - \frac{1^2}{4} - \frac{1^2}{9}} = \sqrt{1 - \frac{1}{4} - \frac{1}{9}} = \sqrt{1 - \frac{13}{36}} = \frac{\sqrt{23}}{6}.$$

NOTE. Figure 4 is a sketch of the domain of f, not a graph of f itself. ■

EXAMPLE 7 Let $w = f(x, y, z) = \ln(x + 2y + z - 3)$. Find the domain and range of f. Calculate $f(1, 2, 3)$.

Solution. $f(x, y, z)$ is defined when $x + 2y + z - 3 > 0$, which occurs when $x + 2y + z > 3$. The equation $x + 2y + z = 3$ is the equation of a plane. Thus the domain of f is the *half space* above (but not including) this plane. The plane is sketched in Figure 5. The range of f is $\mathbb{R}$. Finally, $f(1, 2, 3) = \ln(1 + 4 + 3 - 3) = \ln 5 \approx 1.61$. ■

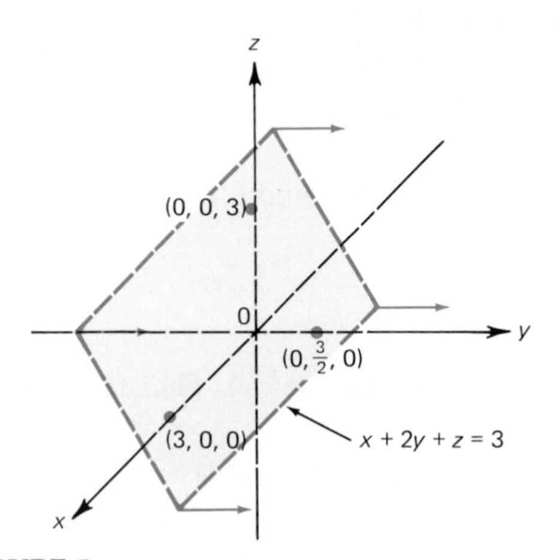

FIGURE 5

We now turn to a discussion of the graph of a function. Recall that the graph of a function of one variable f is the set of all points (x, y) in the plane such that $y = f(x)$. Using this definition as our model, we have the following definition.

Definition 3 GRAPH OF A FUNCTION OF TWO VARIABLES The **graph** of a function f of two variables x and y is the set of all points (x, y, z) in $\mathbb{R}^3$ such that $z = f(x, y)$. The graph of a function of two variables is called a **surface** in $\mathbb{R}^3$.

EXAMPLE 8 Sketch the graph of the function

$$z = f(x, y) = \sqrt{1 - x^2 - y^2}. \tag{4}$$

Solution. We first note that $z \geq 0$. Then squaring both sides of (4), we have $z^2 = 1 - x^2 - y^2$, or $x^2 + y^2 + z^2 = 1$. This equation is, of course, the equation of the unit sphere. However, since $z \geq 0$, the graph of f is the hemisphere sketched in Figure 6. ■

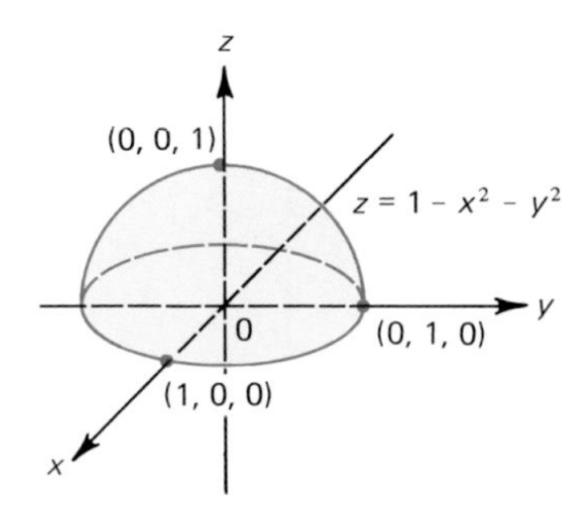

FIGURE 6

It is often very difficult to sketch the graph of a function $z = f(x, y)$ since, except for the quadric surfaces we discussed in Section 3.6, we really do not have a vast "catalog" of surfaces to which to refer. Moreover, the techniques of curve plotting in three dimensions are tedious, to say the least, and plotting points in space will, except in the most trivial of cases, not get us very far. The best we can usually do is to describe **cross sections** of the surface that lie in planes parallel to the coordinate planes. This will give an idea of what the surface looks like. Then if we have access to a computer with graphing capabilities, we can obtain a computer-drawn sketch of the surface. We illustrate this process with two examples.

EXAMPLE 9 Obtain a sketch of the surface

$$z = f(x, y) = x^3 + 3x^2 - y^2 - 9x + 2y - 10. \tag{5}$$

Solution. The xz-plane has the equation $y = 0$. Planes parallel to the xz-plane have the equation $y = c$, where c is a constant. Setting $y = c$ in (5), we have

$$\begin{aligned} z &= x^3 + 3x^2 - 9x - 10 + (-c^2 + 2c) \\ &= x^3 + 3x^2 - 9x - 10 + k, \end{aligned}$$

where $k = -c^2 + 2c$ is a constant. We can obtain the graph of $z = x^3 + 3x^2 - 9x - 10 + k$ by shifting the graph of $z = x^3 + 3x^2 - 9x - 10$ up or down k units. For several values of k, cross sections lying in planes parallel to the xz-plane are given in Figure 7a.

REMARK. The maximum value of the function $g(c) = -c^2 + 2c$ is 1, taken when $c = 1$ (you should verify this). Thus 1 is the largest value k can assume in the parallel cross sections, so $-10 + k \leq -9$, and the "highest" cross section is $z = x^3 + 3x^2 - 9x - 9$.

The yz-plane has the equation $x = 0$. Planes parallel to the yz-plane have the equation $x = c$, where c is a constant. Setting $x = c$ in (5), we obtain

$$z = -y^2 + 2y + (c^3 + 3c^2 - 9c - 10) = -y^2 + 2y + k.$$

For several values of k, cross sections parallel to the yz-plane are drawn in Figure 7b. Note that this k can take on any real value. In Figure 8 we provide a computer-drawn sketch of the surface. ■

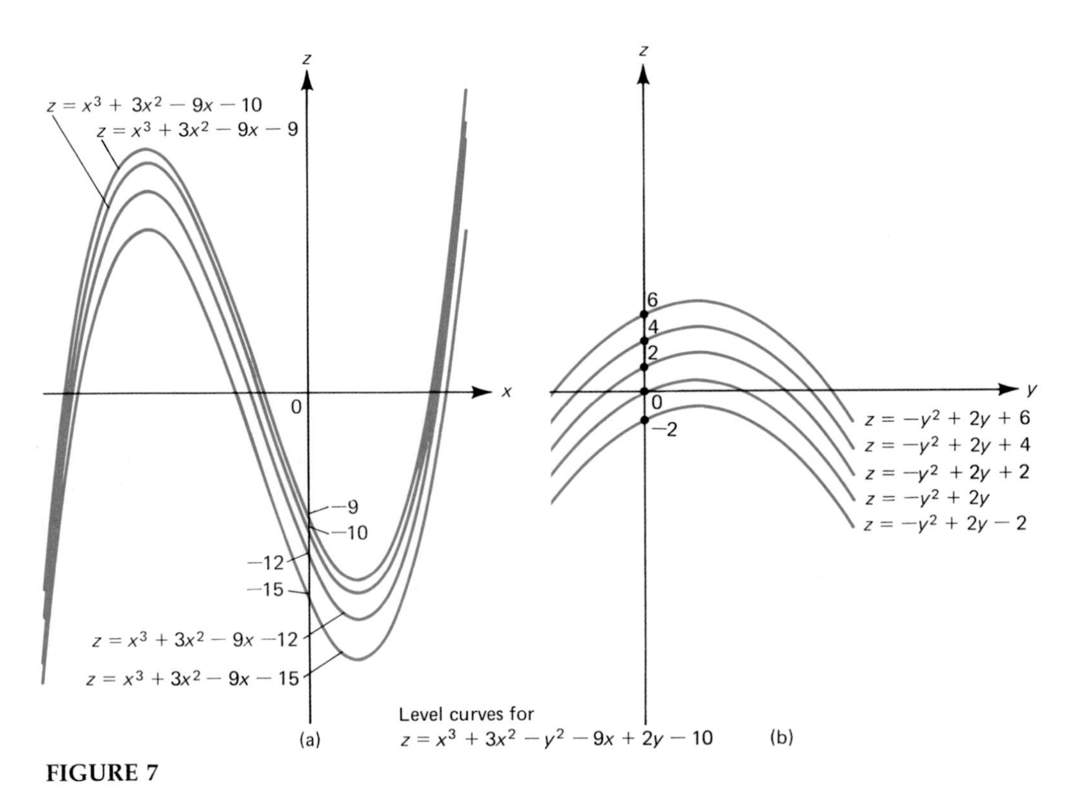

FIGURE 7

EXAMPLE 10 Obtain a sketch of the surface $z = \sin x + e^y$.

Solution. Cross sections lying in planes parallel to the xz-plane have the equation $z = \sin x + e^c = \sin x + k$, where $k > 0$. Some of the cross sections are sketched in Figure 9a. Cross sections lying in planes parallel to the yz-plane have the equation $z = \sin c + e^y = e^y + k$, where $-1 \leq k \leq 1$. Some of these cross sections are sketched in Figure 9b. In Figure 10 we provide a computer-drawn sketch of the surface. ■

REMARK. The curves $y = \sin x$ and $y = e^x$ are easy for us to sketch. Nevertheless, if we combine them as in the last example, the surface is difficult to sketch. For less "obvious" cross sections, graphs can be exceedingly difficult to obtain.

As we have stated, except for the standard quadric surfaces described in Section 3.6, it is extremely difficult to sketch three-dimensional graphs by hand. However, virtually every modern computer (and microcomputer) has graphing capabilities. In Figure 11 we provide microcomputer-drawn sketches of the graphs of four different functions of two variables.

The situation becomes much more complicated when we try to sketch the graph of a function of three variables $w = f(x, y, z)$. We would need *four dimensions* to sketch such a surface. Being human, we are limited to three dimensions, and we see that we have reached the point where our comfortable three-dimensional geometry fails us. We are *not* saying that curves and surfaces in four-dimensional space do not exist; they do—only that we are not able to sketch them.

We have shown that it is usually very difficult to sketch the graph of a function of two variables. Fortunately, there is a way to describe the graph of such a function

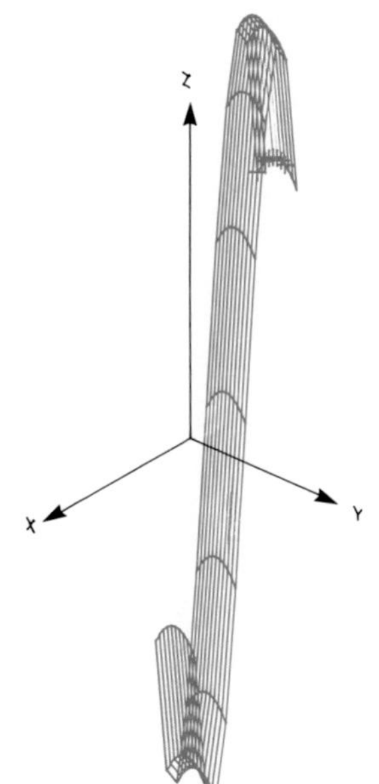

FIGURE 8

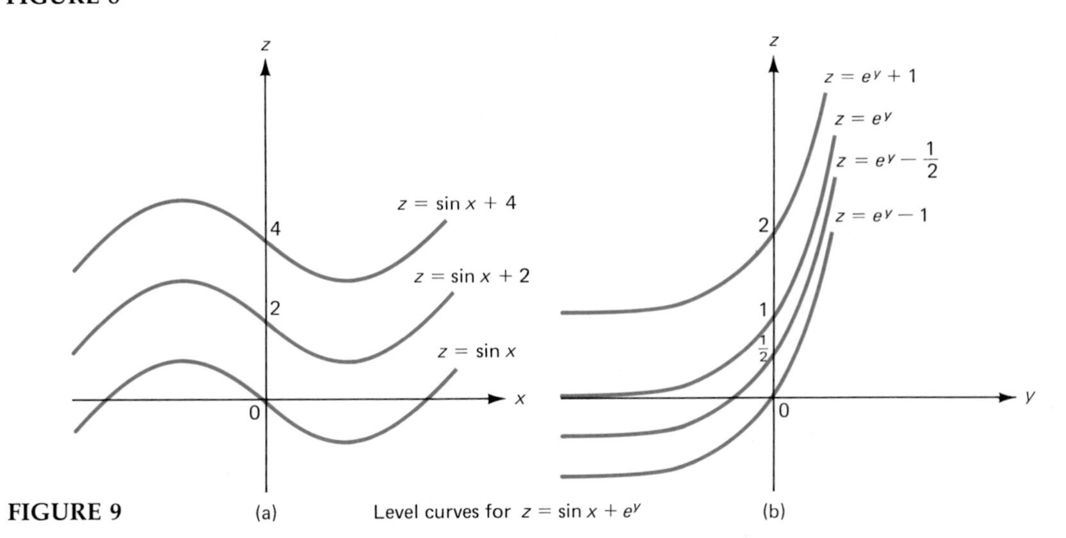

FIGURE 9 Level curves for $z = \sin x + e^y$

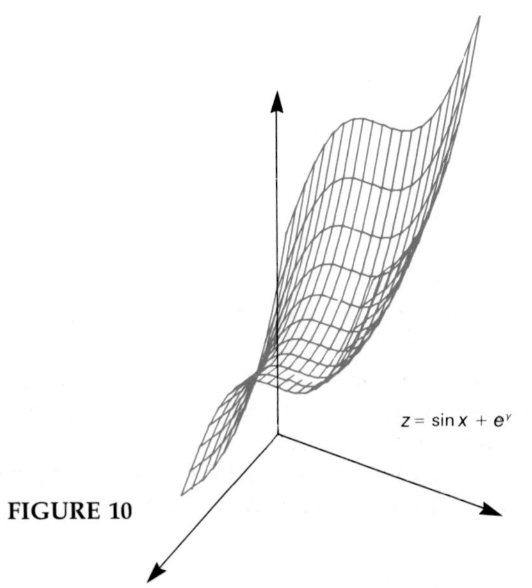

FIGURE 10

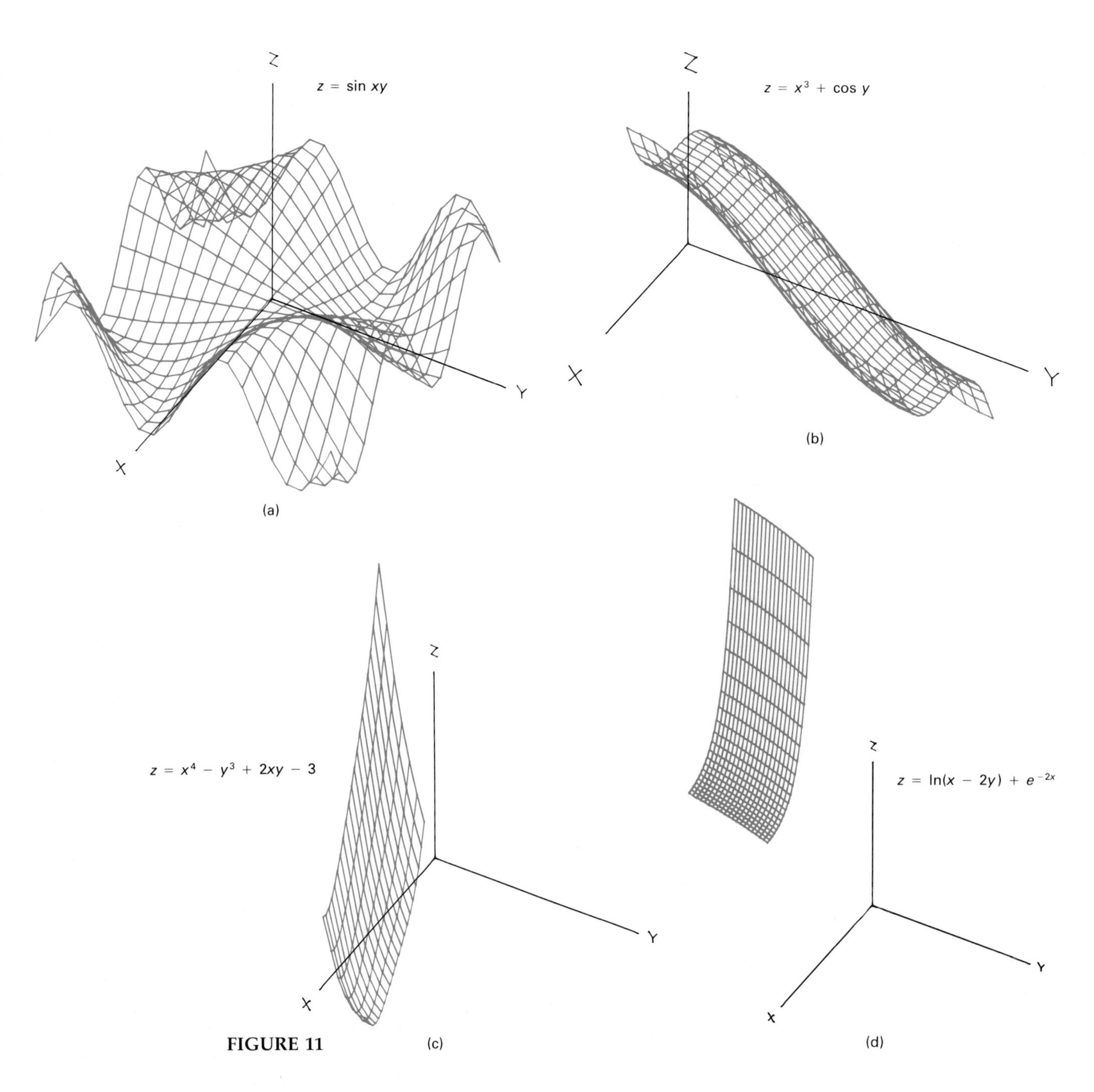

FIGURE 11

in two dimensions. The idea for what we are about to do comes from cartographers (mapmakers). Cartographers have the problem of indicating three-dimensional features (such as mountains and valleys) on a two-dimensional surface. They solve the problem by drawing a **contour map (topographic map),** which is a map in which points of constant elevation are joined to form curves, called **contour curves.** The closer together these contour curves are drawn, the steeper is the terrain. A portion of a typical contour map is sketched in Figure 12.

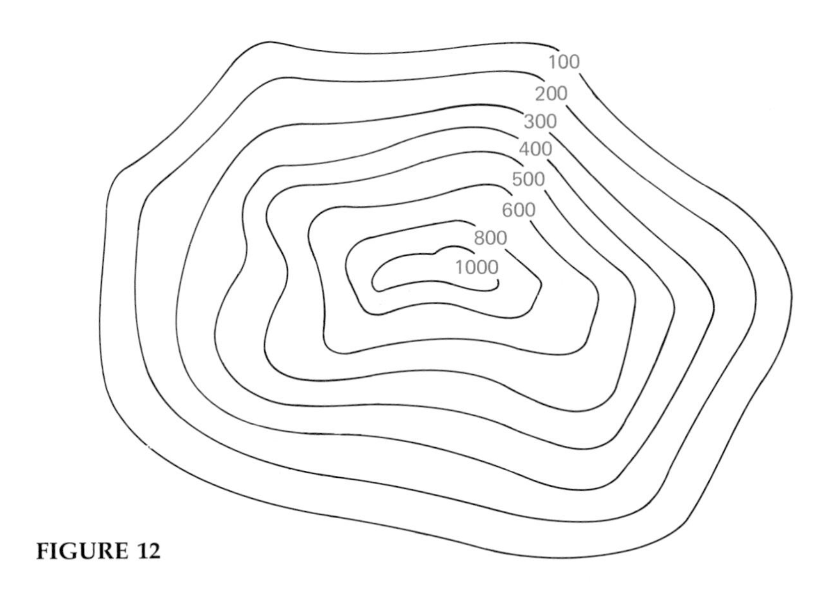

FIGURE 12

LEVEL CURVES

We can use the same idea to depict the function $z = f(x, y)$ graphically. If z is fixed, then the equation $f(x, y) = z$ is the equation of a curve in the xy-plane, called a **level curve.**

We can think of a level curve as the projection of a cross section lying in a plane parallel to the xy-plane. Each value of z gives us such a curve. In other words, a level curve is the projection of the intersection of the surface $z = f(x, y)$ with the plane $z = c$. This idea is best illustrated with some examples.

EXAMPLE 11 Sketch the level curves of $z = x^2 + y^2$.

Solution. If $z > 0$, then $z = a^2$ for some positive number $a > 0$. Hence all level curves are circles of the form $x^2 + y^2 = a^2$. The number a^2 can be thought of as the "elevation" of points on a level curve. Some of these curves are sketched in Figure 13. Each level curve encloses a projection of a "slice" of the actual graph of the function in three dimensions. In this case each circle is the projection onto the xy-plane of a part of the surface in space. Actually, this example is especially simple because we can, without much difficulty, sketch the graph in space. From Section 3.6 the equation $z = x^2 + y^2$ is the equation of an elliptic paraboloid (actually, a circular paraboloid). It is sketched in Figure 14. In this easy case we can see that if we slice this surface parallel to the xy-plane, we obtain the circles whose projections onto the xy-plane are the level curves. In most cases, of course, we will not be able to sketch the graph in space easily, so we will have to rely on our "contour map" sketch in $\mathbb{R}^2$. ■

There are some interesting applications of level curves in the sciences and economics. Three are given below.

(i) Let $T(x, y)$ denote the temperature at a point (x, y) in the xy-plane. The level curves $T(x, y) = c$ are called **isothermal curves.** All points on such a curve have the same temperature.

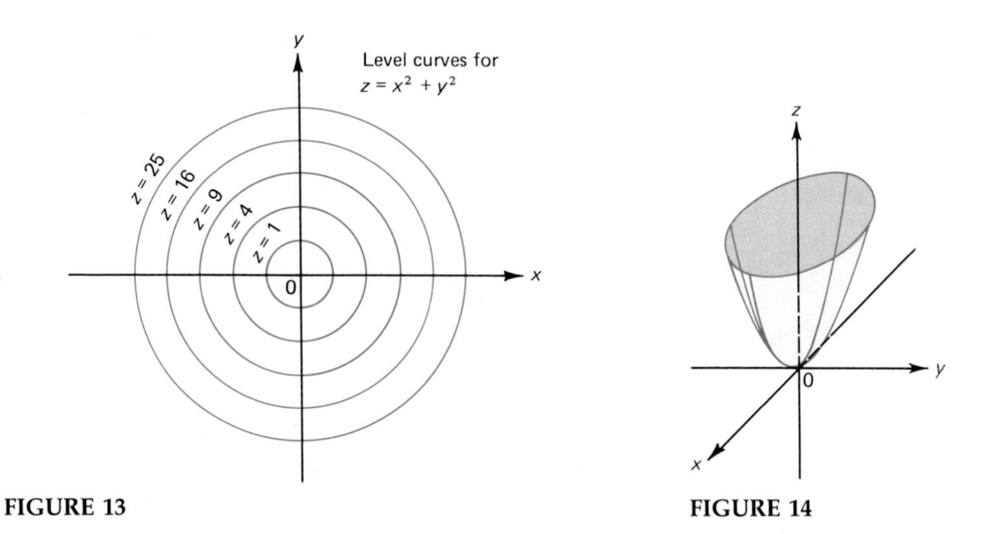

FIGURE 13

FIGURE 14

(ii) Let $V(x, y)$ denote the voltage (or potential) at a point in the xy-plane. The level curves $V(x, y) = c$ are called **equipotential curves.** All points on such a curve have the same voltage.

(iii) A manufacturer makes two products. Let x and y denote the number of units of the first and second products produced during a given year. If $P(x, y)$ denotes the profit the manufacturer receives each year, then the level curves $P(x, y) = c$ are **constant profit curves.** All points on such a curve yield the same profit.

We note that the idea behind level curves can be used to describe functions of three variables of the form $w = f(x, y, z)$. For each fixed w, the equation $f(x, y, z) = w$ is the equation of a surface in space, called a **level surface.** Since surfaces are so difficult to draw, we will not pursue the notion of level surfaces here.

In this section we discussed functions of two and three variables. Little additional work is needed to extend these notions to functions of n variables, where $n > 3$.

FUNCTION OF n VARIABLES

Let Ω be a subset of $\mathbb{R}^n$. A (scalar-valued) **function of n variables** f is a rule that assigns to each vector $\mathbf{x} = (x_1, x_2, \ldots, x_n)$ in Ω a unique real number which we denote $f(\mathbf{x})$ or $f(x_1, x_2, \ldots, x_n)$. The set Ω is called the **domain** of f and is denoted dom f. The set $\{f(\mathbf{x}): \mathbf{x} \in \Omega\}$, which is the set of values the function f takes on, is called the **range** of f.

NOTATION. We will write

$$f: \mathbb{R}^n \to \mathbb{R}$$

to indicate that f is a real-valued function whose domain is a subset of $\mathbb{R}^n$.

EXAMPLE 12 Let $\Omega = \mathbb{R}^5$ and let $f(x_1, x_2, x_3, x_4, x_5) = x_1x_2x_3x_4x_5$. Then the range of f is $\mathbb{R}$. ■

EXAMPLE 13 Let $\Omega = \mathbb{R}^4$ and let $f(x_1, x_2, x_3, x_4) = x_1^2 + x_2^2 + x_3^2 + x_4^2$. Then the range of f is $\mathbb{R}^+$. ■

Usually, as in $\mathbb{R}^2$, when the domain Ω is not given, we take the domain of f to be the largest subset of $\mathbb{R}^n$ for which the expression $f(x)$ makes sense.

EXAMPLE 14 In $\mathbb{R}^4$, let

$$f(\mathbf{x}) = \frac{1}{x_1^2 + x_2^2 + x_3^2 + x_4^2 - 1}.$$

Then $f(\mathbf{x})$ is defined except when $x_1^2 + x_2^2 + x_3^2 + x_4^2 = 1$. This is the equation of the **unit sphere** in $\mathbb{R}^4$. Thus the domain of $f = \{(x_1, x_2, x_3, x_4): x_1^2 + x_2^2 + x_3^2 + x_4^2 \neq 1\}$. It is a bit more difficult to find the range of f. We note that $x_1^2 + x_2^2 + x_3^2 + x_4^2 - 1$ can be made as close to zero—from either side—as desired. Thus $f(\mathbf{x})$ can take on arbitrarily large values. Since $x_1^2 + x_2^2 + x_3^2 + x_4^2 - 1 \geq -1$, $f(\mathbf{x})$ cannot take on negative values larger than -1. In addition, $f(\mathbf{x}) \neq 0$ for any vector $\mathbf{x}$. Finally, if $|\mathbf{x}|$ is large (see Definition 3.7.6 for the definition of the norm, $|\mathbf{x}|$), then $f(\mathbf{x})$ is small and positive. Thus the range of $f = (-\infty, -1] \cup (0, \infty) = \mathbb{R} - (-1, 0]$. ■

As in Example 14, we can define spheres and other geometric objects in $\mathbb{R}^n$, where $n > 3$, by analogy with known objects in $\mathbb{R}^2$ and $\mathbb{R}^3$. However, we cannot, of course, draw these objects.

PROBLEMS 4.1

In Problems 1–45, find the domain and range of the indicated function.

1. $f(x, y) = \sqrt{x^2 + y^2}$
2. $f(x, y) = \sqrt{1 + x + y}$
3. $f(x, y) = \dfrac{x}{y}$
4. $f(x, y) = \sqrt{1 - x^2 - 4y^2}$
5. $f(x, y) = \sqrt{1 - x^2 + 4y^2}$
6. $f(x, y) = \sin(x + y)$
7. $f(x, y) = e^x + e^y$
8. $f(x, y) = \dfrac{1}{(x^2 - y^2)^{3/2}}$
9. $f(x, y) = \tan(x - y)$
10. $f(x, y) = \sqrt{\dfrac{x + y}{x - y}}$
11. $f(x, y) = \sqrt{\dfrac{x - y}{x + y}}$
12. $f(x, y) = \sin^{-1}(x + y)$
13. $f(x, y) = \cos^{-1}(x - y)$
14. $f(x, y) = \dfrac{y}{|x|}$
15. $f(x, y) = \dfrac{x^2 - y^2}{x + y}$
16. $f(x, y) = \ln(1 + x^2 - y^2)$
***17.** $f(x, y) = \left(\dfrac{x}{2y}\right) + \left(\dfrac{2y}{x}\right)$
18. $f(x, y, z) = x + y + z$
19. $f(x, y, z) = \sqrt{x + y + z}$
20. $f(x, y, z) = \dfrac{1}{\sqrt{x^2 + y^2 + z^2}}$
21. $f(x, y, z) = \dfrac{1}{\sqrt{x^2 - y^2 + z^2}}$
22. $f(x, y, z) = \dfrac{1}{\sqrt{x^2 - y^2 - z^2}}$
23. $f(x, y, z) = \sqrt{-x^2 - y^2 - z^2}$
24. $f(x, y, z) = \ln(x - 2y - 3z + 4)$
25. $f(x, y, z) = \dfrac{xy}{z}$

26. $f(x, y, z) = \sin(x + y - z)$
27. $f(x, y, z) = \sin^{-1}(x + y - z)$
28. $f(x, y, z) = \ln(x + y - z)$

29. $f(x, y, z) = \tan^{-1}\left(\frac{x + z}{y}\right)$

30. $f(x, y, z) = e^{xy+z}$

31. $f(x, y, z) = \frac{e^x + e^y}{e^z}$

32. $f(x, y, z) = xyz$

33. $f(x, y, z) = \frac{1}{xyz}$

34. $f(x, y, z) = \frac{x}{y + z}$

35. $f(x, y, z) = \sin x + \cos y + \sin z$
36. $f(x_1, x_2, x_3, x_4) = \sqrt{x_1^2 + x_2^2 + x_3^2 + x_4^2}$
37. $f(x_1, x_2, x_3, x_4) = \sqrt{1 + x_1 + 2x_2 + x_3 - x_4}$
38. $f(x_1, x_2, x_3, x_4) = x_1x_2x_3x_4$

39. $f(x_1, x_2, x_3, x_4) = \frac{x_1x_3}{x_2x_4}$

40. $f(x_1, x_2, x_3, x_4) = \frac{x_1^2 + x_3^2}{x_2^2 - x_4^2}$

41. $f(x_1, x_2, x_3, x_4) = \frac{x_1^2 - x_3^2}{x_2^2 + x_4^2}$

42. $f(x_1, x_2, x_3, x_4, x_5) = \frac{x_1 + x_2 + x_4}{x_3 + x_5}$

43. $f(x_1, x_2, x_3, x_4, x_5) = \frac{x_1^2 - x_2^2 - x_4^2}{x_3^2 + x_5^2}$

44. $f(x_1, x_2, x_3, x_4, x_5) = \frac{x_1 - x_4}{x_2 + x_3 + x_5}$

45. $f(x_1, x_2, x_3, \ldots, x_n) = \sum_{i=1}^{n} x_i^2$

In Problems 46–52, sketch the graph of the given function.

46. $z = 4x^2 + 4y^2$
47. $y = x^2 + 4z^2$
48. $x = 4z^2 - 4y^2$
49. $z = x^2 - 4y^2$
50. $z = \sqrt{x^2 + 4y^2 + 4}$
51. $y = \sqrt{x^2 - 4z^2 + 4}$
52. $x = \sqrt{4 - z^2 - 4y^2}$

In Problems 53–60, describe the level curves of the given function and sketch these curves for the given values of z.

53. $z = \sqrt{1 + x + y}; z = 0, 1, 5, 10$

54. $z = \frac{x}{y}; z = 1, 3, 5, -1, -3$

55. $z = \sqrt{1 - x^2 - 4y^2}; z = 0, \frac{1}{4}, \frac{1}{2}, 1$

56. $z = \sqrt{1 + x^2 - y}; z = 0, 1, 2, 5$

57. $z = \cos^{-1}(x - y); z = 0, \frac{\pi}{6}, \frac{\pi}{3}, \frac{\pi}{2}$

58. $z = \sqrt{\frac{x + y}{x - y}}; z = 0, 1, 2, 5$

59. $z = \tan(x + y); z = 0, 1, -1, \sqrt{3}$
60. $z = \tan^{-1}(x - y^2); z = 0, \pi/6, \pi/4$

61. The temperature T at any point on an object in the plane is given by $T(x, y) = 20 + x^2 + 4y^2$. Sketch the isothermal curves for $T = 50$, $T = 60$, and $T = 70$ degrees.

62. The voltage at a point (x, y) on a metal plate placed in the xy-plane is given by $V(x, y) = \sqrt{1 - 4x^2 - 9y^2}$. Sketch the equipotential curves for $V = 1.0$ V, $V = 0.5$ V, and $V = 0.25$ V.

63. A manufacturer earns $P(x, y) = 100 + 2x^2 + 3y^2$ dollars each year for producing x and y units, respectively, of two products. Sketch the constant profit curves for $P = \$100$, $P = \$200$, and $P = \$1000$.

In Problems 64–73 a function of two variables is given. Match the function to one of the ten computer-drawn graphs given below.

(a)

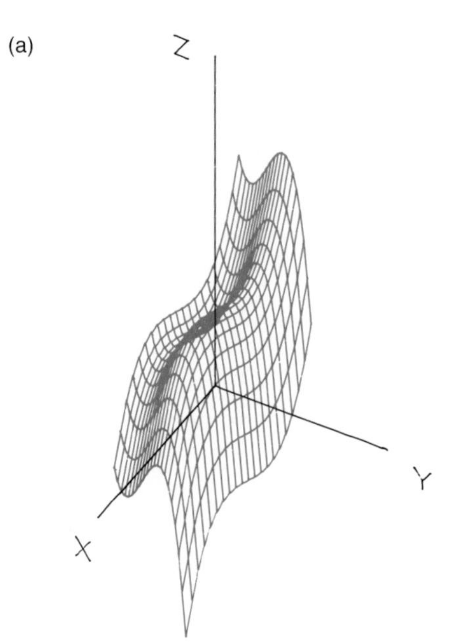

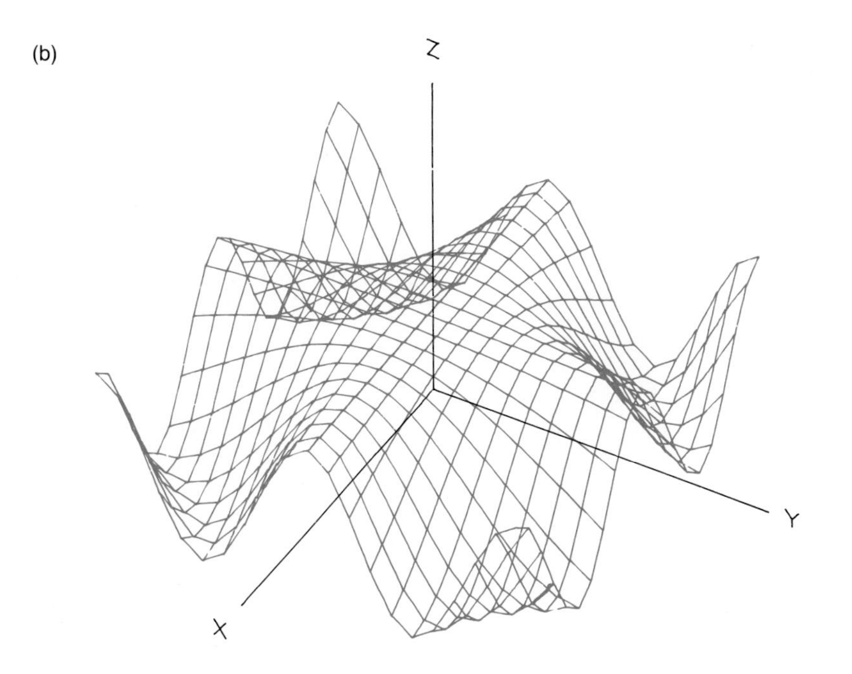
(b)
Z
Y
X

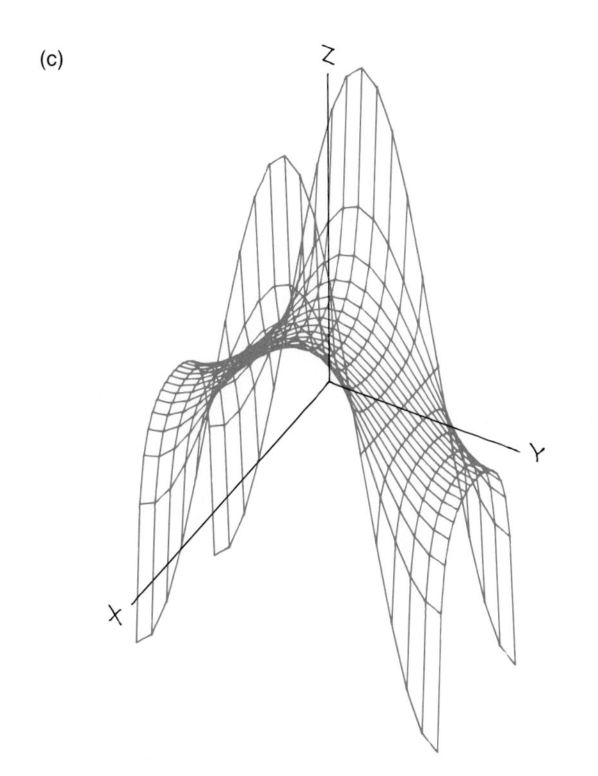
(c)
Z
Y
X

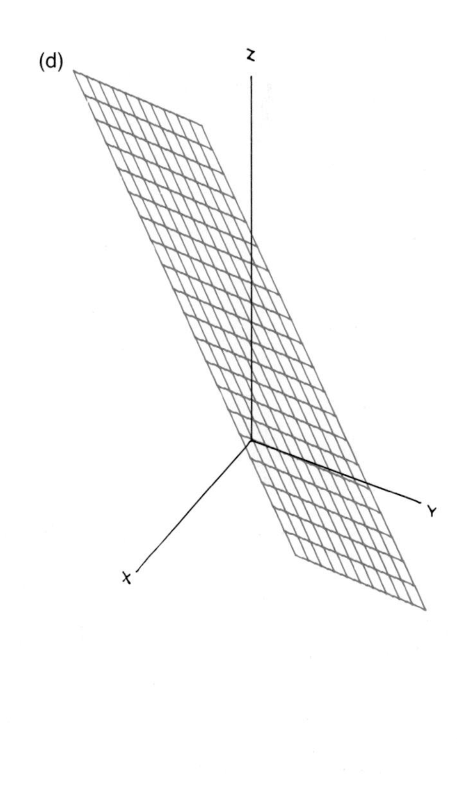
(d)
Z
Y
X

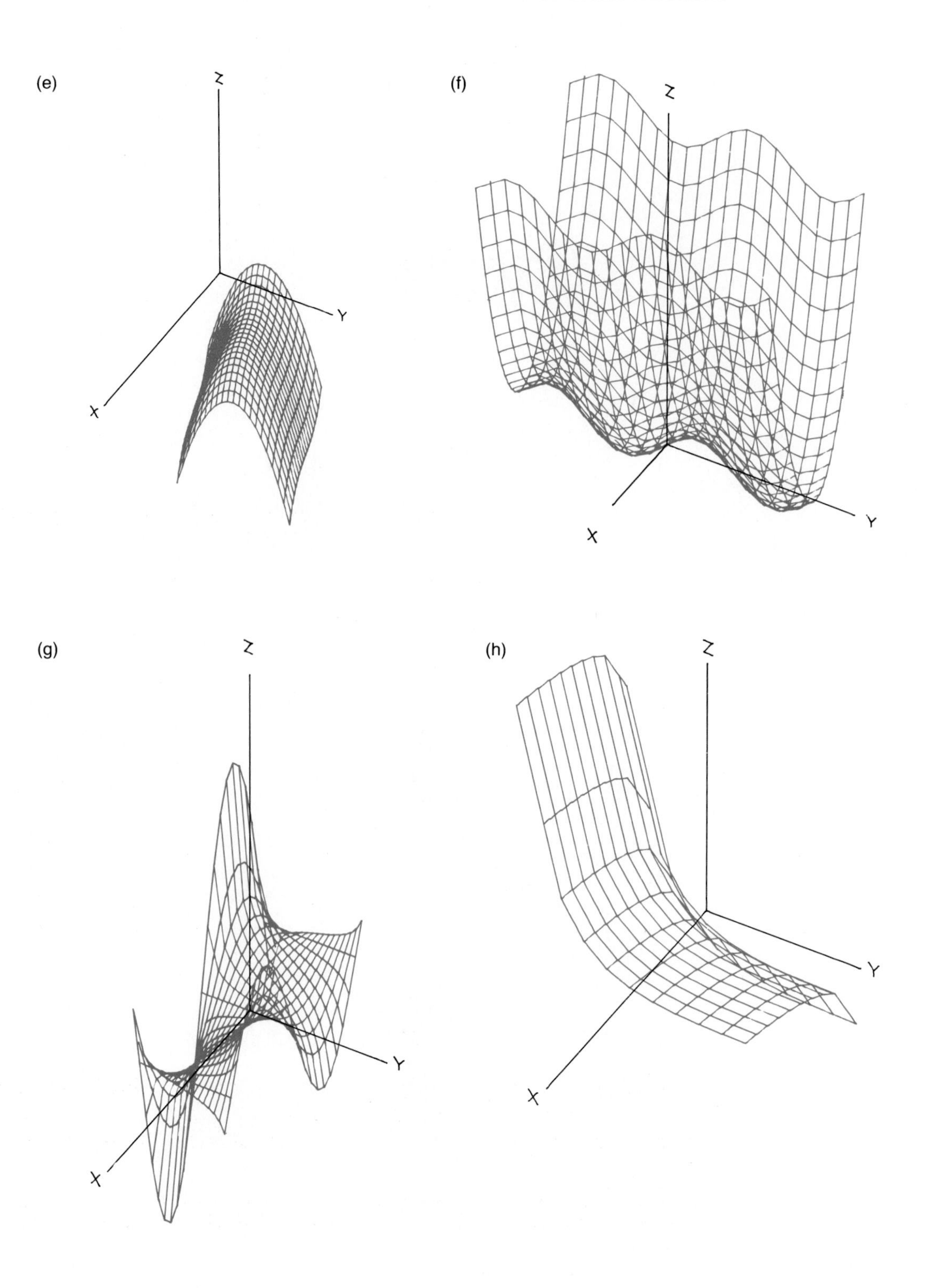
(e)
Z
Y
X
(f)
Z
Y
X
(g)
Z
Y
X
(h)
Z
Y
X

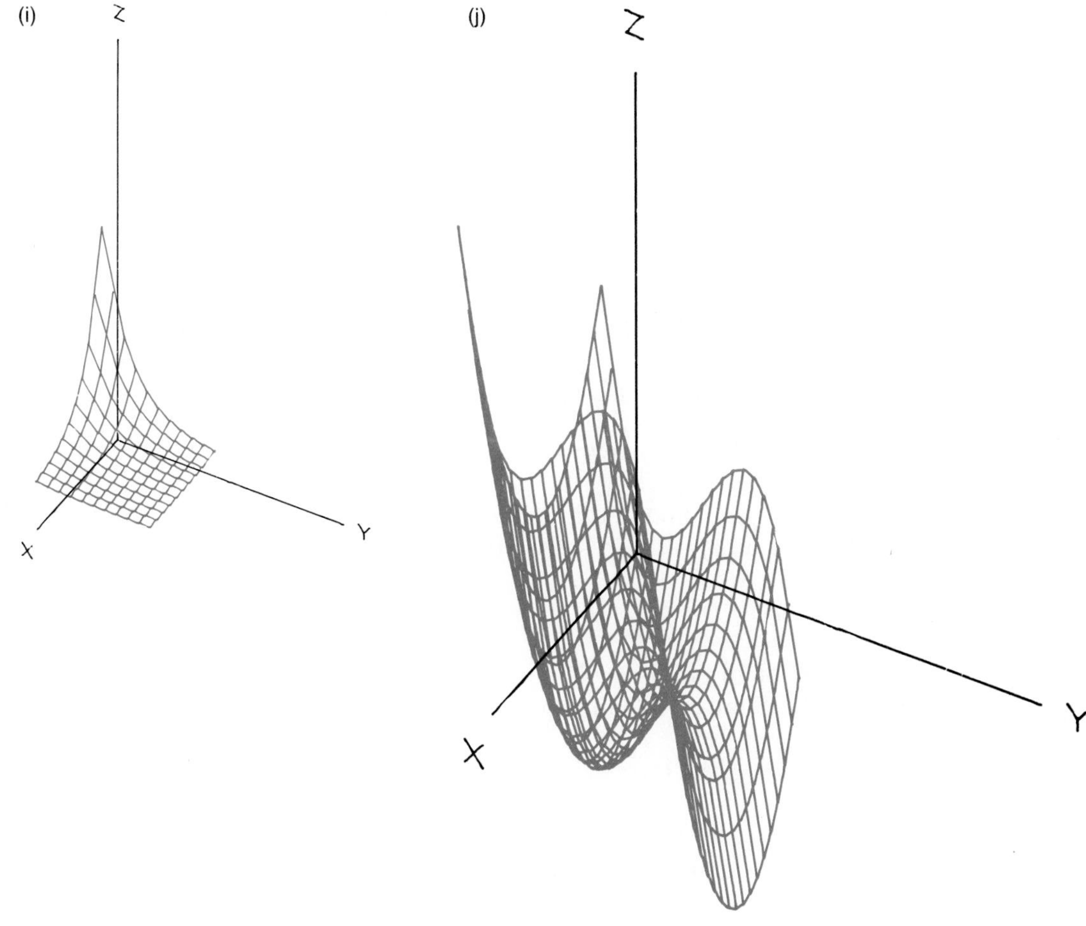

64. $z = x^2 + \sin y$
65. $z = \ln x + e^{-y}$
66. $z = \cos xy$
67. $z = x^3 + y^3 - x^2 + y + 2$
68. $z = x^2 - y^3 + x + 2y - 4$
69. $z = -x^3 - y^2 - x + 2y - 3$
70. $z = e^{-(x+y)}$
71. $z = \sin x \tan y, \ -\pi/2 < y < \pi/2$
72. $z = \sec x \cos y, \ -\pi/2 < x < \pi/2$
73. $z = x - 2y + 4$

4.2 LIMITS AND CONTINUITY

In this section we discuss the fundamental concepts of limits and continuity for functions of two or more variables. Recall that in the definition of a limit,† the notions of an open and closed interval were fundamental. We could say, for example, that x

†The formal definition of a limit of a function of one variable is reproduced here: Suppose that $f(x)$ is defined in a neighborhood of the point x_0 (a finite number) except possibly at the point x_0 itself. Then

$$\lim_{x \to x_0} f(x) = L$$

if for every $\epsilon > 0$, there is a $\delta > 0$, such that if

$$0 < |x - x_0| < \delta, \qquad \text{then} \qquad |f(x) - L| < \epsilon.$$

was close to x_0 if $|x - x_0|$ was sufficiently small or, equivalently, if x was contained in a small open interval (neighborhood) centered at x_0. It is interesting to see how these ideas extend to $\mathbb{R}^2$, $\mathbb{R}^3$, and $\mathbb{R}^n$.

Let (x_0, y_0) be a point in $\mathbb{R}^2$. What do we mean by the equation

$$|(x, y) - (x_0, y_0)| = r? \tag{1}$$

Since (x, y) and (x_0, y_0) are vectors in $\mathbb{R}^2$,

$$|(x, y) - (x_0, y_0)| = |(x - x_0, y - y_0)| \overset{\text{Equation (1.1.1)}}{=} \sqrt{(x - x_0)^2 + (y - y_0)^2}. \tag{2}$$

Inserting (2) into (1) and squaring both sides yields

$$(x - x_0)^2 + (y - y_0)^2 = r^2, \tag{3}$$

which is the equation of the circle of radius r centered at (x_0, y_0). Then the set of points whose coordinates (x, y) satisfy the inequality

$$|(x, y) - (x_0, y_0)| < r \tag{4}$$

is the set of all points in $\mathbb{R}^2$ interior to the circle given by (3). This set is sketched in Figure 1a. Similarly, the inequality

$$|(x, y) - (x_0, y_0)| \leq r \tag{5}$$

describes the set of all points interior to and on the circle given by (3). This set is sketched in Figure 1b. This leads to the following definition.

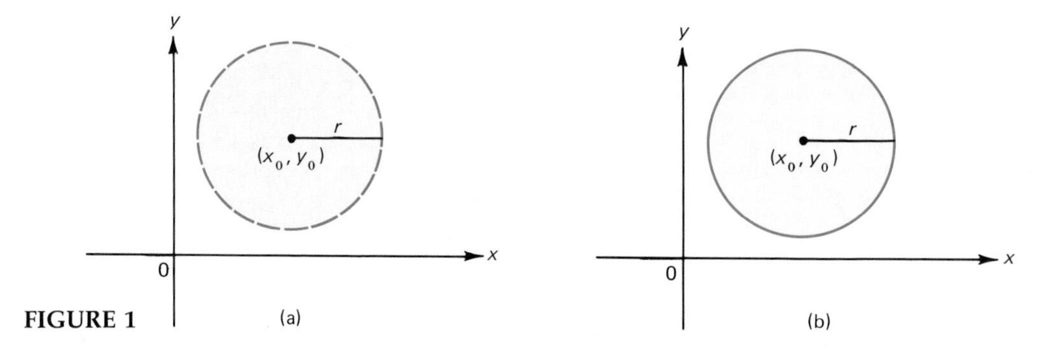

FIGURE 1

Definition 1

(i) The **open disk** D_r centered at (x_0, y_0) with radius r is the subset of $\mathbb{R}^2$ given by

$$\{(x, y): |(x, y) - (x_0, y_0)| < r\}.$$

(ii) The **closed disk** centered at (x_0, y_0) with radius r is the subset of $\mathbb{R}^2$ given by

$$\{(x, y): |(x, y) - (x_0, y_0)| \leq r\}.$$

(iii) The **boundary** of the open or closed disk defined in (i) or (ii) is the circle

$$\{(x, y): |(x, y) - (x_0, y_0)| = r\}.$$

(iv) A **neighborhood** of a point (x_0, y_0) in $\mathbb{R}^2$ is an open disk centered at (x_0, y_0).

REMARK. In this definition the words "open" and "closed" have meanings very similar to their meanings in the terms "open interval" and "closed interval." An open interval does not contain its endpoints. An open disk does not contain any point on its boundary. Similarly, a closed interval contains all its boundary points, as does a closed disk.

With these definitions it is easy to define a limit of a function of two variables. Intuitively, we say that $f(x, y)$ approaches the limit L as (x, y) approaches (x_0, y_0) if $f(x, y)$ gets arbitrarily "close" to L as (x, y) approaches (x_0, y_0) along any **path.**† We define this notion precisely below.

Definition 2 DEFINITION OF A LIMIT Let $f(x, y)$ be defined in a neighborhood of (x_0, y_0) but not necessarily at (x_0, y_0) itself. Then the **limit** of $f(x, y)$ as (x, y) approaches (x_0, y_0) *is* L, written

$$\lim_{(x,y)\to(x_0,y_0)} f(x, y) = L, \tag{6}$$

if for every number $\epsilon > 0$, there is a number $\delta > 0$ such that $|f(x, y) - L| < \epsilon$ for every $(x, y) \neq (x_0, y_0)$ in the open disk centered at (x_0, y_0) with radius δ.

REMARK. This definition is illustrated in Figure 2.

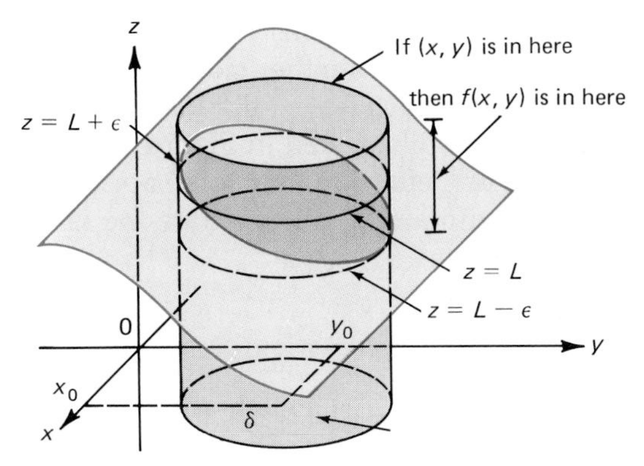

FIGURE 2

EXAMPLE 1 Show, using the definition of a limit, that $\lim_{(x,y)\to(1,2)}(3x + 2y) = 7$.

Solution. Let $\epsilon > 0$ be given. We need to choose a $\delta > 0$ such that $|3x + 2y - 7| < \epsilon$ if $0 < \sqrt{(x - 1)^2 + (y - 2)^2} < \delta$. We start with

†A path is another name for a curve joining (x, y) to (x_0, y_0).

$$|3x + 2y - 7| = |3x - 3 + 2y - 4| \underset{\text{Triangle inequality}}{\leq} |3x - 3| + |2y - 4|$$

$$= |3(x - 1)| + |2(y - 2)| \underset{\text{Explain why.}}{=} 3|x - 1| + 2|y - 2|. \quad \textbf{(7)}$$

Now

$$|x - 1| = \sqrt{(x - 1)^2} \leq \sqrt{(x - 1)^2 + (y - 2)^2}$$

and

$$|y - 2| = \sqrt{(y - 2)^2} \leq \sqrt{(x - 1)^2 + (y - 2)^2}.$$

So from (7)

$$|3x + 2y - 7| \leq 3\sqrt{(x - 1)^2 + (y - 2)^2} + 2\sqrt{(x - 1)^2 + (y - 2)^2}$$
$$= 5\sqrt{(x - 1)^2 + (y - 2)^2}. \quad \textbf{(8)}$$

Now we want $|3x + 2y - 7| < \epsilon$. We choose $\delta = \epsilon/5$. Then from (8)

$$|3x + 2y - 7| \leq 5\sqrt{(x - 1)^2 + (y - 2)^2} < 5\delta = 5\left(\frac{\epsilon}{5}\right) = \epsilon,$$

and the requested limit is shown. ■

In the intuitive definition of a limit which preceded Definition 2, we used the phrase "if $f(x, y)$ gets arbitrarily close to L as (x, y) approaches (x_0, y_0) *along any path.*" Recall that $\lim_{x\to x_0} f(x) = L$ only if $\lim_{x\to x_0^+} f(x) = \lim_{x\to x_0^-} f(x) = L$. That is, the limit exists only if we get the same value from either side of x_0. In $\mathbb{R}^2$ the situation is more complicated because (x, y) can approach (x_0, y_0) not just along two but along an *infinite number* of paths. Some of these are illustrated in Figure 3. Thus the only way we can verify a limit is by making use of Definition 2 or some appropriate limit theorem that can be proven directly from Definition 2. We illustrate the kinds of problems we can encounter in the next two examples.

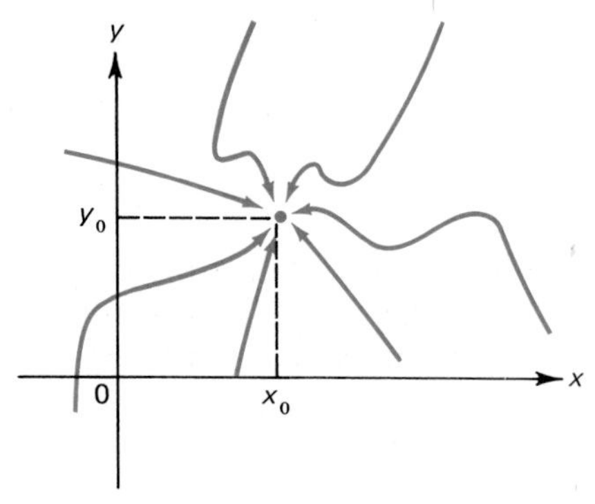

FIGURE 3

EXAMPLE 2 Let $f(x, y) = (y^2 - x^2)/(y^2 + x^2)$ for $(x, y) \neq (0, 0)$. We will show that $\lim_{(x,y)\to(0,0)} f(x, y)$ does not exist. There are an infinite number of approaches to the origin. For example, if we approach along the x-axis, then $y = 0$ and, if the indicated limit exists,

$$\lim_{(x,y)\to(0,0)} \frac{y^2 - x^2}{y^2 + x^2} = \lim_{(x,y)\to(0,0)} \frac{-x^2}{x^2} = \lim_{(x,y)\to(0,0)} -1 = -1.$$

On the other hand, if we approach along the y-axis, then $x = 0$ and

$$\lim_{(x,y)\to(0,0)} \frac{y^2 - x^2}{y^2 + x^2} = \lim_{(x,y)\to(0,0)} \frac{y^2}{y^2} = \lim_{(x,y)\to(0,0)} 1 = 1.$$

Thus we get different answers depending on how we approach the origin. To prove that the limit cannot exist, we note that we have shown that in any open disk centered at the origin, there are points at which f takes on the values $+1$ and -1. Hence f cannot have a limit as $(x, y) \to (0, 0)$. ■

Example 2 leads to the following general rule:

RULE FOR NONEXISTENCE OF A LIMIT

If we get two or more different values for $\lim_{(x,y)\to(x_0,y_0)} f(x, y)$ as we approach (x_0, y_0) along different paths, then $\lim_{(x,y)\to(x_0,y_0)} f(x, y)$ does not exist.

EXAMPLE 3 Let $f(x, y) = xy^2/(x^2 + y^4)$. We will show that $\lim_{(x,y)\to(0,0)} f(x, y)$ does not exist. First, let us approach the origin along a straight line passing through the origin that is not the y-axis. Then $y = mx$ and

$$f(x, y) = \frac{x(m^2x^2)}{x^2 + m^4x^4} = \frac{m^2x}{1 + m^2x^2}$$

which approaches 0 as $x \to 0$. Thus along every straight line, $f(x, y) \to 0$ as $(x, y) \to (0, 0)$. But if we approach $(0, 0)$ along the parabola $x = y^2$, then

$$f(x, y) = \frac{y^2(y^2)}{y^4 + y^4} = \frac{1}{2}$$

so that along this parabola, $f(x, y) \to \frac{1}{2}$ as $(x, y) \to (0, 0)$, and the limit does not exist. ■

EXAMPLE 4 Prove that $\lim_{(x,y)\to(0,0)}[xy^2/(x^2 + y^2)] = 0$.

Solution. It is easy to show that this limit is zero along any straight line passing through the origin. But as we have seen, this is not enough. We must rely on our definition. Let $\epsilon > 0$ be given; then we must show that there is a $\delta > 0$ such that if $0 < \sqrt{x^2 + y^2} < \delta$, then

$$|f(x, y)| = \left|\frac{xy^2}{x^2 + y^2}\right| < \epsilon.$$

But $|x| \le \sqrt{x^2 + y^2}$ and $y^2 \le x^2 + y^2$, so

$$\left|\frac{xy^2}{x^2 + y^2}\right| \le \frac{(\sqrt{x^2 + y^2})(x^2 + y^2)}{x^2 + y^2} = \sqrt{x^2 + y^2}.$$

Hence if we choose $\delta = \epsilon$, we will have $|f(x, y)| < \epsilon$ if $\sqrt{x^2 + y^2} < \delta$. ■

REMARK. This example illustrates the fact that while the existence of different limits for different paths implies that no limit exists, only the definition or an appropriate limit theorem can be used to prove that a limit does exist.

As Examples 1 and 4 illustrate, it is often tedious to calculate limits from the definition. Fortunately, just as in the case of functions of a single variable, there are a number of theorems that greatly facilitate the calculation of limits. As you will recall, one of our important definitions stated that if f is continuous, then $\lim_{x \to x_0} f(x) = f(x_0)$. We now define continuity of a function of two variables.

Definition 3 CONTINUITY

(i) Let $f(x, y)$ be defined at every point (x, y) in a neighborhood of (x_0, y_0). Then f is **continuous** at (x_0, y_0) if all of the following conditions hold:

(a) $f(x_0, y_0)$ exists [i.e., (x_0, y_0) is in the domain of f].
(b) $\lim_{(x,y)\to(x_0,y_0)} f(x, y)$ exists.
(c) $\lim_{(x,y)\to(x_0,y_0)} f(x, y) = f(x_0, y_0)$.

(ii) If one or more of these three conditions fail to hold, then f is said to be **discontinuous** at (x_0, y_0).
(iii) f is **continuous in a subset** S of $\mathbb{R}^2$ if f is continuous at every point (x, y) in S.

REMARK. Condition (c) tells us that if a function f is continuous at (x_0, y_0), then we can calculate $\lim_{(x,y)\to(x_0,y_0)} f(x, y)$ by evaluation of f at (x_0, y_0).

EXAMPLE 5 Let $f(x, y) = xy^2/(x^2 + y^2)$. Then f is discontinuous at (0, 0) because $f(0, 0)$ is not defined. ■

EXAMPLE 6 Let

$$f(x, y) = \begin{cases} \dfrac{xy^2}{x^2 + y^4}, & (x, y) \ne (0, 0) \\ 0, & (x, y) = (0, 0). \end{cases}$$

Here f is defined at (0, 0) but it is still discontinuous there because $\lim_{(x,y)\to(0,0)} f(x, y)$ does not exist (see Example 3). ■

EXAMPLE 7 Let

$$f(x, y) = \begin{cases} \dfrac{xy^2}{x^2 + y^2}, & (x, y) \ne (0, 0) \\ 0, & (x, y) = (0, 0). \end{cases}$$

Here f is continuous at (0, 0), according to Example 4. ■

EXAMPLE 8 Let

$$f(x, y) = \begin{cases} \dfrac{xy^2}{x^2 + y^2}, & (x, y) \neq (0, 0) \\ 1, & (x, y) = (0, 0). \end{cases}$$

Then f is discontinuous at (0, 0) because $\lim_{(x,y)\to(0,0)} f(x, y) = 0 \neq f(0, 0) = 1$, so condition (c) is violated. ■

Naturally, we would like to know what functions are continuous. We can answer this question if we look at the continuous functions of one variable. We start with polynomials.

Definition 4 POLYNOMIAL AND RATIONAL FUNCTION

(i) A **polynomial** $p(x, y)$ in the two variables x and y is a finite sum of terms of the form $Ax^m y^n$, where m and n are nonnegative integers and A is a real number.

(ii) A **rational function** $r(x, y)$ in the two variables x and y is a function that can be written as the quotient of two polynomials: $r(x, y) = p(x, y)/q(x, y)$.

EXAMPLE 9 $p(x, y) = 5x^5y^2 + 12xy^9 - 37x^{82}y^5 + x + 4y - 6$ is a polynomial. ■

EXAMPLE 10 $r(x, y) = \dfrac{8x^3y^7 - 7x^2y^4 + xy - 2y}{1 - 3y^3 + 7x^2y^2 + 18yx^7}$ is a rational function. ■

The limit theorems for functions of one variable can be extended, with minor modifications, to functions of two variables. We will not state them here. However, by using them it is not difficult to prove the following theorem about continuous functions.

Theorem 1

(i) Any polynomial p is continuous at any point in $\mathbb{R}^2$.

(ii) Any rational function $r = p/q$ is continuous at any point (x_0, y_0) for which $q(x_0, y_0) \neq 0$. It is discontinuous when $q(x_0, y_0) = 0$ because it is then not defined at (x_0, y_0).

(iii) If f and g are continuous at (x_0, y_0), then $f + g$, $f - g$, and $f \cdot g$ are continuous at (x_0, y_0).[†]

(iv) If f and g are continuous at (x_0, y_0) and if $g(x_0, y_0) \neq 0$, then f/g is continuous at (x_0, y_0).

(v) If f is continous at (x_0, y_0) and if h is a function of one variable that is continuous at $f(x_0, y_0)$, then the composite function $h \circ f$, defined by $(h \circ f)(x, y) = h(f(x, y))$, is continuous at (x_0, y_0).

[†] $(f + g)(x, y) = f(x, y) + g(x, y)$, just as in the case of functions of one variable. Similarly, $(f \cdot g)(x, y) = [f(x, y)][g(x, y)]$, and $(f/g)(x, y) = f(x, y)/g(x, y)$.

EXAMPLE 11 Calculate $\lim_{(x,y)\to(4,1)}(x^3y^2 - 4xy)/(x + 6xy^3)$.

Solution. $(x^3y^2 - 4xy)/(x + 6xy^3)$ is continuous at (4, 1) so we can calculate the limit by evaluation. We have

$$\lim_{(x,y)\to(4,1)} \frac{x^3y^2 - 4xy}{x + 6xy^3} = \left.\frac{x^3y^2 - 4xy}{x + 6xy^3}\right|_{(4,1)} = \frac{64\cdot 1 - 4\cdot 4\cdot 1}{4 + 6\cdot 4\cdot 1} = \frac{48}{28} = \frac{12}{7}. \blacksquare$$

EXAMPLE 12 Calculate $\lim_{(x,y)\to(\pi/6,\pi/3)} \sin(x + y)$.

Solution. $f(x, y) = x + y$ is continuous, as is $h(x) = \sin x$. Then by (v), $(h \circ f)(x, y) = \sin(x + y)$ is continuous, and therefore

$$\lim_{(x,y)\to(\pi/6,\pi/3)} \sin(x + y) = \sin\left(\frac{\pi}{6} + \frac{\pi}{3}\right) = \sin\left(\frac{\pi}{2}\right) = 1. \blacksquare$$

EXAMPLE 13 For what values of (x, y) is the function

$$f(x, y) = \frac{x^3 + 7xy^5 - 8y^6}{x^2 - y^2}$$

continuous?

Solution. Since f is a rational function in x and y, f is continuous as long as the denominator is not zero, that is, when $y \neq \pm x$. Thus f is continuous at every point (x, y) in $\mathbb{R}^2$ not on one of the lines $y = x$ and $y = -x$. ■

All the ideas in this section can be generalized to functions of three or more variables. A **neighborhood** of radius r of the point (x_0, y_0, z_0) in $\mathbb{R}^3$ consists of all points in $\mathbb{R}^3$ interior to the sphere

$$(x - x_0)^2 + (y - y_0)^2 + (z - z_0)^2 = r^2.$$

That is, the neighborhood (also called an **open ball**) is described by

$$\{(x, y, z): (x - x_0)^2 + (y - y_0)^2 + (z - z_0)^2 < r^2\}.$$

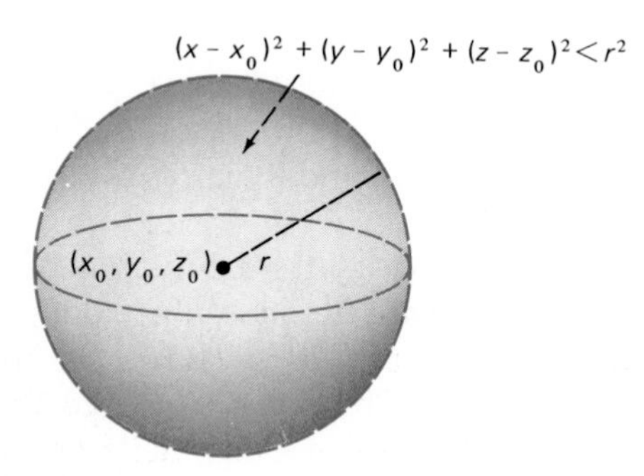

FIGURE 4

A typical neighborhood is sketched in Figure 4. The **closed ball** is described by

$$\{(x, y, z): (x - x_0)^2 + (y - y_0)^2 + (z - z_0)^2 \le r^2\}.$$

With this definition, the definition of the limit $\lim_{(x,y,z)\to(x_0,y_0,z_0)} f(x, y, z) = L$ is analogous to that given in Definition 2. All other definitions, remarks, and theorems given in this section apply to functions of three (or more) variables.

We now extend the notions of limits and continuity to functions of n variables. We begin by defining a neighborhood in $\mathbb{R}^n$.

Definition 5

(i) The **open ball** $B_r(\mathbf{x}_0)$ in $\mathbb{R}^n$ centered at $\mathbf{x}_0$ with radius r is the subset of $\mathbb{R}^n$ given by

$$B_r(\mathbf{x}_0) = \{\mathbf{x} \in \mathbb{R}^n: |\mathbf{x} - \mathbf{x}_0| < r\}. \tag{9}$$

(ii) The **closed ball** $\overline{B}_r(\mathbf{x}_0)$ in $\mathbb{R}^n$ with radius r is the subset of $\mathbb{R}^n$ given by

$$\overline{B}_r(\mathbf{x}_0) = \{\mathbf{x} \in \mathbb{R}^n: |\mathbf{x} - \mathbf{x}_0| \le r\}. \tag{10}$$

(iii) The **boundary** of the open or closed ball defined by (1) or (2) is the *sphere* $S_r(\mathbf{x}_0)$ in $\mathbb{R}^n$ given by

$$S_r(\mathbf{x}_0) = \{\mathbf{x} \in \mathbb{R}^n: |\mathbf{x} - \mathbf{x}_0| = r\}. \tag{11}$$

(iv) A **neighborhood** of a vector $\mathbf{x}_0$ in $\mathbb{R}^n$ is an open ball centered at $\mathbf{x}_0$.

(v) A set Ω in $\mathbb{R}^n$ is **open** if, for every $\mathbf{x}_0 \in \Omega$, there is a neighborhood $B_r(\mathbf{x}_0)$ of $\mathbf{x}_0$ such that $B_r(\mathbf{x}_0) \subset \Omega$.

REMARK 1. The notation in (9), (10), and (11) refers to the norm of a vector defined in Section 3.7 (p. 130).

Thus, for example, if

$$\mathbf{x}_0 = (x_1^{(0)}, x_2^{(0)}, \ldots, x_n^{(0)}) \quad \text{and} \quad \mathbf{x} = (x_1, x_2, \ldots, x_n),$$

then $|\mathbf{x} - \mathbf{x}_0| < r$ means that

$$\sqrt{(x_1 - x_1^{(0)})^2 + (x_2 - x_2^{(0)})^2 + \cdots + (x_n - x_n^{(0)})^2} < r. \tag{12}$$

REMARK 2. It is easy to visualize an open set in $\mathbb{R}^2$. In Figure 5a the set Ω_1 is open because, for every point in the set, it is possible to draw an open disk containing the point that is completely contained in Ω_1. The set Ω_2 in Figure 5b is not open because there are points (on the boundary) for which every neighborhood contains points outside Ω_2.

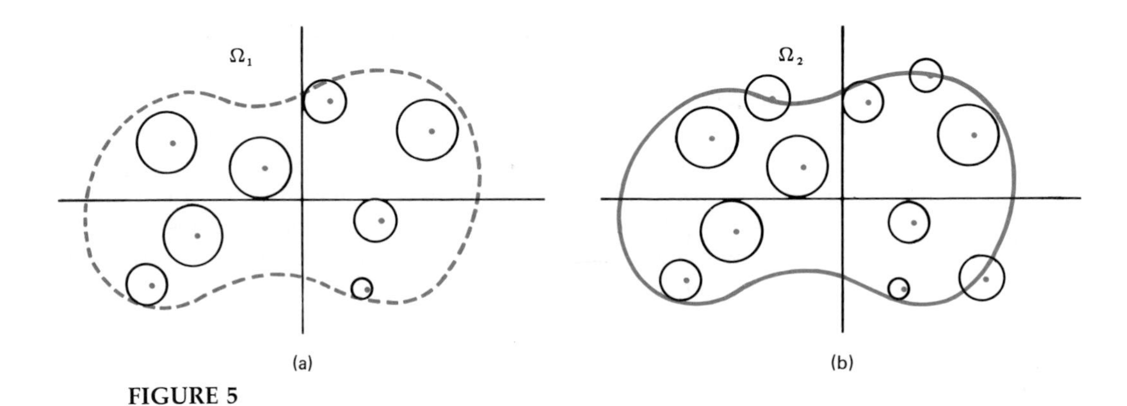

FIGURE 5

Definition 6 LIMIT Let $f\colon \mathbb{R}^n \to \mathbb{R}$ be defined in a neighborhood of $\mathbf{x}_0$ but not necessarily at $\mathbf{x}_0$ itself. Then the **limit of $f(\mathbf{x})$ as $\mathbf{x}$ approaches $\mathbf{x}_0$ is L**, written

$$\lim_{\mathbf{x}\to\mathbf{x}_0} f(\mathbf{x}) = L,$$

if for every number $\epsilon > 0$ there is a number $\delta > 0$ such that $|f(\mathbf{x}) - L| < \epsilon$ whenever $0 < |\mathbf{x} - \mathbf{x}_0| < \delta$.

EXAMPLE 14 Show that

$$\lim_{\mathbf{x}\to(-4,1,0,3)} (x_1 + 2x_2 - x_3 + 3x_4) = 7.$$

Solution. Let $\epsilon > 0$ be given. We need to choose a $\delta > 0$ such that

$$|x_1 + 2x_2 - x_3 + 3x_4 - 7| < \epsilon$$

if

$$0 < \sqrt{(x_1 + 4)^2 + (x_2 - 1)^2 + x_3^2 + (x_4 - 3)^2} < \delta.$$

As in Example 1, we start with the inequality $|x_1 + 2x_2 - x_3 + 3x_4 - 7| < \epsilon$ and work backward to find a suitable value for δ. We use the triangle inequality:

$$\begin{aligned} |x_1 + 2x_2 - x_3 + 3x_4 - 7| &= |x_1 + 4 + 2x_2 - 2 - x_3 + 3x_4 - 9| \\ &\le |x_1 + 4| + |2x_1 - 2| + |x_3| + |3x_4 - 9| \\ &= |x_1 + 4| + 2|x - 1| + |x_3 - 0| + 3|x_4 - 3|. \end{aligned} \quad \textbf{(13)}$$

Now each of the quantities $|x_1 + 4|$, $|x_1 - 1|$, $|x_3|$, and $|x_4 - 3|$ is less than or equal to

$$\sqrt{(x_1 + 4)^2 + (x_2 - 1)^2 + x_3^2 + (x_4 - 3)^2}$$

(explain why). Thus, from (13),

$$|x_1 + 2x_2 - x_3 + 3x_4 - 7| \leq 7\sqrt{(x_1 + 4)^2 + (x_2 - 1)^2 + x_3^2 + (x_4 - 3)^2}$$
$$= 7|(x_1, x_2, x_3, x_4) - (-4, 1, 0, 3)|.$$

Finally, let $\delta = \epsilon/7$; then if $|\mathbf{x} - (-4, 1, 0, 3)| < \delta$, we have

$$|f(\mathbf{x}) - 7| = |x_1 + 2x_2 - x_3 + 3x_4 - 7| \leq 7\delta = 7 \cdot \frac{\epsilon}{7} = \epsilon$$

and, therefore,

$$\lim_{\mathbf{x} \to (-4,1,0,3)} (x_1 + 2x_2 - x_3 + 3x_4) = 7. \quad \blacksquare$$

REMARK. Note the great similarity between the last example and Example 1 (in $\mathbb{R}^2$). This illustrates that, except for our inability to draw things, there is very little difference in the properties of functions of two or three variables and functions of n variables.

EXAMPLE 15 Let

$$f(x_1, x_2, x_3, x_4, x_5) = \frac{x_5^2 - x_4^2 + x_3^2 - x_2^2 + x_1^2}{x_1^2 + x_2^2 + x_3^2 + x_4^2 + x_5^2}$$

for $(x_1, x_2, x_3, x_4, x_5) \neq (0, 0, 0, 0, 0)$. Show that $\lim_{\mathbf{x} \to 0} f(\mathbf{x})$ does not exist.

Solution. We will do this by showing that we get two different answers if we approach the zero vector in two different ways. Let $x_1 = x_2 = x_3 = x_4 = 0$ and let $x_5 \to 0$. That is, we approach zero along the x_5-axis. Then, for $x_5 \neq 0$,

$$f(\mathbf{x}) = \frac{x_5^2}{x_5^2} = 1$$

so that

$$\lim_{(0,0,0,0,x_5) \to \mathbf{0}} f(\mathbf{x}) = 1.$$

Now let $x_1 = x_2 = x_3 = x_5 = 0$ and let $x_4 \to 0$. Then for $x_4 \neq 0$,

$$f(\mathbf{x}) = -\frac{x_4^2}{x_4^2} = -1$$

so that

$$\lim_{(0,0,0,x_4,0) \to \mathbf{0}} f(\mathbf{x}) = -1.$$

Since we get different answers as we approach zero in different ways, we conclude that the limit does not exist.

REMARK. Note the similarity between this example and Example 2. $\blacksquare$

Example 15 enables us to generalize the rule given earlier in this section.

> If we get two or more different values for $\lim_{\mathbf{x}\to\mathbf{x}_0} f(\mathbf{x})$ as we approach $\mathbf{x}_0$ in different ways, then $\lim_{\mathbf{x}\to\mathbf{x}_0} f(\mathbf{x})$ does not exist.

As in $\mathbb{R}^2$ and $\mathbb{R}^3$, we see that the direct computation of limits is quite tedious. Things are simpler once we have defined continuity; for if f is continuous at $\mathbf{x}_0$, then $\lim_{\mathbf{x}\to\mathbf{x}_0} f(\mathbf{x}) = f(\mathbf{x}_0)$.

Definition 7 CONTINUITY

(i) Let $f: \mathbb{R}^n \to \mathbb{R}$ be defined at every point $\mathbf{x}$ in a neighborhood of $\mathbf{x}_0$. Then f is **continuous** at $\mathbf{x}_0$ if all of the following conditions hold:

(a) $f(\mathbf{x}_0)$ exists (i.e., $\mathbf{x}_0$ is in the domain of f).

(b) $\lim_{\mathbf{x}\to\mathbf{x}_0} f(\mathbf{x})$ exists.

(c) $\lim_{\mathbf{x}\to\mathbf{x}_0} f(\mathbf{x}) = f(\mathbf{x}_0)$.

(ii) If one or more of these three conditions fails to hold, then f is said to be **discontinuous** at $\mathbf{x}_0$.

(iii) f is continuous in an open set Ω of $\mathbb{R}^n$ if f is continuous at every point $\mathbf{x}_0$ in Ω.

REMARK. Condition (c) tells us that if a function f is continuous at $\mathbf{x}_0$, then we can calculate $\lim_{\mathbf{x}\to\mathbf{x}_0} f(\mathbf{x})$ by evaluation of f at $\mathbf{x}_0$.

EXAMPLE 16 In $\mathbb{R}^5$,

$$f(x_1, x_2, x_3, x_4, x_5) = \frac{x_5^2 - x_4^2 + x_3^2 - x_2^2 + x_1^2}{x_1^2 + x_2^2 + x_3^2 + x_4^2 + x_5^2}$$

is discontinuous at $\mathbf{0}$ because, as we saw in Example 15, $\lim_{x\to 0}$ does not exist. ■

As in $\mathbb{R}^2$ and $\mathbb{R}^3$, rational functions are continuous when their denominators are not zero.

Definition 8

(i) A **polynomial** $p(\mathbf{x}) = p(x_1, x_2, \ldots, x_n)$ in the n variables $x_1, x_2, \ldots, x_n$ is a finite sum of terms of the form $Ax_1^{m_1}x_2^{m_2}\cdots x_n^{m_n}$ where $m_1, m_2, \ldots, m_n$ are integers and A is a real number. The **degree** of p is the largest value of the sum $m = m_1 + m_2 + \cdots + m_n$ (i.e., the largest sum among the terms constituting $p(\mathbf{x})$).

(ii) A **rational function** $r(\mathbf{x}) = r(x_1, x_2, \ldots, x_n)$ in n variables is a function that can be written as the quotient of two polynomials:

$$r(\mathbf{x}) = \frac{p(\mathbf{x})}{q(\mathbf{x})}.$$

EXAMPLE 17 $p(x_1, x_2, x_3, x_4) = 2x_1^3x_2x_3^2x_4^4 + 5x_1x_2^7x_3^2x_4^3 - 11x_1^4x_2^3x_3^4x_4$ is a polynomial of degree 13, since the sums of the exponents of each of the three terms are $3 + 1 + 2 + 4 = 10$, $1 + 7 + 2 + 3 = 13$, and $4 + 3 + 4 + 1 = 12$, respectively, and 13 is the largest sum. ■

The following theorem, whose proof is omitted (but not difficult), extends the results of Theorem 1. The proof follows directly from an extension to $\mathbb{R}^n$ of the limit theorems of one-variable calculus.

Theorem 2

(i) Any polynomial p is continuous at any point in $\mathbb{R}^n$.

(ii) Any rational function $r = p/q$ is continuous at any point $\mathbf{x}_0$ for which $q(\mathbf{x}_0) \neq 0$. It is discontinuous when $q(\mathbf{x}_0) = 0$ because it is not defined at $\mathbf{x}_0$.

(iii) If f and g are continuous at $\mathbf{x}_0$, then $f + g$, $f - g$, and fg are continuous at $\mathbf{x}_0$.

(iv) If f and g are continuous at $\mathbf{x}_0$ and if $g(\mathbf{x}_0) \neq 0$, then f/g is continuous at $\mathbf{x}_0$.

(v) If f is continuous at $\mathbf{x}_0$ and if h is a function of one variable that is continuous at $f(\mathbf{x}_0)$, then the composite function $h \circ f$, defined by $(h \circ f)(\mathbf{x}_0) = h(f(\mathbf{x}_0))$, is continuous at $\mathbf{x}_0$.

EXAMPLE 18 $p(\mathbf{x}) = x_1^3 x_2^5 x_5^8 - 3x_1 x_2^4 x_4^2 + 5x_1^2 x_2^3 x_3^4 x_4^2 x_5$ is continuous at every $\mathbf{x}$ in $\mathbb{R}^5$. ■

EXAMPLE 19

$$r(\mathbf{x}) = \frac{x_1^2 x_5 x_4^4 + x_1 x_2^2 x_3^3 x_4^5 - x_1^6 x_2 x_5^3}{x_1 - 2x_2 + 3x_3 - 4x_4 + 2x_5 - 6}$$

is continuous at every $\mathbf{x}$ in $\mathbb{R}^5$ except at those $\mathbf{x}$ that satisfy

$$x_1 - 2x_2 + 3x_3 - 4x_4 + 2x_5 = 6. \quad ■ \tag{14}$$

REMARK. The set of vectors in $\mathbb{R}^5$ that satisfy (14) is called a **hyperplane.** In general, a hyperplane H in $\mathbb{R}^n$ is defined by

$$H = \{\mathbf{x}: a_1 x_1 + a_2 x_2 + \cdots + a_n x_n = b\} \tag{15}$$

where $a_1, a_2, \ldots, a_n$ and b are real numbers.

EXAMPLE 20 The function $\sin(x_1^2 + 2x_1 x_4 - x_3^4 x_5^5)$ is continuous at every $\mathbf{x}$ in $\mathbb{R}^5$ by part (v) of Theorem 2 since $\sin x$ is continuous and $x_1^2 + 2x_1 x_4 - x_3^4 x_5^5$ is a polynomial. ■

PROBLEMS 4.2

In Problems 1–5, sketch the indicated region.

1. The open disk centered at (3, 0) with radius 2.
2. The open disk centered at (3, 2) with radius 3.
3. The closed disk centered at (−1, 1) with radius 1.
4. The open ball centered at (1, 0, 0) with radius 1.
5. The closed ball centered at (0, 1, 1) with radius 2.

In Problems 6–11, use Definition 2 to verify the indicated limit.

6. $\lim_{(x,y)\to(1,2)} (3x + y) = 5$
7. $\lim_{(x,y)\to(3,-1)} (x - 7y) = 10$
8. $\lim_{(x,y)\to(5,-2)} (ax + by) = 5a - 2b$

***9.** $\lim_{(x,y)\to(1,1)} \frac{x}{y} = 1$

10. $\lim_{(x,y)\to(0,0)} \frac{2x^2y}{x^2 + y^2} = 0$

11. $\lim_{(x,y)\to(4,1)} (x^2 + 3y^2) = 19$

In Problems 12–21, show that the given limit does not exist.

12. $\lim_{(x,y)\to(0,0)} \frac{x + y}{x - y}$

13. $\lim_{(x,y)\to(0,0)} \frac{xy}{x^2 - y^2}$

14. $\lim_{(x,y)\to(0,0)} \frac{xy}{x^2 + y^2}$

15. $\lim_{(x,y)\to(0,0)} \frac{xy^3}{x^4 + y^4}$

16. $\lim_{(x,y)\to(0,0)} \frac{xy}{x^3 + y^3}$

17. $\lim_{(x,y)\to(0,0)} \frac{(x^2 + y^2)^2}{x^4 + y^4}$

18. $\lim_{(x,y)\to(0,0)} \frac{x^2 - 2y}{y^2 + 2x}$

19. $\lim_{(x,y)\to(0,0)} \frac{ax^2 + by}{cy^2 + dx}, a, b, c, d > 0$

20. $\lim_{(x,y,z)\to(0,0,0)} \frac{xy + 2xz + 3yz}{x^2 + y^2 + z^2}$

21. $\lim_{(x,y,z)\to(0,0,0)} \frac{xyz}{x^3 + y^3 + z^3}$

In Problems 22–25, show that the indicated limit exists and calculate it.

22. $\lim_{(x,y)\to(0,0)} \frac{3xy}{\sqrt{x^2 + y^2}}$

23. $\lim_{(x,y)\to(0,0)} \frac{5x^2y^2}{x^4 + y^2}$

24. $\lim_{(x,y)\to(0,0)} \frac{x^3 + y^3}{x^2 + y^2}$

25. $\lim_{(x,y,z)\to(0,0,0)} \frac{yx^2 + z^3}{x^2 + y^2 + z^2}$

In Problems 26–35, calculate the indicated limit.

26. $\lim_{(x,y)\to(-1,2)} (xy + 4y^2x^3)$

27. $\lim_{(x,y)\to(-1,2)} \frac{4x^3y^2 - 2xy^5 + 7y - 1}{3xy - y^4 + 3x^3}$

28. $\lim_{(x,y)\to(-4,3)} \frac{1 + xy}{1 - xy}$

29. $\lim_{(x,y)\to(\pi,\pi/3)} \ln(x + y)$

30. $\lim_{(x,y)\to(1,2)} \ln(1 + e^{x+y})$

31. $\lim_{(x,y)\to(2,5)} \sinh\left(\frac{x + 1}{y - 2}\right)$

***32.** $\lim_{(x,y)\to(1,1)} \frac{x - y}{x^2 - y^2}$

33. $\lim_{(x,y)\to(2,2)} \frac{x^3 - 2xy + 3x^2 - 2y}{x^2y + 4y^2 - 6x^2 + 24y}$

34. $\lim_{(x,y,z)\to(1,1,3)} \frac{xy^2 - 4xz^2 + 5yz}{3z^2 - 8z^3y^7x^4 + 7x - y + 2}$

35. $\lim_{(x,y,z)\to(4,1,3)} \ln(x - yz + 4x^3y^5z)$

In Problems 36–49, describe the maximum region over which the given function is continuous.

36. $f(x, y) = \sqrt{x - y}$

37. $f(x, y) = \frac{x^3 + 4xy^6 - 7x^4}{x^2 + y^2}$

38. $f(x, y) = \frac{x^3 + 4xy^6 - 7x^4}{x^3 - y^3}$

***39.** $f(x, y) = \frac{xy^3 - 17x^2y^5 + 8x^3y}{xy + 3y - 4x - 12}$

40. $f(x, y) = \ln(3x + 2y + 6)$

41. $f(x, y) = \tan^{-1}(x - y)$

42. $f(x, y) = e^{xy+2}$

43. $f(x, y) = \frac{x^3 - 1 + 3y^5x^2}{1 - xy}$

44. $f(x, y) = \frac{x}{\sqrt{1 - (x^2/4) - y^2}}$

45. $f(x, y, z) = e^{(xy + yz - \sqrt{x})}$

46. $f(x, y, z) = \frac{xyz^2 + yzx^2 - 3x^3yz^5}{x - y + 2z + 4}$

47. $f(x, y, z) = y \ln(xz)$

48. $f(x, y) = \cos^{-1}(x^2 - y)$

49. $f(x, y, z) = \frac{1}{\sqrt{1 - x^2 - y^2 - z^2}}$

50. Find a function $g(x)$ such that the function

$$f(x, y) = \begin{cases} \frac{x^2 - y^2}{x - y} & x \neq y \\ g(x) & x = y \end{cases}$$

is continuous at every point in $\mathbb{R}^2$.

51. Find a number c such that the function

$$f(x, y) = \begin{cases} \dfrac{3xy}{\sqrt{x^2 + y^2}} & (x, y) \neq (0, 0) \\ c & (x, y) = (0, 0) \end{cases}$$

is continuous at the origin.

52. Find a number c such that

$$f(x, y) = \begin{cases} \dfrac{xy}{|x| + |y|} & \text{if } (x, y) \neq (0, 0) \\ c & \text{if } (x, y) = (0, 0) \end{cases}$$

is continuous at the origin.

***53.** Discuss the continuity at the origin of

$$f(x, y, z) = \begin{cases} \dfrac{yz - x^2}{x^2 + y^2 + z^2} & \text{if } (x, y, z) \neq (0, 0, 0) \\ 0 & \text{if } (x, y, z) = (0, 0, 0). \end{cases}$$

In Problems 54–56 use Definition 6 to verify the indicated limit.

54. $\lim_{\mathbf{x} \to (1, -1, 2, 3)} (x_1 - 2x_2 + 3x_3 - x_4) = 6$

55. $\lim_{\mathbf{x} \to (-1, 2, 4, -1, 3)} (2x_1 - 3x_2 + x_3 - 4x_4 + 3x_5) = 9$

56. $\lim_{\mathbf{x} \to (1, -1, 0, 2)} (x_1^2 + x_2^2 + x_3^2 - x_4^2) = -2$

In Problems 57–60 show that the indicated limit does not exist.

57. $\lim_{\mathbf{x} \to \mathbf{0}} \dfrac{x_1 + x_2 + x_3 + x_4}{x_1 - x_2 + x_3 - x_4}$

58. $\lim_{\mathbf{x} \to \mathbf{0}} \dfrac{x_1x_2x_3x_4}{x_1^4 - x_2^4 + x_3^4 - x_4^4}$

59. $\lim_{\mathbf{x} \to \mathbf{0}} \dfrac{x_1x_2x_3x_4}{x_1^4 + x_2^4 + x_3^4 + x_4^4}$

60. $\lim_{\mathbf{x} \to \mathbf{0}} \dfrac{x_1x_4^3 + x_2x_3^3 + x_5x_2^3}{x_1^4 + x_2^4 + x_3^4 + x_4^4 + x_5^4}$

In Problems 61–63 show that the given limit exists and calculate it.

61. $\lim_{\mathbf{x} \to \mathbf{0}} \dfrac{2x_1x_2x_3x_4}{\sqrt{x_1^2 + x_2^2 + x_3^2 + x_4^2}}$

62. $\lim_{\mathbf{x} \to \mathbf{0}} \dfrac{x_1x_2^2 + x_2x_3^2 + x_3x_4^2 + x_4x_1^2}{x_1^2 + x_2^2 + x_3^2 + x_4^2}$

63. $\lim_{\mathbf{x} \to \mathbf{0}} \dfrac{2x_1^2x_2^2 - 3x_2^2x_3^2 + 5x_3x_4^2}{x_1^3 + x_2^3 + x_3^3 + x_4^3}$

In Problems 64–69 calculate the indicated limit.

64. $\lim_{\mathbf{x} \to (1, -1, 2, 3)} (x_1^2 - 4x_2^2 + 5x_3x_4)$

65. $\lim_{\mathbf{x} \to (0, 1, -2, 4)} \dfrac{x_1x_3}{x_2x_4}$

66. $\lim_{\mathbf{x} \to (\pi/2, \pi/4, \pi/3, \pi/6)} \sin(x_1 - 2x_2 + 3x_3 + 5x_4)$

67. $\lim_{\mathbf{x} \to (1, 0, -1, 2, 3)} \ln\left(x_1^2 + x_2^2 + x_3^2 + \dfrac{x_4}{x_5}\right)$

68. $\lim_{\mathbf{x} \to (-2, 3, 1, 4, 6)} \dfrac{x_1 - x_2x_3^3 + x_4^5x_5}{x_1^2 + x_2x_3x_4x_5}$

69. $\lim_{\mathbf{x} \to (-2, 3, 1, 4, 6)} \sqrt{\dfrac{x_1 - x_2x_3^3 + x_4^5x_5}{x_1^2 + x_2x_3x_4x_5}}$

In Problems 70–73 describe the maximum region over which the given function is continuous.

70. $f(x_1, x_2, x_3, x_4) = \sqrt{x_1 + x_2 + x_3 + x_4}$

71. $f(x_1, x_2, x_3, x_4) = \dfrac{x_1x_2}{x_3x_4}$

72. $f(x_1, x_2, x_3, x_4) = \ln(x_1^2 + x_2^2 + x_3^2 + x_4^2)$

73. $f(x_1, x_2, x_3, x_4, x_5) = \dfrac{1}{\sqrt{1 - x_1^2 - x_2^2 - x_3^2 - x_4^2 - x_5^2}}$

74. Find a number c such that the following function is continuous at the origin:

$$f(x_1, x_2, x_3, x_4) = \begin{cases} \dfrac{2(x_1x_2 + x_3x_4)}{\sqrt{x_1^2 + x_2^2 + x_3^2 + x_4^2}} & \mathbf{x} \neq \mathbf{0}, \\ c & \mathbf{x} = \mathbf{0}. \end{cases}$$

4.3 PARTIAL DERIVATIVES

In this section we show one of the ways a function of several variables can be differentiated. The idea is simple. Let $z = f(x, y)$. If we keep one of the variables, say y, fixed, then f can be treated as a function of x only and we can calculate the derivative (if it exists) of f with respect to x. This new function is called the *partial derivative of f with respect to* x and is denoted $\partial f/\partial x$.[†] Before giving a more formal definition, we give an example.

EXAMPLE 1 Let $z = f(x, y) = x^2y + \sin xy^2$. Calculate $\partial f/\partial x$.

Solution. Treating y as if it were a constant, we have

$$\frac{\partial f}{\partial x} = \frac{\partial}{\partial x}(x^2y + \sin xy^2) = \frac{\partial}{\partial x}x^2y + \frac{\partial}{\partial x}\sin xy^2 = 2xy + y^2\cos xy^2. \blacksquare$$

Definition 1 PARTIAL DERIVATIVES Let $z = f(x, y)$.

(i) The **partial derivative of f with respect to x** is the function

$$\frac{\partial z}{\partial x} = \frac{\partial f}{\partial x} = \lim_{\Delta x \to 0} \frac{f(x + \Delta x, y) - f(x, y)}{\Delta x}. \tag{1}$$

The partial derivative $\partial f/\partial x$ is defined at every point (x, y) in the domain of f such that the limit (1) exists.

(ii) The **partial derivative of f with respect to y** is the function

$$\frac{\partial z}{\partial y} = \frac{\partial f}{\partial y} = \lim_{\Delta y \to 0} \frac{f(x, y + \Delta y) - f(x, y)}{\Delta y}. \tag{2}$$

The partial derivative $\partial f/\partial y$ is defined at every point (x, y) in the domain of f such that the limit (2) exists.

REMARK 1. This definition allows us to calculate partial derivatives in the same way we calculate ordinary derivatives by allowing only one of the variables to vary. It also allows us to use all the formulas from one-variable calculus.

REMARK 2. The partial derivatives $\partial f/\partial x$ and $\partial f/\partial y$ give us the rate of change of f as each of the variables x and y change with the other one held fixed. They do *not* tell us how f changes when x and y change simultaneously. We will discuss this different topic in Section 4.5.

REMARK 3. It should be emphasized that while the functions $\partial f/\partial x$ and $\partial f/\partial y$ are computed with one of the variables held constant, each is a function of both variables.

EXAMPLE 2 Let $f(x, y) = \sqrt{x + y^2}$. Calculate $\partial f/\partial x$ and $\partial f/\partial y$.

[†]The symbol ∂ of partial derivatives is not a letter from any alphabet but an invented mathematical symbol, which may be read "partial." Historically, the difference between an ordinary and partial derivative was not recognized at first, and the same symbol d was used for both. The symbol ∂ was introduced in the eighteenth century by the mathematicians Alexis Fontaine dés Bertins (1705–1771), Leonhard Euler (1707–1783), Alexis-Claude Clairaut (1713–1765), and Jean Le Rond d'Alembert (1717–1783) and was used in their development of the theory of partial differentiation.

Solution.

$$\frac{\partial f}{\partial x} = \frac{1}{2\sqrt{x + y^2}} \frac{\partial}{\partial x}(x + y^2) = \frac{1}{2\sqrt{x + y^2}}(1 + 0) = \frac{1}{2\sqrt{x + y^2}}$$

since we are treating y as a constant.

$$\frac{\partial f}{\partial y} = \frac{1}{2\sqrt{x + y^2}} \frac{\partial}{\partial y}(x + y^2) = \frac{1}{2\sqrt{x + y^2}}(0 + 2y) = \frac{y}{\sqrt{x + y^2}}$$

since we are treating x as a constant. ■

EXAMPLE 3 Let $f(x, y) = (1 + x^2 + y^5)^{4/3}$. Calculate $\partial f/\partial x$ and $\partial f/\partial y$ at the point (3, 1).

Solution.

$$\begin{aligned}\frac{\partial f}{\partial x} &= \frac{4}{3}(1 + x^2 + y^5)^{1/3} \frac{\partial}{\partial x}(1 + x^2 + y^5) \\ &= \frac{4}{3}(1 + x^2 + y^5)^{1/3} \cdot 2x = \frac{8x}{3}(1 + x^2 + y^5)^{1/3}.\end{aligned}$$

At (3, 1),

$$\frac{\partial f}{\partial x} = \frac{(8)(3)}{3}(1 + 3^2 + 1^5)^{1/3} = 8\sqrt[3]{11}.$$

Also,

$$\begin{aligned}\frac{\partial f}{\partial y} &= \frac{4}{3}(1 + x^2 + y^5)^{1/3} \frac{\partial}{\partial y}(1 + x^2 + y^5) \\ &= \frac{4}{3}(1 + x^2 + y^5)^{1/3} \cdot 5y^4 = \frac{20y^4}{3}(1 + x^2 + y^5)^{1/3}.\end{aligned}$$

At (3, 1),

$$\frac{\partial f}{\partial y} = \frac{20}{3}\sqrt[3]{11}. \quad ■$$

We now obtain a geometric interpretation of the partial derivative. Let $z = f(x, y)$. As we saw in Section 4.1, this is the equation of a surface in $\mathbb{R}^3$. To obtain $\partial z/\partial x$, we hold y fixed at some constant value y_0. The equation $y = y_0$ is a plane in space parallel to the xz-plane (whose equation is $y = 0$). Thus if y is constant, $\partial z/\partial x$ is the rate of change of f with respect to x as x changes along the curve C, which is at the intersection of the surface $z = f(x, y)$ and the plane $y = y_0$. This is illustrated in Figure 1. To be more precise, if (x_0, y_0, z_0) is a point on the surface $z = f(x, y)$, then $\partial z/\partial x$ evaluated at (x_0, y_0) is the slope of the line tangent to the surface at the point

(x_0, y_0, z_0) that lies in the plane $y = y_0$. Analogously, $\partial z/\partial y$ evaluated at (x_0, y_0) is the slope of the line tangent to the surface at the point (x_0, y_0, z_0) that lies in the plane $x = x_0$ (since x is held fixed in order to calculate $\partial z/\partial y$). This is illustrated in Figure 2.

There are other ways to denote partial derivatives. We will often write

$$f_x = \frac{\partial f}{\partial x} \quad \text{and} \quad f_y = \frac{\partial f}{\partial y}. \tag{3}$$

If f is a function of other variables, say s and t, then we may write $\partial f/\partial s = f_s$ and $\partial f/\partial t = f_t$.

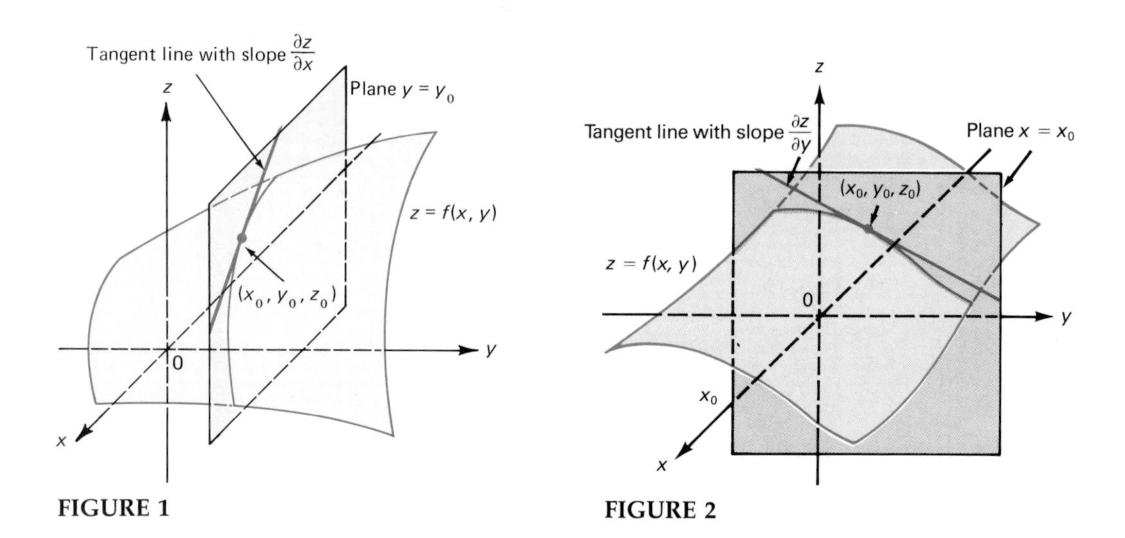

FIGURE 1

FIGURE 2

EXAMPLE 4 The volume of a cone of radius r and height h is given by

$$V = \tfrac{1}{3}\pi r^2 h.$$

Then the rate of change of V with respect to r (with h fixed) is given by

$$\frac{\partial V}{\partial r} = V_r = \frac{\partial}{\partial r}\left(\frac{1}{3}\pi r^2 h\right) = \frac{2}{3}\pi r h,$$

and the rate of change of V with respect to h (with r fixed) is given by

$$\frac{\partial V}{\partial h} = V_h = \frac{\partial}{\partial h}\left(\frac{1}{3}\pi r^2 h\right) = \frac{1}{3}\pi r^2. \quad \blacksquare$$

We now turn to the question of finding partial derivatives of functions of three variables.

Definition 2 PARTIAL DERIVATIVES Let $w = f(x, y, z)$ be defined in a neighborhood of the point (x, y, z).

(i) The **partial derivative of f with respect to x** is the function

$$\frac{\partial w}{\partial x} = \frac{\partial f}{\partial x} = f_x = \lim_{\Delta x \to 0} \frac{f(x + \Delta x, y, z) - f(x, y, z)}{\Delta x}. \tag{4}$$

f_x is defined at every point (x, y, z) in the domain of f at which the limit in (4) exists.

(ii) The **partial derivative of f with respect to y** is the function

$$\frac{\partial w}{\partial y} = \frac{\partial f}{\partial y} = f_y = \lim_{\Delta y \to 0} \frac{f(x, y + \Delta y, z) - f(x, y, z)}{\Delta y}. \tag{5}$$

f_y is defined at every point (x, y, z) in the domain of f at which the limit in (5) exists.

(iii) The **partial derivative of f with respect to z** is the function

$$\frac{\partial w}{\partial z} = \frac{\partial f}{\partial z} = f_z = \lim_{\Delta z \to 0} \frac{f(x, y, z + \Delta z) - f(x, y, z)}{\Delta z}. \tag{6}$$

f_z is defined at any point (x, y, z) in the domain of f at which the limit in (6) exists.

REMARK. As can be seen from Definition 2, to calculate $\partial f/\partial x$, we simply treat y and z as constants and then calculate an ordinary derivative.

EXAMPLE 5 Let $w = f(x, y, z) = xz + e^{y^2 z} + \sqrt{xy^2z^3}$. Calculate $\partial w/\partial x$, $\partial w/\partial y$, and $\partial w/\partial z$.

Solution. To calculate $\partial w/\partial x$, we keep y and z fixed. Then

$$\frac{\partial w}{\partial x} = \frac{\partial f}{\partial x} = f_x = \frac{\partial}{\partial x} xz + \frac{\partial}{\partial x} e^{y^2 z} + \frac{1}{2\sqrt{xy^2z^3}} \frac{\partial}{\partial x}(xy^2z^3)$$

$$= z + 0 + \frac{y^2z^3}{2\sqrt{xy^2z^3}} = z + \frac{y^2z^3}{2\sqrt{xy^2z^3}}.$$

To calculate $\partial w/\partial y$, we keep x and z fixed. Then

$$\frac{\partial w}{\partial y} = \frac{\partial f}{\partial y} = f_y = \frac{\partial}{\partial y} xz + e^{y^2 z} \frac{\partial}{\partial y}(y^2 z) + \frac{1}{2\sqrt{xy^2z^3}} \frac{\partial}{\partial y}(xy^2z^3)$$

$$= 0 + 2yze^{y^2 z} + \frac{2xyz^3}{2\sqrt{xy^2z^3}} = 2yze^{y^2 z} + \frac{xyz^3}{\sqrt{xy^2z^3}}.$$

To calculate $\partial w/\partial z$, we keep x and y fixed. Then

$$\frac{\partial w}{\partial z} = \frac{\partial f}{\partial z} = f_z = x + e^{y^2 z} \frac{\partial}{\partial z} y^2 z + \frac{1}{2\sqrt{xy^2z^3}} \frac{\partial}{\partial z}(xy^2z^3)$$

$$= x + y^2 e^{y^2 z} + \frac{3xy^2z^2}{2\sqrt{xy^2z^3}}. \quad \blacksquare$$

EXAMPLE 6 Let C denote the oxygen consumption (per unit weight) of a fur-bearing animal, let T denote its internal body temperature (in degrees Celsius), let t denote the outside temperature of its fur, and let w denote its weight (in kilograms). It has been experimentally determined that if T is considerably larger than t, then a reasonable model for the oxygen consumption of the animal is given by

$$C = \frac{5(T - t)}{2w^{2/3}}.$$

Calculate (a) C_T, (b) C_t, and (c) C_w.

Solution.

(a) $C_T = \dfrac{\partial C}{\partial T} = \dfrac{\partial}{\partial T}\left(\dfrac{5T}{2w^{2/3}} - \dfrac{5t}{2w^{2/3}}\right) = \dfrac{5}{2w^{2/3}}$

(b) $C_t = -\dfrac{5}{2w^{2/3}}$

(c) $C_w = \dfrac{\partial}{\partial w}\dfrac{5}{2}(T - t)w^{-2/3} = -\dfrac{5}{3}(T - t)w^{-5/3}$

Note that (a) and (b) imply that since $w > 0$, an increase in internal body temperature leads to an increase in oxygen consumption (if t and w do not change) while an increase in fur temperature leads to a decrease in oxygen consumption (with T and w held constant). Does this result make sense intuitively? Furthermore, if T and t are held constant, then since $T > t$, an increase in the animal's weight will lead to a decrease in its oxygen consumption. ■

We know that a differentiable function of one variable is continuous. The situation is more complicated for a function of two or more variables. In fact, there are functions that are discontinuous at a point but for which all partial derivatives exist at that point.

EXAMPLE 7 Let

$$f(x, y) = \begin{cases} \dfrac{xy}{x^2 + y^2} & (xy) \neq (0, 0) \\ 0 & (x, y) = (0, 0). \end{cases}$$

Show that f_x and f_y exist at (0, 0) but that f is not continuous there.

Solution. We first show that $\lim_{(x,y)\to(0,0)} f(x, y)$ does not exist, so that f cannot be continuous at (0, 0). To show that, we first let $(x, y) \to (0, 0)$ along the line $y = x$. Then

$$\frac{xy}{x^2 + y^2} = \frac{x^2}{x^2 + x^2} = \frac{1}{2},$$

so if $\lim_{(x,y)\to(0,0)} f(x, y)$ existed, it would have to equal $\frac{1}{2}$. But if we now let $(x, y) \to (0, 0)$ along the line $y = -x$, we have

$$\frac{xy}{x^2+y^2} = \frac{-x^2}{x^2+x^2} = -\frac{1}{2},$$

so along this line the limit is $-\frac{1}{2}$. Hence the limit does not exist. On the other hand, we have

$$f_x(0, 0) = \lim_{\Delta x \to 0} \frac{f(0+\Delta x, 0) - f(0, 0)}{\Delta x} = \lim_{\Delta x \to 0} \frac{\dfrac{(0+\Delta x)\cdot 0}{\Delta x^2 + 0^2}}{\Delta x}$$

$$= \lim_{\Delta x \to 0} \frac{0}{\Delta x} = \lim_{\Delta x \to 0} 0 = 0.$$

Similarly, $f_y(0, 0) = 0$. Hence both f_x and f_y exist at (0, 0) (and are equal to zero) even though f is not continuous there. ■

In Section 4.5 we will show a relationship between continuity and partial derivatives and show that a certain kind of differentiability does imply continuity.

We now define partial derivatives in $\mathbb{R}^n$.

Definition 3 PARTIAL DERIVATIVES Let $f: \mathbb{R}^n \to \mathbb{R}$. Then the **partial derivative of f with respect to** x_i is the function

$$\frac{\partial f}{\partial x_i} = \lim_{\Delta x_i \to 0} \frac{f(x_1, x_2, \ldots, x_{i-1}, x_i + \Delta x_i, x_{i+1}, \ldots, x_n) - f(x_1, x_2, \ldots, x_i, \ldots, x_n)}{\Delta x_i}. \quad (7)$$

The partial derivative $\partial f/\partial x_i$ is defined at every point $\mathbf{x} = (x_1, x_2, \ldots, x_n)$ in the domain of f such that the limit in (7) exists.

REMARK 1. Equation (7) defines n functions. That is, $\partial f/\partial x_i$ is a function of $(x_1, x_2, \ldots, x_n)$ for $i = 1, 2, \ldots, n$.

REMARK 2. As in the case in which $n = 2$ or 3, $\partial f/\partial x_i$ is computed by treating all variables except the ith one as if they were fixed.

EXAMPLE 8 Let $f(\mathbf{x}) = x_1^2 - x_2^2 + 3x_1x_2x_3 - (x_4/x_1)$. Compute $\partial f/\partial x_i$ for $i = 1, 2, 3, 4$.

Solution.

(a) $\dfrac{\partial f}{\partial x_1} = 2x_1 + 3x_2x_3 + \dfrac{x_4}{x_1^2}$.

(b) $\dfrac{\partial f}{\partial x_2} = -2x_2 + 3x_1x_3$.

(c) $\dfrac{\partial f}{\partial x_3} = 3x_1x_2$.

(d) $\dfrac{\partial f}{\partial x_4} = -\dfrac{1}{x_1}$. ■

NOTATION. We shall write f_i to denote the partial derivative of f with respect to the ith variable. Thus in Example 8 we have

$$f_1 = 2x_1 + 3x_2x_3 + \frac{x_4}{x_1^2}, \qquad f_2 = -2x_3 + 3x_1x_3,$$

and so on.

PROBLEMS 4.3

In Problems 1–12, calculate $\partial z/\partial x$ and $\partial z/\partial y$.

1. $z = x^2y$
2. $z = xy^2$
3. $z = 3e^{xy^3}$
4. $z = \sin(x^2 + y^3)$
5. $z = 4x/y^5$
6. $z = e^y \tan x$
7. $z = \ln(x^3y^5 - 2)$
8. $z = \sqrt{xy + 2y^3}$
9. $z = (x + 5y \sin x)^{4/3}$
10. $z = \sinh(2x - y)$

*11. $z = \sin^{-1}\left(\dfrac{x^2 - y^2}{x^2 + y^2}\right)$

12. $z = \sec xy$

In Problems 13–20, calculate the value of the given partial derivative at the given point.

13. $f(x, y) = x^3 - y^4; f_x(1, -1)$
14. $f(x, y) = \ln(x^2 + y^4); f_y(3, 1)$
15. $f(x, y) = \sin(x + y); f_x(\pi/6, \pi/3)$
16. $f(x, y) = e^{\sqrt{x^2+y}}; f_y(0, 4)$
17. $f(x, y) = \sinh(x - y); f_x(3, 3)$
18. $f(x, y) = \sqrt[3]{x^2y - y^2x^5}; f_y(-2, 4)$
19. $f(x, y) = \dfrac{x^2 - y^2}{x^2 + y^2}; f_y(2, -3)$
20. $f(x, y) = \tan^{-1}\dfrac{y}{x}; f_x(4, 4)$

In Problems 21–29, calculate $\partial w/\partial x$, $\partial w/\partial y$, and $\partial w/\partial z$.

21. $w = xyz$
22. $w = \sqrt{x + y + z}$
23. $w = \dfrac{x + y}{z}$
24. $w = \dfrac{x^2 - y^2 + z^2}{x^2 + y^2 + z^2}$
25. $w = \ln(x^3 + y^2 + z)$
26. $w = e^{x+2y+3x}$
27. $w = \sin(xyz)$
28. $w = \tan^{-1}\dfrac{xz}{y}$
29. $w = \cosh\sqrt{x + 2y + 5z}$

In Problems 30–39, calculate the value of the given partial derivative at the given point.

30. $f(x, y, z) = xyz; f_x(2, 3, 4)$
31. $f(x, y, z) = \sqrt{x + 2y + 3z}; f_y(2, -1, 3)$
32. $f(x, y, z) = \dfrac{x - y}{z}; f_z(-3, -1, 2)$
33. $f(x, y, z) = \sin(z^2 - y^2 + x); f_y(0, 1, 0)$
34. $f(x, y, z) = \ln(x + 2y + 3z); f_z(2, 2, 5)$
35. $f(x, y, z) = \tan^{-1}\dfrac{xy}{z}; f_x(1, 2, -2)$
36. $f(x, y, z) = \dfrac{y^3 - z^5}{x^2y + z}; f_y(4, 0, 1)$
37. $f(x, y, z) = \sqrt{\dfrac{x + y - z}{x + y + z}}; f_z(1, 1, 1)$
38. $f(x, y, z) = e^{xy}(\cosh z - \sinh z); f_z(2, 3, 0)$
39. $f(x, y, z) = \sqrt{x^2 + y^2 + z^2}; f_x(a, b, c)$
40. Let $f(x, y) = \begin{cases} \dfrac{x + y}{x - y} & (x, y) \neq (0, 0) \\ 0 & (x, y) = (0, 0). \end{cases}$
    (a) Show that f is not continuous at (0, 0). [*Hint:* Show that $\lim_{(x,y)\to(0,0)} f(x, y)$ does not exist.]
    (b) Do $f_x(0, 0)$ and $f_y(0, 0)$ exist?
41. A **partial differential equation** is an equation involving partial derivatives. Show that the function $z = f(x, y) = e^{(x+\sqrt{3}y)/4} - 4x - 2y - 4 - 2\sqrt{3}$ satisfies the partial differential equation $\partial z/\partial x + \sqrt{3}\, \partial z/\partial y - z = 4x + 2y$.
42. Show that the function $f(x, y) = z = e^{x+(5/4)y} - \frac{7}{2}e^{y/2} + \frac{7}{2}$ is a solution to the partial differential equation $3f_x - 4f_y + 2f = 7$ that also satisfies $f(x, 0) = e^x$.
43. Find the equation of the line tangent to the surface $z = x^3 - 4y^3$ at the point $(1, -1, 5)$ that (a) lies in the plane $x = 1$; (b) lies in the plane $y = -1$.

44. Find the equation of the line tangent to the surface $z = \tan^{-1}(y/x)$ at the point $(\sqrt{3}, 1, \pi/6)$ that (a) lies in the plane $x = \sqrt{3}$; (b) lies in the plane $y = 1$.

45. Find the equation of the line tangent to the surface $x^2 + 4y^2 + 4z^2 = 9$ that lies on the plane $y = 1$ at the point $(1, 1, 1)$.

46. The ideal gas law states that $P = nRT/V$. Assume that the number of moles n of an ideal gas and the temperature T of the gas are held constant at the values 10 and 20°C (= 293 °K), respectively. What is the rate of change of the pressure, as a function of the volume, when the volume of the gas is 2 liters?

47. For any ideal gas show that $(\partial V/\partial T)(\partial T/\partial P)(\partial P/\partial V) = -1$.

48. The present value of an **annuity** for which B dollars are to be paid every year for t years is given by

$$A_0 = \frac{B}{i}\left[1 - \left(\frac{1}{1+i}\right)^t\right].$$

(a) If t is fixed, how does the present value of the annuity change as the rate of interest changes?

(b) How is the present value changing with respect to i if $B = \$500$ and $i = 6\%$?

(c) If i is fixed, how does the present value of the annuity change as the number of years during which the payments are made is increased?

49. A fur-bearing animal weighing 10 kg has a constant internal body temperature of 23°C. Using the model of Example 6, if the outside temperature is dropping, how is the oxygen consumption of the animal changing when the outside temperature of its fur is 5°C?

50. The cost to a manufacturer of producing x units of product A and y units of product B is given (in dollars) by

$$C(x, y) = \frac{50}{2 + x} + \frac{125}{(3 + y)^2}.$$

Calculate the marginal cost of each of the two products.

51. The revenue received from the manufacturer of Problem 50 is given by

$$R(x, y) = \ln(1 + 50x + 75y) + \sqrt{1 + 40x + 125y}.$$

Calculate the marginal revenue from each of the two products.

52. If a particle is falling in a fluid, then according to **Stokes' law** the velocity of the particle is given by

$$V = \frac{2g}{9}(\rho_P - \rho_f)\frac{r^2}{\eta},$$

where g is the acceleration due to gravity, ρ_P is the density of the particle, ρ_f is the density of the fluid, r is the radius of the particle (in centimeters), and η is the absolute viscosity of the liquid. Calculate V_{ρ_P}, V_{ρ_f}, V_r, and V_η.

53. Let $f(x, y) = g(x)h(y)$, where g and h are differentiable functions of a single variable. Show that the partial derivatives $\partial f/\partial x$ and $\partial f/\partial y$ exist and that

$$\frac{\partial f}{\partial x} = g'(x)h(y) \quad \text{and} \quad \frac{\partial f}{\partial y} = g(x)h'(y).$$

54. Let g, h, and k be differentiable and let $f(x, y, z) = g(x)h(y)k(z)$. Show that $\partial f/\partial x$, $\partial f/\partial y$, and $\partial f/\partial z$ all exist and

$$\frac{\partial f}{\partial x} = g'(x)h(y)k(z), \quad \frac{\partial f}{\partial y} = g(x)h'(y)k(z),$$

$$\frac{\partial f}{\partial z} = g(x)h(y)k'(z).$$

In Problems 55–58, assume that $\partial f/\partial x$ and $\partial g/\partial x$ exist at every point (x, y) in $\mathbb{R}^2$.

55. Prove that $(\partial/\partial x)(f + g)$ exists and

$$\frac{\partial}{\partial x}(f + g) = \frac{\partial f}{\partial x} + \frac{\partial g}{\partial x}.$$

56. Prove that $(\partial/\partial x)(af)$ exists for every constant a and $(\partial/\partial x)(af) = a(\partial f/\partial x)$.

57. Prove that $(\partial/\partial x)(fg)$ exists and is equal to $f(\partial g/\partial x) + g(\partial f/\partial x)$.

58. If $\partial g/\partial x \neq 0$, prove that $(\partial/\partial x)(f/g)$ exists and is equal to $[g(\partial f/\partial x) - f(\partial g/\partial x)]/g^2$.

In Problems 59–65 compute all first-order partial derivatives.

59. $f(x_1, x_2, x_3, x_4) = x_1x_2x_3x_4$

60. $f(x_1, x_2, x_3, x_4) = \sqrt{x_1^2 + x_2^2 + x_3^2 + x_4^2}$

61. $f(x_1, x_2, x_3, x_4) = \dfrac{x_1x_3}{x_2x_4} + e^{x_1x_3/x_2}$

62. $f(x_1, x_2, x_3, x_4, x_5) = x_1^2x_3 - x_2^2x_4 + \sin x_5$

63. $f(x_1, x_2, \ldots, x_n) = \left(\sum_{i=1}^{n} x_i^2\right)^{1/2}$

64. $f(x_1, x_2, \ldots, x_n) = x_1x_2 \cdots x_n$

65. $f(x_1, x_2, \ldots, x_n) = \left(\sum_{i=1}^{n} x_i^2\right)^{-1/2}$

4.4 HIGHER-ORDER PARTIAL DERIVATIVES

We have seen that if $y = f(x)$, then

$$y' = \frac{df}{dx} \quad \text{and} \quad y'' = \frac{d^2f}{dx^2} = \frac{d}{dx}\left(\frac{df}{dx}\right).$$

That is, the second derivative of f is the derivative of the first derivative of f. Analogously, if $z = f(x, y)$, then we can differentiate each of the two "first" partial derivatives $\partial f/\partial x$ and $\partial f/\partial y$ with respect to both x and y to obtain four **second partial derivatives** as follows:

Definition 1 SECOND PARTIAL DERIVATIVES

(i) Differentiate twice with respect to x:

$$\frac{\partial^2 z}{\partial x^2} = \frac{\partial^2 f}{\partial x^2} = f_{xx} = \frac{\partial}{\partial x}\left(\frac{\partial f}{\partial x}\right). \tag{1}$$

(ii) Differentiate first with respect to x and then with respect to y:

$$\frac{\partial^2 z}{\partial y\, \partial x} = \frac{\partial^2 f}{\partial y\, \partial x} = f_{xy} = \frac{\partial}{\partial y}\left(\frac{\partial f}{\partial x}\right). \tag{2}$$

(iii) Differentiate first with respect to y and then with respect to x:

$$\frac{\partial^2 z}{\partial x\, \partial y} = \frac{\partial^2 f}{\partial x\, \partial y} = f_{yx} = \frac{\partial}{\partial x}\left(\frac{\partial f}{\partial y}\right). \tag{3}$$

(iv) Differentiate twice with respect to y:

$$\frac{\partial^2 z}{\partial y^2} = \frac{\partial^2 f}{\partial y^2} = f_{yy} = \frac{\partial}{\partial y}\left(\frac{\partial f}{\partial y}\right). \tag{4}$$

REMARK 1. The derivatives $\partial^2 f/\partial x\,\partial y$ and $\partial^2 f/\partial y\,\partial x$ are called the **mixed second partials.**

REMARK 2. It is much easier to denote the second partials by f_{xx}, f_{xy}, f_{yx}, and f_{yy}. We will therefore use this notation for the remainder of this section. Note that the symbol f_{xy} indicates that we differentiate first with respect to x and then with respect to y.

EXAMPLE 1 Let $z = f(x, y) = x^3y^2 - xy^5$. Calculate the four second partial derivatives.

Solution. We have $f_x = 3x^2y^2 - y^5$ and $f_y = 2x^3y - 5xy^4$.

(a) $f_{xx} = \frac{\partial}{\partial x}(f_x) = 6xy^2$

(b) $f_{xy} = \frac{\partial}{\partial y}(f_x) = 6x^2y - 5y^4$

(c) $f_{yx} = \frac{\partial}{\partial x}(f_y) = 6x^2y - 5y^4$

(d) $f_{yy} = \frac{\partial}{\partial y}(f_y) = 2x^3 - 20xy^3$ ■

In Example 1 we saw that $f_{xy} = f_{yx}$. This result is no accident, as we see by the following theorem whose proof will be given at the end of this section.

Theorem 1 Suppose that f, f_x, f_y, f_{xy}, and f_{yx} are all continuous at (x_0, y_0). Then

$$f_{xy}(x_0, y_0) = f_{yx}(x_0, y_0). \tag{5}$$

This result is often referred to as the **equality of mixed partials.**†

The definition of second partial derivatives and the theorem on the equality of mixed partials are easily extended to functions of three variables. If $w = f(x, y, z)$, then we have the nine second partial derivatives (assuming that they exist):

$$\frac{\partial^2 f}{\partial x^2} = f_{xx} \qquad \frac{\partial^2 f}{\partial y\,\partial x} = f_{xy} \qquad \frac{\partial^2 f}{\partial z\,\partial x} = f_{xz}$$

$$\frac{\partial^2 f}{\partial x\,\partial y} = f_{yx} \qquad \frac{\partial^2 f}{\partial y^2} = f_{yy} \qquad \frac{\partial^2 f}{\partial z\,\partial y} = f_{yz}$$

$$\frac{\partial^2 f}{\partial x\,\partial z} = f_{zx} \qquad \frac{\partial^2 f}{\partial y\,\partial z} = f_{zy} \qquad \frac{\partial^2 f}{\partial z^2} = f_{zz}$$

Theorem 2 If f, f_x, f_y, f_z, and all six mixed partials are continuous at a point (x_0, y_0, z_0), then at that point

$$f_{xy} = f_{yx}; \qquad f_{xz} = f_{zx}; \qquad f_{yz} = f_{zy}.$$

EXAMPLE 2 Let $f(x, y, z) = xy^3 - zx^5 + x^2yz$. Calculate all nine second partial derivatives and show that all three pairs of mixed partials are equal.

Solution. We have

$$f_x = y^3 - 5zx^4 + 2xyz,$$
$$f_y = 3xy^2 + x^2z,$$

†This theorem was first stated by Euler in a 1734 paper devoted to a problem in hydrodynamics.

and

$$f_z = -x^5 + x^2y.$$

Then

$$f_{xx} = -20zx^3 + 2yz, \qquad f_{yy} = 6xy, \qquad f_{zz} = 0,$$

$$f_{xy} = \frac{\partial}{\partial y}(y^3 - 5zx^4 + 2xyz) = 3y^2 + 2xz,$$

$$f_{yx} = \frac{\partial}{\partial x}(3xy^2 + x^2z) = 3y^2 + 2xz,$$

$$f_{xz} = \frac{\partial}{\partial z}(y^3 - 5zx^4 + 2xyz) = -5x^4 + 2xy,$$

$$f_{zx} = \frac{\partial}{\partial x}(-x^5 + x^2y) = -5x^4 + 2xy,$$

$$f_{yz} = \frac{\partial}{\partial z}(3xy^2 + x^2z) = x^2,$$

$$f_{zy} = \frac{\partial}{\partial y}(-x^5 + x^2y) = x^2. \blacksquare$$

We can easily define partial derivatives of orders higher than two. For example,

$$f_{zyx} = \frac{\partial^3 f}{\partial x\, \partial y\, \partial z} = \frac{\partial}{\partial x}\left(\frac{\partial^2 f}{\partial y\, \partial z}\right) = \frac{\partial}{\partial x}(f_{zy}).$$

EXAMPLE 3 Calculate f_{xxx}, f_{xzy}, f_{yxz}, and f_{yxzx} for the function of Example 2.

Solution. We easily obtain the three third partial derivatives:

$$f_{xxx} = \frac{\partial}{\partial x}(f_{xx}) = \frac{\partial}{\partial x}(-20zx^3 + 2yz) = -60zx^2$$

$$f_{xzy} = \frac{\partial}{\partial y}(f_{xz}) = \frac{\partial}{\partial y}(-5x^4 + 2xy) = 2x$$

$$f_{yxz} = \frac{\partial}{\partial z}(f_{yx}) = \frac{\partial}{\partial z}(3y^2 + 2xz) = 2x.$$

Note that $f_{xzy} = f_{yxz}$. This again is no accident and follows from the generalization of Theorem 2 to mixed third partial derivatives. Finally, the fourth partial derivative f_{yxzx} is given by

$$f_{yxzx} = \frac{\partial}{\partial x}(f_{yxz}) = \frac{\partial}{\partial x}(2x) = 2. \blacksquare$$

Higher-order partial derivatives are defined exactly as in $\mathbb{R}^2$ and $\mathbb{R}^3$.

EXAMPLE 4 Let $f(\mathbf{x}) = x_1^2 - x_2^2 + 3x_1x_2x_3 - (x_4/x_1)$. Compute f_{13}, f_{31}, and f_{312}.

Solution. From Example 4.3.8, $f_3 = 3x_1x_2$, so that

$$f_{31} = \frac{\partial^2 f}{\partial x_1\, \partial x_3} = \frac{\partial}{\partial x_1}\left(\frac{\partial f}{\partial x_3}\right) = \frac{\partial}{\partial x_1}(3x_1x_2) = 3x_2.$$

Similarly,

$$f_{13} = \left(\frac{\partial f}{\partial x_1}\right)_3 = \left(2x_1 + 3x_2x_3 - \frac{x_4}{x_1^2}\right)_3 = 3x_2.$$

Finally,

$$f_{312} = (f_{31})_2 = \frac{\partial}{\partial x_2}(3x_2) = 3. \blacksquare$$

REMARK. In Example 4 we found that

$$\frac{\partial^2 f}{\partial x_1\, \partial x_3} = \frac{\partial^2 f}{\partial x_3\, \partial x_1}.$$

This is not a coincidence. In Theorem 1 we stated that $f_{xy} = f_{yx}$ if these partial derivatives are continuous. It is also true that if f is a function of n variables, and if all second-order partial derivatives are continuous, then all higher-order partial derivatives involving differentiation with respect to the same variables are equal. That is, the order in which we take successive partial derivatives makes no difference. The proof of the following theorem is difficult and can be omitted at the first reading.

Theorem 3 Suppose that $f: \mathbb{R}^n \to \mathbb{R}$ and, in an open set Ω, f, f_i, f_j, f_{ij}, and f_{ji} are continuous. Then $f_{ij} = f_{ji}$ at every $\mathbf{x} \in \Omega$.

Proof. We first prove the theorem in the case in which $n = 2$. That is, we show that $f_{12}(\mathbf{x}) = f_{21}(\mathbf{x})$ for every $\mathbf{x}$ in Ω. Let $\mathbf{x} = (x_0, y_0)$. Since Ω is an open set, there is a neighborhood $B_r(\mathbf{x}_0)$ contained in Ω. Let u be a number in the interval $(0, r/\sqrt{2})$ and define

$$g(u) = \frac{1}{u^2}[f(x_0 + u, y_0 + u) - f(x_0, y_0 + u) - f(x_0 + u, y_0) + f(x_0, y_0)]. \quad \textbf{(6)}$$

Note that since $0 < u < r/\sqrt{2}$,

$$|(x_0 + u, y_0 + u) - (x_0, y_0)| = \sqrt{u^2 + u^2} < \sqrt{r^2/2 + r^2/2} = r,$$

so that

$$(x_0 + u, y_0 + u) \in B_r(\mathbf{x}_0) \subset \Omega$$

and $f(x_0 + u, y_0 + u)$ is defined.

Now we define (for u fixed)

$$h(x) = f(x, y_0 + u) - f(x, y_0). \tag{7}$$

Since y_0 and u are fixed, h is a function of x only and

$$h'(x) = f_1(x, y_0 + u) - f_1(x, y_0).$$

Our assumptions on f, f_1, and f_2 imply that h and h' are continous on the interval $[x_0, x_0 + u]$, so we can apply the mean value theorem to obtain a number $c \in (x_0, x_0 + u)$ such that

$$h(x_0 + u) - h(x_0) = h'(c)u = [f_1(c, y_0 + u) - f_1(c, y_0)]u. \tag{8}$$

But then

$$g(u) = \frac{1}{u^2}[h(x_0 + u) - h(x_0)] = \frac{1}{u}[f_1(c, y_0 + u) - f_1(c, y_0)]. \tag{9}$$

Now let

$$\overline{h}(y) = f_1(c, y).$$

Since $\overline{h}'(y) = f_{12}(c, y)$ is continuous, we can again apply the mean value theorem —this time to (9)—to obtain

$$g(u) = \frac{1}{u}[\overline{h}(y_0 + u) - \overline{h}(y_0)] = \frac{1}{u}[\overline{h}'(d)]u = f_{12}(c, d) \tag{10}$$

where $d \in (y_0, y_0 + u)$. Next, we repeat our steps but instead differentiate first with respect to y. This gives us numbers c^* and d^* in $(x_0, x_0 + u)$ and $(y_0, y_0 + u)$, respectively, such that

$$g(u) = f_{21}(c^*, d^*). \tag{11}$$

We now let $u \to 0^+$. Since f_{12} and f_{21} are continuous, we have, combining (10) and (11),

$$\lim_{u\to 0+} g(u) = f_{12}(x_0, y_0) = f_{21}(x_0, y_0).$$

This proves the theorem in the case $n = 2$. If $n > 2$, assume that $i < j$. Let $\mathbf{x}_0 = (x_1^{(0)}, x_2^{(0)}, \ldots, x_n^{(0)})$ and define

$$F(x_i, x_j) = (x_1^{(0)}, x_2^{(0)}, \ldots, x_{i-1}^{(0)}, x_i, x_{i+1}^{(0)}, \ldots, x_{j-1}^{(0)}, x_j, x_{j+1}^{(0)}, \ldots, x_n^{(0)}). \tag{12}$$

That is, only x_i and x_j vary; the other variables stay fixed. But F is a function of two variables only, so that by the part we have already proved,

$$F_{ij}(x_i^{(0)}, x_j^{(0)}) = F_{ji}(x_i^{(0)}, x_j^{(0)}).$$

But

$$F_{ij}(x_i^{(0)}, x_j^{(0)}) = f_{ij}(\mathbf{x}_0) \quad \text{and} \quad F_{ji}(x_i^{(0)}, x_j^{(0)}) = f_{ji}(\mathbf{x}_0).$$

This completes the proof. ■

Let $F: \mathbb{R}^n \to \mathbb{R}$ be a function of n variables. Then the partial derivatives f_1, $f_2, \ldots, f_n$ are called the **first-order partial derivatives** of f. Similarly, the derivatives f_{ij} for $1 \le i, j \le n$ are called the **second-order partial derivatives** of f. Continuing in this manner, we define the **third-order partial derivatives** of f to be derivatives having the form

$$f_{ijk} = \frac{\partial^3 f}{\partial x_k\, \partial x_j\, \partial x_i} \quad \text{for} \quad 1 \le i, j, k \le n.$$

Finally, the **mth-order partial derivatives** are given by

$$f_{i_1 i_2 \cdots i_m} = \frac{\partial^m f}{\partial x_{i_m}\, \partial x_{i_{m-1}} \cdots \partial x_{i_1}}$$

(we emphasize that in this notation $i_1 + i_2 + \cdots + i_m = m$).

EXAMPLE 5 Let $f(x_1, x_2, x_3, x_4) = x_1^2 x_2^3 x_3^4 x_4^5$. Then one fifth-order partial derivative is

$$\begin{aligned} f_{34312}(\mathbf{x}) &= (f_3)_{4312} = (4x_1^2x_2^3x_3^3x_4^5)_{4312} = (20x_1^2x_2^3x_3^3x_4^4)_{312} \\ &= (60x_1^2x_2^3x_3^2x_4^4)_{12} = (120x_1x_2^3x_3^2x_4^4)_2 = 360x_1x_2^2x_3^2x_4^4. \end{aligned}$$ ■

Definition 2

(i) The function $f: \mathbb{R}^n \to \mathbb{R}$ is said to be of **class** $C^{(m)}(\Omega)$ if all partial derivatives of f of orders $\le m$ exist and are continuous in Ω.

(ii) The function $f: \mathbb{R}^n \to \mathbb{R}$ is said to be of **class** $C^{(\infty)}(\Omega)$ if f is of class $C^{(m)}(\Omega)$ for every integer $m \ge 1$.

We will make use of the ideas of $C^{(m)}(\Omega)$ and $C^{(\infty)}(\Omega)$ functions in Section 10.1. Here we simply note that any polynomial in n variables is a $C^{(\infty)}$ function in $\mathbb{R}^n$.

PROBLEMS 4.4

In Problems 1–12, calculate the four second partial derivatives, and show that the mixed partials are equal.

1. $f(x, y) = x^2y$
2. $f(x, y) = xy^2$
3. $f(x, y) = 3e^{xy^3}$
4. $f(x, y) = \sin(x^2 + y^3)$
5. $f(x, y) = \dfrac{4x}{y^5}$
6. $f(x, y) = e^y \tan x$
7. $f(x, y) = \ln(x^3y^5 - 2)$
8. $f(x, y) = \sqrt{xy + 2y^3}$
9. $f(x, y) = (x + 5y \sin x)^{4/3}$

10. $f(x, y) = \sinh(2x - y)$

***11.** $f(x, y) = \sin^{-1}\left(\dfrac{x^2 - y^2}{x^2 + y^2}\right)$ **12.** $f(x, y) = \sec xy$

In Problems 13–21, calculate the nine second partial derivatives, and show that the three pairs of mixed partials are equal.

13. $f(x, y, z) = xyz$ **14.** $f(x, y, z) = x^2y^3z^4$

15. $f(x, y, z) = \dfrac{x + y}{z}$

16. $f(x, y, z) = \sin(x + 2y + z^2)$

17. $f(x, y, z) = \tan^{-1}\dfrac{xz}{y}$ **18.** $f(x, y, z) = \cos xyz$

19. $f(x, y, z) = e^{3xy}\cos z$

20. $f(x, y, z) = \ln(xy + z)$

21. $f(x, y, z) = \cosh\sqrt{x + yz}$

22. How many third partial derivatives are there for a function of (a) two variables; (b) three variables?

23. How many fourth partial derivatives are there for a function of (a) two variables; (b) three variables?

***24.** How many nth partial derivatives are there for a function of (a) two variables; (b) three variables?

In Problems 25–30, calculate the given partial derivative.

25. $f(x, y) = x^2y^3 + 2y;\ f_{xyx}$

26. $f(x, y) = \sin(2xy^4);\ f_{xyy}$

27. $f(x, y) = \ln(3x - 2y);\ f_{yxy}$

28. $f(x, y, z) = x^2y + y^2z - 3\sqrt{xz};\ f_{xyz}$

29. $f(x, y, z) = \cos(x + 2y + 3z);\ f_{zzx}$

30. $f(x, y, z) = e^{xy}\sin z;\ f_{zxyx}$

31. Consider a string that is tightly stretched between two fixed points 0 and L on the x-axis. The string is pulled back and released at a time $t = 0$, causing it to vibrate. One position of the string is sketched in Figure 1. Let $y(x, t)$ denote the height of the string at any time $t \ge 0$ and at any point x in the interval $[0, L]$. It can be shown that $y(x, t)$ satisfies the partial differential equation[†]

$$\frac{\partial^2 y}{\partial t^2} = c^2\frac{\partial^2 y}{\partial x^2}$$

where c is a constant. This equation is called the one-dimensional **wave equation.** Show that

$$y(x, t) = \tfrac{1}{2}[(x - ct)^2 + (x + ct)^2]$$

is a solution to the wave equation, where c is any constant.

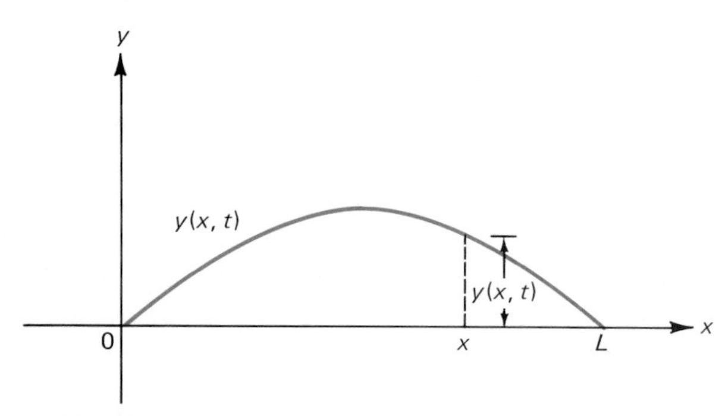

FIGURE 1

32. Consider a cylindrical rod composed of a uniform heat-conducting material of length l and radius r, and assume that heat can enter and leave the rod only through its ends. Let $T(x, t)$ denote the absolute temperature at time t at a point x units along the rod (see Figure 2). Then it can be shown[‡] that T satisfies the partial differential equation

$$\frac{\partial T}{\partial t} = \delta\frac{\partial^2 T}{\partial x^2},$$

where δ is a positive constant called the **diffusivity** of the rod. This equation is called the **heat equation** (or **diffusion equation**). Show that the function

$$T(x, t) = e^{-\alpha^2\delta t}\sin\alpha x$$

satisfies the heat equation for any constant α.

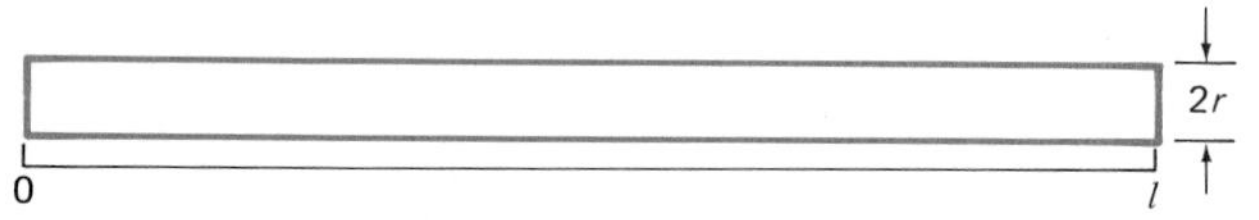

FIGURE 2

33. Show that $T(x, y) = (1/\sqrt{t})e^{-x^2/4\delta t}$ satisfies the heat equation.

34. Find constants α and β such that $T(x, t) = e^{\alpha x + \beta t}$ satisfies the heat equation.

[†]For a derivation of this equation, see W. Derrick and S. Grossman, *Elementary Differential Equations with Applications*, 2nd ed. (Addison-Wesley, Reading, Mass., 1981).

[‡]See Derrick and Grossman, *Elementary Differential Equations.*

35. One of the most important partial differential equations of mathematical physics is **Laplace's equation** *in* $\mathbb{R}^2$, given by

$$\frac{\partial^2 f}{\partial x^2} + \frac{\partial^2 f}{\partial y^2} = 0.$$

Laplace's equation can be used to model steady state heat flow in a closed, bounded region in $\mathbb{R}^2$. Show that the function $f(x, y) = x^2 - y^2$ satisfies Laplace's equation.

36. Show that $f(x, y) = \tan^{-1}(y/x)$ satisfies Laplace's equation.
37. Show that $f(x, y) = \ln(x^2 + y^2)$ satisfies Laplace's equation.
38. Show that $f(x, y) = \sin x \sinh y$ satisfies Laplace's equation.
39. Laplace's equation in $\mathbb{R}^3$ is given by

$$\frac{\partial^2 f}{\partial x^2} + \frac{\partial^2 f}{\partial y^2} + \frac{\partial^2 f}{\partial z^2} = 0.$$

Show that $f(x, y, z) = x^2 + y^2 - 2z^2$ satisfies Laplace's equation in $\mathbb{R}^3$.

40. What is the order of the partial derivative $f_{213214}(\mathbf{x})$?

In Problems 41–46 compute the given higher-order partial derivatives.

41. $f(x_1, x_2, x_3, x_4) = x_1x_2x_3x_4$; f_{13}, f_{11}, f_{134}
42. $f(x_1, x_2, x_3, x_4) = \sqrt{x_1^2 + x_2^2 + x_3^2 + x_4^2}$; f_{12}, f_{33}
43. $f(x_1, x_2, \ldots, x_n) = \left(\sum_{i=1}^{n} x_i^2\right)^{-1/2}$; f_{ij}
44. $f(x_1, x_2) = \sin x_1x_2$; f_{1211}
45. $f(x_1, x_2, x_3) = e^{x_1x_2x_3}$; f_{2211}
46. $f(x_1, x_2, \ldots, x_n) = \sum_{i=1}^{n} x_i^2$; $f_{123\cdots n}$
47. Let f be a function of three variables.
    (a) How many third-order partial derivatives does f have? (Assume that f_{121} and f_{112}, for example, are different third-order partial derivatives.)
    (b) How many fourth-order?
    (c) How many mth-order?

*48. Determine the number of mth-order partial derivatives of a function $f: \mathbb{R}^n \to \mathbb{R}$.

49. Let $p_k: \mathbb{R}^n \to \mathbb{R}$ be a polynomial of degree k. Show that if $p_k^{(k+1)}$ is a partial derivative of p_k of order $k + 1$, then $p_k^{(k+1)}(\mathbf{x}) = 0$ for every $\mathbf{x} \in \mathbb{R}^n$.

4.5 DIFFERENTIABILITY AND THE GRADIENT

In this section we discuss the notion of the differentiability of a function of several variables. There are several ways to introduce this subject and the way we have chosen is designed to illustrate the great similarities between differentiation of functions of one variable and differentiation of functions of several variables.

We begin with a function of one variable,

$$y = f(x).$$

If f is differentiable, then

$$f'(x) = \frac{dy}{dx} = \lim_{\Delta x \to 0} \frac{\Delta y}{\Delta x}. \tag{1}$$

Then if we define the new function $\epsilon(\Delta x)$ by

$$\epsilon(\Delta x) = \frac{\Delta y}{\Delta x} - f'(x), \tag{2}$$

we have

$$\lim_{\Delta x \to 0} \epsilon(\Delta x) = \lim_{\Delta x \to 0} \left(\frac{\Delta y}{\Delta x} - f'(x) \right) = \lim_{\Delta x \to 0} \frac{\Delta y}{\Delta x} - f'(x)$$
$$= f'(x) - f'(x) = 0. \qquad (3)$$

Multiplying both sides of (2) by Δx and rearranging terms, we obtain

$$\Delta y = f'(x)\, \Delta x + \epsilon(\Delta x)\, \Delta x.$$

Note that here Δy depends on both Δx and x. Finally, since $\Delta y = f(x + \Delta x) - f(x)$, we obtain

$$f(x + \Delta x) - f(x) = f'(x)\, \Delta x + \epsilon(\Delta x)\, \Delta x. \qquad (4)$$

Why did we do all this? We did so in order to be able to state the following alternative definition of differentiability of a function f of one variable.

Definition 1 ALTERNATIVE DEFINITION OF DIFFERENTIABILITY OF A FUNCTION OF ONE VARIABLE Let f be a function of one variable. Then f is **differentiable** at a number x if there is a function $f'(x)$ and a function $g(\Delta x)$ such that

$$f(x + \Delta x) - f(x) = f'(x)\, \Delta x + g(\Delta x), \qquad (5)$$

where $\lim_{\Delta x \to 0} [g(\Delta x)/\Delta x] = 0$.

We will soon show how the definition (5) can be extended to a function of two or more variables. First, we give a definition.

Definition 2 DIFFERENTIABILITY OF A FUNCTION OF TWO VARIABLES Let f be a real-valued function of two variables that is defined in a neighborhood of a point (x, y) and such that $f_x(x, y)$ and $f_y(x, y)$ exist. Then f is **differentiable** at (x, y) if there exist functions $\epsilon_1(\Delta x, \Delta y)$ and $\epsilon_2(\Delta x, \Delta y)$ such that

$$f(x + \Delta x, y + \Delta y) - f(x, y) = f_x(x, y)\, \Delta x + f_y(x, y)\, \Delta y + \epsilon_1(\Delta x, \Delta y)\, \Delta x + \epsilon_2(\Delta x, \Delta y)\, \Delta y, \qquad (6)$$

where

$$\lim_{(\Delta x, \Delta y) \to (0,0)} \epsilon_1(\Delta x, \Delta y) = 0 \quad \text{and} \quad \lim_{(\Delta x, \Delta y) \to (0,0)} \epsilon_2(\Delta x, \Delta y) = 0. \qquad (7)$$

EXAMPLE 1 Let $f(x, y) = xy$. Show that f is differentiable at every point (x, y) in $\mathbb{R}^2$.

Solution.

$$\begin{aligned} f(x + \Delta x, y + \Delta y) - f(x, y) &= (x + \Delta x)(y + \Delta y) - xy \\ &= xy + y\, \Delta x + x\, \Delta y + \Delta x\, \Delta y - xy \\ &= y\, \Delta x + x\, \Delta y + \Delta x\, \Delta y \end{aligned}$$

Now $f_x = y$ and $f_y = x$, so we have

$$f(x + \Delta x, y + \Delta y) - f(x, y) = f_x(x, y)\,\Delta x + f_y(x, y)\,\Delta y + \Delta y\,\Delta x + 0 \cdot \Delta y.$$

Setting $\epsilon_1(\Delta x, \Delta y) = \Delta y$ and $\epsilon_2(\Delta x, \Delta y) = 0$, we see that

$$\lim_{(\Delta x, \Delta y) \to (0,0)} \epsilon_1(\Delta x, \Delta y) = \lim_{(\Delta x, \Delta y) \to (0,0)} \epsilon_2(\Delta x, \Delta y) = 0.$$

This result shows that $f(x, y) = xy$ is differentiable at every point in $\mathbb{R}^2$. ■

We now rewrite our definition of differentiability in a more compact form. Since a point (x, y) is a vector in $\mathbb{R}^2$, we will write (as we have done before) $\mathbf{x} = (x, y)$. Then if $z = f(x, y)$, we can simply write

$$z = f(\mathbf{x}).$$

Similarly, if $w = f(x, y, z)$, we may write

$$w = f(\mathbf{x}),$$

where $\mathbf{x}$ is the vector (x, y, z). With this notation we may use the symbol $\mathbf{\Delta x}$ to denote the vector $(\Delta x, \Delta y)$ in $\mathbb{R}^2$ or $(\Delta x, \Delta y, \Delta z)$ in $\mathbb{R}^3$.

Next, we write

$$g(\mathbf{\Delta x}) = \epsilon_1(\Delta x, \Delta y)\,\Delta x + \epsilon_2(\Delta x, \Delta y)\,\Delta y. \tag{8}$$

Note that $(\Delta x, \Delta y) \to (0, 0)$ can be written in the compact form $\mathbf{\Delta x} \to \mathbf{0}$. Then if the conditions (7) hold, we see that

$$|\mathbf{\Delta x}| = \sqrt{\Delta x^2 + \Delta y^2}$$

$$\lim_{\mathbf{\Delta x} \to 0} \frac{|g(\mathbf{\Delta x})|}{|\mathbf{\Delta x}|} \le \lim_{\mathbf{\Delta x} \to 0} |\epsilon_1(\Delta x, \Delta y)| \frac{|\Delta x|}{\sqrt{\Delta x^2 + \Delta y^2}} + \lim_{\mathbf{\Delta x} \to 0} |\epsilon_2(\Delta x, \Delta y)| \frac{|\Delta y|}{\sqrt{\Delta x^2 + \Delta y^2}}$$

$$\frac{|\Delta x|}{\sqrt{\Delta x^2 + \Delta y^2}} \le 1$$

$$\le \lim_{\mathbf{\Delta x} \to 0} |\epsilon_1(\Delta x, \Delta y)| + \lim_{\mathbf{\Delta x} \to 0} |\epsilon_2(\Delta x, \Delta y)| = 0 + 0 = 0.$$

Finally, we have the following important definition.

Definition 3 THE GRADIENT Let f be a function of two variables such that f_x and f_y exist at a point $\mathbf{x} = (x, y)$. Then the **gradient** of f at $\mathbf{x}$, denoted $\nabla f(\mathbf{x})$, is given by

$$\nabla f(\mathbf{x}) = f_x(x, y)\mathbf{i} + f_y(x, y)\mathbf{j}. \tag{9}$$

Note that the gradient of f is a **vector function.** That is, for every point $\mathbf{x}$ in $\mathbb{R}^2$ for which $\nabla f(\mathbf{x})$ is defined, we see that $\nabla f(\mathbf{x})$ is a vector in $\mathbb{R}^2$.

REMARK. The gradient of f is denoted ∇f, which is read "del" f. This symbol, an inverted Greek delta, was first used in the 1850s, although the name "del" first appeared in print only in 1901. The symbol ∇ is also called *nabla*. This name is used because someone once suggested to the Scottish mathematician Peter Guthrie Tait (1831–1901) that ∇ looks like an Assyrian harp, the Assyrian name of which is nabla.[†] Tait, incidentally, was one of the mathematicians who helped carry on Hamilton's development of the theory of quaternions and vectors in the nineteenth century.

EXAMPLE 2 In Example 1, $f(x, y) = xy$, $f_x = y$, and $f_y = x$, so that

$$\nabla f(\mathbf{x}) = \nabla f(x, y) = y\mathbf{i} + x\mathbf{j}. \blacksquare$$

Using this new notation, we observe that

$$\nabla f(\mathbf{x}) \cdot \Delta\mathbf{x} = (f_x\mathbf{i} + f_y\mathbf{j}) \cdot (\Delta x\mathbf{i} + \Delta y\mathbf{j}) = f_x(x, y)\,\Delta x + f_y(x, y)\,\Delta y.$$

Also,

$$f(x + \Delta x, y + \Delta y) = f(\mathbf{x} + \Delta\mathbf{x}).$$

Thus we have the following definition, which is implied by Definition 2.

Definition 4 DIFFERENTIABILITY Let f be a function of two variables that is defined in a neighborhood of a point $\mathbf{x} = (x, y)$. Let $\Delta\mathbf{x} = (\Delta x, \Delta y)$. If $f_x(x, y)$ and $f_y(x, y)$ exist, then f is **differentiable** at $\mathbf{x}$ if there is a function g such that

$$f(\mathbf{x} + \Delta\mathbf{x}) - f(\mathbf{x}) = \nabla f(\mathbf{x}) \cdot \Delta\mathbf{x} + g(\Delta\mathbf{x}), \tag{10}$$

where

$$\lim_{\Delta x \to 0} \frac{g(\Delta\mathbf{x})}{|\Delta\mathbf{x}|} = 0. \tag{11}$$

REMARK. Although formulas (5) and (10) look very similar, there are two fundamental differences. First, f' is a scalar, while ∇f is a vector. Second, $f'(x)\,\Delta x$ is an ordinary product of real numbers, while $\nabla f(\mathbf{x}) \cdot \Delta\mathbf{x}$ is a dot product of vectors.

There are two reasons for giving you this new definition. First, it illustrates that the gradient of a function of two variables is the natural extension of the derivative of a function of one variable. Second, the definition (10), (11) can be used to define the notion of differentiability for a function of three or more variables as well. We will say more about this subject shortly.

One important question remains: What functions are differentiable? A partial answer is given in Theorem 1.

[†]Fortunately, most (but certainly not all) of the mathematical terms currently in use have more to do with the objects they describe. We might further point out that ∇ is also called *atled*, which is delta spelled backward (no kidding).

Theorem 1 Let f, f_x, and f_y be defined and continuous in a neighborhood of $\mathbf{x} = (x, y)$. Then f is differentiable at $\mathbf{x}$.

REMARK 1. It is essential that f_x and f_y be continuous in this theorem. In Example 4.3.6 we saw an example of a function for which f_x and f_y exist at (0, 0) but f itself is not continuous there. As we will see in Theorem 2, if f is differentiable at a point, then it is continuous there. Thus the function in Example 4.3.6 is not differentiable at (0, 0) even though $f_x(0, 0)$ and $f_y(0, 0)$ exist.

REMARK 2. If the hypotheses of Theorem 1 are satisfied, then f is said to be **continuously differentiable** at $\mathbf{x}$.

We will postpone the proof until later in the section.

EXAMPLE 3 Let $z = f(x, y) = \sin xy^2 + e^{x^2y^3}$. Show that f is differentiable and calculate ∇f. Find $\nabla f(1, 1)$.

Solution. $\partial f/\partial x = y^2 \cos xy^2 + 2xy^3e^{x^2y^3}$ and $\partial f/\partial y = 2xy \cos xy^2 + 3x^2y^2e^{x^2y^3}$. Since $\partial f/\partial x$ and $\partial f/\partial y$ are continuous, f is differentiable and

$$\nabla f(x, y) = (y^2 \cos xy^2 + 2xy^3e^{x^2y^3})\mathbf{i} + (2xy \cos xy^2 + 3x^2y^2e^{x^2y^3})\mathbf{j}.$$

At (1, 1), $\nabla f(1, 1) = (\cos 1 + 2e)\mathbf{i} + (2 \cos 1 + 3e)\mathbf{j}$. ■

In Section 4.3 we showed that the existence of all of its partial derivatives at a point does *not* ensure that a function is continuous at that point. However, differentiability (according to Definition 4) does ensure continuity.

Theorem 2 If f is differentiable at $\mathbf{x}_0 = (x_0, y_0)$, then f is continuous at $\mathbf{x}_0$.

Proof. We must show that $\lim_{(x,y)\to(x_0,y_0)} f(x, y) = f(x_0, y_0)$, or, equivalently, $\lim_{\mathbf{x}\to\mathbf{x}_0} f(\mathbf{x}) = f(\mathbf{x}_0)$. But if we define $\Delta\mathbf{x}$ by $\Delta\mathbf{x} = \mathbf{x} - \mathbf{x}_0$, this is the same as showing that

$$\lim_{\Delta\mathbf{x}\to\mathbf{0}} f(\mathbf{x}_0 + \Delta\mathbf{x}) = f(\mathbf{x}_0). \tag{12}$$

Since f is differentiable at $\mathbf{x}_0$,

$$f(\mathbf{x}_0 + \Delta\mathbf{x}) - f(\mathbf{x}_0) = \nabla f(\mathbf{x}_0) \cdot \Delta\mathbf{x} + g(\Delta\mathbf{x}). \tag{13}$$

But as $\Delta\mathbf{x} \to \mathbf{0}$, both terms on the right-hand side of (13) approach zero, so

$$\lim_{\Delta\mathbf{x}\to 0} [f(\mathbf{x}_0 + \Delta\mathbf{x}) - f(\mathbf{x}_0)] = 0,$$

which means that (12) holds and the theorem is proved. ■

The converse to this theorem is false, as it is in one-variable calculus. That is, there are functions that are continuous, but not differentiable, at a given point. For example, the function

$$f(x, y) = \sqrt[3]{x} + \sqrt[3]{y}$$

is continuous at any point (x, y) in $\mathbb{R}^2$. But

$$\nabla f(x, y) = \frac{1}{3x^{2/3}}\mathbf{i} + \frac{1}{3y^{2/3}}\mathbf{j},$$

so f is not differentiable at any point (x, y) for which either x or y is zero. That is, $\nabla f(x, y)$ is not defined on the x- and y-axes. Hence f is not differentiable along these axes.

For functions of one variable, we know that

$$(f + g)' = f' + g' \qquad \text{and} \qquad (\alpha f)' = \alpha f';$$

that is, the derivative of the sum of two functions is the sum of the derivatives of the two functions and the derivative of a scalar multiple of a function is the scalar times the derivative of the function. These results can be extended to the gradient vector.

Theorem 3 Let f and g be differentiable in a neighborhood of $\mathbf{x} = (x, y)$. Then for every scalar α, αf and $f + g$ are differentiable at $\mathbf{x}$, and

(i) $\nabla(\alpha f) = \alpha \nabla f$, and

(ii) $\nabla(f + g) = \nabla f + \nabla g$.

Proof.

(i) From the definition of differentiability (Definition 4), there is a function $h_1(\mathbf{\Delta x})$ such that

$$f(\mathbf{x} + \mathbf{\Delta x}) - f(\mathbf{x}) = \nabla f(\mathbf{x}) \cdot \mathbf{\Delta x} + h_1(\mathbf{\Delta x}),$$

where $\lim_{\mathbf{\Delta x}\to 0} [h_1(\mathbf{\Delta x})/|\mathbf{\Delta x}|] = 0$. Thus $\alpha f(\mathbf{x} + \mathbf{\Delta x}) - \alpha f(\mathbf{x}) = \alpha \nabla f(\mathbf{x}) \cdot \mathbf{\Delta x} + \alpha h_1(\mathbf{\Delta x})$, and

$$\lim_{\mathbf{\Delta x}\to 0} \frac{\alpha h_1(\mathbf{\Delta x})}{|\mathbf{\Delta x}|} = \alpha \lim_{\mathbf{\Delta x}\to 0} \frac{h_1(\mathbf{\Delta x})}{|\mathbf{\Delta x}|} = \alpha 0 = 0.$$

But

$$\alpha \frac{\partial f}{\partial x} = \frac{\partial(\alpha f)}{\partial x}$$

$$\alpha \nabla f(\mathbf{x}) = \alpha(f_x\mathbf{i} + f_y\mathbf{j}) = (\alpha f)_x\mathbf{i} + (\alpha f)_y\mathbf{j} = \nabla(\alpha f).$$

Thus

$$\alpha f(\mathbf{x} + \mathbf{\Delta x}) - \alpha f(\mathbf{x}) = \nabla \alpha f(\mathbf{x}) \cdot \mathbf{\Delta x} + \alpha h_1(\mathbf{\Delta x}),$$

which shows that αf is differentiable and $\nabla(\alpha f) = \alpha \nabla f$.

(ii) As above, there is a function $h_2(\mathbf{\Delta x})$ such that $g(\mathbf{x} + \mathbf{\Delta x}) - g(\mathbf{x}) = \nabla g(\mathbf{x}) \cdot \mathbf{\Delta x} + h_2(\mathbf{\Delta x})$, where $\lim_{\mathbf{\Delta x}\to 0} [h_2(\mathbf{\Delta x})/|\mathbf{\Delta x}|] = 0$. Thus

$$\begin{aligned}(f + g)(\mathbf{x} + \mathbf{\Delta x}) - (f + g)(\mathbf{x}) &= [f(\mathbf{x} + \mathbf{\Delta x}) + g(\mathbf{x} + \mathbf{\Delta x})] - [f(\mathbf{x}) + g(\mathbf{x})] \\ &= [f(\mathbf{x} + \mathbf{\Delta x}) - f(\mathbf{x})] + [g(\mathbf{x} + \mathbf{\Delta x}) - g(\mathbf{x})] \\ &= \nabla f(\mathbf{x}) \cdot \mathbf{\Delta x} + h_1(\mathbf{\Delta x}) + \nabla g(\mathbf{x}) \cdot \mathbf{\Delta x} + h_2(\mathbf{\Delta x}) \\ &= [\nabla f(\mathbf{x}) + \nabla g(\mathbf{x})] \cdot \mathbf{\Delta x} + [h_1(\mathbf{\Delta x}) + h_2(\mathbf{\Delta x})],\end{aligned}$$

where

$$\lim_{\mathbf{\Delta x}\to 0} \frac{[h_1(\mathbf{\Delta x}) + h_2(\mathbf{\Delta x})]}{|\mathbf{\Delta x}|} = 0.$$

To complete the proof, we observe that

$$\begin{aligned}\nabla f(\mathbf{x}) + \nabla g(\mathbf{x}) &= (f_x\mathbf{i} + f_y\mathbf{j}) + (g_x\mathbf{i} + g_y\mathbf{j}) = (f_x + g_x)\mathbf{i} + (f_y + g_y)\mathbf{j} \\ &\qquad \swarrow \frac{\partial}{\partial x}(f + g) = \frac{\partial f}{\partial x} + \frac{\partial g}{\partial x} \\ &= (f + g)_x\mathbf{i} + (f + g)_y\mathbf{j} = \nabla(f + g).\end{aligned}$$

Thus $f + g$ is differentiable and $\nabla(f + g) = \nabla f + \nabla g$. ■

REMARK. Any function that satisfies conditions (i) and (ii) of Theorem 3 is called a **linear mapping** or a **linear operator.** Linear operators play an extremely important role in advanced mathematics.

All the definitions and theorems in this section hold for functions of three or more variables. We give the equivalent results for functions of three variables below.

Definition 5 THE GRADIENT Let f be a scalar function of three variables such that f_x, f_y, and f_z exist at a point $\mathbf{x} = (x, y, z)$. Then the **gradient** of f at $\mathbf{x}$, denoted $\nabla f(\mathbf{x})$, is given by the vector

$$\nabla f(\mathbf{x}) = f_x(x, y, z)\mathbf{i} + f_y(x, y, z)\mathbf{j} + f_z(x, y, z)\mathbf{k}. \quad \textbf{(14)}$$

Definition 6 DIFFERENTIABILITY Let f be a function of three variables that is defined in a neighborhood of $\mathbf{x} = (x, y, z)$, and let $\mathbf{\Delta x} = (\Delta x, \Delta y, \Delta z)$. If $f_x(x, y, z)$, $f_y(x, y, z)$, and $f_z(x, y, z)$ exist, then f is **differentiable** at $\mathbf{x}$ if there is a function g such that

$$f(\mathbf{x} + \mathbf{\Delta x}) - f(\mathbf{x}) = \nabla f \cdot \mathbf{\Delta x} + g(\mathbf{\Delta x}), \quad \textbf{(15)}$$

where

$$\lim_{|\Delta \mathbf{x}| \to 0} \frac{g(\Delta \mathbf{x})}{|\Delta \mathbf{x}|} = 0.$$

Equivalently, we can write

$$f(x + \Delta x, y + \Delta y, z + \Delta z) - f(x, y, z) = f_x(x, y, z)\,\Delta x + f_y(x, y, z)\,\Delta y + f_z(x, y, z)\,\Delta z + g(\Delta x, \Delta y, \Delta z),$$

where

$$\lim_{(\Delta x, \Delta y, \Delta z) \to (0,0,0)} \frac{g(\Delta x, \Delta y, \Delta z)}{\sqrt{\Delta x^2 + \Delta y^2 + \Delta z^2}} = 0.$$

Theorem 1′ If f, f_x, f_y, and f_z exist and are continuous in a neighborhood of $\mathbf{x} = (x, y, z)$, then f is differentiable at $\mathbf{x}$.

Theorem 2′ Let f be a function of three variables that is differentiable at $\mathbf{x}_0$. Then f is continuous at $\mathbf{x}_0$.

EXAMPLE 4 Let $f(x, y, z) = xy^2z^3$. Show that f is differentiable at any point $\mathbf{x}_0$, calculate ∇f, and find $\nabla f(3, -1, 2)$.

Solution. $\partial f/\partial x = y^2z^3$, $\partial f/\partial y = 2xyz^3$, and $\partial f/\partial z = 3xy^2z^2$. Since f, $\partial f/\partial x$, $\partial f/\partial y$, and $\partial f/\partial z$ are all continuous, we know that f is differentiable and that

$$\nabla f = y^2z^3\mathbf{i} + 2xyz^3\mathbf{j} + 3xy^2z^2\mathbf{k}$$

and

$$\nabla f(3, -1, 2) = 8\mathbf{i} - 48\mathbf{j} + 36\mathbf{k}. \blacksquare$$

Theorem 3′ Let f and g be differentiable in a neighborhood of $\mathbf{x} = (x, y, z)$. Then for any scalar α, αf and $f + g$ are differentiable at $\mathbf{x}$, and

(i) $\nabla(\alpha f) = \alpha\,\nabla f$, and
(ii) $\nabla(f + g) = \nabla f + \nabla g$.

We now give a proof of Theorem 1. The proof of Theorem 1′ is similar.

Proof of Theorem 1. We begin by restating the mean value theorem for a function f of one variable.

Mean Value Theorem Let f be continuous on $[a, b]$ and differentiable on (a, b). Then there is a number c in (a, b) such that

$$f(b) - f(a) = f'(c)(b - a).$$

Now we have assumed that f, f_x, and f_y are all continuous in a neighborhood N of $\mathbf{x} = (x, y)$. Choose $\mathbf{\Delta x}$ so small that $\mathbf{x} + \mathbf{\Delta x}$ is in N. Then

$$\Delta f(\mathbf{x}) = f(x + \Delta x, y + \Delta y) - f(x, y)$$

$$= [f(x + \Delta x, y + \Delta y) - \overbrace{f(x + \Delta x, y)] + [f(x + \Delta x, y)}^{\text{This term was added and subtracted}} - f(x, y)]. \quad \textbf{(16)}$$

If $x + \Delta x$ is fixed, then $f(x + \Delta x, y)$ is a function of y that is continuous and differentiable in the interval $[y, y + \Delta y]$. Hence by the mean value theorem there is a number c_2 between y and $y + \Delta y$ such that

$$\begin{aligned} f(x + \Delta x, y + \Delta y) - f(x + \Delta x, y) &= f_y(x + \Delta x, c_2)[(y + \Delta y) - y] \\ &= f_y(x + \Delta x, c_2)\,\Delta y. \end{aligned} \quad \textbf{(17)}$$

Similarly, with y fixed, $f(x, y)$ is a function of x only, and we obtain

$$f(x + \Delta x, y) - f(x, y) = f_x(c_1, y)\,\Delta x, \quad \textbf{(18)}$$

where c_1 is between x and $x + \Delta x$. Thus using (17) and (18) in (16), we have

$$\Delta f(\mathbf{x}) = f_x(c_1, y)\,\Delta x + f_y(x + \Delta x, c_2)\,\Delta y. \quad \textbf{(19)}$$

Now both f_x and f_y are continuous at $\mathbf{x} = (x, y)$, so since c_1 is between x and $x + \Delta x$ and c_2 is between y and $y + \Delta y$, we obtain

$$\lim_{\mathbf{\Delta x} \to \mathbf{0}} f_x(c_1, y) = f_x(x, y) = f_x(\mathbf{x}) \quad \textbf{(20)}$$

and

$$\lim_{\mathbf{\Delta x} \to \mathbf{0}} f_y(x + \Delta x, c_2) = f_y(x, y) = f_y(\mathbf{x}). \quad \textbf{(21)}$$

Let

$$\epsilon_1(\mathbf{\Delta x}) = f_x(c_1, y) - f_x(x, y). \quad \textbf{(22)}$$

From (20) it follows that

$$\lim_{|\mathbf{\Delta x}| \to 0} \epsilon_1(\mathbf{\Delta x}) = 0. \quad \textbf{(23)}$$

Similarly, if

$$\epsilon_2(\mathbf{\Delta x}) = f_y(x + \Delta x, c_2) - f_y(x, y), \quad \textbf{(24)}$$

then

$$\lim_{|\mathbf{\Delta x}| \to 0} \epsilon_2(\mathbf{\Delta x}) = 0. \quad \textbf{(25)}$$

Now define

$$g(\mathbf{\Delta x}) = \epsilon_1(\mathbf{\Delta x})\,\Delta x + \epsilon_2(\mathbf{\Delta x})\,\Delta y. \tag{26}$$

From (23) and (25) it follows that

$$\lim_{|\mathbf{\Delta x}|\to 0} \frac{g(\mathbf{\Delta x})}{|\mathbf{\Delta x}|} = 0. \tag{27}$$

Finally, since

$$f_x(c_1, y) = f_x(x, y) + \epsilon_1(\mathbf{\Delta x}) \qquad \text{From (22)} \tag{28}$$

and

$$f_y(x + \Delta x, c_2) = f_y(x, y) + \epsilon_2(\mathbf{\Delta x}), \qquad \text{From (24)} \tag{29}$$

we may substitute (28) and (29) into (19) to obtain

$$\begin{aligned}\mathbf{\Delta} f(\mathbf{x}) = f(\mathbf{x} + \mathbf{\Delta x}) - f(\mathbf{x}) &= [f_x(\mathbf{x}) + \epsilon_1(\mathbf{\Delta x})]\,\Delta x + [f_y(\mathbf{x}) + \epsilon_2(\mathbf{\Delta x})]\,\Delta y \\ &= f_x(\mathbf{x})\,\Delta x + f_y(\mathbf{x})\,\Delta y + g(\mathbf{\Delta x}) = (f_x\mathbf{i} + f_y\mathbf{j})\cdot(\mathbf{\Delta x}) + g(\mathbf{\Delta x}),\end{aligned}$$

where $\lim_{\mathbf{\Delta x}\to 0} [g(\mathbf{\Delta x})/|\mathbf{\Delta x}|] \to 0$, and the proof is (at last) complete. ■

To define the gradient of a function of n variables, we generalize the formulas (10) and (15).

Definition 7 DIFFERENTIABILITY AND THE GRADIENT Let $f\colon \mathbb{R}^n \to \mathbb{R}$ be defined in a neighborhood of a point $\mathbf{x}$. Let $\mathbf{\Delta x} = (\Delta x_1, \Delta x_2, \ldots, \Delta x_n)$. We say that f is **differentiable** at $\mathbf{x}$ if there exists a vector-valued function $\mathbf{\nabla} f$ and a scalar-valued function g such that

$$f(\mathbf{x} + \mathbf{\Delta x}) - f(\mathbf{x}) = \mathbf{\nabla} f(\mathbf{x}) \cdot \mathbf{\Delta x} + g(\mathbf{\Delta x}) \tag{30}$$

where

$$\lim_{\mathbf{\Delta x}\to 0} \frac{g(\mathbf{\Delta x})}{|\mathbf{\Delta x}|} = 0. \tag{31}$$

The function $\mathbf{\nabla} f$ is called the **gradient of** f.

The following theorem has a proof virtually identical to the proof of Theorem 1.

Theorem 4 Let f and all its first partial derivatives be defined and continuous in a neighborhood of $\mathbf{x} = (x_1, x_2, \ldots, x_n)$. Then f is differentiable at $\mathbf{x}$ and

$$\nabla f(\mathbf{x}) = \left(\frac{\partial f}{\partial x_1}(\mathbf{x}), \frac{\partial f}{\partial x_2}(\mathbf{x}), \ldots, \frac{\partial f}{\partial x_n}(\mathbf{x})\right). \tag{32}$$

In addition, if f is known to be differentiable at $\mathbf{x}$, then $\nabla f(\mathbf{x})$ is given by (32).

EXAMPLE 5 Let $f(\mathbf{x}) = x_1^2 - x_2^2 + 3x_1x_2x_3 - (x_4/x_1)$. Compute ∇f.

Solution. From Example 4.4.4, we have

$$\begin{aligned}\nabla f(\mathbf{x}) &= (f_1(\mathbf{x}), f_2(\mathbf{x}), f_3(\mathbf{x}), f_4(\mathbf{x}))\\ &= \left(2x_1 + 3x_2x_3 + \frac{x_4}{x_1^2}, -2x_2 + 3x_1x_3, 3x_1x_2, -\frac{1}{x_1}\right). \quad \blacksquare\end{aligned}$$

The following theorems generalize Theorems 2 and 3.

Theorem 5 Let $f: \mathbb{R}^n \to \mathbb{R}$ be differentiable at $\mathbf{x}_0$. Then f is continuous at $\mathbf{x}_0$.

Proof. We must show that $\lim_{\mathbf{x}\to\mathbf{x}_0} f(\mathbf{x}) = f(\mathbf{x}_0)$. This is the same as showing that $\lim_{\mathbf{x}\to\mathbf{x}_0} [f(\mathbf{x}) - f(\mathbf{x}_0)] = 0$. But since f is differentiable at $\mathbf{x}_0$,

$$\lim_{\mathbf{x}\to\mathbf{x}_0} [f(\mathbf{x}) - f(\mathbf{x}_0)] = \lim_{\mathbf{x}\to\mathbf{x}_0} [\nabla f(\mathbf{x}_0) \cdot (\mathbf{x} - \mathbf{x}_0) + g(\mathbf{x} - \mathbf{x}_0)] = 0$$

since $\nabla f(\mathbf{x}_0)$ is a constant vector (in terms of the limit) and

$$\lim_{\mathbf{x}\to\mathbf{x}_0} g(\mathbf{x} - \mathbf{x}_0) = \lim_{\mathbf{x}\to\mathbf{x}_0} \frac{g(\mathbf{x} - \mathbf{x}_0)}{|\mathbf{x} - \mathbf{x}_0|} |\mathbf{x} - \mathbf{x}_0| = 0$$

by (31). $\blacksquare$

Theorem 6 Let f and g be differentiable in an open set Ω. Then for every scalar α, αf and $f + g$ are differentiable in Ω and

$$\nabla(f + g) = \nabla f + \nabla g$$

and

$$\nabla(\alpha f) = \alpha \nabla f.$$

REMARK. Theorem 6 shows us that the set of functions that are differentiable in an open set Ω forms a vector space. Moverover, it is then not difficult to show that the gradient operator ∇ is a *linear transformation* from this vector space to the vector space of functions from $\mathbb{R}^n \to \mathbb{R}^n$. We will say much more about this in Section 10.4.

PROBLEMS 4.5

1. Let $f(x, y) = x^2 + y^2$. Show, by using Definition 2, that f is differentiable at any point in $\mathbb{R}^2$.
2. Let $f(x, y) = x^2y^2$. Show, by using Definition 2, that f is differentiable at any point in $\mathbb{R}^2$.
3. Let $f(x, y) =$ be any polynomial in the variables x and y. Show that f is differentiable.

In Problems 4–28, calculate the gradient of the given function. If a point is also given, evaluate the gradient at that point.

4. $f(x, y) = (x + y)^2$
5. $f(x, y) = e^{\sqrt{xy}}$; $(1, 1)$
6. $f(x, y) = \cos(x - y)$; $(\pi/2, \pi/4)$
7. $f(x, y) = \ln(2x - y + 1)$
8. $f(x, y) = \sqrt{x^2 + y^3}$
9. $f(x, y) = \tan^{-1}\dfrac{y}{x}$; $(3, 3)$
10. $f(x, y) = y\tan(y - x)$
11. $f(x, y) = x^2 \sinh y$
12. $f(x, y) = \sec(x + 3y)$; $(0, 1)$
13. $f(x, y) = \dfrac{x - y}{x + y}$; $(3, 1)$
14. $f(x, y) = \dfrac{x^2 - y^2}{x^2 + y^2}$
15. $f(x, y) = \dfrac{e^{x^2} - e^{-y^2}}{3y}$
16. $f(x, y, z) = xyz$; $(1, 2, 3)$
17. $f(x, y, z) = \sin x \cos y \tan z$; $(\pi/6, \pi/4, \pi/3)$
18. $f(x, y, z) = \dfrac{x^2 - y^2 + z^2}{3xy}$; $(1, 2, 0)$
19. $f(x, y, z) = x \ln y - z \ln x$
20. $f(x, y, z) = xy^2 + y^2z^3$; $(2, 3, -1)$
21. $f(x, y, z) = (y - z)e^{x+2y+3z}$; $(-4, -1, 3)$
22. $f(x, y, z) = x \sin y \ln z$; $(1, 0, 1)$
23. $f(x, y, z) = \dfrac{x - z}{\sqrt{1 - y^2 + x^2}}$; $(0, 0, 1)$
24. $f(x, y, z) = x\cosh z - y \sin x$
25. $f(x_1, x_2, x_3, x_4) = x_1x_2x_3x_4$
26. $f(x_1, x_2, \ldots, x_n) = x_1x_2 \cdots x_n$
27. $f(x_1, x_2, x_3, x_4, x_5) = x_1^2 + x_2^2 + x_3^2 + x_4^2 + x_5^2$
28. $f(x_1, x_2, \ldots, x_n) = \left(\sum_{i=1}^{n} x_i^2\right)^{1/2}$
29. Show that if f and g are differentiable functions of three variables, then
$$\nabla(f + g) = \nabla f + \nabla g.$$
30. Show that if f and g are differentiable functions of three variables, then fg is differentiable and
$$\nabla(fg) = f(\nabla g) + g(\nabla f).$$

*31. Show that $\nabla f = \mathbf{0}$ if and only if f is constant.

*32. Show that if $\nabla f = \nabla g$, then there is a constant c for which $f(x, y) = g(x, y) + c$. [*Hint:* Use the result of Problem 31.]

*33. What is the most general function f such that $\nabla f(\mathbf{x}) = \mathbf{x}$ for every $\mathbf{x}$ in $\mathbb{R}^2$?

*34. Let $f(x, y) = \begin{cases} (x^2 + y^2)\sin\dfrac{1}{\sqrt{x^2 + y^2}} & (x, y) \neq (0, 0) \\ 0 & (x, y) = (0, 0). \end{cases}$

(a) Calculate $f_x(0, 0)$ and $f_y(0, 0)$.
(b) Explain why f_x and f_y are *not* continuous at $(0, 0)$.
(c) Show that f is differentiable at $(0, 0)$.

35. Suppose that f is a differentiable function of one variable and g is a differentiable function of three variables. Show that $f \circ g$ is differentiable and $\nabla f \circ g = f'(g)\,\nabla g$.

4.6 THE CHAIN RULES

In this section we derive the chain rule for functions of two and three variables. Let us recall the chain rule for the composition of two functions of one variable:

Let $y = f(u)$ and $u = g(x)$ and assume that f and g are differentiable. Then

$$\frac{dy}{dx} = \frac{dy}{du}\frac{du}{dx} = f'(g(x))g'(x). \tag{1}$$

If $z = f(x, y)$ is a function of two variables, then there are two versions of the chain rule.

Theorem 1 CHAIN RULE Let $z = f(x, y)$ be differentiable and suppose that $x = x(t)$ and $y = y(t)$. Assume further that dx/dt and dy/dt exist and are continuous. Then z can be written as a function of the parameter t, and

$$\frac{dz}{dt} = \frac{\partial z}{\partial x}\frac{dx}{dt} + \frac{\partial z}{\partial y}\frac{dy}{dt} = f_x\frac{dx}{dt} + f_y\frac{dy}{dt}. \tag{2}$$

We can also write this result using our gradient notation. If $\mathbf{g}(t) = x(t)\mathbf{i} + y(t)\mathbf{j}$, then $\mathbf{g}'(t) = (dx/dt)\mathbf{i} + (dy/dt)\mathbf{j}$, and (2) can be written as

$$\frac{d}{dt}f(x(t), y(t)) = (f \circ \mathbf{g})'(t) = [f(\mathbf{g}(t))]' = \nabla f \cdot \mathbf{g}'(t). \tag{3}$$

Theorem 2 CHAIN RULE Let $z = f(x, y)$ be differentiable and suppose that x and y are functions of the two variables r and s. That is, $x = x(r, s)$ and $y = y(r, s)$. Suppose further that $\partial x/\partial r$, $\partial x/\partial s$, $\partial y/\partial r$, and $\partial y/\partial s$ all exist and are continuous. Then z can be written as a function of r and s, and

$$\frac{\partial z}{\partial r} = \frac{\partial z}{\partial x}\frac{\partial x}{\partial r} + \frac{\partial z}{\partial y}\frac{\partial y}{\partial r} \tag{4}$$

$$\frac{\partial z}{\partial s} = \frac{\partial z}{\partial x}\frac{\partial x}{\partial s} + \frac{\partial z}{\partial y}\frac{\partial y}{\partial s}. \tag{5}$$

We will leave the proofs of these theorems until the end of this section.

EXAMPLE 1 Let $z = f(x, y) = xy^2$. Let $x = \cos t$ and $y = \sin t$. Calculate dz/dt.

Solution.

$$\begin{aligned}\frac{dz}{dt} &= \frac{\partial z}{\partial x}\frac{dx}{dt} + \frac{\partial z}{\partial y}\frac{dy}{dt} = y^2(-\sin t) + 2xy(\cos t)\\ &= (\sin^2 t)(-\sin t) + 2(\cos t)(\sin t)(\cos t)\\ &= 2\sin t\cos^2 t - \sin^3 t\end{aligned}$$

We can calculate this result another way. Since $z = xy^2$, we have $z = (\cos t)(\sin^2 t)$. Then

$$\begin{aligned}\frac{dz}{dt} &= (\cos t)(2\sin t\cos t) + (\sin^2 t)(-\sin t)\\ &= 2\sin t\cos^2 t - \sin^3 t. \quad \blacksquare\end{aligned}$$

EXAMPLE 2 Let $z = f(x, y) = \sin xy^2$. Suppose that $x = r/s$ and $y = e^{r-s}$. Calculate $\partial z/\partial r$ and $\partial z/\partial s$.

Solution.

$$\frac{\partial z}{\partial r} = \frac{\partial z}{\partial x}\frac{\partial x}{\partial r} + \frac{\partial z}{\partial y}\frac{\partial y}{\partial r} = (y^2 \cos xy^2)\frac{1}{s} + (2xy \cos xy^2)e^{r-s}$$

$$= \frac{e^{2(r-s)}\cos[(r/s)e^{2(r-s)}]}{s} + \frac{2r}{s}\left\{\cos\left[\frac{r}{s}e^{2(r-s)}\right]\right\}e^{2(r-s)}$$

and

$$\frac{\partial z}{\partial s} = \frac{\partial z}{\partial x}\frac{\partial x}{\partial s} + \frac{\partial z}{\partial y}\frac{\partial y}{\partial s} = (y^2 \cos xy^2)\frac{-r}{s^2} + (2xy \cos xy^2)(-e^{r-s})$$

$$= \frac{-re^{2(r-s)}\cos[(r/s)e^{2(r-s)}]}{s^2} - \frac{2r}{s}\left\{\cos\left[\frac{r}{s}e^{2(r-s)}\right]\right\}e^{2(r-s)}. \quad \blacksquare$$

The chain rules given in Theorem 1 and Theorem 2 can easily be extended to functions of three or more variables.

Theorem 1′ Let $w = f(x, y, z)$ be a differentiable function. If $x = x(t)$, $y = y(t)$, $z = z(t)$, and if dx/dt, dy/dt, and dz/dt exist and are continuous, then

$$\frac{dw}{dt} = \frac{\partial w}{\partial x}\frac{dx}{dt} + \frac{\partial w}{\partial y}\frac{dy}{dt} + \frac{\partial w}{\partial z}\frac{dz}{dt}. \tag{6}$$

Theorem 2′ Let $w = f(x, y, z)$ be a differentiable function and let $x = x(r, s)$, $y = y(r, s)$, and $z = z(r, s)$. Then if all indicated partial derivatives exist and are continuous, we have

$$\frac{\partial w}{\partial r} = \frac{\partial w}{\partial x}\frac{\partial x}{\partial r} + \frac{\partial w}{\partial y}\frac{\partial y}{\partial r} + \frac{\partial w}{\partial z}\frac{\partial z}{\partial r} \tag{7}$$

and

$$\frac{\partial w}{\partial s} = \frac{\partial w}{\partial x}\frac{\partial x}{\partial s} + \frac{\partial w}{\partial y}\frac{\partial y}{\partial s} + \frac{\partial w}{\partial z}\frac{\partial z}{\partial s}. \tag{8}$$

Theorem 3 Let $w = f(x, y, z)$ be a differentiable function and let $x = x(r, s, t)$, $y = y(r, s, t)$, and $z = z(r, s, t)$. Then if all indicated partial derivatives exist and are continuous, we have

$$\begin{aligned}\frac{\partial w}{\partial r} &= \frac{\partial w}{\partial x}\frac{\partial x}{\partial r} + \frac{\partial w}{\partial y}\frac{\partial y}{\partial r} + \frac{\partial w}{\partial z}\frac{\partial z}{\partial r}\\ \frac{\partial w}{\partial s} &= \frac{\partial w}{\partial x}\frac{\partial x}{\partial s} + \frac{\partial w}{\partial y}\frac{\partial y}{\partial s} + \frac{\partial w}{\partial z}\frac{\partial z}{\partial s}\\ \frac{\partial w}{\partial t} &= \frac{\partial w}{\partial x}\frac{\partial x}{\partial t} + \frac{\partial w}{\partial y}\frac{\partial y}{\partial t} + \frac{\partial w}{\partial z}\frac{\partial z}{\partial t}.\end{aligned} \tag{9}$$

In one-variable calculus the chain rule is used to solve "related rates" types of problems. The chain rules for functions of two or more variables are useful in a similar way.

EXAMPLE 3 The radius of a right circular cone is increasing at a rate of 3 cm/sec, while its height is increasing at a rate of 5 cm/sec. How fast is the volume increasing when $r = 15$ cm and $h = 25$ cm?

Solution. We are asked to find dV/dt, where $V = \frac{1}{3}\pi r^2 h$. But

$$\begin{aligned}\frac{dV}{dt} &= \frac{\partial V}{\partial r}\frac{dr}{dt} + \frac{\partial V}{\partial h}\frac{dh}{dt} = \frac{2}{3}\pi rh(3) + \frac{1}{3}\pi r^2(5)\\ &= 2\pi rh + \tfrac{5}{3}\pi r^2 = \pi[2\cdot 15\cdot 25 + \tfrac{5}{3}(15)^2]\\ &= 1125\pi \text{ cm}^3/\text{sec}. \blacksquare\end{aligned}$$

EXAMPLE 4 According to the ideal gas law, the pressure, volume, and absolute temperature of n moles of an ideal gas are related by

$$PV = nRT,$$

where R is a constant. Suppose that the volume of an ideal gas is increasing at a rate of 10 cm^3/min and the pressure is decreasing at a rate of 0.3 N/cm^2/min. How is the temperature of the gas changing when the volume of 5 mol of a gas is 100 cm^3 and the pressure is 2 N/cm^2?

Solution. We have $T = PV/nR$, where $n = 5$. Then

$$\begin{aligned}\frac{dT}{dt} &= \frac{\partial T}{\partial P}\frac{dP}{dt} + \frac{\partial T}{\partial V}\frac{dV}{dt} = \frac{V}{nR}(-0.3) + \frac{P}{nR}(10) = \frac{100}{5R}(-0.3) + \frac{2}{5R}(10)\\ &= \frac{-2}{R}\ {}^\circ\text{K/min}. \blacksquare\end{aligned}$$

We can use the chain rule to compute higher-order partial derivatives.

EXAMPLE 5 Let $z = f(x, y)$, let $x = x(r, s)$ and let $y = y(r, s)$. Compute

$$\frac{\partial^2 z}{\partial r^2}.$$

Solution.

$$\frac{\partial z}{\partial r} = \frac{\partial z}{\partial x}\frac{\partial x}{\partial r} + \frac{\partial z}{\partial y}\frac{\partial y}{\partial r}$$

Then

$$\frac{\partial^2 z}{\partial r^2} = \frac{\partial}{\partial r}\left(\frac{\partial z}{\partial r}\right) = \frac{\partial}{\partial x}\left(\frac{\partial z}{\partial r}\right)\frac{\partial x}{\partial r} + \frac{\partial}{\partial y}\left(\frac{\partial z}{\partial r}\right)\frac{\partial y}{\partial r}$$

and

$$\frac{\partial^2 z}{\partial r^2} = \frac{\partial}{\partial x}\left(\frac{\partial z}{\partial x}\frac{\partial x}{\partial r} + \frac{\partial z}{\partial y}\frac{\partial y}{\partial r}\right)\left(\frac{\partial x}{\partial r}\right) + \frac{\partial}{\partial y}\left(\frac{\partial z}{\partial x}\frac{\partial x}{\partial r} + \frac{\partial z}{\partial y}\frac{\partial y}{\partial r}\right)\left(\frac{\partial y}{\partial r}\right). \quad \blacksquare$$

We next give a proof of Theorem 2. Theorem 1 follows easily from Theorem 2. (Explain why.)

Proof of Theorem 2. We will show that

$$\frac{\partial z}{\partial r} = \frac{\partial z}{\partial x}\frac{\partial x}{\partial r} + \frac{\partial z}{\partial y}\frac{\partial y}{\partial r}.$$

Equation (5) follows in an identical manner. Since $x = x(r, s)$ and $y = y(r, s)$, a change Δr in r will cause a change Δx in x and a change Δy in y. We may therefore write

$$\Delta x = x(r + \Delta r, s) - x(r, s)$$

and

$$\Delta y = y(r + \Delta r, s) - y(r, s).$$

Since z is differentiable, we may write (Definition 4.5.2)

$$\Delta z = \frac{\partial z}{\partial x}\Delta x + \frac{\partial z}{\partial y}\Delta y + \epsilon_1(\Delta x, \Delta y)\,\Delta x + \epsilon_2(\Delta x, \Delta y)\,\Delta y, \tag{10}$$

where

$$\lim_{(\Delta x, \Delta y)\to(0,0)} \epsilon_1(\Delta x, \Delta y) = \lim_{(\Delta x, \Delta y)\to(0,0)} \epsilon_2(\Delta x, \Delta y) = 0.$$

Then dividing both sides of (10) by Δr and taking limits, we obtain

$$\lim_{\Delta r\to 0}\frac{\Delta z}{\Delta r} = \lim_{\Delta r\to 0}\left[\frac{\partial z}{\partial x}\frac{\Delta x}{\Delta r} + \frac{\partial z}{\partial y}\frac{\Delta y}{\Delta r} + \epsilon_1(\Delta x, \Delta y)\frac{\Delta x}{\Delta r} + \epsilon_2(\Delta x, \Delta y)\frac{\Delta y}{\Delta r}\right]. \tag{11}$$

Since x and y are continuous functions of r, we have

$$\lim_{\Delta r\to 0}\Delta x = 0 \quad \text{and} \quad \lim_{\Delta r\to 0}\Delta y = 0,$$

so that

$$\lim_{\Delta r\to 0}\epsilon_1(\Delta x, \Delta y) = \lim_{(\Delta x, \Delta y)\to(0,0)}\epsilon_1(\Delta x, \Delta y) = 0$$

and

$$\lim_{\Delta r\to 0}\epsilon_2(\Delta x, \Delta y) = \lim_{(\Delta x, \Delta y)\to(0,0)}\epsilon_2(\Delta x, \Delta y) = 0.$$

Thus the limits in (11) become

$$\frac{\partial z}{\partial r} = \frac{\partial z}{\partial x}\frac{\partial x}{\partial r} + \frac{\partial z}{\partial y}\frac{\partial y}{\partial r} + 0 \cdot \frac{\partial x}{\partial r} + 0 \cdot \frac{\partial y}{\partial r},$$

and the theorem is proved. ■

We now state and prove one extension of the chain rule in $\mathbb{R}^n$.

Theorem 4 (CHAIN RULE) Let $\mathbf{x}(t)$: $\mathbb{R} \to \mathbb{R}^n$ be differentiable at t_0 and let $f(\mathbf{x})$: $\mathbb{R}^n \to \mathbb{R}$ be differentiable at $\mathbf{x}_0 = \mathbf{x}(t_0)$. Then $f(\mathbf{x}(t))$ is differentiable at t_0 and

$$\frac{d}{dt} f(\mathbf{x}(t_0)) = \nabla f(\mathbf{x}_0) \cdot \mathbf{x}'(t_0). \tag{12}$$

Proof. Since $f(\mathbf{x}(t_0))$ is a function from $\mathbb{R}$ to $\mathbb{R}$, we have

$$\frac{d}{dt} f(\mathbf{x}(t_0)) = \lim_{\Delta t \to 0} \frac{f(\mathbf{x}(t_0 + \Delta t)) - f(\mathbf{x}(t_0))}{\Delta t}. \tag{13}$$

Let $\Delta\mathbf{x}(t_0) = \mathbf{x}(t_0 + \Delta t) - \mathbf{x}(t_0)$. Then, since f is differentiable, we have, using equations (30) and (31) on page 208,

$$f(\mathbf{x}(t_0 + \Delta t) - f(\mathbf{x}(t_0)) = \nabla f(\mathbf{x}(t_0)) \cdot \Delta\mathbf{x}(t_0) + g(\Delta\mathbf{x}(t_0)) \tag{14}$$

where

$$\lim_{|s| \to 0} \frac{g(s)}{|s|} = 0.$$

Let $h(s) = g(s)/s$. Then (14) can be written as

$$f(\mathbf{x}(t_0 + \Delta t) - f(\mathbf{x}(t_0)) = \nabla f(\mathbf{x}(t_0)) \cdot \Delta\mathbf{x}(t_0) + |\Delta\mathbf{x}(t_0)| h(\Delta\mathbf{x}(t_0)) \tag{15}$$

where

$$\lim_{|\Delta\mathbf{x}(t_0)| \to 0} h(\Delta\mathbf{x}(t_0)) = 0. \tag{16}$$

Using (15) in (13), we have

$$\frac{d}{dt} f(\mathbf{x}(t_0)) = \lim_{\Delta t \to 0} \left(\nabla f(\mathbf{x}(t_0)) \cdot \frac{\Delta\mathbf{x}(t_0)}{\Delta t}\right) + \lim_{\Delta t \to 0} \frac{|\Delta\mathbf{x}(t_0)| h(\Delta\mathbf{x}(t_0))}{\Delta t}. \tag{17}$$

Now $\Delta\mathbf{x}(t_0) = \mathbf{x}(t_0 + \Delta t) - \mathbf{x}(t_0)$ and, since each $x_i(t)$ is continuous, $\lim_{\Delta t \to 0} \Delta\mathbf{x}(t) = 0$, so that $\lim_{\Delta t \to 0} |\Delta\mathbf{x}(t)| = 0$. Thus, from (16), $\lim_{\Delta t \to 0} h(\Delta\mathbf{x}(t_0)) = 0$. Also,

$$\lim_{\Delta t \to 0} \frac{x_i(t_0 + \Delta t) - x_i(t_0)}{\Delta t} = x_i'(t_0),$$

so that

$$\lim_{\Delta t \to 0} \frac{|\mathbf{\Delta x}(t_0)|}{\Delta t} =$$

$$\lim_{\Delta t \to 0} \sqrt{\left[\frac{x_1(t_0 + \Delta t) - x_1(t_0)}{\Delta t}\right]^2 + \left[\frac{x_2(t_0 + \Delta t) - x_2(t_0)}{\Delta t}\right]^2 + \cdots + \left[\frac{x_n(t_0 + \Delta t) - x_n(t_0)}{\Delta t}\right]^2}$$
$$= \sqrt{[x_1'(t_0)]^2 + [x_2'(t_0)]^2 + \cdots + [x_n'(t_0)]^2}.$$

Therefore the second limit in (17) is 0. Finally, consider the ith term in the scalar product in (17). We have

$$\lim_{\Delta t \to 0} f_i(\mathbf{x}(t_0))\left[\frac{x_i(t_0 + \Delta t) - x_i(t_0)}{\Delta t}\right] = f_i(\mathbf{x}(t_0)) \lim_{\Delta t \to 0} \left[\frac{x_i(t_0 + \Delta t) - x_i(t_0)}{\Delta t}\right] = f_i(\mathbf{x}(t_0))x_i'(t_0),$$

so that, from (17),

$$\frac{d}{dt} f(\mathbf{x}(t_0)) = \sum_{i=1}^{n} [f_i(\mathbf{x}(t))][x_i'(t)] = \nabla f(\mathbf{x}(t_0)) \cdot \mathbf{x}'(t_0).$$

This completes the proof. ■

EXAMPLE 6 Let $f(\mathbf{x}) = x_1^2 + x_2x_3 - x_4^2$ where $x_1 = t^2$, $x_2 = \sin t$, $x_3 = t^3$, and $x_4 = \ln t$. Compute $(d/dt)\, f(\mathbf{x}(t))$.

Solution.

$$\frac{d}{dt} f(\mathbf{x}(t)) = \nabla f \cdot \mathbf{x}'(t)$$

$$= (2x_1, x_3, x_2, -2x_4) \cdot \left(2t, \cos t, 3t^2, \frac{1}{t}\right)$$

$$= (2t^2, t^3, \sin t, -2 \ln t) \cdot \left(2t, \cos t, 3t^2, \frac{1}{t}\right)$$

$$= 4t^3 + t^3 \cos t + 3t^2 \sin t - \frac{2 \ln t}{t}. \quad ■$$

We will give a more general version of the chain rule in Section 10.4.

We close this section with a discussion of the mean value theorem for functions of n variables. To do so, we first need to define the line segment joining two vectors $\mathbf{x}$ and $\mathbf{y}$ in $\mathbb{R}^n$.

Definition 1 LINE SEGMENT Let $\mathbf{x}$ and $\mathbf{y}$ be two vectors in $\mathbb{R}^n$. Then the **line segment** L joining $\mathbf{x}$ and $\mathbf{y}$ is the set defined by

$$L = \{\mathbf{v}: \mathbf{v} = t\mathbf{x} + (1 - t)\mathbf{y} \text{ for } 0 \leq t \leq 1\}. \tag{18}$$

NOTE. We observe that both **x** and **y** are in L. These vectors are obtained by setting $t = 1$ and $t = 0$ in (18).

Theorem 5 MEAN VALUE THEOREM Let $f: \mathbb{R}^n \to \mathbb{R}$ be differentiable at every point in an open set containing the line segment L joining two vectors **x** and **y** in $\mathbb{R}^n$. Then there is a vector $\mathbf{x}_0$ on L such that

$$f(\mathbf{y}) - f(\mathbf{x}) = \nabla f(\mathbf{x}_0) \cdot (\mathbf{y} - \mathbf{x}). \tag{19}$$

Proof. Let $g(t) = f(t(\mathbf{y} - \mathbf{x}) + \mathbf{x})$. Note that $g(0) = f(\mathbf{x})$, $g(1) = f(\mathbf{y})$, and for $0 < t < 1$, $t(\mathbf{y} - \mathbf{x}) + \mathbf{x} = t\mathbf{y} + (1 - t)\mathbf{x}$ is on L. By the chain rule (Theorem 4),

$$g'(t) = \nabla f(t(\mathbf{y} - \mathbf{x}) + \mathbf{x}) \cdot (\mathbf{y} - \mathbf{x}), \tag{20}$$

since

$$\frac{d}{dt}(t(\mathbf{y} - \mathbf{x}) + \mathbf{x}) = \mathbf{y} - \mathbf{x}$$

because **x** and **y** are constant vectors (see Examples 3.8.10 and 3.8.11 on page 140). Since g is differentiable on $[0, 1]$, the hypotheses of the mean value theorem of one variable are satisfied and there is a number $t_0 \in (0, 1)$ such that

$$g(1) - g(0) = g'(t_0)(1 - 0) = g'(t_0). \tag{21}$$

But, using (20), (21) becomes

$$f(\mathbf{y}) - f(\mathbf{x}) = \nabla f(t_0(\mathbf{y} - \mathbf{x}) + \mathbf{x}) \cdot (\mathbf{y} - \mathbf{x}). \tag{22}$$

Setting $\mathbf{x}_0 = t_0(\mathbf{y} - \mathbf{x}) + \mathbf{x}$ in (22) completes the proof. ■

PROBLEMS 4.6

In Problems 1–11, use the chain rule to calculate dz/dt or dw/dt. Check your answer by first writing z or w as a function of t and then differentiating.

1. $z = xy$, $x = e^t$, $y = e^{2t}$
2. $z = x^2 + y^2$, $x = \cos t$, $y = \sin t$
3. $z = \dfrac{y}{x}$, $x = t^2$, $y = t^3$
4. $z = e^x \sin y$, $x = \sqrt{t}$, $y = \sqrt[3]{t}$
5. $z = \tan^{-1}\dfrac{y}{x}$, $x = \cos 3t$, $y = \sin 5t$
6. $z = \sinh(x - 2y)$, $x = 2t^4$, $y = t^2 + 1$
7. $w = x^2 + y^2 + z^2$, $x = \cos t$, $y = \sin t$, $z = t$
8. $w = xy - yz + zx$, $x = e^t$, $y = e^{2t}$, $z = e^{3t}$

9. $w = \dfrac{x + y}{z}$, $x = t$, $y = t^2$, $z = t^3$

10. $w = \sin(x + 2y + 3z)$, $x = \tan t$, $y = \sec t$, $z = t^5$

11. $w = \ln(2x - 3y + 4z)$, $x = e^t$, $y = \ln t$, $z = \cosh t$

In Problems 12–26, use the chain rule to calculate the indicated partial derivatives.

12. $z = xy$; $x = r + s$; $y = r - s$; $\partial z/\partial r$ and $\partial z/\partial s$

13. $z = x^2 + y^2$; $x = \cos(r + s)$; $y = \sin(r - s)$; $\partial z/\partial r$ and $\partial z/\partial s$

14. $z = \dfrac{y}{x}$; $x = e^r$; $y = e^s$; $\dfrac{\partial z}{\partial r}$ and $\dfrac{\partial z}{\partial s}$

15. $z = \sin \dfrac{y}{x}$; $x = \dfrac{r}{s}$; $y = \dfrac{s}{r}$; $\dfrac{\partial z}{\partial r}$ and $\dfrac{\partial z}{\partial s}$

16. $z = \dfrac{e^{x+y}}{e^{x-y}}$; $x = \ln rs$; $y = \ln \dfrac{r}{s}$; $\dfrac{\partial z}{\partial r}$ and $\dfrac{\partial z}{\partial s}$

17. $z = x^2y^3$; $x = r - s^2$; $y = 2s + r$; $\partial z/\partial r$ and $\partial z/\partial s$

18. $w = x + y + z$; $x = rs$; $y = r + s$; $z = r - s$; $\partial w/\partial r$ and $\partial w/\partial s$

19. $w = \dfrac{xy}{z}$; $x = r$, $y = s$, $z = t$; $\dfrac{\partial w}{\partial r}$, $\dfrac{\partial w}{\partial s}$ and $\dfrac{\partial w}{\partial t}$

20. $w = \dfrac{xy}{z}$; $x = r + s$, $y = t - r$, $z = s + 2t$; $\dfrac{\partial w}{\partial r}$, $\dfrac{\partial w}{\partial s}$ and $\dfrac{\partial w}{\partial t}$

21. $w = \sin xyz$; $x = s^2r$, $y = r^2s$, $z = r - s$; $\partial w/\partial r$ and $\partial w/\partial s$

22. $w = \sinh(x + 2y + 3z)$; $x = \sqrt{r + s}$, $y = \sqrt[3]{s - t}$, $z = \dfrac{1}{r + t}$, $\dfrac{\partial w}{\partial r}$, $\dfrac{\partial w}{\partial s}$ and $\dfrac{\partial w}{\partial t}$

23. $w = x^2y + yz^2$; $x = rst$, $y = \dfrac{rs}{t}$, $z = \dfrac{1}{rst}$; $\dfrac{\partial w}{\partial r}$, $\dfrac{\partial w}{\partial s}$ and $\dfrac{\partial w}{\partial t}$

24. $w = \ln(x + 2y + 3z)$; $x = rt^3 + s$; $y = t - s^5$, $z = e^{r+s}$; $\partial w/\partial r$, $\partial w/\partial s$ and $\partial w/\partial t$

25. $w = e^{xy/z}$; $x = r^2 + t^2$, $y = s^2 - t^2$, $z = r^2 + s^2$; $\partial w/\partial r$, $\partial w/\partial s$ and $\partial w/\partial t$

***26.** $u = xy + w^2 - z^3$; $x = t + r - q$, $y = q^2 + s^2 - t + r$, $z = \dfrac{qr + st}{r^4}$, $w = \dfrac{r - s}{t + q}$; $\dfrac{\partial u}{\partial r}$, $\dfrac{\partial u}{\partial s}$, $\dfrac{\partial u}{\partial t}$ and $\dfrac{\partial u}{\partial q}$

27. The radius of a right circular cone is increasing at a rate of 7 in./min, while its height is decreasing at a rate of 20 in./min. How fast is the volume changing when $r = 45$ in. and $h = 100$ in.? Is the volume increasing or decreasing?

28. The radius of a right circular cylinder is decreasing at a rate of 12 cm/sec, while its height is increasing at a rate of 25 cm/sec. How is the volume changing when $r = 180$ cm and $h = 500$ cm? Is the volume increasing or decreasing?

29. The volume of 10 mol of an ideal gas is decreasing at a rate of 25 cm^3/min and its temperature is increasing at a rate of 1°C/min. How fast is the pressure changing when $V = 1000$ cm^3 and $P = 3$ N/cm^2? Is the pressure increasing or decreasing? Leave your answer in terms of R.

30. The pressure of 8 mol of an ideal gas is decreasing at a rate of 0.4 N/cm^2/min, while the temperature is decreasing at a rate of 0.5 °K/min. How fast is the volume of the gas changing when $V = 1000$ cm^3 and $P = 3$ N/cm^2? Is the volume increasing or decreasing? [*Hint:* Use the value $R = 8.314$ J/mol °K.]

31. The angle A of a triangle ABC is increasing at a rate of 3°/sec, the side AB is increasing at a rate of 1 cm/sec, and the side AC is decreasing at a rate of 2 cm/sec. How fast is the side BC changing when $A = 30°$, $AB = 10$ cm, and $AC = 24$ cm? Is the length of BC increasing or decreasing? [*Hint:* Use the law of cosines and convert to radians.]

32. The wave equation (see Problem 4.4.31) is the partial differential equation

$$\frac{\partial^2 y}{\partial t^2} = c^2 \frac{\partial^2 y}{\partial x^2}.$$

Show that if f is any differentiable function, then

$$y(x, t) = \tfrac{1}{2}[f(x - ct) + f(x + ct)]$$

is a solution to this equation.

33. Let $z = f(x, y)$ be differentiable. Write the expression $(\partial z/\partial x)^2 + (\partial z/\partial y)^2$ in terms of polar coordinates r and θ.

34. Let $w = f(x, y, z)$ be differentiable and let $x = r\cos\theta$, $y = r\sin\theta$, and $z = t$. Calculate $\partial w/\partial r$, $\partial w/\partial\theta$, and $\partial w/\partial t$. (These are cylindrical coordinates.)

35. Let $w = f(x, y, z)$ be differentiable and let $x = \rho \sin \phi \cos \theta$, $y = \rho \sin \phi \sin \theta$, and $z = \rho \cos \phi$. Calculate $\partial w/\partial \rho$, $\partial w/\partial \phi$, and $\partial w/\partial \theta$. (These are spherical coordinates.)

****36.** Laplace's equation (see Problem 4.4.35) is the partial differential equation

$$\frac{\partial^2 f}{\partial x^2} + \frac{\partial^2 f}{\partial y^2} = 0.$$

If we write (x, y) in polar coordinates ($x = r \cos \theta$, $y = r \sin \theta$), show that Laplace's equation becomes

$$\frac{\partial^2 f}{\partial r^2} + \frac{1}{r^2}\frac{\partial^2 f}{\partial \theta^2} + \frac{1}{r}\frac{\partial f}{\partial r} = 0.$$

[*Hint:* Write $\partial^2 f/\partial x^2$ and $\partial^2 f/\partial y^2$ in terms of r and θ, using the result of Example 5.]

37. Show that if $f\colon \mathbb{R}^n \to \mathbb{R}$, then $\nabla f = \mathbf{0}$ if and only if f is constant.

38. Show that if f and $g\colon \mathbb{R}^n \to \mathbb{R}$ and $\nabla f = \nabla g$, then there is a number c such that $f(\mathbf{x}) = g(\mathbf{x}) + c$.

39. Find the most general function $f\colon \mathbb{R}^n \to \mathbb{R}$ such that $\nabla f(\mathbf{x}) = \mathbf{x}$ for every $\mathbf{x} \in \mathbb{R}^n$.

4.7 TANGENT PLANES, NORMAL LINES, AND GRADIENTS

Let $z = f(x, y)$ be a function of two variables. As we have seen, the graph of f is a surface in $\mathbb{R}^3$. More generally, the graph of the equation $F(x, y, z) = 0$ is a surface in $\mathbb{R}^3$. The surface $F(x, y, z) = 0$ is called **differentiable** at a point (x_0, y_0, z_0) if $\partial F/\partial x$, $\partial F/\partial y$, and $\partial F/\partial z$ all exist and are continuous at (x_0, y_0, z_0). In $\mathbb{R}^2$ a differentiable curve has a unique tangent line at each point. In $\mathbb{R}^3$ a differentiable surface in $\mathbb{R}^3$ has a unique tangent plane at each point at which $\partial F/\partial x$, $\partial F/\partial y$, and $\partial F/\partial z$ are not all zero. We will formally define what we mean by a tangent plane to a surface after a bit, although it should be easy enough to visualize (see Figure 1). We note here that not every surface has a tangent plane at every point. For example, the cone $z = \sqrt{x^2 + y^2}$ has no tangent plane at the origin (see Figure 2).

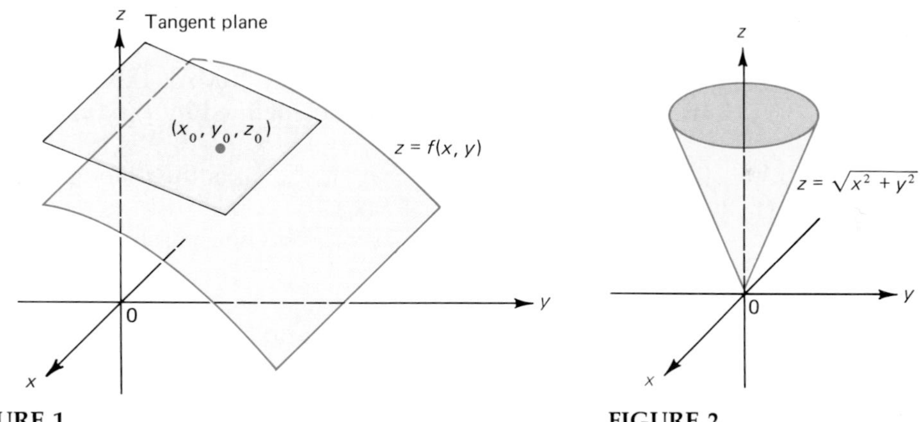

FIGURE 1

FIGURE 2

Assume that the surface S given by $F(x, y, z) = 0$ is differentiable. Let C be any curve lying on S. That is, C can be given parametrically by $\mathbf{g}(t) = x(t)\mathbf{i} + y(t)\mathbf{j} + z(t)\mathbf{k}$. (Recall from Definition 3.8.1, the definition of a curve in $\mathbb{R}^3$.) Then for points on the curve, $F(x, y, z)$ can be written as a function of t, and from the vector form of the chain rule [equation (4.6.3)] we have

$$F'(t) = \nabla F \cdot \mathbf{g}'(t). \tag{1}$$

But since $F(x(t), y(t), z(t)) = 0$ for all t [since $(x(t), y(t), z(t))$ is on S], we see that $F'(t) = 0$ for all t. But $\mathbf{g}'(t)$ is tangent to the curve C for every number t. Thus (1) implies the following:

The gradient of F at a point $\mathbf{x}_0 = (x_0, y_0, z_0)$ on S is orthogonal to the tangent vector at $\mathbf{x}_0$ to any curve C remaining on S and passing through $\mathbf{x}_0$.

This statement is illustrated in Figure 3.

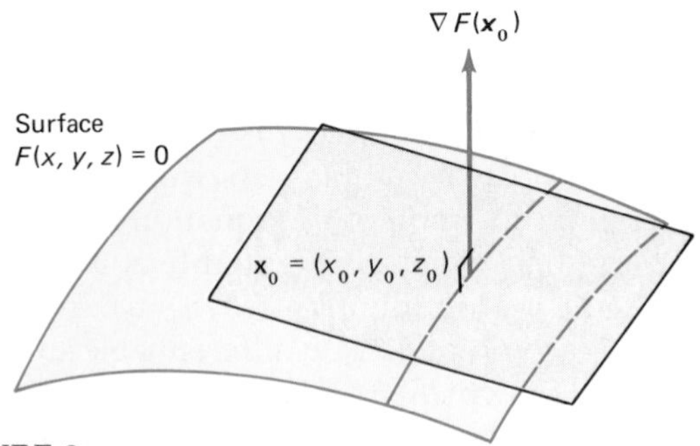

FIGURE 3

Thus if we think of all the vectors tangent to a surface at a point $\mathbf{x}_0$ as constituting a plane, then $\nabla F(\mathbf{x}_0)$ is a *normal* vector to that plane. This motivates the following definition.

Definition 1 TANGENT PLANE AND NORMAL LINE Let F be differentiable at $\mathbf{x}_0 = (x_0, y_0, z_0)$ and let the surface S be defined by $F(x, y, z) = 0$.

(i) The **tangent plane** to S at (x_0, y_0, z_0) is the plane passing through the point (x_0, y_0, z_0) with normal vector $\nabla F(\mathbf{x}_0)$.

(ii) The **normal line** to S at $\mathbf{x}_0$ is the line passing through $\mathbf{x}_0$ having the same direction as $\nabla F(\mathbf{x}_0)$.

EXAMPLE 1 Find the equation of the tangent plane and symmetric equations of the normal line to the ellipsoid $x^2 + (y^2/4) + (z^2/9) = 3$ at the point $(1, 2, 3)$.

Solution. Since $F(x, y, z) = x^2 + (y^2/4) + (z^2/9) - 3 = 0$, we have

$$\nabla F = \frac{\partial F}{\partial x}\mathbf{i} + \frac{\partial F}{\partial y}\mathbf{j} + \frac{\partial F}{\partial z}\mathbf{k} = 2x\mathbf{i} + \frac{y}{2}\mathbf{j} + \frac{2z}{9}\mathbf{k}.$$

Then $\nabla F(1, 2, 3) = 2\mathbf{i} + \mathbf{j} + \frac{2}{3}\mathbf{k}$, and the equation of the tangent plane is

$$2(x - 1) + (y - 2) + \tfrac{2}{3}(z - 3) = 0,$$

or

$$2x + y + \tfrac{2}{3}z = 6.$$

The normal line is given by

$$\frac{x-1}{2} = y - 2 = \tfrac{3}{2}(z - 3). \quad ■$$

The situation is even simpler if we can write the surface in the form $z = f(x, y)$. That is, the surface is the graph of a function of two variables. Then $F(x, y, z) = f(x, y) - z = 0$, so that

$$F_x = f_x \qquad F_y = f_y \qquad F_z = -1,$$

and the normal vector $\mathbf{N}$ to the tangent plane is

$$\mathbf{N} = f_x(x_0, y_0)\mathbf{i} + f_y(x_0, y_0)\mathbf{j} - \mathbf{k}. \qquad \textbf{(2)}$$

REMARK. One interesting consequence of this fact is that if $z = f(x, y)$ and if $\nabla f(x_0, y_0) = \mathbf{0}$, then *the tangent plane to the surface at $(x_0, y_0, f(x_0, y_0))$ is parallel to the xy-plane (i.e., it is horizontal)*. This occurs because at $(x_0, y_0, f(x_0, y_0))$, $\mathbf{N} = (\partial f/\partial x)\mathbf{i} + (\partial f/\partial y)\mathbf{j} - \mathbf{k} = \nabla f - \mathbf{k} = -\mathbf{k}$. Thus the z-axis is normal to the tangent plane.

EXAMPLE 2 Find the tangent plane and normal line to the surface $z = x^3y^5$ at the point (2, 1, 8).

Solution. $\mathbf{N} = (\partial z/\partial x)\mathbf{i} + (\partial z/\partial y)\mathbf{j} - \mathbf{k} = 3x^2y^5\mathbf{i} + 5x^3y^4\mathbf{j} - \mathbf{k} = 12\mathbf{i} + 40\mathbf{j} - \mathbf{k}$ at (2, 1, 8). Then the tangent plane is given by

$$12(x - 2) + 40(y - 1) - (z - 8) = 0,$$

or

$$12x + 40y - z = 56.$$

Symmetric equations of the normal line are

$$\frac{x-2}{12} = \frac{y-1}{40} = \frac{z-8}{-1}. \quad ■$$

We can write the equation of the tangent plane to a surface $z = f(x, y)$ so that it looks like the equation of the tangent line to a curve in $\mathbb{R}^2$. This will further illustrate the connection between the derivative of a function of one variable and the gradient. Recall from Section 3.5 that if P is a point on a plane and $\mathbf{N}$ is a normal vector, then if Q denotes any other point on the plane, the equation of the plane can be written

$$\overrightarrow{PQ} \cdot \mathbf{N} = 0. \qquad \textbf{(3)}$$

In this case, since $z = f(x, y)$, a point on the surface takes the form $(x, y, z) = (x, y, f(x, y))$. Then since $\mathbf{N} = f_x\mathbf{i} + f_y\mathbf{j} - \mathbf{k}$, the equation of the tangent plane at $(x_0, y_0, f(x_0, y_0))$ becomes, using (3),

$$\begin{aligned} 0 &= [(x, y, z) - (x_0, y_0, z_0)] \cdot (f_x, f_y, -1) \\ &= (x - x_0, y - y_0, z - z_0) \cdot (f_x, f_y, -1) \\ &= (x - x_0)f_x + (y - y_0)f_y - (z - z_0). \end{aligned} \tag{4}$$

We can rewrite (4) as

$$z = f(x_0, y_0) + (x - x_0)f_x + (y - y_0)f_y. \tag{5}$$

Denote (x_0, y_0) by $\mathbf{x}_0$ and (x, y) by $\mathbf{x}$. Then (5) can be written as

$$z = f(\mathbf{x}_0) + (\mathbf{x} - \mathbf{x}_0) \cdot \nabla f(\mathbf{x}_0). \tag{6}$$

Recall that if $y = f(x)$ is differentiable at x_0, then the equation of the tangent line to the curve at the point $(x_0, f(x_0))$ is given by

$$\frac{y - f(x_0)}{x - x_0} = f'(x_0),$$

or

$$y = f(x_0) + (x - x_0)f'(x_0). \tag{7}$$

This similarity between (6) and (7) illustrates quite vividly the importance of the gradient vector of a function of several variables as the generalization of the derivative of a function of one variable.

PROBLEMS 4.7

In Problems 1–16, find the equation of the tangent plane and symmetric equations of the normal line to the given surface at the given point.

1. $x^2 + y^2 + z^2 = 1$; $(1, 0, 0)$
2. $x^2 + y^2 + z^2 = 1$; $(0, 1, 0)$
3. $x^2 + y^2 + z^2 = 1$; $(0, 0, 1)$
4. $x^2 - y^2 + z^2 = 1$; $(1, 1, 1)$
5. $\frac{x^2}{a^2} + \frac{y^2}{b^2} + \frac{z^2}{c^2} = 3$; (a, b, c)
6. $\frac{x^2}{a^2} + \frac{y^2}{b^2} + \frac{z^2}{c^2} = 3$; $(-a, b, -c)$
7. $x^{1/2} + y^{1/2} + z^{1/2} = 6$; $(4, 1, 9)$
8. $ax + by + cz = d$; $\left(\frac{1}{a}, \frac{1}{b}, \frac{d-2}{c}\right)$
9. $xyz = 4$; $(1, 2, 2)$
10. $xy^2 - yz^2 + zx^2 = 1$; $(1, 1, 1)$
11. $4x^2 - y^2 - 5z^2 = 15$; $(3, 1, -2)$
12. $xe^y - ye^z = 1$; $(1, 0, 0)$
13. $\sin xy - 2\cos yz = 0$; $(\pi/2, 1, \pi/3)$
14. $x^2 + y^2 + 4x + 2y + 8z = 7$; $(2, -3, -1)$
15. $e^{xyz} = 5$; $(1, 1, \ln 5)$

16. $\sqrt{\dfrac{x+y}{z-1}} = 1$; $(1, 1, 3)$

In Problems 17–24, write the equation of the tangent plane in the form (6) and find the symmetric equations of the normal line to the given surface.

17. $z = xy^2$; $(1, 1, 1)$
18. $z = \ln(x - 2y)$; $(3, 1, 0)$
19. $z = \sin(2x + 5y)$; $(\pi/8, \pi/20, 1)$
20. $z = \sqrt{\dfrac{x+y}{x-y}}$; $(5, 4, 3)$
21. $z = \tan^{-1}\dfrac{y}{x}$; $\left(-2, 2, -\dfrac{\pi}{4}\right)$
22. $z = \sinh xy^2$; $(0, 3, 0)$
23. $z = \sec(x - y)$; $(\pi/2, \pi/6, 2)$
24. $z = e^x \cos y + e^y \cos x$; $(\pi/2, 0, e^{\pi/2})$

*25. Find the two points of intersection of the surface $z = x^2 + y^2$ and the line

$$\frac{x-3}{1} = \frac{y+1}{-1} = \frac{z+2}{-2}.$$

26. At each of the points of intersection found in Problem 25, find the cosine of the angle between the given line and the normal line to the surface.

*27. Show that every line normal to the surface of a sphere passes through the center of the sphere.

28. Show that every line normal to the cone $z^2 = ax^2 + ay^2$ intersects the z-axis.

29. Let f be a differentiable function of one variable and let $z = yf(y/x)$. Show that all tangent planes to the surface defined by this equation have a point in common.

30. The **angle between two surfaces** at a point of intersection is defined to be the angle between their normal lines. Show that if two surfaces $F(x, y, z) = 0$ and $G(x, y, z) = 0$ intersect at right angles at a point $\mathbf{x}_0$, then

$$\nabla F(\mathbf{x}_0) \cdot \nabla G(\mathbf{x}_0) = 0.$$

31. Show that the sum of the squares of the x-, y-, and z-intercepts of any plane tangent to the surface $x^{2/3} + y^{2/3} + z^{2/3} = a^{2/3}$ is constant.

32. The equation $F(x, y) = 0$ defines a curve in $\mathbb{R}^2$. Show that if F is differentiable, then $\nabla F(x, y)$ is normal to the curve at every point.

33. Use the result of Problem 32 to find the equation of the tangent line to a curve $F(x, y) = 0$ at a point (x_0, y_0).

In Problems 34–39, use the results of Problems 32 and 33 to find a normal vector and the equation of the tangent line to the curve at the given point.

34. $xy = 5$; $(1, 5)$
35. $x^2 + xy + y^2 + 3x - 5y = 16$; $(1, -2)$
36. $\dfrac{x+y}{x-y} = 7$; $(4, 3)$
37. $xe^{xy} = 1$; $(1, 0)$
38. $\dfrac{x^2}{4} + \dfrac{y^2}{16} = 1$; $(\sqrt{2}, 2\sqrt{2})$
39. $\tan(x + y) = 1$; $(\pi/4, 0)$

40. Show that at any point (x, y), $y \neq 0$, the curve $x/y = a$ is orthogonal to the curve $x^2 + y^2 = r^2$ for any constants a and r.

4.8 DIRECTIONAL DERIVATIVES AND THE GRADIENT

Let us take another look at the partial derivatives $\partial f/\partial x$ and $\partial f/\partial y$ of the function $z = f(x, y)$. We have

$$\frac{\partial f}{\partial x}(x_0, y_0) = \lim_{\Delta x \to 0} \frac{f(x_0 + \Delta x, y_0) - f(x_0, y_0)}{\Delta x}. \tag{1}$$

This measures the rate of change of f as we approach the point (x_0, y_0) along a vector parallel to the x-axis [since $(x_0 + \Delta x, y_0) - (x_0, y_0) = (\Delta x, 0) = \Delta x\mathbf{i}$]. Similarly

$$\frac{\partial f}{\partial y}(x_0, y_0) = \lim_{\Delta y \to 0} \frac{f(x_0, y_0 + \Delta y) - f(x_0, y_0)}{\Delta y} \tag{2}$$

measures the rate of change of f as we approach the point (x_0, y_0) along a vector parallel to the y-axis.

It is frequently of interest to compute the rate of change of f as we approach (x_0, y_0) along a vector that is not parallel to one of the coordinate axes. The situation is depicted in Figure 1. Suppose that (x, y) approaches the fixed point (x_0, y_0) along the line segment joining them, and let t denote the distance between the two points. We want to determine the relative rate of change in f with respect to a change in t.

Let **u** denote a unit vector with the initial point at (x_0, y_0) and parallel to $\overrightarrow{PQ}$ (see Figure 2). Since **u** and $\overrightarrow{PQ}$ are parallel, there is, by Theorem 1.2.3, a value of t such that

$$\overrightarrow{PQ} = t\mathbf{u}. \tag{3}$$

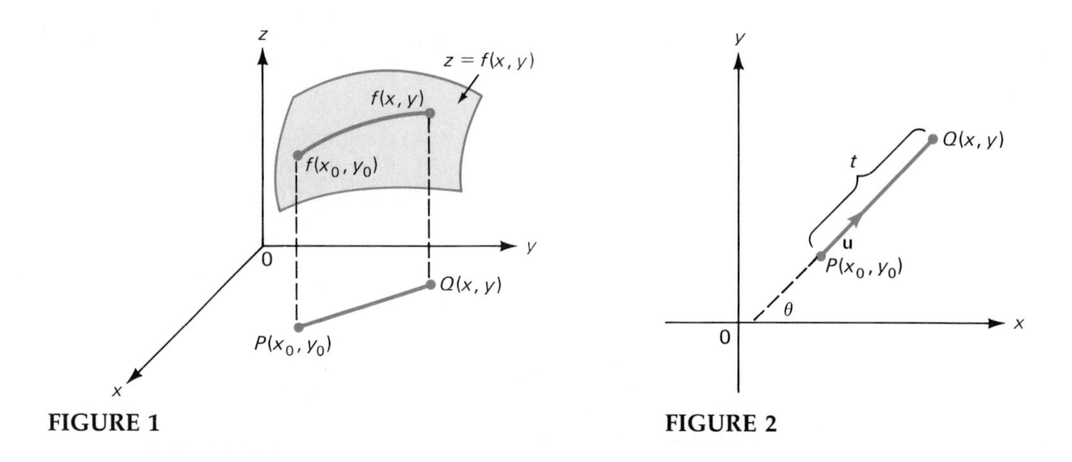

FIGURE 1 **FIGURE 2**

Note that $t > 0$ if **u** and $\overrightarrow{PQ}$ have the same direction and $t < 0$ if **u** and $\overrightarrow{PQ}$ have opposite directions. Now

$$\overrightarrow{PQ} = (x - x_0)\mathbf{i} + (y - y_0)\mathbf{j}, \tag{4}$$

and since **u** is a unit vector, we have

$$\mathbf{u} = \cos\theta\mathbf{i} + \sin\theta\mathbf{j}, \tag{5}$$

where θ is the direction of **u**. Thus inserting (4) and (5) into (3), we have

$$(x - x_0)\mathbf{i} + (y - y_0)\mathbf{j} = t\cos\theta\mathbf{i} + t\sin\theta\mathbf{j},$$

or

$$\begin{aligned} x &= x_0 + t\cos\theta \\ y &= y_0 + t\sin\theta. \end{aligned} \tag{6}$$

The equations (6) are the parametric equations of the line passing through P and Q. Using (6), we have

$$z = f(x, y) = f(x_0 + t\cos\theta,\ y_0 + t\sin\theta). \tag{7}$$

Remember that θ is fixed—it is the direction of approach. Thus $(x, y) \to (x_0, y_0)$ along $\overrightarrow{PQ}$ is equivalent to $t \to 0$ in (7). Hence to compute the instantaneous rate of change of f as $(x, y) \to (x_0, y_0)$ along the vector $\overrightarrow{PQ}$, we need to compute dz/dt. But by the chain rule,

$$\frac{dz}{dt} = \frac{\partial f}{\partial x}(x, y)\frac{dx}{dt} + \frac{\partial f}{\partial y}(x, y)\frac{dy}{dt}$$

or

$$\frac{dz}{dt} = f_x(x, y)\cos\theta + f_y(x, y)\sin\theta, \tag{8}$$

and

$$\frac{dz}{dt} = [f_x(x_0 + t\cos\theta,\ y_0 + t\sin\theta)]\cos\theta + [f_y(x_0 + t\cos\theta,\ y_0 + t\sin\theta)]\sin\theta. \tag{9}$$

If we set $t = 0$ in (9), we obtain the instantaneous rate of change of f in the direction $\overrightarrow{PQ}$ at the point (x_0, y_0). That is,

$$\left.\frac{dz}{dt}\right|_{t=0} = f_x(x_0, y_0)\cos\theta + f_y(x_0, y_0)\sin\theta. \tag{10}$$

But (10) can be written [using (5)] as

$$\left.\frac{dz}{dt}\right|_{t=0} = \nabla f(x_0, y_0) \cdot \mathbf{u}. \tag{11}$$

This leads to the following definition.

Definition 1 DIRECTIONAL DERIVATIVE Let f be differentiable at a point $\mathbf{x}_0 = (x_0, y_0)$ in $\mathbb{R}^2$ and let $\mathbf{u}$ be a unit vector. Then the **directional derivative of f in the direction $\mathbf{u}$**, denoted $f'_{\mathbf{u}}(\mathbf{x}_0)$, is given by

$$f'_{\mathbf{u}}(\mathbf{x}_0) = \nabla f(\mathbf{x}_0) \cdot \mathbf{u}. \tag{12}$$

REMARK 1. Note that if $\mathbf{u} = \mathbf{i}$, then $\nabla f \cdot \mathbf{u} = \partial f/\partial x$ and (12) reduces to the partial derivative $\partial f/\partial x$. Similarly, if $\mathbf{u} = \mathbf{j}$, then (12) reduces to $\partial f/\partial y$.

REMARK 2. Definition 1 makes sense if f is a function of three variables. Then, of course, $\mathbf{u}$ is a unit vector in $\mathbb{R}^3$.

REMARK 3. There is another definition of the directional derivative. It is given by

$$f'_{\mathbf{u}}(\mathbf{x}_0) = \lim_{h\to 0} \frac{f(\mathbf{x}_0 + h\mathbf{u}) - f(\mathbf{x}_0)}{h}. \tag{13}$$

It can be shown that if the limit in (13) exists, it is equal to $\nabla f(\mathbf{x}_0) \cdot \mathbf{u}$ if f is differentiable (see Theorem 2).

EXAMPLE 1 Let $z = f(x, y) = xy^2$. Calculate the directional derivative of f in the direction of the vector $\mathbf{v} = 2\mathbf{i} + 3\mathbf{j}$ at the point $(4, -1)$.

Solution. A unit vector in the direction $\mathbf{v}$ is $\mathbf{u} = (2/\sqrt{13})\mathbf{i} + (3/\sqrt{13})\mathbf{j}$. Also, $\nabla f = y^2\mathbf{i} + 2xy\mathbf{j}$. Thus

$$f'_{\mathbf{u}}(x, y) = \nabla f(\mathbf{x}) \cdot \mathbf{u} = \frac{2y^2}{\sqrt{13}} + \frac{6xy}{\sqrt{13}} = \frac{2y^2 + 6xy}{\sqrt{13}}.$$

At $(4, -1)$, $f'_{\mathbf{u}}(4, -1) = -22/\sqrt{13}$. ■

EXAMPLE 2 Let $f(x, y, z) = x \ln y - e^{xz^3}$. Calculate the directional derivative of f in the direction of the vector $\mathbf{v} = \mathbf{i} - \mathbf{j} + 3\mathbf{k}$. Evaluate this derivative at the point $(-5, 1, -2)$.

Solution. A unit vector in the direction of $\mathbf{v}$ is $\mathbf{u} = (1/\sqrt{11})\mathbf{i} - (1/\sqrt{11})\mathbf{j} + (3/\sqrt{11})\mathbf{k}$, and

$$\nabla f = (\ln y - z^3e^{xz^3})\mathbf{i} + \frac{x}{y}\mathbf{j} - 3xz^2e^{xz^3}\mathbf{k}.$$

Thus

$$f'_{\mathbf{u}}(\mathbf{x}) = \nabla f(\mathbf{x}) \cdot \mathbf{u} = \frac{\ln y - z^3e^{xz^3} - (x/y) - 9xz^2e^{xz^3}}{\sqrt{11}},$$

and at $(-5, 1, -2)$

$$f'_{\mathbf{u}}(-5, 1, -2) = \frac{5 + 188e^{40}}{\sqrt{11}}. \quad ■$$

There is an interesting geometric interpretation of the directional derivative. By Definitions 1.2.5 and 3.2.4 the projection of ∇f on $\mathbf{u}$ is given by

$$\text{Proj}_{\mathbf{u}}\, \nabla f = \frac{\nabla f \cdot \mathbf{u}}{|\mathbf{u}|^2}\mathbf{u},$$

and since $\mathbf{u}$ is a unit vector, the component of ∇f in the direction $\mathbf{u}$ is given by

$$\frac{\nabla f \cdot \mathbf{u}}{|\mathbf{u}|^2} = \nabla f \cdot \mathbf{u}.$$

Thus the *directional derivative of f in the direction* $\mathbf{u}$ *is the component of the gradient of f in the direction* $\mathbf{u}$. This is illustrated in Figure 3.

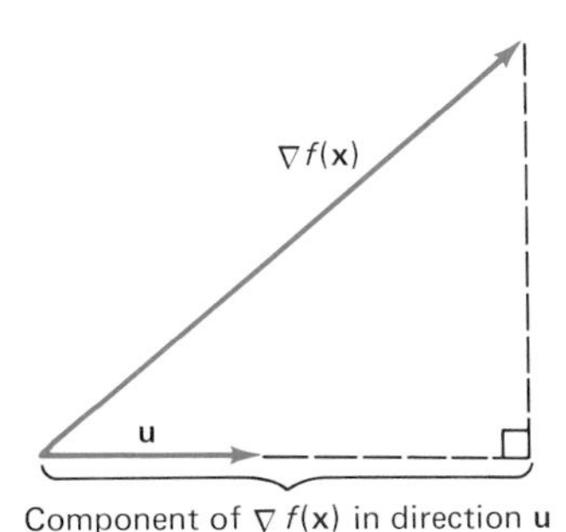

FIGURE 3

We now derive another remarkable property of the gradient. Recall that $\mathbf{u} \cdot \mathbf{v} = |\mathbf{u}|\,|\mathbf{v}| \cos\theta$, where θ is the smallest angle between the vectors $\mathbf{u}$ and $\mathbf{v}$. Thus the directional derivative of f in the direction $\mathbf{u}$ can be written as

$$f'_{\mathbf{u}}(\mathbf{x}) = \nabla f(\mathbf{x}) \cdot \mathbf{u} = |\nabla f(\mathbf{x})|\,|\mathbf{u}| \cos\theta, \tag{14}$$

or since $\mathbf{u}$ is a unit vector,

$$f'_{\mathbf{u}}(\mathbf{x}) = |\nabla f(\mathbf{x})| \cos\theta.$$

Now $\cos\theta = 1$ when $\theta = 0$, which occurs when $\mathbf{u}$ has the direction of ∇f. Similarly, $\cos\theta = -1$ when $\theta = \pi$, which occurs when $\mathbf{u}$ has the direction of $-\nabla f$. Also, $\cos\theta = 0$ when $\theta = \pi/2$. Thus since $-1 \le \cos\theta \le 1$, equation (14) implies the following important result:

Theorem 1 Let f be differentiable. Then f increases most rapidly in the direction of its gradient and decreases most rapidly in the direction opposite to that of its gradient. It changes least in a direction perpendicular to its gradient.

EXAMPLE 3 Consider the sphere $x^2 + y^2 + z^2 = 1$. We can write the upper half of this sphere (i.e., the upper hemisphere) as $z = f(x, y) = \sqrt{1 - x^2 - y^2}$. Then

$$\nabla f = \frac{-x}{\sqrt{1 - x^2 - y^2}}\mathbf{i} + \frac{-y}{\sqrt{1 - x^2 - y^2}}\mathbf{j}.$$

The direction of ∇f is

$$\tan^{-1}\left(\frac{-y/\sqrt{1 - x^2 - y^2}}{-x/\sqrt{1 - x^2 - y^2}}\right) = \tan^{-1}\left(\frac{-y}{-x}\right).$$

But $\tan^{-1}(-y/-x)$ is the direction of the vector $-x\mathbf{i} - y\mathbf{j}$, which points from (x, y) to $(0, 0)$. Thus if we start at a point (x, y, z) on the sphere, the path of steepest ascent (increase) is a great circle passing through the point $(0, 0, 1)$, called the *north pole* of the sphere. ■

EXAMPLE 4 The distribution of voltage on a metal plate is given by

$$V = 50 - x^2 - 4y^2.$$

(a) At the point $(1, -2)$, in what direction does the voltage increase most rapidly?
(b) In what direction does it decrease most rapidly?
(c) What is the magnitude of this increase or decrease?
(d) In what direction does it change least?

Solution. $\nabla V = V_x\mathbf{i} + V_y\mathbf{j} = -2x\mathbf{i} - 8y\mathbf{j}$. At $(1, -2)$, $\nabla V = -2\mathbf{i} + 16\mathbf{j}$.

(a) The voltage increases most rapidly as we move in the direction of $-2\mathbf{i} + 16\mathbf{j}$.
(b) It decreases most rapidly in the direction of $2\mathbf{i} - 16\mathbf{j}$.
(c) The magnitude of the increase or decrease is $\sqrt{2^2 + 16^2} = \sqrt{260}$.
(d) A unit vector perpendicular to ∇V is $(16\mathbf{i} + 2\mathbf{j})/\sqrt{260}$. The voltage changes least in this or the opposite direction. ■

EXAMPLE 5 In Example 4, describe the path of a particle that starts at the point $(1, -2)$ and moves in the direction of greatest voltage increase.

Solution. The path of the particle will be that of the gradient. If the particle follows the path $\mathbf{f}(t) = x(t)\mathbf{i} + y(t)\mathbf{j}$, then since the direction of the path is $\mathbf{f}'(t) = x'(t)\mathbf{i} + y'(t)\mathbf{j}$ and since this direction is also given by $\nabla V = -2x\mathbf{i} - 8y\mathbf{j}$, we must have

$$x'(t) = -2x(t) \qquad \text{and} \qquad y'(t) = -8y(t).$$

The solutions to these differential equations are

$$x(t) = c_1e^{-2t} \qquad \text{and} \qquad y(t) = c_2e^{-8t}.$$

But $x(0) = 1$ and $y(0) = -2$, so

$$x(t) = e^{-2t} \qquad \text{and} \qquad y(t) = -2e^{-8t}.$$

Then since $e^{-8t} = (e^{-2t})^4$, we see that the particle moves along the path

$$y = -2x^4.$$

REMARK. Technically, a direction is a unit vector, so we should choose the direction $(-2x\mathbf{i} - 8y\mathbf{j})/\sqrt{4x^2 + 64y^2}$ in our computations. But this choice would not change the final answer. A method for obtaining the answer by using unit vectors is suggested in Problem 27. ■

One other fact about directional derivatives and gradients should be mentioned here. Since the gradient vector is normal to the curve $f(x, y) = C$, for any constant C, we say that the directional derivative of f in the direction of the gradient is the **normal derivative** of f and is denoted df/dn. We then have, from equation (14),

$$\text{normal derivative} = \frac{df}{dn} = |\nabla f|. \tag{15}$$

EXAMPLE 6 Let $f(x, y) = xy^2$. Calculate the normal derivative. Evaluate df/dn at the point $(3, -2)$.

Solution. $\nabla f = y^2\mathbf{i} + 2xy\mathbf{j}$. Then $df/dn = |\nabla f| = \sqrt{y^4 + 4x^2y^2}$. At $(3, -2)$, $df/dn = \sqrt{16 + 144} = \sqrt{160}$. ■

We now define the directional derivative in $\mathbb{R}^n$.

Definition 2 DIRECTIONAL DERIVATIVE Let $f: \mathbb{R}^n \to \mathbb{R}$ be differentiable at a point $\mathbf{x} = (x_1, x_2, \ldots, x_n)$ in $\mathbb{R}^n$ and let $\mathbf{u}$ be a unit vector in $\mathbb{R}^n$. Then the **directional derivative of f in the direction u**, denoted $f_{\mathbf{u}}'(\mathbf{x})$, is given by

$$f_{\mathbf{u}}'(\mathbf{x}_0) = \nabla f(\mathbf{x}_0) \cdot \mathbf{u}. \tag{16}$$

REMARK. Another, perhaps more common, definition of the directional derivative is

$$f_{\mathbf{u}}'(\mathbf{x}_0) = \lim_{h\to 0} \frac{f(\mathbf{x}_0 + h\mathbf{u}) - f(\mathbf{x}_0)}{h}. \tag{17}$$

Equation (17) illustrates the relationship between the rate of change of f and the direction in which $\mathbf{x} = \mathbf{x}_0 + h\mathbf{u}$ is approaching $\mathbf{x}_0$. For example, if $\mathbf{u} = (0, 0, \ldots, 1, 0, \ldots, 0)$ with a 1 in the ith position, then

$$f(\mathbf{x}_0 + h\mathbf{u}) - f(\mathbf{x}_0) = f(x_1, x_2, \ldots, x_i + h, \ldots, x_n) - f(x_1, x_2, \ldots, x_i, \ldots, x_n)$$

and the limit in (17) defines the partial derivative $\partial f/\partial x_i$. We now show that (17) implies (16).

Theorem 2 Suppose that the directional derivative $f_{\mathbf{u}}'(\mathbf{x})$ is defined by (17). If f is differentiable at $\mathbf{x}$, then f has a directional derivative in every direction and

$$f_{\mathbf{u}}'(\mathbf{x}) = \nabla f(x) \cdot \mathbf{u}.$$

Proof. Since f is differentiable at $\mathbf{x}$, we have

$$f(\mathbf{x} + h\mathbf{u}) - f(\mathbf{x}) = \nabla f(\mathbf{x}) \cdot h\mathbf{u} + g(h\mathbf{u}). \tag{18}$$

Now, from the definition of differentiability,

$$\lim_{|h\mathbf{u}|\to 0} \frac{g(h\mathbf{u})}{|h\mathbf{u}|} = 0.$$

But since $\mathbf{u}$ is a unit vector, $|h\mathbf{u}| = |h|\,|\mathbf{u}| = |h|$. Thus we find that

$$\lim_{h\to 0} \frac{g(h\mathbf{u})}{h} = 0.$$

Then, dividing both sides of (18) by h and taking limits, we obtain

$$\lim_{h\to 0}\frac{f(\mathbf{x}+h\mathbf{u})-f(\mathbf{x})}{h} = \lim_{h\to 0}\frac{\nabla f(\mathbf{x})\cdot h\mathbf{u}}{h} + \lim_{h\to 0}\frac{g(h\mathbf{u})}{h}$$

$$= \lim_{h\to 0}\frac{h(\nabla f(\mathbf{x})\cdot \mathbf{u})}{h} + \lim_{h\to 0}\frac{g(h\mathbf{u})}{h} = \nabla f(\mathbf{x})\cdot\mathbf{u}$$

and the proof is complete. ■

EXAMPLE 7 Let $f(\mathbf{x}) = x_1^2 + x_2x_3^4 - x_4^2$. Compute the directional derivative of f in the direction $(1, 2, 3, 4)$ at the point $(-2, 3, 1, 5)$.

Solution.

$$\nabla f = (2x_1,\ x_3^4,\ 4x_2x_3^3,\ -2x_4)$$
$$= (-4,\ 1,\ 12,\ -10)$$

at $(-2, 3, 1, 5)$. A unit vector in the given direction is $\mathbf{u} = (1/\sqrt{30})(1, 2, 3, 4)$. Thus

$$f_{\mathbf{u}}'(\mathbf{x}) = \nabla f\cdot\mathbf{u} = (-4, 1, 12, 10)\cdot\frac{1}{\sqrt{30}}(1, 2, 3, 4) = \frac{74}{\sqrt{30}}.\ ■$$

In Section 1.2 we showed that the cosine of the angle between two vectors $\mathbf{u}$ and $\mathbf{v}$ in $\mathbb{R}^2$ is given by

$$\cos\theta = \frac{\mathbf{u}\cdot\mathbf{v}}{|\mathbf{u}|\,|\mathbf{v}|}. \tag{19}$$

We can use formula (19) as the *definition*[†] of the angle between two vectors $\mathbf{u}$ and $\mathbf{v}$ in $\mathbb{R}^n$, where θ is always in the interval $[0, \pi]$. Then we have

$$f_{\mathbf{u}}'(\mathbf{x}) = \nabla f(\mathbf{x})\cdot\mathbf{u} = |\nabla f(\mathbf{x})|\,|\mathbf{u}|\cos\theta = |\nabla f(\mathbf{x})|\cos\theta$$

since $\mathbf{u}$ is a unit vector. Thus, as in $\mathbb{R}^2$ and $\mathbb{R}^3$, we see that since $f_{\mathbf{u}}'$ is largest when $\cos\theta = 1$ (so that $\theta = 0$), is smallest when $\cos\theta = -1$, and is 0 when $\cos\theta = 0$ (so that $\theta = \pi/2$). We conclude that

> f increases most rapidly in the direction of its gradient and decreases most rapidly in the direction opposite to that of its gradient. It changes least in a direction orthogonal to its gradient.

[†]It can be shown (as we will do in Problem 9.3.63) that $|\mathbf{u}\cdot\mathbf{v}| \le |\mathbf{u}|\,|\mathbf{v}|$ so that

$$-1 \le \frac{\mathbf{u}\cdot\mathbf{v}}{|\mathbf{u}|\,|\mathbf{v}|} \le 1.$$

This is necessary in order that θ be defined by (19).

We need to say something about the last statement. We say that two vectors in $\mathbb{R}^n$ are *parallel* if one is a scalar multiple of the other. They have the same direction if the scalar is positive and the opposite direction if the scalar is negative. We defined orthogonality in Section 3.7. It is important to note here that even though (for $n > 3$) it is impossible to draw parallel and orthogonal vectors, the ideas developed geometrically in $\mathbb{R}^2$ and $\mathbb{R}^3$ are easily extended to higher dimensions.

PROBLEMS 4.8

In Problems 1–15, calculate the directional derivative of the given function at the given point in the direction of the given vector $\mathbf{v}$.

1. $f(x, y) = xy$ at $(2, 3)$; $\mathbf{v} = \mathbf{i} + 3\mathbf{j}$
2. $f(x, y) = 2x^2 - 3y^2$ at $(1, -1)$; $\mathbf{v} = -\mathbf{i} + 2\mathbf{j}$
3. $f(x, y) = \ln(x + 3y)$ at $(2, 4)$; $\mathbf{v} = \mathbf{i} + \mathbf{j}$
4. $f(x, y) = ax^2 + by^2$ at (c, d); $\mathbf{v} = \alpha\mathbf{i} + \beta\mathbf{j}$
5. $f(x, y) = \tan^{-1}\dfrac{y}{x}$ at $(2, 2)$; $\mathbf{v} = 3\mathbf{i} - 2\mathbf{j}$
6. $f(x, y) = \dfrac{x - y}{x + y}$ at $(4, 3)$; $\mathbf{v} = -\mathbf{i} - 2\mathbf{j}$
7. $f(x, y) = xe^y + ye^x$ at $(1, 2)$; $\mathbf{v} = \mathbf{i} + \mathbf{j}$
8. $f(x, y) = \sin(2x + 3y)$ at $(\pi/12, \pi/9)$; $\mathbf{v} = -2\mathbf{j} + 3\mathbf{j}$
9. $f(x, y, z) = xy + yz + xz$ at $(1, 1, 1)$; $\mathbf{v} = \mathbf{i} + \mathbf{j} + \mathbf{k}$
10. $f(x, y, z) = xy^3z^5$ at $(-3, -1, 2)$; $\mathbf{v} = -\mathbf{i} - 2\mathbf{j} + \mathbf{k}$
11. $f(x, y, z) = \ln(x + 2y + 3z)$ at $(1, 2, 0)$; $\mathbf{v} = 2\mathbf{i} + \mathbf{j} - \mathbf{k}$
12. $f(x, y, z) = xe^{yz}$ at $(2, 0, -4)$; $\mathbf{v} = -\mathbf{i} + 2\mathbf{j} + 5\mathbf{k}$
13. $f(x, y, z) = x^2y^3 + z\sqrt{x}$ at $(1, -2, 3)$; $\mathbf{v} = 5\mathbf{j} + \mathbf{k}$
14. $f(x, y, z) = e^{-(x^2+y^2+z^2)}$ at $(1, 1, 1)$; $\mathbf{v} = \mathbf{i} + 3\mathbf{j} - 5\mathbf{k}$
15. $f(x, y, z) = \dfrac{1}{\sqrt{x^2 + y^2 + z^2}}$ at $(-1, 2, 3)$; $\mathbf{v} = \mathbf{i} - \mathbf{j} + \mathbf{k}$
16. The voltage (potential) at any point on a metal structure is given by
$$v(x, y, z) = \frac{1}{0.02 + \sqrt{x^2 + y^2 + z^2}}.$$
At the point $(1, -1, 2)$, in what direction does the voltage increase most rapidly?
17. The temperature at any point in a solid metal ball centered at the origin is given by
$$T(x, y, z) = 100e^{-(x^2+y^2+z^2)}.$$
(a) Where is the ball hottest?
(b) Show that at any point (x, y, z) on the ball, the direction of greatest increase in temperature is a vector pointing toward the origin.
18. The temperature distribution of a ball centered at the origin is given by
$$T(x, y, z) = \frac{100}{x^2 + y^2 + z^2 + 1}.$$
(a) Where is the ball hottest?
(b) Find the direction of greatest decrease of temperature at the point $(3, -1, 2)$.
(c) Find the direction of greatest increase in temperature. Does this vector point toward the origin?
19. The temperature distribution on a plate is given by
$$T(x, y) = 1 - \frac{x^2}{a^2} - \frac{y^2}{b^2}.$$
Find the path of a heat-seeking particle (i.e., a particle that always moves in the direction of greatest increase in temperature) if it starts at the point (a, b).
20. Find the path of the particle in Problem 19 if it starts at the point $(-a, b)$.
21. The height of a mountain is given by $h(x, y) = 3000 - 2x^2 - y^2$, where the x-axis points north, the y-axis points east, and all distances are measured in meters. Suppose that a mountain climber is at the point $(30, -20, 800)$.
(a) If the climber moves in the southwest direction, will he or she ascend or descend?
(b) In what direction should the climber move so as to ascend most rapidly?

*22. Prove that if $w = f(x, y, z)$ is differentiable at $\mathbf{x}$, then all the first partials exist at $\mathbf{x}$, and

$$\nabla f(\mathbf{x}) = f_x(\mathbf{x})\mathbf{i} + f_y(\mathbf{x})\mathbf{j} + f_z(\mathbf{x})\mathbf{k}.$$

(This result is the converse to Theorem 4.5.1.)

In Problems 23–26, calculate the normal derivative at the given point.

23. $f(x, y) = x + 2y$ at $(1, 4)$
24. $f(x, y) = e^{x+3y}$ at $(1, 0)$
25. $f(x, y) = \tan^{-1}\frac{y}{x}$ at $(-1, -1)$
26. $f(x, y) = \sqrt{\frac{x-y}{x+y}}$ at $(3, 1)$
27. We refer to Example 5.
   (a) Show that a unit vector in the direction of motion is given by

$$\frac{x'}{\sqrt{x'^2 + y'^2}}\mathbf{i} + \frac{y'}{\sqrt{x'^2 + y'^2}}\mathbf{j}.$$

   (b) Show that a unit vector having the direction of the gradient is

$$\frac{-2x}{\sqrt{4x^2 + 64y^2}}\mathbf{i} - \frac{8y}{\sqrt{4x^2 + 64y^2}}\mathbf{j}.$$

   (c) By equating coordinates in (a) and (b), show that

$$\frac{y'(t)}{y(t)} = \frac{4x'(t)}{x(t)}.$$

   (d) Integrate both sides of the equation in (c) to show that $\ln|y'(t)| = \ln[x(t)]^4 + C$.
   (e) Use (d) to show that $y(t) = k[x(t)]^4$, where $k = \pm e^c$.
   (f) Show that $k = -2$ in (e) by using the point $(1, -2)$ on the curve.

In Problems 28–32 find the directional derivative of the given function in the direction of the given vector $\mathbf{v}$ at the point P.

28. $f(x_1, x_2, x_3, x_4) = x_1x_2x_3x_4$; $\mathbf{v} = (2, -1, 4, 6)$; $P = (1, 0, -1, 2)$
29. $f(x_1, x_2, x_3, x_4) = x_1^2 + x_2^2 + x_3^2 + x_4^2$; $\mathbf{v} = (-1, 1, 0, 2)$; $P = (1, 1, -2, 3)$
30. $f(x_1, x_2, x_3, x_4) = \sqrt{x_1^2 + x_2^2 + x_3^2 + x_4^2}$; $\mathbf{v} = (-1, 6, 2, 4)$; $P = (-3, 0, 1, 1)$
31. $f(x_1, x_2, \ldots, x_n) = \Sigma x_i^2$; $\mathbf{v} = (1, 1, \ldots, 1)$; $P = (1, 2, 3, \ldots, n)$
32. $f(x_1, x_2, \ldots, x_n) = x_1x_2 \cdots x_n$; $\mathbf{v} = (1, 1, \ldots, 1)$; $P = (1, 2, \ldots, n)$

4.9 CONSERVATIVE VECTOR FIELDS AND THE GRADIENT (OPTIONAL)

In this brief section we show how the gradient can arise in a fundamental way in physics. Suppose that for every $\mathbf{x}$ in $\mathbb{R}^2$ (or $\mathbb{R}^3$), $\mathbf{F}(\mathbf{x})$ is another vector in $\mathbb{R}^2$ (or $\mathbb{R}^3$). Then $\mathbf{F}$ is really a vector function of a vector. We call such a function a **vector field.**† An important question that can arise is: Does there exist a function $f(\mathbf{x})$ such that $\nabla f(\mathbf{x}) = \mathbf{F}(\mathbf{x})$? That is, is $\mathbf{F}$ the gradient of some function f? In Section 4.11 we answer this question. However, in this section we will assume that there exists a function f such that

$$\mathbf{F} = \nabla f \tag{1}$$

If $\mathbf{F} = -\nabla f$ for some function f, then $\mathbf{F}$ is said to be a **conservative vector field** and f is called a **potential function** for $\mathbf{F}$. The reason for this terminology will be made clear shortly. [If $\mathbf{F} = \nabla f$, then $\mathbf{F} = -\nabla(-f)$, so that the introduction of the minus sign does not cause any problem.]

†We will discuss vector fields in more detail in Section 6.1.

Now let $\mathbf{x}(t) = x(t)\mathbf{i} + y(t)\mathbf{j}$ be a differentiable curve and suppose that a particle of mass m moves along it. Suppose further that the force acting on the particle at any time t is given by $\mathbf{F}(\mathbf{x}(t))$, where $\mathbf{F}$ is assumed to be a conservative vector field. By Newton's second law,

$$\mathbf{F}(\mathbf{x}(t)) = m\mathbf{a}(t) = m\mathbf{x}''(t). \tag{2}$$

But since $\mathbf{F}$ is conservative, (1) implies that $\mathbf{F}(\mathbf{x}) = -\nabla f(\mathbf{x})$ for some differentiable function f. Then we have

$$-\nabla f(\mathbf{x}(t)) = m\mathbf{x}''(t),$$

or

$$m\mathbf{x}'' + \nabla f(\mathbf{x}) = \mathbf{0}. \tag{3}$$

We now take the dot product of both sides of (3) with $\mathbf{x}'$ to obtain

$$m\mathbf{x}' \cdot \mathbf{x}'' + \nabla f(\mathbf{x}) \cdot \mathbf{x}' = 0. \tag{4}$$

But by the product rule,

$$\frac{d}{dx}|\mathbf{x}'(t)|^2 = \frac{d}{dt}(\mathbf{x}'(t) \cdot \mathbf{x}'(t)) = \mathbf{x}'(t) \cdot \mathbf{x}''(t) + \mathbf{x}''(t) \cdot \mathbf{x}'(t) = 2\mathbf{x}'(t) \cdot \mathbf{x}''(t), \tag{5}$$

and by equation (4.6.3),

$$\frac{d}{dt} f(\mathbf{x}(t)) = \nabla f(\mathbf{x}(t)) \cdot \mathbf{x}'(t) \tag{6}$$

Using (5) and (6) in (4), we obtain

$$\frac{d}{dt}\left[\frac{1}{2} m|\mathbf{x}'|^2 + f(\mathbf{x}(t))\right] = 0,$$

which implies that

$$\tfrac{1}{2}m|\mathbf{x}'|^2 + f(\mathbf{x}(t)) = C, \tag{7}$$

where C is a constant. This is one of the versions of the **law of conservation of energy**. The term $\frac{1}{2}m|\mathbf{x}'|^2 = \frac{1}{2}m|\mathbf{v}|^2$ is called the **kinetic energy** of the particle, and the term $f(\mathbf{x}(t))$ is called the **potential energy** of the particle. Equation (7) tells us simply that if the force function $\mathbf{F}$ is conservative, then the total energy of the system is constant and, moreover, the potential function f of $\mathbf{F}$ represents the potential energy of the system.

The **principle of the conservation of energy** states that energy may be transformed from one form to another but cannot be created or destroyed; that is, the total energy is constant. Thus it seems reasonable that force fields in classical physics are conservative (although since the work of Einstein, it has been found that energy can be transformed into mass and vice versa, so that there are forces that are not conservative). One example of a conservative force is given by the force of gravitational attraction. Let m_1 represent the mass of a (relatively) fixed object in space and let m_2 denote the mass of an object moving near the fixed object. Then the magnitude of the gravitational force between the objects is given by[†]

$$|\mathbf{F}| = G\frac{m_1 m_2}{r^2} \tag{8}$$

where r is the distance between the objects and G is a universal constant. If we assume that the first object is at the origin, then we may denote the position of the second object by $\mathbf{x}(t)$, and then since $r = |\mathbf{x}|$, (8) can be written

$$|\mathbf{F}| = \frac{Gm_1 m_2}{|\mathbf{x}|^2}. \tag{9}$$

Also, the force acts toward the origin, that is, in the direction opposite to that of the positive vector $\mathbf{x}$. Therefore

$$\text{direction of } \mathbf{F} = -\frac{\mathbf{x}}{|\mathbf{x}|}, \tag{10}$$

so from (9) and (10),

$$\mathbf{F}(t) = \frac{\alpha\mathbf{x}(t)}{|\mathbf{x}(t)|^3} \tag{11}$$

where $\alpha = -Gm_1m_2$. We now prove that $\mathbf{F}$ is conservative. To show this, we must come up with a function f such that $\mathbf{F} = -\nabla f$. Here we will pull f "out of a hat." In Section 4.11 we will show you how to construct such an f (if one exists). Let

$$f(\mathbf{x}) = \frac{\alpha}{|\mathbf{x}|}.$$

Then $f(\mathbf{x}) = \alpha/\sqrt{x^2 + y^2 + z^2}$ and

$$\begin{aligned} -\nabla f(\mathbf{x}) &= -\alpha\left[\frac{-x}{(x^2 + y^2 + z^2)^{3/2}}\mathbf{i} + \frac{-y}{(x^2 + y^2 + z^2)^{3/2}}\mathbf{j} + \frac{-z}{(x^2 + y^2 + z^2)^{3/2}}\mathbf{k}\right] \\ &= \frac{\alpha}{|\mathbf{x}|^3}(x\mathbf{i} + y\mathbf{j} + z\mathbf{k}) = \frac{\alpha\mathbf{x}}{|\mathbf{x}|^3} = \mathbf{F}. \end{aligned}$$

Thus $\mathbf{F}$ is conservative, so that, with respect to gravitational forces, the law of conservation of energy holds.

[†]This is called *Newton's law of universal gravitation.*

PROBLEMS 4.9

1. Show that the force given by $\mathbf{F}(\mathbf{x}) = -\alpha\mathbf{x}/|\mathbf{x}|^2$ is conservative by finding a potential function for it.
2. Do the same for the force given by $\mathbf{F}(\mathbf{x}) = -\alpha\mathbf{x}/|\mathbf{x}|^4$.
3. If α is a constant, show that the force $\mathbf{F}(\mathbf{x}) = -\alpha\mathbf{x}/|\mathbf{x}|^k$ is conservative.
4. Show that the force $\mathbf{F}(x, y) = y\mathbf{i} + x\mathbf{j}$ is conservative.
5. Show that if $\mathbf{F}$ and $\mathbf{G}$ are conservative, then $\alpha\mathbf{F} + \beta\mathbf{G}$ is also conservative for any constants α and β.

*6. Show that the force $\mathbf{F}(x, y) = y\mathbf{i} - x\mathbf{j}$ is *not* conservative. [*Hint:* Assume that $\mathbf{F} = -\nabla f$, so that $-y = \partial f/\partial x$ and $x = \partial f/\partial y$. Then integrate to obtain a contradiction.]

4.10 THE TOTAL DIFFERENTIAL AND APPROXIMATION

In one-variable calculus the notions of increments and differentials are used to approximate a function. We used the fact that if Δx was small, then

$$f(x + \Delta x) - f(x) = \Delta y \approx f'(x)\,\Delta x. \tag{1}$$

We also defined the differential dy by

$$dy = f'(x)\,dx = f'(x)\,\Delta x \tag{2}$$

(since dx is defined to be equal to Δx). Note that in (2) it is not required that Δx be small.

We now extend these ideas to functions of two or three variables.

Definition 1 INCREMENT AND TOTAL DIFFERENTIAL Let $f = f(\mathbf{x})$ be a function of two or three variables and let $\Delta\mathbf{x} = (\Delta x, \Delta y)$ or $(\Delta x, \Delta y, \Delta z)$.

(i) The **increment of** f, denoted Δf, is defined by

$$\Delta f = f(\mathbf{x} + \Delta\mathbf{x}) - f(\mathbf{x}). \tag{3}$$

(ii) The **total differential of** f, denoted df, is given by

$$df = \nabla f(\mathbf{x}) \cdot \Delta\mathbf{x}. \tag{4}$$

Note that equation (4) is very similar in form to equation (2).

REMARK 1. If f is a function of two variables, then (3) and (4) become

$$\Delta f = f(x + \Delta x, y + \Delta y) - f(x, y), \tag{5}$$

and the total differential is

$$df = f_x(x, y)\,\Delta x + f_y(x, y)\,\Delta y. \tag{6}$$

REMARK 2. If f is a function of three variables, then (3) and (4) become

$$\Delta f = f(x + \Delta x, y + \Delta y, z + \Delta z) - f(x, y, z) \tag{7}$$

and

$$df = f_x(x, y, z)\,\Delta x + f_y(x, y, z)\,\Delta y + f_z(x, y, z)\,\Delta z. \tag{8}$$

REMARK 3. Note that in the definition of the total differential, it is *not* required that $|\mathbf{\Delta x}|$ be small.

From Theorems 4.5.1 and 4.5.1′ and the definition of differentiability, we see that if $|\mathbf{\Delta x}|$ is small and if f is differentiable, then

$$\Delta f \approx df. \tag{9}$$

We can use the relation (9) to approximate functions of several variables in much the same way that we used the relation (1) to approximate the values of functions of one variable.

EXAMPLE 1 Use the total differential to estimate $\sqrt{(2.98)^2 + (4.03)^2}$.

Solution. Let $f(x, y) = \sqrt{x^2 + y^2}$. Then we are asked to calculate $f(2.98, 4.03)$. We know that $f(3, 4) = \sqrt{3^2 + 4^2} = 5$. Thus we need to calculate $f(3 - 0.02, 4 + 0.03)$. Now, at (3, 4),

$$\nabla f(\mathbf{x}) = \frac{x}{\sqrt{x^2 + y^2}}\mathbf{i} + \frac{y}{\sqrt{x^2 + y^2}}\mathbf{j} = \frac{3}{5}\mathbf{i} + \frac{4}{5}\mathbf{j}.$$

Then using (6), we have

$$df = \frac{3}{5}\,\Delta x + \frac{4}{5}\,\Delta y = (0.6)(-0.02) + (0.8)(0.03) = 0.012.$$

Hence

$$f(3 - 0.02, 4 + 0.03) - f(3, 4) = \Delta f \approx df = 0.012,$$

so

$$f(2.98, 4.03) \approx f(3, 4) + 0.012 = 5.012.$$

The exact value of $\sqrt{(2.98)^2 + (4.03)^2}$ is $\sqrt{8.8804 + 16.2409} = \sqrt{25.1213} \approx 5.012115$, so that $\Delta f \approx 0.012115$ and our approximation is very good indeed. ■

EXAMPLE 2 The radius of a cone is measured to be 15 cm and the height of the cone is measured to be 25 cm. There is a maximum error of ± 0.02 cm in the measurement of the radius and ± 0.05 cm in the measurement of the height. (a) What is the approximate volume of the cone? (b) What is the maximum error in the calculation of the volume?

Solution. **(a)** $V = \frac{1}{3}\pi r^2 h \approx \frac{1}{3}\pi(15)^2 25 = 1875\pi \text{ cm}^3 \approx 5890.5 \text{ cm}^3$.

(b) $\nabla V = V_r\mathbf{i} + V_h\mathbf{j} = \frac{2}{3}\pi rh\mathbf{i} + \frac{1}{3}\pi r^2\mathbf{j} = \pi(250\mathbf{i} + 75\mathbf{j})$. Then choosing $\Delta x = 0.02$ and $\Delta y = 0.05$ to find the maximum error, we have

$$\Delta V \approx dV = \nabla V \cdot \Delta \mathbf{x} = \pi[250(0.02) + 75(0.05)] = \pi(5 + 3.75)$$
$$= 8.75\pi \approx 27.5 \text{ cm}^3.$$

Thus the maximum error in the calculation is, approximately, 27.5 cm^3, which means that

$$5890.5 - 27.5 < V < 5890.5 + 27.5,$$

or

$$5863 \text{ cm}^3 < V < 5918 \text{ cm}^3.$$

Note that an error of 27.5 cm^3 is only a **relative error** of $27.5/5890.5 \approx 0.0047$, which is a very small relative error. ■

EXAMPLE 3 A cylindrical tin can has an inside radius of 5 cm and a height of 12 cm. The thickness of the tin is 0.2 cm. Estimate the amount of tin needed to construct the can (including its ends).

Solution. We need to estimate the difference between the "outer" and "inner" volumes of the can. We have $V = \pi r^2 h$. The inner volume is $\pi(5^2)(12) = 300\pi$ cm^3, and the outer volume is $\pi(5.2)^2(12.4)$. The difference is

$$\Delta V = \pi(5.2)^2(12.4) - 300\pi \approx dV.$$

Since $\nabla V = 2\pi rh\mathbf{i} + \pi r^2\mathbf{j} = \pi(120\mathbf{i} + 25\mathbf{j})$, we have

$$dV = \pi(120(0.2) + 25(0.4)) = 34\pi.$$

Thus the amount of tin needed is, approximately, 34π cm^3 ≈ 106.8 cm^3. ■

PROBLEMS 4.10

In Problems 1–12, calculate the total differential df.

1. $f(x, y) = xy^3$

2. $f(x, y) = \tan^{-1}\dfrac{y}{x}$

3. $f(x, y) = \sqrt{\dfrac{x - y}{x + y}}$

4. $f(x, y) = xe^y$

5. $f(x, y) = \ln(2x + 3y)$

6. $f(x, y) = \sin(x - 4y)$

7. $f(x, y, z) = xy^2z^5$

8. $f(x, y, z) = \dfrac{xy}{z}$

9. $f(x, y, z) = \ln(x + 2y + 3z)$

10. $f(x, y, z) = \sec xy - \tan z$

11. $f(x, y, z) = \cosh(xy - z)$

12. $f(x, y, z) = \dfrac{x - z}{y + 3x}$

13. Let $f(x, y) = xy^2$.
(a) Calculate explicitly the difference $\Delta f - df$.
(b) Verify your answer by calculating $\Delta f - df$ at the point (1, 2), where $\Delta x = -0.01$ and $\Delta y = 0.03$.

***14.** Repeat the steps of Problem 13 for the function $f(x, y) = x^3y^2$.

In Problems 15–23, use the total differential to estimate the given number.

15. $3.01/5.99$

16. $19.8\sqrt{65}$

17. $\sqrt{35.6}\,\sqrt[3]{64.08}$

18. $(2.01)^4(3.04)^7 - (2.01)(3.04)^9$

19. $\sqrt{\dfrac{5.02 - 3.96}{5.02 + 3.96}}$

20. $((4.95)^2 + (7.02))^{1/5}$

21. $\dfrac{(3.02)(1.97)}{\sqrt{8.95}}$

22. $\sin\left(\dfrac{11\pi}{24}\right)\cos\left(\dfrac{13\pi}{36}\right)$

23. $(7.92)\sqrt{5.01} - (0.98)^2$

24. The radius and height of a cylinder are, approximately, 10 cm and 20 cm, respectively. The maximum errors in approximation are ± 0.03 cm and ± 0.07 cm.
- **(a)** What is the approximate volume of the cylinder?
- **(b)** What is the approximate maximum error in this calculation?

25. Two sides of a triangular piece of land were measured to be 50 m and 110 m; there was an error of at most ± 0.3 m in each measurement. The angle between the sides was measured to be 60° with an error of at most $\pm 1°$.
- **(a)** Using the law of cosines, find the approximate length of the third side of the triangle.
- **(b)** What is the approximate maximum error of your measurement? [*Hint:* It will be necessary to convert 1° to radians.]

26. How much wood is contained in the sides of a rectangular box with sides of inside measurements 1 m, 1.2 m, and 1.6 m if the thickness of the wood making up the sides is 5 cm (= 0.05 m)?

27. When three resistors are connected in parallel, the total resistance R (measured in ohms) is given by

$$\frac{1}{R} = \frac{1}{r_1} + \frac{1}{r_2} + \frac{1}{r_3},$$

where r_1, r_2, and r_3 are the three separate resistances. Let $r_1 = 6 \pm 0.1\ \Omega$, $r_2 = 8 \pm 0.03\ \Omega$, and $r_3 = 12 \pm 0.15\ \Omega$.
- **(a)** Estimate R.
- **(b)** Find an approximate maximum value for the maximum error in your estimate.

28. The volume of 10 mol of an ideal gas was calculated to be 500 cm³ at a temperature of 40°C (= 313 °K). The maximum error in each measurement n, V, and T was $\frac{1}{2}\%$.
- **(a)** Calculate the approximate pressure of the gas (in newtons per square centimeter).
- **(b)** Find the approximate maximum error in your computation. [*Hint:* Recall that according to the ideal gas law, $PV = nRT$.]

4.11 EXACT VECTOR FIELDS OR HOW TO OBTAIN A FUNCTION FROM ITS GRADIENT

If $z = f(x, y)$ is differentiable, then the gradient of f exists and

$$\nabla f = f_x(x, y)\mathbf{i} + f_y(x, y)\mathbf{j}. \tag{1}$$

We can write (1) as

$$\nabla f = P(x, y)\mathbf{i} + Q(x, y)\mathbf{j}. \tag{2}$$

In this section we will try to reverse this process. This corresponds roughly to finding an antiderivative of a function of one variable. We will ask the following question: Given a *vector field* of the form

$$\mathbf{F}(x, y) = P(x, y)\mathbf{i} + Q(x, y)\mathbf{j}, \tag{3}$$

does there exist a differentiable function f such that $\nabla f = \mathbf{F}$?

Before answering this question in general, we will look at two examples that illustrate what can happen.

EXAMPLE 1 Let $\mathbf{F} = 9x^2y^2\mathbf{i} + (6x^3y + 2y)\mathbf{j}$. Show that there is a function f such that $\nabla f = \mathbf{F}$.

Solution. If there is such an f, then

$$\frac{\partial f}{\partial x} = 9x^2y^2 \quad \text{and} \quad \frac{\partial f}{\partial y} = 6x^3y + 2y. \tag{4}$$

Integrating both sides of the first of these equations *with respect to x,* we obtain

$$f(x, y) = 3x^3y^2 + g(y). \tag{5}$$

The term $g(y)$ is a *constant function of integration* since

$$\frac{\partial}{\partial x}g(y) = 0.$$

Next we differentiate (5) *with respect to* y and use the second equation of (4):

$$\frac{\partial f}{\partial y} = 6x^3y + g'(y) = 6x^3y + 2y.$$

Thus $g'(y) = 2y$ and $g(y) = y^2 + C$. Hence for any constant C,

$$f(x, y) = 3x^3y^2 + y^2 + C$$

is a function that satisfies $\nabla f = \mathbf{F}$. (This result should be checked.) ■

REMARK. When we integrate a function of x and y with respect to x, the "constant" of integration is always a function of y. Analogously, when we integrate a function of x and y with respect to y, the "constant" of integration is always a function of x.

EXAMPLE 2 Show that there is *no* function f that satisfies

$$\nabla f = y\mathbf{i} - x\mathbf{j}.$$

Solution. Suppose that $\nabla f = y\mathbf{i} - x\mathbf{j}$. Then $\partial f/\partial x = y$ and $\partial f/\partial y = -x$. Integration of the first of these equations with respect to x yields

$$f(x, y) = yx + g(y),$$

so that

$$\frac{\partial f}{\partial y} = x + g'(y) = -x.$$

This implies that $g'(y) = -2x$, which is impossible since g and g' are functions of y only. Hence there is no function f whose gradient is $y\mathbf{i} - x\mathbf{j}$. ■

We now state a general result that tells us whether **F** is a gradient of some function f (i.e., whether **F** is conservative). Half its proof is difficult and is omitted.[†] The other half is suggested in Problem 24.

Theorem 1 Let $\mathbf{F}(x, y) = P(x, y)\mathbf{i} + Q(x, y)\mathbf{j}$ and suppose that P, Q, $\partial P/\partial y$, and $\partial Q/\partial x$ are continuous in an open disk D centered at (x, y). Then in D, **F** is the gradient of a function f if and only if

$$\frac{\partial P}{\partial y} = \frac{\partial Q}{\partial x}. \tag{6}$$

Definition 1 EXACT VECTOR FIELD We say that the vector field $P(x, y)\mathbf{i} + Q(x, y)\mathbf{j}$ is **exact** if **F** is the gradient of a function f.

EXAMPLE 3 Show that the vector field

$$\mathbf{F}(x, y) = \left(4x^3y^3 + \frac{1}{x}\right)\mathbf{i} + \left(3x^4y^2 - \frac{1}{y}\right)\mathbf{j}$$

is exact, and find all functions f for which $\nabla f = \mathbf{F}$.

Solution. $P(x, y) = 4x^3y^3 + (1/x)$ and $Q(x, y) = 3x^4y^2 - (1/y)$, so

$$\frac{\partial P}{\partial y} = 12x^3y^2 = \frac{\partial Q}{\partial x}.$$

If $\nabla f = \mathbf{F}$, then $\partial f/\partial x = P$, so

$$f(x, y) = \int\left(4x^3y^3 + \frac{1}{x}\right) dx = x^4y^3 + \ln|x| + g(y).$$

Differentiating with respect to y, we have

$$Q = \frac{\partial f}{\partial y} = 3x^4y^2 + g'(y) = 3x^4y^2 - \frac{1}{y}.$$

Thus $g'(y) = -1/y$, $g(y) = -\ln|y| + C$, and finally,

$$f(x, y) = x^4y^3 + \ln|x| - \ln|y| + C = x^4y^3 + \ln\left|\frac{x}{y}\right| + C.$$

This answer should be checked. Note that the presence of the constant C indicates that the potential function f is not unique. ■

We close this section by pointing out that it is now easy to check whether a force field **F** is conservative. It is necessary only to check whether or not **F** is exact.

[†]For a proof see R. C. Buck, *Advanced Calculus* (McGraw-Hill, New York, 1965), page 425.

PROBLEMS 4.11

In Problems 1–14, test for exactness. If the given vector field is exact, find all functions f for which $\nabla f = \mathbf{F}$.

1. $\mathbf{F}(x, y) = 2xy\mathbf{i} + (x^2 + 1)\mathbf{j}$
2. $\mathbf{F}(x, y) = (4x^3 - ye^{xy})\mathbf{i} + (\tan y - xe^{xy})\mathbf{j}$
3. $\mathbf{F}(x, y) = (4x^2 - 4y^2)\mathbf{i} + (8xy - \ln y)\mathbf{j}$
4. $\mathbf{F}(x, y) = [x\cos(x + y) + \sin(x + y)]\mathbf{i} + x\cos(x + y)]\mathbf{j}$
5. $\mathbf{F}(x, y) = 2x(\cos y)\mathbf{i} + x^2(\sin y)\mathbf{j}$
6. $\mathbf{F}(x, y) = \left[\dfrac{\ln(\ln y)}{x} + \dfrac{2}{3}xy^3\right]\mathbf{i} + \left(\dfrac{\ln x}{y\ln y} + x^2y^2\right)\mathbf{j}$
7. $\mathbf{F}(x, y) = (x - y\cos x)\mathbf{i} - (\sin x)\mathbf{j}$
8. $\mathbf{F}(x, y) = e^{x^2y}\mathbf{i} + e^{x^2y}\mathbf{j}$
9. $\mathbf{F}(x, y) = (3x\ln x + x^5 - y)\mathbf{i} - x\mathbf{j}$
10. $\mathbf{F}(x, y) = \left[\left(\dfrac{1}{x^2}\right) + y^2\right]\mathbf{i} + (2xy)\mathbf{j}$
11. $\mathbf{F}(x, y) = \left(\tan^2 x - \dfrac{y}{x^2 + y^2}\right)\mathbf{i} + \left(\dfrac{x}{x^2 + y^2} - e^y\right)\mathbf{j}$
12. $\mathbf{F}(x, y) = [\cos(x + 3y) - x^4]\mathbf{i} + \left[3\cos(x + 3y) - \left(\dfrac{2}{y}\right)\right]\mathbf{j}$
13. $\mathbf{F}(x, y) = (x^2 + y^2 + 1)\mathbf{i} - (xy + y)\mathbf{j}$
14. $\mathbf{F}(x, y) = \left[-\left(\dfrac{1}{x^3}\right) + 4x^3y\right]\mathbf{i} + (\sin y + \sqrt{y} + x^4)\mathbf{j}$

Consider the vector field

$$\mathbf{F}(x, y, z) = P(x, y, z)\mathbf{i} + Q(x, y, z)\mathbf{j} + R(x, y, z)\mathbf{k}.$$

F is exact (i.e., conservative) if there is a differentiable function f such that

$$\nabla f(x, y, z) = \mathbf{F}(x, y, z).$$

If P, Q, R, $\partial P/\partial y$, $\partial P/\partial z$, $\partial Q/\partial x$, $\partial Q/\partial z$, $\partial R/\partial x$, and $\partial R/\partial y$ are continuous, then, analogously to Theorem 1, **F** is exact if and only if

$$\frac{\partial P}{\partial y} = \frac{\partial Q}{\partial x}, \quad \frac{\partial R}{\partial x} = \frac{\partial P}{\partial z}, \quad \frac{\partial Q}{\partial z} = \frac{\partial R}{\partial y}.$$

In Problems 15–20, use the result above to test for exactness. If **F** is exact, find all functions f for which $\nabla f = \mathbf{F}$.

15. $\mathbf{F} = \mathbf{i} + \mathbf{j} + \mathbf{k}$
16. $\mathbf{F} = yz\mathbf{i} + xz\mathbf{j} + xy\mathbf{k}$
17. $\mathbf{F} = \left[\left(\dfrac{y}{z}\right) + x^2\right]\mathbf{i} + \left[\left(\dfrac{x}{z}\right) - \sin y\right]\mathbf{j} + \left[\cos z - \left(\dfrac{xy}{z^2}\right)\right]\mathbf{k}$
18. $\mathbf{F} = \left(\dfrac{1}{x + 2y + 3z} - 3x\right)\mathbf{i} + \left(\dfrac{2}{x + 2y + 3z} + y^2\right)\mathbf{j} + \left(\dfrac{3}{x + 2y + 3z} - \tan^{-1} z\right)\mathbf{k}$
19. $\mathbf{F} = (e^{yz} + y)\mathbf{i} + (xze^{yz} - x)\mathbf{j} + (xye^{yz} + 2z)\mathbf{k}$
20. $\mathbf{F} = (2xy^3 + x + z)\mathbf{i} + (3x^2y^2 - y)\mathbf{j} + (x + \sin z)\mathbf{k}$

*21. A particle is moving in space with position vector $\mathbf{x}(t)$ and is subjected to a force **F** of magnitude $\alpha/|\mathbf{x}(t)|^3$ (α a constant) directed toward the origin. Show that the force is conservative and find all potential functions for **F**.

*22. Answer the questions in Problem 21 if the magnitude of the force is $\alpha/|\mathbf{x}(t)|^k$ for some positive integer k.

*23. Suppose that a moving particle is subjected to a force of constant magnitude that always points toward the origin. Show that the force is conservative. [*Hint:* If the path is given by $\mathbf{x}(t) = x(t)\mathbf{i} + y(t)\mathbf{j}$, find a unit vector that points toward the origin.]

24. We omitted the proof of Theorem 1 but half of it is easy to prove. Show that if in Theorem 1 $\mathbf{F} = \nabla f$, then

$$\frac{\partial P}{\partial y} = \frac{\partial Q}{\partial x}.$$

[*Hint:* Use Theorem 4.4.1.]

4.12 MAXIMA AND MINIMA FOR A FUNCTION OF TWO VARIABLES

In obtaining maximum and minimum values for a function of one variable, the first basic fact we use is that if f is continuous on a closed bounded interval $[a, b]$, then f takes on a maximum and minimum value in $[a, b]$. We then define a critical point to be a number x at which $f'(x) = 0$ or for which $f(x)$ exists but $f'(x)$ does not. Finally, we define local maxima and minima (which occur at critical points) and give conditions on first and second derivatives that ensure a critical point is a local maximum or minimum.

The theory for functions of two variables is more complicated, but some of the basic ideas are the same. The first thing we need to know is that a function *has* a maximum or minimum. The proof of the following theorem can be found in any advanced calculus book.[†]

Theorem 1 Let f be a function of two variables that is continuous on the closed disk D. Then there exist points $\mathbf{x}_1$ and $\mathbf{x}_2$ in D such that

$$m = f(\mathbf{x}_1) \leq f(\mathbf{x}) \leq f(\mathbf{x}_2) = M \quad (1)$$

for any other point $\mathbf{x}$ in D. The numbers m and M are called, respectively, the **global minimum** and the **global maximum** for the function f over the disk D.

We now define what we mean by local minima and maxima.

Definition 1 DEFINITION OF LOCAL MAXIMA AND MINIMA Let f be defined in a neighborhood of a point $\mathbf{x}_0 = (x_0, y_0)$.

(i) f has a **local maximum** at $\mathbf{x}_0$ if there is a neighborhood N_1 of $\mathbf{x}_0$ such that for every point $\mathbf{x}$ in N_1,

$$f(\mathbf{x}) \leq f(\mathbf{x}_0). \quad (2)$$

(ii) f has a **local minimum** at $\mathbf{x}_0$ if there is a neighborhood N_2 of $\mathbf{x}_0$ such that for every point $\mathbf{x}$ in N_2,

$$f(\mathbf{x}) \geq f(\mathbf{x}_0). \quad (3)$$

(iii) If f has a local maximum or minimum at $\mathbf{x}_0$, then $\mathbf{x}_0$ is called an **extreme point** of f.

A rough sketch of a function with several maxima is given in Figure 1. The following theorem is a natural generalization of the one-variable result.

Theorem 2 Let f be differentiable at $\mathbf{x}_0$ and suppose that $\mathbf{x}_0$ is an extreme point of f. Then

$$\nabla f(\mathbf{x}_0) = \mathbf{0}. \quad (4)$$

Proof. We must show that $f_x(x_0, y_0)$ and $f_y(x_0, y_0) = 0$. To prove that $f_x(x_0, y_0) = 0$, define a function h by

[†]See, for example, Buck, *Advanced Calculus*, page 74.

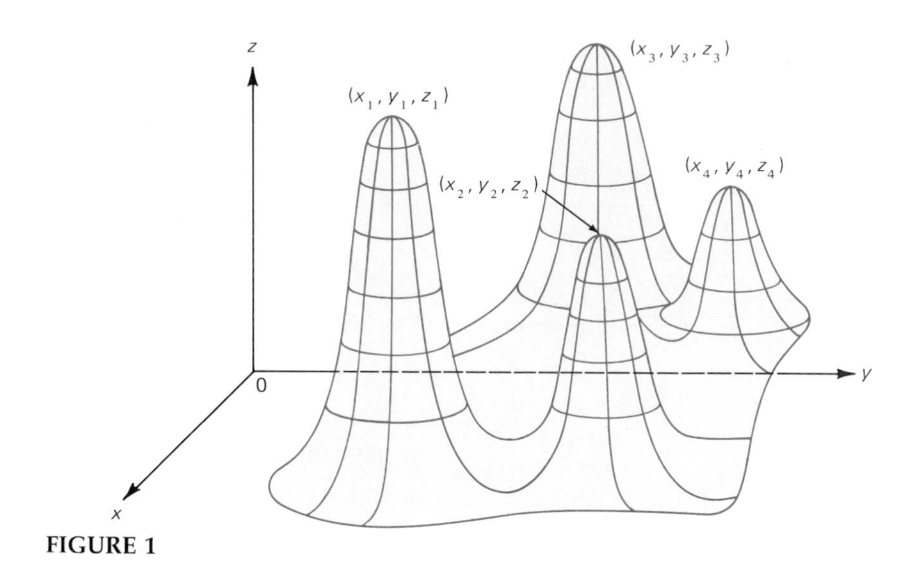

FIGURE 1

$$h(x) = f(x, y_0).$$

Then $h(x_0) = f(x_0, y_0)$ and $h'(x_0) = f_x(x_0, y_0)$. But since f has a local maximum (or minimum) at (x_0, y_0), h has a local maximum (or minimum) at x_0, so that $h'(x_0) = 0$. Thus $f_x(x_0, y_0) = 0$. In a very similar way we can show that $f_y(x_0, y_0) = 0$ by defining $h(y) = f(x_0, y)$, and the theorem is proved. ■

Definition 2 CRITICAL POINT $\mathbf{x}_0$ is a **critical point** of f if f is differentiable at $\mathbf{x}_0$ and $\nabla f(\mathbf{x}_0) = \mathbf{0}$.

REMARK 1. Definitions 1 and 2 and Theorems 1 and 2 hold for functions of three variables with very little change.

REMARK 2. By the remark on page 221, we see that *if $\mathbf{x}_0$ is a critical point of f, then the tangent plane to the surface $z = f(x, y)$ is horizontal* (i.e., parallel to the xy-plane).

REMARK 3. Note that this definition differs from our definition of a critical point of a function of one variable in that, for a function of one variable, x_0 is a critical point if $f(x_0)$ is defined but $f'(x_0)$ does not exist.

As in the case of a function of one variable, the fact that $\mathbf{x}_0$ is a critical point of f does not guarantee that $\mathbf{x}_0$ is an extreme point of f, as Example 3 below illustrates.

EXAMPLE 1 Let $f(x, y) = 1 + x^2 + 3y^2$. Then $\nabla f = 2x\mathbf{i} + 6y\mathbf{j}$, which is zero only when $(x, y) = (0, 0)$. Thus $(0, 0)$ is the only critical point, and it is clearly a local (and global) minimum. This is illustrated in Figure 2. ■

EXAMPLE 2 Let $f(x, y) = 1 - x^2 - 3y^2$. Then $\nabla f = -2x\mathbf{i} - 6y\mathbf{j}$, which is zero only at the origin. In this case $(0, 0)$ is a local (and global) maximum for f. See Figure 3. ■

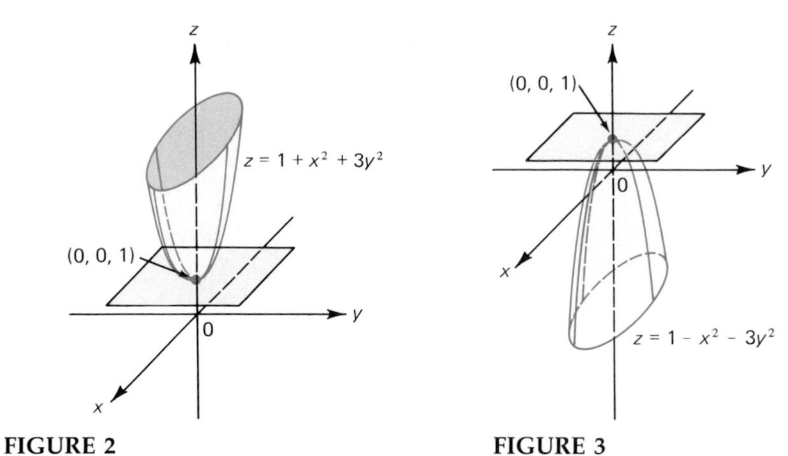

FIGURE 2

FIGURE 3

EXAMPLE 3 Let $f(x, y) = x^2 - y^2$. Then $\nabla f = 2x\mathbf{i} - 2y\mathbf{j}$, which again is zero only at (0, 0). But (0, 0) is *neither* a local maximum nor a local minimum for f. To see this, we simply note that f can take positive and negative values in any neighborhood of (0, 0) since $f(x, y) > 0$ if $|x| > |y|$ and $f(x, y) < 0$ if $|x| < |y|$. This is illustrated by Figure 4. The figure sketched in Figure 4 is called a **hyperbolic paraboloid,** or **saddle surface** (see Figure 3.6.9). ■

We have the following definition.

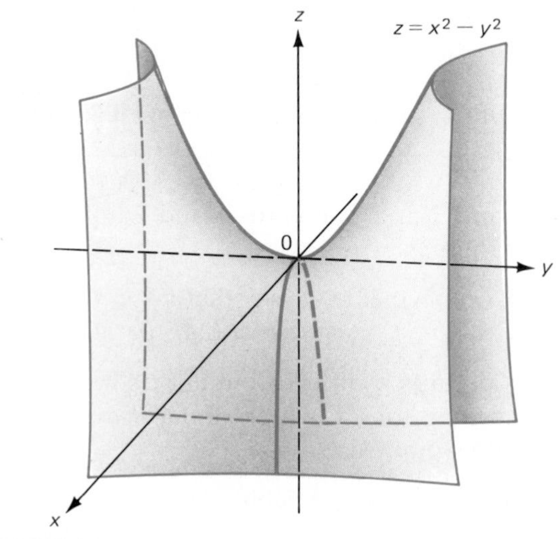

FIGURE 4

Definition 3 SADDLE POINT If $\mathbf{x}_0$ is a critical point of f but is not an extreme point of f, then $\mathbf{x}_0$ is called a **saddle point** of f.

The three examples above illustrate that a critical point of f may be a local maximum, a local minimum, or a saddle point.

EXAMPLE 4 Let $f(x, y) = \sqrt{x^2 + y^2}$. Then it is obvious that f has a local (and global) minimum at $(0, 0)$. But

$$\nabla f = \frac{x}{\sqrt{x^2 + y^2}}\mathbf{i} + \frac{y}{\sqrt{x^2 + y^2}}\mathbf{j},$$

so that $(0, 0)$ is *not* a critical point since $\nabla f(0, 0)$ *does not exist*. We see this clearly in Figure 5. The graph of f (which is a cone) does not have a tangent plane at $(0, 0)$. ■

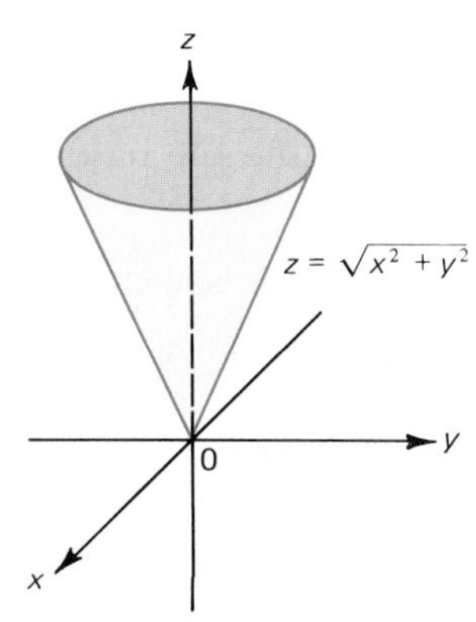

FIGURE 5

Examples 1, 2, and 3 indicate that more is needed to determine whether a critical point is an extreme point (in most cases it will not be at all obvious). You might suspect that the answer has something to do with the signs of the second partial derivatives of f, as in the case of functions of one variable. The following theorem gives the answer. Although we could prove it using results we already know, the proof is very tedious so we will postpone it until our discussion of Taylor's theorem in two variables in Chapter 10 (see Theorem 10.1.2).

Theorem 3 SECOND DERIVATIVES TEST Let f and all its first and second partial derivatives be continuous in a neighborhood of the critical point (x_0, y_0). Let

$$D(x, y) = f_{xx}(x, y)f_{yy}(x, y) - [f_{xy}(x, y)]^2, \tag{5}$$

and let D denote $D(x_0, y_0)$.

(i) If $D > 0$ and $f_{xx}(x_0, y_0) > 0$, then f has a local minimum at (x_0, y_0).
(ii) If $D > 0$ and $f_{xx}(x_0, y_0) < 0$, then f has a local maximum at (x_0, y_0).
(iii) If $D < 0$, then (x_0, y_0) is a saddle point of f.
(iv) If $D = 0$, then any of the preceding is possible.

REMARK 1. Using a 2×2 determinant, we can write (5) as

$$D = \begin{vmatrix} f_{xx} & f_{xy} \\ f_{yx} & f_{yy} \end{vmatrix}.$$

REMARK 2. This theorem does not provide the full answer to the questions of what are the extreme points of f in $\mathbb{R}^2$. It is still necessary to check points at which ∇f does not exist (as in Example 4).

EXAMPLE 5 Let $f(x, y) = 1 + x^2 + 3y^2$. Then as we saw in Example 1, $(0, 0)$ is the only critical point of f. But $f_{xx} = 2$, $f_{yy} = 6$, and $f_{xy} = 0$, so $D(0, 0) = 12$ and $f_{xx} > 0$, which *proves* that f has a local minimum at $(0, 0)$. ■

EXAMPLE 6 Let $f(x, y) = -x^2 - y^2 + 2x + 4y + 5$. Determine the nature of the critical points of f.

Solution. $\nabla f = (-2x + 2)\mathbf{i} + (-2y + 4)\mathbf{j}$. We see that $\nabla f = \mathbf{0}$ when

$$-2x + 2 = 0 \quad \text{and} \quad -2y + 4 = 0,$$

which occurs only at the point $(1, 2)$. Also $f_{xx} = -2$, $f_{yy} = -2$, and $f_{xy} = 0$, so $D = 4$ and $f_{xx} < 0$, which implies that there is a local maximum at $(1, 2)$. At $(1, 2)$, $f = 10$. ■

EXAMPLE 7 Let $f(x, y) = 2x^3 - 24xy + 16y^3$. Determine the nature of the critical points of f.

Solution. We have $\nabla f(x, y) = (6x^2 - 24y)\mathbf{i} + (-24x + 48y^2)\mathbf{j}$. If $\nabla f = \mathbf{0}$, we have

$$6(x^2 - 4y) = 0 \quad \text{and} \quad -24(x - 2y^2) = 0.$$

Clearly, one critical point is $(0, 0)$. To obtain another, we must solve the simultaneous equations

$$x^2 - 4y = 0$$
$$x - 2y^2 = 0.$$

The second equation tells us that $x = 2y^2$. Substituting this expression into the first equation yields

$$4y^4 - 4y = 0,$$

with solutions $y = 0$ and $y = 1$, giving us the critical points $(0, 0)$ and $(2, 1)$. Now $f_{xx} = 12x$, $f_{yy} = 96y$, and $f_{xy} = -24$, so

$$D(x, y) = (12x)(96y) - 24^2 = 1152xy - 576.$$

Then $D(0, 0) = -576$ and $D(2, 1) = 1728$. We find that $(0, 0)$ is a saddle point, and since $f_{xx}(2, 1) > 0$, $(2, 1)$ is a local minimum. ■

EXAMPLE 8 Let $f(x, y) = 4x^2 - 4xy + y^2 + 5$. Determine the nature of the critical points of f.

Solution. $\nabla f = (8x - 4y)\mathbf{i} + (-4x + 2y)\mathbf{j}$. We see that $\nabla f = \mathbf{0}$ when

$$8x - 4y = 0 \quad \text{and} \quad -4x + 2y = 0.$$

This occurs if (x, y) is any point on the line $y = 2x$. Thus in this case there are an *infinite number* of critical points. Now $f_{xx} = 8$, $f_{yy} = 2$, and $f_{xy} = -4$, so $D = 8 \cdot 2 - 16 = 0$. It is impossible to determine the nature of the critical points of x by the second derivatives test. However, since we can find that

$$f(x, y) = (2x - y)^2 + 5,$$

we see that $f(x, y) \geq 5$ and $f(x, y) = 5$ when $y = 2x$. Thus *every point* on the line $y = 2x$ is a local (and global) minimum. ■

EXAMPLE 9 If $f(x, y) = -(4x^2 - 4xy + y^2) + 5$, then following the reasoning of Example 8, it is easy to see that *every point* on the line $y = 2x$ is a local maximum but that $D(x, y) = 0$ at any such point. ■

EXAMPLE 10 Let $f(x, y) = x^3 - y^3$. Then $\nabla f = 3x^2\mathbf{i} - 3y^2\mathbf{j}$, and the only critical point is $(0, 0)$. But $f_{xx} = 6x$, $f_{yy} = -6y$, and $f_{xy} = 0$, so $D(x, y) = -36xy$ and $D(0, 0) = 0$. In this case it is clear that since f can take on positive and negative values in any neighborhood of $(0, 0)$, $(0, 0)$ is a saddle point of f. ■

REMARK. Examples 8, 9, and 10 illustrate that any of the three cases can occur when $D = 0$.

EXAMPLE 11 A rectangular wooden box with an open top is to contain α cubic centimeters, where α is a given positive number. Ignoring the thickness of the wood, how is the box to be constructed so as to use the smallest amount of wood?

Solution. If the dimensions of the box are x, y, and z, then

$$V = xyz = \alpha > 0,$$

so we have $x > 0$, $y > 0$, and $z > 0$. We must minimize the area of the sides of the box, where the area is given by

$$A = xy + 2xz + 2yz$$

(see Figure 6). But $z = \alpha/xy$, so the problem becomes one of minimizing $A(x, y)$, where

$$A(x, y) = xy + 2x\frac{\alpha}{xy} + 2y\frac{\alpha}{xy} = xy + \frac{2\alpha}{y} + \frac{2\alpha}{x}.$$

Now

$$\nabla A(x, y) = \left(y - \frac{2\alpha}{x^2}\right)\mathbf{i} + \left(x - \frac{2\alpha}{y^2}\right)\mathbf{j}.$$

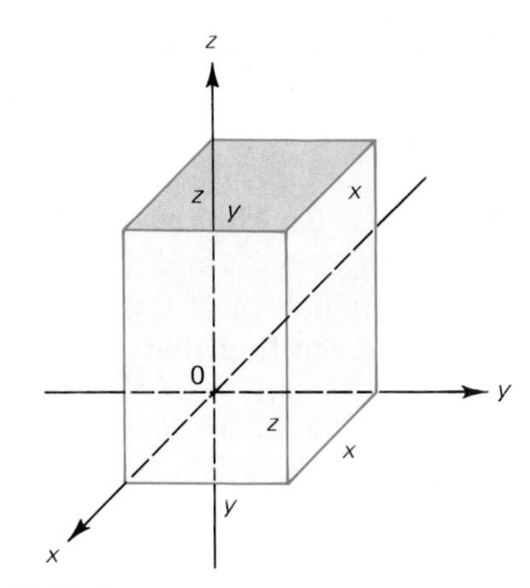

FIGURE 6

This is zero when

$$y = \frac{2\alpha}{x^2} \qquad \text{and} \qquad x = \frac{2\alpha}{y^2}$$

or

$$yx^2 = 2\alpha = xy^2.$$

One obvious point is (0, 0), which is ruled out since $x > 0$ and $y > 0$. Also, (0, 0) is not a critical point since ∇A is not defined at (0, 0). Now $xy^2 = x^2y$ implies that $xy(y - x) = 0$. Since $xy \neq 0$, we have $y = x$, so

$$yx^2 = x^3 = 2\alpha$$

and

$$x = y = \sqrt[3]{2\alpha}.$$

Thus $(\sqrt[3]{2\alpha}, \sqrt[3]{2\alpha})$ is the only critical point.

We can verify that the point $(\sqrt[3]{2\alpha}, \sqrt[3]{2\alpha})$ does indeed give us a local minimum, since $D(\sqrt[3]{2\alpha}, \sqrt[3]{2\alpha}) > 0$ and

$$A_{xx}(\sqrt[3]{2\alpha}, \sqrt[3]{2\alpha}) = \left.\frac{4\alpha}{x^3}\right|_{\sqrt[3]{2\alpha}} = \frac{4\alpha}{2\alpha} = 2,$$

which is greater than 0. Finally, we obtain

$$z = \frac{\alpha}{xy} = \frac{\alpha}{x^2} = \frac{\alpha x}{x^3} = \frac{\alpha x}{2\alpha} = \frac{x}{2}$$

so we can minimize the amount of wood needed by building a box with a square base and a height equal to half its length (or width). Note that since $(\sqrt[3]{2\alpha}, \sqrt[3]{2\alpha})$ is the only critical point, it is easy to check that there is a global minimum there. ■

There is an interesting way that we can use the theory of this section to derive a result which is very useful for statistical analysis and, in fact, any analysis involving the use of a great deal of data. Suppose n data points $(x_1, y_1), (x_2, y_2), \ldots, (x_n, y_n)$ are collected. For example, the x's may represent average tree growth and the y's average daily temperature in a given year in a certain forest. Or x may represent a week's sales and y a week's profit for a certain business. The question arises as to whether we can "fit" these data points to a straight line. That is, is there a straight line which runs "more or less" through the points. If so, then we can write y as a linear function of x, with obvious computational advantages.

The problem is to find the "best" straight line $y = mx + b$ passing through or near these points. Look at Figure 7. If (x_i, y_i) is one of our n points, then, on the line $y = mx + b$, corresponding to x_i we obtain $y_i = mx_i + b$. The "error" ϵ_i between the y value of our actual point and the "approximating" value on the line is given by

$$\epsilon_i = y - mx_i - b. \tag{6}$$

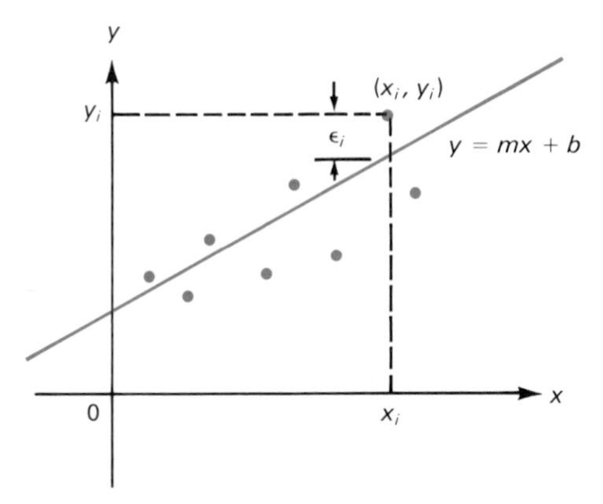

FIGURE 7

One way to choose the approximating line is to use the line which minimizes the sum of the squares of the errors. This is called the **least-squares** criterion for choosing the line. That is, we want to choose m and b such that the function

$$f(m, b) = \epsilon_1^2 + \epsilon_2^2 + \cdots + \epsilon_n^2 = \sum_{i=1}^{n} \epsilon_i^2 = \sum_{i=1}^{n} (y_i - mx_i - b)^2 \tag{7}$$

is a minimum. To do this, we calculate

$$\frac{\partial}{\partial m}(y_i - mx_i - b)^2 = -2x_i(y_i - mx_i - b)$$

and

$$\frac{\partial}{\partial b}(y_i - mx_i - b)^2 = -2(y_i - mx_i - b).$$

Hence

$$\frac{\partial f}{\partial m} = -2 \sum_{i=1}^{n} x_i(y_i - mx_i - b)$$

and

$$\frac{\partial f}{\partial b} = -2 \sum_{i=1}^{n} (y_i - mx_i - b).$$

Setting $\partial f/\partial m = 0$ and $\partial f/\partial b = 0$, we obtain

$$\sum_{i=1}^{n} (x_i y_i - m x_i^2 - b x_i) = 0$$

and

$$\sum_{i=1}^{n} (y_i - mx_i - b) = 0.$$

This leads to the system of two equations in the unknowns m and b

$$\left(\sum_{i=1}^{n} x_i^2\right)m + \left(\sum_{i=1}^{n} x_i\right)b = \sum_{i=1}^{n} x_i y_i \tag{8}$$

and

$$\left(\sum_{i=1}^{n} x_i\right)m + nb = \sum_{i=1}^{n} y_i. \tag{9}$$

Here we have used the fact that

$$\sum_{i=1}^{n} b = nb.$$

The system (8) and (9) is not hard to solve for m and b. To do so, we multiply both sides of (8) by n and both sides of (9) by

$$\sum_{i=1}^{n} x_i$$

and then subtract to obtain finally

$$m = \frac{n \sum_{i=1}^{n} x_i y_i - \left[\sum_{i=1}^{n} x_i\right]\left[\sum_{i=1}^{n} y_i\right]}{n \sum_{i=1}^{n} x_i^2 - \left[\sum_{i=1}^{n} x_i\right]^2} \tag{10}$$

and

$$b = \frac{\left[\sum_{i=1}^{n} x_i^2\right]\left[\sum_{i=1}^{n} y_i\right] - \left[\sum_{i=1}^{n} x_i\right]\left[\sum_{i=1}^{n} x_i y_i\right]}{n \sum_{i=1}^{n} x_i^2 - \left[\sum_{i=1}^{n} x_i\right]^2}. \tag{11}$$

We will leave it to you to check that the numbers m and b given in (10) and (11) do indeed provide a minimum. The line $y = mx + b$ given by (10) and (11) is called the **regression line** for the n points.

REMARK. Equations (10) and (11) make sense only if

$$n \sum_{i=1}^{n} x_i^2 - \left(\sum_{i=1}^{n} x_i\right)^2 \neq 0.$$

But, in fact,

$$n \sum_{i=1}^{n} x_i^2 - \left(\sum_{i=1}^{n} x_i\right)^2 \geq 0$$

and is equal to zero only when all the x_i's are equal (in which case the regression line is the vertical line $x = x_i$). We will not prove this fact.

EXAMPLE 12 Find the regression line for the points (1, 2), (2, 4), and (5, 5).

Solution. We tabulate some appropriate values in Table 1. Then, from (10),

$$m = \frac{3(35) - (8)(11)}{3(30) - 8^2} = \frac{17}{26} \approx 0.654$$

and

$$b = \frac{(30)(11) - 8(35)}{26} = \frac{50}{26} \approx 1.923.$$

Thus the regression line is

$$y = 0.654x + 1.923.$$

This is all illustrated in Figure 8.

TABLE 1

i	x_i	y_i	x_i^2	x_iy_i
1	1	2	1	2
2	2	4	4	8
3	5	5	25	25
$\sum_{i=1}^{3}$	8	11	30	35

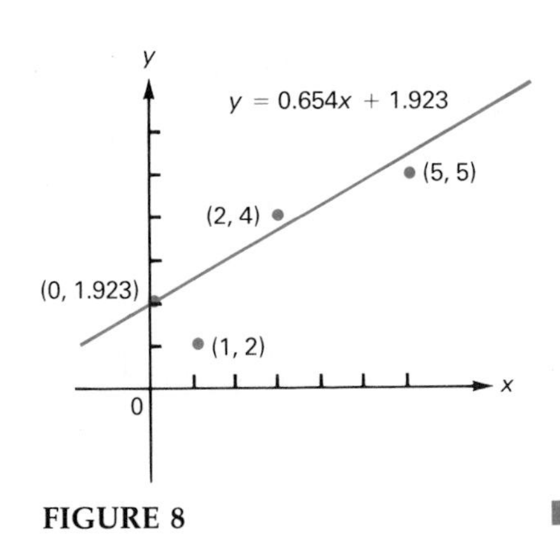

FIGURE 8

PROBLEMS 4.12

In Problems 1–18, determine the nature of the critical points of the given function.

1. $f(x, y) = 7x^2 - 8xy + 3y^2 + 1$
2. $f(x, y) = x^2 + y^3 - 3xy$
3. $f(x, y) = x^2 + 3y^2 + 4x - 6y + 3$
4. $f(x, y) = x^2 + y^2 + 4xy + 6y - 3$
5. $f(x, y) = x^2 + y^2 + 4x - 2y + 3$
6. $f(x, y) = xy^2 + x^2y - 3xy$
7. $f(x, y) = x^3 + 3xy^2 + 3y^2 - 15x + 2$
8. $f(x, y) = x^3 + y^3 - 3xy$
9. $f(x, y) = \frac{1}{y} - \frac{1}{x} - 4x + y$
10. $f(x, y) = \frac{1}{x} + \frac{2}{y} + 2x + y + 1$
11. $f(x, y) = x^2 - xy + y^2 + 2x + 2y$

*12. $f(x, y) = xy + \frac{8}{x} + \frac{1}{y}$

13. $f(x, y) = (4 - x - y)xy$

*14. $f(x, y) = \sin x + \sin y + \sin(x + y)$

15. $f(x, y) = 2x^2 + y^2 + \frac{2}{x^2y}$
16. $f(x, y) = 4x^2 + 12xy + 9y^2 + 25$
17. $f(x, y) = x^{25} - y^{25}$
18. $f(x, y) = \tan xy$
19. Find three positive numbers whose sum is 50 and such that their product is a maximum.
20. Find three positive numbers x, y, and z whose sum is 50 such that the product xy^2z^3 is a maximum.
21. Find three numbers whose sum is 50 and the sum of whose squares is a minimum.
22. Use the methods of this section to find the minimum distance from the point $(1, -1, 2)$ to the plane $x + 2y - z = 4$. [*Hint:* Express in terms of x and y the distance between $(1, -1, 2)$ and a point (x, y, z) on the plane.]
23. Find the minimum distance between the point $(2, 0, 1)$ and the plane $-x + y + z = 8$.
24. What is the maximum volume of an open-top rectangular box that can be built from α square meters of wood? Assume that any size board can be obtained without waste.

*25. Find the dimensions of the rectangular box of maximum volume that can be inscribed in the ellipsoid

$$\frac{x^2}{a^2} + \frac{y^2}{b^2} + \frac{z^2}{c^2} = 1$$

whose faces are parallel to the coordinate planes.

26. Show that the rectangular box inscribed in a sphere that encloses the greatest volume has the dimensions of a cube.
27. A company uses two types of raw material, I and II, for its product. If it uses x units of I and y units of II, it can produce U units of the finished item, where

$$U(x, y) = 8xy + 32x + 40y - 4x^2 - 6y^2.$$

Each unit of I costs \$10 and each unit of II costs \$4. Each unit of the product can be sold for \$40. How can the company maximize its profits?

28. Find the regression line for the points (1, 1), (2, 3), and (3, 6). Sketch the line and the points.

29. Find the regression line for the points $(-1, 3)$, $(1, 2)$, $(2, 0)$, and $(4, -2)$. Sketch the line and the points.

30. Find the regression line for the points $(1, 4)$, $(3, -2)$, $(5, 8)$, and $(7, 3)$. Sketch the line and the points.

31. Show that if n points lie on the same straight line L, then the regression line for the n points is L.

***32.** Show that the function $e^{(x^2+y^2)/y} + (11/x) + (5/y) + \sin x^2y$ has a global minimum in the first quadrant.

4.13 CONSTRAINED MAXIMA AND MINIMA —LAGRANGE MULTIPLIERS

In the previous section we saw how to find the maximum and minimum of a function of two variables by taking gradients and applying a second derivative test. It often happens that there are side conditions (or **constraints**) attached to a problem. For example, we have been asked to find the shortest distance from a point (x_0, y_0) to a line $y = mx + b$. We could write this problem as follows:

Minimize the function: $z = \sqrt{(x - x_0)^2 + (y - y_0)^2}$

subject to the constraint: $y - mx - b = 0$.

As another example, suppose that a region of space containing a sphere is heated and a function $w = T(x, y, z)$ gives the temperature of every point of the region. Then if the sphere is given by $x^2 + y^2 + z^2 = r^2$, and if we wish to find the hottest point on the sphere, we have the following problem:

Maximize: $w = T(x, y, z)$

subject to the constraint: $x^2 + y^2 + z^2 - r^2 = 0$.

We now generalize these two examples. Let f and g be functions of two variables. Then we can formulate a **constrained maximization** (or **minimization**) problem as follows:

Maximize (or minimize): $z = f(x, y)$ **(1)**

subject to the constraint: $g(x, y) = 0$. **(2)**

If f and g are functions of three variables, we have the following problem:

Maximize (or minimize): $w = f(x, y, z)$ **(3)**

subject to the constraint: $g(x, y, z) = 0$. **(4)**

We now develop a method for dealing with problems of the type (1), (2), or (3), (4). Let C be a curve in $\mathbb{R}^2$ or $\mathbb{R}^3$ given parametrically by the differentiable function $\mathbf{F}(t)$.

That is, C is given by

$$\mathbf{F}(t) = x(t)\mathbf{i} + y(t)\mathbf{j} \qquad (\text{in } \mathbb{R}^2)$$

or

$$\mathbf{F}(t) = x(t)\mathbf{i} + y(t)\mathbf{j} + z(t)\mathbf{k} \qquad (\text{in } \mathbb{R}^3).$$

Let $f(\mathbf{x})$ denote the function (of two or three variables) that is to be maximized.

Theorem 1 Suppose that f is differentiable at a point $\mathbf{x}_0$ and that among all points on a curve C, f takes its maximum (or minimum) value at $\mathbf{x}_0$. Then $\nabla f(\mathbf{x}_0)$ is orthogonal to C at $\mathbf{x}_0$. That is, since $\mathbf{F}'(t)$ is tangent to C, if $\mathbf{x}_0 = \mathbf{F}(t_0)$, then

$$\nabla f(\mathbf{x}_0) \cdot \mathbf{F}'(t_0) = 0. \tag{5}$$

Proof. For $\mathbf{x}$ on C, $\mathbf{x} = \mathbf{F}(t)$, so that the composite function $f(\mathbf{F}(t))$ has a maximum (or minimum) at t_0. Therefore its derivative at t_0 is 0. By the chain rule,

$$\frac{d}{dt} f(\mathbf{F}(t)) = \nabla f(\mathbf{F}(t)) \cdot \mathbf{F}'(t),$$

and at t_0

$$0 = \nabla f(\mathbf{F}(t_0)) \cdot \mathbf{F}'(t_0) = \nabla f(\mathbf{x}_0) \cdot \mathbf{F}'(t_0),$$

and the theorem is proved. ■

For f as a function of two variables, the result of Theorem 1 is illustrated in Figure 1.

We can use Theorem 1 to make an interesting geometric observation.[†] Suppose that, subject to the constraint $g(x, y) = 0$, f takes its maximum (or minimum) at the point $\mathbf{x}_0 = (x_0, y_0)$. The equation $g(x, y) = 0$ determines a curve C in the xy-plane, and

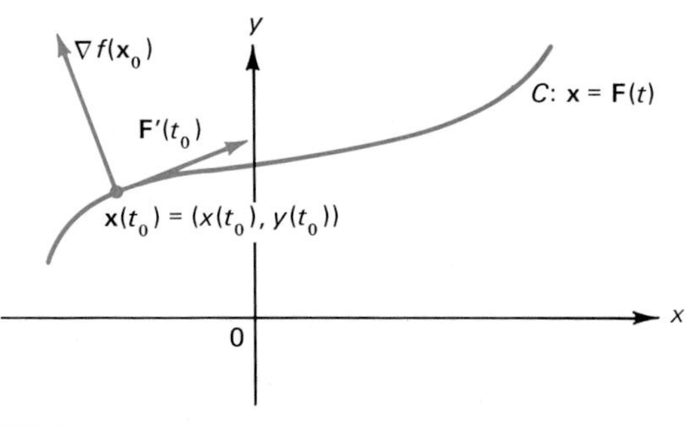

FIGURE 1

[†]This observation was first made by the French mathematician Joseph Louis Lagrange (1736–1813). See the accompanying biographical sketch.

1736–
1813

Joseph Louis Lagrange

Joseph Louis Lagrange
The Granger Collection

Joseph Louis Lagrange was one of the two greatest mathematicians of the eighteenth century—the other being Leonhard Euler. Born in 1736 in Turin, Italy, Lagrange was the youngest of eleven children of French and Italian parents and the only one to survive to adulthood. Educated in Turin, Joseph Louis became a professor of mathematics in the military academy there when he was still quite young.

Lagrange's early publications established his reputation. When Euler left his post at the court of Frederick the Great in Berlin in 1766, he recommended that Lagrange be appointed his successor. Accepting Euler's advice, Frederick wrote to Lagrange that "the greatest king in Europe" wished to invite to his court "the greatest mathematician in Europe." Lagrange accepted and remained in Berlin for twenty years. Afterwards, he accepted a post at the École Polytechnique in France.

Lagrange had a deep influence on nineteenth and twentieth century mathematics. He is perhaps best known as the first great mathematician to attempt to make calculus mathematically rigorous. His major work in this area was his 1797 paper "Théorie des fonctions analytiques contenant les principes du calcul différentiel." In this work, Lagrange tried to make calculus more logical—rather than more useful. His key idea was to represent a function $f(x)$ by a Taylor series. For example, we can write $1/(1 - x) = 1 + x + x^2 + \cdots + x^n + \cdots$ (a result that can be obtained by long division). Lagrange multiplied the coefficient of x^n by $n!$ and called the result the nth *derived function* of $1/(1 - x)$ at $x = 0$. This is the origin of the word *derivative.* The notation $f'(x)$, $f''(x)$, . . . was first used by Lagrange.

Lagrange is known for much else as well. Beginning in the 1750s, he invented the calculus of variations. He made significant contributions to ordinary differential equations, partial differential equations, numerical analysis, number theory, and algebra. In 1788 he published his *Mécanique Analytique*, which contained the equations of motion of a dynamical system. Today these equations are known as *Lagrange's equations.*

Lagrange lived in France during the French revolution. In 1790 he was placed on a committee to reform weights and measures and later became the head of a related committee that, in 1799, recommended the adoption of the system that we know today as the *metric system.* Despite his work for the revolution, however, Lagrange was disgusted by its cruelties. After the great French chemist Lavoisier was guillotined, Lagrange exclaimed, "It took the mob only a moment to remove his head; a century will not suffice to reproduce it."

In his later years, Lagrange was often lonely and depressed. When he was 56, the 17-year-old daughter of his friend the astronomer P. C. Lemonier was so moved by his unhappiness that she proposed to him. The resulting marriage apparently turned out to be ideal for both.

Perhaps the greatest tribute to Lagrange was given by Napoleon Bonaparte, who said, "Lagrange is the lofty pyramid of the mathematical sciences."

by Theorem 1, $\nabla f(x_0, y_0)$ is orthogonal to C at (x_0, y_0). But from Section 4.7, $\nabla g(x_0, y_0)$ is also orthogonal to C at (x_0, y_0). Thus we see that

$$\nabla g(x_0, y_0) \text{ and } \nabla f(x_0, y_0) \text{ are parallel.}$$

Hence there is a number λ such that

$$\nabla f(x_0, y_0) = \lambda \nabla g(x_0, y_0).$$

We can extend this observation to the following rule, which applies equally well to functions of three or more variables:

If, subject to the constraint $g(\mathbf{x}) = 0$, f takes its maximum (or minimum) value at a point $\mathbf{x}_0$, then there is a number λ such that

$$\nabla f(\mathbf{x}_0) = \lambda \nabla g(\mathbf{x}_0). \tag{6}$$

The number λ is called a **Lagrange multiplier.** We will illustrate the Lagrange multiplier technique with a number of examples.

EXAMPLE 1 Find the maximum and minimum values of $f(x, y) = xy^2$ subject to the condition $x^2 + y^2 = 1$.

Solution. We have $g(x, y) = x^2 + y^2 - 1 = 0$. Then

$$\nabla f = y^2\mathbf{i} + 2xy\mathbf{j} \quad \text{and} \quad \nabla g = 2x\mathbf{i} + 2y\mathbf{j}.$$

At a maximizing or minimizing point we have

$$\nabla f = \lambda \nabla g,$$

or

$$y^2\mathbf{i} + 2xy\mathbf{j} = 2x\lambda\mathbf{i} + 2y\lambda\mathbf{j},$$

which leads to the equations

$$y^2 = 2x\lambda$$
$$2xy = 2y\lambda.$$

Multiplying the first equation by y and the second by x, we obtain

$$y^3 = 2xy\lambda$$
$$2x^2y = 2xy\lambda,$$

or

$$y^3 = 2x^2y.$$

But $x^2 = 1 - y^2$, so

$$y^3 = 2(1 - y^2)y = 2y - 2y^3,$$

or

$$3y^3 = 2y.$$

The solutions to this last equation are $y = 0$ and $y = \pm\sqrt{\frac{2}{3}}$. This leads to the six points

$$(1, 0), (-1, 0), \left(\frac{1}{\sqrt{3}}, \sqrt{\frac{2}{3}}\right), \left(-\frac{1}{\sqrt{3}}, \sqrt{\frac{2}{3}}\right), \left(\frac{1}{\sqrt{3}}, -\sqrt{\frac{2}{3}}\right), \left(-\frac{1}{\sqrt{3}}, -\sqrt{\frac{2}{3}}\right).$$

Evaluating $f(x, y) = xy^2$ at these points, we have

$$f(1, 0) = f(-1, 0) = 0, \qquad f\left(\frac{1}{\sqrt{3}}, \sqrt{\frac{2}{3}}\right) = f\left(\frac{1}{\sqrt{3}}, -\sqrt{\frac{2}{3}}\right) = \frac{2}{3\sqrt{3}},$$

and

$$f\left(-\frac{1}{\sqrt{3}}, \sqrt{\frac{2}{3}}\right) = f\left(-\frac{1}{\sqrt{3}}, -\sqrt{\frac{2}{3}}\right) = -\frac{2}{3\sqrt{3}}.$$

Thus the maximum value of f is $2/(3\sqrt{3})$ and the minimum value of f is $-2/(3\sqrt{3})$. Note that there is neither a maximum nor a minimum at $(1, 0)$ and $(-1, 0)$ even though $\nabla f(1, 0) = 0\,\nabla g(1, 0)$ and $\nabla f(-1, 0) = 0\,\nabla g(-1, 0) = \mathbf{0}$. ■

EXAMPLE 2 Find the points on the sphere $x^2 + y^2 + z^2 = 1$ closest to and farthest from the point $(1, 2, 3)$.

Solution. We wish to minimize and maximize $f(x, y, z) = (x - 1)^2 + (y - 2)^2 + (z - 3)^2$ subject to $g(x, y, z) = x^2 + y^2 + z^2 - 1 = 0$. We have $\nabla f(x, y, z) = 2(x - 1)\mathbf{i} + 2(y - 2)\mathbf{j} + 2(z - 3)\mathbf{k}$, and $\nabla g(x, y, z) = 2x\mathbf{i} + 2y\mathbf{j} + 2z\mathbf{k}$. Condition (6) implies that at a maximizing point, $\nabla f = \lambda \nabla g$, so

$$2(x - 1) = 2x\lambda$$
$$2(y - 2) = 2y\lambda$$
$$2(z - 3) = 2z\lambda.$$

If $\lambda \neq 1$, then we find that

$$x - 1 = x\lambda, \qquad \text{or} \qquad x - x\lambda = 1, \qquad \text{or} \qquad x(1 - \lambda) = 1,$$

and

$$x = \frac{1}{1 - \lambda}.$$

Similarly, we obtain

$$y = \frac{2}{1-\lambda} \quad \text{and} \quad z = \frac{3}{1-\lambda}.$$

Then

$$1 = x^2 + y^2 + z^2 = \frac{1}{(1-\lambda)^2}(1^2 + 2^2 + 3^2) = \frac{14}{(1-\lambda)^2}$$

so that

$$(1-\lambda)^2 = 14, \qquad (1-\lambda) = \pm\sqrt{14},$$

and

$$\lambda = 1 \pm \sqrt{14}.$$

If $\lambda = 1 + \sqrt{14}$, then $(x, y, z) = (-1/\sqrt{14}, -2/\sqrt{14}, -3/\sqrt{14})$. If $\lambda = 1 - \sqrt{14}$, then $(x, y, z) = (1/\sqrt{14}, 2/\sqrt{14}, 3/\sqrt{14})$. Finally, evaluation shows us that f is maximized at $(-1/\sqrt{14}, -2/\sqrt{14}, -3/\sqrt{14})$ and that f is minimized at $(1/\sqrt{14}, 2/\sqrt{14}, 3/\sqrt{14})$. ■

We note that we can use Lagrange multipliers in $\mathbb{R}^3$ if there are two or more constraint equations. Suppose, for example, we wish to maximize (or minimize) $w = f(x, y, z)$ subject to the constraints

$$g(x, y, z) = 0 \tag{7}$$

and

$$h(x, y, z) = 0. \tag{8}$$

Each of the equations (7) and (8) represents a surface in $\mathbb{R}^3$, and their intersection forms a curve in $\mathbb{R}^3$. By an argument very similar to the one we used earlier (but applied in $\mathbb{R}^3$ instead of $\mathbb{R}^2$), we find that if f is maximized (or minimized) at (x_0, y_0, z_0), then $\nabla f(x_0, y_0, z_0)$ is in the plane determined by $\nabla g(x_0, y_0, z_0)$ and $\nabla h(x_0, y_0, z_0)$. Thus there are numbers λ and μ such that

$$\nabla f(x_0, y_0, z_0) = \lambda \nabla g(x_0, y_0, z_0) + \mu \nabla h(x_0, y_0, z_0) \tag{9}$$

(see Problem 3.5.35). Formula (9) is the generalization of formula (6) in the case of two constraints.

EXAMPLE 3 Find the maximum value of $w = xyz$ among all points (x, y, z) lying on the line of intersection of planes $x + y + z = 30$ and $x + y - z = 0$.

Solution. Setting $f(x, y, z) = xyz$, $g(x, y, z) = x + y + z - 30$, and $h(x, y, z) = x + y - z$, we obtain

$$\nabla f = yz\mathbf{i} + xz\mathbf{j} + xy\mathbf{k}$$
$$\nabla g = \mathbf{i} + \mathbf{j} + \mathbf{k}$$
$$\nabla h = \mathbf{i} + \mathbf{j} - \mathbf{k},$$

and using equation (9) to obtain the maximum, we obtain the equations

$$yz = \lambda + \mu$$
$$xz = \lambda + \mu$$
$$xy = \lambda - \mu.$$

Multiplying the three equations by x, y, and z, respectively, we find that

$$xyz = (\lambda + \mu)x$$
$$xyz = (\lambda + \mu)y$$
$$xyz = (\lambda - \mu)z.$$

If $\lambda + \mu = 0$, then $yz = 0$ and $xyz = 0$, which is not a maximum value since xyz can be positive (for example, $x = 8$, $y = 7$, $z = 15$ is in the constraint set and $xyz = 840$). Thus we can divide the first two equations by $\lambda + \mu$ to find that

$$x = y.$$

Since $x + y - z = 0$, we have $2x - z = 0$, or $z = 2x$. But then

$$30 = x + y + z = x + x + 2x = 4x,$$

or

$$x = \tfrac{15}{2}.$$

Then

$$y = \tfrac{15}{2}, \qquad z = 15,$$

and the maximum value of xyz occurs at $(\frac{15}{2}, \frac{15}{2}, 15)$ and is equal to $(\frac{15}{2})(\frac{15}{2})15 = 843\frac{3}{4}$. ■

We conclude this section with two observations. First, while the outlined steps make the method of Lagrange multipliers seem easy, it should be noted that solving three nonlinear equations in three unknowns or four such equations in four unknowns often entails very involved algebraic manipulations. Second, no method is given for determining whether a solution found actually yields a maximum, a minimum, or neither. Fortunately, in many practical applications the existence of a maximum or a minimum can readily be inferred from the nature of the particular problem.

PROBLEMS 4.13

1. Use Lagrange multipliers to find the minimum distance from the point (1, 2) to the line $2x + 3y = 5$.
2. Use Lagrange multipliers to find the minimum distance from the point (3, −2) to the line $y = 2 - x$.
3. Use Lagrange multipliers to find the minimum distance from the point (1, −1, 2) to the plane $x + y - z = 3$.
4. Use Lagrange multipliers to find the minimum distance from the point (3, 0, 1) to the plane $2x - y + 4z = 5$.
5. Use Lagrange multipliers to find the minimum distance from the plane $ax + by + cz = d$ to the origin.

*6. Find the maximum and minimum values of $x^2 + y^2$ subject to the condition $x^3 + y^3 = 6xy$.

7. Find the maximum and minimum values of $2x^2 + xy + y^2 - 2y$ subject to the condition $y = 2x - 1$.
8. Find the maximum and minimum values of $x^2 + y^2 + z^2$ subject to the condition $z^2 = x^2 - 1$.
9. Find the maximum and minimum values of $x^3 + y^3 + z^3$ if (x, y, z) lies on the sphere $x^2 + y^2 + z^2 = 4$.
10. Find the maximum and minimum values of $x + y + z$ if (x, y, z) lies on the sphere $x^2 + y^2 + z^2 = 1$.
11. Find the maximum and minimum values of xyz if (x, y, z) is on the ellipsoid $x^2 + (y^2/4) + (z^2/9) = 1$.
12. Solve Problem 11 if (x, y, z) is on the ellipsoid $(x^2/a^2) + (y^2/b^2) + (z^2/c^2) = 1$.

*13. Minimize the function $x^2 + y^2 + z^2$ for (x, y, z) on the planes $3x - y + z = 6$ and $x + 2y + 2z = 2$.

14. Find the minimum values of $x^3 + y^3 + z^3$ for (x, y, z) on the planes $x + y + z = 2$ and $x + y - z = 3$.
15. Find the maximum and minimum distances from the origin to a point on the ellipse $(x^2/a^2) + (y^2/b^2) = 1$.
16. Find the maximum and minimum distances from the origin to a point on the ellipsoid $(x^2/a^2) + (y^2/b^2) + (z^2/c^2) = 1$.

*17. Find the maximum value of $x_1 + x_2 + x_3 + x_4$ subject to $x_1^2 + x_2^2 + x_3^2 + x_4^2 = 1$. [*Hint:* Use the obvious generalization of Lagrange multipliers to functions of four variables.]

*18. Find the maximum value of $x_1 + x_2 + \cdots + x_n$ subject to $x_1^2 + x_2^2 + \cdots + x_n^2 = 1$.

19. Using Lagrange multipliers, show that among all rectangles with the same perimeter, the square encloses the greatest area.

*20. Show that among all triangles having the same perimeter, the equilateral triangle has the greatest area. [*Hint:* The area is $\sqrt{s(s-a)(s-b)(s-c)}$, where a, b, and c are the sides and $s = (a + b + c)/2$.]

21. Find the maximum and minimum values of xyz subject to $x^2 + z^2 = 1$ and $x = y$.
22. Find the volume of the largest rectangular parallelepiped that can be inscribed in the ellipsoid $x^2 + 4y^2 + 9z^2 = 9$.
23. A silo is in the shape of a cylinder topped with a cone (see Figure 2). The radius of each is 6 m, and the total surface area is 200 m^2 (excluding the base). What are the heights of the cylinder and cone that maximize the volume enclosed by the silo?

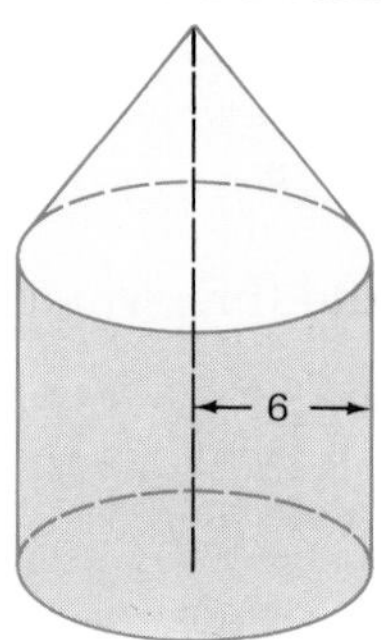

FIGURE 2

24. Prove the **general arithmetic-geometric inequality**: If $p + q + r = 1$, then $x^p y^q z^r \le px + qy + rz$ with equality if and only if $x = y = z$.
25. The base of an open-top rectangular box costs \$3 per square meter to construct, while the sides cost only \$1 per square meter. Find the dimensions of the box of greatest volume that can be constructed for \$36.
26. A manufacturing company has three plants I, II, and III, which produce x, y, and z units, respectively, of a certain product. The annual revenue from this production is given by

$$R(x, y, z) = 6xyz^2 - 400{,}000x - 400{,}000y - 400{,}000z.$$

If the company is to produce 1000 units annually, how should it allocate production so as to maximize profits?

27. A firm has \$250,000 to spend on labor and raw materials. The output of the firm is αxy, where α is a constant and x and y are, respectively, the quantities of labor and raw materials consumed. The unit price of hiring labor is \$5000, and the unit price of raw materials is \$2500. Find the ratio of x to y that maximizes output.
28. The temperature of a point (x, y, z) on the unit sphere is given by $T(x, y, z) = xy + yz$. What is the hottest point on the sphere?

REVIEW EXERCISES FOR CHAPTER FOUR

In Exercises 1–6, find the domain and range of the indicated function.

1. $f(x, y) = \sqrt{x^2 - y^2}$
2. $f(x, y) = \dfrac{1}{\sqrt{x^2 + y^2}}$
3. $f(x, y) = \cos(x + 3y)$
4. $f(x, y, z) = \sqrt{1 - x^2 - y^2 - z^2}$
5. $f(x, y, z) = \dfrac{1}{\sqrt{x^2 + y^2 + z^2 - 1}}$
6. $f(x, y) = \ln(x - y + 4z - 3)$

In Exercises 7–10, describe the level curves of the given function and sketch these curves for the given values of z.

7. $z = \sqrt{1 - x - y};\ z = 0, 1, 3, 8$
8. $z = \sqrt{1 - y^2 + x};\ z = 0, 2, 4, 7$
9. $z = \ln(x - 3y);\ z = 0, 1, 2, 3$
10. $z = \dfrac{x^2 + y^2}{x^2 - y^2};\ z = 1, 3, 6$
11. Sketch the open disk centered at $(-1, 2)$ with radius 4.
12. Sketch the closed ball centered at $(1, 2, 3)$ with radius 2.
13. Show that $\lim_{(x,y)\to(0,0)} \dfrac{xy}{y^2 - x^2}$ does not exist.
14. Show that $\lim_{(x,y)\to(0,0)} \dfrac{y^2 - 2x}{y^2 + 2x}$ does not exist.
15. Show that $\lim_{(x,y)\to(0,0)} \dfrac{4xy^3}{x^2 + y^4} = 0$.
16. Show that $\lim_{(x,y,z)\to(0,0,0)} \dfrac{zy^2 + x^3}{x^2 + y^2 + z^2} = 0$.
17. Calculate $\lim_{(x,y)\to(1,-2)} \dfrac{1 + x^2y}{2 - y}$.
18. Calculate $\lim_{(x,y)\to(1,4)} \ln(1 + x + \sqrt{y})$.
19. Calculate $\lim_{(x,y,z)\to(2,-1,1)} \dfrac{x^2 - yz^3}{1 + xyz - 2y^5}$.
20. Find the maximum region over which the function $f(x, y) = \ln(1 - 2x + 3y)$ is continuous.
21. Find the maximum region over which the function $f(x, y, z) = \ln(x - y - z + 4)$ is continuous.
22. Find the maximum region over which the function
$$f(x, y, z) = \frac{1}{\sqrt{1 - x^2 + y^2 - z^2}}$$
is continuous.
23. Find a number c such that the function
$$f(x, y) = \begin{cases} \dfrac{-2xy}{\sqrt{x^2 + y^2}}, & (x, y) \neq (0, 0) \\ c, & (x, y) = (0, 0) \end{cases}$$
is continuous at the origin.

In Exercises 24–35, calculate all first partial derivatives.

24. $f(x, y) = \dfrac{y}{x}$
25. $f(x, y) = \cos(x - 3y)$
26. $f(x, y) = \dfrac{1}{\sqrt{x^2 - y^2}}$
27. $f(x, y) = \tan^{-1}\dfrac{y}{1 + x}$
28. $f(x, y, z) = \ln(x - y + 4z)$
29. $f(x, y, z) = \dfrac{1}{\sqrt{x^2 + y^2 + z^2}}$
30. $f(x, y) = \cosh\dfrac{y}{x^2}$
31. $f(x, y, z) = \sec\dfrac{x - y}{z}$
32. $f(x, y, z) = (x^2y - y^3z^5 + x\sqrt{z})^{2/3}$
33. $f(x, y, z) = \dfrac{x^2 - y^3}{y^3 + z^4}$
34. $f(x, y, z, w) = \dfrac{x - z + w}{y + 2w - x}$
35. $f(x, y, z, w) = e^{(x-w)/(y+z)}$

In Exercises 36–41, calculate all second partial derivatives and show that all pairs of mixed partials are equal.

36. $f(x, y) = xy^3$
37. $f(x, y) = \tan^{-1}\dfrac{y}{x}$
38. $f(x, y) = \sqrt{x^2 - y^2}$

39. $f(x, y) = \dfrac{x + y}{x - y}$

40. $f(x, y, z) = \ln(2 - 3x + 4y - 7z)$

41. $f(x, y, z) = \dfrac{1}{\sqrt{1 - x^2 - y^2 - z^2}}$

42. Let $f(x, y, z) = x^2y^3 - zx^5$. Calculate f_{yzx}.

43. Let $f(x, y, z) = (x - y)/z$. Calculate f_{zxxyz}.

In Exercises 44–51, calculate the gradient of the given function at the given point.

44. $f(x, y) = x^2 - y^3$; $(1, 2)$

45. $f(x, y) = \tan^{-1}\dfrac{y}{x}$; $(-1, -1)$

46. $f(x, y) = \dfrac{x - y}{x + y}$; $(3, 2)$

47. $f(x, y) = \cos(x - 2y)$; $\left(\dfrac{\pi}{2}, \dfrac{\pi}{6}\right)$

48. $f(x, y, z) = xy + yz^3$; $(1, 2, -1)$

49. $f(x, y, z) = \dfrac{x - y}{3z}$; $(2, 1, 4)$

50. $f(x, y, z) = \dfrac{1}{\sqrt{x^2 + y^2 + z^2}}$; (a, b, c)

51. $f(x, y, z) = e^{-(x^2+y^3+z^4)}$; $(0, -1, 1)$

In Exercises 52–60, use the chain rule to calculate the indicated derivative.

52. $z = 2xy$; $x = \cos t$, $y = \sin t$; dz/dt

53. $z \sin^{-1}\dfrac{y}{x}$; $x = 1 + t$; $y = t^2$; $\dfrac{dz}{dt}$

54. $w = \ln(1 - x - 2y + 3z)$; $x = e^t \sin t$, $y = e^t \cos t$, $z = t^2$; dw/dt

55. $z = \dfrac{y}{x}$; $x = r - s$; $y = r + s$; $\dfrac{\partial z}{\partial s}$

56. $z = xy^3$; $x = \dfrac{r}{s}$; $y = \dfrac{s^2}{r}$; $\dfrac{\partial z}{\partial r}$

57. $z = \sin(x - y)$; $x = e^{r+s}$, $y = e^{r-s}$; $\partial z/\partial s$

58. $w = xyz$; $x = rs$, $y = \dfrac{r}{s}$, $z = s^2r^3$; $\dfrac{\partial w}{\partial r}$ and $\dfrac{\partial w}{\partial s}$

59. $w = x^3y + y^3z$; $x = rst$; $y = \dfrac{rs}{t}$, $z = \dfrac{rt}{s}$; $\dfrac{\partial w}{\partial s}$ and $\dfrac{\partial w}{\partial t}$

60. $w = \ln(x + 2y + 5z)$; $x = e^{r+s+t}$, $y = \sqrt{rst^2}$, $z = \dfrac{1}{\sqrt{r + s + t}}$; $\dfrac{\partial w}{\partial r}$ and $\dfrac{\partial w}{\partial t}$

In Exercises 61–66, find the equation of the tangent plane and symmetric equations of the normal line to the given surface at the given point.

61. $x^2 + y^2 + z^2 = 3$; $(1, 1, 1)$

62. $x^{1/2} + y^{3/2} + z^{1/2} = 3$; $(1, 0, 4)$

63. $3x - y + 5z = 15$; $(-1, 2, 4)$

64. $xy^2 - yz^3 = 0$; $(1, 1, 1)$

65. $xyz = 6$; $(-2, 1, -3)$

66. $\sqrt{\dfrac{x - y}{y + z}} = \dfrac{1}{2}$; $(2, 1, 3)$

In Exercises 67–72, calculate the directional derivative of the given function at the given point, in the direction of the given vector **v**.

67. $f(x, y) = \dfrac{y}{x}$ at $(1, 2)$; $\mathbf{v} = \mathbf{i} - \mathbf{j}$

68. $f(x, y) = 3x^2 - 4xy$ at $(3, -1)$; $\mathbf{v} = 2\mathbf{i} + 5\mathbf{j}$

69. $f(x, y) = \tan^{-1}\dfrac{y}{x}$ at $(1, -1)$; $\mathbf{v} = -3\mathbf{i} + 2\mathbf{j}$

70. $f(x, y, z) = xy^2 - zy^3$ at $(1, 2, 3)$; $\mathbf{v} = \mathbf{i} - \mathbf{j} + 2\mathbf{k}$

71. $f(x, y, z) = \dfrac{1}{\sqrt{x^2 + y^2 + z^2}}$ at $(1, -1, 2)$; $\mathbf{v} = -2\mathbf{i} + \mathbf{j} - 3\mathbf{k}$

72. $f(x, y, z) = e^{-(x+y^2-xz)}$ at $(1, 0, -1)$; $\mathbf{v} = 2\mathbf{i} + 5\mathbf{j} + \mathbf{k}$

73. Show that the force $\mathbf{f}(\mathbf{x}) = -3\mathbf{x}/|\mathbf{x}|^5$ is conservative and find a potential function for **f**.

74. Show that the force $\mathbf{f}(x, y) = y^2\mathbf{i} - x^2\mathbf{j}$ is not conservative.

In Exercises 75–80, calculate the total differential df.

75. $f(x, y) = x^3y^2$

76. $f(x, y) = \cos^{-1}\dfrac{y}{x}$

77. $f(x, y) = \sqrt{\dfrac{x + 1}{y - 1}}$

78. $f(x, y, z) = xy^5z^3$

79. $f(x, y, z) = \ln(x - y + 4z)$

80. $f(x, y) = \left(\dfrac{x - y}{y + z}\right)^{1/3}$

In Exercises 81–83, estimate the given number by using the total differential.

81. $4.03/6.97$
82. $\sqrt{(4.97)^2 + (12.02)^2}$
83. $\sqrt{3.97}(10.05 - 1.03)^{3/2}$

84. How much wood is contained in the sides of a rectangular box with sides of inside measurements 1.5 m, 1.3 m, and 2 m if the thickness of the wood making up the sides is 3 cm (= 0.03 m)?

In Exercises 85–90, test for exactness. If the vector field is exact, find all functions f for which $\nabla f = \mathbf{F}$.

85. $\mathbf{F}(x, y) = (y^2 + 1)\mathbf{i} + (2xy)\mathbf{j}$
86. $\mathbf{F}(x, y) = -(\sin y)\mathbf{i} + (y - x\cos y)\mathbf{j}$

87. $\mathbf{F}(x, y) = (2xy)\mathbf{i} + \left(\dfrac{1}{y^2} - x^2\right)\mathbf{j}$

88. $\mathbf{F}(x, y) = (\sin x + \sqrt{x} + y^4)\mathbf{i} + \left(\dfrac{-1}{y^3} + 4y^3x\right)\mathbf{j}$

89. $\mathbf{F}(x, y, z) = (y^3z^5 + 3x)\mathbf{i} + (-4\sqrt{y} + 3xy^2z^5)\mathbf{j} + (\sin z + 5xy^3z^4)\mathbf{k}$

90. $\mathbf{F}(x, y, z) = \left(x + \dfrac{1}{z}\right)\mathbf{i} + \left(y - \dfrac{1}{z}\right)\mathbf{j} + \left(z + \dfrac{y - x}{z^2}\right)\mathbf{k}$

In Exercises 91–96, determine the nature of the critical points of the given function.

91. $f(x, y) = 6x^2 + 14y^2 - 16xy + 2$
92. $f(x, y) = x^5 - y^5$

93. $f(x, y) = \dfrac{1}{y} + \dfrac{2}{x} + 2y + x + 4$

94. $f(x, y) = 49 - 16x^2 + 24xy - 9y^2$

95. $f(x, y) = x^2 + y^2 + \dfrac{2}{xy^2}$

96. $f(x, y) = \cot xy$

97. Find the minimum distance from the point $(2, -1, 4)$ to the plane $x - y + 3z = 7$.

98. What is the smallest amount of wood needed to build an open-top rectangular box enclosing a volume of 25 m^3?

99. What is the maximum volume of an open-top rectangular box that can be built from 10 m^2 of wood?

100. Show that there is no maximum or minimum value of xy^3z^2 if (x, y, z) lies on the plane $x - y + 2z = 2$.

101. What are the dimensions of the rectangular parallelepiped of maximum volume that can be inscribed in the ellipsoid $x^2 + 9y^2 + 4z^2 = 36$?

102. Minimize the function $x^2 + y^2 + z^2$ for (x, y, z) on the planes $2x + y + z = 2$ and $x - y - 3z = 4$.

In Exercises 103–106 find the domain and range of the given function.

103. $f(x_1, x_2, x_3, x_4) = \sqrt{x_1^4 + x_2^4 + x_3^4 + x_4^4}$

104. $f(x_1, x_2, x_3, x_4) = \dfrac{x_1x_2}{x_3 - x_4}$

105. $f(x_1, x_2, x_3, x_4) = \dfrac{x_1^2 + x_2^2}{x_3^2 + x_4^2}$

106. $f(x_1, x_2, x_3, x_4) = \sin(x_1 + x_2 - x_3) + \cos x_4$

107. Verify directly from Definition 4.2.6 that

$$\lim_{\mathbf{x}\to(2,-1,1,3)} (2x_1 - 5x_2 + x_3 - 3x_4) = 1.$$

108. Show that

$$\lim_{\mathbf{x}\to 0} \frac{x_1x_2 + x_3x_4}{x_1^2 + x_2^2 + x_3^2 + x_4^2}$$

does not exist.

109. Compute

$$\lim_{\mathbf{x}\to(1,-1,3,2)} \left(x_1^2 - 4x_2^2 + \frac{3x_3x_4}{x_2}\right).$$

110. Compute

$$\lim_{\mathbf{x}\to(1,0,2,3)} \left[(e^{x_1 - 2x_2 + x_3 - x_4}) \cos\frac{\pi x_3}{6x_1}\right].$$

111. Compute all first-order partial derivatives of

$$f(x_1, x_2, x_3, x_4) = x_1^3x_2^2x_3 - \ln(x_1 + 2x_2 - x_4) + \frac{x_4^5}{x_1}.$$

112. Do the same as in Exercise 111 for

$$f(x_1, x_2, x_3, x_4) = \frac{x_1 - x_3}{x_2 + x_4} + \sin(x_1x_3) - \tan(\sqrt{x_2x_4}).$$

113. For the function of Exercise 111 compute f_{12}, f_{31}, and f_{124}.

114. Compute the derivative of $\mathbf{f}(t) = (t^3, \cos t, \ln t, e^{2t})$.

115. Compute the derivative of $\mathbf{f}(t) = (\sqrt{t}, t^{5/3}, (1/t), t^5 - 2t + 2)$.

116. Compute the unit tangent vector and the unit normal vector to the curve $\mathbf{f}(t) = (t^2, t^4, t^6, t^8)$ at $t = 1$.

117. Compute the length of the curve $\mathbf{f}(t) = (t^2, t^2, t, t)$ for $t \in [0, 1]$.

118. Compute the gradient of the function in Exercise 103.

119. Compute the gradient of the function in Exercise 104.

120. Find the directional derivative of $f(x_1, x_2, x_3, x_4) = x_1^2 + x_2^2 + x_3^2 + x_4^2$ in the direction $\mathbf{v} = (2, -1, 4, 1)$ at the point $(3, 1, -1, 4)$.

5 Multiple Integration

In your previous calculus course you studied the definite integral of a function of one variable. In this chapter we show how the notion of the definite integral can be extended to functions of several variables. In particular, we will discuss the double integral of a function of two variables and the triple integral of a function of three variables.

Multiple integrals were developed essentially by the great Swiss mathematician Leonhard Euler[†]. Euler had a clear conception of the double integral over a bounded region enclosed by arcs, and in a paper published in 1769,[‡] he gave a procedure for evaluating double integrals by repeated integration. We will discuss Euler's technique in Section 5.2. Multiple integrals were first used by Newton in his work (which appeared in the *Principia*) on the gravitational attraction exerted by spheres on particles. It should be mentioned, however, that Newton had only a geometric interpretation of a multiple integral. The more precise analytical definitions had to await the work of Euler and, later, Lagrange.

[†]See the accompanying biographical sketch.
[‡]*Novi Comm. Acad. Sci. Petrop.*, **14,** 72–103 (1769).

1707–
1783

Leonhard Euler

Leonhard Euler
The Granger Collection

Leonhard Euler (pronounced "oiler") was born in Basil, Switzerland. His father was a clergyman who hoped that his son would follow him into the ministry. The father was adept at mathematics, however, and together with Johann Bernoulli, instructed young Leonhard in that subject. Euler also studied theology, astronomy, physics, medicine, and several Eastern languages.

In 1727 Euler applied to and was accepted for a chair of medicine and physiology at the St. Petersburg Academy. The day Euler arrived in Russia, however, Catherine I—founder of the Academy—died, and the Academy was plunged into turmoil. By 1730 Euler was pursuing his mathematical career from the chair of natural philosophy. Accepting an invitation from Frederick the Great, Euler went to Berlin in 1741 to head the Prussian Academy. Twenty-five years later, he returned to St. Petersburg, where he died in 1783 at the age of 76.

The most prolific writer in the history of mathematics, Euler found new results in virtually every branch of pure and applied mathematics. Although German was his native language, he wrote mostly in Latin and occasionally in French. His amazing productivity did not decline even when he became totally blind in 1766. During his lifetime, Euler published 530 books and papers. When he died, he left so many unpublished manuscripts that the St. Petersburg Academy was still publishing his work in its *Proceedings* almost half a century later. Euler's work enriched such diverse areas as hydraulics, celestial mechanics, lunar theory, and the theory of music, as well as mathematics.

Euler had a phenomenal memory. As a young man he memorized the entire *Aeneid* by Virgil (in Latin) and many years later could still recite the entire work. He was able to solve astonishingly complex mathematical problems in his head and is said to have solved, again in his head, problems in astronomy that stymied Newton. The French academician François Arago once commented that Euler could calculate without effort "just as men breathe, as eagles sustain themselves in the air."

Euler wrote in a mathematical language that is largely in use today. Among many symbols first used by him are:

$f(x)$	for functional notation
e	for the base of the natural logarithm
Σ	for the summation sign
i	to denote $\sqrt{-1}$

Euler's textbooks were models of clarity. His texts included the *Introductio in analysin infinitorum* (1748), his *Institutiones calculi differentialis* (1755), and the three-volume *Institutiones calculi integralis* (1768–74). This and others of his works served as models for many of today's mathematics textbooks.

It is said that Euler did for mathematical analysis what Euclid did for geometry. It is no wonder that so many later mathematicians expressed their debt to him.

We begin our discussion of multiple integration with an introduction to the double integral. Our development will closely parallel the development of the definite integral of a one-variable function.

5.1 VOLUME UNDER A SURFACE AND THE DOUBLE INTEGRAL

Study of the definite integral usually begins by calculating the area under a curve $y = f(x)$ (and above the x-axis) for x in the interval $[a, b]$. It is initially assumed that, on $[a, b]$, $f(x) \geq 0$. We carry out a similar development here by obtaining an expression which represents a volume in $\mathbb{R}^3$.

We begin by considering an especially simple case. Let R denote the rectangle in $\mathbb{R}^2$ given by

$$R = \{(x, y): a \leq x \leq b \text{ and } c \leq y \leq d\}. \tag{1}$$

This rectangle is sketched in Figure 1. Let $z = f(x, y)$ be a continuous function that is nonnegative over R. That is, $f(x, y) \geq 0$ for every (x, y) in R. We now ask: What is the volume "under" the surface $z = f(x, y)$ and "over" the rectangle R? The volume requested is sketched in Figure 2.

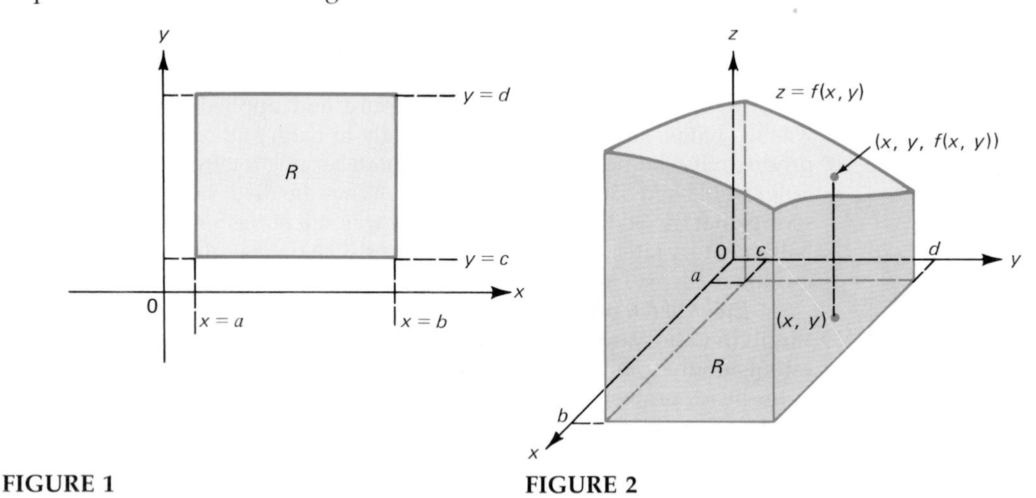

FIGURE 1 **FIGURE 2**

We will calculate this volume in much the same way the area under a curve was calculated. We begin by "partitioning" the rectangle.

Step 1. Form a **regular partition** (i.e., all subintervals have the same length) of the intervals $[a, b]$ and $[c, d]$:

$$a = x_0 < x_1 < x_2 < \cdots < x_{n-1} < x_n = b, \tag{2}$$

$$c = y_0 < y_1 < y_2 < \cdots < y_{m-1} < y_m = d. \tag{3}$$

We then define

$$\Delta x = x_i - x_{i-1} = \frac{b - a}{n} \tag{4}$$

$$\Delta y = y_j - y_{j-1} = \frac{d - c}{m}, \tag{5}$$

and define the subrectangles R_{ij} by

$$R_{ij} = \{(x, y): x_{i-1} \le x \le x_i \text{ and } y_{j-1} \le y \le y_j\} \tag{6}$$

for $i = 1, 2, \ldots, n$ and $j = 1, 2, \ldots, m$. This is sketched in Figure 3. Note that there are nm subrectangles R_{ij} covering the rectangle R.

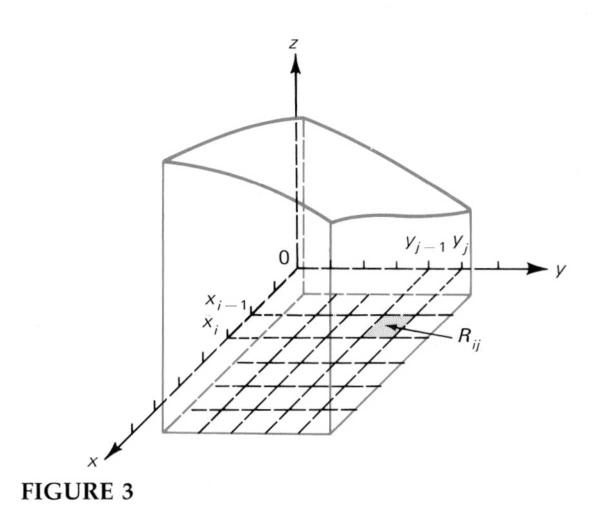

FIGURE 3

Step 2. Estimate the volume under the surface and over each subrectangle.

Let (x_i^*, y_j^*) be a point in R_{ij}. Then the volume V_{ij} under the surface and over R_{ij} is approximated by

$$V_{ij} \approx f(x_i^*, y_j^*)\, \Delta x\, \Delta y = f(x_i^*, y_j^*)\, \Delta A, \tag{7}$$

where $\Delta A = \Delta x\, \Delta y$ is the area of R_{ij}. The expression on the right-hand side of (7) is simply the volume of the parallelepiped (three-dimensional box) with base R_{ij} and height $f(x_i^*, y_j^*)$. Unless $f(x, y)$ is a plane parallel to the xy-plane, the expression $f(x_i^*, y_j^*)\, \Delta A$ will not in general be equal to the volume under the surface S. But if Δx and Δy are small, the approximation will be a good one. The difference between the actual V_{ij} and the approximate volume given in (7) is illustrated in Figure 4.

Step 3. Add up the approximate volumes to obtain an approximation to the total volume sought.

The total volume is

$$V = V_{11} + V_{12} + \cdots + V_{1m} + V_{21} + V_{22} + \cdots + V_{2m} + \cdots + V_{n1} + V_{n2} + \cdots + V_{nm}. \tag{8}$$

To simplify notation, we use the summation sign Σ. Since we are summing over two variables i and j, we need two such signs:

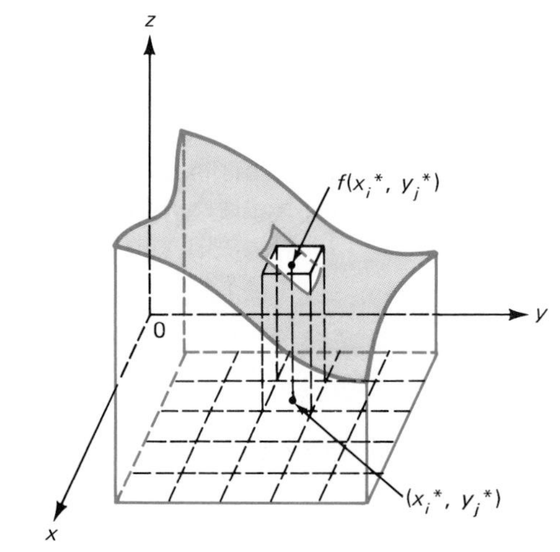

FIGURE 4

$$V = \sum_{i=1}^{n} \sum_{j=1}^{m} V_{ij}. \tag{9}$$

The expression in (9) is called a **double sum.** If we "write out"[†] the expression in (9), we obtain the expression in (8). Then combining (7) and (9), we have

$$V \approx \sum_{i=1}^{n} \sum_{j=1}^{m} f(x_i^*, y_j^*)\,\Delta A. \tag{10}$$

Step 4. Take a limit as both Δx and Δy approach zero.
To indicate that this is happening, we define

$$\Delta s = \sqrt{(\Delta x)^2 + (\Delta y)^2}.$$

Geometrically, Δs is the length of a diagonal of the rectangle R_{ij} whose sides have lengths Δx and Δy (see Figure 5). As $\Delta s \to 0$, the number of subrectangles R_{ij} increases without bound and the area of each R_{ij} approaches zero. This implies that the volume approximation given by (7) is getting closer and closer to the "true" volume over R_{ij}. Thus the approximation (10) gets better and better as $\Delta s \to 0$, enabling us to write

$$V = \lim_{\Delta s \to 0} \sum_{i=1}^{n} \sum_{j=1}^{m} f(x_i^*, y_j^*)\,\Delta A. \tag{11}$$

[†]This writing out is done by summing over j first and then over i. For example,

$$\sum_{i=1}^{3} \sum_{j=1}^{4} a_{ij} = \sum_{i=1}^{3} (a_{i1} + a_{i2} + a_{i3} + a_{i4})$$
$$= a_{11} + a_{12} + a_{13} + a_{14} + a_{21} + a_{22} + a_{23} + a_{24} + a_{31} + a_{32} + a_{33} + a_{34}.$$

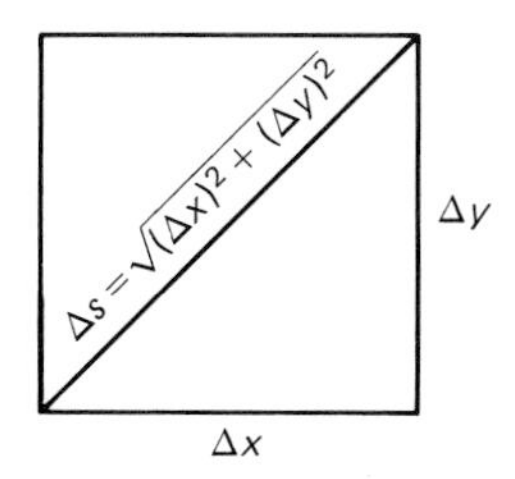

FIGURE 5

EXAMPLE 1 Calculate the volume under the plane $z = x + 2y$ and over the rectangle $R = \{(x, y): 1 \leq x \leq 2 \text{ and } 3 \leq y \leq 5\}$.

Solution. The solid whose volume we wish to calculate is sketched in Figure 6.

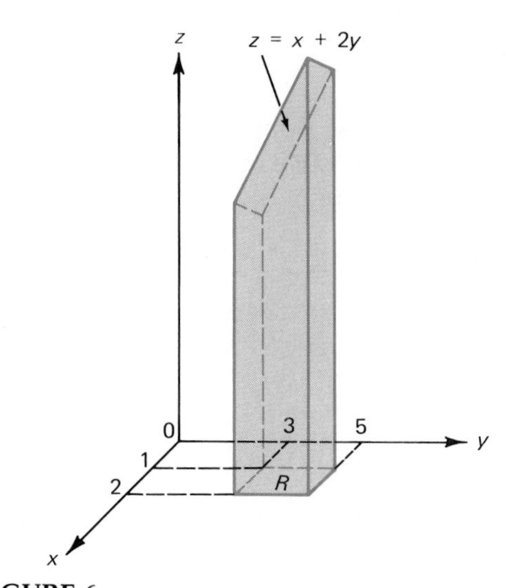

FIGURE 6

Step 1. For simplicity, we partition each of the intervals [1, 2] and [3, 5] into n subintervals of equal length (i.e., $m = n$):

$$1 = x_0 < x_1 < \cdots < x_n = 2$$

$$3 = y_0 < y_1 < \cdots < y_n = 5,$$

where

$$x_i = 1 + \frac{i}{n}, \qquad \Delta x = \frac{1}{n}$$

and

$$y_j = 3 + \frac{2j}{n}, \qquad \Delta y = \frac{2}{n}.$$

Step 2. Then choosing $x_i^* = x_i$ and $y_j^* = y_j$, we obtain

$$\begin{aligned} V_{ij} &\approx f(x_i^*, y_j^*)\,\Delta A = (x_i + 2y_j)\,\Delta x\,\Delta y \\ &= \left[\left(1 + \frac{i}{n}\right) + 2\left(3 + \frac{2j}{n}\right)\right]\frac{1}{n}\cdot\frac{2}{n} \\ &= \left(7 + \frac{i}{n} + \frac{4j}{n}\right)\frac{2}{n^2}. \end{aligned}$$

Step 3. $$\begin{aligned} V = \sum_{i=1}^{n}\sum_{j=1}^{n} V_{ij} &\approx \sum_{i=1}^{n}\sum_{j=1}^{n}\left(\frac{14}{n^2} + \frac{2i}{n^3} + \frac{8j}{n^3}\right) \\ &= \underbrace{\sum_{i=1}^{n}\sum_{j=1}^{n}\frac{14}{n^2}}_{①} + \underbrace{\sum_{i=1}^{n}\sum_{j=1}^{n}\frac{2i}{n^3}}_{②} + \underbrace{\sum_{i=1}^{n}\sum_{j=1}^{n}\frac{8j}{n^3}}_{③}. \end{aligned}$$

It is not difficult to evaluate each of these double sums. There are n^2 terms in each sum. Since $14/n^2$ does not depend on i or j, we evaluate the sum ① by simply adding up the term $14/n^2$ a total of n^2 times. Thus

$$\sum_{i=1}^{n}\sum_{j=1}^{n}\frac{14}{n^2} = n^2\left(\frac{14}{n^2}\right) = 14.$$

Next, if we set $i = 1$ in ③, then we have $\Sigma_{j=1}^{n} 8j/n^3$. Similarly, setting $i = 2, 3, 4, \ldots, n$ in ③ yields $\Sigma_{j=1}^{n} 8j/n^3$. Thus in ③ we obtain the term $(\Sigma_{j=1}^{n} 8j/n^3)$ n times. But

$$\begin{aligned} \sum_{j=1}^{n}\frac{8j}{n^3} &= \frac{8}{n^3}\sum_{j=1}^{n} j = \frac{8}{n^3}(1 + 2 + \cdots + n) \\ &= \frac{8}{n^3}\left[\frac{n(n+1)}{2}\right] \qquad \text{see Example 1 in Appendix 1} \\ &= \frac{4(n+1)}{n^2}. \end{aligned}$$

Thus

$$\sum_{i=1}^{n}\sum_{j=1}^{n}\frac{8j}{n^3} = n\left\{\sum_{j=1}^{n}\frac{8j}{n^3}\right\} = n\left[\frac{4(n+1)}{n^2}\right] = \frac{4(n+1)}{n}.$$

To calculate ②, we use the same argument as in ③:

$$\sum_{j=1}^{n}\frac{2i}{n^3} = n\left(\frac{2i}{n^3}\right) = \frac{2i}{n^2},$$

so that

$$\sum_{i=1}^{n}\sum_{j=1}^{n}\frac{2i}{n^3} = \sum_{i=1}^{n}\frac{2i}{n^2} = \frac{2}{n^2}\sum_{i=1}^{n} i = \frac{2}{n^2}\left[\frac{n(n+1)}{2}\right] = \frac{n+1}{n}.$$

Finally, we have

$$\sum_{i=1}^{n}\sum_{j=1}^{n} V_{ij} \approx 14 + \frac{4(n+1)}{n} + \frac{n+1}{n}.$$

Step 4. Now as $\Delta s \to 0$, both Δx and Δy approach 0, so $n = (b - a)/\Delta x \to \infty$. Thus

$$V = \lim_{\Delta s \to 0} \sum_{i=1}^{n}\sum_{j=1}^{n} f(x_i^*, y_j^*)\,\Delta A = \lim_{n\to\infty} \sum_{i=1}^{n}\sum_{j=1}^{n} f(x_i^*, y_j^*)\,\Delta A$$

$$= \lim_{n\to\infty} \left[14 + 4\left(\frac{n+1}{n}\right) + \frac{n+1}{n}\right] = 14 + 4 + 1 = 19. \ \blacksquare$$

The calculation we just made was very tedious. Instead of making other calculations like this one, we will define the double integral and, in Section 5.2, show how double integrals can be easily calculated.

Definition 1 THE DOUBLE INTEGRAL Let $z = f(x, y)$ and let the rectangle R be given by (1). Let $\Delta A = \Delta x\,\Delta y$. Suppose that

$$\lim_{\Delta s \to 0} \sum_{i=1}^{n}\sum_{j=1}^{m} f(x_i^*, y_j^*)\,dA$$

exists and is independent of the way in which the points (x_i^*, y_j^*) are chosen. Then the **double integral of f over R**, written $\iint_R f(x, y)\,dA$, is defined by

$$\iint_R f(x, y)\,dA = \lim_{\Delta s \to 0} \sum_{i=1}^{n}\sum_{j=1}^{m} f(x_i^*, y_j^*)\,\Delta A. \tag{12}$$

If the limit in (12) exists, then the function f is said to be **integrable** over R.

We observe that this definition says nothing about volumes. For example, if $f(x, y)$ takes on negative values in R, then the limit in (12) will not represent the volume under the surface $z = f(x, y)$. However, the limit in (12) may still exist, and in that case f will be integrable over R.

NOTE. $\iint_R f(x, y)\,dA$ is a number, not a function. This is analogous to the fact that the definite integral $\int_a^b f(x)\,dx$ is a number. We will not encounter indefinite double integrals in this book.

As we already stated, we will not calculate any other double integrals in this section but will wait until Section 5.2 to see how these calculations can be made simple. We should note, however, that the result of Example 1 can now be restated as

$$\iint_R (x + 2y)\,dA = 19,$$

where R is the rectangle $\{(x, y)\colon 1 \le x \le 2 \text{ and } 3 \le y \le 5\}$.

What functions are integrable over a rectangle R? The following theorem gives a partial answer.

Theorem 1 EXISTENCE OF THE DOUBLE INTEGRAL OVER A RECTANGLE If f is continuous on R, then f is integrable over R.

We will not give proofs of the theorems stated in this section regarding double integrals. The proofs of all these theorems can be found in any standard advanced calculus text.[†]

We now turn to the question of defining double integrals over regions in $\mathbb{R}^2$ that are not rectangular. We will denote a region in $\mathbb{R}^2$ by Ω. The two types of regions in which we will be most interested are illustrated in Figure 7. In this figure g_1, g_2, h_1, and h_2 denote continuous functions. A more general region Ω is sketched in Figure 8. We assume that the region is bounded. This means that there is a number M such that for every (x, y) in Ω, $|(x, y)| = \sqrt{x^2 + y^2} \leq M$. Since Ω is bounded, we can draw a rectangle R around it. Let f be defined over Ω. We then define a new function F by

$$F(x, y) = \begin{cases} f(x, y), & \text{for } (x, y) \text{ in } \Omega \\ 0, & \text{for } (x, y) \text{ in } R \text{ but not in } \Omega. \end{cases} \tag{13}$$

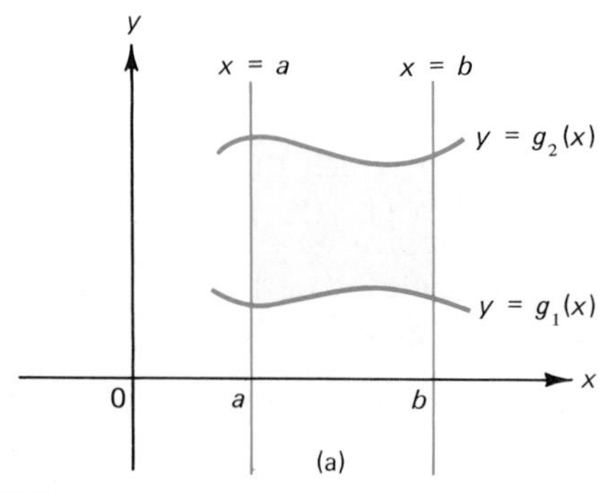

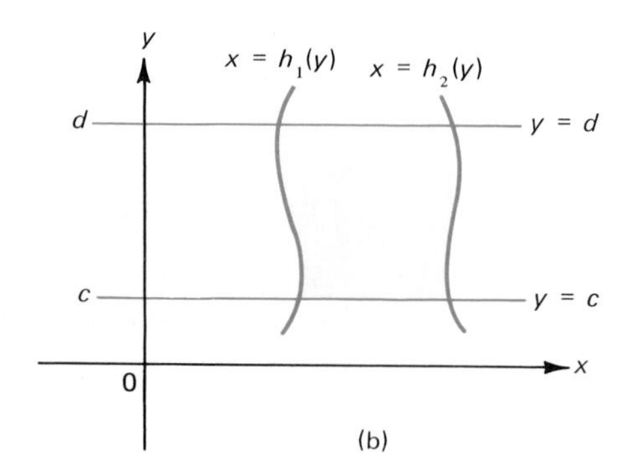

FIGURE 7

Definition 2 INTEGRABILITY OVER A REGION Let f be defined for (x, y) in Ω and let F be defined by (13). Then we write

$$\iint_{\Omega} f(x, y)\, dA = \iint_{R} F(x, y)\, dA \tag{14}$$

if the integral on the right exists. In this case we say that f is **integrable** over Ω.

REMARK. If we divide R into nm subrectangles, as in Figure 9, then we can see what is happening. For each subrectangle R_{ij} that lies entirely in Ω, $F = f$, so the volume of the "parallelepiped" above R_{ij} is given by

$$V_{ij} \approx f(x_i^*, y_j^*)\, \Delta x\, \Delta y = F(x_i^*, y_j^*)\, \Delta x\, \Delta y.$$

[†]See, for example, R. C. Buck, *Advanced Calculus* (McGraw-Hill, New York, 1965).

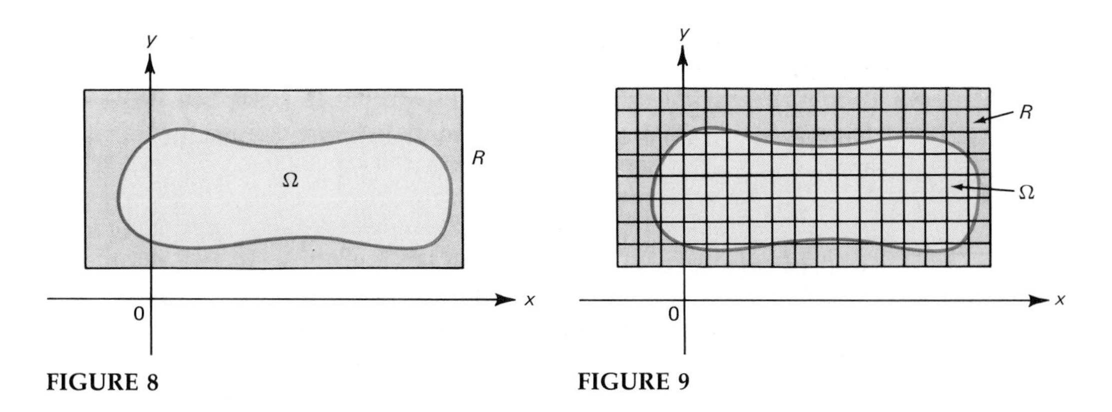

FIGURE 8

FIGURE 9

However, if R_{ij} is in R but not in Ω, then $F = 0$, so

$$V_{ij} \approx F(x_i^*, y_j^*)\,\Delta x\,\Delta y = 0.$$

Finally, if R_{ij} is partly in Ω and partly outside of Ω, then there is no real problem since, as $\Delta s \to 0$, the sum of the volumes above these rectangles (along the boundary of Ω) will approach zero—unless the boundary of Ω is very complicated indeed. Thus we see that the limit of the sum of the volumes of the "parallelepipeds" above R is the same as the limit of the sum of the volumes of the "parallelepipeds" above Ω. This should help explain the "reasonableness" of expression (14).

Theorem 2 EXISTENCE OF THE DOUBLE INTEGRAL OVER A MORE GENERAL REGION Let Ω be one of the regions depicted in Figure 7 where the functions g_1 and g_2 or h_1 and h_2 are continuous. Let F be defined by (13). If f is continuous over Ω, then f is integrable over Ω and its integral is given by (14).

REMARK 1. There are some regions Ω that are so complicated that it is possible to find functions continuous but not integrable over Ω. We will not concern ourselves with such regions in this book.

REMARK 2. *If f is nonnegative and integrable over Ω, then*

$$\iint_\Omega f(x, y)\, dA$$

is defined as the volume under the surface $z = f(x, y)$ and over the region Ω.

REMARK 3. *If the function $f(x, y) = 1$ is integrable over Ω, then*

$$\iint_\Omega 1\, dA = \iint_\Omega dA \tag{15}$$

is equal to the area of the region Ω. To see this, note that

$$V_{ij} \approx f(x_i^*, y_j^*)\,\Delta A = \Delta A,$$

so the double integral (15) is the limit of the sum of areas of rectangles in Ω.

We close this section by stating five theorems about double integrals. Each one is analogous to a theorem about definite integrals.

Theorem 3 If f is integrable over Ω, then for any constant c, cf is integrable over Ω, and

$$\iint_{\Omega} cf(x, y)\, dA = c \iint_{\Omega} f(x, y)\, dA. \tag{16}$$

Theorem 4 If f and g are integrable over Ω, then $f + g$ is integrable over Ω, and

$$\iint_{\Omega} [f(x, y) + g(x, y)]\, dA = \iint_{\Omega} f(x, y)\, dA + \iint_{\Omega} g(x, y)\, dA. \tag{17}$$

Theorem 5 If f is integrable over Ω_1 and Ω_2, where Ω_1 and Ω_2 have no points in common except perhaps those of their common boundary, then f is integrable over $\Omega = \Omega_1 \cup \Omega_2$, and

$$\iint_{\Omega} f(x, y)\, dA = \iint_{\Omega_1} f(x, y)\, dA + \iint_{\Omega_2} f(x, y)\, dA.$$

A typical region Ω is depicted in Figure 10.

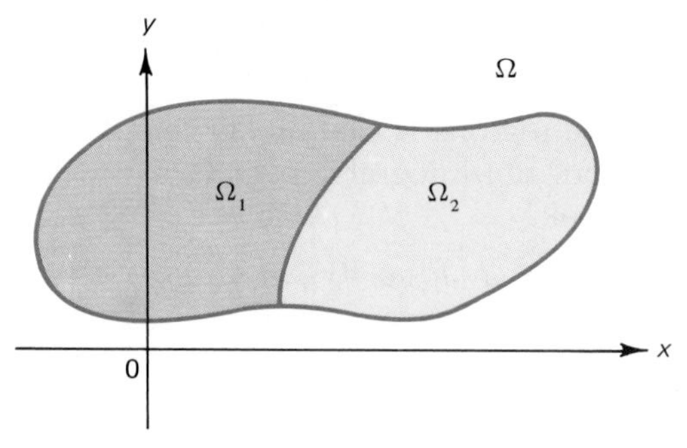

FIGURE 10

Theorem 6 If f and g are integrable over Ω and $f(x, y) \le g(x, y)$ for every (x, y) in Ω, then

$$\iint_{\Omega} f(x, y)\, dA \le \iint_{\Omega} g(x, y)\, dA \tag{18}$$

Theorem 7 Let f be integrable over Ω. Suppose that there exist constants m and M such that

$$m \le f(x, y) \le M \tag{19}$$

for every (x, y) in Ω. If A_Ω denotes the area of Ω, then

$$mA_\Omega \le \iint_\Omega f(x, y)\, dA \le MA_\Omega. \tag{20}$$

Theorem 7 can be useful for estimating double integrals.

EXAMPLE 2 Let Ω be the rectangle $\{(x, y): a \le x \le b \text{ and } c \le y \le d\}$. Find upper and lower bounds for

$$\iint_\Omega \sin(x - 3y^3)\, dA.$$

Solution. Since $-1 \le \sin(x - 3y^3) \le 1$, and since $A_\Omega = (b - a)(d - c)$, we have, using (20),

$$-(b - a)(d - c) \le \iint_\Omega \sin(x - 3y^3)\, dA \le (b - a)(d - c). \blacksquare$$

EXAMPLE 3 Let Ω be the disk $\{(x, y): x^2 + y^2 \le 1\}$. Find upper and lower bounds for

$$\iint_\Omega \frac{1}{1 + x^2 + y^2}\, dA.$$

Solution. Since $0 \le x^2 + y^2 \le 1$ in Ω, we easily see that

$$\frac{1}{2} \le \frac{1}{1 + x^2 + y^2} \le 1.$$

Since the area of the disk is π, we have

$$\frac{\pi}{2} \le \iint_\Omega \frac{1}{1 + x^2 + y^2}\, dA \le \pi.$$

In fact, it can be shown (see Problem 5.4.25) that the value of the integral is $\pi \ln 2 \approx 0.693\pi$. $\blacksquare$

PROBLEMS 5.1

In Problems 1–8, let Ω denote the rectangle $\{(x, y): 0 \le x \le 3 \text{ and } 1 \le y \le 2\}$. Use the technique employed in Example 1 to calculate the given double integral. Use Theorem 3 and/or 4 where appropriate.

1. $\iint_\Omega (2x + 3y)\, dA$
2. $\iint_\Omega (x - y)\, dA$
3. $\iint_\Omega (y - x)\, dA$
4. $\iint_\Omega (ax + by + c)\, dA$
5. $\iint_\Omega (x^2 + y^2)\, dA$
6. $\iint_\Omega (x^2 - y^2)\, dA$
[*Hint:* Use formula (2) in Appendix 1.
7. $\iint_\Omega (2x^2 + 3y^2)\, dA$
8. $\iint_\Omega (ax^2 + by)\, dA$

In Problems 9–14, let Ω denote the rectangle $\{(x, y): -1 \le x \le 0 \text{ and } -2 \le y \le 3\}$. Calculate the double integral.

9. $\iint_\Omega (x + y)\, dA$
10. $\iint_\Omega (3x - y)\, dA$
11. $\iint_\Omega (y - 2x)\, dA$
12. $\iint_\Omega (x^2 + 2y^2)\, dA$
13. $\iint_\Omega (y^2 - x^2)\, dA$
14. $\iint_\Omega (3x^2 - 5y^2)\, dA$

In Problems 15–19, use Theorem 7 to obtain upper and lower bounds for the given integral.

15. $\iint_{\Omega}(x^5y^2 + xy)\,dA$, where Ω is the rectangle $\{(x, y)\colon 0 \le x \le 1 \text{ and } 1 \le y \le 2\}$.

16. $\iint_{\Omega} e^{-(x^2+y^2)}\,dA$, where Ω is the disk $x^2 + y^2 \le 4$.

***17.** $\iint_{\Omega}[(x - y)/(4 - x^2 - y^2)]\,dA$, where Ω is the disk $x^2 + y^2 \le 1$.

18. $\iint_{\Omega} \cos(\sqrt{|x|} - \sqrt{|y|})\,dA$, where Ω is the region of Problem 17.

19. $\iint_{\Omega} \ln(1 + x + y)\,dA$, where Ω is the region bounded by the lines $y = x$, $y = 1 - x$, and the x-axis.

***20.** Let Ω be one of the regions depicted in Figure 7. Which is greater:

$$\iint_{\Omega} e^{(x^2+y^2)}\,dA \qquad \text{or} \qquad \iint_{\Omega} (x^2 + y^2)\,dA?$$

5.2 THE CALCULATION OF DOUBLE INTEGRALS

In this section we derive an easy method for calculating $\iint_{\Omega} f(x, y)\,dx\,dy$, where Ω is one of the regions depicted in Figure 5.1.7.

We begin, as in Section 5.1, by considering

$$\iint_R f(x, y)\,dA, \tag{1}$$

where R is the rectangle

$$R = \{(x, y)\colon a \le x \le b \text{ and } c \le y \le d\}. \tag{2}$$

If $z = f(x, y) \ge 0$ for (x, y) in R, then the double integral in (1) is the volume under the surface $z = f(x, y)$ and over the rectangle R in the xy-plane. We now calculate this volume by partitioning the x-axis taking "slices" parallel to the yz-plane. This is illustrated in Figure 1. We can approximate the volume by adding up the volumes of the various "slices." The face of each "slice" lies in the plane $x = x_i$, and the volume of the ith slice is approximately equal to the area of its face times its thickness Δx. What is the area of the face? If x is fixed, then $z = f(x, y)$ can be thought of as a curve lying in the plane $x = x_i$. Thus the area of the ith face is the area bounded by this curve, the y-axis, and the lines $y = c$ and $y = d$. This area is sketched in Figure 2. If $f(x_i, y)$ is a continuous function of y, then the area of the ith face, denoted A_i, is given by

$$A_i = \int_c^d f(x_i, y)\,dy.$$

By treating x_i as a constant, we can compute A_i as an ordinary definite integral, where the variable is y. Note, too, that $A(x) = \int_c^d f(x, y)\,dy$ is a function of x only and can therefore be integrated. Then the volume of the ith slice is approximated by

$$V_i \approx \left\{\int_c^d f(x_i, y)\,dy\right\} \Delta x$$

so that, adding up these "subvolumes" and taking the limit as Δx approaches zero, we obtain

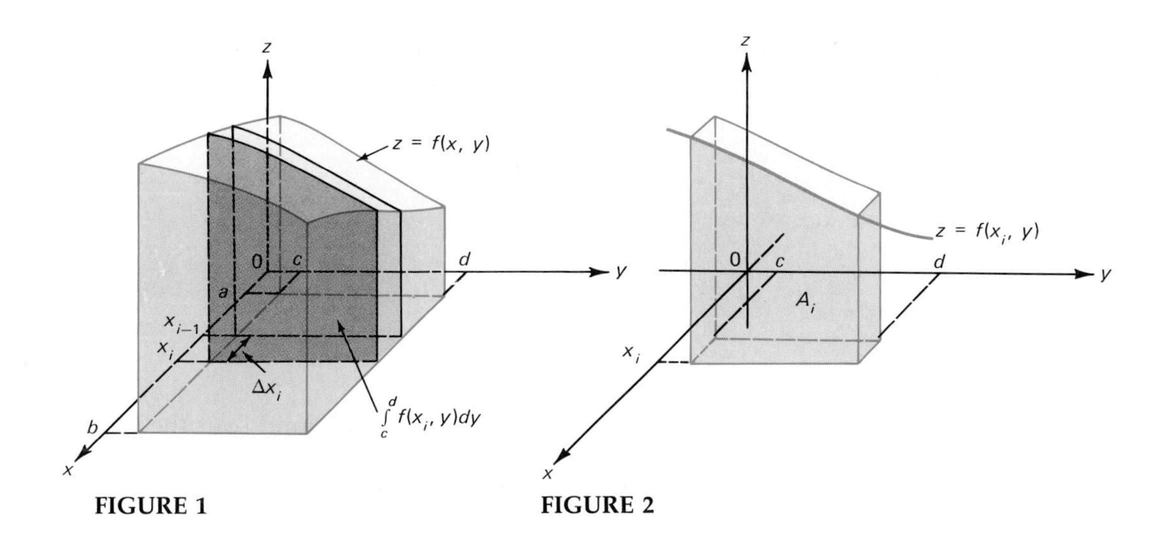

FIGURE 1 **FIGURE 2**

$$V = \int_a^b \left\{ \int_c^d f(x, y)\, dy \right\} dx = \int_a^b A(x)\, dx. \tag{3}$$

Definition 1 REPEATED INTEGRAL The expression in (3) is called a **repeated integral** or **iterated integral.** Since we also have

$$V = \iint_R f(x, y)\, dA,$$

we obtain

$$\iint_R f(x, y)\, dA = \int_a^b \left\{ \int_c^d f(x, y)\, dy \right\} dx. \tag{4}$$

REMARK 1. Usually we will write equation (4) without braces. We then have

$$\iint_R f(x, y)\, dA = \int_a^b \int_c^d f(x, y)\, dy\, dx. \tag{5}$$

REMARK 2. We should emphasize that the first integration in $\int_a^b \int_c^d f(x, y)\, dy\, dx$ is performed by treating x as a constant.

Similarly, if we instead begin by partitioning the y-axis, we find that the area of the face of a "slice" lying in the plane $y = y_i$ is given by

$$A_i = \int_a^b f(x, y_i)\, dx,$$

where now A_i is an integral in the variable x. Thus as before,

$$V = \int_c^d \left\{ \int_a^b f(x, y)\, dx \right\} dy, \tag{6}$$

and

$$\iint_R f(x, y)\, dA = \int_c^d \int_a^b f(x, y)\, dx\, dy. \tag{7}$$

EXAMPLE 1 Calculate the volume under the plane $z = x + 2y$ and over the rectangle

$$R = \{(x, y) \colon 1 \le x \le 2 \text{ and } 3 \le y \le 5\}.$$

Solution. We calculated this volume in Example 5.1.1. Using equation (5), we have

$$\begin{aligned} V &= \iint_R (x + 2y)\, dA = \int_1^2 \left[\int_3^5 (x + 2y)\, dy \right] dx^{\dagger} \\ &= \int_1^2 \left[(xy + y^2) \Big|_{y=3}^{y=5} \right] dx = \int_1^2 [(5x + 25) - (3x + 9)]\, dx \\ &= \int_1^2 (2x + 16)\, dx = (x^2 + 16x) \Big|_1^2 = 19. \end{aligned}$$

Similarly, using equation (7), we have

$$\begin{aligned} V &= \int_3^5 \left\{ \int_1^2 (x + 2y)\, dx \right\} dy = \int_3^5 \left[\left(\frac{x^2}{2} + 2yx \right) \Big|_{x=1}^{x=2} \right] dy \\ &= \int_3^5 \left\{ (2 + 4y) - \left(\frac{1}{2} + 2y \right) \right\} dy = \int_3^5 \left(2y + \frac{3}{2} \right) dy \\ &= \left(y^2 + \frac{3}{2} y \right) \Big|_3^5 = 19. \blacksquare \end{aligned}$$

EXAMPLE 2 Calculate the volume of the region beneath the surface $z = xy^2 + y^3$ and over the rectangle $R = \{(x, y) \colon 0 \le x \le 2 \text{ and } 1 \le y \le 3\}$.

Solution. A computer-drawn sketch of this region is given in Figure 3. Using equation (5), we have

---

†Remember, in computing the bracketed integral, we treat x as a constant.

$$V = \int_0^2 \int_1^3 (xy^2 + y^3)\, dy\, dx = \int_0^2 \left[\left(\frac{xy^3}{3} + \frac{y^4}{4} \right) \Big|_1^3 \right] dx$$
$$= \int_0^2 \left[\left(9x + \frac{81}{4} \right) - \left(\frac{x}{3} + \frac{1}{4} \right) \right] dx = \int_0^2 \left(\frac{26}{3} x + 20 \right) dx$$
$$= \left(\frac{13x^2}{3} + 20x \right) \Big|_0^2 = \frac{52}{3} + 40 = \frac{172}{3}.$$

You should verify that the same answer is obtained by using equation (7). ■

We now extend our results to more general regions. Let

$$\Omega = \{(x, y): a \le x \le b \text{ and } g_1(x) \le y \le g_2(x)\}. \tag{8}$$

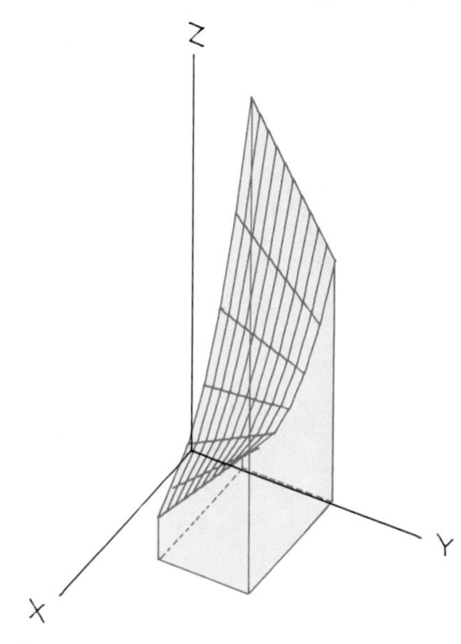

$Z = XY^2 + Y^3$

FIGURE 3

This region is sketched in Figure 4. We assume that for every x in $[a, b]$,

$$g_1(x) \le g_2(x). \tag{9}$$

If we partition the x-axis as before, then we obtain slices lying in the planes $x = x_i$, a typical one of which is sketched in Figure 5. Then

$$A_i = \int_{g_1(x_i)}^{g_2(x_i)} f(x_i, y)\, dy, \qquad V_i \approx \left\{ \int_{g_1(x_i)}^{g_2(x_i)} f(x_i, y)\, dy \right\} \Delta x,$$

and

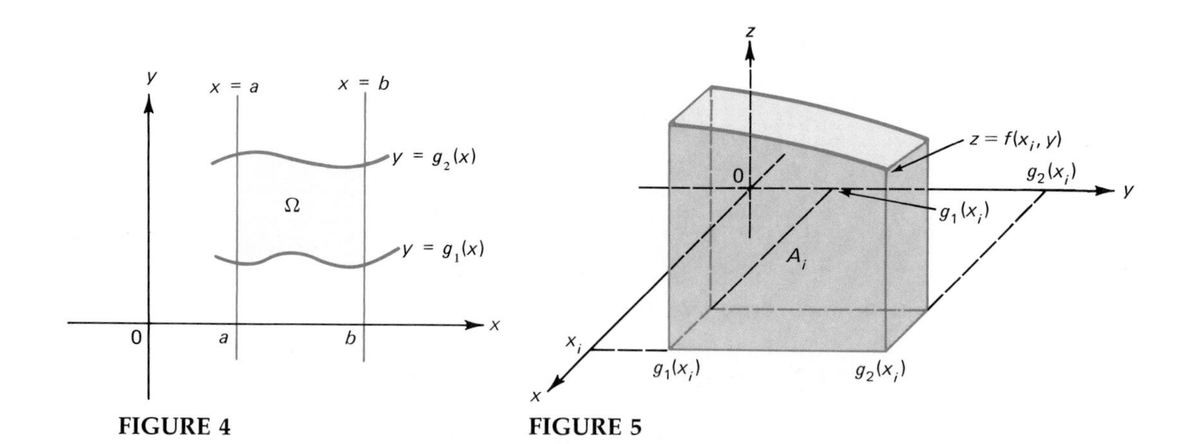

FIGURE 4 **FIGURE 5**

$$V = \iint_{\Omega} f(x, y)\, dA = \int_a^b \int_{g_1(x)}^{g_2(x)} f(x, y)\, dy\, dx. \tag{10}$$

Similarly, let

$$\Omega = \{(x, y): h_1(y) \leq x \leq h_2(y) \text{ and } c \leq y \leq d\} \tag{11}$$

(see Figure 6). Then

$$V = \int_c^d \int_{h_1(y)}^{h_2(y)} f(x, y)\, dx\, dy. \tag{12}$$

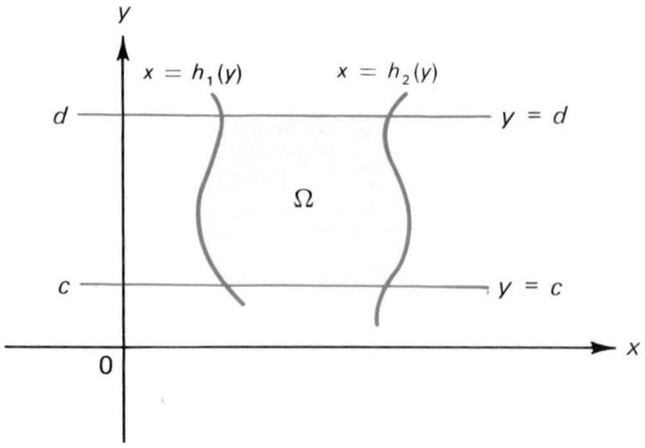

FIGURE 6

We summarize these results in the following theorem.

Theorem 1 Let f be continuous over a region Ω given by equation (8) or (11).

(i) If Ω is of the form (8), where g_1 and g_2 are continuous, then

$$\iint_{\Omega} f(x, y)\, dA = \int_a^b \int_{g_1(x)}^{g_2(x)} f(x, y)\, dy\, dx.$$

(ii) If Ω is of the form (11), where h_1 and h_2 are continuous, then

$$\iint_{\Omega} f(x, y)\, dA = \int_c^d \int_{h_1(y)}^{h_2(y)} f(x, y)\, dx\, dy.$$

REMARK 1. We have not actually proved this theorem here but have merely indicated why it should be so. A rigorous proof can be found in any advanced calculus text.

REMARK 2. Note that this theorem says nothing about volume. It can be used to calculate any double integral if the hypotheses of the theorem are satisfied and if each function being integrated has an antiderivative that can be written in terms of elementary functions.

REMARK 3. Many regions are of the form (8) or (11). In addition, almost all regions that arise in practical applications can be broken into a finite number of regions of the form (8) or (11), and the integration can be carried out using Theorem 5.1.5.

EXAMPLE 3 Find the volume of the solid under the surface $z = x^2 + y^2$ and lying above the region

$$\Omega = \{(x, y)\colon 0 \le x \le 1 \text{ and } x^2 \le y \le \sqrt{x}\}.$$

Solution. Ω is sketched in Figure 7. We see that $0 \le x \le 1$ and $x^2 \le y \le \sqrt{x}$. Then using (10), we have

$$V = \int_0^1 \int_{x^2}^{\sqrt{x}} (x^2 + y^2)\, dy\, dx = \int_0^1 \left\{ \left(x^2 y + \frac{y^3}{3} \right) \bigg|_{x^2}^{\sqrt{x}} \right\} dx$$

$$= \int_0^1 \left\{ \left(x^2\sqrt{x} + \frac{(\sqrt{x})^3}{3} \right) - \left(x^2 \cdot x^2 + \frac{(x^2)^3}{3} \right) \right\} dx$$

$$= \int_0^1 \left(x^{5/2} + \frac{x^{3/2}}{3} - x^4 - \frac{x^6}{3} \right) dx$$

$$= \left(\frac{2x^{7/2}}{7} + \frac{2x^{5/2}}{15} - \frac{x^5}{5} - \frac{x^7}{21} \right) \bigg|_0^1 = \frac{2}{7} + \frac{2}{15} - \frac{1}{5} - \frac{1}{21} = \frac{18}{105}.$$

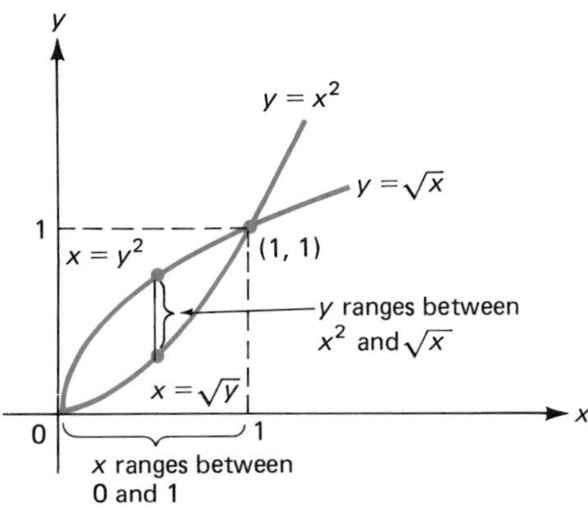

FIGURE 7

We can calculate this integral in another way. We note that x varies between the curves $x = y^2$ and $x = \sqrt{y}$. Then using (12), since $0 \le y \le 1$ and $y^2 \le x \le \sqrt{y}$, we have

$$V = \int_0^1 \int_{y^2}^{\sqrt{y}} (x^2 + y^2)\, dx\, dy,$$

which is equal to 18/105. ■

EXAMPLE 4 Let $f(x, y) = x^2 y$. Calculate the integral of f over the region bounded by the x-axis and the semicircle $x^2 + y^2 = 4$, $y \ge 0$.

Solution. The region of integration is sketched in Figure 8. Using equation (8), we see that $0 \le y \le \sqrt{4 - x^2}$, $-2 \le x \le 2$, so that, integrating first with respect to y, we obtain

$$\iint_\Omega x^2 y\, dA = \int_{-2}^{2} \int_0^{\sqrt{4-x^2}} x^2 y\, dy\, dx$$

$$= \int_{-2}^{2} \left\{ \frac{x^2 y^2}{2} \Big|_0^{\sqrt{4-x^2}} \right\} dx = \int_{-2}^{2} \frac{x^2(4 - x^2)}{2}\, dx$$

$$= \int_{-2}^{2} \left(2x^2 - \frac{x^4}{2}\right) dx = \left(\frac{2x^3}{3} - \frac{x^5}{10}\right)\Big|_{-2}^{2} = \frac{64}{15}.$$

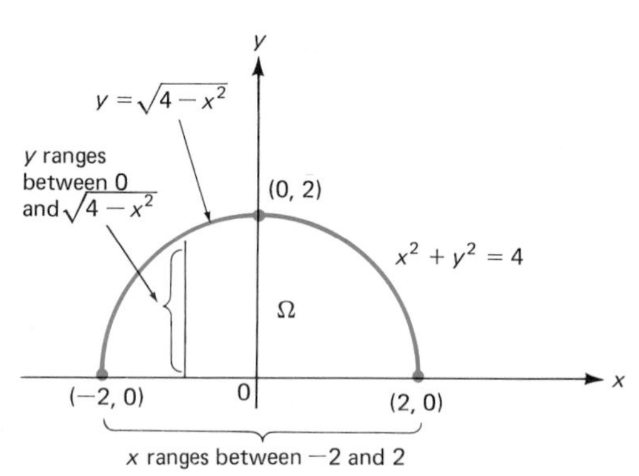

FIGURE 8

We can also use equation (11) and integrate first with respect to x. Then $-\sqrt{4 - y^2} \le x \le \sqrt{4 - y^2}$, $0 \le y \le 2$, and

$$V = \int_0^2 \int_{-\sqrt{4-y^2}}^{\sqrt{4-y^2}} x^2 y\, dx\, dy = \int_0^2 \left(\frac{x^3 y}{3} \Big|_{-\sqrt{4-y^2}}^{\sqrt{4-y^2}} \right) dy$$

$$= \int_0^2 \frac{2}{3}(4 - y^2)^{3/2}\, y\, dy = \frac{-2}{15}(4 - y^2)^{5/2}\Big|_0^2 = \frac{64}{15}. \quad ■$$

REVERSING THE ORDER OF INTEGRATION

EXAMPLE 5 Evaluate $\int_1^2 \int_1^{x^2} (x/y)\, dy\, dx$.

Solution.

$$\int_1^2 \int_1^{x^2} \frac{x}{y}\, dy\, dx = \int_1^2 \left\{ x \ln y \Big|_1^{x^2} \right\} dx = \int_1^2 x \ln x^2\, dx = \int_1^2 2x \ln x\, dx$$

It is necessary to use integration by parts to complete the problem. Setting $u = \ln x$ and $dv = 2x\, dx$, we have $du = (1/x)\, dx$, $v = x^2$, and

$$\int_1^2 2x \ln x\, dx = x^2 \ln x \Big|_1^2 - \int_1^2 x\, dx = 4 \ln 2 - \frac{x^2}{2}\Big|_1^2 = 4 \ln 2 - \frac{3}{2}.$$

There is an easier way to calculate the double integral. We simply **reverse the order of integration.** The region of integration is sketched in Figure 9. If we want to integrate first with respect to x, we note that we can describe the region by

$$\Omega = \{(x, y): \sqrt{y} \le x \le 2 \text{ and } 1 \le y \le 4\}.$$

Then

$$\int_1^2 \int_1^{x^2} \frac{x}{y}\, dy\, dx = \iint_\Omega \frac{x}{y}\, dA = \int_1^4 \int_{\sqrt{y}}^2 \frac{x}{y}\, dx\, dy = \int_1^4 \left\{ \frac{x^2}{2y}\Big|_{\sqrt{y}}^2 \right\} dy$$

$$= \int_1^4 \left(\frac{2}{y} - \frac{1}{2}\right) dy = \left(2 \ln y - \frac{y}{2}\right)\Big|_1^4 = 2 \ln 4 - \frac{3}{2}$$

$$= 4 \ln 2 - \frac{3}{2}.$$

Note that in this case it is easier to integrate first with respect to x. ■

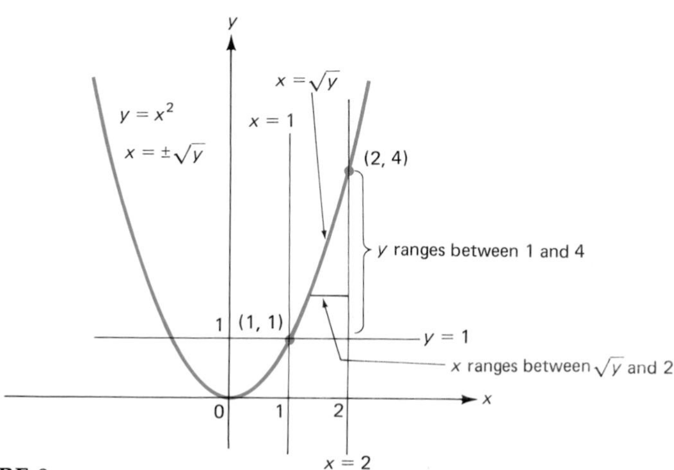

FIGURE 9

The technique used in Example 5 suggests the following:

When changing the order of integration, first sketch the region of integration in the xy-plane.

REMARK. Why is it legitimate to reverse the order of integration? There is a theorem[†] that asserts the following: Suppose that the region Ω can be written as

$$\Omega = \{(x, y): a \le x \le b, g_1(x) \le y \le g_2(x)\} = \{(x, y): c \le y \le d, h_1(y) \le x \le h_2(y)\}.$$

Then if f is continuous on Ω,

$$\iint_\Omega f(x, y)\, dA = \int_a^b \int_{g_1(x)}^{g_2(x)} f(x, y)\, dy\, dx = \int_c^d \int_{h_1(y)}^{h_2(y)} f(x, y)\, dx\, dy.$$

We will not prove this result here.

EXAMPLE 6 Compute $\int_0^2 \int_y^2 e^{x^2}\, dx\, dy$.

Solution. The region of integration is sketched in Figure 10. We first observe that the double integral cannot be evaluated directly since it is impossible to find an antiderivative for e^{x^2}. Instead, we reverse the order of integration. From Figure 10 we see that Ω can be written as $0 \le y \le x$, $0 \le x \le 2$, so

$$\int_0^2 \int_y^2 e^{x^2}\, dx\, dy = \iint_\Omega e^{x^2}\, dA = \int_0^2 \int_0^x e^{x^2}\, dy\, dx = \int_0^2 \left(ye^{x^2} \Big|_{y=0}^{y=x} \right) dx = \int_0^2 xe^{x^2}\, dx$$
$$= \frac{1}{2} e^{x^2} \Big|_0^2 = \frac{1}{2}(e^4 - 1). \blacksquare$$

EXAMPLE 7 Reverse the order of integration in the iterated integral $\int_0^1 \int_{\sqrt{y}}^2 f(x, y)\, dx\, dy$.

Solution. The region of integration is sketched in Figure 11. This region is divided into two subregions Ω_1 and Ω_2. What happens if we integrate first with respect

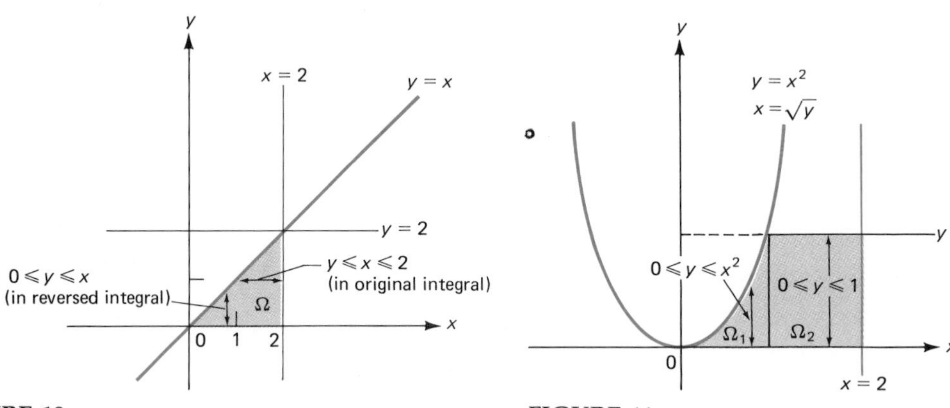

FIGURE 10 **FIGURE 11**

[†]This result is a special case of **Fubini's theorem,** which is proved in most advanced calculus books.

to y? In Ω_1, $0 \le y \le x^2$. In Ω_2, $0 \le y \le 1$. Thus

Theorem 5.1.5

$$\int_0^1 \int_{\sqrt{y}}^2 f(x, y)\, dx\, dy = \iint_\Omega f(x, y)\, dA = \iint_{\Omega_1} f(x, y)\, dA + \iint_{\Omega_2} f(x, y)\, dA$$
$$= \int_0^1 \int_0^{x^2} f(x, y)\, dy\, dx + \int_1^2 \int_0^1 f(x, y)\, dy\, dx. \blacksquare$$

EXAMPLE 8 Find the volume in the first octant bounded by the three coordinate planes and the surface $z = 1/(1 + x + 3y)^3$.

Solution. The solid here extends over the infinite region $\{(x, y): 0 \le x \le \infty$ and $0 \le y \le \infty\}$. Thus

$$V = \int_0^\infty \int_0^\infty \frac{1}{(1 + x + 3y)^3}\, dx\, dy = \int_0^\infty \lim_{N\to\infty}\left(-\frac{1}{2(1 + x + 3y)^2}\bigg|_0^N\right) dy$$
$$= \int_0^\infty \frac{1}{2(1 + 3y)^2}\, dy = \lim_{N\to\infty}\left(-\frac{1}{6(1 + 3y)}\right)\bigg|_0^N = \frac{1}{6}.$$

Note that improper double integrals can be treated in the same way that we treat improper "single" integrals. ■

EXAMPLE 9 Find the volume of the solid bounded by the coordinate planes and the plane $2x + y + z = 2$.

Solution. We have $z = 2 - 2x - y$ and this expression must be integrated over the region in the xy-plane bounded by the line $2x + y = 2$ (obtained when $z = 0$) and the x- and y-axes. See Figure 12. We therefore have

$$V = \int_0^1 \int_0^{2-2x} (2 - 2x - y)\, dy\, dx = \int_0^1 \left\{\left(2y - 2xy - \frac{y^2}{2}\right)\bigg|_0^{2-2x}\right\} dx$$
$$= \int_0^1 \left\{2(2 - 2x) - 2x(2 - 2x) - \frac{(2 - 2x)^2}{2}\right\} dx$$
$$= \int_0^1 (2x^2 - 4x + 2)\, dx = \left(\frac{2x^3}{3} - 2x^2 + 2x\right)\bigg|_0^1 = \frac{2}{3}. \blacksquare$$

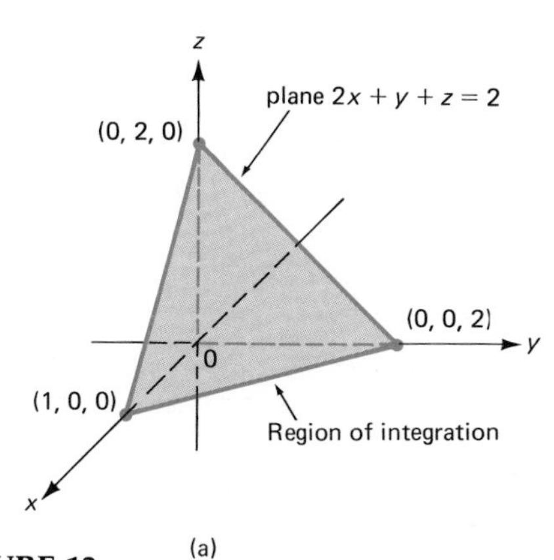

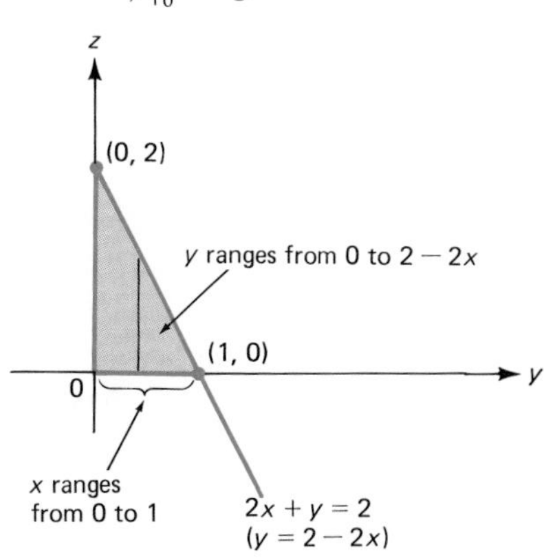

FIGURE 12

EXAMPLE 10 Find the volume of the solid bounded by the circular paraboloid $x = y^2 + z^2$ and the plane $x = 1$.

Solution. The solid is sketched in Figure 13. There are many ways to obtain its volume. Perhaps the best way to do so is to get another perspective on the picture. In Figure 14 we redraw Figure 13 with x as the vertical axis. The volume of the indicated element is $(1 - x)\,\Delta y\,\Delta z$, where $x = y^2 + z^2$. We can take advantage of symmetry to write

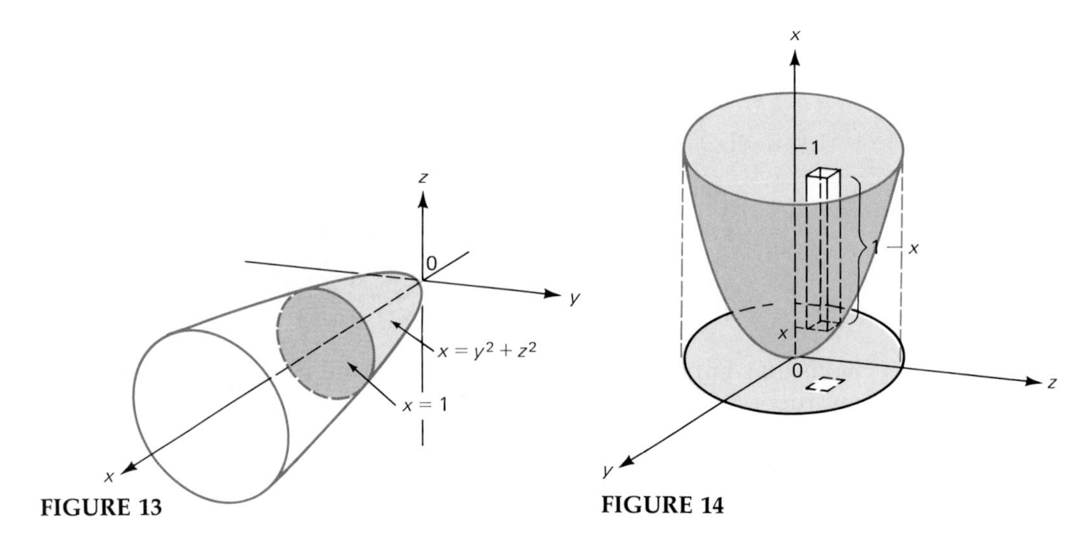

FIGURE 13

FIGURE 14

$$
\begin{aligned}
V &= 4\int_0^1 \int_0^{\sqrt{1-z^2}} (1 - x)\,dy\,dz = 4\int_0^1 \int_0^{\sqrt{1-z^2}} (1 - y^2 - z^2)\,dy\,dz \\
&= 4\int_0^1 \left(y - \frac{y^3}{3} - yz^2\right)\Bigg|_{y=0}^{y=\sqrt{1-z^2}} dz \\
&= 4\int_0^1 \left\{(1 - z^2)^{1/2} - \frac{(1 - z^2)^{3/2}}{3} - z^2(1 - z^2)^{1/2}\right\} dz.
\end{aligned}
$$

We can integrate this by making the substitution $z = \sin\theta$. Then

$$
\begin{aligned}
V &= 4\int_0^{\pi/2} \left(\cos\theta - \frac{\cos^3\theta}{3} - \sin^2\theta\cos\theta\right)\cos\theta\,d\theta \\
&= 4\int_0^{\pi/2} \Bigg[\cos\theta\underbrace{(1 - \sin^2\theta)}_{=\cos^2\theta} - \frac{\cos^3\theta}{3}\Bigg]\cos\theta\,d\theta \\
&= \frac{8}{3}\int_0^{\pi/2} \cos^4\theta\,d\theta = \frac{2}{3}\int_0^{\pi/2} (1 + \cos 2\theta)^2\,d\theta \\
&= \frac{2}{3}\int_0^{\pi/2} (1 + 2\cos 2\theta + \cos^2 2\theta)\,d\theta
\end{aligned}
$$

$$= \frac{2}{3}\int_0^{\pi/2}\left(1 + 2\cos 2\theta + \frac{1}{2} + \frac{1}{2}\cos 4\theta\right) d\theta$$

$$= \frac{2}{3}\cdot\frac{3}{2}\cdot\frac{\pi}{2} = \frac{\pi}{2}. \quad ■$$

REMARK. We will derive a much easier way to obtain this answer in Section 5.4 (see Example 5.4.7).

PROBLEMS 5.2

In Problems 1–23, evaluate the given double integral.

1. $\int_0^1\int_0^2 xy^2\,dx\,dy$
2. $\int_{-1}^3\int_2^4 (x^2 - y^3)\,dy\,dx$
3. $\int_2^5\int_0^4 e^{(x-y)}\,dx\,dy$
4. $\int_0^1\int_{x^2}^x x^3y\,dy\,dx$
5. $\int_2^4\int_{1+y}^{2+3y} (x - y^2)\,dx\,dy$
6. $\int_{\pi/4}^{\pi/3}\int_{\sin x}^{\cos x} (x + 2y)\,dy\,dx$
7. $\int_0^3\int_{-\sqrt{9-y^2}}^{\sqrt{9-y^2}} x^2y\,dx\,dy$
8. $\int_1^2\int_{y^5}^{3y^5} \frac{1}{x}\,dx\,dy$
9. $\iint_\Omega (x^2 + y^2)\,dA$, where $\Omega = \{(x, y)\colon 1 \le x \le 2$ and $-1 \le y \le 1\}$.
10. $\iint_\Omega 2xy\,dA$, where $\Omega = \{(x, y)\colon 0 \le x \le 4$ and $1 \le y \le 3\}$.
11. $\iint_\Omega (x - y)^2\,dA$, where $\Omega = \{(x, y)\colon -2 \le x \le 2$ and $0 \le y \le 1\}$.
12. $\iint_\Omega \sin(2x + 3y)\,dA$, where $\Omega = \{(x, y)\colon 0 \le x \le \pi/6$ and $0 \le y \le \pi/18\}$.
13. $\iint_\Omega xe^{(x^2+y)}\,dA$, where Ω is the region of Problem 10.
14. $\iint_\Omega (x - y^2)\,dA$, where Ω is the region in the first quadrant bounded by the x-axis, the y-axis, and the unit circle.
15. $\iint_\Omega (x^2 + y)\,dA$, where Ω is the region of Problem 14.
16. $\iint_\Omega (x^3 - y^3)\,dA$, where Ω is the region of Problem 14.
17. $\iint_\Omega (x + 2y)\,dA$, where Ω is the triangular region bounded by the lines $y = x$, $y = 1 - x$, and the y-axis.
18. $\iint_\Omega e^{x+2y}\,dA$, where Ω is the region of Problem 17.
19. $\iint_\Omega (x^2 + y)\,dA$, where Ω is the region in the first quadrant between the parabolas $y = x^2$ and $y = 1 - x^2$.
20. $\iint_\Omega (1/\sqrt{y})\,dA$, where Ω is the region of Problem 19.
21. $\iint_\Omega (y/\sqrt{x^2 + y^2})\,dA$, where $\Omega = \{(x, y)\colon 1 \le x \le y$ and $1 \le y \le 2\}$.
22. $\iint_\Omega [e^{-y}/(1 + x^2)]\,dA$, where Ω is the first quadrant.
23. $\iint_\Omega (x + y)e^{-(x+y)}\,dA$, where Ω is the first quadrant.

In Problems 24–33, (a) sketch the region over which the integral is taken. Then (b) change the order of integration, and (c) evaluate the given integral.

24. $\int_0^2\int_{-1}^3 dx\,dy$
25. $\int_0^4\int_{-5}^8 (x + y)\,dy\,dx$
26. $\int_2^4\int_1^y \frac{y^3}{x^3}\,dx\,dy$
27. $\int_0^1\int_0^x dy\,dx$
28. $\int_0^1\int_x^1 dy\,dx$
29. $\int_0^{\pi/2}\int_0^{\cos y} y\,dx\,dy$
30. $\int_0^2\int_0^{\sqrt{4-y^2}} (4 - x^2)^{3/2}\,dx\,dy$
31. $\int_0^1\int_{\sqrt{x}}^{\sqrt[3]{x}} (1 + y^6)\,dy\,dx$
32. $\int_0^1\int_{\sqrt{y}}^1 \sqrt{3 - x^3}\,dx\,dy$
33. $\int_0^\infty\int_x^\infty \frac{1}{(1 + y^2)^{7/5}}\,dy\,dx$
34. Show that if both integrals exist, then
$$\int_0^\infty\int_0^x f(x, y)\,dy\,dx = \int_0^\infty\int_y^\infty f(x, y)\,dx\,dy.$$
[*Hint:* Draw a picture.]

In Problems 35–44, find the volume of the given solid.

35. The solid bounded by the plane $x + y + z = 3$ and the three coordinate planes.
*36. The solid bounded by the planes $x = 0$, $z = 0$, $x + 2y + z = 6$, and $x - 2y + z = 6$.

37. The solid bounded by the cylinders $x^2 + y^2 = 4$ and $y^2 + z^2 = 4$.

38. The solid bounded by the cylinder $x^2 + z^2 = 1$ and the planes $y = 0$ and $y = 2$.

39. The ellipsoid $x^2 + 4y^2 + 9z^2 = 36$.

40. The solid bounded above by the sphere $x^2 + y^2 + z^2 = 9$ and below by the plane $z = \sqrt{5}$.

41. The solid bounded by the planes $y = 0$, $y = x$, and the cylinder $x + z^2 = 2$.

***42.** The solid bounded by the parabolic cylinder $x = z^2$ and the planes $y = 1$, $y = 5$, $z = 1$, and $x = 0$.

***43.** The solid bounded by the paraboloid $y = x^2 + z^2$ and the plane $x + y = 3$.

44. The solid bounded by the surface $z = e^{-(x+y)}$ and the three coordinate planes.

***45.** Use a double integral to find the area of each of the regions bounded by the x-axis and the curves $y = x^3 + 1$ and $y = 3 - x^2$.

46. Use a double integral to find the area in the first quadrant bounded by the curves $y = x^{1/m}$ and $y = x^{1/n}$, where m and n are positive and $n > m$.

47. Let $f(x, y) = g(x)h(y)$, where g and h are continuous. Let Ω be the rectangle $\{(x, y): a \le x \le b$ and $c \le y \le d\}$. Show that

$$\iint_\Omega f(x, y)\, dA = \left\{\int_a^b g(x)\, dx\right\} \cdot \left\{\int_c^d h(y)\, dy\right\}.$$

48. Sketch the solid whose volume is given by

$$V = \int_1^3 \int_0^2 (x + 3y)\, dx\, dy.$$

***49.** Sketch the solid whose volume is given by

$$V = \int_0^1 \int_{x^2}^{\sqrt{x}} \sqrt{x^2 + y^2}\, dy\, dx.$$

5.3 DENSITY, MASS, AND CENTER OF MASS (OPTIONAL)

Let $\rho(x, y)$ denote the density of a plane object (like a thin lamina, for example). Suppose that the object occupies a region Ω in the xy-plane. Then the mass of a small rectangle of sides Δx and Δy centered at the point (x, y) is approximated by

$$\rho(x, y)\, \Delta x\, \Delta y = \rho(x, y)\, \Delta A \tag{1}$$

and the total mass of the object is

$$\mu = \iint_\Omega \rho(x, y)\, dA. \tag{2}$$

Compare this formula with the formula for the mass of an object lying along the x-axis with density $\rho(x)$:

$$\mu(x) = \int \rho(x) dx.$$

In one-variable calculus we saw how to calculate the first moment and center of mass of an object around the x- and y-axes.[†] For example, we may define

$$M_y = \int_a^b x\rho(x)\, dx \tag{3}$$

[†]See *Calculus* or *Calculus of One Variable*, Sections 9.4 and 9.5. See, in particular, equation (9.4.7) on p. 552.

to be the first moment about the y-axis when we had a system of masses distributed along the x-axis. Similarly, we calculated the x-coordinate of the center of mass of the object to be

$$\bar{x} = \frac{\int_b^a x\rho(x)\,dx}{\int_a^b \rho(x)\,dx} = \frac{\text{first moment about } y\text{-axis}}{\text{mass}} = \frac{M_y}{\mu}. \tag{4}$$

When we calculated the centroid of a plane region, we found that it was necessary to assume that the region had a constant area density ρ. However, by using double integrals, we can get away from this restriction. Consider the plane region whose mass is given by (2). Then we define

$$M_y = \text{first moment around } y\text{-axis} = \iint_\Omega x\rho(x, y)\,dA. \tag{5}$$

Look at Figure 1. The first moment about the y-axis of a small rectangle centered at (x, y) is given by

$$x_i\rho(x_i^*, y_j^*)\,\Delta x\,\Delta y, \tag{6}$$

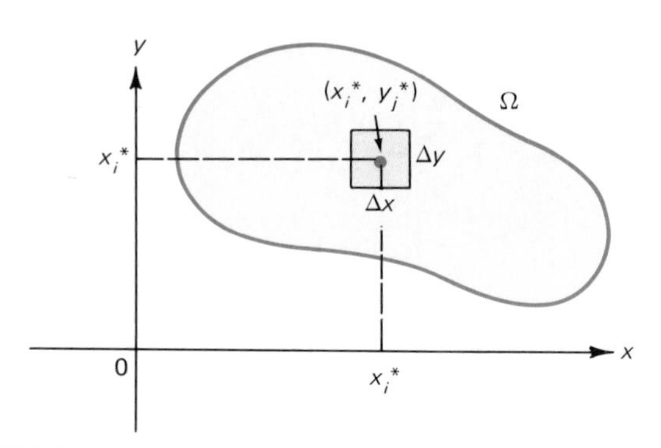

FIGURE 1

and if we add up these moments for all such "subrectangles" and take a limit, we arrive at equation (5). Finally, we define the **center of mass** of the plane region to be the point $(\bar{x}, \bar{y})$, where

$$\bar{x} = \frac{M_y}{\mu} = \frac{\iint_\Omega x\rho(x, y)\,dA}{\iint_\Omega \rho(x, y)\,dA} \tag{7}$$

and

$$\bar{y} = \frac{M_x}{\mu} = \frac{\iint_\Omega y\rho(x, y)\,dA}{\iint_\Omega \rho(x, y)\,dA}. \tag{8}$$

If ρ is constant, the center of mass of a region is called the **centroid** of that region.

EXAMPLE 1 A plane lamina has the shape of the triangle bounded by the lines $y = x$, $y = 2 - x$, and the x-axis. Its density function is given by $\rho(x, y) = 1 + 2x + y$. Distance is measured in meters, and mass is measured in kilograms. Find the mass and center of mass of the lamina.

Solution. The region is sketched in Figure 2. The mass is given by

$$\mu = \iint_{\Omega} \rho(x, y)\, dA = \int_0^1 \int_y^{2-y} (1 + 2x + y)\, dx\, dy$$

$$= \int_0^1 \left\{ (x + x^2 + xy) \Big|_y^{2-y} \right\} dy = \int_0^1 (6 - 4y - 2y^2)\, dy$$

$$= \left(6y - 2y^2 - \frac{2y^3}{3}\right)\Big|_0^1 = \frac{10}{3}\ \text{kg}.$$

Then

$$M_y = \int_0^1 \int_y^{2-y} x(1 + 2x + y)\, dx\, dy$$

$$= \int_0^1 \int_y^{2-y} (x + 2x^2 + xy)\, dx\, dy = \int_0^1 \left\{ \left(\frac{x^2}{2} + \frac{2x^3}{3} + \frac{x^2 y}{2}\right) \Big|_y^{2-y} \right\} dy$$

$$= \int_0^1 \left(\frac{22}{3} - 8y + 2y^2 - \frac{4y^3}{3}\right) dy$$

$$= \left(\frac{22}{3} y - 4y^2 + \frac{2y^3}{3} - \frac{y^4}{3}\right)\Big|_0^1 = \frac{11}{3}\ \text{kg} \cdot \text{m}.$$

$$M_x = \int_0^1 \int_y^{2-y} y(1 + 2x + y)\, dx\, dy = \int_0^1 \left\{ (xy + x^2 y + xy^2) \Big|_y^{2-y} \right\} dy$$

$$= \int_0^1 (6y - 4y^2 - 2y^3)\, dy = \left(3y^2 - \frac{4y^3}{3} - \frac{y^4}{2}\right)\Big|_0^1 = \frac{7}{6}\ \text{kg} \cdot \text{m}.$$

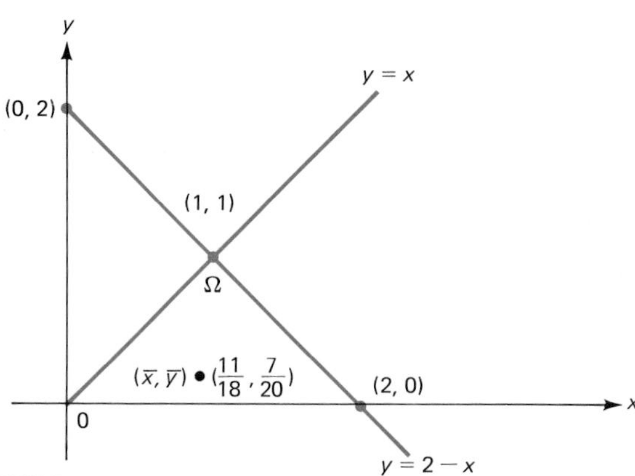

FIGURE 2

Thus

$$\bar{x} = \frac{M_y}{\mu} = \frac{11/3}{10/3} = \frac{11}{10}\text{ m} \quad \text{and} \quad \bar{y} = \frac{M_x}{\mu} = \frac{7/6}{10/3} = \frac{7}{20}\text{ m.} \quad \blacksquare$$

The following result shows how centroids can be used to compute volumes of revolution.

Theorem 1 FIRST THEOREM OF PAPPUS[†] Suppose that the plane region Ω is revolved about a line L in the xy-plane that does not intersect it. Then the volume generated is equal to the product of the area of Ω and the length of the circumference of the circle traced by the centroid of Ω.

Proof. We construct a coordinate system, placing the y-axis so that it coincides with the line L and Ω is in the first quadrant. The situation is then as depicted in Figure 3. We form a regular partition of the region Ω and choose the point (x_i^*, y_j^*) to be a point in the subrectangle Ω_{ij}. If Ω_{ij} is revolved about L (the y-axis), it forms a ring, as depicted in Figure 4. The volume of the ring is, approximately,

$$\begin{aligned} V_{ij} &\approx (\text{circumference of circle with radius } x_i^*) \times (\text{thickness of ring}) \\ &\quad \times (\text{height of ring}) \\ &= 2\pi x_i^* \, \Delta x \, \Delta y = 2\pi x_i^* \, \Delta A, \end{aligned}$$

and using a familiar limiting argument, we have

$$V = \lim_{\substack{\Delta x \to 0 \\ \Delta y \to 0}} \sum_{i=1}^{n} \sum_{j=1}^{m} V_{ij} = \iint_{\Omega} 2\pi x \, dA = 2\pi \iint_{\Omega} x \, dA. \tag{9}$$

Now we can think of Ω as a thin lamina with constant density $\rho = 1$. Then from (7) (with $\rho = 1$),

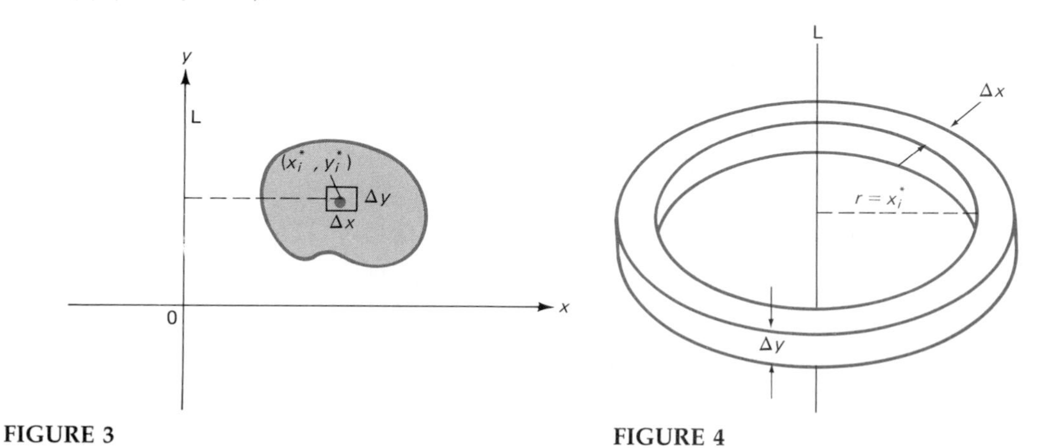

FIGURE 3

FIGURE 4

[†]See the accompanying biographical sketch.

Born c. 290 A.D.

Pappus of Alexandria

Born during the reign of the Roman emperor Diocletian (A.D. 284–305), Pappus of Alexandria lived approximately 600 years after the time of Euclid and Archimedes and devoted much of his life to revitalizing interest in the traditional study of Greek geometry.

Pappus' greatest work was his *Mathematical Collection,* written between A.D. 320 and 340. This work is significant for three reasons. First, in it are collected works of more than 30 different mathematicians of antiquity. We owe much of our knowledge of Greek geometry to the *Collection.* Second, it provides alternative proofs of the results of the greatest of the Greeks, including Euclid and Archimedes. Third, the *Collection* contains a variety of discoveries not found in any earlier work.

The *Collection* comprises eight books, each one containing a variety of interesting and sometimes amusing results. In Book V, for example, Pappus showed that if two regular polygons have equal perimeters, then the one with the greater number of sides has the larger area. Pappus used this result to suggest the great wisdom of bees, as bees construct their hives using hexagonal (6-sided) cells, rather than square or triangular ones.

Book VII contains the theorem proved in this section. It is also the book that is the most important in the history of mathematics in that it contains the *Treasury of Analysis.* The *Treasury* is a collection of mathematical facts that, together with Euclid's *Elements,* claimed to contain the material the practicing mathematician in the fourth century needed to know. Although mathematicians wrote in Greek for another thousand or so years, no follower wrote a work of equal significance.

$$\bar{x} = \frac{\iint_\Omega x\,dA}{\iint_\Omega dA} \overset{\text{From (9)}}{=} \frac{V/2\pi}{\iint_\Omega dA} \overset{\text{Formula (5.1.15)}}{=} \frac{V/2\pi}{\text{area of }\Omega}$$

or

$$\begin{aligned} V &= (2\pi\bar{x})(\text{area of }\Omega) \\ &= \begin{pmatrix}\text{length of the circumference of the circle} \\ \text{traced by the centroid of }\Omega\end{pmatrix} \times (\text{area of }\Omega). \quad \blacksquare \end{aligned}$$

EXAMPLE 2 Use the first theorem of Pappus to calculate the volume of the torus generated by rotating the circle $(x - a)^2 + y^2 = r^2$ $(r < a)$ about the y-axis.

Solution. The circle and the torus are sketched in Figure 5. The area of the circle

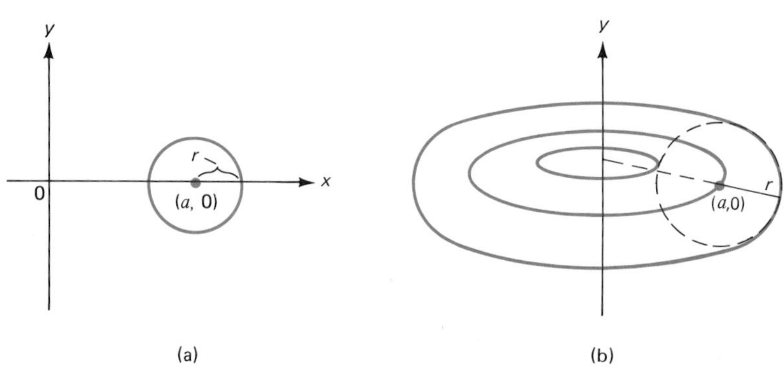

FIGURE 5

is πr^2. The radius of the circle traced by the centroid $(a, 0)$ is a, and the circumference is $2\pi a$. Thus

$$V = (2\pi a)\pi r^2 = 2\pi^2 ar^2. \quad \blacksquare$$

PROBLEMS 5.3

In Problems 1–12, find the mass and center of mass of an object that lies in the given region with the given area density function.

1. $\Omega = \{(x, y): 1 \le x \le 2, -1 \le y \le 1\}$; $\rho(x, y) = x^2 + y^2$.
2. $\Omega = \{(x, y): 0 \le x \le 4, 1 \le y \le 3\}$; $\rho(x, y) = 2xy$.
3. $\Omega = \{(x, y): 0 \le x \le \pi/6, 0 \le y \le \pi/18\}$; $\rho(x, y) = \sin(2x + 3y)$.
4. $\Omega = \{(x, y): -2 \le x \le 2, 0 \le y \le 1\}$; $\rho(x, y) = (x - y)^2$.
5. Ω is the region of Problem 2; $\rho(x, y) = xe^{x-y}$.
6. Ω is the quarter of the unit circle lying in the first quadrant; $\rho(x, y) = x + y^2$.
7. Ω is the region of Problem 6; $\rho(x, y) = x^2 + y$.
8. Ω is the region of Problem 6; $\rho(x, y) = x^3 + y^3$.
9. Ω is the triangular region bounded by the lines $y = x$, $y = 1 - x$, and the x-axis; $\rho(x, y) = x + 2y$.
10. Ω is the region of Problem 9; $\rho(x, y) = e^{x+2y}$.
11. Ω is the first quadrant; $\rho(x, y) = e^{-y}/(1 + x)^3$.
12. Ω is the first quadrant; $\rho(x, y) = (x + y)e^{-(x+y)}$.
13. Use the first theorem of Pappus to calculate the volume of the torus generated by rotating the unit circle about the line $y = 4 - x$.
14. Use the first theorem of Pappus to calculate the volume of the solid generated by rotating the triangle with vertices $(-1, 2)$, $(1, 2)$, and $(0, 4)$ about the x-axis.
15. Use the first theorem of Pappus to calculate the volume of the "elliptical torus" generated by rotating the ellipse $(x^2/a^2) + (y^2/b^2) = 1$ about the line $y = 3a$. Assume that $3a > b > 0$.

5.4 DOUBLE INTEGRALS IN POLAR COORDINATES

In this section we will see how to evaluate double integrals of functions in the form $z = f(r, \theta)$, where r and θ denote the polar coordinates of a point in the plane.

Let $z = f(r, \theta)$ and let Ω denote the "polar rectangle"

$$\theta_1 \le \theta \le \theta_2, \qquad r_1 \le r \le r_2. \tag{1}$$

This region is sketched in Figure 1. We will calculate the volume of the solid between the surface $z = f(r, \theta)$ and the region Ω. We partition Ω by small "polar rectangles"

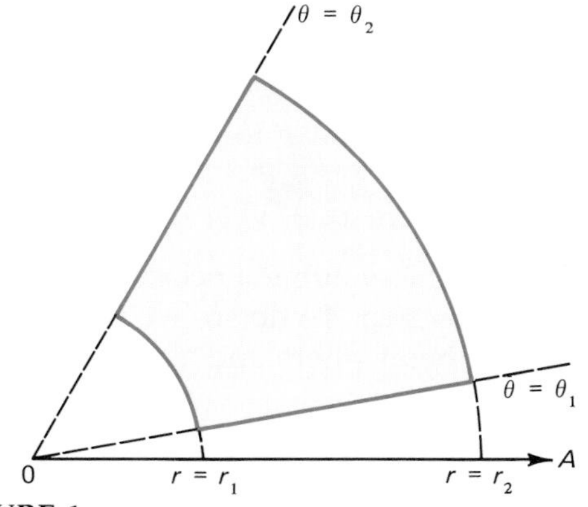

FIGURE 1

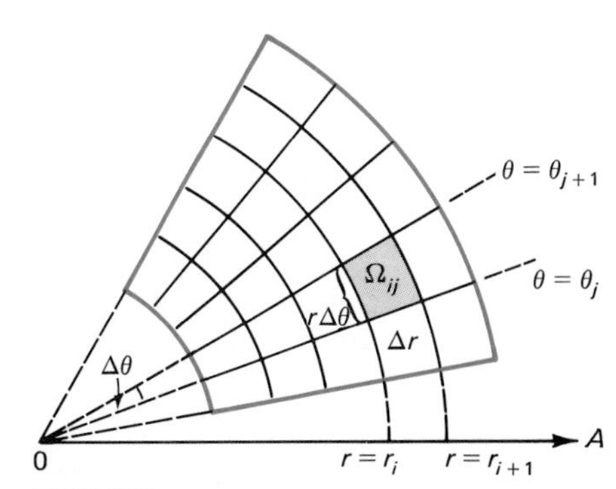

FIGURE 2

and calculate the volume over such a region (see Figure 2). The volume of the part of the solid over the region Ω_{ij} is given, approximately, by

$$f(r_i, \theta_j)A_{ij}, \tag{2}$$

where A_{ij} is the area of Ω_{ij}. Recall that if $r = f(\theta)$, then the area bounded by the lines $\theta = \alpha$, $\theta = \beta$, and the curve $r = f(\theta)$ is given by[†]

$$A = \int_\alpha^\beta \frac{1}{2}[f(\theta)]^2\, d\theta. \tag{3}$$

Thus

$$\begin{aligned}
A_{ij} &= \frac{1}{2}\int_{\theta_j}^{\theta_{j+1}} (r_{i+1}^2 - r_i^2)\, d\theta \\
&= \frac{1}{2}(r_{i+1}^2 - r_i^2)\, \theta \Big|_{\theta_j}^{\theta_{j+1}} = \frac{1}{2}(r_{i+1}^2 - r_i^2)(\theta_{j+1} - \theta_j) \\
&= \frac{1}{2}(r_{i+1} + r_i)(r_{i+1} - r_i)(\theta_{j+1} - \theta_j) \\
&= \frac{1}{2}(r_{i+1} + r_i)\, \Delta r\, \Delta\theta.
\end{aligned}$$

But if Δr is small, then $r_{i+1} \approx r_i$, and we have

$$A_{ij} \approx \tfrac{1}{2}(2r_i)\, \Delta r\, \Delta\theta = r_i\, \Delta r\, \Delta\theta.$$

Then

$$V_{ij} \approx f(r_i, \theta_j)A_{ij} \approx f(r_i, \theta_j)r_i\, \Delta r\, \Delta\theta,$$

so that, adding up the individual volumes and taking a limit, we obtain

$$V = \iint_\Omega f(r, \theta)r\, dr\, d\theta. \tag{4}$$

NOTE. Do not forget the extra r in the above formula.

EXAMPLE 1 Find the volume enclosed by the sphere $x^2 + y^2 + z^2 = a^2$.

Solution. We will calculate the volume enclosed by the hemisphere $z = \sqrt{a^2 - x^2 - y^2}$ and then multiply by two. To do so, we first note that this volume is the volume of the solid under the hemisphere and above the disk $x^2 + y^2 \le a^2$. We

---

[†]See *Calculus* or *Calculus of One Variable*, p. 636.

use polar coordinates, since in polar coordinates $x^2 + y^2 = (r\cos\theta)^2 + (r\sin\theta^2) = r^2$. On the disk, $0 \le r \le a$ and $0 \le \theta \le 2\pi$ (see Figure 3). Then $z = \sqrt{a^2 - (x^2 + y^2)} = \sqrt{a^2 - r^2}$, so by (4)

$$V = \int_0^{2\pi}\int_0^a \sqrt{a^2 - r^2}\, r\, dr\, d\theta = \int_0^{2\pi}\left\{-\frac{1}{3}(a^2 - r^2)^{3/2}\Big|_0^a\right\} d\theta$$

$$= \int_0^{2\pi} \frac{1}{3}a^3\, d\theta = \frac{2\pi a^3}{3}.$$

Thus the volume of the sphere is $2(2\pi a^3/3) = (4/3)\pi a^3$. ■

EXAMPLE 2 Find the volume of the solid bounded above by the surface $z = 3 + r$ and below by the region enclosed by the cardioid $r = 1 + \sin\theta$.

Solution. The cardioid is sketched in Figure 4 and can be described by

$$\Omega = \{(r, \theta): 0 \le \theta \le 2\pi \text{ and } 0 \le r \le 1 + \sin\theta\}.$$

Then, from (4),

$$V = \int_0^{2\pi}\int_0^{1+\sin\theta} (3 + r)r\, dr\, d\theta = \int_0^{2\pi}\left\{\left(\frac{3r^2}{2} + \frac{r^3}{3}\right)\Big|_0^{1+\sin\theta}\right\} d\theta$$

$$= \int_0^{2\pi}\left\{\frac{3}{2}(1 + 2\sin\theta + \sin^2\theta) + \frac{1}{3}(1 + 3\sin\theta + 3\sin^2\theta + \sin^3\theta)\right\} d\theta$$

$$= \int_0^{2\pi}\left(\frac{11}{6} + 4\sin\theta + \frac{5}{2}\sin^2\theta + \frac{\sin^3\theta}{3}\right) d\theta$$

$$= \int_0^{2\pi}\left\{\frac{11}{6} + 4\sin\theta + \frac{5}{4}(1 - \cos 2\theta) + \frac{\sin\theta}{3}(1 - \cos^2\theta)\right\} d\theta$$

$$= \left(\frac{11}{6}\theta - 4\cos\theta + \frac{5}{4}\theta - \frac{5}{8}\sin 2\theta - \frac{\cos\theta}{3} + \frac{\cos^3\theta}{9}\right)\Big|_0^{2\pi} = \frac{37}{6}\pi. \ ■$$

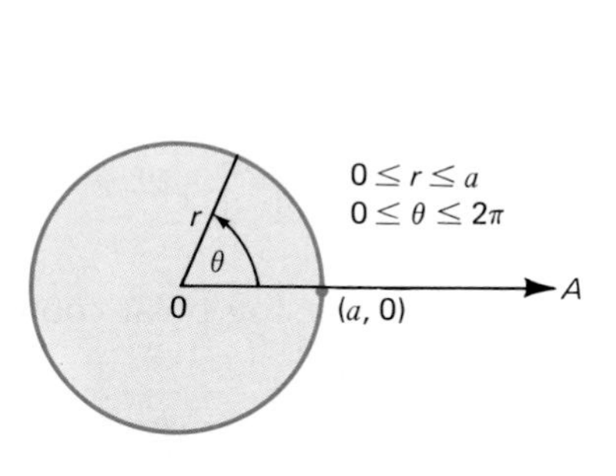

FIGURE 3

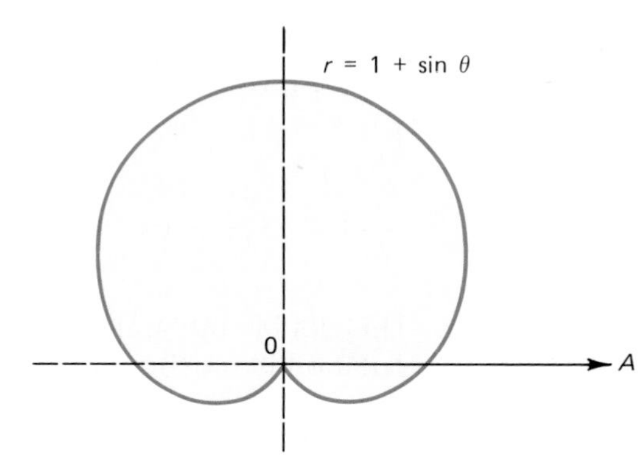

FIGURE 4

REMARK. We can also use this technique to calculate areas. If $f(r, \theta) = 1$, then

$$\iint_{\Omega} r\,dr\,d\theta = \text{area of } \Omega. \tag{5}$$

(See Remark 3 following Theorem 5.1.2.)

EXAMPLE 3 Calculate the area enclosed by the cardioid $r = 1 + \sin\theta$.

Solution.

$$\begin{aligned} A &= \int_0^{2\pi}\int_0^{1+\sin\theta} r\,dr\,d\theta = \int_0^{2\pi}\left\{\frac{r^2}{2}\Big|_0^{1+\sin\theta}\right\}d\theta \\ &= \frac{1}{2}\int_0^{2\pi}(1 + 2\sin\theta + \sin^2\theta)\,d\theta = \frac{1}{2}\int_0^{2\pi}\left(1 + 2\sin\theta + \frac{1-\cos 2\theta}{2}\right)d\theta \\ &= \frac{1}{2}\left(\frac{3\theta}{2} - 2\cos\theta - \frac{\sin 2\theta}{4}\right)\Big|_0^{2\pi} = \frac{3\pi}{2}. \quad \blacksquare \end{aligned}$$

As we will see, it is often very useful to write a double integral in terms of polar coordinates. Let $z = f(x, y)$ be a function defined over a region Ω. Then using polar coordinates, we can write

$$z = f(r\cos\theta, r\sin\theta), \tag{6}$$

and we can also describe Ω in terms of polar coordinates. The volume of the solid under f and over Ω is the same whether we use rectangular or polar coordinates. Thus writing the volume in both rectangular and polar coordinates, we obtain the useful **change-of-variables formula**

$$\iint_{\Omega} f(x, y)\,dA = \iint_{\Omega} f(r\cos\theta, r\sin\theta)r\,dr\,d\theta. \tag{7}$$

EXAMPLE 4 The density at any point on a semicircular plane lamina is proportional to the square of the distance from the point to the center of the circle. Find the mass of the lamina.

Solution. We have $\rho(x, y) = \alpha(x^2 + y^2) = \alpha r^2$ and $\Omega = \{(r, \theta): 0 \le r \le a$ and $0 \le \theta \le \pi\}$, where a is the radius of the circle. Then

$$\mu = \int_0^{\pi}\int_0^{a}(\alpha r^2)r\,dr\,d\theta = \int_0^{\pi}\left\{\frac{\alpha r^4}{4}\Big|_0^a\right\}d\theta = \frac{\alpha a^4}{4}\int_0^{\pi}d\theta = \frac{\alpha\pi a^4}{4}.$$

This double integral can be computed without using polar coordinates, but the computation is much more tedious. Try it! $\blacksquare$

EXAMPLE 5 Find the volume of the solid bounded by the xy-plane, the cylinder $x^2 + y^2 = 4$, and the paraboloid $z = 2(x^2 + y^2)$.

Solution. The volume requested is the volume under the surface $z = 2(x^2 + y^2) = 2r^2$ and above the circle $x^2 + y^2 = 4$. Thus

$$V = \int_0^{2\pi}\int_0^2 2r^2 \cdot r\,dr\,d\theta = \int_0^{2\pi}\left\{\frac{r^4}{2}\bigg|_0^2\right\}d\theta = 16\pi. \blacksquare$$

EXAMPLE 6 In probability theory one of the most important integrals that is encountered is the integral

$$\int_{-\infty}^{\infty} e^{-x^2}\,dx.$$

We now show how a combination of double integrals and polar coordinates can be used to evaluate it. Let

$$I = \int_0^{\infty} e^{-x^2}\,dx.$$

Then by symmetry $\int_{-\infty}^{\infty} e^{-x^2}\,dx = 2I$. Thus we need only to evaluate I. But since any dummy variable can be used in a definite integral, we also have

$$I = \int_0^{\infty} e^{-y^2}\,dy.$$

Thus

$$I^2 = \left(\int_0^{\infty} e^{-x^2}\,dx\right)\left(\int_0^{\infty} e^{-y^2}\,dy\right),$$

and from the result of Problem 5.2.47,†

$$I^2 = \int_0^{\infty}\int_0^{\infty} e^{-x^2}e^{-y^2}\,dx\,dy = \int_0^{\infty}\int_0^{\infty} e^{-(x^2+y^2)}\,dx\,dy = \iint_{\Omega} e^{-(x^2+y^2)}\,dA,$$

where Ω denotes the first quadrant. In polar coordinates the first quadrant can be written as

$$\Omega = \left\{(r, \theta): 0 \le r < \infty \text{ and } 0 \le \theta \le \frac{\pi}{2}\right\}.$$

Thus since $x^2 + y^2 = r^2$, we obtain

---

†Problem 5.2.47 really does not apply to improper integrals such as this one. The answer we obtain is correct, but additional theory is needed to justify this next step. The needed result is best left to a course in advanced calculus.

$$I^2 = \int_0^{\pi/2} \int_0^{\infty} e^{-r^2} r\, dr\, d\theta = \int_0^{\pi/2} \left(\lim_{N\to\infty} \int_0^N e^{-r^2} r\, dr \right) d\theta$$

$$= \int_0^{\pi/2} \left(\lim_{N\to\infty} -\frac{1}{2} e^{-r^2} \Big|_0^N \right) d\theta = \frac{1}{2} \int_0^{\pi/2} d\theta = \frac{\pi}{4}.$$

Hence $I^2 = \pi/4$, so $I = \sqrt{\pi}/2$, and

$$\int_{-\infty}^{\infty} e^{-x^2}\, dx = 2I = \sqrt{\pi}.$$

By making the substitution $u = x/\sqrt{2}$, we can show that

$$\int_{-\infty}^{\infty} e^{-x^2/2}\, dx = \sqrt{2\pi}.$$

The function $\rho(x) = (1/\sqrt{2\pi})e^{-x^2/2}$ is called the **density function for the unit normal distribution.** We have just shown that $\int_{-\infty}^{\infty} \rho(x)\, dx = 1.$ ■

EXAMPLE 7 Find the volume of the solid bounded by the circular paraboloid $x = y^2 + z^2$ and the plane $x = 1$.

Solution. This problem is the same as the one in Example 5.2.10. In that example we found that

$$V = \iint_{\Omega} (1 - y^2 - z^2)\, dy\, dz,$$

where Ω is the circle $y^2 + z^2 = 1$. But this integral is easily evaluated by using polar coordinates (with y and z in place of x and y in the polar coordinate formulas). We have

$$V = \int_0^{2\pi} \int_0^1 (1 - r^2) r\, dr\, d\theta$$

$$= \int_0^{2\pi} \left\{ \left(\frac{r^2}{2} - \frac{r^4}{4} \right) \Big|_0^1 \right\} d\theta = \int_0^{2\pi} \frac{1}{4}\, d\theta = \frac{\pi}{2}. \quad ■$$

PROBLEMS 5.4

In Problems 1–5, calculate the volume under the given surface that lies over the given region Ω.

1. $z = r$; Ω is the circle of radius a.
2. $z = r^n$; n is a positive integer; Ω is the circle of radius a.
3. $z = 3 - r$; Ω is the circle $r = 2 \cos \theta$.
4. $z = r^2$; Ω is the cardioid $r = 4(1 - \cos \theta)$.
5. $z = r^3$; Ω is the region enclosed by the spiral of Archimedes $r = a\theta$ and the polar axis for θ between 0 and 2π.

In Problems 6–15, calculate the area of the region enclosed by the given curve or curves.

6. $r = 1 - \cos \theta$
7. $r = 4(1 + \cos \theta)$
8. $r = 1 + 2 \cos \theta$ (outer loop)
9. $r = 3 - 2 \sin \theta$
10. $r^2 = \cos 2\theta$
11. $r^2 = 4 \sin 2\theta$
12. $r = a + b \sin \theta$, $a > b > 0$
13. $r = \tan \theta$ and the line $\theta = \pi/4$

14. Outside the circle $r = 6$ and inside the cardioid $r = 4(1 + \sin\theta)$.
15. Inside the cardioid $r = 2(1 + \cos\theta)$ but outside the circle $r = 2$.
16. Find the volume of the solid bounded above by the sphere $x^2 + y^2 + z^2 = 4a^2$, below by the xy-plane, and on the sides by the cylinder $x^2 + y^2 = a^2$.
17. Find the area of the region interior to the curve $(x^2 + y^2)^3 = 9y^2$.
18. Find the volume of the solid bounded by the cone $x^2 + y^2 = z^2$ and the cylinder $x^2 + y^2 = 4y$.
19. Find the volume of the solid bounded by the cone $z^2 = x^2 + y^2$ and the paraboloid $2z = x^2 + y^2$.
20. Find the volume of the solid bounded by the cylinder $x^2 + y^2 = 9$ and the hyperboloid $x^2 + y^2 - z^2 = 1$.
21. Find the volume of the solid centered at the origin that is bounded above by the surface $z = e^{-(x^2+y^2)}$ and below by the unit circle.
22. Find the centroid of the region bounded by $r = \cos\theta + 2\sin\theta$.
23. Find the centroid of the region bounded by the limaçon $r = 3 + \sin\theta$.
24. Find the centroid of the region bounded by the limaçon $r = a + b\cos\theta$, $a > b > 0$.
25. Show that $\iint_\Omega 1/(1 + x^2 + y^2)\, dA = \pi \ln 2$, where Ω is the unit disk.

5.5 SURFACE AREA

In one-variable calculus we may derive a formula for the length of a plane curve [given by $y = f(x)$] by showing that the length of a small "piece" of the curve is approximately equal to $\sqrt{1 + [f'(x)]^2}\,\Delta x$. We now define the area of a surface $z = f(x, y)$ that lies over a region Ω in the xy-plane. We define it by analogy with the arc length formula

Definition 1 LATERAL SURFACE AREA Let f be continuous with continuous partial derivatives in the region Ω in the xy-plane. Then the **lateral surface area** σ of the graph of f over Ω is defined by

$$\sigma = \iint_\Omega \sqrt{1 + f_x^2(x, y) + f_y^2(x, y)}\, dA. \tag{1}$$

REMARK 1. The assumption that f is continuously differentiable over Ω ensures that the integral in (1) exists.

REMARK 2. We will show why formula (1) makes intuitive sense at the end of this section.

REMARK 3. We emphasize that formula (1) is a definition. A definition is not something that we have to prove, of course, but it is something that we have to live with. It will be comforting, therefore, to use our definition to evaluate areas where we know what we want the answer to be.

EXAMPLE 1 Calculate the lateral surface area cut from the cylinder $y^2 + z^2 = 9$ by the planes $x = 0$ and $x = 4$.

Solution. The surface is a right circular cylinder with height 4 and radius 3 whose axis lies along the x-axis. Thus

$$S = 2\pi rh = 24\pi.$$

We now solve the problem by using formula (1). The surface area consists of two equal parts, one for $z > 0$ and one for $z < 0$. We calculate the surface area for $z > 0$ and multiply the result by 2. We have, for $z > 0$, $z = f(x, y) = \sqrt{9 - y^2}$, so $f_x = 0$, $f_y = -y/\sqrt{9 - y^2}$, and $f_y^{\,2} = y^2/(9 - y^2)$. Thus since $0 \le x \le 4$ and $-3 \le y \le 3$, we have

$$\begin{aligned} S &= 2\int_0^4 \int_{-3}^3 \sqrt{1 + \frac{y^2}{9 - y^2}}\, dy\, dx = 4\int_0^4 \int_0^3 \sqrt{\frac{9}{9 - y^2}}\, dy\, dx \\ &= 12\int_0^4 \left(\int_0^3 \frac{dy}{\sqrt{9 - y^2}}\right) dx. \end{aligned}$$

Setting $y = 3 \sin \theta$, we have

$$S = 12\int_0^4 \left(\int_0^{\pi/2} \frac{3\cos\theta}{3\cos\theta}\, d\theta\right) dx = 12\int_0^4 \frac{\pi}{2}\, dx = 12(4)\left(\frac{\pi}{2}\right) = 24\pi. \quad \blacksquare$$

We now compute surface area in problems where we don't know the answer in advance.

EXAMPLE 2 Calculate the lateral surface area cut from the cylinder $y^2 + z^2 = 9$ by the planes $x = 0$, $x = 1$, $y = 0$, and $y = 2$.

Solution. The required area is sketched in Figure 1. It consists of two equal parts: one for $z > 0$ and one for $z < 0$. We will calculate the surface area for $z > 0$ and multiply the result by 2. For $z > 0$, we have $z = \sqrt{9 - y^2}$. Ω is the rectangle $\{(x, y)$: $0 \le x \le 1$ and $0 \le y \le 2\}$. Then

$$f_x = 0 \quad \text{and} \quad f_y = -\frac{y}{\sqrt{9 - y^2}}.$$

So

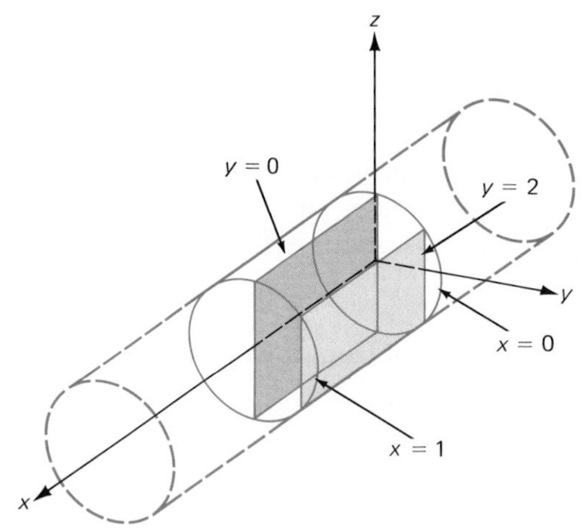

FIGURE 1

$$\sigma = \int_0^2 \int_0^1 \sqrt{1 + 0^2 + \left(\frac{-y}{\sqrt{9-y^2}}\right)^2}\, dx\, dy = \int_0^2 \int_0^1 \sqrt{1 + \frac{y^2}{9-y^2}}\, dx\, dy$$
$$= 3 \int_0^2 \int_0^1 \frac{1}{\sqrt{9-y^2}}\, dx\, dy = 3 \int_0^2 \left(\frac{x}{\sqrt{9-y^2}}\Big|_0^1\right) dy = 3 \int_0^2 \frac{dy}{\sqrt{9-y^2}}$$
$$= 3 \sin^{-1} \frac{y}{3}\Big|_0^2 = 3 \sin^{-1} \frac{2}{3}.$$

Thus the total surface area is $6 \sin^{-1} \frac{2}{3} \approx 4.38$. ■

EXAMPLE 3 Find the lateral surface area of the circular paraboloid $z = x^2 + y^2$ between the xy-plane and the plane $z = 9$.

Solution. The surface area requested is sketched in Figure 2. The region Ω is the disk $x^2 + y^2 \leq 9$. We have

$$f_x = 2x \qquad \text{and} \qquad f_y = 2y,$$

so

$$\sigma = \iint_\Omega \sqrt{1 + 4x^2 + 4y^2}\, dA.$$

Clearly, this problem calls for the use of polar coordinates. We have

$$\sigma = \int_0^{2\pi} \int_0^3 \sqrt{1 + 4r^2}\, r\, dr\, d\theta = \int_0^{2\pi} \left\{\frac{1}{12}(1 + 4r^2)^{3/2}\Big|_0^3\right\} d\theta$$
$$= \frac{\pi}{6}(37^{3/2} - 1) \approx 117.3. \; ■$$

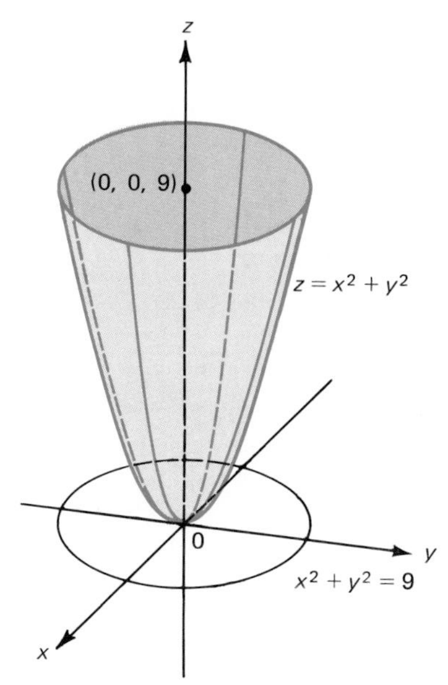

FIGURE 2

EXAMPLE 4 Calculate the area of the part of the surface $z = x^3 + y^4$ that lies over the square $\{(x, y): 0 \le x \le 1, 0 \le y \le 1\}$.

Solution. $f_x = 3x^2$ and $f_y = 4y^3$, so

$$\sigma = \int_0^1 \int_0^1 \sqrt{1 + 9x^4 + 16y^6}\, dx\, dy.$$

However, this is as far as we can go unless we resort to numerical techniques to approximate this double integral. As with ordinary definite integrals, many double integrals cannot be integrated in terms of functions that we know. However, there are a great number of techniques for approximating a double integral numerically. ■

Derivation of the Formula for Surface Area. We begin by calculating the surface area $\Delta\sigma$ over a rectangle ΔR with sides Δx and Δy. The situation is depicted in Figure 3, in which it is assumed that $f(x, y) > 0$ for (x, y) in Ω. We assume that f has continuous partial derivatives over R. If Δx and Δy are small, then the region $PQSR$ in space has, approximately, the shape of a parallelogram. Thus by equation (3.4.6) on p. 108,

$$\Delta\sigma \approx \text{area of parallelogram} = |\overrightarrow{PQ} \times \overrightarrow{PR}|. \tag{2}$$

Now

$$\begin{aligned}\overrightarrow{PQ} &= (x + \Delta x, y, f(x + \Delta x, y)) - (x, y, f(x, y)) \\ &= (\Delta x, 0, f(x + \Delta x, y) - f(x, y)).\end{aligned}$$

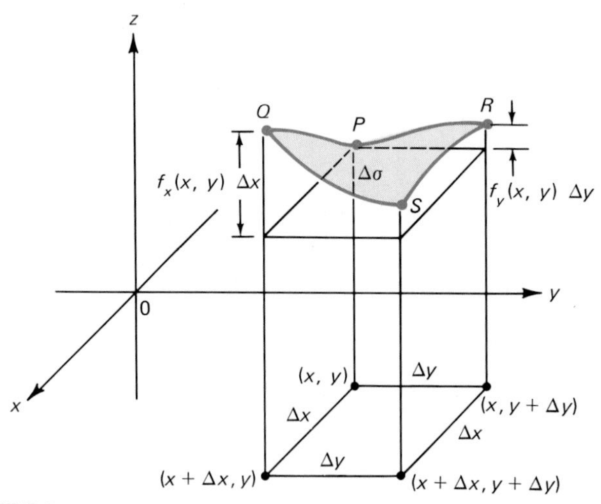

FIGURE 3

But if Δx is small, then

$$\frac{f(x + \Delta x, y) - f(x, y)}{\Delta x} \approx f_x(x, y),$$

so

$$f(x + \Delta x, y) - f(x, y) \approx f_x(x, y)\,\Delta x$$

and

$$\overrightarrow{PQ} \approx (\Delta x, 0, f_x(x, y)\,\Delta x). \tag{3}$$

Similarly,

$$\begin{aligned}\overrightarrow{PR} &= (x, y + \Delta y, f(x, y + \Delta y)) - (x, y, f(x, y)) \\ &= (0, \Delta y, f(x, y + \Delta y) - f(x, y)),\end{aligned}$$

and if Δy is small,

$$\overrightarrow{PR} \approx (0, \Delta y, f_y(x, y)\,\Delta y). \tag{4}$$

Thus from (3) and (4),

$$\begin{aligned}\overrightarrow{PQ} \times \overrightarrow{PQ} &\approx \begin{vmatrix} \mathbf{i} & \mathbf{j} & \mathbf{k} \\ \Delta x & 0 & f_x(x, y)\,\Delta x \\ 0 & \Delta y & f_y(x, y)\,\Delta y \end{vmatrix} \\ &= -f_x(x, y)\,\Delta x\,\Delta y\,\mathbf{i} - f_y(x, y)\,\Delta x\,\Delta y\,\mathbf{j} + \Delta x\,\Delta y\,\mathbf{k} \\ &= (-f_x(x, y)\mathbf{i} - f_y(x, y)\mathbf{j} + \mathbf{k})\,\Delta x\,\Delta y,\end{aligned}$$

so that from (2)

$$\Delta\sigma \approx \sqrt{f_x^2(x, y) + f_y^2(x, y) + 1}\;\overbrace{\Delta x\,\Delta y}^{=\,\Delta A}. \tag{5}$$

Finally, adding up the surface area over rectangles that partition Ω and taking a limit yields

$$\sigma = \iint_{\Omega} \sqrt{1 + f_x^2(x, y) + f_y^2(x, y)}\,dA. \;\blacksquare \tag{1}$$

REMARK. We caution that equation (1) has *not* been proved. However, if f has continuous partial derivatives, it certainly is plausible.

PROBLEMS 5.5

In Problems 1–9, find the area of the part of the surface that lies over the given region.

1. $z = x + 2y$; $\Omega = \{(x, y): 0 \le x \le y, 0 \le y \le 2\}$
2. $z = 4x + 7y$; Ω = region between $y = x^2$ and $y = x^5$
3. $z = ax + by$; Ω = upper half of unit circle
4. $z = y^2$; $\Omega = \{(x, y): 0 \le x \le 2, 0 \le y \le 4\}$
***5.** $z = 3 + x^{2/3}$; $\Omega = \{(x, y): -1 \le x \le 1, 1 \le y \le 2\}$
6. $z = (x^4/4) + (1/8x^2)$; $\Omega = \{(x, y): 1 \le x \le 2, 0 \le y \le 5\}$
7. $z = \frac{1}{3}(y^2 + 2)^{3/2}$; $\Omega = \{(x, y): -4 \le x \le 7, 0 \le y \le 3\}$
8. $z = 2 \ln(1 + y)$; $\Omega = \{(x, y): 0 \le x \le 2, 0 \le y \le 1\}$
***9.** $(z + 1)^2 = 4x^3$; $\Omega = \{(x, y): 0 \le x \le 1, 0 \le y \le 2\}$

***10.** Calculate the lateral surface area of the cylinder $y^{2/3} + z^{2/3} = 1$ for x in the interval $[0, 2]$.
***11.** Find the surface area of the hemisphere $x^2 + y^2 + z^2 = a^2$, $z \ge 0$.
***12.** Find the surface area of the part of the sphere $x^2 + y^2 + z^2 = a^2$ that is also inside the cylinder $x^2 + y^2 = ay$.
***13.** Find the area of the surface in the first octant cut from the cylinder $x^2 + y^2 = 16$ by the plane $y = z$.
14. Find the area of the portion of the sphere $x^2 + y^2 + z^2 = 16z$ lying within the circular paraboloid $z = x^2 + y^2$.
15. Find the area of the surface cut from the hyperbolic paraboloid $4z = x^2 - y^2$ by the cylinder $x^2 + y^2 = 16$.
***16.** Let $z = f(x, y)$ be the equation of a plane (i.e., $z = ax + by + c$). Show that over the region Ω, the area of the plane is given by

$$\sigma = \iint_\Omega \sec \gamma \, dA,$$

where γ is the angle between the normal vector **N** to the plane and the positive z-axis. [*Hint:* Show, using the dot product, that

$$\cos \gamma = \frac{\mathbf{N} \cdot \mathbf{k}}{|\mathbf{N}|} = \frac{1}{\sqrt{1 + a^2 + b^2}}.]$$

In Problems 17–20, find a double integral that represents the area of the given surface over the given region. Do *not* try to evaluate the integral.

17. $z = x^3 + y^3$; Ω is the unit circle.
18. $z = \ln(x + 2y)$; $\Omega = \{(x, y): 0 \le x \le 1, 0 \le y \le 4\}$.
19. $z = \sqrt{1 + x + y}$; Ω is the triangle bounded by $y = x$, $y = 4 - x$, and the y-axis.
20. $z = e^{x-y}$; Ω is the ellipse $4x^2 + 9y^2 = 36$.
***21.** Find a double integral that represents the surface area of the ellipsoid $(x^2/a^2) + (y^2/b^2) + (z^2/c^2) = 1$.

5.6 THE TRIPLE INTEGRAL

In this section we discuss the idea behind the triple integral of a function of three variables $f(x, y, z)$ over a region S in $\mathbb{R}^3$. This is really a simple extension of the double integral. For that reason we will omit a number of technical details.

We start with a parallelepiped π in $\mathbb{R}^3$, which can be written as

$$\pi = \{(x, y, z): a_1 \le x \le a_2, b_1 \le y \le b_2, c_1 \le z \le c_2\} \quad \textbf{(1)}$$

and is sketched in Figure 1. We construct regular partitions of the three intervals $[a_1, a_2]$, $[b_1, b_2]$, and $[c_1, c_2]$:

$$a_1 = x_0 < x_1 < \cdots < x_n = a_2$$
$$b_1 = y_0 < y_1 < \cdots < y_m = b_2$$
$$c_1 = z_0 < z_1 < \cdots < z_p = c_2,$$

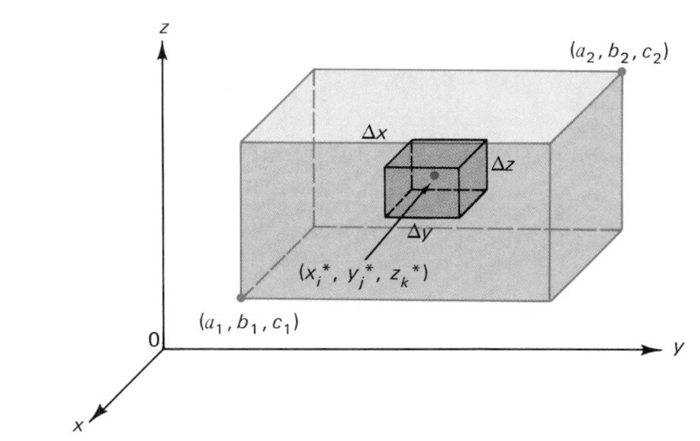

FIGURE 1

to obtain *nmp* "boxes." The volume of a typical box B_{ijk} is given by

$$\Delta V = \Delta x\, \Delta y\, \Delta z, \tag{2}$$

where $\Delta x = x_i - x_{i-1}$, $\Delta y = y_j - y_{j-1}$, and $\Delta z = z_k - z_{k-1}$. We then form the sum

$$\sum_{i=1}^{n} \sum_{j=1}^{m} \sum_{k=1}^{p} f(x_i^*, y_j^*, z_k^*)\, \Delta V, \tag{3}$$

where (x_i^*, y_j^*, z_k^*) is in B_{ijk}.

We now define

$$\Delta u = \sqrt{\Delta x^2 + \Delta y^2 + \Delta z^2}.$$

Geometrically, Δu is the length of a diagonal of a rectangular solid with sides having lengths Δx, Δy, and Δz, respectively. We see that as $\Delta u \to 0$, Δx, Δy, and Δz approach 0, and the volume of each box tends to zero. We then take the limit as $\Delta u \to 0$.

Definition 1 THE TRIPLE INTEGRAL Let $w = f(x, y, z)$ and let the parallelepiped π be given by (1). Suppose that

$$\lim_{\Delta u \to 0} \sum_{i=1}^{n} \sum_{j=1}^{m} \sum_{k=1}^{p} f(x_i^*, y_j^*, z_k^*)\, \Delta x\, \Delta y\, \Delta z$$

exists and is independent of the way in which the points (x_i^*, y_j^*, z_k^*) are chosen. Then the **triple integral** of f over π, written $\iiint_\pi f(x, y, z)\, dV$, is defined by

$$\iiint_\pi f(x, y, z)\, dV = \lim_{\Delta u \to 0} \sum_{i=1}^{n} \sum_{j=1}^{m} \sum_{k=1}^{p} f(x_i^*, y_j^*, z_k^*)\, \Delta V \tag{4}$$

As with double integrals, we can write triple integrals as iterated (or repeated) integrals. If π is defined by (1), we have

$$\iiint_{\pi} f(x, y, z)\, dV = \int_{a_1}^{a_2} \int_{b_1}^{b_2} \int_{c_1}^{c_2} f(x, y, z)\, dz\, dy\, dx. \tag{5}$$

EXAMPLE 1 Evaluate $\iiint_{\pi} xy \cos yz\, dV$, where π is the parallelepiped

$$\left\{(x, y, z)\colon 0 \le x \le 1, 0 \le y \le 1, 0 \le z \le \frac{\pi}{2}\right\}.$$

Solution.

$$\begin{aligned}\iiint_{\pi} xy \cos yz\, dV &= \int_0^1 \int_0^1 \int_0^{\pi/2} xy \cos yz\, dz\, dy\, dx \\ &= \int_0^1 \int_0^1 \left\{xy \cdot \frac{1}{y} \sin yz \Big|_0^{\pi/2}\right\} dy\, dx = \int_0^1 \int_0^1 x \sin\left(\frac{\pi}{2} y\right) dy\, dx \\ &= \int_0^1 \left\{-\frac{2}{\pi} x \cos \frac{\pi}{2} y \Big|_0^1\right\} dx = \int_0^1 \frac{2}{\pi} x\, dx = \frac{x^2}{\pi}\Big|_0^1 = \frac{1}{\pi}. \quad \blacksquare\end{aligned}$$

THE TRIPLE INTEGRAL OVER A MORE GENERAL REGION

We now define the triple integral over a more general region S. We assume that S is bounded. Then we can enclose S in a parallelepiped π and define a new function F by

$$F(x, y, z) = \begin{cases} f(x, y, z), & \text{if } (x, y, z) \text{ is in } S \\ 0, & \text{if } (x, y, z) \text{ is in } \pi \text{ but not in } S. \end{cases}$$

We then define

$$\iiint_{S} f(x, y, z)\, dV = \iiint_{\pi} F(x, y, z)\, dV. \tag{6}$$

REMARK 1. If f is continuous over S and if S is a region of the type we will discuss below, then $\iiint_S f(x, y, z)\, dV$ will exist. The proof of this fact is beyond the scope of this text but can be found in any advanced calculus text.

REMARK 2. If $f \ge 0$ on S, then the triple integral $\iiint_S f(x, y, z)\, dV$ represents the "volume" in four-dimensional space $\mathbb{R}^4$ of the region bounded above by f and below by S. We cannot, of course, draw this volume, but otherwise the theory of volumes carries over to four (and more) dimensions.

Now let S take the form

$$S = \{(x, y, z)\colon a_1 \le x \le a_2, g_1(x) \le y \le g_2(x), h_1(x, y) \le z \le h_2(x, y)\}. \tag{7}$$

What does such a solid look like? We first note that the equations $z = h_1(x, y)$ and $z = h_2(x, y)$ are the equations of surfaces in $\mathbb{R}^3$. The equations $y = g_1(x)$ and $y = g_2(x)$ are equations of cylinders in $\mathbb{R}^3$, and the equations $x = a_1$ and $x = a_2$ are equations of

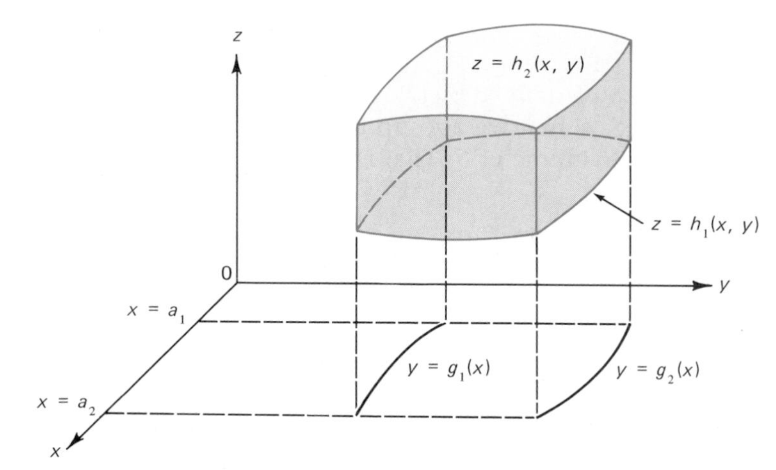

FIGURE 2

planes in $\mathbb{R}^3$. The solid S is sketched in Figure 2. We assume that g_1, g_2, h_1, and h_2 are continuous. If f is continuous, then $\iiint_S f(x, y, z)\, dV$ will exist and

$$\iiint_S f(x, y, z)\, dV = \int_{a_1}^{a_2} \int_{g_1(x)}^{g_2(x)} \int_{h_1(x, y)}^{h_2(x, y)} f(x, y, z)\, dz\, dy\, dx. \tag{8}$$

Note the similarity between equation (8) and equation (5.2.10).

EXAMPLE 2 Evaluate $\iiint_S 2x^3y^2z\, dV$, where S is the region

$$\{(x, y, z) \colon 0 \le x \le 1,\ x^2 \le y \le x,\ x - y \le z \le x + y\}.$$

Solution.

$$\begin{aligned}
\iiint_S 2x^3y^2z\, dV &= \int_0^1 \int_{x^2}^{x} \int_{x-y}^{x+y} 2x^3y^2z\, dz\, dy\, dx \\
&= \int_0^1 \int_{x^2}^{x} \left\{ x^3y^2z^2 \Big|_{x-y}^{x+y} \right\} dy\, dx \\
&= \int_0^1 \int_{x^2}^{x} x^3y^2[(x + y)^2 - (x - y)^2]\, dy\, dx \\
&= \int_0^1 \int_{x^2}^{x} 4x^4y^3\, dy\, dx = \int_0^1 \left\{ x^4y^4 \Big|_{x^2}^{x} \right\} dx \\
&= \int_0^1 (x^8 - x^{12})\, dx = \frac{1}{9} - \frac{1}{13} = \frac{4}{117}. \quad ■
\end{aligned}$$

Many of the applications we saw for the double integral can be extended to the triple integral. We present three of them below.

I. VOLUME

Let the region S be defined by (7). Then, since $\Delta V = \Delta x\, \Delta y\, \Delta z$ represents the volume of a "box" in S, when we add up the volumes of these boxes and take a limit, we obtain the total volume of S. That is,

$$\text{volume of } S = \iiint_S dV. \tag{9}$$

EXAMPLE 3 Calculate the volume of the region of Example 2.

Solution.

$$V = \int_0^1 \int_{x^2}^x \int_{x-y}^{x+y} dz\, dy\, dx = \int_0^1 \int_{x^2}^x \left\{ z \Big|_{x-y}^{x+y} \right\} dy\, dx$$

$$= \int_0^1 \int_{x^2}^x 2y\, dy\, dx = \int_0^1 \left\{ y^2 \Big|_{x^2}^x \right\} dx = \int_0^1 (x^2 - x^4)\, dx$$

$$= \frac{1}{3} - \frac{1}{5} = \frac{2}{15}. \blacksquare$$

EXAMPLE 4 Find the volume of the tetrahedron formed by the planes $x = 0$, $y = 0$, $z = 0$, and $x + (y/2) + (z/4) = 1$.

Solution. The tetrahedron is sketched in Figure 3. We see that z ranges from 0 to the plane $x + (y/2) + (z/4) = 1$, or $z = 4[1 - x - (y/2)]$. This last plane intersects the xy-plane in a line whose equation (obtained by setting $z = 0$) is given by $0 = 1 - x - (y/2)$ or $y = 2(1 - x)$, so that y ranges from 0 to $2(1 - x)$ (see Figure 4). Finally, this line intersects the x-axis at the point $(1, 0, 0)$, so that x ranges from 0 to 1, and we have

$$V = \int_0^1 \int_0^{2(1-x)} \int_0^{4(1-x-y/2)} dz\, dy\, dx = \int_0^1 \int_0^{2(1-x)} 4\left(1 - x - \frac{y}{2}\right) dy\, dx$$

$$= 4 \int_0^1 \left\{ \left(y(1 - x) - \frac{y^2}{4} \right) \Big|_0^{2(1-x)} \right\} dx = 4 \int_0^1 [2(1 - x)^2 - (1 - x)^2]\, dx$$

$$= -\frac{4}{3}(1 - x)^3 \Big|_0^1 = \frac{4}{3}.$$

REMARK. It was not necessary to integrate in the order z, then y, then x. We could have written, for example,

$$0 \le x \le 1 - \frac{y}{2} - \frac{z}{4}.$$

The intersection of this plane with the yz-plane is the line $0 = 1 - (y/2) - (z/4)$, or $z = 4[1 - (y/2)]$. The intersection of this line with the y-axis occurs at the

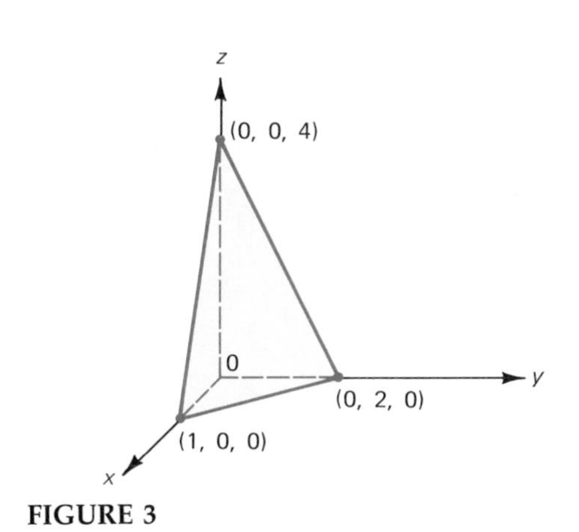

FIGURE 3

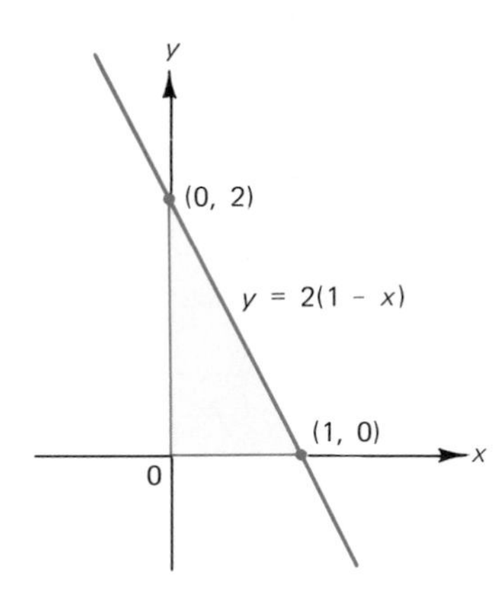

FIGURE 4

point (0, 2, 0). Thus

$$\begin{aligned} V &= \int_0^2 \int_0^{4(1-y/2)} \int_0^{1-y/2-z/4} dx\, dz\, dy \\ &= \int_0^2 \int_0^{4(1-y/2)} \left(1 - \frac{y}{2} - \frac{z}{4}\right) dz\, dy = \int_0^2 \left\{ \left[\left(1 - \frac{y}{2}\right) z - \frac{z^2}{8} \right]\Bigg|_0^{4(1-y/2)} \right\} dy \\ &= \int_0^2 \left\{ 4\left(1 - \frac{y}{2}\right)^2 - \frac{16[1-(y/2)]^2}{8} \right\} dy = 2\int_0^2 \left(1 - \frac{y}{2}\right)^2 dy \\ &= -\frac{4}{3}(1-x)^3 \Big|_0^1 = \frac{4}{3}. \end{aligned}$$

We could also integrate in any of four other orders (xyz, yzx, yxz, and zxy) to obtain the same result. ■

II. DENSITY AND MASS

Let the function $\rho(x, y, z)$ denote the density (in kilograms per cubic meter, say) of a solid S in $\mathbb{R}^3$. Then for a "box" of sides Δx, Δy, and Δz, the approximate mass of the box will be equal to $\rho(x_i^*, y_j^*, z_k^*)\,\Delta x\,\Delta y\,\Delta z = \rho(x_i^*, y_j^*, z_k^*)\,\Delta V$ if Δx, Δy, and Δz are small. We then obtain

$$\text{total mass of } S = \mu(S) = \iiint_S \rho(x, y, z)\, dV \qquad \textbf{(10)}$$

EXAMPLE 5 The density of the solid of Example 2 is given by $\rho(x, y, z) = x + 2y + 4z$ kg/m^3. Calculate the total mass of the solid.

Solution.

$$\begin{aligned}\mu(S) &= \int_0^1 \int_{x^2}^x \int_{x-y}^{x+y} (x + 2y + 4z)\, dz\, dy\, dx = \int_0^1 \int_{x^2}^x \left\{[(x + 2y)z + 2z^2]\Big|_{x-y}^{x+y}\right\} dy\, dx \\ &= \int_0^1 \int_{x^2}^x (10xy + 4y^2)\, dy\, dx = \int_0^1 \left\{\left(5xy^2 + \frac{4y^3}{3}\right)\Big|_{x^2}^{x}\right\} dx \\ &= \int_0^1 \left(5x^3 - 5x^5 + \frac{4}{3}x^3 - \frac{4}{3}x^6\right) dx = \int_0^1 \left(\frac{19}{3}x^3 - 5x^5 - \frac{4}{3}x^6\right) dx \\ &= \frac{19}{12} - \frac{5}{6} - \frac{4}{21} = \tfrac{47}{84} \text{ kg.} \quad \blacksquare\end{aligned}$$

III. FIRST MOMENTS AND CENTER OF MASS

In $\mathbb{R}^3$ we use the symbol M_{yz} to denote the first moment with respect to the yz-plane. Similarly, M_{xz} denotes the first moment with respect to the xz-plane, and M_{xy} denotes the first moment with respect to the xy-plane. Since the distance from a point (x, y, z) to the yz-plane is x, and so on, we may use familiar reasoning to obtain

$$M_{yz} = \iiint_S x\rho(x, y, z)\, dV \tag{11}$$

$$M_{xz} = \iiint_S y\rho(x, y, z)\, dV \tag{12}$$

$$M_{xy} = \iiint_S z\rho(x, y, z)\, dV. \tag{13}$$

The **center of mass** of S is then given by

$$(\overline{x}, \overline{y}, \overline{z}) = \left(\frac{M_{yz}}{\mu}, \frac{M_{xz}}{\mu}, \frac{M_{xy}}{\mu}\right), \tag{14}$$

where μ denotes the mass of S. If $\rho(x, y, z)$ is constant, then as in $\mathbb{R}^1$ and $\mathbb{R}^2$, the center of mass is called the **centroid.**

EXAMPLE 6 Find the center of mass of the solid in Example 5.

Solution. We have already found that $\mu = \frac{47}{84}$. We next calculate the moments.

$$M_{yz} = \int_0^1 \int_{x^2}^x \int_{x-y}^{x+y} x(x + 2y + 4z)\, dz\, dy\, dx$$

$$= \int_0^1 \int_{x^2}^x \left\{ ((x^2 + 2xy)z + 2xz^2)\Big|_{x-y}^{x+y} \right\} dy\, dx$$

$$= \int_0^1 \int_{x^2}^x (10x^2y + 4xy^2)\, dy\, dx = \int_0^1 \left\{ \left(5x^2y^2 + \frac{4xy^3}{3}\right)\Big|_{x^2}^{x} \right\} dx$$

$$= \int_0^1 \left(5x^4 - 5x^6 + \frac{4}{3}x^4 - \frac{4}{3}x^7\right) dx = 1 - \frac{5}{7} + \frac{4}{15} - \frac{1}{6} = \frac{27}{70}$$

$$M_{xz} = \int_0^1 \int_{x^2}^x \int_{x-y}^{x+y} y(x + 2y + 4z)\, dz\, dy\, dx$$

$$= \int_0^1 \int_{x^2}^x \left\{ [(xy + 2y^2)z + 2yz^2]\Big|_{x-y}^{x+y} \right\} dy\, dx$$

$$= \int_0^1 \int_{x^2}^x (10xy^2 + 4y^3)\, dy\, dx = \int_0^1 \left\{ \left(\frac{10}{3}xy^3 + y^4\right)\Big|_{x^2}^{x} \right\} dx$$

$$= \int_0^1 \left(\frac{10}{3}x^4 - \frac{10}{3}x^7 + x^4 - x^8\right) dx = \frac{2}{3} - \frac{5}{12} + \frac{1}{5} - \frac{1}{9} = \frac{61}{180}$$

$$M_{xy} = \int_0^1 \int_{x^2}^x \int_{x-y}^{x+y} z(x + 2y + 4z)\, dz\, dy\, dx$$

$$= \int_0^1 \int_{x^2}^x \left\{ \left[(x + 2y)\frac{z^2}{2} + \frac{4z^3}{3} \right]\Big|_{x-y}^{x+y} \right\} dy\, dz$$

$$= \int_0^1 \int_{x^2}^x \left[(x + 2y)(2xy) + \frac{4}{3}(6x^2y + 2y^3) \right] dy\, dx$$

$$= \int_0^1 \int_{x^2}^x \left(10x^2y + 4xy^2 + \frac{8}{3}y^3\right) dy\, dx$$

$$= \int_0^1 \left\{ \left(5x^2y^2 + \frac{4}{3}xy^3 + \frac{2}{3}y^4\right)\Big|_{x^2}^{x} \right\} dx$$

$$= \int_0^1 \left(5x^4 - 5x^6 + \frac{4}{3}x^4 - \frac{4}{3}x^7 + \frac{2}{3}x^4 - \frac{2}{3}x^8\right) dx$$

$$= \int_0^1 \left(7x^4 - 5x^6 - \frac{4}{3}x^7 - \frac{2}{3}x^8\right) dx = \frac{7}{5} - \frac{5}{7} - \frac{1}{6} - \frac{2}{27} = \frac{841}{1890}$$

Thus

$$(\bar{x}, \bar{y}, \bar{z}) = \left(\frac{M_{yz}}{\mu}, \frac{M_{xz}}{\mu}, \frac{M_{xy}}{\mu}\right) = \left(\frac{27/70}{47/84}, \frac{61/180}{47/84}, \frac{841/1890}{47/84}\right)$$

$$= \left(\frac{162}{235}, \frac{427}{705}, \frac{1682}{2115}\right) \approx (0.689, 0.606, 0.795).$$

Here distances are measured in meters. ■

PROBLEMS 5.6

In Problems 1–7, evaluate the repeated triple integral.

1. $\int_0^1 \int_0^y \int_0^x y\,dz\,dx\,dy$

2. $\int_0^2 \int_{-z}^{z} \int_{y-z}^{y+z} 2xz\,dx\,dy\,dz$

3. $\int_{a_1}^{a_2} \int_{b_1}^{b_2} \int_{c_1}^{c_2} dy\,dx\,dz$

4. $\int_0^{\pi/2} \int_0^{\pi/2} \int_0^z \sin\left(\frac{x}{z}\right) dx\,dz\,dy$

5. $\int_1^2 \int_{1-y}^{1+y} \int_0^{yz} 6xyz\,dx\,dz\,dy$

6. $\int_0^1 \int_0^{\sqrt{1-x^2}} \int_0^x yz\,dz\,dy\,dx$

7. $\int_0^1 \int_{-\sqrt{1-x^2}}^{\sqrt{1-x^2}} \int_{-\sqrt{1-x^2-y^2}}^{\sqrt{1-x^2-y^2}} z^2\,dz\,dy\,dx$ [*Hint:* Use polar coordinates.]

***8.** Change the order of integration in Problem 1 and write the integral in the form (a) $\int_?^? \int_?^? \int_?^? y\,dx\,dy\,dz$; (b) $\int_?^? \int_?^? \int_?^? y\,dy\,dz\,dx$. [*Hint:* Sketch the region in Problem 1 from the given limits and then find the new limits directly from the figure.]

***9.** Write the integral of Problem 6 in the form (a) $\int_?^? \int_?^? \int_?^? yz\,dx\,dy\,dz$; (b) $\int_?^? \int_?^? \int_?^? yz\,dy\,dz\,dx$.

***10.** Write the integral of Problem 7 in the form $\int_?^? \int_?^? \int_?^? z^2\,dx\,dz\,dy$.

In Problems 11–18, find the volume of the given solid.

11. The tetrahedron with vertices at the points (0, 0, 0), (1, 0, 0), (0, 1, 0), and (0, 0, 1).

12. The tetrahedron with vertices at the points (0, 0, 0), $(a, 0, 0)$, $(0, b, 0)$, and $(0, 0, c)$.

13. The solid in the first octant bounded by the cylinder $x^2 + z^2 = 9$, the plane $x + y = 4$, and the three coordinate planes.

***14.** The solid bounded by the planes $x - 2y + 4z = 4$, $-2x + 3y - z = 6$, $x = 0$, and $y = 0$.

15. The solid bounded above by the sphere $x^2 + y^2 + z^2 = 16$ and below by the plane $z = 2$.

16. The solid bounded by the parabolic cylinder $z = 5 - x^2$ and the planes $z = y$ and $z = 2y$ that lies in the half space $y \geq 0$.

17. The solid bounded by the ellipsoid $(x^2/a^2) + (y^2/b^2) + (z^2/c^2) = 1$.

18. The solid bounded by the elliptic cylinder $9x^2 + y^2 = 9$ and the planes $z = 0$ and $x + y + 9z = 9$.

In Problems 19–23, find the mass and center of mass of the given solid having the given density.

19. the solid of Problem 11; $\rho(x, y, z) = x$

20. the solid of Problem 12; $\rho(x, y, z) = x^2 + y^2 + z^2$

21. the solid of Problem 13; $\rho(x, y, z) = z$

22. the solid of Problem 15; $\rho(x, y, z) = x + 2y + 4z$

23. the solid of Problem 17; $\rho(x, y, z) = \alpha x^2 + \beta y^2 + \gamma z^2$

24. Find the mass of a cube whose side has a length of k units if its density at any point is proportional to the distance from a given face of the cube.

25. Find the centroid of the tetrahedron of Problem 12.

26. Find the centroid of the ellipsoid of Problem 17.

27. Find the centroid of the solid of Problem 16.

28. Find the centroid of the solid of Problem 18.

29. The solid S lies in the first octant and is bounded by the planes $z = 0$, $y = 1$, $x = y$, and the hyperboloid $z = xy$. Its density is given by $\rho(x, y, z) = 1 + 2z$. (a) Find the center of mass of S. (b) Show that the center of mass lies inside S.

5.7 THE TRIPLE INTEGRAL IN CYLINDRICAL AND SPHERICAL COORDINATES

In this section we show how triple integrals can be written by using cylindrical and spherical coordinates.

I. CYLINDRICAL COORDINATES

Recall from Section 3.9 that the cylindrical coordinates of a point in $\mathbb{R}^3$ are (r, θ, z), where r and θ are the polar coordinates of the projection of the point onto the xy-plane and z is the usual z-coordinate. In order to calculate an integral of a region given in cylindrical coordinates, we go through a procedure very similar to the one we used to write double integrals in polar coordinates in Section 5.4.

Consider the "cylindrical parallelepiped" given by

$$\pi_c = \{(r, \theta, z): r_1 \le r \le r_2;\ \theta_1 \le \theta \le \theta_2;\ z_1 \le z \le z_2\}.$$

This solid is sketched in Figure 1. If we partition the z-axis for z in $[z_1, z_2]$, we obtain "slices." For a fixed z the area of the face of a slice of π_c is, according to equation (5.4.5), given by

$$A_i = \int_{\theta_1}^{\theta_2}\int_{r_1}^{r_2} r\,dr\,d\theta. \tag{1}$$

The volume of a slice is, by (1), given by

$$V_i = A_i\,\Delta z = \left\{\int_{\theta_1}^{\theta_2}\int_{r_1}^{r_2} r\,dr\,d\theta\right\}\Delta z.$$

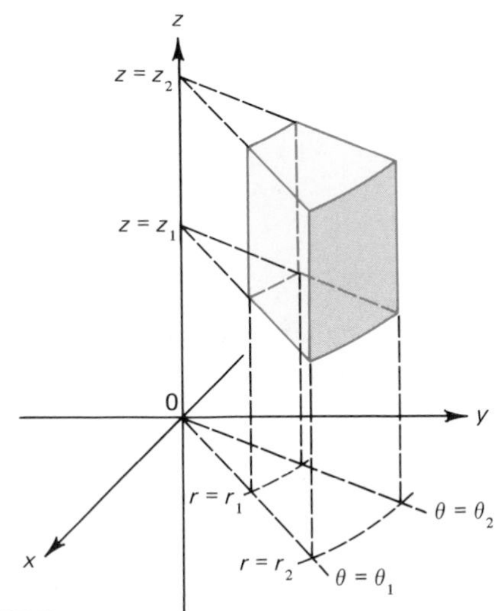

FIGURE 1

Then adding these volumes and taking the limit as before, we obtain

$$\text{volume of } \pi_c = \int_{z_1}^{z_2}\int_{\theta_1}^{\theta_2}\int_{r_1}^{r_2} r\,dr\,d\theta\,dz. \tag{2}$$

In general, let the region S be given in cylindrical coordinates by

$$S = \{(r, \theta, z): \theta_1 \le \theta \le \theta_2,\ 0 \le g_1(\theta) \le r \le g_2(\theta),\ h_1(r, \theta) \le z \le h_2(r, \theta)\}, \tag{3}$$

and let f be a function of r, θ, and z. Then the triple integral of f over S is given by

$$\iiint\limits_S f = \int_{\theta_1}^{\theta_2}\int_{g_1(\theta)}^{g_2(\theta)}\int_{h_1(r,\theta)}^{h_2(r,\theta)} f(r, \theta, z)r\,dz\,dr\,d\theta. \tag{4}$$

EXAMPLE 1 Find the mass of a solid bounded by the cylinder $r = \sin\theta$, the planes $z = 0$, $\theta = 0$, $\theta = \pi/3$, and the cone $z = r$, if the density is given by $\rho(r, \theta, z) = 4r$.

Solution. We first note that the solid may be written as

$$S = \{(r, \theta, z)\colon 0 \le \theta \le \frac{\pi}{3},\ 0 \le r \le \sin\theta,\ 0 \le z \le r\}.$$

Then

$$\begin{aligned}
\mu &= \int_0^{\pi/3}\int_0^{\sin\theta}\int_0^{r} (4r)r\,dz\,dr\,d\theta \\
&= \int_0^{\pi/3}\int_0^{\sin\theta}\left\{4r^2z\Big|_0^r\right\}dr\,d\theta = \int_0^{\pi/3}\int_0^{\sin\theta} 4r^3\,dr\,d\theta \\
&= \int_0^{\pi/3}\left\{r^4\Big|_0^{\sin\theta}\right\}d\theta = \int_0^{\pi/3}\sin^4\theta\,d\theta = \int_0^{\pi/3}\left(\frac{1-\cos 2\theta}{2}\right)^2 d\theta \\
&= \frac{1}{4}\int_0^{\pi/3}\left(1 - 2\cos 2\theta + \frac{1+\cos 4\theta}{2}\right)d\theta \\
&= \frac{1}{4}\left(\frac{3\theta}{2} - \sin 2\theta + \frac{\sin 4\theta}{8}\right)\Big|_0^{\pi/3} = \frac{1}{4}\left(\frac{\pi}{2} - \frac{\sqrt{3}}{2} - \frac{\sqrt{3}}{16}\right) \\
&= \frac{\pi}{8} - \frac{9\sqrt{3}}{64}. \quad \blacksquare
\end{aligned}$$

The formula for conversion from rectangular to cylindrical coordinates is virtually identical to formula (5.4.7). Let S be a region in $\mathbb{R}^3$. Since $x = r\cos\theta$, $y = r\sin\theta$, and $z = z$, we have

$$\underbrace{\iiint_S f(x, y, z)\,dV}_{\text{Given in rectangular coordinates}} = \underbrace{\iiint_S f(r\cos\theta, r\sin\theta, z)r\,dz\,dr\,d\theta}_{\text{Given in cylindrical coordinates}}. \tag{5}$$

REMARK. Again, do not forget the extra r when you convert to cylindrical coordinates.

EXAMPLE 2 Find the mass of the solid bounded by the paraboloid $z = x^2 + y^2$ and the plane $z = 4$ if the density at any point is proportional to the distance from the point to the z-axis.

Solution. The solid is sketched in Figure 2. The density is given by $\rho(x, y, z) = \alpha\sqrt{x^2 + y^2}$, where α is a constant of proportionality. Thus since the solid may be written as

$$S = \{(x, y, z)\colon -2 \le x \le 2,\ -\sqrt{4 - x^2} \le y \le \sqrt{4 - x^2},\ x^2 + y^2 \le z \le 4\},$$

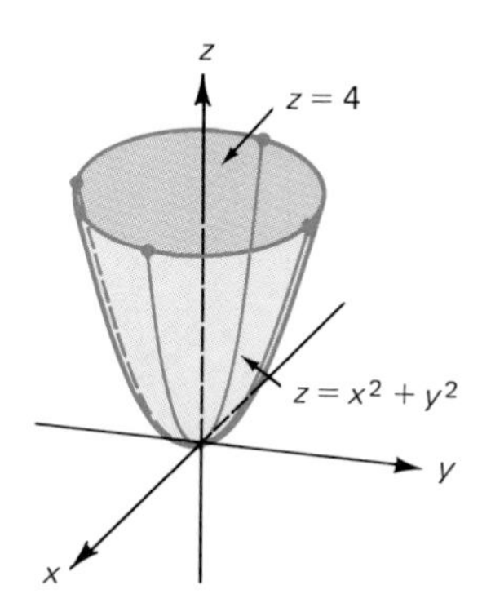

FIGURE 2

we have

$$\mu = \iiint_S \alpha\sqrt{x^2 + y^2}\, dV = \int_{-2}^{2} \int_{-\sqrt{4-x^2}}^{\sqrt{4-x^2}} \int_{x^2+y^2}^{4} \alpha\sqrt{x^2 + y^2}\, dz\, dy\, dx.$$

We write this expression in cylindrical coordinates, using the fact that $\alpha\sqrt{x^2 + y^2} = \alpha r$. We note that the largest value of r is 2, since at the "top" of the solid, $r^2 = x^2 + y^2 = 4$. Then

$$\mu = \int_0^{2\pi} \int_0^{2} \int_{r^2}^{4} (\alpha r) r\, dz\, dr\, d\theta = \int_0^{2\pi} \int_0^{2} \alpha r^2(4 - r^2)\, dr\, d\theta$$

$$= \alpha \int_0^{2\pi} \left\{ \left(\frac{4r^3}{3} - \frac{r^5}{5} \right) \Big|_0^2 \right\} d\theta = \frac{64\alpha}{15} \int_0^{2\pi} d\theta = \frac{128}{15}\pi\alpha. \quad \blacksquare$$

II. SPHERICAL COORDINATES

Recall from Section 3.9 that a point P in $\mathbb{R}^3$ can be written in the spherical coordinates (ρ, θ, φ), where $\rho \geq 0$, $0 \leq \theta < 2\pi$, $0 \leq \varphi \leq \pi$. Here ρ is the distance between the point and the origin, θ is the same as in cylindrical coordinates, and φ is the angle between $\overrightarrow{0P}$ and the positive z-axis. Consider the "spherical parallelepiped"

$$\pi_s = \{(\rho, \theta, \varphi): \rho_1 \leq \rho \leq \rho_2,\ \theta_1 \leq \theta \leq \theta_2,\ \varphi_1 \leq \varphi \leq \varphi_2\}. \tag{6}$$

This solid is sketched in Figure 3. To approximate the volume of π_s, we partition the intervals $[\rho_1, \rho_2]$, $[\theta_1, \theta_2]$, and $[\varphi_1, \varphi_2]$. This partition gives us a number of "spherical boxes," one of which is sketched in Figure 4. The length of an arc of a circle is given by

$$L = r\theta, \tag{7}$$

where r is the radius of the circle and θ is the angle that "cuts off" the arc. In Figure 4 one side of the spherical box is $\Delta\rho$. Since $\rho = \rho_i$ is the equation of a sphere, we find, from (7), that the length of a second side is $\rho_i\Delta\varphi$. Finally, the length of the third side

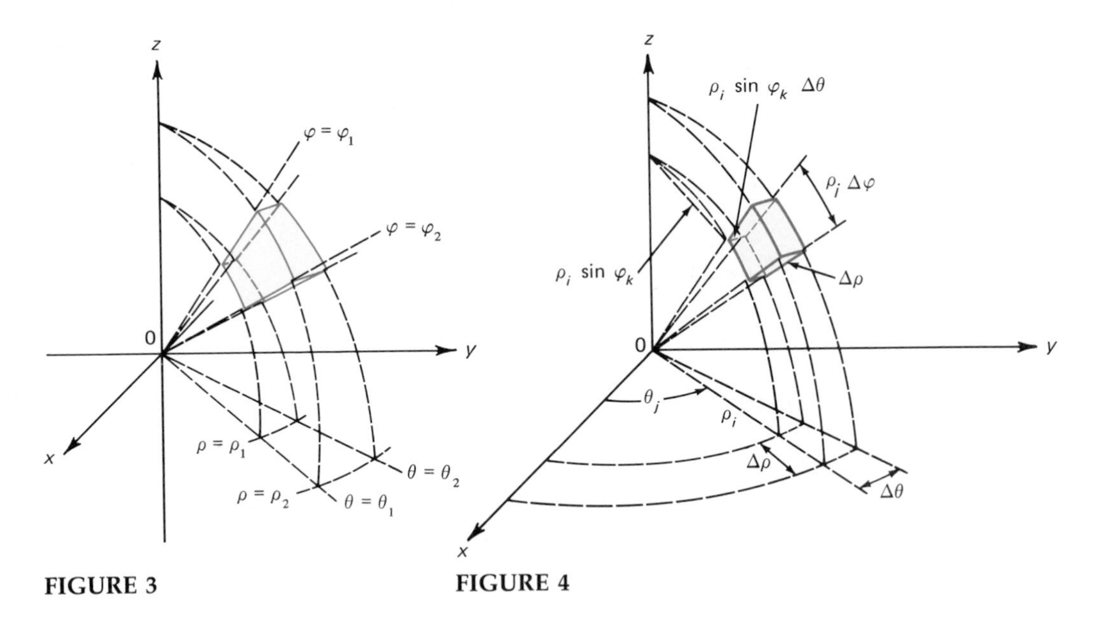

FIGURE 3 FIGURE 4

is $\rho_i \sin \varphi_k \Delta\theta$. This follows from equation (7) and equation (3.9.5). Thus the volume of the box in Figure 4 is given, approximately, by

$$V_i \approx (\Delta\rho)(\rho_i \, \Delta\varphi)(\rho_i \sin \varphi_k \, \Delta\theta),$$

and using a familiar argument, we have

$$\text{volume of } \pi_s = \iiint\limits_{\pi_s} \rho^2 \sin \varphi \, d\rho \, d\varphi \, d\theta = \int_{\theta_1}^{\theta_2} \int_{\varphi_1}^{\varphi_2} \int_{\rho_1}^{\rho_2} \rho^2 \sin \varphi \, d\rho \, d\varphi \, d\theta. \tag{8}$$

If f is a function of the variables ρ, θ, and φ, we have

$$\iiint\limits_{\pi_s} f = \iiint\limits_{\pi_s} f(\rho, \theta, \varphi)\rho^2 \sin \varphi \, d\rho \, d\varphi \, d\theta. \tag{9}$$

More generally, let the region S be defined in spherical coordinates by

$$S = \{(\rho, \theta, \varphi)\colon \theta_1 \le \theta \le \theta_2, \, g_1(\theta) \le \varphi \le g_2(\theta), \, h_1(\theta, \varphi) \le \rho \le h_2(\theta, \varphi)\}. \tag{10}$$

Then

$$\iiint\limits_{S} f = \int_{\theta_1}^{\theta_2} \int_{g_1(\theta)}^{g_2(\theta)} \int_{h_1(\theta, \varphi)}^{h_2(\theta, \varphi)} f(\rho, \theta, \varphi)\rho^2 \sin \varphi \, d\rho \, d\varphi \, d\theta. \tag{11}$$

Recall that to convert from rectangular to spherical coordinates, we have the formulas [see equations (3.9.6), (3.9.7), and (3.9.8)]

$$x = \rho \sin \varphi \cos \theta, \qquad y = \rho \sin \varphi \sin \theta, \qquad z = \rho \cos \varphi.$$

Thus to convert from a triple integral in rectangular coordinates to a triple integral in spherical coordinates, we have

$$\underbrace{\iiint_S f(x, y, z)\, dV}_{\text{Given in rectangular coordinates}} = \underbrace{\iiint_S f(\rho \sin \varphi \cos \theta, \rho \sin \varphi \sin \theta, \rho \cos \varphi)\rho^2 \sin \varphi \, d\rho \, d\varphi \, d\theta.}_{\text{Given in spherical coordinates}}$$

EXAMPLE 3 Calculate the volume enclosed by the sphere $x^2 + y^2 + z^2 = a^2$.

Solution. In rectangular coordinates, since

$$S = \{(x, y, z): -a \le x \le a, -\sqrt{a^2 - x^2} \le y \le \sqrt{a^2 - x^2},$$
$$-\sqrt{a^2 - x^2 - y^2} \le z \le \sqrt{a^2 - x^2 - y^2}\},$$

we can write the volume as

$$V = \int_{-a}^{a} \int_{-\sqrt{a^2-x^2}}^{\sqrt{a^2-x^2}} \int_{-\sqrt{a^2-x^2-y^2}}^{\sqrt{a^2-x^2-y^2}} dz\, dy\, dx.$$

This expression is very tedious to calculate. So instead, we note that the region enclosed by a sphere can be represented in spherical coordinates as

$$S = \{(\rho, \theta, \varphi): 0 \le \rho \le a, 0 \le \theta \le 2\pi, 0 \le \varphi \le \pi\}.$$

Thus

$$V = \int_0^{2\pi} \int_0^{\pi} \int_0^{a} \rho^2 \sin \varphi \, d\rho \, d\varphi \, d\theta = \int_0^{2\pi} \int_0^{\pi} \frac{a^3}{3} \sin \varphi \, d\varphi \, d\theta$$
$$= \frac{a^3}{3} \int_0^{2\pi} \left\{ -\cos \varphi \Big|_0^{\pi} \right\} d\theta = \frac{2a^3}{3} \int_0^{2\pi} d\theta = \frac{4\pi a^3}{3}. \blacksquare$$

EXAMPLE 4 Find the mass of the sphere in Example 3 if its density at a point is proportional to the distance from the point to the origin.

Solution. We have density $= \alpha\sqrt{x^2 + y^2 + z^2} = \alpha\rho$. Thus

$$\mu = \int_0^{2\pi} \int_0^{\pi} \int_0^{a} \alpha\rho(\rho^2 \sin \varphi)\, d\rho \, d\varphi \, d\theta = \frac{\alpha a^4}{4} \int_0^{2\pi} \int_0^{\pi} \sin \varphi \, d\varphi \, d\theta$$
$$= \frac{\alpha a^4}{4}(4\pi) = \pi a^4 \alpha. \blacksquare$$

EXAMPLE 5 Find the centroid of the "ice-cream-cone-shaped" region below the sphere $x^2 + y^2 + z^2 = z$ and above the cone $z^2 = x^2 + y^2$.

Solution. The region is sketched in Figure 5. Since $x^2 + y^2 + z^2 = \rho^2$ and $z = \rho \cos \varphi$, the sphere $\{(x, y, z): x^2 + y^2 + z^2 = z\}$ can be written in spherical

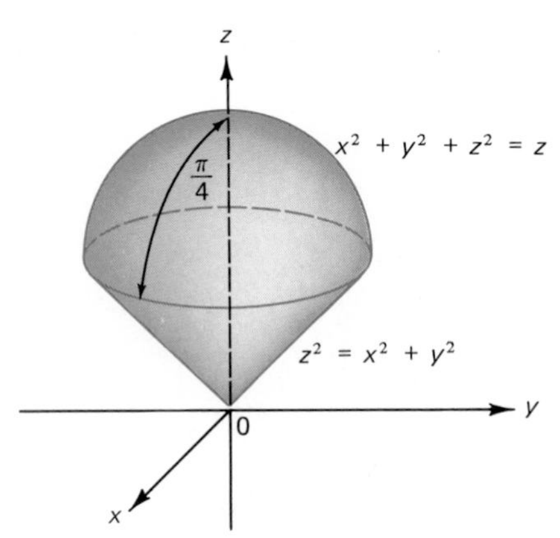

FIGURE 5

coordinates as

$$\rho^2 = \rho \cos \varphi, \qquad \text{or} \qquad \rho = \cos \varphi.$$

Since $x^2 + y^2 = \rho^2 \sin^2 \varphi \cos^2 \theta + \rho^2 \sin^2 \varphi \sin^2 \theta = \rho^2 \sin^2 \varphi$, the equation of the cone is

$$\rho^2 \cos^2 \varphi = \rho^2 \sin^2 \varphi, \qquad \text{or} \qquad \cos^2 \varphi = \sin^2 \varphi,$$

and since $0 \leq \varphi \leq \pi$, the equation is

$$\varphi = \frac{\pi}{4}.$$

Thus, assuming that the density of the region is the constant k, we have

$$\begin{aligned}
\mu &= \int_0^{2\pi} \int_0^{\pi/4} \int_0^{\cos \varphi} k\rho^2 \sin \varphi \, d\rho \, d\varphi \, d\theta \\
&= k \int_0^{2\pi} \int_0^{\pi/4} \sin \varphi \left\{ \frac{\rho^3}{3} \bigg|_0^{\cos \varphi} \right\} d\varphi \, d\theta = k \int_0^{2\pi} \int_0^{\pi/4} \frac{\cos^3 \varphi}{3} \sin \varphi \, d\varphi \, d\theta \\
&= k \int_0^{2\pi} \left\{ - \frac{\cos^4 \varphi}{12} \bigg|_0^{\pi/4} \right\} d\theta = \frac{1}{12} k \int_0^{2\pi} \frac{3}{4} d\theta = \frac{\pi k}{8}.
\end{aligned}$$

Since $x = \rho \sin \varphi \cos \theta$, we have

$$\begin{aligned}
M_{yz} &= k \int_0^{2\pi} \int_0^{\pi/4} \int_0^{\cos \varphi} (\rho \sin \varphi \cos \theta) \rho^2 \sin \varphi \, d\rho \, d\varphi \, d\theta \\
&= k \int_0^{2\pi} \cos \theta \, d\theta \int_0^{\pi/4} \int_0^{\cos \varphi} \rho^3 \sin^2 \varphi \, d\rho \, d\varphi = 0,
\end{aligned}$$

since

$$\int_0^{2\pi} \cos\theta \, d\theta = 0$$

(an unsurprising result because of symmetry). Similarly, since $y = \rho \sin\varphi \sin\theta$,

$$M_{xz} = k \int_0^{2\pi} \int_0^{\pi/4} \int_0^{\cos\varphi} (\rho \sin\varphi \sin\theta)\rho^2 \sin\varphi \, d\rho \, d\varphi \, d\theta = 0$$

(again because of symmetry). Finally, since $z = \rho \cos\varphi$,

$$\begin{aligned} M_{xy} &= k \int_0^{2\pi} \int_0^{\pi/4} \int_0^{\cos\varphi} (\rho \cos\varphi)\rho^2 \sin\varphi \, d\rho \, d\varphi \, d\theta \\ &= \frac{k}{4} \int_0^{2\pi} \int_0^{\pi/4} \cos^5\varphi \sin\varphi \, d\varphi \, d\theta = \frac{k}{4} \int_0^{2\pi} \left\{ -\frac{\cos^6\varphi}{6}\Big|_0^{\pi/4} \right\} d\theta \\ &= \frac{1}{24} \cdot \frac{7}{8} k \int_0^{2\pi} d\theta = \frac{7\pi k}{96}. \end{aligned}$$

Thus $\bar{x} = \bar{y} = 0$ and $\bar{z} = M_{xy}/\mu = (7\pi k/96)/(\pi k/8) = 7/12$. ■

PROBLEMS 5.7

Solve Problems 1–11 by using cylindrical coordinates.

1. Find the volume of the region inside both the sphere $x^2 + y^2 + z^2 = 4$ and the cylinder $(x - 1)^2 + y^2 = 1$.
2. Find the centroid of the region of Problem 1.
*3. Suppose the density of the region of Problem 1 is proportional to the square of the distance to the xy-plane and is measured in kilograms per cubic meter. Find the center of mass of the region.
4. Find the volume of the solid bounded above by the paraboloid $z = 4 - x^2 - y^2$ and below by the xy-plane.
5. Find the center of mass of the solid of Problem 4 if the density is proportional to the distance to the xy-plane.
6. Find the volume of the solid bounded by the plane $z = y$ and the paraboloid $z = x^2 + y^2$.
7. Find the centroid of the region of Problem 6.
8. Find the volume of the solid bounded by the two cones $z^2 = x^2 + y^2$ and $z^2 = 16x^2 + 16y^2$ between $z = 0$ and $z = 2$.
9. Find the center of mass of the solid in Problem 8 if the density at any point is proportional to the distance to the z-axis.
10. Find the volume of the solid bounded by the hyperboloid of two sheets $z^2 - x^2 - y^2 = 1$ and the cone $z^2 = x^2 + y^2$ for $0 \le z \le a$.
11. Evaluate $\displaystyle\int_0^1 \int_0^{\sqrt{1-x^2}} \int_{\sqrt{x^2+y^2}}^{\sqrt{2-x^2-y^2}} z^3 \, dz \, dy \, dx$.

Solve Problems 12–21 by using spherical coordinates.

12. Find the mass of the unit sphere if the density at any point is proportional to the distance to the boundary of the sphere.
13. Find the volume of the solid inside the sphere $x^2 + y^2 + z^2 = 4$ and outside the cone $z^2 = x^2 + y^2$.
14. Find the center of mass of the solid of Problem 12 if the density at a point is proportional to the distance from the origin.
15. A solid fills the space between two concentric spheres centered at the origin. The radii of the spheres are a and b, where $0 < a < b$. Find the mass of the solid if the density at each point is inversely proportional to its distance from the origin.
16. Find the center of mass of the object of Problem 13 if the density at a point is proportional to the square of the distance from the origin.

17. Find the volume of one of the smaller wedges cut from the unit sphere by two planes that meet at a diameter at an angle of $\pi/6$.

18. Find the mass of a wedge cut from a sphere of radius a by two planes that meet at a diameter at an angle of b radians if the density at any point on the wedge is proportional to the distance to that diameter.

19. Evaluate

$$\int_{-3}^{3}\int_{-\sqrt{9-x^2}}^{\sqrt{9-x^2}}\int_{-\sqrt{9-x^2-y^2}}^{\sqrt{9-x^2-y^2}} (x^2+y^2+z^2)^{3/2}\,dz\,dy\,dx.$$

20. Evaluate

$$\int_{0}^{1}\int_{0}^{\sqrt{1-x^2}}\int_{0}^{\sqrt{1-x^2-y^2}} (1/\sqrt{x^2+y^2+z^2})\,dz\,dy\,dx.$$

21. Evaluate

$$\int_{0}^{a}\int_{0}^{\sqrt{a^2-x^2}}\int_{0}^{\sqrt{a^2-x^2-y^2}} \frac{z^3}{(x^2+y^2)^{1/2}}\,dz\,dy\,dx.$$

***22.** A "sphere" in $\mathbb{R}^4$ has the equation $x^2+y^2+z^2+w^2=a^2$.

(a) Explain why it is reasonable that the volume of this "sphere" be given by

$$V = \int_{-a}^{a}\int_{-\sqrt{a^2-x^2}}^{\sqrt{a^2-x^2}}\int_{-\sqrt{a^2-x^2-y^2}}^{\sqrt{a^2-x^2-y^2}}\int_{-\sqrt{a^2-x^2-y^2-z^2}}^{\sqrt{a^2-x^2-y^2-z^2}} dw\,dz\,dy\,dx$$

$$= 16\int_{0}^{a}\int_{0}^{\sqrt{a^2-x^2}}\int_{0}^{\sqrt{a^2-x^2-y^2}}\int_{0}^{\sqrt{a^2-x^2-y^2-z^2}} dw\,dz\,dy\,dx.$$

(b) Using spherical coordinates, evaluate the integral in (a). [*Hint:* The first integral is easy and reduces the "quadruple" integral to a triple integral.]

REVIEW EXERCISES FOR CHAPTER FIVE

In Exercises 1–8, evaluate the integral.

1. $\int_0^1\int_0^2 x^2y\,dx\,dy$

2. $\int_0^1\int_{x^2}^{x} xy^3\,dy\,dx$

3. $\int_2^4\int_{1+y}^{2+5y} (x-y^2)\,dx\,dy$

4. $\int_0^4\int_{-\sqrt{16-x^2}}^{\sqrt{16-x^2}} 4y\,dy\,dx$

5. $\iint_\Omega (y-x^2)\,dx\,dy$, where $\Omega=\{(x,y): -3\le x\le 3, 0\le y\le 2\}$.

6. $\iint_\Omega (y-x^2)\,dx\,dy$, where Ω is the region in the first quadrant bounded by the x-axis, the y-axis, and the circle $x^2+y^2=4$.

7. $\iint_\Omega (x+y^2)\,dx\,dy$, where Ω is the region in the first quadrant bounded by the x-axis, the y-axis, and the unit circle.

8. $\iint_\Omega (2x+y)e^{-(x+y)}\,dx\,dy$, where Ω is the first quadrant.

9. Find upper and lower bounds for $\iint_\Omega \sin(x^5+y^{12})\,dA$, where Ω is the region in the first quadrant bounded by the x-axis, the y-axis, and the circle $x^2+y^2=4$.

10. Change the order of integration of $\int_2^5\int_1^x 3x^2y\,dy\,dx$ and evaluate.

11. Change the order of integration of $\int_0^3\int_0^{\sqrt{9-x^2}} (9-y^2)^{3/2}\,dy\,dx$ and evaluate.

12. Change the order of integration of $\int_0^\infty\int_x^\infty f(x,y)\,dy\,dx$.

13. Find the volume of the solid bounded by the plane $x+2y+3z=6$ and the three coordinate planes.

14. Find the volume of the solid bounded by the cylinder $y^2+z^2=4$ and the planes $x=0$ and $x=3$.

15. Find the volume enclosed by the ellipsoid $4x^2+y^2+25z^2=100$.

16. Find the volume of the solid bounded by the paraboloid $x=y^2+z^2$ and the plane $y=x-2$.

17. Find the center of mass of the unit disk if the density at a point is proportional to the distance to the boundary of the disk.

18. Find the center of mass of the region $\Omega=\{(x,y): 0\le x\le 3, 1\le y\le 5\}$ if $\rho(x,y)=3x^2y$.

19. Find the center of mass of the triangular region bounded by the lines $y=x$, $y=2-x$, and the y-axis if $\rho(x,y)=3xy+y^3$.

20. Calculate the volume under the surface $z=r^5$ and over the circle of radius 6 centered at the origin.

21. Calculate the area of the region enclosed by the curve $r=2(1+\sin\theta)$.

22. Calculate the area of the region inside the cardioid $r=3(1+\sin\theta)$ and outside the circle $r=2$.

23. Find the volume of the solid bounded by the cone $z^2=x^2+y^2$ and the paraboloid $4z=x^2+y^2$.

24. Find the centroid of the region bounded by the right-hand loop of the lemniscate $r^2=4\cos 2\theta$.

25. Find the area of the part of the plane $z=3x+5y$ over the region bounded by the curves $x=y^2$ and $x=y^3$.

26. Find the area of the part of the surface $z=(y^4/4)+(1/8y^2)$ over the rectangle $\{(x,y): 0\le x\le 3, 2\le y\le 4\}$.

27. Find the surface area of the hemisphere $x^2 + y^2 + z^2 = 16$, $z \geq 0$.

28. Evaluate $\int_0^1 \int_0^y \int_0^x x^2 \, dz \, dx \, dy$.

29. Evaluate $\int_1^2 \int_{2-x}^{2+x} \int_0^{xz} 12xyz \, dy \, dz \, dx$.

30. Change the order of integration in Exercise 28 and write the integral in the following forms:
(a) $\int_?^? \int_?^? \int_?^? x^2 \, dx \, dy \, dz$ (b) $\int_?^? \int_?^? \int_?^? x^2 \, dy \, dz \, dx$

31. Find the volume of the tetrahedron with vertices at (0, 0, 0), (2, 0, 0), (0, 1, 0), and (0, 0, 3).

32. Find the volume of the solid bounded by the parabolic cylinder $z = 4 - y^2$ and the planes $z = x$ and $z = 2x$ that lies in the half plane $z \geq 0$.

33. Find the mass of the solid of Exercise 31 if the density function is $\rho(x, y, z) = 2x + y^2$.

34. Find the center of mass of the solid bounded by the ellipsoid $4x^2 + y^2 + 25z^2 = 100$ if the density function is $\rho(x, y, z) = z^2$.

35. Find the centroid of the solid in Exercise 32.

36. Find the volume of the region inside both the sphere $x^2 + y^2 + z^2 = a^2$ and the cylinder $x^2 + [y - (a/2)]^2 = (a/2)^2$.

37. Find the centroid of the solid of Exercise 36.

38. The density of the region of Exercise 36 is proportional to the square of the distance to the xy-plane. Find the center of mass of the region.

39. Find the volume of the solid bounded by the plane $z = x$ and the paraboloid $z = x^2 + y^2$.

40. Evaluate $\displaystyle\int_{-2}^{2} \int_{-\sqrt{4-x^2}}^{\sqrt{4-x^2}} \int_{\sqrt{x^2+y^2}}^{\sqrt{4-x^2-y^2}} z^2 \, dz \, dy \, dx$.

41. Find the volume of the solid inside the sphere $x^2 + y^2 + z^2 = 9$ and outside the cone $z^2 = x^2 + y^2$.

42. Find the center of mass of the solid in Exercise 41 if the density at a point is proportional to its distance to the origin.

43. Find the volume of a wedge cut from a sphere of radius 2 by two planes that meet at a diameter at an angle of $\pi/3$.

44. Evaluate $\displaystyle\int_0^2 \int_0^{\sqrt{4-x^2}} \int_0^{\sqrt{4-x^2-y^2}} \frac{z^2}{(x^2 + y^2)^{1/2}} \, dz \, dy \, dx$.

6 Introduction to Vector Analysis

6.1 VECTOR FIELDS

We wrote about vector fields briefly in our discussion of potential functions and the gradient in Section 4.9. The notion of a vector field is central to the subject of **vector analysis.** We provide formal definitions now.

Definition 1 VECTOR FIELD IN $\mathbb{R}^2$ Let Ω be a region in $\mathbb{R}^2$. Then $\mathbf{F}$ is a **vector field** in $\mathbb{R}^2$ if $\mathbf{F}$ assigns to every $\mathbf{x}$ in Ω a unique vector $\mathbf{F}(\mathbf{x})$ in $\mathbb{R}^2$.

REMARK. Simply put, a vector field in $\mathbb{R}^2$ is a function whose domain is a subset of $\mathbb{R}^2$ and whose range is a subset of $\mathbb{R}^2$. Two vector fields in $\mathbb{R}^2$ are sketched in Figure 1. The meaning of this sketch is that to every point $\mathbf{x}$ in Ω a unique vector $\mathbf{F}(\mathbf{x})$ is assigned. That is, the function value $\mathbf{F}(\mathbf{x})$ is represented by an arrow with $\mathbf{x}$ at the "tail" of the arrow.

EXAMPLE 1 **(Gradient Field in $\mathbb{R}^2$)** Let $z = f(x, y)$ be a differentiable function. Then $\mathbf{F} = \nabla f = f_x\mathbf{i} + f_y\mathbf{j}$ is a vector field. ■

Definition 1 can be extended, in an obvious way, to $\mathbb{R}^3$.

Definition 2 VECTOR FIELD IN $\mathbb{R}^3$ Let S be a region in $\mathbb{R}^3$. Then $\mathbf{F}$ is a **vector field** in $\mathbb{R}^3$ if $\mathbf{F}$ assigns to each vector $\mathbf{x}$ in S a unique vector $\mathbf{F}(\mathbf{x})$ in $\mathbb{R}^3$.

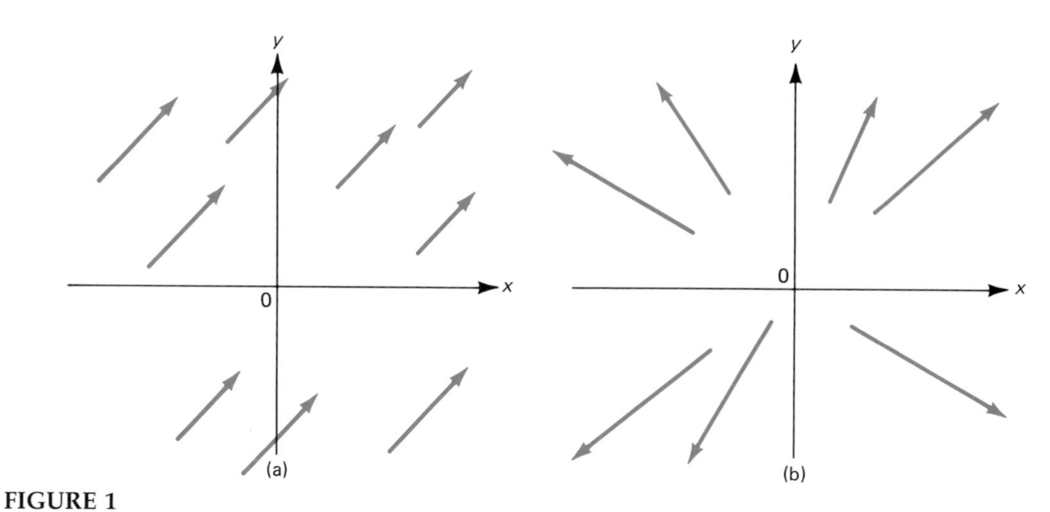

FIGURE 1

REMARK. A vector field in $\mathbb{R}^3$ is a function whose domain and range are subsets of $\mathbb{R}^3$.

Two vector fields in $\mathbb{R}^3$ are sketched in Figure 2.

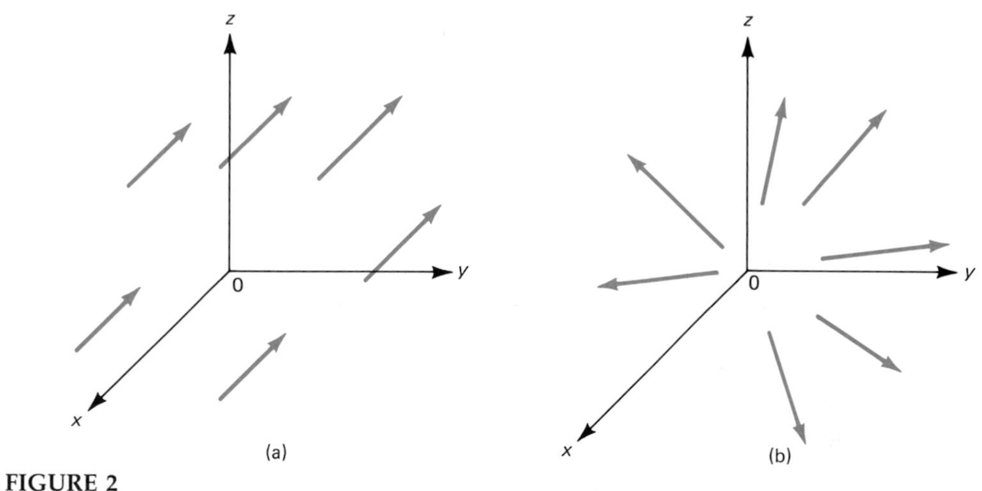

FIGURE 2

EXAMPLE 2 **(Gradient Field in $\mathbb{R}^3$)** If $u = f(x, y, z)$ is differentiable, then $\mathbf{F} = \nabla f = f_x\mathbf{i} + f_y\mathbf{j} + f_z\mathbf{k}$ is a vector field. It is called a **gradient vector field.** ■

EXAMPLE 3 From Section 4.9, if $\mathbf{F} = -\nabla f$ for some differentiable function $z = f(x, y)$ or $w = f(x, y, z)$, then $\mathbf{F}$ is called a **conservative vector field.** ■

EXAMPLE 4 **(Gravitational Field)** Let m_1 represent the mass of a (relatively) fixed object in space and let m_2 denote the mass of an object moving near the fixed object. Then the magnitude of the gravitational force between the objects is given by (see page 234)

$$|\mathbf{F}| = G\frac{m_1 m_2}{r^2} \tag{1}$$

where r is the distance between the objects and G is a universal constant. If we assume that the first object is at the origin, then we may denote the position of the second object by $\mathbf{x}(t)$, and then, since $r = |\mathbf{x}|$, (1) can be written

$$|\mathbf{F}| = \frac{Gm_1 m_2}{|\mathbf{x}|^2}. \tag{2}$$

Also, the force acts toward the origin, that is, in the direction opposite to that of the position vector $\mathbf{x}$. Therefore

$$\text{direction of } \mathbf{F} = -\frac{\mathbf{x}}{|\mathbf{x}|} \tag{3}$$

so from (2) and (3),

$$\mathbf{F}(t) = \frac{\alpha\mathbf{x}(t)}{|\mathbf{x}(t)|^3} \tag{4}$$

where $\alpha = -Gm_1 m_2$.

The vector field (4) is called a **gravitational field.** As we saw in Section 4.9, such a field is conservative. We can sketch this vector field without much difficulty because for every $\mathbf{x} \neq \mathbf{0} \in \mathbb{R}^3$, $\mathbf{F}(\mathbf{x})$ points toward the origin. The sketch appears in Figure 3. ■

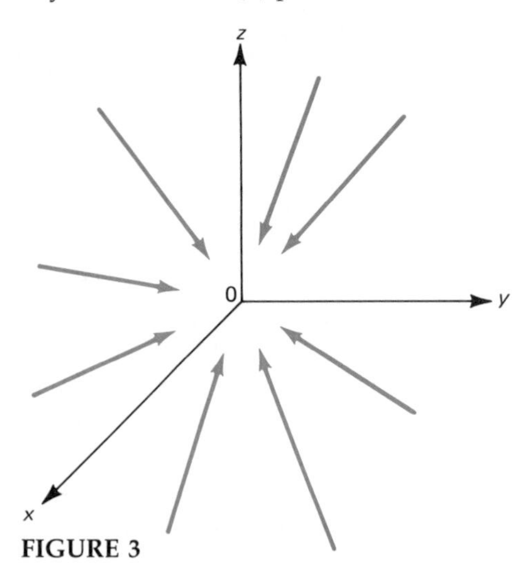

FIGURE 3

Vector fields arise in a great number of physical applications. There are, for example, mechanical force fields, magnetic fields, electric fields, velocity fields, and direction fields.

EXAMPLE 5 Let $\mathbf{F}$ be the vector field in $\mathbb{R}^2$ given by

$$\mathbf{F} = \frac{2(x^2 - y^2 - 1)}{[(x+1)^2 + y^2][(x-1)^2 + y^2]}\mathbf{i} + \frac{4xy}{[(x+1)^2 + y^2][(x-1)^2 + y^2]}\mathbf{j}. \qquad (5)$$

It can be shown that **F** is the electric field in the plane caused by two infinite straight wires that are perpendicular to the xy-plane and that pass through the points (1, 0) and (−1, 0). It is assumed that the wires are uniformly charged with electricity and that the charge of one is opposite to the charge of the other. This field is sketched in Figure 4. (Also, see Problem 25.) ■

EXAMPLE 6 **(Velocity Field of a Rotating Body)** Suppose a body is rotating in space around an axis of rotation (see Figure 5). We define a vector **w** as follows: **w** has the direction of the axis of rotation such that if we look from the initial point of **w** to its terminal point, the motion appears clockwise. We now define

$$\textbf{angular speed } \omega = \frac{\text{tangential speed of a point } \mathbf{x}(t)}{\text{distance from } \mathbf{x} \text{ to the axis of rotation}}. \qquad (6)$$

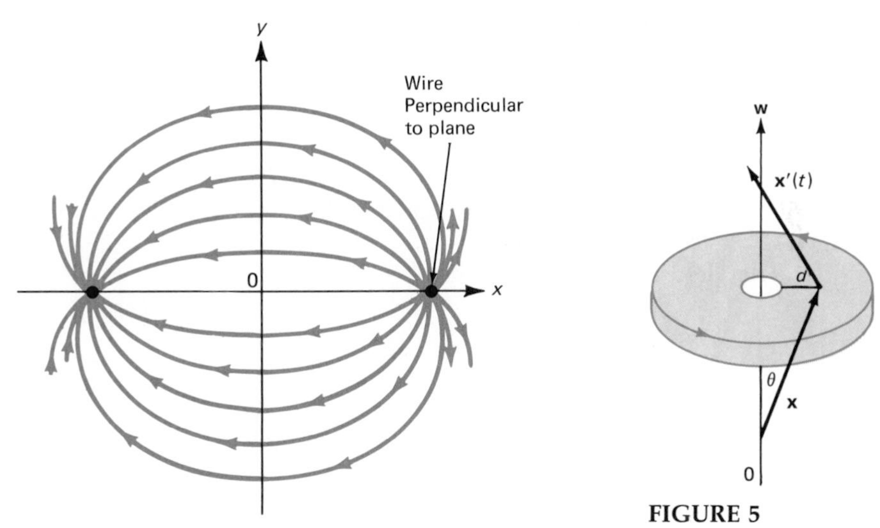

FIGURE 4

FIGURE 5

Note that if $\mathbf{x}(t)$ represents the position vector, then the velocity vector is $\mathbf{x}'(t)$ and its speed is $|\mathbf{x}'(t)|$. Thus if d denotes the distance of a point **x** from the axis of rotation, we have, from (6),

$$\omega = \frac{|\mathbf{x}'(t)|}{d}. \qquad (7)$$

Then we define the magnitude of **w** to be ω. That is,

$$|\mathbf{w}| = \omega. \qquad (8)$$

Let θ denote the angle between **x** and **w**. Then from Figure 5

$$\sin\theta = \frac{d}{|\mathbf{x}|}$$

or

$$d = |\mathbf{x}| \sin\theta,$$

and

$$\omega d = |\mathbf{x}|\, \omega \sin\theta = |\mathbf{x}|\, |\mathbf{w}| \sin\theta = |\mathbf{x} \times \mathbf{w}|.$$

Thus from (7)

$$|\mathbf{x}'(t)| = |\mathbf{x} \times \mathbf{w}|. \tag{9}$$

Now, again from Figure 5, we see that the velocity vector $\mathbf{x}'$ is perpendicular to both $\mathbf{x}$ and $\mathbf{w}$, so that the direction of $\mathbf{v} = \mathbf{x}'$ is the direction of $\mathbf{w} \times \mathbf{x}$. Combining this result with (9), we obtain

$$\mathbf{v} = \mathbf{w} \times \mathbf{x}.$$

We can write this **velocity field** as

$$\mathbf{v}(x, y, z) = \mathbf{w} \times (x\mathbf{i} + y\mathbf{j} + z\mathbf{k}). \tag{10}$$

This field is sketched in Figure 6. ■

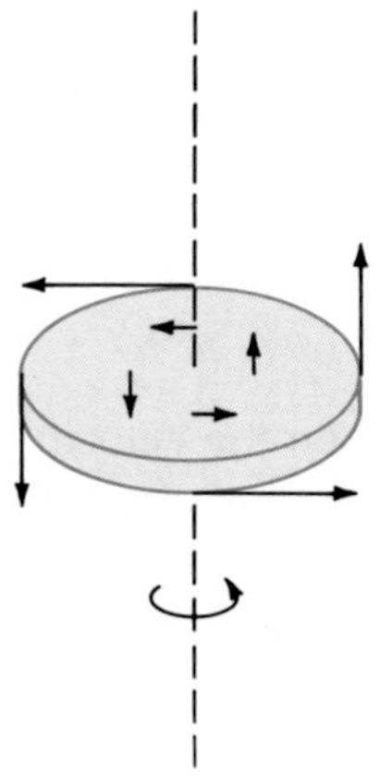

FIGURE 6

EXAMPLE 7 Suppose that the axis of rotation in Example 6 is the y-axis and that $\mathbf{w}$ points in the positive y-direction. Then

$$\mathbf{w} = \omega\mathbf{j} \quad \text{and} \quad \mathbf{w} \times \mathbf{x} = \begin{vmatrix} \mathbf{i} & \mathbf{j} & \mathbf{k} \\ 0 & \omega & 0 \\ x & y & z \end{vmatrix} = \omega z\mathbf{i} - \omega x\mathbf{k} = \omega(z\mathbf{i} - x\mathbf{k}). \quad \blacksquare$$

NOTE. In Examples 6 and 7, ω is not necessarily constant. If ω is constant, the body is said to have **constant angular speed.**

PROBLEMS 6.1

In Problems 1–21, compute the gradient vector field of the given function.

1. $f(x, y) = (x + y)^2$
2. $f(x, y) = e^{\sqrt{xy}}$
3. $f(x, y) = \cos(x - y)$
4. $f(x, y) = \ln(2x - y + 1)$
5. $f(x, y) = \sqrt{x^2 + y^3}$
6. $f(x, y) = \tan^{-1} \dfrac{y}{x}$
7. $f(x, y) = y \tan(y - x)$
8. $f(x, y) = x^2 \sinh y$
9. $f(x, y) = \sec(x + 3y)$
10. $f(x, y) = \dfrac{x - y}{x + y}$
11. $f(x, y) = \dfrac{x^2 - y^2}{x^2 + y^2}$
12. $f(x, y) = \dfrac{e^{x^2} - e^{-y^2}}{3y}$
13. $f(x, y, z) = xyz$
14. $f(x, y, z) = \sin x \cos y \tan z$
15. $f(x, y, z) = \dfrac{x^2 - y^2 + z^2}{3xy}$
16. $f(x, y, z) = x \ln y - z \ln x$
17. $f(x, y, z) = xy^2 + y^2z^3$
18. $f(x, y, z) = (y - z)e^{x+2y+3z}$
19. $f(x, y, z) = x \sin y \ln z$
20. $f(x, y, z) = \dfrac{x - z}{\sqrt{1 - y^2 + x^2}}$
21. $f(x, y, z) = x \cosh z - y \sin x$
22. Define a gravitational field in $\mathbb{R}^2$ and show that it is conservative.
23. Find the velocity field of a rotating body whose axis of rotation is the z-axis with $\mathbf{w}$ pointing in the direction of the positive z-axis. Assume that ω is constant.
24. Answer the question of Problem 23 after replacing the z-axis with the x-axis.
25. Show that the force field given in Example 5 is the gradient of the function

$$f(x, y) = \ln\frac{\sqrt{(x - 1)^2 + y^2}}{\sqrt{(x + 1)^2 + y^2}}.$$

6.2 WORK, LINE INTEGRALS IN THE PLANE, AND INDEPENDENCE OF PATH

In Section 1.3 we showed that if a constant force $\mathbf{F}$ is applied to a particle that moves along a vector $\mathbf{d}$, then the work done by the force on the particle is given by

$$W = \mathbf{F} \cdot \mathbf{d}. \tag{1}$$

We now calculate the work done when a particle moves along a curve C. In doing so, we will define an important concept in applied mathematics—the *line integral.*

Suppose that a curve in the plane is given parametrically by

$$C: \quad \mathbf{x}(t) = f(t)\mathbf{i} + g(t)\mathbf{j}, \tag{2}$$

or

$$C: \quad \mathbf{x}(t) = x(t)\mathbf{i} + y(t)\mathbf{j}. \tag{2'}$$

If a force is applied to a particle moving along C, then such a force will have magnitude and direction, so the force will be a vector function of the vector $\mathbf{x}(t)$. That is, the force will be a vector field. We write

$$\mathbf{F}(\mathbf{x}) = \mathbf{F}(x, y) = P(x, y)\mathbf{i} + Q(x, y)\mathbf{j} \tag{3}$$

where P and Q are scalar-valued functions. The problem is to determine the work done when a particle moves on C from a point $\mathbf{x}(a)$ to a point $\mathbf{x}(b)$ subject to the force $\mathbf{F}$ given by (3). We will assume in our discussion that the curve C is *smooth* or *piecewise smooth*. By that we mean that the functions $f(t)$ and $g(t)$ in (2) are continuously differentiable or that f' and g' exist and are piecewise continuous.[†]

A typical curve C is sketched in Figure 1. Let

$$W(t) = \text{work done in moving from } \mathbf{x}(a) \text{ to } \mathbf{x}(t). \tag{4}$$

Then the work done in moving from $\mathbf{x}(t)$ to $\mathbf{x}(t + \Delta t)$ is given by

$$W(t + \Delta t) - W(t). \tag{5}$$

Now if Δt is small, then the part of the curve between $\mathbf{x}(t)$ and $\mathbf{x}(t + \Delta t)$ is "close" to a straight line and so can be approximated by the vector

$$\mathbf{x}(t + \Delta t) - \mathbf{x}(t). \tag{6}$$

If Δt is small and if $\mathbf{F}(\mathbf{x})$ is continuous, then the force applied between $\mathbf{x}(t)$ and $\mathbf{x}(t + \Delta t)$ is approximately equal to $\mathbf{F}(\mathbf{x}(t))$. Thus by (1), if Δt is small, then

$$W(t + \Delta t) - W(t) \approx \mathbf{F}(\mathbf{x}(t)) \cdot [\mathbf{x}(t + \Delta t) - \mathbf{x}(t)]. \tag{7}$$

We divide both sides of (7) by Δt and take the limit as $\Delta t \to 0$ to obtain

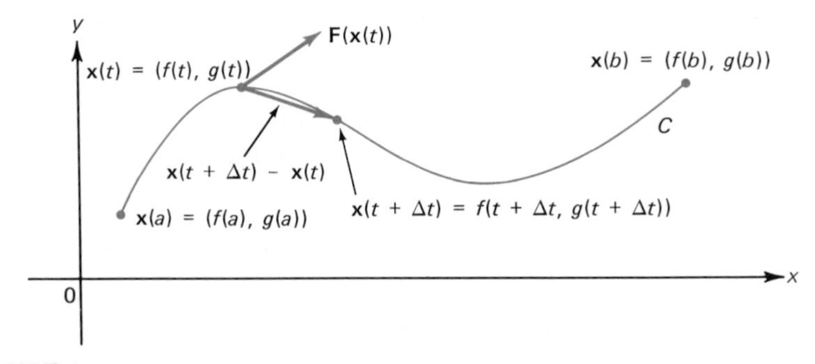

FIGURE 1

[†]See p. 122 in *Calculus* or *Calculus of One Variable* for a definition of piecewise continuity.

$$\lim_{\Delta t \to 0} \frac{W(t + \Delta t) - W(t)}{\Delta t} = \lim_{\Delta t \to 0} \left\{ \mathbf{F}(\mathbf{x}(t)) \cdot \frac{[\mathbf{x}(t + \Delta t) - \mathbf{x}(t)]}{\Delta t} \right\}$$

or

$$W'(t) = \mathbf{F}(\mathbf{x}(t)) \cdot \mathbf{x}'(t). \tag{8}$$

But

$$W(a) = 0$$

and

$$W(b) = \text{total work done on the particle.}$$

Thus

$$W = \text{total work done} = W(b) - W(a) = \int_a^b W'(t)\, dt,$$

or

$$W = \int_a^b \mathbf{F}(\mathbf{x}(t)) \cdot \mathbf{x}'(t)\, dt. \tag{9}$$

We write equation (9) as

$$W = \int_C \mathbf{F}(\mathbf{x}) \cdot \mathbf{dx}. \tag{10}$$

The symbol $\int_C$ is read "the integral along the curve C." The integral in (10) is called a *line integral of* **F** *over* C.

REMARK 1. If C lies along the x-axis, then C is given by $\mathbf{x}(t) = x(t)\mathbf{i} + 0\mathbf{j}$[†] and $\mathbf{x}'(t) = x'(t)\mathbf{i}$, so that (10) becomes

$$\int_{[a,\,b]} F(x)\, dx = \int_a^b F(x)\, dx,$$

which is our usual definite integral.

REMARK 2. Since **F** is given by (3), we can write equation (9) as

$$W = \int_a^b [P(x(t), y(t))\, x'(t) + Q(x(t), y(t))\, y'(t)]\, dt. \tag{11}$$

[†]We have substituted $x(t)$ for $f(t)$ and 0 for $g(t)$ here.

EXAMPLE 1 A particle is moving along the parabola $y = x^2$ subject to a force given by the vector field $2xy\mathbf{i} + (x^2 + y^2)\mathbf{j}$. How much work is done in moving from the point (1, 1) to the point (3, 9) if forces are measured in newtons and distances are measured in meters?

Solution. The curve C is given parametrically by

$$\mathbf{x}(t) = t\mathbf{i} + t^2\mathbf{j} \qquad \text{between} \qquad t = 1 \quad \text{and} \quad t = 3.$$

We therefore have $f(t) = t$ and $g(t) = t^2$. Then $P = 2xy = 2t^3$ and $Q = x^2 + y^2 = t^2 + t^4$. Also, $f'(t) = 1$ and $g'(t) = 2t$, so by (11)

$$W = \int_1^3 [(2t^3)1 + (t^2 + t^4)(2t)]\,dt = \int_1^3 (4t^3 + 2t^5)\,dt = 322\tfrac{2}{3}\text{ J.} \quad ■$$

EXAMPLE 2 Two electrical charges of like polarity (i.e., both positive or both negative) will repel each other. If a charge of α coulombs is placed at the origin and a charge of 1 coulomb of the same polarity is at the point (x, y), then the force of repulsion is given by

$$F(x, y) = \frac{\alpha x}{(x^2 + y^2)^{3/2}}\mathbf{i} + \frac{\alpha y}{(x^2 + y^2)^{3/2}}\mathbf{j}.$$

How much work is done by the force on the 1-coulomb charge as the charge moves on the straight line from (1, 0) to (3, −2)?

Solution. The straight line is the line $y = 1 - x$, or parametrically,

$$\mathbf{x}(t) = t\mathbf{i} + (1 - t)\mathbf{j} \qquad \text{for} \qquad 1 \le t \le 3.$$

Then

$$P = \frac{\alpha t}{[(t^2 + (1 - t)^2]^{3/2}} = \frac{\alpha t}{(2t^2 - 2t + 1)^{3/2}}$$

and

$$Q = \frac{\alpha(1 - t)}{[t^2 + (1 - t)^2]^{3/2}} = \frac{\alpha(1 - t)}{(2t^2 - 2t + 1)^{3/2}}.$$

Then since $f'(t) = 1$ and $g'(t) = -1$,

$$\begin{aligned} W &= \alpha \int_1^3 \left\{\frac{t}{(2t^2 - 2t + 1)^{3/2}}(1) + \frac{(1 - t)}{(2t^2 - 2t + 1)^{3/2}}(-1)\right\} dt \\ &= \alpha \int_1^3 \frac{2t - 1}{(2t^2 - 2t + 1)^{3/2}}\,dt = \frac{\alpha}{2}\int_1^3 \frac{4t - 2}{(2t^2 - 2t + 1)^{3/2}}\,dt \\ &= -\alpha(2t^2 - 2t + 1)^{-1/2}\Big|_1^3 = \alpha\left(1 - \frac{1}{\sqrt{13}}\right). \quad ■ \end{aligned}$$

EXAMPLE 3 How much work is done by the force on the charge in Example 2 if it moves in the counterclockwise direction along the semicircle $x^2 + y^2 = a^2$, $y \geq 0$?

Solution. Here

$$\mathbf{x}(t) = a(\cos t)\mathbf{i} + a(\sin t)\mathbf{j}, \qquad 0 \leq t \leq \pi.$$

Then

$$P = \frac{\alpha x}{(x^2 + y^2)^{3/2}} = \frac{a\alpha \cos t}{(a^2 \cos^2 t + a^2 \sin^2 t)^{3/2}} = \frac{a\alpha \cos t}{a^3} = \frac{\alpha \cos t}{a^2}$$

and

$$Q = \frac{\alpha y}{(x^2 + y^2)^{3/2}} = \frac{\alpha \sin t}{a^2}.$$

Since $f'(t) = -a \sin t$ and $g'(t) = a \cos t$,

$$W = \frac{\alpha}{a} \int_0^{\pi} [\cos t(-\sin t) + \sin t(\cos t)]\, dt = 0.$$

This result is no surprise since in this example the force **F** and the direction of the curve $\mathbf{x}'$ are orthogonal. Thus the component of the force in the direction of motion is zero, which means that the force does no work. ■

We used the notion of work to motivate the discussion of the line integral. We now give a general definition of the line integral in the plane.

Definition 1 LINE INTEGRAL IN THE PLANE Let P and Q be continuous on a set S containing the smooth (or piecewise smooth) curve C given by

$$C: \quad \mathbf{x}(t) = f(t)\mathbf{i} + g(t)\mathbf{j}, \qquad t \in [a, b].$$

Let the vector field **F** be given by

$$\mathbf{F}(x, y) = P(x, y)\mathbf{i} + Q(x, y)\mathbf{j}.$$

Then the **line integral** of **F** over C is given by

$$\int_C \mathbf{F}(\mathbf{x}) \cdot d\mathbf{x} = \int_a^b [P(f(t), g(t))f'(t) + Q(f(t), g(t))g'(t)]\, dt. \qquad \mathbf{(12)}$$

REMARK. If C is piecewise smooth but not smooth, then C is made up of a number of "sections," each of which is smooth. Since C is continuous, these sections are joined. Some typical piecewise smooth curves are sketched in Figure 2. If C is made up of the n smooth curves $C_1, C_2, \ldots, C_n$, then

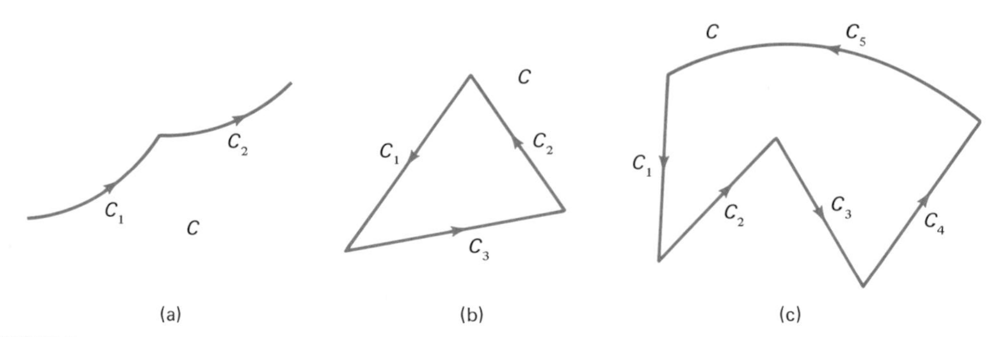

FIGURE 2

$$\int_C \mathbf{F(x)} \cdot \mathbf{dx} = \int_{C_1} \mathbf{F(x)} \cdot \mathbf{dx} + \int_{C_2} \mathbf{F(x)} \cdot \mathbf{dx} + \cdots + \int_{C_n} \mathbf{F(x)} \cdot \mathbf{dx}. \tag{13}$$

EXAMPLE 4 Calculate $\int_C \mathbf{F(x)} \cdot \mathbf{dx}$, where $\mathbf{F}(x, y) = xy\mathbf{i} + ye^x\mathbf{j}$ and C is the rectangle joining the points (0, 0), (2, 0), (2, 1), and (0, 1) if C is traversed in the counterclockwise direction.

Solution. The rectangle is sketched in Figure 3, and it is made up of four smooth curves (straight lines). We have

$$\begin{aligned}
C_1:&\quad \mathbf{x}(t) = t\mathbf{i}, && 0 \le t \le 2\\
C_2:&\quad \mathbf{x}(t) = 2\mathbf{i} + t\mathbf{j}, && 0 \le t \le 1\\
C_3:&\quad \mathbf{x}(t) = (2 - t)\mathbf{i} + \mathbf{j}, && 0 \le t \le 2\\
C_4:&\quad \mathbf{x}(t) = (1 - t)\mathbf{j}, && 0 \le t \le 1.
\end{aligned}$$

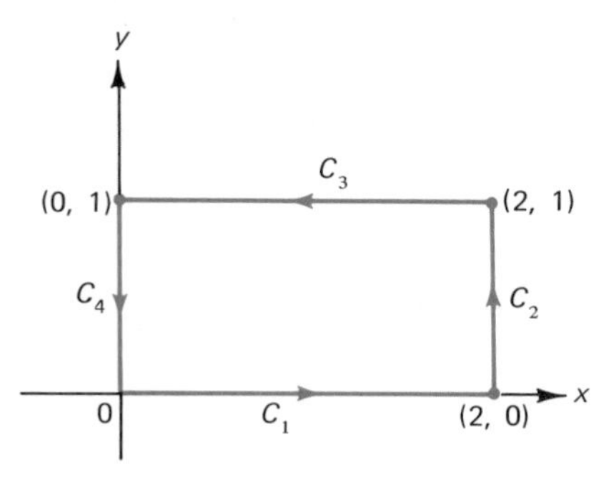

FIGURE 3

Note, for example, that on C_3, $t = 0$ corresponds to the point $2\mathbf{i} + \mathbf{j} = (2, 1)$ and $t = 2$ corresponds to $(2 - 2)\mathbf{i} + \mathbf{j} = (0, 1)$. Thus as t increases, we do move along C_3 in the direction indicated by the arrow in Figure 3. This illustrates why our parametrization of the rectangle is correct.

Now

$$\int_C \mathbf{F(x)} \cdot \mathbf{dx} = \int_{C_1} \mathbf{F(x)} \cdot \mathbf{dx} + \int_{C_2} \mathbf{F(x)} \cdot \mathbf{dx} + \int_{C_3} \mathbf{F(x)} \cdot \mathbf{dx} + \int_{C_4} \mathbf{F(x)} \cdot \mathbf{dx}.$$

On C_1, $x = t$ and $y = 0$, so that $xy = 0$, $ye^x = 0$, and

$$\int_{C_1} \mathbf{F(x)} \cdot \mathbf{dx} = 0.$$

On C_2, $x = 2$, $y = t$, $x' = 0$, $y' = 1$, and $ye^x = te^2$, so that

$$\int_{C_2} \mathbf{F(x)} \cdot \mathbf{dx} = \int_0^1 te^2\, dt = \frac{e^2}{2}.$$

On C_3, $x = (2 - t)$, $y = 1$, $x' = -1$, $y' = 0$, and $xy = 2 - t$, so that

$$\int_{C_3} \mathbf{F(x)} \cdot \mathbf{dx} = \int_0^2 (2 - t)(-1)\, dt = -2.$$

On C_4, $x = 0$, $y = 1 - t$, $x' = 0$, $y' = -1$, $xy = 0$, and $ye^x = 1 - t$, so that

$$\int_{C_4} \mathbf{F(x)} \cdot \mathbf{dx} = \int_0^1 (1 - t)(-1)\, dt = -\frac{1}{2}.$$

Thus

$$\int_{C} \mathbf{F(x)} \cdot \mathbf{dx} = 0 + \frac{e^2}{2} - 2 - \frac{1}{2} = \frac{e^2}{2} - \frac{5}{2} \approx 1.2. \quad \blacksquare$$

There are certain conditions under which the calculation of a line integral becomes very easy. We first illustrate what we have in mind with an example.

EXAMPLE 5 Let $\mathbf{F}(x, y) = y\mathbf{i} + x\mathbf{j}$. Calculate $\int_C \mathbf{F(x)} \cdot \mathbf{dx}$, where C is as follows:

(a) The straight line from (0, 0) to (1, 1).
(b) The parabola $y = x^2$ from (0, 0) to (1, 1).
(c) The curve $\mathbf{x}(t) = t^{3/2}\mathbf{i} + t^5\mathbf{j}$ from (0, 0) to (1, 1).

Solution. **(a)** C is given by $\mathbf{x}(t) = t\mathbf{i} + t\mathbf{j}$. Then $\mathbf{x}'(t) = \mathbf{i} + \mathbf{j}$ and $\mathbf{F} \cdot \mathbf{x}' = (t\mathbf{i} + t\mathbf{j}) \cdot (\mathbf{i} + \mathbf{j}) = 2t$, so

$$\int_C \mathbf{F(x)} \cdot \mathbf{dx} = \int_0^1 2t\, dt = 1.$$

(b) C is given by $\mathbf{x}(t) = t\mathbf{i} + t^2\mathbf{j}$. Then $\mathbf{x}'(t) = \mathbf{i} + 2t\mathbf{j}$ and $\mathbf{F} \cdot \mathbf{x}' = (t^2\mathbf{i} + t\mathbf{j}) \cdot (\mathbf{i} + 2t\mathbf{j}) = 3t^2$, so

$$\int_C \mathbf{F(x)} \cdot \mathbf{dx} = \int_0^1 3t^2\, dt = 1.$$

(c) Here $\mathbf{x}'(t) = \frac{3}{2}\sqrt{t}\mathbf{i} + 5t^4\mathbf{j}$ and $\mathbf{F}(x, y) = t^5\mathbf{i} + t^{3/2}\mathbf{j}$, so $\mathbf{F} \cdot \mathbf{x}' = (t^5\mathbf{i} + t^{3/2}\mathbf{j}) \cdot (\frac{3}{2}\sqrt{t}\mathbf{i} + 5t^4\mathbf{j}) = \frac{3}{2}t^{11/2} + 5t^{11/2} = \frac{13}{2}t^{11/2}$. Then

$$\int_C \mathbf{F} \cdot \mathbf{x}'\, dt = \int_0^1 \tfrac{13}{2} t^{11/2}\, dt = 1. \blacksquare$$

Before stating our next result, we make several definitions.

Definition 2 CONNECTED SET A set Ω in $\mathbb{R}^2$ is **connected** if any two points in Ω can be joined by a piecewise smooth curve lying entirely in Ω.

Definition 3 OPEN SET A set Ω in $\mathbb{R}^2$ is **open** if for every point $\mathbf{x} \in \Omega$, there is an open disk D with center at $\mathbf{x}$ that is contained in Ω.

Definition 4 REGION A **region** Ω in $\mathbb{R}^2$ is an open, connected set.

These definitions are illustrated in Figure 4.

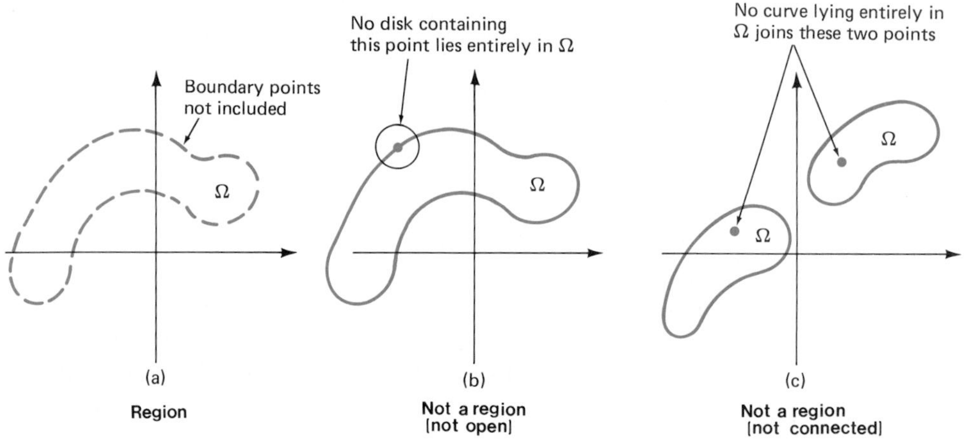

FIGURE 4

INDEPENDENCE OF PATH

In Example 5 we saw that on three very different curves, we obtained the same answer as we moved between the points (0, 0) and (1, 1). In fact, as we will show in a moment, we will get the same answer if we integrate along any piecewise smooth curve C joining these two points. When this happens, we say that the line integral is **independent of the path.** A condition that ensures that a line integral is independent of the path over which it is integrated is given below.

Theorem 1 Let $\mathbf{F}$ be continuous in a region Ω in $\mathbb{R}^2$. Then $\mathbf{F}$ is the gradient of a differentiable function f if and only if for any piecewise smooth curve C lying in Ω, the line integral $\int_C \mathbf{F}(\mathbf{x}) \cdot \mathbf{dx}$ is independent of the path.

Proof. We first assume that $\mathbf{F}(\mathbf{x}) = \nabla f$ for some differentiable function f. Recall the vector form of the chain rule:

$$\frac{d}{dt} f(\mathbf{x}(t)) = \nabla f(\mathbf{x}(t)) \cdot \mathbf{x}'(t) \qquad \textbf{(14)}$$

[see equation (4.6.3)]. Now suppose that C is given by $\mathbf{x}(t)$: $a \le t \le b$, $\mathbf{x}(a) = \mathbf{x}_0$, and $\mathbf{x}(b) = \mathbf{x}_1$. We will assume that C is smooth. Otherwise, we could write the line integral in the form (13) and treat the integral over each smooth curve C_i separately. Using (14), we have

$$\int_C \mathbf{F}(\mathbf{x}) \cdot \mathbf{dx} = \int_a^b \mathbf{F}(\mathbf{x}(t)) \cdot \mathbf{x}'(t)\,dt \overset{\mathbf{F} = \nabla f}{=} \int_a^b \nabla f(\mathbf{x}(t)) \cdot \mathbf{x}'(t)\,dt$$

$$= \int_a^b \frac{d}{dt} f(\mathbf{x}(t))\,dt = f(\mathbf{x}(b)) - f(\mathbf{x}(a)) = f(\mathbf{x}_1) - f(\mathbf{x}_0).$$

This proves that the line integral is independent of the path since $f(\mathbf{x}_1) - f(\mathbf{x}_0)$ does not depend on the particular curve chosen.

We now assume that $\int_C \mathbf{F}(\mathbf{x}) \cdot \mathbf{dx}$ is independent of the path and prove that $\mathbf{F} = \nabla f$ for some differentiable function f. Let $\mathbf{x}_0$ be a fixed point in Ω and let $\mathbf{x}$ be any other point in Ω. Since Ω is connected, there is at least one piecewise smooth path C joining $\mathbf{x}_0$ and $\mathbf{x}$, with C wholly contained in Ω. We define a function f by

$$f(\mathbf{x}) = \int_C \mathbf{F}(\mathbf{x}) \cdot \mathbf{dx}.$$

This function is well defined because, by hypothesis, $\int_C \mathbf{F}(\mathbf{x}) \cdot \mathbf{dx}$ is the same no matter what path is chosen between $\mathbf{x}_0$ and $\mathbf{x}$. Write $\mathbf{x}_0 = (x_0, y_0)$ and $\mathbf{x} = (x, y)$. Since Ω is open, there is an open disk D centered at (x, y) that is contained in Ω. Choose $\Delta x > 0$ such that $(x + \Delta x, y) \in D$, and let C_1 be the horizontal line segment joining (x, y) to $(x + \Delta x, y)$. The situation is depicted in Figure 5. We see that $C \cup C_1$ is a path joining (x_0, y_0) to $(x + \Delta x, y)$, so that

$$f(x + \Delta x, y) = \int_C \mathbf{F}(\mathbf{x}) \cdot \mathbf{dx} + \int_{C_1} \mathbf{F}(\mathbf{x}) \cdot \mathbf{dx} = f(x, y) + \int_{C_1} \mathbf{F}(\mathbf{x}) \cdot \mathbf{dx}. \qquad (15)$$

A parametrization for C_1 (which is a horizontal line) is

$$\mathbf{x}(t) = (x + t\,\Delta x, y), \qquad 0 \le t \le 1$$

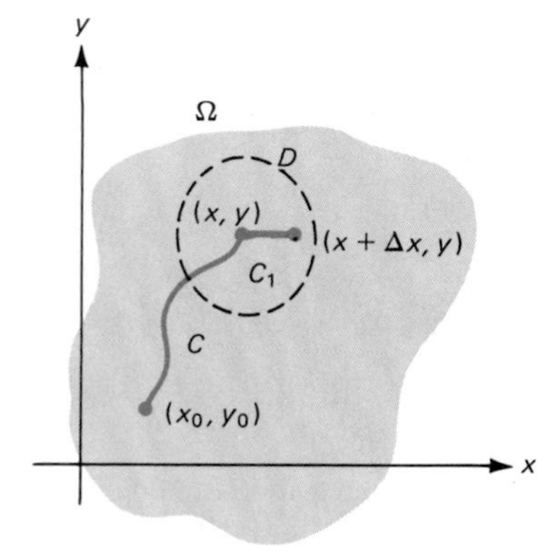

FIGURE 5

and

$$\mathbf{x}'(t) = (\Delta x, 0).$$

Suppose that $\mathbf{F}(\mathbf{x}) = P(x, y)\mathbf{i} + Q(x, y)\mathbf{j}$. Then

$$\mathbf{F}(\mathbf{x}) \cdot \mathbf{dx} = P(x, y)\, \Delta x. \tag{16}$$

From (15) and (16) we compute

$$f(x + \Delta x, y) - f(x, y) = \int_{C_1} \mathbf{F}(\mathbf{x}) \cdot \mathbf{dx} = \int_0^1 P(x + t\,\Delta x, y)\, \Delta x\, dt,$$

so that

$$\frac{f(x + \Delta x, y) - f(x, y)}{\Delta x} = \int_0^1 P(x + t\,\Delta x, y)\, dt. \tag{17}$$

In (17), $P(x + t\,\Delta x, y)$ is a function of one variable (t), so we may apply the mean value theorem for integrals[†] to see that there is a number $\bar{x}$ with $x < \bar{x} < x + \Delta x$ such that

$$\int_0^1 P(x + t\,\Delta x, y)\, dt = P(\bar{x}, y) \int_0^1 dt = P(\bar{x}, y).$$

Thus taking the limits as $\Delta x \to 0$ on both sides of (17), we have

$$\frac{\partial f(\mathbf{x})}{\partial x} = \lim_{\Delta x \to 0} \frac{f(x + \Delta x, y) - f(x, y)}{\Delta x} = \lim_{\Delta x \to 0} P(\bar{x}, y) \overset{\substack{P \text{ is continuous and } \bar{x} \to x \text{ as } \Delta x \to 0 \\ \downarrow}}{=} P(x, y).$$

In a similar manner, we can show that $(\partial f/\partial y)(\mathbf{x}) = Q(x, y)$. This shows that $\mathbf{F} = \nabla f$, and the proof is complete. ■

REMARK. If **F** is the gradient of a differentiable function f, then **F** is said to be **exact.**

In proving this theorem, we also proved the following corollary.

Corollary 1 Suppose that **F** is continuous in a region Ω in $\mathbb{R}^2$ and $\mathbf{F} = \nabla f$ for some differentiable function f. Then for any piecewise smooth curve C in Ω starting at the point $\mathbf{x}_0$ and ending at the point $\mathbf{x}_1$,

$$\int_C \mathbf{F}(\mathbf{x}) \cdot \mathbf{dx} = f(\mathbf{x}_1) - f(\mathbf{x}_0); \tag{18}$$

that is, the value of the integral depends only on the endpoints of the path.

[†]The mean value theorem for integrals states that if f and g are continuous on $[a, b]$ and $g(x)$ is never zero on (a, b), then there exists a number c in (a, b) such that $\int_a^b f(x)g(x)\,dx = f(c)\int_a^b g(x)\,dx$. For a proof see *Calculus* or *Calculus of One Variable,* p. 343.

REMARK 1. This corollary is really the line integral analog of the fundamental theorem of calculus. It says that we can evaluate the line integral of a gradient field by evaluating at two points the function for which **F** is the gradient.

REMARK 2. In Theorem 1 it is important that **F** be continuous on Ω, not only on C. (See Problem 29.)

EXAMPLE 5 (Continued). Since $y\mathbf{i} + x\mathbf{j} = \nabla(xy)$, we immediately find that for any curve C starting at (0, 0) and ending at (1, 1)

$$\int_C \mathbf{F}(\mathbf{x}) \cdot \mathbf{dx} = xy\big|_{(1,1)} - xy\big|_{(0,0)} = 1 - 0 = 1. \blacksquare$$

Using Theorem 4.11.1 we have the following obvious corollary to Theorem 1.

Corollary 2 Let $\mathbf{F}(x, y) = P(x, y)\mathbf{i} + Q(x, y)\mathbf{j}$. Then if P, Q, $\partial P/\partial y$, and $\partial Q/\partial x$ are all continuous on a region containing C and if $\partial P/\partial y = \partial Q/\partial x$, then $\int_C \mathbf{F}(\mathbf{x}) \cdot \mathbf{dx}$ is independent of the path.

EXAMPLE 6 Let $\mathbf{F}(x, y) = [4x^3y^3 + (1/x)]\mathbf{i} + [3x^4y^2 - (1/y)]\mathbf{j}$. Calculate $\int_C \mathbf{F}(\mathbf{x}) \cdot \mathbf{dx}$ for any smooth curve C, not crossing the x- or y-axis, starting at (1, 1) and ending at (2, 3).

Solution. We showed in Example 4.11.3 that **F** is exact and that $\mathbf{F} = \nabla f$, where $f(x, y) = x^4y^3 + \ln|x/y|$. Thus $\int_C \mathbf{F}(\mathbf{x}) \cdot \mathbf{dx}$ is independent of the path, and

$$\int_C \mathbf{F}(\mathbf{x}) \cdot \mathbf{dx} = f(2, 3) - f(1, 1) = (16)(27) + \ln\left(\frac{2}{3}\right) - 1 = 431 + \ln\left(\frac{2}{3}\right). \blacksquare$$

EXAMPLE 7 Let $\mathbf{F}(\mathbf{x}) = x\mathbf{i} + (x - y)\mathbf{j}$. Let C_1 be the part of the curve $y = x^2$ that connects (0, 0) to (1, 1), and let C_2 be the part of the curve $y = x^3$ that connects these two points. Compute $\int_{C_1} \mathbf{F}(\mathbf{x}) \cdot \mathbf{dx}$ and $\int_{C_2} \mathbf{F}(\mathbf{x}) \cdot \mathbf{dx}$.

Solution. Along C_1, $\mathbf{x}(t) = t\mathbf{i} + t^2\mathbf{j}$ and $\mathbf{x}'(t) = \mathbf{i} + 2t\mathbf{j}$ for $0 \le t \le 1$, so that

$$\int_{C_1} \mathbf{F}(\mathbf{x}) \cdot \mathbf{dx} = \int_0^1 [t \cdot 1 + (t - t^2)2t]\, dt = \int_0^1 (t + 2t^2 - 2t^3)\, dt$$

$$= \left(\frac{t^2}{2} + \frac{2t^3}{3} - \frac{t^4}{2}\right)\Bigg|_0^1 = \frac{2}{3}.$$

Along C_2, $\mathbf{x}(t) = t\mathbf{i} + t^3\mathbf{j}$ and $\mathbf{x}' = \mathbf{i} + 3t^2\mathbf{j}$ for $0 \le t \le 1$, so

$$\int_{C_2} \mathbf{F}(\mathbf{x}) \cdot \mathbf{dx} = \int_0^1 [t \cdot 1 + (t - t^3)(3t^2)]\, dt = \int_0^1 (t + 3t^3 - 3t^5)\, dt$$

$$= \left(\frac{t^2}{2} + \frac{3}{4}t^4 - \frac{1}{2}t^6\right)\Bigg|_0^1 = \frac{3}{4}.$$

We see that $\int_C \mathbf{F}(\mathbf{x}) \cdot \mathbf{dx}$ is *not* independent of the path. Note that here $\partial P/\partial y = 0$ and $\partial Q/\partial x = 1$, so **F** is *not* exact. $\blacksquare$

There is another important consequence of Theorem 1. Let C be a closed curve (i.e., $\mathbf{x}_0 = \mathbf{x}_1$).

If $\mathbf{F}$ is continuous in a region Ω and if $\mathbf{F}$ is the gradient of a differentiable function f, then for any closed curve C lying in Ω,

$$\int_C \mathbf{F}(\mathbf{x}) \cdot d\mathbf{x} = f(\mathbf{x}_1) - f(\mathbf{x}_0) = 0.$$

There is an interesting physical interpretation of this fact:

The work done by a conservative force field as it moves a particle completely around a closed path is zero.[†]

This also follows from the more general fact that if $\mathbf{F}$ is conservative, then the work done in moving an object between two points depends only on the points, and not on the path taken between the points.

PROBLEMS 6.2

In Problems 1–14, calculate $\int_C \mathbf{F}(\mathbf{x}) \cdot d\mathbf{x}$.

1. $\mathbf{F}(x, y) = x^2\mathbf{i} + y^2\mathbf{j}$; C is the straight line from (0, 0) to (2, 4).
2. $\mathbf{F}(x, y) = x^2\mathbf{i} + y^2\mathbf{j}$; C is the parabola $y = x^2$ from (0, 0) to (2, 4).
3. $\mathbf{F}(x, y) = xy\mathbf{i} + (y - x)\mathbf{j}$; C is the line $y = 2x - 4$ from (1, −2) to (2, 0).
4. $\mathbf{F}(x, y) = xy\mathbf{i} + (y - x)\mathbf{j}$; C is the curve $y = \sqrt{x}$ from (0, 0) to (1, 1).
5. $\mathbf{F}(x, y) = xy\mathbf{i} + (y - x)\mathbf{j}$; C is the unit circle in the counterclockwise direction.
6. $\mathbf{F}(x, y) = xy\mathbf{i} + (y - x)\mathbf{j}$; C is the triangle joining the points (0, 0), (0, 1), and (1, 0) in the counterclockwise direction.
7. $\mathbf{F}(x, y) = xy\mathbf{i} + (y - x)\mathbf{j}$; C is the triangle joining the points (0, 0), (1, 0), and (1, 1) in the counterclockwise direction.
8. $\mathbf{F}(x, y) = e^x\mathbf{i} + e^y\mathbf{j}$; C is the curve of Problem 4.
9. $\mathbf{F}(x, y) = (x^2 + 2y)\mathbf{i} - y^2\mathbf{j}$; C is the part of the ellipse $x^2 + 9y^2 = 9$ joining the points (0, −1) and (0, 1) in the clockwise direction.
10. $\mathbf{F}(x, y) = (\cos x)\mathbf{i} - (\sin y)\mathbf{j}$; C is the curve of Problem 9.
11. $\mathbf{F}(x, y) = e^{x+y}\mathbf{i} + e^{x-y}\mathbf{j}$; C is the curve of Problem 6.
12. $\mathbf{F}(x, y) = e^{x+y}\mathbf{i} + e^{x-y}\mathbf{j}$; C is the curve of Problem 7.
13. $\mathbf{F}(x, y) = (y/x^2)\mathbf{i} + (x/y^2)\mathbf{j}$; C is the straight line segment from (2, 1) to (4, 6).
14. $\mathbf{F}(x, y) = (\ln x)\mathbf{i} + (\ln y)\mathbf{j}$; C is the curve $\mathbf{x}(t) = 2t\mathbf{i} + t^3\mathbf{j}$ for $1 \le t \le 4$.

In Problems 15–19 forces are given in newtons and distances are given in meters.

15. Calculate the work done when a force field $\mathbf{F}(x, y) = x^3\mathbf{i} + xy\mathbf{j}$ moves a particle from the point (0, 1) to the point $(1, e^{\pi/2})$ along the curve $\mathbf{x}(t) = (\sin t)\mathbf{i} + e^t\mathbf{j}$.
16. Calculate the work done when the force field $\mathbf{F}(x, y) = xy\mathbf{i} + (2x^3 - y)\mathbf{j}$ moves a particle around the unit circle in the counterclockwise direction.
17. What is the work done if the particle in Problem 16 is moved in the clockwise direction?
18. Calculate the work done by the force field $\mathbf{F}(x, y) = -y^2x\mathbf{i} + 2x\mathbf{j}$ when a particle is moved around the ellipse $(x^2/a^2) + (y^2/b^2) = 1$ in the counterclockwise direction.

[†]This is true in general when the closed path is contained in a region over which the force field is continuous.

19. Calculate the work done by the force field $\mathbf{F}(x, y) = 2xy\mathbf{i} + y^2\mathbf{j}$ when a particle is moved around the triangle of Problem 7.

20. What is the work done on the 1-coulomb particle of Example 2 if it moves along the line $y = 2x - 3$ from $(1, -1)$ to $(2, 1)$?

In Problems 21–28, show that **F** is exact and use Theorem 1 to calculate $\int_C \mathbf{F}(\mathbf{x}) \cdot \mathbf{dx}$, where C is any smooth curve starting at $\mathbf{x}_0$ and ending at $\mathbf{x}_1$.

21. $\mathbf{F}(x, y) = 2xy\mathbf{i} + (x^2 + 1)\mathbf{j}$; $\mathbf{x}_0 = (0, 1)$, $\mathbf{x}_1 = (2, 3)$

22. $\mathbf{F}(x, y) = (4x^2 - 4y^2)\mathbf{i} + (\ln y - 8xy)\mathbf{j}$; $\mathbf{x}_0 = (-1, 1)$, $\mathbf{x}_1 = (4, e)$

23. $\mathbf{F}(x, y) = [x\cos(x + y) + \sin(x + y)]\mathbf{i} + x\cos(x + y)\mathbf{j}$; $\mathbf{x}_0 = (0, 0)$, $\mathbf{x}_1 = (\pi/6, \pi/3)$

24. $\mathbf{F}(x, y) = \left[\dfrac{1}{x^2} + y^2\right]\mathbf{i} - 2xy\mathbf{j}$; $\mathbf{x}_0 = (1, 4)$, $\mathbf{x}_1 = (3, 2)$

25. $\mathbf{F}(x, y) = 2x(\cos y)\mathbf{i} - x^2(\sin y)\mathbf{j}$; $\mathbf{x}_0 = (0, \pi/2)$, $\mathbf{x}_1 = (\pi/2, 0)$

26. $\mathbf{F}(x, y) = (2xy^3 - 2)\mathbf{i} + (3x^2y^2 + \cos y)\mathbf{j}$; $\mathbf{x}_0 = (1, 0)$, $\mathbf{x}_1 = (0, -\pi)$

27. $\mathbf{F}(x, y) = e^y\mathbf{i} + xe^y\mathbf{j}$; $\mathbf{x}_0 = (0, 0)$, $\mathbf{x}_1 = (5, 7)$

28. $\mathbf{F}(x, y) = (\cosh x)(\cosh y)\mathbf{i} + (\sinh x)(\sinh y)\mathbf{j}$; $\mathbf{x}_0 = (0, 0)$, $\mathbf{x}_1 = (1, 2)$

***29.** Let the force field **F** be given by

$$\mathbf{F}(x, y) = \frac{y}{x^2 + y^2}\mathbf{i} - \frac{x}{x^2 + y^2}\mathbf{j}.$$

(a) Show that **F** is exact.

(b) Calculate $\int_C \mathbf{F}(\mathbf{x}) \cdot \mathbf{dx}$, where C is the unit circle in the counterclockwise direction.

(c) Explain why $\int_C \mathbf{F}(\mathbf{x}) \cdot \mathbf{dx} \neq 0$. Does this result contradict the last fact stated in this section?

6.3 GREEN'S THEOREM IN THE PLANE

In this section we state a result that gives an important relationship between line integrals and double integrals.

Let Ω be a region in the plane (a typical region is sketched in Figure 1). The curve (indicated by the arrows) that goes around the edge of Ω in the direction that keeps Ω on the left (sometimes called the *counterclockwise* direction) is called the **boundary of** Ω and is denoted $\partial\Omega$. Let

$$\mathbf{F}(x, y) = P(x, y)\mathbf{i} + Q(x, y)\mathbf{j}.$$

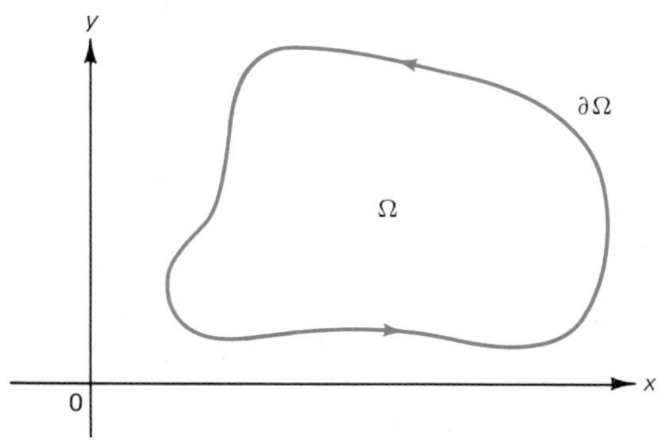

FIGURE 1

If the curve $\partial\Omega$ is given by

$$\partial\Omega: \quad \mathbf{x}(t) = x(t)\mathbf{i} + y(t)\mathbf{j},$$

we can write

$$\mathbf{F}(\mathbf{x}) \cdot \mathbf{dx} = P\,dx + Q\,dy. \tag{1}$$

We then denote the line integral of **F** around $\partial\Omega$ by

$$\oint_{\partial\Omega} P\,dx + Q\,dy. \tag{2}$$

The symbol $\oint_{\partial\Omega}$ indicates that $\partial\Omega$ is a closed curve around which we integrate in the counterclockwise direction (the direction of the arrow).

Definition 1 SIMPLE CURVE A curve C is called **simple** if it does not cross itself. That is, suppose C is given by

$$C: \quad \mathbf{x}(t) = f(t)\mathbf{i} + g(t)\mathbf{j}, \qquad t \in [a, b].$$

Then C is simple if and only if $\mathbf{x}(t_1) \neq \mathbf{x}(t_2)$ whenever $t_1 \neq t_2$.

This notion is illustrated in Figure 2.

Definition 2 SIMPLY CONNECTED REGION A region Ω in the xy-plane is called **simply connected** if it has the following property: If C is a simple closed curve contained in Ω, then every point in the region enclosed by C is also in Ω. Intuitively, a region is simply connected if it has no holes.

We illustrate this definition in Figure 3.

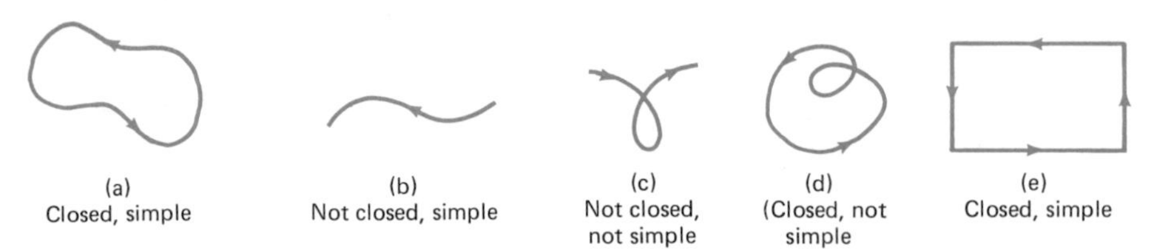

FIGURE 2

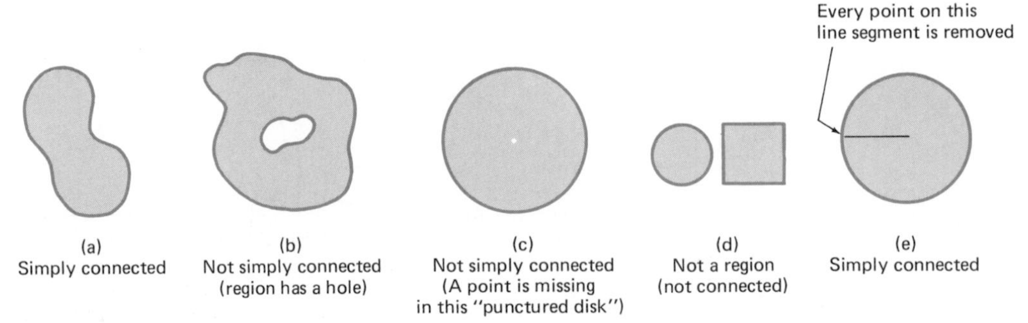

FIGURE 3

We are now ready to state the principal result of this section.

Theorem 1 GREEN'S THEOREM[†] IN THE PLANE Let Ω be a simply connected region in the xy-plane bounded by a piecewise smooth curve $\partial\Omega$. Let P and Q be continuous with continuous first partials in an open disk containing Ω. Then

$$\oint_{\partial\Omega} P\,dx + Q\,dy = \iint_{\Omega} \left(\frac{\partial Q}{\partial x} - \frac{\partial P}{\partial y}\right) dx\,dy. \tag{3}$$

REMARK. This theorem shows how the line integral of a function around the boundary of a region is related to a double integral over that region.

Partial Proof of Green's Theorem. We prove Green's theorem in the case in which Ω takes the simple form given in Figure 4. The region Ω can be written

$$\{(x, y)\colon a \leq x \leq b,\ g_1(x) \leq y \leq g_2(x)\}, \tag{4}$$

and

$$\{(x, y)\colon c \leq y \leq d,\ h_1(y) \leq x \leq h_2(y)\}. \tag{5}$$

We first calculate

$$\begin{aligned}\iint_{\Omega} \frac{\partial P}{\partial y}\,dx\,dy &= \int_a^b \left\{\int_{g_1(x)}^{g_2(x)} \frac{\partial P}{\partial y}\,dy\right\} dx = \int_a^b \left\{P(x, y)\Big|_{y=g_1(x)}^{y=g_2(x)}\right\} dx \\ &= \int_a^b [P(x, g_2(x)) - P(x, g_1(x))]\,dx.\end{aligned} \tag{6}$$

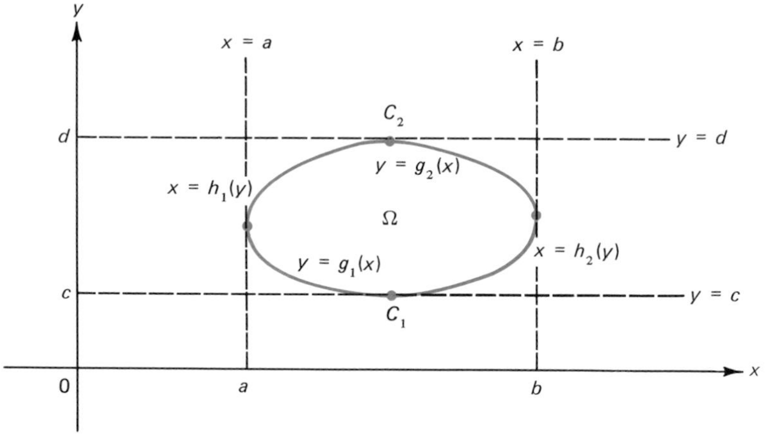

FIGURE 4

[†]Named after George Green (1793–1841), a British mathematician and physicist who wrote an essay in 1828 on electricity and magnetism that contained this important theorem. Green was the self-educated son of a baker. His 1828 essay was published for private circulation. It was largely overlooked until it was rediscovered by Lord Kelvin in 1846. The theorem was independently discovered by the Russian mathematician Michel Ostrogradski (1801–1861), and to this day the theorem is known in the Soviet Union as *Ostrogradski's theorem*.

Now

$$\oint_{\partial\Omega} P\,dx = \int_{C_1} P\,dx + \int_{C_2} P\,dx, \tag{7}$$

where C_1 is the graph of $g_1(x)$ from $x = a$ to $x = b$ and C_2 is the graph of $g_2(x)$ from $x = b$ to $x = a$ (note the order). Since on C_1, $y = g_1(x)$, we have

$$\int_{C_1} P(x, y)\,dx = \int_a^b P(x, g_1(x))\,dx. \tag{8}$$

Similarly,

$$\int_{C_2} P(x, y)\,dx = \int_b^a P(x, g_2(x))\,dx = -\int_a^b P(x, g_2(x))\,dx. \tag{9}$$

Thus

$$\begin{aligned}\oint_{\partial\Omega} P\,dx &= \int_a^b P(x, g_1(x))\,dx - \int_a^b P(x, g_2(x))\,dx \\ &= -\int_a^b [P(x, g_2(x)) - P(x, g_1(x))]\,dx. \end{aligned} \tag{10}$$

Comparing (6) and (10), we find that

$$\iint_{\Omega} -\frac{\partial P}{\partial y}\,dx\,dy = \oint_{\partial\Omega} P\,dx. \tag{11}$$

Similarly, using the representation (5), we have

$$\begin{aligned}\iint_{\Omega} \frac{\partial Q}{\partial x}\,dx\,dy &= \int_c^d \left\{\int_{h_1(y)}^{h_2(y)} \frac{\partial Q}{\partial x}\,dx\right\} dy \\ &= \int_c^d [Q(h_2(y), y) - Q(h_1(y), y)]\,dy,\end{aligned}$$

which by analogous reasoning yields the equation

$$\iint_{\Omega} \frac{\partial Q}{\partial x}\,dx\,dy = \oint_{\partial\Omega} Q\,dy. \tag{12}$$

Adding (11) and (12) completes the proof of the theorem in the special case that Ω can be written in the form (4) and (5).† ■

†For a proof of Green's theorem for more general regions, see R. C. Buck, *Advanced Calculus* (McGraw-Hill, New York, 1965), page 408.

REMARK. The relationship between independence of path and Green's theorem is given in the diagram below. Here $\mathbf{F}(x, y) = P(x, y)\mathbf{i} + Q(x, y)\mathbf{j}$.

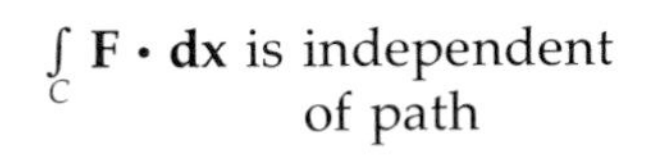

$\int_C \mathbf{F} \cdot \mathbf{dx}$ is independent of path

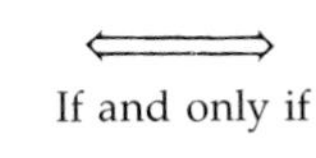

If and only if

there is an f such that $\nabla f = \mathbf{F}$ (i.e. $\mathbf{F}$ is exact)

$\oint_C \mathbf{F} \cdot \mathbf{dx} = 0$ around all simple closed curves $\Leftrightarrow$ The hypotheses of Green's theorem apply to the region and

$$\frac{\partial P}{\partial y} = \frac{\partial Q}{\partial x}$$

EXAMPLE 1 Evaluate $\oint_{\partial\Omega} xy\,dx + (x - y)\,dy$, where Ω is the rectangle $\{(x, y)\colon 0 \le x \le 1, 1 \le y \le 3\}$.

Solution. $P(x, y) = xy$, $Q(x, y) = x - y$, $\partial Q/\partial x = 1$, and $\partial P/\partial y = x$, so

$$\oint_{\partial\Omega} xy\,dx + (x - y)\,dy = \int_0^1 \int_1^3 (1 - x)\,dy\,dx = \int_0^1 \left\{(1 - x)y\Big|_1^3\right\} dx$$

$$= \int_0^1 2(1 - x)\,dx = 1. \quad \blacksquare$$

EXAMPLE 2 Evaluate $\oint_C (x^3 + y^3)\,dx + (2y^3 - x^3)\,dy$, where C is the unit circle.

Solution. We first note that $C = \partial\Omega$, where Ω is the unit disk. Next, we have $(\partial Q/\partial x) - (\partial P/\partial y) = -3x^2 - 3y^2 = -3(x^2 + y^2)$. Thus

$$\oint_C (x^3 + y^3)\,dx + (2y^3 - x^3)\,dy = -3 \iint_{\substack{\text{unit}\\ \text{disk}}} (x^2 + y^2)\,dx\,dy$$

Converting to polar coordinates

$$\downarrow$$

$$= -3 \int_0^{2\pi} \int_0^1 r^2 \cdot r\,dr\,d\theta$$

$$= -3 \int_0^{2\pi} \left\{\frac{r^4}{4}\Big|_0^1\right\} d\theta$$

$$= -\frac{3}{4} \int_0^{2\pi} d\theta = -\frac{3\pi}{2}. \quad \blacksquare$$

Green's theorem can be useful for calculating area. Recall that

$$\text{area enclosed by } \Omega = \iint_\Omega dA. \tag{13}$$

But by Green's theorem,

$$\iint_{\Omega} dA = \oint_{\partial\Omega} x\,dy = \oint_{\partial\Omega} (-y)\,dx = \frac{1}{2}\oint_{\partial\Omega} [(-y)\,dx + x\,dy] \tag{14}$$

(explain why). Any of the line integrals in (14) can be used to calculate area.

EXAMPLE 3 Use Green's theorem to calculate the area enclosed by the ellipse $(x^2/a^2) + (y^2/b^2) = 1$.

Solution. The ellipse can be written parametrically as

$$\mathbf{x}(t) = a(\cos t)\mathbf{i} + b(\sin t)\mathbf{j}, \qquad 0 \le t \le 2\pi.$$

Then using the first line integral in (14), we obtain

$$\begin{aligned} A &= \oint_{\partial\Omega} x\,dy = \int_0^{2\pi} (a\cos t)\frac{d}{dt}(b\sin t)\,dt \\ &= \int_0^{2\pi} (a\cos t)b\cos t\,dt = ab\int_0^{2\pi} \cos^2 t\,dt \\ &= \frac{ab}{2}\int_0^{2\pi} (1 + \cos 2t)\,dt = \pi ab. \end{aligned}$$

Note how much easier this calculation is than the direct evaluation of $\iint_A dx\,dy$, where A denotes the area enclosed by the ellipse. ■

There are two very interesting and important vector interpretations of Green's theorem.

Definition 3 CURL AND DIVERGENCE Let $\mathbf{F}(x, y) = P(x, y)\mathbf{i} + Q(x, y)\mathbf{j}$ be a vector field in the plane.

(i) The **curl** of $\mathbf{F}$ is given by

$$\text{curl } \mathbf{F} = \frac{\partial Q}{\partial x} - \frac{\partial P}{\partial y}. \tag{15}$$

(ii) The **divergence** of $\mathbf{F}$, denoted div $\mathbf{F}$, is given by

$$\text{div } \mathbf{F} = \frac{\partial P}{\partial x} + \frac{\partial Q}{\partial y}. \tag{16}$$

Before explaining these terms more fully, we will write Green's theorem in two equivalent vector forms.

First, recall from equation (2.6.2) that if $\mathbf{T}$ denotes the unit tangent vector to a curve $\mathbf{x}(t) = x(t)\mathbf{i} + y(t)\mathbf{j}$ and if s denotes the parameter of arc length, then

$$\mathbf{T} = \frac{\mathbf{dx}}{ds}$$

or

$$\mathbf{dx} = \mathbf{T}\, ds. \tag{17}$$

Now let $\mathbf{T}$ denote the unit tangent vector to $\partial\Omega$. Then if $\mathbf{F} = P\mathbf{i} + Q\mathbf{j}$, we can write, using (17),

$$\oint_{\partial\Omega} P\, dx + Q\, dy = \oint_{\partial\Omega} \mathbf{F} \cdot \mathbf{dx} = \oint_{\partial\Omega} \mathbf{F} \cdot \mathbf{T}\, ds. \tag{18}$$

Then applying Green's theorem and using (15) and (18), we obtain the following theorem.

Theorem 2 FIRST VECTOR FORM OF GREEN'S THEOREM[†] Under the hypotheses of Theorem 1,

$$\oint_{\partial\Omega} \mathbf{F} \cdot \mathbf{T}\, ds = \iint_{\Omega} \text{curl}\, \mathbf{F}\, dx\, dy. \tag{19}$$

Now from the expression

$$\oint_{\partial\Omega} P\, dx + Q\, dy = \iint_{\Omega} \left(\frac{\partial Q}{\partial x} - \frac{\partial P}{\partial y}\right) dx\, dy,$$

we may replace P by $-Q$ and Q by P to obtain

$$\oint_{\partial\Omega} -Q\, dx + P\, dy = \iint_{\Omega} \left(\frac{\partial P}{\partial x} + \frac{\partial Q}{\partial y}\right) dx\, dy. \tag{20}$$

If $dx\,\mathbf{i} + dy\,\mathbf{j}$ represents the vector $\mathbf{T}\, ds$, then $dy\,\mathbf{i} - dx\,\mathbf{j}$ represents the vector $\mathbf{n}\, ds$, where $\mathbf{n}$ is the unit normal vector to the curve. This is easy to see since $(dx\,\mathbf{i} + dy\,\mathbf{j}) \cdot (dy\,\mathbf{i} - dx\,\mathbf{j}) = 0$ and both $\mathbf{T}\, ds$ and $\mathbf{n}\, ds$ have the same magnitude. Thus the left-hand side of (20) can be written

$$\oint_{\partial\Omega} -Q\, dx + P\, dy = \oint_{\partial\Omega} \mathbf{F} \cdot \mathbf{n}\, ds. \tag{21}$$

Using (16) and (21) in (20) yields the next theorem.

[†]This form of Green's theorem is sometimes called **Stokes's theorem in the plane.**

Theorem 3 SECOND VECTOR FORM OF GREEN'S THEOREM[†] Under the hypotheses of Theorem 1,

$$\oint_{\partial\Omega} \mathbf{F} \cdot \mathbf{n}\, ds = \iint_{\Omega} \operatorname{div} \mathbf{F}\, dx\, dy. \tag{22}$$

CIRCULATION AND FLUX

There are interesting physical interpretations of the two vector forms of Green's theorem. Let $\mathbf{F}(x, y)$ denote the direction and rate of flow of a fluid at a point (x, y) in the plane. The integral

$$\oint_{\partial\Omega} \mathbf{F} \cdot \mathbf{T}\, ds$$

is the integral of the component of the flow in the direction tangent to the boundary of Ω and is called the **circulation** of $\mathbf{F}$ around the boundary of Ω (see Figure 5). If Ω is small, then curl $\mathbf{F}$ is nearly a constant, and by (19) we have

$$\begin{aligned}\oint_{\partial\Omega} \mathbf{F} \cdot \mathbf{T}\, ds &= \iint_{\Omega} \operatorname{curl} \mathbf{F}\, dx\, dy \approx \operatorname{curl} \mathbf{F}(x, y) \iint_{\Omega} dx\, dy \\ &= [\operatorname{curl} \mathbf{F}(x, y)](\text{area of } \Omega).\end{aligned} \tag{23}$$

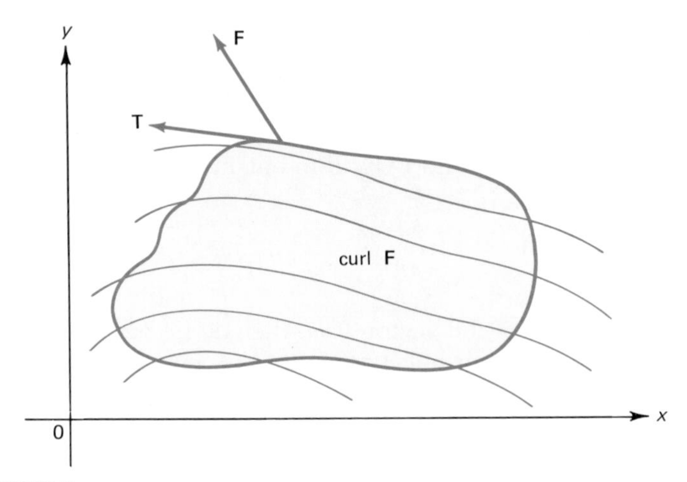

FIGURE 5

From (23) it appears that the curl represents the circulation per unit area at the point (x, y). If curl $\mathbf{F} = 0$ for every (x, y) in Ω, then the fluid flow $\mathbf{F}$ is called **irrotational.**

Now $\oint_{\partial\Omega} \mathbf{F} \cdot \mathbf{n}\, ds$ is the component of flow in the direction of the outward normal to $\partial\Omega$ and is called the **flux** across $\partial\Omega$ (see Figure 6). The flux is the rate at which fluid

[†]This form of Green's theorem is sometimes called the **divergence theorem in the plane.**

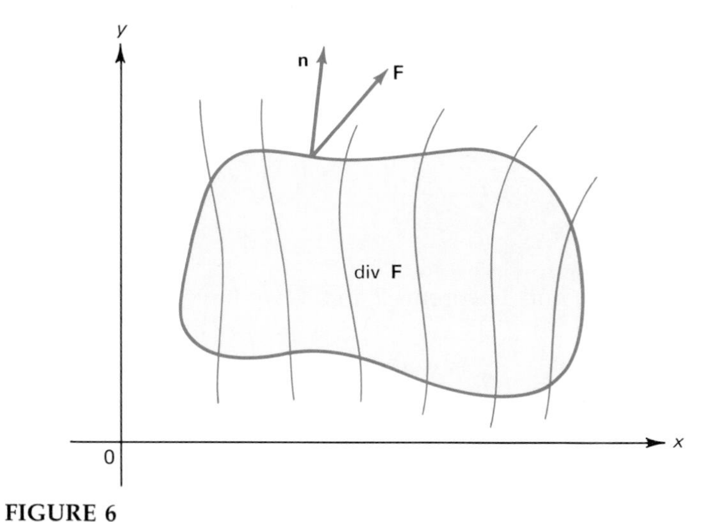

FIGURE 6

is flowing across the boundary of Ω from inside Ω. If Ω is small, then using (22), we have

$$\oint_{\partial\Omega} \mathbf{F} \cdot \mathbf{n}\, ds = \iint_{\Omega} \text{div } \mathbf{F}\, dx\, dy \approx [\text{div } \mathbf{F}(x, y)](\text{area of } \Omega). \tag{24}$$

Thus the divergence of $\mathbf{F}$ represents the net rate of flow away from (x, y). If div $\mathbf{F} = 0$ for every (x, y) in Ω, then the flow $\mathbf{F}$ is called **incompressible,** because if div $\mathbf{F} = 0$, then the net flow into the region is 0. That is, fluid is not accumulating. The accumulation of fluid in a fixed region would cause it to become compressed.

Before leaving this section we note that there are other ways to write the curl and divergence, and therefore there are other ways to write Green's theorem in the plane. Very loosely speaking, we can think of the gradient symbol ∇ as the vector function $(\partial/\partial x, \partial/\partial y)$. Then if $\mathbf{F} = (P, Q)$,

$$\boldsymbol{\nabla} \times \mathbf{F} = \begin{vmatrix} \mathbf{i} & \mathbf{j} & \mathbf{k} \\ \dfrac{\partial}{\partial x} & \dfrac{\partial}{\partial y} & 0 \\ P & Q & 0 \end{vmatrix} = \left(\frac{\partial Q}{\partial x} - \frac{\partial P}{\partial y}\right)\mathbf{k} = \text{curl } \mathbf{F}\, \mathbf{k}.$$

Thus curl $\mathbf{F} = (\text{curl } \mathbf{F}\, \mathbf{k}) \cdot \mathbf{k} = (\boldsymbol{\nabla} \times \mathbf{F}) \cdot \mathbf{k}$ (since $\mathbf{k} \cdot \mathbf{k} = 1$). For this reason curl $\mathbf{F}$ is often written (with the $\mathbf{k}$ omitted) as

$$\text{curl } \mathbf{F} = \boldsymbol{\nabla} \times \mathbf{F}. \tag{25}$$

Similarly,

$$\nabla \cdot \mathbf{F} = \left(\frac{\partial}{\partial x}, \frac{\partial}{\partial y}\right) \cdot (P, Q) = \frac{\partial P}{\partial x} + \frac{\partial Q}{\partial y} = \text{div } \mathbf{F},$$

and we write

$$\text{div } \mathbf{F} = \nabla \cdot \mathbf{F}. \tag{26}$$

Then using (25), (26), and Theorems 2 and 3, we have the following forms of Green's theorem:

$$\oint_{\partial\Omega} \mathbf{F} \cdot \mathbf{T}\, ds = \iint_{\Omega} (\nabla \times \mathbf{F}) \cdot \mathbf{k}\, dx\, dy \tag{27}$$

and

$$\oint_{\partial\Omega} \mathbf{F} \cdot \mathbf{n}\, ds = \iint_{\Omega} \nabla \cdot \mathbf{F}\, dx\, dy. \tag{28}$$

PROBLEMS 6.3

In Problems 1–15, find the line integral by using Green's theorem.

1. $\oint_{\partial\Omega} 3y\, dx + 5x\, dy$; $\Omega = \{(x, y): 0 \le x \le 1, 0 \le y \le 1\}$.
2. $\oint_{\partial\Omega} ay\, dx + bx\, dy$; Ω is the region of Problem 1.
3. $\oint_{\partial\Omega} e^x \cos y\, dx + e^x \sin y\, dy$; Ω is the region enclosed by the triangle with vertices at (0, 0), (1, 0), and (0, 1).
4. The integral of Problem 3, where Ω is the region enclosed by the triangle with vertices at (0, 0), (1, 1), and (1, 0).
5. The integral of Problem 3, where Ω is the region enclosed by the rectangle with vertices at (0, 0), (2, 0), (2, 1), and (0, 1).
6. $\oint_{\partial\Omega} 2xy\, dx + x^2\, dy$; Ω is the unit disk.
7. $\oint_{\partial\Omega}(x^2 + y^2)\, dx - 2xy\, dy$; Ω is the unit disk.
8. $\oint_{\partial\Omega}(1/y)\, dx + (1/x)\, dy$; Ω is the region bounded by the lines $y = 1$ and $x = 16$ and the curve $y = \sqrt{x}$.
9. $\oint_{\partial\Omega} \cos y\, dx + \cos x\, dy$; Ω is the region enclosed by the rectangle $\{(x, y): 0 \le x \le \pi/4, 0 \le y \le \pi/3\}$.
10. $\oint_{\partial\Omega} x^2y\, dx - xy^2\, dy$; Ω is the disk $x^2 + y^2 \le 9$.
11. $\oint_{\partial\Omega} y \ln x\, dy$; $\Omega = \{(x, y): 1 \le y \le 3, e^y \le x \le e^{y^3}\}$.
12. $\oint_{\partial\Omega}\sqrt{1 + y^2}\, dx$; $\Omega = \{(x, y): -1 \le y \le 1, y^2 \le x \le 1\}$.
13. $_{\partial\Omega}\, ay\, dx + bx\, dy$; Ω is a region of the type (4), (5).
14. $\oint_{\partial\Omega} e^x \sin y\, dx + e^x \cos y\, dy$; Ω is the region enclosed by the ellipse $(x^2/a^2) + (y^2/b^2) = 1$.
15. $\oint_{\partial\Omega}(-4x/\sqrt{1 + y^2})\, dx + (2x^2y/(1 + y^2)^{3/2})\, dy$; Ω is a region of the type (4), (5).
16. Use one of the line integrals in (14) to calculate the area of the circle $\mathbf{x}(t) = a(\cos t)\mathbf{i} + a(\sin t)\mathbf{j}$.
17. Use Green's theorem to calculate the area enclosed by the triangle with vertices at (a_1, b_1), (a_2, b_2), and (a_3, b_3), assuming that the three points are not collinear.
18. Use Green's theorem to calculate the area of the quadrilateral with vertices at (0, 0), (2, 1), (−1, 3), and (4, 4).
19. Use Green's theorem to calculate the area of the quadrilateral with vertices at (a_1, b_1), (a_2, b_2), (a_3, b_3), and (a_4, b_4), assuming that no point is within the triangle whose vertices are the other three points and that this order yields a simple polygon.

In Problems 20–26, calculate (a) curl $\mathbf{F}$, (b) $\oint_{\partial\Omega} \mathbf{F} \cdot \mathbf{T}\, ds$, (c) div $\mathbf{F}$, and (d) $\oint_{\partial\Omega} \mathbf{F} \cdot \mathbf{n}\, ds$.

20. $\mathbf{F}(x, y) = x^2\mathbf{i} + y^2\mathbf{j}$; Ω is the region of Problem 1.
21. $\mathbf{F}(x, y) = y^2\mathbf{i} + x^2\mathbf{j}$; Ω is the region of Problem 1.
22. $\mathbf{F}(x, y) = ay\mathbf{i} + bx\mathbf{j}$; Ω is the region of Problem 1.

23. $\mathbf{F}(x, y) = y^3\mathbf{i} + x^3\mathbf{j}$; Ω is the unit disk.
24. $\mathbf{F}(x, y) = x\mathbf{i} + y\mathbf{j}$; Ω is the unit disk.
25. $\mathbf{F}(x, y) = y\mathbf{i} - x\mathbf{j}$; Ω is the unit disk.
26. $\mathbf{F}(x, y) = xy\mathbf{i} + (y^2 - x^2)\mathbf{j}$; Ω is the region of Problem 3.

27. Let $\partial\Omega$ be the ellipse $(x^2/a^2) + (y^2/b^2) = 1$. Let $\mathbf{F}(x, y)$ be the vector field $-x\mathbf{i} - y\mathbf{j}$ that, at any point (x, y), points toward the origin. Show that $\oint_{\partial\Omega} \mathbf{F} \cdot \mathbf{T}\, ds = 0$.

*28. Let Ω be the disk $x^2 + y^2 \le a^2$. Show that $\oint_{\partial\Omega} \alpha\sqrt{x^2 + y^2}\, dx + \beta\sqrt{x^2 + y^2}\, dy = 0$.

29. Let Ω be as in Problem 28 and suppose that g is continuously differentiable. Show that $\oint_{\partial\Omega} \alpha g(x^2 + y^2)\, dx + \beta g(x^2 + y^2)\, dy = 0$.

30. Show that the vector flow $\mathbf{F}(x, y) = x\mathbf{i} + y\mathbf{j}$ is irrotational.

31. Show that the vector flow $\mathbf{F}(x, y) = \sin x\, e^x\mathbf{i} + y^{5/2}\mathbf{j}$ is irrotational.

32. Show that for any continuously differentiable functions f and g, the vector flow $\mathbf{F}(x, y) = f(x)\mathbf{i} + g(y)\mathbf{j}$ is irrotational.

33. Show that the vector flow $\mathbf{F}(x, y) = y\sqrt{x^2 + y^2}\mathbf{i} - x\sqrt{x^2 + y^2}\mathbf{j}$ is incompressible.

34. Let g be as in Problem 29. Show that the vector flow $\mathbf{F}(x, y) = -yg(x^2 + y^2)\mathbf{i} + xg(x^2 + y^2)\mathbf{j}$ is incompressible.

35. Let $\mathbf{F}(x, y) = y/(x^2 + y^2)\mathbf{i} - x/(x^2 + y^2)\mathbf{j}$.
    (a) Show that curl $\mathbf{F} = 0$.
    (b) Show that $\oint_C \mathbf{F} \cdot \mathbf{T}\, ds \ne 0$ if C is the unit circle oriented counterclockwise.
    (c) Explain why the results of (a) and (b) do not contradict Theorem 2.

6.4 LINE INTEGRALS IN SPACE

In this section we begin a development that will extend the results of the last two sections to $\mathbb{R}^3$. We start with the notion of a line integral in space, which is the natural generalization of a line integral in the plane.

Suppose that a curve C in space is given parametrically as

$$C: \quad \mathbf{x}(t) = f(t)\mathbf{i} + g(t)\mathbf{j} + h(t)\mathbf{k}, \tag{1}$$

or

$$C: \quad \mathbf{x}(t) = x(t)\mathbf{i} + y(t)\mathbf{j} + z(t)\mathbf{k}. \tag{2}$$

As in Section 6.2, we assume that a force $\mathbf{F}$ is applied to a particle moving along the curve. Now $\mathbf{F} = \mathbf{F}(x, y, z)$ is a vector field in $\mathbb{R}^3$, and we can write

$$\mathbf{F}(\mathbf{x}) = \mathbf{F}(x, y, z) = P(x, y, z)\mathbf{i} + Q(x, y, z)\mathbf{j} + R(x, y, z)\mathbf{k}. \tag{3}$$

As before, we will assume that the curve C is *smooth* or *piecewise smooth.* That is, the functions $f(t)$, $g(t)$, and $h(t)$ are continuously differentiable, or f', g', and h' exist and are piecewise continuous.

Following the development in Section 6.2 [see equation (6.2.7)], we find that the work done in moving along the curve from $\mathbf{x}(t)$ to $\mathbf{x}(t + \Delta t)$ is, if Δt is small, given approximately by

$$W(t + \Delta t) - W(t) \approx \mathbf{F}(\mathbf{x}(t)) \cdot [\mathbf{x}(t + \Delta t) - \mathbf{x}(t)], \tag{4}$$

so that, as before,

$$W'(t) = \mathbf{F}(\mathbf{x}(t)) \cdot \mathbf{x}'(t) \tag{5}$$

and

$$W = \text{total work done} = \int_a^b \mathbf{F}(\mathbf{x}(t)) \cdot \mathbf{x}'(t)\, dt, \tag{6}$$

which we write as

$$W = \int_C \mathbf{F}(\mathbf{x}) \cdot \mathbf{dx}. \tag{7}$$

We see that there is no essential difference between the formula for work in $\mathbb{R}^2$ and the formula for work in $\mathbb{R}^3$.

EXAMPLE 1 A particle moves along the elliptical helix (see Example 3.8.1) $\mathbf{x}(t) = (\cos t)\mathbf{i} + 4(\sin t)\mathbf{j} + t\mathbf{k}$. It is subject to a force given by the vector field $x^2\mathbf{i} + y^2\mathbf{j} + 2xyz\mathbf{k}$. How much work is done in moving from the point (1, 0, 0) to the point $(0, 4, \pi/2)$ if forces are measured in newtons and distances are measured in meters?

Solution. The curve C is given by

$$\mathbf{x}(t) = (\cos t)\mathbf{i} + (4 \sin t)\mathbf{j} + t\mathbf{k}, \qquad 0 \le t \le \frac{\pi}{2}.$$

Then

$$\mathbf{x}'(t) = (-\sin t)\mathbf{i} + 4(\cos t)\mathbf{j} + \mathbf{k}$$
$$\mathbf{F}(\mathbf{x}(t)) = (\cos^2 t)\mathbf{i} + 16(\sin^2 t)\mathbf{j} + 8t(\sin t)(\cos t)\mathbf{k}$$

and

$$\mathbf{F}(\mathbf{x}(t)) \cdot \mathbf{x}'(t) = -\sin t \cos^2 t + 64 \cos t \sin^2 t + 8t \sin t \cos t.$$

Thus from (6)

$$W = \int_0^{\pi/2} (-\sin t \cos^2 t + 64 \cos t \sin^2 t + 8t \sin t \cos t)\, dt$$
$$= \left(\frac{\cos^3 t}{3} + \frac{64 \sin^3 t}{3}\right)\Bigg|_0^{\pi/2} + \int_0^{\pi/2} 8t \sin t \cos t\, dt = 21 + \int_0^{\pi/2} 8t \sin t \cos t\, dt.$$

Now

$$\int 8t \sin t \cos t\, dt = 4 \int t \sin 2t\, dt.$$

We integrate by parts. Let $u = t$ and $dv = \sin 2t\, dt$; then $du = dt$, $v = (-\cos 2t)/2$, and

$$\int_0^{\pi/2} 8t \sin t \cos t \, dt = -4t \frac{\cos 2t}{2}\bigg|_0^{\pi/2} + 4\int_0^{\pi/2} \frac{\cos 2t}{2}\, dt = \pi.$$

Thus

$$W = 21 + \pi \text{ joules.} \blacksquare$$

A **region** S in space is, as in $\mathbb{R}^2$, an open, connected set. The region is **simply connected** if it has no holes.

We now define the line integral in space.

Definition 1 LINE INTEGRAL IN SPACE Suppose that P, Q, and R are continuous on a region S in $\mathbb{R}^3$ containing the smooth or piecewise smooth curve C given by

$$C: \ \mathbf{x}(t) = f(t)\mathbf{i} + g(t)\mathbf{j} + h(t)\mathbf{k}, \qquad t \in [a, b].$$

Let the vector field $\mathbf{F}$ be given by

$$\mathbf{F}(x, y, z) = P(x, y, z)\mathbf{i} + Q(x, y, z)\mathbf{j} + R(x, y, z)\mathbf{k}.$$

Then the **line integral** of $\mathbf{F}$ over C is given by

$$\int_C \mathbf{F}(\mathbf{x}) \cdot d\mathbf{x} = \int_a^b [P(f(t), g(t), h(t))f'(t) + Q(f(t), g(t), h(t))g'(t) + R(f(t), g(t), h(t))h'(t)]\, dt. \quad \text{(8)}$$

As in the plane, the computation of a line integral in space is very easy if the integral is **independent of the path.** The following theorem generalizes Theorem 6.2.1. Its proof is essentially identical to the proof of that theorem.

Theorem 1 Let $\mathbf{F}$ be continuous on a region S in $\mathbb{R}^3$. Then $\mathbf{F}$ is the gradient of a differentiable function f if and only if the following conditions hold:

(i) For any piecewise smooth curve C lying in S, the line integral

$$\int_C \mathbf{F}(\mathbf{x}) \cdot d\mathbf{x}$$

is independent of the path.

(ii) For any piecewise smooth curve C in S starting at the point $\mathbf{x}_0$ and ending at the point $\mathbf{x}_1$,

$$\int_C \mathbf{F}(\mathbf{x}) \cdot d\mathbf{x} = f(\mathbf{x}_1) - f(\mathbf{x}_0). \quad \text{(9)}$$

EXAMPLE 2 Let $\mathbf{F}(x, y, z) = y^2z^3\mathbf{i} + 2xyz^3\mathbf{j} + 3xy^2z^2\mathbf{k}$. Compute $\int_C \mathbf{F}(\mathbf{x}) \cdot d\mathbf{x}$, where C is any curve starting at (1, 2, 3) and ending at (−1, 4, 1).

Solution. In Example 4.5.4 we saw that $\mathbf{F} = \nabla f$, where $f = xy^2z^3$. Thus from (9)

$$\int_C \mathbf{F}(\mathbf{x}) \cdot d\mathbf{x} = f(-1, 4, 1) - f(1, 2, 3) = -16 - 108 = -124. \blacksquare$$

In Section 4.11 we saw that $\mathbf{F}(x, y) = P(x, y)\mathbf{i} + Q(x, y)\mathbf{j}$ is the gradient of a function f if $\partial P/\partial y = \partial Q/\partial x$. If $\mathbf{F}(x, y, z) = P(x, y, z)\mathbf{i} + Q(x, y, z)\mathbf{j} + R(x, y, z)\mathbf{k}$, then there is a condition that can be used to check whether there is a differentiable function f such that $\mathbf{F} = \nabla f$. We give this result without proof.[†]

Theorem 2 Let $\mathbf{F}(x, y, z) = P(x, y, z)\mathbf{i} + Q(x, y, z)\mathbf{j} + R(x, y, z)\mathbf{k}$ and suppose that $P, Q, R, \partial P/\partial y, \partial P/\partial z, \partial Q/\partial x, \partial Q/\partial z, \partial R/\partial x$, and $\partial R/\partial y$ are continuous in a simply connected region. Then $\mathbf{F}$ is the gradient of a differentiable function f if and only if

$$\frac{\partial P}{\partial y} = \frac{\partial Q}{\partial x}; \quad \frac{\partial R}{\partial x} = \frac{\partial P}{\partial z}; \quad \frac{\partial Q}{\partial z} = \frac{\partial R}{\partial y}. \tag{10}$$

If $\mathbf{F}$ is the gradient of a differentiable function f, $\mathbf{F}$ is said to be **exact.**

EXAMPLE 3 Let $\mathbf{F}(x, y, z) = yz\mathbf{i} + xz\mathbf{j} + xy\mathbf{k}$. Then $\partial P/\partial y = z = \partial Q/\partial x$, $\partial R/\partial x = y = \partial P/\partial z$, and $\partial Q/\partial z = x = \partial R/\partial y$. Thus $\mathbf{F}$ is exact, and there is a differentiable function f such that $\mathbf{F} = \nabla f$, so that

$$f_x = yz, \quad f_y = xz, \quad f_z = xy.$$

Then

$$f(x, y, z) = \int f_x\, dx = \int yz\, dx = xyz + g(y, z)$$

for some differentiable function g of y and z only. (This means that g is a constant function *with respect to* x.) Hence

$$f_y = xz + g_y(y, z) = xz,$$

so $g_y(y, z) = 0$ and $g(y, z) = h(z)$, where h is a differentiable function of z only. Then

$$f(x, y, z) = xyz + h(z)$$

and

$$f_z(x, y, z) = xy + h'(z) = xy,$$

[†]For a proof, see R. C. Buck, *Advanced Calculus* (McGraw-Hill, New York, 1965), page 424.

so $h'(z) = 0$ and $h(z) = C$, a constant. Finally, we have

$$f(x, y, z) = xyz + C,$$

and it is easily verified that $\nabla f = \mathbf{F}$. ■

EXAMPLE 4 Let $\mathbf{F}(x, y, z) = xy\mathbf{i} + yz\mathbf{j} + xz\mathbf{k}$. Then $\partial P/\partial y = x$ and $\partial Q/\partial x = 0$, so $\mathbf{F}$ is not exact. ■

Once we know how to determine whether a vector field is exact, we can use the following fact, which follows immediately from Theorem 1.

The work done by a conservative force field as it moves a particle completely around a closed path in $\mathbb{R}^3$ is zero.

Before leaving this section we note that line integrals in space are often written in a different way. If $\mathbf{F}(x, y, z) = P(x, y, z)\mathbf{i} + Q(x, y, z)\mathbf{j} + R(x, y, z)\mathbf{k}$, and if we write $\mathbf{dx} = dx\,\mathbf{i} + dy\,\mathbf{j} + dz\,\mathbf{k}$, then

$$\int_C \mathbf{F}(\mathbf{x}) \cdot \mathbf{dx} = \int_C (P\,dx + Q\,dy + R\,dz). \qquad \textbf{(11)}$$

We emphasize that formula (11) does not tell us anything new; it simply provides a different notation. For example, in Example 2

$$\int_C \mathbf{F}(\mathbf{x}) \cdot \mathbf{dx} = \int_C (y^2z^3\,dx + 2xyz^3\,dy + 3xy^2z^2\,dz),$$

and we still need formula (8) to evaluate the integral.

PROBLEMS 6.4

In Problems 1–4, evaluate $\int_C \mathbf{F}(\mathbf{x}) \cdot \mathbf{dx}$.

1. $\mathbf{F}(x, y, z) = x\mathbf{i} + y\mathbf{j} + z\mathbf{k}$; C is the curve $\mathbf{x}(t) = t\mathbf{i} + t^2\mathbf{j} + t^3\mathbf{k}$ from (0, 0, 0) to (1, 1, 1).
2. $\mathbf{F}(x, y, z) = 2xz\mathbf{i} - xy\mathbf{j} + yz^2\mathbf{k}$; C is the curve of Problem 1.
3. $\mathbf{F}(x, y, z) = x^2\mathbf{i} + y^2\mathbf{j} + z^2\mathbf{k}$; C is the helix $\mathbf{x}(t) = (\cos t)\mathbf{i} + (\sin t)\mathbf{j} + t\mathbf{k}$ from (1, 0, 0) to $(0, 1, \pi/2)$.
4. $\mathbf{F}(x, y, z) = yz\mathbf{i} + xz\mathbf{j} + xy\mathbf{k}$; C is the curve of Problem 3.

In Problems 5–7, show that $\mathbf{F}$ is exact and use Theorem 1 to evaluate $\int_C \mathbf{F}(\mathbf{x}) \cdot \mathbf{dx}$ when C is any piecewise smooth curve joining the two given points.

5. $\mathbf{F}(x, y, z) = y^2z^4\mathbf{i} + 2xyz^4\mathbf{j} + 4xy^2z^3\mathbf{k}$ from (0, 0, 0) to (3, 2, 1).
6. $\mathbf{F}(x, y, z) = [(y/z) + x^2]\mathbf{i} + [(x/z) - \sin y]\mathbf{j} + [\cos z - (xy/z^2)]\mathbf{k}$ from $(0, \pi/3, \pi/2)$ to $(1, \pi/2, \pi)$.
7. $\mathbf{F}(x, y, z) = [1/(x + 2y + 3z) - 3x]\mathbf{i} + [2/(x + 2y + 3z) + y^2]\mathbf{j} + [3/(x + 2y + 3z)]\mathbf{k}$ from (1, 0, 0) to (2, 3, 4).
8. Compute $\int_{(2,1,2)}^{(-1,0,4)} \mathbf{F}(\mathbf{x}) \cdot \mathbf{dx}$, where $\mathbf{F}(\mathbf{x}) = (yz + 2)\mathbf{i} + (xz - 3)\mathbf{j} + (xy + 5)\mathbf{k}$.
9. Suppose that $\mathbf{F}(\mathbf{x}) = P(x, y, z)\mathbf{i} + Q(x, y, z)\mathbf{j} + R(x, y, z)\mathbf{k}$ is continuously differentiable and the gradient of a differentiable function. Show that

$$\frac{\partial P}{\partial y} = \frac{\partial Q}{\partial x}, \qquad \frac{\partial P}{\partial z} = \frac{\partial R}{\partial x}, \qquad \frac{\partial Q}{\partial x} = \frac{\partial R}{\partial y}.$$

6.5 SURFACE INTEGRALS

In Section 6.3 we discussed Green's theorem, which gave a relationship between a line integral over a closed curve and a double integral over a region enclosed by that curve. In this section we discuss the notion of a surface integral. This will enable us, in Section 6.7, to extend Green's theorem to integrals over regions in space.

Recall from Section 3.6 that a surface in space is defined as the set of points satisfying the equation

$$G(x, y, z) = 0. \tag{1}$$

In this section we will begin by considering surfaces that can be written in the form

$$z = f(x, y) \tag{2}$$

for some function f. Note that (2) is a special case of (1), as can be seen by defining $G(x, y, z) = z - f(x, y)$ so that $z = f(x, y)$ is equivalent to $G(x, y, z) = 0$.

Definition 1 SMOOTH SURFACE A surface is called **smooth** at a point (x_0, y_0, z_0) if $\partial f/\partial x$ and $\partial f/\partial y$ are continuous at (x_0, y_0). If the surface is smooth at all points in the domain of f, we speak of it as a **smooth surface.** That is, if f is continuously differentiable, then the surface is smooth.

A surface integral is very much like a double integral. Suppose that the surface $z = f(x, y)$ is smooth for (x, y) in a bounded region Ω in the xy-plane. Since $f(x, y)$ is continuous on Ω, we find from Definitions 5.1.1 and 5.1.2 and Theorem 5.1.1, that

$$\iint_{\Omega} f(x, y)\, dA = \lim_{\Delta s \to 0} \sum_{i=1}^{n} \sum_{j=1}^{m} f(x_i^*, y_j^*)\, \Delta x\, \Delta y \tag{3}$$

exists, where $\Delta s = \sqrt{\Delta x^2 + \Delta y^2}$ and the limit is independent of the way in which the points (x_i^*, y_j^*) are chosen in the rectangle R_{ij}. We stress that in (3) the quantity $\Delta x\, \Delta y$ represents the *area* of the rectangle R_{ij}.

The double integral (3) is an integral over a region in the plane. A *surface integral,* which we will soon define, is an integral over a surface in space. Suppose we wish to integrate the function $F(x, y, z)$ over the surface S given by $z = f(x, y)$ where $(x, y) \in \Omega$ and, as before, Ω is a bounded region in the xy-plane. Such a surface is sketched in Figure 1.

We partition Ω into rectangles (and parts of rectangles) as before. This procedure provides a partition of S into mn "subsurfaces" S_{ij}, where

$$S_{ij} = \{(x, y, z): z = f(x, y) \text{ and } (x, y) \in R_{ij}\}. \tag{4}$$

We choose a point in each S_{ij}. Such a point will have the form (x_i^*, y_j^*, z_{ij}^*), where $z_{ij}^* = f(x_i^*, y_j^*)$ and $(x_i^*, y_j^*) \in R_{ij}$. We let $\Delta\sigma_{ij}$ denote the surface area of S_{ij}. This is

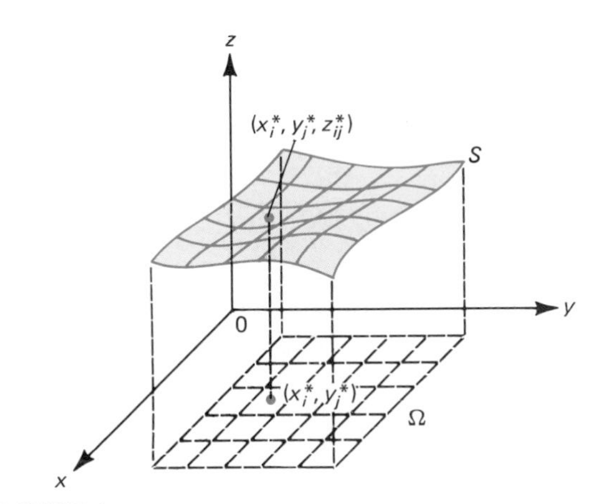

FIGURE 1

analogous to the notation $\Delta x \, \Delta y$ as the area of the rectangle R_{ij}. Then we write the double sum

$$\sum_{i=1}^{n}\sum_{j=1}^{m} F(x_i^*, y_j^*, z_{ij}^*) \, \Delta\sigma_{ij}$$

and consider

$$\lim_{\Delta s \to 0} \sum_{i=1}^{n}\sum_{j=1}^{m} F(x_i^*, y_j^*, z_{ij}^*) \, \Delta\sigma_{ij}. \tag{5}$$

Definition 2 INTEGRAL OVER A SURFACE Suppose the limit in (5) exists and is independent of the way the surface S is partitioned and the way in which the points (x_i^*, y_j^*, z_{ij}^*) are chosen in S_{ij}. Then F is said to be **integrable** over S and the **surface integral** of F over S, denoted $\iint_S F(x, y, z) \, d\sigma$, is given by

$$\iint_S F(x, y, z) \, d\sigma = \lim_{\Delta s \to 0} \sum_{i=1}^{n}\sum_{j=1}^{m} F(x_i^*, y_j^*, z_{ij}^*) \, \Delta\sigma_{ij}. \tag{6}$$

There are three central questions that are raised by this definition.

(i) Under what conditions does the surface integral of F over S exist?
(ii) How can we compute the surface integral if it does exist?
(iii) What is a surface integral good for?

We will answer the first two questions now and will answer the third one by giving some applications of the surface integral later in the section.

The key to evaluating the limit in (6) is to note that, from formula (5.5.5),

$$\Delta\sigma_{ij} \approx \sqrt{f_x^2(x_i^*, y_j^*) + f_y^2(x_i^*, y_j^*) + 1} \, \Delta x \, \Delta y. \tag{7}$$

Then inserting (7) into the limit in (6) and noting that $z_{ij}^* = f(x_i^*, y_j^*)$, we have

$$\lim_{\Delta s \to 0} \sum_{i=1}^{n} \sum_{j=1}^{m} F(x_i^*, y_j^*, z_{ij}^*) \, \Delta\sigma_{ij}$$

$$= \lim_{\Delta s \to 0} \sum_{i=1}^{n} \sum_{j=1}^{m} F(x_i^*, y_j^*, f(x_i^*, y_j^*)) \sqrt{f_x^2(x_i^*, y_j^*) + f_y^2(x_i^*, y_j^*) + 1} \, \Delta x \, \Delta y. \quad \textbf{(8)}$$

Now if $z = f(x, y)$ is a smooth surface over Ω, then f_x and f_y are continuous over Ω, so that $\sqrt{f_x^2 + f_y^2 + 1}$ is also continuous over Ω. Furthermore, if $F(x, y, z)$ is continuous for (x, y, z) on S, then by Theorem 5.1.1 and Definitions 5.1.1 and 5.1.2,

$$\iint_\Omega F(x, y, f(x, y)) \sqrt{f_x^2(x, y) + f_y^2(x, y) + 1} \, dA$$

exists and is equal to the right-hand limit in (8). We therefore have the following important result.

Theorem 1 Let $S: z = f(x, y)$ be a smooth surface for (x, y) in the bounded region Ω in the xy-plane. Then if F is continuous on S, F is integrable over S, and

$$\iint_S F(x, y, z) \, d\sigma = \iint_\Omega F(x, y, f(x, y)) \sqrt{f_x^2(x, y) + f_y^2(x, y) + 1} \, dA. \quad \textbf{(9)}$$

REMARK 1. Using (9), we can compute a surface integral over S by transforming it into an ordinary double integral over the region Ω that is the projection of S into the xy-plane.

REMARK 2. If $F(x, y, z) = 1$ in (9), then (9) reduces to

$$\sigma = \iint_S d\sigma = \iint_\Omega \sqrt{f_x^2(x, y) + f_y^2(x, y) + 1} \, dA. \quad \textbf{(10)}$$

This is the formula for surface area given in Section 5.5 on page 299.

EXAMPLE 1 Compute $\iint_S (x + 2y + 3z) \, d\sigma$, where S is the part of the plane $2x - y + z = 3$, that lies above the triangular region Ω in the xy-plane bounded by the x- and y-axes and the line $y = 1 - 2x$.

Solution. We have $z = f(x, y) = 3 - 2x + y$, so that $f_x = -2$, $f_y = 1$, and $\sqrt{f_x^2(x, y) + f_y^2(x, y) + 1} = \sqrt{6}$. The region Ω is sketched in Figure 2. Then from (9)

$$I = \iint_S (x + 2y + 3z) \, d\sigma = \sqrt{6} \iint_\Omega [x + 2y + 3(3 - 2x + y)] \, dA$$

$$= \sqrt{6} \int_0^{1/2} \int_0^{1-2x} (9 - 5x + 5y) \, dy \, dx = \sqrt{6} \int_0^{1/2} \left(9y - 5xy + \frac{5y^2}{2}\right)\Bigg|_0^{1-2x} dx$$

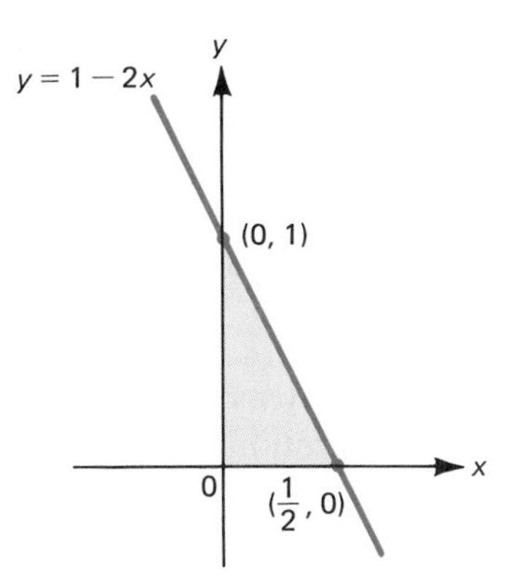

FIGURE 2

$$= \sqrt{6}\int_0^{1/2}\left[9(1-2x) - 5x(1-2x) + \frac{5}{2}(1-2x)^2\right]dx$$

$$= \sqrt{6}\int_0^{1/2}\left(20x^2 - 33x + \frac{23}{2}\right)dx = \frac{59}{24}\sqrt{6}. \quad \blacksquare$$

EXAMPLE 2 Compute $\iint_S (x^2 + y^2 + 3z^2)\,d\sigma$, where S is the part of the circular paraboloid $z = x^2 + y^2$ with $x^2 + y^2 \le 9$.

Solution. Here $f(x, y) = x^2 + y^2$, so that $f_x = 2x$, $f_y = 2y$, and $d\sigma = \sqrt{1 + 4x^2 + 4y^2}\,dx\,dy$. Thus

$$I = \iint_S (x^2 + y^2 + 3z^2)\,d\sigma = \iint_\Omega [x^2 + y^2 + 3(x^2 + y^2)^2]\sqrt{1 + 4x^2 + 4y^2}\,dA,$$

where Ω is the disk (in the xy-plane) $x^2 + y^2 \le 9$. The problem is greatly simplified by the use of polar coordinates. We have, using $x^2 + y^2 = r^2$,

$$I = \int_0^{2\pi}\int_0^3 [r^2 + 3(r^2)^2]\sqrt{1 + 4r^2}\,r\,dr\,d\theta$$

$$= \int_0^{2\pi}\int_0^3 (r^3 + 3r^5)\sqrt{1 + 4r^2}\,dr\,d\theta$$

$$= 2\pi\int_0^3 (r^3 + 3r^5)\sqrt{1 + 4r^2}\,dr.$$

There are several ways to complete the evaluation of this integral. One way is to integrate by parts twice (start with $u = r^2 + 3r^4$). Another way is to make the substitution $r = \frac{1}{2}\tan\varphi$. The result is

$$I = 2\pi\left\{21(37)^{3/2} + \frac{1}{120}[1 - 55(37)^{5/2}] + \frac{1}{280}(37^{7/2} - 1)\right\} \approx 12{,}629.4. \quad \blacksquare$$

EXAMPLE 3 Evaluate $\iint_S (x + y + z)\, d\sigma$, where S is the part of the surface $z = x^3 + y^4$ lying over the square $\{(x, y): 0 \le x \le 1, 0 \le y \le 1\}$.

Solution. We have $f_x = 3x^2$, $f_y = 4y^3$, and $d\sigma = \sqrt{1 + 9x^4 + 16y^6}\, dx\, dy$, so

$$\iint_S (x + y + z)\, d\sigma = \int_0^1 \int_0^1 (x + x^3 + y + y^4)\sqrt{1 + 9x^4 + 16y^6}\, dx\, dy.$$

This is as far as we can go because there is no way to compute this integral directly. The best we can do is to use numerical integration to approximate the answer. ■

REMARK. Example 3 is typical. Like the computations of arc length and surface area, the direct evaluation of a surface integral will often be impossible since it will involve integrals of functions for which antiderivatives cannot be found.

There are many applications for which it is necessary to compute a surface integral. For example, suppose that for (x, y) in a region Ω in the xy-plane, $z = f(x, y)$ is the equation of a thin metallic surface in space. Suppose further that the density of the surface varies and is given by $\rho(x, y, z)$ for (x, y, z) on the surface. Our problem is to compute the total mass of the sheet. We do so by first partitioning Ω in the usual way. This procedure leads to a partition of the metallic surface into subsurfaces, as before. We again denote this subsurface S_{ij}:

$$S_{ij} = \{(x, y, z): z = f(x, y) \text{ and } (x, y) \in R_{ij}\}.$$

If $\Delta\sigma_{ij}$ denotes the surface area of S_{ij}, if $\Delta\sigma_{ij}$ is small, and if $\rho(x, y, z)$ is continuous, then $\rho(x, y, z)$ is approximately constant on S_{ij}. If (x_i^*, y_j^*) is a point in R_{ij}, then $(x_i^*, y_j^*, z_{ij}^*) = (x_i^*, y_j^*, f(x_i^*, y_j^*))$ is a point on S_{ij}, and

$$\rho(x, y, z) \approx \rho(x_i^*, y_j^*, f(x_i^*, y_j^*)) \tag{11}$$

for (x, y, z) on S_{ij}. Now if area density is constant, then the mass μ of an object with area σ and density ρ is given by

$$\mu = \rho\sigma. \tag{12}$$

Thus if μ_{ij} denotes the mass of S_{ij}, then combining (11) and (12), we have

$$\mu_{ij} \approx \rho(x_i^*, y_j^*, f(x_i^*, y_j^*))\, \Delta\sigma_{ij}, \tag{13}$$

and adding up the masses of the mn "subsurfaces" S_{ij}, we obtain

$$\mu \approx \sum_{i=1}^{n} \sum_{j=1}^{m} \rho(x_i^*, y_j^*, f(x_i^*, y_j^*))\, \Delta\sigma_{ij}. \tag{14}$$

Finally, we let $\Delta s \to 0$, as before. This gives us

$$\mu = \lim_{\Delta s \to 0} \sum_{i=1}^{n} \sum_{j=1}^{m} \rho(x_i^*, y_j^*, f(x_i^*, y_j^*))\, \Delta\sigma_{ij} = \iint_S \rho(x, y, f(x, y))\, d\sigma. \tag{15}$$

EXAMPLE 4 A metallic dome has the shape of a hemisphere centered at the origin with radius 4 m. Its area density at a point (x, y, z) in space is given by $\rho(x, y, z) = 25 - x^2 - y^2$ kilograms per square meter. Find the total mass of the dome.

Solution. The equation of the hemisphere is $x^2 + y^2 + z^2 = 16$ with $z \geq 0$. Thus $z = f(x, y) = \sqrt{16 - x^2 - y^2}$ (for $x^2 + y^2 \leq 16$),

$$f_x = \frac{-x}{\sqrt{16 - x^2 - y^2}} \qquad f_y = \frac{-y}{\sqrt{16 - x^2 - y^2}}$$

and

$$\begin{aligned}\sqrt{f_x^2 + f_y^2 + 1} &= \sqrt{\frac{x^2}{16 - x^2 - y^2} + \frac{y^2}{16 - x^2 - y^2} + 1} \\ &= \sqrt{\frac{x^2 + y^2 + (16 - x^2 - y^2)}{16 - x^2 - y^2}} = \frac{4}{\sqrt{16 - x^2 - y^2}}.\end{aligned}$$

Then using polar coordinates, we have

$$\begin{aligned}\mu &= \iint_S \rho(x, y, z)\, d\sigma \overset{\downarrow}{=} \int_0^{2\pi} \int_0^4 (25 - r^2)\left(\frac{4}{\sqrt{16 - r^2}}\right) r\, dr\, d\theta \\ &= 2\pi \int_0^4 (25 - r^2)\left(\frac{4}{\sqrt{16 - r^2}}\right) r\, dr\, d\theta.\end{aligned}$$

(Ω is the circle centered at (0, 0) with radius 4)

Now let $16 - r^2 = u^2$, so that $25 - r^2 = 9 + u^2$ and $r\, dr = -u\, du$. Then the integral becomes

$$\begin{aligned}2\pi \int_4^0 (9 + u^2)\left(\frac{4}{u}\right)(-u\, du) &= 8\pi \int_0^4 (9 + u^2)\, du \\ &= (8\pi)\left(\frac{172}{3}\right) = \frac{1376\pi}{3} \approx 1441 \text{ kg.} \quad \blacksquare\end{aligned}$$

We now derive another way to represent a surface integral. Let $z = f(x, y)$ be a smooth surface in space for $(x, y) \in \Omega$. As before, the surface can be written

$$G(x, y, z) = 0,$$

where

$$G(x, y, z) = z - f(x, y). \tag{16}$$

Definition 3 OUTWARD UNIT NORMAL VECTOR We know from Section 4.7 that the gradient vector ∇G is orthogonal or normal to the surface at every point on the surface. Then if $\nabla G \neq 0$,

$$\mathbf{n} = \frac{\nabla G}{|\nabla G|} \tag{17}$$

is a unit normal vector. It is called the **outward unit normal vector.** In general, if $G(x, y, z) = 0$ is any surface in space, then at any point at which ∇G exists and is nonzero, the vector defined by (17) is called the outward unit normal vector.

REMARK 1. There are two unit vectors normal to a smooth surface at any point. Which one is the outward normal vector? This is a difficult question to answer because, depending on how the surface is defined, we can obtain either one. To see why, note that the surface $G(x, y, z) = 0$ can also be written as $-G(x, y, z) = 0$. In the first case $\mathbf{n} = \nabla G|\nabla G|$, while in the second $\mathbf{n} = \nabla(-G)/|\nabla(-G)| = -\nabla G/|-\nabla G| = -\nabla G/|\nabla G|$. The second "outward" normal is the negative of the first and so points in the opposite direction. To resolve this difficulty, we must define the *orientation* of a surface. This topic is discussed in advanced calculus books. For our purposes we will write closed surfaces so that **n** points outside the surface. We will say more about the direction of **n** for other kinds of surfaces in Section 6.7.

REMARK 2. We can think of the direction of **n** in the following way. If $G(x, y, z) = 0$, then the outward normal vector **n** points from the surface $G(x, y, z) = 0$ toward the surface $G(x, y, z) = \epsilon$ for every $\epsilon > 0$. This occurs because ∇G points in the direction of increasing G (see Theorem 4.8.1). For example, the unit outward normal vector **n** to the unit sphere $x^2 + y^2 + z^2 - 1 = 0$ points toward the sphere of radius 2: $x^2 + y^2 + z^2 - 1 = 3$. Evidently, **n** points away from the center of the sphere.

Now from (16), $\nabla G = G_x\mathbf{i} + G_y\mathbf{j} + G_z\mathbf{k} = -f_x\mathbf{i} - f_y\mathbf{j} + \mathbf{k}$, so that

$$\mathbf{n} = \frac{\nabla G}{|\nabla G|} = \frac{-f_x\mathbf{i} - f_y\mathbf{j} + \mathbf{k}}{\sqrt{f_x^2 + f_y^2 + 1}}. \tag{18}$$

We now define the angle γ to be the acute angle between **n** and the positive z-axis. This angle is depicted in Figure 3. Since **k** is a unit vector having the direction of the positive z-axis, we have, from Theorem 3.2.3 on p. 94,

$$\cos\gamma = \frac{\mathbf{n}\cdot\mathbf{k}}{|\mathbf{n}|\,|\mathbf{k}|} = \frac{1}{\sqrt{f_x^2 + f_y^2 + 1}} > 0 \tag{19}$$

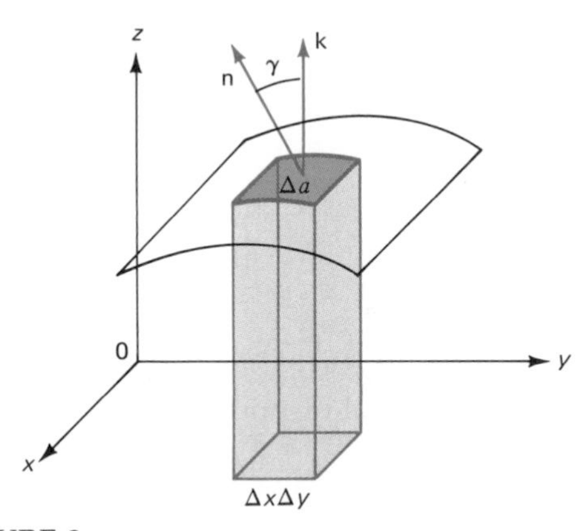

FIGURE 3

and

$$\sec \gamma = \sqrt{f_x^2 + f_y^2 + 1}. \tag{20}$$

Hence we have

$$\iint_S F(x, y, z)\, d\sigma = \iint_\Omega F(x, y, f(x, y)) \sec \gamma(x, y)\, dx\, dy. \tag{21}$$

EXAMPLE 5 Compute the surface integral of Example 1 by using formula (21).

Solution. Here S is the plane $2x - y + z = 3$. From Section 3.5 we know that $\mathbf{N} = 2\mathbf{i} - \mathbf{j} + \mathbf{k}$ is normal to this surface, so $\mathbf{n} = \mathbf{N}/|\mathbf{N}| = (1/\sqrt{6})(2\mathbf{i} - \mathbf{j} + \mathbf{k})$ is the outward unit normal vector. Then $\cos \gamma = \mathbf{n} \cdot \mathbf{k} = 1/\sqrt{6}$ and $\sec \gamma = \sqrt{6}$, so

$$\iint_S (x + 2y + 3z)\, d\sigma = \iint_\Omega [(x + 2y) + 3(3 - 2x + y)]\sqrt{6}\, dx\, dy,$$

and we can complete the integration as before. ■

In all the preceding discussion we assumed that the surface was written as $z = f(x, y)$. Suppose instead that the surface is given by

$$S\colon x = g(y, z), \tag{22}$$

where $(y, z) \in \Omega$, a region in the yz-plane, or

$$S\colon y = h(x, z), \tag{23}$$

where $(x, z) \in \Omega$, a region in the xz-plane. Let α and β be defined as the acute angles between the unit outward normal to S and the positive x-axis and y-axis, respectively (see Figure 4). Then we have the following:

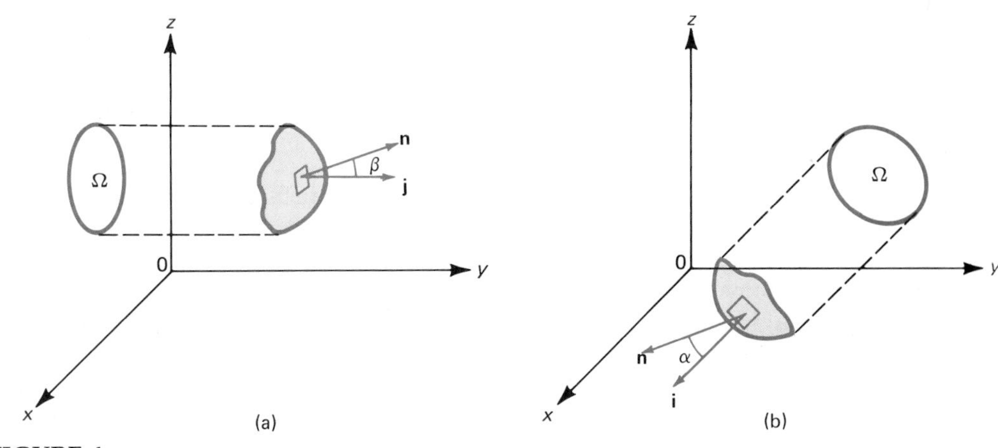

FIGURE 4

(i) If the surface is given by (22), then,

$$\iint_S F(x, y, z)\, d\sigma = \iint_\Omega F(g(y, z), y, z)\sqrt{g_y^2 + g_z^2 + 1}\, dy\, dz \tag{24}$$

$$= \iint_\Omega F(g(y, z), y, x) \sec\alpha\, dy\, dz.$$

(ii) If the surface is given by (23), then

$$\iint_S F(x, y, z)\, d\sigma = \iint_\Omega F(x, h(x, z), z)\sqrt{h_x^2 + h_z^2 + 1}\, dx\, dz \tag{25}$$

$$= \iint_\Omega F(x, h(x, z), z) \sec\beta\, dx\, dz.$$

EXAMPLE 6 Compute $\iint_S y\, d\sigma$, where S is the hyperbolic paraboloid $x = y^2 - (z^2/2)$ for $0 \le y \le \sqrt{2}$ and $0 \le z \le 1$.

Solution. Here $g(y, z) = y^2 - (z^2/2)$, $g_y = 2y$, $g_z = -z$, and $\sqrt{g_y^2 + g_z^2 + 1} = \sqrt{4y^2 + z^2 + 1}$, so

$$I = \iint_S y\, d\sigma = \int_0^1 \int_0^{\sqrt{2}} y\sqrt{4y^2 + z^2 + 1}\, dy\, dz = \frac{1}{12}\int_0^1 (4y^2 + z^2 + 1)^{3/2}\Big|_{y=0}^{y=\sqrt{2}}\, dz$$

$$= \frac{1}{12}\left[\int_0^1 (z^2 + 9)^{3/2}\, dz - \int_0^1 (z^2 + 1)^{3/2}\, dz\right].$$

These integrals can be integrated by making the substitutions $z = 3\tan\theta$ and $z = \tan\theta$, respectively. We then obtain

$$\iint_S y\, d\sigma = \frac{1}{12}\left[\frac{47}{8}\sqrt{10} + \frac{243}{8}\ln\left(\frac{1 + \sqrt{10}}{9}\right) - \frac{19}{32}\sqrt{2} - \frac{3}{32}\ln(1 + \sqrt{2})\right]. \blacksquare$$

FLUX

We now give another application of the surface integral and represent the integral in yet another way. Suppose that a smooth surface is immersed in a continuous vector field. For example, this could be a permeable surface immersed in a fluid or a conductor surrounded by an electric field. We denote the vector field by $\mathbf{v}(x, y, z)$ and assume that it is a flow field; that is, we assume that something is moving. For example, the fluid could be in motion. As a particle moves through the fluid, it traces out a **streamline** which is determined by **lines of force.** In a flow field a line of force is an imaginary line that, at a given point, has the same direction as the flow field at that point. One streamline is illustrated in Figure 5. As the particle moves from P_1 to P_2 to P_3, it traces out a path that is determined by the vectors $\mathbf{v}(x, y, z)$ at each point through which it passes. That is, $\mathbf{v}$ is tangent to the streamline at every point. Thus if the particle is at the point (x_0, y_0, z_0), then $\mathbf{v}(x_0, y_0, z_0)$ determines where the particle will be in the next instant of time. Loosely speaking, the *flux* of a vector field is determined by the number of lines of force that cut through the surface in a direction normal to the surface. For an electric field the flux density is proportional to the number of lines of force that cut through the surface.

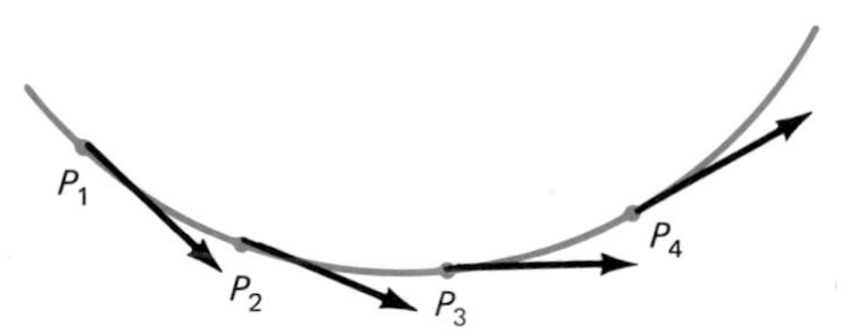

FIGURE 5

A typical surface immersed in a vector field is sketched in Figure 6a. In Figure 6b we have sketched the flux across a small subsurface. The flux across a point on the surface can be thought of as the component of the vector field in the direction normal to the surface at that point. From Definition 3.2.4, this **normal component,** as it is called, of the vector field is given by

$$\frac{\mathbf{v} \cdot \mathbf{n}}{|\mathbf{n}|} = \mathbf{v} \cdot \mathbf{n}$$

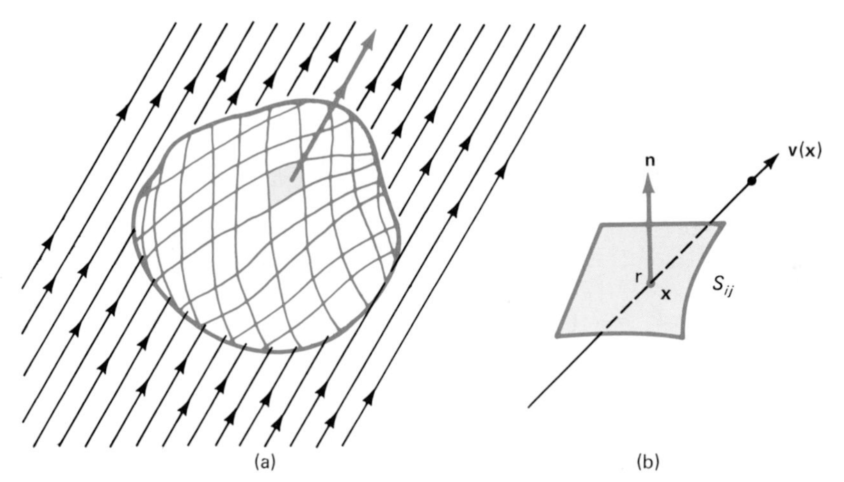

FIGURE 6

since $|\mathbf{n}| = 1$. If the subsurface S_{ij} is very small, then because $\mathbf{v}$ is continuous, $\mathbf{v} \cdot \mathbf{n}$ will vary little over S_{ij} and the flux across S_{ij} will be approximately equal to $\mathbf{v}(x_i^*, y_j^*, f(x_i^*, y_j^*)) \cdot \mathbf{n}$ times the surface area of S_{ij}. Thus

$$\text{flux across } S_{ij} \approx \mathbf{v}(x_i^*, y_j^*, f(x_i^*, y_j^*)) \cdot \mathbf{n}\, \Delta\sigma_{ij}. \tag{26}$$

Then

$$\text{total flux} \approx \sum_{i=1}^{n} \sum_{j=1}^{m} \mathbf{v}(x_i^*, y_j^*, f(x_i^*, y_j^*)) \cdot \mathbf{n}\, \Delta\sigma_{ij}. \tag{27}$$

This result leads to the following formal definition.

Definition 4 FLUX OF A VECTOR FIELD OVER A SURFACE Let S be a smooth surface with the vector field $\mathbf{v}(x, y, z)$ continuous on S. Then the **flux** of $\mathbf{v}$ over S is defined by

$$\text{flux of } \mathbf{v} \text{ over } S = \iint_S \mathbf{v} \cdot \mathbf{n}\, d\sigma. \tag{28}$$

The following theorem tells us how flux can be computed.

Theorem 2 Let $S\colon z = f(x, y)$ for $(x, y) \in \Omega$ be a smooth surface and let $\mathbf{n}(x, y, z)$ be its outward unit normal vector. Let $\mathbf{v} = \mathbf{v}(x, y, z)$ be a vector field continuous on S. If

$$\mathbf{v}(x, y, z) = P(x, y, z)\mathbf{i} + Q(x, y, z)\mathbf{j} + R(x, y, z)\mathbf{k}, \tag{29}$$

then

$$\text{flux of } \mathbf{v} \text{ over } S = \iint_S \mathbf{v} \cdot \mathbf{n}\, d\sigma = \iint_\Omega (-Pf_x - Qf_y + R)\, dx\, dy. \tag{30}$$

Proof. Using (18) and (19), we find that

$$\begin{aligned} \mathbf{v} \cdot \mathbf{n} &= (P\mathbf{i} + Q\mathbf{j} + R\mathbf{k}) \cdot \left(\frac{-f_x\mathbf{i} - f_y\mathbf{j} + \mathbf{k}}{\sqrt{f_x^2 + f_y^2 + 1}} \right) \\ &= \frac{1}{\sqrt{f_x^2 + f_y^2 + 1}} (-Pf_x - Qf_y + R) \\ &= \frac{-Pf_x - Qf_y + R}{\sec \gamma}. \end{aligned} \tag{31}$$

Now from (21)

$$\iint_S \mathbf{v} \cdot \mathbf{n}\, d\sigma = \iint_\Omega \mathbf{v} \cdot \mathbf{n} \sec \gamma\, dx\, dy. \tag{32}$$

But from (31)

$$\mathbf{v} \cdot \mathbf{n} \sec \gamma = -Pf_x - Qf_y + R. \tag{33}$$

Inserting (33) into the right-hand integral in (32) completes the proof of the theorem. ■

EXAMPLE 7 Find the flux across the conical surface $z = \sqrt{x^2 + y^2}$, $x^2 + y^2 \neq 0$, where $x^2 + y^2 \leq 1$, $x \geq 0$, $y \geq 0$, if the velocity field is given by $\mathbf{v} = x^2\mathbf{i} + y^2\mathbf{j} + z\mathbf{k}$.

Solution. We have $f_x = x/\sqrt{x^2 + y^2}$ and $f_y = y/\sqrt{x^2 + y^2}$, so

$$\text{flux} = \iint_\Omega (-Pf_x - Qf_y + R)\,dx\,dy$$

$$= \iint_\Omega \left(-\frac{x^3}{\sqrt{x^2+y^2}} - \frac{y^3}{\sqrt{x^2+y^2}} + \sqrt{x^2+y^2}\right) dx\,dy$$

Using polar coordinates

$$= \int_0^{\pi/2}\int_0^1 \left(-\frac{r^3\cos^3\theta}{r} - \frac{r^3\sin^3\theta}{r} + r\right) r\,dr\,d\theta$$

$$= \int_0^{\pi/2}\int_0^1 [-r^3(\cos^3\theta + \sin^3\theta) + r^2]\,dr\,d\theta$$

$$= \int_0^{\pi/2} \left\{\left[-\frac{r^4}{4}(\cos^3\theta + \sin^3\theta) + \frac{r^3}{3}\right]\Bigg|_0^1\right\} d\theta$$

$$= \int_0^{\pi/2} \frac{1}{3}\,d\theta - \frac{1}{4}\int_0^{\pi/2} (\cos^3\theta + \sin^3\theta)\,d\theta = \frac{\pi}{6} - \frac{1}{3}. \blacksquare$$

EXAMPLE 8 Compute the flux of the vector field $\mathbf{v}(x, y, z) = x^2y\mathbf{i} - 2yz\mathbf{j} + x^3y^2\mathbf{k}$ over the surface of the rectangular solid S:

$$0 \leq x \leq 1, \qquad 0 \leq y \leq 2, \qquad 0 \leq z \leq 3.$$

Solution. The surface S is sketched in Figure 7. This surface is not smooth, but it is piecewise smooth. That is, it is smooth on each face. So we compute the surface integral over each face and add up the results. The work is summarized in Table 1. For example, consider the face $x = 1$. This face is in a plane parallel to the yz-plane, so that $\Delta\sigma_{ij} = \Delta y\,\Delta z$ and

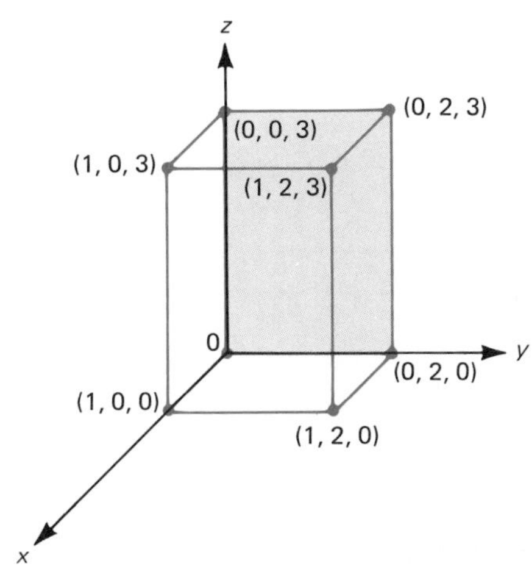

FIGURE 7

TABLE 1

Face	Outward unit normal $\mathbf{n}$	$\mathbf{v} \cdot \mathbf{n}$	$\iint_{\text{face}} (\mathbf{v} \cdot \mathbf{n})\, d\sigma$ = flux along face
$x = 0$	$-\mathbf{i}$	$-x^2y = 0$	0, since $x = 0$
$x = 1$	$\mathbf{i}$	$x^2y = y$	6
$y = 0$	$-\mathbf{j}$	$2yz = 0$	0, since $y = 0$
$y = 2$	$\mathbf{j}$	$-2yz = -4z$	-18
$z = 0$	$-\mathbf{k}$	$-x^3y^2$	$-\frac{2}{3}$
$z = 3$	$\mathbf{k}$	x^3y^2	$\frac{2}{3}$

$$\iint_{x=1} (\mathbf{v} \cdot \mathbf{n})\, d\sigma = \int_0^3 \int_0^2 y\, dy\, dz = 6.$$

The surface $y = 2$ is in a plane parallel to the xz-plane, so, analogously, $\Delta\sigma_{ij} = \Delta x\, \Delta z$ and

$$\iint_{y=2} (\mathbf{v} \cdot \mathbf{n})\, d\sigma = \int_0^3 \int_0^1 -4z\, dx\, dz = -18.$$

Similarly,

$$\iint_{z=0} (\mathbf{v} \cdot \mathbf{n})\, d\sigma = \int_0^2 \int_0^1 -x^3y^2\, dx\, dy = -\frac{2}{3}$$

and

$$\iint_{z=3} (\mathbf{v} \cdot \mathbf{n})\, d\sigma = \int_0^2 \int_0^1 x^3y^2\, dx\, dy = \frac{2}{3}.$$

Thus the total flux is $6 - 18 - \frac{2}{3} + \frac{2}{3} = -12$. ■

Formulas (21), (24), and (25) gives us a way to represent the **differential** $d\sigma$. We have

$$d\sigma = \sec \gamma\, dx\, dy, \tag{34}$$

$$d\sigma = \sec \beta\, dx\, dz, \tag{35}$$

and

$$d\sigma = \sec \alpha\, dy\, dz. \tag{36}$$

We have already seen [formula (19)] that

$$\cos \gamma = \mathbf{n} \cdot \mathbf{k}. \tag{37}$$

Similarly, it is easy to show that

$$\cos \beta = \mathbf{n} \cdot \mathbf{j} \tag{38}$$

and

$$\cos \alpha = \mathbf{n} \cdot \mathbf{i}. \tag{39}$$

Thus

$$\mathbf{n} = (\cos \alpha)\mathbf{i} + (\cos \beta)\mathbf{j} + (\cos \gamma)\mathbf{k}, \tag{40}$$

and if $\mathbf{v} = P\mathbf{i} + Q\mathbf{j} + R\mathbf{k}$, we can write the flux as

$$\iint_S \mathbf{v} \cdot \mathbf{n}\, d\sigma = \iint_S (P \cos \alpha + Q \cos \beta + R \cos \gamma)\, d\sigma. \tag{41}$$

PROBLEMS 6.5

In Problems 1–17, evaluate the surface integral over the given surface.†

1. $\iint_S x\, d\sigma$, where S: $z = x^2$, $0 \le x \le 1$, $0 \le y \le 2$.
2. $\iint_S y\, d\sigma$, where S is as in Problem 1.
3. $\iint_S x^2\, d\sigma$, where S is as in Problem 1.
4. $\iint_S (x^2 - 2y^2)\, d\sigma$, where S is as in Problem 1.
5. $\iint_S \sqrt{1 + 4z}\, d\sigma$, where S is as in Problem 1.

*6. $\iint_S x\, d\sigma$, where S is the hemisphere $x^2 + y^2 + z^2 = 4$, $z \ge 0$, $x^2 + y^2 \le 4$.

*7. $\iint_S xy\, d\sigma$, where S is as in Problem 6.

8. $\iint_S (x + y)\, d\sigma$, where S is the plane $x + 2y - 3z = 4$; $0 \le x \le 1$, $1 \le y \le 2$.
9. $\iint_S yz\, d\sigma$, where S is as in Problem 8.
10. $\iint_S z^2\, d\sigma$, where S is as in Problem 8.

*11. $\iint_S (x^2 + y^2 + z^2)\, d\sigma$, where S is the part of the plane $x - y = 4$ that lies inside the cylinder $y^2 + z^2 = 4$.

12. $\iint_S \cos z\, d\sigma$, where S is the plane $2x + 3y + z = 1$ for $0 \le x \le 1$ and $-1 \le y \le 2$.

*13. $\iint_S z\, d\sigma$, where S is the tetrahedron bounded by the coordinate planes and the plane $4x + 8y + 2z = 16$.

14. $\iint_S |x|\, d\sigma$, where S is the hemisphere $x^2 + y^2 + z^2 = 4$; $x \ge 0$, $y^2 + z^2 \le 4$.
15. $\iint_S z\, d\sigma$, where S is the surface of Problem 14.
16. $\iint_S z^2\, d\sigma$, where S is the hemisphere $x^2 + y^2 + z^2 = 9$, $y \ge 0$, $x^2 + z^2 \le 9$.
17. $\iint_S x^2\, d\sigma$, where S is the surface of Problem 16.
18. Find the mass of a triangular metallic sheet with corners at (1, 0, 0), (0, 1, 0), and (0, 0, 1) if its density is constant.
19. Find the mass of the sheet of Problem 18 if its density is proportional to x^2.
20. Find the mass of a metallic sheet in the shape of the hemisphere $x^2 + y^2 + z^2 = 9$, $z \ge 0$, $x^2 + y^2 \le 9$, if its density is proportional to its distance from the origin.
21. **(a)** Show that $\iint_S(x^2 + y^2)\, d\sigma$ over the hemisphere $x^2 + y^2 + z^2 = r^2$, $z \ge 0$, $x^2 + y^2 \le r^2$, is equal to $\frac{4}{3}\pi r^4$.
    (b) Show that $\iint_S x^2\, d\sigma = \iint_S y^2\, d\sigma = \iint_S z^2\, d\sigma = \frac{1}{3}\iint_S(x^2 + y^2 + z^2)\, d\sigma$.
    (c) Explain why the last integral is equal to $\frac{2}{3}\pi r^4$, without performing any integration.
    (d) Use (c) to explain why $\iint_S(x^2 + y^2)\, d\sigma = \frac{4}{3}\pi r^4$, without performing any integration.

In Problems 22–30, compute the flux $\iint_S \mathbf{v} \cdot \mathbf{n}\, d\sigma$ for the given surface lying in the given vector field.

22. S: $z = xy$; $0 \le x \le 1$, $0 \le y \le 2$; $\mathbf{v} = x^2y\mathbf{i} - z\mathbf{j}$
23. S: $z = 4 - x - y$; $x \ge 0$, $y \ge 0$, $z \ge 0$; $\mathbf{v} = -3x\mathbf{i} - y\mathbf{j} + 3z\mathbf{k}$
24. S: $x^2 + y^2 + z^2 = 1$; $z \ge 0$; $\mathbf{v} = x\mathbf{i} + y\mathbf{j} + z\mathbf{k}$
25. S: $x^2 + y^2 + z^2 = 1$; $y \ge 0$; $\mathbf{v} = x\mathbf{i} + y\mathbf{j} + z\mathbf{k}$
26. S: $x^2 + y^2 + z^2 = 1$; $x \le 0$; $\mathbf{v} = x\mathbf{i} + y\mathbf{j} + z\mathbf{k}$
27. S: $z = \sqrt{x^2 + y^2}$; $x^2 + y^2 \le 1$; $\mathbf{v} = x\mathbf{i} - y\mathbf{j} + xy\mathbf{k}$

†Note that there may be two answers if it is not clear which of the two normals points outward.

28. $S: x = \sqrt{y^2 + z^2};\ y^2 + z^2 \leq 1;\ \mathbf{v} = y\mathbf{i} - z\mathbf{j} + yz\mathbf{k}$

***29.** S: region bounded by $y = 1$ and $y = \sqrt{x^2 + z^2}$; $x^2 + z^2 \leq 1$; $\mathbf{v} = x\mathbf{i} - z\mathbf{j} + xz\mathbf{k}$

30. S: unit sphere; $\mathbf{v} = x\mathbf{i} + y\mathbf{j} + z\mathbf{k}$

31. Show that if $\mathbf{v} = a\mathbf{i} + b\mathbf{j} + c\mathbf{k}$, where a, b, and c are constants, then $\iint_S \mathbf{v} \cdot \mathbf{n}\, d\sigma = 0$, where S is the sphere $x^2 + y^2 + z^2 = r^2$.

32. Find the flux of the vector field $\mathbf{v} = xz\mathbf{i} + y^2\mathbf{j} - xy^3z^2\mathbf{k}$ over the rectangular solid of Example 8.

33. Find the flux of the vector field of Problem 32 over the tetrahedron formed by the coordinate planes and the plane $x + y + z = 1$.

34. If all integrals exist, show that $\iint_S [F(x, y, z) + G(x, y, z)]\, d\sigma = \iint_S F(x, y, z)\, d\sigma + \iint_S G(x, y, z)\, d\sigma$.

6.6 DIVERGENCE AND CURL OF A VECTOR FIELD IN $\mathbb{R}^3$

In Section 6.3 we defined the divergence of $F(x, y) = P(x, y)\mathbf{i} + Q(x, y)\mathbf{j}$ by

$$\text{div}\ \mathbf{F} = \frac{\partial P}{\partial x} + \frac{\partial Q}{\partial y} \tag{1}$$

and the curl of $\mathbf{F}$ by

$$\text{curl}\ \mathbf{F} = \frac{\partial Q}{\partial x} - \frac{\partial P}{\partial y}. \tag{2}$$

We also wrote

$$\text{div}\ \mathbf{F} = \boldsymbol{\nabla} \cdot \mathbf{F} \tag{3}$$

and

$$\text{curl}\ \mathbf{F} = \boldsymbol{\nabla} \times \mathbf{F}, \tag{4}$$

where $\boldsymbol{\nabla}$, the gradient symbol, is regarded as the vector $(\partial/\partial x, \partial/\partial y)$. Moreover, we used both div $\mathbf{F}$ and curl $\mathbf{F}$ to give alternative versions of Green's theorem in the plane.

In this section we define the divergence and curl of a vector field in $\mathbb{R}^3$, and in Sections 6.7 and 6.8 we will show how these are used in the statement of two very important theorems about surface integrals.

Let the function $F(x, y, z)$ be given. Then the gradient of F is given by

$$\boldsymbol{\nabla} F = \frac{\partial F}{\partial x}\mathbf{i} + \frac{\partial F}{\partial y}\mathbf{j} + \frac{\partial F}{\partial z}\mathbf{k}. \tag{5}$$

We can think of the gradient as a function that takes a differentiable function of (x, y, z) into a vector field in $\mathbb{R}^3$. We write this function, symbolically, as

$$\boldsymbol{\nabla} = \frac{\partial}{\partial x}\mathbf{i} + \frac{\partial}{\partial y}\mathbf{j} + \frac{\partial}{\partial z}\mathbf{k}. \tag{6}$$

Of course, the "thing" written in (6) has no real meaning in itself. However, it is a useful device for writing things down. For example, (5) can be written as

$$\nabla F = \left(\frac{\partial}{\partial x}\mathbf{i} + \frac{\partial}{\partial y}\mathbf{j} + \frac{\partial}{\partial z}\mathbf{k}\right)F = \frac{\partial F}{\partial x}\mathbf{i} + \frac{\partial F}{\partial y}\mathbf{j} + \frac{\partial F}{\partial z}\mathbf{k}.$$

We now define the divergence and curl of a vector field $\mathbf{F}$ in $\mathbb{R}^3$ given by

$$\mathbf{F}(x, y, z) = P(x, y, z)\mathbf{i} + Q(x, y, z)\mathbf{j} + R(x, y, z)\mathbf{k} \tag{7}$$

Definition 1 DIVERGENCE AND CURL Let the vector field $\mathbf{F}$ be given by (7), where P, Q, and R are differentiable. Then the **divergence** of $\mathbf{F}$ (div $\mathbf{F}$) and **curl** of $\mathbf{F}$ (curl $\mathbf{F}$) are given by

$$\text{div } \mathbf{F} = \frac{\partial P}{\partial x} + \frac{\partial Q}{\partial y} + \frac{\partial R}{\partial z} \tag{8}$$

and

$$\text{curl } \mathbf{F} = \left(\frac{\partial R}{\partial y} - \frac{\partial Q}{\partial z}\right)\mathbf{i} + \left(\frac{\partial P}{\partial z} - \frac{\partial R}{\partial x}\right)\mathbf{j} + \left(\frac{\partial Q}{\partial x} - \frac{\partial P}{\partial y}\right)\mathbf{k}. \tag{9}$$

NOTE. div $\mathbf{F}$ is a scalar function and curl $\mathbf{F}$ is a vector field.

Before giving examples of divergence and curl, we derive an easy way to remember how to compute them.

OBSERVATION. Let ∇ be given by (6) and let the differentiable vector field $\mathbf{F}$ be given by (7). Then

$$\text{(i)}\quad \text{div } \mathbf{F} = \nabla \cdot \mathbf{F} \tag{10}$$

and

$$\text{(ii)}\quad \text{curl } \mathbf{F} = \nabla \times \mathbf{F}. \tag{11}$$

Proof.

$$\text{(i)}\quad \nabla \cdot \mathbf{F} = \left(\frac{\partial}{\partial x}\mathbf{i} + \frac{\partial}{\partial y}\mathbf{j} + \frac{\partial}{\partial z}\mathbf{k}\right) \cdot (P\mathbf{i} + Q\mathbf{j} + R\mathbf{k})$$

$$= \frac{\partial P}{\partial x} + \frac{\partial Q}{\partial y} + \frac{\partial R}{\partial z} = \text{div } \mathbf{F}$$

$$\text{(ii)}\quad \nabla \times \mathbf{F} = \begin{vmatrix} \mathbf{i} & \mathbf{j} & \mathbf{k} \\ \dfrac{\partial}{\partial x} & \dfrac{\partial}{\partial y} & \dfrac{\partial}{\partial z} \\ P & Q & R \end{vmatrix}$$

$$= \left(\frac{\partial R}{\partial y} - \frac{\partial Q}{\partial z}\right)\mathbf{i} + \left(\frac{\partial P}{\partial z} - \frac{\partial R}{\partial x}\right)\mathbf{j} + \left(\frac{\partial Q}{\partial x} - \frac{\partial P}{\partial y}\right)\mathbf{k} = \text{curl } \mathbf{F}. \quad ■$$

EXAMPLE 1 Compute the divergence and curl of $\mathbf{F}(x, y, z) = xy\mathbf{i} + (z^2 - 2y)\mathbf{j} + \cos yz\mathbf{k}$.

Solution.

$$\begin{aligned}\operatorname{div} \mathbf{F} &= \frac{\partial}{\partial x}(xy) + \frac{\partial}{\partial y}(z^2 - 2y) + \frac{\partial}{\partial z}(\cos yz) \\ &= y - 2 - y \sin yz\end{aligned}$$

and

$$\begin{aligned}\operatorname{curl} \mathbf{F} &= \begin{vmatrix} \mathbf{i} & \mathbf{j} & \mathbf{k} \\ \dfrac{\partial}{\partial x} & \dfrac{\partial}{\partial y} & \dfrac{\partial}{\partial z} \\ xy & z^2 - 2y & \cos yz \end{vmatrix} \\ &= \left[\frac{\partial}{\partial y}\cos yz - \frac{\partial}{\partial z}(z^2 - 2y)\right]\mathbf{i} + \left(\frac{\partial}{\partial z}xy - \frac{\partial}{\partial x}\cos yz\right)\mathbf{j} \\ &\quad + \left[\frac{\partial}{\partial x}(z^2 - 2y) - \frac{\partial}{\partial y}xy\right]\mathbf{k} \\ &= (-z \sin yz - 2z)\mathbf{i} - x\mathbf{k}. \quad \blacksquare\end{aligned}$$

EXAMPLE 2 Compute the divergence and curl of

$$\mathbf{F}(x, y, z) = yz\mathbf{i} + xz\mathbf{j} + xy\mathbf{k}.$$

Solution.

$$\operatorname{div} \mathbf{F} = \frac{\partial}{\partial x}yz + \frac{\partial}{\partial y}xz + \frac{\partial}{\partial z}xy = 0$$

$$\begin{aligned}\operatorname{curl} \mathbf{F} &= \begin{vmatrix} \mathbf{i} & \mathbf{j} & \mathbf{k} \\ \dfrac{\partial}{\partial x} & \dfrac{\partial}{\partial y} & \dfrac{\partial}{\partial z} \\ yz & xz & xy \end{vmatrix} \\ &= \left(\frac{\partial}{\partial y}yx - \frac{\partial}{\partial z}xz\right)\mathbf{i} + \left(\frac{\partial}{\partial z}yz - \frac{\partial}{\partial x}xy\right)\mathbf{j} + \left(\frac{\partial}{\partial x}xz - \frac{\partial}{\partial y}yz\right)\mathbf{k} \\ &= (x - x)\mathbf{i} + (y - y)\mathbf{j} + (z - z)\mathbf{k} = \mathbf{0}. \quad \blacksquare\end{aligned}$$

As in $\mathbb{R}^2$, the curl of a vector field **F** represents the circulation per unit area at the point (x, y, z). If $\operatorname{curl} \mathbf{F} = \mathbf{0}$ for every (x, y, z) in some region W in $\mathbb{R}^3$, then the fluid flow **F** is called **irrotational.** The divergence of **F** at a point (x, y, z) represents the net rate of flow away from (x, y, z). If $\operatorname{div} \mathbf{F} = 0$ for every (x, y, z) in W, then the flow **F** is called **incompressible.** The vector field in Example 2 is both irrotational and incompressible.

Recall from Theorem 6.4.2 that the vector field $\mathbf{F} = P\mathbf{i} + Q\mathbf{j} + R\mathbf{k}$ is the gradient of a function f if and only if

$$\frac{\partial P}{\partial y} = \frac{\partial Q}{\partial x}; \quad \frac{\partial R}{\partial x} = \frac{\partial P}{\partial z}; \quad \frac{\partial Q}{\partial z} = \frac{\partial R}{\partial y}. \tag{12}$$

Using (12) and (9), we obtain the following interesting result.

Theorem 1 In a simply connected region, the differentiable vector field $\mathbf{F}$ is the gradient of a function f if and only if curl $\mathbf{F} = \mathbf{0}$.

PROBLEMS 6.6

In Problems 1–10, compute the divergence and curl of the given vector field.

1. $\mathbf{F}(x, y, z) = x^2\mathbf{i} + y^2\mathbf{j} + z^2\mathbf{k}$
2. $\mathbf{F}(x, y, z) = (\sin y)\mathbf{i} + (\sin z)\mathbf{j} + (\sin x)\mathbf{k}$
3. $\mathbf{F}(x, y, z) = a\mathbf{i} + b\mathbf{j} + c\mathbf{k}$; a, b, c constants
4. $\mathbf{F}(x, y, z) = \sqrt{1 + x^2 + y^2}\mathbf{i} + \sqrt{1 + x^2 + y^2}\mathbf{j} + z^4\mathbf{k}$
5. $\mathbf{F}(x, y, z) = xy\mathbf{i} + yz\mathbf{j} + xz\mathbf{k}$
6. $\mathbf{F}(x, y, z) = (y^2 + z^2)\mathbf{i} + (x^2 + z^2)\mathbf{j} + (x^2 + y^2)\mathbf{k}$
7. $\mathbf{F}(x, y, z) = e^{yz}\mathbf{i} + e^{xz}\mathbf{j} + e^{xy}\mathbf{k}$
8. $\mathbf{F}(x, y, z) = e^{xy}\mathbf{i} + e^{yz}\mathbf{j} + e^{xz}\mathbf{k}$
9. $\mathbf{F}(x, y, z) = \frac{x}{y}\mathbf{i} + \frac{y}{z}\mathbf{j} + \frac{z}{x}\mathbf{k}$
10. $\mathbf{F}(x, y, z) = \sqrt{y + z}\mathbf{i} + \sqrt{x + z}\mathbf{j} + \sqrt{x + y}\mathbf{k}$
11. Let f, g, and h be differentiable functions of two variables. Show that the vector field $\mathbf{F}(x, y, z) = f(y, z)\mathbf{i} + g(x, z)\mathbf{j} + h(x, y)\mathbf{k}$ is incompressible.

In Problems 12–18, assume that all given functions are sufficiently differentiable.

12. Show that div($\mathbf{F} + \mathbf{G}$) = div $\mathbf{F}$ + div $\mathbf{G}$.
13. If $f = f(x, y, z)$ is a scalar function, show that div $(f\mathbf{F}) = f$ div $\mathbf{F} + \nabla f \cdot \mathbf{F}$.
14. Show that curl($\mathbf{F} + \mathbf{G}$) = curl $\mathbf{F}$ + curl $\mathbf{G}$.
15. If f is as in Problem 13, show that curl($f\mathbf{F}$) = f curl $\mathbf{F} + \nabla f \times \mathbf{F}$.
16. If f is as in Problem 13, show that curl grad f = curl(∇f) = $\mathbf{0}$.
17. Show that div curl $\mathbf{F} = 0$.
18. Show that div($\mathbf{F} \times \mathbf{G}$) = $\mathbf{G} \cdot$ curl $\mathbf{F} - \mathbf{F} \cdot$ curl $\mathbf{G}$.

The **Laplacian** of a twice-differentiable scalar function $f = f(x, y, z)$, denoted $\nabla^2 f$, is defined by

$$\text{Laplacian of } f = \nabla^2 f = \frac{\partial^2 f}{\partial x^2} + \frac{\partial^2 f}{\partial y^2} + \frac{\partial^2 f}{\partial x^2}. \tag{13}$$

In Problems 19–22, compute $\nabla^2 f$.

19. $f(x, y, z) = xyz$
20. $f(x, y, z) = x^2 + y^2 + z^2$
21. $f(x, y, z) = \dfrac{1}{\sqrt{x^2 + y^2 + z^2}}$
22. $f(x, y, z) = 2x^2 + 5y^2 + 3z^2$
23. A function that satisfies the equation $\nabla^2 f = 0$, called **Laplace's equation,** is called **harmonic.** Which of the functions in Problems 19–22 are harmonic?
24. Show that $\nabla^2 f$ = div (grad f).
25. The **Laplacian** of a vector field $\mathbf{F}$ is given by $\nabla^2\mathbf{F} = \nabla^2 P\mathbf{i} + \nabla^2 Q\mathbf{j} + \nabla^2 R\mathbf{k}$, where $\mathbf{F} = P\mathbf{i} + Q\mathbf{j} + R\mathbf{k}$. Show that curl curl $\mathbf{F} = \nabla$ div $\mathbf{F} - \nabla^2 \mathbf{F}$.
26. Let $\mathbf{F}$ be the vector field of Problem 1. Verify that div curl $\mathbf{F} = 0$.
27. Let f be the function given in Problem 21. Verify that curl grad $f = \mathbf{0}$.
28. It is true, although we will not attempt to prove it, that if div $\mathbf{F} = 0$, then $\mathbf{F}$ = curl $\mathbf{G}$ for some vector field $\mathbf{G}$. Let

$$\mathbf{F} = x\mathbf{i} + \frac{y}{2}\mathbf{j} - \frac{3}{2}z\mathbf{k}.$$

   (a) Verify that div $\mathbf{F} = 0$.
   (b) Find a vector field $\mathbf{G}$ such that $\mathbf{F}$ = curl $\mathbf{G}$.
29. If curl $\mathbf{F} = \mathbf{0}$, then $\mathbf{F} = \nabla f$ for some function f. Let $\mathbf{F} = 4xyz\mathbf{i} + 2x^2z\mathbf{j} + 2x^2y\mathbf{k}$.
   (a) Verify that curl $\mathbf{F} = \mathbf{0}$.
   (b) Find a function f such that $\mathbf{F} = \nabla f$.

6.7 STOKES'S THEOREM

In Section 6.3 we discussed Green's theorem in the plane, one vector form of which was given by (see Theorem 6.3.2)

$$\oint_{\partial\Omega} \mathbf{F} \cdot \mathbf{T}\, dx = \iint_{\Omega} \text{curl } \mathbf{F}\, dx\, dy, \tag{1}$$

where $\partial\Omega$ is the piecewise smooth boundary of a region Ω in the xy-plane. We now generalize this result.

If C is a closed curve enclosing the region Ω in the plane, then by *traversing C in the positive sense* we mean moving around C so that Ω is always on the left. This motion corresponds to moving in a counterclockwise direction (see Figure 1).

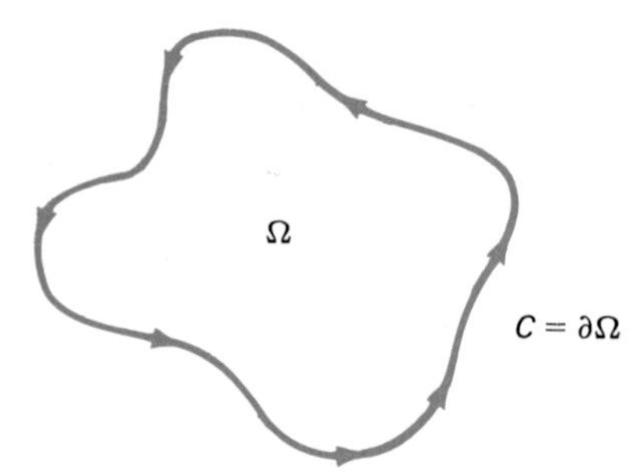

FIGURE 1

REMARK. Like the definition of the outward unit normal vector, the words "counterclockwise" and "left" are ambiguous when applied to closed curves in $\mathbb{R}^3$. To see why, picture a circle in a plane in $\mathbb{R}^3$. You can look at the circle from either side of the plane. The counterclockwise direction from one side would appear as the clockwise direction from the other side. (Take a large piece of translucent paper, place it between yourself and a friend, and ask him or her to draw a circle in the counterclockwise direction. It will appear clockwise to you.) Thus the "positive sense" in which a curve is traversed is arbitrary. However, once we have chosen a positive or counterclockwise direction, the outward unit normal vector can be unambiguously defined: At any point on a smooth surface, we choose a tangent vector $\mathbf{u}$ lying in the unique tangent plane at that point. We then obtain a second tangent vector $\mathbf{v}$ by rotating $\mathbf{u}$ 90° in the counterclockwise direction while remaining in the tangent plane. Finally, we define the direction of $\mathbf{n}$ to be the direction of $\mathbf{u} \times \mathbf{v}$ (so that the vectors $\mathbf{u}$, $\mathbf{v}$, and $\mathbf{n}$ form a right-handed system).

Let $S: z = f(x, y)$ be a smooth surface for $(x, y) \in \Omega$, where Ω is bounded. We assume that the boundary of S, denoted ∂S, is a piecewise smooth, simple closed curve in $\mathbb{R}^3$. The positive direction on ∂S corresponds to the positive direction of $\partial\Omega$, where $\partial\Omega$ is the projection of ∂S into the xy-plane. This orientation is illustrated in Figure 2.

We now state the second major result in vector calculus (the first was Green's theorem).

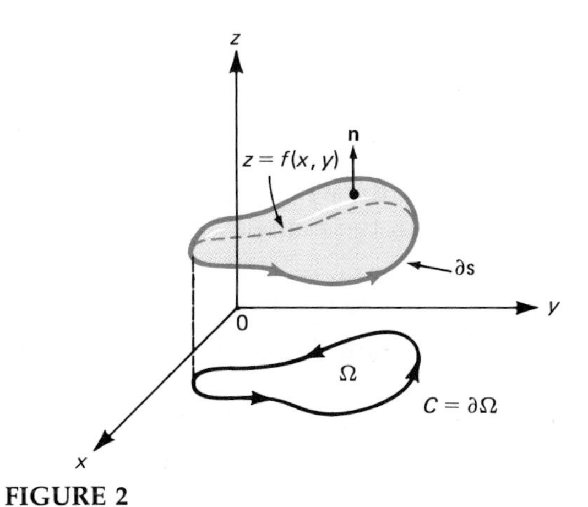

FIGURE 2

Theorem 1 STOKES'S THEOREM[†] Let $\mathbf{F}(x, y, z) = P(x, y, z)\mathbf{i} + Q(x, y, z)\mathbf{j} + R(x, y, z)\mathbf{k}$ be continuously differentiable on a bounded region W in space that contains the smooth surface S, and let ∂S be a piecewise smooth, simple closed curve traversed in the positive sense. Then

$$\oint_{\partial S} \mathbf{F} \cdot \mathbf{T}\, ds = \iint_S \text{curl } \mathbf{F} \cdot \mathbf{n}\, d\sigma, \tag{2}$$

where $\mathbf{T}$ is the unit tangent vector to the curve ∂S at a point (x, y, z) and $\mathbf{n}$ is the outward unit normal vector to the surface S at the point (x, y, z).

The proof of this theorem is difficult and is left to the end of this section.

REMARK 1. Note the similarity between Stokes's theorem and the first vector form of Green's theorem.

REMARK 2. If ∂S is given parametrically by $\mathbf{x}(t) = x(t)\mathbf{i} + y(t)\mathbf{j} + z(t)\mathbf{k}$, then by Theorem 3.8.3(v),

$$\mathbf{T} = \frac{\mathbf{dx}}{ds}, \quad \text{or} \quad \mathbf{dx} = \mathbf{T}\, ds. \tag{3}$$

Thus Stokes's theorem can be written as

$$\oint_{\partial S} \mathbf{F} \cdot \mathbf{dx} = \iint_S \text{curl } \mathbf{F} \cdot \mathbf{n}\, d\sigma. \tag{4}$$

REMARK 3. Using the notation given in equation (6.4.11), we can also write Stokes's theorem as

[†]Named after the English mathematician and physicist Sir G. G. Stokes (1819–1903). Stokes is also known as one of the first to discuss the notion of uniform convergence (in 1848).

$$\oint_{\partial S} P\,dx + Q\,dy + R\,dz = \iint_S \text{curl}\,\mathbf{F} \cdot \mathbf{n}\,d\sigma. \tag{5}$$

EXAMPLE 1 Use Stokes's theorem to evaluate $\oint_C \mathbf{F} \cdot \mathbf{dx}$, where $\mathbf{F}(x, y, z) = (z - 2y)\mathbf{i} + (3x - 4y)\mathbf{j} + (z + 3y)\mathbf{k}$ and C is the unit circle in the plane $z = 2$.

Solution. C is given parametrically by $\mathbf{x}(t) = (\cos t)\mathbf{i} + (\sin t)\mathbf{j} + 2\mathbf{k}$. Clearly, C bounds the unit disk $x^2 + y^2 \leq 1$, $z = 2$ in the plane $z = 2$. (It bounds many surfaces—for example, a hemisphere with this circle as its base. But the disk is the simplest one to use.) Then

$$\oint_C \mathbf{F} \cdot \mathbf{dx} = \iint_S \text{curl}\,\mathbf{F} \cdot \mathbf{n}\,d\sigma.$$

We compute

$$\text{curl}\,\mathbf{F} = \begin{vmatrix} \mathbf{i} & \mathbf{j} & \mathbf{k} \\ \dfrac{\partial}{\partial x} & \dfrac{\partial}{\partial y} & \dfrac{\partial}{\partial z} \\ z - 2y & 3x - 4y & z + 3y \end{vmatrix} = 3\mathbf{i} + \mathbf{j} + 5\mathbf{k},$$

and $\mathbf{n} = \mathbf{k}$ (since $\mathbf{k}$ is normal to any vector lying in a plane parallel to the xy-plane). Thus

$$\text{curl}\,\mathbf{F} \cdot \mathbf{n} = 5,$$

so that

$$\oint_C \mathbf{F} \cdot \mathbf{dx} = 5 \iint_S d\sigma.$$

But $\iint_S d\sigma$ = the surface area of S = the area of the unit disk $= \pi$. Thus

$$\oint_C \mathbf{F} \cdot \mathbf{dx} = 5\pi. \quad \blacksquare$$

EXAMPLE 2 Compute $\oint_C \mathbf{F} \cdot \mathbf{dx}$, where $\mathbf{F}$ is as in Example 1 and C is the boundary of the triangle joining the points (1, 0, 0), (0, 1, 0), and (0, 0, 1).

Solution. We already have found that curl $\mathbf{F} = 3\mathbf{i} + \mathbf{j} + 5\mathbf{k}$. The curve C lies in a plane. Two vectors on the plane are $\mathbf{i} - \mathbf{j}$ and $\mathbf{j} - \mathbf{k}$ (explain why). A normal vector to the plane is therefore given by

$$\mathbf{N} = (\mathbf{i} - \mathbf{j}) \times (\mathbf{j} - \mathbf{k}) = \begin{vmatrix} \mathbf{i} & \mathbf{j} & \mathbf{k} \\ 1 & -1 & 0 \\ 0 & 1 & -1 \end{vmatrix} = \mathbf{i} + \mathbf{j} + \mathbf{k},$$

so that $\mathbf{n} = (1/\sqrt{3})(\mathbf{i} + \mathbf{j} + \mathbf{k})$ is the required outward unit normal vector. Thus curl $\mathbf{F} \cdot \mathbf{n} = 9/\sqrt{3} = 3\sqrt{3}$, and we have

$$\oint_C \mathbf{F} \cdot \mathbf{dx} = 3\sqrt{3} \iint_S d\sigma = 3\sqrt{3} \times (\text{area of the triangle}).$$

The triangle is an equilateral triangle with the length of one side given by

$$\ell = \text{distance between } (1, 0, 0) \text{ and } (0, 1, 0) = \sqrt{2}.$$

Then the area of the triangle is easily computed to be $\sqrt{3}/2$. Thus

$$\oint_C \mathbf{F} \cdot \mathbf{dx} = (3\sqrt{3})\left(\frac{\sqrt{3}}{2}\right) = \frac{9}{2}. \blacksquare$$

REMARK. In the preceding example tangent vectors to the surface (a plane) are vectors lying on the plane. The vector $\mathbf{j} - \mathbf{k}$ is obtained from the vector $\mathbf{i} - \mathbf{j}$ by rotating $\mathbf{i} - \mathbf{j}$ 120° in the counterclockwise direction when viewed from the side of the plane that does not contain the origin. This procedure leads to $\mathbf{n} = (1/\sqrt{3})(\mathbf{i} + \mathbf{j} + \mathbf{k})$. If we chose the "other" counterclockwise direction, then we would obtain $\mathbf{n} = -(1/\sqrt{3})(\mathbf{i} + \mathbf{j} + \mathbf{k})$ and $\oint_C \mathbf{F} \cdot \mathbf{dx} = -\frac{9}{2}$. Which answer is correct? Both, because the answer depends on our choice of a counterclockwise direction. Without defining the orientation of a surface, this is the best we can do. There will always be two possible answers unless either the counterclockwise direction or the direction of the outward unit normal vector is specified in advance. The vector $\mathbf{n}$ can be distinguished from the vector $-\mathbf{n}$ by the following property:

> Let $\mathbf{n}$ be the outward unit normal vector to a surface. S. Then, on the boundary of S, $\mathbf{T} \times \mathbf{n}$ points off the edge of the surface.

We will not worry any more about this problem in this text.

EXAMPLE 3 Verify Stokes's theorem for the surface $z = x^2 + y^2$, $x^2 + y^2 \le 1$, $\mathbf{F}(x, y, z) = y^2\mathbf{i} + x\mathbf{j} + z^2\mathbf{k}$.

Solution. The surface is a circular paraboloid with boundary curve given by the circle $x^2 + y^2 = 1$, $z = 1$ (see Figure 3). This curve can be given parametrically by

$$\mathbf{x}(t) = (\cos t)\mathbf{i} + (\sin t)\mathbf{j} + \mathbf{k}, \qquad 0 \le t \le 2\pi. \tag{6}$$

Thus $\mathbf{x}'(t) = -\sin t\,\mathbf{i} + \cos t\,\mathbf{j}$ and $\mathbf{F} \cdot \mathbf{dx} = (-\sin^3 t + \cos^2 t)\,dt$, so that

$$\oint_{\partial S} \mathbf{F} \cdot \mathbf{dx} = \int_0^{2\pi} (-\sin^3 t + \cos^2 t)\,dt = \pi.$$

Now

$$\text{curl } \mathbf{F} = \begin{vmatrix} \mathbf{i} & \mathbf{j} & \mathbf{k} \\ \dfrac{\partial}{\partial x} & \dfrac{\partial}{\partial y} & \dfrac{\partial}{\partial z} \\ y^2 & x & z^2 \end{vmatrix} = (1 - 2y)\mathbf{k},$$

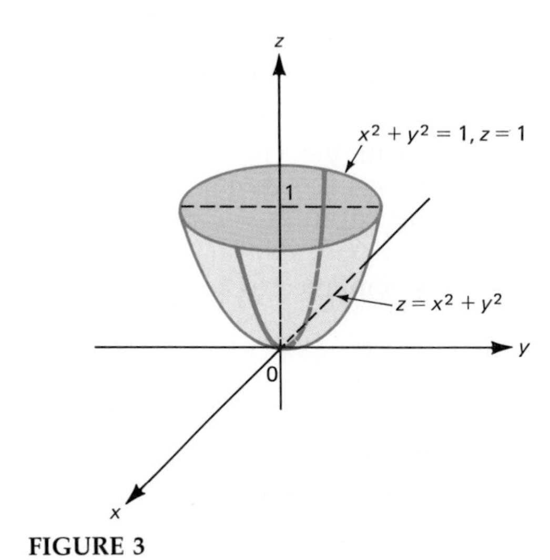

FIGURE 3

so that using formula (6.5.30)

$$\iint_S \text{curl } \mathbf{F} \cdot \mathbf{n}\, d\sigma = \iint_\Omega (1 - 2y)\, dx\, dy.$$

Using polar coordinates, we have $y = r \sin \theta$ and

$$\iint_S \text{curl } \mathbf{F} \cdot \mathbf{n}\, d\sigma = \int_0^{2\pi}\int_0^1 (1 - 2r \sin \theta) r\, dr\, d\theta$$

$$= \int_0^{2\pi}\int_0^1 (r - 2r^2 \sin \theta)\, dr\, d\theta = \int_0^{2\pi}\left\{\left(\frac{r^2}{2} - \frac{2r^3}{3} \sin \theta\right)\Big|_0^1\right\} d\theta$$

$$= \int_0^{2\pi}\left(\frac{1}{2} - \frac{2}{3} \sin \theta\right) d\theta = \pi. \quad \blacksquare$$

From Theorem 6.6.2, in a simply connected region, **F** is the gradient of a function f if and only if curl $\mathbf{F} = \mathbf{0}$. This result provides another proof of the fact that

$$\oint_C \mathbf{F} \cdot \mathbf{dx} = 0$$

if **F** is the gradient of a function f, since

$$\oint_C \mathbf{F} \cdot \mathbf{dx} = \iint_S \text{curl } \mathbf{F} \cdot \mathbf{n}\, d\sigma = 0,$$

where S is any smooth surface whose boundary is C.

We can combine several results to obtain the following theorem.

Theorem 2 Let $\mathbf{F} = P(x, y, z)\mathbf{i} + Q(x, y, z)\mathbf{j} + R(x, y, z)\mathbf{k}$ be continuously differentiable on a simply connected region S in space. Then the following conditions are equivalent. That is, if one is true, all are true.

(i) $\mathbf{F}$ is the gradient of a differentiable function f.

(ii) $\dfrac{\partial P}{\partial y} = \dfrac{\partial Q}{\partial x}$; $\dfrac{\partial R}{\partial x} = \dfrac{\partial P}{\partial z}$; and $\dfrac{\partial Q}{\partial z} = \dfrac{\partial R}{\partial y}$.

(iii) $\oint_C \mathbf{F} \cdot \mathbf{dx} = 0$ for every piecewise smooth, simple closed curve C lying in S.

(iv) $\oint_C \mathbf{F} \cdot \mathbf{dx}$ is independent of path.

(v) Curl $\mathbf{F} = \mathbf{0}$.

There are many interesting physical results that follow from Stokes's theorem. One of them is given here. Suppose a steady current is flowing in a wire with an electric current density given by the vector field $\mathbf{i}$. It is well known in physics that such a current will set up a magnetic field, which is usually denoted $\mathbf{B}$. In Figure 4 we see a collection of compass needles near a wire carrying (a) no current and (b) a very strong current. A special case of one of the famous Maxwell equations[†] states that

$$\text{curl } \mathbf{B} = \mathbf{i}.^{\ddagger} \tag{7}$$

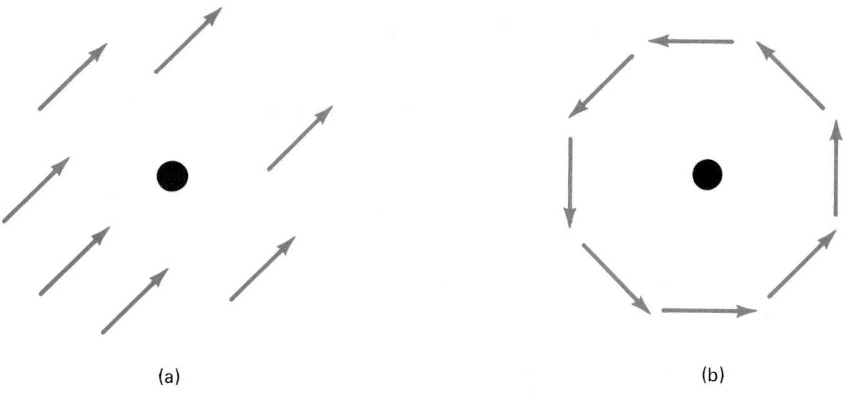

FIGURE 4

From this equation we can deduce an important physical law. Let S be a surface with smooth boundary ∂S. Then $\oint_{\partial S} \mathbf{B} \cdot \mathbf{T}\, ds$ is defined as the **circulation** of the magnetic field around ∂S. By Stokes's theorem

$$\oint_{\partial S} \mathbf{B} \cdot \mathbf{T}\, ds = \iint_S \text{curl } \mathbf{B} \cdot \mathbf{n}\, d\sigma.$$

[†] James Maxwell (1831–1879) was a British physicist. He formulated four equations, known as **Maxwell's equations,** which were supposed to explain all electromagnetic phenomena.

[‡] The vector $\mathbf{i}$ here is not to be confused with the unit vector $\mathbf{i} = (1, 0, 0)$.

But by (7), $\iint_S \text{curl}\,\mathbf{B} \cdot \mathbf{n}\, d\sigma = \iint_S \mathbf{i} \cdot \mathbf{n}\, d\sigma$, so

$$\oint_{\partial S} \mathbf{B} \cdot \mathbf{T}\, ds = \iint_S \mathbf{i} \cdot \mathbf{n}\, d\sigma. \tag{8}$$

In other words, (8) states that *the total current flowing through an electric field in a surface S is equal to the circulation of the magnetic field induced by* $\mathbf{i}$ *around the boundary of S.* This important result is known as **Ampère's law.**

We close this section with a proof of Stokes's theorem.

Proof of Stokes's Theorem (Optional). We have $\mathbf{F} = P\mathbf{i} + Q\mathbf{j} + R\mathbf{k}$ and

$$\text{curl}\,\mathbf{F} = \left(\frac{\partial R}{\partial y} - \frac{\partial Q}{\partial z}\right)\mathbf{i} + \left(\frac{\partial P}{\partial z} - \frac{\partial R}{\partial x}\right)\mathbf{j} + \left(\frac{\partial Q}{\partial x} - \frac{\partial P}{\partial y}\right)\mathbf{k}. \tag{9}$$

The surface S is given by S: $z = f(x, y)$ for (x, y) in Ω. Then by Theorem 6.5.2,

$$\iint_S \text{curl}\,\mathbf{F} \cdot \mathbf{n}\, d\sigma = \iint_\Omega \left[\left(\frac{\partial R}{\partial y} - \frac{\partial Q}{\partial z}\right)\left(-\frac{\partial f}{\partial x}\right) + \left(\frac{\partial P}{\partial z} - \frac{\partial R}{\partial x}\right)\left(-\frac{\partial f}{\partial y}\right) + \left(\frac{\partial Q}{\partial x} - \frac{\partial P}{\partial y}\right)\right] dx\, dy. \tag{10}$$

Suppose that $\mathbf{x}(t) = x(t)\mathbf{i} + y(t)\mathbf{j}$ for $t_0 \le t \le t_1$ parametrizes $\partial\Omega$. Then ∂S is parametrized by

$$\mathbf{x}(t) = x(t)\mathbf{i} + y(t)\mathbf{j} + f(x(t), y(t))\mathbf{k}, \qquad t_0 \le t \le t_1.$$

Thus

$$\oint_{\partial S} \mathbf{F} \cdot \mathbf{T}\, ds = \oint_{\partial S} \mathbf{F} \cdot d\mathbf{x} = \int_{t_0}^{t_1} \left(P\frac{dx}{dt} + Q\frac{dy}{dt} + R\frac{dz}{dt}\right) dt. \tag{11}$$

By the chain rule,

$$\frac{dz}{dt} = \frac{\partial z}{\partial x}\frac{dx}{dt} + \frac{\partial z}{\partial y}\frac{dy}{dt}. \tag{12}$$

Substituting (12) into (11) yields

$$\begin{aligned}\oint_{\partial S} \mathbf{F} \cdot \mathbf{T}\, ds &= \int_{t_0}^{t_1} \left[\left(P + R\frac{\partial z}{\partial x}\right)\frac{dx}{dt} + \left(Q + R\frac{\partial z}{\partial y}\right)\frac{dy}{dt}\right] dt \\ &= \oint_{\partial\Omega} \left(P + R\frac{\partial z}{\partial x}\right) dx + \left(Q + R\frac{\partial z}{\partial y}\right) dy. \end{aligned} \tag{13}$$

We apply Green's theorem to (13) to obtain

$$\oint_{\partial S} \mathbf{F}\cdot\mathbf{T}\,ds = \iint_{\Omega} \left\{ \frac{\partial[Q + R(\partial z/\partial y)]}{\partial x} - \frac{\partial[P + R(\partial z/\partial x)]}{\partial y} \right\} dx\,dy. \tag{14}$$

We now apply the chain rule again. This is a bit complicated. For example,

$$\frac{\partial}{\partial x} Q(x, y, z) = \frac{\partial Q}{\partial x} + \frac{\partial Q}{\partial y}\frac{\partial y}{\partial x} + \frac{\partial Q}{\partial z}\frac{\partial z}{\partial x} = \frac{\partial Q}{\partial x} + \frac{\partial Q}{\partial z}\frac{\partial z}{\partial x}$$

since y is not a function of x (so that $\partial y/\partial x = 0$). Also

$$\begin{aligned}\frac{\partial}{\partial x}\left(R\frac{\partial z}{\partial y}\right) &= \left[\frac{\partial}{\partial x} R(x, y, z)\right]\frac{\partial z}{\partial y} + R\frac{\partial}{\partial x}\left\{\frac{\partial[z(x, y)]}{\partial y}\right\} \\ &= \frac{\partial R}{\partial x}\frac{\partial z}{\partial y} + \frac{\partial R}{\partial z}\frac{\partial z}{\partial x}\frac{\partial z}{\partial y} + R\frac{\partial^2 z}{\partial x\,\partial y} + R\frac{\partial^2 z}{\partial y^2}\frac{\partial y}{\partial x}\end{aligned}$$

and the last term is zero because y is not a function of x. Differentiating inside the right-hand integral in (14), we obtain

$$\begin{aligned}\oint_{\partial S} \mathbf{F}\cdot\mathbf{T}\,ds = \iint_{\Omega} \Bigg[&\left(\frac{\partial Q}{\partial x} + \frac{\partial Q}{\partial z}\frac{\partial z}{\partial x} + \frac{\partial R}{\partial x}\frac{\partial z}{\partial y} + \frac{\partial R}{\partial z}\frac{\partial z}{\partial x}\frac{\partial z}{\partial y} + R\frac{\partial^2 z}{\partial x\,\partial y}\right) \\ &- \left(\frac{\partial P}{\partial y} + \frac{\partial P}{\partial z}\frac{\partial z}{\partial y} + \frac{\partial R}{\partial y}\frac{\partial z}{\partial x} + \frac{\partial R}{\partial z}\frac{\partial z}{\partial y}\frac{\partial z}{\partial x} + R\frac{\partial^2 z}{\partial y\,\partial x}\right)\Bigg]\,dz\,dy \end{aligned} \tag{15}$$

After terms cancel

$$= \iint_{\Omega} \left(\frac{\partial Q}{\partial x} + \frac{\partial Q}{\partial z}\frac{\partial z}{\partial x} + \frac{\partial R}{\partial x}\frac{\partial z}{\partial y} - \frac{\partial P}{\partial y} - \frac{\partial P}{\partial z}\frac{\partial z}{\partial y} - \frac{\partial R}{\partial y}\frac{\partial z}{\partial x}\right) dx\,dy.$$

After rearranging and noting that $\partial f/\partial x = \partial z/\partial x$ and $\partial f/\partial y = \partial z/\partial y$, we see that the integrals in (10) and (15) are identical. Thus

$$\iint_S \operatorname{curl}\mathbf{F}\cdot\mathbf{n}\,d\sigma = \oint_{\partial S} \mathbf{F}\cdot\mathbf{T}\,ds. \ \blacksquare$$

NOTE. We have proved this theorem only under the simplifying assumption that the surface can be written in the form $z = f(x, y)$. The proof for more general surfaces is best left to an advanced calculus book.

PROBLEMS 6.7

In Problems 1–8, evaluate the line integral by using Stokes's theorem.

1. $\oint_C \mathbf{F}\cdot d\mathbf{x}$, where $\mathbf{F}(x, y, z) = (x + y)\mathbf{i} + (z - 2x + y)\mathbf{j} + (y - z)\mathbf{k}$ and C is the unit circle in the plane $z = 5$.
2. $\oint_C \mathbf{F}\cdot d\mathbf{x}$, where $\mathbf{F}(x, y, z) = ax\mathbf{i} + by\mathbf{j} + cz\mathbf{k}$ and C is the curve of Problem 1.
3. $\oint_C \mathbf{F}\cdot d\mathbf{x}$, where $\mathbf{F}$ is as in Example 1 and C is the boundary of the triangle joining the points (2, 0, 0), (0, 2, 0), and (0, 0, 2).

4. $\oint_C \mathbf{F} \cdot \mathbf{dx}$, where $\mathbf{F}$ is as in Example 2 and C is the boundary of the triangle joining the points $(d, 0, 0)$, $(0, d, 0)$, and $(0, 0, d)$.

5. $\oint_C y^2x^3\,dx + 2xyz^3\,dy + 3xy^2z^2\,dz$, where C is given parametrically by $\mathbf{x}(t) = 2\cos t\mathbf{i} + 3\mathbf{j} + 2\sin t\mathbf{k}$, $0 \le t \le 2\pi$.

6. $\oint_C \mathbf{F} \cdot \mathbf{dx}$, where $\mathbf{F} = 2y(x - z)\mathbf{i} + (x^2 + z^2)\mathbf{j} + y^3\mathbf{k}$, C is the square $0 \le x \le 3$, $0 \le y \le 3$, $z = 4$.

7. $\oint_C e^x\,dx + x\sin y\,dy + (y^2 - x^2)\,dz$, where C is the equilateral triangle formed by the intersection of the plane $x + y + z = 3$ with the three coordinate planes.

8. $\oint_C \mathbf{F} \cdot \mathbf{T}\,ds$, where $\mathbf{F} = -3y\mathbf{i} + 3x\mathbf{j} + \mathbf{k}$ and C is the circle $x^2 + y^2 = 1$, $z = 3$.

9. Show that Green's theorem is really a special case of Stokes's theorem.

In Problems 10–13, verify Stokes's theorem for the given S and $\mathbf{F}$.

10. $\mathbf{F} = z^2\mathbf{i} + x^2\mathbf{j} + y^2\mathbf{k}$; S is the part of the plane $x + y + z = 1$ lying in the first octant (i.e., $x \ge 0$, $y \ge 0$, $z \ge 0$).

11. $\mathbf{F} = y^2\mathbf{i} + z^2\mathbf{j} + x^2\mathbf{k}$ and S is the part of the plane $x + 2y + 3z = 6$ lying in the first octant.

12. $\mathbf{F} = 2y\mathbf{i} + x^2\mathbf{j} + 3x\mathbf{k}$; S is the hemisphere $x^2 + y^2 + z^2 = 16$, $z \ge 0$.

13. $\mathbf{F} = y\mathbf{i} - x\mathbf{j} - z\mathbf{k}$; S is the circle $x^2 + y^2 \le 9$, $z = 2$.

14. Let S be a sphere. Use Stokes's theorem to show that $\iint_S \text{curl}\,\mathbf{F} \cdot \mathbf{n}\,d\sigma = 0$.

***15.** Let S be a smooth closed surface bounding a simply connected region. Show that $\iint_S \text{curl}\,\mathbf{F} \cdot \mathbf{n}\,d\sigma = 0$ for any continuously differentiable vector field $\mathbf{F}$.

***16.** Let $\mathbf{E}$ be an electric field and let $\mathbf{B}(t)$ be a magnetic field in space induced by $\mathbf{E}$. One of Maxwell's equations states that $\text{curl}\,\mathbf{E} = -\partial\mathbf{B}/\partial t$. Let S be a surface in space with smooth boundary C. We define

$$\text{voltage drop around } C = \oint_C \mathbf{E} \cdot \mathbf{dx}.$$

Prove **Faraday's law,** which states that the voltage drop around C is equal to the time rate of decrease of the magnetic flux through S. [*Warning:* At one point it is necessary to interchange the order of differentiation and integration. A proof that this manipulation is "legal" requires techniques from advanced calculus; so just assume that it can be done.]

6.8 THE DIVERGENCE THEOREM

In Section 6.3 we gave the second vector form of Green's theorem (Theorem 6.3.3):

$$\oint_{\partial\Omega} \mathbf{F} \cdot \mathbf{n}\,ds = \iint_{\Omega} \text{div}\,\mathbf{F}\,dx\,dy, \tag{1}$$

where $\partial\Omega$ is the piecewise smooth boundary of a region Ω in the xy-plane. In this section we extend this theorem to obtain the third major result in vector integral calculus: the **divergence theorem.**

Green's theorem gives us a relationship between a line integral in $\mathbb{R}^2$ around a closed curve and a double integral over the region in $\mathbb{R}^2$ enclosed by that curve. Stokes's theorem shows a relationship between a line integral in $\mathbb{R}^3$ around a closed curve and a surface integral over a surface that has the closed curve as a boundary. As we will see, the divergence theorem gives us a relationship between a surface integral over a closed surface and a triple integral over the solid bounded by that surface.

Let S be a surface that forms the complete boundary of a solid W in space. A typical region and its boundary are sketched in Figure 1. We will assume that S is smooth or piecewise smooth. This assumption ensures that $\mathbf{n}(x, y, z)$, the outward unit normal to S, is continuous or piecewise continuous as a function of (x, y, z).

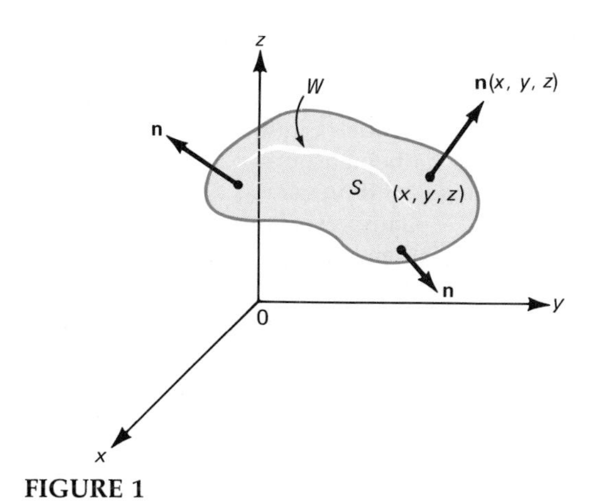

FIGURE 1

Theorem 1 THE DIVERGENCE THEOREM† Let W be a solid in $\mathbb{R}^3$ totally bounded by the smooth or piecewise smooth surface S. Let $\mathbf{F}$ be a smooth vector field on W, and let $\mathbf{n}$ denote the outward unit normal to S. Then

$$\iint_S \mathbf{F} \cdot \mathbf{n}\, d\sigma = \iiint_W \operatorname{div} \mathbf{F}\, dx\, dy\, dz. \tag{2}$$

REMARK 1. Just as Green's theorem transforms a line integral to an ordinary double integral, the divergence theorem transforms a surface integral to an ordinary triple integral.

REMARK 2. If $\mathbf{F} = P\mathbf{i} + Q\mathbf{j} + R\mathbf{k}$, then using equation (6.5.41), we can write the divergence theorem in the form

$$\iint_S (P \cos \alpha + Q \cos \beta + R \cos \gamma)\, d\sigma = \iiint_W \operatorname{div} \mathbf{F}\, dx\, dy\, dz, \tag{3}$$

where α, β, and γ are the acute angles that $\mathbf{n}$ makes with the positive coordinate axes.

Proof of the Divergence Theorem. In the proof of Green's theorem we assumed (see Figure 6.3.4) that the region Ω could be enclosed by two curves—an upper curve $y = g_2(x)$ and a lower curve $y = g_1(x)$ or a left curve $x = h_1(y)$ and a right curve $x = h_2(y)$. This is equivalent to saying that any line parallel to the x- or y-axis crosses $\partial\Omega$ in at most two points. In order to prove the divergence theorem, we assume much the same thing. That is, we assume that W can be enclosed by an upper surface and a lower surface so that any line parallel to one of the coordinate axes crosses S in at most two points. This assumption is not

†This theorem is also known as Gauss's theorem, named after the German mathematician Carl Friedrich Gauss (1777–1855). See the accompanying biographical sketch.

1777–

1855

Carl Friedrich Gauss

Carl Friedrich Gauss
The Granger Collection

The greatest mathematician of the nineteenth century, Carl Friedrich Gauss is considered one of the three greatest mathematicians of all time—the others being Archimedes and Newton.

Gauss was born in Brunswick, Germany, in 1777. His father, a hard-working laborer who was exceptionally stubborn and did not believe in formal education, did what he could to keep Gauss from appropriate schooling. Fortunately for Carl (and for mathematics), his mother, while uneducated herself, encouraged her son in his studies and took considerable pride in his achievements until her death at the age of 97.

Gauss was a child prodigy. At the age of three, he found an error in his father's bookkeeping. A famous story tells of Carl, age 10, as a student in the local Brunswick school. The teacher there was known to assign tasks to keep his pupils busy. One day he asked his students to add the numbers from 1 to 100. Almost at once, Carl placed his slate face down with the words, "There it is." Afterwards, the teacher found that Gauss was the only one with the correct answer, 5050. Gauss had noticed that the numbers could be arranged in 50 pairs, each with the sum 101 ($1 + 100$, $2 + 99$, and so on) and $50 \times 101 = 5050$. Later in life, Gauss joked that he could add before he could speak.

When Gauss was 15, the Duke of Brunswick noticed him and became his patron. The Duke helped him enter Brunswick College in 1795 and, three years later, to enter the university at Göttingen. Undecided between careers in mathematics and philosophy, Gauss chose mathematics after two remarkable discoveries. First, he invented the method of least squares a decade before the result was published by Legendre. Second, a month before his 19th birthday, he solved a problem whose solution had been sought for more than two thousand years. Gauss showed how to construct, using compass and ruler, a regular polygon with the number of sides not a multiple of 2, 3, or 5. On March 30, 1796, the day of this discovery, he began a diary, which contained as its first entry rules for construction of a 17-sided regular polygon. The diary, which contains 146 statements of results in only 19 pages, is one of the most important documents in the history of mathematics.

After a short period at Göttingen, Gauss went to the University of Helmstädt and, in 1798 at the age of 20, wrote his now famous doctoral dissertation. In it he gave the first mathematically rigorous proof of the fundamental theorem of algebra—that every polynomial of degree n, has, counting multiplicities, exactly n roots. Many mathematicians, including Euler, Newton, and Lagrange, had attempted to prove this result.

Gauss made a great number of discoveries in physics as well as in mathematics. For example, in 1801 he used a new procedure to calculate, from very little data, the orbit of the planetoid Ceres. In 1833, he invented the electromagnetic telegraph with his colleague Wilhelm Weber (1804–1891). While he did brilliant work in astronomy and electricity, however, it was Gauss's mathematical output that was astonishing. He made fundamental contributions to algebra and geometry. In 1811, he discovered a result that led to the development of complex variable theory by Cauchy. He is encountered in courses in matrix theory in the Gauss–Jordan method of elimination. Students of numerical analysis study Gaussian quadrature—a technique for numerical integration.

Gauss became a professor of mathematics at Göttingen in 1807 and remained in that post until his death in 1855. Even after his death, his mathematical spirit remained to haunt nineteenth century mathematicians. Often it turned out that an important new result was discovered earlier by Gauss and could be found in his unpublished notes.

In his mathematical writings, Gauss was a perfectionist and is probably the last mathematician who knew everything in his subject. Claiming that a cathedral was not a cathedral until the last piece of scaffolding was removed, he endeavored to make each of his published works complete, concise, and polished. He used a seal that pictured a tree carrying only a few fruit together with the motto *pauca sed matura* (few, but ripe). But Gauss also believed that mathematics must reflect the real world. At his death, Gauss was honored by a commemorative medal on which was inscribed "George V. King of Hanover to the Prince of Mathematicians."

necessary; smoothness of S is sufficient. But it allows us to give a reasonably simple proof.

From (3) it is sufficient to prove that

$$\iint_S P \cos \alpha \, d\sigma = \iiint_W \frac{\partial P}{\partial x} \, dx \, dy \, dz, \tag{4}$$

$$\iint_S Q \cos \beta \, d\sigma = \iiint_W \frac{\partial Q}{\partial y} \, dx \, dy \, dz, \tag{5}$$

and

$$\iint_S R \cos \gamma \, d\sigma = \iiint_W \frac{\partial R}{\partial z} \, dx \, dy \, dz. \tag{6}$$

The proofs of (4), (5), and (6) are virtually identical, so we will prove (6) only and leave (4) and (5) to you. By the assumption stated above, we can think of S as "walnut shaped." That is, S consists of two surfaces: an upper surface S_2: $z = f_2(x, y)$ and a lower surface S_1: $z = f_1(x, y)$. This is depicted in Figure 2. We then have

$$\iint_S R \cos \gamma \, d\sigma = \iint_{S_1} R \cos \gamma \, d\sigma + \iint_{S_2} R \cos \gamma \, d\sigma.$$

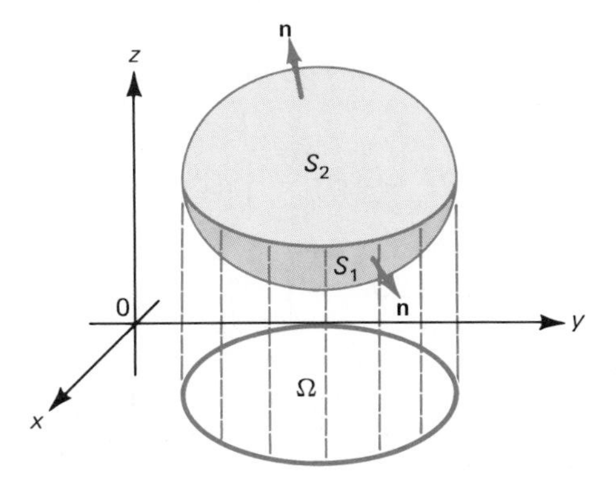

FIGURE 2

Now from equation (6.5.21),

$$\iint_{S_2} R(x, y, z) \cos \gamma \, d\sigma = \iint_{\Omega} R(x, y, f_2(x, y)) \cos \gamma \sec \gamma \, dx \, dy$$

$$= \iint_{\Omega} R(x, y, f_2(x, y)) \, dx \, dy, \tag{7}$$

where Ω is the projection of S_2 into the xy-plane. On S_1 the situation is a bit different because the outer normal $\mathbf{n}$ on S_1 points downward and we assumed in deriving equation (6.5.21), that $\mathbf{n}$ points upward (so that the angle γ between $\mathbf{n}$ and the positive z-axis is an acute angle). We solve this problem by using $-\mathbf{n}$ in (6.5.21) so that on S_1 we have

$$\iint_{S_1} R(x, y, z) \cos \gamma \, d\sigma = -\iint_{\Omega} R(x, y, f_1(x, y)) \cos \gamma \sec \gamma \, dx \, dy$$

$$= -\iint_{\Omega} R(x, y, f_1(x, y)) \, dx \, dy.$$

Thus

$$\iint_{S} R \cos \gamma \, ds = \iint_{\Omega} [R(x, y, f_2(x, y)) - R(x, y, f_1(x, y)] \, dx \, dy$$

$$= \iint_{\Omega} \left(\int_{f_1(x,y)}^{f_2(x,y)} \frac{\partial R}{\partial z} \, dz \right) dx \, dy = \iiint_{W} \frac{\partial R}{\partial z} \, dz \, dx \, dy,$$

and under the restrictions already mentioned, this completes the proof. ■

EXAMPLE 1 Compute $\iint_S \mathbf{F} \cdot \mathbf{n} \, d\sigma$, where $\mathbf{F}(x, y, z) = x^2\mathbf{i} + 2y\mathbf{j} + 4z^2\mathbf{k}$ and S is the surface of the cylinder $x^2 + y^2 \le 4$, $0 \le z \le 2$.

Solution. div $\mathbf{F} = 2x + 2 + 8z$, so

$$\iint_{S} \mathbf{F} \cdot \mathbf{n} \, ds = \iiint_{W} (2x + 2 + 8z) \, dz \, dx \, dy$$

$$= \iint_{x^2+y^2 \le 4} \int_0^2 (2x + 2 + 8z) \, dz \, dx \, dy$$

$$= \iint_{x^2+y^2 \le 4} \left[(2xz + 2z + 4z^2) \Big|_{z=0}^{z=2} \right] dx \, dy$$

$$= \iint_{x^2+y^2 \le 4} (4x + 20) \, dx \, dy = \int_0^{2\pi} \int_0^2 (4r \cos \theta + 20) r \, dr \, d\theta$$

$$= 4 \int_0^{2\pi} \int_0^2 (r^2 \cos \theta + 5r) \, dr \, d\theta = 4 \int_0^{2\pi} \left(\frac{8}{3} \cos \theta + 10 \right) d\theta = 80\pi. \quad ■$$

EXAMPLE 2 Compute $\iint_S \mathbf{F} \cdot \mathbf{n} \, d\sigma$, where $\mathbf{F} = 2xy\mathbf{i} + 3y\mathbf{j} + 2z\mathbf{k}$ and S is the boundary of the solid bounded by the three coordinate planes and the plane $x + y + z = 1$.

Solution. div $\mathbf{F} = 2y + 3 + 2 = 2y + 5$, so

$$\iint_S \mathbf{F} \cdot \mathbf{n}\, d\sigma = \iiint_W (2y + 5)\, dx\, dy\, dz,$$

$$\int_0^1 \int_0^{1-y} \int_0^{1-x-y} (2y + 5)\, dz\, dx\, dy = \int_0^1 \int_0^{1-y} (2y + 5)(1 - x - y)\, dx\, dy$$

$$= \int_0^1 \left[-(2y + 5) \frac{(1 - x - y)^2}{2} \Bigg|_{x=0}^{x=1-y} \right] dy$$

$$= \frac{1}{2} \int_0^1 (2y + 5)(1 - y)^2\, dy = \frac{11}{12}. \blacksquare$$

EXAMPLE 3 Compute $\iint_S \mathbf{F} \cdot \mathbf{n}\, d\sigma$, where $\mathbf{F} = (5x + \sin y \tan z)\mathbf{i} + (y^2 - e^{x-2\cos z^3})\mathbf{j} + (4xy)^{3/5}\mathbf{k}$ and S is the boundary of the solid bounded by the parabolic cylinder $y = 9 - x^2$, the plane $y + z = 1$, and the xy- and xz-planes.

Solution. The solid is sketched in Figure 3. It would be difficult to compute this surface integral directly, but with the divergence theorem it becomes relatively easy. Since div $\mathbf{F} = 5 + 2y$, we have

$$\iint_S \mathbf{F} \cdot \mathbf{n}\, d\sigma = \iiint_W (5 + 2y)\, dx\, dy\, dz$$

$$= \int_0^1 \int_{-\sqrt{9-y}}^{\sqrt{9-y}} \int_0^{1-y} (5 + 2y)\, dz\, dx\, dy$$

$$= 2 \int_0^1 \int_0^{\sqrt{9-y}} (5 + 2y)(1 - y)\, dx\, dy$$

$$= 2 \int_0^1 \int_0^{\sqrt{9-y}} (5 - 3y - 2y^2)\, dx\, dy$$

$$= 2 \int_0^1 (5 - 3y - 2y^2) \sqrt{9 - y}\, dy \approx 16.66. \blacksquare$$

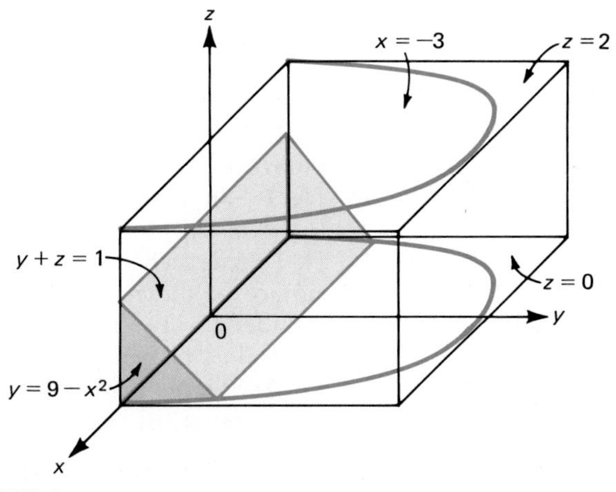

FIGURE 3

At the end of Section 6.3 we stated that the divergence of **F** at a point (x, y) represents the net rate of flow away from (x, y). We can see this more clearly in light of the divergence theorem. Let us divide this region W into a set of small subregions W_i with boundaries S_i. Since div **F** is continuous, if W_i is sufficiently small, div **F** is approximately constant, and

$$\iiint_{W_i} \text{div}\,\mathbf{F}\,dx\,dy\,dz \approx \text{div}\,\mathbf{F}\iiint_{W_i} dx\,dy\,dz$$
$$= (\text{div}\,\mathbf{F})\text{ volume } W_i. \tag{8}$$

But

$$\iiint_{W_i} \text{div}\,\mathbf{F}\,dx\,dy\,dz = \iint_{S_i} \mathbf{F}\cdot\mathbf{n}\,d\sigma, \tag{9}$$

which is the flux of **F** through S_i. Thus by (8) and (9), at a point (x, y, z)

$$\text{div}\,\mathbf{F}(x, y, z) \approx \frac{\iiint_{W_i} \text{div}\,\mathbf{F}\,dx\,dy\,dz}{\text{volume of } W_i} = \frac{\iint_{S_i} \mathbf{F}\cdot\mathbf{n}\,d\sigma}{\text{volume of } W}.$$

This last term can be thought of as *flux per unit volume* or net average rate of flow away from (x, y, z).

PROBLEMS 6.8

In Problems 1–15, evaluate the surface integral by using the divergence theorem.

1. $\iint_S \mathbf{F}\cdot\mathbf{n}\,d\sigma$, where $\mathbf{F} = x\mathbf{i} + y\mathbf{j} + z\mathbf{k}$ and S is the unit sphere.
2. $\iint_S \mathbf{F}\cdot\mathbf{n}\,d\sigma$, where $\mathbf{F} = x^2\mathbf{i} + y^2\mathbf{j} + z^2\mathbf{k}$ and S is the unit sphere.
3. $\iint_S \mathbf{F}\cdot\mathbf{n}\,d\sigma$, where $\mathbf{F}$ is as in Problem 1 and S is the cylinder $x^2 + y^2 \le 4$, $0 \le z \le 3$.
4. $\iint_S \mathbf{F}\cdot\mathbf{n}\,d\sigma$, where $\mathbf{F} = y\mathbf{i} + z\mathbf{j} + x\mathbf{k}$ and S is as in Problem 3.
5. $\iint_S \mathbf{F}\cdot\mathbf{n}\,d\sigma$, where $\mathbf{F} = (y^2 + z^2)^{3/2}\mathbf{i} + [\sin(x^2 - z^5)^{4/3}]\mathbf{j} + e^{x^2-y^3}\mathbf{k}$ and S is the ellipsoid
$$\frac{x^2}{a^2} + \frac{y^2}{b^2} + \frac{z^2}{c^2} = 1, \qquad abc \ne 0.$$
6. $\iint_S \mathbf{F}\cdot\mathbf{n}\,d\sigma$, where $\mathbf{F} = x\mathbf{i} + y\mathbf{j} + z\mathbf{k}$ and S is the surface of the unit cube $0 \le x \le 1$, $0 \le y \le 1$, $0 \le z \le 1$.
7. $\iint_S \mathbf{F}\cdot\mathbf{n}\,d\sigma$, where $\mathbf{F} = x^2\mathbf{i} + y^2\mathbf{j} - xy\mathbf{k}$ and S is as in Problem 6.
8. $\iint_S \mathbf{F}\cdot\mathbf{n}\,d\sigma$, where $\mathbf{F} = xyz\mathbf{i} + yz\mathbf{j} + z\mathbf{k}$ and S is as in Problem 6.
9. $\iint_S \mathbf{F}\cdot\mathbf{n}\,d\sigma$, where $\mathbf{F} = 2x\mathbf{i} + 3y\mathbf{j} + z\mathbf{k}$ and S is the boundary of the hemisphere $x^2 + y^2 + z^2 = 9$, $z \ge 0$.
10. $\iint_S \mathbf{F}\cdot\mathbf{n}\,d\sigma$, where $\mathbf{F} = x^2\mathbf{i} + y^2\mathbf{j} + z^2\mathbf{k}$ and S is as in Problem 9.
11. $\iint_S \mathbf{F}\cdot\mathbf{n}\,d\sigma$, where $\mathbf{F} = xy\mathbf{i} + y^2\mathbf{j} + yz\mathbf{k}$ and S is the boundary of the tetrahedron with vertices at $(0, 0, 0)$, $(1, 0, 0)$, $(0, 1, 0)$, and $(0, 0, 1)$.
12. $\iint_S \mathbf{F}\cdot\mathbf{n}\,d\sigma$, where $\mathbf{F} = y^2\mathbf{i} + x^2\mathbf{j} + z^2\mathbf{k}$ and S is as in Problem 11.
13. $\iint_S \mathbf{F}\cdot\mathbf{n}\,d\sigma$, where $\mathbf{F} = x(1 - \sin y)\mathbf{i} + (y - \cos y)\mathbf{j} + z\mathbf{k}$ and S is as in Problem 11.
14. $\iint_S \mathbf{F}\cdot\mathbf{n}\,d\sigma$, where $\mathbf{F} = x\mathbf{i} + y\mathbf{j} + z\mathbf{k}$ and S is the surface of the region bounded by the parabolic cylinder $z = 1 - y^2$, the plane $x + z = 2$, and the xy- and yz-planes.
15. $\iint_S \mathbf{F}\cdot\mathbf{n}\,d\sigma$, where $\mathbf{F} = (x^2 + e^{y\cos z})\mathbf{i} + (xy - \tan z^{1/3})\mathbf{j} + (x - y^{3/5})^{2/9}\mathbf{k}$ and S is as in Problem 14.

16. If $\mathbf{F}$ is twice continuously differentiable, prove that $\iint_S \text{curl } \mathbf{F} \cdot \mathbf{n}\, d\sigma = 0$ for any smooth closed surface S.

17. Show that $\iint_S \mathbf{F} \cdot \mathbf{n}\, d\sigma = 0$ if $\mathbf{F}$ is a constant vector field and S is a closed smooth surface.

18. If $\mathbf{F} = x\mathbf{i} + y\mathbf{j} + z\mathbf{k}$ and W is a solid with closed smooth boundary S, show that

$$\text{volume of } W = \frac{1}{3}\iint_S \mathbf{F} \cdot \mathbf{n}\, d\sigma.$$

19. Suppose that a vector field $\mathbf{v}$ is tangent to a closed smooth surface S, where S is the boundary of a solid W. Show that $\iiint_W \text{div } \mathbf{v}\, dx\, dy\, dz = 0$.

20. One of Maxwell's equations states that an electric field $\mathbf{E}$ in space satisfies div $\mathbf{E} = (1/\epsilon_0)\rho$, where ϵ_0 is a constant and ρ is the charge density. Show that the flux of the displacement $D = \epsilon_0\mathbf{E}$ across a closed surface is equal to the charge q inside the surface. This last result is known as **Gauss's law.**

6.9 CHANGING VARIABLES IN MULTIPLE INTEGRALS AND THE JACOBIAN

In many places in this and the preceding chapter we found it useful to evaluate a double integral by first converting to polar coordinates. In Section 5.4 we proved that [see equation (5.4.7) on p. 296]

$$\iint_\Omega f(x, y)\, dx\, dy = \iint_\Omega f(\cos\theta, r \sin\theta) r\, dr\, d\theta. \tag{1}$$

In Section 5.7 we showed how to convert an ordinary triple integral from an integral in Cartesian coordinates to an integral in cyclindrical or spherical coordinates.

In this section we show how, under certain conditions, it is possible to convert from one set of coordinates to another in double and triple integrals. Much of what we do is a generalization of the formula for changing variables in ordinary definite integrals. Let us recall that formula now. If $x = g(u)$ is a differentiable one-to-one function, then

$$\int_b^a f(x)\, dx = \int_c^d f(g(u))\, g'(u)\, du, \tag{2}$$

where $c = g^{-1}(a)$ and $d = g^{-1}(b)$.

Now suppose we wish to change the variables of integration in the double integral

$$\iint_\Omega f(x, y)\, dx\, dy. \tag{3}$$

We assume that the new variables are called u and v and that they are related to the old variables x and y by the relations

$$x = g(u, v) \qquad \text{and} \qquad y = h(u, v). \tag{4}$$

The functional relationship described by (4) is called a **mapping** from the uv-plane into the xy-plane. We will assume that there is a region Σ in the uv-plane that gets mapped **onto** Ω by the mapping described by (4) and the mapping is one-to-one. That is:

(i) For every $(x, y) \in \Omega$, there is a $(u, v) \in \Sigma$ such that $x = g(u, v)$ and $y = h(u, v)$.
(ii) If $g(u_1, v_1) = g(u_2, v_2)$ and $h(u_1, v_1) = h(u_2, v_2)$, then $u_1 = u_2$ and $v_1 = v_2$.

Condition (ii) is the natural extension of the definition of one-to-one for functions of two variables. We illustrate what is going on in Figure 1.

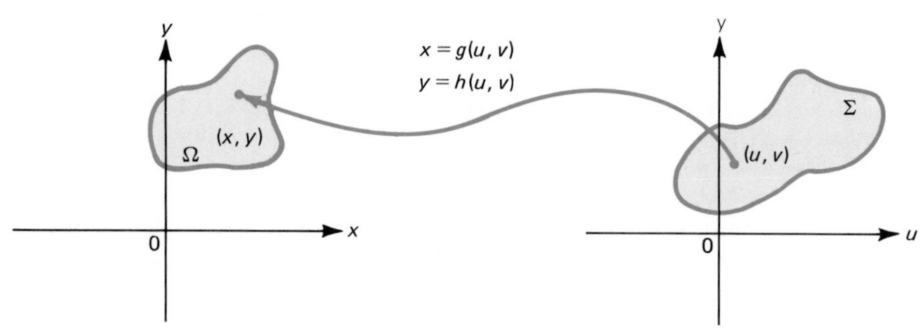

FIGURE 1

EXAMPLE 1 Let $x = u^2 - v^2$ and $y = 3uv$. We will compute the image of the vertical line $u = k$ in the uv-plane under the mapping. Substituting k for u in the equations given above, we have

$$x = k^2 - v^2 \quad \text{and} \quad y = 3kv.$$

Thus $v = y/3k$ and

$$x = k^2 - \frac{y^2}{9k^2}. \tag{5}$$

For every $k \neq 0$, equation (5) is the equation of a parabola in the xy-plane. Moreover, if $v = c$, a constant, then $x = u^2 - c^2$, $y = 3uc$, $u = y/3c$, and

$$x = \frac{y^2}{9c^2} - c^2, \tag{6}$$

which is also a parabola for $c \neq 0$. Thus the functions given above map straight lines into parabolas. Some of these curves are sketched in Figure 2. ■

Definition 1 JACOBIAN† Let $x = g(u, v)$ and $y = h(u, v)$ be differentiable. Then the **Jacobian** of x and y with respect to u and v, denoted $\partial(x, y)/\partial(u, v)$, is given by

†Named after the German mathematician Carl Gustav Jacob Jacobi (1804–1851). See the accompanying biographical sketch.

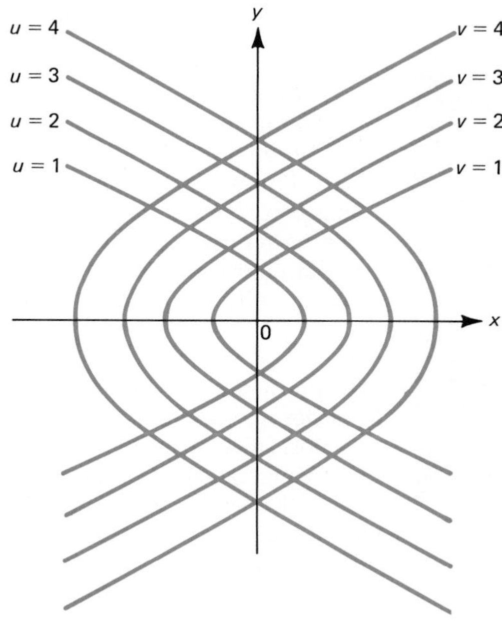

FIGURE 2

$$\frac{\partial(x, y)}{\partial(u, v)} = \frac{\partial x}{\partial u}\frac{\partial y}{\partial v} - \frac{\partial x}{\partial v}\frac{\partial y}{\partial u} = \begin{vmatrix} \dfrac{\partial x}{\partial u} & \dfrac{\partial x}{\partial v} \\ \dfrac{\partial y}{\partial u} & \dfrac{\partial y}{\partial v} \end{vmatrix}. \tag{7}$$

EXAMPLE 2 If $x = u^2 - v^2$ and $y = 3uv$ as in Example 1, then $\partial x/\partial u = 2u$, $\partial x/\partial v = -2v$, $\partial y/\partial u = 3v$, and $\partial y/\partial v = 3u$, so

$$\begin{vmatrix} \dfrac{\partial x}{\partial u} & \dfrac{\partial x}{\partial v} \\ \dfrac{\partial y}{\partial u} & \dfrac{\partial y}{\partial v} \end{vmatrix} = \begin{vmatrix} 2u & -2v \\ 3v & 3u \end{vmatrix} = 6(u^2 + v^2). \blacksquare$$

EXAMPLE 3 Let $x = r \cos \theta$ and $y = r \sin \theta$. Then

$$\frac{\partial x}{\partial r} = \cos \theta, \qquad \frac{\partial x}{\partial \theta} = -r \sin \theta, \qquad \frac{\partial y}{\partial r} = \sin \theta, \qquad \frac{\partial y}{\partial \theta} = r \cos \theta,$$

so

$$\begin{aligned} \frac{\partial(x, y)}{\partial(u, v)} &= \begin{vmatrix} \cos \theta & -r \sin \theta \\ \sin \theta & r \cos \theta \end{vmatrix} \\ &= r \cos^2 \theta + r \sin^2 \theta = r(\sin^2 \theta + \cos^2 \theta) = r. \blacksquare \end{aligned}$$

1804–
1851

Carl Gustav Jacob Jacobi

Carl Gustav Jacob Jacobi
Historical Pictures Service, Chicago

The son of a prosperous banker, Carl Gustav Jacob Jacobi was born in Potsdam, Germany, in 1804. He was educated at the University of Berlin, where he received his doctorate in 1825. In 1827, he was appointed Extraordinary Professor of Mathematics at the University of Königsberg. Jacobi taught at Königsberg until 1842, when he returned to Berlin under a pension from the Prussian government. He remained in Berlin until his death in 1851.

A prolific writer of mathematical treatises, Jacobi was best known in his time for his results in the theory of elliptic functions. Today, however, he is most remembered for his work on determinants. He was one of the two most creative developers of determinant theory, the other being Cauchy. In 1829, Jacobi published a paper on algebra that contained the notation for the Jacobian that we use today. In 1841 he published an extensive treatise titled *De determinantibus functionalibus,* which was devoted to results about the Jacobian. Jacobi showed the relationship between the Jacobian of functions of several variables and the derivative of a function of one variable. He also showed that n functions of n variables are linearly independent if and only if their Jacobian is not identically zero.

In addition to being a fine mathematician, Jacobi was considered the greatest teacher of mathematics of his generation. He inspired and influenced an astonishing number of students. To dissuade his students from mastering great amounts of mathematics before setting off to do their own research, Jacobi often remarked, "Your father would never have married, and you would not be born, if he had insisted on knowing all the girls in the world before marrying one."

Jacobi believed strongly in research in pure mathematics and frequently defended it against the claim that research should always be applicable to something. He once said, "The real end of science is the honor of the human mind."

To obtain our main result we need assumptions (i) and (ii) cited earlier. We need also to assume the following:

(iii) $C_1 = \partial\Omega$ and $C_2 = \partial\Sigma$ are simple closed curves, and as (u, v) moves once about C_2 in the positive direction, $(x, y) = (g(u, v), h(u, v))$ moves once around C_1 in the positive or negative direction.

(iv) All second order partial derivatives are continuous.

Theorem 1 CHANGE OF VARIABLES IN A DOUBLE INTEGRAL If assumptions (i), (ii), (iii), and (iv) hold, then

$$\iint_{\Omega} f(x, y)\, dx\, dy = \pm\iint_{\Sigma} f[g(u, v), h(u, v)]\frac{\partial(x, y)}{\partial(u, v)}\, du\, dv. \tag{8}$$

The plus (minus) sign is taken if as (u, v) moves around C_2 in the positive direction $(x, y) = (g(u, v), h(u, v))$ moves around C_1 in the positive (negative) direction.

Proof. Define $F(x, y) = \int_{x_0}^{x} f(t, y)\, dt$. By the fundamental theorem of calculus, F is continuous and

$$\frac{\partial F}{\partial x} = f(x, y). \tag{9}$$

Then by Green's theorem

$$\iint_{\Omega} f(x, y)\, dx\, dy = \iint_{\Omega} \frac{\partial F}{\partial x}\, dx\, dy = \oint_{C_1} F\, dy. \tag{10}$$

We wish to write $\oint_{C_1} F\, dy$ as a line integral in the uv-plane. Let $\mathbf{u} = u(t)\mathbf{i} + v(t)\mathbf{j}$ be a parametric representation of C_2 in the uv-plane for $a \le t \le b$. Then by assumption (iii), $\mathbf{x} = x(t)\mathbf{i} + y(t)\mathbf{j} = g(u(t), v(t))\mathbf{i} + h(u(t), v(t))\mathbf{j}$, $a \le t \le b$, is a parametric representation for C_1. The only difference is that this representation may traverse C_1 in either the positive or the negative direction. Thus

$$\oint_{C_1} F(x, y)\, dy = \int_a^b F(g(u(t), v(t)), h(u(t), v(t))) \frac{dy}{dt}\, dt. \tag{11}$$

Now by the chain rule and the fact that $y = h(u, v)$,

$$\frac{dy}{dt} = \frac{\partial h}{\partial u} u'(t) + \frac{\partial h}{\partial v} v'(t). \tag{12}$$

To simplify notation, let $\overline{F}(u, v) = F(g(u, v), h(u, v))$. Then if we substitute (12) into (11), we obtain

$$\begin{aligned} \oint_{C_1} F(x, y)\, dy &= \int_a^b \overline{F}(u(t), v(t)) \left[\frac{\partial h}{\partial u} u'(t) + \frac{\partial h}{\partial v} v'(t)\right] dt \\ &= \int_a^b \left[\overline{F} \frac{\partial h}{\partial u} u'(t) + \overline{F} \frac{\partial h}{\partial v} v'(t)\right] dt. \end{aligned} \tag{13}$$

Using the definition of the line integral, we can write (13) as

$$\oint_{C_1} F(x, y)\, dy = \pm \oint_{C_1} \overline{F} \frac{\partial h}{\partial u}\, du + \overline{F} \frac{\partial h}{\partial v}\, dv, \tag{14}$$

where the $\pm$ depends on whether (x, y) traverses C_1 in the same direction as, or in the direction opposite to, that in which (u, v) traverses C_2 as t goes from a to b. Let

$$P = \overline{F} \frac{\partial h}{\partial u} \quad \text{and} \quad Q = \overline{F} \frac{\partial h}{\partial v}. \tag{15}$$

Then by Green's theorem

$$\oint_{C_2} P\,du + Q\,dv = \iint_{\Sigma} \left(\frac{\partial Q}{\partial u} - \frac{\partial P}{\partial v}\right) du\,dv. \tag{16}$$

But from the product rule

$$\frac{\partial Q}{\partial u} = \frac{\partial \overline{F}}{\partial u}\frac{\partial h}{\partial v} + \overline{F}\frac{\partial^2 h}{\partial u\,\partial v} \tag{17}$$

and

$$\frac{\partial P}{\partial v} = \frac{\partial \overline{F}}{\partial v}\frac{\partial h}{\partial u} + \overline{F}\frac{\partial^2 h}{\partial v\,\partial u}, \tag{18}$$

so that from (14)–(18),

Equal because mixed partials are continuous

$$\begin{aligned}\oint_{C_1} F\,dy &= \pm\iint_{\Sigma} \left(\frac{\partial \overline{F}}{\partial u}\frac{\partial h}{\partial v} + \overline{F}\frac{\partial^2 h}{\partial u\,\partial v} - \frac{\partial \overline{F}}{\partial v}\frac{\partial h}{\partial u} - \overline{F}\frac{\partial^2 h}{\partial v\,\partial u}\right) du\,dv \\ &= \pm\iint_{\Sigma} \left(\frac{\partial \overline{F}}{\partial u}\frac{\partial h}{\partial v} - \frac{\partial \overline{F}}{\partial v}\frac{\partial h}{\partial u}\right) du\,dv.\end{aligned} \tag{19}$$

But

$$\frac{\partial \overline{F}}{\partial u} = \frac{\partial F}{\partial x}\frac{\partial x}{\partial u} + \frac{\partial F}{\partial y}\frac{\partial y}{\partial u} = \frac{\partial \overline{F}}{\partial x}\frac{\partial g}{\partial u} + \frac{\partial \overline{F}}{\partial y}\frac{\partial h}{\partial u},$$

and similarly for $\partial \overline{F}/\partial v$. Thus

$$\begin{aligned}\oint_{C_1} F\,dy &= \pm\iint_{\Sigma} \left\{\left(\frac{\partial \overline{F}}{\partial x}\frac{\partial g}{\partial u} + \frac{\partial \overline{F}}{\partial y}\frac{\partial h}{\partial u}\right)\frac{\partial h}{\partial v} - \left(\frac{\partial \overline{F}}{\partial x}\frac{\partial g}{\partial v} + \frac{\partial \overline{F}}{\partial y}\frac{\partial h}{\partial v}\right)\frac{\partial h}{\partial u}\right\} du\,dv \\ &= \pm\iint_{\Sigma} \frac{\partial \overline{F}}{\partial x}\left(\frac{\partial g}{\partial u}\frac{\partial h}{\partial v} - \frac{\partial g}{\partial v}\frac{\partial h}{\partial u}\right) du\,dv.\end{aligned}$$

But $\partial \overline{F}/\partial x = f(g(u, v), h(u, v))$ and $(\partial g/\partial u)(\partial h/\partial v) - (\partial g/\partial v)(\partial h/\partial u) = \partial(x, y)/\partial(u, v)$. Thus

$$\oint_{C_1} F\,dy = \pm\iint_{\Sigma} f(g(u, v), h(u, v))\frac{\partial(x, y)}{\partial(u, v)}\,du\,dv. \tag{20}$$

Combining (10) with (20) completes the proof. ■

REMARK. Conditions (i) and (ii) given on page 388 are often difficult to verify. It can be shown that these conditions hold if all partial derivatives are continuous in Ω and the Jacobian $\partial(x, y)/\partial(u, v)$ is not zero on Ω.

EXAMPLE 4 We can use Theorem 1 to obtain the polar coordinate formula (1) very easily. For if $x = r\cos\theta$ and $y = r\sin\theta$, then $\partial(x, y)/\partial(r, \theta) = r$ by Example 3, and (1) follows immediately from (8). ■

EXAMPLE 5 Let Ω be the region in the upper half of the xy-plane bounded by the parabolas $y^2 = 9 - 9x$ and $y^2 = 9 + 9x$ and by the x-axis. Compute $\iint_\Omega (x + y)\,dx\,dy$ by making the change of variables $x = u^2 - v^2$, $y = 3uv$.

Solution. We saw in Example 1 that this mapping takes straight lines in the uv-plane into parabolas in the xy-plane. For example, if $y^2 = 9 - 9x$, then $y^2 = 9u^2v^2 = 9 - 9x$, or $u^2v^2 = 1 - x = 1 - u^2 + v^2$, or $u^2 + u^2v^2 = 1 + v^2 = u^2(1 + v^2)$, and $u^2 = 1$, so $u = \pm 1$. Similarly, if $y^2 = 9 + 9x$, then $v = \pm 1$. Since, in Ω, $y \geq 0$, we must have $uv \geq 0$, so that u and v have the same sign. We will choose $u = v = 1$ for reasons to be made clear shortly. Note also that if $u = 0$, then $y = 0$ and $x = -v^2 \leq 0$, so that the positive v-axis in the uv-plane is mapped into the negative x-axis in the xy-plane. Similarly, the positive u-axis in the uv-plane is mapped into the positive x-axis in the xy-plane. The situation is sketched in Figure 3. The reason we chose $u = v = 1$ is that moving around Σ in the positive direction corresponds to moving around Ω in the positive direction. [Try it: Take the path $v = 0$ (the u-axis) to $u = 1$ to $v = 1$ to $u = 0$.] Thus the integral around the "parabolic" region given in the problem can be reduced to an integral around a square. Also, as we computed in Example 2,

$$\frac{\partial(x, y)}{\partial(u, v)} = 6(u^2 + v^2).$$

Finally, $x + y = u^2 + 3uv - v^2$, so

$$\iint_\Omega (x + y)\,dx\,dy = 6\int_0^1\int_0^1 (u^2 + 3uv - v^2)(u^2 + v^2)\,du\,dv$$

$$= 6\int_0^1\int_0^1 (u^4 + 3u^3v + 3uv^3 - v^4)\,du\,dv = \frac{9}{2}. \quad ■$$

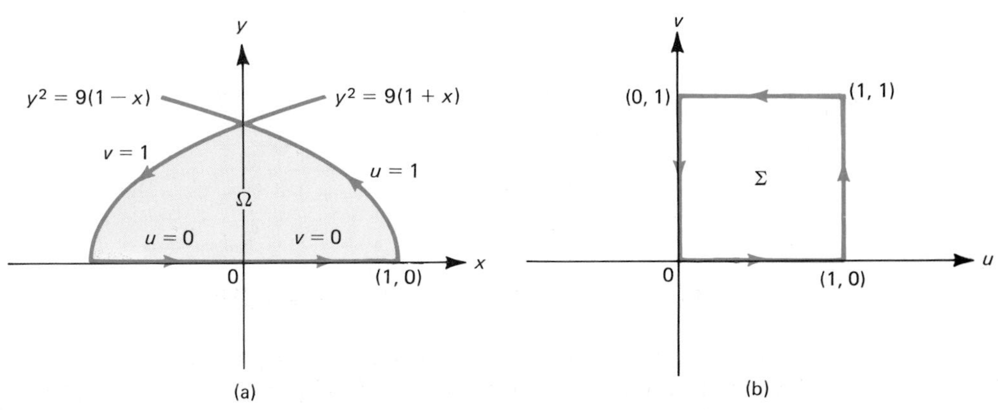

FIGURE 3

For triple integrals there is a result analogous to the one we have proven for double integrals. We will state this result without proof.

Let

$$x = g(u, v, w) \qquad y = h(u, v, w), \qquad z = j(u, v, w). \tag{21}$$

We define the **Jacobian** of the transformation from a region U in a uvw-space to a region W in xyz-space as

$$\text{Jacobian} = \frac{\partial(x, y, z)}{\partial(u, v, w)} = \begin{vmatrix} \frac{\partial x}{\partial u} & \frac{\partial x}{\partial v} & \frac{\partial x}{\partial w} \\ \frac{\partial y}{\partial u} & \frac{\partial y}{\partial v} & \frac{\partial y}{\partial w} \\ \frac{\partial z}{\partial u} & \frac{\partial z}{\partial v} & \frac{\partial z}{\partial w} \end{vmatrix}. \tag{22}$$

Then under hypotheses similar to the ones made in Theorem 1, we have the following theorem.

Theorem 2 CHANGE OF VARIABLES IN A TRIPLE INTEGRAL

$$\iiint_W F(x, y, z)\, dx\, dy\, dz =$$

$$\pm \iiint_U F(g(u, v, w), h(u, v, w), j(u, v, w)) \frac{\partial(x, y, z)}{\partial(u, v, w)}\, du\, dv\, dw$$

EXAMPLE 6 Let $x = r \cos \theta$, $y = r \sin \theta$, and $z = z$. These are cyclindrical coordinates. Then

$$\frac{\partial(x, y, z)}{\partial(r, \theta, z)} = \begin{vmatrix} \frac{\partial x}{\partial r} & \frac{\partial x}{\partial \theta} & \frac{\partial x}{\partial z} \\ \frac{\partial y}{\partial r} & \frac{\partial y}{\partial \theta} & \frac{\partial y}{\partial z} \\ \frac{\partial z}{\partial r} & \frac{\partial z}{\partial \theta} & \frac{\partial z}{\partial z} \end{vmatrix} = \begin{vmatrix} \cos \theta & -r \sin \theta & 0 \\ \sin \theta & r \cos \theta & 0 \\ 0 & 0 & 1 \end{vmatrix} = r(\cos^2\theta + \sin^2 \theta) = r,$$

so

$$\iiint_W f(x, y, z)\, dx\, dy\, dz = \iiint_U f(r \cos \theta, r \sin \theta, z) r\, dz\, dr\, d\theta.$$

This is equation 5.7.5. ■

EXAMPLE 7 Let $x = \rho \sin \varphi \cos \theta$, $y = \rho \sin \varphi \sin \theta$, and $z = \rho \cos \varphi$. These are spherical coordinates (see Section 3.9). Then

$$\frac{\partial(x, y, z)}{\partial(\rho, \varphi, \theta)} = \begin{vmatrix} \frac{\partial x}{\partial \rho} & \frac{\partial x}{\partial \varphi} & \frac{\partial x}{\partial \theta} \\ \frac{\partial y}{\partial \rho} & \frac{\partial y}{\partial \varphi} & \frac{\partial y}{\partial \theta} \\ \frac{\partial z}{\partial \rho} & \frac{\partial z}{\partial \varphi} & \frac{\partial z}{\partial \theta} \end{vmatrix} = \begin{vmatrix} \sin \varphi \cos \theta & \rho \cos \varphi \cos \theta & -\rho \sin \varphi \sin \theta \\ \sin \varphi \sin \theta & \rho \cos \varphi \sin \theta & \rho \sin \varphi \cos \theta \\ \cos \varphi & -\rho \sin \varphi & 0 \end{vmatrix}$$

Expanding in the last row
↓

$$= \cos \varphi \begin{vmatrix} \rho \cos \varphi \cos \theta & -\rho \sin \varphi \sin \theta \\ \rho \cos \varphi \sin \theta & \rho \sin \varphi \cos \theta \end{vmatrix} + \rho \sin \theta \begin{vmatrix} \sin \varphi \cos \theta & -\rho \sin \varphi \sin \theta \\ \sin \varphi \sin \theta & \rho \sin \varphi \cos \theta \end{vmatrix}$$

$$= \cos \varphi(\rho^2 \cos \varphi \sin \varphi \cos^2 \theta + \rho^2 \sin \varphi \cos \varphi \sin^2 \theta) + \rho \sin \varphi(\rho \sin^2 \varphi \cos^2 \theta + \rho \sin^2 \varphi \sin^2 \theta)$$

$$= \cos \varphi(\rho^2 \sin \varphi \cos \varphi) + \rho \sin \varphi(\rho \sin^2 \varphi)$$

$$= \rho^2 \sin \varphi(\cos^2 \varphi + \sin^2 \varphi) = \rho^2 \sin \varphi.$$

Thus

$$\iiint_W f(x, y, z)\, dx\, dy\, dz = \iiint_U f(\rho \sin \varphi \cos \theta, \rho \sin \varphi \sin \theta, \rho \cos \theta)\rho^2 \sin \varphi\, d\rho\, d\varphi\, d\theta.$$

This is formula (5.7.12). ■

PROBLEMS 6.9

In Problems 1–20, compute the Jacobian of the given transformation.

1. $x = u + v$, $y = u - v$
2. $x = u^2 - v^2$, $y = u^2 + v^2$
3. $x = u^2 - v^2$, $y = 2uv$
4. $x = \sin u$, $y = \cos v$
5. $x = u + 3v - 1$, $y = 2u + 4v + 6$
6. $x = v - 2u$, $y = u + 2v$
7. $x = au + bv$, $y = bu - av$
8. $x = e^v$, $y = e^u$
9. $x = ue^v$, $y = ve^u$

*10. $x = u^v$, $y = v^u$

11. $x = \ln(u + v)$, $y = \ln uv$
12. $x = \tan u$, $y = \sec v$
13. $x = u \sec v$, $y = v \csc u$
14. $x = u \ln v$, $y = v \ln u$
15. $x = u + v + w$, $y = u - v - w$, $z = -u + v + w$
16. $x = au + bv + cw$, $y = au - bv - cw$, $z = -au + bv + cw$
17. $x = u^2 + v^2 + w^2$, $y = u + v + w$, $z = uvw$
18. $x = u \sin v$, $y = v \cos w$, $z = w \sin u$
19. $x = e^u$, $y = e^v$, $z = e^w$
20. $x = u \ln(v + w)$, $y = v \ln(u + w)$, $z = w \ln(u + v)$

In Problems 21–25, transform the integral in (x, y) to an integral in (u, v) by using the given transformation. You need not evaluate the integral.

21. $\int_0^1\int_y^1 xy\,dx\,dy$; $x = u - v$, $y = u + v$.
22. $\iint_\Omega e^{(x+y)/(x-y)}\,dx\,dy$, where Ω is the region in the first quadrant between the lines $x + y = 1$ and $x + y = 2$; $x = u + v$, $y = u - v$.
23. $\iint_\Omega y\,dx\,dy$, where Ω is the region $7x^2 + 6\sqrt{3}x(y - 1) + 13(y - 1)^2 \le 16$; use the transformation $x = \sqrt{3}u + (\frac{1}{2})v$, $y = 1 - u + (\sqrt{3}/2)v$.
24. $\int_0^1\int_0^x (x^2 + y^2)\,dx\,dy$; $x = v$, $y = u$.
*25. $\iint_\Omega (y - x)\,dy\,dx$, Ω is the region bounded by $y = 2$, $y = x$, and $x = -y^2$; $x = v - u^2$, $y = u + v$.
26. Let Ω be the region in the first quadrant of the xy-plane bounded by the hyperbolas $xy = 1$, $xy = 2$, and the lines $x = y$ and $x = 4y$. Compute $\iint_\Omega x^2y^2\,dx\,dy$ by setting $x = u$ and $y = u/v$.
27. Let W be the solid enclosed by the ellipsoid $(x^2/a^2) + (y^2/b^2) + (z^2/c^2) = 1$. Then

$$\text{volume of } W = \iiint_W dx\,dy\,dz.$$

Compute this volume by making the transformation $x = au$, $y = bv$, $z = cw$. [*Hint:* In uvw-space you'll obtain a sphere.]
28. Compute $\iiint_W (xy + xz + yz)\,dx\,dy\,dz$, where W is the region of Problem 27.

REVIEW EXERCISES FOR CHAPTER SIX

In Exercises 1–6, compute the gradient vector field of the given function.

1. $f(x, y) = (x + y)^3$
2. $f(x, y) = \sin(x + 2y)$
3. $f(x, y) = \sqrt{xy}$
4. $f(x, y) = \dfrac{x + y}{x - y}$
5. $f(x, y, z) = x^2 + y^2 + z^2$
6. $f(x, y, z) = xyz$

In Exercises 7 and 8, sketch the given vector field.

7. $\mathbf{F}(x, y) = (x^2 + y^2)\mathbf{i} - 2xy\mathbf{j}$
8. $\mathbf{F}(x, y, z) = x\mathbf{i} - y\mathbf{j} - z\mathbf{k}$

In Exercises 9–11, calculate $\int_C \mathbf{F}(\mathbf{x}) \cdot \mathbf{dx}$.

9. $\mathbf{F}(x, y) = x^2\mathbf{i} + y^2\mathbf{j}$; C is the curve $y = x^{3/2}$ from $(0, 0)$ to $(1, 1)$.
10. $\mathbf{F}(x, y) = x^2y\mathbf{i} - xy^2\mathbf{j}$; C is the unit circle in the counterclockwise direction.
11. $\mathbf{F}(x, y) = 3xy\mathbf{i} - y\mathbf{j}$; C is the triangle joining the points $(0, 0)$, $(1, 1)$, and $(0, 1)$ in the counterclockwise direction.

12. Calculate the work done when the force field $\mathbf{F}(x, y) = x^2y\mathbf{i} + (y^3 + x^3)\mathbf{j}$ moves a particle around the unit circle in the counterclockwise direction.
13. Calculate the work done when the force field $\mathbf{F}(x, y) = 3(x - y)\mathbf{i} + x^5\mathbf{j}$ moves a particle around the triangle of Exercise 11.
14. Show that $\mathbf{F}(x, y) = 3x^2y^2\mathbf{i} + 2x^3y\mathbf{j}$ is exact, and calculate $\int_C \mathbf{F}(\mathbf{x}) \cdot \mathbf{dx}$, where C starts at $(1, 2)$ and ends at $(3, -1)$.
15. Show that $\mathbf{F}(x, y) = e^{xy}(1 + xy)\mathbf{i} + x^2e^{xy}\mathbf{j}$ is exact, and calculate $\int_C \mathbf{F}(\mathbf{x}) \cdot \mathbf{dx}$, where C starts at $(-1, 0)$ and ends at $(0, 1)$.
16. Evaluate $\oint_{\partial\Omega} 2y\,dx + 4x\,dy$, where $\Omega = \{(x, y): 0 \le x \le 2, 0 \le y \le 2\}$.
17. Evaluate $\oint_{\partial\Omega} x^2y\,dx + xy^2\,dy$, where Ω is the region enclosed by the triangle of Exercise 11.
18. Evaluate $\oint_{\partial\Omega}(x^3 + y^3)\,dx + (x^2y^2)\,dy$, where Ω is the unit disk.
19. Evaluate $\oint_{\partial\Omega}\sqrt{1 + x^2}\,dy$; $\Omega = \{(x, y): -1 \le x \le 1, x^2 \le y \le 1\}$.
20. Evaluate

$$\oint_{\partial\Omega} \frac{2x^2 + y^2 - xy}{(x^2 + y^2)^{1/2}}\,dx + \frac{xy - x^2 - 2y^2}{(x^2 + y^2)^{1/2}}\,dy,$$

where Ω is a region of the type (6.3.4) or (6.3.5) that does not contain the origin.
21. Let $\mathbf{F}(x, y) = xy^2\mathbf{i} + x^2y\mathbf{j}$ and let Ω denote the disk of radius 2 centered at $(0, 0)$. Calculate (a) curl $\mathbf{F}$, (b) $\oint_{\partial\Omega} \mathbf{F} \cdot \mathbf{T}\,ds$, (c) div $\mathbf{F}$, and (d) $\oint_{\partial\Omega} \mathbf{F} \cdot \mathbf{n}\,ds$.
22. Do the same for $\mathbf{F}(x, y) = y^2\mathbf{i} - x^2\mathbf{j}$, where Ω is the region enclosed by the triangle of Exercise 11.
23. Show that the vector flow $\mathbf{F}(x, y) = (\cos x^2)\mathbf{i} + e^y\mathbf{j}$ is irrotational.
24. Show that the vector flow $\mathbf{F}(x, y) = -(x^2y + y^3)\mathbf{i} + (x^3 + xy^2)\mathbf{j}$ is incompressible.

In Exercises 25 and 26, evaluate $\int_C \mathbf{F}(\mathbf{x}) \cdot \mathbf{dx}$.

25. $\mathbf{F}(x, y, z) = x\mathbf{i} + y\mathbf{j} + z\mathbf{k}$; C is the curve $\mathbf{x}(t) = t^3\mathbf{i} + t^2\mathbf{j} + t\mathbf{k}$ from $(0, 0, 0)$ to $(1, 1, 1)$.
26. $\mathbf{F}(x, y, z) = x^2\mathbf{i} + y^2\mathbf{j} + z^2\mathbf{k}$; C is the helix $\mathbf{x}(t) = (\sin t)\mathbf{i} + (\cos t)\mathbf{j} + 2t\mathbf{k}$ from $(0, 1, 0)$ to $(1, 0, \pi)$.

27. Show that $\mathbf{F}(x, y, z) = -(y/z)\mathbf{i} - (x/z)\mathbf{j} + (xy/z^2)\mathbf{k}$ is exact, and use that fact to evaluate $\int_C \mathbf{F}(\mathbf{x}) \cdot d\mathbf{x}$, where C is a piecewise smooth curve joining the points $(1, 1, 1)$ and $(2, -1, 3)$ and not crossing the xy-plane.

In Exercises 28–33, evaluate the surface integral over the given surface.

28. $\iint_S y\, d\sigma$, where $S: z = y^2$, $0 \le x \le 2$, $0 \le y \le 1$.
29. $\iint_S (x^2 + y^2)\, d\sigma$, where S is as in Exercise 28.
30. $\iint_S y^2\, d\sigma$, where S is the hemisphere $x^2 + y^2 + z^2 = 9$, $z \ge 0$, $x^2 + y^2 \le 9$.
31. $\iint_S xz\, d\sigma$, where S is as in Exercise 30.
32. $\iint_S y\, d\sigma$, where S is the boundary of the tetrahedron bounded by the coordinate planes and the plane $x + y + z = 1$.
33. $\iint_S (x^2 + y^2 + z^2)\, d\sigma$, where S is the part of the plane $y - x = 3$ that lies inside the cylinder $y^2 + z^2 = 9$.
34. Find the mass of a triangular metallic sheet with corners at $(1, 0, 0)$, $(0, 1, 0)$, and $(0, 0, 1)$ if its density is proportional to y^2.
35. Find the mass of a metallic sheet in the shape of the hemisphere $x^2 + y^2 + z^2 = 1$, $y \ge 0$, $x^2 + z^2 \le 1$, if its density is proportional to the distance from the y-axis.

In Exercises 36–39, compute the flux $\iint_S \mathbf{v} \cdot \mathbf{n}\, d\sigma$ for the given surface lying in the given vector field.

36. $S: z = 2xy$; $0 \le x \le 1$, $0 \le y \le 4$; $\mathbf{v} = xy^2\mathbf{i} - 2z\mathbf{j}$
37. $S: x^2 + y^2 + z^2 = 1$; $z \ge 0$; $\mathbf{v} = x\mathbf{i} + 2y\mathbf{j} + 3z\mathbf{k}$
38. $S: y = \sqrt{x^2 + z^2}$; $x^2 + z^2 \le 4$; $\mathbf{v} = x\mathbf{i} - xz\mathbf{j} + z\mathbf{k}$
39. S: unit sphere; $\mathbf{v} = x^2\mathbf{i} + y^2\mathbf{j} + z^2\mathbf{k}$

In Exercises 39–45, compute the divergence and curl of the given vector field.

40. $\mathbf{F}(x, y, z) = x\mathbf{i} + y\mathbf{j} + z\mathbf{k}$
41. $\mathbf{F}(x, y, z) = (x - y)\mathbf{i} + (y - z)\mathbf{j} + (z - x)\mathbf{k}$
42. $\mathbf{F}(x, y, z) = yz\mathbf{i} + xz\mathbf{j} + xy\mathbf{k}$
43. $\mathbf{F}(x, y, z) = \ln x\mathbf{i} + \ln y\mathbf{j} + \ln z\mathbf{k}$
44. $\mathbf{F}(x, y, z) = e^{yz}\mathbf{i} + e^{xz}\mathbf{j} + e^{xy}\mathbf{k}$
45. $\mathbf{F}(x, y, z) = \cos y\mathbf{i} + \cos x\mathbf{j} + \cos z\mathbf{k}$

In Exercises 46–48, evaluate the line integral by using Stokes's theorem.

46. $\oint_C \mathbf{F} \cdot d\mathbf{x}$, where $\mathbf{F}(x, y, z) = (x + 2y)\mathbf{i} + (y - 3z)\mathbf{j} + (z - x)\mathbf{k}$ and C is the unit circle in the plane $z = 2$.
47. $\oint_C \mathbf{F} \cdot d\mathbf{x}$, where $\mathbf{F} = x\mathbf{i} + y\mathbf{j} + z\mathbf{k}$ and C is the boundary of the triangle joining the points $(1, 0, 0)$, $(0, 1, 0)$, and $(0, 0, 1)$.
48. $\oint_C \mathbf{F} \cdot d\mathbf{x}$, where $\mathbf{F} = y^2\mathbf{i} + x^2\mathbf{j} + z^2\mathbf{k}$ and C is the boundary of the part of the plane $x + y + z = 1$ lying in the first octant.

49. Compute $\iint_S \text{curl}\, \mathbf{F} \cdot \mathbf{n}\, d\sigma$, where S is the unit sphere and $\mathbf{F} = e^{xy}\mathbf{i} + \tan^{-1} z\mathbf{j} + (x + y + z)^{7/3} z^2\mathbf{k}$.

In Exercises 50–55, evaluate the surface integral by using the divergence theorem.

50. $\iint_S \mathbf{F} \cdot \mathbf{n}\, d\sigma$, where $\mathbf{F} = x\mathbf{i} + 2y\mathbf{j} + 3z\mathbf{k}$ and S is the unit sphere.
51. $\iint_S \mathbf{F} \cdot \mathbf{n}\, d\sigma$, where $\mathbf{F} = ax\mathbf{i} + by\mathbf{j} + cz\mathbf{k}$ and S is the unit sphere.
52. $\iint_S \mathbf{F} \cdot \mathbf{n}\, d\sigma$, where $\mathbf{F}$ is as in Exercise 50 and S is the cylinder $x^2 + y^2 = 9$, $0 \le z \le 6$.
53. $\iint_S \mathbf{F} \cdot \mathbf{n}\, d\sigma$, where $\mathbf{F} = (y^3 - z)\mathbf{i} + x^2 e^z\mathbf{j} + \sin xy\mathbf{k}$ and S is the ellipsoid $(x^2/4) + (y^2/16) + (z^2/25) = 1$.
54. $\iint_S \mathbf{F} \cdot \mathbf{n}\, d\sigma$, where $\mathbf{F}$ is as in Exercise 50 and S is the boundary of the unit cube.
55. $\iint_S \mathbf{F} \cdot \mathbf{n}\, d\sigma$, where $\mathbf{F} = xz\mathbf{i} + xy\mathbf{j} + xyz\mathbf{k}$ and S is the unit cube in the first octant.

In Exercises 56–63, compute the Jacobian of the given transformation.

56. $x = u + 2v$, $y = 2u - v$
57. $x = u^3 - v^3$, $y = u^3 + v^3$
58. $x = u \ln v$, $y = v \ln u$
59. $x = ve^u$, $y = ue^v$
60. $x = \dfrac{u}{v}$, $y = \dfrac{v}{u}$
61. $x = v \tan u$, $y = \tan uv$
62. $x = u + v + w$, $y = u - 2v + 3w$, $z = -2u + v - 5w$
63. $x = vw$, $y = uw$, $z = uv$

64. Transform the integral $\int_0^1 \int_x^1 xy\, dx\, dy$ by making the transformation $x = u + v$, $y = u - v$.
65. Transform $\iint_\Omega e^{(x-y)/(x+y)}\, dx\, dy$, where Ω is the region in the first quadrant between the lines $x + y = 2$ and $x + y = 3$, by making the transformation $u = x - y$ and $v = x + y$. Then evaluate the integral.

7 Matrices and Linear Systems of Equations

We first discussed the space $\mathbb{R}^n$ in Section 3.7. In this chapter and in Chapters 8 and 9 we shall investigate properties of this space. An important tool for carrying out algebraic operations in $\mathbb{R}^n$ is the *matrix*. Therefore, we begin by defining a matrix.

7.1 MATRICES

MATRIX

An $\boldsymbol{m} \times \boldsymbol{n}$ **matrix**[†] A is a rectangular array of mn numbers arranged in m rows and n columns:[‡]

$$A = \begin{pmatrix} a_{11} & a_{12} & \cdots & a_{1j} & \cdots & a_{1n} \\ a_{21} & a_{22} & \cdots & a_{2j} & \cdots & a_{2n} \\ \vdots & \vdots & & \vdots & & \vdots \\ a_{i1} & a_{i2} & \cdots & a_{ij} & \cdots & a_{in} \\ \vdots & \vdots & & \vdots & & \vdots \\ a_{m1} & a_{m2} & \cdots & a_{mj} & \cdots & a_{mn} \end{pmatrix}. \tag{1}$$

[†]The term "matrix" was first used in 1850 by the British mathematician James Joseph Sylvester (1814–1897) to distinguish matrices from determinants (which we shall discuss in Chapter 8). In fact, the term "matrix" was intended to mean "mother of determinants."

[‡]As with vectors, we shall always assume, unless stated otherwise, that the numbers in a matrix are real.

The ijth **component** of A, denoted a_{ij}, is the number appearing in the ith row and jth column of A. We will sometimes write the matrix A as $A = (a_{ij})$. Usually, matrices will be denoted by capital letters.

If A is an $m \times n$ matrix with $m = n$, then A is a **square matrix.** An $m \times n$ matrix with all components equal to zero is called the $m \times n$ **zero matrix.**

An $m \times n$ matrix is said to have the **size** $m \times n$. Two matrices $A = (a_{ij})$ and $B = (b_{ij})$ are **equal** if (i) they have the same size and (ii) corresponding components are equal.

EXAMPLE 1 Five matrices of different sizes are given below:

(i) $A = \begin{pmatrix} 1 & 3 \\ 4 & 2 \end{pmatrix}$, 2×2 (square) **(ii)** $A = \begin{pmatrix} -1 & 3 \\ 4 & 0 \\ 1 & -2 \end{pmatrix}$, 3×2

(iii) $\begin{pmatrix} -1 & 4 & 1 \\ 3 & 0 & 2 \end{pmatrix}$, 2×3 **(iv)** $\begin{pmatrix} 1 & 6 & -2 \\ 3 & 1 & 4 \\ 2 & -6 & 5 \end{pmatrix}$, 3×3 (square)

(v) $\begin{pmatrix} 0 & 0 & 0 & 0 \\ 0 & 0 & 0 & 0 \end{pmatrix}$, 2×4 zero matrix ■

Each vector is a special kind of matrix. Thus, for example, the n-component row vector $(a_1, a_2, \ldots, a_n)$ is a $1 \times n$ matrix whereas the n-component column vector $\begin{pmatrix} a_1 \\ a_2 \\ \vdots \\ a_n \end{pmatrix}$ is an $n \times 1$ matrix.

Matrices, like vectors, arise in a great number of practical situations. For example, we saw in Section 3.7 (Example 3.7.9 on p. 129) how the vector $\begin{pmatrix} 30 \\ 20 \\ 40 \\ 10 \end{pmatrix}$ could represent order quantities for four different products used by one manufacturer. Suppose that there were five different plants. Then the 4×5 matrix

$$Q = \begin{pmatrix} 30 & 20 & 15 & 16 & 25 \\ 20 & 10 & 20 & 25 & 22 \\ 40 & 22 & 18 & 20 & 13 \\ 10 & 40 & 50 & 35 & 45 \end{pmatrix}$$

could represent the orders for the four products in each of the five plants. We can see, for example, that plant 4 orders 25 units of the second product and plant 2 orders 40 units of the fourth product.

Matrices, like vectors, can be added and multiplied by scalars.[†]

Definition 1 ADDITION OF MATRICES Let $A = (a_{ij})$ and $B = (b_{ij})$ be two $m \times n$ matrices. Then the sum of A and B is the $m \times n$ matrix $A + B$ given by

$$A + B = (a_{ij} + b_{ij}) = \begin{pmatrix} a_{11} + b_{11} & a_{12} + b_{12} & \cdots & a_{1n} + b_{1n} \\ a_{21} + b_{21} & a_{22} + b_{22} & \cdots & a_{2n} + b_{2n} \\ \vdots & \vdots & & \vdots \\ a_{m1} + b_{m1} & a_{m2} + b_{m2} & \cdots & a_{mn} + b_{mn} \end{pmatrix}$$

That is, $A + B$ is the $m \times n$ matrix obtained by adding the corresponding components of A and B.

WARNING: The sum of two matrices is defined only when both matrices have the same size. Thus, for example, it is not possible to add together the matrices $\begin{pmatrix} 1 & 2 & 3 \\ 4 & 5 & 6 \end{pmatrix}$ and $\begin{pmatrix} -1 & 0 \\ 2 & -5 \\ 4 & 7 \end{pmatrix}$.

EXAMPLE 2

$$\begin{pmatrix} 2 & 4 & -6 & 7 \\ 1 & 3 & 2 & 1 \\ -4 & 3 & -5 & 5 \end{pmatrix} + \begin{pmatrix} 0 & 1 & 6 & -2 \\ 2 & 3 & 4 & 3 \\ -2 & 1 & 4 & 4 \end{pmatrix} = \begin{pmatrix} 2 & 5 & 0 & 5 \\ 3 & 6 & 6 & 4 \\ -6 & 4 & -1 & 9 \end{pmatrix}. \blacksquare$$

Definition 2 MULTIPLICATION OF A MATRIX BY A SCALAR If $A = (a_{ij})$ is an $m \times n$ matrix and if α is a scalar, then the $m \times n$ matrix αA is given by

$$\alpha A = (\alpha a_{ij}) = \begin{pmatrix} \alpha a_{11} & \alpha a_{12} & \cdots & \alpha a_{1n} \\ \alpha a_{21} & \alpha a_{22} & \cdots & \alpha a_{2n} \\ \vdots & \vdots & & \vdots \\ \alpha a_{m1} & \alpha a_{m2} & \cdots & \alpha a_{mn} \end{pmatrix} \quad (2)$$

In other words, $\alpha A = (\alpha a_{ij})$ is the matrix obtained by multiplying each component of A by α.

[†]The algebra of matrices, that is, the rules by which matrices can be added and multiplied, was developed by the English mathematician Arthur Cayley (1821–1895) in 1857. Matrices arose with Cayley in connection with linear transformations of the type

$$x' = ax + by,$$
$$y' = cx + dy,$$

where a, b, c, d are real numbers, and which may be thought of as mapping the point (x, y) into the point (x', y'). Clearly, the above transformation is completely determined by the four coefficients a, b, c, d, and so the transformation can be symbolized by the square array

$$\begin{pmatrix} a & b \\ c & d \end{pmatrix},$$

which we have called a (*square*) *matrix*. We shall discuss linear transformations in Section 9.6.

EXAMPLE 3 Let $A = \begin{pmatrix} 1 & -3 & 4 & 2 \\ 3 & 1 & 4 & 6 \\ -2 & 3 & 5 & 7 \end{pmatrix}$. Then $2A = \begin{pmatrix} 2 & -6 & 8 & 4 \\ 6 & 2 & 8 & 12 \\ -4 & 6 & 10 & 14 \end{pmatrix}$,

$$-3A = \begin{pmatrix} -3 & 9 & -12 & -6 \\ -9 & -3 & -12 & -18 \\ 6 & -9 & -15 & -21 \end{pmatrix}, \text{ and } 0A = \begin{pmatrix} 0 & 0 & 0 & 0 \\ 0 & 0 & 0 & 0 \\ 0 & 0 & 0 & 0 \end{pmatrix}. \ ■$$

EXAMPLE 4 Let $A = \begin{pmatrix} 1 & 2 & 4 \\ -7 & 3 & -2 \end{pmatrix}$ and $B = \begin{pmatrix} 4 & 0 & 5 \\ 1 & -3 & 6 \end{pmatrix}$. Calculate $-2A + 3B$.

Solution. $-2A + 3B = (-2)\begin{pmatrix} 1 & 2 & 4 \\ -7 & 3 & -2 \end{pmatrix} + (3)\begin{pmatrix} 4 & 0 & 5 \\ 1 & -3 & 6 \end{pmatrix} =$

$$\begin{pmatrix} -2 & -4 & -8 \\ 14 & -6 & 4 \end{pmatrix} + \begin{pmatrix} 12 & 0 & 15 \\ 3 & -9 & 18 \end{pmatrix} = \begin{pmatrix} 10 & -4 & 7 \\ 17 & -15 & 22 \end{pmatrix}. \ ■$$

The next theorem is similar to Theorem 3.7.1. Its proof is left as an exercise (see Problems 21–24).

Theorem 1 Let A, B, and C be $m \times n$ matrices and let α be a scalar. Then:

(i) $A + 0 = A$
(ii) $0A = 0$
(iii) $A + B = B + A$ **(commutative law for matrix addition)**
(iv) $(A + B) + C = A + (B + C)$ **(associative law for matrix addition)**
(v) $\alpha(A + B) = \alpha A + \alpha B$ **(distributive law for scalar multiplication)**
(vi) $1A = A$

REMARK. The zero in part (i) of the theorem is the $m \times n$ zero matrix. In part (ii) the zero on the left is a scalar while the zero on the right is the $m \times n$ zero matrix.

EXAMPLE 5 To illustrate the associative law we note that

$$\left[\begin{pmatrix} 1 & 4 & -2 \\ 3 & -1 & 0 \end{pmatrix} + \begin{pmatrix} 2 & -2 & 3 \\ 1 & -1 & 5 \end{pmatrix}\right] + \begin{pmatrix} 3 & -1 & 2 \\ 0 & 1 & 4 \end{pmatrix} = \begin{pmatrix} 3 & 2 & 1 \\ 4 & -2 & 5 \end{pmatrix}$$

$$+ \begin{pmatrix} 3 & -1 & 2 \\ 0 & 1 & 4 \end{pmatrix} = \begin{pmatrix} 6 & 1 & 3 \\ 4 & -1 & 9 \end{pmatrix}$$

Similarly,

$$\begin{pmatrix} 1 & 4 & -2 \\ 3 & -1 & 0 \end{pmatrix} + \left[\begin{pmatrix} 2 & -2 & 3 \\ 1 & -1 & 5 \end{pmatrix} + \begin{pmatrix} 3 & -1 & 2 \\ 0 & 1 & 4 \end{pmatrix}\right] = \begin{pmatrix} 1 & 4 & -2 \\ 3 & -1 & 0 \end{pmatrix}$$

$$+ \begin{pmatrix} 5 & -3 & 5 \\ 1 & 0 & 9 \end{pmatrix} = \begin{pmatrix} 6 & 1 & 3 \\ 4 & -1 & 9 \end{pmatrix}. \ ■$$

As with vectors, the associative law for matrix addition enables us to define the sum of three or more matrices.

PROBLEMS 7.1

In Problems 1–12 perform the indicated computation with $A = \begin{pmatrix} 1 & 3 \\ 2 & 5 \\ -1 & 2 \end{pmatrix}$, $B = \begin{pmatrix} -2 & 0 \\ 1 & 4 \\ -7 & 5 \end{pmatrix}$, and $C = \begin{pmatrix} -1 & 1 \\ 4 & 6 \\ -7 & 3 \end{pmatrix}$.

1. $3A$
2. $A + B$
3. $A - C$
4. $2C - 5A$
5. $0B$ (0 is the scalar zero)
6. $-7A + 3B$
7. $A + B + C$
8. $C - A - B$
9. $2A - 3B + 4C$
10. $7C - B + 2A$
11. Find a matrix D such that $2A + B - D$ is the 3×2 zero matrix.
12. Find a matrix E such that $A + 2B - 3C + E$ is the 3×2 zero matrix.

In Problems 13–20 perform the indicated computation with $A = \begin{pmatrix} 1 & -1 & 2 \\ 3 & 4 & 5 \\ 0 & 1 & -1 \end{pmatrix}$, $B = \begin{pmatrix} 0 & 2 & 1 \\ 3 & 0 & 5 \\ 7 & -6 & 0 \end{pmatrix}$, and $C = \begin{pmatrix} 0 & 0 & 2 \\ 3 & 1 & 0 \\ 0 & -2 & 4 \end{pmatrix}$.

13. $A - 2B$
14. $3A - C$
15. $A + B + C$
16. $2A - B + 2C$
17. $C - A - B$
18. $4C - 2B + 3A$
19. Find a matrix D such that $A + B + C + D$ is the 3×3 zero matrix.
20. Find a matrix E such that $3C - 2B + 8A - 4E$ is the 3×3 zero matrix.

21. Let $A = (a_{ij})$ be an $m \times n$ matrix and let $\bar{0}$ denote the $m \times n$ zero matrix. Use Definitions 1 and 2 to show that $0A = \bar{0}$ and $\bar{0} + A = A$. Similarly, show that $1A = A$.

22. Let $A = (a_{ij})$ and $B = (b_{ij})$ be $m \times n$ matrices. Compute $A + B$ and $B + A$ and show that they are equal.

23. If α is a scalar and A and B are as in Problem 22, compute $\alpha(A + B)$ and $\alpha A + \alpha B$ and show that they are equal.

24. If $A = (a_{ij})$, $B = (b_{ij})$, and $C = (c_{ij})$ are $m \times n$ matrices, compute $(A + B) + C$ and $A + (B + C)$ and show that they are equal.

25. Consider the "graph" joining the four points in the figure. Construct a 4×4 matrix having the property that $a_{ij} = 0$ if point i is not connected (joined by a line) to point j and $a_{ij} = 1$ if point i is connected to point j.

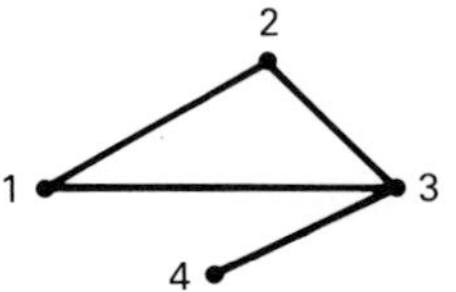

26. Do the same (this time constructing a 5×5 matrix) for the accompanying graph.

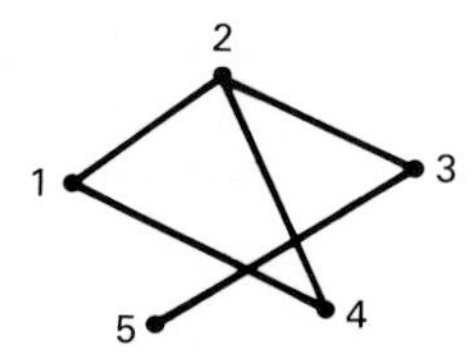

7.2 MATRIX PRODUCTS

In this section we see how two matrices can be multiplied together. Quite obviously, we could define the product of two $m \times n$ matrices $A = (a_{ij})$ and $B = (b_{ij})$ to be the $m \times n$ matrix whose ijth component is $a_{ij}b_{ij}$. However, for just about all the important applications involving matrices, another kind of product is needed. It comes as the generalization of the scalar product.

Definition 1 PRODUCT OF TWO MATRICES Let $A = (a_{ij})$ be an $m \times n$ matrix whose ith row is denoted $\mathbf{a}_i$. Let $B = (b_{ij})$ be an $n \times p$ matrix whose jth column is denoted $\mathbf{b}_j$. Then the product of A and B is an $m \times p$ matrix $C = (c_{ij})$, where

$$c_{ij} = \mathbf{a}_i \cdot \mathbf{b}_j. \tag{1}$$

That is, the ijth element of AB is the scalar product of the ith row of A ($\mathbf{a}_i$) and the jth column of B ($\mathbf{b}_j$). If we write this out, we obtain

$$c_{ij} = a_{i1}b_{1j} + a_{i2}b_{2j} + \cdots + a_{in}b_{nj}. \tag{2}$$

■ WARNING: Two matrices can be multiplied together only if the number of columns of the first is equal to the number of rows of the second. Otherwise the vectors $\mathbf{a}_i$ and $\mathbf{b}_j$ will have different numbers of components and the scalar product in equation (1) will not be defined.

EXAMPLE 1 If $A = \begin{pmatrix} 1 & 3 \\ -2 & 4 \end{pmatrix}$ and $B = \begin{pmatrix} 3 & -2 \\ 5 & 6 \end{pmatrix}$, calculate AB and BA.

Solution. Let $C = (c_{ij}) = AB$. Then $c_{11} = \mathbf{a}_1 \cdot \mathbf{b}_1 = (1 \;\; 3) \cdot \begin{pmatrix} 3 \\ 5 \end{pmatrix} = 3 + 15 = 18$; $c_{12} = \mathbf{a}_1 \cdot \mathbf{b}_2 = (1 \;\; 3) \cdot \begin{pmatrix} -2 \\ 6 \end{pmatrix} = -2 + 18 = 16$; $c_{21} = (-2 \;\; 4) \cdot \begin{pmatrix} 3 \\ 5 \end{pmatrix} = -6 + 20 = 14$; and $c_{22} = (-2 \;\; 4) \cdot \begin{pmatrix} -2 \\ 6 \end{pmatrix} = 4 + 24 = 28$. Thus $C = AB = \begin{pmatrix} 18 & 16 \\ 14 & 28 \end{pmatrix}$. Similarly, leaving out the intermediate steps, we see that

$$C' = BA = \begin{pmatrix} 3 & -2 \\ 5 & 6 \end{pmatrix}\begin{pmatrix} 1 & 3 \\ -2 & 4 \end{pmatrix} = \begin{pmatrix} 3+4 & 9-8 \\ 5-12 & 15+24 \end{pmatrix} = \begin{pmatrix} 7 & 1 \\ -7 & 39 \end{pmatrix} \quad \blacksquare$$

REMARK. Example 1 illustrates an important fact: **Matrix products do not, in general, commute.** That is, $AB \neq BA$ in general. It sometimes happens that $AB = BA$, but this will be the exception, not the rule. In fact, as the next example illustrates, it may occur that AB is defined while BA is not. Thus we must be careful of *order* when multiplying two matrices together.

EXAMPLE 2 Let $A = \begin{pmatrix} 2 & 0 & -3 \\ 4 & 1 & 5 \end{pmatrix}$ and $B = \begin{pmatrix} 7 & -1 & 4 & 7 \\ 2 & 5 & 0 & -4 \\ -3 & 1 & 2 & 3 \end{pmatrix}$. Calculate AB.

Solution. We first note that A is a 2×3 matrix and B is a 3×4 matrix. Hence the number of columns of A equals the number of rows of B. The product AB is therefore defined and is a 2×4 matrix. Let $AB = C = (c_{ij})$. Then

$$c_{11} = (2 \;\; 0 \;\; -3) \cdot \begin{pmatrix} 7 \\ 2 \\ -3 \end{pmatrix} = 23 \qquad c_{12} = (2 \;\; 0 \;\; -3) \cdot \begin{pmatrix} -1 \\ 5 \\ 1 \end{pmatrix} = -5$$

$$c_{13} = (2 \;\; 0 \;\; -3) \cdot \begin{pmatrix} 4 \\ 0 \\ 2 \end{pmatrix} = 2 \qquad c_{14} = (2 \;\; 0 \;\; -3) \cdot \begin{pmatrix} 7 \\ -4 \\ 3 \end{pmatrix} = 5$$

$$c_{21} = (4 \quad 1 \quad 5) \cdot \begin{pmatrix} 7 \\ 2 \\ -3 \end{pmatrix} = 15 \qquad c_{22} = (4 \quad 1 \quad 5) \cdot \begin{pmatrix} -1 \\ 5 \\ 1 \end{pmatrix} = 6$$

$$c_{23} = (4 \quad 1 \quad 5) \cdot \begin{pmatrix} 4 \\ 0 \\ 2 \end{pmatrix} = 26 \qquad c_{24} = (4 \quad 1 \quad 5) \cdot \begin{pmatrix} 7 \\ -4 \\ 3 \end{pmatrix} = 39$$

Hence $AB = \begin{pmatrix} 23 & -5 & 2 & 5 \\ 15 & 6 & 26 & 39 \end{pmatrix}$.

Note that the product BA is *not* defined since the number of columns of B (four) is not equal to the number of rows of A (two). ■

EXAMPLE 3 **(Direct and Indirect Contact with a Contagious Disease)** In this example we show how matrix multiplication can be used to model the spread of a contagious disease. Suppose that four individuals have contracted such a disease. This group has contacts with six people in a second group. We can represent these contacts, called *direct contacts*, by a 4×6 matrix. An example of such a matrix is given below:

DIRECT CONTACT MATRIX: First and second groups

$$A = \begin{pmatrix} 0 & 1 & 0 & 0 & 1 & 0 \\ 1 & 0 & 0 & 1 & 0 & 1 \\ 0 & 0 & 0 & 1 & 1 & 0 \\ 1 & 0 & 0 & 0 & 0 & 1 \end{pmatrix}$$

Here we set $a_{ij} = 1$ if the ith person in the first group has made contact with the jth person in the second group. For example, the 1 in the 2,4 position means that the second person in the first (infected) group has been in contact with the fourth person in the second group. Now suppose that a third group of five people has had a variety of direct contact with individuals of the second group. We can also represent this by a matrix.

DIRECT CONTACT MATRIX: Second and third groups.

$$B = \begin{pmatrix} 0 & 0 & 1 & 0 & 1 \\ 0 & 0 & 0 & 1 & 0 \\ 0 & 1 & 0 & 0 & 0 \\ 1 & 0 & 0 & 0 & 1 \\ 0 & 0 & 0 & 1 & 0 \\ 0 & 0 & 1 & 0 & 0 \end{pmatrix}$$

Note that $b_{64} = 0$, which means that the sixth person in the second group has had no contact with the fourth person in the third group.

The *indirect* or *second-order* contacts between the individuals in the first and third groups is represented by the 4×5 matrix $C = AB$. To see this, observe that a person in group 3 can be infected from someone in group 2 who, in turn, has been infected by someone in group 1. For example, since $a_{24} = 1$ and $b_{45} = 1$, we see that, indirectly, the fifth person in group 3 has contact (through the fourth person in group 2) with the second person in group 1. The total number of indirect contacts between the second person in group 1 and the fifth person in group 3 is given by

$$\begin{aligned} c_{25} &= a_{21}b_{15} + a_{22}b_{25} + a_{23}b_{35} + a_{24}b_{45} + a_{25}b_{55} + a_{26}b_{65} \\ &= 1 \cdot 1 + 0 \cdot 0 + 0 \cdot 0 + 1 \cdot 1 + 0 \cdot 0 + 1 \cdot 0 = 2 \end{aligned}$$

INDIRECT CONTACT MATRIX: First and third groups:

$$C = AB = \begin{pmatrix} 0 & 0 & 0 & 2 & 0 \\ 1 & 0 & 2 & 0 & 2 \\ 1 & 0 & 0 & 1 & 1 \\ 0 & 0 & 2 & 0 & 1 \end{pmatrix}$$

We observe that only the second person in Group 3 has no indirect contacts with the disease. The fifth person in this group has $2 + 1 + 1 = 4$ indirect contacts. ■

We have seen that for matrix multiplication the commutative law does not hold. The next theorem shows that the associative law does hold.

Theorem 1 ASSOCIATIVE LAW FOR MATRIX MULTIPLICATION Let $A = (a_{ij})$ be an $n \times m$ matrix, $B = (b_{ij})$ an $m \times p$ matrix, and $C = (c_{ij})$ a $p \times q$ matrix. Then the **associative law**

$$A(BC) = (AB)C \tag{3}$$

holds and ABC, defined by either side of (3), is an $n \times q$ matrix.

The proof of this theorem is not difficult, but it is somewhat tedious. For that reason let us defer it until the end of the section.

EXAMPLE 4 Verify the associative law for $A = \begin{pmatrix} 1 & -3 \\ 0 & 2 \end{pmatrix}$, $B = \begin{pmatrix} 2 & -1 & 4 \\ 3 & 1 & 5 \end{pmatrix}$, and $C = \begin{pmatrix} 0 & -2 & 1 \\ 4 & 3 & 2 \\ -5 & 0 & 6 \end{pmatrix}$.

Solution. We first note that A is 2×2, B is 2×3, and C is 3×3. Hence all products used in the statement of the associative law are defined and the resulting product will be a 2×3 matrix. We then calculate

$$AB = \begin{pmatrix} 1 & -3 \\ 0 & 2 \end{pmatrix}\begin{pmatrix} 2 & -1 & 4 \\ 3 & 1 & 5 \end{pmatrix} = \begin{pmatrix} -7 & -4 & -11 \\ 6 & 2 & 10 \end{pmatrix}$$

$$(AB)C = \begin{pmatrix} -7 & -4 & -11 \\ 6 & 2 & 10 \end{pmatrix}\begin{pmatrix} 0 & -2 & 1 \\ 4 & 3 & 2 \\ -5 & 0 & 6 \end{pmatrix} = \begin{pmatrix} 39 & 2 & -81 \\ -42 & -6 & 70 \end{pmatrix}$$

Similarly,

$$BC = \begin{pmatrix} 2 & -1 & 4 \\ 3 & 1 & 5 \end{pmatrix}\begin{pmatrix} 0 & -2 & 1 \\ 4 & 3 & 2 \\ -5 & 0 & 6 \end{pmatrix} = \begin{pmatrix} -24 & -7 & 24 \\ -21 & -3 & 35 \end{pmatrix}$$

$$A(BC) = \begin{pmatrix} 1 & -3 \\ 0 & 2 \end{pmatrix}\begin{pmatrix} -24 & -7 & 24 \\ -21 & -3 & 35 \end{pmatrix} = \begin{pmatrix} 39 & 2 & -81 \\ -42 & -6 & 70 \end{pmatrix}$$

Thus $(AB)C = A(BC)$. ■

From now on we shall write the product of three matrices simply as ABC. We can do this because $(AB)C = A(BC)$; thus we get the same answer no matter how the multiplication is carried out (provided that we do not commute any of the matrices).

The associative law can be extended to longer products. For example, suppose that AB, BC, and CD are defined. Then

$$ABCD = A(B(CD)) = ((AB)C)D = A(BC)D = (AB)(CD) \tag{4}$$

There are two distributive laws for matrix multiplication.

Theorem 2 DISTRIBUTIVE LAWS FOR MATRIX MULTIPLICATION If all the following sums and products are defined, then

$$A(B + C) = AB + AC \tag{5}$$

and

$$(A + B)C = AC + BC. \tag{6}$$

Proofs of Theorems 1 and 2. ASSOCIATIVE LAWS Since A is $n \times m$ and B is $m \times p$, AB is $n \times p$. Thus $(AB)C = (n \times p) \times (p \times q)$ is an $n \times q$ matrix. Similarly BC is $m \times q$ and $A(BC)$ is $n \times q$ so that $(AB)C$ and $A(BC)$ are both of the same size. We must show that the ijth component of $(AB)C$ equals the ijth component of $A(BC)$. Define $D = (d_{ij}) = AB$. Then $d_{ij} = \sum_{k=1}^{m} a_{ik}b_{kj}$. The ijth component of $(AB)C = DC$ is $\sum_{l=1}^{p} d_{il}c_{lj} = \sum_{l=1}^{p}\left(\sum_{k=1}^{m} a_{ik}b_{kl}\right)c_{lj} = \sum_{k=1}^{m}\sum_{l=1}^{p} a_{ik}b_{kl}c_{lj}$. Next we define

$E = (e_{ij}) = BC$. Then $e_{ij} = \sum_{l=1}^{p} b_{il}c_{lj}$ and the ijth component of $A(BC) = AE$ is $\sum_{k=1}^{m} a_{ik}e_{kj} = \sum_{k=1}^{m}\sum_{l=1}^{p} a_{ik}b_{kl}c_{lj}$. Thus the ijth component of $(AB)C$ is equal to the ijth component of $A(BC)$. This proves the associative law. ■

DISTRIBUTIVE LAWS We prove the first distributive law (5). The proof of the second one (6) is virtually identical and is therefore omitted. Let A be $n \times m$ and let B and C be $m \times p$. Then the kjth component of $B + C$ is $b_{kj} + c_{kj}$ and the ijth component of $A(B + C)$ is $\sum_{k=1}^{m} a_{ik}(b_{kj} + c_{kj}) = \sum_{k=1}^{m} a_{ik}b_{kj} + \sum_{k=1}^{m} a_{ik}c_{kj} = ij$th component of AB plus the ijth component of AC and this proves (5). ■

PROBLEMS 7.2

In Problems 1–15 perform the indicated computation.

1. $\begin{pmatrix} 2 & 3 \\ -1 & 2 \end{pmatrix}\begin{pmatrix} 4 & 1 \\ 0 & 6 \end{pmatrix}$

2. $\begin{pmatrix} 3 & -2 \\ 1 & 4 \end{pmatrix}\begin{pmatrix} -5 & 6 \\ 1 & 3 \end{pmatrix}$

3. $\begin{pmatrix} 1 & -1 \\ 1 & 1 \end{pmatrix}\begin{pmatrix} -1 & 0 \\ 2 & 3 \end{pmatrix}$

4. $\begin{pmatrix} -5 & 6 \\ 1 & 3 \end{pmatrix}\begin{pmatrix} 3 & -2 \\ 1 & 4 \end{pmatrix}$

5. $\begin{pmatrix} -4 & 5 & 1 \\ 0 & 4 & 2 \end{pmatrix}\begin{pmatrix} 3 & -1 & 1 \\ 5 & 6 & 4 \\ 0 & 1 & 2 \end{pmatrix}$

6. $\begin{pmatrix} 7 & 1 & 4 \\ 2 & -3 & 5 \end{pmatrix}\begin{pmatrix} 1 & 6 \\ 0 & 4 \\ -2 & 3 \end{pmatrix}$

7. $\begin{pmatrix} 1 & 6 \\ 0 & 4 \\ -2 & 3 \end{pmatrix}\begin{pmatrix} 7 & 1 & 4 \\ 2 & -3 & 5 \end{pmatrix}$

8. $\begin{pmatrix} 1 & 4 & -2 \\ 3 & 0 & 4 \end{pmatrix}\begin{pmatrix} 0 & 1 \\ 2 & 3 \end{pmatrix}$

9. $\begin{pmatrix} 1 & 4 & 6 \\ -2 & 3 & 5 \\ 1 & 0 & 4 \end{pmatrix}\begin{pmatrix} 2 & -3 & 5 \\ 1 & 0 & 6 \\ 2 & 3 & 1 \end{pmatrix}$

10. $\begin{pmatrix} 2 & -3 & 5 \\ 1 & 0 & 6 \\ 2 & 3 & 1 \end{pmatrix}\begin{pmatrix} 1 & 4 & 6 \\ -2 & 3 & 5 \\ 1 & 0 & 4 \end{pmatrix}$

11. $(1 \quad 4 \quad 0 \quad 2)\begin{pmatrix} 3 & -6 \\ 2 & 4 \\ 1 & 0 \\ -2 & 3 \end{pmatrix}$

12. $\begin{pmatrix} 1 \\ 3 \\ 5 \end{pmatrix}(2 \quad -1 \quad 4)$

13. $\begin{pmatrix} 3 \\ -1 \\ 10 \\ 2 \end{pmatrix}(1 \quad 5 \quad -3 \quad 8)$

14. $\begin{pmatrix} 1 & 0 & 0 \\ 0 & 1 & 0 \\ 0 & 0 & 1 \end{pmatrix}\begin{pmatrix} 3 & -2 & 1 \\ 4 & 0 & 6 \\ 5 & 1 & 9 \end{pmatrix}$

15. $\begin{pmatrix} a & b & c \\ d & e & f \\ g & h & j \end{pmatrix}\begin{pmatrix} 1 & 0 & 0 \\ 0 & 1 & 0 \\ 0 & 0 & 1 \end{pmatrix}$, where $a, b, c, d, e, f, g, h, j$ are real numbers.

16. Find a matrix $A = \begin{pmatrix} a & b \\ c & d \end{pmatrix}$ such that $A\begin{pmatrix} 2 & 3 \\ 1 & 2 \end{pmatrix} = \begin{pmatrix} 1 & 0 \\ 0 & 1 \end{pmatrix}$.

***17.** Let a_{11}, a_{12}, a_{21}, and a_{22} be given real numbers such that $a_{11}a_{22} - a_{12}a_{21} \neq 0$. Find numbers b_{11}, b_{12}, b_{21}, and b_{22} such that $\begin{pmatrix} a_{11} & a_{12} \\ a_{21} & a_{22} \end{pmatrix}\begin{pmatrix} b_{11} & b_{12} \\ b_{21} & b_{22} \end{pmatrix} = \begin{pmatrix} 1 & 0 \\ 0 & 1 \end{pmatrix}$.

18. Verify the associative law for multiplication for the matrices $A = \begin{pmatrix} 2 & -1 & 4 \\ 1 & 0 & 6 \end{pmatrix}$, $B = \begin{pmatrix} 1 & 0 & 1 \\ 2 & -1 & 2 \\ 3 & -2 & 0 \end{pmatrix}$, and $C = \begin{pmatrix} 1 & 6 \\ -2 & 4 \\ 0 & 5 \end{pmatrix}$.

19. As in Example 3, suppose that a group of people have contracted a contagious disease. These persons have contacts with a second group who in turn have contacts with a third group. Let $A = \begin{pmatrix} 1 & 0 & 1 & 0 \\ 0 & 1 & 1 & 0 \\ 1 & 0 & 0 & 1 \end{pmatrix}$ represent the contacts between the contagious group and the members of group 2, and let

$$B = \begin{pmatrix} 1 & 0 & 1 & 0 & 0 \\ 0 & 0 & 0 & 1 & 0 \\ 1 & 1 & 0 & 0 & 0 \\ 0 & 0 & 1 & 0 & 1 \end{pmatrix}$$

represent the contacts between groups 2 and 3. (a) How many people are in each group? (b) Find the matrix of indirect contacts between groups 1 and 3.

20. Answer the questions of Problem 19 for $A = \begin{pmatrix} 1 & 0 & 1 & 1 & 0 \\ 0 & 1 & 0 & 1 & 1 \end{pmatrix}$ and

$$B = \begin{pmatrix} 1 & 0 & 0 & 0 & 0 & 0 & 1 \\ 0 & 1 & 0 & 1 & 0 & 0 & 0 \\ 1 & 1 & 0 & 0 & 1 & 1 & 1 \\ 0 & 0 & 0 & 1 & 1 & 0 & 1 \\ 0 & 1 & 0 & 0 & 0 & 0 & 0 \end{pmatrix}.$$

21. A company pays its executives a salary and gives them shares of its stock as an annual bonus. Last year, the president of the company received $80,000 and 50 shares of stock, each of the three vice-presidents was paid $45,000 and 20 shares of stock, and the treasurer was paid $40,000 and 10 shares of stock.

(a) Express the payments to the executives in money and stock by means of a 2×3 matrix.

(b) Express the number of executives of each rank by means of a column vector.

(c) Use matrix multiplication to calculate the total amount of money and the total number of shares of stock the company paid these executives last year.

22. Sales, unit gross profits, and unit taxes for sales of a large corporation are given in the table below. Find a matrix that shows total profits and taxes in each of the four months.

23. Let A be a square matrix. Then A^2 is defined simply as AA. Calculate $\begin{pmatrix} 2 & -1 \\ 4 & 6 \end{pmatrix}^2$.

24. Calculate A^2, where $A = \begin{pmatrix} 1 & -2 & 4 \\ 2 & 0 & 3 \\ 1 & 1 & 5 \end{pmatrix}$.

25. Calculate A^3, where $A = \begin{pmatrix} -1 & 2 \\ 3 & 4 \end{pmatrix}$.

26. Calculate A^2, A^3, A^4, and A^5, where

$$A = \begin{pmatrix} 0 & 1 & 0 & 0 \\ 0 & 0 & 1 & 0 \\ 0 & 0 & 0 & 1 \\ 0 & 0 & 0 & 0 \end{pmatrix}.$$

27. Calculate A^2, A^3, A^4, and A^5, where

$$A = \begin{pmatrix} 0 & 1 & 0 & 0 & 0 \\ 0 & 0 & 1 & 0 & 0 \\ 0 & 0 & 0 & 1 & 0 \\ 0 & 0 & 0 & 0 & 1 \\ 0 & 0 & 0 & 0 & 0 \end{pmatrix}.$$

28. An $n \times n$ matrix A has the property that AB is the zero matrix for any $n \times n$ matrix B. Prove that A is the zero matrix.

29. A **probability matrix** is a square matrix having two properties: (i) every component is nonnegative (≥ 0) and (ii) the sum of the elements in each row is 1. The following are probability matrices:

	Product					
	Sales of Item					
Month	i	ii	iii	Item	Unit Profit (in hundreds of dollars)	Unit Taxes (in hundreds of dollars)
January	4	2	20	i	3.5	1.5
February	6	1	9	ii	2.75	2
March	5	3	12	iii	1.5	0.6
April	8	2.5	20			

$$P = \begin{pmatrix} \frac{1}{3} & \frac{1}{3} & \frac{1}{3} \\ \frac{1}{4} & \frac{1}{2} & \frac{1}{4} \\ 0 & 0 & 1 \end{pmatrix} \quad \text{and} \quad Q = \begin{pmatrix} \frac{1}{6} & \frac{1}{6} & \frac{2}{3} \\ 0 & 1 & 0 \\ \frac{1}{5} & \frac{1}{5} & \frac{3}{5} \end{pmatrix}$$

Show that PQ is a probability matrix.

***30.** Let P be a probability matrix. Show that P^2 is a probability matrix.

****31.** Let P and Q be probability matrices of the same size. Prove that PQ is a probability matrix.

32. Prove formula (4) by using the associative law [equation (3)].

***33.** A round robin tennis tournament can be organized in the following way. Each of the n players plays all the others, and the results are recorded in an $n \times n$ matrix R as follows:

$$R_{ij} = \begin{cases} 1 & \text{if the } i\text{th player beats the } j\text{th player} \\ 0 & \text{if the } i\text{th player loses to the } j\text{th player} \\ 0 & \text{if } i = j \end{cases}$$

The ith player is then assigned the score

$$S_i = \sum_{j=1}^{n} R_{ij} + \frac{1}{2}\sum_{j=1}^{n} (R^2)_{ij}{}^{\dagger}$$

(a) In a tournament between four players

$$R = \begin{pmatrix} 0 & 1 & 0 & 0 \\ 0 & 0 & 1 & 1 \\ 1 & 0 & 0 & 0 \\ 1 & 0 & 1 & 0 \end{pmatrix}.$$

Rank the players according to their scores.

(b) Interpret the meaning of the score.

34. Let O be the $m \times n$ zero matrix and let A be an $n \times p$ matrix. Show that $OA = O_1$, where O_1 is the $m \times p$ zero matrix.

35. Verify the distributive law [equation (5)] for the matrices

$$A = \begin{pmatrix} 1 & 2 & 4 \\ 3 & -1 & 0 \end{pmatrix} \quad B = \begin{pmatrix} 2 & 7 \\ -1 & 4 \\ 6 & 0 \end{pmatrix} \quad C = \begin{pmatrix} -1 & 2 \\ 3 & 7 \\ 4 & 1 \end{pmatrix}.$$

7.3 LINEAR SYSTEMS OF EQUATIONS

In this section we describe a method for finding all solutions (if any) to a system of m linear equations in n unknowns. To see what can happen, we begin by looking at a system of two equations in two unknowns:

$$\begin{aligned} a_{11}x_1 + a_{12}x_2 &= b_1, \\ a_{21}x_1 + a_{22}x_2 &= b_2 \end{aligned} \tag{1}$$

where a_{11}, a_{12}, a_{21}, a_{22}, b_1, and b_2 are given numbers. Each of these equations is the equation of a straight line (in the x_1x_2-plane instead of the xy-plane). The slope of the first line is $-a_{11}/a_{12}$; the slope of the second line is $-a_{21}/a_{22}$ (if $a_{12} \neq 0$ and $a_{22} \neq 0$). A **solution** to system (1) is a pair of numbers, denoted by the vector (x_1, x_2), that satisfies (1). Three questions naturally arise.

(i) Does the system (1) have any solutions?
(ii) If so, how many?
(iii) What are the solutions?

It is easy to answer the first two questions by sketching the graphs of the two lines. There are three possible cases, all of which are illustrated in Figure 1. If the lines are not parallel

---

$^{\dagger}(R^2)_{ij}$ is the ijth component of the matrix R^2.

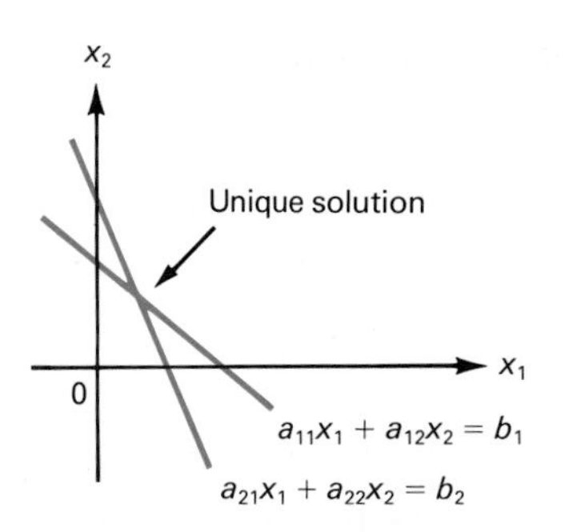

(a) Lines intersecting at one point

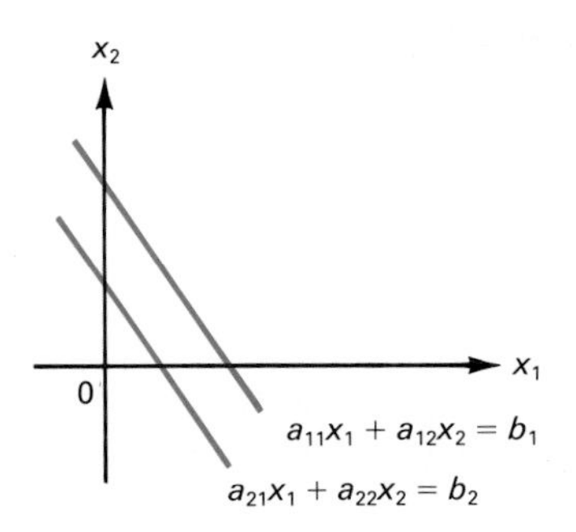

(b) Parallel lines

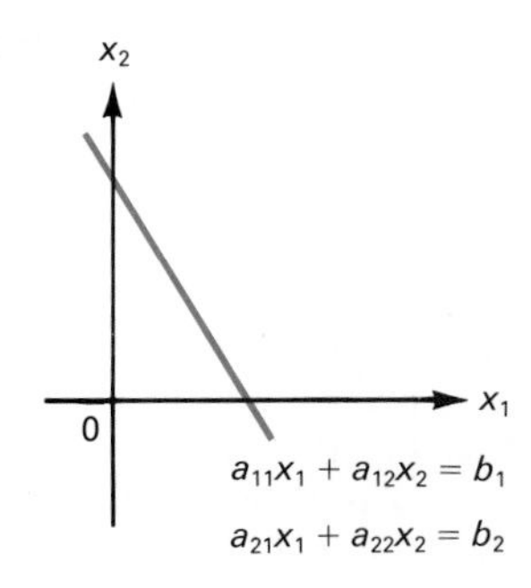

(c) Coincident lines

FIGURE 1

(i.e., if they have different slopes), then they intersect at exactly one point and the system (1) has a unique solution. If they are parallel and not coincident, then they have no points in common, and the system has no solution. Finally, if the lines are coincident, then there are an infinite number of solutions.

We shall soon see that the same is true for a system of m linear equations in n unknowns. That is, there is either a unique solution, no solution, or an infinite number of solutions. Before describing a general method for determining these solutions (if there are any), we look at some simple examples.

EXAMPLE 1 Solve the system

$$\begin{aligned} 2x_1 + 4x_2 + 6x_3 &= 18 \\ 4x_1 + 5x_2 + 6x_3 &= 24 \\ 3x_1 + x_2 - 2x_3 &= 4 \end{aligned} \tag{2}$$

Solution. Here we seek three numbers x_1, x_2, and x_3 such that the three equations in (2) are satisfied. Our method of solution will be to simplify the equations so that solutions can be readily identified. We begin by dividing the first equation by 2. This gives us

$$\begin{aligned} x_1 + 2x_2 + 3x_3 &= 9 \\ 4x_1 + 5x_2 + 6x_3 &= 24 \\ 3x_1 + x_2 - 2x_3 &= 4 \end{aligned} \tag{3}$$

When we add two equations together, we obtain a third, valid equation. This equation may replace either of the two equations used to obtain it in the system. We begin simplifying system (3) by multiplying both sides of the first equation in (3) by -4 and adding this new equation to the second equation. This gives us

$$\begin{array}{r} -4x_1 - 8x_2 - 12x_3 = -36 \\ 4x_1 + 5x_2 + 6x_3 = 24 \\ \hline -3x_2 - 6x_3 = -12 \end{array}$$

The equation $-3x_2 - 6x_3 = -12$ is our new second equation, and the system is now as follows:

$$\begin{aligned} x_1 + 2x_2 + 3x_3 &= 9 \\ -3x_2 - 6x_3 &= -12 \\ 3x_1 + x_2 - 2x_3 &= 4 \end{aligned}$$

We then multiply the first equation by -3 and add it to the third equation:

$$\begin{aligned} x_1 + 2x_2 + 3x_3 &= 9 \\ -3x_2 - 6x_3 &= -12 \\ -5x_2 - 11x_3 &= -23 \end{aligned} \tag{4}$$

Note that in system (4) the variable x_1 has been eliminated from the second and third equations. Next we divide the second equation by -3:

$$\begin{aligned} x_1 + 2x_2 + 3x_3 &= 9 \\ x_2 + 2x_3 &= 4 \\ -5x_2 - 11x_3 &= -23 \end{aligned}$$

We multiply the second equation by -2 and add it to the first and then multiply the second equation by 5 and add it to the third:

$$\begin{aligned} x_1 - x_3 &= 1 \\ x_2 + 2x_3 &= 4 \\ -x_3 &= -3 \end{aligned}$$

We multiply the third equation by -1:

$$\begin{aligned} x_1 - x_3 &= 1 \\ x_2 + 2x_3 &= 4 \\ x_3 &= 3 \end{aligned}$$

Finally, we add the third equation to the first and then multiply the third equation by -2 and add it to the second to obtain the following system, which is equivalent to system (2):

$$\begin{aligned} x_1 &= 4 \\ x_2 &= -2 \\ x_3 &= 3 \end{aligned}$$

This is the unique solution to the system. We write it as the vector $(4, -2, 3)$. The method we used here is called **Gauss–Jordan elimination.**[†]

REMARK. Solving a system of linear equations is often a complicated process, and errors can creep in. However, the final answer is easy to check. Just insert the results into the original equations to verify that they work. In this example, with $x_1 = 4$, $x_2 = -2$ and $x_3 = 3$, we obtain

$$\begin{aligned} 2x_1 + 4x_2 + 6x_3 &= 2(4) + 4(-2) + 6(3) = 8 - 8 + 18 = 18 \\ 4x_1 + 5x_2 + 6x_3 &= 4(4) + 5(-2) + 6(3) = 16 - 10 + 18 = 24 \\ 3x_1 + x_2 - 2x_3 &= 3(4) + (-2) - 2(3) = 12 - 2 - 6 = 4 \end{aligned}$$

Thus equations (2) hold, and our answer is correct. ■

Before going on to another example, let us summarize what we have done in this example:

(i) We divided to make the coefficient of x_1 in the first equation equal to 1.
(ii) We "eliminated" the x_1 terms in the second and third equations. That is, we made the coefficients of these terms equal to zero by multiplying the first equation by appropriate numbers and then adding it to the second and third equations, respectively.
(iii) We divided to make the coefficient of the x_2 term in the second equation equal to 1 and then proceeded to use the second equation to eliminate the x_2 terms in the first and third equations.
(iv) We divided to make the coefficient of the x_3 term in the third equation equal to 1 and then proceeded to use the third equation to eliminate the x_3 terms in the first and second equations.

We emphasize that, at every step, we obtained systems that were equivalent. That is, each system had the same set of solutions as the one that preceded it.

Before solving other systems of equations, we introduce notation that makes it easier to write down each step in our procedure. The coefficients of the variables x_1, x_2, x_3 in system (2) can be written as the entries of a matrix A, called the **coefficient matrix** of the system:

$$A = \begin{pmatrix} 2 & 4 & 6 \\ 4 & 5 & 6 \\ 3 & 1 & -2 \end{pmatrix}$$

Using matrix notation, system (2) can be written as the following **augmented matrix:**

$$\left(\begin{array}{ccc|c} 2 & 4 & 6 & 18 \\ 4 & 5 & 6 & 24 \\ 3 & 1 & -2 & 4 \end{array}\right) \tag{5}$$

[†]Named after the great German mathematician Carl Friedrich Gauss (1777–1855) and the French mathematician Camille Jordan (1838–1922). See the biographical sketch of Gauss on p. 382.

For example, the first row in the augmented matrix (5) is read $2x_1 + 4x_2 + 6x_3 = 18$. Note that each row of the augmented matrix corresponds to one of the equations in the system.

We now introduce some terminology. We have seen that multiplying (or dividing) the sides of an equation by a nonzero number gives us a new, valid equation. Moreover, adding a multiple of one equation to another equation in a system gives us another valid equation. Finally, if we interchange two equations in a system of equations, we obtain an equivalent system. These three operations, when applied to the rows of the augmented matrix representation of a system of equations, are called **elementary row operations.**

To sum up, the three elementary row operations applied to the augmented matrix representation of a system of equations are:

ELEMENTARY ROW OPERATIONS

(i) Multiply (or divide) one row by a nonzero number.
(ii) Add a multiple of one row to another row.
(iii) Interchange two rows.

The process of applying elementary row operations to simplify an augmented matrix is called **row reduction.**

NOTATION:

(i) $M_i(c)$ stands for "multiply the ith row of a matrix by the number c."
(ii) $A_{i,j}(c)$ stands for "multiply the ith row by c and add it to the jth row."
(iii) $P_{i,j}$ stands for "interchange (permute) rows i and j."
(iv) $A \to B$ indicates that the augmented matrices A and B are equivalent; that is, the systems they represent have the same solution.

In Example 1 we saw that by using the elementary row operations (i) and (ii) several times we could obtain a system in which the solutions to the system were given explicitly. We now repeat the steps in Example 1, using the notation just introduced:

$$\left(\begin{array}{ccc|c} 2 & 4 & 6 & 18 \\ 4 & 5 & 6 & 24 \\ 3 & 1 & -2 & 4 \end{array}\right) \xrightarrow{M_1(\frac{1}{2})} \left(\begin{array}{ccc|c} 1 & 2 & 3 & 9 \\ 4 & 5 & 6 & 24 \\ 3 & 1 & -2 & 4 \end{array}\right) \xrightarrow[A_{1,3}(-3)]{A_{1,2}(-4)} \left(\begin{array}{ccc|c} 1 & 2 & 3 & 9 \\ 0 & -3 & -6 & -12 \\ 0 & -5 & -11 & -23 \end{array}\right)$$

$$\xrightarrow{M_2(-\frac{1}{3})} \left(\begin{array}{ccc|c} 1 & 2 & 3 & 9 \\ 0 & 1 & 2 & 4 \\ 0 & -5 & -11 & -23 \end{array}\right) \xrightarrow[A_{2,3}(5)]{A_{2,1}(-2)} \left(\begin{array}{ccc|c} 1 & 0 & -1 & 1 \\ 0 & 1 & 2 & 4 \\ 0 & 0 & -1 & -3 \end{array}\right)$$

$$\xrightarrow{M_3(-1)} \left(\begin{array}{ccc|c} 1 & 0 & -1 & 1 \\ 0 & 1 & 2 & 4 \\ 0 & 0 & 1 & 3 \end{array}\right) \xrightarrow[A_{3,2}(-2)]{A_{3,1}(1)} \left(\begin{array}{ccc|c} 1 & 0 & 0 & 4 \\ 0 & 1 & 0 & -2 \\ 0 & 0 & 1 & 3 \end{array}\right)$$

Again we can easily "see" the solution $x_1 = 4$, $x_2 = -2$, $x_3 = 3$. ■

EXAMPLE 2 Solve the following system:

$$\begin{aligned} 2x_1 + 4x_2 + 6x_3 &= 18 \\ 4x_1 + 5x_2 + 6x_3 &= 24 \\ 2x_1 + 7x_2 + 12x_3 &= 30 \end{aligned}$$

Solution. We proceed as in Example 1, first writing the system as an augmented matrix:

$$\left(\begin{array}{ccc|c} 2 & 4 & 6 & 18 \\ 4 & 5 & 6 & 24 \\ 2 & 7 & 12 & 30 \end{array}\right)$$

We then obtain, successively,

$$\xrightarrow{M_1(\frac{1}{2})} \left(\begin{array}{ccc|c} 1 & 2 & 3 & 9 \\ 4 & 5 & 6 & 24 \\ 2 & 7 & 12 & 30 \end{array}\right) \xrightarrow[A_{1,3}(-2)]{A_{1,2}(-4)} \left(\begin{array}{ccc|c} 1 & 2 & 3 & 9 \\ 0 & -3 & -6 & -12 \\ 0 & 3 & 6 & 12 \end{array}\right)$$

$$\xrightarrow{M_2(-\frac{1}{3})} \left(\begin{array}{ccc|c} 1 & 2 & 3 & 9 \\ 0 & 1 & 2 & 4 \\ 0 & 3 & 6 & 12 \end{array}\right) \xrightarrow[A_{2,3}(-3)]{A_{2,1}(-2)} \left(\begin{array}{ccc|c} 1 & 0 & -1 & 1 \\ 0 & 1 & 2 & 4 \\ 0 & 0 & 0 & 0 \end{array}\right)$$

This is equivalent to the system of equations

$$\begin{aligned} x_1 - x_3 &= 1 \\ x_2 + 2x_3 &= 4 \end{aligned}$$

This is as far as we can go. There are now only two equations in the three unknowns x_1, x_2, x_3 and there are an infinite number of solutions. To see this, let x_3 be chosen. Then $x_2 = 4 - 2x_3$ and $x_1 = 1 + x_3$. This will be a solution for any number x_3. We write these solutions in the form $(1 + x_3, 4 - 2x_3, x_3)$. For example, if $x_3 = 0$ we obtain the solution $(1, 4, 0)$. For $x_3 = 10$ we obtain the solution $(11, -16, 10)$. ■

EXAMPLE 3 Solve the system

$$\begin{aligned} 2x_1 + 4x_2 + 6x_3 &= 18 \\ 4x_1 + 5x_2 + 6x_3 &= 24 \\ 2x_1 + 7x_2 + 12x_3 &= 40 \end{aligned} \tag{6}$$

Solution. We use the augmented-matrix form and proceed exactly as in Example 2 to obtain, successively, the following systems. (Note how, in each step, we use either elementary row operation (i) or (ii).)

$$\begin{pmatrix} 2 & 4 & 6 & | & 18 \\ 4 & 5 & 6 & | & 24 \\ 2 & 7 & 12 & | & 40 \end{pmatrix} \xrightarrow{M_1(\frac{1}{2})} \begin{pmatrix} 1 & 2 & 3 & | & 9 \\ 4 & 5 & 6 & | & 24 \\ 2 & 7 & 12 & | & 40 \end{pmatrix}$$

$$\xrightarrow[A_{1,3}(-2)]{A_{1,2}(-4)} \begin{pmatrix} 1 & 2 & 3 & | & 9 \\ 0 & -3 & -6 & | & -12 \\ 0 & 3 & 6 & | & 22 \end{pmatrix} \xrightarrow{M_2(-\frac{1}{3})} \begin{pmatrix} 1 & 2 & 3 & | & 9 \\ 0 & 1 & 2 & | & 4 \\ 0 & 3 & 6 & | & 22 \end{pmatrix}$$

$$\xrightarrow[A_{2,3}(-3)]{A_{2,1}(-2)} \begin{pmatrix} 1 & 0 & -1 & | & 1 \\ 0 & 1 & 2 & | & 4 \\ 0 & 0 & 0 & | & 10 \end{pmatrix} \xrightarrow{M_3(\frac{1}{10})} \begin{pmatrix} 1 & 0 & -1 & | & 1 \\ 0 & 1 & 2 & | & 4 \\ 0 & 0 & 0 & | & 1 \end{pmatrix}$$

The last equation now reads $0x_1 + 0x_2 + 0x_3 = 1$, which is impossible since $0 \neq 1$. Thus system (6) has *no* solution. ■

Let us take another look at these three examples. In Example 1 we began with the matrix

$$A_1 = \begin{pmatrix} 2 & 4 & 6 \\ 4 & 5 & 6 \\ 3 & 1 & -2 \end{pmatrix}.$$

In the process of row reduction A_1 was "reduced" to the matrix

$$R_1 = \begin{pmatrix} 1 & 0 & 0 \\ 0 & 1 & 0 \\ 0 & 0 & 1 \end{pmatrix}.$$

In Example 2 we started with

$$A_2 = \begin{pmatrix} 2 & 4 & 6 \\ 4 & 5 & 6 \\ 2 & 7 & 12 \end{pmatrix}$$

and ended up with

$$R_2 = \begin{pmatrix} 1 & 0 & -1 \\ 0 & 1 & 2 \\ 0 & 0 & 0 \end{pmatrix}.$$

In Example 3 we began with

$$A_3 = \begin{pmatrix} 2 & 4 & 6 \\ 4 & 5 & 6 \\ 2 & 7 & 12 \end{pmatrix}$$

and again ended up with

$$R_3 = \begin{pmatrix} 1 & 0 & -1 \\ 0 & 1 & 2 \\ 0 & 0 & 0 \end{pmatrix}.$$

The matrices R_1, R_2, and R_3 are called the *reduced row echelon forms* of the matrices A_1, A_2, and A_3, respectively. We have the following definition:

Definition 1 REDUCED ROW ECHELON FORM A matrix is in **reduced row echelon form** if the following four conditions hold:

- **(i)** Any row (if any) consisting entirely of zeros appears at the bottom of the matrix.
- **(ii)** The first nonzero number (starting from the left) in any row not consisting entirely of zeros is 1.
- **(iii)** If two successive rows do not consist entirely of zeros, then the first 1 in the lower row occurs farther to the right than the first 1 in the higher row.
- **(iv)** Any column containing the first 1 in a row has zeros everywhere else.

EXAMPLE 4 The following matrices are in reduced row echelon form:

(a) $\begin{pmatrix} 1 & 0 & 0 \\ 0 & 1 & 0 \\ 0 & 0 & 1 \end{pmatrix}$ **(b)** $\begin{pmatrix} 1 & 0 & 0 & 0 \\ 0 & 1 & 0 & 0 \\ 0 & 0 & 0 & 1 \end{pmatrix}$ **(c)** $\begin{pmatrix} 1 & 0 & 0 & 5 \\ 0 & 0 & 1 & 2 \end{pmatrix}$

(d) $\begin{pmatrix} 1 & 0 \\ 0 & 1 \end{pmatrix}$ **(e)** $\begin{pmatrix} 1 & 0 & 2 & 5 \\ 0 & 1 & 3 & 6 \\ 0 & 0 & 0 & 0 \end{pmatrix}$ ■

EXAMPLE 5 The following matrices are not in reduced row echelon form.

(a) $\begin{pmatrix} 0 & 0 & 0 \\ 0 & 1 & 0 \\ 0 & 0 & 1 \end{pmatrix}$ Condition (i) is violated.

(b) $\begin{pmatrix} 1 & 0 & 0 \\ 0 & 2 & 0 \\ 0 & 0 & 1 \end{pmatrix}$ Condition (ii) is violated.

(c) $\begin{pmatrix} 0 & 1 & 0 \\ 1 & 0 & 0 \\ 0 & 0 & 1 \end{pmatrix}$ Condition (iii) is violated.

(d) $\begin{pmatrix} 1 & 0 & 0 \\ 0 & 1 & 2 \\ 0 & 0 & 1 \end{pmatrix}$ Condition (iv) is violated. ■

Definition 2 ROW ECHELON FORM A matrix is in **row echelon form** if conditions (i), (ii), and (iii) of Definition 1 hold.

EXAMPLE 6 The following matrices are in row echelon form:

$$\textbf{(a)}\ \begin{pmatrix} 1 & 2 & 3 \\ 0 & 1 & 5 \\ 0 & 0 & 1 \end{pmatrix} \qquad \textbf{(b)}\ \begin{pmatrix} 1 & -1 & 6 & 4 \\ 0 & 1 & 2 & -8 \\ 0 & 0 & 0 & 1 \end{pmatrix}$$

$$\textbf{(c)}\ \begin{pmatrix} 1 & 0 & 2 & 5 \\ 0 & 0 & 1 & 2 \end{pmatrix} \qquad \textbf{(d)}\ \begin{pmatrix} 1 & 2 \\ 0 & 1 \end{pmatrix} \qquad \textbf{(e)}\ \begin{pmatrix} 1 & 3 & 2 & 5 \\ 0 & 1 & 3 & 6 \\ 0 & 0 & 0 & 0 \end{pmatrix} \blacksquare$$

REMARK 1. The difference between row echelon form and reduced row echelon form should be clear from the examples. In row echelon form, all the numbers below the first 1 in a row are zero. In reduced row echelon form, all the numbers above and below the first 1 in a row are zero. Thus reduced row echelon form is more exclusive. That is, every matrix in reduced row echelon form is in row echelon form, but not conversely.

REMARK 2. We can always reduce a matrix to reduced row echelon form or row echelon form by performing elementary row operations. We saw this reduction to reduced row echelon form in Examples 1, 2, and 3.

As we saw in Examples 1, 2, and 3, there is a strong connection between the reduced row echelon form of a matrix and the existence of a unique solution to the system. In Example 1, the reduced row echelon form of the *coefficient matrix* (that is, the first three columns of the augmented matrix) had a 1 in each row and there was a unique solution. In Examples 2 and 3, the reduced row echelon form of the coefficient matrix had a row of zeros and the system had either no solution or an infinite number of solutions. This turns out always to be true in any system with the same number of equations as unknowns. But before turning to the general case, let us discuss the usefulness of the row echelon form of a matrix. It is possible to solve the system in Example 1 by reducing the coefficient matrix to its row echelon form.

EXAMPLE 7 Solve the system of Example 1 by reducing the coefficient matrix to row echelon form.

Solution. We begin as before:

$$\left(\begin{array}{ccc|c} 2 & 4 & 6 & 18 \\ 4 & 5 & 6 & 24 \\ 3 & 1 & -2 & 4 \end{array}\right) \xrightarrow{M_1(\frac{1}{2})} \left(\begin{array}{ccc|c} 1 & 2 & 3 & 9 \\ 4 & 5 & 6 & 24 \\ 3 & 1 & -2 & 4 \end{array}\right) \xrightarrow[A_{1,3}(-3)]{A_{1,2}(-4)}$$

$$\left(\begin{array}{ccc|c} 1 & 2 & 3 & 9 \\ 0 & -3 & -6 & -12 \\ 0 & -5 & -11 & -23 \end{array}\right) \xrightarrow{M_2(-\frac{1}{3})} \left(\begin{array}{ccc|c} 1 & 2 & 3 & 9 \\ 0 & 1 & 2 & 4 \\ 0 & -5 & -11 & -23 \end{array}\right)$$

So far, this process is identical to our earlier one. Now, however, we only make zero the number (-5) below the first 1 in the second row:

$$\xrightarrow{A_{2,3}(5)} \left(\begin{array}{ccc|c} 1 & 2 & 3 & 9 \\ 0 & 1 & 2 & 4 \\ 0 & 0 & -1 & -3 \end{array}\right) \xrightarrow{M_3(-1)} \left(\begin{array}{ccc|c} 1 & 2 & 3 & 9 \\ 0 & 1 & 2 & 4 \\ 0 & 0 & 1 & 3 \end{array}\right)$$

The augmented matrix of the system (and the coefficient matrix) are now in row echelon form and we immediately see that $x_3 = 3$. We then use **back substitution** to solve for x_2 and then x_1. The second equation reads $x_2 + 2x_3 = 4$. Thus $x_2 + 2(3) = 4$ and $x_2 = -2$. Similarly, from the first equation we obtain $x_1 + 2(-2) + 3(3) = 9$ or $x_1 = 4$. Thus we again obtain the solution $(4, -2, 3)$. The method of solution just employed is called **Gaussian elimination.** ■

We therefore have two methods for solving our sample systems of equations:

(i) GAUSS-JORDAN ELIMINATION
Row reduce the coefficient matrix to reduced row echelon form.
(ii) GAUSSIAN ELIMINATION
Row reduce the coefficient matrix to row echelon form, solve for the last unknown, and then use back substitution to solve for the other unknowns.

Which method is more useful? It depends. In solving systems of equations on a computer, Gaussian elimination is the preferred method because it involves fewer elementary row operations. On the other hand, there are times when it is essential to obtain the reduced row echelon form of a matrix (one of these is discussed in Section 7.5). In these cases Gauss-Jordan elimination is the preferred method.

We now turn to the solution of a general system of m equations in n unknowns. Because of our need to do so in Section 7.5, we shall be solving most of the systems by Gauss-Jordan elimination. Keep in mind, however, that Gaussian elimination is sometimes the preferred approach.

The general $m \times n$ system of m linear equations in n unknowns is given by

$$\begin{aligned} a_{11}x_1 + a_{12}x_2 + a_{13}x_3 + \cdots + a_{1n}x_n &= b_1 \\ a_{21}x_1 + a_{22}x_2 + a_{23}x_3 + \cdots + a_{2n}x_n &= b_2 \\ a_{31}x_1 + a_{32}x_2 + a_{33}x_3 + \cdots + a_{3n}x_n &= b_3 \\ \vdots \qquad \vdots \qquad \vdots \qquad \qquad \vdots \quad &\quad \vdots \\ a_{m1}x_1 + a_{m2}x_2 + a_{m3}x_3 + \cdots + a_{mn}x_n &= b_m \end{aligned} \tag{7}$$

In system (7) all the a's and b's are given real numbers. The problem is to find all sets of n numbers, denoted $(x_1, x_2, x_3, \ldots, x_n)$, that satisfy each of the m equations in (7). The number a_{ij} is the coefficient of the variable x_j in that ith equation.

We solve system (7) by writing the system as an augmented matrix and row-reducing the matrix to its reduced row echelon form. We start by dividing the first row by a_{11} (elementary row operation (i)). If $a_{11} = 0$ then we rearrange[†] the equations so

[†]To rearrange a system of equations we simply write the same equations in a different order. For example, the first equation can become the fourth equation, the third equation can become the second equation, and so on. This is a sequence of elementary row operations (iii).

that, with rearrangement, the new $a_{11} \neq 0$. We then use the first equation to eliminate the x_1 term in each of the other equations (using elementary row operation (ii)). Then the new second equation is divided by the new a_{22} term and the new, new second equation is used to eliminate the x_2 terms in all the other equations. The process is continued until one of three situations occurs:

(i) The last nonzero[‡] equation reads $x_n = c$ for some constant c. Then there is either a unique solution or an infinite number of solutions to the system.
(ii) The last nonzero equation reads $x_j + a'_{i,j+1}x_{j+1} + \cdots + a'_{i,j+k}x_n = c$ for some constant c where at least two of the a's are nonzero. That is, the last equation is a linear equation in two or more of the variables. Then there are an infinite number of solutions.
(iii) The last equation reads $0 = c$, where $c \neq 0$. Then there is no solution. In this case the system is called **inconsistent.** In cases (i) and (ii) the system is called **consistent.**

EXAMPLE 8 Solve the system

$$\begin{aligned} x_1 + 3x_2 - 5x_3 + x_4 &= 4 \\ 2x_1 + 5x_2 - 2x_3 + 4x_4 &= 6 \end{aligned}$$

Solution. We write this system as an augmented matrix and row reduce:

$$\left(\begin{array}{cccc|c} 1 & 3 & -5 & 1 & 4 \\ 2 & 5 & -2 & 4 & 6 \end{array}\right) \xrightarrow{A_{1,2}(-2)} \left(\begin{array}{cccc|c} 1 & 3 & -5 & 1 & 4 \\ 0 & -1 & 8 & 2 & -2 \end{array}\right) \xrightarrow{M_2(-1)}$$

$$\left(\begin{array}{cccc|c} 1 & 3 & -5 & 1 & 4 \\ 0 & 1 & -8 & -2 & 2 \end{array}\right) \xrightarrow{A_{2,1}(-3)} \left(\begin{array}{cccc|c} 1 & 0 & 19 & 7 & -2 \\ 0 & 1 & -8 & -2 & 2 \end{array}\right)$$

This is as far as we can go. The coefficient matrix is in reduced row echelon form—case (ii) above. There are evidently an infinite number of solutions. The variables x_3 and x_4 can be chosen arbitrarily. Then $x_2 = 2 + 8x_3 + 2x_4$ and $x_1 = -2 - 19x_3 - 7x_4$. All solutions are, therefore, represented by $(-2 - 19x_3 - 7x_4, 2 + 8x_3 + 2x_4, x_3, x_4)$. For example, if $x_3 = 1$ and $x_4 = 2$, we obtain the solution $(-35, 14, 1, 2)$. ■

As you will see if you do a lot of system solving, the computations can become very messy. It is a good rule of thumb to use a calculator whenever the fractions become unpleasant. It should be noted, however, that if computations are carried out on a computer or calculator, "round-off" errors can be introduced. This problem is not discussed in this text.[§]

[‡]The "zero equation" is the equation $0 = 0$.

[§]See, for example, Section 8.1 of S. I. Grossman, *Elementary Linear Algebra,* second edition, (Wadsworth, Belmont, CA, 1984).

HOMOGENEOUS SYSTEM OF EQUATIONS

A linear system in which all the constant terms—(the b_i's in (7))—are zero is called **homogeneous.**

EXAMPLE 9 Solve the following homogeneous system:

$$\begin{aligned} 2x_1 + 4x_2 + 6x_3 &= 0 \\ 4x_1 + 5x_2 + 6x_3 &= 0 \\ 3x_1 + x_2 - 2x_3 &= 0 \end{aligned}$$

Solution. This is the homogeneous version of the system in Example 7.3.1. Reducing successively, we obtain (after dividing the first equation by 2)

$$\left(\begin{array}{ccc|c} 1 & 2 & 3 & 0 \\ 4 & 5 & 6 & 0 \\ 3 & 1 & -2 & 0 \end{array}\right) \xrightarrow[A_{1,3}(-3)]{A_{1,2}(-4)} \left(\begin{array}{ccc|c} 1 & 2 & 3 & 0 \\ 0 & -3 & -6 & 0 \\ 0 & -5 & -11 & 0 \end{array}\right) \xrightarrow{M_2(-\frac{1}{3})} \left(\begin{array}{ccc|c} 1 & 2 & 3 & 0 \\ 0 & 1 & 2 & 0 \\ 0 & -5 & -11 & 0 \end{array}\right)$$

$$\xrightarrow[A_{2,3}(5)]{A_{2,1}(-2)} \left(\begin{array}{ccc|c} 1 & 0 & -1 & 0 \\ 0 & 1 & 2 & 0 \\ 0 & 0 & -1 & 0 \end{array}\right) \xrightarrow{M_3(-1)} \left(\begin{array}{ccc|c} 1 & 0 & -1 & 0 \\ 0 & 1 & 2 & 0 \\ 0 & 0 & 1 & 0 \end{array}\right) \xrightarrow[A_{3,2}(-2)]{A_{3,1}(1)} \left(\begin{array}{ccc|c} 1 & 0 & 0 & 0 \\ 0 & 1 & 0 & 0 \\ 0 & 0 & 1 & 0 \end{array}\right)$$

Thus the system has the unique solution (0, 0, 0). This solution is called the **trivial** or **zero** solution. ■

EXAMPLE 10 Solve the following homogeneous system:

$$\begin{aligned} x_1 + 2x_2 - x_3 &= 0 \\ 3x_1 - 3x_2 + 2x_3 &= 0 \\ -x_1 - 11x_2 + 6x_3 &= 0 \end{aligned}$$

Solution. Using Gauss-Jordan elimination we obtain, successively,

$$\left(\begin{array}{ccc|c} 1 & 2 & -1 & 0 \\ 3 & -3 & 2 & 0 \\ -1 & -11 & 6 & 0 \end{array}\right) \xrightarrow[A_{1,3}(1)]{A_{1,2}(-3)} \left(\begin{array}{ccc|c} 1 & 2 & -1 & 0 \\ 0 & -9 & 5 & 0 \\ 0 & -9 & 5 & 0 \end{array}\right)$$

$$\xrightarrow{M_2(-\frac{1}{9})} \left(\begin{array}{ccc|c} 1 & 2 & -1 & 0 \\ 0 & 1 & -\frac{5}{9} & 0 \\ 0 & -9 & 5 & 0 \end{array}\right) \xrightarrow[A_{2,3}(9)]{A_{2,1}(-2)} \left(\begin{array}{ccc|c} 1 & 0 & \frac{1}{9} & 0 \\ 0 & 1 & -\frac{5}{9} & 0 \\ 0 & 0 & 0 & 0 \end{array}\right)$$

The augmented matrix is now in reduced row echelon form and, evidently, there are an infinite number of solutions given by $(-\frac{1}{9}x_3, \frac{5}{9}x_3, x_3)$. If $x_3 = 0$, for example, we obtain the trivial solution. If $x_3 = 1$ we obtain the solution $(-\frac{1}{9}, \frac{5}{9}, 1)$. ■

EXAMPLE 11 Solve the system

$$\begin{aligned} x_1 + x_2 - x_3 &= 0 \\ 4x_1 - 2x_2 + 7x_3 &= 0 \end{aligned}$$

Solution. Row-reducing, we obtain

$$\left(\begin{array}{ccc|c} 1 & 1 & -1 & 0 \\ 4 & -2 & 7 & 0 \end{array}\right) \xrightarrow{A_{1,2}(-4)} \left(\begin{array}{ccc|c} 1 & 1 & -1 & 0 \\ 0 & -6 & 11 & 0 \end{array}\right)$$

$$\xrightarrow{M_2(-\frac{1}{6})} \left(\begin{array}{ccc|c} 1 & 1 & -1 & 0 \\ 0 & 1 & -\frac{11}{6} & 0 \end{array}\right) \xrightarrow{A_{2,1}(-1)} \left(\begin{array}{ccc|c} 1 & 0 & \frac{5}{6} & 0 \\ 0 & 1 & -\frac{11}{6} & 0 \end{array}\right)$$

Thus there are an infinite number of solutions given by $(-\frac{5}{6}x_3, \frac{11}{6}x_3, x_3)$. This is not surprising since the system contains three unknowns and only two equations. ■

We can generalize Example 10 to obtain the following useful result:

Theorem 1 A homogeneous system with more unknowns than equations has an infinite number of solutions.

Proof. Suppose, in system (7), that $n > m$ and $b_1 = b_2 = \cdots = b_m = 0$ so that the system is homogeneous with more unknowns than equations. Evidently the trivial solution is a solution. Suppose that it is the only one. Then row reduction would lead us to the system

$$\begin{aligned} x_1 \qquad\qquad &= 0 \\ x_2 \qquad &= 0 \\ &\vdots \\ x_n &= 0 \end{aligned}$$

and, possibly, additional equations of the form $0 = 0$. But this system has at least as many equations as unknowns. Since row reduction does not change either the number of equations or the number of unknowns, we have a contradiction of our assumption that there were more unknowns than equations. ■

We close this section with three examples illustrating how a system of linear equations can arise in a practical situation.

EXAMPLE 12 A model that is often used in economics is the **Leontief input-output model.**[†] Suppose an economic system has n industries. There are two kinds of demands on each industry. First there is the *external* demand from outside the system. If the system is

[†]Named after American economist Wassily W. Leontief. This model was used in his pioneering paper "Qualitative Input and Output Relations in the Economic System of the United States" in *Review of Economic Statistics* **18,** 105–125 (1936). An updated version of this model appears in Leontief's book *Input-Output Analysis* (Oxford University Press, New York, 1966). Leontief won the Nobel Prize in economics in 1973 for his development of input-output analysis.

a country, for example, then the external demand could be from another country. Second there is the demand placed on one industry by another industry in the same system. In the United States, for example, there is a demand on the output of the steel industry by the automobile industry.

Let e_i represent the external demand placed on the ith industry. Let a_{ij} represent the internal demand placed on the ith industry by the jth industry. More precisely, a_{ij} represents the number of units of the output of industry i needed to produce one unit of the output of industry j. Let x_i represent the output of industry i. Now we assume that the output of each industry is equal to its demand (that is, there is no overproduction). The total demand is equal to the sum of the internal and external demands. To calculate the internal demand on industry 2, for example, we note that $a_{21}x_1$ is the demand on industry 2 made by industry 1. Thus the total internal demand on industry 2 is $a_{21}x_1 + a_{22}x_2 + \cdots + a_{2n}x_n$.

We are led to the following system of equations obtained by equating the total demand with the output of each industry:

$$\begin{aligned} a_{11}x_1 + a_{12}x_2 + \cdots + a_{1n}x_n + e_1 &= x_1 \\ a_{21}x_1 + a_{22}x_2 + \cdots + a_{2n}x_n + e_2 &= x_2 \\ \vdots \quad\quad \vdots \quad\quad\quad\quad \vdots \quad\quad \vdots \quad &\quad \vdots \\ a_{n1}x_1 + a_{n2}x_2 + \cdots + a_{nn}x_n + e_n &= x_n \end{aligned} \tag{8}$$

Or, rewriting (8) so it looks like system (7), we get

$$\begin{aligned} (1 - a_{11})x_1 - a_{12}x_2 - \cdots - a_{1n}x_n &= e_1 \\ -a_{21}x_1 + (1 - a_{22})x_2 - \cdots - a_{2n}x_n &= e_2 \\ \vdots \quad\quad\quad\quad \vdots \quad\quad\quad\quad \vdots \quad &\quad \vdots \\ -a_{n1}x_1 - a_{n2}x_2 - \cdots + (1 - a_{nn})x_n &= e_n \end{aligned} \tag{9}$$

System (9) of n equations in n unknowns is very important in economic analysis.

EXAMPLE 13 In an economic system of three industries, suppose that the external demands are, respectively, 10, 25, and 20. Suppose that $a_{11} = 0.2$, $a_{12} = 0.5$, $a_{13} = 0.15$, $a_{21} = 0.4$, $a_{22} = 0.1$, $a_{23} = 0.3$, $a_{31} = 0.25$, $a_{32} = 0.5$, and $a_{33} = 0.15$. Find the output in each industry such that supply exactly equals demand.

Solution. Here $n = 3$, $1 - a_{11} = 0.8$, $1 - a_{22} = 0.9$, and $1 - a_{33} = 0.85$. Then system (9) is

$$\begin{aligned} 0.8x_1 - 0.5x_2 - 0.15x_3 &= 10 \\ -0.4x_1 + 0.9x_2 - 0.3x_3 &= 25 \\ -0.25x_1 - 0.5x_2 + 0.85x_3 &= 20 \end{aligned}$$

Solving this system by using a calculator, we obtain successively (using five-decimal-place accuracy and Gauss-Jordan elimination)

$$\left(\begin{array}{ccc|c} 0.8 & -0.5 & -0.15 & 10 \\ -0.4 & 0.9 & -0.3 & 25 \\ -0.25 & -0.5 & 0.85 & 20 \end{array}\right) \xrightarrow{M_1(\frac{1}{0.8})} \left(\begin{array}{ccc|c} 1 & -0.625 & -0.1875 & 12.5 \\ -0.4 & 0.9 & -0.3 & 25 \\ -0.25 & -0.5 & 0.85 & 20 \end{array}\right)$$

$$\xrightarrow[A_{1,3}(0.25)]{A_{1,2}(0.4)} \left(\begin{array}{ccc|c} 1 & -0.625 & -0.1875 & 12.5 \\ 0 & 0.65 & -0.375 & 30 \\ 0 & -0.65625 & 0.80313 & 23.125 \end{array}\right)$$

$$\xrightarrow{M_2(\frac{1}{0.65})} \left(\begin{array}{ccc|c} 1 & -0.625 & -0.1875 & 12.5 \\ 0 & 1 & -0.57692 & 46.15385 \\ 0 & -0.65625 & 0.80313 & 23.125 \end{array}\right)$$

$$\xrightarrow[A_{2,3}(0.65625)]{A_{2,1}(0.625)} \left(\begin{array}{ccc|c} 1 & 0 & -0.54808 & 41.34616 \\ 0 & 1 & -0.57692 & 46.15385 \\ 0 & 0 & 0.42453 & 53.41346 \end{array}\right)$$

$$\xrightarrow{M_3(1/0.42453)} \left(\begin{array}{ccc|c} 1 & 0 & -0.54808 & 41.34616 \\ 0 & 1 & -0.57692 & 46.15385 \\ 0 & 0 & 1 & 125.81787 \end{array}\right)$$

$$\xrightarrow[A_{3,2}(0.57692)]{A_{3,1}(0.54808)} \left(\begin{array}{ccc|c} 1 & 0 & 0 & 110.30442 \\ 0 & 1 & 0 & 118.74070 \\ 0 & 0 & 1 & 125.81787 \end{array}\right)$$

We conclude that the outputs needed for supply to equal demand are, approximately, $x_1 = 110$, $x_2 = 119$, and $x_3 = 126$. ■

EXAMPLE 14 A State Fish and Game Department supplies three types of food to a lake that supports three species of fish. Each fish of Species 1 consumes, each week, an average of one unit of Food 1, one unit of Food 2, and two units of Food 3. Each fish of Species 2 consumes, each week, an average of three units of Food 1, four units of Food 2, and five units of Food 3. For a fish of Species 3, the average weekly consumption is two units of Food 1, one unit of Food 2, and five units of Food 3. Each week 25,000 units of Food 1, 20,000 units of Food 2, and 55,000 units of Food 3 are supplied to the lake. If we assume that all food is eaten, how many fish of each species can coexist in the lake?

Solution. We let x_1, x_2, and x_3 denote the numbers of fish of the three species being supported by the lake environment. Using the information in the problem, we see that x_1 fish of Species 1 consume x_1 units of Food 1, x_2 fish of Species 2 consume $3x_2$ units of Food 1, and x_3 fish of Species 3 consume $2x_3$ units of Food 1. Thus $x_1 + 3x_2 + 2x_3 = 25{,}000$ = total weekly supply of Food 1. Obtaining a similar equation for each of the other two foods, we are led to the following system:

$$\begin{aligned} x_1 + 3x_2 + 2x_3 &= 25{,}000 \\ x_1 + 4x_2 + x_3 &= 20{,}000 \\ 2x_1 + 5x_2 + 5x_3 &= 55{,}000 \end{aligned}$$

Upon solving, we obtain

$$\left(\begin{array}{ccc|c} 1 & 3 & 2 & 25{,}000 \\ 1 & 4 & 1 & 20{,}000 \\ 2 & 5 & 5 & 55{,}000 \end{array}\right)$$

$$\xrightarrow[A_{1,3}(-2)]{A_{1,2}(-1)} \left(\begin{array}{ccc|c} 1 & 3 & 2 & 25{,}000 \\ 0 & 1 & -1 & -5{,}000 \\ 0 & -1 & 1 & 5{,}000 \end{array}\right) \xrightarrow[A_{2,3}(1)]{A_{2,1}(-3)} \left(\begin{array}{ccc|c} 1 & 0 & 5 & 40{,}000 \\ 0 & 1 & -1 & -5{,}000 \\ 0 & 0 & 0 & 0 \end{array}\right)$$

Thus, if x_3 is chosen arbitrarily, we have an infinite number of solutions given by $(40{,}000 - 5x_3, x_3 - 5{,}000, x_3)$. Of course, we must have $x_1 \geq 0$, $x_2 \geq 0$ and $x_3 \geq 0$. Since $x_2 = x_3 - 5{,}000 \geq 0$, we have $x_3 \geq 5{,}000$. This means that $0 \leq x_1 \leq 40{,}000 - 5(5{,}000) = 15{,}000$. Finally, since $40{,}000 - 5x_3 \geq 0$, we see that $x_3 \leq 8{,}000$. This means that the populations that can be supported by the lake with all food consumed are

$$x_1 = 40{,}000 - 5x_3$$

$$x_2 = x_3 - 5{,}000$$

$$5{,}000 \leq x_3 \leq 8{,}000$$

For example, if $x_3 = 6{,}000$, then $x_1 = 10{,}000$ and $x_2 = 1{,}000$. ■

PROBLEMS 7.3

In Problems 1–36, use Gauss-Jordan or Gaussian elimination to find all solutions, if any, to the given systems.

1. $\begin{aligned} x_1 - 3x_2 &= 4 \\ -4x_1 + 2x_2 &= 6 \end{aligned}$

2. $\begin{aligned} 2x_1 - x_2 &= -3 \\ 5x_1 + 7x_2 &= 4 \end{aligned}$

3. $\begin{aligned} 2x_1 - 8x_2 &= 5 \\ -3x_1 + 12x_2 &= 8 \end{aligned}$

✓**4.** $\begin{aligned} 2x_1 - 8x_2 &= 6 \\ -3x_1 + 12x_2 &= -9 \end{aligned}$

5. $\begin{aligned} 6x_1 + x_2 &= 3 \\ -4x_1 - x_2 &= 8 \end{aligned}$

6. $\begin{aligned} 3x_1 + x_2 &= 0 \\ 2x_1 - 3x_2 &= 0 \end{aligned}$

7. $\begin{aligned} x_1 - 2x_2 + 3x_3 &= 11 \\ 4x_1 + x_2 - x_3 &= 4 \\ 2x_1 - x_2 + 3x_3 &= 10 \end{aligned}$

8. $\begin{aligned} -2x_1 + x_2 + 6x_3 &= 18 \\ 5x_1 + 8x_3 &= -16 \\ 3x_1 + 2x_2 - 10x_3 &= -3 \end{aligned}$

9. $\begin{aligned} 3x_1 + 6x_2 - 6x_3 &= 9 \\ 2x_1 - 5x_2 + 4x_3 &= 6 \\ -x_1 + 16x_2 - 14x_3 &= -3 \end{aligned}$

10. $\begin{aligned} 3x_1 + 6x_2 - 6x_3 &= 9 \\ 2x_1 - 5x_2 + 4x_3 &= 6 \\ 5x_1 + 28x_2 - 26x_3 &= -8 \end{aligned}$

11. $\begin{aligned} x_1 + x_2 - x_3 &= 7 \\ 4x_1 - x_2 + 5x_3 &= 4 \\ 2x_1 + 2x_2 - 3x_3 &= 0 \end{aligned}$

12. $\begin{aligned} x_1 + x_2 - x_3 &= 7 \\ 4x_1 - x_2 + 5x_3 &= 4 \\ 6x_1 + x_2 + 3x_3 &= 18 \end{aligned}$

13. $\begin{aligned} x_1 + x_2 - x_3 &= 7 \\ 4x_1 - x_2 + 5x_3 &= 4 \\ 6x_1 + x_2 + 3x_3 &= 20 \end{aligned}$

14. $\begin{aligned} x_1 - 2x_2 + 3x_3 &= 0 \\ 4x_1 + x_2 - x_3 &= 0 \\ 2x_1 - x_2 + 3x_3 &= 0 \end{aligned}$

15. $\begin{aligned} x_1 + x_2 - x_3 &= 0 \\ 4x_1 - x_2 + 5x_3 &= 0 \\ 6x_1 + x_2 + 3x_3 &= 0 \end{aligned}$

16. $\begin{aligned} 2x_2 + 5x_3 &= 6 \\ x_1 - 2x_3 &= 4 \\ 2x_1 + 4x_2 &= -2 \end{aligned}$

17. $\begin{aligned} x_1 + 2x_2 - x_3 &= 4 \\ 3x_1 + 4x_2 - 2x_3 &= 7 \end{aligned}$

✓**18.** $\begin{aligned} x_1 + 2x_2 - 4x_3 &= 4 \\ -2x_1 - 4x_2 + 8x_3 &= -8 \end{aligned}$

19. $\begin{aligned} x_1 + 2x_2 - 4x_3 &= 4 \\ -2x_1 - 4x_2 + 8x_3 &= -9 \end{aligned}$

20. $\begin{aligned} x_1 + 2x_2 - x_3 + x_4 &= 7 \\ 3x_1 + 6x_2 - 3x_3 + 3x_4 &= 21 \end{aligned}$

✓**21.** $\begin{aligned} 2x_1 + 6x_2 - 4x_3 + 2x_4 &= 4 \\ x_1 - x_3 + x_4 &= 5 \\ -3x_1 + 2x_2 - 2x_3 &= -2 \end{aligned}$

22. $\begin{aligned} x_1 - 2x_2 + x_3 + x_4 &= 2 \\ 3x_1 \quad + 2x_3 - 2x_4 &= -8 \\ 4x_2 - x_3 - x_4 &= 1 \\ -x_1 + 6x_2 - 2x_3 \quad &= 7 \end{aligned}$

23. $\begin{aligned} x_1 - 2x_2 + x_3 + x_4 &= 2 \\ 3x_1 \quad + 2x_3 - 2x_4 &= -8 \\ 4x_2 - x_3 - x_4 &= 1 \\ 5x_1 \quad + 3x_3 - x_4 &= -3 \end{aligned}$

24. $\begin{aligned} x_1 - 2x_2 + x_3 + x_4 &= 2 \\ 3x_1 \quad + 2x_3 - 2x_4 &= -8 \\ 4x_2 - x_3 - x_4 &= 1 \\ 5x_1 \quad + 3x_3 - x_4 &= 0 \end{aligned}$

25. $\begin{aligned} x_1 + x_2 &= 4 \\ 2x_1 - 3x_2 &= 7 \\ 3x_1 + 2x_2 &= 8 \end{aligned}$

26. $\begin{aligned} x_1 + x_2 &= 4 \\ 2x_1 - 3x_2 &= 7 \\ 3x_1 - 2x_2 &= 11 \end{aligned}$

27. $\begin{aligned} x_1 + x_2 - x_3 &= 0 \\ 2x_1 - 4x_2 + 3x_3 &= 0 \\ 3x_1 + 7x_2 - x_3 &= 0 \end{aligned}$

28. $\begin{aligned} x_1 + x_2 - x_3 &= 0 \\ 2x_1 - 4x_2 + 3x_3 &= 0 \\ -x_1 - 7x_2 + 6x_3 &= 0 \end{aligned}$

29. $\begin{aligned} x_1 + x_2 - x_3 &= 0 \\ 2x_1 - 4x_2 + 3x_3 &= 0 \\ -5x_1 + 13x_2 - 10x_3 &= 0 \end{aligned}$

30. $\begin{aligned} 2x_1 + 3x_2 - x_3 &= 0 \\ 6x_1 - 5x_2 + 7x_3 &= 0 \end{aligned}$

31. $\begin{aligned} 4x_1 - x_2 &= 0 \\ 7x_1 + 3x_2 &= 0 \\ -8x_1 + 6x_2 &= 0 \end{aligned}$

32. $\begin{aligned} x_1 - x_2 + 7x_3 - x_4 &= 0 \\ 2x_1 + 3x_2 - 8x_3 + x_4 &= 0 \end{aligned}$

33. $\begin{aligned} x_1 - 2x_2 + x_3 + x_4 &= 0 \\ 3x_1 \quad + 2x_3 - 2x_4 &= 0 \\ 4x_2 - x_3 - x_4 &= 0 \\ 5x_1 \quad + 3x_3 - x_4 &= 0 \end{aligned}$

34. $\begin{aligned} -2x_1 \quad + 7x_4 &= 0 \\ x_1 + 2x_2 - x_3 + 4x_4 &= 0 \\ 3x_1 \quad - x_3 + 5x_4 &= 0 \\ 4x_1 + 2x_2 + 3x_3 \quad &= 0 \end{aligned}$

35. $\begin{aligned} 2x_1 - x_2 &= 0 \\ 3x_1 + 5x_2 &= 0 \\ 7x_1 - 3x_2 &= 0 \\ -2x_1 + 3x_2 &= 0 \end{aligned}$

36. $\begin{aligned} x_1 - 3x_2 &= 0 \\ -2x_1 + 6x_2 &= 0 \\ 4x_1 - 12x_2 &= 0 \end{aligned}$

In Problems 37–45, determine whether the given matrix is in row echelon form (but not reduced row echelon form), reduced row echelon form, or neither.

37. $\begin{pmatrix} 1 & 1 & 0 \\ 0 & 1 & 1 \\ 0 & 0 & 1 \end{pmatrix}$

38. $\begin{pmatrix} 2 & 0 & 0 \\ 0 & 1 & 0 \\ 0 & 0 & -1 \end{pmatrix}$

39. $\begin{pmatrix} 1 & 0 & 1 & 0 \\ 0 & 1 & 1 & 0 \\ 0 & 0 & 0 & 0 \end{pmatrix}$

40. $\begin{pmatrix} 1 & 0 & 0 & 0 \\ 0 & 0 & 1 & 0 \\ 0 & 0 & 0 & 1 \end{pmatrix}$

41. $\begin{pmatrix} 0 & 1 & 0 & 0 \\ 1 & 0 & 0 & 0 \\ 0 & 0 & 0 & 0 \end{pmatrix}$

42. $\begin{pmatrix} 1 & 0 & 1 & 2 \\ 0 & 1 & 3 & 4 \end{pmatrix}$

43. $\begin{pmatrix} 1 & 0 \\ 0 & 1 \\ 0 & 0 \end{pmatrix}$

44. $\begin{pmatrix} 1 & 0 & 0 \\ 0 & 0 & 0 \\ 0 & 0 & 1 \end{pmatrix}$

45. $\begin{pmatrix} 1 & 0 & 0 & 4 \\ 0 & 1 & 0 & 5 \\ 0 & 1 & 1 & 6 \end{pmatrix}$

In Problems 46–51 use the elementary row operations to reduce the given matrices to row echelon form and reduced row echelon form.

46. $\begin{pmatrix} 1 & 1 \\ 2 & 3 \end{pmatrix}$

47. $\begin{pmatrix} -1 & 6 \\ 4 & 2 \end{pmatrix}$

48. $\begin{pmatrix} 1 & -1 & 1 \\ 2 & 4 & 3 \\ 5 & 6 & -2 \end{pmatrix}$

49. $\begin{pmatrix} 2 & -4 & 8 \\ 3 & 5 & 8 \\ -6 & 0 & 4 \end{pmatrix}$

50. $\begin{pmatrix} 2 & -4 & -2 \\ 3 & 1 & 6 \end{pmatrix}$

51. $\begin{pmatrix} 2 & -7 \\ 3 & 5 \\ 4 & -3 \end{pmatrix}$

52. In the Leontief input-output model of Example 12 suppose that there are three industries. Suppose further that $e_1 = 10$, $e_2 = 15$, $e_3 = 30$, $a_{11} = \frac{1}{3}$, $a_{12} = \frac{1}{2}$, $a_{13} = \frac{1}{6}$, $a_{21} = \frac{1}{4}$, $a_{22} = \frac{1}{4}$, $a_{23} = \frac{1}{8}$, $a_{31} = \frac{1}{12}$, $a_{32} = \frac{1}{3}$, and $a_{33} = \frac{1}{6}$. Find the output of each industry such that supply exactly equals demand.

53. In Example 14 assume that there are 15,000 units of the first food, 10,000 units of the second, and 35,000 units of the third supplied to the lake each week. Assuming that all three foods are consumed, what populations of the three species can coexist in the lake? Is there a unique solution?

54. The Sunrise Porcelain Company manufactures ceramic cups and saucers. For each cup or saucer a worker measures a fixed amount of material and puts it into a forming machine, from which it is automatically glazed and dried. On the average, a worker needs 3 minutes to get the process started for a cup and 2 minutes for a saucer. The material for a cup costs 25¢ and the material for a saucer costs 20¢. If $44 is allocated daily for production of cups and saucers, how many of each can be manufactured in an 8-hour work day if a worker is working every minute and exactly $44 is spent on materials?

55. Answer the question of Problem 54 if the materials for a cup and saucer cost 15¢ and 10¢, respectively, and $24 is spent in an 8-hour day.

56. Answer the question of Problem 55 if $25 is spent in an 8-hour day.

57. An ice-cream shop sells only ice-cream sodas and milk shakes. It puts 1 ounce of syrup and 4 ounces of ice cream in an ice-cream soda, and 1 ounce of syrup and 3 ounces of ice cream in a milk shake. If the store used 4 gallons of ice cream and 5 quarts of syrup in a day, how many ice-cream sodas and milk shakes did it sell? [*Hint:* 1 quart = 32 ounces; 1 gallon = 128 ounces]

58. A traveler just returned from Europe spent $30 a day for housing in England, $20 a day in France and $20 a day in Spain. For food the traveler spent $20 a day in England, $30 a day in France, and $20 a day in Spain. The traveler spent $10 a day in each country for incidental expenses. The traveler's records of the trip indicate a total of $340 spent for housing, $320 for food, and $140 for incidental expenses while traveling in these countries. Calculate the number of days the traveler spent in each of the countries or show that the records must be incorrect, because the amounts spent are incompatible with each other.

59. An investor remarks to a stockbroker that all her stock holdings are in three companies, Eastern Airlines, Hilton Hotels, and McDonald's, and that two days ago the value of her stocks went down $350 but yesterday the value increased by $600. The broker recalls that two days ago the price of Eastern Airlines stock dropped by $1 a share, Hilton Hotels dropped $1.50, but the price of McDonald's stock rose by $0.50. The broker also remembers that yesterday the price of Eastern Airlines stock rose $1.50, there was a further drop of $0.50 a share in Hilton Hotels' stock, and McDonald's stock rose $1. Show that the broker does not have enough information to calculate the number of shares the investor owns of each company's stock, but that when the investor says that she owns 200 shares of McDonald's stock, the broker can calculate the number of shares of Eastern Airlines and Hilton Hotels.

60. An intelligence agent knows that 60 aircraft, consisting of fighter planes and bombers, are stationed at a certain secret airfield. The agent wishes to determine how many of the aircraft are fighter planes and how many are bombers. There is a type of rocket carried by both sorts of planes; the fighter carries six of these rockets, the bomber only two. The agent learns that 250 rockets are requird to arm every plane at this airfield. Furthermore, the agent overhears a remark that there are twice as many fighter planes as bombers at the base (that is, the number of fighter planes minus twice the number of bombers equals zero). Calculate the number of fighter planes and bombers at the airfield or show that the agent's information must be incorrect, because it is inconsistent.

61. Consider the following system:

$$
\begin{aligned}
2x_1 - x_2 + 3x_3 &= a \\
3x_1 + x_2 - 5x_3 &= b \\
-5x_1 - 5x_2 + 21x_3 &= c
\end{aligned}
$$

Show that the system is inconsistent if $c \neq 2a - 3b$.

62. Consider the system

$$
\begin{aligned}
2x_1 + 3x_2 - x_3 &= a \\
x_1 - x_2 + 3x_3 &= b \\
3x_1 + 7x_2 - 5x_3 &= c
\end{aligned}
$$

Find conditions on a, b, and c such that the system is consistent.

***63.** Consider the general system of three linear equations in three unknowns:

$$
\begin{aligned}
a_{11}x_1 + a_{12}x_2 + a_{13}x_3 &= b_1 \\
a_{21}x_1 + a_{22}x_2 + a_{23}x_3 &= b_2 \\
a_{31}x_1 + a_{32}x_2 + a_{33}x_3 &= b_3
\end{aligned}
$$

Find conditions on the coefficients a_{ij} such that the system has a unique solution.

64. Solve the following system using a hand calculator and carrying five decimal places of accuracy:

$$
\begin{aligned}
2x_2 - x_3 - 4x_4 &= 2 \\
x_1 - x_2 + 5x_3 + 2x_4 &= -4 \\
3x_1 + 3x_2 - 7x_3 - x_4 &= 4 \\
-x_1 - 2x_2 + 3x_3 &= -7
\end{aligned}
$$

65. Do the same for the system

$$
\begin{aligned}
3.8x_1 + 1.6x_2 + 0.9x_3 &= 3.72 \\
-0.7x_1 + 5.4x_2 + 1.6x_3 &= 3.16 \\
1.5x_1 + 1.1x_2 - 3.2x_3 &= 43.78
\end{aligned}
$$

7.4 MATRICES AND LINEAR SYSTEMS OF EQUATIONS

In Section 7.3 we discussed the following systems of m equations in n unknowns:

$$\begin{aligned} a_{11}x_1 + a_{12}x_2 + \cdots + a_{1n}x_n &= b_1 \\ a_{21}x_1 + a_{22}x_2 + \cdots + a_{2n}x_n &= b_2 \\ \vdots \qquad \vdots \qquad\qquad \vdots \quad &\quad \vdots \\ a_{m1}x_1 + a_{m_2}x_2 + \cdots + a_{mn}x_n &= b_m \end{aligned} \tag{1}$$

We define the matrix

$$A = \begin{pmatrix} a_{11} & a_{12} & \cdots & a_{1n} \\ a_{21} & a_{22} & \cdots & a_{2n} \\ \vdots & \vdots & & \vdots \\ a_{m1} & a_{m2} & \cdots & a_{mn} \end{pmatrix},$$

the vector $\mathbf{x} = \begin{pmatrix} x_1 \\ x_2 \\ \vdots \\ x_n \end{pmatrix}$, and the vector $\mathbf{b} = \begin{pmatrix} b_1 \\ b_2 \\ \vdots \\ b_m \end{pmatrix}$. Since A is an $m \times n$ matrix and $\mathbf{x}$ is an $n \times 1$ matrix, the matrix product $A\mathbf{x}$ is defined as an $m \times 1$ matrix. It is not difficult to see that system (1) can be written as

$$A\mathbf{x} = \mathbf{b} \tag{2}$$

EXAMPLE 1 Consider the following system:

$$\begin{aligned} 2x_1 + 4x_2 + 6x_3 &= 18 \\ 4x_1 + 5x_2 + 6x_3 &= 24 \\ 3x_1 + x_2 - 2x_3 &= 4 \end{aligned} \tag{3}$$

(See Example 7.3.1.) This can be written in the form $A\mathbf{x} = \mathbf{b}$ with $A = \begin{pmatrix} 2 & 4 & 6 \\ 4 & 5 & 6 \\ 3 & 1 & -2 \end{pmatrix}$, $\mathbf{x} = \begin{pmatrix} x_1 \\ x_2 \\ x_3 \end{pmatrix}$, and $\mathbf{b} = \begin{pmatrix} 18 \\ 24 \\ 4 \end{pmatrix}$. ■

It is obviously easier to write out system (1) in the form $A\mathbf{x} = \mathbf{b}$. There are many other advantages, too. In Section 7.5 we shall see how a square system can be solved almost at once if we know a matrix called the *inverse* of A. Even without that, as we saw in Section 7.3, computations are much easier to write down by using an augmented matrix.

In this last example it is important to note that the last system of equations can be written as

$$Ix = \mathbf{s} \tag{4}$$

where $I = \begin{pmatrix} 1 & 0 & 0 \\ 0 & 1 & 0 \\ 0 & 0 & 1 \end{pmatrix}$ and $\mathbf{s}$ is the solution vector $\begin{pmatrix} 4 \\ -2 \\ 3 \end{pmatrix}$. We shall be making use of this fact in Section 7.5.

PROBLEMS 7.4

In Problems 1–6 write the given system in the form $A\mathbf{x} = \mathbf{b}$.

1. $\begin{aligned} 2x_1 - x_2 &= 3 \\ 4x_1 + 5x_2 &= 7 \end{aligned}$

2. $\begin{aligned} x_1 - x_2 + 3x_3 &= 11 \\ 4x_1 + x_2 - x_3 &= -4 \\ 2x_1 - x_2 + 3x_3 &= 10 \end{aligned}$

3. $\begin{aligned} 3x_1 + 6x_2 - 7x_3 &= 0 \\ 2x_1 - x_2 + 3x_3 &= 1 \end{aligned}$

4. $\begin{aligned} 4x_1 - x_2 + x_3 - x_4 &= -7 \\ 3x_1 + x_2 - 5x_3 + 6x_4 &= 8 \\ 2x_1 - x_2 + x_3 &= 9 \end{aligned}$

5. $\begin{aligned} x_2 - x_3 &= 7 \\ x_1 + x_3 &= 2 \\ 3x_1 + 2x_2 + x_3 &= -5 \end{aligned}$

6. $\begin{aligned} 2x_1 + 3x_2 - x_3 &= 0 \\ -4x_1 + 2x_2 + x_3 &= 0 \\ 7x_1 + 3x_2 - 9x_3 &= 0 \end{aligned}$

7.5 THE INVERSE OF A SQUARE MATRIX

In this section we define two kinds of matrices that are central to matrix theory. We begin with a simple example. Let $A = \begin{pmatrix} 2 & 5 \\ 1 & 3 \end{pmatrix}$ and $B = \begin{pmatrix} 3 & -5 \\ -1 & 2 \end{pmatrix}$. Then an easy computation shows that $AB = BA = \begin{pmatrix} 1 & 0 \\ 0 & 1 \end{pmatrix}$. The matrix $\begin{pmatrix} 1 & 0 \\ 0 & 1 \end{pmatrix}$ is called the 2×2 *identity matrix* and is denoted I or I_2. The matrix B is called the *inverse* of A and is written A^{-1}.

Definition 1 IDENTITY MATRIX Then $n \times n$ **identity matrix** is the $n \times n$ matrix with ones down the **main diagonal**† and zeros everywhere else. That is,

$$I_n = (b_{ij}) \quad \text{where} \quad b_{ij} = \begin{cases} 1 & \text{if } i = j \\ 0 & \text{if } i \neq j \end{cases} \tag{1}$$

†The main diagonal of $A = (a_{ij})$ consists of the components a_{11}, a_{22}, a_{33}, and so on. Unless otherwise stated, we shall refer to the main diagonal simply as the **diagonal.**

EXAMPLE 1 $I_3 = \begin{pmatrix} 1 & 0 & 0 \\ 0 & 1 & 0 \\ 0 & 0 & 1 \end{pmatrix}$ and $I_5 = \begin{pmatrix} 1 & 0 & 0 & 0 & 0 \\ 0 & 1 & 0 & 0 & 0 \\ 0 & 0 & 1 & 0 & 0 \\ 0 & 0 & 0 & 1 & 0 \\ 0 & 0 & 0 & 0 & 1 \end{pmatrix}$ ■

Theorem 1 Let A be a square $n \times n$ matrix. Then

$$AI_n = I_nA = A.$$

That is, I_n commutes with every $n \times n$ matrix and leaves it unchanged after multiplication on the left or right.

REMARK. I_n functions for $n \times n$ matrices the way the number 1 functions for real numbers (since $1 \cdot a = a \cdot 1 = a$ for every real number a).

Proof. Let c_{ij} be the ijth element of AI_n. Then

$$c_{ij} = a_{i1}b_{1j} + a_{i2}b_{2j} + \cdots + a_{ij}b_{jj} + \cdots + a_{in}b_{nj}.$$

But, from (1), this sum is equal to a_{ij}. Thus $AI_n = A$. In a similar fashion we can show that $I_nA = A$, and this proves the theorem. ■

NOTATION: From now on we shall write an identity matrix simply as I, since if A is $n \times n$, the products IA and AI are defined only if I is also $n \times n$.

Definition 2 THE INVERSE OF A MATRIX Let A and B be $n \times n$ matrices. Suppose that

$$AB = BA = I$$

Then B is called the **inverse** of A and is written as A^{-1}. We then have

$$AA^{-1} = A^{-1}A = I.$$

If A has an inverse, then A is said to be **invertible.**

REMARK 1. From this definition it immediately follows that $(A^{-1})^{-1} = A$ if A is invertible.

REMARK 2. This definition does *not* state that every square matrix has an inverse. In fact there are many square matrices that have no inverse. (See, for instance, Example 3 below.)

In Definition 2 we defined *the* inverse of a matrix. This statement suggests that inverses are unique. This is indeed the case, as the following theorem shows.

Theorem 2 If a square matrix A is invertible, then its inverse is unique.

Proof. Suppose B and C are two inverses for A. We can show that $B = C$. By definition, we have $AB = BA = I$ and $AC = CA = I$. Then $B(AC) = BI = B$ and $(BA)C = IC = C$. But $B(AC) = (BA)C$ by the associative law of matrix multiplication. Hence $B = C$ and the theorem is proved. ■

Another important fact about inverses is given below.

Theorem 3 Let A and B be invertible $n \times n$ matrices. Then AB is invertible and

$$(AB)^{-1} = B^{-1}A^{-1}.$$

Proof. To prove this result, we refer to Definition 2. That is, $B^{-1}A^{-1} = (AB)^{-1}$ if and only if $B^{-1}A^{-1}(AB) = (AB)(B^{-1}A^{-1}) = I$. But this follows since

$$(B^{-1}A^{-1})(AB) \overset{\text{Equation (4) on page 406}}{=} B^{-1}(A^{-1}A)B = B^{-1}IB = B^{-1}B = I$$

and

$$(AB)(B^{-1}A^{-1}) = A(BB^{-1})A^{-1} = AIA^{-1} = AA^{-1} = I. \quad ■$$

Consider the system of n equations in n unknowns

$$A\mathbf{x} = \mathbf{b},$$

and suppose that A is invertible. Then

$$A^{-1}A\mathbf{x} = A^{-1}\mathbf{b} \quad \text{We multiplied on the left by } A^{-1}.$$

$$I\mathbf{x} = A^{-1}\mathbf{b} \quad A^{-1}A = I$$

$$\mathbf{x} = A^{-1}\mathbf{b} \quad I\mathbf{x} = \mathbf{x}$$

That is,

If A is invertible, the system $A\mathbf{x} = \mathbf{b}$ has the unique solution $\mathbf{x} = A^{-1}\mathbf{b}$. **(2)**

This is one of the reasons we study matrix inverses.

Two basic questions come to mind once we have defined the inverse of a matrix:

(i) What matrices do have inverses?
(ii) If a matrix has an inverse, how can we compute it?

We answer both questions in this section. Rather than starting by giving you what seems to be a set of arbitrary rules, we look first at what happens in the 2×2 case.

EXAMPLE 2 Let $A = \begin{pmatrix} 2 & -3 \\ -4 & 5 \end{pmatrix}$. Compute A^{-1} if it exists.

Solution. Suppose that A^{-1} exists. We write $A^{-1} = \begin{pmatrix} x & y \\ z & w \end{pmatrix}$ and use the fact that $AA^{-1} = I$. Then

$$AA^{-1} = \begin{pmatrix} 2 & -3 \\ -4 & 5 \end{pmatrix}\begin{pmatrix} x & y \\ z & w \end{pmatrix} = \begin{pmatrix} 2x - 3z & 2y - 3w \\ -4x + 5z & -4y + 5w \end{pmatrix} = \begin{pmatrix} 1 & 0 \\ 0 & 1 \end{pmatrix}$$

The last two matrices can be equal only if each of their corresponding components are equal. This means that

$$2x \qquad - 3z \qquad = 1 \tag{3}$$

$$2y \qquad - 3w = 0 \tag{4}$$

$$-4x \qquad + 5z \qquad = 0 \tag{5}$$

$$-4y \qquad + 5w = 1 \tag{6}$$

This is a system of four equations in four unknowns. Note that there are two equations involving x and z only—equations (3) and (5)—and two equations involving y and w only—equations (4) and (6). We write these two systems in augmented matrix form:

$$\left(\begin{array}{cc|c} 2 & -3 & 1 \\ -4 & 5 & 0 \end{array}\right) \tag{7}$$

$$\left(\begin{array}{cc|c} 2 & -3 & 0 \\ -4 & 5 & 1 \end{array}\right). \tag{8}$$

Now, we know from Section 7.3 that if system (7) (in the variables x and z) has a unique solution, then Gauss-Jordan elimination of (7) will result in

$$\left(\begin{array}{cc|c} 1 & 0 & x \\ 0 & 1 & z \end{array}\right)$$

where (x, z) is the unique pair of numbers that satisfies $2x - 3y = 1$ and $-4x + 5z = 0$. Similarly, row reduction of (8) will result in

$$\left(\begin{array}{cc|c} 1 & 0 & y \\ 0 & 1 & w \end{array}\right)$$

where (y, w) is the unique pair of numbers that satisfies $2y - 3w = 0$ and $-4y + 5w = 1$.

Since the coefficient matrices in (7) and (8) are the same, we can perform the row reductions on the two augmented matrices simultaneously, by considering the new augmented matrix

$$\left(\begin{array}{cc|cc} 2 & -3 & 1 & 0 \\ -4 & 5 & 0 & 1 \end{array}\right). \tag{9}$$

If A^{-1} is invertible, then the system defined by (3), (4), (5), and (6) has a unique solution and, by what we said above, Gauss-Jordan elimination will result in

$$\left(\begin{array}{cc|cc} 1 & 0 & x & y \\ 0 & 1 & z & w \end{array}\right).$$

We now carry out the computation, noting that the matrix on the left in (9) is A and the matrix on the right in (9) is I:

$$\left(\begin{array}{cc|cc} 2 & -3 & 1 & 0 \\ -4 & 5 & 0 & 1 \end{array}\right) \xrightarrow{M_1(\frac{1}{2})} \left(\begin{array}{cc|cc} 1 & -\frac{3}{2} & \frac{1}{2} & 0 \\ -4 & 5 & 0 & 1 \end{array}\right)$$

$$\xrightarrow{A_{1,2}(4)} \left(\begin{array}{cc|cc} 1 & -\frac{3}{2} & \frac{1}{2} & 0 \\ 0 & -1 & 2 & 1 \end{array}\right)$$

$$\xrightarrow{M_2(-1)} \left(\begin{array}{cc|cc} 1 & -\frac{3}{2} & \frac{1}{2} & 0 \\ 0 & 1 & -2 & -1 \end{array}\right)$$

$$\xrightarrow{A_{2,1}(\frac{3}{2})} \left(\begin{array}{cc|cc} 1 & 0 & -\frac{5}{2} & -\frac{3}{2} \\ 0 & 1 & -2 & -1 \end{array}\right).$$

Thus $x = -\frac{5}{2}$, $y = -\frac{3}{2}$, $z = -2$, $w = -1$, and $A^{-1} = \begin{pmatrix} -\frac{5}{2} & -\frac{3}{2} \\ -2 & -1 \end{pmatrix}$. We still must check our answer. We have

$$AA^{-1} = \begin{pmatrix} 2 & -3 \\ -4 & 5 \end{pmatrix}\begin{pmatrix} -\frac{5}{2} & -\frac{3}{2} \\ -2 & -1 \end{pmatrix} = \begin{pmatrix} 1 & 0 \\ 0 & 1 \end{pmatrix}$$

and

$$A^{-1}A = \begin{pmatrix} -\frac{5}{2} & -\frac{3}{2} \\ -2 & -1 \end{pmatrix}\begin{pmatrix} 2 & -3 \\ -4 & 5 \end{pmatrix} = \begin{pmatrix} 1 & 0 \\ 0 & 1 \end{pmatrix}.$$

Thus A is invertible and $A^{-1} = \begin{pmatrix} -\frac{5}{2} & -\frac{3}{2} \\ -2 & -1 \end{pmatrix}$. ■

EXAMPLE 3 Let $A = \begin{pmatrix} 1 & 2 \\ -2 & -4 \end{pmatrix}$. Calculate A^{-1} if it exists.

Solution. If $A^{-1} = \begin{pmatrix} x & y \\ z & w \end{pmatrix}$ exists, then

$$AA^{-1} = \begin{pmatrix} 1 & 2 \\ -2 & -4 \end{pmatrix}\begin{pmatrix} x & y \\ z & w \end{pmatrix} = \begin{pmatrix} x + 2z & y + 2w \\ -2x - 4z & -2y - 4w \end{pmatrix} = \begin{pmatrix} 1 & 0 \\ 0 & 1 \end{pmatrix}.$$

This leads to the system

$$\begin{aligned} x \qquad\quad + 2z \qquad\quad &= 1 \\ y \qquad\quad + 2w &= 0 \\ -2x \qquad\quad - 4z \qquad\quad &= 0 \\ -2y \qquad\quad - 4w &= 1. \end{aligned} \tag{10}$$

Using the same reasoning as in Example 2, we can write this system in the augmented matrix form $(A|I)$ and row reduce.

$$\left(\begin{array}{rr|rr} 1 & 2 & 1 & 0 \\ -2 & -4 & 0 & 1 \end{array}\right) \xrightarrow{A_{1,2}(2)} \left(\begin{array}{rr|rr} 1 & 2 & 1 & 0 \\ 0 & 0 & 2 & 1 \end{array}\right)$$

This is as far as we can go. The last line reads $0 = 2$ or $0 = 1$, depending on which of the two systems of equations (in x and z or in y and w) is being solved. Thus system (10) is inconsistent and A is not invertible. ■

The last two examples illustrate a procedure that always works when you are trying to find the inverse of a matrix.

PROCEDURE FOR COMPUTING THE INVERSE OF A SQUARE MATRIX A

(i) Write the augmented matrix $(A|I)$.
(ii) Use row reduction to reduce the matrix A to its reduced row echelon form.
(iii) Decide if A is invertible.
- **(a)** If A can be reduced to the identity matrix I, then A^{-1} will be the matrix to the right of the vertical bar.
- **(b)** If the row reduction of A leads to a row of zeros to the left of the vertical bar, then A is not invertible.

REMARK. We can rephrase (a) and (b) as follows.

A square matrix A is invertible if and only if its reduced row echelon form is the identity matrix.

We now give a formula that is useful for the computation of the inverse of a 2×2 matrix.

Let $A = \begin{pmatrix} a_{11} & a_{12} \\ a_{21} & a_{22} \end{pmatrix}$. Then we define

$$\text{Determinant of } A = a_{11}a_{22} - a_{12}a_{21} \tag{11}$$

We abbreviate the determinant of A by $\det A$.

Theorem 4 Let A be a 2×2 matrix. Then, if $\det A \neq 0$, A is invertible, and

$$A^{-1} = \frac{1}{\det A}\begin{pmatrix} a_{22} & -a_{12} \\ -a_{21} & a_{11} \end{pmatrix} \tag{12}$$

Proof. Let $B = (1/\det A)\begin{pmatrix} a_{22} & -a_{12} \\ -a_{21} & a_{11} \end{pmatrix}$. Then

$$BA = \frac{1}{\det A}\begin{pmatrix} a_{22} & -a_{12} \\ -a_{21} & a_{11} \end{pmatrix}\begin{pmatrix} a_{11} & a_{12} \\ a_{21} & a_{22} \end{pmatrix}$$

$$= \frac{1}{a_{11}a_{22} - a_{12}a_{21}}\begin{pmatrix} a_{22}a_{11} - a_{12}a_{21} & 0 \\ 0 & -a_{21}a_{12} + a_{11}a_{22} \end{pmatrix} = \begin{pmatrix} 1 & 0 \\ 0 & 1 \end{pmatrix} = I$$

Similarly $AB = I$, which shows that A is invertible and that $B = A^{-1}$. ■

REMARK. This formula can be obtained directly by applying our procedure for computing an inverse. You are asked to do this in Problem 46.

EXAMPLE 4 Let $A = \begin{pmatrix} 2 & -4 \\ 1 & 3 \end{pmatrix}$. Calculate A^{-1} if it exists.

Solution. We find that $\det A = (2)(3) - (-4)(1) = 10$; hence A^{-1} exists. From equation (12), we get

$$A^{-1} = \frac{1}{10}\begin{pmatrix} 3 & 4 \\ -1 & 2 \end{pmatrix} = \begin{pmatrix} \frac{3}{10} & \frac{4}{10} \\ -\frac{1}{10} & \frac{2}{10} \end{pmatrix}.$$

Check.

$$A^{-1}A = \frac{1}{10}\begin{pmatrix} 3 & 4 \\ -1 & 2 \end{pmatrix}\begin{pmatrix} 2 & -4 \\ 1 & 3 \end{pmatrix} = \frac{1}{10}\begin{pmatrix} 10 & 0 \\ 0 & 10 \end{pmatrix} = \begin{pmatrix} 1 & 0 \\ 0 & 1 \end{pmatrix}$$

and

$$AA^{-1} = \begin{pmatrix} 2 & -4 \\ 1 & 3 \end{pmatrix}\begin{pmatrix} \frac{3}{10} & \frac{4}{10} \\ -\frac{1}{10} & \frac{2}{10} \end{pmatrix} = \begin{pmatrix} 1 & 0 \\ 0 & 1 \end{pmatrix}.$$ ■

EXAMPLE 5 Let $A = \begin{pmatrix} 1 & 2 \\ -2 & -4 \end{pmatrix}$. Calculate A^{-1} if it exists.

Solution. We find that $\det A = (1)(-4) - (2)(-2) = -4 + 4 = 0$, so that A^{-1} does not exist, as we saw in Example 3. ■

The procedure described above works for $n \times n$ matrices where $n > 2$. We illustrate this with a number of examples.

EXAMPLE 6 Let $A = \begin{pmatrix} 2 & 4 & 6 \\ 4 & 5 & 6 \\ 3 & 1 & -2 \end{pmatrix}$ (see Example 7.3.1). Calculate A^{-1} if it exists.

Solution. We first put I next to A in an augmented matrix form

$$\left(\begin{array}{ccc|ccc} 2 & 4 & 6 & 1 & 0 & 0 \\ 4 & 5 & 6 & 0 & 1 & 0 \\ 3 & 1 & -2 & 0 & 0 & 1 \end{array}\right)$$

and then carry out the row reduction.

$$\xrightarrow{M_1(\frac{1}{2})} \left(\begin{array}{ccc|ccc} 1 & 2 & 3 & \frac{1}{2} & 0 & 0 \\ 4 & 5 & 6 & 0 & 1 & 0 \\ 3 & 1 & -2 & 0 & 0 & 1 \end{array}\right) \xrightarrow[A_{1,3}(-3)]{A_{1,2}(-4)} \left(\begin{array}{ccc|ccc} 1 & 2 & 3 & \frac{1}{2} & 0 & 0 \\ 0 & -3 & -6 & -2 & 1 & 0 \\ 0 & -5 & -11 & -\frac{3}{2} & 0 & 1 \end{array}\right)$$

$$\xrightarrow{M_2(-\frac{1}{3})} \left(\begin{array}{ccc|ccc} 1 & 2 & 3 & \frac{1}{2} & 0 & 0 \\ 0 & 1 & 2 & \frac{2}{3} & -\frac{1}{3} & 0 \\ 0 & -5 & -11 & -\frac{3}{2} & 0 & 1 \end{array}\right) \xrightarrow[A_{2,3}(5)]{A_{2,1}(-2)} \left(\begin{array}{ccc|ccc} 1 & 0 & -1 & -\frac{5}{6} & \frac{2}{3} & 0 \\ 0 & 1 & 2 & \frac{2}{3} & -\frac{1}{3} & 0 \\ 0 & 0 & -1 & \frac{11}{6} & -\frac{5}{3} & 1 \end{array}\right)$$

$$\xrightarrow{M_3(-1)} \left(\begin{array}{ccc|ccc} 1 & 0 & -1 & -\frac{5}{6} & \frac{2}{3} & 0 \\ 0 & 1 & 2 & \frac{2}{3} & -\frac{1}{3} & 0 \\ 0 & 0 & 1 & -\frac{11}{6} & \frac{5}{3} & -1 \end{array}\right) \xrightarrow[A_{3,2}(-2)]{A_{3,1}(1)} \left(\begin{array}{ccc|ccc} 1 & 0 & 0 & -\frac{8}{3} & \frac{7}{3} & -1 \\ 0 & 1 & 0 & \frac{13}{3} & -\frac{11}{3} & 2 \\ 0 & 0 & 1 & -\frac{11}{6} & \frac{5}{3} & -1 \end{array}\right)$$

Since A has now been reduced to I, we have

$$A^{-1} = \begin{pmatrix} -\frac{8}{3} & \frac{7}{3} & -1 \\ \frac{13}{3} & -\frac{11}{3} & 2 \\ -\frac{11}{6} & \frac{5}{3} & -1 \end{pmatrix} = \frac{1}{6}\begin{pmatrix} -16 & 14 & -6 \\ 26 & -22 & 12 \\ -11 & 10 & -6 \end{pmatrix}$$

We factor out $\frac{1}{6}$ to make computations easier.

Check. $A^{-1}A = \frac{1}{6}\begin{pmatrix} -16 & 14 & -6 \\ 26 & -22 & 12 \\ -11 & 10 & -6 \end{pmatrix}\begin{pmatrix} 2 & 4 & 6 \\ 4 & 5 & 6 \\ 3 & 1 & -2 \end{pmatrix} = \frac{1}{6}\begin{pmatrix} 6 & 0 & 0 \\ 0 & 6 & 0 \\ 0 & 0 & 6 \end{pmatrix} = I.$

We can also verify that $AA^{-1} = I$. ■

■ WARNING: It is easy to make numerical errors in computing A^{-1}. Therefore it is essential to check the computations by verifying that $A^{-1}A = I$.

EXAMPLE 7 Let $A = \begin{pmatrix} 2 & 4 & 3 \\ 0 & 1 & -1 \\ 3 & 5 & 7 \end{pmatrix}$. Calculate A^{-1} if it exists.

Solution. Proceeding as in Example 6 we obtain, successively, the following augmented matrices:

$$\left(\begin{array}{ccc|ccc} 2 & 4 & 3 & 1 & 0 & 0 \\ 0 & 1 & -1 & 0 & 1 & 0 \\ 3 & 5 & 7 & 0 & 0 & 1 \end{array}\right) \xrightarrow{M_1(\frac{1}{2})} \left(\begin{array}{ccc|ccc} 1 & 2 & \frac{3}{2} & \frac{1}{2} & 0 & 0 \\ 0 & 1 & -1 & 0 & 1 & 0 \\ 3 & 5 & 7 & 0 & 0 & 1 \end{array}\right)$$

$$\xrightarrow{A_{1,3}(-3)} \left(\begin{array}{ccc|ccc} 1 & 2 & \frac{3}{2} & \frac{1}{2} & 0 & 0 \\ 0 & 1 & -1 & 0 & 1 & 0 \\ 0 & -1 & \frac{5}{2} & -\frac{3}{2} & 0 & 1 \end{array}\right) \xrightarrow[A_{2,3}(1)]{A_{2,1}(-2)} \left(\begin{array}{ccc|ccc} 1 & 0 & \frac{7}{2} & \frac{1}{2} & -2 & 0 \\ 0 & 1 & -1 & 0 & 1 & 0 \\ 0 & 0 & \frac{3}{2} & -\frac{3}{2} & 1 & 1 \end{array}\right)$$

$$\xrightarrow{M_3(\frac{2}{3})} \left(\begin{array}{ccc|ccc} 1 & 0 & \frac{7}{2} & \frac{1}{2} & -2 & 0 \\ 0 & 1 & -1 & 0 & 1 & 0 \\ 0 & 0 & 1 & -1 & \frac{2}{3} & \frac{2}{3} \end{array}\right) \xrightarrow[A_{3,2}(1)]{A_{3,1}(-\frac{7}{2})} \left(\begin{array}{ccc|ccc} 1 & 0 & 0 & 4 & -\frac{13}{3} & -\frac{7}{3} \\ 0 & 1 & 0 & -1 & \frac{5}{3} & \frac{2}{3} \\ 0 & 0 & 1 & -1 & \frac{2}{3} & \frac{2}{3} \end{array}\right)$$

Thus

$$A^{-1} = \begin{pmatrix} 4 & -\frac{13}{3} & -\frac{7}{3} \\ -1 & \frac{5}{3} & \frac{2}{3} \\ -1 & \frac{2}{3} & \frac{2}{3} \end{pmatrix}$$

Check. $A^{-1}A = \begin{pmatrix} 4 & -\frac{13}{3} & -\frac{7}{3} \\ -1 & \frac{5}{3} & \frac{2}{3} \\ -1 & \frac{2}{3} & \frac{2}{3} \end{pmatrix} \begin{pmatrix} 2 & 4 & 3 \\ 0 & 1 & -1 \\ 3 & 5 & 7 \end{pmatrix} = \begin{pmatrix} 1 & 0 & 0 \\ 0 & 1 & 0 \\ 0 & 0 & 1 \end{pmatrix}$ ■

EXAMPLE 8 Let $A = \begin{pmatrix} 1 & -3 & 4 \\ 2 & -5 & 7 \\ 0 & -1 & 1 \end{pmatrix}$. Calculate A^{-1} if it exists.

Solution. Proceeding as before we obtain, successively,

$$\left(\begin{array}{ccc|ccc} 1 & -3 & 4 & 1 & 0 & 0 \\ 2 & -5 & 7 & 0 & 1 & 0 \\ 0 & -1 & 1 & 0 & 0 & 1 \end{array}\right) \xrightarrow{A_{1,2}(-2)} \left(\begin{array}{ccc|ccc} 1 & -3 & 4 & 1 & 0 & 0 \\ 0 & 1 & -1 & -2 & 1 & 0 \\ 0 & -1 & 1 & 0 & 0 & 1 \end{array}\right)$$

$$\xrightarrow[A_{2,3}(1)]{A_{2,1}(3)} \left(\begin{array}{ccc|ccc} 1 & 0 & 1 & -5 & 3 & 0 \\ 0 & 1 & -1 & -2 & 1 & 0 \\ 0 & 0 & 0 & -2 & 1 & 1 \end{array}\right)$$

This is as far as we can go. The matrix A *cannot* be reduced to the identity matrix and we can conclude that A is *not* invertible. ■

There is another way to see the result of the last example. Let $\mathbf{b}$ be any 3-vector and consider the system $A\mathbf{x} = \mathbf{b}$. If we tried to solve this by Gaussian elimination, we would end up with an equation that reads $0 = c \neq 0$ as in Example 3, or $0 = 0$. This is case (ii) or (iii) of Section 7.3 (see page 419). That is, the system either has no solution or it has an infinite number of solutions. The one possibility ruled out is the case in which the system has a unique solution. But if A^{-1} existed, then there would be a unique solution given by $\mathbf{x} = A^{-1}\mathbf{b}$. We are left to conclude that:

If in the row reduction of A we end up with a row of zeros, then A is *not* invertible.

Definition 3 ROW-EQUIVALENT MATRICES Suppose that by elementary row operations we can transform the matrix A into the matrix B. Then A and B are said to be **row equivalent.**

In this chapter we have introduced a number of concepts that seem to be related. The following important theorem states that four of these are equivalent. (Two statements are **equivalent** if each one implies the other. That is, if the first is true, then the second is true, and if the second is true, then the first is true.) In Problem 38 you are asked to prove the theorem. In Chapters 8 and 9 we will add parts to the theorem.

Theorem 5 SUMMING-UP THEOREM—VIEW 1 Let A be an $n \times n$ matrix. Then the following four statements are equivalent. (That is, if one is true, all are true).

(i) A is invertible.
(ii) The system $A\mathbf{x} = \mathbf{b}$ has a unique solution for every n-vector $\mathbf{b}$.
(iii) The only solution to the homogeneous system $A\mathbf{x} = \mathbf{0}$ is the trivial solution ($\mathbf{x} = \mathbf{0}$).
(iv) A is row equivalent to the identity matrix I_n; that is, the reduced row echelon form of A is I_n.

Corollary. If A is invertible, then the unique solution to the system $A\mathbf{x} = \mathbf{b}$ is given by $\mathbf{x} = A^{-1}\mathbf{b}$.

EXAMPLE 9 Solve the system

$$\begin{aligned} 2x_1 + 4x_2 + 3x_3 &= 6 \\ x_2 - x_3 &= -4 \\ 3x_1 + 5x_2 + 7x_3 &= 7 \end{aligned}$$

Solution. This system can be written as $A\mathbf{x} = \mathbf{b}$, where $A = \begin{pmatrix} 2 & 4 & 3 \\ 0 & 1 & -1 \\ 3 & 5 & 7 \end{pmatrix}$ and $\mathbf{b} = \begin{pmatrix} 6 \\ -4 \\ 7 \end{pmatrix}$. In Example 7 we found that A^{-1} exists and

$$A^{-1} = \begin{pmatrix} 4 & -\frac{13}{3} & -\frac{7}{3} \\ -1 & \frac{5}{3} & \frac{2}{3} \\ -1 & \frac{2}{3} & \frac{2}{3} \end{pmatrix}$$

Thus the unique solution is given by

$$\mathbf{x} = \begin{pmatrix} x_1 \\ x_2 \\ x_3 \end{pmatrix} = A^{-1}\mathbf{b} = \begin{pmatrix} 4 & -\frac{13}{3} & -\frac{7}{3} \\ -1 & \frac{5}{3} & \frac{2}{3} \\ -1 & \frac{2}{3} & \frac{2}{3} \end{pmatrix} \begin{pmatrix} 6 \\ -4 \\ 7 \end{pmatrix} = \begin{pmatrix} 25 \\ -8 \\ -4 \end{pmatrix}. \blacksquare$$

EXAMPLE 10 In the Leontief input-output model described in Example 7.3.12 we obtained the system

$$\begin{aligned} a_{11}x_1 + a_{12}x_2 + \cdots + a_{1n}x_n + e_1 &= x_1 \\ a_{21}x_1 + a_{22}x_2 + \cdots + a_{2n}x_n + e_2 &= x_2 \\ \vdots \qquad \vdots \qquad\qquad \vdots \qquad \vdots \;\; &\quad \vdots \\ a_{n1}x_1 + a_{n2}x_2 + \cdots + a_{nn}x_n + e_n &= x_n \end{aligned} \tag{13}$$

which can be written as

$$A\mathbf{x} + \mathbf{e} = \mathbf{x} = I\mathbf{x}$$

or

$$(I - A)\mathbf{x} = \mathbf{e} \tag{14}$$

The matrix A of internal demands is called the **technology matrix,** and the matrix $I - A$ is called the **Leontief matrix.** If the Leontief matrix is invertible, then systems (13) and (14) have unique solutions.

Leontief used his model to analyze the 1958 American economy.† He divided the economy into 81 sectors and grouped them into six families of related sectors. For simplicity, we treat each family of sectors as a single sector so we can treat the American economy as an economy with six industries. These industries are listed in Table 1.

TABLE 1

Sector	Examples
Final nonmetal (FN)	Furniture, processed food
Final metal (FM)	Household appliances, motor vehicles
Basic metal (BM)	Machine-shop products, mining
Basic nonmetal (BN)	Agriculture, printing
Energy (E)	Petroleum, coal
Services (S)	Amusements, real estate

The input-output table, Table 2, gives internal demands in 1958 based on Leontief's figures. The units in the table are millions of dollars. Thus, for example, the number 0.173 in the 6,5 position means that in order to produce \$1 million worth of energy, it is necessary to provide \$0.173 million = \$173,000 worth of services. Similarly, the 0.037 in the 4,2 position means that in order to produce \$1 million worth of final metal, it is necessary to expend \$0.037 million = \$37,000 on basic nonmetal products.

†*Scientific American* (April 1965), pp. 26–27.

TABLE 2 INTERNAL DEMANDS IN 1958 U.S. ECONOMY

	FN	*FM*	*BM*	*BN*	*E*	*S*
FN	0.170	0.004	0	0.029	0	0.008
FM	0.003	0.295	0.018	0.002	0.004	0.016
BM	0.025	0.173	0.460	0.007	0.011	0.007
BN	0.348	0.037	0.021	0.403	0.011	0.048
E	0.007	0.001	0.039	0.025	0.358	0.025
S	0.120	0.074	0.104	0.123	0.173	0.234

Finally, Leontief estimated the following external demands on the 1958 American economy (in millions of dollars).

TABLE 3 EXTERNAL DEMANDS ON 1958 U.S. ECONOMY (MILLIONS OF DOLLARS)

FN	$99,640
FM	$75,548
BM	$14,444
BN	$33,501
E	$23,527
S	$263,985

In order to run the American economy in 1958 and meet all external demands, how many units in each of the six sectors had to be produced?

Solution. The technology matrix is given by

$$A = \begin{pmatrix} 0.170 & 0.004 & 0 & 0.029 & 0 & 0.008 \\ 0.003 & 0.295 & 0.018 & 0.002 & 0.004 & 0.016 \\ 0.025 & 0.173 & 0.460 & 0.007 & 0.011 & 0.007 \\ 0.348 & 0.037 & 0.021 & 0.403 & 0.011 & 0.048 \\ 0.007 & 0.001 & 0.039 & 0.025 & 0.358 & 0.025 \\ 0.120 & 0.074 & 0.104 & 0.123 & 0.173 & 0.234 \end{pmatrix}$$

and

$$\mathbf{e} = \begin{pmatrix} 99{,}640 \\ 75{,}548 \\ 14{,}444 \\ 33{,}501 \\ 23{,}527 \\ 263{,}985 \end{pmatrix}$$

To obtain the Leontief matrix, we subtract to obtain

$$I - A = \begin{pmatrix} 1 & 0 & 0 & 0 & 0 & 0 \\ 0 & 1 & 0 & 0 & 0 & 0 \\ 0 & 0 & 1 & 0 & 0 & 0 \\ 0 & 0 & 0 & 1 & 0 & 0 \\ 0 & 0 & 0 & 0 & 1 & 0 \\ 0 & 0 & 0 & 0 & 0 & 1 \end{pmatrix}$$

$$= \begin{pmatrix} 0.170 & 0.004 & 0 & 0.029 & 0 & 0.008 \\ 0.003 & 0.295 & 0.018 & 0.002 & 0.004 & 0.016 \\ 0.025 & 0.173 & 0.460 & 0.007 & 0.011 & 0.007 \\ 0.348 & 0.037 & 0.021 & 0.403 & 0.011 & 0.048 \\ 0.007 & 0.001 & 0.039 & 0.025 & 0.358 & 0.025 \\ 0.120 & 0.074 & 0.104 & 0.123 & 0.173 & 0.234 \end{pmatrix}$$

$$= \begin{pmatrix} 0.830 & -0.004 & 0 & -0.029 & 0 & -0.008 \\ -0.003 & 0.705 & -0.018 & -0.002 & -0.004 & -0.016 \\ -0.025 & -0.173 & 0.540 & -0.007 & -0.011 & -0.007 \\ -0.348 & -0.037 & -0.021 & 0.597 & -0.011 & -0.048 \\ -0.007 & -0.001 & -0.039 & -0.025 & 0.642 & -0.025 \\ -0.120 & -0.074 & -0.104 & -0.123 & -0.173 & 0.766 \end{pmatrix}$$

The computation of the inverse of a 6×6 matrix is a tedious affair. Carrying three decimal places on a calculator, we obtain the matrix below. Intermediate steps are omitted.

$$(I - A)^{-1} = \begin{pmatrix} 1.234 & 0.014 & 0.006 & 0.064 & 0.007 & 0.018 \\ 0.017 & 1.436 & 0.057 & 0.012 & 0.020 & 0.032 \\ 0.071 & 0.465 & 1.877 & 0.019 & 0.045 & 0.031 \\ 0.751 & 0.134 & 0.100 & 1.740 & 0.066 & 0.124 \\ 0.060 & 0.045 & 0.130 & 0.082 & 1.578 & 0.059 \\ 0.339 & 0.236 & 0.307 & 0.312 & 0.376 & 1.349 \end{pmatrix}$$

Therefore the "ideal" output vector is given by

$$\mathbf{x} = (I - A)^{-1}\mathbf{e} = \begin{pmatrix} 1.234 & 0.014 & 0.006 & 0.064 & 0.007 & 0.018 \\ 0.017 & 1.436 & 0.057 & 0.012 & 0.020 & 0.032 \\ 0.071 & 0.465 & 1.877 & 0.019 & 0.045 & 0.031 \\ 0.751 & 0.134 & 0.100 & 1.740 & 0.066 & 0.124 \\ 0.060 & 0.045 & 0.130 & 0.082 & 1.578 & 0.059 \\ 0.339 & 0.236 & 0.307 & 0.312 & 0.376 & 1.349 \end{pmatrix} \begin{pmatrix} 99{,}640 \\ 75{,}548 \\ 14{,}444 \\ 33{,}501 \\ 23{,}527 \\ 263{,}985 \end{pmatrix}$$

$$= \begin{pmatrix} 131{,}161 \\ 120{,}324 \\ 79{,}194 \\ 178{,}936 \\ 66{,}703 \\ 426{,}542 \end{pmatrix}$$

This means that it would require 131,161 units ($131,161 million worth) of final nonmetal products, 120,324 units of final metal products, 79,194 units of basic metal products, 178,936 units of basic nonmetal products, 66,703 units of energy and 426,542 service units to run the U.S. economy and meet the external demands in 1958. ■

PROBLEMS 7.5

In Problems 1–15 determine whether the given matrix is invertible. If it is, calculate the inverse.

1. $\begin{pmatrix} 2 & 1 \\ 3 & 2 \end{pmatrix}$ **2.** $\begin{pmatrix} -1 & 6 \\ 2 & -12 \end{pmatrix}$

3. $\begin{pmatrix} 0 & 1 \\ 1 & 0 \end{pmatrix}$ **4.** $\begin{pmatrix} 1 & 1 \\ 3 & 3 \end{pmatrix}$

5. $\begin{pmatrix} a & a \\ b & b \end{pmatrix}$ **6.** $\begin{pmatrix} 1 & 1 & 1 \\ 0 & 2 & 3 \\ 5 & 5 & 1 \end{pmatrix}$

7. $\begin{pmatrix} 3 & 2 & 1 \\ 0 & 2 & 2 \\ 0 & 0 & -1 \end{pmatrix}$ **8.** $\begin{pmatrix} 1 & 1 & 1 \\ 0 & 1 & 1 \\ 0 & 0 & 1 \end{pmatrix}$

9. $\begin{pmatrix} 1 & 6 & 2 \\ -2 & 3 & 5 \\ 7 & 12 & -4 \end{pmatrix}$ **10.** $\begin{pmatrix} 3 & 1 & 0 \\ 1 & -1 & 2 \\ 1 & 1 & 1 \end{pmatrix}$

11. $\begin{pmatrix} 2 & -1 & 4 \\ -1 & 0 & 5 \\ 19 & -7 & 3 \end{pmatrix}$ **12.** $\begin{pmatrix} 1 & 2 & 3 \\ 1 & 1 & 2 \\ 0 & 1 & 2 \end{pmatrix}$

13. $\begin{pmatrix} 1 & 1 & 1 & 1 \\ 1 & 2 & -1 & 2 \\ 1 & -1 & 2 & 1 \\ 1 & 3 & 3 & 2 \end{pmatrix}$ **14.** $\begin{pmatrix} 1 & 0 & 2 & 3 \\ -1 & 1 & 0 & 4 \\ 2 & 1 & -1 & 3 \\ -1 & 0 & 5 & 7 \end{pmatrix}$

15. $\begin{pmatrix} 1 & -3 & 0 & -2 \\ 3 & -12 & -2 & -6 \\ -2 & 10 & 2 & 5 \\ -1 & 6 & 1 & 3 \end{pmatrix}$

16. Show that if A, B, and C are invertible matrices, then ABC is invertible and $(ABC)^{-1} = C^{-1}B^{-1}A^{-1}$.

17. If $A_1, A_2, \ldots, A_m$ are invertible $n \times n$ matrices, show that $A_1A_2 \cdots A_m$ is invertible and calculate its inverse.

18. Show that the matrix $\begin{pmatrix} 3 & 4 \\ -2 & -3 \end{pmatrix}$ is equal to its own inverse.

19. Show that the matrix $\begin{pmatrix} a_{11} & a_{12} \\ a_{21} & a_{22} \end{pmatrix}$ is equal to its own inverse if $A = \pm I$ or if $a_{11} = -a_{22}$ and $a_{21}a_{12} = 1 - a_{11}^2$.

20. Find the output vector $\mathbf{x}$ in the Leontief input–output model if $n = 3$, $\mathbf{e} = \begin{pmatrix} 30 \\ 20 \\ 40 \end{pmatrix}$, and $A = \begin{pmatrix} \frac{1}{5} & \frac{1}{5} & 0 \\ \frac{2}{5} & \frac{2}{5} & \frac{3}{5} \\ \frac{1}{5} & \frac{1}{10} & \frac{2}{5} \end{pmatrix}$.

***21.** Suppose that A is $n \times m$ and B is $m \times n$ so that AB is $n \times n$. Show that AB is not invertible if $n > m$. [*Hint:* Show that there is a nonzero vector $\mathbf{x}$ such that $AB\mathbf{x} = \mathbf{0}$ and then apply Theorem 5.]

***22.** Use the methods of this section to find the inverses of the following matrices with complex entries (see Appendix 3).

(a) $\begin{pmatrix} i & 2 \\ 1 & -i \end{pmatrix}$ **(b)** $\begin{pmatrix} 1-i & 0 \\ 0 & 1+i \end{pmatrix}$

(c) $\begin{pmatrix} 1 & i & 0 \\ -i & 0 & 1 \\ 0 & 1+i & 1-i \end{pmatrix}$

23. Show that for every real number θ the matrix $\begin{pmatrix} \sin\theta & \cos\theta & 0 \\ \cos\theta & -\sin\theta & 0 \\ 0 & 0 & 1 \end{pmatrix}$ is invertible and find its inverse.

24. Calculate the inverse of $A = \begin{pmatrix} 2 & 0 & 0 \\ 0 & 3 & 0 \\ 0 & 0 & 4 \end{pmatrix}$.

25. A square matrix $A = (a_{ij})$ is called **diagonal** if all its elements off the main diagonal are zero. That is, $a_{ij} = 0$ if $i \neq j$. (The matrix of Problem 24 is diagonal.) Show that a diagonal matrix is invertible if and only if each of its diagonal components is nonzero.

26. Let

$$A = \begin{pmatrix} a_{11} & 0 & \cdots & 0 \\ 0 & a_{22} & \cdots & 0 \\ & & \ddots & \\ 0 & 0 & \cdots & a_{nn} \end{pmatrix}$$

be a diagonal matrix such that each of its diagonal components is nonzero. Calculate A^{-1}.

27. Calculate the inverse of $A = \begin{pmatrix} 2 & 1 & -1 \\ 0 & 3 & 4 \\ 0 & 0 & 5 \end{pmatrix}$.

28. Show that the matrix $A = \begin{pmatrix} 1 & 0 & 0 \\ -2 & 0 & 0 \\ 4 & 6 & 1 \end{pmatrix}$ is not invertible.

***29.** A square matrix is called **upper (lower) triangular** if all its elements below (above) the main diagonal are zero. (The matrix of Problem 27 is upper triangular and the matrix of Problem 28 is lower triangular.) Show that an upper or lower triangular matrix is invertible if and only if each of its diagonal elements is nonzero.

***30.** Show that the inverse of an invertible upper triangular matrix is upper triangular. [*Hint:* First prove the result for a 3×3 matrix.]

In Problems 31 and 32, show that the matrix given is not invertible by finding a nonzero vector $\mathbf{x}$ such that $A\mathbf{x} = \mathbf{0}$.

31. $\begin{pmatrix} 2 & -1 \\ -4 & 2 \end{pmatrix}$ **32.** $\begin{pmatrix} 1 & -1 & 3 \\ 0 & 4 & -2 \\ 2 & -6 & 8 \end{pmatrix}$

33. A factory for the construction of quality furniture has two divisions: a machine shop where the parts of the furniture are fabricated, and an assembly and finishing division where the parts are put together into the finished product. Suppose there are 12 employees in the machine shop and 20 in the assembly and finishing division and that each employee works an 8-hour day. Suppose further that the factory produces only two products: chairs and tables. A chair requires $\frac{384}{17}$ hours of machine shop time and $\frac{480}{17}$ hours of assembly and finishing time. A table requires $\frac{240}{17}$ hours of machine shop time and $\frac{640}{17}$ hours of assembly and finishing time. Assuming that there is an unlimited demand for these products and that the manufacturer wishes to keep all employees busy, how many chairs and how many tables can this factory produce each day?

34. A witch's magic cupboard contains 10 oz of ground four-leaf clovers and 14 oz of powdered mandrake root. The cupboard will replenish itself automatically provided she uses up exactly all her supplies. A batch of love potion requires $3\frac{1}{13}$ oz of ground four-leaf clovers and $2\frac{2}{13}$ oz of powdered mandrake root. One recipe of a well-known (to witches) cure for the common cold requires $5\frac{5}{13}$ oz of four-leaf clovers and $10\frac{10}{13}$ oz of mandrake root. How much of the love potion and the cold remedy should the witch make in order to use up the supply in the cupboard exactly?

35. A farmer feeds cattle a mixture of two types of feed. One standard unit of type A feed supplies a steer with 10% of its minimum daily requirement of protein and 15% of its requirement of carbohydrates. Type B feed contains 12% of the requirement of protein and 8% of the requirement of carbohydrates in a standard unit. If the farmer wishes to feed the cattle exactly 100% of their minimum daily requirement of protein and carbohydrates, how many units of each type of feed should she give a steer each day?

36. A much simplified version of an input-output table for the 1958 Israeli economy divides that economy into three sectors—agriculture, manufacturing, and energy—with the following result.[†]

	Agriculture	Manufacturing	Energy
Agriculture	0.293	0	0
Manufacturing	0.014	0.207	0.017
Energy	0.044	0.010	0.216

[†]Wassily Leontief, *Input-Output Economics* (Oxford University Press, New York, 1966), 54–57.

(a) How many units of agricultural production are required to produce one unit of agricultural output?
(b) How many units of agricultural production are required to produce 200,000 units of agricultural output?
(c) How many units of agricultural product go into the production of 50,000 units of energy?
(d) How many units of energy go into the production of 50,000 units of agricultural products?

37. Continuing Problem 36, exports (in thousands of Israeli pounds) in 1958 were as follows:

Agriculture	13,213
Manufacturing	17,597
Energy	1,786

(a) Compute the technology and Leontief matrices.
(b) Determine the number of Israeli pounds worth of agricultural products, manufactured goods, and energy required to run this model of the Israeli economy and export the stated value of products.

38. Prove Theorem 5. [*Hint:* Explain why (i) implies (ii) implies (iii) implies (iv) implies (i).]

In Problems 39–45 compute the row echelon form of the given matrix and use it to determine directly whether the given matrix is invertible.

39. The matrix of Problem 1.
40. The matrix of Problem 4.
41. The matrix of Problem 7.
42. The matrix of Problem 9.
43. The matrix of Problem 11.
44. The matrix of Problem 13.
45. The matrix of Problem 14.
46. Let $A = \begin{pmatrix} a_{11} & a_{12} \\ a_{21} & a_{22} \end{pmatrix}$ and assume that $a_{11}a_{22} - a_{12}a_{21} \neq 0$. Derive formula (12) by row reducing the augmented matrix $\left(\begin{array}{cc|cc} a_{11} & a_{12} & 1 & 0 \\ a_{21} & a_{22} & 0 & 1 \end{array}\right)$.

7.6 THE TRANSPOSE OF A MATRIX

Corresponding to every matrix is another matrix that, as we shall see in Chapter 8, has properties very similar to those of the original matrix.

Definition 1 TRANSPOSE Let $A = (a_{ij})$ be an $m \times n$ matrix. Then the **transpose** *of* A, written A^t, is the $n \times m$ matrix obtained by interchanging the rows and columns of A. Succinctly, we may write $A^t = (a_{ji})$. In other words,

$$\text{if } A = \begin{pmatrix} a_{11} & a_{12} & \cdots & a_{1n} \\ a_{21} & a_{22} & \cdots & a_{2n} \\ \vdots & \vdots & & \vdots \\ a_{m1} & a_{m2} & \cdots & a_{mn} \end{pmatrix}, \quad \text{then} \quad A^t = \begin{pmatrix} a_{11} & a_{21} & \cdots & a_{m1} \\ a_{12} & a_{22} & \cdots & a_{m2} \\ \vdots & \vdots & & \vdots \\ a_{1n} & a_{2n} & \cdots & a_{mn} \end{pmatrix}. \tag{1}$$

Simply put, the ith row of A is the ith column of A^t and the jth column of A is the jth row of A^t.

EXAMPLE 1 Find the transposes of the matrices

$$A = \begin{pmatrix} 2 & 3 \\ 1 & 4 \end{pmatrix} \qquad B = \begin{pmatrix} 2 & 3 & 1 \\ -1 & 4 & 6 \end{pmatrix} \qquad C = \begin{pmatrix} 1 & 2 & -6 \\ 2 & -3 & 4 \\ 0 & 1 & 2 \\ 2 & -1 & 5 \end{pmatrix}$$

Solution. Interchanging the rows and columns of each matrix, we obtain

$$A^t = \begin{pmatrix} 2 & 1 \\ 3 & 4 \end{pmatrix} \qquad B^t = \begin{pmatrix} 2 & -1 \\ 3 & 4 \\ 1 & 6 \end{pmatrix} \qquad C^t = \begin{pmatrix} 1 & 2 & 0 & 2 \\ 2 & -3 & 1 & -1 \\ -6 & 4 & 2 & 5 \end{pmatrix}$$

Note, for example, that 4 is the component in row 2 and column 3 of C while 4 is the component in row 3 and column 2 of C^t. That is, the 2,3 element of C is the 3,2 element of C^t. ■

Theorem 1 Suppose $A = (a_{ij})$ is an $n \times m$ matrix and $B = (b_{ij})$ is an $m \times p$ matrix. Then:

(i) $(A^t)^t = A$. **(2)**

(ii) $(AB)^t = B^tA^t$ **(3)**

(iii) If A and B are $n \times m$, then $(A + B)^t = A^t + B^t$. **(4)**

Proof.

(i) This follows directly from the definition of the transpose.

(ii) First we note that AB is an $n \times p$ matrix, so $(AB)^t$ is $p \times n$. Also, B^t is $p \times m$ and A^t is $m \times n$, so B^tA^t is $p \times n$. Thus both matrices in equation (3) have the same size. Now the ijth element of AB is $\sum_{k=1}^{m} a_{ik}b_{kj}$ and this is the jith element of $(AB)^t$. Let $C = B^t$ and $D = A^t$. Then the ijth element c_{ij} of C is b_{ji} and the ijth element d_{ij} of D is a_{ji}. Thus the jith element of CD = the jith element of $B^tA^t = \sum_{k=1}^{m} c_{jk}d_{ki} = \sum_{k=1}^{m} b_{kj}a_{ik} = \sum_{k=1}^{m} a_{ik}b_{kj}$ = the jith element of $(AB)^t$. This completes the proof of part (ii).

(iii) This part is left as an exercise (see Problem 11). ■

The transpose plays an important role in matrix theory. We shall see in succeeding chapters that A and A^t have many properties in common. Since columns of A^t are rows of A, we shall be able to use facts about the transpose to conclude that just about anything which is true about the rows of a matrix is true about its columns. We conclude this section with an important definition.

Definition 2 SYMMETRIC MATRIX The $n \times n$ (square) matrix A is called **symmetric** if $A^t = A$.

EXAMPLE 2 The following four matrices are symmetric:

$$I \qquad A = \begin{pmatrix} 1 & 2 \\ 2 & 3 \end{pmatrix} \qquad B = \begin{pmatrix} 1 & -4 & 2 \\ -4 & 7 & 5 \\ 2 & 5 & 0 \end{pmatrix} \qquad C = \begin{pmatrix} -1 & 2 & 4 & 6 \\ 2 & 7 & 3 & 5 \\ 4 & 3 & 8 & 0 \\ 6 & 5 & 0 & -4 \end{pmatrix}$$ ■

PROBLEMS 7.6

In Problems 1–10 find the transpose of the given matrix.

1. $\begin{pmatrix} -1 & 4 \\ 6 & 5 \end{pmatrix}$

2. $\begin{pmatrix} 3 & 0 \\ 1 & 2 \end{pmatrix}$

3. $\begin{pmatrix} 2 & 3 \\ -1 & 2 \\ 1 & 4 \end{pmatrix}$

4. $\begin{pmatrix} 2 & -1 & 0 \\ 1 & 5 & 6 \end{pmatrix}$

5. $\begin{pmatrix} 1 & 2 & 3 \\ -1 & 0 & 4 \\ 1 & 5 & 5 \end{pmatrix}$

6. $\begin{pmatrix} 1 & 2 & 3 \\ 2 & 4 & -5 \\ 3 & -5 & 7 \end{pmatrix}$

7. $\begin{pmatrix} 1 & 0 & 1 & 0 \\ 0 & 1 & 0 & 1 \end{pmatrix}$

8. $\begin{pmatrix} 2 & -1 \\ 2 & 4 \\ 1 & 6 \\ 1 & 5 \end{pmatrix}$

9. $\begin{pmatrix} a & b & c \\ d & e & f \\ g & h & i \end{pmatrix}$

10. $\begin{pmatrix} 0 & 0 & 0 \\ 0 & 0 & 0 \end{pmatrix}$

11. Let A and B be $n \times m$ matrices. Show, using Definition 1, that $(A + B)^t = A^t + B^t$.

12. Find numbers α and β such that $\begin{pmatrix} 2 & \alpha & 3 \\ 5 & -6 & 2 \\ \beta & 2 & 4 \end{pmatrix}$ is symmetric.

13. If A and B are symmetric $n \times n$ matrices, prove that $A + B$ is symmetric.

14. If A and B are symmetric $n \times n$ matrices, show that $(AB)^t = BA$.

15. For any matrix A, show that the product matrix AA^t is defined and is a symmetric matrix.

16. Show that every diagonal matrix (see Problem 7.5.25) is symmetric.

17. Show that the transpose of every upper triangular matrix (see Problem 7.5.29) is lower triangular.

18. A square matrix is called **skew-symmetric** if $A^t = -A$ (that is, $a_{ij} = -a_{ji}$). Which of the following matrices are skew-symmetric?

(a) $\begin{pmatrix} 1 & -6 \\ 6 & 0 \end{pmatrix}$ **(b)** $\begin{pmatrix} 0 & -6 \\ 6 & 0 \end{pmatrix}$

(c) $\begin{pmatrix} 2 & -2 & -2 \\ 2 & 2 & -2 \\ 2 & 2 & 2 \end{pmatrix}$ **(d)** $\begin{pmatrix} 0 & 1 & -1 \\ -1 & 0 & 2 \\ 1 & -2 & 0 \end{pmatrix}$

19. Let A and B be $n \times n$ skew-symmetric matrices. Show that $A + B$ is skew-symmetric.

20. If A is skew-symmetric, show that every component on the main diagonal of A is zero.

21. If A and B are skew-symmetric $n \times n$ matrices, show that $(AB)^t = BA$, so that AB is symmetric if and only if A and B commute.

***22.** Let $A = \begin{pmatrix} a_{11} & a_{12} \\ a_{21} & a_{22} \end{pmatrix}$ be a matrix with nonnegative entries having the properties that (i) $a_{11}^2 + a_{21}^2 = 1$ and $a_{12}^2 + a_{22}^2 = 1$ and (ii) $\begin{pmatrix} a_{11} \\ a_{21} \end{pmatrix} \cdot \begin{pmatrix} a_{12} \\ a_{22} \end{pmatrix} = 0$. Show that A is invertible and that $A^{-1} = A^t$.

REVIEW EXERCISES FOR CHAPTER SEVEN

In Exercises 1–14, find all solutions (if any) to the given systems.

1. $\begin{aligned} 3x_1 + 6x_2 &= 9 \\ -2x_1 + 3x_2 &= 4 \end{aligned}$

2. $\begin{aligned} 3x_1 + 6x_2 &= 9 \\ 2x_1 + 4x_2 &= 6 \end{aligned}$

3. $\begin{aligned} 3x_1 - 6x_2 &= 9 \\ -2x_1 + 4x_2 &= 6 \end{aligned}$

4. $\begin{aligned} x_1 + x_2 + x_3 &= 2 \\ 2x_1 - x_2 + 2x_3 &= 4 \\ -3x_1 + 2x_2 + 3x_3 &= 8 \end{aligned}$

5. $\begin{aligned} x_1 + x_2 + x_3 &= 0 \\ 2x_1 - x_2 + 2x_3 &= 0 \\ -3x_1 + 2x_2 + 3x_3 &= 0 \end{aligned}$

6. $\begin{aligned} x_1 + x_2 + x_3 &= 2 \\ 2x_1 - x_2 + 2x_3 &= 4 \\ -x_1 + 4x_2 + x_3 &= 2 \end{aligned}$

7. $\begin{aligned} x_1 + x_2 + x_3 &= 2 \\ 2x_1 - x_2 + 2x_3 &= 4 \\ -x_1 + 4x_2 + x_3 &= 3 \end{aligned}$

8. $\begin{aligned} x_1 + x_2 + x_3 &= 0 \\ 2x_1 - x_2 + 2x_3 &= 0 \\ -x_1 + 4x_2 + x_3 &= 0 \end{aligned}$

9. $\begin{aligned} 2x_1 + x_2 - 3x_3 &= 0 \\ 4x_1 - x_2 + x_3 &= 0 \end{aligned}$

10. $\begin{aligned} x_1 + x_2 &= 0 \\ 2x_1 + x_2 &= 0 \\ 3x_1 + x_2 &= 0 \end{aligned}$

11. $\begin{aligned} x_1 + x_2 &= 1 \\ 2x_1 + x_2 &= 3 \\ 3x_1 + x_2 &= 4 \end{aligned}$

12. $\begin{aligned} x_1 + x_2 + x_3 + x_4 &= 4 \\ 2x_1 - 3x_2 - x_3 + 4x_4 &= 7 \\ -2x_1 + 4x_2 + x_3 - 2x_4 &= 1 \\ 5x_1 - x_2 + 2x_3 + x_4 &= -1 \end{aligned}$

13. $\begin{aligned} x_1 + x_2 + x_3 + x_4 &= 0 \\ 2x_1 - 3x_2 - x_3 + 4x_4 &= 0 \\ -2x_1 + 4x_2 + x_3 - 2x_4 &= 0 \\ 5x_1 - x_2 + 2x_3 + x_4 &= 0 \end{aligned}$

14. $\begin{aligned} x_1 + x_2 + x_3 + x_4 &= 0 \\ 2x_1 - 3x_2 - x_3 + 4x_4 &= 0 \\ -2x_1 + 4x_2 + x_3 - 2x_4 &= 0 \end{aligned}$

In Exercises 15–19 determine whether the given matrix is in row echelon form (but not reduced row echelon form), reduced row echelon form, or neither.

15. $\begin{pmatrix} 1 & 0 & 0 & 0 \\ 0 & 1 & 0 & 2 \\ 0 & 0 & 1 & 3 \end{pmatrix}$

16. $\begin{pmatrix} 1 & 8 & 1 & 0 \\ 0 & 1 & 5 & -7 \\ 0 & 0 & 1 & 4 \end{pmatrix}$

17. $\begin{pmatrix} 1 & 0 \\ 0 & 3 \\ 0 & 0 \end{pmatrix}$

18. $\begin{pmatrix} 1 & 0 & 2 & 0 \\ 0 & 1 & 3 & 0 \end{pmatrix}$

19. $\begin{pmatrix} 1 & 1 & 1 & 1 \\ 0 & 1 & 1 & 1 \end{pmatrix}$

In Exercises 20 and 21, reduce the matrix to row echelon form and reduced row echelon form.

20. $\begin{pmatrix} 2 & 8 & -2 \\ 1 & 0 & -6 \end{pmatrix}$

21. $\begin{pmatrix} 1 & -1 & 2 & 4 \\ -1 & 2 & 0 & 3 \\ 2 & 3 & -1 & 1 \end{pmatrix}$

In Exercises 22–29 perform the indicated computations.

22. $3\begin{pmatrix} -2 & 1 \\ 0 & 4 \\ 2 & 3 \end{pmatrix}$

23. $\begin{pmatrix} 1 & 0 & 3 \\ 2 & -1 & 6 \end{pmatrix} + \begin{pmatrix} 2 & 0 & 4 \\ -2 & 5 & 8 \end{pmatrix}$

24. $5\begin{pmatrix} 2 & 1 & 3 \\ -1 & 2 & 4 \\ -6 & 1 & 5 \end{pmatrix} - 3\begin{pmatrix} -2 & 1 & 4 \\ 5 & 0 & 7 \\ 2 & -1 & 3 \end{pmatrix}$

25. $\begin{pmatrix} 2 & 3 \\ -1 & 4 \end{pmatrix}\begin{pmatrix} 5 & -1 \\ 2 & 7 \end{pmatrix}$

26. $\begin{pmatrix} 2 & 3 & 1 & 5 \\ 0 & 6 & 2 & 4 \end{pmatrix}\begin{pmatrix} 5 & 7 & 1 \\ 2 & 0 & 3 \\ 1 & 0 & 0 \\ 0 & 5 & 6 \end{pmatrix}$

27. $\begin{pmatrix} 2 & 3 & 5 \\ -1 & 6 & 4 \\ 1 & 0 & 6 \end{pmatrix}\begin{pmatrix} 0 & -1 & 2 \\ 3 & 1 & 2 \\ -7 & 3 & 5 \end{pmatrix}$

28. $\begin{pmatrix} 1 & 0 & 3 & -1 & 5 \\ 2 & 1 & 6 & 2 & 5 \end{pmatrix}\begin{pmatrix} 7 & 1 \\ 2 & 3 \\ -1 & 0 \\ 5 & 6 \\ 2 & 3 \end{pmatrix}$

29. $\begin{pmatrix} 1 & -1 & 2 \\ 3 & 5 & 6 \\ 2 & 4 & -1 \end{pmatrix}\begin{pmatrix} 2 \\ 1 \\ 3 \end{pmatrix}$

30. Verify the associative law of matrix multiplication for the matrices

$$A = \begin{pmatrix} 2 & 3 & 1 \\ 0 & 4 & 6 \end{pmatrix}, \quad B = \begin{pmatrix} 1 & 0 & 2 \\ 0 & 3 & 3 \\ 5 & 1 & -1 \end{pmatrix}, \text{ and } C = \begin{pmatrix} 5 & 6 \\ -1 & 2 \\ 0 & 1 \end{pmatrix}.$$

In Exercises 31–35 calculate the row echelon form and the inverse of the given matrix (if the inverse exists).

31. $\begin{pmatrix} 2 & 3 \\ -1 & 4 \end{pmatrix}$

32. $\begin{pmatrix} -1 & 2 \\ 2 & -4 \end{pmatrix}$

33. $\begin{pmatrix} 1 & 2 & 0 \\ 2 & 1 & -1 \\ 3 & 1 & 1 \end{pmatrix}$

34. $\begin{pmatrix} -1 & 2 & 0 \\ 4 & 1 & -3 \\ 2 & 5 & -3 \end{pmatrix}$

35. $\begin{pmatrix} 2 & 0 & 4 \\ -1 & 3 & 1 \\ 0 & 1 & 2 \end{pmatrix}$

In Exercises 36–38 first write the system in the form $A\mathbf{x} = \mathbf{b}$, then calculate A^{-1}, and finally, use matrix multiplication to obtain the solution vector.

36. $\begin{aligned} x_1 - 3x_2 &= 4 \\ 2x_1 + 5x_2 &= 7 \end{aligned}$

37. $\begin{aligned} x_1 + 2x_2 \phantom{{}- x_3} &= 3 \\ 2x_1 + x_2 - x_3 &= -1 \\ 3x_1 + x_2 + x_3 &= 7 \end{aligned}$

38. $\begin{aligned} 2x_1 \phantom{{}+ 3x_2} + 4x_3 &= 7 \\ -x_1 + 3x_2 + x_3 &= -4 \\ x_2 + 2x_3 &= 5 \end{aligned}$

In Exercises 39–44 calculate the transpose of the given matrix and determine whether the matrix is symmetric or skew-symmetric.†

39. $\begin{pmatrix} 2 & 3 & 1 \\ -1 & 0 & 2 \end{pmatrix}$

40. $\begin{pmatrix} 4 & 6 \\ 6 & 4 \end{pmatrix}$

√ **41.** $\begin{pmatrix} 2 & 3 & 1 \\ 3 & -6 & -5 \\ 1 & -5 & 9 \end{pmatrix}$

42. $\begin{pmatrix} 0 & 5 & 6 \\ -5 & 0 & 4 \\ -6 & -4 & 0 \end{pmatrix}$

43. $\begin{pmatrix} 1 & -1 & 4 & 6 \\ -1 & 2 & 5 & 7 \\ 4 & 5 & 3 & -8 \\ 6 & 7 & -8 & 9 \end{pmatrix}$

44. $\begin{pmatrix} 0 & 1 & -1 & 1 \\ -1 & 0 & 1 & -2 \\ 1 & 1 & 0 & 1 \\ 1 & -2 & -1 & 0 \end{pmatrix}$

Skew symmetric $A^t = -A$

†From Problem 7.6.18 we have that A is skew-symmetric if $A^t = -A$.

8 Determinants

8.1 DEFINITIONS

Let $A = \begin{pmatrix} a_{11} & a_{12} \\ a_{21} & a_{22} \end{pmatrix}$ be a 2×2 matrix. In Section 7.5 (page 433) we defined the determinant of A by

$$\det A = a_{11}a_{22} - a_{12}a_{21} \tag{1}$$

We shall often denote $\det A$ by

$$|A| = \begin{vmatrix} a_{11} & a_{12} \\ a_{21} & a_{22} \end{vmatrix} \tag{2}$$

We showed that A is invertible if $\det A \neq 0$. As we shall see, this important theorem is valid for $n \times n$ matrices.

In this chapter we shall develop some of the basic properties of determinants and see how they can be used to calculate inverses and solve systems of n linear equations in n unknowns.

We shall define the determinant of an $n \times n$ matrix *inductively*. In other words, we use our knowledge of a 2×2 determinant to define a 3×3 determinant, use this to define a 4×4 determinant, and so on. We start by defining a 3×3 determinant.†

Definition 1 3×3 DETERMINANT Let $A = \begin{pmatrix} a_{11} & a_{12} & a_{13} \\ a_{21} & a_{22} & a_{23} \\ a_{31} & a_{32} & a_{33} \end{pmatrix}$. Then

$$\det A = |A| = a_{11}\begin{vmatrix} a_{22} & a_{23} \\ a_{32} & a_{33} \end{vmatrix} - a_{12}\begin{vmatrix} a_{21} & a_{23} \\ a_{31} & a_{33} \end{vmatrix} + a_{13}\begin{vmatrix} a_{21} & a_{22} \\ a_{31} & a_{32} \end{vmatrix} \quad (3)$$

Note the minus sign before the second term on the right side of (3).

EXAMPLE 1

Let $A = \begin{pmatrix} 3 & 5 & 2 \\ 4 & 2 & 3 \\ -1 & 2 & 4 \end{pmatrix}$. Calculate $|A|$.

Solution.

$$|A| = \begin{vmatrix} 3 & 5 & 2 \\ 4 & 2 & 3 \\ -1 & 2 & 4 \end{vmatrix} = 3\begin{vmatrix} 2 & 3 \\ 2 & 4 \end{vmatrix} - 5\begin{vmatrix} 4 & 3 \\ -1 & 4 \end{vmatrix} + 2\begin{vmatrix} 4 & 2 \\ -1 & 2 \end{vmatrix}$$

$$= 3 \cdot 2 - 5 \cdot 19 + 2 \cdot 10 = -69 \quad \blacksquare$$

EXAMPLE 2 Calculate $\begin{vmatrix} 2 & -3 & 5 \\ 1 & 0 & 4 \\ 3 & -3 & 9 \end{vmatrix}$.

Solution.

$$\begin{vmatrix} 2 & -3 & 5 \\ 1 & 0 & 4 \\ 3 & -3 & 9 \end{vmatrix} = 2\begin{vmatrix} 0 & 4 \\ -3 & 9 \end{vmatrix} - (-3)\begin{vmatrix} 1 & 4 \\ 3 & 9 \end{vmatrix} + 5\begin{vmatrix} 1 & 0 \\ 3 & -3 \end{vmatrix}$$

$$= 2 \cdot 12 + 3(-3) + 5(-3) = 0 \quad \blacksquare$$

There is a simpler method for calculating 3×3 determinants. From equation (3) we have

$$\begin{vmatrix} a_{11} & a_{12} & a_{13} \\ a_{21} & a_{22} & a_{23} \\ a_{31} & a_{32} & a_{33} \end{vmatrix} = a_{11}(a_{22}a_{33} - a_{23}a_{32}) - a_{12}(a_{21}a_{33} - a_{23}a_{31}) + a_{13}(a_{21}a_{32} - a_{22}a_{31})$$

†There are several ways to define a determinant and this is one of them. It is important to realize that "det" is a function which assigns a *number* to a *square* matrix.

or

$$|A| = a_{11}a_{22}a_{33} + a_{12}a_{23}a_{31} + a_{13}a_{21}a_{32} - a_{13}a_{22}a_{31} - a_{12}a_{21}a_{33} - a_{11}a_{32}a_{23}. \qquad \textbf{(4)}$$

We write A and adjoin to it its first two columns

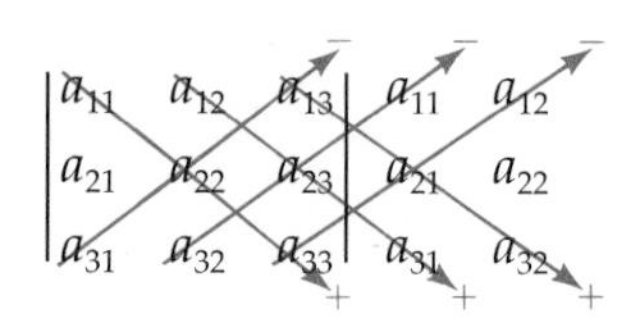

We then calculate the six products, put minus signs before the products with arrows pointing upward, and add. This gives the sum in equation (4).

EXAMPLE 3 Calculate $\begin{vmatrix} 3 & 5 & 2 \\ 4 & 2 & 3 \\ -1 & 2 & 4 \end{vmatrix}$ by using this new method.

Solution. Writing $\left|\begin{matrix} 3 & 5 & 2 \\ 4 & 2 & 3 \\ -1 & 2 & 4 \end{matrix}\right|\begin{matrix} 3 & 5 \\ 4 & 2 \\ -1 & 2 \end{matrix}$ and multiplying as indicated, we obtain

$$\begin{aligned} |A| &= (3)(2)(4) + (5)(3)(-1) + (2)(4)(2) - (-1)(2)(2) - 2(3)(3) - (4)(4)(5) \\ &= 24 - 15 + 16 + 4 - 18 - 80 = -69. \blacksquare \end{aligned}$$

■ WARNING: The method given above will *not* work for $n \times n$ determinants if $n \neq 3$. If you try something analogous for 4×4 or higher-order determinants, you will get the wrong answer.

Before defining $n \times n$ determinants, we first note that in equation (3), $\begin{pmatrix} a_{22} & a_{23} \\ a_{32} & a_{33} \end{pmatrix}$ is the matrix obtained by deleting the first row and first column of A; $\begin{pmatrix} a_{21} & a_{23} \\ a_{31} & a_{33} \end{pmatrix}$ is the matrix obtained by deleting the first row and second column of A; and $\begin{pmatrix} a_{21} & a_{22} \\ a_{31} & a_{32} \end{pmatrix}$ is the matrix obtained by deleting the first row and third column of A. If we denote these three matrices by M_{11}, M_{12}, and M_{13}, respectively, and if $A_{11} = \det M_{11}$, $A_{12} = -\det M_{12}$, and $A_{13} = \det M_{13}$, then equation (3) can be written

$$\det A = |A| = a_{11}A_{11} + a_{12}A_{12} + a_{13}A_{13} \qquad \textbf{(5)}$$

Definition 2 MINOR Let A be an $n \times n$ matrix and let M_{ij} be the $(n-1) \times (n-1)$ matrix obtained from A by deleting the ith row and jth column of A. M_{ij} is called the ***ij*th minor** of A.

EXAMPLE 4 Let $A = \begin{pmatrix} 2 & -1 & 4 \\ 0 & 1 & 5 \\ 6 & 3 & -4 \end{pmatrix}$. Find M_{13} and M_{32}.

Solution. Deleting the first row and third column of A, we obtain $M_{13} = \begin{pmatrix} 0 & 1 \\ 6 & 3 \end{pmatrix}$. Similarly, by eliminating the third row and second column we obtain $M_{32} = \begin{pmatrix} 2 & 4 \\ 0 & 5 \end{pmatrix}$. ■

EXAMPLE 5 Let $A = \begin{pmatrix} 1 & -3 & 5 & 6 \\ 2 & 4 & 0 & 3 \\ 1 & 5 & 9 & -2 \\ 4 & 0 & 2 & 7 \end{pmatrix}$. Find M_{32} and M_{24}.

Solution. Deleting the third row and second column of A, we find that $M_{32} = \begin{pmatrix} 1 & 5 & 6 \\ 2 & 0 & 3 \\ 4 & 2 & 7 \end{pmatrix}$; similarly, $M_{24} = \begin{pmatrix} 1 & -3 & 5 \\ 1 & 5 & 9 \\ 4 & 0 & 2 \end{pmatrix}$. ■

Definition 3 COFACTOR Let A be an $n \times n$ matrix. The ***ij*th cofactor** of A, denoted A_{ij}, is given by

$$A_{ij} = (-1)^{i+j}|M_{ij}|. \tag{6}$$

That is, the ijth cofactor of A is obtained by taking the determinant of the ijth minor and multiplying it by $(-1)^{i+j}$. Note that

$$(-1)^{i+j} = \begin{cases} 1 & \text{if } i + j \text{ is even} \\ -1 & \text{if } i + j \text{ is odd} \end{cases}$$

REMARK. Definition 3 makes sense because we are going to define an $n \times n$ determinant with the assumption that we already know what an $(n - 1) \times (n - 1)$ determinant is.

EXAMPLE 6 In Example 5 we have

$$A_{32} = (-1)^{3+2}|M_{32}| = -\begin{vmatrix} 1 & 5 & 6 \\ 2 & 0 & 3 \\ 4 & 2 & 7 \end{vmatrix} = -8$$

and

$$A_{24} = (-1)^{2+4}\begin{vmatrix} 1 & -3 & 5 \\ 1 & 5 & 9 \\ 4 & 0 & 2 \end{vmatrix} = -192. \ ■$$

We now consider the general $n \times n$ matrix. Here

$$A = \begin{pmatrix} a_{11} & a_{12} & \cdots & a_{1n} \\ a_{21} & a_{22} & \cdots & a_{2n} \\ \vdots & \vdots & & \vdots \\ a_{n1} & a_{n2} & \cdots & a_{nn} \end{pmatrix}. \tag{7}$$

Definition 4 $n \times n$ DETERMINANT Let A be an $n \times n$ matrix with $n \geq 3$. Then the determinant of A, written det A or $|A|$, is given by

$$\det A = |A| = a_{11}A_{11} + a_{12}A_{12} + a_{13}A_{13} + \cdots + a_{1n}A_{1n}$$
$$= \sum_{k=1}^{n} a_{1k}A_{1k}. \tag{8}$$

The expression on the right side of (8) is called an **expansion by cofactors.**

In equation (8) we defined the determinant by expanding by cofactors using components of A in the first row. We shall see in the next section (Theorem 8.2.1) that we get the same answer if we expand by cofactors in any row or column.

EXAMPLE 7 Calculate det A, where

$$A = \begin{pmatrix} 1 & 3 & 5 & 2 \\ 0 & -1 & 3 & 4 \\ 2 & 1 & 9 & 6 \\ 3 & 2 & 4 & 8 \end{pmatrix}.$$

Solution.

$$\begin{vmatrix} 1 & 3 & 5 & 2 \\ 0 & -1 & 3 & 4 \\ 2 & 1 & 9 & 6 \\ 3 & 2 & 4 & 8 \end{vmatrix} = a_{11}A_{11} + a_{12}A_{12} + a_{13}A_{13} + a_{14}A_{14}$$

$$= 1\begin{vmatrix} -1 & 3 & 4 \\ 1 & 9 & 6 \\ 2 & 4 & 8 \end{vmatrix} - 3\begin{vmatrix} 0 & 3 & 4 \\ 2 & 9 & 6 \\ 3 & 4 & 8 \end{vmatrix} + 5\begin{vmatrix} 0 & -1 & 4 \\ 2 & 1 & 6 \\ 3 & 2 & 8 \end{vmatrix} - 2\begin{vmatrix} 0 & -1 & 3 \\ 2 & 1 & 9 \\ 3 & 2 & 4 \end{vmatrix}$$

$$= 1(-92) - 3(-70) + 5(2) - 2(-16) = 160 \quad \blacksquare$$

It is clear that calculating the determinant of an $n \times n$ matrix can be tedious. To calculate a 4×4 determinant, we must calculate four 3×3 determinants. To calculate a 5×5 determinant, we must calculate five 4×4 determinants—which is the same as calculating twenty 3×3 determinants. Fortunately, there are techniques for greatly simplifying these computations. Some of these methods are discussed in the next section. There are, however, some matrices whose determinants can easily be calculated.

Definition 5 A square matrix is called **upper triangular** if all its components below the diagonal are zero. It is **lower triangular** if all its components above the diagonal are zero. A matrix is called **diagonal** if all its elements not on the diagonal are zero; that is, $A = (a_{ij})$ is upper triangular if $a_{ij} = 0$ for $i > j$, lower triangular if $a_{ij} = 0$ for $i < j$, and diagonal if $a_{ij} = 0$ for $i \neq j$. Note that a diagonal matrix is both upper and lower triangular.

EXAMPLE 8 The matrices $A = \begin{pmatrix} 2 & 1 & 7 \\ 0 & 2 & -5 \\ 0 & 0 & 1 \end{pmatrix}$ and $B = \begin{pmatrix} -2 & 3 & 0 & 1 \\ 0 & 0 & 2 & 4 \\ 0 & 0 & 1 & 3 \\ 0 & 0 & 0 & -2 \end{pmatrix}$ are upper triangular; $C = \begin{pmatrix} 5 & 0 & 0 \\ 2 & 3 & 0 \\ -1 & 2 & 4 \end{pmatrix}$ and $D = \begin{pmatrix} 0 & 0 \\ 1 & 0 \end{pmatrix}$ are lower triangular; I and $E = \begin{pmatrix} 2 & 0 & 0 \\ 0 & -7 & 0 \\ 0 & 0 & -4 \end{pmatrix}$ are diagonal. ■

EXAMPLE 9 Let

$$A = \begin{pmatrix} a_{11} & 0 & 0 & 0 \\ a_{21} & a_{22} & 0 & 0 \\ a_{31} & a_{32} & a_{33} & 0 \\ a_{41} & a_{42} & a_{43} & a_{44} \end{pmatrix}$$

be lower triangular. Compute det A.

Solution.

$$\begin{aligned} \det A &= a_{11}A_{11} + 0A_{12} + 0A_{13} + 0A_{14} = a_{11}A_{11} \\ &= a_{11} \begin{vmatrix} a_{22} & 0 & 0 \\ a_{32} & a_{33} & 0 \\ a_{42} & a_{43} & a_{44} \end{vmatrix} \\ &= a_{11}a_{22} \begin{vmatrix} a_{33} & 0 \\ a_{43} & a_{44} \end{vmatrix} \\ &= a_{11}a_{22}a_{33}a_{44} \quad ■ \end{aligned}$$

Example 9 can easily be generalized to prove the following:

Theorem 1 Let $A = (a_{ij})$ be an upper[†] or lower triangular $n \times n$ matrix. Then

$$\det A = a_{11}a_{22}a_{33} \cdots a_{nn} \tag{9}$$

[†]The proof for the upper triangular case is more difficult at this stage, but it will be just the same once we know that det A can be evaluated by expanding in any column (Theorem 8.2.1).

That is,

> The determinant of a triangular matrix equals the product of its diagonal components.

EXAMPLE 10 The determinants of the six matrices in Example 8 are $|A| = 2 \cdot 2 \cdot 1 = 4$; $|B| = (-2)(0)(1)(-2) = 0$; $|C| = 5 \cdot 3 \cdot 4 = 60$; $|D| = 0$; $|I| = 1$; $|E| = (2)(-7)(-4) = 56$. ■

PROBLEMS 8.1

In Problems 1–10 calculate the determinant.

1. $\begin{vmatrix} 1 & 0 & 3 \\ 0 & 1 & 4 \\ 2 & 1 & 0 \end{vmatrix}$

2. $\begin{vmatrix} -1 & 1 & 0 \\ 2 & 1 & 4 \\ 1 & 5 & 6 \end{vmatrix}$

3. $\begin{vmatrix} 3 & -1 & 4 \\ 6 & 3 & 5 \\ 2 & -1 & 6 \end{vmatrix}$

4. $\begin{vmatrix} -1 & 0 & 6 \\ 0 & 2 & 4 \\ 1 & 2 & -3 \end{vmatrix}$

5. $\begin{vmatrix} -2 & 3 & 1 \\ 4 & 6 & 5 \\ 0 & 2 & 1 \end{vmatrix}$

6. $\begin{vmatrix} 5 & -2 & 1 \\ 6 & 0 & 3 \\ -2 & 1 & 4 \end{vmatrix}$

7. $\begin{vmatrix} 2 & 0 & 3 & 1 \\ 0 & 1 & 4 & 2 \\ 0 & 0 & 1 & 5 \\ 1 & 2 & 3 & 0 \end{vmatrix}$

8. $\begin{vmatrix} -3 & 0 & 0 & 0 \\ -4 & 7 & 0 & 0 \\ 5 & 8 & -1 & 0 \\ 2 & 3 & 0 & 6 \end{vmatrix}$

9. $\begin{vmatrix} -2 & 0 & 0 & 7 \\ 1 & 2 & -1 & 4 \\ 3 & 0 & -1 & 5 \\ 4 & 2 & 3 & 0 \end{vmatrix}$

10. $\begin{vmatrix} 2 & 3 & -1 & 4 & 5 \\ 0 & 1 & 7 & 8 & 2 \\ 0 & 0 & 4 & -1 & 5 \\ 0 & 0 & 0 & -2 & 8 \\ 0 & 0 & 0 & 0 & 6 \end{vmatrix}$

11. Show that if A and B are diagonal $n \times n$ matrices, then $\det AB = \det A \det B$.

***12.** Show that if A and B are lower triangular matrices, then $\det AB = \det A \det B$.

13. Show that, in general, it is not true that $\det(A + B) = \det A + \det B$.

14. Show that if A is triangular, then $\det A \neq 0$ if and only if all the diagonal components of A are nonzero.

15. Prove Theorem 1 for a lower triangular matrix.

***16.** We say that the vectors $\begin{pmatrix} 1 \\ 0 \end{pmatrix}$ and $\begin{pmatrix} 0 \\ 1 \end{pmatrix}$ *generate the area* 1 in the plane since if we construct a square with three of its vertices at (0, 0), (1, 0), and (0, 1), we see that the area is 1. (See Figure 1a.) More generally, if $\begin{pmatrix} x_1 \\ y_1 \end{pmatrix}$ and $\begin{pmatrix} x_2 \\ y_2 \end{pmatrix}$ are not collinear, then they generate an area defined to be the area of the parallelogram with three of its four vertices at (0, 0), (x_1, y_1), and (x_2, y_2). (See Figure 1b.)

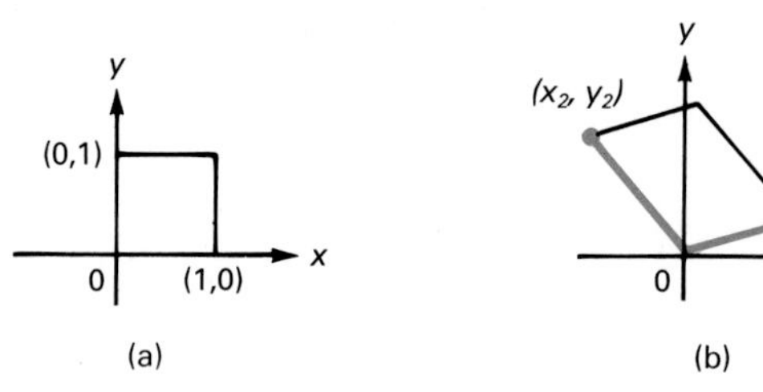

FIGURE 1

Let A be a 2×2 matrix. If k denotes the area generated by $\begin{pmatrix} x_1 \\ y_1 \end{pmatrix}$ and $\begin{pmatrix} x_2 \\ y_2 \end{pmatrix}$, where $\begin{pmatrix} x_1 \\ y_1 \end{pmatrix} = A\begin{pmatrix} 1 \\ 0 \end{pmatrix}$ and $\begin{pmatrix} x_2 \\ y_2 \end{pmatrix} = A\begin{pmatrix} 0 \\ 1 \end{pmatrix}$, show that $k = |\det A|$.

****17.** Let $\mathbf{u}_1$ and $\mathbf{u}_2$ be two 2-vectors and let $\mathbf{v}_1 = A\mathbf{u}_1$ and $\mathbf{v}_2 = A\mathbf{u}_2$. Show that

$$\text{(area generated by } \mathbf{v}_1 \text{ and } \mathbf{v}_2) = \text{(area generated by } \mathbf{u}_1 \text{ and } \mathbf{u}_2)\ |\det A|$$

This provides a geometric interpretation of the determinant.

8.2 PROPERTIES OF DETERMINANTS

Determinants have many properties that can make computations easier. We begin to describe these properties by stating a theorem from which everything else follows. The proof of this theorem is difficult and is given in Appendix 4.

Theorem 1 BASIC THEOREM Let

$$A = \begin{pmatrix} a_{11} & a_{12} & \cdots & a_{1n} \\ a_{21} & a_{22} & \cdots & a_{2n} \\ \vdots & \vdots & & \vdots \\ a_{n1} & a_{n2} & \cdots & a_{nn} \end{pmatrix}$$

be an $n \times n$ matrix. Then

$$\det A = a_{i1}A_{i1} + a_{i2}A_{i2} + \cdots + a_{in}A_{in} = \sum_{k=1}^{n} a_{ik}A_{ik} \tag{1}$$

for $i = 1, 2, \ldots, n$. That is, we can calculate det A by expanding by cofactors in *any* row of A. Furthermore:

$$\det A = a_{1j}A_{1j} + a_{2j}A_{2j} + \cdots + a_{nj}A_{nj} = \sum_{k=1}^{n} a_{kj}A_{kj} \tag{2}$$

Since the jth column of A is $\begin{pmatrix} a_{1j} \\ a_{2j} \\ \vdots \\ a_{nj} \end{pmatrix}$, equation (2) indicates that we can calculate det A by expanding by cofactors in any column of A.

EXAMPLE 1 For $A = \begin{pmatrix} 3 & 5 & 2 \\ 4 & 2 & 3 \\ -1 & 2 & 4 \end{pmatrix}$, we saw in Example 8.1.1 that det $A = -69$. Expanding in the second row we obtain

$$\begin{aligned} \det A &= 4A_{21} + 2A_{22} + 3A_{23} \\ &= 4(-1)^{2+1}\begin{vmatrix} 5 & 2 \\ 2 & 4 \end{vmatrix} + 2(-1)^{2+2}\begin{vmatrix} 3 & 2 \\ -1 & 4 \end{vmatrix} + 3(-1)^{2+3}\begin{vmatrix} 3 & 5 \\ -1 & 2 \end{vmatrix} \\ &= -4(16) + 2(14) - 3(11) = -69. \end{aligned}$$

Similarly, if we expand in the third column, say, we obtain

$$\begin{aligned}\det A &= 2A_{13} + 3A_{23} + 4A_{33}\\ &= 2(-1)^{1+3}\begin{vmatrix} 4 & 2\\ -1 & 2\end{vmatrix} + 3(-1)^{2+3}\begin{vmatrix} 3 & 5\\ -1 & 2\end{vmatrix} + 4(-1)^{3+3}\begin{vmatrix} 3 & 5\\ 4 & 2\end{vmatrix}\\ &= 2(10) - 3(11) + 4(-14) = -69.\end{aligned}$$

You should verify that we get the same answer if we expand in the third row or the first or second column. ■

We now list and prove some additional properties of determinants. In each case we assume that A is an $n \times n$ matrix.† We shall see that these properties can be used to reduce greatly the work involved in evaluating a determinant.

Property 1 If any row or column of A is the zero vector, then $\det A = 0$.

Proof. Suppose the ith row of A contains all zeroes. That is, $a_{ij} = 0$ for $j = 1, 2, \ldots, n$. Then $\det A = a_{i1}A_{i1} + a_{i2}A_{i2} + \cdots + a_{in}A_{in} = 0 + 0 + \cdots + 0 = 0$. The same proof works if the jth column is the zero vector. ■

EXAMPLE 2 It is easy to verify that

$$\begin{vmatrix} 2 & 3 & 5\\ 0 & 0 & 0\\ 1 & -2 & 4\end{vmatrix} = 0 \quad \text{and} \quad \begin{vmatrix} -1 & 3 & 0 & 1\\ 4 & 2 & 0 & 5\\ -1 & 6 & 0 & 4\\ 2 & 1 & 0 & 1\end{vmatrix} = 0. \ ■$$

Property 2 If the ith row or the jth column of A is multiplied by the constant c, then $\det A$ is multiplied by c. That is, if we call this new matrix B, then

$$|B| = \begin{vmatrix} a_{11} & a_{12} & \cdots & a_{1n}\\ a_{21} & a_{22} & \cdots & a_{2n}\\ \vdots & \vdots & & \vdots\\ ca_{i1} & ca_{i2} & \cdots & ca_{in}\\ \vdots & \vdots & & \vdots\\ a_{n1} & a_{n2} & \cdots & a_{nn}\end{vmatrix} = c\begin{vmatrix} a_{11} & a_{12} & \cdots & a_{1n}\\ a_{21} & a_{22} & \cdots & a_{2n}\\ \vdots & \vdots & & \vdots\\ a_{i1} & a_{i2} & \cdots & a_{in}\\ \vdots & \vdots & & \vdots\\ a_{n1} & a_{n2} & \cdots & a_{nn}\end{vmatrix} = c|A|. \tag{3}$$

Proof. To prove (3) we expand in the ith row of A to obtain

$$\begin{aligned}\det B &= ca_{i1}A_{i1} + ca_{i2}A_{i2} + \cdots + ca_{in}A_{in}\\ &= c(a_{i1}A_{i1} + a_{i2}A_{i2} + \cdots + a_{in}A_{in}) = c \det A\end{aligned}$$

†The proofs of these properties are given in terms of either the rows (for columns) of a matrix. Using Theorem 1 the same properties can be proved for columns (or rows).

A similar proof works for columns. ■

EXAMPLE 3 Let $A = \begin{pmatrix} 1 & -1 & 2 \\ 3 & 1 & 4 \\ 0 & -2 & 5 \end{pmatrix}$. Then $\det A = 16$. If we multiply the second row by 4, we have $B = \begin{pmatrix} 1 & -1 & 2 \\ 12 & 4 & 16 \\ 0 & -2 & 5 \end{pmatrix}$ and $\det B = 64 = 4 \det A$. If the third column is multiplied by -3, we obtain $C = \begin{pmatrix} 1 & -1 & -6 \\ 3 & 1 & -12 \\ 0 & -2 & -15 \end{pmatrix}$ and $\det C = -48 = -3 \det A$. ■

REMARK. Using Property 2 we can prove (see Problem 28) the following interesting fact: For any scalar α and $n \times n$ matrix A, $\det \alpha A = \alpha^n \det A$.

Property 3 Let

$$A = \begin{pmatrix} a_{11} & a_{12} & \cdots & a_{1j} & \cdots & a_{1n} \\ a_{21} & a_{22} & \cdots & a_{2j} & \cdots & a_{2n} \\ \vdots & \vdots & & \vdots & & \vdots \\ a_{n1} & a_{n2} & \cdots & a_{nj} & \cdots & a_{nn} \end{pmatrix}, \quad B = \begin{pmatrix} a_{11} & a_{12} & \cdots & \alpha_{1j} & \cdots & a_{1n} \\ a_{21} & a_{22} & \cdots & \alpha_{2j} & \cdots & a_{2n} \\ \vdots & \vdots & & \vdots & & \vdots \\ a_{n1} & a_{n2} & \cdots & \alpha_{nj} & \cdots & a_{nn} \end{pmatrix},$$

and

$$C = \begin{pmatrix} a_{11} & a_{12} & \cdots & a_{1j} + \alpha_{1j} & \cdots & a_{1n} \\ a_{21} & a_{22} & \cdots & a_{2j} + \alpha_{2j} & \cdots & a_{2n} \\ \vdots & \vdots & & \vdots & & \vdots \\ a_{n1} & a_{n2} & \cdots & a_{nj} + \alpha_{nj} & \cdots & a_{nn} \end{pmatrix}.$$

Then

$$\det C = \det A + \det B. \tag{4}$$

In other words, suppose that A, B, and C are identical except for the jth column and that the jth column of C is the sum of the jth columns of A and B. Then $\det C = \det A + \det B$. The same statement is true for rows.

Proof. We expand $\det C$ in the jth column to obtain

$$\begin{aligned} \det C &= (a_{1j} + \alpha_{1j})A_{1j} + (a_{2j} + \alpha_{2j})A_{2j} + \cdots + (a_{nj} + \alpha_{nj})A_{nj} \\ &= (a_{1j}A_{1j} + a_{2j}A_{2j} + \cdots + a_{nj}A_{nj}) \\ &\quad + (\alpha_{1j}A_{1j} + \alpha_{2j}A_{2j} + \cdots + \alpha_{nj}A_{nj}) = \det A + \det B. \end{aligned}$$ ■

EXAMPLE 4 Let $A = \begin{pmatrix} 1 & -1 & 2 \\ 3 & 1 & 4 \\ 0 & -2 & 5 \end{pmatrix}$, $B = \begin{pmatrix} 1 & -6 & 2 \\ 3 & 2 & 4 \\ 0 & 4 & 5 \end{pmatrix}$, and $C = \begin{pmatrix} 1 & -1-6 & 2 \\ 3 & 1+2 & 4 \\ 0 & -2+4 & 5 \end{pmatrix} = \begin{pmatrix} 1 & -7 & 2 \\ 3 & 3 & 4 \\ 0 & 2 & 5 \end{pmatrix}$. Then $\det A = 16$, $\det B = 108$, and $\det C = 124 = \det A + \det B$. ■

Property 4 Interchanging any two rows (or columns) of A has the effect of multiplying $\det A$ by -1.

Proof. We prove the statement for rows and assume first that two adjacent rows are interchanged. That is, we assume that the ith and $(i+1)$st rows are interchanged. Let

$$A = \begin{pmatrix} a_{11} & a_{12} & \cdots & a_{1n} \\ a_{21} & a_{22} & \cdots & a_{2n} \\ \vdots & \vdots & & \vdots \\ a_{i1} & a_{i2} & \cdots & a_{in} \\ a_{i+1,1} & a_{i+1,2} & \cdots & a_{i+1,n} \\ \vdots & \vdots & & \vdots \\ a_{n1} & a_{n2} & \cdots & a_{nn} \end{pmatrix} \quad \text{and } B = \begin{pmatrix} a_{11} & a_{12} & \cdots & a_{1n} \\ a_{21} & a_{22} & \cdots & a_{2n} \\ \vdots & \vdots & & \vdots \\ a_{i+1,1} & a_{i+1,2} & \cdots & a_{i+1,n} \\ a_{i1} & a_{i2} & \cdots & a_{in} \\ \vdots & \vdots & & \vdots \\ a_{n1} & a_{n2} & \cdots & a_{nn} \end{pmatrix}.$$

Then, expanding $\det A$ in its ith row and $\det B$ in its $(i+1)$st row, we obtain

$$\begin{aligned} \det A &= a_{i1}A_{i1} + a_{i2}A_{i2} + \cdots + a_{in}A_{in} \\ \det B &= a_{i1}B_{i+1,1} + a_{i2}B_{i+1,2} + \cdots + a_{in}B_{i+1,n}. \end{aligned} \tag{5}$$

Here $A_{ij} = (-1)^{i+j}|M_{ij}|$, where M_{ij} is obtained by crossing off the ith row and jth column of A. Notice now that if we cross off the $(i+1)$st row and jth column of B, we obtain the same M_{ij}. Thus

$$B_{i+1,j} = (-1)^{i+1+j}|M_{ij}| = -(-1)^{i+j}|M_{ij}| = -A_{ij}$$

so that, from equations (5), $\det B = -\det A$.

Now suppose that $i < j$ and that the ith and jth rows are to be interchanged. We can do this by interchanging adjacent rows several times. It will take $j - i$ interchanges to move row j into the ith row. Then row i will be in the $(i+1)$st row, and it will take an additional $j - i - 1$ interchanges to move row i into the jth row. To illustrate, we interchange rows 2 and 6:[†]

[†]Note that all the numbers here refer to rows.

$$\begin{matrix} 1 & & 1 & & 1 & & 1 & & 1 & & 1 & & 1 & & 1 \\ 2 & & 2 & & 2 & & 2 & & 6 & & 6 & & 6 & & 6 \\ 3 & & 3 & & 3 & & 6 & & 2 & & 3 & & 3 & & 3 \\ 4 & \to & 4 & \to & 6 & \to & 3 & \to & 3 & \to & 2 & \to & 4 & \to & 4 \\ 5 & & 6 & & 4 & & 4 & & 4 & & 4 & & 2 & & 5 \\ 6 & & 5 & & 5 & & 5 & & 5 & & 5 & & 5 & & 2 \\ 7 & & 7 & & 7 & & 7 & & 7 & & 7 & & 7 & & 7 \end{matrix}$$

$6 - 2 = 4$ interchanges to move the 6 into the 2 position; $6 - 2 - 1 = 3$ interchanges to get the 2 into the 6 position

Finally, the total number of interchanges of adjacent rows is $(j - i) + (j - i - 1) = 2j - 2i - 1$, which is odd. Thus det A is multiplied by -1 an odd number of times, which is what we needed to show. ■

EXAMPLE 5 Let $A = \begin{pmatrix} 1 & -1 & 2 \\ 3 & 1 & 4 \\ 0 & -2 & 5 \end{pmatrix}$. By interchanging the first and third rows we obtain $B = \begin{pmatrix} 0 & -2 & 5 \\ 3 & 1 & 4 \\ 1 & -1 & 2 \end{pmatrix}$. By interchanging the first and second columns of A we obtain $C = \begin{pmatrix} -1 & 1 & 2 \\ 1 & 3 & 4 \\ -2 & 0 & 5 \end{pmatrix}$. Then, by direct calculation, we find that det $A = 16$ and det B = det $C = -16$. ■

Property 5 If A has two equal rows or columns, then det $A = 0$.

Proof. Suppose the ith and jth rows of A are equal. By interchanging these rows we get a matrix B having the property that det $B = -$det A (from Property 4). But since row i = row j, interchanging them gives us the same matrix. Thus $A = B$ and det A = det $B = -$det A. Thus 2 det $A = 0$, which can happen only if det $A = 0$. ■

EXAMPLE 6 By direct calculation we can verify that for $A = \begin{pmatrix} 1 & -1 & 2 \\ 5 & 7 & 3 \\ 1 & -1 & 2 \end{pmatrix}$, which has two equal rows, and $B = \begin{pmatrix} 5 & 2 & 2 \\ 3 & -1 & -1 \\ -2 & 4 & 4 \end{pmatrix}$, which has two equal columns, det A = det $B = 0$. ■

Property 6 If one row (column) of A is a constant multiple of another row (column), then det $A = 0$.

Proof. Let $(a_{j1}, a_{j2}, \ldots, a_{jn}) = c(a_{i1}, a_{i2}, \ldots, a_{in})$. Then, from Property 2,

$$\det A = c \begin{vmatrix} a_{11} & a_{12} & \cdots & a_{1n} \\ a_{21} & a_{22} & \cdots & a_{2n} \\ \vdots & \vdots & & \vdots \\ a_{i1} & a_{i2} & \cdots & a_{in} \\ \vdots & \vdots & & \vdots \\ a_{i1} & a_{i2} & \cdots & a_{in} \\ \vdots & \vdots & & \vdots \\ a_{n1} & a_{n2} & \cdots & a_{nn} \end{vmatrix} = 0 \qquad \text{From Property 5.}$$

(jth row $\rightarrow$ the second row $a_{i1}\ a_{i2}\ \cdots\ a_{in}$.) ■

EXAMPLE 7 $\begin{vmatrix} 2 & -3 & 5 \\ 1 & 7 & 2 \\ -4 & 6 & -10 \end{vmatrix} = 0$ since the third row is -2 times the first row. ■

EXAMPLE 8 $\begin{vmatrix} 2 & 4 & 1 & 12 \\ -1 & 1 & 0 & 3 \\ 0 & -1 & 9 & -3 \\ 7 & 3 & 6 & 9 \end{vmatrix} = 0$ since the fourth column is three times the second column. ■

Property 7 If a multiple of one row (column) of A is added to another row (column) of A, then the determinant is unchanged.

Proof. Let B be the matrix obtained by adding c times the ith row of A to the jth row of A. Then

$$\det B = \begin{vmatrix} a_{11} & a_{12} & \cdots & a_{1n} \\ a_{21} & a_{22} & \cdots & a_{2n} \\ \vdots & \vdots & & \vdots \\ a_{i1} & a_{i2} & \cdots & a_{in} \\ \vdots & \vdots & & \vdots \\ a_{j1} + ca_{i1} & a_{j2} + ca_{i2} & \cdots & a_{jn} + ca_{in} \\ \vdots & \vdots & & \vdots \\ a_{n1} & a_{n2} & \cdots & a_{nn} \end{vmatrix}$$

$$\overset{\text{From Property 3}}{=} \begin{vmatrix} a_{11} & a_{12} & \cdots & a_{1n} \\ a_{21} & a_{22} & \cdots & a_{2n} \\ \vdots & \vdots & & \vdots \\ a_{i1} & a_{i2} & \cdots & a_{in} \\ \vdots & \vdots & & \vdots \\ a_{j1} & a_{j2} & \cdots & a_{jn} \\ \vdots & \vdots & & \vdots \\ a_{n1} & a_{n2} & \cdots & a_{nn} \end{vmatrix} + \begin{vmatrix} a_{11} & a_{12} & \cdots & a_{1n} \\ a_{21} & a_{22} & \cdots & a_{2n} \\ \vdots & \vdots & & \vdots \\ a_{i1} & a_{i2} & \cdots & a_{in} \\ \vdots & \vdots & & \vdots \\ ca_{i1} & ca_{i2} & \cdots & ca_{in} \\ \vdots & \vdots & & \vdots \\ a_{n1} & a_{n2} & \cdots & a_{nn} \end{vmatrix}$$

$$= \det A + 0 = \det A. \quad ■$$

(From Property 6 — the 0)

EXAMPLE 9 Let $A = \begin{pmatrix} 1 & -1 & 2 \\ 3 & 1 & 4 \\ 0 & -2 & 5 \end{pmatrix}$. Then det $A = 16$. If we multiply the third row by 4 and add it to the second row, we obtain a new matrix B given by

$$B = \begin{pmatrix} 1 & -1 & 2 \\ 3+4(0) & 1+4(-2) & 4+5(4) \\ 0 & -2 & 5 \end{pmatrix} = \begin{pmatrix} 1 & -1 & 2 \\ 3 & -7 & 24 \\ 0 & -2 & 5 \end{pmatrix}$$

and det $B = 16 =$ det A. ■

The properties discussed above make it much easier to evaluate high-order determinants. We simply row-reduce the determinant, using Property 7, until the determinant is in an easily evaluated form. The most common goal will be to use Property 7 repeatedly until either (i) the new determinant has a row (column) of zeros or one row (column) a multiple of another row (column)—in which case the determinant is zero—or (ii) the new matrix is triangular so that its determinant is the product of its diagonal elements.

EXAMPLE 10 Calculate

$$|A| = \begin{vmatrix} 1 & 3 & 5 & 2 \\ 0 & -1 & 3 & 4 \\ 2 & 1 & 9 & 6 \\ 3 & 2 & 4 & 8 \end{vmatrix}$$

Solution. (See Example 8.1.7.)

There is already a zero in the first column, so it is simplest to reduce other elements in the first column to zero. We then continue to reduce, aiming for a triangular matrix:

Multiply the first row by -2 and add it to the third row; and multiply the first row by -3 and add it to the fourth row:

$$|A| = \begin{vmatrix} 1 & 3 & 5 & 2 \\ 0 & -1 & 3 & 4 \\ 0 & -5 & -1 & 2 \\ 0 & -7 & -11 & 2 \end{vmatrix}$$

Multiply the second row by -5 and -7 and add it to the third and fourth rows, respectively:

$$= \begin{vmatrix} 1 & 3 & 5 & 2 \\ 0 & -1 & 3 & 4 \\ 0 & 0 & -16 & -18 \\ 0 & 0 & -32 & -26 \end{vmatrix}$$

Factor out -16 from the third row (using Property 2):

$$= -16 \begin{vmatrix} 1 & 3 & 5 & 2 \\ 0 & -1 & 3 & 4 \\ 0 & 0 & 1 & \frac{9}{8} \\ 0 & 0 & -32 & -26 \end{vmatrix}$$

Multiply the third row by 32 and add it to the fourth row:

$$= -16\begin{vmatrix} 1 & 3 & 5 & 2 \\ 0 & -1 & 3 & 4 \\ 0 & 0 & 1 & \frac{9}{8} \\ 0 & 0 & 0 & 10 \end{vmatrix}$$

Now we have an upper triangular matrix and $|A| = -16(1)(-1)(1)(10) = (-16)(-10) = 160$. ■

EXAMPLE 11 Calculate

$$|A| = \begin{vmatrix} -2 & 1 & 0 & 4 \\ 3 & -1 & 5 & 2 \\ -2 & 7 & 3 & 1 \\ 3 & -7 & 2 & 5 \end{vmatrix}$$

Solution. There are a number of ways to proceed here, and it is not apparent which way will get us the answer most quickly. However, since there is already one zero in the first row, we begin our reduction in that row by making the first and fourth components in the first row equal to zero.

Multiply the second column by 2 and -4 and add it to the first and fourth columns, respectively:

$$|A| = \begin{vmatrix} 0 & 1 & 0 & 0 \\ 1 & -1 & 5 & 6 \\ 12 & 7 & 3 & -27 \\ -11 & -7 & 2 & 33 \end{vmatrix}$$

Interchange the first two columns:

$$= -\begin{vmatrix} 1 & 0 & 0 & 0 \\ -1 & 1 & 5 & 6 \\ 7 & 12 & 3 & -27 \\ -7 & -11 & 2 & 33 \end{vmatrix}$$

Multiply the second column by -5 and -6 and it to the third and fourth columns, respectively:

$$= -\begin{vmatrix} 1 & 0 & 0 & 0 \\ -1 & 1 & 0 & 0 \\ 7 & 12 & -57 & -99 \\ -7 & -11 & 57 & 99 \end{vmatrix}$$

Since the fourth column is now a multiple of the third column (column 4 $= \frac{99}{57} \times$ column 3), we see that $|A| = 0$. ■

There are three additional facts about determinants that will be very useful to us.

Theorem 2 Let A be an $n \times n$ matrix. Then

$$a_{i1}A_{j1} + a_{i2}A_{j2} + \cdots + a_{in}A_{jn} = 0 \quad \text{if } i \neq j. \tag{6}$$

NOTE. From Theorem 1 the sum in equation (6) equals det A if $i = j$.

Proof. Let

$$B = \begin{pmatrix} a_{11} & a_{12} & \cdots & a_{1n} \\ a_{21} & a_{22} & \cdots & a_{2n} \\ \vdots & \vdots & & \vdots \\ a_{i1} & a_{i2} & \cdots & a_{in} \\ \vdots & \vdots & & \vdots \\ a_{i1} & a_{i2} & \cdots & a_{in} \\ \vdots & \vdots & & \vdots \\ a_{n1} & a_{n2} & \cdots & a_{nn} \end{pmatrix} \quad \leftarrow j\text{th row}$$

Then, since two rows of B are equal, det $B = 0$. But $B = A$ except in the jth row. Thus if we calculate det B by expanding in the jth row of B, we obtain the sum in (6) and the theorem is proved. Note that when we expand in the jth row, the jth row is deleted in computing the cofactors of B. Thus $B_{jk} = A_{jk}$ for $k = 1, 2, \ldots, n$. ■

Theorem 3 Let A be an $n \times n$ matrix. Then

$$\det A = \det A^t. \tag{7}$$

Proof. This proof uses mathematical induction. If you are unfamiliar with this important method of proof, refer to Appendix 1. We first prove the theorem in the case $n = 2$. If

$$|A| = \begin{vmatrix} a_{11} & a_{12} \\ a_{21} & a_{22} \end{vmatrix} = a_{11}a_{22} - a_{12}a_{21}$$

then

$$|A^t| = \begin{vmatrix} a_{11} & a_{21} \\ a_{12} & a_{22} \end{vmatrix} = a_{11}a_{22} - a_{21}a_{12} = |A|.$$

So the theorem is true for $n = 2$. Next we assume the theorem to be true for $(n - 1) \times (n - 1)$ matrices and prove it for $n \times n$ matrices. This will prove the theorem. Let $B = A^t$. Then

$$|A| = \begin{vmatrix} a_{11} & a_{12} & \cdots & a_{1n} \\ a_{21} & a_{22} & \cdots & a_{2n} \\ \vdots & \vdots & & \vdots \\ a_{n1} & a_{n2} & \cdots & a_{nn} \end{vmatrix} \quad \text{and} \quad |A^t| = |B| = \begin{vmatrix} a_{11} & a_{21} & \cdots & a_{n1} \\ a_{12} & a_{22} & \cdots & a_{n2} \\ \vdots & \vdots & & \vdots \\ a_{1n} & a_{2n} & \cdots & a_{nn} \end{vmatrix}$$

We expand $|A|$ in the first row and expand $|B|$ in the first column. This gives us

$$|A| = a_{11}A_{11} + a_{12}A_{12} + \cdots + a_{1n}A_{1n}$$
$$|B| = a_{11}B_{11} + a_{12}B_{21} + \cdots + a_{1n}B_{n1}$$

We need to show that $A_{1k} = B_{k1}$ for $k = 1, 2, \ldots, n$. But $A_{1k} = (-1)^{1+k}|M_{1k}|$ and $B_{k1} = (-1)^{k+1}|N_{k1}|$, where M_{1k} is the $1k$th minor of A and N_{k1} is the $k1$st minor of B. Then

$$|M_{1k}| = \begin{vmatrix} a_{21} & a_{22} & \cdots & a_{2,k-1} & a_{2,k+1} & \cdots & a_{2n} \\ a_{31} & a_{32} & \cdots & a_{3,k-1} & a_{3,k+1} & \cdots & a_{3n} \\ \vdots & \vdots & & \vdots & \vdots & & \vdots \\ a_{n1} & a_{n2} & \cdots & a_{n,k-1} & a_{n,k+1} & \cdots & a_{nn} \end{vmatrix}$$

and

$$|N_{k1}| = \begin{vmatrix} a_{21} & a_{31} & \cdots & a_{n1} \\ a_{22} & a_{32} & \cdots & a_{n2} \\ \vdots & \vdots & & \vdots \\ a_{2,k-1} & a_{3,k-1} & \cdots & a_{n,k-1} \\ a_{2,k+1} & a_{3,k+1} & \cdots & a_{n,k+1} \\ \vdots & \vdots & & \vdots \\ a_{2n} & a_{3n} & \cdots & a_{nn} \end{vmatrix}$$

Clearly $M_{1k} = N_{k1}^t$, and since both are $(n-1) \times (n-1)$ matrices, the induction hypothesis tells us that $|M_{1k}| = |N_{k1}|$. Thus $A_{1k} = B_{k1}$ and the proof is complete. ■

EXAMPLE 12 Let $A = \begin{pmatrix} 1 & -1 & 2 \\ 3 & 1 & 4 \\ 0 & -2 & 5 \end{pmatrix}$. Then $A^t = \begin{pmatrix} 1 & 3 & 0 \\ -1 & 1 & -2 \\ 2 & 4 & 5 \end{pmatrix}$, and it is easy to verify that $|A| = |A^t| = 16$. ■

Theorem 4 Let A and B be $n \times n$ matrices. Then

$$\det AB = \det A \det B \tag{8}$$

That is:

The determinant of the product is the product of the determinants.

REMARK. The proof of this theorem is not conceptually difficult, but, as you might imagine from having worked with matrix products, it is extremely cumbersome. For that reason we shall simply prove the theorem in the case that A and B are 2×2 matrices. Let $A = \begin{pmatrix} a_{11} & a_{12} \\ a_{21} & a_{22} \end{pmatrix}$ and $B = \begin{pmatrix} b_{11} & b_{12} \\ b_{21} & b_{22} \end{pmatrix}$. Then

$$\det A \det B = (a_{11}a_{22} - a_{12}a_{21})(b_{11}b_{22} - b_{12}b_{21})$$
$$= a_{11}a_{22}b_{11}b_{22} - a_{11}a_{22}b_{12}b_{21} - a_{12}a_{21}b_{11}b_{22} + a_{12}a_{21}b_{12}b_{21}$$

and

$$AB = \begin{pmatrix} a_{11}b_{11} + a_{12}b_{21} & a_{11}b_{12} + a_{12}b_{22} \\ a_{21}b_{11} + a_{22}b_{21} & a_{21}b_{12} + a_{22}b_{22} \end{pmatrix}.$$

Hence

$$\begin{aligned} \det AB &= (a_{11}b_{11} + a_{12}b_{21})(a_{21}b_{12} + a_{22}b_{22}) - (a_{11}b_{12} + a_{12}b_{22})(a_{21}b_{11} + a_{22}b_{21}) \\ &= a_{11}b_{11}a_{21}b_{12} + a_{11}b_{11}a_{22}b_{22} + a_{12}b_{21}a_{21}b_{12} + a_{12}b_{21}a_{22}b_{22} \\ &\quad - a_{11}b_{12}a_{21}b_{11} - a_{11}b_{12}a_{22}b_{21} - a_{12}b_{22}a_{21}b_{11} - a_{12}b_{22}a_{22}b_{21} \\ &= a_{11}b_{11}a_{22}b_{22} + a_{12}b_{21}a_{21}b_{12} - a_{11}b_{12}a_{22}b_{21} - a_{12}b_{22}a_{21}b_{11} \\ &= \det A \det B. \ \blacksquare \end{aligned}$$

EXAMPLE 13 Verify equation (8) for $A = \begin{pmatrix} 1 & -1 & 2 \\ 3 & 1 & 4 \\ 0 & -2 & 5 \end{pmatrix}$ and $B = \begin{pmatrix} 1 & -2 & 3 \\ 0 & -1 & 4 \\ 2 & 0 & -2 \end{pmatrix}$.

Solution. $\det A = 16$ and $\det B = -8$. We calculate

$$AB = \begin{pmatrix} 1 & -1 & 2 \\ 3 & 1 & 4 \\ 0 & -2 & 5 \end{pmatrix}\begin{pmatrix} 1 & -2 & 3 \\ 0 & -1 & 4 \\ 2 & 0 & -2 \end{pmatrix} = \begin{pmatrix} 5 & -1 & -5 \\ 11 & -7 & 5 \\ 10 & 2 & -18 \end{pmatrix}$$

and $\det AB = -128 = (16)(-8) = \det A \det B$. ■

PROBLEMS 8.2

In Problems 1–20 evaluate the determinant by using the methods of this section.

1. $\begin{vmatrix} 3 & -5 \\ 2 & 6 \end{vmatrix}$

2. $\begin{vmatrix} 4 & 1 \\ 0 & -3 \end{vmatrix}$

3. $\begin{vmatrix} -1 & 0 & 2 \\ 3 & 1 & 4 \\ 2 & 0 & -6 \end{vmatrix}$

4. $\begin{vmatrix} 2 & 1 & -1 \\ 3 & -2 & 0 \\ 5 & 1 & 6 \end{vmatrix}$

5. $\begin{vmatrix} -3 & 2 & 4 \\ 1 & -1 & 2 \\ -1 & 4 & 0 \end{vmatrix}$

6. $\begin{vmatrix} 0 & -2 & 3 \\ 1 & 2 & -3 \\ 4 & 0 & 5 \end{vmatrix}$

7. $\begin{vmatrix} -2 & 3 & 6 \\ 4 & 1 & 8 \\ -2 & 0 & 0 \end{vmatrix}$

8. $\begin{vmatrix} 2 & -1 & 3 \\ 4 & 0 & 6 \\ 5 & -2 & 3 \end{vmatrix}$

9. $\begin{vmatrix} 1 & -1 & 2 & 4 \\ 0 & -3 & 5 & 6 \\ 1 & 4 & 0 & 3 \\ 0 & 5 & -6 & 7 \end{vmatrix}$

10. $\begin{vmatrix} 2 & -3 & 1 & 4 \\ 0 & -2 & 0 & 0 \\ 3 & 7 & -1 & 2 \\ 4 & 1 & -3 & 8 \end{vmatrix}$

11. $\begin{vmatrix} 1 & 1 & -1 & 0 \\ -3 & 4 & 6 & 0 \\ 2 & 5 & -1 & 3 \\ 4 & 0 & 3 & 0 \end{vmatrix}$

12. $\begin{vmatrix} 3 & -1 & 2 & 1 \\ 4 & 3 & 1 & -2 \\ -1 & 0 & 2 & 3 \\ 6 & 2 & 5 & 2 \end{vmatrix}$

13. $\begin{vmatrix} 2 & 0 & 0 & 0 \\ 0 & 0 & 3 & 0 \\ 0 & -1 & 0 & 0 \\ 0 & 0 & 0 & 4 \end{vmatrix}$ **14.** $\begin{vmatrix} 0 & a & 0 & 0 \\ b & 0 & 0 & 0 \\ 0 & 0 & 0 & c \\ 0 & 0 & d & 0 \end{vmatrix}$

15. $\begin{vmatrix} 1 & 2 & 0 & 0 \\ 3 & -2 & 0 & 0 \\ 0 & 0 & 1 & -5 \\ 0 & 0 & 7 & 2 \end{vmatrix}$ **16.** $\begin{vmatrix} a & b & 0 & 0 \\ c & d & 0 & 0 \\ 0 & 0 & a & -b \\ 0 & 0 & c & d \end{vmatrix}$

17. $\begin{vmatrix} 2 & -1 & 0 & 4 & 1 \\ 3 & 1 & -1 & 2 & 0 \\ 3 & 2 & -2 & 5 & 1 \\ 0 & 0 & 4 & -1 & 6 \\ 3 & 2 & 1 & -1 & 1 \end{vmatrix}$ **18.** $\begin{vmatrix} 1 & -1 & 2 & 0 & 0 \\ 3 & 1 & 4 & 0 & 0 \\ 2 & -1 & 5 & 0 & 0 \\ 0 & 0 & 0 & 2 & 3 \\ 0 & 0 & 0 & -1 & 4 \end{vmatrix}$

19. $\begin{vmatrix} a & 0 & 0 & 0 & 0 \\ 0 & 0 & b & 0 & 0 \\ 0 & 0 & 0 & 0 & c \\ 0 & 0 & 0 & d & 0 \\ 0 & e & 0 & 0 & 0 \end{vmatrix}$ **20.** $\begin{vmatrix} 2 & 5 & -6 & 8 & 0 \\ 0 & 1 & -7 & 6 & 0 \\ 0 & 0 & 0 & 4 & 0 \\ 0 & 2 & 1 & 5 & 1 \\ 4 & -1 & 5 & 3 & 0 \end{vmatrix}$

In Problems 21 – 27 compute the determinant assuming that

$$\begin{vmatrix} a_{11} & a_{12} & a_{13} \\ a_{21} & a_{22} & a_{23} \\ a_{31} & a_{32} & a_{33} \end{vmatrix} = 8$$

21. $\begin{vmatrix} a_{31} & a_{32} & a_{33} \\ a_{21} & a_{22} & a_{23} \\ a_{11} & a_{12} & a_{13} \end{vmatrix}$ **22.** $\begin{vmatrix} a_{31} & a_{32} & a_{33} \\ a_{11} & a_{12} & a_{13} \\ a_{21} & a_{22} & a_{23} \end{vmatrix}$

23. $\begin{vmatrix} a_{11} & a_{12} & a_{13} \\ 2a_{21} & 2a_{22} & 2a_{23} \\ a_{31} & a_{32} & a_{33} \end{vmatrix}$

24. $\begin{vmatrix} -3a_{11} & -3a_{12} & -3a_{13} \\ 2a_{21} & 2a_{22} & 2a_{23} \\ 5a_{31} & 5a_{32} & 5a_{33} \end{vmatrix}$ **25.** $\begin{vmatrix} a_{11} & 2a_{13} & a_{12} \\ a_{21} & 2a_{23} & a_{22} \\ a_{31} & 2a_{33} & a_{32} \end{vmatrix}$

26. $\begin{vmatrix} a_{11} - a_{12} & a_{12} & a_{13} \\ a_{21} - a_{22} & a_{22} & a_{23} \\ a_{31} - a_{32} & a_{32} & a_{33} \end{vmatrix}$

27. $\begin{vmatrix} 2a_{11} - 3a_{21} & 2a_{12} - 3a_{22} & 2a_{13} - 3a_{23} \\ a_{31} & a_{32} & a_{33} \\ a_{21} & a_{22} & a_{23} \end{vmatrix}$

28. Using Property 2, show that if α is a number and A is an $n \times n$ matrix, then $\det \alpha A = \alpha^n \det A$.

***29.** Show that

$$\begin{vmatrix} 1 + x_1 & x_2 & x_3 & \cdots & x_n \\ x_1 & 1 + x_2 & x_3 & \cdots & x_n \\ x_1 & x_2 & 1 + x_3 & \cdots & x_n \\ \vdots & \vdots & \vdots & & \vdots \\ x_1 & x_2 & x_3 & \cdots & 1 + x_n \end{vmatrix} = 1 + x_1 + x_2 + \cdots + x_n$$

***30.** A matrix is **skew-symmetric** if $A^t = -A$. If A is an $n \times n$ skew-symmetric matrix, show that $\det A = (-1)^n \det A$.

31. Using the result of Problem 30, show that if A is a skew-symmetric $n \times n$ matrix and n is odd, then $\det A = 0$.

32. A matrix A is called **orthogonal** if A is invertible and $A^{-1} = A^t$. Show that if A is orthogonal, then $\det A = \pm 1$.

****33.** Let Δ denote the triangle in the plane with vertices at (x_1, y_1), (x_2, y_2), and (x_3, y_3). Show that the area of the triangle is given by

$$\text{Area of } \Delta = \pm\tfrac{1}{2} \begin{vmatrix} 1 & x_1 & y_1 \\ 1 & x_2 & y_2 \\ 1 & x_3 & y_3 \end{vmatrix}$$

Under what circumstances will this determinant equal zero?

****34.** Three lines, no two of which are parallel, determine a triangle in the plane. Suppose that the lines are given by

$$a_{11}x + a_{12}y + a_{13} = 0$$
$$a_{21}x + a_{22}y + a_{23} = 0$$
$$a_{31}x + a_{32}y + a_{33} = 0$$

Show that the area determined by the lines is

$$\frac{\pm 1}{2A_{13}A_{23}A_{33}} \begin{vmatrix} A_{11} & A_{12} & A_{13} \\ A_{21} & A_{22} & A_{23} \\ A_{31} & A_{32} & A_{33} \end{vmatrix}$$

35. The 3×3 **Vandermonde**[†] **determinant** is given by

$$D_3 = \begin{vmatrix} 1 & 1 & 1 \\ a_1 & a_2 & a_3 \\ a_1^2 & a_2^2 & a_3^2 \end{vmatrix}$$

[†] A. T. Vandermonde (1735–1796) was a French mathematician.

Show that $D_3 = (a_2 - a_1)(a_3 - a_1)(a_3 - a_2)$.

36. $D_4 = \begin{vmatrix} 1 & 1 & 1 & 1 \\ a_1 & a_2 & a_3 & a_4 \\ a_1^2 & a_2^2 & a_3^2 & a_4^2 \\ a_1^3 & a_2^3 & a_3^3 & a_4^3 \end{vmatrix}$ is the 4×4 Vandermonde determinant. Show that $D_4 = (a_2 - a_1)(a_3 - a_1) \times (a_4 - a_1)(a_3 - a_2)(a_4 - a_2)(a_4 - a_3)$.

****37. (a)** Define the $n \times n$ Vandermonde determinant D_n.

(b) Show that $D_n = \prod_{\substack{i=1 \\ j>i}}^{n} (a_j - a_i)$, where Π stands for the word "product." Note that the product in Problem 36 can be written $\prod_{\substack{i=1 \\ j>i}}^{4} (a_j - a_i)$.

8.3 DETERMINANTS AND INVERSES

In this section we shall see how matrix inverses can be calculated by using determinants. We begin with a simple result.

Theorem 1 If A is invertible, then $\det A \neq 0$ and

$$\det A^{-1} = \frac{1}{\det A}. \qquad (1)$$

Proof. From Theorems 8.2.4, page 464, and 8.1.1, page 453, we have

$$1 = \det I = \det AA^{-1} = \det A \det A^{-1} \qquad (2)$$

If $\det A$ were equal to zero, then equation (2) would read $1 = 0$. Thus $\det A \neq 0$ and $\det A^{-1} = 1/\det A$. ■

Before using determinants to calculate inverses, we need to define the *adjoint* of a matrix $A = (a_{ij})$. Let $B = (A_{ij})$ be the matrix of cofactors of A. (Remember that a cofactor, defined on page 451, is a number.) Then

$$B = \begin{pmatrix} A_{11} & A_{12} & \cdots & A_{1n} \\ A_{21} & A_{22} & \cdots & A_{2n} \\ \vdots & \vdots & & \vdots \\ A_{n1} & A_{n2} & \cdots & A_{nn} \end{pmatrix}. \qquad (3)$$

Definition 1 THE ADJOINT Let A be an $n \times n$ matrix and let B, given by (3), denote the matrix of its cofactors. Then the **adjoint** *of* A, written adj A, is the transpose of the $n \times n$ matrix B; that is,

$$\text{adj } A = B^t = \begin{pmatrix} A_{11} & A_{21} & \cdots & A_{n1} \\ A_{12} & A_{22} & \cdots & A_{n2} \\ \vdots & \vdots & & \vdots \\ A_{1n} & A_{2n} & \cdots & A_{nn} \end{pmatrix}. \qquad (4)$$

EXAMPLE 1 Let $A = \begin{pmatrix} 2 & 4 & 3 \\ 0 & 1 & -1 \\ 3 & 5 & 7 \end{pmatrix}$. Compute adj A.

Solution. We have $A_{11} = \begin{vmatrix} 1 & -1 \\ 5 & 7 \end{vmatrix} = 12$, $A_{12} = -\begin{vmatrix} 0 & -1 \\ 3 & 7 \end{vmatrix} = -3$, $A_{13} = -3$, $A_{21} = -13$, $A_{22} = 5$, $A_{23} = 2$, $A_{31} = -7$, $A_{32} = 2$, and $A_{33} = 2$. Thus $B = \begin{pmatrix} 12 & -3 & -3 \\ -13 & 5 & 2 \\ -7 & 2 & 2 \end{pmatrix}$ and adj $A = B^t = \begin{pmatrix} 12 & -13 & -7 \\ -3 & 5 & 2 \\ -3 & 2 & 2 \end{pmatrix}$. ■

EXAMPLE 2 Let

$$A = \begin{pmatrix} 1 & -3 & 0 & -2 \\ 3 & -12 & -2 & -6 \\ -2 & 10 & 2 & 5 \\ -1 & 6 & 1 & 3 \end{pmatrix}.$$

Calculate adj A.

Solution. This is more tedious since we have to compute sixteen 3×3 determinants. For example, we have $A_{12} = -\begin{vmatrix} 3 & -2 & -6 \\ -2 & 2 & 5 \\ -1 & 1 & 3 \end{vmatrix} = -1$, $A_{24} = \begin{vmatrix} 1 & -3 & 0 \\ -2 & 10 & 2 \\ -1 & 6 & 1 \end{vmatrix} = -2$, and $A_{43} = -\begin{vmatrix} 1 & -3 & -2 \\ 3 & -12 & -6 \\ -2 & 10 & 5 \end{vmatrix} = 3$. Completing these calculations, we find that

$$B = \begin{pmatrix} 0 & -1 & 0 & 2 \\ -1 & 1 & -1 & -2 \\ 0 & 2 & -3 & -3 \\ -2 & -2 & 3 & 2 \end{pmatrix}$$

and

$$\text{adj } A = B^t = \begin{pmatrix} 0 & -1 & 0 & -2 \\ -1 & 1 & 2 & -2 \\ 0 & -1 & -3 & 3 \\ 2 & -2 & -3 & 2 \end{pmatrix}. \quad ■$$

EXAMPLE 3 Let $A = \begin{pmatrix} a_{11} & a_{12} \\ a_{21} & a_{22} \end{pmatrix}$. Then adj $A = \begin{pmatrix} A_{11} & A_{21} \\ A_{12} & A_{22} \end{pmatrix} = \begin{pmatrix} a_{22} & -a_{12} \\ -a_{21} & a_{11} \end{pmatrix}$. ■

■ WARNING: In taking the adjoint of a matrix, do not forget to transpose the matrix of cofactors.

Theorem 2 Let A be an $n \times n$ matrix. Then

$$(A)(\text{adj } A) = (\text{adj } A)A = \begin{pmatrix} \det A & 0 & 0 & \cdots & 0 \\ 0 & \det A & 0 & \cdots & 0 \\ 0 & 0 & \det A & \cdots & 0 \\ \vdots & \vdots & \vdots & & \vdots \\ 0 & 0 & 0 & \cdots & \det A \end{pmatrix} = (\det A)I. \quad (5)$$

Proof. Let $C = (c_{ij}) = (A)(\text{adj } A)$. Then

$$C = \begin{pmatrix} a_{11} & a_{12} & \cdots & a_{1n} \\ a_{21} & a_{22} & \cdots & a_{2n} \\ \vdots & \vdots & & \vdots \\ a_{n1} & a_{n2} & \cdots & a_{nn} \end{pmatrix} \begin{pmatrix} A_{11} & A_{21} & \cdots & A_{n1} \\ A_{12} & A_{22} & \cdots & A_{n2} \\ \vdots & \vdots & & \vdots \\ A_{1n} & A_{2n} & \cdots & A_{nn} \end{pmatrix}. \quad (6)$$

We have

$$c_{ij} = (i\text{th row of } A) \cdot (j\text{th column of adj } A)$$

$$= (a_{i1} \quad a_{i2} \quad \cdots \quad a_{in}) \cdot \begin{pmatrix} A_{j1} \\ A_{j2} \\ \vdots \\ A_{jn} \end{pmatrix}.$$

Thus

$$c_{ij} = a_{i1}A_{j1} + a_{i2}A_{j2} + \cdots + a_{in}A_{jn}. \quad (7)$$

Now if $i = j$, the sum in (7) equals $a_{i1}A_{i1} + a_{i2}A_{i2} + \cdots + a_{in}A_{in}$, which is the expansion of $\det A$ in the ith row of A. On the other hand, if $i \neq j$, then from Theorem 8.2.2 the sum in (7) equals zero. Thus

$$c_{ij} = \begin{cases} \det A & \text{if } i = j \\ 0 & \text{if } i \neq j \end{cases}$$

Thus $C = (\det A)I$. A similar proof shows that $(\text{adj } A)A = C$. ■

We can now state the main result.

Theorem 3 Let A be an $n \times n$ matrix. Then A is invertible if and only if $\det A \neq 0$. If $\det A \neq 0$, then

$$A^{-1} = \frac{1}{\det A} \text{adj } A. \quad (8)$$

Note that Theorem 7.5.4 for 2×2 matrices is a special case of this theorem.

Proof. If A is invertible, then $\det A \neq 0$ by Theorem 1. If $\det A \neq 0$, then from Theorem 2,

$$(A)\left(\frac{1}{\det A}\operatorname{adj} A\right) = \frac{1}{\det A}[A(\operatorname{adj} A)] = \frac{1}{\det A}(\det A)I = I.$$

Similarly,

$$\left(\frac{1}{\det A}\operatorname{adj} A\right)(A) = I$$

and the proof is complete. ■

EXAMPLE 4 Let $A = \begin{pmatrix} 2 & 4 & 3 \\ 0 & 1 & -1 \\ 3 & 5 & 7 \end{pmatrix}$. Determine whether A is invertible, and, if it is, calculate A^{-1}.

Solution. Since $\det A = 3 \neq 0$, we see that A is invertible. From Example 1,

$$\operatorname{adj} A = \begin{pmatrix} 12 & -13 & -7 \\ -3 & 5 & 2 \\ -3 & 2 & 2 \end{pmatrix}.$$

Thus

$$A^{-1} = \frac{1}{3}\begin{pmatrix} 12 & -13 & -7 \\ -3 & 5 & 2 \\ -3 & 2 & 2 \end{pmatrix} = \begin{pmatrix} 4 & -\frac{13}{3} & -\frac{7}{3} \\ -1 & \frac{5}{3} & \frac{2}{3} \\ -1 & \frac{2}{3} & \frac{2}{3} \end{pmatrix}$$

Check.

$$A^{-1}A = \frac{1}{3}\begin{pmatrix} 12 & -13 & -7 \\ -3 & 5 & 2 \\ -3 & 2 & 2 \end{pmatrix}\begin{pmatrix} 2 & 4 & 3 \\ 0 & 1 & -1 \\ 3 & 5 & 7 \end{pmatrix} = \frac{1}{3}\begin{pmatrix} 3 & 0 & 0 \\ 0 & 3 & 0 \\ 0 & 0 & 3 \end{pmatrix} = I$$ ■

EXAMPLE 5 Let

$$A = \begin{pmatrix} 1 & -3 & 0 & -2 \\ 3 & -12 & -2 & -6 \\ -2 & 10 & 2 & 5 \\ -1 & 6 & 1 & 3 \end{pmatrix}$$

Determine whether A is invertible and, if so, calculate A^{-1}.

Solution. Using properties of determinants, we compute

$$\begin{vmatrix} 1 & -3 & 0 & -2 \\ 3 & -12 & -2 & -6 \\ -2 & 10 & 2 & 5 \\ -1 & 6 & 1 & 3 \end{vmatrix}$$

Multiply the first column by 3 and 2 and add it to the second and fourth columns, respectively.

$$= \begin{vmatrix} 1 & 0 & 0 & 0 \\ 3 & -3 & -2 & 0 \\ -2 & 4 & 2 & 1 \\ -1 & 3 & 1 & 1 \end{vmatrix}$$

Expand in the first row.

$$= \begin{vmatrix} -3 & -2 & 0 \\ 4 & 2 & 1 \\ 3 & 1 & 1 \end{vmatrix} = -1$$

Thus $\det A = -1 \neq 0$ and A^{-1} exists. By Example 2, we have

$$\operatorname{adj} A = \begin{pmatrix} 0 & -1 & 0 & -2 \\ -1 & 1 & 2 & -2 \\ 0 & -1 & -3 & 3 \\ 2 & -2 & -3 & 3 \end{pmatrix}$$

Thus

$$A^{-1} = \frac{1}{-1}\begin{pmatrix} 0 & -1 & 0 & -2 \\ -1 & 1 & 2 & -2 \\ 0 & -1 & -3 & 3 \\ 2 & -2 & -3 & 2 \end{pmatrix} = \begin{pmatrix} 0 & 1 & 0 & 2 \\ 1 & -1 & -2 & 2 \\ 0 & 1 & 3 & -3 \\ -2 & 2 & 3 & -2 \end{pmatrix}$$

Check.

$$AA^{-1} = \begin{pmatrix} 1 & -3 & 0 & -2 \\ 3 & -12 & -2 & -6 \\ -2 & 10 & 2 & 5 \\ -1 & 6 & 1 & 3 \end{pmatrix}\begin{pmatrix} 0 & 1 & 0 & 2 \\ 1 & -1 & -2 & 2 \\ 0 & 1 & 3 & -3 \\ -2 & 2 & 3 & -2 \end{pmatrix}$$

$$= \begin{pmatrix} 1 & 0 & 0 & 0 \\ 0 & 1 & 0 & 0 \\ 0 & 0 & 1 & 0 \\ 0 & 0 & 0 & 1 \end{pmatrix} \blacksquare$$

NOTE. As you may have noticed, if $n > 3$ it is generally easier to compute A^{-1} by row reduction then by using adj A since, even for the 4×4 case, it is necessary to calculate 17 determinants (16 for the adjoint plus det A). Nevertheless, Theorem 3 is very important since, before you do any row reduction, the calculation of det A (if it can be done easily) will tell you whether or not A^{-1} exists. However, we stress:

Do not use determinants to compute an inverse if $n > 3$.

Theorem 3 has a useful corollary whose proof is left as an exercise.

Corollary 1 Let A be an $n \times n$ matrix. Then the homogeneous system $A\mathbf{v} = \mathbf{0}$ has an infinite number of solutions if and only if $\det A = 0$.

EXAMPLE 6 In Example 7.3.10 on page 420 we used row reduction to show that the following homogeneous system has an infinite number of solutions:

$$\begin{aligned} x_1 + 2x_2 - x_3 &= 0 \\ 3x_1 - 3x_2 + 2x_3 &= 0 \\ -x_1 - 11x_2 + 6x_3 &= 0 \end{aligned}$$

We can use determinants to show, with much less work, that there are an infinite number of solutions:

$$A = \begin{pmatrix} 1 & 2 & -1 \\ 3 & -3 & 2 \\ -1 & -11 & 6 \end{pmatrix}$$

and $\det A = 0$. The result then follows from corollary 1. ■

In Section 7.5 (page 437) we gave the first form of the summing-up theorem. Using Theorem 3, we can extend this central result.

Theorem 4 SUMMING-UP THEOREM—VIEW 2 Let A be an $n \times n$ matrix. Then each of the following five statements implies the other four (that is, if one is true, all are true):

(i) A is invertible.
(ii) The system $A\mathbf{x} = \mathbf{b}$ has a unique solution for every n-vector $\mathbf{b}$.
(iii) The only solution to the homogeneous system $A\mathbf{x} = \mathbf{0}$ is the trivial solution ($\mathbf{x} = \mathbf{0}$).
(iv) A is row equivalent to the identity matrix I_n; that is, the reduced row echelon form of A is I_n.
(v) $\det A \neq 0$.

PROBLEMS 8.3

In Problems 1–12 use the methods of this section to determine whether the given matrix is invertible. If so, compute the inverse.

1. $\begin{pmatrix} 3 & 2 \\ 1 & 2 \end{pmatrix}$

2. $\begin{pmatrix} 3 & 6 \\ -4 & -8 \end{pmatrix}$

3. $\begin{pmatrix} 0 & 1 \\ 1 & 0 \end{pmatrix}$

4. $\begin{pmatrix} 1 & 1 & 1 \\ 0 & 2 & 3 \\ 5 & 5 & 1 \end{pmatrix}$

5. $\begin{pmatrix} 3 & 2 & 1 \\ 0 & 2 & 2 \\ 0 & 1 & -1 \end{pmatrix}$

6. $\begin{pmatrix} 1 & 1 & 1 \\ 0 & 1 & 1 \\ 0 & 0 & 1 \end{pmatrix}$

7. $\begin{pmatrix} 1 & 2 & 3 \\ 1 & 1 & 2 \\ 0 & 1 & 2 \end{pmatrix}$

8. $\begin{pmatrix} 3 & 1 & 0 \\ 1 & -1 & 2 \\ 1 & 1 & 1 \end{pmatrix}$

9. $\begin{pmatrix} 2 & -1 & 4 \\ -1 & 0 & 5 \\ 19 & -7 & 3 \end{pmatrix}$

10. $\begin{pmatrix} 1 & 6 & 2 \\ -2 & 3 & 5 \\ 7 & 12 & -4 \end{pmatrix}$

11. $\begin{pmatrix} 1 & 1 & 1 & 1 \\ 1 & 2 & -1 & 2 \\ 1 & -1 & 2 & 1 \\ 1 & 3 & 3 & 2 \end{pmatrix}$

12. $\begin{pmatrix} 1 & -3 & 0 & -2 \\ 3 & -12 & -2 & -6 \\ -2 & 10 & 2 & 5 \\ -1 & 6 & 1 & 3 \end{pmatrix}$

In Problems 13–16 use determinants to determine whether the given homogeneous system has a unique solution (the trivial solution) or an infinite number of solutions.

13.
$$\begin{aligned} 2x_1 + x_2 &= 0 \\ -x_1 + 6x_2 &= 0 \\ 4x_1 + 2x_2 + x_3 &= 0 \end{aligned}$$

14.
$$\begin{aligned} 3x_1 - x_2 - x_3 &= 0 \\ 4x_1 + 2x_2 + x_3 &= 0 \\ 2x_1 - 4x_2 - 3x_3 &= 0 \end{aligned}$$

15.
$$\begin{aligned} x_1 + x_2 + x_3 &= 0 \\ -2x_1 - 3x_2 + 4x_3 &= 0 \\ x_1 + 7x_3 &= 0 \end{aligned}$$

16.
$$\begin{aligned} 2x_1 - x_2 - x_3 &= 0 \\ x_1 + x_2 + 5x_3 &= 0 \\ 3x_1 + 2x_2 - x_3 &= 0 \end{aligned}$$

17. Show that an $n \times n$ matrix A is invertible if and only if A^t is invertible.

18. For $A = \begin{pmatrix} 1 & 1 \\ 2 & 5 \end{pmatrix}$, verify that $\det A^{-1} = 1/\det A$.

19. For $A = \begin{pmatrix} 1 & -1 & 3 \\ 4 & 1 & 6 \\ 2 & 0 & -2 \end{pmatrix}$, verify that $\det A^{-1} = 1/\det A$.

20. For what values of α is the matrix $\begin{pmatrix} \alpha & -3 \\ 4 & 1-\alpha \end{pmatrix}$ not invertible?

21. For what values of α does the matrix
$$\begin{pmatrix} -\alpha & \alpha - 1 & \alpha + 1 \\ 1 & 2 & 3 \\ 2 - \alpha & \alpha + 3 & \alpha + 7 \end{pmatrix}$$
not have an inverse?

22. Suppose that the $n \times n$ matrix A is not invertible. Show that $(A)(\text{adj } A)$ is the zero matrix.

8.4 CRAMER'S RULE (OPTIONAL)

In this section we examine an old method for solving systems with the same number of unknowns as equations. Consider the system of n equations in n unknowns

$$\begin{aligned} a_{11}x_1 + a_{12}x_2 + \cdots + a_{1n}x_n &= b_1 \\ a_{21}x_1 + a_{22}x_2 + \cdots + a_{2n}x_n &= b_2 \\ \vdots \qquad\quad \vdots \qquad\qquad\quad \vdots \quad &\quad \vdots \\ a_{n1}x_1 + a_{n2}x_2 + \cdots + a_{nn}x_n &= b_n \end{aligned} \tag{1}$$

which can be written in the form

$$A\mathbf{x} = \mathbf{b}. \tag{2}$$

We suppose that $\det A \neq 0$. Then system (2) has a unique solution given by $\mathbf{x} = A^{-1}\mathbf{b}$. We can develop a method for finding that solution without row reduction and without computing A^{-1}.

Let $D = \det A$. We define n new matrices:

$$A_1 = \begin{pmatrix} b_1 & a_{12} & \cdots & a_{1n} \\ b_2 & a_{22} & \cdots & a_{2n} \\ \vdots & \vdots & & \vdots \\ b_n & a_{n2} & \cdots & a_{nn} \end{pmatrix}, \quad A_2 = \begin{pmatrix} a_{11} & b_1 & \cdots & a_{1n} \\ a_{21} & b_2 & \cdots & a_{2n} \\ \vdots & \vdots & & \vdots \\ a_{n1} & b_n & \cdots & a_{nn} \end{pmatrix}, \ldots,$$

$$A_n = \begin{pmatrix} a_{11} & a_{12} & \cdots & b_1 \\ a_{21} & a_{22} & \cdots & b_2 \\ \vdots & \vdots & & \vdots \\ a_{n1} & a_{n2} & \cdots & b_n \end{pmatrix}$$

That is, A_i is the matrix obtained by replacing the ith column of A with $\mathbf{b}$. Finally, let $D_1 = \det A_1, D_2 = \det A_2, \ldots, D_n = \det A_n$.

Theorem 1 CRAMER'S RULE[†] Let A be an $n \times n$ matrix and suppose that $\det A \neq 0$. Then the unique solution to the system $A\mathbf{x} = \mathbf{b}$ is given by

$$x_1 = \frac{D_1}{D}, \; x_2 = \frac{D_2}{D}, \ldots, x_i = \frac{D_i}{D}, \ldots, x_n = \frac{D_n}{D} \tag{3}$$

Proof. The solution to $A\mathbf{x} = \mathbf{b}$ is $\mathbf{x} = A^{-1}\mathbf{b}$. But

$$A^{-1}\mathbf{b} = \frac{1}{D}(\text{adj } A)\mathbf{b} = \frac{1}{D}\begin{pmatrix} A_{11} & A_{21} & \cdots & A_{n1} \\ A_{12} & A_{22} & \cdots & A_{n2} \\ \vdots & \vdots & & \vdots \\ A_{1n} & A_{2n} & \cdots & A_{nn} \end{pmatrix}\begin{pmatrix} b_1 \\ b_2 \\ \vdots \\ b_n \end{pmatrix} \tag{4}$$

Now $(\text{adj } A)\mathbf{b}$ is an n-vector, the jth component of which is

[†]Named for the Swiss mathematician Gabriel Cramer (1704–1752). Cramer published the rule in 1750 in his *Introduction to the Analysis of Lines of Algebraic Curves.* Actually, there is much evidence to suggest that the rule was known as early as 1729 to Colin Maclaurin (1698–1746), who was probably the most outstanding British mathematician in the years following the death of Newton.

$$(A_{1j}\ A_{2j}\ \cdots\ A_{nj}) \cdot \begin{pmatrix} b_1 \\ b_2 \\ \vdots \\ b_n \end{pmatrix} = b_1A_{1j} + b_2A_{2j} + \cdots + b_nA_{nj} \tag{5}$$

Consider the matrix A_j:

$$A_j = \begin{pmatrix} a_{11} & a_{12} & \cdots & b_1 & \cdots & a_{1n} \\ a_{21} & a_{22} & \cdots & b_2 & \cdots & a_{2n} \\ \vdots & \vdots & & \vdots & & \vdots \\ a_{n1} & a_{n2} & \cdots & b_n & \cdots & a_{nn} \end{pmatrix} \tag{6}$$

$\uparrow$ jth column

If we expand the determinant of A_j in its jth column, we obtain

$$D_j = b_1 \text{ (cofactor of } b_1) + b_2 \text{ (cofactor of } b_2) + \cdots + b_n \text{ (cofactor of } b_n) \tag{7}$$

But to find the cofactor of b_i, say, we delete the ith row and jth column of A_j (since b_i is in the jth column of A_j). But the jth column of A_j is $\mathbf{b}$ and, with this deleted, we simply have the ijth minor, M_{ij}, of A. Thus

$$\text{Cofactor of } b_i \text{ in } A_j = A_{ij}$$

so that (7) becomes

$$D_j = b_1A_{1j} + b_2A_{2j} + \cdots + b_nA_{nj} \tag{8}$$

But this is the same as the right side of (5). Thus the ith component of (adj A)$\mathbf{b}$ is D_i and we have

$$\mathbf{x} = \begin{pmatrix} x_1 \\ x_2 \\ \vdots \\ x_n \end{pmatrix} = A^{-1}\mathbf{b} = \frac{1}{D}(\text{adj } A)\mathbf{b} = \frac{1}{D}\begin{pmatrix} D_1 \\ D_2 \\ \vdots \\ D_n \end{pmatrix} = \begin{pmatrix} D_1/D \\ D_2/D \\ \vdots \\ D_n/D \end{pmatrix}$$

and the proof is complete. ■

EXAMPLE 1 Solve, using Cramer's rule, the system

$$\begin{aligned} 2x_1 + 4x_2 + 6x_3 &= 18 \\ 4x_1 + 5x_2 + 6x_3 &= 24 \\ 3x_1 + x_2 - 2x_3 &= 4 \end{aligned} \tag{9}$$

Solution. We have solved this before—using row reduction in Example 7.3.1. We could also solve it by calculating A^{-1} (Example 7.5.6) and then finding $A^{-1}\mathbf{b}$. We now solve it by using Cramer's rule. First we have

$$D = \begin{vmatrix} 2 & 4 & 6 \\ 4 & 5 & 6 \\ 3 & 1 & -2 \end{vmatrix} = 6 \neq 0,$$

so that system (9) has a unique solution. Then $D_1 = \begin{vmatrix} 18 & 4 & 6 \\ 24 & 5 & 6 \\ 4 & 1 & -2 \end{vmatrix} = 24$, $D_2 = \begin{vmatrix} 2 & 18 & 6 \\ 4 & 24 & 6 \\ 3 & 4 & -2 \end{vmatrix} = -12$ and $D_3 = \begin{vmatrix} 2 & 4 & 18 \\ 4 & 5 & 24 \\ 3 & 1 & 4 \end{vmatrix} = 18$. Hence $x_1 = \dfrac{D_1}{D} = \dfrac{24}{6} = 4$, $x_2 = \dfrac{D_2}{D} = -\dfrac{12}{6} = -2$ and $x_3 = \dfrac{D_3}{D} = \dfrac{18}{6} = 3$. ■

EXAMPLE 2 Show that the system

$$\begin{aligned} x_1 + 3x_2 + 5x_3 + 2x_4 &= 2 \\ -x_2 + 3x_3 + 4x_4 &= 0 \\ 2x_1 + x_2 + 9x_3 + 6x_4 &= -3 \\ 3x_1 + 2x_2 + 4x_3 + 8x_4 &= -1 \end{aligned} \tag{10}$$

has a unique solution and find it by using Cramer's rule.

Solution. We saw in Example 8.2.10 that

$$|A| = \begin{vmatrix} 1 & 3 & 5 & 2 \\ 0 & -1 & 3 & 4 \\ 2 & 1 & 9 & 6 \\ 3 & 2 & 4 & 8 \end{vmatrix} = 160 \neq 0$$

Thus the system has a unique solution. To find it we compute: $D_1 = -464$; $D_2 = 280$; $D_3 = -56$; $D_4 = 112$. Thus $x_1 = D_1/D = -464/160$, $x_2 = D_2/D = 280/160$, $x_3 = D_3/D = -56/160$, and $x_4 = D_4/D = 112/160$. These solutions can be verified by direct substitution into system (10). ■

REMARK. Cramer's rule is discussed here because it occasionally appears in science textbooks and because it is sometimes useful in proving an interesting result (see, for example, Problem 10). However it should be stressed that when faced with the necessity of solving a system of equations, you should almost always use some other method. Gaussian elimination is frequently one of the fastest methods.

PROBLEMS 8.4

In Problems 1–9 solve the given system by using Cramer's rule.

1. $\begin{aligned} 2x_1 + 3x_2 &= -1 \\ -7x_1 + 4x_2 &= 47 \end{aligned}$

2. $\begin{aligned} 3x_1 - x_2 &= 0 \\ 4x_1 + 2x_2 &= 5 \end{aligned}$

3. $\begin{aligned} 2x_1 + x_2 + x_3 &= 6 \\ 3x_1 - 2x_2 - 3x_3 &= 5 \\ 8x_1 + 2x_2 + 5x_3 &= 11 \end{aligned}$

4. $\begin{aligned} x_1 + x_2 + x_3 &= 8 \\ 4x_2 - x_3 &= -2 \\ 3x_1 - x_2 + 2x_3 &= 0 \end{aligned}$

5. $\begin{aligned} 2x_1 + 2x_2 + x_3 &= 7 \\ x_1 + 2x_2 - x_3 &= 0 \\ -x_1 + x_2 + 3x_3 &= 1 \end{aligned}$

6. $\begin{aligned} 2x_1 + 5x_2 - x_3 &= -1 \\ 4x_1 + x_2 + 3x_3 &= 3 \\ -2x_1 + 2x_2 &= 0 \end{aligned}$

7. $\begin{aligned} 2x_1 + x_2 - x_3 &= 4 \\ x_1 + x_3 &= 2 \\ -x_2 + 5x_3 &= 1 \end{aligned}$

8. $\begin{aligned} x_1 + x_2 + x_3 + x_4 &= 6 \\ 2x_1 - x_3 - x_4 &= 4 \\ 3x_3 + 6x_4 &= 3 \\ x_1 - x_4 &= 5 \end{aligned}$

9. $\begin{aligned} x_1 - x_4 &= 7 \\ 2x_2 + x_3 &= 2 \\ 4x_1 - x_2 &= -3 \\ 3x_3 - 5x_4 &= 2 \end{aligned}$

***10.** Consider the triangle in Figure 1.

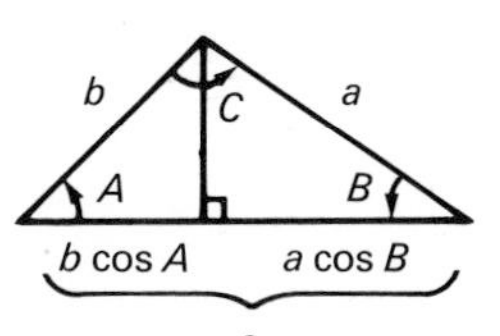

FIGURE 1

(a) Show, using elementary trigonometry, that

$$\begin{aligned} c\cos A \qquad\qquad + a\cos C &= b \\ b\cos A + a\cos B \qquad\qquad &= c \\ c\cos B + b\cos C &= a \end{aligned}$$

(b) If the system of part (*a*) is thought of as a system of three equations in the three unknowns $\cos A$, $\cos B$, and $\cos C$, show that the determinant of the system is nonzero.
(c) Use Cramer's rule to solve for $\cos C$.
(d) Use part (*c*) to prove the **law of cosines**: $c^2 = a^2 + b^2 - 2ab\cos C$.

REVIEW EXERCISES FOR CHAPTER EIGHT

In Exercises 1–8 calculate the determinant.

1. $\begin{vmatrix} -1 & 2 \\ 0 & 4 \end{vmatrix}$

2. $\begin{vmatrix} -3 & 5 \\ -7 & 4 \end{vmatrix}$

3. $\begin{vmatrix} 1 & -2 & 3 \\ 0 & 4 & 5 \\ 0 & 0 & 6 \end{vmatrix}$

4. $\begin{vmatrix} 5 & 0 & 0 \\ 6 & 2 & 0 \\ 10 & 100 & 6 \end{vmatrix}$

5. $\begin{vmatrix} 1 & -1 & 2 \\ 3 & 4 & 2 \\ -2 & 3 & 4 \end{vmatrix}$

6. $\begin{vmatrix} 3 & 1 & -2 \\ 4 & 0 & 5 \\ -6 & 1 & 3 \end{vmatrix}$

7. $\begin{vmatrix} 1 & -1 & 2 & 3 \\ 4 & 0 & 2 & 5 \\ -1 & 2 & 3 & 7 \\ 5 & 1 & 0 & 4 \end{vmatrix}$

8. $\begin{vmatrix} 3 & 15 & 17 & 19 \\ 0 & 2 & 21 & 60 \\ 0 & 0 & 1 & 50 \\ 0 & 0 & 0 & -1 \end{vmatrix}$

In Exercises 9–14 use determinants to calculate the inverse (if one exists).

9. $\begin{pmatrix} -3 & 4 \\ 2 & 1 \end{pmatrix}$

10. $\begin{pmatrix} 3 & -5 & 7 \\ 0 & 2 & 4 \\ 0 & 0 & -3 \end{pmatrix}$

11. $\begin{pmatrix} 1 & -1 & 2 \\ 3 & 1 & 4 \\ 5 & -1 & 8 \end{pmatrix}$

12. $\begin{pmatrix} 1 & 1 & 1 \\ 1 & 0 & 1 \\ 0 & 1 & 1 \end{pmatrix}$

13. $\begin{pmatrix} 2 & 1 & 0 & 0 \\ 0 & -1 & 3 & 0 \\ 1 & 0 & 0 & -2 \\ 3 & 0 & -1 & 0 \end{pmatrix}$

14. $\begin{pmatrix} 3 & -1 & 2 & 4 \\ 1 & 1 & 0 & 3 \\ -2 & 4 & 1 & 5 \\ 6 & -4 & 1 & 2 \end{pmatrix}$

In Exercises 15–18 solve the system by using Cramer's rule.

15. $\begin{aligned} 2x_1 - x_2 &= 3 \\ 3x_1 + 2x_2 &= 5 \end{aligned}$

16. $\begin{aligned} x_1 - x_2 + x_3 &= 7 \\ 2x_1 - 5x_3 &= 4 \\ 3x_2 - x_3 &= 2 \end{aligned}$

17. $\begin{aligned} 2x_1 + 3x_2 - x_3 &= 5 \\ -x_1 + 2x_2 + 3x_3 &= 0 \\ 4x_1 - x_2 + x_3 &= -1 \end{aligned}$

18. $\begin{aligned} x_1 - x_3 + x_4 &= 7 \\ 2x_2 + 2x_3 - 3x_4 &= -1 \\ 4x_1 - x_2 - x_3 &= 0 \\ -2x_1 + x_2 + 4x_3 &= 2 \end{aligned}$

9 Vector Spaces and Linear Transformations

9.1 VECTOR SPACES

As we saw in Section 3.7, the set $\mathbb{R}^n$ has a number of nice properties. We can add two vectors in $\mathbb{R}^n$ and obtain another vector in $\mathbb{R}^n$. Under addition, vectors in $\mathbb{R}^n$ obey the commutative and associative laws. If $\mathbf{x} \in \mathbb{R}^n$, then $\mathbf{x} + \mathbf{0} = \mathbf{x}$ and $\mathbf{x} + (-\mathbf{x}) = \mathbf{0}$. We can multiply vectors in $\mathbb{R}^n$ by scalars, and we can show that this scalar multiplication satisfies a distributive law.

The set $\mathbb{R}^n$ is called a *vector space*. Loosely speaking, we can say that a vector space is a set of objects that obey the rules given in the preceding paragraph.

In this chapter we make a seemingly great leap from the concrete world of solving equations and dealing with easily visualized vectors to the world of abstract vector spaces. There is a great advantage in doing so. Once we have established a fact about vector spaces in general, we can apply that fact to *every* vector space. Otherwise, we would have to prove that fact again and again, once for each new vector space we encounter (and there is an endless supply of them). But, as you will see, the abstract theorems we shall prove are really no more difficult than the ones already encountered.

Definition 1 REAL VECTOR SPACE A **real vector space** V is a set of objects, called **vectors,** together with two operations called **addition** and **scalar multiplication** that satisfy the ten axioms listed below.

NOTATION: If $\mathbf{x}$ and $\mathbf{y}$ are in V and if α is a real number, then we write $\mathbf{x} + \mathbf{y}$ for the sum of $\mathbf{x}$ and $\mathbf{y}$ and $\alpha\mathbf{x}$ for the scalar product of α and $\mathbf{x}$.

Before we list the properties satisfied by vectors in a vector space, two things should be mentioned. First, it may be helpful to think of $\mathbb{R}^2$ or $\mathbb{R}^3$ when dealing with a vector space, but often a vector space appears to be very different from these comfortable spaces. (We shall see this shortly.) Second, Definition 1 gives a definition of a *real* vector space. The word "real" means that the scalars we use are real numbers. It would be just as easy to define a *complex* vector space by using complex numbers instead of real ones. This book deals primarily with real vector spaces, but other sets of scalars present little difficulty. In a few instances we will refer to complex numbers. For that reason, a review of the elementary properties of complex numbers is given in Appendix 3.

AXIOMS OF A VECTOR SPACE

(i) If $\mathbf{x} \in V$ and $\mathbf{y} \in V$, then $\mathbf{x} + \mathbf{y} \in V$ **(closure under addition).**

(ii) For all $\mathbf{x}$, $\mathbf{y}$, and $\mathbf{z}$ in V, $(\mathbf{x} + \mathbf{y}) + \mathbf{z} = \mathbf{x} + (\mathbf{y} + \mathbf{z})$ **(associative law of vector addition).**

(iii) There is a vector $\mathbf{0} \in V$ such that for all $\mathbf{x} \in V$, $\mathbf{x} + \mathbf{0} = \mathbf{0} + \mathbf{x} = \mathbf{x}$ ($\mathbf{0}$ is called the **additive identity**).

(iv) If $\mathbf{x} \in V$, there is a vector $-\mathbf{x}$ in V such that $\mathbf{x} + (-\mathbf{x}) = \mathbf{0}$ ($-\mathbf{x}$ is called the **additive inverse** of $\mathbf{x}$).

(v) If $\mathbf{x}$ and $\mathbf{y}$ are in V, then $\mathbf{x} + \mathbf{y} = \mathbf{y} + \mathbf{x}$ **(commutative law of vector addition).**

(vi) If $\mathbf{x} \in V$ and α is a scalar (real number), then $\alpha\mathbf{x} \in V$ **(closure under scalar multiplication).**

(vii) If $\mathbf{x}$ and $\mathbf{y}$ are in V and α is a scalar, then $\alpha(\mathbf{x} + \mathbf{y}) = \alpha\mathbf{x} + \alpha\mathbf{y}$ **(first distributive law).**

(viii) If $\mathbf{x} \in V$ and α and β are scalars, then $(\alpha + \beta)\mathbf{x} = \alpha\mathbf{x} + \beta\mathbf{x}$ **(second distributive law).**

(ix) If $\mathbf{x} \in V$ and α and β are scalars, then $\alpha(\beta\mathbf{x}) = \alpha\beta\mathbf{x}$ **(associative law of scalar multiplication).**

(x) For every vector $\mathbf{x} \in V$, $1\mathbf{x} = \mathbf{x}$ (the scalar 1 is called a **multiplicative identity**).

REMARK. It is important to keep in mind the two requirements that make a set a vector space. The most obvious are the "things" in the set, the things we call *vectors.* Second, and equally important, are the two operations on the vectors. Without defining the operations of addition and scalar multiplication, no set is a vector space.

EXAMPLE 1 THE SPACE $\mathbb{R}^n$ Let $V = \mathbb{R}^n = \{(x_1, x_2, \ldots, x_n): x_i \in \mathbb{R} \text{ for } i = 1, 2, \ldots, n\}$.

From Section 3.7 (see Theorem 3.7.1) we see that V satisfies all the axioms of a vector space if we take the set of scalars to be $\mathbb{R}$ and the operations of vector addition and scalar multiplication as defined on page 126. ■

EXAMPLE 2 Let $V = \{0\}$. That is, V consists of the single number 0. Since $0 + 0 = 1 \cdot 0 = 0 + (0 + 0) = (0 + 0) + 0 = 0$, we see that V is a vector space. It is often referred to as a **trivial** vector space. We also obtain a trivial vector space by taking $V = \{\mathbf{0}\}$ where $\mathbf{0}$ is the zero vector in $\mathbb{R}^n$. ■

EXAMPLE 3 Let $V = \{1\}$. That is, V consists of the single number 1. This is *not* a vector space since it violates axiom (i)—the closure axiom. To see this we simply note that $1 + 1 = 2 \notin V$. Here we are assuming the usual addition and multiplication of real numbers. ■

EXAMPLE 4 Let $V = \{(x, y): y = mx$, where m is a fixed real number and x is an arbitrary real number$\}$. That is, V consists of all points lying on the line $y = mx$ passing through the origin with slope m. Suppose that (x_1, y_1) and (x_2, y_2) are in V. Then $y_1 = mx_1$, $y_2 = mx_2$, and

$$\begin{aligned}(x_1, y_1) + (x_2, y_2) &= (x_1, mx_1) + (x_2, mx_2) = (x_1 + x_2, mx_1 + mx_2)\\ &= (x_1 + x_2, m(x_1 + x_2)) \in V.\end{aligned}$$

Thus axiom (i) is satisfied. Axioms (ii), (iii), and (v) are obvious. Further,

$$-(x, mx) = (-x, -mx) = (-x, m(-x)) \in V$$

and

$$(x, mx) + (-x, m(-x)) = (0, 0) = \mathbf{0}$$

so that axiom (iv) is satisfied. The other axioms are easily verified, and we see that the set of points in the plane lying on a straight line passing through the origin constitutes a vector space. ■

EXAMPLE 5 Let $V = \{(x, y): y = 2x + 1, \mathbf{x} \in \mathbb{R}\}$. That is, V is the set of points lying on the line $y = 2x + 1$. V is *not* a vector space because closure is violated, as in Example 3. To see this, let us suppose that (x_1, y_1) and (x_2, y_2) are in V. Then

$$(x_1, y_1) + (x_2, y_2) = (x_1 + x_2, y_1 + y_2).$$

If this last vector were in V, we would have

$$y_1 + y_2 = 2(x_1 + x_2) + 1 = 2x_1 + 2x_2 + 1.$$

But $y_1 = 2x_1 + 1$ and $y_2 = 2x_2 + 1$ so that

$$y_1 + y_2 = (2x_1 + 1) + (2x_2 + 1) = 2x_1 + 2x_2 + 2.$$

Hence we conclude that

$$(x_1 + x_2, y_1 + y_2) \notin V \quad \text{if} \quad (x_1, y_1) \in V \quad \text{and} \quad (x_2, y_2) \in V. \ ■$$

EXAMPLE 6 Let $V = \{(x, y, z): ax + by + cz = 0\}$. That is, V is the set of points in $\mathbb{R}^3$ lying on the plane passing through the origin with normal vector (a, b, c). Suppose (x_1, y_1, z_1) and (x_2, y_2, z_2) are in V. Then $(x_1, y_1, z_1) + (x_2, y_2, z_2) = (x_1 + x_2, y_1 + y_2, z_1 + z_2) \in V$ because $a(x_1 + x_2) + b(y_1 + y_2) + c(z_1 + z_2) = (ax_1 + by_1 + cz_1) + (ax_2 + by_2 + cz_2) =$

$0 + 0 = 0$; hence axiom (i) is satisfied. The other axioms are easily verified. Thus the set of points lying on a plane in $\mathbb{R}^3$ that passes through the origin comprises a vector space. ■

EXAMPLE 7 Let P_n denote the set of polynomials with real coefficients of degree less than or equal to n. If $p \in P_n$, then

$$p(x) = a_n x^n + a_{n-1}x^{n-1} + \cdots + a_1 x + a_0$$

where each a_i is real. The sum $p(x) + q(x)$ is defined as in elementary algebra. If $q(x) = b_n x^n + b_{n-1}x^{n-1} + \cdots + b_1 x + b_0$, then

$$\begin{aligned} p(x) + q(x) &= (a_n + b_n)x^n + (a_{n-1} + b_{n-1})x^{n-1} \\ &\quad + \cdots + (a_1 + b_1)x + (a_0 + b_0). \end{aligned} \tag{1}$$

Let us check the ten axioms, one at a time, to verify that P_n is indeed a vector space:

Axiom (i). If $p(x) \in P_n$ and $q(x) \in P_n$, then $p(x) + q(x)$, given by (1) is a polynomial of degree n; so $p(x) + q(x) \in P_n$, and the closure rule holds.

Axiom (ii). If $p(x)$ and $q(x)$ are as above and $r(x) = c_n x^n + c_{n-1}x^{n-1} + \cdots + c_1 x + c_0$, then, using the rule (1) twice,

$$\begin{aligned} [p(x) + q(x)] + r(x) &= (a_n + b_n)x^n + (a_{n-1} + b_{n-1})x^{n-1} + \cdots + (a_1 + b_1)x \\ &\quad + (a_0 + b_0) + c_n x^n + c_{n-1}x^{n-1} + \cdots + c_1 x + c_0 \\ &= (a_n + b_n + c_n)x^n + (a_{n-1} + b_{n-1} + c_{n-1})x^{n-1} \\ &\quad + \cdots + (a_1 + b_1 + c_1)x + (a_0 + b_0 + c_0). \end{aligned}$$

Similarly,

$$\begin{aligned} p(x) + [q(x) + r(x)] &= a_n x^n + a_{n-1}x^{n-1} + \cdots + a_1 x + a_0 + (b_n + c_n)x^n \\ &\quad + (b_{n-1} + c_{n-1})x^{n-1} + \cdots + (b_1 + c_1)x + (b_0 + c_0) \\ &= (a_n + b_n + c_n)x^n + (a_{n-1} + b_{n-1} + c_{n-1})x^{n-1} \\ &\quad + \cdots + (a_1 + b_1 + c_1)x + (a_0 + b_0 + c_0). \end{aligned}$$

Thus

$$(p + q) + r = p + (q + r),$$

and the associative law of vector addition holds. Here we have made repeated use of the fact that if a, b and c are real numbers, then

$$a + (b + c) = (a + b) + c = a + b + c.$$

Axiom (iii). If we define the zero polynomial by $\mathbf{0} = 0x^n + 0x^{n-1} + \cdots + 0x + 0$, then clearly $\mathbf{0} \in P_n$, and axiom (iii) is satisfied.

Axiom (iv). We define $-p(x)$ by $-p(x) = -a_nx^n - a_{n-1}x^{n-1} - \cdots - a_1x - a_0$. Then by (1),

$$\begin{aligned} p(x) + [-p(x)] &= [a_n + (-a_n)]x^n + [a_{n-1} + (-a_{n-1})]x^{n-1} \\ &\quad + \cdots + [a_1 + (-a_1)]x + [a_0 + (-a_0)] \\ &= 0x^n + 0x^{n-1} + \cdots + 0x + 0 = \mathbf{0}. \end{aligned}$$

Axiom (v). This follows from (1) and the fact that if a and b are real numbers, then $a + b = b + a$ (that is, real numbers commute under addition).

Axiom (vi). If α is a real number, then

$$\begin{aligned} \alpha p(x) &= \alpha(a_nx^n + a_{n-1}x^{n-1} + \cdots + a_1x + a_0) = \alpha a_nx^n + \alpha a_{n-1}x^{n-1} \\ &\quad + \cdots + \alpha a_1x + \alpha a_0 \in P_n. \end{aligned}$$

Axiom (vii).

$$\begin{aligned} \alpha(p(x) + q(x)) &= \alpha(a_n + b_n)x^n + \alpha(a_{n-1} + b_{n-1})x^{n-1} + \cdots + \alpha(a_1 + b_1)x \\ &\quad + \alpha(a_0 + b_0) \\ &= \alpha a_nx^n + \alpha a_{n-1}x^{n-1} + \cdots + \alpha a_1x + \alpha a_0 + \alpha b_nx^n \\ &\quad + \alpha b_{n-1}x^{n-1} + \cdots + \alpha b_1x + \alpha b_0 = \alpha p(x) + \alpha q(x). \end{aligned}$$

Here we have used the fact that if a, b, and c are real numbers, then $a(b + c) = ab + ac$.

Axiom (viii).

$$\begin{aligned} (\alpha + \beta)p(x) &= (\alpha + \beta)a_nx^n + (\alpha + \beta)a_{n-1}x^{n-1} + \cdots + (\alpha + \beta)a_1x + (\alpha + \beta)a_0 \\ &= \alpha a_nx^n + \alpha a_{n-1}x^{n-1} + \cdots + \alpha a_1x + \alpha a_0 \\ &\quad + \beta a_nx^n + \beta a_{n-1}x^{n-1} + \cdots + \beta a_1x + \beta a_0 \\ &= \alpha p(x) + \beta p(x). \end{aligned}$$

Here we have used the fact that if a, b, and c are real numbers, then

$$(a + b)c = ac + bc.$$

Axiom (ix).

$$\begin{aligned} \alpha(\beta p(x)) &= \alpha(\beta a_nx^n + \beta a_{n-1}x^{n-1} + \cdots + \beta a_1x + \beta a_0) \\ &= (\alpha\beta a_nx^n + \alpha\beta a_{n-1}x^{n-1} + \cdots + \alpha\beta a_1x + \alpha\beta a_0) = \alpha\beta p(x). \end{aligned}$$

Axiom (x).

$$\begin{aligned} 1p(x) &= 1(a_nx^n + a_{n-1}x^{n-1} + \cdots + a_1x + a_0) \\ &= 1a_nx^n + 1a_{n-1}x^{n-1} + \cdots + 1a_1x + 1a_0 \\ &= a_nx^n + a_{n-1}x^{n-1} + \cdots + a_1x + a_0 = p(x). \end{aligned}$$

Thus the ten axioms are satisfied, and P_n is a vector space. ■

In Example 7 we worked through every axiom. Only such a comprehensive check can prove that a set is a vector space. However, many of the axioms are fairly easy to check, and so, in the remainder of the examples in this section, we will check only those axioms that might present some difficulty.

EXAMPLE 8 Let $C[0, 1]$ denote the set of real-valued continuous functions defined on the interval $[0, 1]$. We define $(f + g)x = f(x) + g(x)$ and $(\alpha f)(x) = \alpha[f(x)]$. Since the sum of continuous functions is continuous, axiom (i) is satisfied and the other axioms are easily verified with $\mathbf{0}$ = the zero function and $(-f)(x) = -f(x)$. ■

EXAMPLE 9 Let M_{34} denote the set of 3×4 matrices with real components. Then with the usual sum and scalar multiplication of matrices, it is again easy to verify that M_{34} is a vector space with $\mathbf{0}$ being the 3×4 zero matrix. If $A = (a_{ij})$ is in M_{34}, then $-A = (-a_{ij})$ is also in M_{34}. ■

EXAMPLE 10 In an identical manner we see that M_{mn}, the set of $m \times n$ matrices with real components, forms a vector space for any integers m and n. ■

EXAMPLE 11 Let S_3 denote the set of invertible 3×3 matrices. Define the "sum" $A \oplus B$ by $A \oplus B = AB$. If A and B are invertible, then AB is invertible (by Theorem 7.5.3) so that axiom (i) is satisfied. Axiom (ii) is simply the associative law for matrix multiplication (Theorem 7.2.1); axioms (iii) and (iv) are satisfied with $\mathbf{0} = I_3$ and $-A = A^{-1}$. Axiom (v) fails, however, since, in general, $AB \neq BA$ so that S_3 is not a vector space. ■

EXAMPLE 12 Let $V = \{(x, y): y \geq 0\}$. V consists of the points in $\mathbb{R}^2$ in the upper half plane (the first two quadrants). If $y_1 \geq 0$ and $y_2 \geq 0$, then $y_1 + y_2 \geq 0$; hence if $(x_1, y_1) \in V$ and $(x_2, y_2) \in V$, then $(x_1 + x_2, y_1 + y_2) \in V$. V is not a vector space, however, since the vector $(1, 1)$, for example, does not have an inverse in V because $(-1, -1) \notin V$. Moreover, axiom (vi) fails since if $(x, y) \in V$, then $\alpha(x, y) \notin V$ if $\alpha < 0$. ■

EXAMPLE 13 THE SPACE $\mathbb{C}^n$ Let $\mathbb{C}^n = \{(c_1, c_2, \ldots, c_n): c_i$ is a complex number for $i = 1, 2, \ldots, n\}$ and the set of scalars is the set of complex numbers. It is easy to verify that $\mathbb{C}^n$, too, is a vector space since the only difference between a complex and real space is that the scalars are complex numbers. ■

As these examples suggest, there are many different kinds of vector spaces and many kinds of sets that are *not* vector spaces. Before leaving this section let us prove some elementary results about vector spaces.

Theorem 1 Let V be a vector space. Then:

(i) $\alpha\mathbf{0} = \mathbf{0}$ for every real number α.
(ii) $0 \cdot \mathbf{x} = \mathbf{0}$ for every $\mathbf{x} \in V$.
(iii) If $\alpha\mathbf{x} = \mathbf{0}$, then $\alpha = 0$ or $\mathbf{x} = \mathbf{0}$ (or both).
(iv) $(-1)\mathbf{x} = -\mathbf{x}$ for every $\mathbf{x} \in V$.

Proof.

(i) By axiom (iii), $\mathbf{0} + \mathbf{0} = \mathbf{0}$; and from axiom (vii),

$$\alpha(\mathbf{0} + \mathbf{0}) = \alpha\mathbf{0} + \alpha\mathbf{0} = \alpha\mathbf{0}. \tag{2}$$

Adding $-\alpha\mathbf{0}$ to both sides of the last equation in (2) and using the associative law (axiom ii), we obtain

$$\begin{aligned}[\alpha\mathbf{0} + \alpha\mathbf{0}] + (-\alpha\mathbf{0}) &= \alpha\mathbf{0} + (-\alpha\mathbf{0})\\ \alpha\mathbf{0} + [\alpha\mathbf{0} + (-\alpha\mathbf{0})] &= \mathbf{0}\\ \alpha\mathbf{0} + \mathbf{0} &= \mathbf{0}\\ \alpha\mathbf{0} &= \mathbf{0}.\end{aligned}$$

(ii) Essentially the same proof as used in part (i) works. We start with $0 + 0 = 0$ and use axiom (viii) to see that $0\mathbf{x} = (0 + 0)\mathbf{x} = 0\mathbf{x} + 0\mathbf{x}$ or $0\mathbf{x} + (-0\mathbf{x}) = 0\mathbf{x} + [0\mathbf{x} + (-0\mathbf{x})]$ or $\mathbf{0} = 0\mathbf{x} + \mathbf{0} = 0\mathbf{x}$.

(iii) Let $\alpha\mathbf{x} = \mathbf{0}$. If $\alpha \neq 0$, we multiply both sides of the equation by $1/\alpha$ to obtain $(1/\alpha)(\alpha\mathbf{x}) = (1/\alpha)\mathbf{0} = \mathbf{0}$ (by part i). But $(1/\alpha)(\alpha\mathbf{x}) = 1\mathbf{x} = \mathbf{x}$ (by axiom ix), so $\mathbf{x} = \mathbf{0}$.

(iv) We start with the fact that $1 + (-1) = 0$. Then, using part (ii), we obtain

$$\mathbf{0} = 0\mathbf{x} = [1 + (-1)]\mathbf{x} = 1\mathbf{x} + (-1)\mathbf{x} = \mathbf{x} + (-1)\mathbf{x}. \tag{3}$$

We add $-\mathbf{x}$ to both sides of (3) to obtain

$$\begin{aligned}\mathbf{0} + (-\mathbf{x}) &= \mathbf{x} + (-1)\mathbf{x} + (-\mathbf{x}) = \mathbf{x} + (-\mathbf{x}) + (-1)\mathbf{x}\\ &= \mathbf{0} + (-1)\mathbf{x} = (-1)\mathbf{x}.\end{aligned}$$

Thus $-\mathbf{x} = (-1)\mathbf{x}$. Note that we were able to reverse the order of addition in the preceding equation by using the commutative law (axiom v). ■

REMARK. Part (iii) of Theorem 1 is not as obvious as it seems. There are objects which have the property that $xy = 0$ does not imply that either x or y is zero. As an example, we look at the multiplication of 2×2 matrices. If $A = \begin{pmatrix} 0 & 1 \\ 0 & 0 \end{pmatrix}$ and $B = \begin{pmatrix} 0 & -2 \\ 0 & 0 \end{pmatrix}$, then neither A nor B is zero although, as is easily verified, the product $AB = 0$, the zero matrix.

PROBLEMS 9.1

In Problems 1–20 determine whether the given set is a vector space. If it is not, list the axioms that do not hold.

1. The set of diagonal $n \times n$ matrices under the usual matrix addition and the usual scalar multiplication.
2. The set of diagonal $n \times n$ matrices under multiplication (that is, $A + B = AB$).
3. $\{(x, y): y \leq 0; x, y \text{ real}\}$ with the usual addition and scalar multiplication of vectors.
4. The vectors in the plane lying in the first quadrant.
5. The set of vectors in $\mathbb{R}^3$ in the form (x, x, x).

6. The set of polynomials of degree 4 under the operations of Example 7.
7. The set of symmetric matrices (see page 444) under the usual addition and scalar multiplication.
8. The set of 2×2 matrices of the form $\begin{pmatrix} 0 & a \\ b & 0 \end{pmatrix}$ under the usual addition and scalar multiplication.
9. The set of matrices of the form $\begin{pmatrix} 1 & \alpha \\ \beta & 1 \end{pmatrix}$ with the matrix operations of addition and scalar multiplication.
10. The set consisting of the single vector (0, 0) under the usual operations in $\mathbb{R}^2$.
11. The set of polynomials of degree $\leq n$ with zero constant term.
12. The set of polynomials of degree $\leq n$ with positive constant term a_0.
13. The set of continuous functions in [0, 1] with $f(0) = 0$ and $f(1) = 0$ under the operations of Example 8.
14. The set of points in $\mathbb{R}^3$ lying on a line passing through the origin.
15. The set of points in $\mathbb{R}^3$ lying on the line $x = t + 1$, $y = 2t$, $z = t - 1$.
16. $\mathbb{R}^2$ with addition defined by $(x_1, y_1) + (x_2, y_2) = (x_1 + x_2 + 1, y_1 + y_2 + 1)$ and ordinary scalar multiplication.
17. The set of Problem 16 with scalar multiplication defined by $\alpha(x, y) = (\alpha + \alpha x - 1, \alpha + \alpha y - 1)$.
18. The set consisting of one object with addition defined by *object* + *object* = *object* and scalar multiplication defined by α(*object*) = *object*.
19. The set of differentiable functions defined on [0, 1] with the operations of Example 8.

*20. The set of real numbers of the form $a + b\sqrt{2}$, where a and b are rational numbers, under the usual addition of real numbers and with scalar multiplication defined only for rational scalars; i.e., here the scalars are the rational numbers.

21. Show that in a vector space the additive identity element is unique (i.e., there is only one identity).
22. Show that in a vector space each vector has a unique additive inverse.
23. If $\mathbf{x}$ and $\mathbf{y}$ are vectors in a vector space V, show that there is a unique vector $\mathbf{z} \in V$ such that $\mathbf{x} + \mathbf{z} = \mathbf{y}$.
24. Show that the set of positive real numbers forms a vector space under the operations $x + y = xy$ and $\alpha x = x^\alpha$, for α real.

9.2 SUBSPACES

From Example 9.1.1, page 479, we know that $\mathbb{R}^2 = \{(x, y): x \in \mathbb{R} \text{ and } y \in \mathbb{R}\}$ is a vector space. In Example 9.1.4, page 480, we saw that $V = \{(x, y): y = mx\}$ is also a vector space. Moreover, it is clear that $V \subset \mathbb{R}^2$. That is, $\mathbb{R}^2$ has a subset that is also a vector space. In fact, all vector spaces have subsets that are also vector spaces. We shall examine these important subsets in this section.

Definition 1 SUBSPACE Let H be a nonempty subset of a vector space V, and suppose that H is itself a vector space under the operations of addition and scalar multiplication defined on V. Then H is said to be a **subspace** of V.

We shall encounter many examples of subspaces in this chapter. But first we prove a result that makes it relatively easy to determine whether a subset of V is indeed a subspace of V.

Theorem 1 A nonempty subset H of the vector space V is a subspace of V if the two closure rules hold:

RULES FOR CHECKING WHETHER A SUBSET IS A SUBSPACE

(i) If $\mathbf{x} \in H$ and $\mathbf{y} \in H$, then $\mathbf{x} + \mathbf{y} \in H$.

(ii) If $\mathbf{x} \in H$, then $\alpha\mathbf{x} \in H$ for every scalar α.

Proof. To show that H is a vector space, we must show that axioms (i) to (x) on page 479 hold under the operations of vector addition and scalar multiplication defined in V. The two closure operations, axioms (i) and (vi), hold by hypothesis. Since vectors in H are also in V, the associative, commutative, distributive, and multiplicative identity laws [axioms (ii), (v), (vii), (viii), (ix), and (x)] hold. Let $\mathbf{x} \in H$. Then $0\mathbf{x} \in H$ by hypothesis (ii). But by Theorem 9.1.1 (ii), $0\mathbf{x} = \mathbf{0}$. Thus $\mathbf{0} \in H$ and axiom (iii) holds. Finally, by part (ii), $(-1)\mathbf{x} \in H$ for every $\mathbf{x} \in H$. By Theorem 9.1.1 (iv), $-\mathbf{x} = (-1)\mathbf{x} \in H$ so that axiom (iv) also holds and the proof is complete. ■

This theorem shows that to test whether H is a subspace of V, it is only necessary to verify that

$\mathbf{x} + \mathbf{y}$ and $\alpha\mathbf{x}$ are in H when $\mathbf{x}$ and $\mathbf{y}$ are in H and α is a scalar.

The preceding proof contains a fact that is important enough to mention explicitly:

Every subspace of a vector space V contains $\mathbf{0}$. **(1)**

This fact often makes it easy to see that a particular subset of V is *not* a vector space. That is, if a subset does not contain $\mathbf{0}$, then it is not a subspace.

We now give some examples of subspaces.

EXAMPLE 1 For any vector space V, the subset $\{\mathbf{0}\}$ consisting of the zero vector alone is a subspace since $\mathbf{0} + \mathbf{0} = \mathbf{0}$ and $\alpha\mathbf{0} = \mathbf{0}$ for every real number α (part (i) of Theorem 9.1.1). It is called the **trivial subspace**. ■

EXAMPLE 2 V is a subspace of itself for every vector space V. ■

PROPER SUBSPACE

The first two examples show that every vector space V contains two subspaces $\{\mathbf{0}\}$ and V (unless, of course, $V = \{\mathbf{0}\}$). It is more interesting to find other subspaces. Subspaces other than $\{\mathbf{0}\}$ and $\mathbf{V}$ are called **proper subspaces**.

EXAMPLE 3 Let $H = \{(x, y): y = mx\}$ (see Example 9.1.4). Then, as we have already mentioned, H is a subspace of $\mathbb{R}^2$. As we shall see in Section 9.4 the sets of vectors lying on straight lines through the origin are the only proper subspaces of $\mathbb{R}^2$. ■

EXAMPLE 4 Let $H = \{(x, y, z): x = at, y = bt, \text{ and } z = ct; a, b, c, t \text{ real}\}$. Then H consists of the vectors in $\mathbb{R}^3$ lying on a straight line passing through the origin. We verify that H is a subspace of $\mathbb{R}^3$. Let $\mathbf{x} = (at_1, bt_1, ct_1) \in H$ and $\mathbf{y} = (at_2, bt_2, ct_2) \in H$. Then $\mathbf{x} + \mathbf{y} = (a(t_1 + t_2), b(t_1 + t_2), c(t_1 + t_2)) \in H$ and $\alpha\mathbf{x} = (a(\alpha t_1), b(\alpha t_2), c(\alpha t_3)) \in H$. Thus H is a subspace of $\mathbb{R}^3$. ■

EXAMPLE 5 Let $\pi = \{(x, y, z): ax + by + cz = 0; a, b, c \text{ real}\}$. Then, as we saw in Example 9.1.6, π is a vector space; thus π is a subspace of $\mathbb{R}^3$. Here the letter π is used to indicate that the set is a plane. ■

We shall prove in Section 9.4 (Example 9.4.10) that sets of vectors lying on lines and planes through the origin are the only proper subspaces of $\mathbb{R}^3$. However we may observe here that no line or plane in $\mathbb{R}^3$ that does not pass through the origin is a subspace of $\mathbb{R}^3$ since such a line or plane does not contain the zero vector (0, 0, 0).

Before studying more examples, we note that **not every vector space has proper subspaces.**

EXAMPLE 6 Let H be a subspace of $\mathbb{R}$.† If $H \neq \{\mathbf{0}\}$, then H contains a nonzero real number α. Then, by axiom (vi), $1 = (1/\alpha)\alpha \in H$ and $\beta 1 = \beta \in H$ for every real number β. Thus if H is not the trivial subspace, then $H = \mathbb{R}$. That is, $\mathbb{R}$ has *no* proper subspace. ■

EXAMPLE 7 If P_n denotes the vector space of polynomials of degree $\leq n$ (Example 9.1.7), and if $0 < m < n$, then P_m is a proper subspace of P_n, as is easily verified. For example, suppose that $m = 2$ and $n = 5$. Then

$$(a_0 + a_1x + a_2x^2) + (b_0 + b_1x + b_2x^2) = (a_0 + b_0) + (a_1 + b_1)x + (a_2 + b_2)x^2 \in P_2$$

and

$$\alpha(a_0 + a_1x + a_2x^2) = \alpha a_0 + \alpha a_1x + \alpha a_2x^2 \in P_2.$$

This shows that the sum and scalar multiples of polynomials in P_2 are again in P_2 so the two closure rules hold and P_2 is a subspace of P_5. ■

EXAMPLE 8 Let M_{mn} (Example 9.1.10) denote the vector space of $m \times n$ matrices with real components and let $H = \{A \in M_{mn}: a_{11} = 0\}$. By the definition of matrix addition and scalar multiplication it is clear that the two closure axioms hold, so that H is a subspace. ■

EXAMPLE 9 Let $V = M_{nn}$ (the $n \times n$ matrices) and let $H = \{A \in M_{nn}: A \text{ is invertible}\}$. Then H is not a subspace since the $n \times n$ zero matrix is not in H. ■

EXAMPLE 10 Let $V = M_{33}$ and let $H = \{A \in M_{33}: A \text{ is lower triangular}\}$. If

$$A = \begin{pmatrix} a_1 & 0 & 0 \\ a_2 & a_3 & 0 \\ a_4 & a_5 & a_6 \end{pmatrix} \quad \text{and} \quad B = \begin{pmatrix} b_1 & 0 & 0 \\ b_2 & b_3 & 0 \\ b_4 & b_5 & b_6 \end{pmatrix}$$

then

†Note that $\mathbb{R}$ is a vector space over itself; that is, $\mathbb{R}$ is a vector space where the scalars are taken to be the reals, and addition is the customary addition of real numbers. This is Example 9.1.1 with $n = 1$.

$$A + B = \begin{pmatrix} a_1 + b_1 & 0 & 0 \\ a_2 + b_2 & a_3 + b_3 & 0 \\ a_4 + b_4 & a_5 + b_5 & a_6 + b_6 \end{pmatrix} \quad \text{and} \quad \alpha A = \begin{pmatrix} \alpha a_1 & 0 & 0 \\ \alpha a_2 & \alpha a_3 & 0 \\ \alpha a_4 & \alpha a_5 & \alpha a_6 \end{pmatrix}$$

are lower triangular. Thus the two closure rules hold, and H is a subspace of M_{33}. ■

EXAMPLE 11 $P_n[0, 1]^\dagger \subset C[0, 1]$ (see Example 9.1.8) because every polynomial is continuous and P_n is a vector space for every integer n, so that each $P_n[0, 1]$ is a subspace of $C[0, 1]$. ■

EXAMPLE 12 Let $C'[0, 1]$ denote the set of functions with continuous first derivatives defined on $[0, 1]$. Since every differentiable function is continuous, we have $C'[0, 1] \subset C[0, 1]$. Since the sum and scalar multiple of two differentiable functions are differentiable, we see that $C'[0, 1]$ is a subspace of $C[0, 1]$. It is a proper subspace because not every continuous function is differentiable. ■

EXAMPLE 13 If $f \in C[0, 1]$, then $\int_0^1 f(x)\,dx$ exists. Let $H = \{f \in C[0, 1]\colon \int_0^1 f(x)\,dx = 0\}$. If $f \in H$ and $g \in H$, then $\int_0^1 [f(x) + g(x)]\,dx = \int_0^1 f(x)\,dx + \int_0^1 g(x)\,dx = 0 + 0 = 0$ and $\int_0^1 \alpha f(x)\,dx = \alpha \int_0^1 f(x)\,dx = 0$. Thus $f + g$ and αf are in H for every real number α. This shows that H is a proper subspace of $C[0, 1]$. ■

As the last three examples illustrate, a vector space can have a great number and variety of proper subspaces. Before leaving this section, we prove an interesting fact about subspaces.

Theorem 2 Let H_1 and H_2 be subspaces of a vector space V. Then $H_1 \cap H_2$ is a subspace of V.

Proof. Let $\mathbf{x}_1 \in H_1 \cap H_2$ and $\mathbf{x}_2 \in H_1 \cap H_2$. Then, since H_1 and H_2 are subspaces, $\mathbf{x}_1 + \mathbf{x}_2 \in H_1$ and $\mathbf{x}_1 + \mathbf{x}_2 \in H_2$. This means that $\mathbf{x}_1 + \mathbf{x}_2 \in H_1 \cap H_2$. Similarly, $\alpha\mathbf{x}_1 \in H_1 \cap H_2$. Thus the two closure axioms are satisfied, and $H_1 \cap H_2$ is a subspace.[‡] ■

EXAMPLE 14 In $\mathbb{R}^3$, let $H_1 = \{(x, y, z)\colon 2x - y - z = 0\}$ and $H_2 = \{(x, y, z)\colon x + 2y + 3z = 0\}$. Then H_1 and H_2 consist of vectors lying on planes through the origin and are, by Example 5, subspaces of $\mathbb{R}^3$. $H_1 \cap H_2$ is the intersection of the two planes which we compute as in Section 7.3:

$$\begin{aligned} x + 2y + 3z &= 0 \\ 2x - y - z &= 0 \end{aligned}$$

or, row-reducing,

$$\left(\begin{array}{ccc|c} 1 & 2 & 3 & 0 \\ 2 & -1 & -1 & 0 \end{array}\right) \xrightarrow{A_{1,2}(-2)} \left(\begin{array}{ccc|c} 1 & 2 & 3 & 0 \\ 0 & -5 & -7 & 0 \end{array}\right) \xrightarrow{M_2(-\frac{1}{5})}$$

$$\left(\begin{array}{ccc|c} 1 & 2 & 3 & 0 \\ 0 & 1 & \frac{7}{5} & 0 \end{array}\right) \xrightarrow{A_{2,1}(-2)} \left(\begin{array}{ccc|c} 1 & 0 & \frac{1}{5} & 0 \\ 0 & 1 & \frac{7}{5} & 0 \end{array}\right).$$

† $P_n[0, 1]$ denotes the set of polynomials of degree $\leq n$ defined on the interval $[0, 1]$.

‡ Note, in particular, that as $\mathbf{0} \in H_1$ and $\mathbf{0} \in H_2$, we have $\mathbf{0} \in H_1 \cap H_2$.

Thus all solutions to the homogeneous system are given by $(-\frac{1}{5}z, -\frac{7}{5}z, z)$. Setting $z = t$, we obtain the parametric equations of a line L in $\mathbb{R}^3$: $x = -\frac{1}{5}t$, $y = -\frac{7}{5}t$, $z = t$. As we saw in Example 4, the set of vectors on L constitutes a subspace of $\mathbb{R}^3$. ■

PROBLEMS 9.2

In Problems 1–20 determine whether the given subset H of the vector space V is a subspace of V.

1. $V = \mathbb{R}^2$; $H = \{(x, y): y \geq 0\}$
2. $V = \mathbb{R}^2$; $H = \{(x, y): x = y\}$
3. $V = \mathbb{R}^3$; $H =$ the xy-plane
4. $V = \mathbb{R}^2$; $H = \{(x, y): x^2 + y^2 \leq 1\}$
5. $V = M_{nn}$; $H = \{D \in M_{nn}: D \text{ is diagonal}\}$
6. $V = M_{nn}$; $H = \{T \in M_{nn}: T \text{ is upper triangular}\}$
7. $V = M_{nn}$; $H = \{S \in M_{nn}: S \text{ is symmetric}\}$
8. $V = M_{mn}$; $H = \{A \in M_{mn}: a_{ij} = 0\}$
9. $V = M_{22}$; $H = \left\{A \in M_{22}: A = \begin{pmatrix} a & b \\ -b & c \end{pmatrix}\right\}$
10. $V = M_{22}$; $H = \left\{A \in M_{22}: A = \begin{pmatrix} a & 1+a \\ 0 & 0 \end{pmatrix}\right\}$
11. $V = M_{22}$; $H = \left\{A \in M_{22}: A = \begin{pmatrix} 0 & a \\ b & 0 \end{pmatrix}\right\}$
12. $V = P_4$; $H = \{p \in P_4: \deg p = 4\}$
13. $V = P_4$; $H = \{p \in P_4: p(0) = 0\}$
14. $V = P_n$; $H = \{p \in P_n: p(0) = 0\}$
15. $V = P_n$; $H = \{p \in P_n: p(0) = 1\}$
16. $V = C[0, 1]$; $H = \{f \in C[0, 1]: f(0) = f(1) = 0\}$
17. $V = C[0, 1]$; $H = \{f \in C[0, 1]: f(0) = 2\}$
18. $V = C'[0, 1]$; $H = \{f \in C'[0, 1]: f'(0) = 0\}$
19. $V = C[a, b]$, where a and b are real numbers and $a < b$; $H = \{f \in C[a, b]: \int_a^b f(x)\,dx = 0\}$
20. $V = C[a, b]$; $H = \{f \in C[a, b]: \int_a^b f(x)\,dx = 1\}$
21. Let $V = M_{22}$; let $H_1 = \{A \in M_{22}: a_{11} = 0\}$ and $H_2 = \left\{A \in M_{22}: A = \begin{pmatrix} -b & a \\ a & b \end{pmatrix}\right\}$.
    (a) Show that H_1 and H_2 are subspaces.
    (b) Describe the subset $H = H_1 \cap H_2$, and show that it is a subspace.
22. If $V = C[0, 1]$, let H_1 denote the subspace of Example 11 and H_2 denote the subspace of Example 12. Describe the set $H_1 \cap H_2$, and show that it is a subspace.
23. Let A be an $n \times m$ matrix and let $H = \{\mathbf{x} \in \mathbb{R}^m: A\mathbf{x} = \mathbf{0}\}$. Show that H is a subspace of $\mathbb{R}^m$. H is called the *kernel* of the matrix A.
24. In Problem 23 let $H = \{\mathbf{x} \in \mathbb{R}^m: A\mathbf{x} \neq 0\}$. Show that H is not a subspace of $\mathbb{R}^m$.
25. Let $H = \{(x, y, z, w): ax + by + cz + dw = 0\}$, where a, b, c, and d are real numbers not all zero. Show that H is a proper subspace of $\mathbb{R}^4$. H is called a *hyperplane* in $\mathbb{R}^4$.
26. Let $H = \{(x_1, x_2, \ldots, x_n): a_1x_1 + a_2x_2 + \cdots + a_nx_n = 0\}$, where $a_1, a_2, \ldots, a_n$ are real numbers not all zero. Show that H is a proper subspace of $\mathbb{R}^n$. H, as in Problem 25, is called a **hyperplane** in $\mathbb{R}^n$.
27. Let H_1 and H_2 be subspaces of a vector space V. Let $H_1 + H_2 = \{\mathbf{v}: \mathbf{v} = \mathbf{v}_1 + \mathbf{v}_2 \text{ with } \mathbf{v}_1 \in H_1 \text{ and } \mathbf{v}_2 \in H_2\}$. Show that $H_1 + H_2$ is a subspace of V.
28. Let $\mathbf{v}_1$ and $\mathbf{v}_2$ be two vectors in $\mathbb{R}^2$. Show that $H = \{\mathbf{v}: \mathbf{v} = a\mathbf{v}_1 + b\mathbf{v}_2;\ a, b \text{ real}\}$ is a subspace of $\mathbb{R}^2$.

*29. In Problem 28 show that if $\mathbf{v}_1$ and $\mathbf{v}_2$ are not collinear, then $H = \mathbb{R}^2$.

*30. Let $\mathbf{v}_1, \mathbf{v}_2, \ldots, \mathbf{v}_n$ be arbitrary vectors in a vector space V. Let $H = \{\mathbf{v} \in V: \mathbf{v} = a_1\mathbf{v}_1 + a_2\mathbf{v}_2 + \cdots + a_n\mathbf{v}_n$, where $a_1, a_2, \ldots, a_n$ are scalars$\}$. Show that H is a subspace of V. H is called the subspace *spanned* by the vectors $\mathbf{v}_1, \mathbf{v}_2, \ldots, \mathbf{v}_n$.

9.3 LINEAR INDEPENDENCE, LINEAR COMBINATION, AND SPAN

In the study of linear algebra, one of the central ideas is that of the linear dependence or independence of vectors. In this section we shall define what we mean by linear independence. In the next section we shall show how this notion is used to define the dimension of a vector space.

Consider the vectors $\mathbf{v}_1 = \begin{pmatrix}1\\2\end{pmatrix}$ and $\mathbf{v}_2 = \begin{pmatrix}2\\4\end{pmatrix}$. We see that $\mathbf{v}_2 = 2\mathbf{v}_1$ or, writing this equation in another way, that

$$2\mathbf{v}_1 - \mathbf{v}_2 = \mathbf{0}. \tag{1}$$

Is there an equation relating the vectors $\mathbf{v}_1 = \begin{pmatrix}1\\2\\3\end{pmatrix}$, $\mathbf{v}_2 = \begin{pmatrix}-4\\1\\5\end{pmatrix}$, and $\mathbf{v}_3 = \begin{pmatrix}-5\\8\\19\end{pmatrix}$? This question is more difficult to answer at first glance. It is easy to verify, however, that $\mathbf{v}_3 = 3\mathbf{v}_1 + 2\mathbf{v}_2$, or, rewriting, that

$$3\mathbf{v}_1 + 2\mathbf{v}_2 - \mathbf{v}_3 = \mathbf{0}. \tag{2}$$

It appears that the two vectors in equation (1) and the three vectors in (2) are more closely related than an arbitrary pair of 2-vectors or an arbitrary triple of 3-vectors. In each case we say that the vectors are *linearly dependent*. We have the following important definitions:

Definition 1 LINEARLY DEPENDENT VECTORS The set of vectors $\mathbf{v}_1, \mathbf{v}_2, \ldots, \mathbf{v}_n$ in a vector space V is **linearly dependent** if there exist scalars $c_1, c_2, \ldots, c_n$ *not all zero* such that

$$c_1\mathbf{v}_1 + c_2\mathbf{v}_2 + \cdots + c_n\mathbf{v}_n = \mathbf{0}. \tag{3}$$

With this definition we see that the vectors in equation (1) $[c_1 = 2,\ c_2 = -1]$ and equation (2) $[c_1 = 3,\ c_2 = 2,\ c_3 = -1]$ are linearly dependent.

Definition 2 LINEARLY INDEPENDENT VECTORS The set of vectors $\mathbf{v}_1, \mathbf{v}_2, \ldots, \mathbf{v}_n$ in a vector space V is **linearly independent** if it is not linearly dependent.

Putting this another way, $\mathbf{v}_1, \mathbf{v}_2, \ldots, \mathbf{v}_n$ are linearly independent if the equation $c_1\mathbf{v}_1 + c_2\mathbf{v}_2 + \cdots + c_n\mathbf{v}_n = \mathbf{0}$ holds only for $c_1 = c_2 = \cdots = c_n = 0$.

How do we determine whether a set of vectors is linearly dependent or independent? The case for two vectors is easy.

Theorem 1 Two vectors are linearly dependent if and only if one is a scalar multiple of the other.

Proof. First suppose that $\mathbf{v}_2 = c\mathbf{v}_1$ for some scalar $c \neq 0$. Then $c\mathbf{v}_1 - \mathbf{v}_2 = \mathbf{0}$ and $\mathbf{v}_1$ and $\mathbf{v}_2$ are linearly dependent. On the other hand, suppose that $\mathbf{v}_1$ and $\mathbf{v}_2$ are dependent. Then there are constants c_1 and c_2, not both zero, such that $c_1\mathbf{v}_1 + c_2\mathbf{v}_2 = \mathbf{0}$. If $c_1 \neq 0$, then, dividing by c_1, we obtain $\mathbf{v}_1 + (c_2/c_1)\mathbf{v}_2 = \mathbf{0}$ or

$$\mathbf{v}_1 = \left(-\frac{c_2}{c_1}\right)\mathbf{v}_2.$$

That is, $\mathbf{v}_1$ is a scalar multiple of $\mathbf{v}_2$. If $c_1 = 0$ then $c_2 \neq 0$ and hence $\mathbf{v}_2 = \mathbf{0} = 0\mathbf{v}_1$. ■

■ WARNING: Theorem 1 does *not* say that three or more vectors are linearly independent if no one of them is a multiple of any other. For example, (1, 1), (1, 0) and (0, 1) in $\mathbb{R}^2$ are linearly dependent because $(1, 1) - (1, 0) - (0, 1) = (0, 0)$. However no one of the three vectors is a multiple of any other one of the three vectors. **Theorem 1 can be used only when considering exactly two vectors.**

EXAMPLE 1 The vectors $\mathbf{v}_1 = \begin{pmatrix} 2 \\ -1 \\ 0 \\ 3 \end{pmatrix}$ and $\mathbf{v}_2 = \begin{pmatrix} -6 \\ 3 \\ 0 \\ -9 \end{pmatrix}$ are linearly dependent since $\mathbf{v}_2 = -3\mathbf{v}_1$. ■

EXAMPLE 2 The vectors $\begin{pmatrix} 1 \\ 2 \\ 4 \end{pmatrix}$ and $\begin{pmatrix} 2 \\ 5 \\ -3 \end{pmatrix}$ are linearly independent; if they were not, we would have

$$\begin{pmatrix} 2 \\ 5 \\ -3 \end{pmatrix} = c\begin{pmatrix} 1 \\ 2 \\ 4 \end{pmatrix} = \begin{pmatrix} c \\ 2c \\ 4c \end{pmatrix}.$$

Then $2 = c$, $5 = 2c$, and $-3 = 4c$, which is clearly impossible for any number c. ■

There are several techniques for determining whether a set of vectors is linearly independent. Let us examine two of these techniques here.

EXAMPLE 3 Determine whether the vectors $\begin{pmatrix} 1 \\ -2 \\ 3 \end{pmatrix}$, $\begin{pmatrix} 2 \\ -2 \\ 0 \end{pmatrix}$, and $\begin{pmatrix} 0 \\ 1 \\ 7 \end{pmatrix}$ are linearly dependent or independent.

Solution. Suppose that

$$c_1\begin{pmatrix} 1 \\ -2 \\ 3 \end{pmatrix} + c_2\begin{pmatrix} 2 \\ -2 \\ 0 \end{pmatrix} + c_3\begin{pmatrix} 0 \\ 1 \\ 7 \end{pmatrix} = \mathbf{0} = \begin{pmatrix} 0 \\ 0 \\ 0 \end{pmatrix}.$$

Then, multiplying through and adding, we have

$$\begin{pmatrix} c_1 + 2c_2 \\ -2c_1 - 2c_2 + c_3 \\ 3c_1 + 7c_3 \end{pmatrix} = \begin{pmatrix} 0 \\ 0 \\ 0 \end{pmatrix}.$$

This yields a system of three equations in the three unknowns c_1, c_2, and c_3:

$$\begin{aligned} c_1 + 2c_2 \qquad &= 0, \\ -2c_1 - 2c_2 + c_3 &= 0, \\ 3c_1 \qquad + 7c_3 &= 0. \end{aligned} \tag{4}$$

Thus the vectors will be linearly dependent if and only if system (4) has nontrivial solutions. We write system (4) using an augmented matrix and then row reduce:

$$\left(\begin{array}{rrr|r} 1 & 2 & 0 & 0 \\ -2 & -2 & 1 & 0 \\ 3 & 0 & 7 & 0 \end{array}\right) \xrightarrow[A_{1,3}(-3)]{A_{1,2}(2)} \left(\begin{array}{rrr|r} 1 & 2 & 0 & 0 \\ 0 & 2 & 1 & 0 \\ 0 & -6 & 7 & 0 \end{array}\right)$$

$$\xrightarrow{M_2(\frac{1}{2})} \left(\begin{array}{rrr|r} 1 & 2 & 0 & 0 \\ 0 & 1 & \frac{1}{2} & 0 \\ 0 & -6 & 7 & 0 \end{array}\right) \xrightarrow[A_{2,3}(6)]{A_{2,1}(-2)} \left(\begin{array}{rrr|r} 1 & 0 & -1 & 0 \\ 0 & 1 & \frac{1}{2} & 0 \\ 0 & 0 & 10 & 0 \end{array}\right)$$

$$\xrightarrow{M_3(\frac{1}{10})} \left(\begin{array}{rrr|r} 1 & 0 & -1 & 0 \\ 0 & 1 & \frac{1}{2} & 0 \\ 0 & 0 & 1 & 0 \end{array}\right) \xrightarrow[A_{3,2}(-\frac{1}{2})]{A_{3,1}(1)} \left(\begin{array}{rrr|r} 1 & 0 & 0 & 0 \\ 0 & 1 & 0 & 0 \\ 0 & 0 & 1 & 0 \end{array}\right)$$

The last system of equations reads $c_1 = 0$, $c_2 = 0$, $c_3 = 0$. Hence (4) has no nontrivial solutions and the given vectors are linearly independent. ■

There is a much faster way to solve this problem. We know from Corollary 8.3.1 on page 472 that the homogeneous system (4) has a unique solution if and only if the determinant of the coefficient matrix is nonzero. But

$$\det A = \begin{vmatrix} 1 & 2 & 0 \\ -2 & -2 & 1 \\ 3 & 0 & 7 \end{vmatrix} = 20 \neq 0.$$

Thus the system (4) has a unique solution, the trivial solution, and the vectors are linearly independent. ■

■ **WARNING:** Using determinants to establish linear independence or dependence works only if the resulting system of equations has the same number of equations as unknowns. Otherwise the underlying matrix is not square, and the determinant is not defined.

EXAMPLE 4 Determine whether the vectors $\begin{pmatrix} 1 \\ -3 \\ 0 \end{pmatrix}$, $\begin{pmatrix} 3 \\ 0 \\ 4 \end{pmatrix}$, and $\begin{pmatrix} 11 \\ -6 \\ 12 \end{pmatrix}$ are linearly dependent or independent.

Solution. The equation $c_1\begin{pmatrix} 1 \\ -3 \\ 0 \end{pmatrix} + c_2\begin{pmatrix} 3 \\ 0 \\ 4 \end{pmatrix} + c_3\begin{pmatrix} 11 \\ -6 \\ 12 \end{pmatrix} = \begin{pmatrix} 0 \\ 0 \\ 0 \end{pmatrix}$ leads to the homogeneous system

$$
\begin{aligned}
c_1 + 3c_2 + 11c_3 &= 0 \\
-3c_1 \qquad - 6c_3 &= 0 \\
4c_2 + 12c_3 &= 0.
\end{aligned}
\tag{5}
$$

Here

$$\det A = \begin{vmatrix} 1 & 3 & 11 \\ -3 & 0 & -6 \\ 0 & 4 & 12 \end{vmatrix} = 0.$$

So the homogeneous system (5) has an infinite number of solutions, and the vectors are linearly dependent. If we wish to find the equation relating these vectors, we write system (5) in augmented matrix form and row reduce to obtain, successively,

$$\left(\begin{array}{ccc|c} 1 & 3 & 11 & 0 \\ -3 & 0 & -6 & 0 \\ 0 & 4 & 12 & 0 \end{array}\right) \xrightarrow{A_{1,2}(3)} \left(\begin{array}{ccc|c} 1 & 3 & 11 & 0 \\ 0 & 9 & 27 & 0 \\ 0 & 4 & 12 & 0 \end{array}\right)$$

$$\xrightarrow{M_2(\frac{1}{9})} \left(\begin{array}{ccc|c} 1 & 3 & 11 & 0 \\ 0 & 1 & 3 & 0 \\ 0 & 4 & 12 & 0 \end{array}\right) \xrightarrow[A_{2,3}(-4)]{A_{2,1}(-3)} \left(\begin{array}{ccc|c} 1 & 0 & 2 & 0 \\ 0 & 1 & 3 & 0 \\ 0 & 0 & 0 & 0 \end{array}\right).$$

We can stop here. The system now reads

$$
\begin{aligned}
c_1 \qquad + 2c_3 &= 0, \\
c_2 + 3c_3 &= 0.
\end{aligned}
$$

If, for example, we choose $c_3 = 1$, we have $c_2 = -3$ and $c_1 = -2$; so that, as is easily verified,

$$-2\begin{pmatrix} 1 \\ -3 \\ 0 \end{pmatrix} - 3\begin{pmatrix} 3 \\ 0 \\ 4 \end{pmatrix} + \begin{pmatrix} 11 \\ -6 \\ 12 \end{pmatrix} = \begin{pmatrix} 0 \\ 0 \\ 0 \end{pmatrix}. \blacksquare$$

The following useful result follows immediately from Theorem 7.3.1 on p. 421 (see Problem 27).

Theorem 2 A set of n vectors in $\mathbb{R}^m$ is always linearly dependent if $n > m$.

EXAMPLE 5 The vectors

$$\begin{pmatrix} 2 \\ -3 \\ 4 \end{pmatrix}, \quad \begin{pmatrix} 4 \\ 7 \\ -6 \end{pmatrix}, \quad \begin{pmatrix} 18 \\ -11 \\ 4 \end{pmatrix}, \quad \text{and} \quad \begin{pmatrix} 2 \\ -7 \\ 3 \end{pmatrix}$$

are linearly dependent since they constitute a set of four vectors in $\mathbb{R}^3$. ■

There is a very important (and obvious) corollary to Theorem 2.

Corollary 1 A set of linearly independent n-vectors contains at most n vectors.

NOTE. We can rephrase the corollary as follows: If we have n linearly independent n-vectors, then we cannot add any more vectors without making the set linearly dependent.

The following important result links the notions of linear independence, invertibility, and determinant:

Theorem 3 Let A be an $n \times n$ matrix. Then the following three statements are equivalent (i.e., if one is true, all three are true):

(i) A is invertible;
(ii) $\det A \neq 0$;
(iii) the columns of A (considered as n-vectors) are linearly independent.

Proof. We know, from Theorem 8.3.3, that statements (i) and (ii) are equivalent. We shall show the equivalence of (i) and (iii). Suppose that A is invertible and let $\mathbf{a}_1, \mathbf{a}_2, \ldots, \mathbf{a}_n$ denote the columns of A. Suppose that

$$c_1\mathbf{a}_1 + c_2\mathbf{a}_2 + \cdots + c_n\mathbf{a}_n = \mathbf{0}. \tag{6}$$

System (6) can be written in the form $A\mathbf{c} = \mathbf{0}$ where

$$\mathbf{c} = \begin{pmatrix} c_1 \\ c_2 \\ \vdots \\ c_n \end{pmatrix}.$$

Then $\mathbf{c} = A^{-1}\mathbf{0} = \mathbf{0}$, which shows that $c_1 = c_2 = \cdots = c_n = 0$, so that the vectors $\mathbf{a}_1, \mathbf{a}_2, \ldots, \mathbf{a}_n$ are independent. Thus statement (i) implies statement (iii).

Now suppose that the columns of A are linearly independent. Let

$$\mathbf{a}_1 = \begin{pmatrix} a_{11} \\ a_{21} \\ \vdots \\ a_{n1} \end{pmatrix}, \quad \mathbf{a}_2 = \begin{pmatrix} a_{12} \\ a_{22} \\ \vdots \\ a_{n2} \end{pmatrix},$$

and so on. If we write out system (6), we obtain

$$\begin{aligned} a_{11}c_1 + a_{12}c_2 + \cdots + a_{1n}c_n &= 0 \\ a_{21}c_1 + a_{22}c_2 + \cdots + a_{2n}c_n &= 0 \\ \vdots \qquad \vdots \qquad\qquad \vdots \quad &\ \ \vdots \\ a_{n1}c_1 + a_{n2}c_2 + \cdots + a_{nn}c_n &= 0. \end{aligned} \tag{7}$$

If A were not row equivalent to I_n, then row reduction of the augmented matrix associated with (7) would leave us with a row of zeros. But if, say, the last row is zero, then the last equation reads $0 = 0$. That is, we can choose a value for c_n arbitrarily and obtain an infinite number of solutions. But since the vectors $\mathbf{a}_1, \mathbf{a}_2, \ldots, \mathbf{a}_n$ are linearly independent, the only solution to (6) and (7) is $c_1 = c_2 = \cdots = c_n = 0$. This contradiction shows that A is row equivalent to I_n. Then, by part (i) of Theorem 7.5.5, A is invertible. This completes the proof. ■

Using Theorem 3, we can extend our summing-up theorem, last seen in Section 8.3 (page 472).

Theorem 4 SUMMING-UP THEOREM—VIEW 3 Let A be an $n \times n$ matrix. Then each of the following six statements implies the other five. (That is, if one is true, all are true.)

- **(i)** A is invertible.
- **(ii)** The system $A\mathbf{x} = \mathbf{b}$ has a unique solution for every n-vector $\mathbf{b}$.
- **(iii)** The only solution to the homogeneous system $A\mathbf{x} = \mathbf{0}$ is the trivial solution ($\mathbf{x} = \mathbf{0}$).
- **(iv)** A is row equivalent to the identity matrix I_n; that is, the reduced row echelon form of A is I_n.
- **(v)** $\det A \neq 0$.
- **(vi)** The columns of A (considered as n-vectors) are linearly independent.

All the examples given so far in this section have been in $\mathbb{R}^n$. We now turn to some other vector spaces.

EXAMPLE 6 In M_{23} let $A_1 = \begin{pmatrix} 1 & 0 & 2 \\ 3 & 1 & -1 \end{pmatrix}$, $A_2 = \begin{pmatrix} -1 & 1 & 4 \\ 2 & 3 & 0 \end{pmatrix}$, and $A_3 = \begin{pmatrix} -1 & 0 & 1 \\ 1 & 2 & 1 \end{pmatrix}$. Determine whether A_1, A_2, and A_3 are linearly dependent or independent.

Solution. Suppose that $c_1A_1 + c_2A_2 + c_3A_3 = 0$. Then

$$\begin{pmatrix} 0 & 0 & 0 \\ 0 & 0 & 0 \end{pmatrix} = c_1\begin{pmatrix} 1 & 0 & 2 \\ 3 & 1 & -1 \end{pmatrix} + c_2\begin{pmatrix} -1 & 1 & 4 \\ 2 & 3 & 0 \end{pmatrix} + c_3\begin{pmatrix} -1 & 0 & 1 \\ 1 & 2 & 1 \end{pmatrix}$$

$$= \begin{pmatrix} c_1 - c_2 - c_3 & c_2 & 2c_1 + 4c_2 + c_3 \\ 3c_1 + 2c_2 + c_3 & c_1 + 3c_2 + 2c_3 & -c_1 + c_3 \end{pmatrix}.$$

This gives us a homogeneous system of six equations in the three unknowns c_1, c_2, and c_3, and it is quite easy to verify that the only solution is $c_1 = c_2 = c_3 = 0$. Thus the three matrices are linearly independent. ■

EXAMPLE 7 In P_3, determine whether the polynomials 1, x, x^2, and x^3 are linearly dependent or independent.

Solution. Suppose that $c_1 + c_2x + c_3x^2 + c_4x^3 = 0$. This must hold for every real number x. In particular, if $x = 0$, we obtain $c_1 = 0$. Then, setting $x = 1, -1, 2$, we obtain, successively,

$$\begin{aligned} c_2 + c_3 + c_4 &= 0, \\ -c_2 + c_3 - c_4 &= 0, \\ 2c_2 + 4c_3 + 8c_4 &= 0. \end{aligned}$$

The determinant of this homogeneous system is

$$\begin{vmatrix} 1 & 1 & 1 \\ -1 & 1 & -1 \\ 2 & 4 & 8 \end{vmatrix} = 12 \neq 0,$$

so that the system has the unique solution $c_2 = c_3 = c_4 = 0$ and the four polynomials are linearly independent. We can see this in another way. It is a basic law from algebra that a polynomial of degree n has at most n roots. Thus, we know that any polynomial of degree 3 has at most three real roots. But if $c_1 + c_2x + c_3x^2 + c_4x^3 = 0$ for some nonzero constants c_1, c_2, c_3, and c_4 and for every real number x, then we have constructed a cubic polynomial for which every real number is a root. This is impossible.

We can solve this problem by another method. If $c_1 + c_2x + c_3x^2 + c_4x^3 = 0$ for every x, then as before, $c_1 = 0$. Then, differentiation yields

$$c_2 + 2c_3x + 3c_4x^2 = 0.$$

Setting $x = 0$ again, we conclude that $c_2 = 0$. Another differentiation gives us

$$2c_3 + 6c_4x = 0,$$

so that $c_3 = 0$. Finally, a third differentiation results in

$$6c_4 = 0 \qquad \text{or} \qquad c_4 = 0.$$

Thus $c_1 = c_2 = c_3 = c_4 = 0$, and the polynomials are linearly independent. ■

EXAMPLE 8 In P_2 determine whether the polynomials $x - 2x^2$, $x^2 - 4x$, and $-7x + 8x^2$ are linearly dependent or independent.

Solution. Let $c_1(x - 2x^2) + c_2(x^2 - 4x) + c_3(-7x + 8x^2) = 0$. Then, rearranging terms, we obtain

$$\begin{aligned} (c_1 - 4c_2 - 7c_3)x &= 0, \\ (-2c_1 + c_2 + 8c_3)x^2 &= 0. \end{aligned}$$

These equations hold for every x if and only if

$$c_1 - 4c_2 - 7c_3 = 0$$

and

$$-2c_1 + c_2 + 8c_3 = 0.$$

But it is easy to show that this system of two equations in three unknowns has an infinite number of solutions. If we solve this system, we obtain, successively,

$$\left(\begin{array}{rrr|r} 1 & -4 & -7 & 0 \\ -2 & 1 & 8 & 0 \end{array}\right) \xrightarrow{A_{1,2}(2)} \left(\begin{array}{rrr|r} 1 & -4 & -7 & 0 \\ 0 & -7 & -6 & 0 \end{array}\right)$$

$$\xrightarrow{M_2(-\frac{1}{7})} \left(\begin{array}{rrr|r} 1 & -4 & -7 & 0 \\ 0 & 1 & \frac{6}{7} & 0 \end{array}\right) \xrightarrow{A_{2,1}(4)} \left(\begin{array}{rrr|r} 1 & 0 & -\frac{25}{7} & 0 \\ 0 & 1 & \frac{6}{7} & 0 \end{array}\right).$$

Thus c_3 can be chosen arbitrarily and $c_1 = \frac{25}{7}c_3$ and $c_2 = -\frac{6}{7}c_3$. If $c_3 = 7$, for example, then $c_1 = 25$, $c_2 = -6$, and we have $25(x - 2x^2) - 6(x^2 - 4x) + 7(-7x + 8x^2) = 0$. This shows that the polynomials are linearly dependent. ■

We now define a new concept that is closely related to the notion of linear dependence.

Definition 3 LINEAR COMBINATION Let $\mathbf{v}_1, \mathbf{v}_2, \ldots, \mathbf{v}_n$ be vectors in a vector space V. Then any expression of the form

$$a_1\mathbf{v}_1 + a_2\mathbf{v}_2 + \cdots + a_n\mathbf{v}_n, \tag{8}$$

where $a_1, a_2, \ldots, a_n$ are scalars, is called a **linear combination** of $\mathbf{v}_1, \mathbf{v}_2, \ldots, \mathbf{v}_n$.

EXAMPLE 9 In $\mathbb{R}^3$, $\begin{pmatrix} -7 \\ 7 \\ 7 \end{pmatrix}$ is a linear combination of $\begin{pmatrix} -1 \\ 2 \\ 4 \end{pmatrix}$ and $\begin{pmatrix} 5 \\ -3 \\ 1 \end{pmatrix}$ since $\begin{pmatrix} -7 \\ 7 \\ 7 \end{pmatrix} = 2\begin{pmatrix} -1 \\ 2 \\ 4 \end{pmatrix} - \begin{pmatrix} 5 \\ -3 \\ 1 \end{pmatrix}$. ■

EXAMPLE 10 In M_{23}, $\begin{pmatrix} -3 & 2 & 8 \\ -1 & 9 & 3 \end{pmatrix} = 3\begin{pmatrix} -1 & 0 & 4 \\ 1 & 1 & 5 \end{pmatrix} + 2\begin{pmatrix} 0 & 1 & -2 \\ -2 & 3 & -6 \end{pmatrix}$, which shows that $\begin{pmatrix} -3 & 2 & 8 \\ -1 & 9 & 3 \end{pmatrix}$ is a linear combination of $\begin{pmatrix} -1 & 0 & 4 \\ 1 & 1 & 5 \end{pmatrix}$ and $\begin{pmatrix} 0 & 1 & -2 \\ -2 & 3 & -6 \end{pmatrix}$. ■

EXAMPLE 11 In P_n every polynomial can be written as a linear combination of the "monomials" $1, x, x^2, \ldots, x^n$. ■

Definition 4 SPAN OF A VECTOR SPACE The vectors $\mathbf{v}_1, \mathbf{v}_2, \ldots, \mathbf{v}_n$ in a vector space V are said to **span** V if every vector in V can be written as a linear combination of them. That is, for every $\mathbf{v} \in V$ there are scalars $a_1, a_2, \ldots, a_n$ such that

$$\mathbf{v} = a_1\mathbf{v}_1 + a_2\mathbf{v}_2 + \cdots + a_n\mathbf{v}_n \tag{9}$$

EXAMPLE 12 We saw in Section 1.1 that the vectors $\mathbf{i} = \begin{pmatrix} 1 \\ 0 \end{pmatrix}$ and $\mathbf{j} = \begin{pmatrix} 0 \\ 1 \end{pmatrix}$ span $\mathbb{R}^2$. In Section 3.1 we saw that $\mathbf{i} = \begin{pmatrix} 1 \\ 0 \\ 0 \end{pmatrix}$, $\mathbf{j} = \begin{pmatrix} 0 \\ 1 \\ 0 \end{pmatrix}$, and $\mathbf{k} = \begin{pmatrix} 0 \\ 0 \\ 1 \end{pmatrix}$ span $\mathbb{R}^3$. ■

In fact, it is not difficult to find sets of vectors that span $\mathbb{R}^n$, as the following theorem states. The proof is given as part of Theorem 9.4.5.

Theorem 5 Any set of n linearly independent vectors in $\mathbb{R}^n$ spans $\mathbb{R}^n$.

EXAMPLE 13 The three vectors

$$\begin{pmatrix} 1 \\ -2 \\ 3 \end{pmatrix}, \quad \begin{pmatrix} 2 \\ -2 \\ 0 \end{pmatrix}, \quad \text{and} \quad \begin{pmatrix} 0 \\ 1 \\ 7 \end{pmatrix}$$

are linearly independent (by Example 3), so by Theorem 4, they span $\mathbb{R}^3$. ■

EXAMPLE 14 From Example 11 it follows that the monomials $1, x, x^2, \ldots, x^n$ span P_n. ■

EXAMPLE 15 Since $\begin{pmatrix} a & b \\ c & d \end{pmatrix} = a\begin{pmatrix} 1 & 0 \\ 0 & 0 \end{pmatrix} + b\begin{pmatrix} 0 & 1 \\ 0 & 0 \end{pmatrix} + c\begin{pmatrix} 0 & 0 \\ 1 & 0 \end{pmatrix} + d\begin{pmatrix} 0 & 0 \\ 0 & 1 \end{pmatrix}$, we see that $\begin{pmatrix} 1 & 0 \\ 0 & 0 \end{pmatrix}$, $\begin{pmatrix} 0 & 1 \\ 0 & 0 \end{pmatrix}$, $\begin{pmatrix} 0 & 0 \\ 1 & 0 \end{pmatrix}$, and $\begin{pmatrix} 0 & 0 \\ 0 & 1 \end{pmatrix}$ span M_{22}. ■

EXAMPLE 16 Let P denote the vector space of polynomials of any degree. Then no *finite* set of polynomials spans P. To see this suppose that $p_1, p_2, \ldots, p_m$ are polynomials. Let p_k be the polynomial of largest degree in this set and let $N = \deg p_k$. Then it is clear that the polynomial $p(x) = x^{N+1}$ cannot be written as a linear combination of $p_1, p_2, \ldots, p_m$. ■

We now turn to another way of finding subspaces of a vector space V.

Definition 5 SPAN OF A SET OF VECTORS Let $\mathbf{v}_1, \mathbf{v}_2, \ldots, \mathbf{v}_n$ be n vectors in a vector space V. The **span** of $\{\mathbf{v}_1, \mathbf{v}_2, \ldots, \mathbf{v}_n\}$ is the set of linear combinations of $\mathbf{v}_1, \mathbf{v}_2, \ldots, \mathbf{v}_n$. That is,

$$\text{span}\,\{\mathbf{v}_1, \mathbf{v}_2, \ldots, \mathbf{v}_n\} = \{\mathbf{v}: \mathbf{v} = a_1\mathbf{v}_1 + a_2\mathbf{v}_2 + \cdots + a_n\mathbf{v}_n\} \qquad \textbf{(10)}$$

where $a_1, a_2, \ldots, a_n$ are scalars.

Theorem 6 Span $\{\mathbf{v}_1, \mathbf{v}_2, \ldots, \mathbf{v}_n\}$ is a subspace of V.

Proof. The proof is easy and is left as an exercise (see Problem 48). ■

EXAMPLE 17 Let $\mathbf{v}_1 = (2, -1, 4)$ and $\mathbf{v}_2 = (4, 1, 6)$. Then $H = \text{span}\,\{\mathbf{v}_1, \mathbf{v}_2\} = \{\mathbf{v}: \mathbf{v} = a_1(2, -1, 4) + a_2(4, 1, 6)\}$. What does H look like? If $\mathbf{v} = (x, y, z) \in H$, then we have $x = 2a_1 + 4a_2$, $y = -a_1 + a_2$, and $z = 4a_1 + 6a_2$. We can view these equations as a system of three equations in the two unknowns, a_1 and a_2, and solve this system in the usual way:

$$\left(\begin{array}{rr|l} -1 & 1 & y \\ 2 & 4 & x \\ 4 & 6 & z \end{array}\right) \xrightarrow{M_1(-1)} \left(\begin{array}{rr|l} 1 & -1 & -y \\ 2 & 4 & x \\ 4 & 6 & z \end{array}\right) \xrightarrow[A_{1,3}(-4)]{A_{1,2}(-2)} \left(\begin{array}{rr|c} 1 & -1 & -y \\ 0 & 6 & x + 2y \\ 0 & 10 & z + 4y \end{array}\right)$$

$$\xrightarrow{M_2(\frac{1}{6})} \left(\begin{array}{rr|c} 1 & -1 & -y \\ 0 & 1 & (x + 2y)/6 \\ 0 & 10 & z + 4y \end{array}\right) \xrightarrow[A_{2,3}(-10)]{A_{2,1}(1)} \left(\begin{array}{rr|c} 1 & 0 & x/6 - 2y/3 \\ 0 & 1 & x/6 + y/3 \\ 0 & 0 & -5x/3 + 2y/3 + z \end{array}\right)$$

We see that the system has a solution only if $-5x/3 + 2y/3 + z = 0$; or, multiplying through by -3, if

$$5x - 2y - 3z = 0. \tag{11}$$

Equation (11) is the equation of a plane in $\mathbb{R}^3$ passing through the origin. ■

The last example can be generalized to prove the following interesting fact:

> *The span of two nonzero vectors in $\mathbb{R}^3$ that are not parallel is a plane passing through the origin.*

For a suggested proof see Problems 49 and 50.

We can give a geometric interpretation of this result. Look at the vectors in Figure 1. We know (from Section 1.1) the geometric interpretation of the vectors $2\mathbf{u}$, $-\mathbf{u}$, and $\mathbf{u} + \mathbf{v}$, for example. Using these, we see that any other vector in the plane of $\mathbf{u}$ and $\mathbf{v}$ can be obtained as a linear combination of $\mathbf{u}$ and $\mathbf{v}$. Figure 2 shows how in four different situations a third vector $\mathbf{w}$ in the plane of $\mathbf{u}$ and $\mathbf{v}$ can be written as $\alpha\mathbf{u} + \beta\mathbf{v}$ for appropriate choices of the numbers α and β.

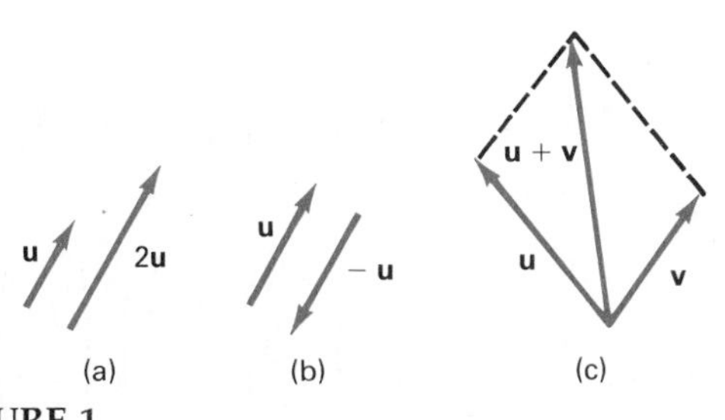

FIGURE 1

We close this section by citing a useful result. Its proof is not difficult and is left as an exercise (see Problem 51).

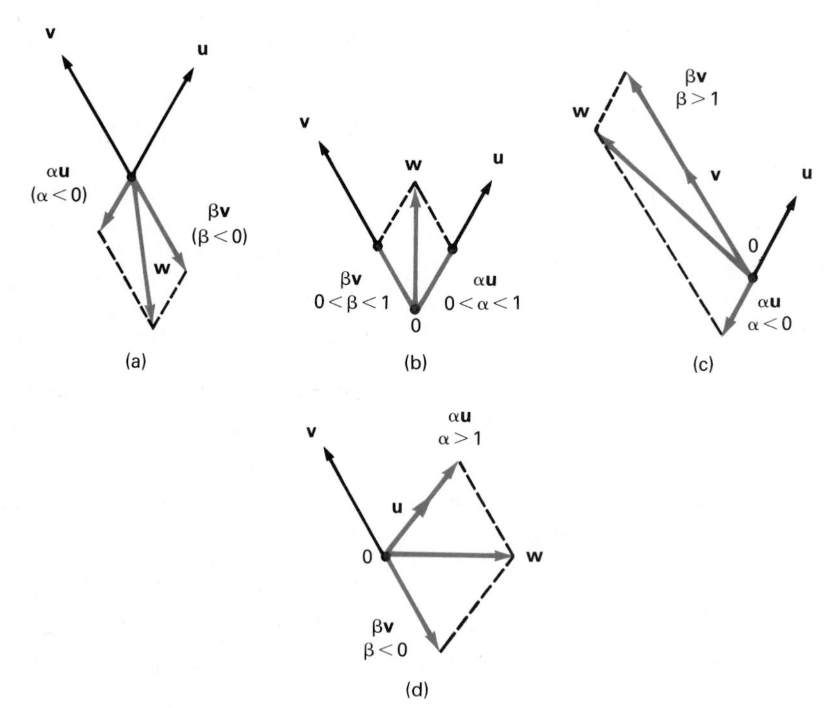

FIGURE 2

Theorem 7 Let $\mathbf{v}_1, \mathbf{v}_2, \ldots, \mathbf{v}_n, \mathbf{v}_{n+1}$ be $n + 1$ vectors in a vector space V. If $\mathbf{v}_1, \mathbf{v}_2, \ldots, \mathbf{v}_n$ span V, then $\mathbf{v}_1, \mathbf{v}_2, \ldots, \mathbf{v}_n, \mathbf{v}_{n+1}$ also span V. That is, the addition of one (or more) vectors to a spanning set yields another spanning set.

PROBLEMS 9.3

In Problems 1–22 determine whether the given set of vectors is linearly dependent or independent.

1. $\begin{pmatrix}1\\2\end{pmatrix}; \begin{pmatrix}-1\\-3\end{pmatrix}$

2. $\begin{pmatrix}2\\-1\\4\end{pmatrix}; \begin{pmatrix}4\\-2\\7\end{pmatrix}$

3. $\begin{pmatrix}2\\-1\\4\end{pmatrix}; \begin{pmatrix}4\\-2\\8\end{pmatrix}$

4. $\begin{pmatrix}-2\\3\end{pmatrix}; \begin{pmatrix}4\\7\end{pmatrix}$

5. $\begin{pmatrix}-3\\2\end{pmatrix}; \begin{pmatrix}1\\10\end{pmatrix}; \begin{pmatrix}4\\-5\end{pmatrix}$

6. $\begin{pmatrix}1\\0\\1\end{pmatrix}; \begin{pmatrix}0\\1\\1\end{pmatrix}; \begin{pmatrix}1\\1\\0\end{pmatrix}$

7. $\begin{pmatrix}1\\0\\0\end{pmatrix}; \begin{pmatrix}0\\1\\0\end{pmatrix}; \begin{pmatrix}0\\0\\1\end{pmatrix}$

8. $\begin{pmatrix}-3\\4\\2\end{pmatrix}; \begin{pmatrix}7\\-1\\3\end{pmatrix}; \begin{pmatrix}1\\2\\8\end{pmatrix}$

9. $\begin{pmatrix}-3\\4\\2\end{pmatrix}; \begin{pmatrix}7\\-1\\3\end{pmatrix}; \begin{pmatrix}1\\1\\8\end{pmatrix}$

10. $\begin{pmatrix}1\\-2\\1\\1\end{pmatrix}; \begin{pmatrix}3\\0\\2\\-2\end{pmatrix}; \begin{pmatrix}0\\4\\-1\\-1\end{pmatrix}; \begin{pmatrix}5\\0\\3\\-1\end{pmatrix}$

11. $\begin{pmatrix}1\\-2\\1\\1\end{pmatrix}; \begin{pmatrix}3\\0\\2\\-2\end{pmatrix}; \begin{pmatrix}0\\4\\-1\\1\end{pmatrix}; \begin{pmatrix}5\\0\\3\\-1\end{pmatrix}$

12. $\begin{pmatrix}1\\-1\\2\end{pmatrix}; \begin{pmatrix}4\\0\\0\end{pmatrix}; \begin{pmatrix}-2\\3\\5\end{pmatrix}; \begin{pmatrix}7\\1\\2\end{pmatrix}$

13. In P_2: $1 - x, x$

14. In P_2: $-x, x^2 - 2x, 3x + 5x^2$

15. In P_2: $1 - x, 1 + x, x^2$

16. In P_3: $x, x^2 - x, x^3 - x$

17. In P_3: $2x, x^3 - 3, 1 + x - 4x^3, x^3 + 18x - 9$

18. In M_{22}: $\begin{pmatrix} 2 & -1 \\ 4 & 0 \end{pmatrix}, \begin{pmatrix} 0 & -3 \\ 1 & 5 \end{pmatrix}, \begin{pmatrix} 4 & 1 \\ 7 & -5 \end{pmatrix}$

19. In M_{22}: $\begin{pmatrix} 1 & -1 \\ 0 & 6 \end{pmatrix}, \begin{pmatrix} -1 & 0 \\ 3 & 1 \end{pmatrix}, \begin{pmatrix} 1 & 1 \\ -1 & 2 \end{pmatrix}, \begin{pmatrix} 0 & 1 \\ 1 & 0 \end{pmatrix}$

20. In M_{22}: $\begin{pmatrix} -1 & 0 \\ 1 & 2 \end{pmatrix}, \begin{pmatrix} 2 & 3 \\ 7 & -4 \end{pmatrix}, \begin{pmatrix} 8 & -5 \\ 7 & 6 \end{pmatrix}, \begin{pmatrix} 4 & -1 \\ 2 & 3 \end{pmatrix}, \begin{pmatrix} 2 & 3 \\ -1 & 4 \end{pmatrix}$

***21.** In $C[0, 1]$: $\sin x, \cos x$

***22.** In $C[0, 1]$: $x, \sqrt{x}, \sqrt[3]{x}$

23. Determine a condition on the numbers a, b, c, and d such that the vectors $\begin{pmatrix} a \\ b \end{pmatrix}$ and $\begin{pmatrix} c \\ d \end{pmatrix}$ are linearly dependent.

24. Find a condition on the numbers a_{ij} such that the vectors $\begin{pmatrix} a_{11} \\ a_{12} \\ a_{13} \end{pmatrix}$, $\begin{pmatrix} a_{21} \\ a_{22} \\ a_{23} \end{pmatrix}$, and $\begin{pmatrix} a_{31} \\ a_{32} \\ a_{33} \end{pmatrix}$ are linearly dependent.

25. For what value(s) of α will the vectors $\begin{pmatrix} 1 \\ 2 \\ 3 \end{pmatrix}$, $\begin{pmatrix} 2 \\ -1 \\ 4 \end{pmatrix}$, $\begin{pmatrix} 3 \\ \alpha \\ 4 \end{pmatrix}$ be linearly dependent?

26. For what value(s) of α are the vectors $\begin{pmatrix} 2 \\ -3 \\ 1 \end{pmatrix}$, $\begin{pmatrix} -4 \\ 6 \\ -2 \end{pmatrix}$, $\begin{pmatrix} \alpha \\ 1 \\ 2 \end{pmatrix}$ linearly dependent? [*Hint:* Look carefully.]

27. Prove Theorem 2. [*Hint:* Write out the expression $c_1\mathbf{v}_1 + c_2\mathbf{v}_2 + \cdots + c_n\mathbf{v}_n = \mathbf{0}$ in terms of components and use Theorem 7.3.1.]

28. Prove that if the vectors $\mathbf{v}_1, \mathbf{v}_2, \ldots, \mathbf{v}_n$ are linearly dependent m-vectors and if $\mathbf{v}_{n+1}$ is any other m-vector, then the set $\mathbf{v}_1, \mathbf{v}_2, \ldots, \mathbf{v}_n, \mathbf{v}_{n+1}$ is linearly dependent.

29. Show that if $\mathbf{v}_1, \mathbf{v}_2, \ldots, \mathbf{v}_n$ $(n \geq 2)$ are linearly independent, then so too are $\mathbf{v}_1, \mathbf{v}_2, \ldots, \mathbf{v}_k$, where $k < n$.

30. Show that if the nonzero vectors $\mathbf{v}_1$ and $\mathbf{v}_2$ are orthogonal (that is, $\mathbf{v}_1 \cdot \mathbf{v}_2 = 0$) then the set $\{\mathbf{v}_1, \mathbf{v}_2\}$ is linearly independent.

***31.** Suppose that $\mathbf{v}_1$ is orthogonal to $\mathbf{v}_2$ and $\mathbf{v}_3$ and that $\mathbf{v}_2$ is orthogonal to $\mathbf{v}_3$. If $\mathbf{v}_1$, $\mathbf{v}_2$, and $\mathbf{v}_3$ are nonzero, show that the set $\{\mathbf{v}_1, \mathbf{v}_2, \mathbf{v}_3\}$ is linearly independent.

32. Let A be a square matrix whose columns are the vectors $\mathbf{v}_1, \mathbf{v}_2, \ldots, \mathbf{v}_n$. Show that $\mathbf{v}_1, \mathbf{v}_2, \ldots, \mathbf{v}_n$ are linearly independent if and only if the row echelon form of A does not contain a row of zeros.

In Problems 33–39 determine whether the given set of vectors spans the given vector space.

33. In $\mathbb{R}^3$: $(1, -1, 2), (1, 1, 2), (0, 0, 1)$

34. In $\mathbb{R}^3$: $(1, -1, 2), (-1, 1, 2), (0, 0, 1)$

35. In P_2: $1 - x, 3 - x^2$

36. In P_2: $1 - x, 3 - x^2, x$

37. In M_{22}: $\begin{pmatrix} 2 & 1 \\ 0 & 0 \end{pmatrix}, \begin{pmatrix} 0 & 0 \\ 2 & 1 \end{pmatrix}, \begin{pmatrix} 3 & -1 \\ 0 & 0 \end{pmatrix}, \begin{pmatrix} 0 & 0 \\ 3 & 1 \end{pmatrix}$

38. In M_{22}: $\begin{pmatrix} 1 & 0 \\ 1 & 0 \end{pmatrix}, \begin{pmatrix} 1 & 2 \\ 0 & 0 \end{pmatrix}, \begin{pmatrix} 4 & -1 \\ 3 & 0 \end{pmatrix}, \begin{pmatrix} -2 & 5 \\ 6 & 0 \end{pmatrix}$

39. In M_{23}: $\begin{pmatrix} 1 & 0 & 0 \\ 0 & 0 & 0 \end{pmatrix}, \begin{pmatrix} 0 & 1 & 0 \\ 0 & 0 & 0 \end{pmatrix}, \begin{pmatrix} 0 & 0 & 1 \\ 0 & 0 & 0 \end{pmatrix}, \begin{pmatrix} 0 & 0 & 0 \\ 1 & 0 & 0 \end{pmatrix}, \begin{pmatrix} 0 & 0 & 0 \\ 0 & 1 & 0 \end{pmatrix}, \begin{pmatrix} 0 & 0 & 0 \\ 0 & 0 & 1 \end{pmatrix}$

40. Show that any four polynomials in P_2 are linearly dependent.

41. Show that two polynomials cannot span P_2.

***42.** Show that any $n + 2$ polynomials in P_n are linearly dependent.

***43.** If $p_1, p_2, \ldots, p_m$ span P_n, show that $m \geq n + 1$.

44. Show that any seven matrices in M_{32} are linearly dependent.

***45.** Prove that any $mn + 1$ matrices in M_{mn} are linearly dependent.

46. Let S_1 and S_2 be two finite, linearly independent sets in a vector space V. Show that $S_1 \cap S_2$ is a linearly independent set.

47. Show that the infinite set $\{1, x, x^2, x^3, \ldots\}$ spans P, the vector space of polynomials.

48. Show that if $\mathbf{u}$ and $\mathbf{v}$ are in span $\{\mathbf{v}_1, \mathbf{v}_2, \ldots, \mathbf{v}_n\}$, then $\mathbf{u} + \mathbf{v}$ and $\alpha\mathbf{u}$ are in span $\{\mathbf{v}_1, \mathbf{v}_2, \ldots, \mathbf{v}_n\}$ [*Hint:* Using the definition of span, write $\mathbf{u} + \mathbf{v}$ and $\alpha\mathbf{u}$ as linear combinations of $\mathbf{v}_1, \mathbf{v}_2, \ldots, \mathbf{v}_n$.]

49. Let $\mathbf{v}_1 = (x_1, y_1, z_1)$ and $\mathbf{v}_2 = (x_2, y_2, z_2)$ be in $\mathbb{R}^3$. Show that if $\mathbf{v}_2 = c\mathbf{v}_1$, then span $\{\mathbf{v}_1, \mathbf{v}_2\}$ is a line passing through the origin.

****50.** In Problem 49 assume that $\mathbf{v}_1$ and $\mathbf{v}_2$ are independent. Show that $H = \text{span}\{\mathbf{v}_1, \mathbf{v}_2\}$ is a plane passing through the origin. What is the equation of that plane? [*Hint:* If $(x, y, z) \in H$, write $\mathbf{v} = a_1\mathbf{v}_1 + a_2\mathbf{v}_2$ and find a condition relating x, y, and z such that the resulting 3×2 system has a solution.]

51. Prove Theorem 7. [*Hint:* If $\mathbf{v} \in V$, write $\mathbf{v}$ as a linear combination of $\mathbf{v}_1, \mathbf{v}_2, \ldots, \mathbf{v}_n, \mathbf{v}_{n+1}$ with the coefficient of $\mathbf{v}_{n+1}$ equal to zero.]

52. Let H be a subspace of V containing $\mathbf{v}_1, \mathbf{v}_2, \ldots, \mathbf{v}_n$. Show that span $\{\mathbf{v}_1, \mathbf{v}_2, \ldots, \mathbf{v}_n\} \subseteq H$. That is, span $\{\mathbf{v}_1, \mathbf{v}_2, \ldots, \mathbf{v}_n\}$ is the *smallest* subspace of V containing $\mathbf{v}_1, \mathbf{v}_2, \ldots, \mathbf{v}_n$.

53. Show that any subset of a set of linearly independent vectors is linearly independent.

54. Let $\{\mathbf{v}_1, \mathbf{v}_2, \ldots, \mathbf{v}_n\}$ be a linearly independent set. Show that the vectors $\mathbf{v}_1, \mathbf{v}_1 + \mathbf{v}_2, \mathbf{v}_1 + \mathbf{v}_2 + \mathbf{v}_3, \ldots, \mathbf{v}_1 + \mathbf{v}_2 + \cdots + \mathbf{v}_n$ are linearly independent.

55. Show that M_{22} can be spanned by invertible matrices.

56. Let $\{\mathbf{v}_1, \mathbf{v}_2, \ldots, \mathbf{v}_n\}$ be a set of vectors having the property that the set $\{\mathbf{v}_i, \mathbf{v}_j\}$ is linearly dependent when $i \neq j$. Show that each vector in the set is a multiple of a single vector in the set.

57. Let $S = \{\mathbf{v}_1, \mathbf{v}_2, \ldots, \mathbf{v}_n\}$ be a linearly dependent set in a vector space V. Show that at least one of the vectors in S can be written as a linear combination of the vectors that precede it. That is, show that there is an integer $k \leq n$ and scalars, $a_1, a_2, \ldots, a_{k-1}$ such that $\mathbf{v}_k = a_1\mathbf{v}_1 + a_2\mathbf{v}_2 + \cdots + a_{k-1}\mathbf{v}_{k-1}$.

58. Let f and g be in $C'[0, 1]$. Then the **Wronskian** of f and g is defined by

$$W(f, g)(x) = \begin{vmatrix} f(x) & g(x) \\ f'(x) & g'(x) \end{vmatrix}.$$

Show that f and g are linearly independent if $W(f, g)(x) \neq 0$ for every $x \in [0, 1]$.

59. Let $\{\mathbf{v}_1, \mathbf{v}_2, \ldots, \mathbf{v}_n\}$ be a linearly independent set and suppose that $\mathbf{v} \notin \text{span}\{\mathbf{v}_1, \mathbf{v}_2, \ldots, \mathbf{v}_n\}$. Show that $\{\mathbf{v}_1, \mathbf{v}_2, \ldots, \mathbf{v}_n, \mathbf{v}\}$ is a linearly independent set.

60. Find a set of three linearly independent vectors in $\mathbb{R}^3$ that contains the vectors $\begin{pmatrix} 2 \\ 1 \\ 2 \end{pmatrix}$ and $\begin{pmatrix} -1 \\ 3 \\ 4 \end{pmatrix}$. [*Hint:* Find a vector $\mathbf{v} \notin \text{span}\left\{\begin{pmatrix} 2 \\ 1 \\ 2 \end{pmatrix}, \begin{pmatrix} -1 \\ 3 \\ 4 \end{pmatrix}\right\}$.]

61. Find a set of three linearly independent vectors in P_2 that contains the polynomials $1 - x^2$ and $1 + x^2$.

***62.** Show that in P_n the polynomials $1, x, x^2, \ldots, x^n$ are linearly independent.

***63.** Prove the **Cauchy-Schwartz inequality** in $\mathbb{R}^n$: $|\mathbf{u} \cdot \mathbf{v}| \leq |\mathbf{u}|\,|\mathbf{v}|$.

64. Show that, in Problem 63, $|\mathbf{u} \cdot \mathbf{v}| = |\mathbf{u}|\,|\mathbf{v}|$ if and only if $\mathbf{u} = \lambda\mathbf{v}$ for some real number λ.

65. Using the result of Problem 64, prove that if $|\mathbf{u} + \mathbf{v}| = |\mathbf{u}| + |\mathbf{v}|$, then $\mathbf{u}$ and $\mathbf{v}$ are linearly dependent.

66. Using the result of Problem 63, prove the **triangle inequality:** $|\mathbf{u} + \mathbf{v}| \leq |\mathbf{u}| + |\mathbf{v}|$. [*Hint:* Expand $|\mathbf{u} + \mathbf{v}|^2$.]

9.4 BASIS AND DIMENSION

We have seen that in $\mathbb{R}^2$ it is convenient to write vectors in terms of the vectors $\mathbf{i} = \begin{pmatrix} 1 \\ 0 \end{pmatrix}$ and $\mathbf{j} = \begin{pmatrix} 0 \\ 1 \end{pmatrix}$. In $\mathbb{R}^3$ we wrote vectors in terms of $\begin{pmatrix} 1 \\ 0 \\ 0 \end{pmatrix}$, $\begin{pmatrix} 0 \\ 1 \\ 0 \end{pmatrix}$, and $\begin{pmatrix} 0 \\ 0 \\ 1 \end{pmatrix}$. We now generalize this idea.

Definition 1 BASIS A set of vectors $\{\mathbf{v}_1, \mathbf{v}_2, \ldots, \mathbf{v}_n\}$ forms a **basis** for V if

(i) $\{\mathbf{v}_1, \mathbf{v}_2, \ldots, \mathbf{v}_n\}$ is linearly independent, and
(ii) $\{\mathbf{v}_1, \mathbf{v}_2, \ldots, \mathbf{v}_n\}$ spans V.

We have already seen quite a few examples of bases. In Theorem 9.3.5, for instance, we saw that any set of n linearly independent vectors in $\mathbb{R}^n$ spans $\mathbb{R}^n$. Thus,

Every set of n linearly independent vectors in $\mathbb{R}^n$ is a basis in $\mathbb{R}^n$.

In $\mathbb{R}^n$ we define

$$\mathbf{e}_1 = \begin{pmatrix} 1 \\ 0 \\ 0 \\ \vdots \\ 0 \end{pmatrix}, \quad \mathbf{e}_2 = \begin{pmatrix} 0 \\ 1 \\ 0 \\ \vdots \\ 0 \end{pmatrix}, \quad \mathbf{e}_3 = \begin{pmatrix} 0 \\ 0 \\ 1 \\ \vdots \\ 0 \end{pmatrix}, \ldots, \mathbf{e}_n = \begin{pmatrix} 0 \\ 0 \\ 0 \\ \vdots \\ 1 \end{pmatrix}.$$

Then since the $\mathbf{e}_i$'s are the columns of the identity matrix (which has determinant 1), $\{\mathbf{e}_1, \mathbf{e}_2, \ldots, \mathbf{e}_n\}$ is linearly independent and therefore constitutes a basis in $\mathbb{R}^n$. This special basis is called the **standard basis** in $\mathbb{R}^n$. We shall now find bases for some other spaces.

EXAMPLE 1 By Example 9.3.7, the polynomials 1, x, x^2, x^3 are linearly independent in P_3. By Example 9.3.11 these polynomials span P_3. Thus $\{1, x, x^2, x^3\}$ is a basis for P_3. In general, the monomials $\{1, x, x^2, x^3, \ldots, x^n\}$ constitute a basis for P_n. This is called the **standard basis** for P_n. ■

EXAMPLE 2 We saw in Example 9.3.15 that $\begin{pmatrix} 1 & 0 \\ 0 & 0 \end{pmatrix}$, $\begin{pmatrix} 0 & 1 \\ 0 & 0 \end{pmatrix}$, $\begin{pmatrix} 0 & 0 \\ 1 & 0 \end{pmatrix}$, and $\begin{pmatrix} 0 & 0 \\ 0 & 1 \end{pmatrix}$ span M_{22}. If $c_1\begin{pmatrix} 1 & 0 \\ 0 & 0 \end{pmatrix} + c_2\begin{pmatrix} 0 & 1 \\ 0 & 0 \end{pmatrix} + c_3\begin{pmatrix} 0 & 0 \\ 1 & 0 \end{pmatrix} + c_4\begin{pmatrix} 0 & 0 \\ 0 & 1 \end{pmatrix} = \begin{pmatrix} 0 & 0 \\ 0 & 0 \end{pmatrix}$, then, obviously, $c_1 = c_2 = c_3 = c_4 = 0$. Thus these four matrices are linearly independent and form a basis for M_{22}. This set of matrices is called the **standard basis** for M_{22}. ■

EXAMPLE 3 Find a basis for the set of vectors lying on the plane

$$\pi = \left\{ \begin{pmatrix} x \\ y \\ z \end{pmatrix} : 2x - y + 3z = 0 \right\}$$

Solution. We saw in Example 9.1.6 that π is a vector space. If $(x, y, z) \in \pi$, then $y = 2x + 3z$ and each vector in π can be written

$$\begin{pmatrix} x \\ y \\ z \end{pmatrix} = \begin{pmatrix} x \\ 2x + 3z \\ 3z \end{pmatrix} = \begin{pmatrix} x \\ 2x \\ 0 \end{pmatrix} + \begin{pmatrix} 0 \\ 3z \\ z \end{pmatrix} = x\begin{pmatrix} 1 \\ 2 \\ 0 \end{pmatrix} + z\begin{pmatrix} 0 \\ 3 \\ 1 \end{pmatrix}.$$

This shows that $\begin{pmatrix} 1 \\ 2 \\ 0 \end{pmatrix}$ and $\begin{pmatrix} 0 \\ 3 \\ 1 \end{pmatrix}$ span π. Since neither of these vectors is a multiple of the other, they are independent, and, therefore, they form a basis for π. ■

If $\mathbf{v}_1, \mathbf{v}_2, \ldots, \mathbf{v}_n$ is a basis for V, then any vector $\mathbf{v} \in V$ can be written $\mathbf{v} = c_1\mathbf{v}_1 + c_2\mathbf{v}_2 + \cdots + c_n\mathbf{v}_n$. Can it be written in another way as a linear combination of the $\mathbf{v}_i$'s? The answer is *no*.

Theorem 1 If $\{\mathbf{v}_1, \mathbf{v}_2, \ldots, \mathbf{v}_n\}$ is a basis for V and if $\mathbf{v} \in V$, then there exists a *unique* set of scalars $c_1, c_2, \ldots, c_n$ such that $\mathbf{v} = c_1\mathbf{v}_1 + c_2\mathbf{v}_2 + \cdots + c_n\mathbf{v}_n$.

Proof. At least one such set of scalars exists because $\{\mathbf{v}_1, \mathbf{v}_2, \ldots, \mathbf{v}_n\}$ spans V. Suppose then that $\mathbf{v}$ can be written in two ways as a linear combination of the basis vectors. That is, suppose that

$$\mathbf{v} = c_1\mathbf{v}_1 + c_2\mathbf{v}_2 + \cdots + c_n\mathbf{v}_n = d_1\mathbf{v}_1 + d_2\mathbf{v}_2 + \cdots + d_n\mathbf{v}_n.$$

Then, subtracting, we obtain the equation

$$(c_1 - d_1)\mathbf{v}_1 + (c_2 - d_2)\mathbf{v}_2 + \cdots + (c_n - d_n)\mathbf{v}_n = \mathbf{0}.$$

But, since the $\mathbf{v}_i$'s are linearly independent, this equation can hold only if $c_1 - d_1 = c_2 - d_2 = \cdots = c_n - d_n = 0$. Thus $c_1 = d_1, c_2 = d_2, \ldots, c_n = d_n$, and the theorem is proved. ■

We have seen that a vector space may have many bases. A question naturally arises: Do all bases contain the same number of vectors? In $\mathbb{R}^3$, the answer is certainly yes. To see this we note that any three linearly independent vectors in $\mathbb{R}^3$ form a basis. But fewer than three vectors cannot form a basis since, as we saw in Section 9.3, the span of two linearly independent vectors in $\mathbb{R}^3$ is a plane in $\mathbb{R}^3$—and a plane is not all of $\mathbb{R}^3$. Similarly a set of four or more vectors in $\mathbb{R}^3$ cannot be linearly independent; for if the first three vectors in the set are linearly independent, then they form a basis and, therefore, all other vectors in the set can be written as a linear combination of the first three. Thus all bases in $\mathbb{R}^3$ contain three vectors. The next theorem tells us that the answer to the question posed above is *yes* for any vector space that has a finite basis.

Theorem 2 If $\{\mathbf{u}_1, \mathbf{u}_2, \ldots, \mathbf{u}_m\}$ and $\{\mathbf{v}_1, \mathbf{v}_2, \ldots, \mathbf{v}_n\}$ are bases for the vector space V, then $m = n$; that is, **any two bases in a vector space V have the same number of vectors.**

Proof†. Let $S_1 = \{\mathbf{u}_1, \ldots, \mathbf{u}_m\}$ and $S_2 = \{\mathbf{v}_1, \ldots, \mathbf{v}_n\}$ be two bases for V. We must show that $m = n$. We prove this by showing that if $m > n$, then S_1 is a linearly dependent set, which contradicts the hypothesis that S_1 is a basis. This will show that $m \leq n$. The same proof will then show that $n \leq m$, and this will prove the theorem. Hence all we must show is that if $m > n$, then S_1 is dependent. Since S_2 constitutes a basis, we can write each $\mathbf{u}_i$ as a linear combination of the $\mathbf{v}_i$'s. We have

$$\begin{aligned}
\mathbf{u}_1 &= a_{11}\mathbf{v}_1 + a_{12}\mathbf{v}_2 + \cdots + a_{1n}\mathbf{v}_n \\
\mathbf{u}_2 &= a_{21}\mathbf{v}_1 + a_{22}\mathbf{v}_2 + \cdots + a_{2n}\mathbf{v}_n \\
&\ \ \vdots \\
\mathbf{u}_m &= a_{m1}\mathbf{v}_1 + a_{m2}\mathbf{v}_2 + \cdots + a_{mn}\mathbf{v}_n
\end{aligned} \tag{1}$$

†This proof is given for vector spaces with bases containing a finite number of vectors.

To show that S_1 is dependent, we must find scalars $c_1, c_2, \ldots, c_m$, not all zero, such that

$$c_1\mathbf{u}_1 + c_2\mathbf{u}_2 + \cdots + c_m\mathbf{u}_m = \mathbf{0}. \tag{2}$$

Inserting (1) into (2), we obtain

$$c_1(a_{11}\mathbf{v}_1 + a_{12}\mathbf{v}_2 + \cdots + a_{1n}\mathbf{v}_n) + c_2(a_{21}\mathbf{v}_1 + a_{22}\mathbf{v}_2 + \cdots + a_{2n}\mathbf{v}_n) + \cdots + c_m(a_{m1}\mathbf{v}_1 + a_{m2}\mathbf{v}_2 + \cdots + a_{mn}\mathbf{v}_n) = \mathbf{0}. \tag{3}$$

Equation (3) can be rewritten as

$$(a_{11}c_1 + a_{21}c_2 + \cdots + a_{m1}c_m)\mathbf{v}_1 + (a_{12}c_1 + a_{22}c_2 + \cdots + a_{m2}c_m)\mathbf{v}_2 + \cdots + (a_{1n}c_1 + a_{2n}c_2 + \cdots + a_{mn}c_m)\mathbf{v}_n = \mathbf{0}. \tag{4}$$

But, since $\mathbf{v}_1, \mathbf{v}_2, \ldots, \mathbf{v}_n$ are linearly independent, we must have

$$\begin{aligned} a_{11}c_1 + a_{21}c_2 + \cdots + a_{m1}c_m &= 0 \\ a_{12}c_1 + a_{22}c_2 + \cdots + a_{m2}c_m &= 0 \\ \vdots \qquad \vdots \qquad\quad \vdots \quad &\ \ \vdots \\ a_{1n}c_1 + a_{2n}c_2 + \cdots + a_{mn}c_m &= 0. \end{aligned} \tag{5}$$

System (5) is a homogeneous system of n equations in the m unknowns $c_1, c_2, \ldots, c_m$ and, since $m > n$, Theorem 7.3.1 tells us that the system has an infinite number of solutions. Thus there are scalars $c_1, c_2, \ldots, c_m$, not all zero, such that (2) is satisfied and therefore S_1 is a linearly dependent set. This contradiction proves that $m \leq n$ and, by exchanging the roles of S_1 and S_2, we can show that $n \leq m$, and the proof is complete. ■

With this theorem we can define one of the central concepts in linear algebra.

Definition 2 DIMENSION The **dimension** of a vector space V is the number of vectors in a basis of V. If this number is finite, then V is called a **finite dimensional vector space**. Otherwise V is called an **infinite dimensional vector space**. If $V = \{\mathbf{0}\}$, then V is said to be **zero dimensional**.

NOTATION: We write the dimension of V as dim V.

REMARK. We have not proved that every nontrivial vector space has a basis. A proof of this fact is beyond the scope of this text. But we do not need this fact in order for Definition 2 to make sense; for *if* V has a finite basis, then V is finite dimensional. Otherwise V is infinite dimensional. Thus in order to show that V is infinite dimensional, it is only necessary to show that V does not have a finite basis. We can do this by showing that V contains an infinite number of linearly independent vectors (see Example 7 below). It is not necessary to construct an infinite basis for V.

EXAMPLE 4 Since n linearly independent vectors in $\mathbb{R}^n$ comprise a basis, we see that

$$\dim \mathbb{R}^n = n. \quad ■$$

EXAMPLE 5 By Example 1 and Problem 9.3.62, the polynomials $\{1, x, x^2, \ldots, x^n\}$ constitute a basis in P_n. Thus $\dim P_n = n + 1$. ■

EXAMPLE 6 In M_{mn} let A_{ij} be the $m \times n$ matrix with a 1 in the ijth position and a zero everywhere else. It is easy to show that the A_{ij} for $i = 1, 2, \ldots, m$ and $j = 1, 2, \ldots, n$ form a basis for M_{mn}. Thus $\dim M_{mn} = mn$. ■

EXAMPLE 7 In Example 9.3.16, we saw that no finite set of polynomials spans P. Thus P has no finite basis and is, therefore, an infinite dimensional vector space. ■

There are a number of theorems that tell us something about the dimension of a vector space.

Theorem 3 Suppose that $\dim V = n$. If $\mathbf{u}_1, \mathbf{u}_2, \ldots, \mathbf{u}_m$ is a set of m linearly independent vectors in V, then $m \le n$.

Proof. Let $\mathbf{v}_1, \mathbf{v}_2, \ldots, \mathbf{v}_n$ be a basis for V. If $m > n$, then, as in proof of Theorem 2, we can find constants $c_1, c_2, \ldots, c_m$ not all zero such that equation (2) is satisfied. This would contradict the linear independence of the $\mathbf{u}_i$'s. Thus $m \le n$. ■

Theorem 4 Let H be a subspace of the finite dimensional vector space V. Then H is finite dimensional and

$$\dim H \le \dim V. \tag{6}$$

Proof. Let $\dim V = n$. Any set of linearly independent vectors in H is also a linearly independent set in V. By Theorem 3, any linearly independent set in H can contain at most n vectors. Hence H is finite dimensional. Moreover, since any basis in H is a linearly independent set, we see that $\dim H \le n$. ■

Theorem 4 has some interesting consequences. We give three of them here.

EXAMPLE 8 Let $P[0, 1]$ denote the set of polynomials defined on the interval $[0, 1]$. Then $P[0, 1] \subset C[0, 1]$. If $C[0, 1]$ were finite dimensional, then $P[0, 1]$ would be finite dimensional also. But, by Example 7, this is not the case. Hence $C[0, 1]$ is infinite dimensional. Similarly, since $P[0, 1] \subset C'[0, 1]$ (since every polynomial is differentiable), we also see that $C'[0, 1]$ is infinite dimensional. In general,

Any vector space containing an infinite dimensional subspace is infinite dimensional.

■

EXAMPLE 9 We show that the only proper subspaces of $\mathbb{R}^2$ are straight lines passing through the origin. First, we note that any *proper* subspace of $\mathbb{R}^2$ must have dimension one (explain why). Let $\mathbf{v} = (a, b)$ be a basis for this subspace, which we denote by H. If $\mathbf{u} = (x, y)$ is in H, then $\mathbf{u}$ can be written as a linear combination of $\mathbf{v}$. That is, there is a number t such that $\mathbf{u} = t\mathbf{v}$, or

$$(x, y) = t(a, b) = (ta, tb).$$

But, then,

$$x = ta \qquad \text{and} \qquad y = tb.$$

So

$$t = \frac{x}{a} \qquad \text{and} \qquad y = \frac{x}{a}\, b = \frac{b}{a}\, x.$$

Therefore, we have shown that the subspace H consists of all points $(x, y) \in \mathbb{R}^2$ whose coordinates satisfy the equation $y = \left(\frac{b}{a}\right)x$. This is the equation of a straight line through the origin. ■

EXAMPLE 10 We can use Theorem 4 to find *all* subspaces of $\mathbb{R}^3$. Let H be a subspace of $\mathbb{R}^3$. Then there are four possibilities: $H = \{\mathbf{0}\}$; $\dim H = 1$, $\dim H = 2$, and $\dim H = 3$. If $\dim H = 3$, then H contains a basis of three linearly independent vectors $\mathbf{v}_1, \mathbf{v}_2, \mathbf{v}_3$ in $\mathbb{R}^3$. But then $\mathbf{v}_1, \mathbf{v}_2, \mathbf{v}_3$ also form a basis for $\mathbb{R}^3$. Thus $H = \text{span}\,\{\mathbf{v}_1, \mathbf{v}_2, \mathbf{v}_3\} = \mathbb{R}^3$. Hence the only way to get a *proper* subspace of $\mathbb{R}^3$ is to have $\dim H = 1$ or $\dim H = 2$. If $\dim H = 1$, then H has a basis consisting of one vector $\mathbf{v} = (a, b, c)$. Let $\mathbf{x}$ be in H. Then $\mathbf{x} = t(a, b, c)$ for some real number t (since (a, b, c) spans H). If $\mathbf{x} = (x, y, z)$, this means that $x = at$, $y = bt$, $z = ct$. But these are the equations of a line in $\mathbb{R}^3$ passing through the origin with direction vector (a, b, c).

Now suppose $\dim H = 2$ and let $\mathbf{v}_1 = (a_1, b_1, c_1)$ and $\mathbf{v}_2 = (a_2, b_2, c_2)$ be a basis for H. If $(x, y, z) \in H$, then there exist real numbers s and t such that $(x, y, z) = s(a_1, b_1, c_1) + t(a_2, b_2, c_2)$. Then

$$\begin{aligned} x &= sa_1 + ta_2 \\ y &= sb_1 + tb_2 \\ z &= sc_1 + tb_2. \end{aligned} \tag{7}$$

Let $\mathbf{v}_3 = \mathbf{v}_1 \times \mathbf{v}_2 = (\alpha, \beta, \gamma)$. Then, from Theorem 3.4.2 part (vi), we have $\mathbf{v}_3 \cdot \mathbf{v}_1 = 0$ and $\mathbf{v}_3 \cdot \mathbf{v}_2 = 0$. Now, we calculate

$$\begin{aligned} \alpha x + \beta y + \gamma z &= \alpha(sa_1 + ta_2) + \beta(sb_1 + tb_2) + \gamma(sc_1 + tc_2) \\ &= (\alpha a_1 + \beta b_1 + \gamma c_1)s + (\alpha a_2 + \beta b_2 + \gamma c_2)t \\ &= (\mathbf{v}_3 \cdot \mathbf{v}_1)s + (\mathbf{v}_3 \cdot \mathbf{v}_2)t = 0 \end{aligned}$$

Thus if $(x, y, z) \in H$, then $\alpha x + \beta y + \gamma z = 0$, which shows that H is a plane passing through the origin with normal vector $\mathbf{v}_3 = \mathbf{v}_1 \times \mathbf{v}_2$. Therefore we have proved that

The only proper subspaces of $\mathbb{R}^3$ are sets of vectors lying on lines and planes passing through the origin.

■

EXAMPLE 11 Let A be an $m \times n$ matrix and let $S = \{\mathbf{x} \in \mathbb{R}^n; A\mathbf{x} = \mathbf{0}\}$. Let $\mathbf{x}_1 \in S$ and $\mathbf{x}_2 \in S$; then $A(\mathbf{x}_1 + \mathbf{x}_2) = A\mathbf{x}_1 + A\mathbf{x}_2 = \mathbf{0} + \mathbf{0} = \mathbf{0}$ and $A(\alpha\mathbf{x}_1) = \alpha(A\mathbf{x}_1) = \alpha\mathbf{0} = \mathbf{0}$, so that S is a subspace of $\mathbb{R}^n$ and $\dim S \le n$. S is called the **solution space** of the homogeneous system $A\mathbf{x} = \mathbf{0}$. It is also called the **kernel** of the matrix A. ■

EXAMPLE 12 Find a basis for (and the dimension of) the solution space S of the following homogeneous system:

$$\begin{aligned} x + 2y - z &= 0 \\ 2x - y + 3z &= 0 \end{aligned}$$

Solution. Here $A = \begin{pmatrix} 1 & 2 & -1 \\ 2 & -1 & 3 \end{pmatrix}$. Since A is a 2×3 matrix, S is a subspace of $\mathbb{R}^3$. Row-reducing, we find, successively,

$$\left(\begin{array}{ccc|c} 1 & 2 & -1 & 0 \\ 2 & -1 & 3 & 0 \end{array}\right) \xrightarrow{A_{1,2}(-2)} \left(\begin{array}{ccc|c} 1 & 2 & -1 & 0 \\ 0 & -5 & 5 & 0 \end{array}\right)$$

$$\xrightarrow{M_2(-\frac{1}{5})} \left(\begin{array}{ccc|c} 1 & 2 & -1 & 0 \\ 0 & 1 & -1 & 0 \end{array}\right) \xrightarrow{A_{2,1}(-2)} \left(\begin{array}{ccc|c} 1 & 0 & 1 & 0 \\ 0 & 1 & -1 & 0 \end{array}\right).$$

Then $y = z$ and $x = -z$, so that all solutions are of the form $\begin{pmatrix} -z \\ z \\ z \end{pmatrix}$. Thus $\begin{pmatrix} -1 \\ 1 \\ 1 \end{pmatrix}$ is a basis for S and $\dim S = 1$. Note that S is the set of vectors lying on the straight line $x = -t$, $y = t$, $z = t$. ■

EXAMPLE 13 Find a basis for the solution space S of the system

$$\begin{aligned} 2x - y + 3z &= 0 \\ 4x - 2y + 6z &= 0 \\ -6x + 3y - 9z &= 0. \end{aligned}$$

Solution. Row-reducing as above, we obtain

$$\left(\begin{array}{ccc|c} 2 & -1 & 3 & 0 \\ 4 & -2 & 6 & 0 \\ -6 & 3 & -9 & 0 \end{array}\right) \xrightarrow[A_{1,3}(3)]{A_{1,2}(-2)} \left(\begin{array}{ccc|c} 2 & -1 & 3 & 0 \\ 0 & 0 & 0 & 0 \\ 0 & 0 & 0 & 0 \end{array}\right).$$

giving the single equation $2x - y + 3z = 0$. S is a plane and, by Example 3, a basis is given by $\begin{pmatrix} 1 \\ 2 \\ 0 \end{pmatrix}$ and $\begin{pmatrix} 0 \\ 3 \\ 1 \end{pmatrix}$ and $\dim S = 2$. Note that we have shown that any solution to the homogeneous equation can be written as

$$c_1 \begin{pmatrix} 1 \\ 2 \\ 0 \end{pmatrix} + c_2 \begin{pmatrix} 0 \\ 3 \\ 1 \end{pmatrix}.$$

For example, if $c_1 = 2$ and $c_2 = -3$, we obtain the solution

$$\mathbf{x} = 2\begin{pmatrix}1\\2\\0\end{pmatrix} - 3\begin{pmatrix}0\\3\\1\end{pmatrix} = \begin{pmatrix}2\\4\\0\end{pmatrix} + \begin{pmatrix}0\\-9\\-3\end{pmatrix} = \begin{pmatrix}2\\-5\\-3\end{pmatrix}. \blacksquare$$

Before leaving this section we prove a result that is very useful in finding bases in a vector space. We have seen that n linearly independent vectors in $\mathbb{R}^n$ comprise a basis for $\mathbb{R}^n$. This fact holds in *any* finite dimensional vector space.

Theorem 5 Any n linearly independent vectors in a vector space V of dimension n constitute a basis for V.

Proof. Let $\mathbf{v}_1, \mathbf{v}_2, \ldots, \mathbf{v}_n$ be the n vectors. If they span V, then they constitute a basis. If they do not, then there is a vector $\mathbf{u} \in V$ such that $\mathbf{u} \notin$ span $\{\mathbf{v}_1, \mathbf{v}_2, \ldots, \mathbf{v}_n\}$. This means that the $n + 1$ vectors $\mathbf{v}_1, \mathbf{v}_2, \ldots, \mathbf{v}_n, \mathbf{u}$ are linearly independent. To see this note that if

$$c_1\mathbf{v}_1 + c_2\mathbf{v}_2 + \cdots + c_n\mathbf{v}_n + c_{n+1}\mathbf{u} = 0, \tag{8}$$

then $c_{n+1} = 0$ for, if not, we could write $\mathbf{u}$ as a linear combination of $\mathbf{v}_1, \mathbf{v}_2, \ldots, \mathbf{v}_n$ by dividing equation (8) by c_{n+1} and putting all terms except $\mathbf{u}$ on the right-hand side. But if $c_{n+1} = 0$, then (8) reads

$$c_1\mathbf{v}_1 + c_2\mathbf{v}_2 + \cdots + c_n\mathbf{v}_n = 0,$$

which means that $c_1 = c_2 = \cdots = c_n = 0$ since the $\mathbf{v}_i$'s are linearly independent. Now let $W =$ span $\{\mathbf{v}_1, \mathbf{v}_2, \ldots, \mathbf{v}_n, \mathbf{u}\}$. Then as all the vectors in braces are in V, W is a subspace of V. Since $\mathbf{v}_1, \mathbf{v}_2, \ldots, \mathbf{v}_n, \mathbf{u}$ are linearly independent, they form a basis for W. Thus dim $W = n + 1$. But from Theorem 4, dim $W \leq n$. This contradiction shows that there is *no* vector $\mathbf{u} \in V$ such that $\mathbf{u} \notin$ span $\{\mathbf{v}_1, \mathbf{v}_2, \ldots, \mathbf{v}_n\}$. Thus $\mathbf{v}_1, \mathbf{v}_2, \ldots, \mathbf{v}_n$ span V and therefore constitute a basis for V. ■

PROBLEMS 9.4

In Problems 1–10 determine whether the given set of vectors is a basis for the given vector space.

1. In P_2: $1 - x^2, x$
2. In P_2: $-3x, 1 + x^2, x^2 - 5$
3. In P_2: $x^2 - 1, x^2 - 2, x^2 - 3$
4. In P_3: $1, 1 + x, 1 + x^2, 1 + x^3$
5. In P_3: $3, x^3 - 4x + 6, x^2$
6. In M_{22}: $\begin{pmatrix}3 & 1\\0 & 0\end{pmatrix}, \begin{pmatrix}3 & 2\\0 & 0\end{pmatrix}, \begin{pmatrix}-5 & 1\\0 & 6\end{pmatrix}, \begin{pmatrix}0 & 1\\0 & -7\end{pmatrix}$
7. In M_{22}: $\begin{pmatrix}a & 0\\0 & 0\end{pmatrix}, \begin{pmatrix}0 & b\\0 & 0\end{pmatrix}, \begin{pmatrix}0 & 0\\c & 0\end{pmatrix}, \begin{pmatrix}0 & 0\\0 & d\end{pmatrix}$, where $abcd \neq 0$
8. In M_{22}: $\begin{pmatrix}-1 & 0\\3 & 1\end{pmatrix}, \begin{pmatrix}2 & 1\\1 & 4\end{pmatrix}, \begin{pmatrix}-6 & 1\\5 & 8\end{pmatrix}, \begin{pmatrix}7 & -2\\1 & 0\end{pmatrix}, \begin{pmatrix}0 & 1\\0 & 0\end{pmatrix}$
9. $H = \{(x, y) \in \mathbb{R}^2: x + y = 0\}$; $(1, -1)$
10. $H = \{(x, y) \in \mathbb{R}^2: x + y = 0\}$; $(1, -1), (-3, 3)$

11. Find a basis in $\mathbb{R}^3$ for the set of vectors in the plane $2x - y - z = 0$.

12. Find a basis in $\mathbb{R}^3$ for the set of vectors in the plane $3x - 2y + 6z = 0$.

13. Find a basis in $\mathbb{R}^3$ for the set of vectors on the line $x/2 = y/3 = z/4$.

14. Find a basis in $\mathbb{R}^3$ for the set of vectors on the line $x = 3t$, $y = -2t$, $z = t$.

15. Show that one-dimensional subspaces of $\mathbb{R}^4$ have the form $x_1 = at$, $x_2 = bt$, $x_3 = ct$, $x_4 = dt$ where a, b, c and d are real numbers, not all 0.

16. In $\mathbb{R}^4$ let $H = \{(x, y, z, w): ax + by + cz + dw = 0\}$, where $abcd \neq 0$.

(a) Show that H is a subspace of $\mathbb{R}^4$.

(b) Find a basis for H.

(c) What is dim H?

***17.** In $\mathbb{R}^n$ a **hyperplane** is a subspace of dimension $n - 1$. If H is a hyperplane in $\mathbb{R}^n$ show that

$$H = \{(x_1, x_2, \ldots, x_n): a_1x_1 + a_2x_2 + \cdots + a_nx_n = 0\}$$

where $a_1, a_2, \ldots, a_n$ are fixed real numbers, not all of which are zero.

18. In $\mathbb{R}^5$ find a basis for the hyperplane

$$H = \{(x_1, x_2, x_3, x_4, x_5): 2x_1 - 3x_2 + x_3 + 4x_4 - x_5 = 0\}$$

In Problems 19–23 find a basis for the solution space of the given homogeneous system.

19. $\begin{aligned} x - y &= 0 \\ -2x + 2y &= 0 \end{aligned}$

20. $\begin{aligned} x - 2y &= 0 \\ 3x + y &= 0 \end{aligned}$

21. $\begin{aligned} x - y - z &= 0 \\ 2x - y + z &= 0 \end{aligned}$

22. $\begin{aligned} x - 3y + z &= 0 \\ -2x + 2y - 3z &= 0 \\ 4x - 8y + 5z &= 0 \end{aligned}$

23. $\begin{aligned} 2x - 6y + 4z &= 0 \\ -x + 3y - 2z &= 0 \\ -3x + 9y - 6z &= 0 \end{aligned}$

24. Find a basis for D_3, the vector space of diagonal 3×3 matrices. What is the dimension of D_3?

25. What is the dimension of D_n, the space of diagonal $n \times n$ matrices?

26. Let S_{nn} denote the vector space of symmetric $n \times n$ matrices. Show that S_{nn} is a subspace of M_{nn} and that dim $S_{nn} = [n(n + 1)]/2$.

27. Suppose that $\mathbf{v}_1, \mathbf{v}_2, \ldots, \mathbf{v}_m$ are linearly independent vectors in a vector space V of dimension n and $m < n$. Show that $\{\mathbf{v}_1, \mathbf{v}_2, \ldots, \mathbf{v}_m\}$ can be enlarged to a basis for V. That is, there exist vectors $\mathbf{v}_{m+1}, \mathbf{v}_{m+2}, \ldots, \mathbf{v}_n$ such that $\{\mathbf{v}_1, \mathbf{v}_2, \ldots, \mathbf{v}_n\}$ is a basis. [*Hint:* Look at the proof of Theorem 5.]

28. Let $\{\mathbf{v}_1, \mathbf{v}_2, \ldots, \mathbf{v}_n\}$ be a basis for V. Let $\mathbf{u}_1 = \mathbf{v}_1$, $\mathbf{u}_2 = \mathbf{v}_1 + \mathbf{v}_2$, $\mathbf{u}_3 = \mathbf{v}_1 + \mathbf{v}_2 + \mathbf{v}_3, \ldots, \mathbf{u}_n = \mathbf{v}_1 + \mathbf{v}_2 + \cdots + \mathbf{v}_n$. Show that $\{\mathbf{u}_1, \mathbf{u}_2, \ldots, \mathbf{u}_n\}$ is also a basis for V.

29. Show that if $\{\mathbf{v}_1, \mathbf{v}_2, \ldots, \mathbf{v}_n\}$ spans V, then dim $V \leq n$. [*Hint:* Use the result of Problem 9.3.57.]

30. Let H and K be subspaces of V such that $H \subseteq K$ and dim H = dim $K < \infty$. Show that $H = K$.

31. Let H and K be subspaces of V and define $H + K = \{\mathbf{h} + \mathbf{k}: \mathbf{h} \in H \text{ and } \mathbf{k} \in K\}$.

(a) Show that $H + K$ is a subspace of V.

(b) If $H \cap K = \{\mathbf{0}\}$, show that dim $(H + K)$ = dim H + dim K.

***32.** If H is a subspace of the finite dimensional vector space V, show that there exists a unique subspace K of V such that **(a)** $H \cap K = \{\mathbf{0}\}$ and **(b)** $H + K = V$.

33. Show that two vectors $\mathbf{v}_1$ and $\mathbf{v}_2$ in $\mathbb{R}^2$ with endpoints at the origin are collinear if and only if dim span $\{\mathbf{v}_1, \mathbf{v}_2\} = 1$.

34. Show that three vectors $\mathbf{v}_1$, $\mathbf{v}_2$, and $\mathbf{v}_3$ in $\mathbb{R}^3$ with endpoints at the origin are coplanar if and only if dim span $\{\mathbf{v}_1, \mathbf{v}_2, \mathbf{v}_3\} \leq 2$.

35. Show that any n vectors which span an n-dimensional space V form a basis for V. [*Hint:* Show that if the n vectors are not linearly independent, then dim $V < n$.]

***36.** Show that every subspace of a finite dimensional vector space has a basis.

(SKIP)

9.5 CHANGE OF BASIS (Optional)

In $\mathbb{R}^2$ we wrote vectors in terms of the "standard" basis $\mathbf{i} = \begin{pmatrix} 1 \\ 0 \end{pmatrix}$, $\mathbf{j} = \begin{pmatrix} 0 \\ 1 \end{pmatrix}$. In $\mathbb{R}^n$ we defined the standard basis $\{\mathbf{e}_1, \mathbf{e}_2, \ldots, \mathbf{e}_n\}$. In P_n we defined the standard basis to be $\{1, x, x^2, \ldots, x^n\}$. These bases are most commonly used because it is relatively easy to work with them. But it sometimes happens that another basis is more convenient.

There are obviously many bases to choose from since in an n-dimensional vector space *any* n linearly independent vectors form a basis. In this section we shall see how to change from one basis to another by computing a certain matrix.

We start with a simple example. Let $\mathbf{u}_1 = \begin{pmatrix} 1 \\ 0 \end{pmatrix}$ and $\mathbf{u}_2 = \begin{pmatrix} 0 \\ 1 \end{pmatrix}$. Then $B_1 = \{\mathbf{u}_1, \mathbf{u}_2\}$ is the standard basis in $\mathbb{R}^2$. Let $\mathbf{v}_1 = \begin{pmatrix} 1 \\ 3 \end{pmatrix}$ and $\mathbf{v}_2 = \begin{pmatrix} -1 \\ 2 \end{pmatrix}$. Since $\mathbf{v}_1$ and $\mathbf{v}_2$ are linearly independent (because $\mathbf{v}_1$ is not a multiple of $\mathbf{v}_2$), $B_2 = \{\mathbf{v}_1, \mathbf{v}_2\}$ is a second basis in $\mathbb{R}^2$. Let $x = \begin{pmatrix} x_1 \\ x_2 \end{pmatrix}$ be a vector in $\mathbb{R}^2$. This notation means that

$$\mathbf{x} = \begin{pmatrix} x_1 \\ x_2 \end{pmatrix} = x_1\begin{pmatrix} 1 \\ 0 \end{pmatrix} + x_2\begin{pmatrix} 0 \\ 1 \end{pmatrix} = x_1\mathbf{u}_1 + x_2\mathbf{u}_2.$$

That is, $\mathbf{x}$ is written in terms of the vectors in the basis B_1. To emphasize this fact, we write

$$(\mathbf{x})_{B_1} = \begin{pmatrix} x_1 \\ x_2 \end{pmatrix}.$$

Since B_2 is another basis in $\mathbb{R}^2$, there are scalars c_1 and c_2 such that

$$\mathbf{x} = c_1\mathbf{v}_1 + c_2\mathbf{v}_2. \tag{1}$$

Once these scalars are found, we write

$$(\mathbf{x})_{B_2} = \begin{pmatrix} c_1 \\ c_2 \end{pmatrix}$$

to indicate that $\mathbf{x}$ is now expressed in terms of the vectors in B_2. To find the numbers c_1 and c_2, we write the old basis vectors ($\mathbf{u}_1$ and $\mathbf{u}_2$) in terms of the new basis vectors ($\mathbf{v}_1$ and $\mathbf{v}_2$). It is easy to verify that

$$\mathbf{u}_1 = \begin{pmatrix} 1 \\ 0 \end{pmatrix} = \tfrac{2}{5}\begin{pmatrix} 1 \\ 3 \end{pmatrix} - \tfrac{3}{5}\begin{pmatrix} -1 \\ 2 \end{pmatrix} = \tfrac{2}{5}\mathbf{v}_1 - \tfrac{3}{5}\mathbf{v}_2 \tag{2}$$

and

$$\mathbf{u}_2 = \begin{pmatrix} 0 \\ 1 \end{pmatrix} = \tfrac{1}{5}\begin{pmatrix} 1 \\ 3 \end{pmatrix} + \tfrac{1}{5}\begin{pmatrix} -1 \\ 2 \end{pmatrix} = \tfrac{1}{5}\mathbf{v}_1 + \tfrac{1}{5}\mathbf{v}_2. \tag{3}$$

That is,

$$(\mathbf{u}_1)_{B_2} = \begin{pmatrix} \frac{2}{5} \\ -\frac{3}{5} \end{pmatrix} \qquad \text{and} \qquad (\mathbf{u}_2)_{B_2} = \begin{pmatrix} \frac{1}{5} \\ \frac{1}{5} \end{pmatrix}.$$

Then

$$\mathbf{x} = x_1\mathbf{u}_1 + x_2\mathbf{u}_2 \overset{\text{From (2) and (3)}}{=} x_1(\tfrac{2}{5}\mathbf{v}_1 - \tfrac{3}{5}\mathbf{v}_2) + x_2(\tfrac{1}{5}\mathbf{v}_1 + \tfrac{1}{5}\mathbf{v}_2)$$
$$= (\tfrac{2}{5}x_1 + \tfrac{1}{5}x_2)\mathbf{v}_1 + (-\tfrac{3}{5}x_1 + \tfrac{1}{5}x_2)\mathbf{v}_2.$$

Thus, from (1),

$$c_1 = \tfrac{2}{5}x_1 + \tfrac{1}{5}x_2$$
$$c_2 = -\tfrac{3}{5}x_1 + \tfrac{1}{5}x_2$$

or

$$(\mathbf{x})_{B_2} = \begin{pmatrix} c_1 \\ c_2 \end{pmatrix} = \begin{pmatrix} \tfrac{2}{5}x_1 + \tfrac{1}{5}x_2 \\ -\tfrac{3}{5}x_1 + \tfrac{1}{5}x_2 \end{pmatrix} = \begin{pmatrix} \tfrac{2}{5} & \tfrac{1}{5} \\ -\tfrac{3}{5} & \tfrac{1}{5} \end{pmatrix}\begin{pmatrix} x_1 \\ x_2 \end{pmatrix}$$

For example, if $(\mathbf{x})_{B_1} = \begin{pmatrix} 3 \\ -4 \end{pmatrix}$, then

$$(\mathbf{x})_{B_2} = \begin{pmatrix} \tfrac{2}{5} & \tfrac{1}{5} \\ -\tfrac{3}{5} & \tfrac{1}{5} \end{pmatrix}\begin{pmatrix} 3 \\ -4 \end{pmatrix} = \begin{pmatrix} \tfrac{2}{5} \\ -\tfrac{13}{5} \end{pmatrix}.$$

Check. $\tfrac{2}{5}\mathbf{v}_1 - \tfrac{13}{5}\mathbf{v}_2 = \tfrac{2}{5}\begin{pmatrix} 1 \\ 3 \end{pmatrix} - \tfrac{13}{5}\begin{pmatrix} -1 \\ 2 \end{pmatrix} = \begin{pmatrix} \tfrac{2}{5} + \tfrac{13}{5} \\ \tfrac{6}{5} - \tfrac{26}{5} \end{pmatrix} = \begin{pmatrix} 3 \\ -4 \end{pmatrix} = 3\begin{pmatrix} 1 \\ 0 \end{pmatrix} - 4\begin{pmatrix} 0 \\ 1 \end{pmatrix}$
$= 3\mathbf{u}_1 - 4\mathbf{u}_2.$

The matrix $A = \begin{pmatrix} \tfrac{2}{5} & \tfrac{1}{5} \\ -\tfrac{3}{5} & \tfrac{1}{5} \end{pmatrix}$ is called the **transition matrix** from B_1 to B_2, and we have shown that

$$(\mathbf{x})_{B_2} = A(\mathbf{x})_{B_1}. \tag{4}$$

This example can be easily generalized, but first we need to extend our notation. Let $B_1 = \{\mathbf{u}_1, \mathbf{u}_2, \ldots, \mathbf{u}_n\}$ and $B_2 = \{\mathbf{v}_1, \mathbf{v}_2, \ldots, \mathbf{v}_n\}$ be two bases for an n-dimensional real vector space V. Let $\mathbf{x} \in V$. Then $\mathbf{x}$ can be written in terms of both bases:

$$\mathbf{x} = b_1\mathbf{u}_1 + b_2\mathbf{u}_2 + \cdots + b_n\mathbf{u}_n \tag{5}$$

and

$$\mathbf{x} = c_1\mathbf{v}_1 + c_2\mathbf{v}_2 + \cdots + c_n\mathbf{v}_n \tag{6}$$

where b_i and c_i are real numbers. We then write $(\mathbf{x})_{B_1} = \begin{pmatrix} b_1 \\ b_2 \\ \vdots \\ b_n \end{pmatrix}$ to denote the representation of $\mathbf{x}$ in terms of the basis B_1. This is unambiguous because the coefficients b_i in (5) are unique by Theorem 9.4.1. Likewise $(\mathbf{x})_{B_2} = \begin{pmatrix} c_1 \\ c_2 \\ \vdots \\ c_n \end{pmatrix}$ has a similar meaning. Suppose that $\mathbf{w}_1 = a_1\mathbf{u}_1 + a_2\mathbf{u}_2 + \cdots + a_n\mathbf{u}_n$ and $\mathbf{w}_2 = b_1\mathbf{u}_1 + b_2\mathbf{u}_2 + \cdots + b_n\mathbf{u}_n$. Then $\mathbf{w}_1 + \mathbf{w}_2 = (a_1 + b_1)\mathbf{u}_1 + (a_2 + b_2)\mathbf{u}_2 + \cdots + (a_n + b_n)\mathbf{u}_n$, so that

$$(\mathbf{w}_1 + \mathbf{w}_2)_{B_1} = (\mathbf{w}_1)_{B_1} + (\mathbf{w}_2)_{B_1}$$

That is, in the new notation we can add vectors just as we add vectors in $\mathbb{R}^n$. Moreover, it is easy to show that

$$\alpha(\mathbf{w})_{B_1} = (\alpha\mathbf{w})_{B_1}$$

Now, since B_2 is a basis, each $\mathbf{u}_j$ in B_1 can be written as a linear combination of the $\mathbf{v}_i$'s. Thus there exists a unique set of scalars $a_{1j}, a_{2j}, \ldots, a_{nj}$ such that for $j = 1, 2, \ldots, n$

$$\mathbf{u}_j = a_{1j}\mathbf{v}_1 + a_{2j}\mathbf{v}_2 + \cdots + a_{nj}\mathbf{v}_n \tag{7}$$

or

$$(\mathbf{u}_j)_{B_2} = \begin{pmatrix} a_{1j} \\ a_{2j} \\ \vdots \\ a_{nj} \end{pmatrix}. \tag{8}$$

Definition 1 TRANSITION MATRIX The $n \times n$ matrix A whose columns are given by (8) is called the **transition matrix** from basis B_1 to basis B_2. That is,

$$A = \begin{pmatrix} a_{11} & a_{12} & a_{13} & \cdots & a_{1n} \\ a_{21} & a_{22} & a_{23} & \cdots & a_{2n} \\ \vdots & \vdots & \vdots & & \vdots \\ a_{n1} & a_{n2} & a_{n3} & \cdots & a_{nn} \end{pmatrix}. \tag{9}$$

$$\begin{matrix} \uparrow & \uparrow & \uparrow & & \uparrow \\ (\mathbf{u}_1)_{B_2} & (\mathbf{u}_2)_{B_2} & (\mathbf{u}_3)_{B_2} & \cdots & (\mathbf{u}_n)_{B_2} \end{matrix}$$

Theorem 1 Let B_1 and B_2 be bases for a vector space V. Let A be the transition matrix from B_1 to B_2. Then, for every $\mathbf{x} \in V$,

$$(\mathbf{x})_{B_2} = A(\mathbf{x})_{B_1} \tag{10}$$

Proof. We use the representation of **x** given in (5) and (6):

$$\mathbf{x} \overset{\text{From (5)}}{=} b_1\mathbf{u}_1 + b_2\mathbf{u}_2 + \cdots + b_n\mathbf{u}_n$$

$$\overset{\text{From (7)}}{=} b_1(a_{11}\mathbf{v}_1 + a_{21}\mathbf{v}_2 + \cdots + a_{n1}\mathbf{v}_n) + b_2(a_{12}\mathbf{v}_1 + a_{22}\mathbf{v}_2 + \cdots + a_{n2}\mathbf{v}_n)$$
$$+ \cdots + b_n(a_{1n}\mathbf{v}_1 + a_{2n}\mathbf{v}_2 + \cdots + a_{nn}\mathbf{v}_n)$$
$$= (a_{11}b_1 + a_{12}b_2 + \cdots + a_{1n}b_n)\mathbf{v}_1 + (a_{21}b_1 + a_{22}b_2 + \cdots + a_{2n}b_n)\mathbf{v}_2 + \cdots$$
$$+ (a_{n1}b_1 + a_{n2}b_2 + \cdots + a_{nn}b_n)\mathbf{v}_n$$

$$\overset{\text{From (6)}}{=} c_1\mathbf{v}_1 + c_2\mathbf{v}_2 + \cdots + c_n\mathbf{v}_n. \tag{11}$$

Thus

$$(\mathbf{x})_{B_2} = \begin{pmatrix} c_1 \\ c_2 \\ \vdots \\ c_n \end{pmatrix} \overset{\text{From (11)}}{=} \begin{pmatrix} a_{11}b_1 + a_{12}b_2 + \cdots + a_{1n}b_n \\ a_{21}b_1 + a_{22}b_2 + \cdots + a_{2n}b_n \\ \vdots \\ a_{n1}b_1 + a_{n2}b_2 + \cdots + a_{nn}b_n \end{pmatrix}$$
$$= \begin{pmatrix} a_{11} & a_{12} & \cdots & a_{1n} \\ a_{21} & a_{22} & \cdots & a_{2n} \\ \vdots & \vdots & & \vdots \\ a_{n1} & a_{n2} & \cdots & a_{nn} \end{pmatrix} \begin{pmatrix} b_1 \\ b_2 \\ \vdots \\ b_n \end{pmatrix} = A(\mathbf{x})_{B_1}. \blacksquare \tag{12}$$

Before doing any further examples, we prove a theorem that is very useful for computations.

Theorem 2 If A is the transition matrix from B_1 to B_2, then A^{-1} is the transition matrix from B_2 to B_1.

Proof. Let C be the transition matrix from B_2 to B_1. Then, from (10), we have

$$(\mathbf{x})_{B_1} = C(\mathbf{x})_{B_2}. \tag{13}$$

But $(\mathbf{x})_{B_2} = A(\mathbf{x})_{B_1}$, and substituting this into (13) yields

$$(\mathbf{x})_{B_1} = CA(\mathbf{x})_{B_1}. \tag{14}$$

We leave it as an exercise (see Problem 39) to show that (14) can hold for every **x** in V only if $CA = I$ and $AC = I$. Thus $C = A^{-1}$ and the theorem is proven. ■

REMARK. This theorem makes it especially easy to find the transition matrix from the standard basis $B_1 = \{\mathbf{e}_1, \mathbf{e}_2, \ldots, \mathbf{e}_n\}$ in $\mathbb{R}^n$ to any other basis in $\mathbb{R}^n$. Let $B_2 = \{\mathbf{v}_1, \mathbf{v}_2, \ldots, \mathbf{v}_n\}$ be any other basis. Let C be the matrix whose columns are the vectors $\mathbf{v}_1, \mathbf{v}_2, \ldots, \mathbf{v}_n$. Then C is the transition matrix from B_2 to B_1 since each vector $\mathbf{v}_i$ is already written in terms of the standard basis. For example,

$$\begin{pmatrix} 1 \\ 3 \\ -2 \\ 4 \end{pmatrix}_{B_1} = \begin{pmatrix} 1 \\ 3 \\ -2 \\ 4 \end{pmatrix} = 1\begin{pmatrix} 1 \\ 0 \\ 0 \\ 0 \end{pmatrix} + 3\begin{pmatrix} 0 \\ 1 \\ 0 \\ 0 \end{pmatrix} - 2\begin{pmatrix} 0 \\ 0 \\ 1 \\ 0 \end{pmatrix} + 4\begin{pmatrix} 0 \\ 0 \\ 0 \\ 1 \end{pmatrix}.$$

Thus the transition matrix from B_1 to B_2 is C^{-1}.

PROCEDURE FOR FINDING THE TRANSITION MATRIX FROM STANDARD BASIS TO BASIS $B_2 = \{\mathbf{v}_1, \mathbf{v}_2, \ldots, \mathbf{v}_n\}$

(i) Write the matrix C whose columns are $\mathbf{v}_1, \mathbf{v}_2, \ldots, \mathbf{v}_n$.

(ii) Compute C^{-1}. This is the required transition matrix.

EXAMPLE 1 In $\mathbb{R}^3$ let $B_1 = \{\mathbf{i}, \mathbf{j}, \mathbf{k}\}$ and let $B_2 = \left\{\begin{pmatrix} 1 \\ 0 \\ 2 \end{pmatrix}, \begin{pmatrix} 3 \\ -1 \\ 0 \end{pmatrix}, \begin{pmatrix} 0 \\ 1 \\ -2 \end{pmatrix}\right\}$. If $\mathbf{x} = \begin{pmatrix} x \\ y \\ z \end{pmatrix} \in \mathbb{R}^3$, write $\mathbf{x}$ in terms of the vectors in B_2.

Solution. We first verify that B_2 is a basis. This is evident since $\begin{vmatrix} 1 & 3 & 0 \\ 0 & -1 & 1 \\ 2 & 0 & -2 \end{vmatrix} = 8 \neq 0$. Since $\mathbf{u}_1 = \begin{pmatrix} 1 \\ 0 \\ 0 \end{pmatrix}$, $\mathbf{u}_2 = \begin{pmatrix} 0 \\ 1 \\ 0 \end{pmatrix}$, and $\mathbf{u}_3 = \begin{pmatrix} 0 \\ 0 \\ 1 \end{pmatrix}$, we immediately see that the transition matrix, C, from B_2 to B_1 is given by

$$C = \begin{pmatrix} 1 & 3 & 0 \\ 0 & -1 & 1 \\ 2 & 0 & -2 \end{pmatrix}.$$

Thus, from Theorem 2, the transition matrix A from B_1 to B_2 is

$$A = C^{-1} = \tfrac{1}{8}\begin{pmatrix} 2 & 6 & 3 \\ 2 & -2 & -1 \\ 2 & 6 & -1 \end{pmatrix}.$$

For example, if $(\mathbf{x})_{B_1} = \begin{pmatrix} 1 \\ -2 \\ 4 \end{pmatrix}$, then

$$(\mathbf{x})_{B_2} = \frac{1}{8}\begin{pmatrix} 2 & 6 & 3 \\ 2 & -2 & -1 \\ 2 & 6 & -1 \end{pmatrix}\begin{pmatrix} 1 \\ -2 \\ 4 \end{pmatrix} = \frac{1}{8}\begin{pmatrix} 2 \\ 2 \\ -14 \end{pmatrix} = \begin{pmatrix} \frac{1}{4} \\ \frac{1}{4} \\ -\frac{7}{4} \end{pmatrix}.$$

As a check, note that

$$\frac{1}{4}\begin{pmatrix} 1 \\ 0 \\ 2 \end{pmatrix} + \frac{1}{4}\begin{pmatrix} 3 \\ -1 \\ 0 \end{pmatrix} - \frac{7}{4}\begin{pmatrix} 0 \\ 1 \\ -2 \end{pmatrix} = \begin{pmatrix} 1 \\ -2 \\ 4 \end{pmatrix} = 1\begin{pmatrix} 1 \\ 0 \\ 0 \end{pmatrix} - 2\begin{pmatrix} 0 \\ 1 \\ 0 \end{pmatrix} + 4\begin{pmatrix} 0 \\ 0 \\ 1 \end{pmatrix}. \blacksquare$$

EXAMPLE 2 In P_2 the standard basis is $B_1 = \{1, x, x^2\}$. Another basis is $B_2 = \{4x - 1, 2x^2 - x, 3x^2 + 3\}$. If $p = a_0 + a_1x + a_2x^2$, write p in terms of the polynomials in B_2.

Solution. We first verify that B_2 is a basis. If $c_1(4x - 1) + c_2(2x^2 - x) + c_3(3x^2 + 3) = 0$ for all x, then, rearranging terms, we obtain

$$(-c_1 + 3c_3)1 + (4c_1 - c_2)x + (2c_2 + 3c_3)x^2 = 0.$$

But, since $\{1, x, x^2\}$ is a linearly independent set, we must have

$$\begin{aligned} -c_1 \quad\quad + 3c_3 &= 0 \\ 4c_1 - c_2 \quad\quad &= 0 \\ 2c_2 + 3c_3 &= 0 \end{aligned}$$

The determinant of this homogeneous system is $\begin{vmatrix} -1 & 0 & 3 \\ 4 & -1 & 0 \\ 0 & 2 & 3 \end{vmatrix} = 27 \neq 0$, which means that $c_1 = c_2 = c_3 = 0$ is the only solution. Now $(4x - 1)_{B_1} = \begin{pmatrix} -1 \\ 4 \\ 0 \end{pmatrix}$, $(2x^2 - x)_{B_1} = \begin{pmatrix} 0 \\ -1 \\ 2 \end{pmatrix}$, and $(3 + 3x^2)_{B_1} = \begin{pmatrix} 3 \\ 0 \\ 3 \end{pmatrix}$. Hence

$$C = \begin{pmatrix} -1 & 0 & 3 \\ 4 & -1 & 0 \\ 0 & 2 & 3 \end{pmatrix}$$

is the transition matrix from B_2 to B_1 so that

$$A = C^{-1} = \frac{1}{27}\begin{pmatrix} -3 & 6 & 3 \\ -12 & -3 & 12 \\ 8 & 2 & 1 \end{pmatrix}$$

is the transition matrix from B_1 to B_2. Since $(a_0 + a_1x + a_2x^2)_{B_1} = \begin{pmatrix} a_0 \\ a_1 \\ a_2 \end{pmatrix}$, we have

$$(a_0 + a_1x + a_2x^2)_{B_2} = \frac{1}{27}\begin{pmatrix} -3 & 6 & 3 \\ -12 & -3 & 12 \\ 8 & 2 & 1 \end{pmatrix}\begin{pmatrix} a_0 \\ a_1 \\ a_2 \end{pmatrix}$$
$$= \begin{pmatrix} \frac{1}{27}[-3a_0 + 6a_1 + 3a_2] \\ \frac{1}{27}[-12a_0 - 3a_1 + 12a_2] \\ \frac{1}{27}[8a_0 + 2a_1 + a_2] \end{pmatrix}.$$

For example, if $p(x) = 5x^2 - 3x + 4$, then

$$(5x^2 - 3x + 4)_{B_2} = \frac{1}{27}\begin{pmatrix} -3 & 6 & 3 \\ -12 & -3 & 12 \\ 8 & 2 & 1 \end{pmatrix}\begin{pmatrix} 4 \\ -3 \\ 5 \end{pmatrix} = \begin{pmatrix} -\frac{15}{27} \\ \frac{21}{27} \\ \frac{31}{27} \end{pmatrix}$$

or

Check this

$$5x^2 - 3x + 4 = -\tfrac{15}{27}(4x - 1) + \tfrac{21}{27}(2x^2 - x) + \tfrac{31}{27}(3x^2 + 3). \blacksquare$$

EXAMPLE 3 Let $B_1 = \left\{\begin{pmatrix} 3 \\ 1 \end{pmatrix}, \begin{pmatrix} 2 \\ -1 \end{pmatrix}\right\}$ and $B_2 = \left\{\begin{pmatrix} 2 \\ 4 \end{pmatrix}, \begin{pmatrix} -5 \\ 3 \end{pmatrix}\right\}$ be two bases in $\mathbb{R}^2$. If $(\mathbf{x})_{B_1} = \begin{pmatrix} b_1 \\ b_2 \end{pmatrix}$, write $\mathbf{x}$ in terms of the vectors in B_2.

Solution. This problem is a bit more difficult because neither basis is the standard basis. We must write the vectors in B_1 as linear combinations of the vectors in B_2. That is, we must find constants a_{11}, a_{21}, a_{12}, a_{22} such that

$$\begin{pmatrix} 3 \\ 1 \end{pmatrix} = a_{11}\begin{pmatrix} 2 \\ 4 \end{pmatrix} + a_{21}\begin{pmatrix} -5 \\ 3 \end{pmatrix} \quad \text{and} \quad \begin{pmatrix} 2 \\ -1 \end{pmatrix} = a_{12}\begin{pmatrix} 2 \\ 4 \end{pmatrix} + a_{22}\begin{pmatrix} -5 \\ 3 \end{pmatrix}.$$

This leads to the following systems:

$$\begin{aligned} 2a_{11} - 5a_{21} &= 3 \\ 4a_{11} + 3a_{21} &= 1 \end{aligned} \quad \text{and} \quad \begin{aligned} 2a_{12} - 5a_{22} &= 2 \\ 4a_{12} + 3a_{22} &= -1 \end{aligned}$$

The solutions are $a_{11} = \frac{7}{13}$, $a_{21} = -\frac{5}{13}$, $a_{12} = \frac{1}{26}$, and $a_{22} = -\frac{5}{13}$. Thus

$$A = \frac{1}{26}\begin{pmatrix} 14 & 1 \\ -10 & -10 \end{pmatrix}$$

and

$$(\mathbf{x})_{B_2} = \frac{1}{26}\begin{pmatrix} 14 & 1 \\ -10 & -10 \end{pmatrix}\begin{pmatrix} b_1 \\ b_2 \end{pmatrix} = \begin{pmatrix} \frac{1}{26}(14b_1 + b_2) \\ -\frac{10}{26}(b_1 + b_2) \end{pmatrix}.$$

For example, let $\mathbf{x} \overset{\text{In standard basis}}{\downarrow} = \begin{pmatrix} 7 \\ 4 \end{pmatrix}$. Then

$$\begin{pmatrix} 7 \\ 4 \end{pmatrix}_{B_1} = b_1\begin{pmatrix} 3 \\ 1 \end{pmatrix} + b_2\begin{pmatrix} 2 \\ -1 \end{pmatrix} = 3\begin{pmatrix} 3 \\ 1 \end{pmatrix} - \begin{pmatrix} 2 \\ -1 \end{pmatrix}.$$

So that

$$\begin{pmatrix} 7 \\ 4 \end{pmatrix}_{B_1} = \begin{pmatrix} 3 \\ -1 \end{pmatrix}$$

and

$$\begin{pmatrix} 7 \\ 4 \end{pmatrix}_{B_2} = \frac{1}{26}\begin{pmatrix} 14 & 1 \\ -10 & -10 \end{pmatrix}\begin{pmatrix} 3 \\ -1 \end{pmatrix} = \begin{pmatrix} \frac{41}{26} \\ -\frac{20}{26} \end{pmatrix}.$$

That is

$$\begin{pmatrix} 7 \\ 4 \end{pmatrix} \overset{\text{Check!}}{\downarrow}= \frac{41}{26}\begin{pmatrix} 2 \\ 4 \end{pmatrix} - \frac{20}{26}\begin{pmatrix} -5 \\ 3 \end{pmatrix}. \quad ■$$

Using the notation of this section we can derive a convenient way to determine whether a given set of vectors in any finite dimensional real vector space is linearly dependent or independent.

Theorem 3 Let $B_1 = \{\mathbf{v}_1, \mathbf{v}_2, \ldots, \mathbf{v}_n\}$ be a basis for the n-dimensional vector space V. Suppose that

$$(\mathbf{x}_1)_{B_1} = \begin{pmatrix} a_{11} \\ a_{21} \\ \vdots \\ a_{n1} \end{pmatrix}, (\mathbf{x}_2)_{B_1} = \begin{pmatrix} a_{12} \\ a_{22} \\ \vdots \\ a_{n2} \end{pmatrix}, \ldots, (\mathbf{x}_n)_{B_1} = \begin{pmatrix} a_{1n} \\ a_{21} \\ \vdots \\ a_{nn} \end{pmatrix}$$

Let

$$A = \begin{pmatrix} a_{11} & a_{12} & \cdots & a_{1n} \\ a_{21} & a_{22} & \cdots & a_{2n} \\ \vdots & \vdots & & \vdots \\ a_{n1} & a_{n2} & \cdots & a_{nn} \end{pmatrix}.$$

Then $\mathbf{x}_1, \mathbf{x}_2, \ldots, \mathbf{x}_n$ are linearly independent if and only if $\det A \neq 0$.

Proof. Let $\mathbf{a}_1, \mathbf{a}_2, \ldots, \mathbf{a}_n$ denote the columns of A. Suppose that

$$c_1\mathbf{x}_1 + c_2\mathbf{x}_2 + \cdots + c_n\mathbf{x}_n = \mathbf{0} \tag{15}$$

Then, using the addition defined on page 513, we may write (15) as

$$(c_1\mathbf{a}_1 + c_2\mathbf{a}_2 + \cdots + c_n\mathbf{a}_n)_{B_1} = (\mathbf{0})_{B_1} \tag{16}$$

Equation (16) gives two representations of the zero vector in V in terms of the basis vectors in B_1. Since the representation of a vector in terms of basis vectors is unique (by Theorem 9.4.1) we conclude that

$$c_1\mathbf{a}_1 + c_2\mathbf{a}_2 + \cdots + c_n\mathbf{a}_n = \mathbf{0} \tag{17}$$

where the zero on the right-hand side is the zero vector in $\mathbb{R}^n$. But this proves the theorem, since Equation (17) involves the columns of A, which are linearly independent if and only if $\det A \neq 0$. ■

EXAMPLE 4 In P_2, determine whether the polynomials $3 - x$, $2 + x^2$, and $4 + 5x - 2x^2$ are linearly dependent or independent.

Solution. Using the basis $B_1 = \{1, x, x^2\}$, we have $(3 - x)_{B_1} = \begin{pmatrix} 3 \\ -1 \\ 0 \end{pmatrix}$, $(2 + x^2)_{B_1} = \begin{pmatrix} 2 \\ 0 \\ 1 \end{pmatrix}$, and $(4 + 5x - 2x^2)_{B_1} = \begin{pmatrix} 4 \\ 5 \\ -2 \end{pmatrix}$. Then $\det A = \begin{vmatrix} 3 & 2 & 4 \\ -1 & 0 & 5 \\ 0 & 1 & -2 \end{vmatrix} = -23 \neq 0$, so the polynomials are independent. ■

EXAMPLE 5 In M_{22}, determine whether the matrices $\begin{pmatrix} 1 & 2 \\ 3 & 6 \end{pmatrix}$, $\begin{pmatrix} -1 & 3 \\ -1 & 1 \end{pmatrix}$, $\begin{pmatrix} 2 & -1 \\ 0 & 1 \end{pmatrix}$, and $\begin{pmatrix} 1 & 4 \\ 4 & 9 \end{pmatrix}$ are linearly dependent or independent.

Solution. Using the standard basis $B_1 = \left\{\begin{pmatrix} 1 & 0 \\ 0 & 0 \end{pmatrix}, \begin{pmatrix} 0 & 1 \\ 0 & 0 \end{pmatrix}, \begin{pmatrix} 0 & 0 \\ 1 & 0 \end{pmatrix}, \begin{pmatrix} 0 & 0 \\ 0 & 1 \end{pmatrix}\right\}$ we obtain

$$\det A = \begin{vmatrix} 1 & -1 & 2 & 1 \\ 2 & 3 & -1 & 4 \\ 3 & -1 & 0 & 4 \\ 6 & 1 & 1 & 9 \end{vmatrix} = 0$$

so the matrices are dependent. Note that $\det A = 0$ because the fourth row of A is the sum of the first three rows of A. Note also that

$$-29\begin{pmatrix} 1 & 2 \\ 3 & 6 \end{pmatrix} - 7\begin{pmatrix} -1 & 3 \\ -1 & 1 \end{pmatrix} + \begin{pmatrix} 2 & -1 \\ 0 & 1 \end{pmatrix} + 20\begin{pmatrix} 1 & 4 \\ 4 & 9 \end{pmatrix} = \begin{pmatrix} 0 & 0 \\ 0 & 0 \end{pmatrix},$$

which illustrates that the four matrices are linearly dependent. ■

PROBLEMS 9.5

In Problems 1–5 write $\begin{pmatrix} x \\ y \end{pmatrix} \in \mathbb{R}^2$ in terms of the given basis.

1. $\begin{pmatrix} 1 \\ 1 \end{pmatrix}, \begin{pmatrix} 1 \\ -1 \end{pmatrix}$

2. $\begin{pmatrix} 2 \\ -3 \end{pmatrix}, \begin{pmatrix} 3 \\ -2 \end{pmatrix}$

3. $\begin{pmatrix} 5 \\ 7 \end{pmatrix}\begin{pmatrix} 3 \\ -4 \end{pmatrix}$

4. $\begin{pmatrix} -1 \\ -2 \end{pmatrix}, \begin{pmatrix} -1 \\ 2 \end{pmatrix}$

5. $\begin{pmatrix} a \\ c \end{pmatrix}, \begin{pmatrix} b \\ d \end{pmatrix}$, where $ad - bc \neq 0$

In Problems 6–10 write $\begin{pmatrix} x \\ y \\ z \end{pmatrix} \in \mathbb{R}^3$ in terms of the given basis.

6. $\begin{pmatrix} 1 \\ 0 \\ 0 \end{pmatrix}, \begin{pmatrix} 0 \\ 0 \\ 1 \end{pmatrix}, \begin{pmatrix} 1 \\ 1 \\ 1 \end{pmatrix}$

7. $\begin{pmatrix} 1 \\ 0 \\ 0 \end{pmatrix}, \begin{pmatrix} 1 \\ 1 \\ 0 \end{pmatrix}, \begin{pmatrix} 1 \\ 1 \\ 1 \end{pmatrix}$

8. $\begin{pmatrix} 1 \\ 0 \\ -1 \end{pmatrix}, \begin{pmatrix} -1 \\ 1 \\ 0 \end{pmatrix}, \begin{pmatrix} 0 \\ 1 \\ 1 \end{pmatrix}$

9. $\begin{pmatrix} 2 \\ 1 \\ 3 \end{pmatrix}, \begin{pmatrix} -1 \\ 4 \\ 5 \end{pmatrix}, \begin{pmatrix} 3 \\ -2 \\ -4 \end{pmatrix}$

10. $\begin{pmatrix} a \\ 0 \\ 0 \end{pmatrix}, \begin{pmatrix} b \\ d \\ 0 \end{pmatrix}, \begin{pmatrix} c \\ e \\ f \end{pmatrix}$, where $adf \neq 0$

In Problems 11–13 write the polynomial $a_0 + a_1x + a_2x^2$ in P_2 in terms of the given basis.

11. $1, x - 1, x^2 - 1$

12. $6, 2 + 3x, 3 + 4x + 5x^2$

13. $x + 1, x - 1, x^2 - 1$

14. In M_{22} write the matrix $\begin{pmatrix} 2 & -1 \\ 4 & 6 \end{pmatrix}$ in terms of the basis $\left\{\begin{pmatrix} 1 & 1 \\ -1 & 0 \end{pmatrix}, \begin{pmatrix} 2 & 0 \\ 3 & 1 \end{pmatrix}, \begin{pmatrix} 0 & 1 \\ -1 & 0 \end{pmatrix}, \begin{pmatrix} 0 & -2 \\ 0 & 4 \end{pmatrix}\right\}$.

15. In P_3 write the polynomial $2x^3 - 3x^2 + 5x - 6$ in terms of the basis polynomials $1, 1 + x, x + x^2, x^2 + x^3$.

16. In P_3 write the polynomial $4x^2 - x + 5$ in terms of the basis polynomials $1, 1 - x, (1 - x)^2, (1 - x)^3$.

17. In $\mathbb{R}^2$ suppose that $\mathbf{x} = \begin{pmatrix} 2 \\ -1 \end{pmatrix}_{B_1}$, where $B_1 = \left\{\begin{pmatrix} 1 \\ 1 \end{pmatrix}, \begin{pmatrix} 2 \\ 3 \end{pmatrix}\right\}$. Write $\mathbf{x}$ in terms of the basis $B_2 = \left\{\begin{pmatrix} 0 \\ 3 \end{pmatrix}, \begin{pmatrix} 5 \\ -1 \end{pmatrix}\right\}$.

18. In $\mathbb{R}^2$, $\mathbf{x} = \begin{pmatrix} 4 \\ -1 \end{pmatrix}_{B_1}$, where $B_1 = \left\{\begin{pmatrix} 2 \\ -5 \end{pmatrix}, \begin{pmatrix} 7 \\ 3 \end{pmatrix}\right\}$.

Write $\mathbf{x}$ in terms of $B_2 = \left\{\begin{pmatrix} -2 \\ 1 \end{pmatrix}, \begin{pmatrix} -3 \\ 2 \end{pmatrix}\right\}$.

19. In $\mathbb{R}^3$, $\mathbf{x} = \begin{pmatrix} 2 \\ -1 \\ 4 \end{pmatrix}_{B_1}$, where $B_1 = \left\{\begin{pmatrix} 1 \\ -1 \\ 0 \end{pmatrix}, \begin{pmatrix} 0 \\ 1 \\ -1 \end{pmatrix}, \begin{pmatrix} 1 \\ 0 \\ 1 \end{pmatrix}\right\}$. Write $\mathbf{x}$ in terms of $B_2 = \left\{\begin{pmatrix} 3 \\ 0 \\ 0 \end{pmatrix}, \begin{pmatrix} 1 \\ 2 \\ -1 \end{pmatrix}, \begin{pmatrix} 0 \\ 1 \\ 5 \end{pmatrix}\right\}$.

20. In P_2, $\mathbf{x} = \begin{pmatrix} 2 \\ 1 \\ 3 \end{pmatrix}_{B_1}$, where $B_1 = \{1 - x, 3x, x^2 - x - 1\}$. Write $\mathbf{x}$ in terms of $B_2 = \{3 - 2x, 1 + x, x + x^2\}$.

In Problems 21–28 use Theorem 3 to determine whether the given set of vectors is linearly dependent or independent.

21. In P_2: $2 + 3x + 5x^2, 1 - 2x + x^2, -1 + 6x^2$

22. In P_2: $-3 + x^2, 2 - x + 4x^2, 4 + 2x$

23. In P_2: $x + 4x^2, -2 + 2x, 2 + x + 12x^2$

24. In P_2: $-2 + 4x - 2x^2, 3 + x, 6 + 8x$

25. In P_3: $1 + x^2, -1 - 3x + 4x^2 + 5x^3, 2 + 5x - 6x^3, 4 + 6x + 3x^2 + 7x^3$

26. In M_{22}: $\begin{pmatrix} 2 & 0 \\ 3 & 4 \end{pmatrix}, \begin{pmatrix} -3 & -2 \\ 7 & 1 \end{pmatrix}, \begin{pmatrix} 1 & 0 \\ -1 & -3 \end{pmatrix}, \begin{pmatrix} 11 & 2 \\ -5 & -5 \end{pmatrix}$

27. In M_{22}: $\begin{pmatrix} 1 & -3 \\ 2 & 4 \end{pmatrix}, \begin{pmatrix} 1 & 4 \\ 5 & 0 \end{pmatrix}, \begin{pmatrix} -1 & 6 \\ -1 & 3 \end{pmatrix}, \begin{pmatrix} 0 & 0 \\ 3 & 0 \end{pmatrix}$

28. In M_{22}: $\begin{pmatrix} a & 0 \\ 0 & 0 \end{pmatrix}, \begin{pmatrix} b & c \\ 0 & 0 \end{pmatrix}, \begin{pmatrix} d & e \\ f & 0 \end{pmatrix}, \begin{pmatrix} g & h \\ j & k \end{pmatrix}$, where $acfk \neq 0$

29. In P_n, let $p_1, p_2, \ldots, p_{n+1}$ be $n + 1$ polynomials such that $p_i(0) = 0$ for $i = 1, 2, \ldots, n + 1$. Show that the polynomials are linearly dependent.

***30.** In Problem 29 suppose that $p_i^{(j)} = 0$ for $i = 1, 2, \ldots, n + 1$, and for some j with $1 \leq j \leq n$, and $p_i^{(j)}$ denotes the jth derivative of p_i. Show that the polynomials are linearly dependent in P_n.

31. In M_{mn} let $A_1, A_2, \ldots, A_{mn}$ be mn matrices each of whose components in the 1, 1 position is zero. Show that the matrices are linearly dependent.

***32.** Suppose the x- and y-axes in the plane are rotated counterclockwise through an angle of θ (measured in degrees or radians). This gives us new axes which we denote (x', y'). What are the x- and y-coordinates of the now rotated basis vectors **i** and **j**?

33. Show that the "change of coordinates" matrix in Problem 32 is given by

$$A^{-1} = \begin{pmatrix} \cos\theta & \sin\theta \\ -\sin\theta & \cos\theta \end{pmatrix}.$$

34. If, in Problems 32 and 33, $\theta = \pi/6 = 30°$, write the vector $\begin{pmatrix} -4 \\ 3 \end{pmatrix}$ in terms of the new coordinate axes x' and y'.

35. If $\theta = \pi/4 = 45°$, write $\begin{pmatrix} 2 \\ -7 \end{pmatrix}$ in terms of the new coordinate axes.

36. If $\theta = 2\pi/3 = 120°$, write $\begin{pmatrix} 4 \\ 5 \end{pmatrix}$ in terms of the new coordinate axes.

37. Let $C = (c_{ij})$ be an $n \times n$ invertible matrix and let $B_1 = \{\mathbf{v}_1, \mathbf{v}_2, \ldots, \mathbf{v}_n\}$ be a basis for a vector space V. Let

$$\mathbf{c}_1 = \begin{pmatrix} c_{11} \\ c_{21} \\ \vdots \\ c_{n1} \end{pmatrix}_{B_1}, \mathbf{c}_2 = \begin{pmatrix} c_{12} \\ c_{22} \\ \vdots \\ c_{n2} \end{pmatrix}_{B_1}, \ldots, \mathbf{c}_n = \begin{pmatrix} c_{1n} \\ c_{2n} \\ \vdots \\ c_{nn} \end{pmatrix}_{B_1}$$

Show that $B_2 = \{\mathbf{c}_1, \mathbf{c}_2, \ldots, \mathbf{c}_n\}$ is a basis for V.

38. Show that $(\mathbf{x})_{B_1} = CA(\mathbf{x})_{B_1}$ for every $\mathbf{x}$ in a vector space V if and only if $CA = I$ and $AC = I$. [*Hint:* Let $\mathbf{x}_i$ be the ith vector in B_1. Then $(\mathbf{x}_i)_{B_1}$ has a 1 in the ith position and a 0 everywhere else. What can you say about $CA(\mathbf{x}_i)_{B_1}$?]

9.6 LINEAR TRANSFORMATIONS

In this section we discuss a special class of functions, called *linear transformations,* which occur with great frequency in linear alebra and other branches of mathematics. They are also important in a wide variety of applications. Before defining a linear transformation, let us study two simple examples to see what can happen.

EXAMPLE 1 In $\mathbb{R}^2$, define a function T by the formula

$$T\begin{pmatrix} x \\ y \end{pmatrix} = \begin{pmatrix} x \\ -y \end{pmatrix}.$$

Geometrically, T takes a vector in $\mathbb{R}^2$ and reflects it about the x-axis. This is illustrated in Figure 1. Moreover, let $\mathbf{u} = \begin{pmatrix} x_1 \\ y_1 \end{pmatrix}$ and $\mathbf{v} = \begin{pmatrix} x_2 \\ y_2 \end{pmatrix}$. Then

$$T\mathbf{u} = T\begin{pmatrix} x_1 \\ y_1 \end{pmatrix} = \begin{pmatrix} x_1 \\ -y_1 \end{pmatrix}$$

$$T\mathbf{v} = T\begin{pmatrix} x_2 \\ y_2 \end{pmatrix} = \begin{pmatrix} x_2 \\ -y_2 \end{pmatrix}$$

$$\mathbf{u} + \mathbf{v} = \begin{pmatrix} x_1 + x_2 \\ y_1 + y_2 \end{pmatrix}$$

and

$$T(\mathbf{u} + \mathbf{v}) = \begin{pmatrix} x_1 + x_2 \\ -(y_1 + y_2) \end{pmatrix} = \begin{pmatrix} x_1 + x_2 \\ -y_1 - y_2 \end{pmatrix} = \begin{pmatrix} x_1 \\ -y_1 \end{pmatrix} + \begin{pmatrix} x_2 \\ -y_2 \end{pmatrix}$$
$$= T\mathbf{u} + T\mathbf{v}.$$

Also, if α is a scalar,

$$T(\alpha\mathbf{u}) = T\begin{pmatrix} \alpha x_1 \\ \alpha y_1 \end{pmatrix} = \begin{pmatrix} \alpha x_1 \\ -\alpha y_1 \end{pmatrix} = \alpha\begin{pmatrix} x_1 \\ -y_1 \end{pmatrix} = \alpha T\mathbf{u}.$$

Once we have given our basic definition, we shall see that T is a linear transformation from $\mathbb{R}^2$ into $\mathbb{R}^2$.

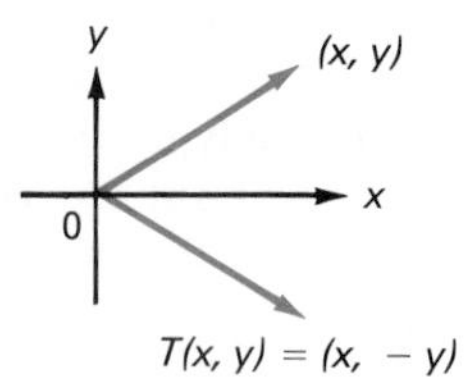

FIGURE 1 ■

EXAMPLE 2 A manufacturer makes four different products, each of which requires three raw materials. We denote the four products P_1, P_2, P_3, and P_4 and the raw materials R_1, R_2, and R_3. The accompanying table gives the number of units of each raw material required to manufacture one unit of each product.

		Needed to product one unit of			
		P_1	P_2	P_3	P_4
Number of units of raw material	R_1	2	1	3	4
	R_2	4	2	2	1
	R_3	3	3	1	2

A natural question arises: If certain numbers of the four products are produced, how many units of each raw material are needed? We let p_1, p_2, p_3, and p_4 denote the number of items of the four products manufactured and let r_1, r_2, and r_3 denote the number of units of the three raw materials needed. Then, we define

$$\mathbf{p} = \begin{pmatrix} p_1 \\ p_2 \\ p_3 \\ p_4 \end{pmatrix}, \quad \mathbf{r} = \begin{pmatrix} r_1 \\ r_2 \\ r_3 \end{pmatrix}, \quad A = \begin{pmatrix} 2 & 1 & 3 & 4 \\ 4 & 2 & 2 & 1 \\ 3 & 3 & 1 & 2 \end{pmatrix}$$

For example, suppose $\mathbf{p} = \begin{pmatrix} 10 \\ 30 \\ 20 \\ 50 \end{pmatrix}$. How many units of R_1 are needed to produce these numbers of units of the four products? From the table, we find that

$$\begin{aligned} r_1 &= p_1 \cdot 2 + p_2 \cdot 1 + p_3 \cdot 3 + p_4 \cdot 4 \\ &= 10 \cdot 2 + 30 \cdot 1 + 20 \cdot 3 + 50 \cdot 4 = 310 \text{ units} \end{aligned}$$

Similarly,

$$r_2 = 10 \cdot 4 + 30 \cdot 2 + 20 \cdot 2 + 50 \cdot 1 = 190 \text{ units}$$

and

$$r_3 = 10 \cdot 3 + 30 \cdot 3 + 20 \cdot 1 + 50 \cdot 2 = 240 \text{ units}$$

In general, we see that

$$\begin{pmatrix} 2 & 1 & 3 & 4 \\ 4 & 2 & 2 & 1 \\ 3 & 3 & 1 & 2 \end{pmatrix} \begin{pmatrix} p_1 \\ p_2 \\ p_3 \\ p_4 \end{pmatrix} = \begin{pmatrix} r_1 \\ r_2 \\ r_3 \end{pmatrix}$$

or

$$\mathbf{r} = A\mathbf{p}.$$

We can look at this in another way. If $\mathbf{p}$ is called the *production vector* and $\mathbf{r}$ the *raw material vector*, we define the function T by $\mathbf{r} = T\mathbf{p} = A\mathbf{p}$. That is, T is the function that "transforms" the production vector into the raw material vector. It is defined by ordinary matrix multiplication. As we shall see, this function is also a linear transformation. ■

Before defining a linear transformation, let us say a bit about functions. In Section 7.4 (page 427) we wrote a linear system of equations as

$$A\mathbf{x} = \mathbf{b} \tag{1}$$

where A is an $m \times n$ matrix, $\mathbf{x}$ is an n-vector (a vector in $\mathbb{R}^n$), and $\mathbf{b} \in \mathbb{R}^m$. In Section 7.4 we were interested in finding $\mathbf{x}$ when A and $\mathbf{b}$ were known. However, we can look at equation (1) in another way. Suppose A is given. Then equation (1) says: Give me an $\mathbf{x}$ in $\mathbb{R}^n$ and I'll give you a $\mathbf{b}$ in $\mathbb{R}^m$. That is, A represents a function with domain $\mathbb{R}^n$ and range in $\mathbb{R}^m$.

As another example, consider the space $C'[0, 1]$ consisting of all real-valued functions that are differentiable with continuous derivatives on the interval $[0, 1]$. If

$f \in C[0, 1]$, then we can define a function D by $Df = f'$. That is, D takes a differentiable function and gives you its (continuous) derivative. For example, $D(x^2) = 2x$, $D(\ln x) = 1/x$ and $D(\sin x) = \cos x$.

The point of these two examples is to show that functions may take many different forms. One special kind of function is defined below.

Definition 1 LINEAR TRANSFORMATION Let V and W be vector spaces. A **linear transformation** T from V into W is a function that assigns to each vector $\mathbf{v} \in V$ a unique vector $T\mathbf{v} \in W$ and that satisfies, for each $\mathbf{u}$ and $\mathbf{v}$ in V and each scalar α,

$$T(\mathbf{u} + \mathbf{v}) = T\mathbf{u} + T\mathbf{v} \tag{2}$$

and

$$T(\alpha\mathbf{v}) = \alpha T\mathbf{v} \tag{3}$$

NOTATION: We write $T: V \to W$ to indicate that T takes V into W.

TERMINOLOGY: Linear transformations are often called **linear operators**. Functions that satisfy (2) and (3) are called **linear functions**.

EXAMPLE 3 Let $T: \mathbb{R}^2 \to \mathbb{R}^3$ be defined by $T\begin{pmatrix} x \\ y \end{pmatrix} = \begin{pmatrix} x + y \\ x - y \\ 3y \end{pmatrix}$. For example, $T\begin{pmatrix} 2 \\ -3 \end{pmatrix} = \begin{pmatrix} -1 \\ 5 \\ -9 \end{pmatrix}$. Then

$$T\left[\begin{pmatrix} x_1 \\ y_1 \end{pmatrix} + \begin{pmatrix} x_2 \\ y_2 \end{pmatrix}\right] = T\begin{pmatrix} x_1 + x_2 \\ y_1 + y_2 \end{pmatrix} = \begin{pmatrix} x_1 + x_2 + y_1 + y_2 \\ x_1 + x_2 - y_1 - y_2 \\ 3y_1 + 3y_2 \end{pmatrix}$$

$$= \begin{pmatrix} x_1 + y_1 \\ x_1 - y_1 \\ 3y_1 \end{pmatrix} + \begin{pmatrix} x_2 + y_2 \\ x_2 - y_2 \\ 3y_2 \end{pmatrix}.$$

But

$$\begin{pmatrix} x_1 + y_1 \\ x_1 - y_1 \\ 3y_1 \end{pmatrix} = T\begin{pmatrix} x_1 \\ y_1 \end{pmatrix} \quad \text{and} \quad \begin{pmatrix} x_2 + y_2 \\ x_2 - y_2 \\ 3y_2 \end{pmatrix} = T\begin{pmatrix} x_2 \\ y_2 \end{pmatrix}.$$

Thus

$$T\left[\begin{pmatrix} x_1 \\ y_1 \end{pmatrix} + \begin{pmatrix} x_2 \\ y_2 \end{pmatrix}\right] = T\begin{pmatrix} x_1 \\ y_1 \end{pmatrix} + T\begin{pmatrix} x_2 \\ y_2 \end{pmatrix}.$$

Similarly,

$$T\left[\alpha\begin{pmatrix}x\\y\end{pmatrix}\right] = T\begin{pmatrix}\alpha x\\ \alpha y\end{pmatrix} = \begin{pmatrix}\alpha x + \alpha y\\ \alpha x - \alpha y\\ 3\alpha y\end{pmatrix} = \alpha\begin{pmatrix}x + y\\ x - y\\ 3y\end{pmatrix} = \alpha T\begin{pmatrix}x\\y\end{pmatrix}.$$

Thus T is a linear transformation. ■

EXAMPLE 4 Let V and W be vector spaces and define $T: V \to W$ by $T\mathbf{v} = \mathbf{0}$ for every $\mathbf{v}$ in V. Then $T(\mathbf{v}_1 + \mathbf{v}_2) = \mathbf{0} = \mathbf{0} + \mathbf{0} = T\mathbf{v}_1 + T\mathbf{v}_2$ and $T(\alpha\mathbf{v}) = \mathbf{0} = \alpha\mathbf{0} = \alpha T\mathbf{v}$. Here T is called the **zero transformation**. ■

EXAMPLE 5 Let V be a vector space and define $I: V \to V$ by $I\mathbf{v} = \mathbf{v}$ for every $\mathbf{v}$ in V. Here I is obviously a linear transformation. It is called the **identity transformation** or **identity operator**. ■

EXAMPLE 6 Let $T: \mathbb{R}^2 \to \mathbb{R}^2$ be defined by $T\begin{pmatrix}x\\y\end{pmatrix} = \begin{pmatrix}-x\\y\end{pmatrix}$. We saw in Example 1 that T is linear. ■

EXAMPLE 7 Let A be an $m \times n$ matrix and define $T: \mathbb{R}^n \to \mathbb{R}^m$ by $T\mathbf{x} = A\mathbf{x}$. It is not difficult to verify (see Problem 39) that $A(\mathbf{x} + \mathbf{y}) = A\mathbf{x} + A\mathbf{y}$ and $A(\alpha\mathbf{x}) = \alpha A\mathbf{x}$ if $\mathbf{x}$ and $\mathbf{y}$ are in $\mathbb{R}^n$ and that T is a linear transformation. Thus: **Every $m \times n$ matrix A gives rise to a linear transformation from $\mathbb{R}^n$ into $\mathbb{R}^m$.** In Section 9.9 we shall see that a certain converse is true: *Every linear transformation between finite dimensional vector spaces can be represented by a matrix.* ■

EXAMPLE 8 Suppose the vector $\mathbf{v} = \begin{pmatrix}x\\y\end{pmatrix}$ in the xy-plane is rotated through an angle of θ (measured in degrees or radians) in the counterclockwise direction. Call the new rotated vector $\mathbf{v}' = \begin{pmatrix}x'\\y'\end{pmatrix}$. Then, as in Figure 2, if r denotes the length of $\mathbf{v}$ (which is unchanged by rotation),

$$x = r\cos\alpha \qquad y = r\sin\alpha$$

$$x' = r\cos(\theta + \alpha) \qquad y' = r\sin(\theta + \alpha)^{\dagger}$$

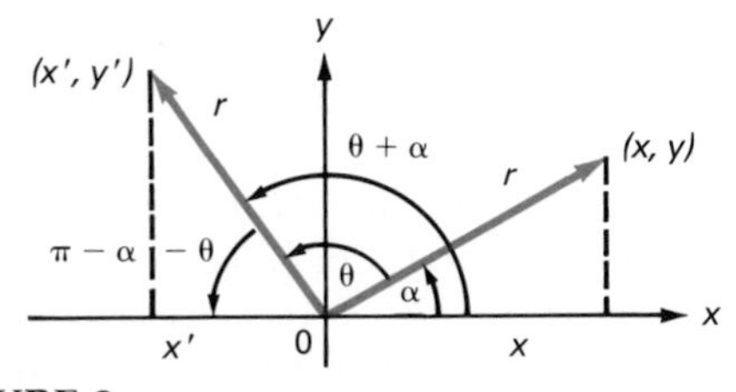

FIGURE 2

†These follow from the standard definitions of cos θ and sin θ as the x and y coordinates of a point on the unit circle. If (x, y) is a point on the circle centered at the origin of radius r, then $x = r\cos\phi$ and $y = r\sin\phi$, where ϕ is the angle the vector (x, y) makes with the positive x-axis.

But $r\cos(\theta+\alpha) = r\cos\theta\cos\alpha - r\sin\theta\sin\alpha$, so that

$$x' = x\cos\theta - y\sin\theta. \tag{4}$$

Similarly, $r\sin(\theta+\alpha) = r\sin\theta\cos\alpha + r\cos\theta\sin\alpha$ or

$$y' = x\sin\theta + y\cos\theta. \tag{5}$$

Let

$$A_\theta = \begin{pmatrix} \cos\theta & -\sin\theta \\ \sin\theta & \cos\theta \end{pmatrix} \tag{6}$$

Then, from (4) and (5), we see that $A_\theta\begin{pmatrix} x \\ y \end{pmatrix} = \begin{pmatrix} x' \\ y' \end{pmatrix}$. The linear transformation $T\colon \mathbb{R}^2 \to \mathbb{R}^2$ defined by $T\mathbf{v} = A_\theta\mathbf{v}$, where A_θ is given by (6), is called a **rotation transformation**. ■

EXAMPLE 9 Let $D\colon C'[0, 1] \to C[0, 1]$ be defined by $Df = f'$. Since $(f+g)' = f' + g'$ and $(\alpha f)' = \alpha f'$ if f and g are differentiable, we see that D is linear. D is called a **differential operator**. ■

EXAMPLE 10 Let $T\colon \mathbb{R}^3 \to \mathbb{R}^3$ be defined by $T\begin{pmatrix} x \\ y \\ z \end{pmatrix} = \begin{pmatrix} x \\ y \\ 0 \end{pmatrix}$. Then T is the projection operator taking a vector in space and projecting it into the xy-plane. Similarly, $T\begin{pmatrix} x \\ y \\ z \end{pmatrix} = \begin{pmatrix} x \\ 0 \\ z \end{pmatrix}$ projects a vector in space into the xz-plane. ■

EXAMPLE 11 Define $T\colon M_{mn} \to M_{nm}$ by $T(A) = A^t$. Since $(A+B)^t = A^t + B^t$ and $(\alpha A)^t = \alpha A^t$, we see that T, called the **transpose operator**, is a linear transformation. ■

EXAMPLE 12 Let $J\colon C[0, 1] \to \mathbb{R}$ be defined by $Jf = \int_0^1 f(x)\,dx$. Since $\int_0^1 [f(x) + g(x)\,dx = \int_0^1 f(x)\,dx + \int_0^1 g(x)\,dx$ and $\int_0^1 \alpha f(x)\,dx = \alpha\int_0^1 f(x)\,dx$ if f and g are continuous, we see that J is linear. For example, $T(x^3) = \frac{1}{4}$. J is called an **integral operator**. ■

■ WARNING: Not every function that looks linear actually is linear. For example, define $T\colon \mathbb{R} \to \mathbb{R}$ by $Tx = 2x + 3$. Then $\{(x, Tx)\colon x \in \mathbb{R}\}$ is a straight line in the xy-plane. But T is not linear since $T(x+y) = 2(x+y) + 3 = 2x + 2y + 3$ and $Tx + Ty = (2x+3) + (2y+3) = 2x + 2y + 6$. The only linear functions from $\mathbb{R}$ to $\mathbb{R}$ are functions of the form $f(x) = mx$ for some real number m. Thus among all straight-line functions, the only ones that are linear are those that pass through the origin.

EXAMPLE 13 Let $T\colon C[0, 1] \to \mathbb{R}$ be defined by $Tf = f(0) + 1$. Then T is not linear. To see this, we compute

$$T[f+g] = (f+g)(0) + 1 = f(0) + g(0) + 1$$

and

$$Tf + Tg = [f(0) + 1] + [g(0) + 1] = f(0) + g(0) + 2.$$

This provides another example of a transformation that might look linear but in fact is not. ■

We provide two more examples of functions that are not linear.

EXAMPLE 14 (**The norm function**) Let $N: \mathbb{R}^2 \to \mathbb{R}$ be defined by $N(\mathbf{v}) = |\mathbf{v}|$. It is generally not true that $|\mathbf{u} + \mathbf{v}| = |\mathbf{u}| + |\mathbf{v}|$.

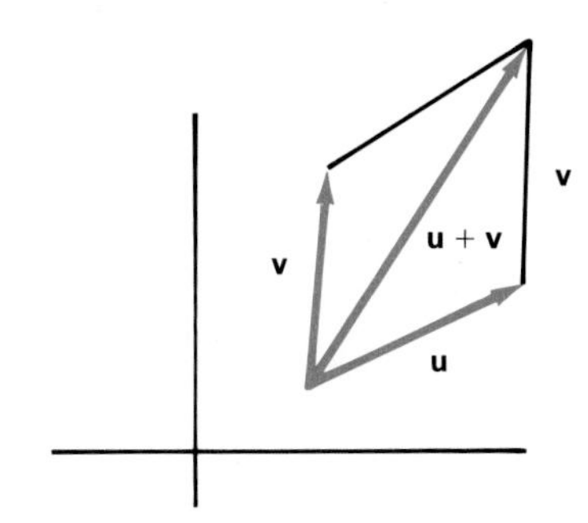

FIGURE 3

In fact, unless **u** and **v** lie on the same straight line, $|\mathbf{u} + \mathbf{v}| < |\mathbf{u}| + |\mathbf{v}|$ (see Figure 3). Thus N is not a linear transformation. ■

EXAMPLE 15 Let $T: \mathbb{R}^2 \to \mathbb{R}^2$ be given by $T\begin{pmatrix} x \\ y \end{pmatrix} = \begin{pmatrix} 2 \\ y \end{pmatrix}$. Then, for example,

$$T\begin{pmatrix} 1 \\ 5 \end{pmatrix} = \begin{pmatrix} 2 \\ 5 \end{pmatrix},$$

$$T\begin{pmatrix} 3 \\ 4 \end{pmatrix} = \begin{pmatrix} 2 \\ 4 \end{pmatrix},$$

and

$$T\begin{pmatrix} 1 \\ 5 \end{pmatrix} + T\begin{pmatrix} 3 \\ 4 \end{pmatrix} = \begin{pmatrix} 4 \\ 9 \end{pmatrix}.$$

But

$$T\left[\begin{pmatrix}1\\5\end{pmatrix}+\begin{pmatrix}3\\4\end{pmatrix}\right]=T\begin{pmatrix}4\\9\end{pmatrix}=\begin{pmatrix}2\\9\end{pmatrix}\neq\begin{pmatrix}4\\9\end{pmatrix}=T\begin{pmatrix}1\\5\end{pmatrix}+T\begin{pmatrix}3\\4\end{pmatrix}.$$

Thus T is not linear. As added proof, note that $T\left[\alpha\begin{pmatrix}x\\y\end{pmatrix}\right]=T\begin{pmatrix}\alpha x\\\alpha y\end{pmatrix}=\begin{pmatrix}2\\\alpha y\end{pmatrix}$. But $\alpha T\begin{pmatrix}x\\y\end{pmatrix}=\alpha\begin{pmatrix}2\\y\end{pmatrix}=\begin{pmatrix}2\alpha\\\alpha y\end{pmatrix}\neq\begin{pmatrix}2\\\alpha y\end{pmatrix}=T\left[\alpha\begin{pmatrix}x\\y\end{pmatrix}\right]$ unless $\alpha=1$. ■

PROBLEMS 9.6

In Problems 1–29 determine whether the given transformation from V to W is linear.

1. $T: \mathbb{R}^2 \to \mathbb{R}^2$; $T\begin{pmatrix}x\\y\end{pmatrix}=\begin{pmatrix}x\\0\end{pmatrix}$
2. $T: \mathbb{R}^2 \to \mathbb{R}^2$; $T\begin{pmatrix}x\\y\end{pmatrix}=\begin{pmatrix}1\\y\end{pmatrix}$
3. $T: \mathbb{R}^3 \to \mathbb{R}^2$; $T\begin{pmatrix}x\\y\\z\end{pmatrix}=\begin{pmatrix}x\\y\end{pmatrix}$
4. $T: \mathbb{R}^3 \to \mathbb{R}^2$; $T\begin{pmatrix}x\\y\\z\end{pmatrix}=\begin{pmatrix}0\\y\end{pmatrix}$
5. $T: \mathbb{R}^3 \to \mathbb{R}^2$; $T\begin{pmatrix}x\\y\\z\end{pmatrix}=\begin{pmatrix}1\\z\end{pmatrix}$
6. $T: \mathbb{R}^2 \to \mathbb{R}^2$; $T\begin{pmatrix}x\\y\end{pmatrix}=\begin{pmatrix}x^2\\y^2\end{pmatrix}$
7. $T: \mathbb{R}^2 \to \mathbb{R}^2$; $T\begin{pmatrix}x\\y\end{pmatrix}=\begin{pmatrix}y\\x\end{pmatrix}$
8. $T: \mathbb{R}^2 \to \mathbb{R}^2$; $T\begin{pmatrix}x\\y\end{pmatrix}=\begin{pmatrix}x+y\\x-y\end{pmatrix}$
9. $T: \mathbb{R}^2 \to \mathbb{R}$; $T\begin{pmatrix}x\\y\end{pmatrix}=xy$
10. $T: \mathbb{R}^n \to \mathbb{R}$; $T\begin{pmatrix}x_1\\x_2\\\vdots\\x_n\end{pmatrix}=x_1+x_2+\cdots+x_n$
11. $T: \mathbb{R} \to \mathbb{R}^n$; $T(x)=\begin{pmatrix}x\\x\\\vdots\\x\end{pmatrix}$
12. $T: \mathbb{R}^4 \to \mathbb{R}^2$; $T\begin{pmatrix}x\\y\\z\\w\end{pmatrix}=\begin{pmatrix}x+z\\y+w\end{pmatrix}$
13. $T: \mathbb{R}^4 \to \mathbb{R}^2$; $T\begin{pmatrix}x\\y\\z\\w\end{pmatrix}=\begin{pmatrix}xz\\yw\end{pmatrix}$
14. $T: M_{33} \to M_{33}$; $T(A)=AB$, where $B=\begin{pmatrix}1&0&2\\3&-1&1\\-2&4&5\end{pmatrix}$
15. $T: M_{nn} \to M_{nn}$; $T(A)=A^tA$
16. $T: M_{mn} \to M_{mp}$; $T(A)=AB$, where B is a fixed $n \times p$ matrix
17. $T: D_n \to D_n$; $T(D)=D^2$ (D_n is the set of $n \times n$ diagonal matrices)
18. $T: D_n \to D_n$; $T(D)=I+D$
19. $T: P_2 \to P_1$; $T(a_0+a_1x+a_2x^2)=a_0+a_1x$
20. $T: P_2 \to P_1$; $T(a_0+a_1x+a_2x^2)=a_1+a_2x$
21. $T: \mathbb{R} \to P_n$; $T(a)=a+ax+ax^2+\cdots+ax^n$
22. $T: P_2 \to P_4$; $T(p(x))=[p(x)]^2$
23. $T: C[0, 1] \to C[0, 1]$; $Tf(x)=f^2(x)\ [=f(x)\cdot f(x)]$
24. $T: C[0, 1] \to C[0, 1]$; $Tf(x)=f(x)+1$
25. $T: C[0, 1] \to \mathbb{R}$; $Tf=\int_0^1 f(x)g(x)\,dx$, where g is a fixed function in $C[0, 1]$

26. $T: C'[0, 1] \to C[0, 1]$; $Tf = (fg)'$, where g is a fixed function in $C'[0, 1]$
27. $T: C[0, 1] \to C[1, 2]$; $Tf(x) = f(x - 1)$
28. $T: C[0, 1] \to \mathbb{R}$; $Tf = f(\frac{1}{2})$
29. $T: M_{nn} \to \mathbb{R}$; $T(A) = \det A$

30. Let $T: \mathbb{R}^2 \to \mathbb{R}^2$ be given by $T(x, y) = (-x, -y)$. Describe T geometrically.
31. Let T be a linear transformation from $\mathbb{R}^2 \to \mathbb{R}^3$ such that $T\begin{pmatrix} 1 \\ 0 \end{pmatrix} = \begin{pmatrix} 1 \\ 2 \\ 3 \end{pmatrix}$ and $T\begin{pmatrix} 0 \\ 1 \end{pmatrix} = \begin{pmatrix} -4 \\ 0 \\ 5 \end{pmatrix}$. Find: **(a)** $T\begin{pmatrix} 2 \\ 4 \end{pmatrix}$ and **(b)** $T\begin{pmatrix} -3 \\ 7 \end{pmatrix}$.

32. In Example 8: **(a)** Find the rotation matrix A_θ when $\theta = \pi/6$. **(b)** What happens to the vector $\begin{pmatrix} -3 \\ 4 \end{pmatrix}$ if it is rotated through an angle of $\pi/6$ in the counterclockwise direction?

33. Let $A_\theta = \begin{pmatrix} \cos\theta & -\sin\theta & 0 \\ \sin\theta & \cos\theta & 0 \\ 0 & 0 & 1 \end{pmatrix}$. Describe geometrically the linear transformation $T: \mathbb{R}^3 \to \mathbb{R}^3$ given by $T\mathbf{x} = A_\theta\mathbf{x}$.
34. Answer the questions in Problem 33 for $A_\theta = \begin{pmatrix} \cos\theta & 0 & -\sin\theta \\ 0 & 1 & 0 \\ \sin\theta & 0 & \cos\theta \end{pmatrix}$.
35. Suppose that, in a real vector space V, T satisfies $T(\mathbf{x} + \mathbf{y}) = T\mathbf{x} + T\mathbf{y}$ and $T(\alpha\mathbf{x}) = \alpha T\mathbf{x}$ for $\alpha \geq 0$. Show that T is linear.
36. Find a linear transformation $T: M_{33} \to M_{22}$.
37. If T is a linear transformation from V to W, show that $T(\mathbf{x} - \mathbf{y}) = T\mathbf{x} - T\mathbf{y}$.
38. If T is a linear transformation from V to W, show that $T\mathbf{0} = \mathbf{0}$. Are the two zero vectors here the same?
39. Let A be an $m \times n$ matrix and let $\mathbf{x}$ and $\mathbf{y}$ be in $\mathbb{R}^n$. By writing out components, show that $A(\mathbf{x} + \mathbf{y}) = A\mathbf{x} + A\mathbf{y}$ and $A(\alpha\mathbf{x}) = \alpha A\mathbf{x}$, for every scalar α.

9.7 PROPERTIES OF LINEAR TRANSFORMATIONS: RANGE AND KERNEL

In this section we develop some of the basic properties of linear transformations.

Theorem 1 Let $T: V \to W$ be a linear transformation. Then for all vectors $\mathbf{u}, \mathbf{v}, \mathbf{v}_1, \mathbf{v}_2, \ldots, \mathbf{v}_n$ in V and all scalars $\alpha_1, \alpha_2, \ldots, \alpha_n$:

(i) $T(\mathbf{0}) = \mathbf{0}$
(ii) $T(\mathbf{u} - \mathbf{v}) = T\mathbf{u} - T\mathbf{v}$
(iii) $T(\alpha_1\mathbf{v}_1 + \alpha_2\mathbf{v}_2 + \cdots + \alpha_n\mathbf{v}_n) = \alpha_1 T\mathbf{v}_1 + \alpha_2 T\mathbf{v}_2 + \cdots + \alpha_n T\mathbf{v}_n$

NOTE. In part (i) the $\mathbf{0}$ on the left is the zero vector in V while the $\mathbf{0}$ on the right is the zero vector in W.

Proof.

(i) $T(\mathbf{0}) = T(\mathbf{0} + \mathbf{0}) = T(\mathbf{0}) + T(\mathbf{0})$. Thus $\mathbf{0} = T(\mathbf{0}) - T(\mathbf{0}) = T(\mathbf{0}) + T(\mathbf{0}) - T(\mathbf{0}) = T(\mathbf{0})$.
(ii) $T(\mathbf{u} - \mathbf{v}) = T[\mathbf{u} + (-1)\mathbf{v}] = T\mathbf{u} + T[(-1)\mathbf{v}] = T\mathbf{u} + (-1)T\mathbf{v} = T\mathbf{u} - T\mathbf{v}$.
(iii) We prove this part by induction (see Appendix 1). For $n = 2$, we get $T(\alpha_1\mathbf{v}_1 + \alpha_2\mathbf{v}_2) = T(\alpha_1\mathbf{v}_1) + T(\alpha_2\mathbf{v}_2) = \alpha_1 T\mathbf{v}_1 + \alpha_2 T\mathbf{v}_2$. Thus the equation holds for $n = 2$. We assume that it holds for $n = k$ and prove it

for $n = k + 1$: $T(\alpha_1\mathbf{v}_1 + \alpha_2\mathbf{v}_2 + \cdots + \alpha_k\mathbf{v}_k + \alpha_{k+1}\mathbf{v}_{k+1}) = T(\alpha_1\mathbf{v}_1 + \alpha_2\mathbf{v}_2 + \cdots + \alpha_k\mathbf{v}_k) + T(\alpha_{k+1}\mathbf{v}_{k+1})$, and using the equation in part (iii) for $n = k$, this is equal to $(\alpha_1 T\mathbf{v}_1 + \alpha_2 T\mathbf{v}_2 + \alpha_k T\mathbf{v}_k) + \alpha_{k+1}T\mathbf{v}_{k+1}$, which is what we wanted to show. This completes the proof. ■

REMARK. Note that part (ii) of Theorem 1 is a special case of part (iii).

An important fact about linear transformations is that **they are completely determined by what they do to basis vectors**.

Theorem 2 Let V be a finite dimensional vector space with basis $B = \{\mathbf{v}_1, \mathbf{v}_2, \ldots, \mathbf{v}_n\}$. Let $\mathbf{w}_1, \mathbf{w}_2, \ldots, \mathbf{w}_n$ be n vectors in W. Suppose that T_1 and T_2 are two linear transformations from V to W such that $T_1\mathbf{v}_i = T_2\mathbf{v}_i = \mathbf{w}_i$ for $i = 1, 2, \ldots, n$. Then for any vector $\mathbf{v} \in V$, $T_1\mathbf{v} = T_2\mathbf{v}$. That is, $T_1 = T_2$.

Proof. Since B is a basis for V, there exists a unique set of scalars $\alpha_1, \alpha_2, \ldots, \alpha_n$ such that $\mathbf{v} = \alpha_1\mathbf{v}_1 + \alpha_2\mathbf{v}_2 + \cdots + \alpha_n\mathbf{v}_n$. Then, from part (iii) of Theorem 1,

$$\begin{aligned} T_1\mathbf{v} = T_1(\alpha_1\mathbf{v}_1 + \alpha_2\mathbf{v}_2 + \cdots + \alpha_n\mathbf{v}_n) &= \alpha_1 T_1\mathbf{v}_1 + \alpha_2 T_1\mathbf{v}_2 + \cdots + \alpha_n T_1\mathbf{v}_n \\ &= \alpha_1\mathbf{w}_1 + \alpha_2\mathbf{w}_2 + \cdots + \alpha_n\mathbf{w}_n \end{aligned}$$

Similarly,

$$\begin{aligned} T_2\mathbf{v} = T_2(\alpha_1\mathbf{v}_1 + \alpha_2\mathbf{v}_2 + \cdots + \alpha_n\mathbf{v}_n) &= \alpha_1 T_2\mathbf{v}_1 + \alpha_2 T_2\mathbf{v}_2 + \cdots + \alpha_n T_2\mathbf{v}_n \\ &= \alpha_1\mathbf{w}_1 + \alpha_2\mathbf{w}_2 + \cdots + \alpha_n\mathbf{w}_n \end{aligned}$$

Thus $T_1\mathbf{v} = T_2\mathbf{v}$. ■

Theorem 2 tells us that if $T: V \to W$ and V is finite dimensional, then we need to know only what T does to basis vectors in V to determine T completely. To see this let $\mathbf{v}_1, \mathbf{v}_2, \ldots, \mathbf{v}_n$ be a basis in V and let $\mathbf{v}$ be another vector in V. Then as in the proof of Theorem 2,

$$T\mathbf{v} = \alpha_1 T\mathbf{v}_1 + \alpha_2 T\mathbf{v}_2 + \cdots + \alpha_n T\mathbf{v}_n$$

Thus we can compute $T\mathbf{v}$ for any vector $\mathbf{v} \in V$ if we know $T\mathbf{v}_1, T\mathbf{v}_2, \ldots, T\mathbf{v}_n$.

EXAMPLE 1 Let T be a linear transformation from $\mathbb{R}^3$ into $\mathbb{R}^2$ and suppose that $T\begin{pmatrix}1\\0\\0\end{pmatrix} = \begin{pmatrix}2\\3\end{pmatrix}$, $T\begin{pmatrix}0\\1\\0\end{pmatrix} = \begin{pmatrix}-1\\4\end{pmatrix}$, and $T\begin{pmatrix}0\\0\\1\end{pmatrix} = \begin{pmatrix}5\\-3\end{pmatrix}$. Compute $T\begin{pmatrix}3\\-4\\5\end{pmatrix}$.

Solution. We have $\begin{pmatrix}3\\-4\\5\end{pmatrix} = 3\begin{pmatrix}1\\0\\0\end{pmatrix} - 4\begin{pmatrix}0\\1\\0\end{pmatrix} + 5\begin{pmatrix}0\\0\\1\end{pmatrix}$. Thus

$$T\begin{pmatrix}3\\-4\\5\end{pmatrix} = 3T\begin{pmatrix}1\\0\\0\end{pmatrix} - 4T\begin{pmatrix}0\\1\\0\end{pmatrix} + 5T\begin{pmatrix}0\\0\\1\end{pmatrix}$$

$$= 3\begin{pmatrix}2\\3\end{pmatrix} - 4\begin{pmatrix}-1\\4\end{pmatrix} + 5\begin{pmatrix}5\\-3\end{pmatrix} = \begin{pmatrix}6\\9\end{pmatrix} + \begin{pmatrix}4\\-16\end{pmatrix} + \begin{pmatrix}25\\-15\end{pmatrix} = \begin{pmatrix}35\\-22\end{pmatrix}. \blacksquare$$

Another question arises: If $\mathbf{w}_1, \mathbf{w}_2, \ldots, \mathbf{w}_n$ are n vectors in W, does there exist a linear transformation T such that $T\mathbf{v}_i = \mathbf{w}_i$ for $i = 1, 2, \ldots, n$? The answer is yes, as the next theorem shows.

Theorem 3 Let V be a finite dimensional vector space with basis $B = \{\mathbf{v}_1, \mathbf{v}_2, \ldots, \mathbf{v}_n\}$. Let W be a vector space containing the n vectors $\mathbf{w}_1, \mathbf{w}_2, \ldots, \mathbf{w}_n$. Then there exists a unique linear transformation $T: V \to W$ such that $T\mathbf{v}_i = \mathbf{w}_i$ for $i = 1, 2, \ldots, n$.

Proof. Define a function T as follows:

(i) $T\mathbf{v}_i = \mathbf{w}_i$
(ii) If $\mathbf{v} = \alpha_1\mathbf{v}_1 + \alpha_2\mathbf{v}_2 + \cdots + \alpha_n\mathbf{v}_n$, then

$$T\mathbf{v} = \alpha_1\mathbf{w}_1 + \alpha_2\mathbf{w}_2 + \cdots + \alpha_n\mathbf{w}_n. \tag{1}$$

Because B is a basis for V, T is defined for every $\mathbf{v} \in V$; and since W is a vector space, $T\mathbf{v} \in W$. Thus it only remains to show that T is linear. But this follows directly from equation (1). For if $\mathbf{u} = \alpha_1\mathbf{v}_1 + \alpha_2\mathbf{v}_2 + \cdots + \alpha_n\mathbf{v}_n$ and $\mathbf{v} = \beta_1\mathbf{v}_1 + \beta_2\mathbf{v}_2 + \cdots + \beta_n\mathbf{v}_n$, then

$$\begin{aligned} T(\mathbf{u} + \mathbf{v}) &= T[(\alpha_1 + \beta_1)\mathbf{v}_1 + (\alpha_2 + \beta_2)\mathbf{v}_2 + \cdots + (\alpha_n + \beta_n)\mathbf{v}_n] \\ &= (\alpha_1 + \beta_1)\mathbf{w}_1 + (\alpha_2 + \beta_2)\mathbf{w}_2 + \cdots + (\alpha_n + \beta_n)\mathbf{w}_n \\ &= (\alpha_1\mathbf{w}_1 + \alpha_2\mathbf{w}_2 + \cdots + \alpha_n\mathbf{w}_n) + (\beta_1\mathbf{w}_1 + \beta_2\mathbf{w}_2 + \cdots + \beta_n\mathbf{w}_n) \\ &= T\mathbf{u} + T\mathbf{v}. \end{aligned}$$

Similarly $T(\alpha\mathbf{v}) = \alpha T\mathbf{v}$, so T is linear. The uniqueness of T follows from Theorem 2, and the theorem is proved. ■

REMARK. In Theorems 2 and 3 the vectors $\mathbf{w}_1, \mathbf{w}_2, \ldots, \mathbf{w}_n$ need not be distinct. Moreover, we emphasize that the theorems are true if V is any finite dimensional vector space, not just $\mathbb{R}^n$. Note also that W does not have to be finite dimensional.

EXAMPLE 2 Find a linear transformation from $\mathbb{R}^2$ into the plane

$$W = \left\{\begin{pmatrix}x\\y\\z\end{pmatrix}: 2x - y + 3z = 0\right\}.$$

Solution. From Example 9.4.3 on p. 503, we know that W is a two-dimensional subspace of $\mathbb{R}^3$ with basis vectors $\mathbf{w}_1 = \begin{pmatrix}1\\2\\0\end{pmatrix}$ and $\mathbf{w}_2 = \begin{pmatrix}0\\3\\1\end{pmatrix}$. Using the standard basis in $\mathbb{R}^2$, $\mathbf{v}_1 = \begin{pmatrix}1\\0\end{pmatrix}$ and $\mathbf{v}_2 = \begin{pmatrix}0\\1\end{pmatrix}$, we define the linear transformation T by $T\begin{pmatrix}1\\0\end{pmatrix} = \begin{pmatrix}1\\2\\0\end{pmatrix}$ and $T\begin{pmatrix}0\\1\end{pmatrix} = \begin{pmatrix}0\\3\\1\end{pmatrix}$. Then, as the discussion following Theorem 2 shows, T is completely determined; that is,

$$T\begin{pmatrix}x\\y\end{pmatrix} = xT\begin{pmatrix}1\\0\end{pmatrix} + yT\begin{pmatrix}0\\1\end{pmatrix} = \begin{pmatrix}x\\2x\\0\end{pmatrix} + \begin{pmatrix}0\\3y\\y\end{pmatrix} = \begin{pmatrix}x\\2x+3y\\y\end{pmatrix}.$$

For example,

$$T\begin{pmatrix}5\\-7\end{pmatrix} = T\left[5\begin{pmatrix}1\\0\end{pmatrix} - 7\begin{pmatrix}0\\1\end{pmatrix}\right] = 5T\begin{pmatrix}1\\0\end{pmatrix} - 7T\begin{pmatrix}0\\1\end{pmatrix} = 5\begin{pmatrix}1\\2\\0\end{pmatrix} - 7\begin{pmatrix}0\\3\\1\end{pmatrix} = \begin{pmatrix}5\\-11\\-7\end{pmatrix}. \blacksquare$$

We now turn to two important definitions in the theory of linear transformations.

Definition 1 KERNEL AND RANGE OF A LINEAR TRANSFORMATION Let V and W be vector spaces and let $T: V \to W$ be a linear transformation. Then

(i) The **kernel** of T, denoted ker T, is given by

$$\ker T = \{\mathbf{v} \in V: T\mathbf{v} = \mathbf{0}\}. \tag{2}$$

(ii) The **range** of T, denoted range T, is given by

$$\text{Range } T = \{\mathbf{w} \in W: \mathbf{w} = T\mathbf{v} \text{ for some } \mathbf{v} \in V\}. \tag{3}$$

REMARK 1. Note that ker T is nonempty because, by Theorem 1, $T(\mathbf{0}) = \mathbf{0}$ so that $\mathbf{0} \in \ker T$ for any linear transformation T. We shall be interested in finding other vectors in V that get "mapped to zero." Again note that when we write $T(\mathbf{0}) = \mathbf{0}$, the $\mathbf{0}$ on the left is in V and the $\mathbf{0}$ on the right is in W.

REMARK 2. Range T is simply the set of "images" of vectors in V under the transformation T. In fact, if $\mathbf{w} = T\mathbf{v}$, we shall say that $\mathbf{w}$ is the **image** of $\mathbf{v}$ under T.

Before giving examples of kernels and ranges, we prove a theorem that will be very useful.

Theorem 4 If $T: V \to W$ is a linear transformation, then

(i) $\ker T$ is a subspace of V, and
(ii) range T is a subspace of W.

Proof.

(i) Let $\mathbf{u}$ and $\mathbf{v}$ be in $\ker T$; then $T(\mathbf{u} + \mathbf{v}) = T\mathbf{u} + T\mathbf{v} = \mathbf{0} + \mathbf{0} = \mathbf{0}$ and $T(\alpha\mathbf{u}) = \alpha T\mathbf{u} = \alpha\mathbf{0} = \mathbf{0}$ so that $\mathbf{u} + \mathbf{v}$ and $\alpha\mathbf{u}$ are in $\ker T$.
(ii) Let $\mathbf{w}$ and $\mathbf{x}$ be in range T. Then $\mathbf{w} = T\mathbf{u}$ and $\mathbf{x} = T\mathbf{v}$ for two vectors $\mathbf{u}$ and $\mathbf{v}$ in V. This means that $T(\mathbf{u} + \mathbf{v}) = T\mathbf{u} + T\mathbf{v} = \mathbf{w} + \mathbf{x}$ and $T(\alpha\mathbf{u}) = \alpha T\mathbf{u} = \alpha\mathbf{w}$. Thus $\mathbf{w} + \mathbf{x}$ and $\alpha\mathbf{w}$ are in range T. ■

EXAMPLE 3 Let $T\mathbf{v} = \mathbf{0}$ for every $\mathbf{v} \in V$. (T is the zero transformation.) Then $\ker T = V$ and range $T = \{\mathbf{0}\}$. ■

EXAMPLE 4 Let $T\mathbf{v} = \mathbf{v}$ for every $\mathbf{v} \in V$. (T is the identity transformation.) Then $\ker T = \{\mathbf{0}\}$ and range $T = V$. ■

The zero and identity transformations provide two extremes. In the first, everything is in the kernel. In the second, only the zero vector is in the kernel. The cases in between are more interesting.

EXAMPLE 5 Let $T: \mathbb{R}^3 \to \mathbb{R}^3$ be defined by $T\begin{pmatrix} x \\ y \\ z \end{pmatrix} = \begin{pmatrix} x \\ y \\ 0 \end{pmatrix}$. That is (see Example 9.6.10), T is the projection operator from $\mathbb{R}^3$ into the xy-plane. If $T\begin{pmatrix} x \\ y \\ z \end{pmatrix} = \begin{pmatrix} x \\ y \\ 0 \end{pmatrix} = \mathbf{0} = \begin{pmatrix} 0 \\ 0 \\ 0 \end{pmatrix}$, then $x = y = 0$. Thus $\ker T = \{(x, y, z): x = y = 0\} =$ the z-axis, and range $T = \{(x, y, z): z = 0\} =$ the xy-plane. Note that $\dim \ker T = 1$ and $\dim \text{range } T = 2$. ■

Definition 2 NULLITY AND RANK OF A LINEAR TRANSFORMATION If T is a linear transformation from V to W, then we define

$$\textbf{Nullity of } T = \nu(T) = \dim \ker T \tag{4}$$

$$\textbf{Rank of } T = \rho(T) = \dim \text{range } T. \tag{5}$$

In Section 9.8 we define the rank and nullity of a matrix and show how they are related to the rank and nullity of a linear transformation. Note that in the last example $\nu(T) = 1$ and $\rho(T) = 2$.

EXAMPLE 6 Let $T: \mathbb{R}^n \to \mathbb{R}^m$ be given by $T\mathbf{v} = A\mathbf{v}$, where A is an $m \times n$ matrix. Then $\ker T = \{\mathbf{v}: A\mathbf{v} = \mathbf{0}\}$ and range $T = \{\mathbf{w}: A\mathbf{v} = \mathbf{w}$ for some $\mathbf{v} \in \mathbb{R}^n\}$. These two subspaces are called, respectively, the **kernel** and **range** of the matrix A. We shall discuss them in great detail in Section 9.8. ■

EXAMPLE 7 Let $V = M_{mn}$ and define $T: M_{mn} \to M_{mn}$ by $T(A) = A^t$ (see Example 9.6.11). If $TA = A^t = 0$, then A^t is the $n \times m$ zero matrix so that A is the $m \times n$ zero matrix. Thus ker $T = \{\mathbf{0}\}$ and, clearly, range $T = M_{nm}$. This means that $\nu(T) = 0$ and $\rho(T) = nm$. ■

EXAMPLE 8 Let $T: P_3 \to P_2$ be defined by $T(p) = T(a_0 + a_1x + a_2x^2 + a_3x^3) = a_0 + a_1x + a_2x^2$. Then if $T(p) = 0$, $a_0 + a_1x + a_2x^2 = 0$ for every x, which implies that $a_0 = a_1 = a_2 = 0$. Thus ker $T = \{p \in P_3: p(x) = a_3x^3\}$ and range $T = P_2$, $\nu(T) = 1$, and $\rho(T) = 3$. ■

EXAMPLE 9 Let $V = C[0, 1]$ and define $J: C[0, 1] \to \mathbb{R}$ by $Jf = \int_0^1 f(x)\,dx$ (see Example 9.6.12). Then ker $J = \{f \in C[0, 1]: \int_0^1 f(x)\,dx = 0\}$. Let α be a real number. Then the constant function $f(x) = \alpha$ for $x \in [0, 1]$ is in $C[0, 1]$ and $\int_0^1 \alpha\,dx = \alpha$. Since this is true for every real number α, we have range $J = \mathbb{R}$. ■

In the next section we shall see how to find the range and kernel of a matrix. In Section 9.9 we shall see how every linear transformation from one finite dimensional vector space to another can be represented by a matrix. This will enable us to compute the kernel and range of any linear transformation between finite dimensional vector spaces by finding the kernel and range of a corresponding matrix.

PROBLEMS 9.7

In Problems 1–10 find the kernel, range, rank, and nullity of the given linear transformation.

1. $T: \mathbb{R}^2 \to \mathbb{R}^2;\ T\begin{pmatrix} x \\ y \end{pmatrix} = \begin{pmatrix} x \\ 0 \end{pmatrix}$
2. $T: \mathbb{R}^3 \to \mathbb{R}^2;\ T\begin{pmatrix} x \\ y \\ z \end{pmatrix} = \begin{pmatrix} z \\ y \end{pmatrix}$
3. $T: \mathbb{R}^2 \to \mathbb{R};\ T\begin{pmatrix} x \\ y \end{pmatrix} = x + y$
4. $T: \mathbb{R}^4 \to \mathbb{R}^2;\ T\begin{pmatrix} x \\ y \\ z \\ w \end{pmatrix} = \begin{pmatrix} x + z \\ y + w \end{pmatrix}$
5. $T: M_{22} \to M_{22}$: $T(A) = AB$, where $B = \begin{pmatrix} 1 & 2 \\ 0 & 1 \end{pmatrix}$
6. $T: \mathbb{R} \to P_3$: $T(a) = a + ax + ax^2 + ax^3$
*7. $T: M_{nn} \to M_{nn}$: $T(A) = A^t + A$
8. $T: C'[0, 1] \to C[0, 1]$; $Tf = f'$
9. $T: C[0, 1] \to \mathbb{R}$; $Tf = f(\frac{1}{2})$
10. $T: \mathbb{R}^2 \to \mathbb{R}^2$: T is rotation through an angle of $\pi/3$
11. Let $T: V \to W$ be a linear transformation, let $\{\mathbf{v}_1, \mathbf{v}_2, \ldots, \mathbf{v}_n\}$ be a basis for V, and suppose that $T\mathbf{v}_i = \mathbf{0}$ for $i = 1, 2, \ldots, n$. Show that T is the zero transformation.
12. In Problem 11, suppose that $W = V$ and $T\mathbf{v}_i = \mathbf{v}_i$ for $i = 1, 2, \ldots, n$. Show that T is the identity operator.
13. Let $T: V \to \mathbb{R}^3$. Prove that range T is either **(a)** $\{\mathbf{0}\}$, **(b)** a line through the origin, **(c)** a plane through the origin, or **(d)** $\mathbb{R}^3$.
14. Let $T: \mathbb{R}^3 \to V$. Show that ker T is one of four spaces listed in Problem 13.
15. Find all linear transformations from $\mathbb{R}^2$ into $\mathbb{R}^2$ such that the line $y = 0$ is carried into the line $x = 0$.
16. Find all linear transformations from $\mathbb{R}^2$ into $\mathbb{R}^2$ that carry the line $y = ax$ into the line $y = bx$.
17. Find a linear transformation T from $\mathbb{R}^3 \to \mathbb{R}^3$ such that ker $T = \{(x, y, z): 2x - y + z = 0\}$.
18. Find a linear transformation T from $\mathbb{R}^3 \to \mathbb{R}^3$ such that range $T = \{(x, y, z): 2x - y + z = 0\}$.
19. Let $T: M_{nn} \to M_{nn}$ be defined by $TA = A - A^t$. Show that ker T = {symmetric $n \times n$ matrices} and range of T = {skew-symmetric $n \times n$ matrices}.
*20. Let $T: C'[0, 1] \to C[0, 1]$ be defined by $Tf(x) = xf'(x)$. Find the kernel and range of T.

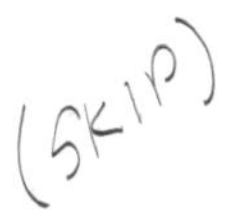

9.8 THE RANK AND NULLITY OF A MATRIX

Let A be an $m \times n$ matrix and let

$$N_A = \{\mathbf{x} \in \mathbb{R}^n\colon A\mathbf{x} = \mathbf{0}\}. \tag{1}$$

Then, as we saw in Example 9.4.11, N_A is a subspace of $\mathbb{R}^n$. Moreover, since T_A: $\mathbb{R}^n \to \mathbb{R}^m$ defined by $T\mathbf{x} = A\mathbf{x}$ is a linear operator, we see that $N_A = \ker T_A$.

Definition 1 KERNEL AND NULLITY OF A MATRIX N_A is called the **kernel** of A and $\nu(A) = \dim N_A$ is called the **nullity** of A. If N_A contains only the zero vector, then $\nu(A) = 0$.

REMARK. In Section 9.7 we defined the rank and nullity of a linear transformation. Example 9.6.7 showed that if A is an $m \times n$ matrix, then the transformation T defined by $T\mathbf{x} = A\mathbf{x}$ is a linear transformation from $\mathbb{R}^n \to \mathbb{R}^m$. We can see that $\nu(T) = \nu(A)$, and this shows that the nullity of a matrix is a special case of the nullity of a linear transformation. The same fact holds for the rank of a matrix (which we define shortly).

EXAMPLE 1 Let $A = \begin{pmatrix} 1 & 2 & -1 \\ 2 & -1 & 3 \end{pmatrix}$. Then, as we saw in Example 9.4.12, N_A is spanned by $\begin{pmatrix} -1 \\ 1 \\ 1 \end{pmatrix}$ and $\nu(A) = 1$. ■

EXAMPLE 2 Let $A = \begin{pmatrix} 2 & -1 & 3 \\ 4 & -2 & 6 \\ -6 & 3 & -9 \end{pmatrix}$. Then, by Example 9.4.13, $\left\{\begin{pmatrix} 1 \\ 2 \\ 0 \end{pmatrix}, \begin{pmatrix} 0 \\ 3 \\ 1 \end{pmatrix}\right\}$ is a basis for N_A and $\nu(A) = 2$. ■

Let A be an $m \times n$ matrix and let

$$R_A = \{\mathbf{y} \in \mathbb{R}^m\colon A\mathbf{x} = \mathbf{y} \text{ for some } \mathbf{x} \in \mathbb{R}^n\}. \tag{2}$$

Then R_A is a subspace of $\mathbb{R}^m$ since $R_A = \text{range } T_A$, where $T_A\mathbf{x} = A\mathbf{x}$.

Definition 2 RANGE AND RANK OF A MATRIX R_A is called the **range** of A and $\rho(A) = \dim R_A$ is called the **rank** of A.

Before giving examples, we shall give two definitions and a theorem that make the calculation of rank relatively easy.

Definition 3 ROW AND COLUMN SPACE OF A MATRIX If A is an $m \times n$ matrix, let $\{\mathbf{r}_1, \mathbf{r}_2, \ldots, \mathbf{r}_m\}$ denote the rows of A and $\{\mathbf{c}_1, \mathbf{c}_2, \ldots, \mathbf{c}_n\}$, the columns of A. Then we define

$$RS_A = \textbf{row space } \textit{of } A = \text{span}\,\{\mathbf{r}_1, \mathbf{r}_2, \ldots, \mathbf{r}_m\} \qquad \textbf{(3)}$$

and

$$CS_A = \textbf{column space } \textit{of } A = \text{span}\,\{\mathbf{c}_1, \mathbf{c}_2, \ldots, \mathbf{c}_n\}. \qquad \textbf{(4)}$$

NOTE. RS_A is a subspace of $\mathbb{R}^n$ and CS_A is a subspace of $\mathbb{R}^m$.

We have introduced a lot of notation in just two pages. Let us stop for a moment to illustrate these ideas with an example.

EXAMPLE 3 Let $A = \begin{pmatrix} 1 & 2 & -1 \\ 2 & -1 & 3 \end{pmatrix}$. A is a 2×3 matrix.

(a) *The kernel of* $A = N_A = \{\mathbf{x} \in \mathbb{R}^3\colon A\mathbf{x} = \mathbf{0}\}$. As we saw in Example 1, $N_A = \text{span}\left\{\begin{pmatrix} -1 \\ 1 \\ 1 \end{pmatrix}\right\}$.

(b) *The nullity of* $A = \nu(A) = \dim N_A = 1$.

(c) *The range of* $A = R_A = \{\mathbf{y} \in \mathbb{R}^2\colon A\mathbf{x} = \mathbf{y} \text{ for some } \mathbf{x} \in \mathbb{R}^3\}$. Let $\mathbf{y} = \begin{pmatrix} y_1 \\ y_2 \end{pmatrix}$ be in $\mathbb{R}^2$.

Then, if $\mathbf{y} \in R_A$, there is an $\mathbf{x} \in \mathbb{R}^3$ such that $A\mathbf{x} = \mathbf{y}$. Writing $\mathbf{x} = \begin{pmatrix} x_1 \\ x_2 \\ x_3 \end{pmatrix}$, we have

$$\begin{pmatrix} 1 & 2 & -1 \\ 2 & -1 & 3 \end{pmatrix}\begin{pmatrix} x_1 \\ x_2 \\ x_3 \end{pmatrix} = \begin{pmatrix} y_1 \\ y_2 \end{pmatrix}$$

or

$$\begin{aligned} x_1 + 2x_2 - x_3 &= y_1 \\ 2x_1 - x_2 + 3x_3 &= y_2. \end{aligned}$$

Row-reducing this system, we have

$$\left(\begin{array}{ccc|c} 1 & 2 & -1 & y_1 \\ 2 & -1 & 3 & y_2 \end{array}\right) \xrightarrow{A_{1,2}(-2)} \left(\begin{array}{ccc|c} 1 & 2 & -1 & y_1 \\ 0 & -5 & 5 & y_2 - 2y_1 \end{array}\right)$$

$$\xrightarrow{M_2(-\frac{1}{5})} \left(\begin{array}{ccc|c} 1 & 2 & -1 & y_1 \\ 0 & 1 & -1 & \dfrac{2y_1 - y_2}{5} \end{array}\right) \xrightarrow{A_{2,1}(-2)} \left(\begin{array}{ccc|c} 1 & 0 & 1 & \dfrac{y_1 + 2y_2}{5} \\ 0 & 1 & -1 & \dfrac{2y_1 - y_2}{5} \end{array}\right)$$

Thus if x_3 is chosen arbitrarily, we see that

$$x_1 = -x_3 + \frac{y_1 + 2y_2}{5} \quad \text{and} \quad x_2 = x_3 + \frac{2y_1 - y_2}{5}.$$

That is, for every $\mathbf{y} = \begin{pmatrix} y_1 \\ y_2 \end{pmatrix} \in \mathbb{R}^2$, there are an infinite number of vectors $\mathbf{x} \in \mathbb{R}^3$ such that $A\mathbf{x} = \mathbf{y}$. Thus $R_A = \mathbb{R}^2$. Note, for example, that if $\mathbf{y} = \begin{pmatrix} 2 \\ -3 \end{pmatrix}$, then, choosing $x_3 = 0$ (the simplest choice), we have

$$x_1 = \frac{2 + 2(-3)}{5} = -\tfrac{4}{5} \quad \text{and} \quad x_2 = \frac{2(2) - (-3)}{5} = \tfrac{7}{5}$$

and

$$A\mathbf{x} = \begin{pmatrix} 1 & 2 & -1 \\ 2 & -1 & 3 \end{pmatrix} \begin{pmatrix} -\frac{4}{5} \\ \frac{7}{5} \\ 0 \end{pmatrix} = \begin{pmatrix} \frac{10}{5} \\ -\frac{15}{5} \end{pmatrix} = \begin{pmatrix} 2 \\ -3 \end{pmatrix} = \mathbf{y}.$$

(d) *The rank of* $A = \rho(A) = \dim R_A = \dim \mathbb{R}^2 = 2$.

(e) *The row space of* $A = RS_A = \text{span}\,\{(1, 2, -1), (2, -1, 3)\}$. Since these two vectors are linearly independent, we see that S_A is a two-dimensional subspace of $\mathbb{R}^3$. From Example 9.4.10, we observe that RS_A is a plane passing through the origin.

(f) *The column space of* $A = CS_A = \text{span}\left\{\begin{pmatrix} 1 \\ 2 \end{pmatrix}, \begin{pmatrix} 2 \\ -1 \end{pmatrix}, \begin{pmatrix} -1 \\ 3 \end{pmatrix}\right\} = \mathbb{R}^2$ since $\begin{pmatrix} 1 \\ 2 \end{pmatrix}$ and $\begin{pmatrix} 2 \\ -1 \end{pmatrix}$, being linearly independent, comprise a basis for $\mathbb{R}^2$. ■

In Example 2 we may observe that $R_A = CS_A = \mathbb{R}^2$ and $\dim RS_A = \dim CS_A = \dim R_A = \rho(A) = 2$. This is no coincidence.

Theorem 1 If A is an $m \times n$ matrix, then

(i) $CS_A = R_A$[†]
(ii) $\dim RS_A = \dim CS_A = \dim R_A = \rho(A)$.

The proof of this theorem is not difficult, but it is quite long. For that reason, we omit it.[‡]

[†]Do not be confused by this notation. It indicates that the column space of A is the same as the range of A. It does *not* say that the column space of A equals the row space of A. This last statement is usually false. It is always false if A is not a square matrix because then CS_A, the column space of A, contains m-vectors and RS_A, the row space of A, contains n-vectors.

[‡]For a proof, see S. I. Grossman, *Elementary Linear Algebra*, second edition, (Wadsworth, Belmont, Ca., 1984), p. 251.

EXAMPLE 4 Find a basis for R_A and determine the rank of $A = \begin{pmatrix} 2 & -1 & 3 \\ 4 & -2 & 6 \\ -6 & 3 & -9 \end{pmatrix}$.

Solution. Since $\mathbf{c}_2 = -\frac{1}{2}\mathbf{c}_1$ and $\mathbf{c}_3 = \frac{3}{2}\mathbf{c}_1$, we see that $\rho(A) = 1$. Thus any column in CS_A is a basis for CS_A $(=R_A)$. For example, $\begin{pmatrix} 2 \\ 4 \\ -6 \end{pmatrix}$ is a basis for R_A. ■

The following theorem will simplify our computations.

Theorem 2 If A is row (or column) equivalent to B, then $\rho(A) = \rho(B)$ and $\nu(A) = \nu(B)$.

Proof. Recall from Definition 7.5.3, page 437, that A is row equivalent to B if A can be "reduced" to B by elementary row operations. The definition for "column equivalent" is similar. Now interchanging rows of A leaves the same number of linearly independent rows as does multiplying any row by a nonzero constant. Suppose that $\mathbf{r}_1, \mathbf{r}_2, \ldots, \mathbf{r}_m$ are linearly independent. Consider the set $S = \{\mathbf{r}_1, \mathbf{r}_2, \ldots, \mathbf{r}_{i-1}, \mathbf{r}_i + \alpha\mathbf{r}_j, \mathbf{r}_{i+1}, \ldots, \mathbf{r}_j, \ldots, \mathbf{r}_m\}$. Suppose that $c_1\mathbf{r}_1 + c_2\mathbf{r}_2 + \cdots + c_{i-1}\mathbf{r}_{i-1} + c_i(\mathbf{r}_i + \alpha\mathbf{r}_j) + c_{i+1}\mathbf{r}_{i+1} + \cdots + c_j\mathbf{r}_j + \cdots + c_m\mathbf{r}_m = \mathbf{0}$. Then

$$c_1\mathbf{r}_1 + c_2\mathbf{r}_2 + \cdots + c_{i-1}\mathbf{r}_{i-1} + c_i\mathbf{r}_i + c_{i+1}\mathbf{r}_{i+1} + \cdots + (\alpha c_i + c_j)\mathbf{r}_j + \cdots + c_m\mathbf{r}_m = \mathbf{0}.$$

By independence, we have $c_1 = c_2 = \cdots = c_{i-1} = c_i = c_{i+1} = \cdots = \alpha c_i + c_j = \cdots = c_m = 0$. Moreover, $-c_j = \alpha c_i = \alpha 0 = 0$. This implies that S is a linearly independent set. Therefore we have shown that if we add a constant multiple of one row to another, we do not change the rank. Thus $\rho(A) = \rho(B)$. The other part is easier. The equation $A\mathbf{x} = \mathbf{0}$ is a homogeneous system of m equations in n unknowns. If B is obtained from A by row reduction, then, as we saw in Section 7.3, the solutions to the system are unchanged. Thus if $A\mathbf{x}_1 = \mathbf{0}$, then $B\mathbf{x}_1 = \mathbf{0}$ and vice versa. This means that $N_A = N_B$ so that $\nu(A) = \nu(B)$. ■

EXAMPLE 5 Determine the rank of $A = \begin{pmatrix} 1 & -1 & 3 \\ 2 & 0 & 4 \\ -1 & -3 & 1 \end{pmatrix}$.

Solution. We row reduce to obtain a simpler matrix:

$$\begin{pmatrix} 1 & -1 & 3 \\ 2 & 0 & 4 \\ -1 & -3 & 1 \end{pmatrix} \xrightarrow[A_{1,3}(1)]{A_{1,2}(-2)} \begin{pmatrix} 1 & -1 & 3 \\ 0 & 2 & -2 \\ 0 & -4 & 4 \end{pmatrix}$$

$$\xrightarrow{M_2(\frac{1}{2})} \begin{pmatrix} 1 & -1 & 3 \\ 0 & 1 & -1 \\ 0 & -4 & 4 \end{pmatrix} \xrightarrow{A_{2,3}(4)} \begin{pmatrix} 1 & -1 & 3 \\ 0 & 1 & -1 \\ 0 & 0 & 0 \end{pmatrix} = B$$

Since B has two independent rows, we have $\rho(B) = \rho(A) = 2$. ■

The next theorem gives the relationship between rank and nullity.

Theorem 3 Let A be an $m \times n$ matrix. Then

$$\rho(A) + \nu(A) = n. \tag{5}$$

Proof. We assume that $k = \rho(A)$ and that the first k columns of A are linearly independent. Let $\mathbf{c}_i$ $(i > k)$ denote any other column of A. Since $\mathbf{c}_1, \mathbf{c}_2, \ldots, \mathbf{c}_k$ form a basis for CS_A, we have, for some scalars $a_1, a_2, \ldots, a_k$,

$$\mathbf{c}_i = a_1\mathbf{c}_1 + a_2\mathbf{c}_2 + \cdots + a_k\mathbf{c}_k. \tag{6}$$

Thus, by adding $-a_1\mathbf{c}_1, -a_2\mathbf{c}_2, \ldots, -a_k\mathbf{c}_k$ successively to the ith column of A, we obtain a new $m \times n$ matrix B with $\rho(B) = \rho(A)$ and $\nu(B) = \nu(A)$ with the ith column of $B = \mathbf{0}$. We do this to all other columns of A (except the first k) to obtain the matrix

$$D = \begin{pmatrix} a_{11} & a_{12} & \cdots & a_{1k} & 0 & 0 & \cdots & 0 \\ a_{21} & a_{22} & \cdots & a_{2k} & 0 & 0 & \cdots & 0 \\ \vdots & \vdots & & \vdots & \vdots & \vdots & & \vdots \\ a_{m1} & a_{m2} & \cdots & a_{mk} & 0 & 0 & \cdots & 0 \end{pmatrix} \tag{7}$$

where $\rho(D) = \rho(A)$ and $\nu(D) = \nu(A)$. By possibly rearranging the rows of D we can assume that the first k rows of D are independent. Then we do the same thing to the rows, (i.e., add multiples of the first k rows to the last $m - k$ rows) to obtain a new matrix:

$$F = \begin{pmatrix} a_{11} & a_{12} & \cdots & a_{1k} & 0 & \cdots & 0 \\ a_{21} & a_{22} & \cdots & a_{2k} & 0 & \cdots & 0 \\ \vdots & \vdots & & \vdots & \vdots & & \vdots \\ a_{k1} & a_{k2} & \cdots & a_{kk} & 0 & \cdots & 0 \\ 0 & 0 & \cdots & 0 & 0 & \cdots & 0 \\ \vdots & \vdots & & \vdots & \vdots & & \vdots \\ 0 & 0 & \cdots & 0 & 0 & \cdots & 0 \end{pmatrix}$$

where $\rho(F) = \rho(A)$ and $\nu(F) = \nu(A)$. It is now obvious that if $i > k$, then $F\mathbf{e}_i = \mathbf{0}$,[†] so $E_k = \{\mathbf{e}_{k+1}, \mathbf{e}_{k+2}, \ldots, \mathbf{e}_n\}$ is a linearly independent set of $n - k$ vectors in N_F. We now show that E_k spans N_F. Let the vector $\mathbf{x} \in N_F$ have the form

[†] Recall that $\mathbf{e}_i$ is the vector with a 1 in the ith position and a zero everywhere else.

$$\mathbf{x} = \begin{pmatrix} x_1 \\ x_2 \\ \vdots \\ x_k \\ \vdots \\ x_n \end{pmatrix}.$$

Then

$$\mathbf{0} = F\mathbf{x} = \begin{pmatrix} a_{11}x_1 + a_{12}x_2 + \cdots + a_{1k}x_k \\ a_{21}x_1 + a_{22}x_2 + \cdots + a_{2k}x_k \\ \vdots \qquad \vdots \qquad\qquad \vdots \\ a_{k1}x_1 + a_{k2}x_2 + \cdots + a_{kk}x_k \\ 0 \\ \vdots \\ 0 \end{pmatrix} = \begin{pmatrix} 0 \\ 0 \\ \vdots \\ 0 \end{pmatrix}.$$

The determinant of the matrix of the $k \times k$ homogeneous system described above is nonzero, since the rows of this matrix are linearly independent. Thus all the solutions to the system have $x_1 = x_2 = \cdots = x_k = 0$. Thus $\mathbf{x}$ has the form

$$(0, 0, \ldots, 0, x_{k+1}, x_{k+2}, \ldots, x_n) = x_{k+1}\mathbf{e}_{k+1} + x_{k+2}\mathbf{e}_{k+2} + \cdots + x_n\mathbf{e}_n$$

This means that E_k spans N_F so that $\nu(F) = n - k = n - \rho(F)$. This completes the proof. ■

EXAMPLE 6 For $A = \begin{pmatrix} 1 & 2 & -1 \\ 2 & -1 & 3 \end{pmatrix}$ we calculated (in Examples 1 and 3) that $\rho(A) = 2$ and $\nu(A) = 1$; this illustrates that $\rho(A) + \nu(A) = n\ (=3)$. ■

EXAMPLE 7 For $A = \begin{pmatrix} 1 & -1 & 3 \\ 2 & 0 & 4 \\ -1 & -3 & 1 \end{pmatrix}$ calculate $\nu(A)$.

Solution. In Example 5 we found that $\rho(A) = 2$. Thus $\nu(A) = 3 - 2 = 1$. ■

We next show how the notion of rank can be used to obtain information about solutions to a linear system of equations. Again we consider the system of m equations in n unknowns

$$\begin{aligned} a_{11}x_1 + a_{12}x_2 + \cdots + a_{1n}x_n &= b_1 \\ a_{21}x_1 + a_{22}x_2 + \cdots + a_{2n}x_n &= b_2 \\ \vdots \qquad \vdots \qquad\qquad \vdots \quad &\quad \vdots \\ a_{m1}x_1 + a_{m2}x_2 + \cdots + a_{mn}x_n &= b_m. \end{aligned} \tag{8}$$

which we write as $A\mathbf{x} = \mathbf{b}$. We use the symbol $(A, \mathbf{b})$ to denote the $m \times (n + 1)$

augmented matrix obtained (as in Section 7.3) by adjoining the vector $\mathbf{b}$ to A.

Theorem 4 The system $A\mathbf{x} = \mathbf{b}$ has at least one solution if and only if A and the augmented matrix $(A, \mathbf{b})$ have the same rank.

Proof. If $\mathbf{c}_1, \mathbf{c}_2, \ldots, \mathbf{c}_n$ are the columns of A, then we can write system (8) as

$$x_1\mathbf{c}_1 + x_2\mathbf{c}_2 + \cdots + x_n\mathbf{c}_n = \mathbf{b}. \tag{9}$$

System (9) has a solution if and only if $\mathbf{b}$ can be written as a linear combination of the columns of A. That is, to have a solution we must have $\mathbf{b} \in CS_A$. If $\mathbf{b} \in CS_A$, then $(A, \mathbf{b})$ has the same number of linearly independent columns as A so that A and $(A, \mathbf{b})$ have the same rank. If $\mathbf{b} \notin C_A$, then $\rho(A, \mathbf{b}) = \rho(A) + 1$ and the system has no solutions. This completes the proof. ■

EXAMPLE 8 Determine whether the system

$$\begin{aligned} 2x_1 + 4x_2 + 6x_3 &= 18 \\ 4x_1 + 5x_2 + 6x_3 &= 24 \\ 2x_1 + 7x_2 + 12x_3 &= 40 \end{aligned}$$

has solutions.

Solution. Let $A = \begin{pmatrix} 2 & 4 & 6 \\ 4 & 5 & 6 \\ 2 & 7 & 12 \end{pmatrix}$. Then we row reduce to obtain, successively,

$$\xrightarrow{M_1(\frac{1}{2})} \begin{pmatrix} 1 & 2 & 3 \\ 4 & 5 & 6 \\ 2 & 7 & 12 \end{pmatrix} \xrightarrow[A_{1,3}(-2)]{A_{1,2}(-4)} \begin{pmatrix} 1 & 2 & 3 \\ 0 & -3 & -6 \\ 0 & 3 & 6 \end{pmatrix}$$

$$\xrightarrow{M_2(-\frac{1}{3})} \begin{pmatrix} 1 & 2 & 3 \\ 0 & 1 & 2 \\ 0 & 3 & 6 \end{pmatrix} \xrightarrow[A_{2,3}(-3)]{A_{2,1}(-2)} \begin{pmatrix} 1 & 0 & -1 \\ 0 & 1 & 2 \\ 0 & 0 & 0 \end{pmatrix}.$$

Thus $\rho(A) = 2$. Similarly, we row reduce $(A, \mathbf{b})$ to obtain

$$\left(\begin{array}{ccc|c} 2 & 4 & 6 & 18 \\ 4 & 5 & 6 & 24 \\ 2 & 7 & 12 & 40 \end{array}\right) \xrightarrow{M_1(\frac{1}{2})} \left(\begin{array}{ccc|c} 1 & 2 & 3 & 9 \\ 4 & 5 & 6 & 24 \\ 2 & 7 & 12 & 40 \end{array}\right)$$

$$\xrightarrow[A_{1,3}(-2)]{A_{1,2}(-4)} \left(\begin{array}{ccc|c} 1 & 2 & 3 & 9 \\ 0 & -3 & -6 & -12 \\ 0 & 3 & 6 & 22 \end{array}\right)$$

$$\xrightarrow{M_2(-\frac{1}{3})} \left(\begin{array}{ccc|c} 1 & 2 & 3 & 9 \\ 0 & 1 & 2 & 4 \\ 0 & 3 & 6 & 22 \end{array}\right) \xrightarrow[A_{2,3}(-3)]{A_{2,1}(-2)} \left(\begin{array}{ccc|c} 1 & 0 & -1 & 1 \\ 0 & 1 & 2 & 4 \\ 0 & 0 & 0 & 10 \end{array}\right).$$

It is easy to see that the last three columns of the last matrix are linearly independent. Thus $\rho(A, \mathbf{b}) = 3$ and there are no solutions to the system. ■

EXAMPLE 9 Determine whether the system

$$\begin{aligned} x_1 - x_2 + 2x_3 &= 4 \\ 2x_1 + x_2 - 3x_3 &= -2 \\ 4x_1 - x_2 + x_3 &= 6 \end{aligned}$$

has solutions.

Solution. Let $A = \begin{pmatrix} 1 & -1 & 2 \\ 2 & 1 & -3 \\ 4 & -1 & 1 \end{pmatrix}$. Then $\det A = 0$, so $\rho(A) < 3$. Since the first column is not a multiple of the second, we see that the first two columns are linearly independent; hence $\rho(A) = 2$. To compute $\rho(A, \mathbf{b})$, we row reduce:

$$\left(\begin{array}{rrr|r} 1 & -1 & 2 & 4 \\ 2 & 1 & -3 & -2 \\ 4 & -1 & 1 & 6 \end{array}\right) \xrightarrow[A_{1,3}(-4)]{A_{1,2}(-2)} \left(\begin{array}{rrr|r} 1 & -1 & 2 & 4 \\ 0 & 3 & -7 & -10 \\ 0 & 3 & -7 & -10 \end{array}\right)$$

We see that $\rho(A, \mathbf{b}) = 2$ and there are an infinite number of solutions to the system. (If there were a unique solution, we would have $\det A \neq 0$. ■

PROBLEMS 9.8

In Problems 1–15 find the rank and nullity of the given matrix.

1. $\begin{pmatrix} 1 & 2 \\ 3 & 4 \end{pmatrix}$

2. $\begin{pmatrix} 1 & -1 & 2 \\ 3 & 1 & 0 \end{pmatrix}$

3. $\begin{pmatrix} -1 & 3 & 2 \\ 2 & -6 & -4 \end{pmatrix}$

4. $\begin{pmatrix} 1 & -1 & 2 \\ 3 & 1 & 4 \\ -1 & 0 & 4 \end{pmatrix}$

5. $\begin{pmatrix} 1 & -1 & 2 \\ 3 & 1 & 4 \\ 5 & -1 & 8 \end{pmatrix}$

6. $\begin{pmatrix} -1 & 2 & 1 \\ 2 & -4 & -2 \\ -3 & 6 & 3 \end{pmatrix}$

7. $\begin{pmatrix} 1 & -1 & 2 & 3 \\ 0 & 1 & 4 & 3 \\ 1 & 0 & 6 & 6 \end{pmatrix}$

8. $\begin{pmatrix} 1 & -1 & 2 & 3 \\ 0 & 1 & 4 & 3 \\ 1 & 0 & 6 & 5 \end{pmatrix}$

9. $\begin{pmatrix} 2 & 3 \\ -1 & 1 \\ 4 & 7 \end{pmatrix}$

10. $\begin{pmatrix} 1 & -1 & 2 & 3 \\ 0 & 1 & 0 & 1 \\ 1 & 0 & 1 & 0 \\ 0 & 0 & 0 & 1 \end{pmatrix}$

11. $\begin{pmatrix} 1 & -1 & 2 & 1 \\ -1 & 0 & 1 & 2 \\ 1 & -2 & 5 & 4 \\ 2 & -1 & 1 & -1 \end{pmatrix}$

12. $\begin{pmatrix} 1 & -1 & 2 & 3 \\ -2 & 2 & -4 & -6 \\ 2 & -2 & 4 & 6 \\ 3 & -3 & 6 & 9 \end{pmatrix}$

13. $\begin{pmatrix} -1 & -1 & 0 & 0 \\ 0 & 0 & 2 & 3 \\ 4 & 0 & -2 & 1 \\ 3 & -1 & 0 & 4 \end{pmatrix}$

14. $\begin{pmatrix} 3 & 0 & 0 \\ 0 & 0 & 0 \\ 0 & 0 & 6 \end{pmatrix}$

15. $\begin{pmatrix} 1 & 2 & 3 \\ 0 & 0 & 4 \\ 0 & 0 & 6 \end{pmatrix}$

In Problems 16–22 find a basis for the range and kernel of the indicated matrix.

16. The matrix of Problem 2
17. The matrix of Problem 5
18. The matrix of Problem 6
19. The matrix of Problem 8

20. The matrix of Problem 11

21. The matrix of Problem 12

22. The matrix of Problem 13

In Problems 23–26 use Theorem 4 to determine whether the given system has any solutions.

23.
$$\begin{aligned} x_1 + x_2 - x_3 &= 7 \\ 4x_1 - x_2 + 5x_3 &= 4 \\ 6x_1 + x_2 + 3x_3 &= 20 \end{aligned}$$

24.
$$\begin{aligned} x_1 + x_2 - x_3 &= 7 \\ 4x_1 - x_2 + 5x_3 &= 4 \\ 6x_1 + x_2 + 3x_3 &= 18 \end{aligned}$$

25.
$$\begin{aligned} x_1 - 2x_2 + x_3 + x_4 &= 2 \\ 3x_1 \qquad + 2x_3 - 2x_4 &= -8 \\ 4x_2 - x_3 - x_4 &= 1 \\ 5x_1 \qquad + 3x_3 - x_4 &= -3 \end{aligned}$$

26.
$$\begin{aligned} x_1 - 2x_2 + x_3 + x_4 &= 2 \\ 3x_1 \qquad + 2x_3 - 2x_4 &= -8 \\ 4x_2 - x_3 - x_4 &= 1 \\ 5x_1 \qquad + 3x_3 - x_4 &= 0 \end{aligned}$$

27. Show that the rank of a diagonal matrix is equal to the number of nonzero components on the diagonal.

28. Let A be an upper triangular $n \times n$ matrix with zeros on the diagonal. Show that $\rho(A) < n$.

29. Show that for any matrix A, $\rho(A) = \rho(A^t)$.

30. Show that if A is an $m \times n$ matrix and $m < n$, then **(a)** $\rho(A) \leq m$ and **(b)** $\nu(A) \geq n - m$.

31. Let A be an $m \times n$ matrix and let B and C be invertible $m \times m$ and $n \times n$ matrices, respectively. Prove that $\rho(A) = \rho(BA) = \rho(AC)$. That is, multiplying a matrix by an invertible matrix does not change its rank.

32. Let A and B be $m \times n$ and $n \times p$ matrices, respectively. Show that $\rho(AB) \leq \min(\rho(A), \rho(B))$.

33. Let A be a 5×7 matrix with rank 5. Show that the linear system $A\mathbf{x} = \mathbf{b}$ has at least one solution for every 5-vector $\mathbf{b}$.

***34.** Let A and B be $m \times n$ matrices. Show that if $\rho(A) = \rho(B)$, then there exist invertible matrices C and D such that $B = CAD$.

35. If $B = CAD$, where C and D are invertible, prove that $\rho(A) = \rho(B)$.

36. Suppose that any k rows of A are linearly independent while any $k + 1$ rows of A are linearly dependent. Show that $\rho(A) = k$.

37. If A is an $n \times n$ matrix, show that $\rho(A) < n$ if and only if there is a vector $\mathbf{x} \in \mathbb{R}^n$ such that $\mathbf{x} \neq \mathbf{0}$ and $A\mathbf{x} = \mathbf{0}$.

38. Let A be an $m \times n$ matrix. Suppose that for every $\mathbf{y} \in \mathbb{R}^m$ there is an $\mathbf{x} \in \mathbb{R}^n$ such that $A\mathbf{x} = \mathbf{y}$. Show that $\rho(A) = m$.

9.9 THE MATRIX REPRESENTATION OF A LINEAR TRANSFORMATION

If A is an $m \times n$ matrix and $T: \mathbb{R}^n \to \mathbb{R}^m$ is defined by $T\mathbf{x} = A\mathbf{x}$ then, as we saw in Example 9.6.7, T is a linear transformation. We shall now see that for *every* linear transformation from $\mathbb{R}^n$ into $\mathbb{R}^m$, there exists an $m \times n$ matrix A such that $T\mathbf{x} = A\mathbf{x}$ for every $\mathbf{x} \in \mathbb{R}^n$. This fact is extremely useful. As we saw in Definitions 9.8.1 and 9.8.2, if $T\mathbf{x} = A\mathbf{x}$, then $\ker T = N_A$ and range $T = R_A$. Moreover, $\nu(T) = \dim \ker T = \nu(A)$ and $\rho(T) = \dim \text{range } T = \rho(A)$. Thus we can determine the kernel, range, nullity, and rank of a linear transformation from $\mathbb{R}^n \to \mathbb{R}^m$ by determining the kernel and range space of a corresponding matrix. Moreover, once we know that $T\mathbf{x} = A\mathbf{x}$, we can evaluate $T\mathbf{x}$ for any $\mathbf{x}$ in $\mathbb{R}^n$ by simple matrix multiplication.

But this is not all. As we shall see, any linear transformation between finite dimensional vector spaces can be represented by a matrix.

Theorem 1 Let $T: \mathbb{R}^n \to \mathbb{R}^m$ be a linear transformation. Then there exists a unique $m \times n$ matrix A_T such that

$$T\mathbf{x} = A_T\mathbf{x} \quad \text{for every } \mathbf{x} \in \mathbb{R}^n. \tag{1}$$

Proof. Let $\mathbf{w}_1 = T\mathbf{e}_1, \mathbf{w}_2 = T\mathbf{e}_2, \ldots, \mathbf{w}_n = T\mathbf{e}_n$. Let A_T be the matrix whose columns are $\mathbf{w}_1, \mathbf{w}_2, \ldots, \mathbf{w}_n$. If

$$\mathbf{w}_i = \begin{pmatrix} a_{1i} \\ a_{2i} \\ \vdots \\ a_{mi} \end{pmatrix} \qquad \text{for } i = 1, 2, \ldots, n$$

then

$$A_T\mathbf{e}_i = \begin{pmatrix} a_{11} & a_{12} & \cdots & a_{1i} & \cdots & a_{1n} \\ a_{21} & a_{22} & \cdots & a_{2i} & \cdots & a_{2n} \\ \vdots & \vdots & & \vdots & & \vdots \\ a_{m1} & a_{m2} & & a_{mi} & & a_{mn} \end{pmatrix} \begin{pmatrix} 0 \\ 0 \\ \vdots \\ 1 \\ 0 \\ \vdots \\ 0 \end{pmatrix} = \begin{pmatrix} a_{1i} \\ a_{2i} \\ \vdots \\ a_{mi} \end{pmatrix} = \mathbf{w}_i.$$

(1 in the ith position)

Thus $A_T\mathbf{e}_i = \mathbf{w}_i$ for $i = 1, 2, \ldots, n$. If $\mathbf{x} \in \mathbb{R}^n$, there exists a unique set of real numbers $c_1, c_2, \ldots, c_n$ such that $\mathbf{x} = c_1\mathbf{e}_1 + c_2\mathbf{e}_2 + \cdots + c_n\mathbf{e}_n$. Then $T\mathbf{x} = c_1T\mathbf{e}_1 + c_2T\mathbf{e}_2 + \cdots + c_nT\mathbf{e}_n = c_1\mathbf{w}_1 + c_2\mathbf{w}_2 + \cdots + c_n\mathbf{w}_n$. But $A_T\mathbf{x} = A_T(c_1\mathbf{e}_1 + c_2\mathbf{e}_2 + \cdots + c_n\mathbf{e}_n) = c_1A_T\mathbf{e}_1 + c_2A_T\mathbf{e}_2 + \cdots + c_nA_T\mathbf{e}_n = c_1\mathbf{w}_1 + c_2\mathbf{w}_2 + \cdots + c_n\mathbf{w}_n$. Thus $T\mathbf{x} = A_T\mathbf{x}$. We can now show that A_T is unique. Suppose that $T\mathbf{x} = A_T\mathbf{x}$ and $T\mathbf{x} = B_T\mathbf{x}$ for every $\mathbf{x} \in \mathbb{R}^n$. Then $A_T\mathbf{x} = B_T\mathbf{x}$ or, setting $C_T = A_T - B_T$, we have $C_T\mathbf{x} = \mathbf{0}$ for every $\mathbf{x} \in \mathbb{R}^n$. In particular, $C_T\mathbf{e}_i = \mathbf{0}$ for $i = 1, 2, \ldots, n$. But, as we see from the proof of the first part of the theorem, $C_T\mathbf{e}_i$ is the ith column of C_T. Thus each of the n columns of C_T is the m-zero vector and $C_T = 0$, the $m \times n$ zero matrix. This shows that $A_T = B_T$ and the theorem is proved. ■

REMARK 1. In this theorem we assumed that every vector in $\mathbb{R}^n$ and $\mathbb{R}^m$ is written in terms of the standard basis vectors in those spaces. If we choose other bases for $\mathbb{R}^n$ and $\mathbb{R}^m$ we shall, of course, get a different matrix A_T. See, for instance, Example 9.5.1.

REMARK 2. The proof of the theorem shows us that A_T is easily obtained as the matrix whose columns are the vectors $T\mathbf{e}_i$.

Definition 1 TRANSFORMATION MATRIX The matrix A_T in Theorem 1 is called the **transformation matrix** corresponding to T.

In Section 9.7 we defined the range, rank, kernel, and nullity of a linear transformation. In Section 9.8 we defined the range, rank, kernel, and nullity of a matrix. The proof of the following theorem follows easily from Theorem 1 and is left as an exercise (see Problem 35).

Theorem 2 Let A_T be the transformation matrix corresponding to the linear transformation $T: \mathbb{R}^n \to \mathbb{R}^m$. Then

(i) range $T = R_{A_T} = CS_{A_T}$
(ii) $\rho(T) = \rho(A_T)$
(iii) ker $T = N_{A_T}$
(iv) $\nu(T) = \nu(A_T)$

EXAMPLE 1 Find the transformation matrix A_T corresponding to the projection of a vector in $\mathbb{R}^3$ onto the xy-plane.

Solution. Here $T\begin{pmatrix} x \\ y \\ z \end{pmatrix} = \begin{pmatrix} x \\ y \\ 0 \end{pmatrix}$. In particular, $T\begin{pmatrix} 1 \\ 0 \\ 0 \end{pmatrix} = \begin{pmatrix} 1 \\ 0 \\ 0 \end{pmatrix}$, $T\begin{pmatrix} 0 \\ 1 \\ 0 \end{pmatrix} = \begin{pmatrix} 0 \\ 1 \\ 0 \end{pmatrix}$, and $T\begin{pmatrix} 0 \\ 0 \\ 1 \end{pmatrix} = \begin{pmatrix} 0 \\ 0 \\ 0 \end{pmatrix}$. Thus $A_T = \begin{pmatrix} 1 & 0 & 0 \\ 0 & 1 & 0 \\ 0 & 0 & 0 \end{pmatrix}$. Note that $A_T\begin{pmatrix} x \\ y \\ z \end{pmatrix} = \begin{pmatrix} 1 & 0 & 0 \\ 0 & 1 & 0 \\ 0 & 0 & 0 \end{pmatrix}\begin{pmatrix} x \\ y \\ z \end{pmatrix} = \begin{pmatrix} x \\ y \\ 0 \end{pmatrix}$. ■

EXAMPLE 2 Let $T: \mathbb{R}^3 \to \mathbb{R}^4$ be defined by

$$T\begin{pmatrix} x \\ y \\ z \end{pmatrix} = \begin{pmatrix} x - y \\ y + z \\ 2x - y - z \\ -x + y + 2z \end{pmatrix}.$$

Find A_T, ker T, range T, $\nu(T)$, and $\rho(T)$.

Solution. $T\begin{pmatrix} 1 \\ 0 \\ 0 \end{pmatrix} = \begin{pmatrix} 1 \\ 0 \\ 2 \\ -1 \end{pmatrix}$, $T\begin{pmatrix} 0 \\ 1 \\ 0 \end{pmatrix} = \begin{pmatrix} -1 \\ 1 \\ -1 \\ 1 \end{pmatrix}$, and $T\begin{pmatrix} 0 \\ 0 \\ 1 \end{pmatrix} = \begin{pmatrix} 0 \\ 1 \\ -1 \\ 2 \end{pmatrix}$. Thus

$$A_T = \begin{pmatrix} 1 & -1 & 0 \\ 0 & 1 & 1 \\ 2 & -1 & -1 \\ -1 & 1 & 2 \end{pmatrix}.$$

Note (as a check) that

$$\begin{pmatrix} 1 & -1 & 0 \\ 0 & 1 & 1 \\ 2 & -1 & -1 \\ -1 & 1 & 2 \end{pmatrix}\begin{pmatrix} x \\ y \\ z \end{pmatrix} = \begin{pmatrix} x - y \\ x + z \\ 2x - y - z \\ -x + y + 2z \end{pmatrix}.$$

Next we compute the kernel and range of A. Row-reducing, we obtain

$$\begin{pmatrix} 1 & -1 & 0 \\ 0 & 1 & 1 \\ 2 & -1 & -1 \\ -1 & 1 & 2 \end{pmatrix} \xrightarrow[A_{1,4}(1)]{A_{1,3}(-2)} \begin{pmatrix} 1 & -1 & 0 \\ 0 & 1 & 1 \\ 0 & 1 & -1 \\ 0 & 0 & 2 \end{pmatrix}$$

$$\xrightarrow[A_{2,3}(-1)]{A_{2,1}(1)} \begin{pmatrix} 1 & 0 & 1 \\ 0 & 1 & 1 \\ 0 & 0 & -2 \\ 0 & 0 & 2 \end{pmatrix} \xrightarrow{A_{3,4}(1)} \begin{pmatrix} 1 & 0 & 1 \\ 0 & 1 & 1 \\ 0 & 0 & -2 \\ 0 & 0 & 0 \end{pmatrix}$$

Thus $\rho(A) = 3$ and $\nu(A) \overset{\text{since } \rho(A)+\nu(A)=3}{=} 3 - 3 = 0$. This means that $\ker T = \{\mathbf{0}\}$, range $T = \text{span}\left\{\begin{pmatrix} 1 \\ 0 \\ 2 \\ -1 \end{pmatrix}, \begin{pmatrix} -1 \\ 1 \\ -1 \\ 1 \end{pmatrix}, \begin{pmatrix} 0 \\ 1 \\ -1 \\ 2 \end{pmatrix}\right\}$, $\nu(T) = 0$, and $\rho(T) = 3$. ■

EXAMPLE 3 Let $T: \mathbb{R}^3 \to \mathbb{R}^3$ be defined by $T\begin{pmatrix} x \\ y \\ z \end{pmatrix} = \begin{pmatrix} 2x - y + 3z \\ 4x - 2y + 6z \\ -6x + 3y - 9z \end{pmatrix}$. Find A_T, $\ker T$, range T, $\nu(T)$, and $\rho(T)$.

Solution. Since $T\begin{pmatrix} 1 \\ 0 \\ 0 \end{pmatrix} = \begin{pmatrix} 2 \\ 4 \\ -6 \end{pmatrix}$, $T\begin{pmatrix} 0 \\ 1 \\ 0 \end{pmatrix} = \begin{pmatrix} -1 \\ -2 \\ 3 \end{pmatrix}$, and $T\begin{pmatrix} 0 \\ 0 \\ 1 \end{pmatrix} = \begin{pmatrix} 3 \\ 6 \\ -9 \end{pmatrix}$, we have

$$A_T = \begin{pmatrix} 2 & -1 & 3 \\ 4 & -2 & 6 \\ -6 & 3 & -9 \end{pmatrix}.$$

From Example 9.8.4 we see that $\rho(A) \overset{\text{Theorem 2(ii)}}{=} \rho(T) = 1$ and range $T = \text{span}\left\{\begin{pmatrix} 2 \\ 4 \\ -6 \end{pmatrix}\right\}$. Then $\nu(T) = 2$. To find $N_A \overset{\text{Theorem 2(iii)}}{=} \ker T$, we row-reduce to solve the system $A\mathbf{x} = \mathbf{0}$:

$$\left(\begin{array}{ccc|c} 2 & -1 & 3 & 0 \\ 4 & -2 & 6 & 0 \\ -6 & 3 & -9 & 0 \end{array}\right) \xrightarrow[A_{1,3}(3)]{A_{1,2}(-2)} \left(\begin{array}{ccc|c} 2 & -1 & 3 & 0 \\ 0 & 0 & 0 & 0 \\ 0 & 0 & 0 & 0 \end{array}\right).$$

This means that $\begin{pmatrix} x \\ y \\ z \end{pmatrix} \in N_A$ if $2x - y + 3z = 0$ or $y = 2x + 3z$. First setting $x = 1$, $z = 0$ and then $x = 0$, $z = 1$, we obtain a basis for N_A: $\ker T = N_A = \text{span}\left\{\begin{pmatrix} 1 \\ 2 \\ 0 \end{pmatrix}, \begin{pmatrix} 0 \\ 3 \\ 1 \end{pmatrix}\right\}$. ■

EXAMPLE 4 It is easy to verify that if T is the zero transformation from $\mathbb{R}^n \to \mathbb{R}^m$, then A_T is the $m \times n$ zero matrix. Similarly, if T is the identity transformation from $\mathbb{R}^n \to \mathbb{R}^n$, then $A_T = I_n$. ■

EXAMPLE 5 We saw in Example 9.6.8 that if T is the function which rotates every vector in $\mathbb{R}^2$ through an angle of θ, then $A_T = \begin{pmatrix} \cos\theta & -\sin\theta \\ \sin\theta & \cos\theta \end{pmatrix}$. ■

We now generalize the notion of matrix representation to arbitrary finite dimensional vector spaces.

Theorem 3 Let V be a real n-dimensional vector space, W be a real m-dimensional vector space, and $T: V \to W$ be a linear transformation. Let $B_1 = \{\mathbf{v}_1, \mathbf{v}_2, \ldots, \mathbf{v}_n\}$ be a basis for V and let $B_2 = \{\mathbf{w}_1, \mathbf{w}_2, \ldots, \mathbf{w}_m\}$ be a basis for W. Then there is a unique $m \times n$ matrix A_T such that

$$(T\mathbf{x})_{B_2} = (A_T\mathbf{x})_{B_2} = A_T(\mathbf{x})_{B_1}. \tag{2}$$

REMARK 1. The notation in (2) is the notation of Section 9.5. If $\mathbf{x} \in V = c_1\mathbf{v}_1 + c_2\mathbf{v}_2 + \cdots + c_n\mathbf{v}_n$, then $(\mathbf{x})_{B_1} = \begin{pmatrix} c_1 \\ c_2 \\ \vdots \\ c_n \end{pmatrix}$. Let $\mathbf{c} = \begin{pmatrix} c_1 \\ c_2 \\ \vdots \\ c_n \end{pmatrix}$. Then $A_T\mathbf{c}$ is an m-vector that we denote $\mathbf{d} = \begin{pmatrix} d_1 \\ d_2 \\ \vdots \\ d_m \end{pmatrix}$. Equation (2) says that

$$(T\mathbf{x})_{B_2} = \begin{pmatrix} d_1 \\ d_2 \\ \vdots \\ d_m \end{pmatrix}.$$

That is,

$$T\mathbf{x} = d_1\mathbf{w}_1 + d_2\mathbf{w}_2 + \cdots + d_m\mathbf{w}_m.$$

REMARK 2. As in Theorem 1, the uniqueness of A_T is relative to the bases B_1 and B_2. If we change the bases, we change A_T.

Proof. Let $T\mathbf{v}_1 = \mathbf{y}_1, T\mathbf{v}_2 = \mathbf{y}_2, \ldots, T\mathbf{v}_n = \mathbf{y}_n$. Since $\mathbf{y}_i \in W$, we have, for $i = 1, 2, \ldots, n$,

$$\mathbf{y}_i = a_{1i}\mathbf{w}_1 + a_{2i}\mathbf{w}_2 + \cdots + a_{mi}\mathbf{w}_m$$

for some (unique) set of scalars $a_{1i}, a_{2i}, \ldots, a_{mi}$, and we write

$$(\mathbf{y}_1)_{B_2} = \begin{pmatrix} a_{11} \\ a_{21} \\ \vdots \\ a_{m1} \end{pmatrix}, \ (\mathbf{y}_2)_{B_2} = \begin{pmatrix} a_{12} \\ a_{22} \\ \vdots \\ a_{m2} \end{pmatrix}, \ldots, (\mathbf{y}_n)_{B_2} = \begin{pmatrix} a_{1n} \\ a_{2n} \\ \vdots \\ a_{mn} \end{pmatrix}.$$

We now define

$$A_T = \begin{pmatrix} a_{11} & a_{21} & \cdots & a_{1n} \\ a_{21} & a_{22} & \cdots & a_{2n} \\ \vdots & \vdots & & \vdots \\ a_{m1} & a_{m2} & \cdots & a_{mn} \end{pmatrix}.$$

Since

$$(\mathbf{v}_1)_{B_1} = \begin{pmatrix} 1 \\ 0 \\ \vdots \\ 0 \end{pmatrix}, \ (\mathbf{v}_2)_{B_1} = \begin{pmatrix} 0 \\ 1 \\ 0 \\ \vdots \\ 0 \end{pmatrix}, \ldots, (\mathbf{v}_n)_{B_1} = \begin{pmatrix} 0 \\ 0 \\ \vdots \\ 1 \end{pmatrix},$$

we have, as in the proof of Theorem 1,

$$A_T\mathbf{v}_i = \begin{pmatrix} a_{1i} \\ a_{2i} \\ \vdots \\ a_{mi} \end{pmatrix} = (\mathbf{y}_i)_{B_2}.$$

If $\mathbf{x}$ is in V, then

$$\mathbf{x} = c_1\mathbf{v}_1 + c_2\mathbf{v}_2 + \cdots + c_n\mathbf{v}_n,$$

$$(\mathbf{x})_{B_1} = \begin{pmatrix} c_1 \\ c_2 \\ \vdots \\ c_n \end{pmatrix},$$

and

$$(A_T(\mathbf{x})_{B_1})_{B_2} = \begin{pmatrix} a_{11} & a_{12} & \cdots & a_{1n} \\ a_{21} & a_{22} & \cdots & a_{2n} \\ \vdots & \vdots & & \vdots \\ a_{m1} & a_{m2} & \cdots & a_{mn} \end{pmatrix} \begin{pmatrix} c_1 \\ c_2 \\ \vdots \\ c_n \end{pmatrix}$$

$$= \begin{pmatrix} a_{11}c_1 + a_{12}c_2 + \cdots + a_{1n}c_n \\ a_{21}c_1 + a_{22}c_2 + \cdots + a_{2n}c_n \\ \vdots \quad\quad \vdots \quad\quad\quad \vdots \\ a_{m1}c_1 + a_{m2}c_2 + \cdots + a_{mn}c_n \end{pmatrix}$$

$$= c_1\begin{pmatrix} a_{11} \\ a_{21} \\ \vdots \\ a_{m1} \end{pmatrix} + c_2\begin{pmatrix} a_{12} \\ a_{22} \\ \vdots \\ a_{m2} \end{pmatrix} + \cdots + c_n\begin{pmatrix} a_{1n} \\ a_{2n} \\ \vdots \\ a_{mn} \end{pmatrix}$$

$$= c_1(\mathbf{y}_1)_{B_2} + c_2(\mathbf{y}_2)_{B_2} + \cdots + c_n(\mathbf{y}_n)_{B_2}.$$

Similarly, $T\mathbf{x} = T(c_1\mathbf{v}_1 + c_2\mathbf{v}_2 + \cdots + c_n\mathbf{v}_n) = c_1T\mathbf{v}_1 + c_2T\mathbf{v}_2 + \cdots + c_nT\mathbf{v}_n = c_1\mathbf{y}_1 + c_2\mathbf{y}_2 + \cdots + c_n\mathbf{y}_n$. Thus $(T\mathbf{x})_{B_2} = A_T(\mathbf{x})_{B_1}$. The proof of uniqueness is exactly as in the proof of uniqueness in Theorem 1. ■

The following useful result follows immediately from Theorem 9.8.4 and generalizes Theorem 2. Its proof is left as an exercise (see Problem 36).

Theorem 4 Let V and W be finite dimensional vector spaces with $\dim V = n$. Let $T: V \to W$ be a linear transformation and let A_T be a matrix representation of T. Then

(i) $\rho(T) = \rho(A_T)$
(ii) $\nu(T) = \nu(A_T)$
(iii) $\nu(T) + \rho(T) = n$.

EXAMPLE 6 Let $T: P_2 \to P_3$ be defined by $(Tp)(x) = xp(x)$. Find A_T, and use it to determine the kernel and range of T.

Solution. Using the standard basis $B_1 = \{1, x, x^2\}$ in P_2 and $B_2 = \{1, x, x^2, x^3\}$ in P_3, we have

$$T(1) = x = \begin{pmatrix} 0 \\ 1 \\ 0 \\ 0 \end{pmatrix}_{B_2}, \quad T(x) = x^2 = \begin{pmatrix} 0 \\ 0 \\ 1 \\ 0 \end{pmatrix}_{B_2}, \text{ and } T(x^2) = x^3 = \begin{pmatrix} 0 \\ 0 \\ 0 \\ 1 \end{pmatrix}_{B_2}.$$

Thus

$$A_T = \begin{pmatrix} 0 & 0 & 0 \\ 1 & 0 & 0 \\ 0 & 1 & 0 \\ 0 & 0 & 1 \end{pmatrix}.$$

Clearly $\rho(A) = 3$ and a basis for R_A is $\left\{\begin{pmatrix} 0 \\ 1 \\ 0 \\ 0 \end{pmatrix}, \begin{pmatrix} 0 \\ 0 \\ 1 \\ 0 \end{pmatrix}, \begin{pmatrix} 0 \\ 0 \\ 0 \\ 1 \end{pmatrix}\right\}$. Therefore, range $T =$ span $\{x, x^2, x^3\}$. Since $\nu(A) = 3 - \rho(A) = 0$, we see that $\ker T = \{0\}$. ■

EXAMPLE 7 Let $T: P_3 \to P_2$ be defined by $T(a_0 + a_1x + a_2x^2 + a_3x^3) = a_1 + a_2x^2$. Compute A_T, and use it to find the kernel and range of T.

Solution. Using the standard bases $B_1 = \{1, x, x^2, x^3\}$ in P_3 and $B_2 = \{1, x, x^2\}$ in P_2, we immediately see that $T(1) = \begin{pmatrix}0\\0\\0\end{pmatrix}_{B_2}$, $T(x) = \begin{pmatrix}1\\0\\0\end{pmatrix}_{B_2}$, $T(x^2) = \begin{pmatrix}0\\0\\1\end{pmatrix}_{B_2}$, and $T(x^3) = \begin{pmatrix}0\\0\\0\end{pmatrix}_{B_2}$. Thus $A_T = \begin{pmatrix}0&1&0&0\\0&0&0&0\\0&0&1&0\end{pmatrix}$. Clearly $\rho(A) = 2$ and a basis for R_A is $\left\{\begin{pmatrix}1\\0\\0\end{pmatrix}\begin{pmatrix}0\\0\\1\end{pmatrix}\right\}$ so that range T = span $\{1, x^2\}$. Then $\nu(A) = 4 - 2 = 2$; and if $A_T\begin{pmatrix}a_0\\a_1\\a_2\\a_3\end{pmatrix} = \begin{pmatrix}0\\0\\0\end{pmatrix}$, then $a_1 = 0$ and $a_2 = 0$. Hence a_0 and a_3 are arbitrary and a basis for N_A is $\left\{\begin{pmatrix}1\\0\\0\\0\end{pmatrix}, \begin{pmatrix}0\\0\\0\\1\end{pmatrix}\right\}$, so that a basis for ker T is $\{1, x^3\}$. ■

In all the examples of this section we have obtained the matrix A_T by using the standard basis in each vector space. However, Theorem 3 holds for any bases in V and W. You are asked to compute A for nonstandard bases in Problems 9, 10, 11, 12, and 20.

PROBLEMS 9.9

In Problems 1–30 find the matrix representation A_T of the linear transformation T, ker T, range T, $\nu(T)$, and $\rho(T)$. Unless otherwise stated, assume that B_1 and B_2 are standard bases.

1. $T: \mathbb{R}^2 \to \mathbb{R}^2$; $T\begin{pmatrix}x\\y\end{pmatrix} = \begin{pmatrix}x - 2y\\-x + y\end{pmatrix}$

2. $T: \mathbb{R}^2 \to \mathbb{R}^3$; $T\begin{pmatrix}x\\y\end{pmatrix} = \begin{pmatrix}x + y\\x - y\\2x + 3y\end{pmatrix}$

3. $T: \mathbb{R}^3 \to \mathbb{R}^2$; $T\begin{pmatrix}x\\y\\z\end{pmatrix} = \begin{pmatrix}x - y + z\\-2x + 2y - 2z\end{pmatrix}$

4. $T: \mathbb{R}^2 \to \mathbb{R}^2$; $T\begin{pmatrix}x\\y\end{pmatrix} = \begin{pmatrix}ax + by\\cx + dy\end{pmatrix}$

5. $T: \mathbb{R}^3 \to \mathbb{R}^3$; $T\begin{pmatrix}x\\y\\z\end{pmatrix} = \begin{pmatrix}x - y + 2z\\3x + y + 4z\\5x - y + 8z\end{pmatrix}$

6. $T: \mathbb{R}^3 \to \mathbb{R}^3$; $T\begin{pmatrix}x\\y\\z\end{pmatrix} = \begin{pmatrix}-x + 2y + z\\2x - 4y - 2z\\-3x + 6y + 3z\end{pmatrix}$

7. $T: \mathbb{R}^4 \to \mathbb{R}^3$; $T\begin{pmatrix}x\\y\\z\\w\end{pmatrix} = \begin{pmatrix}x - y + 2z + 3w\\y + 4z + 3w\\x + 6z + 6w\end{pmatrix}$

8. $T: \mathbb{R}^4 \to \mathbb{R}^4$; $T\begin{pmatrix}x\\y\\z\\w\end{pmatrix} = \begin{pmatrix}x - y + 2z + w\\-x + z + 2w\\x - 2y + 5z + 4w\\2x - y + z - w\end{pmatrix}$

* 9. $T: \mathbb{R}^2 \to \mathbb{R}^2;\ T\begin{pmatrix} x \\ y \end{pmatrix} = \begin{pmatrix} x - 2y \\ 2x + y \end{pmatrix};$

$B_1 = B_2 = \left\{\begin{pmatrix} 1 \\ -2 \end{pmatrix}, \begin{pmatrix} 3 \\ 2 \end{pmatrix}\right\}$

*10. $T: \mathbb{R}^2 \to \mathbb{R}^2;\ T\begin{pmatrix} x \\ y \end{pmatrix} = \begin{pmatrix} 4x - y \\ 3x + 2y \end{pmatrix};$

$B_1 = B_2 = \left\{\begin{pmatrix} -1 \\ 1 \end{pmatrix}, \begin{pmatrix} 4 \\ 3 \end{pmatrix}\right\}$

*11. $T: \mathbb{R}^3 \to \mathbb{R}^2;\ T\begin{pmatrix} x \\ y \\ z \end{pmatrix} = \begin{pmatrix} 2x + y + z \\ y - 3z \end{pmatrix};$

$B_1 = \left\{\begin{pmatrix} 1 \\ 0 \\ 1 \end{pmatrix}, \begin{pmatrix} 1 \\ 1 \\ 0 \end{pmatrix}, \begin{pmatrix} 1 \\ 1 \\ 1 \end{pmatrix}\right\};\ B_2 = \left\{\begin{pmatrix} 1 \\ -1 \end{pmatrix}, \begin{pmatrix} 2 \\ 3 \end{pmatrix}\right\}$

*12. $T: \mathbb{R}^2 \to \mathbb{R}^3;\ T\begin{pmatrix} x \\ y \end{pmatrix} = \begin{pmatrix} x - y \\ 2x + y \\ y \end{pmatrix};$

$B_1 = \left\{\begin{pmatrix} 2 \\ 1 \end{pmatrix}, \begin{pmatrix} 1 \\ 2 \end{pmatrix}\right\};\ B_2 = \left\{\begin{pmatrix} 1 \\ -1 \\ 0 \end{pmatrix}, \begin{pmatrix} 0 \\ 2 \\ 0 \end{pmatrix}, \begin{pmatrix} 0 \\ 2 \\ 5 \end{pmatrix}\right\}$

13. $T: P_2 \to P_3;\ T(a_0 + a_1x + a_2x^2) = a_1 - a_1x + a_0x^3$
14. $T: \mathbb{R} \to P_3;\ T(a) = a + ax + ax^2 + ax^3$
15. $T: P_3 \to \mathbb{R};\ T(a_0 + a_1x + a_2x^2 + a_3x^3) = a_2$
16. $T: P_3 \to P_1;\ T(a_0 + a_1x + a_2x^2 + a_3x^3) = (a_1 + a_3)x - a_2$
17. $T: P_3 \to P_2;\ T(a_0 + a_1x + a_2x^2 + a_3x^3) = (a_0 - a_1 + 2a_2 + 3a_3) + (a_1 + 4a_2 + 3a_3)x + (a_0 + 6a_2 + 5a_3)x^2$
18. $T: M_{22} \to M_{22};\ T\begin{pmatrix} a & b \\ c & d \end{pmatrix} = \begin{pmatrix} a - b + 2c + d & -a + c + 2d \\ a - 2b + 5c + 4d & 2a - b + c - d \end{pmatrix}$
19. $T: M_{22} \to M_{22};\ T\begin{pmatrix} a & b \\ c & d \end{pmatrix} = \begin{pmatrix} a + b + c + d & a + b + c \\ a + b & a \end{pmatrix}$

*20. $T: P_2 \to P_3;\ T[p(x)] = xp(x);\ B_1 = \{1, x, x^2\};$ $B_2 = \{1, (1 + x), (1 + x)^2, (1 + x)^3\}$
21. $D: P_4 \to P_3;\ Dp(x) = p'(x)$
22. $T: P_4 \to P_4;\ Tp(x) = xp'(x) - p(x)$
23. $D: P_n \to P_{n-1};\ Dp(x) = p'(x)$
24. $D: P_4 \to P_2;\ Dp(x) = p''(x)$

*25. $T: P_4 \to P_4;\ Tp(x) = p''(x) + xp'(x) + 2p(x)$
26. $D: P_n \to P_{n-k};\ Dp(x) = p^{(k)}(x)$

*27. $T: P_n \to P_n;\ Tp(x) = x^n p^{(n)}(x) + x^{n-1}p^{(n-1)}(x) + \cdots + xp'(x) + p(x)$
28. $J: P_n \to \mathbb{R};\ Jp = \int_0^1 p(x)\,dx$
29. $T: \mathbb{R}^3 \to P_2;\ T\begin{pmatrix} a \\ b \\ c \end{pmatrix} = a + bx + cx^2$
30. $T: P_3 \to \mathbb{R}^3;\ T(a_0 + a_1x + a_2x^2 + a_3x^3) = \begin{pmatrix} a_3 - a_2 \\ a_1 + a_3 \\ a_2 - a_1 \end{pmatrix}$
31. Let $T: M_{mn} \to M_{nm}$ be given by $TA = A^t$. Find A_T with respect to the standard bases in M_{mn} and M_{nm}.

*32. Let $T: \mathbb{C}^2 \to \mathbb{C}^2$ be given by $T\begin{pmatrix} x \\ y \end{pmatrix} = \begin{pmatrix} x + iy \\ (1 + i)y - x \end{pmatrix}$. Find A_T.
33. Let $V = \text{span}\,\{1, \sin x, \cos x\}$. Find A_D, where $D: V \to V$ is defined by $Df(x) = f'(x)$. Find range D and ker D.
34. Answer the questions of Problems 33 given $V = \text{span}\,\{e^x, xe^x, x^2e^x\}$.
35. Prove Theorem 2.
36. Prove Theorem 4.

9.10 EIGENVALUES AND EIGENVECTORS

In this section we introduce one of the central ideas in matrix theory: the notion of eigenvalues and eigenvectors. Eigenvalues are used in many applications, one of which is discussed in Section 9.11. In Section 12.6 we use eigenvalues and eigenvectors to obtain information about systems of differential equations.

Before giving the definitions, we remind you that the vector space $\mathbb{C}^n$ consists of vectors of the form $(z_1, z_2, \ldots, z_n)$ where each z_i is a complex number (see Example 9.1.13).

Let $T: V \to V$ be a linear transformation. In a great variety of applications (one of which is given in the next section), it is useful to find a vector $\mathbf{v}$ in V such that $T\mathbf{v}$ and $\mathbf{v}$ are parallel. That is, we seek a vector $\mathbf{v}$ and a scalar λ such that

$$T\mathbf{v} = \lambda\mathbf{v}. \tag{1}$$

If $\mathbf{v} \neq \mathbf{0}$ and λ satisfy (1), then λ is called an *eigenvalue* of T and $\mathbf{v}$ is called an *eigenvector* of T corresponding to the eigenvalue λ. If V is finite dimensional, then T can be represented by a matrix A_T. For that reason we shall discuss eigenvalues and eigenvectors of $n \times n$ matrices.

Definition 1 EIGENVALUE AND EIGENVECTOR Let A be an $n \times n$ matrix with real[†] components. The number λ (real or complex) is called an **eigenvalue** of A if there is a *nonzero* vector $\mathbf{v}$ in $\mathbb{C}^n$ such that

$$A\mathbf{v} = \lambda\mathbf{v}. \tag{2}$$

The vector $\mathbf{v} \neq \mathbf{0}$ is called an **eigenvector** *of A corresponding to the eigenvalue* λ.

NOTE. The word *eigen* is the German word for "own" or "proper." Eigenvalues are also called **proper values** or **characteristic values** and eigenvectors are called **proper vectors** or **characteristic vectors**.

REMARK. We shall see (for instance, in Example 6) that a matrix with real components can have complex eigenvalues and eigenvectors. That is why, in the definition, we have asserted that $\mathbf{v} \in \mathbb{C}^n$. We shall not be using many facts about complex numbers in this book. For a discussion of those few facts we do need, see Appendix 3.

EXAMPLE 1 Let $A = \begin{pmatrix} 10 & -18 \\ 6 & -11 \end{pmatrix}$. Then $A\begin{pmatrix} 2 \\ 1 \end{pmatrix} = \begin{pmatrix} 10 & -18 \\ 6 & -11 \end{pmatrix}\begin{pmatrix} 2 \\ 1 \end{pmatrix} = \begin{pmatrix} 2 \\ 1 \end{pmatrix}$. Thus $\lambda_1 = 1$ is an eigenvalue of A with corresponding eigenvector $\mathbf{v}_1 = \begin{pmatrix} 2 \\ 1 \end{pmatrix}$. Similarly, $A\begin{pmatrix} 3 \\ 2 \end{pmatrix} = \begin{pmatrix} 10 & -18 \\ 6 & -11 \end{pmatrix}\begin{pmatrix} 3 \\ 2 \end{pmatrix} = \begin{pmatrix} -6 \\ -4 \end{pmatrix} = -2\begin{pmatrix} 3 \\ 2 \end{pmatrix}$, so that $\lambda_2 = -2$ is an eigenvalue of A with corresponding eigenvector $\mathbf{v}_2 = \begin{pmatrix} 3 \\ 2 \end{pmatrix}$. As we soon shall see, these are the only eigenvalues of A. ■

EXAMPLE 2 Let $A = I$. Then for any $\mathbf{v} \in \mathbb{C}^n$, $A\mathbf{v} = I\mathbf{v} = \mathbf{v}$. Thus 1 is the only eigenvalue of A and every $\mathbf{v} \neq \mathbf{0} \in \mathbb{C}^n$ is an eigenvector of I. ■

We shall compute the eigenvalues and eigenvectors of many matrices in this section. But first we need to prove some facts that will simplify our computations.

Suppose that λ is an eigenvalue of A. Then there exists a nonzero vector

[†]This definition is also valid if A has complex components; but as the matrices we shall be dealing with will, for the most part, have real components, the definition is sufficient for our purposes.

$\mathbf{v} = \begin{pmatrix} x_1 \\ x_2 \\ \vdots \\ x_n \end{pmatrix} \neq \mathbf{0}$ such that $A\mathbf{v} = \lambda\mathbf{v} = \lambda I\mathbf{v}$. Rewriting this, we have

$$(A - \lambda I)\mathbf{v} = \mathbf{0}. \tag{3}$$

If A is an $n \times n$ matrix, equation (3) is a homogeneous system of n equations in the unknowns $x_1, x_2, \ldots, x_n$. Since, by assumption, the system has nontrivial solutions, we conclude that $\det(A - \lambda I) = 0$. Conversely, if $\det(A - \lambda I) = 0$, then equation (3) has nontrivial solutions, and λ is an eigenvalue of A. On the other hand, if $\det(A - \lambda I) \neq 0$, then (3) has only the solution $\mathbf{v} = \mathbf{0}$ so that λ is *not* an eigenvalue of A. Summing up these facts, we have the following theorem:

Theorem 1 Let A be an $n \times n$ matrix. Then λ is an eigenvalue of A if and only if

$$p(\lambda) = \det(A - \lambda I) = 0. \tag{4}$$

Definition 2 CHARACTERISTIC EQUATION AND POLYNOMIAL Equation (4) is called the **characteristic equation** of A; $p(\lambda)$ is called the **characteristic polynomial** of A.

As will become apparent in the examples, $p(\lambda)$ is a polynomial of degree n in λ. For example, if $A = \begin{pmatrix} a & b \\ c & d \end{pmatrix}$, then $A - \lambda I = \begin{pmatrix} a & b \\ c & d \end{pmatrix} - \begin{pmatrix} \lambda & 0 \\ 0 & \lambda \end{pmatrix} = \begin{pmatrix} a - \lambda & b \\ c & d - \lambda \end{pmatrix}$ and $p(\lambda) = \det(A - \lambda I) = (a - \lambda)(d - \lambda) - bc = \lambda^2 - (a + d)\lambda + (ad - bc)$.

By the fundamental theorem of algebra, any polynomial of degree n with real or complex coefficients has exactly n roots (counting multiplicities). By this we mean, for example, that the polynomial $(\lambda - 1)^5$ has five roots, all equal to the number 1. Since any eigenvalue of A is a root of the characteristic equation of A, we conclude that

Counting multiplicities, every $n \times n$ matrix has exactly n eigenvalues.

Theorem 2 Let λ be an eigenvalue of the $n \times n$ matrix A and let $E_\lambda = \{\mathbf{v}: A\mathbf{v} = \lambda\mathbf{v}\}$. Then E_λ is a subspace of $\mathbb{C}^n$.

Proof. If $A\mathbf{v} = \lambda\mathbf{v}$, then $(A - \lambda I)\mathbf{v} = \mathbf{0}$. Thus E_λ is the kernel of the matrix $A - \lambda I$, which, by Example 9.4.11, is a subspace[†] of $\mathbb{C}^n$. ■

Definition 3 EIGENSPACE Let λ be an eigenvalue of A. The subspace E_λ is called the **eigenspace**[‡] of A corresponding to the eigenvalue λ.

We now prove another useful result.

[†]In Example 9.4.11 we saw that ker A is a subspace of $\mathbb{R}^n$ if A is a real matrix. The extension of this result to $\mathbb{C}^n$ presents no difficulties.

[‡]Note that $\mathbf{0} \in E_\lambda$ since E_λ is a subspace.

Theorem 3 Let A be an $n \times n$ matrix and let $\lambda_1, \lambda_2, \ldots, \lambda_m$ be distinct eigenvalues of A with corresponding eigenvectors $\mathbf{v}_1, \mathbf{v}_2, \ldots, \mathbf{v}_m$. Then $\mathbf{v}_1, \mathbf{v}_2, \ldots, \mathbf{v}_m$ are linearly independent. That is, **eigenvectors corresponding to distinct eigenvalues are linearly independent.**

Proof. We prove this by mathematical induction. We start with $m = 2$. Suppose that

$$c_1\mathbf{v}_1 + c_2\mathbf{v}_2 = \mathbf{0}. \tag{5}$$

Then, multiplying both sides of (5) by A, we have $\mathbf{0} = A(c_1\mathbf{v}_1 + c_2\mathbf{v}_2) = c_1A\mathbf{v}_1 + c_2A\mathbf{v}_2$ or (since $A\mathbf{v}_i = \lambda_i\mathbf{v}_i$ for $i = 1, 2$)

$$c_1\lambda_1\mathbf{v}_1 + c_2\lambda_2\mathbf{v}_2 = \mathbf{0}. \tag{6}$$

We then multiply (5) by λ_1 and subtract it from (6) to obtain

$$(c_1\lambda_1\mathbf{v}_1 + c_2\lambda_2\mathbf{v}_2) - (c_1\lambda_1\mathbf{v}_1 + c_2\lambda_1\mathbf{v}_2) = \mathbf{0}$$

or

$$c_2(\lambda_2 - \lambda_1)\mathbf{v}_2 = \mathbf{0}.$$

Since $\mathbf{v}_2 \neq \mathbf{0}$ (by the definition of an eigenvector) and since $\lambda_2 \neq \lambda_1$, we conclude that $c_2 = 0$. Then inserting $c_2 = 0$ in (5), we see that $c_1 = 0$, which proves the theorem in the case $m = 2$. Now suppose that the theorem is true for $m = k$. That is, we assume that any k eigenvectors corresponding to distinct eigenvalues are linearly independent. We prove the theorem for $m = k + 1$. So we assume that

$$c_1\mathbf{v}_1 + c_2\mathbf{v}_2 + \cdots + c_k\mathbf{v}_k + c_{k+1}\mathbf{v}_{k+1} = \mathbf{0}. \tag{7}$$

Then, multiplying both sides of (7) by A and using the fact that $A\mathbf{v}_i = \lambda_i\mathbf{v}_i$, we obtain

$$c_1\lambda_1\mathbf{v}_1 + c_2\lambda_2\mathbf{v}_2 + \cdots + c_k\lambda_k\mathbf{v}_k + c_{k+1}\lambda_{k+1}\mathbf{v}_{k+1} = \mathbf{0}. \tag{8}$$

We multiply both sides of (7) by λ_{k+1} and subtract it from (8):

$$c_1(\lambda_1 - \lambda_{k+1})\mathbf{v}_1 + c_2(\lambda_2 - \lambda_{k+1})\mathbf{v}_2 + \cdots + c_k(\lambda_k - \lambda_{k+1})\mathbf{v}_k = \mathbf{0}.$$

But, by the induction assumption, $\mathbf{v}_1, \mathbf{v}_2, \ldots, \mathbf{v}_k$ are linearly independent. Thus $c_1(\lambda_1 - \lambda_{k+1}) = c_2(\lambda_2 - \lambda_{k+1}) = \cdots = c_k(\lambda_k - \lambda_{k+1}) = 0$; and, since $\lambda_i \neq \lambda_{k+1}$ for $i = 1, 2, \ldots, k$, we conclude that $c_1 = c_2 = \cdots = c_k = 0$. But, from (7), this means that $c_{k+1} = 0$. Thus the theorem is true for $m = k + 1$ and the proof is complete. ■

If

$$A = \begin{pmatrix} a_{11} & a_{12} & \cdots & a_{1n} \\ a_{21} & a_{22} & \cdots & a_{2n} \\ \vdots & \vdots & & \vdots \\ a_{n1} & a_{n2} & \cdots & a_{nn} \end{pmatrix}$$

then

$$p(\lambda) = \det(A - \lambda I) = \begin{vmatrix} a_{11} - \lambda & a_{12} & \cdots & a_{1n} \\ a_{21} & a_{22} - \lambda & \cdots & a_{2n} \\ \vdots & \vdots & & \vdots \\ a_{n1} & a_{n2} & \cdots & a_{nn} - \lambda \end{vmatrix}$$

and $p(\lambda) = 0$ can be written in the form

$$p(\lambda) = \lambda^n + b_{n-1}\lambda^{n-1} + \cdots + b_1\lambda + b_0 = 0. \tag{9}$$

Equation (9) has n roots, some of which may be repeated. If $\lambda_1, \lambda_2, \ldots, \lambda_m$ are the distinct roots of (9) with multiplicities $r_1, r_2, \ldots, r_m$, respectively, then (9) may be factored to obtain

$$p(\lambda) = (\lambda - \lambda_1)^{r_1}(\lambda - \lambda_2)^{r_2} \cdots (\lambda - \lambda_m)^{r_m} = 0. \tag{10}$$

ALGEBRAIC MULTIPLICITY

The numbers $r_1, r_2, \ldots, r_m$ are called the **algebraic multiplicities** of the eigenvalues $\lambda_1, \lambda_2, \ldots, \lambda_m$ respectively.

We now calculate eigenvalues and corresponding eigenspaces. We do this in a three-step procedure:

PROCEDURE FOR COMPUTING EIGENVALUES AND EIGENVECTORS

(i) Find $p(\lambda) = \det(A - \lambda I)$.

(ii) Find the roots $\lambda_1, \lambda_2, \ldots, \lambda_m$ of $p(\lambda) = 0$.

(iii) Corresponding to each eigenvalue λ_i, solve the homogeneous system $(A - \lambda_i I)\mathbf{v} = \mathbf{0}$.

REMARK. Step (ii) is usually the most difficult one.

EXAMPLE 3 Let $A = \begin{pmatrix} 4 & 2 \\ 3 & 3 \end{pmatrix}$. Then $\det(A - \lambda I) = \begin{vmatrix} 4 - \lambda & 2 \\ 3 & 3 - \lambda \end{vmatrix} = (4 - \lambda)(3 - \lambda) - 6 = \lambda^2 - 7\lambda + 6 = (\lambda - 1)(\lambda - 6) = 0$. Thus the eigenvalues of A are $\lambda_1 = 1$ and $\lambda_2 = 6$. For $\lambda_1 = 1$, we solve $(A - I)\mathbf{v} = \mathbf{0}$ or $\begin{pmatrix} 3 & 2 \\ 3 & 2 \end{pmatrix}\begin{pmatrix} x_1 \\ x_2 \end{pmatrix} = \begin{pmatrix} 0 \\ 0 \end{pmatrix}$. Any eigenvector corresponding to $\lambda_1 = 1$ satisfies $3x_1 + 2x_2 = 0$. One such eigenvector is $\mathbf{v}_1 = \begin{pmatrix} 2 \\ -3 \end{pmatrix}$.

Thus $E_1 = \text{span}\left\{\begin{pmatrix} 2 \\ -3 \end{pmatrix}\right\}$. Similarly, the equation $(A - 6I)\mathbf{v} = \mathbf{0}$ means that $\begin{pmatrix} -2 & 2 \\ 3 & -3 \end{pmatrix}\begin{pmatrix} x_1 \\ x_2 \end{pmatrix} = \begin{pmatrix} 0 \\ 0 \end{pmatrix}$ or $x_1 = x_2$. Thus $\mathbf{v}_2 = \begin{pmatrix} 1 \\ 1 \end{pmatrix}$ is an eigenvector corresponding to $\lambda_2 = 6$ and $E_6 = \text{span}\left\{\begin{pmatrix} 1 \\ 1 \end{pmatrix}\right\}$. Note that $\mathbf{v}_1$ and $\mathbf{v}_2$ are linearly independent since neither one is a multiple of the other. ■

EXAMPLE 4 Let $A = \begin{pmatrix} 1 & -1 & 4 \\ 3 & 2 & -1 \\ 2 & 1 & -1 \end{pmatrix}$. Then

$$\det(A - \lambda I) = \begin{vmatrix} 1-\lambda & -1 & 4 \\ 3 & 2-\lambda & -1 \\ 2 & 1 & -1-\lambda \end{vmatrix}$$
$$= -(\lambda^3 - 2\lambda^2 - 5\lambda + 6) = -(\lambda - 1)(\lambda + 2)(\lambda - 3) = 0.$$

Thus the eigenvalues of A are $\lambda_1 = 1$, $\lambda_2 = -2$, and $\lambda_3 = 3$. Corresponding to $\lambda_1 = 1$ we have

$$(A - I)\mathbf{v} = \begin{pmatrix} 0 & -1 & 4 \\ 3 & 1 & -1 \\ 2 & 1 & -2 \end{pmatrix}\begin{pmatrix} x_1 \\ x_2 \\ x_3 \end{pmatrix} = \begin{pmatrix} 0 \\ 0 \\ 0 \end{pmatrix}.$$

Solving by row reduction, we obtain, successively,

$$\left(\begin{array}{ccc|c} 0 & -1 & 4 & 0 \\ 3 & 1 & -1 & 0 \\ 2 & 1 & -2 & 0 \end{array}\right) \xrightarrow[A_{1,3}(1)]{A_{1,2}(1)} \left(\begin{array}{ccc|c} 0 & -1 & 4 & 0 \\ 3 & 0 & 3 & 0 \\ 2 & 0 & 2 & 0 \end{array}\right) \xrightarrow{M_2(\frac{1}{3})}$$
$$\left(\begin{array}{ccc|c} 0 & -1 & 4 & 0 \\ 1 & 0 & 1 & 0 \\ 2 & 0 & 2 & 0 \end{array}\right) \xrightarrow{A_{2,3}(-2)} \left(\begin{array}{ccc|c} 0 & -1 & 4 & 0 \\ 1 & 0 & 1 & 0 \\ 0 & 0 & 0 & 0 \end{array}\right).$$

Thus $x_1 = -x_3$, $x_2 = 4x_3$, an eigenvector is $\mathbf{v}_1 = \begin{pmatrix} -1 \\ 4 \\ 1 \end{pmatrix}$, and $E_1 = \text{span}\left\{\begin{pmatrix} -1 \\ 4 \\ 1 \end{pmatrix}\right\}$.

For $\lambda_2 = -2$, we have $[A - (-2I)]\mathbf{v} = (A + 2I)\mathbf{v} = \mathbf{0}$ or $\begin{pmatrix} 3 & -1 & 4 \\ 3 & 4 & -1 \\ 2 & 1 & 1 \end{pmatrix}\begin{pmatrix} x_1 \\ x_2 \\ x_3 \end{pmatrix} = \begin{pmatrix} 0 \\ 0 \\ 0 \end{pmatrix}$.

This leads to

$$\left(\begin{array}{rrr|r} 3 & -1 & 4 & 0 \\ 3 & 4 & -1 & 0 \\ 2 & 1 & 1 & 0 \end{array}\right) \xrightarrow[A_{1,3}(1)]{A_{1,2}(4)} \left(\begin{array}{rrr|r} 3 & -1 & 4 & 0 \\ 15 & 0 & 15 & 0 \\ 5 & 0 & 5 & 0 \end{array}\right) \xrightarrow{M_2(\frac{1}{15})}$$

$$\left(\begin{array}{rrr|r} 3 & -1 & 4 & 0 \\ 1 & 0 & 1 & 0 \\ 5 & 0 & 5 & 0 \end{array}\right) \xrightarrow[A_{2,3}(-5)]{A_{2,1}(-4)} \left(\begin{array}{rrr|r} -1 & -1 & 0 & 0 \\ 1 & 0 & 1 & 0 \\ 0 & 0 & 0 & 0 \end{array}\right).$$

Thus $x_2 = -x_1$, $x_3 = -x_1$, and an eigenvector is $\mathbf{v}_2 = \begin{pmatrix} 1 \\ -1 \\ -1 \end{pmatrix}$. Thus $E_{-2} = \text{span}\left\{\begin{pmatrix} 1 \\ -1 \\ -1 \end{pmatrix}\right\}$. Finally, for $\lambda_3 = 3$, we have

$$(A - 3I)\mathbf{v} = \begin{pmatrix} -2 & -1 & 4 \\ 3 & -1 & -1 \\ 2 & 1 & -4 \end{pmatrix}\begin{pmatrix} x_1 \\ x_2 \\ x_3 \end{pmatrix} = \begin{pmatrix} 0 \\ 0 \\ 0 \end{pmatrix}$$

and

$$\left(\begin{array}{rrr|r} -2 & -1 & 4 & 0 \\ 3 & -1 & -1 & 0 \\ 2 & 1 & -4 & 0 \end{array}\right) \xrightarrow[A_{1,3}(1)]{A_{1,2}(-1)} \left(\begin{array}{rrr|r} -2 & -1 & 4 & 0 \\ 5 & 0 & -5 & 0 \\ 0 & 0 & 0 & 0 \end{array}\right)$$

$$\xrightarrow{M_2(\frac{1}{5})} \left(\begin{array}{rrr|r} -2 & -1 & 4 & 0 \\ 1 & 0 & -1 & 0 \\ 0 & 0 & 0 & 0 \end{array}\right) \xrightarrow{A_{2,1}(4)} \left(\begin{array}{rrr|r} 2 & -1 & 0 & 0 \\ 1 & 0 & -1 & 0 \\ 0 & 0 & 0 & 0 \end{array}\right).$$

Hence $x_3 = x_1$, $x_2 = 2x_1$, and $\mathbf{v}_3 = \begin{pmatrix} 1 \\ 2 \\ 1 \end{pmatrix}$ so that $E_3 = \text{span}\left\{\begin{pmatrix} 1 \\ 2 \\ 1 \end{pmatrix}\right\}$.

REMARK. In this and every other example, there is always an infinite number of choices for each eigenvector. We arbitrarily choose a simple one by setting one or more of the x_i's equal to a convenient number. Here we have set one of the x_i's equal to 1. ■

EXAMPLE 5 Let $A = \begin{pmatrix} 2 & -1 \\ -4 & 2 \end{pmatrix}$. Then $\det(A - \lambda I) = \begin{vmatrix} 2-\lambda & -1 \\ -4 & 2-\lambda \end{vmatrix} = \lambda^2 - 4\lambda = \lambda(\lambda - 4)$. Thus the eigenvalues are $\lambda_1 = 0$ and $\lambda_2 = 4$. The eigenspace corresponding to zero is simply the kernel of A. We calculate $\begin{pmatrix} 2 & -1 \\ -4 & 2 \end{pmatrix}\begin{pmatrix} x_1 \\ x_2 \end{pmatrix} = \begin{pmatrix} 0 \\ 0 \end{pmatrix}$ or $2x_1 = x_2$ and an eigenvector is $\mathbf{v}_1 = \begin{pmatrix}1\\2\end{pmatrix}$. Thus $\ker A = E_0 = \text{span}\left\{\begin{pmatrix}1\\2\end{pmatrix}\right\}$. Corresponding to $\lambda_2 = 4$ we have $\begin{pmatrix}-2 & -1\\-4 & -2\end{pmatrix}\begin{pmatrix}x_1\\x_2\end{pmatrix} = \begin{pmatrix}0\\0\end{pmatrix}$, so $E_4 = \text{span}\left\{\begin{pmatrix}1\\-2\end{pmatrix}\right\}$. ■

EXAMPLE 6 Let $A = \begin{pmatrix}3 & -5\\1 & -1\end{pmatrix}$. Then $\det(A - \lambda I) = \begin{vmatrix}3-\lambda & -5\\1 & -1-\lambda\end{vmatrix} = \lambda^2 - 2\lambda + 2 = 0$ and

$$\lambda = \frac{-(-2) \pm \sqrt{4 - 4(1)(2)}}{2} = \frac{2 \pm \sqrt{-4}}{2} = \frac{2 \pm 2i}{2} = 1 \pm i.$$

Thus $\lambda_1 = 1 + i$ and $\lambda_2 = 1 - i$. We compute

$$[A - (1+i)I]\mathbf{v} = \begin{pmatrix}2-i & -5\\1 & -2-i\end{pmatrix}^{\dagger}\begin{pmatrix}x_1\\x_2\end{pmatrix} = \begin{pmatrix}0\\0\end{pmatrix}$$

and we obtain $(2 - i)x_1 - 5x_2 = 0$ and $x_1 + (-2 - i)x_2 = 0$. Thus $x_1 = (2 + i)x_2$, which yields the eigenvector $\mathbf{v}_1 = \begin{pmatrix}2+i\\1\end{pmatrix}$ and $E_{1+i} = \text{span}\left\{\begin{pmatrix}2+i\\1\end{pmatrix}\right\}$. Similarly, $[A - (1-i)I]\mathbf{v} = \begin{pmatrix}2+i & -5\\1 & -2+i\end{pmatrix}\begin{pmatrix}x_1\\x_2\end{pmatrix} = \begin{pmatrix}0\\0\end{pmatrix}$ or $x_1 + (-2 + i)x_2 = 0$, which yields $x_1 = (2 - i)x_2$, $\mathbf{v}_2 = \begin{pmatrix}2-i\\1\end{pmatrix}$, and $E_{1-i} = \text{span}\left\{\begin{pmatrix}2-i\\1\end{pmatrix}\right\}$. ■

REMARK 1. Example 6 illustrates that a real matrix may have complex eigenvalues and eigenvectors. Some texts define eigenvalues of real matrices to be the *real* roots of the characteristic equation. By this definition the matrix of the last example has *no* eigenvalues. This might make the computations simpler, but it also greatly reduces the usefulness of the theory of eigenvalues and eigenvectors. We shall see a significant illustration of the use of complex eigenvalues in Section 12.6.

REMARK 2. Note that $\lambda_2 = 1 - i$ is the complex conjugate of $\lambda_1 = 1 + i$. Also, the components of $\mathbf{v}_2$ are complex conjugates of the components of $\mathbf{v}_1$. This is no coincidence. In Problem 33 you are asked to prove that **the eigenvalues of a real matrix occur in complex conjugate pairs** and that **the corresponding eigenvector pairs are complex conjugates of each other.**

EXAMPLE 7 Let $A = \begin{pmatrix}4 & 0\\0 & 4\end{pmatrix}$. Then $\det(A - \lambda I) = \begin{vmatrix}4-\lambda & 0\\0 & 4-\lambda\end{vmatrix} = (\lambda - 4)^2 = 0$; hence $\lambda = 4$ is an eigenvalue of algebraic multiplicity 2. It is obvious that $A\mathbf{v} = 4\mathbf{v}$ for every vector $\mathbf{v} \in \mathbb{R}^2$, so that $E_4 = \mathbb{R}^2 = \text{span}\left\{\begin{pmatrix}1\\0\end{pmatrix}, \begin{pmatrix}0\\1\end{pmatrix}\right\}$. ■

---

†Note that the columns of this matrix are linearly dependent because $\begin{pmatrix}-5\\-2-i\end{pmatrix} = (-2-i)\begin{pmatrix}2-i\\1\end{pmatrix}$.

EXAMPLE 8 Let $A = \begin{pmatrix} 4 & 1 \\ 0 & 4 \end{pmatrix}$. Then $\det(A - \lambda I) = \begin{vmatrix} 4-\lambda & 1 \\ 0 & 4-\lambda \end{vmatrix} = (\lambda - 4)^2 = 0$; thus $\lambda = 4$ is again an eigenvalue of algebraic multiplicity 2. But this time we have $(A - 4I)\mathbf{v} = \begin{pmatrix} 0 & 1 \\ 0 & 0 \end{pmatrix}\begin{pmatrix} x_1 \\ x_2 \end{pmatrix} = \begin{pmatrix} x_2 \\ 0 \end{pmatrix}$. Thus $x_2 = 0$, $\mathbf{v}_1 = \begin{pmatrix} 1 \\ 0 \end{pmatrix}$ is an eigenvector, and $E_4 = \text{span}\left\{\begin{pmatrix} 1 \\ 0 \end{pmatrix}\right\}$. ■

EXAMPLE 9 Let $A = \begin{pmatrix} 3 & 2 & 4 \\ 2 & 0 & 2 \\ 4 & 2 & 3 \end{pmatrix}$. Then $\det(A - \lambda I) = \begin{vmatrix} 3-\lambda & 2 & 4 \\ 2 & -\lambda & 2 \\ 4 & 2 & 3-\lambda \end{vmatrix} = -\lambda^3 + 6\lambda^2 + 15\lambda + 8 = -(\lambda + 1)^2(\lambda - 8) = 0$, so that the eigenvalues are $\lambda_1 = 8$ and $\lambda_2 = -1$ (with algebraic multiplicity 2). For $\lambda_1 = 8$, we obtain

$$(A - 8I)\mathbf{v} = \begin{pmatrix} -5 & 2 & 4 \\ 2 & -8 & 2 \\ 4 & 2 & -5 \end{pmatrix}\begin{pmatrix} x_1 \\ x_2 \\ x_3 \end{pmatrix} = \begin{pmatrix} 0 \\ 0 \\ 0 \end{pmatrix}$$

or, row-reducing,

$$\left(\begin{array}{rrr|r} -5 & 2 & 4 & 0 \\ 2 & -8 & 2 & 0 \\ 4 & 2 & -5 & 0 \end{array}\right) \xrightarrow[A_{1,3}(-1)]{A_{1,2}(4)} \left(\begin{array}{rrr|r} -5 & 2 & 4 & 0 \\ -18 & 0 & 18 & 0 \\ 9 & 0 & -9 & 0 \end{array}\right) \xrightarrow{M_2(\frac{1}{18})}$$

$$\left(\begin{array}{rrr|r} -5 & 2 & 4 & 0 \\ -1 & 0 & 1 & 0 \\ 9 & 0 & -9 & 0 \end{array}\right) \xrightarrow{A_{2,3}(9)} \left(\begin{array}{rrr|r} 0 & 2 & -1 & 0 \\ -1 & 0 & 1 & 0 \\ 0 & 0 & 0 & 0 \end{array}\right).$$

Hence $x_3 = 2x_2$ and $x_1 = x_3$, we obtain the eigenvector $\mathbf{v}_1 = \begin{pmatrix} 2 \\ 1 \\ 2 \end{pmatrix}$, and $E_8 = \text{span}\left\{\begin{pmatrix} 2 \\ 1 \\ 2 \end{pmatrix}\right\}$. For $\lambda_2 = -1$, we have $(A + I)\mathbf{v} = \begin{pmatrix} 4 & 2 & 4 \\ 2 & 1 & 2 \\ 4 & 2 & 4 \end{pmatrix}\begin{pmatrix} x_1 \\ x_2 \\ x_3 \end{pmatrix} = \begin{pmatrix} 0 \\ 0 \\ 0 \end{pmatrix}$, which gives us the single equation $2x_1 + x_2 + 2x_3 = 0$ or $x_2 = -2x_1 - 2x_3$. If $x_1 = 1$ and $x_3 = 0$, we obtain $\mathbf{v}_2 = \begin{pmatrix} 1 \\ -2 \\ 0 \end{pmatrix}$. If $x_1 = 0$ and $x_3 = 1$, we obtain $\mathbf{v}_3 = \begin{pmatrix} 0 \\ -2 \\ 1 \end{pmatrix}$. Thus $E_{-1} = \text{span}\left\{\begin{pmatrix} 1 \\ -2 \\ 0 \end{pmatrix}, \begin{pmatrix} 0 \\ -2 \\ 1 \end{pmatrix}\right\}$. There are other convenient choices for eigenvectors. For example, $\mathbf{v} = \begin{pmatrix} 1 \\ 0 \\ -1 \end{pmatrix}$ is in E_{-1} since $\mathbf{v} = \mathbf{v}_2 - \mathbf{v}_3$. ■

EXAMPLE 10 Let $A = \begin{pmatrix} -5 & -5 & -9 \\ 8 & 9 & 18 \\ -2 & -3 & -7 \end{pmatrix}$. Then $\det(A - \lambda I) = \begin{vmatrix} -5-\lambda & -5 & -9 \\ 8 & 9-\lambda & 18 \\ -2 & -3 & -7-\lambda \end{vmatrix} = -\lambda^3 - 3\lambda^2 - 3\lambda - 1 = -(\lambda + 1)^3 = 0$. Thus $\lambda = -1$ is an eigenvalue of algebraic multiplicity 3. To compute E_{-1}, we set $(A + I)\mathbf{v} = \begin{pmatrix} -4 & -5 & -9 \\ 8 & 10 & 18 \\ -2 & -3 & -6 \end{pmatrix}\begin{pmatrix} x_1 \\ x_2 \\ x_3 \end{pmatrix} = \begin{pmatrix} 0 \\ 0 \\ 0 \end{pmatrix}$ and row reduce to obtain, successively,

$$\left(\begin{array}{ccc|c} -4 & -5 & -9 & 0 \\ 8 & 10 & 18 & 0 \\ -2 & -3 & -6 & 0 \end{array}\right) \xrightarrow[A_{3,2}(4)]{A_{3,1}(-2)} \left(\begin{array}{ccc|c} 0 & 1 & 3 & 0 \\ 0 & -2 & -6 & 0 \\ -2 & -3 & -6 & 0 \end{array}\right) \xrightarrow[A_{1,3}(3)]{A_{1,2}(2)} \left(\begin{array}{ccc|c} 0 & 1 & 3 & 0 \\ 0 & 0 & 0 & 0 \\ -2 & 0 & 3 & 0 \end{array}\right)$$

This yields $x_2 = -3x_3$ and $2x_1 = 3x_3$. Setting $x_3 = 2$ we obtain only one linearly independent eigenvector: $\mathbf{v}_1 = \begin{pmatrix} 3 \\ -6 \\ 2 \end{pmatrix}$. Thus $E_{-1} = \text{span}\left\{\begin{pmatrix} 3 \\ -6 \\ 2 \end{pmatrix}\right\}$. ■

EXAMPLE 11 Let $A = \begin{pmatrix} -1 & -3 & -9 \\ 0 & 5 & 18 \\ 0 & -2 & -7 \end{pmatrix}$. Then $\det(A - \lambda I) = \begin{vmatrix} -1-\lambda & -3 & -9 \\ 0 & 5-\lambda & 18 \\ 0 & -2 & -7-\lambda \end{vmatrix} = -(\lambda + 1)^3 = 0$. Thus, as in Example 10, $\lambda = -1$ is an eigenvalue of algebraic multiplicity 3. To find E_{-1}, we compute $(A + I)\mathbf{v} = \begin{pmatrix} 0 & -3 & -9 \\ 0 & 6 & 18 \\ 0 & -2 & -6 \end{pmatrix}\begin{pmatrix} x_1 \\ x_2 \\ x_3 \end{pmatrix} = \begin{pmatrix} 0 \\ 0 \\ 0 \end{pmatrix}$. Thus $-2x_2 - 6x_3 = 0$ or $x_2 = -3x_3$, and x_1 is arbitrary. Setting $x_1 = 0$, $x_3 = 1$, we obtain $\mathbf{v}_1 = \begin{pmatrix} 0 \\ -3 \\ 1 \end{pmatrix}$. Setting $x_1 = 1$, $x_3 = 1$ yields $\mathbf{v}_2 = \begin{pmatrix} 1 \\ -3 \\ 1 \end{pmatrix}$. Thus $E_{-1} = \text{span}\left\{\begin{pmatrix} 0 \\ -3 \\ 1 \end{pmatrix}, \begin{pmatrix} 1 \\ -3 \\ 1 \end{pmatrix}\right\}$. ■

In each of the last five examples we found an eigenvalue with an algebraic multiplicity of 2 or more. But, as we saw in Examples 8, 10, and 11, the number of linearly independent eigenvectors is not necessarily equal to the algebraic multiplicity of the eigenvalue (as was the case in Examples 7 and 9). This observation leads to the following definition:

Definition 4 GEOMETRIC MULTIPLICITY Let λ be an eigenvalue of the matrix A. Then the **geometric multiplicity** of λ is the dimension of the eigenspace corresponding to λ (which is the nullity of the matrix $A - \lambda I$). That is,

$$\text{Geometric multiplicity of } \lambda = \dim E_\lambda = \nu(A - \lambda I)$$

In Examples 7 and 9 we saw that for the eigenvalues of algebraic multiplicity 2, the geometric multiplicities were also 2. In Example 8 the geometric multiplicity of $\lambda = 4$ was 1 while the algebraic multiplicity was 2. In Example 10 the algebraic multiplicity was 3 and the geometric multiplicity was 1. In Example 11 the algebraic multiplicity was 3 and the geometric multiplicity was 2. These examples illustrate the fact that if the algebraic multiplicity of λ is greater than 1, then we cannot predict the geometric multiplicity of λ without additional information.

If A is a 2×2 matrix and λ is an eigenvalue with algebraic multiplicity 2, then the geometric multiplicity of λ is ≤ 2 since there can be at most two linearly independent vectors in a two-dimensional space. Let A be a 3×3 matrix having two eigenvalues λ_1 and λ_2 with algebraic multiplicities 1 and 2, respectively. Then the geometric multiplicity of λ_2 is ≤ 2 because otherwise we would have at least four linearly independent vectors in a three-dimensional space. Intuitively, it seems that the geometric multiplicity of an eigenvalue is always less than or equal to its algebraic multiplicity. The proof of the following theorem is not difficult if additional facts about determinants are proved. Since this would take us too far afield, we omit the proof.

Theorem 4 Let λ be an eigenvalue of A. Then,

Geometric multiplicity of $\lambda \leq$ algebraic multiplicity of λ

NOTE. The geometric multiplicity of an eigenvalue is never zero. This follows from Definition 1, which states that if λ is an eigenvalue, then there exists a *nonzero* eigenvector corresponding to λ.

In the rest of this chapter an important problem for us will be to determine whether a given $n \times n$ matrix does or does not have n linearly independent eigenvectors. From what we have already discussed in this section, the following theorem is apparent:

Theorem 5 Let A be an $n \times n$ matrix. Then A has n linearly independent eigenvectors if and only if the geometric multiplicity of every eigenvalue is equal to its algebraic multiplicity. In particular, A has n linearly independent eigenvectors if all the eigenvalues are distinct (since then the algebraic multiplicity of every eigenvalue is 1).

Theorem 5 is useful only after we have computed the eigenvalues of A. However, there is one case in which there are always n linearly independent eigenvectors. The proof of the following theorem is omitted.[†]

Theorem 6 An $n \times n$ symmetric matrix has n linearly independent eigenvectors. Moreover, each eigenvalue of a symmetric matrix is real.

Suppose that 0 is an eigenvalue of A. Then $\det (A - 0I) = \det A = 0$. Conversely, if $\det A = 0$, then 0 is an eigenvalue of A. Thus 0 is an eigenvalue of A if and only if $\det A = 0$. This fact allows us to extend, for the last time, our Summing-up Theorem, last seen in Section 9.3 (page 495).

[†]For a proof of Theorem 6 see Section 7.4 in S. I. Grossman, *Elementary Linear Algebra,* second edition, (Wadsworth, Belmont, CA, 1984), p. 302.

Theorem 7 SUMMING-UP THEOREM—VIEW 4 Let A be an $n \times n$ matrix. Then each of the following seven statements implies the other six. That is, if one is true, all are true.

(i) A is invertible.
(ii) The system $A\mathbf{x} = \mathbf{b}$ has a unique solution for every n-vector $\mathbf{b}$.
(iii) The only solution to the homogeneous system $A\mathbf{x} = \mathbf{0}$ is the trivial solution ($\mathbf{x} = \mathbf{0}$).
(iv) A is row equivalent to the identity matrix I_n; that is, the reduced row echelon form of A is I_n.
(v) $\det A \neq 0$.
(vi) The columns of A (considered as n-vectors) are linearly independent.
(vii) 0 is not an eigenvalue of A.

PROBLEMS 9.10

In Problems 1–20 calculate the eigenvalues and eigenspaces of the given matrix. If the algebraic multiplicity of an eigenvalue is greater than 1, calculate its geometric multiplicity.

1. $\begin{pmatrix} -2 & -2 \\ -5 & 1 \end{pmatrix}$ **2.** $\begin{pmatrix} -12 & 7 \\ -7 & 2 \end{pmatrix}$

3. $\begin{pmatrix} 2 & -1 \\ 5 & -2 \end{pmatrix}$ **4.** $\begin{pmatrix} -3 & 0 \\ 0 & -3 \end{pmatrix}$

5. $\begin{pmatrix} -3 & 2 \\ 0 & -3 \end{pmatrix}$ **6.** $\begin{pmatrix} 3 & 2 \\ -5 & 1 \end{pmatrix}$

7. $\begin{pmatrix} 1 & -1 & 0 \\ -1 & 2 & -1 \\ 0 & -1 & 1 \end{pmatrix}$ **8.** $\begin{pmatrix} 1 & 1 & -2 \\ -1 & 2 & 1 \\ 0 & 1 & -1 \end{pmatrix}$

9. $\begin{pmatrix} 5 & 4 & 2 \\ 4 & 5 & 2 \\ 2 & 2 & 2 \end{pmatrix}$ **10.** $\begin{pmatrix} 1 & 2 & 2 \\ 0 & 2 & 1 \\ -1 & 2 & 2 \end{pmatrix}$

11. $\begin{pmatrix} 0 & 1 & 0 \\ 0 & 0 & 1 \\ 1 & -3 & 3 \end{pmatrix}$ **12.** $\begin{pmatrix} -3 & -7 & -5 \\ 2 & 4 & 3 \\ 1 & 2 & 2 \end{pmatrix}$

13. $\begin{pmatrix} 1 & -1 & -1 \\ 1 & -1 & 0 \\ 1 & 0 & -1 \end{pmatrix}$ **14.** $\begin{pmatrix} 7 & -2 & -4 \\ 3 & 0 & -2 \\ 6 & -2 & -3 \end{pmatrix}$

15. $\begin{pmatrix} 4 & 6 & 6 \\ 1 & 3 & 2 \\ -1 & -5 & -2 \end{pmatrix}$ **16.** $\begin{pmatrix} 4 & 1 & 0 & 1 \\ 2 & 3 & 0 & 1 \\ -2 & 1 & 2 & -3 \\ 2 & -1 & 0 & 5 \end{pmatrix}$

17. $\begin{pmatrix} a & 0 & 0 & 0 \\ 0 & a & 0 & 0 \\ 0 & 0 & a & 0 \\ 0 & 0 & 0 & a \end{pmatrix}$ **18.** $\begin{pmatrix} a & b & 0 & 0 \\ 0 & a & 0 & 0 \\ 0 & 0 & a & 0 \\ 0 & 0 & 0 & a \end{pmatrix}; b \neq 0$

19. $\begin{pmatrix} a & b & 0 & 0 \\ 0 & a & c & 0 \\ 0 & 0 & a & 0 \\ 0 & 0 & 0 & a \end{pmatrix}; bc \neq 0$

20. $\begin{pmatrix} a & b & 0 & 0 \\ 0 & a & c & 0 \\ 0 & 0 & a & d \\ 0 & 0 & 0 & a \end{pmatrix}; bcd \neq 0$

21. Show that for any real numbers a and b, the matrix $A = \begin{pmatrix} a & b \\ -b & a \end{pmatrix}$ has the eigenvectors $\begin{pmatrix} 1 \\ i \end{pmatrix}$ and $\begin{pmatrix} 1 \\ -i \end{pmatrix}$.

In Problems 22–28 assume that the matrix A has the eigenvalues $\lambda_1, \lambda_2, \ldots, \lambda_k$.

22. Show that the eigenvalues of A^t are $\lambda_1, \lambda_2, \ldots, \lambda_k$.
23. Show that the eigenvalues of αA are $\alpha\lambda_1, \alpha\lambda_2, \ldots, \alpha\lambda_k$.
24. Show that A^{-1} exists if and only if $\lambda_1\lambda_2 \cdots \lambda_k \neq 0$.
***25.** If A^{-1} exists, show that the eigenvalues of A^{-1} are $1/\lambda_1, 1/\lambda_2, \ldots, 1/\lambda_k$.
26. Show that the matrix $A - \alpha I$ has the eigenvalues $\lambda_1 - \alpha, \lambda_2 - \alpha, \ldots, \lambda_k - \alpha$.
***27.** Show that the eigenvalues of A^2 are $\lambda_1^2, \lambda_2^2, \ldots, \lambda_k^2$.

***28.** Show that the eigenvalues of A^m are $\lambda_1{}^m$, $\lambda_2{}^m$, $\ldots, \lambda_k{}^m$ for $m = 1, 2, 3, \ldots$.

29. Let λ be an eigenvalue of A with corresponding eigenvector $\mathbf{v}$. Let $p(\lambda) = a_0 + a_1\lambda + a_2\lambda^2 + \cdots + a_n\lambda^n$. Define the matrix $p(A)$ by $p(A) = a_0I + a_1A + a_2A^2 + \cdots + a_nA^n$. Show that $p(A)\mathbf{v} = p(\lambda)\mathbf{v}$.

30. Using the result of Problem 29, show that if $\lambda_1, \lambda_2, \ldots, \lambda_k$ are eigenvalues of A, then $p(\lambda_1), p(\lambda_2), \ldots, p(\lambda_k)$ are eigenvalues of $p(A)$.

31. Show that if A is an upper triangular matrix, then the eigenvalues of A are the diagonal components of A.

32. Let $A_1 = \begin{pmatrix} 2 & 0 & 0 & 0 \\ 0 & 2 & 0 & 0 \\ 0 & 0 & 2 & 0 \\ 0 & 0 & 0 & 2 \end{pmatrix}$, $A_2 = \begin{pmatrix} 2 & 1 & 0 & 0 \\ 0 & 2 & 0 & 0 \\ 0 & 0 & 2 & 0 \\ 0 & 0 & 0 & 2 \end{pmatrix}$, $A_3 = \begin{pmatrix} 2 & 1 & 0 & 0 \\ 0 & 2 & 1 & 0 \\ 0 & 0 & 2 & 0 \\ 0 & 0 & 0 & 2 \end{pmatrix}$, and $A_4 = \begin{pmatrix} 2 & 1 & 0 & 0 \\ 0 & 2 & 1 & 0 \\ 0 & 0 & 2 & 1 \\ 0 & 0 & 0 & 2 \end{pmatrix}$. Show that, for each matrix, $\lambda = 2$ is an eigenvalue of algebraic multiplicity 4. In each ease, compute the geometric multiplicity of $\lambda = 2$.

***33.** Let A be a real $n \times n$ matrix. Show that if λ_1 is a complex eigenvalue of A with eigenvector $\mathbf{v}_1$, then $\overline{\lambda}_1$ is an eigenvalue of A with eigenvector $\overline{\mathbf{v}}_1$.

***34.** A **probability matrix** is an $n \times n$ matrix having two properties:

(i) $a_{ij} \geq 0$ for every i and j.

(ii) The sum of the components in every column is 1.

Prove that 1 is an eigenvalue of every probability matrix.

(SKIP)

9.11 IF TIME PERMITS: A MODEL OF POPULATION GROWTH

In this section we show how the theory of eigenvalues and eigenvectors can be used to analyze a model of the growth of a bird population.[†] We begin by discussing a simple model of population growth. We assume that a certain species grows at a constant rate; that is, the population of the species after one time period (which could be an hour, a week, a month, a year, etc.) is a constant multiple of the population in the previous time period. One way this could happen, for example, is that each generation is distinct and each organism produces r offspring and then dies. If p_n denotes the population after the nth time period, we would have

$$p_n = rp_{n-1}.$$

For example, this model might describe a bacteria population where, at a given time, an organism splits into two separate organisms. Then $r = 2$. Let p_0 denote the initial population. Then $p_1 = rp_0$, $p_2 = rp_1 = r(rp_0) = r^2p_0$, $p_3 = rp_2 = r(r^2p_0) = r^3p_0$, and so on, so that

$$p_n = r^np_0. \tag{1}$$

From this model we see that the population increases without bound if $r > 1$ and decreases to zero if $r < 1$. If $r = 1$ the population remains at the constant value p_0.

[†]The material in this section is based on a paper by D. Cooke: "A 2 × 2 Matrix Model of Population Growth," *Mathematical Gazette* **61**(416): 120–123.

This model is, evidently, very simplistic. One obvious objection is that the number of offspring produced depends, in many cases, on the ages of the adults. For example, in a human population the average female adult over 50 would certainly produce fewer children than the average 21-year-old female. To deal with this difficulty, we introduce a model that allows for age groupings with different fertility rates.

We now look at a model of population growth for a species of birds. In this bird population we assume that the number of female birds equals the number of males. Let $p_{j,n-1}$ denote the population of juvenile (immature) females in the $(n-1)$st year and let $p_{a,n-1}$ denote the number of adult females in the $(n-1)$st year. Some of the juvenile birds will die during the year. We assume that a certain proportion α of the juvenile birds survive to become adults in the spring of the nth year. Later in the spring each surviving female bird produces eggs, which hatch to produce, on the average, k juvenile female birds in the following spring. Adults also die, and the proportion of adults that survive from one spring to the next is β.

This constant survival rate of birds is not just a simplistic assumption. It appears to be the case with most of the natural bird populations that have been studied. This means that the adult survival rate of many bird species is independent of age. Perhaps few birds in the wild survive long enough to exhibit the effects of old age. Moreover, in many species the number of offspring seems to be uninfluenced by the age of the mother.

In the notation introduced above, $p_{j,n}$ and $p_{a,n}$ represent, respectively, the populations of juvenile and adult females in the nth year. Putting together all the information given, we arrive at the following 2×2 system:

$$\begin{aligned} p_{j,n} &= \phantom{\alpha p_{j,n-1} + {}} k p_{a,n-1} \\ p_{a,n} &= \alpha p_{j,n-1} + \beta p_{a,n-1} \end{aligned} \tag{2}$$

or

$$\mathbf{p}_n = A\mathbf{p}_{n-1} \tag{3}$$

where $\mathbf{p}_n = \begin{pmatrix} p_{j,n} \\ p_{a,n} \end{pmatrix}$ and $A = \begin{pmatrix} 0 & k \\ \alpha & \beta \end{pmatrix}$. It is clear, from (3), that $\mathbf{p}_1 = A\mathbf{p}_0$, $\mathbf{p}_2 = A\mathbf{p}_1 = A(A\mathbf{p}_0) = A^2\mathbf{p}_0, \ldots$, and so on. Hence

$$\mathbf{p}_n = A^n\mathbf{p}_0 \tag{4}$$

where $\mathbf{p}_0$ is the vector of initial populations of juvenile and adult females.

Equation (4) is like equation (1), but now we are able to distinguish between the survival rates of juvenile and adult birds.

EXAMPLE 1 Let $A = \begin{pmatrix} 0 & 2 \\ 0.3 & 0.5 \end{pmatrix}$. This means that each adult female produces two female offspring and, since the number of males is assumed equal to the number of females, at least four eggs—and probably many more, since losses among fledglings are likely to be

TABLE 1

Year n	No. of juveniles $p_{j,n}$	No. of adults $p_{a,n}$	Total female population T_n in nth year	$p_{j,n}/p_{a,n}$ †	T_n/T_{n-1} †
0	0	10	10	0	—
1	20	5	25	4.00	2.50
2	10	8	18	1.18	0.74
3	17	7	24	2.34	1.31
4	14	8	22	1.66	0.96
5	17	8	25	2.00	1.13
10	22	12	34	1.87	1.06
11	24	12	36	1.88	1.07
12	25	13	38	1.88	1.06
20	42	22	64	1.88	1.06

†The figures in these columns were obtained before the numbers in the previous columns were rounded. Thus, for example, in year 2, $p_{j,2}/p_{a,2} = 10/8.5 \approx 1.176470588 \approx 1.18$.

high. From the model, it is apparent that α and β lie in the interval $[0, 1]$. Since juvenile birds are not as likely as adults to survive, we must have $\alpha < \beta$.

In Table 1 we assume that, initially, there are 10 female (and 10 male) adults and no juveniles. The computations were done on a computer, but the work would not be too onerous if done on a hand calculator. For example, $\mathbf{p}_1 = \begin{pmatrix} 0 & 2 \\ 0.3 & 0.5 \end{pmatrix}\begin{pmatrix} 0 \\ 10 \end{pmatrix} = \begin{pmatrix} 20 \\ 5 \end{pmatrix}$, so that $p_{j,1} = 20$, $p_{a,1} = 5$, the total female population after one year is 25, and the ratio of juvenile to adult females is 4 to 1. In the second year, $\mathbf{p}_2 = \begin{pmatrix} 0 & 2 \\ 0.3 & 0.5 \end{pmatrix}\begin{pmatrix} 20 \\ 5 \end{pmatrix} = \begin{pmatrix} 10 \\ 8.5 \end{pmatrix}$, which we round down to $\begin{pmatrix} 10 \\ 8 \end{pmatrix}$ since we cannot have $8\frac{1}{2}$ adult birds. Table 1 tabulates the ratios $p_{j,n}/p_{a,n}$ and the ratios T_n/T_{n-1} of the total number of females in successive years. ■

In Table 1 the ratio $p_{j,n}/p_{a,n}$ seems to be approaching the constant 1.88 while the total population seems to be increasing at a constant rate of 6 percent a year. Let us see if we can determine why this is the case.

First, we return to the general case of equation (4). Suppose that A has the real distinct eigenvalues λ_1 and λ_2 with corresponding eigenvectors $\mathbf{v}_1$ and $\mathbf{v}_2$. Since $\mathbf{v}_1$ and $\mathbf{v}_2$ are linearly independent, we can write

$$\mathbf{p}_0 = a_1\mathbf{v}_1 + a_2\mathbf{v}_2 \tag{5}$$

for some real numbers a_1 and a_2. Then (4) becomes

$$\mathbf{p}_n = A^n(a_1\mathbf{v}_1 + a_2\mathbf{v}_2). \tag{6}$$

But $A\mathbf{v}_1 = \lambda_1\mathbf{v}_1$ and $A^2\mathbf{v}_1 = A(A\mathbf{v}_1) = A(\lambda_1\mathbf{v}_1) = \lambda_1 A\mathbf{v}_1 = \lambda_1(\lambda_1\mathbf{v}_1) = \lambda_1^2\mathbf{v}_1$. Thus we can see that $A^n\mathbf{v}_1 = \lambda_1^n\mathbf{v}_1$, $A^n\mathbf{v}_2 = \lambda_2^n\mathbf{v}_2$, and, from (6),

$$\mathbf{p}_n = a_1\lambda_1^n\mathbf{v}_1 + a_2\lambda_2^n\mathbf{v}_2. \tag{7}$$

The characteristic equation of A is $\begin{vmatrix} -\lambda & k \\ \alpha & \beta - \lambda \end{vmatrix} = \lambda^2 - \beta\lambda - k\alpha = 0$ or $\lambda = (\beta \pm \sqrt{\beta^2 + 4\alpha k})/2$. By assumption, $k > 0$, $0 < \alpha < 1$, and $0 < \beta < 1$. Hence $4\alpha k > 0$ and $\beta^2 + 4\alpha k > 0$, which means that the eigenvalues are, indeed, real and distinct and that one eigenvalue λ_1 is positive, one λ_2 is negative, and $|\lambda_1| > |\lambda_2|$. We can write (7) as

$$\mathbf{p}_n = \lambda_1^n\left[a_1\mathbf{v}_1 + \left(\frac{\lambda_2}{\lambda_1}\right)^n a_2\mathbf{v}_2\right]. \tag{8}$$

Since $|\lambda_2/\lambda_1| < 1$, it is apparent that $(\lambda_2/\lambda_1)^n$ gets very small as n gets large. Thus, for n large,

$$\mathbf{p}_n \approx a_1\lambda_1^n\mathbf{v}_1. \tag{9}$$

This means that, in the long run, the age distribution stabilizes and is proportional to $\mathbf{v}_1$. Each age group will change by a factor of λ_1 each year. Thus—in the long run—equation (4) acts just like equation (1). In the short term—that is, before "stability" is reached—the numbers oscillate. The magnitude of this oscillation depends on the magnitude of λ_2/λ_1 (which is negative, thus explaining the oscillation).

EXAMPLE 1 (continued) For $A = \begin{pmatrix} 0 & 2 \\ 0.3 & 0.5 \end{pmatrix}$, we have $\lambda^2 - 0.5\lambda - 0.6 = 0$ or $\lambda = (0.5 \pm \sqrt{0.25 + 2.4})/2 = (0.5 \pm \sqrt{2.65})/2$, so that $\lambda_1 \approx 1.06$ and $\lambda_2 \approx -0.56$. This explains the 6 percent increase in population noted in the last column of Table 1. Corresponding to the eigenvalue $\lambda_1 = 1.06$, we compute $(A - 1.06I)\mathbf{v}_1 = \begin{pmatrix} -1.06 & 2 \\ 0.3 & -0.56 \end{pmatrix}\begin{pmatrix} x_1 \\ x_2 \end{pmatrix} = \begin{pmatrix} 0 \\ 0 \end{pmatrix}$ or $1.06x_1 = 2x_2$, so that $\mathbf{v}_1 = \begin{pmatrix} 1 \\ 0.53 \end{pmatrix}$ is an eigenvector. Similarly, $(A + 0.56)\mathbf{v}_2 = \begin{pmatrix} 0.56 & 2 \\ 0.3 & 1.06 \end{pmatrix}\begin{pmatrix} x_1 \\ x_2 \end{pmatrix} = \begin{pmatrix} 0 \\ 0 \end{pmatrix}$, so that $0.56x_1 + 2x_2 = 0$ and $\mathbf{v}_2 = \begin{pmatrix} 1 \\ -0.28 \end{pmatrix}$ is a second eigenvector. Note that in $\mathbf{v}_1$ we have $1/0.53 \approx 1.88$. This explains the ratio $p_{j,n}/p_{a,n}$ in the fifth column of the table. ■

It is remarkable just how much information is available from a simple computation of eigenvalues. For example, whether a population will ultimately increase or decrease may be a critical question. It will increase if $\lambda_1 > 1$, and the condition for that is $(\beta + \sqrt{\beta^2 + 4\alpha k})/2 > 1$ or $\sqrt{\beta^2 + 4\alpha k} > 2 - \beta$ or $\beta^2 + 4\alpha k > (2 - \beta)^2 = 4 - 4\beta + \beta^2$. This leads to $4\alpha k > 4 - 4\beta$ or

$$k > \frac{1 - \beta}{\alpha}. \tag{10}$$

In Example 1 we had $\beta = 0.5$, $\alpha = 0.3$; thus (10) is satisfied when $k > 0.5/0.3 \approx 1.67$.

Before we close this section we indicate two limitations of this model:

(i) Birth and death rates often change from year to year and are particularly dependent on the weather. This model assumes a constant environment.

(ii) Ecologists have found that for many species birth and death rates vary with the size of the population. In particular, a population cannot grow when it reaches a certain size due to the effects of limited food resources and overcrowding. Clearly a population cannot grow indefinitely at a constant rate. Otherwise that population would overrun the earth.

PROBLEMS 9.11

In Problems 1–3 find the numbers of juvenile and adult female birds after 1, 2, 5, 10, 19, and 20 years. Then find the long-term ratios of $p_{j,n}$ to $p_{a,n}$ and T_n to T_{n-1}. [*Hint:* Use equations (7) and (9) and a hand calculator and round to three decimals.]

1. $\mathbf{p}_0 = \begin{pmatrix} 0 \\ 12 \end{pmatrix}$; $k = 3$, $\alpha = 0.4$, $\beta = 0.6$

2. $\mathbf{p}_0 = \begin{pmatrix} 0 \\ 15 \end{pmatrix}$; $k = 1$, $\alpha = 0.3$, $\beta = 0.4$

3. $\mathbf{p}_0 = \begin{pmatrix} 0 \\ 20 \end{pmatrix}$; $k = 4$, $\alpha = 0.7$, $\beta = 0.8$

4. Show that $\alpha = \beta$ and $\alpha > \frac{1}{2}$, then the bird population will always increase in the long run if at least one female offspring on the average is produced by each female adult.

5. Show that, in the long run, the ratio $p_{j,n}/p_{a,n}$ approaches the limiting value k/λ_1.

6. Suppose we divide the adult birds into two age groups; those 1–5 years old and those more than 5 years old. Assume that the survival rate for birds in the first group is β while in the second group it is γ (and $\beta > \gamma$). Assume that the birds in the first group are equally divided as to age. (That is, if there are 100 birds in the group, then 20 are 1 year old, 20 are 2 years old, and so on.) Formulate a 3×3 matrix model for this situation.

(SKIP)

9.12 SIMILAR MATRICES AND DIAGONALIZATION

In this section we describe an interesting and useful relationship that can hold between two matrices.

Definition 1 SIMILAR MATRICES Two $n \times n$ matrices A and B are said to be **similar** if there exists an invertible $n \times n$ matrix C such that

$$B = C^{-1}AC. \qquad (1)$$

The function defined by (1) which takes the matrix A into the matrix B is called a **similarity transformation**.

NOTE. $C^{-1}(A_1 + A_2)C = C^{-1}A_1C + C^{-1}A_2C$ and $C^{-1}(\alpha A)C = \alpha C^{-1}AC$, so that the function defined by (1) is, in fact, a linear transformation. This explains the use of the word "transformation" in Definition 1.

The purpose of this section is to show (i) that similar matrices have several important properties in common and (ii) that most matrices are similar to diagonal matrices.

EXAMPLE 1 Let $A = \begin{pmatrix} 2 & 1 \\ 0 & -1 \end{pmatrix}$, $B = \begin{pmatrix} 4 & -2 \\ 5 & -3 \end{pmatrix}$, and $C = \begin{pmatrix} 2 & -1 \\ -1 & 1 \end{pmatrix}$. Then $CB = \begin{pmatrix} 2 & -1 \\ -1 & 1 \end{pmatrix}\begin{pmatrix} 4 & -2 \\ 5 & -3 \end{pmatrix} = \begin{pmatrix} 3 & -1 \\ 1 & -1 \end{pmatrix}$ and $AC = \begin{pmatrix} 2 & 1 \\ 0 & -1 \end{pmatrix}\begin{pmatrix} 2 & -1 \\ -1 & 1 \end{pmatrix} = \begin{pmatrix} 3 & -1 \\ 1 & -1 \end{pmatrix}$. Thus $CB = AC$. Since $\det C = 1 \neq 0$, C is invertible; and since $CB = AC$, we have $C^{-1}CB = C^{-1}AC$ or $B = C^{-1}AC$. This shows that A and B are similar. ■

EXAMPLE 2 Let $D = \begin{pmatrix} 1 & 0 & 0 \\ 0 & -1 & 0 \\ 0 & 0 & 2 \end{pmatrix}$, $A = \begin{pmatrix} -6 & -3 & -25 \\ 2 & 1 & 8 \\ 2 & 2 & 7 \end{pmatrix}$, and $C = \begin{pmatrix} 2 & 4 & 3 \\ 0 & 1 & -1 \\ 3 & 5 & 7 \end{pmatrix}$. C is invertible because $\det C = 3 \neq 0$. We then compute:

$$CA = \begin{pmatrix} 2 & 4 & 3 \\ 0 & 1 & -1 \\ 3 & 5 & 7 \end{pmatrix}\begin{pmatrix} -6 & -3 & -25 \\ 2 & 1 & 8 \\ 2 & 2 & 7 \end{pmatrix} = \begin{pmatrix} 2 & 4 & 3 \\ 0 & -1 & 1 \\ 6 & 10 & 14 \end{pmatrix}$$

$$DC = \begin{pmatrix} 1 & 0 & 0 \\ 0 & -1 & 0 \\ 0 & 0 & 2 \end{pmatrix}\begin{pmatrix} 2 & 4 & 3 \\ 0 & 1 & -1 \\ 3 & 5 & 7 \end{pmatrix} = \begin{pmatrix} 2 & 4 & 3 \\ 0 & -1 & 1 \\ 6 & 10 & 14 \end{pmatrix}$$

Thus $CA = DC$ and $A = C^{-1}DC$, so A and D are similar. ■

NOTE. In Examples 1 and 2 it was not necessary to compute C^{-1}. It was only necessary to know that C is nonsingular.

Theorem 1 If A and B are similar $n \times n$ matrices, then A and B have the same characteristic equation and, therefore, have the same eigenvalues.

Proof. Since A and B are similar, $B = C^{-1}AC$ and

$$\begin{aligned} \det(B - \lambda I) &= \det(C^{-1}AC - \lambda I) = \det[C^{-1}AC - C^{-1}(\lambda I)C] \\ &= \det[C^{-1}(A - \lambda I)C] = \det(C^{-1}) \det(A - \lambda I) \det(C) \\ &= \det(C^{-1}) \det(C) \det(A - \lambda I) = \det(C^{-1}C) \det(A - \lambda I) \\ &= \det I \det(A - \lambda I) = \det(A - \lambda I). \end{aligned}$$

This means that A and B have the same characteristic equation and, since eigenvalues are roots of the characteristic equation, they have the same eigenvalues. ■

EXAMPLE 3 In Example 2 it is obvious that the eigenvalues of $D = \begin{pmatrix} 1 & 0 & 0 \\ 0 & -1 & 0 \\ 0 & 0 & 2 \end{pmatrix}$ are 1, −1, and 2.

Thus these are the eigenvalues of $A = \begin{pmatrix} -6 & -3 & -25 \\ 2 & 1 & 8 \\ 2 & 2 & 7 \end{pmatrix}$. Check this by verifying that $\det(A - I) = \det(A + I) = \det(A - 2I) = 0$. ■

In a variety of applications it is useful to "diagonalize" a matrix A—that is, to find a diagonal matrix similar to A.

Definition 2 DIAGONALIZABLE MATRIX An $n \times n$ matrix A is **diagonalizable** if there is a diagonal matrix D such that A is similar to D.

REMARK. If D is a diagonal matrix, then its eigenvalues are its diagonal components. If A is similar to D, then A and D have the same eigenvalues (by Theorem 1). Putting these two facts together, we observe that if A is diagonalizable, then A is similar to a diagonal matrix whose diagonal components are the eigenvalues of A.

The next theorem tells us when a matrix is diagonalizable.

Theorem 2 An $n \times n$ matrix A is diagonalizable if and only if it has n linearly independent eigenvectors. In that case, the diagonal matrix D similar to A is given by

$$D = \begin{pmatrix} \lambda_1 & 0 & 0 & \cdots & 0 \\ 0 & \lambda_2 & 0 & \cdots & 0 \\ 0 & 0 & \lambda_3 & \cdots & 0 \\ \vdots & \vdots & \vdots & & \vdots \\ 0 & 0 & 0 & \cdots & \lambda_n \end{pmatrix} \tag{2}$$

where $\lambda_1, \lambda_2, \ldots, \lambda_n$ are the eigenvalues of A. If C is a matrix whose columns are linearly independent eigenvectors of A, then

$$D = C^{-1}AC. \tag{3}$$

Proof. We first assume that A has n linearly independent eigenvectors $\mathbf{v}_1, \mathbf{v}_2, \ldots, \mathbf{v}_n$ corresponding to the (not necessarily distinct) eigenvalues $\lambda_1, \lambda_2, \ldots, \lambda_n$. Let

$$\mathbf{v}_1 = \begin{pmatrix} c_{11} \\ c_{21} \\ \vdots \\ c_{n1} \end{pmatrix}, \mathbf{v}_2 = \begin{pmatrix} c_{12} \\ c_{22} \\ \vdots \\ c_{n2} \end{pmatrix}, \ldots, \mathbf{v}_n = \begin{pmatrix} c_{1n} \\ c_{2n} \\ \vdots \\ c_{nn} \end{pmatrix}$$

and let

$$C = \begin{pmatrix} c_{11} & c_{12} & \cdots & c_{1n} \\ c_{21} & c_{22} & \cdots & c_{2n} \\ \vdots & \vdots & & \vdots \\ c_{n1} & c_{n2} & \cdots & c_{nn} \end{pmatrix}.$$

Then C is invertible since its columns are linearly independent. Now

$$AC = \begin{pmatrix} a_{11} & a_{12} & \cdots & a_{1n} \\ a_{21} & a_{22} & \cdots & a_{2n} \\ \vdots & \vdots & & \vdots \\ a_{n1} & a_{n2} & \cdots & a_{nn} \end{pmatrix} \begin{pmatrix} c_{11} & c_{12} & \cdots & c_{1n} \\ c_{21} & c_{22} & \cdots & c_{2n} \\ \vdots & \vdots & & \vdots \\ c_{n1} & c_{n2} & \cdots & c_{nn} \end{pmatrix}$$

and we see that the ith column of AC is $A\begin{pmatrix} c_{1i} \\ c_{2i} \\ \vdots \\ c_{ni} \end{pmatrix} = A\mathbf{v}_i = \lambda_i\mathbf{v}_i$. Thus AC is the matrix whose ith column is $\lambda_i\mathbf{v}_i$ and

$$AC = \begin{pmatrix} \lambda_1 c_{11} & \lambda_2 c_{12} & \cdots & \lambda_n c_{1n} \\ \lambda_1 c_{21} & \lambda_2 c_{22} & \cdots & \lambda_n c_{2n} \\ \vdots & \vdots & & \vdots \\ \lambda_1 c_{n1} & \lambda_2 c_{n2} & \cdots & \lambda_n c_{nn} \end{pmatrix}$$

But

$$CD = \begin{pmatrix} c_{11} & c_{12} & \cdots & c_{1n} \\ c_{21} & c_{22} & \cdots & c_{2n} \\ \vdots & \vdots & & \vdots \\ c_{n1} & c_{n2} & \cdots & c_{nn} \end{pmatrix} \begin{pmatrix} \lambda_1 & 0 & \cdots & 0 \\ 0 & \lambda_2 & \cdots & 0 \\ \vdots & \vdots & & \vdots \\ 0 & 0 & \cdots & \lambda_n \end{pmatrix} = \begin{pmatrix} \lambda_1 c_{11} & \lambda_2 c_{12} & \cdots & \lambda_n c_{1n} \\ \lambda_1 c_{21} & \lambda_2 c_{22} & \cdots & \lambda_n c_{2n} \\ \vdots & \vdots & & \vdots \\ \lambda_1 c_{n1} & \lambda_2 c_{n2} & \cdots & \lambda_n c_{nn} \end{pmatrix}.$$

Thus

$$AC = CD \tag{4}$$

and, since C is invertible, we can multiply both sides of (4) on the left by C^{-1} to obtain

$$D = C^{-1}AC \tag{5}$$

This proves that if A has n linearly independent eigenvectors, then A is diagonalizable. Conversely, suppose that A is diagonalizable. That is, suppose that (5) holds for some invertible matrix C. Let $\mathbf{v}_1, \mathbf{v}_2, \ldots, \mathbf{v}_n$ be the columns of C. Then $AC = CD$ and, reversing the arguments above, we immediately see that $A\mathbf{v}_i = \lambda_i\mathbf{v}_i$ for $i = 1, 2, \ldots, n$. Thus $\mathbf{v}_1, \mathbf{v}_2, \ldots, \mathbf{v}_n$ are eigenvectors of A and are linearly independent because C is invertible. ■

NOTATION: To indicate that D is a diagonal matrix with diagonal components $\lambda_1, \lambda_2, \ldots, \lambda_n$, we write $D = \text{diag}\,(\lambda_1, \lambda_2, \ldots, \lambda_n)$.

Theorem 2 has a useful corollary that follows immediately from Theorem 9.10.3:

Corollary If the $n \times n$ matrix A has n distinct eigenvalues, then A is diagonalizable.

REMARK. If the real coefficients of a polynomial of degree n are picked at random, then, with probability 1, the polynomial will have n distinct roots. It is not difficult to see, intuitively, why this is so. If $n = 2$, for example, then the equation $\lambda^2 + a\lambda + b = 0$ has equal roots if and only if $a^2 = 4b$—a highly unlikely event if a and b are chosen at random. We can, of course, write down polynomials having roots of algebraic multiplicity greater than 1, but these polynomials are exceptional. Thus, without attempting to be mathematically precise, we can accurately say that *most* polynomials have distinct roots. Hence *most* matrices have distinct eigenvalues and, as we stated at the beginning of the section, *most* matrices are diagonalizable.

EXAMPLE 4 Let $A = \begin{pmatrix} 4 & 2 \\ 3 & 3 \end{pmatrix}$. In Example 9.10.3 we found the two linearly independent eigenvectors $\mathbf{v}_1 = \begin{pmatrix} 2 \\ -3 \end{pmatrix}$ and $\mathbf{v}_2 = \begin{pmatrix} 1 \\ 1 \end{pmatrix}$. Then, setting $C = \begin{pmatrix} 2 & 1 \\ -3 & 1 \end{pmatrix}$, we find that

$$C^{-1}AC = \frac{1}{5}\begin{pmatrix} 1 & -1 \\ 3 & 2 \end{pmatrix}\begin{pmatrix} 4 & 2 \\ 3 & 3 \end{pmatrix}\begin{pmatrix} 2 & 1 \\ -3 & 1 \end{pmatrix}$$

$$= \frac{1}{5}\begin{pmatrix} 1 & -1 \\ 3 & 2 \end{pmatrix}\begin{pmatrix} 2 & 6 \\ -3 & 6 \end{pmatrix} = \frac{1}{5}\begin{pmatrix} 5 & 0 \\ 0 & 30 \end{pmatrix} = \begin{pmatrix} 1 & 0 \\ 0 & 6 \end{pmatrix}$$

which is the matrix whose diagonal components are the eigenvalues of A. ■

EXAMPLE 5 Let $A = \begin{pmatrix} 1 & -1 & 4 \\ 3 & 2 & -1 \\ 2 & 1 & -1 \end{pmatrix}$. In Example 9.10.4 we computed the three linearly independent eigenvectors $\mathbf{v}_1 = \begin{pmatrix} -1 \\ 4 \\ 1 \end{pmatrix}$, $\mathbf{v}_2 = \begin{pmatrix} 1 \\ -1 \\ -1 \end{pmatrix}$, and $\mathbf{v}_3 = \begin{pmatrix} 1 \\ 2 \\ 1 \end{pmatrix}$. Then $C = \begin{pmatrix} -1 & 1 & 1 \\ 4 & -1 & 2 \\ 1 & -1 & 1 \end{pmatrix}$

and

$$C^{-1}AC = -\frac{1}{6}\begin{pmatrix} 1 & -2 & 3 \\ -2 & -2 & 6 \\ -3 & 0 & -3 \end{pmatrix}\begin{pmatrix} 1 & -1 & 4 \\ 3 & 2 & -1 \\ 2 & 1 & -1 \end{pmatrix}\begin{pmatrix} -1 & 1 & 1 \\ 4 & -1 & 2 \\ 1 & -1 & 1 \end{pmatrix}$$

$$= -\frac{1}{6}\begin{pmatrix} 1 & -2 & 3 \\ -2 & -2 & 6 \\ -3 & 0 & -3 \end{pmatrix}\begin{pmatrix} -1 & -2 & 3 \\ 4 & 2 & 6 \\ 1 & 2 & 3 \end{pmatrix}$$

$$= -\frac{1}{6}\begin{pmatrix} -6 & 0 & 0 \\ 0 & 12 & 0 \\ 0 & 0 & -18 \end{pmatrix} = \begin{pmatrix} 1 & 0 & 0 \\ 0 & -2 & 0 \\ 0 & 0 & 3 \end{pmatrix}$$

with eigenvalues 1, -2, and 3. ■

REMARK. Since there are an infinite number of ways to choose an eigenvector, there are an infinite number of ways to choose the diagonalizing matrix C. The only advice is to choose the eigenvectors and matrix C that are, arithmetically, the easiest to work with. This usually means that you should insert as many zeros and ones as possible.

EXAMPLE 6 Let $A = \begin{pmatrix} 3 & 2 & 4 \\ 2 & 0 & 2 \\ 4 & 2 & 3 \end{pmatrix}$. Then, from Example 9.10.9 we have the three linearly independent eigenvectors $\mathbf{v}_1 = \begin{pmatrix} 2 \\ 1 \\ 2 \end{pmatrix}$, $\mathbf{v}_2 = \begin{pmatrix} 1 \\ -2 \\ 0 \end{pmatrix}$, and $\mathbf{v}_3 = \begin{pmatrix} 0 \\ -2 \\ 1 \end{pmatrix}$. Setting $C = \begin{pmatrix} 2 & 1 & 0 \\ 1 & -2 & -2 \\ 2 & 0 & 1 \end{pmatrix}$, we obtain

$$C^{-1}AC = -\frac{1}{9}\begin{pmatrix} -2 & -1 & -2 \\ -5 & 2 & 4 \\ 4 & 2 & -5 \end{pmatrix}\begin{pmatrix} 3 & 2 & 4 \\ 2 & 0 & 2 \\ 4 & 2 & 3 \end{pmatrix}\begin{pmatrix} 2 & 1 & 0 \\ 1 & -2 & -2 \\ 2 & 0 & 1 \end{pmatrix}$$

$$= -\frac{1}{9}\begin{pmatrix} -2 & -1 & -2 \\ -5 & 2 & 4 \\ 4 & 2 & -5 \end{pmatrix}\begin{pmatrix} 16 & -1 & 0 \\ 8 & 2 & 2 \\ 16 & 0 & -1 \end{pmatrix}$$

$$= -\frac{1}{9}\begin{pmatrix} -72 & 0 & 0 \\ 0 & 9 & 0 \\ 0 & 0 & 9 \end{pmatrix} = \begin{pmatrix} 8 & 0 & 0 \\ 0 & -1 & 0 \\ 0 & 0 & -1 \end{pmatrix}.$$

This example illustrates that A is diagonalizable even though its eigenvalues are not distinct. ■

EXAMPLE 7 Let $A = \begin{pmatrix} 4 & 1 \\ 0 & 4 \end{pmatrix}$. In Example 9.10.8 we saw that A did *not* have two linearly independent eigenvectors. Suppose that A were diagonalizable (in contradiction to Theorem 2). Then $D = \begin{pmatrix} 4 & 0 \\ 0 & 4 \end{pmatrix}$ and there would be an invertible matrix C such that $C^{-1}AC = D$. Multiplying this equation on the left by C and on the right by C^{-1}, we find that $A = CDC^{-1} = C\begin{pmatrix} 4 & 0 \\ 0 & 4 \end{pmatrix}C^{-1} = C(4I)C^{-1} = 4CIC^{-1} = 4CC^{-1} = 4I = \begin{pmatrix} 4 & 0 \\ 0 & 4 \end{pmatrix} = D$. But $A \neq D$, so no such C exists. ■

We have seen that many matrices are similar to diagonal matrices. However, one question remains: What do we do if A is not diagonalizable? The answer to this question involves a discussion of the *Jordan canonical form* of a matrix, a subject not in this text.[†]

[†]This topic is discussed in Section 7.6 of S. I. Grossman, *Elementary Linear Algebra*, Second Edition (Wadsworth, Belmont, CA 1984).

PROBLEMS 9.12

In Problems 1–15 determine whether the given matrix A is diagonalizable. If it is, find a matrix C such that $C^{-1}AC = D$.

1. $\begin{pmatrix} -2 & -2 \\ -5 & 1 \end{pmatrix}$

2. $\begin{pmatrix} 3 & -1 \\ -2 & 4 \end{pmatrix}$

3. $\begin{pmatrix} 2 & -1 \\ 5 & -2 \end{pmatrix}$

4. $\begin{pmatrix} 3 & -5 \\ 1 & -1 \end{pmatrix}$

5. $\begin{pmatrix} 3 & 2 \\ -5 & 1 \end{pmatrix}$

6. $\begin{pmatrix} 1 & -1 & 0 \\ -1 & 2 & -1 \\ 0 & -1 & 1 \end{pmatrix}$

7. $\begin{pmatrix} 1 & 1 & -2 \\ -1 & 2 & 1 \\ 0 & 1 & -1 \end{pmatrix}$

8. $\begin{pmatrix} 2 & 1 & 0 \\ 0 & 0 & 1 \\ 0 & 0 & 0 \end{pmatrix}$

9. $\begin{pmatrix} 3 & 0 & 0 \\ 0 & 0 & 1 \\ 0 & 0 & 2 \end{pmatrix}$

10. $\begin{pmatrix} 3 & -1 & -1 \\ 1 & 1 & -1 \\ 1 & -1 & 1 \end{pmatrix}$

11. $\begin{pmatrix} 7 & -2 & -4 \\ 3 & 0 & -2 \\ 6 & -2 & -3 \end{pmatrix}$

12. $\begin{pmatrix} 4 & 6 & 6 \\ 1 & 3 & 2 \\ -1 & -5 & -2 \end{pmatrix}$

13. $\begin{pmatrix} -3 & -7 & -5 \\ 2 & 4 & 3 \\ 1 & 2 & 2 \end{pmatrix}$

14. $\begin{pmatrix} -2 & -2 & 0 & 0 \\ -5 & 1 & 0 & 0 \\ 0 & 0 & 2 & -1 \\ 0 & 0 & 5 & -2 \end{pmatrix}$

15. $\begin{pmatrix} 4 & 1 & 0 & 1 \\ 2 & 3 & 0 & 1 \\ -2 & 1 & 2 & -3 \\ 2 & -1 & 0 & 5 \end{pmatrix}$

16. Show that if A is similar to B and B is similar to C, then A is similar to C.

***17.** Let T be a linear transformation from $\mathbb{R}^n \to \mathbb{R}^n$. Let B be the matrix representation of T with respect to a basis B_1, and let C be the matrix representation of T with respect to a basis B_2. Show that B is similar to C by showing that $B = ACA^{-1}$, where A is the transition matrix from B_2 to B_1 (see Theorem 9.5.1).

18. If A is similar to B, show that $\rho(A) = \rho(B)$ and $\nu(A) = \nu(B)$. [*Hint:* First prove that if C is invertible, then $\nu(CA) = \nu(A)$ by showing that $\mathbf{x} \in N_A$ if and only if $\mathbf{x} \in N_{CA}$. Next prove that $\rho(AC) = \rho(A)$ by showing that $R_A = R_{AC}$. Conclude that $\rho(AC) = \rho(CA) = \rho(A)$. Finally, use the fact that C^{-1} is invertible to show that $\rho(C^{-1}AC) = \rho(A)$.]

19. If A is similar to B, show that A^n is similar to B^n for any positive integer n.

20. If A is similar to B, show that $\det A = \det B$.

21. Let $D = \begin{pmatrix} 1 & 0 \\ 0 & -1 \end{pmatrix}$. Compute D^{20}.

22. Let $A = \begin{pmatrix} 3 & -4 \\ 2 & -3 \end{pmatrix}$. Compute A^{20}. [*Hint:* Find a C such that $A = CDC^{-1}$, where D is diagonal and show that $A^{20} = CD^{20}C^{-1}$.]

23. Suppose that $C^{-1}AC = D$. Show that for any integer n, $A^n = CD^nC^{-1}$. This gives an easy way to compute powers of a diagonalizable matrix.

24. Use the result of Problem 23 and Example 6 to compute A^{10}, where $A = \begin{pmatrix} 3 & 2 & 4 \\ 2 & 0 & 2 \\ 4 & 2 & 3 \end{pmatrix}$.

***25.** Let A be an $n \times n$ matrix whose characteristic equation is $(\lambda - c)^n = 0$. Show that A is diagonalizable if and only if $A = cI$.

26. If A is diagonalizable, show that $\det A = \lambda_1\lambda_2 \cdots \lambda_n$, where $\lambda_1, \lambda_2, \ldots, \lambda_n$ are the eigenvalues of A.

***27.** Let A and B be real $n \times n$ matrices with distinct eigenvalues. Prove that $AB = BA$ if and only if A and B have the same eigenvectors.

REVIEW EXERCISES FOR CHAPTER NINE

In Exercises 1–10 determine whether the given set is a vector space. If so, determine its dimension. If it is finite dimensional, find a basis for it.

1. The vectors (x, y, z) in $\mathbb{R}^3$ satisfying $x + 2y - z = 0$

2. The vectors (x, y, z) in $\mathbb{R}^3$ satisfying $x + 2y - z \leq 0$

3. The vectors (x, y, z, w) in $\mathbb{R}^4$ satisfying $x + y + z + w = 0$

4. The vectors in $\mathbb{R}^3$ satisfying $x - 2 = y + 3 = z - 4$

5. The set of upper triangular $n \times n$ matrices under the operations of matrix addition and scalar multiplication.

6. The set of polynomials of degree ≤ 5

7. The set of polynomials of degree 5
8. The set of 3×2 matrices $A = (a_{ij})$, with $a_{12} = 0$, under the operations of matrix addition and scalar multiplication
9. The set in Exercise 8 except that $a_{12} = 1$
10. The set $S = \{f \in C[0, 2]: f(2) = 0\}$

In Exercises 11–14 determine whether the given set of vectors is linearly dependent or independent.

11. In P_3: $1, 2 - x^2, 3 - x, 7x^2 - 8x$
12. In P_3: $1, 2 + x^3, 3 - x, 7x^2 - 8x$
13. In M_{22}: $\begin{pmatrix} 1 & -1 \\ 0 & 0 \end{pmatrix}, \begin{pmatrix} 1 & 1 \\ 0 & 0 \end{pmatrix}, \begin{pmatrix} 0 & 0 \\ 1 & 1 \end{pmatrix}, \begin{pmatrix} 0 & 0 \\ 1 & -1 \end{pmatrix}$
14. In M_{22}: $\begin{pmatrix} 1 & 1 \\ 0 & 0 \end{pmatrix}, \begin{pmatrix} 1 & -1 \\ 0 & 0 \end{pmatrix}, \begin{pmatrix} 0 & 0 \\ 1 & 1 \end{pmatrix}, \begin{pmatrix} 0 & 0 \\ 1 & -1 \end{pmatrix}$

In Exercises 15–20 find a basis for the given vector space and determine its dimension.

15. The vectors in $\mathbb{R}^3$ lying on the plane $2x + 3y - 4z = 0$
16. $H = \{(x, y) \in \mathbb{R}^2: 2x - 3y = 0\}$
17. $\{\mathbf{v} \in \mathbb{R}^4: 3x - y - z + w = 0\}$
18. $\{p \in P_3: p(0) = 0\}$
19. The set of diagonal 4×4 matrices
20. M_{32}

In Exercises 21–23 write the given vector in terms of the new given basis vectors.

21. In $\mathbb{R}^2$: $\mathbf{x} = \begin{pmatrix} 2 \\ -1 \end{pmatrix}$; $\begin{pmatrix} 1 \\ 2 \end{pmatrix}, \begin{pmatrix} -1 \\ 2 \end{pmatrix}$
22. In $\mathbb{R}^3$: $\mathbf{x} = \begin{pmatrix} -3 \\ 4 \\ 2 \end{pmatrix}$; $\begin{pmatrix} 1 \\ 0 \\ 1 \end{pmatrix}, \begin{pmatrix} 1 \\ 1 \\ 0 \end{pmatrix}, \begin{pmatrix} 0 \\ 2 \\ 3 \end{pmatrix}$
23. In P_2: $\mathbf{x} = 4 + x^2$; $1 + x^2, 1 + x, 1$

In Exercises 24–29 determine whether the given transformation from V to W is linear.

24. $T: \mathbb{R}^2 \to \mathbb{R}^2$; $T(x, y) = (0, -y)$
25. $T: \mathbb{R}^3 \to \mathbb{R}^3$; $T(x, y, z) = (1, y, z)$
26. $T: \mathbb{R}^2 \to \mathbb{R}^2$; $T(x, y) = x/y$
27. $T: P_1 \to P_2$; $(Tp)(x) = xp(x)$
28. $T: P_2 \to P_2$; $(Tp)(x) = 1 + p(x)$
29. $T: C[0, 1] \to C[0, 1]$; $Tf(x) = f(1)$

In Exercises 30–35 find the kernel, range, nullity, and rank of the given matrix.

30. $A = \begin{pmatrix} 1 & -2 \\ -2 & 4 \end{pmatrix}$
31. $A = \begin{pmatrix} 1 & -1 & 3 \\ 2 & 0 & 4 \\ 0 & -2 & 2 \end{pmatrix}$
32. $A = \begin{pmatrix} 1 & -1 & 2 \\ 0 & 1 & 4 \\ 1 & -1 & 0 \end{pmatrix}$
33. $A = \begin{pmatrix} 2 & 4 & -2 \\ -1 & -2 & 1 \end{pmatrix}$
34. $A = \begin{pmatrix} 2 & 3 \\ -1 & 2 \\ 4 & 6 \end{pmatrix}$
35. $A = \begin{pmatrix} 1 & -1 & 2 & 3 \\ 0 & 1 & -1 & 0 \\ 1 & -2 & 3 & 3 \\ 2 & -3 & 5 & 6 \end{pmatrix}$

In Exercises 36–40 find the matrix representation of the given linear transformation and find the kernel, range, nullity, and rank of the transformation.

36. $T: \mathbb{R}^2 \to \mathbb{R}^2$; $T(x, y) = (0, -y)$
37. $T: \mathbb{R}^3 \to \mathbb{R}^2$; $T(x, y, z) = (y, z)$
38. $T: \mathbb{R}^4 \to \mathbb{R}^2$; $T(x, y, z, w) = (x - 2z, 2y + 3w)$
39. $T: P_3 \to P_4$; $(Tp)(x) = xp(x)$
40. $T: M_{22} \to M_{22}$; $TA = AB$, where $B = \begin{pmatrix} -1 & 0 \\ 1 & 2 \end{pmatrix}$

In Exercises 41–46 calculate the eigenvalues and eigenspaces of the given matrix.

41. $\begin{pmatrix} -8 & 12 \\ -6 & 10 \end{pmatrix}$
42. $\begin{pmatrix} 2 & 5 \\ 0 & 2 \end{pmatrix}$
43. $\begin{pmatrix} 1 & 0 & 0 \\ 3 & 7 & 0 \\ -2 & 4 & -5 \end{pmatrix}$
44. $\begin{pmatrix} 1 & -1 & 0 \\ 1 & 2 & 1 \\ -2 & 1 & -1 \end{pmatrix}$
45. $\begin{pmatrix} 5 & -2 & 0 & 0 \\ 4 & -1 & 0 & 0 \\ 0 & 0 & 3 & -1 \\ 0 & 0 & 2 & 3 \end{pmatrix}$
46. $\begin{pmatrix} -2 & 1 & 0 \\ 0 & -2 & 1 \\ 0 & 0 & -2 \end{pmatrix}$

In Exercises 47–55 determine whether the given matrix A is diagonalizable. If it is, find a matrix C such that $C^{-1}AC = D$.

47. $\begin{pmatrix} -18 & -15 \\ 20 & 17 \end{pmatrix}$

48. $\begin{pmatrix} \frac{17}{2} & \frac{9}{2} \\ -15 & -8 \end{pmatrix}$

49. $\begin{pmatrix} 1 & 1 & 1 \\ -1 & -1 & 0 \\ -1 & 0 & -1 \end{pmatrix}$

50. $\begin{pmatrix} 4 & 2 & 0 \\ 2 & 4 & 0 \\ 0 & 0 & -3 \end{pmatrix}$

51. $\begin{pmatrix} -3 & 2 & 1 \\ -7 & 4 & 2 \\ -5 & 3 & 2 \end{pmatrix}$

52. $\begin{pmatrix} 8 & 0 & 12 \\ 0 & -2 & 0 \\ 12 & 0 & -2 \end{pmatrix}$

53. $\begin{pmatrix} 2 & 2 & 0 \\ 2 & 2 & 0 \\ 0 & 0 & -3 \end{pmatrix}$

54. $\begin{pmatrix} 4 & 2 & -2 & 2 \\ 1 & 3 & 1 & -1 \\ 0 & 0 & 2 & 0 \\ 1 & 1 & -3 & 5 \end{pmatrix}$

55. $\begin{pmatrix} 3 & 4 & -4 & 0 \\ 0 & -1 & 0 & 0 \\ 0 & 0 & -1 & 0 \\ 0 & -4 & 4 & 3 \end{pmatrix}$

10 Calculus in $\mathbb{R}^n$

In Chapters 3 and 4 we discussed vector functions and functions of several variables first in $\mathbb{R}^2$ and $\mathbb{R}^3$ and then in $\mathbb{R}^n$. In this chapter we go further and discuss functions with domain in $\mathbb{R}^n$ and range in $\mathbb{R}^m$. We begin by extending Taylor's theorem to n variables.

10.1 TAYLOR'S THEOREM IN n VARIABLES

In your one-variable calculus course you may have studied Taylor's theorem. This important result is stated below. If this topic is unfamiliar, please read the review material in Sections 13.1–13.3 before going on in this section.

TAYLOR'S THEOREM FOR A FUNCTION OF ONE VARIABLE

Let the function f and its first $n + 1$ derivatives be continuous on the interval $[x_0, x]$. Then

$$f(x) = f(x_0) + \frac{f'(x_0)}{1!}(x - x_0) + \frac{f''(x_0)}{2!}(x - x_0)^2 + \cdots + \frac{f^{(n)}(x_0)}{n!}(x - x_0)^n + R_n(x) \tag{1}$$

where

$$R_n(x) = \frac{f^{(n+1)}(c)}{(n+1)!}(x - x_0)^{n+1} \tag{2}$$

for some number c in $[x_0, x]$. We can rewrite (1) as

$$f(x) = p_n(x) + R_n(x). \tag{3}$$

Before continuing, we list four common Taylor polynomials. These are obtained in Sections 13.1 and 13.3:

$$e^x \approx 1 + x + \frac{x^2}{2!} + \frac{x^3}{3!} + \cdots + \frac{x^n}{n!} \tag{4}$$

$$\sin x \approx x - \frac{x^3}{3!} + \frac{x^5}{5!} - \cdots + \frac{(-1)^n x^{2n+1}}{(2n+1)!} \tag{5}$$

$$\cos x \approx 1 - \frac{x^2}{2!} + \frac{x^4}{4!} - \cdots + \frac{(-1)^n x^{2n}}{(2n)!} \tag{6}$$

$$\ln(1 + x) \approx x - \frac{x^2}{2} + \frac{x^3}{3} - \cdots + \frac{(-1)^{n+1} x^n}{n} \tag{7}$$

In this section we shall extend Taylor's theorem to a function of n variables. More precisely, we shall indicate how a function $f(\mathbf{x})$ can be approximated by an nth-degree polynomial $P_n(x)$ with a remainder term similar to (2).

The result we want is obtained by repeated application of the chain rule (Theorem 4.6.4 on p. 215). We assume that the function $f\colon \mathbb{R}^n \to \mathbb{R}$ is defined on an open set Ω and that the line segment joining $\mathbf{x}$ and $\mathbf{x}_0$ in $\mathbb{R}^n$ is in Ω (see Definition 4.6.1 on p. 216). We define the vector $\mathbf{h}$ by

$$\mathbf{h} = \mathbf{x} - \mathbf{x}_0. \tag{8}$$

Keep in mind that $\mathbf{h} = (h_1, h_2, \ldots, h_n)$ is an n-vector. Let

$$g(t) = f(\mathbf{x}_0 + t\mathbf{h}). \tag{9}$$

Note that g is defined for $t \in [0, 1]$ because $\{\mathbf{x}\colon \mathbf{x} = \mathbf{x}_0 + t\mathbf{h}, 0 \le t \le 1\}$ is the line segment joining $\mathbf{x}_0$ and $\mathbf{x}$. We assume that f is of class $C^{(m)}(\Omega)$. Then g has m continuous derivatives, which we now compute. We first note that

$$\frac{d}{dt}(\mathbf{x}_0 + t\mathbf{h}) = \mathbf{h}$$

since $\mathbf{h}$ is assumed to be constant (see Example 3.8.11 on p. 140). Thus, by the chain rule,

$$g'(t) = \frac{d}{dt} f(\mathbf{x}_0 + t\mathbf{h}) = \nabla f(\mathbf{x}_0 + t\mathbf{h}) \cdot \mathbf{h}. \tag{10}$$

Writing out the terms in the scalar product in (10), we have

$$g'(t) = \sum_{i=1}^{n} f_i(\mathbf{x}_0 + t\mathbf{h})h_i. \tag{11}$$

The idea now is to compute higher-order derivatives of g and then to apply Taylor's theorem of one variable to g. Consider the function $f_i(\mathbf{x}_0 + t\mathbf{h})$. We can apply the chain rule again to obtain

$$\frac{d}{dt} f_i(\mathbf{x}_0 + t\mathbf{h}) = \nabla f_i(\mathbf{x}_0 + t\mathbf{h}) \cdot \mathbf{h}. \tag{12}$$

But $\nabla f_i = (f_{i1}, f_{i2}, \ldots, f_{in})$, so that, from (12),

$$\frac{d}{dt} f_i(\mathbf{x}_0 + t\mathbf{h}) = \sum_{j=1}^{n} f_{ij}(\mathbf{x}_0 + t\mathbf{h})\mathbf{h}_j. \tag{13}$$

Note that the subscripts on f represent partial derivatives, whereas those on the vector **h** represent the components of **h**. Now, from (11),

$$g''(t) = \sum_{i=1}^{n} \frac{d}{dt} f_i(\mathbf{x}_0 + t\mathbf{h})h_i. \tag{14}$$

So inserting (13) into (14) yields

$$g''(t) = \sum_{i=1}^{n} \left[\sum_{j=1}^{n} f_{ij}(\mathbf{x}_0 + t\mathbf{h})h_j \right] h_i. \tag{15}$$

This is getting complicated, but if you have come this far, you can probably see the pattern. We have

$$g'''(t) = \sum_{i=1}^{n} \left\{ \sum_{j=1}^{n} \left[\sum_{k=1}^{n} f_{ijk}(\mathbf{x}_0 + t\mathbf{h})h_k \right] h_j \right\} h_i \tag{16}$$

which we write in the abbreviated form

$$g'''(t) = \sum_{i,j,k=1}^{n} f_{ijk}(\mathbf{x}_0 + t\mathbf{h})h_i h_j h_k. \tag{17}$$

Finally, we have

$$g^{(m)}(t) = \sum_{i_1,i_2,\ldots,i_m=1}^{n} f_{i_1 i_2 i_3 \cdots i_m}(\mathbf{x}_0 + t\mathbf{h})h_{i_1}h_{i_2}\cdots h_{i_m}. \tag{18}$$

Formula (18) is formidable, but it is useful in deriving Taylor's theorem for n variables.

Theorem 1 Let $f: \mathbb{R}^n \to \mathbb{R}$ be of class $C^{(m+1)}(\Omega)$ and let the line segment joining $\mathbf{x}_0$ and $\mathbf{x}$ be in Ω. Then if $\mathbf{x} = (x_1, x_2, \ldots, x_n)$ and $\mathbf{x}_0 = (x_1^{(0)}, x_2^{(0)}, \ldots, x_n^{(0)})$,

$$\begin{aligned} \mathbf{f}(\mathbf{x}) &= f(\mathbf{x}_0) + \sum_{i=1}^{n} f_i(\mathbf{x}_0)(x_i - x_i^{(0)}) + \frac{1}{2!}\sum_{i=1}^{n}\sum_{j=1}^{n} f_{ij}(\mathbf{x}_0)(x_i - x_i^{(0)})(x_j - x_j^{(0)}) \\ &+ \frac{1}{3!}\sum_{i,j,k=1}^{n} f_{ijk}(\mathbf{x}_0)(x_i - x_i^{(0)})(x_j - x_j^{(0)})(x_k - x_k^{(0)}) + \cdots \\ &+ \frac{1}{m!}\sum_{i_1,i_2,\ldots,i_m=1}^{n} f_{i_1 i_2 \cdots i_m}(\mathbf{x}_0)(x_{i_1} - x_{i_1}^{(0)})(x_{i_2} - x_{i_2}^{(0)}) \cdots (x_{i_m} - x_{i_m}^{(0)}) + R_m(\mathbf{x}) \end{aligned} \tag{19}$$

where

$$\begin{aligned} R_m(\mathbf{x}) &= \frac{1}{(m+1)!} \\ &\times \sum_{i_1,i_2,\ldots,i_{m+1}=1}^{n} f_{i_1 i_2 \cdots i_{m+1}}(\mathbf{x}_0 + c\mathbf{h})(x_{i_1} - x_{i_1}^{(0)})(x_{i_2} - x_{i_2}^{(0)}) \cdots (x_{i_{m+1}} - x_{i_{m+1}}^{(0)}) \end{aligned} \tag{20}$$

for some number c in $(0, 1)$.

Proof. Since $g(t)$ has $m + 1$ continuous derivatives on $[0, 1]$ we have, from Taylor's theorem

$$g(1) = g(0) + g'(0) + \frac{1}{2!}g''(0) + \cdots + \frac{1}{m!}g^{(m)}(0) + \frac{1}{(m+1)!}g^{(m+1)}(c) \tag{21}$$

for some c in $(0, 1)$. But, from (8) and (9),

$$g(1) = f(\mathbf{x}_0 + \mathbf{h}) = f(\mathbf{x}_0 + \mathbf{x} - \mathbf{x}_0) = f(\mathbf{x}) \quad \text{and} \quad g(0) = f(\mathbf{x}_0).$$

Also, $h_i = (x_i - x_i^{(0)})$, so inserting (11), (15), (17), and (18) into (21) gives the desired result. ■

NOTATION: We shall denote the polynomial given in (19) by $p_m(\mathbf{x})$.

Taylor's theorem is not terribly difficult to apply, but it does become tedious when computing higher-order terms of a function of n variables. We shall therefore keep our examples simple.

EXAMPLE 1 Compute $p_3(\mathbf{x})$ around $\mathbf{x}_0 = \mathbf{0}$ for the function $f(x, y) = x^2 y$.

Solution. $f(\mathbf{x}_0) = f(0, 0) = 0$. The second term is (19) is

$$\sum_{i=1}^{n} f_i(\mathbf{0})(x_i - 0).$$

Here, $x_1 = x$ and $x_2 = y$. Also, $f_1 = 2xy$ and $f_2 = x^2$, so that $f_1(\mathbf{0}) = f_2(\mathbf{0}) = 0$. The third term is

$$\frac{1}{2!}\sum_{i=1}^{2}\sum_{j=1}^{2} f_{ij}(\mathbf{0})x_i x_j. \tag{22}$$

We compute this methodically, starting by setting $i = 1$ in (22). Then we compute $f_{11} = 2y$ and $f_{12} = 2x$, both of which are zero at (0, 0). Setting $i = 2$, we find that $f_{21}(\mathbf{0}) = f_{22}(\mathbf{0}) = 0$, so that the third term is 0. Finally, the fourth term in $p_3(\mathbf{x})$ is

$$\frac{1}{3!}\sum_{i=1}^{2}\sum_{j=1}^{2}\sum_{k=1}^{2} f_{ijk}(\mathbf{0})x_i x_j x_k. \tag{23}$$

Setting $i = 1$, we have

$$\sum_{j=1}^{2}\sum_{k=1}^{2} f_{1jk}(\mathbf{0})x_1 x_j x_k. \tag{24}$$

Setting $j = 1$ in (24) gives us

$$\sum_{k=1}^{2} f_{11k}(\mathbf{0})x^2 x_k. \tag{25}$$

Now $f_{111} = 0$ and $f_{112} = 2$, so that (25) reduces to $2x^2x_2 = 2x^2y$. Next, setting $j = 2$ in (24) yields

$$\sum_{k=1}^{2} f_{12k}(\mathbf{0})xyx_k. \tag{26}$$

But $f_{121} = 2$ and $f_{122} = 0$, so that (26) also reduces to $2x^2y$. Thus for $i = 1$ in (23), we obtain the sum $4x^2y$. Setting $i = 2$ in (23) yields

$$\sum_{j=1}^{2}\sum_{k=1}^{2} f_{2jk}(\mathbf{0})yx_j x_k. \tag{27}$$

Now $f_{211} = 2$, $f_{212} = 0$, $f_{221} = 0$, and $f_{222} = 0$, so that (27) reduces to $2x^2y$. Finally, (23) becomes $p_3 = (1/3!)(4x^2y + 2x^2y) = x^2y$. All this work has led to an unsurprising result. The polynomial x^2y is of degree 3 (see Definition 4.2.8 on p. 175), so that it is equal to its own third-degree Taylor polynomial. Note that all fourth-order derivatives of p_3 are zero so that $R_3 \equiv 0$. ■

REMARK. It is true that if $f(\mathbf{x})$ is a polynomial of degree m, then $f(\mathbf{x}) = p_m(\mathbf{x})$. We will not attempt to prove this fact. The proof is not conceptually difficult, but it involves computations like those carried out in Example 1 (see Problems 17, 18, and 19).

EXAMPLE 2 Compute $p_2(\mathbf{x})$ for $\mathbf{x}_0 = \mathbf{0}$ and $f(\mathbf{x}) = e^{x_1+x_2+\cdots+x_n}$.

Solution. $f(\mathbf{x}_0) = e^0 = 1$. The second term in (19) is

$$\sum_{i=1}^{n} f_i(\mathbf{0})x_i. \tag{28}$$

Now $f_i(x) = e^{x_1+x_2+\cdots+x_n}$, so that $f_i(\mathbf{0}) = 1$ and (28) becomes

$$\sum_{i=1}^{n} x_i = x_1 + x_2 + \cdots + x_n.$$

The third term in (19) is

$$\frac{1}{2!}\sum_{i=1}^{n}\sum_{j=1}^{n} f_{ij}(\mathbf{0})x_i x_j. \tag{29}$$

Again $f_{ij}(\mathbf{0}) = 1$ and (29) becomes

$$\frac{1}{2!}\sum_{i=1}^{n}\sum_{j=1}^{n} x_i x_j = \frac{1}{2!}\sum_{i=1}^{n} x_i \sum_{j=1}^{n} x_j = \frac{1}{2!}\left(\sum_{i=1}^{n} x_i\right)^2 = \frac{1}{2!}(x_1 + x_2 + \cdots + x_n)^2.$$

Thus

$$p_2(\mathbf{x}) = 1 + (x_1 + x_2 + \cdots + x_n) + \frac{1}{2!}(x_1 + x_2 + \cdots + x_n)^2.$$

We can obtain this answer immediately from formula (4), for if $x = x_1 + x_2 + \cdots + x_n$, then

$$e^{x_1+x_2+\cdots+x_n} \approx 1 + (x_1 + x_2 + \cdots + x_n) + \frac{(x_1 + x_2 + \cdots + x_n)^2}{2!} + \cdots + \frac{(x_1 + x_2 + \cdots + x_n)^k}{k!}. \blacksquare$$

TAYLOR POLYNOMIALS AND THE BINOMIAL THEOREM

Before doing any more examples, we illustrate how the computation of Taylor polynomials can be made a bit easier. We first suppose that $n = 2$ so that $\mathbf{x} = (x, y)$ and $\mathbf{x}_0 = (x_0, y_0)$. Let us compute $p_2(x, y)$ with $x_1 = x$, $x_2 = y$, $x_1^{(0)} = x_0$ and $x_2^{(0)} = y_0$. Also $f_1(\mathbf{x}_0) = f_x(x_0, y_0)$ and $f_2(\mathbf{x}_0) = f_y(x_0, y_0)$. Equation (19) then becomes

$$\begin{aligned} f(x, y) &= f(x_0, y_0) + f_x(x_0, y_0)(x - x_0) + f_y(x_0, y_0)(y - y_0) \\ &\quad + \frac{1}{2!}[f_{xx}(x_0, y_0)(x - x_0)^2 + 2f_{xy}(x_0, y_0)(x - x_0)(y - y_0) + f_{yy}(x, y)(y - y_0)^2] \\ &\quad + R_3(x, y). \end{aligned} \tag{30}$$

Here we have used the fact that $f_{xy} = f_{yx}$.

The linear term is easy to remember. What about the quadratic term? Recall that

$$(a + b)^2 = a^2 + 2ab + b^2. \tag{31}$$

Compare this expression with the quadratic terms in brackets in equation (30)!

What about the cubic term? The binomial theorem states that (see Appendix 2)

$$(a + b)^3 = a^3 + 3a^2b + 3ab^2 + b^3. \tag{32}$$

If we compute the third degree term in (19), again with $n = 2$, we obtain

$$\frac{1}{3!}[f_{xxx}(x_0, y_0)(x - x_0)^3 + 3f_{xxy}(x_0, y_0)(x - x_0)^2(y - y_0) + 3f_{xyy}(x_0, y_0)(x - x_0)(y - y_0)^2 + f_{yyy}(y - y_0)^3]. \tag{33}$$

Compare expressions (32) and (33).

These results for the quadratic and cubic terms suggest the following rule, which we offer without proof:

If $n = 2$, the kth degree term in the Taylor polynomial of f at (x_0, y_0) is obtained as follows:

Step (i) Expand $(a + b)^k$ by the binomial theorem.

Step (ii) A typical term in the expansion is ([see equation (2) in Appendix 2])

$$\binom{k}{j} a^{k-j} b^j.$$

Write the term

$$\binom{k}{j} f_{\underbrace{xx\cdots x}_{k-j \text{ times}}\underbrace{yy\cdots y}_{j \text{ times}}}(x - x_0)^{k-j}(y - y_0)^j. \tag{34}$$

There will be $k + 1$ such terms (for $j = 0, 1, 2, \ldots, k$)

Step (iii) The kth degree term in the Taylor polynomial is

$$\frac{1}{k!}[\text{sum of the } k + 1 \text{ terms obtained in step (ii)}]$$

NOTE. This procedure can be extended to $n > 2$. For example,

$$(a + b + c)^2 = a^2 + 2ab + 2ac + b^2 + 2bc + c^2.$$

Then, as can be verified, the quadratic term in (19) at $\mathbf{x}_0 = (x_0, y_0, z_0)$ for $n = 3$ is

$$\frac{1}{2!}[f_{xx}(x - x_0)^2 + 2f_{xy}(x - x_0)(y - y_0) + 2f_{xz}(x - x_0)(z - z_0) + f_{yy}(y - y_0)^2 + 2f_{yz}(y - y_0)(z - z_0) + f_{zz}(z - z_0)^2]. \tag{35}$$

In (35), f_{xx} denotes $f_{xx}(x_0, y_0, z_0)$, f_{xz} denotes $f_{xz}(x_0, y_0, z_0)$, and so on.

After a few computations, it becomes apparent that computing a Taylor polynomial is not much more difficult than computing expressions of the form $(x_1 + x_2 + \cdots + x_n)^k$.

EXAMPLE 3 Compute $p_2(\mathbf{x})$ for $f(\mathbf{x}) = \sin(2x + y)$ at $\mathbf{x}_0 = (\pi/6, \pi/3)$.

Solution. We see that

$$f(x_0, y_0) = \sin\left(\frac{\pi}{3} + \frac{\pi}{3}\right) = \frac{\sqrt{3}}{2}.$$

$f_x(x, y) = 2\cos(2x + y)$ and $f_y(x, y) = \cos(2x + y)$, so that

$$f_x\left(\frac{\pi}{6}, \frac{\pi}{3}\right) = -1, \qquad f_y\left(\frac{\pi}{6}, \frac{\pi}{3}\right) = -\frac{1}{2}$$

Next

$$f_{xx}(x, y) = -4\sin(2x + y) \quad \text{and} \quad f_{xx}\left(\frac{\pi}{6}, \frac{\pi}{3}\right) = -2\sqrt{3}$$

$$f_{xy}(x, y) = -2\sin(2x + y) \quad \text{and} \quad 2f_{xy}\left(\frac{\pi}{6}, \frac{\pi}{3}\right) = -2\sqrt{3}$$

$$f_{yy}(y, y) = -\sin(2x + y) \quad \text{and} \quad f_{yy}\left(\frac{\pi}{6}, \frac{\pi}{3}\right) = -\frac{\sqrt{3}}{2}$$

Then, from (30), we have

$$\begin{aligned}
p_2(x, y) &= \frac{\sqrt{3}}{2} - \left(x - \frac{\pi}{6}\right) - \frac{1}{2}\left(y - \frac{\pi}{3}\right) + \frac{1}{2}\left[-2\sqrt{3}\left(x - \frac{\pi}{6}\right)^2\right. \\
&\quad \left. - 2\sqrt{3}\left(x - \frac{\pi}{6}\right)\left(y - \frac{\pi}{3}\right) - \frac{\sqrt{3}}{2}\left(y - \frac{\pi}{3}\right)^2\right]. \\
&= \frac{\sqrt{3}}{2} - \left(x - \frac{\pi}{6}\right) - \frac{1}{2}\left(y - \frac{\pi}{3}\right) - \sqrt{3}\left(x - \frac{\pi}{6}\right)^2 \\
&\quad - \sqrt{3}\left(x - \frac{\pi}{6}\right)\left(y - \frac{\pi}{3}\right) - \frac{\sqrt{3}}{4}\left(y - \frac{\pi}{3}\right)^2. \quad \blacksquare
\end{aligned}$$

EXAMPLE 4 Compute $p_2(\mathbf{x})$ for $f(\mathbf{x}) = \ln(1 + 2x + 5y)$ at $\mathbf{x}_0 = (0, 0)$.

Solution. As in Example 2, there are two ways to solve this problem. The easier method is to use formula (7). Substituting $2x + 5y$ for x in (7), we obtain

$$\ln(1 + 2x + 5y) \approx 2x + 5y - \frac{(2x + 5y)^2}{2} + \frac{(2x + 5y)^3}{3} - \cdots + \frac{(-1)^{n+1}(2x + 5y)^n}{n}.$$

Thus

$$p_2(\mathbf{x}) = 2x + 5y - \frac{(2x + 5y)^2}{2} = 2x + 5y - 2x^2 - 10xy - \frac{25}{2}y^2.$$

We next compute $p_2(\mathbf{x})$ the "hard" way to illustrate further the accuracy of formula (30). We have

$$f(x_0, y_0) = f(0, 0) = \ln 1 = 0$$

$$f_x(x, y) = \frac{2}{1 + 2x + 5y} \quad \text{and} \quad f_x(0, 0) = 2$$

$$f_y(x, y) = \frac{5}{1 + 2x + 5y} \quad \text{and} \quad f_y(0, 0) = 5$$

$$f_{xx}(x, y) = -\frac{4}{(1 + 2x + 5y)^2} \quad \text{and} \quad f_{xx}(0, 0) = -4$$

$$f_{xy}(x, y) = -\frac{10}{(1 + 2x + 5y)^2} \quad \text{and} \quad 2f_{xy}(0, 0) = -20$$

$$f_{yy}(x, y) = -\frac{25}{(1 + 2x + 5y)^2} \quad \text{and} \quad f_{yy}(0, 0) = -25.$$

Then, from (30),

$$\begin{aligned} p_2(x, y) &= 2x + 5y + \frac{1}{2!}(-4x^2 - 20xy - 25y^2) \\ &= 2x + 5y - 2x^2 - 10xy - (25/2)y^2. \ \blacksquare \end{aligned}$$

Once we know how to compute Taylor polynomials of n variables, we can use them to approximate functions as is done Section 13.3. The accuracy of the approximation depends, of course, on the size of the remainder term R_m given by (20). It is evident that if all the $(m + 1)$st-order partial derivatives of f are bounded in some interval, then $p_m(\mathbf{x})$ is a good approximation to $f(\mathbf{x})$ if $\mathbf{x}$ is close to $\mathbf{x}_0$, and a bound on the error $|f(\mathbf{x}) - R_m(\mathbf{x})|$ can be computed. We will not give any examples of this here.

We also note that Taylor's theorem in n variables has many practical applications. For example, it is a useful tool in the study of qualitative properties of systems of ordinary differential equations.

In Section 4.12 we stated the second derivatives test for a function of two variables (see Theorem 4.12.3). Using Taylor's theorem in two variables, we can prove it.

Theorem 2 SECOND DERIVATIVES TEST Let f and all its first and second partial derivatives be continuous in a neighborhood of the critical point (x_0, y_0). Let

$$D(x, y) = f_{xx}(x, y)f_{yy}(x, y) - [f_{xy}(x, y)]^2 \tag{36}$$

and let D denote $D(x_0, y_0)$.

(i) If $D > 0$ and $f_{xx}(x_0, y_0) > 0$, then f has a local minimum at (x_0, y_0).
(ii) If $D > 0$ and $f_{xx}(x_0, y_0) < 0$, then f has a local maximum at (x_0, y_0).
(iii) If $D < 0$, then (x_0, y_0) is a saddle point of f.
(iv) If $D = 0$, then any of the preceding alternatives is possible.

Proof. Because this is a very long proof we shall give it in steps. We begin by making the simplifying assumption that $f \in C^3(\Omega)$ where Ω is a neighborhood of (x_0, y_0). This assumption is not necessary but it does make things easier. We shall assume, also to make things simpler, that $(x_0, y_0) = (0, 0)$.

step 1. By Taylor's theorem and formula (30),

$$f(x, y) = f(0, 0) + f_x(0, 0)x + f_y(0, 0)y$$
$$+ \tfrac{1}{2}[f_{xx}(0, 0)x^2 + f_{xy}(0, 0)xy + f_{yx}(0, 0)xy + f_{yy}(0, 0)y^2] + R_2(x, y). \quad \textbf{(37)}$$

Now, since $(0, 0)$ is a critical point, $f_x(0, 0) = f_y(0, 0) = 0$ and, since f_{xy} and f_{yx} are continuous, $f_{xy}(0, 0) = f_{yx}(0, 0)$. Also, by (20), $R_2(x, y)$ contains terms of the form x^3, x^2y, xy^2, and y^3 (since $f \in C^3(\Omega)$). A key step in the proof is to observe that if x and y are very small, then third-order terms in x and y (like x^3, x^2y, xy^2, and y^3) are considerably smaller than the second-order terms x^2, xy, and y^2. For example, if $x = y = 0.1$, then $x^3 = x^2y = xy^2 = y^3 = 0.001$, while $x^2 = xy = y^2 = 0.01$. The difference is a factor of 10. Thus, for x and y small enough, $R_2(x, y)$ is negligible compared to the other terms in (31) and we can write

$$f(x, y) \approx f(0, 0) + \tfrac{1}{2}[f_{xx}(0, 0)x^2 + 2f_{xy}(0, 0)xy + f_{yy}(0, 0)y^2]. \quad \textbf{(38)}$$

Let $A = f_{xx}(0, 0)$, $B = f_{xy}(0, 0)$, and $C = f_{yy}(0, 0)$. Then (38) becomes

$$f(x, y) - f(0, 0) \approx \tfrac{1}{2}(Ax^2 + 2Bxy + Cy^2). \quad \textbf{(39)}$$

step 2. We now show that if $D = AC - B^2 > 0$ and if $A > 0$, then $(Ax^2 + 2Bxy + Cy^2) > 0$ for all vectors $(x, y) \neq (0, 0)$. Using (39), this will show that $f(x, y) > f(0, 0)$ if x and y are sufficiently small, which means that $(0, 0)$ is a local minimum. This will prove part (i). We have

$$Ax^2 + 2Bxy + Cy^2 = A\left(x^2 + \frac{2B}{A}xy + \frac{C}{A}y^2\right)$$

Completing the square ↘

$$= A\left[\left(x + \frac{B}{A}y\right)^2 + \left(\frac{C}{A} - \frac{B^2}{A^2}\right)y^2\right]$$
$$= A\left[\left(x + \frac{B}{A}y\right)^2 + \frac{AC - B^2}{A^2}y^2\right]. \quad \textbf{(40)}$$

If $AC - B^2 > 0$ and if $A > 0$, then the last expression in (40) is nonnegative; it is positive if $(x, y) \neq (0, 0)$. This proves part (i). The proof of part (ii) follows from (40) because if $A < 0$ and $AC - B^2 > 0$, then $Ax^2 + 2Bxy + Cy^2 < 0$ for $(x, y) \neq (0, 0)$ and, from (39), $f(x, y) < f(0, 0)$ if x and y are sufficiently small.

step 3. We prove part (iii). We assume that $A = f_{xx}(0, 0) > 0$. A similar argument works in the case $A < 0$. We must show that we can find values of x and y as small as we like that make $Ax^2 + 2Bxy + Cy^2$ positive or negative. Choose y small and let $x = (-B/A)y$. Then, from (40),

$$Ax^2 + 2Bxy + Cy^2 = A\left(\frac{AC - B^2}{A^2}y^2\right) < 0 \qquad \text{if } A > 0,\ AC - B^2 < 0.$$

Now, choose $y = 0$ but $x \neq 0$. Then $Ax^2 + 2Bxy + Cy^2 = Ax^2 > 0$. Thus there are vectors (x, y) arbitrarily close to $(0, 0)$ such that $f(x, y) > f(0, 0)$ and $f(x, y) < f(0, 0)$. Hence $(0, 0)$ is a saddle point.

step 4. We prove part (iv) by means of examples. Let

$$f(x, y) = x^2 + 2xy + y^2 = (x + y)^2.$$

Then, since $f(x, y) \geq 0$, $(0, 0)$ is a local minimum (which is, of course, not unique). But $f_{xx} = f_{xy} = f_{yy} = 2$, so that $D = 2^2 - 2 \cdot 2 = 0$. Similarly, $D = 0$ when $f(x, y) = -x^2 - 2xy - y^2 = -(x + y)^2$ and $(0, 0)$ is a local maximum. Finally, let $f(x, y) = x^3 - y^3$. Then $f_{xx}(0, 0) = f_{xy}(0, 0) = f_{yy}(0, 0) = 0$, so that $D = 0$. However, $(0, 0)$ is a critical point and $x^3 - y^3 > 0$ if $x > y$ and $x^3 - y^3 < 0$ if $x < y$. Thus $(0, 0)$ is a saddle point. These three examples show that if $D = 0$, then a critical point can be a local maximum, a local minimum, or a saddle point. ■

PROBLEMS 10.1

The **linearization** of a function $f: \mathbb{R}^n \to \mathbb{R}$ at $\mathbf{x}_0$ is the first-degree Taylor polynomial that approximates that function at $\mathbf{x}_0$. In Problems 1–7 find the linearization of the given function.

1. $\sin(x + y)$; $\mathbf{x}_0 = \mathbf{0}$
2. $\cos(3x - 2y)$; $\mathbf{x}_0 = \mathbf{0}$
3. $e^{x_1 - 4x_2 + x_3}$; $\mathbf{x}_0 = \mathbf{0}$
4. $\ln(1 + x_1 + x_2 + \cdots + x_n)$; $\mathbf{x}_0 = \mathbf{0}$
5. $\sqrt{x + y}$; $\mathbf{x}_0 = (2, 2)$
6. $\sqrt{x_1 + x_2 + x_3 + x_4}$; $\mathbf{x}_0 = (1, 1, 1, 1)$
7. $\dfrac{x}{y} + \dfrac{y}{x}$; $\mathbf{x}_0 = (2, 1)$

In Problems 8–16 find the Taylor polynomial of degree m that approximates the given function at the given point.

8. $f(x, y) = \ln(x + 2y)$; $\mathbf{x}_0 = (1, 0)$; $m = 2$
9. $f(x, y) = \sin(x^2 + y^2)$; $\mathbf{x}_0 = (0, 0)$; $m = 2$
10. $f(x_1, x_2, x_3, x_4) = \sin(x_1 + x_2 + x_3 + x_4)$;
$\mathbf{x}_0 = \left(\dfrac{\pi}{8}, \dfrac{\pi}{8}, \dfrac{\pi}{8}, \dfrac{\pi}{8}\right)$; $m = 2$
11. $f(x, y, z) = \sin xyz$; $\mathbf{x}_0 = \mathbf{0}$; $m = 9$
12. $f(x_1, x_2, \ldots, x_n) = \cos(x_1 x_2 \cdots x_n)$; $\mathbf{x}_0 = \mathbf{0}$; $m = 4n$
13. $f(x, y) = e^x \sin y$; $\mathbf{x}_0 = \mathbf{0}$; $m = 2$
14. $f(x, y) = \sin x \cos y$; $\mathbf{x}_0 = \mathbf{0}$; $m = 2$
15. $f(x, y) = e^x \sin y$; $\mathbf{x}_0 = \left(2, \dfrac{\pi}{4}\right)$; $m = 2$
16. $f(x, y) = \sin x \cos y$; $\mathbf{x}_0 = \left(\dfrac{\pi}{2}, \pi\right)$; $m = 3$
17. Verify that the second-degree Taylor polynomial of xy around $\mathbf{0}$ is xy.
18. Verify that the fifth-degree Taylor polynomial of x^2yz^2 around $\mathbf{0}$ is x^2yz^2.
***19.** **(a)** Show that if $f(x, y)$ is a polynomial of degree m, then its mth-degree Taylor polynomial $p_m(\mathbf{x})$ around $\mathbf{0}$ is equal to f.
(b) What can you say about the mth-degree Taylor polynomial around a value $\mathbf{x}_0 \neq \mathbf{0}$?
20. Write out the third-degree Taylor polynomial for a function of three variables. [*Hint:* First compute $(a + b + c)^3$.]
21. Write out the fifth-degree term in the Taylor polynomial for a function of two variables.

10.2 INVERSE FUNCTIONS AND THE IMPLICIT FUNCTION THEOREM: I

In this section we discuss one of the most interesting and important theorems in calculus: the implicit function theorem. We use a special case of the theorem each time we say that $f(x, y) = 0$ defines y as a function of x. The result is discussed here in the setting of functions from $\mathbb{R}^{n+1} \to \mathbb{R}$. We shall return to the implicit function theorem in a more general setting in Section 10.5. We begin by looking at functions from $\mathbb{R}$ to $\mathbb{R}$.

Let $y = f(x)$ be a function of one variable. We know that f has an inverse function defined on its range if f is one-to-one[†]. But a differentiable f is one-to-one in a neighborhood of a point x_0 if $f'(x_0) \neq 0$. Summarizing, we have

Theorem 1 Let Ω be an open set in $\mathbb{R}$ and let $f: \mathbb{R} \to \mathbb{R}$ have domain Ω. Suppose that f is differentiable and that $f'(x) \neq 0$ for every $x \in \Omega$. Then there exists a differentiable function $g: \mathbb{R} \to \mathbb{R}$ with domain $f(\Omega)$ such that $(g \circ f)(x) = x$ for every $x \in \Omega$ and $(f \circ g)(y) = y$ for every $y \in f(\Omega)$. In this case g is called the **inverse** of f and we write $g = f^{-1}$.

We can look at the problem of finding inverse functions in another way. Suppose that $y = f(x)$. Let $F(x, y) = f(x) - y$. Then $y = f(x)$ is equivalent to

$$F(x, y) = 0. \tag{1}$$

More generally, equation (1) gives y *implicitly* as a function of x and x implicitly as a function of y. If F takes the special form $f(x) - y$, then writing x explicitly as a function of y is equivalent to finding an inverse function for f—so that $x = f^{-1}(y)$.

But suppose again that $F(x, y) = y - f(x)$. Now $\partial F/\partial x = -f'(x)$ and if $f'(x) \neq 0$, then f has an inverse function and x can be determined explicitly in terms of y. This leads us to suspect that in (1) we can write x as a function of y if $\partial F/\partial x \neq 0$ and y as a function of x if $\partial F/\partial y \neq 0$. This is the basic idea behind the very deep results we are about to discuss. But first we shall give two examples.

EXAMPLE 1 Let $F(x, y) = x^2 + y^2 - 1 = 0$. This is the equation of the unit circle in $\mathbb{R}^2$. If we make no restrictions on x or y, then neither variable can be written in terms of the other since

$$x = \pm\sqrt{1 - y^2} \qquad \text{and} \qquad y = \pm\sqrt{1 - x^2} \tag{2}$$

and neither expression in (2) is a function. However, if we specify $x > 0$ (so that we have only the right semicircle in Figure 1a), then $x = \sqrt{1 - y^2}$, and we have written x explicitly as a function of y. Note that $(\partial F/\partial x)(x, y) = 2x \neq 0$ if $x > 0$. There are three other regions depicted in Figure 1 in which one of the variables can be written as a function of the other. In each case one of the partial derivatives of F is nonzero over the region. ■

[†]See Section 6.1 in *Calculus* or *Calculus, Part I.*

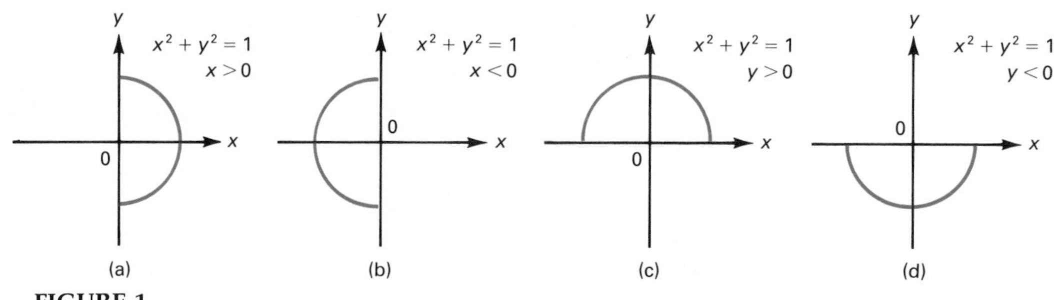

FIGURE 1

EXAMPLE 2 Let $F(x, y) = x^2y^5 + e^{2x+y} + \sqrt{3x + y} = 0$. Again, we would like to write y as a function of x or x as a function of y. Unfortunately, if you try to do this, you will quickly find it an impossible task. However, we can calculate

$$\frac{\partial F}{\partial x}(x, y) = 2xy^5 + 2e^{2x+y} + \frac{3}{2\sqrt{3x + y}}$$

and

$$\frac{\partial F}{\partial y}(x, y) = 5x^2y^4 + e^{2x+y} + \frac{1}{2\sqrt{3x + y}}.$$

In the region $\Omega = \{(x, y): x > 0, y > 0\}$ both $\partial F/\partial x$ and $\partial F/\partial y$ are nonzero. Then, if our intuition is correct, there are functions f and g such that $y = f(x)$ and $x = g(y)$ in Ω. Note that there is a difference between asserting the existence of the functions f and g and being able to write them explicitly. In Example 1 we could write them; in this example we cannot. ■

We now generalize the ideas discussed above to higher dimensions. Let $f: \mathbb{R}^n \to \mathbb{R}$ and write

$$x_{n+1} = f(x_1, x_2, \ldots, x_n). \tag{3}$$

In (3) we have written x_{n+1} explicity in terms of the variables $x_1, x_2, \ldots, x_n$. If we define $F: \mathbb{R}^{n+1} \to \mathbb{R}$ by

$$F(x_1, x_2, \ldots, x_n, x_{n+1}) = x_{n+1} - f(x_1, x_2, \ldots, x_n),$$

then (3) can be written

$$F(x_1, x_2, \ldots, x_n, x_{n+1}) = 0. \tag{4}$$

More generally, if $F: \mathbb{R}^{n+1} \to \mathbb{R}$ is a given function, we ask whether the $(n + 1)$st variable x_{n+1} can be written as a function of the other n variables if equation (4) holds.

EXAMPLE 3 Let $F: \mathbb{R}^{n+1} \to \mathbb{R}$ be given by

$$F(x_1, x_2, \ldots, x_n, x_{n+1}) = x_1^2 + x_2^2 + \cdots + x_n^2 + x_{n+1}^2 - 1.$$

If $F(x_1, x_2, \ldots, x_n, x_{n+1}) = 0$, then, as in Example 1, no one of the variables can be written as a function of the others if we make no restrictions. However, if we specify $x_{n+1} > 0$, say, then we can write

$$x_{n+1} = \sqrt{1 - x_1^2 - x_2^2 - \cdots - x_n^2}.$$

Note that $\partial f/\partial x_{n+1} = 2x_{n+1} \neq 0$ in the open set $\{\mathbf{x} \in \mathbb{R}^{n+1}: x_{n+1} > 0\}$. ■

We are now ready to state the first form of the implicit function theorem. The proofs of this theorem and its generalization (given in Section 10.5) are far more difficult than any proofs encountered in this text. For that reason they are omitted.†

Theorem 2 IMPLICIT FUNCTION THEOREM—FIRST FORM Let $F: \mathbb{R}^{n+1} \to \mathbb{R}$ be defined on an open set Ω and let

$$\mathbf{x}_0 = (x_1^{(0)}, x_2^{(0)}, \ldots, x_n^{(0)}, x_{n+1}^{(0)})$$

be in Ω. Suppose that the partial derivatives $(\partial F/\partial x_i)(\mathbf{x})$ are continuous in a neighborhood of $\mathbf{x}_0$ for $i = 1, 2, \ldots, n + 1$. Suppose further that

$$F(\mathbf{x}_0) = 0 \quad \text{and} \quad \frac{\partial F}{\partial x_{n+1}}(\mathbf{x}_0) \neq 0.$$

Then there is an open set $\overline{\Omega}$ in $\mathbb{R}^n$ that contains $\overline{\mathbf{x}}_0 = (x_1^{(0)}, x_2^{(0)}, \ldots, x_n^{(0)})$ and a unique continuous function $f: \mathbb{R}^n \to \mathbb{R}$ such that $x_{n+1}^{(0)} = f(\overline{\mathbf{x}}_0)$ and $F(\mathbf{x}, f(\mathbf{x})) = 0$ for all $\mathbf{x}$ in $\overline{\Omega}$. Furthermore, f has continuous first partial derivatives on $\overline{\Omega}$ given by

$$\frac{\partial f}{\partial x_i}(\mathbf{x}) = \frac{-\dfrac{\partial F}{\partial x_i}(\mathbf{x}, f(\mathbf{x}))}{\dfrac{\partial F}{\partial x_{n+1}}(\mathbf{x}, f(\mathbf{x}))} \quad \text{for} \quad 1 \leq i \leq n. \tag{5}$$

REMARK 1. In light of this theorem we say that $F(\mathbf{x}, x_{n+1}) = 0$ defines x_{n+1} **implicitly** as a function of $\mathbf{x} = (x_1, x_2, \ldots, x_n)$ near $\overline{\mathbf{x}}_0 = (x_1^{(0)}, x_2^{(0)}, \ldots, x_n^{(0)})$. In this case we write $x_{n+1} = f(\mathbf{x})$.

REMARK 2. It is important to emphasize that the implicit function theorem is a **local** result. That is, we can guarantee that x_{n+1} can be written as a function of $x_1, x_2, \ldots, x_n$ only in some open set in $\mathbb{R}^n$. This open set could be very small or it could be all of $\mathbb{R}^n$. The theorem does not tell you how big $\overline{\Omega}$ is. All we know is that $x_{n+1} = f(\mathbf{x})$ for $\mathbf{x}$ "near" $\mathbf{x}_0$. Nevertheless, even this limited information can be very useful.

---

†Reasonably comprehensible proofs can be found in C. H. Edwards, Jr., *Advanced Calculus in Several Variables* (Academic Press, New York, 1973).

REMARK 3. Although the implicit function theorem is, as we have stated, difficult to prove, the condition $(\partial F/\partial x_{n+1})(\mathbf{x}_0) \neq 0$ and equation (5) are very natural. For suppose that $F(x_1, x_2, \ldots, x_n, f(x_1, x_2, \ldots, x_n)) = 0$. Then, using the chain rule to compute $(\partial f/\partial x_i)$, we have

$$\frac{\partial F}{\partial x_i} + \frac{\partial F}{\partial x_{n+1}}\frac{\partial f}{\partial x_i} = 0,$$

which shows that $\partial f/\partial x_i$ is given by (5) if $\partial F/\partial x_{n+1} \neq 0$.

EXAMPLE 4 Let $F(x, y, z, w) = x^2 + y^4 + z^6 + w^2 - 1 = 0$. Find a region over which w can be written as a function f of x, y, and z and compute $\partial f/\partial x$, $\partial f/\partial y$, and $\partial f/\partial z$.

Solution. We have $\partial F/\partial w = 2w$, which is nonzero on two different open sets in $\mathbb{R}^4$: $\Omega_1 = \{(x, y, z, w): w > 0\}$ and $\Omega_2 = \{(x, y, z, w): w < 0\}$. We will choose Ω_1; the analysis for Ω_2 is similar. In Ω_1 we can write

$$w = f(x, y, z) = \sqrt{1 - x^2 - y^4 - z^6},$$

so that

$$\frac{\partial f}{\partial x} = \frac{-x}{\sqrt{1 - x^2 - y^4 - z^6}} \qquad \frac{\partial f}{\partial y} = \frac{-2y^3}{\sqrt{1 - x^2 - y^4 - z^6}}$$

and

$$\frac{\partial f}{\partial z} = \frac{-3z^5}{\sqrt{1 - x^2 - y^4 - z^6}}.$$

Here we have no need for formula (5), since we have found f explicitly. However, we can use our explicit representation to verify formula (5) in this case. We have

$$\frac{\partial f}{\partial x} = \frac{-\partial F/\partial x}{\partial F/\partial w} = \frac{-2x}{2w} = \frac{-x}{w} = \frac{-x}{\sqrt{1 - x^2 - y^4 - z^6}}$$

$$\frac{\partial f}{\partial y} = \frac{-\partial F/\partial y}{\partial F/\partial w} = \frac{-4y^3}{2w} = \frac{-2y^3}{w} = \frac{-2y^3}{\sqrt{1 - x^2 - y^4 - z^6}}$$

and

$$\frac{\partial f}{\partial z} = \frac{-\partial F/\partial z}{\partial F/\partial w} = \frac{-6z^5}{2w} = \frac{-3z^5}{w} = \frac{-3z^5}{\sqrt{1 - x^2 - y^4 - z^6}}. \quad ■$$

EXAMPLE 5 Let $F(x, y, z) = x^2y^3 + 4xyz^8 - 5xyz^4 + z^2 + 1 = 0$. Show that z can be written as a function of x and y in a neighborhood of the point $(2, -1, 1)$ and compute $\partial z/\partial x$ and $\partial z/\partial y$ at that point.

Solution. First we note that $F(2, -1, 1) = -4 - 8 + 10 + 2 = 0$ so that the hypotheses of the implicit function theorem are verified. In this problem it is pretty clear that we cannot write z explicitly as a function of x and y. (There is, according to a famous theorem proved by a French mathematician named Galois, no explicit formula for solving all eighth-degree polynomials.) However, we can compute

$$\frac{\partial F}{\partial z} = 32xyz^7 - 20xyz^3 + 2z$$

and, at $(2, -1, 1)$, $\partial F/\partial z = -22 \neq 0$; so that for (x, y) in a neighborhood of $(2, -1)$ there is unique function f such that $z = f(x, y)$ and $F(x, y, f(x, y)) = 0$.[(Note that $f(2, -1) = 1$.)] Then

$$\frac{\partial z}{\partial x} = \frac{-\partial F/\partial x}{\partial F/\partial z} = \frac{-2xy^3 - 4yz^8 + 5yz^4}{-22} = \frac{3}{-22} = -\frac{3}{22}$$

and

$$\frac{\partial z}{\partial y} = \frac{-\partial F/\partial y}{\partial F/\partial z} = \frac{-3x^2y^2 - 4xz^8 + 5xz^4}{-22} = \frac{-10}{-22} = \frac{5}{11}$$

at the point $(2, -1, 1)$. ■

REMARK. We emphasize again that the implicit function theorem enables us to compute partial derivatives of the implicitly defined function f even though we may not be able to write f explicitly.

EXAMPLE 6 Let $F(x, y) = x - y^5 = 0$. Then, clearly, $y = x^{1/5}$ for every $x \in \mathbb{R}$. However, $\partial F/\partial y = -5y^4 = 0$ at $(0, 0)$, so that the hypotheses of the implicit function theorem are not satisfied near $(0, 0)$. This example shows that functions may be implicitly defined in cases where the implicit function theorem cannot be used. ■

PROBLEMS 10.2

In Problems 1–9 a function F from $\mathbb{R}^{n+1} \to \mathbb{R}$ is given. Show that $F(x_1, x_2, \ldots, x_n, x_{n+1}) = 0$ and that x_{n+1} can be written as $f(x_1, x_2, \ldots, x_n)$ in a neighborhood of the given point $(x_1, x_2, \ldots, x_{n+1})$. Then compute $\partial f/\partial x_i$ for $i = 1, 2, \ldots, n$ at $(x_1, x_2, \ldots, x_n)$.

1. $F(x, y) = x^2 + xy + y^2 - 3$; $(1, 1)$
2. $F(x, y) = x^3 + x^2y^5 - 2\sqrt{xy}$; $(1, 1)$
3. $F(x, y) = e^{x+y} - \sin\left[\frac{\pi}{2}(x - 3y^2)\right] + \frac{y}{x} - e + 1$; $(1, 0)$
4. $F(x, y, z) = xy^2 + yz^2 + x^2z - 3$; $(1, 1, 1)$
5. $F(x, y, z) = x + y + z - \sin(xyz)$; $(0, 0, 0)$
6. $F(x, y, z) = e^{x+2y+3z} - 1$; $(0, 0, 0)$
7. $F(x, y, z) = xe^z - ze^y + (xz/y) - 1$; $(1, 3, 0)$
8. $F(x, y, z) = 1 - e^z\cos(y - x)$; $(3\pi/2, 3\pi/2, 0)$
9. $F(x_1, x_2, x_3, \ldots, x_n, x_{n+1}) = \sum_{i=1}^{n+1}(x_i)^4 - n - 1$; $(1, 1, 1, \ldots, 1)$

In Problems 10–17 find formulas for $\partial z/\partial x$ and $\partial z/\partial y$, and state where they are valid.

10. $x^2 + y^2 + z^2 - 1 = 0$
11. $x^4 + y^4 + z^4 - 1 = 0$
12. $xy + xz + yz^5 - 2 = 0$
13. $x^3 \sin xyz + y^3 \cos xyz = 0$
14. $z^3e^{xy} - (xy/z) - 3 = 0$
15. $e^{xy}\ln(z/x) + \cos(x^5 - 4y/z) = 0$

16. $\sinh\left(\dfrac{x - 2y}{3z}\right) + \cosh\left(\dfrac{4x}{z^5}\right) = 0$

17. $e^{\sqrt{zx}} - 3x^2y^3z^4 + \sin\left(\dfrac{5z}{x + y}\right) = 0$

18. Let $F: \mathbb{R}^2 \to \mathbb{R}$ be continuously differentiable with $F(x_0, y_0) = 0$ and $(\partial F/\partial y)(x_0, y_0) > 0$. Show that there are intervals (a, b) and (c, d) such that $x_0 \in (a, b)$, $y_0 \in (c, d)$, $F(x, c) < 0$ for all $x \in (a, b)$, $F(x, d) > 0$ for all $x \in (a, b)$, and $(\partial F/\partial x)(x, y) > 0$ for all $x \in (a, b)$ and $y \in (c, d)$. [*Hint:* Use the continuity of F and $\partial F/\partial y$.]

19. Let $\bar{x}$ be in $[a, b]$ and define the function $F_{\bar{x}}(y) = F(\bar{x}, y)$. Show that there is a unique number $\bar{y}$ in $[c, d]$ such that $F_{\bar{x}}(\bar{y}) = F(\bar{x}, \bar{y}) = 0$. [*Hint:* Use the intermediate value theorem.]

20. Use the results of Problems 18 and 19 to prove the following special case of the implicit function theorem: Let $F: \mathbb{R}^2 \to \mathbb{R}$ be continuously differentiable and let $(x_0, y_0) \in \mathbb{R}^2$ be a point at which $F(x_0, y_0) = 0$ and $(\partial F/\partial y)(x_0, y_0) \neq 0$. Then there exists a closed interval $[a, b]$ containing x_0 and a function $f: \mathbb{R} \to \mathbb{R}$ with domain $[a, b]$ such that $y = f(x)$ satisfies $F(x, f(x)) = 0$ in a neighborhood of (x_0, y_0).

10.3 FUNCTIONS FROM $\mathbb{R}^n$ TO $\mathbb{R}^m$

In Section 4.1 we described functions from $\mathbb{R}^n$ to $\mathbb{R}$ and in Section 3.8 we discussed functions from $\mathbb{R}$ to $\mathbb{R}^n$. In this section we begin our study of functions from $\mathbb{R}^n$ to $\mathbb{R}^m$ where $n \geq 1$ and $m \geq 1$.

Definition 1 FUNCTIONS FROM $\mathbb{R}^n$ TO $\mathbb{R}^m$ Let Ω be a subset of $\mathbb{R}^n$. Then a **function** or **mapping f** from $\mathbb{R}^n$ to $\mathbb{R}^m$ is a rule that assigns to each $\mathbf{x} = (x_1, x_2, \ldots, x_n)$ in Ω a unique vector $\mathbf{f}(\mathbf{x})$ in $\mathbb{R}^m$. The set Ω is called the **domain** of **f**. The set $\{\mathbf{f}(\mathbf{x}): \mathbf{x} \in \Omega\}$ is called the **range** of **f**.

NOTATION: We will write $\mathbf{f}: \mathbb{R}^n \to \mathbb{R}^m$ and, since $\mathbf{f}(\mathbf{x})$ is a subset of $\mathbb{R}^m$, we will represent it in terms of its **component functions.** That is,

$$\begin{aligned} \mathbf{f}(\mathbf{x}) &= (f_1(\mathbf{x}), f_2(\mathbf{x}), \ldots, f_m(\mathbf{x})) \\ &= (f_1(x_1, x_2, \ldots, x_n), f_2(x_1, x_2, \ldots, x_n), \ldots, f_m(x_1, x_2, \ldots, x_n)). \end{aligned} \tag{1}$$

Note that each of the m component functions is a mapping from $\mathbb{R}^n \to \mathbb{R}$.

EXAMPLE 1 Let $\mathbf{f}: \mathbb{R}^2 \to \mathbb{R}^2$ be defined by

$$\mathbf{f}(x, y) = (x + 2y, -3x + 4y).$$

For example, the image of (1, 2) under this mapping is (5, 5) and the image of (−4, 2) is (0, 20). Note that **f** can be represented by a matrix:

$$\mathbf{f}(x, y) = \begin{pmatrix} 1 & 2 \\ -3 & 4 \end{pmatrix}\begin{pmatrix} x \\ y \end{pmatrix}.$$

That is, **f** is a linear transformation from $\mathbb{R}^2 \to \mathbb{R}^2$. Here the component functions are $f_1(x, y) = x + 2y$ and $f_2(x, y) = -3x + 4y$. ■

EXAMPLE 2 Let $\mathbf{f}$: $\mathbb{R}^2 \to \mathbb{R}^2$ be defined by $\mathbf{f}(r, \theta) = (r \cos \theta, r \sin \theta)$. This is called the **polar coordinate mapping.** To have an idea how this mapping works, consider the rectangle $R = \{(r, \theta): 0 \le r \le 1, 0 \le \theta \le \pi\}$. This rectangle is sketched in Figure 1a. Keep in mind that r and θ are just numbers here and that the vector (r, θ) is an ordinary vector in $\mathbb{R}^2$. The image of R under the mapping is the semicircular disk S pictured in Figure 1b and given by $S = \{(x, y): x = r \cos \theta, y = r \sin \theta, 0 \le r \le 1, 0 \le \theta \le \pi\}$. ■

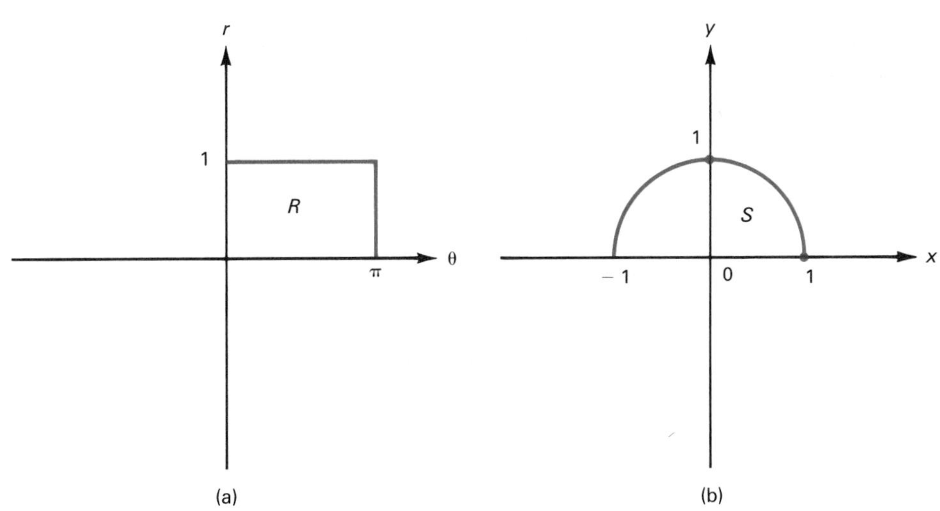

FIGURE 1

EXAMPLE 3 Let $\mathbf{f}$: $\mathbb{R}^3 \to \mathbb{R}^4$ be given by $\mathbf{f}(x, y, z) = (xy, xz, y + z, xyz)$. Thus, for example, $\mathbf{f}(1, 2, 3) = (2, 3, 5, 6)$ and $\mathbf{f}(-1, 2, 0) = (-2, 0, 2, 0)$. Here the component functions are $f_1(x, y, z) = xy$, $f_2(x, y, z) = xz$, $f_3(x, y, z) = y + z$, and $f_4(x, y, z) = xyz$. ■

EXAMPLE 4 As we saw in Section 9.6, a mapping $\mathbf{f}$: $\mathbb{R}^n \to \mathbb{R}^m$ is called a **linear transformation** if $\mathbf{f}(\mathbf{x} + \mathbf{y}) = \mathbf{f}(\mathbf{x}) + \mathbf{f}(\mathbf{y})$ and $\mathbf{f}(\alpha\mathbf{x}) = \alpha\mathbf{f}(\mathbf{x})$ for any scalar α. If $\mathbf{f}$ is a linear transformation from $\mathbb{R}^n \to \mathbb{R}^m$, then from Section 9.9 there exists an $m \times n$ matrix A such that $\mathbf{f}(\mathbf{x}) = A\mathbf{x}$ for every $\mathbf{x}$ in $\mathbb{R}^n$. ■

Functions from $\mathbb{R}^n \to \mathbb{R}^m$ have many interesting properties. In Chapter 9 we discussed some of the properties of linear transformations. But most mappings from $\mathbb{R}^n \to \mathbb{R}^m$ are not linear. In this section we shall discuss some basic properties of these functions, including the notions of limits and continuity. In Section 10.4 we shall address the more difficult concept of differentiation.

Definition 2 Let $\mathbf{f}$: $\mathbb{R}^n \to \mathbb{R}^m$ and $\mathbf{g}$: $\mathbb{R}^n \to \mathbb{R}^m$ have domains Ω_1 and Ω_2, respectively. Then we define the functions of $\alpha\mathbf{f}$ and $\mathbf{f} + \mathbf{g}$ by

(i) $(\alpha\mathbf{f})(\mathbf{x}) = \alpha\mathbf{f}(\mathbf{x})$ with domain Ω_1, and
(ii) $(\mathbf{f} + \mathbf{g})(\mathbf{x}) = f(\mathbf{x}) + g(\mathbf{x})$ with domain $\Omega_1 \cap \Omega_2$.

EXAMPLE 5 Let **f** and **g**: $\mathbb{R}^3 \to \mathbb{R}^2$ be given by

$$\mathbf{f}(x, y, z) = (\sqrt{1 - x^2 - y^2 - z^2}, xy^2z^3) \qquad \text{and} \qquad \mathbf{g}(x, y, z) = \left(\frac{1}{x + y + z}, \sqrt{x}\right).$$

Compute $-4\mathbf{f}$ and $\mathbf{f} + \mathbf{g}$ and determine their respective domains.

Solution. The domain of $\mathbf{f} = \{(x, y, z): x^2 + y^2 + z^2 \leq 1\}$. This is the closed unit ball. Thus, $-4\mathbf{f}(x, y, z) = (-4\sqrt{1 - x^2 - y^2 - z^2}, -4xy^2z^3)$ with the closed unit ball as its domain. In addition,

$$(\mathbf{f} + \mathbf{g})(x, y, z) = \left(\sqrt{1 - x^2 - y^2 - z^2} + \frac{1}{x + y + z}, xy^2z^3 + \sqrt{x}\right).$$

The domain of $\mathbf{g} = \{(x, y, z): x + y + z \neq 0 \text{ and } x \geq 0\}$. This is the set of points in the half-space $x \geq 0$ that are not on the plane $x + y + z = 0$. Hence the domain of $\mathbf{f} + \mathbf{g}$ consists of those points in the closed unit half ball $\{(x, y, z): x^2 + y^2 + z^2 \leq 1, x \geq 0\}$ that do not lie on the plane $x + y + z = 0$. ■

Definition 3 COMPOSITE FUNCTION Let **f**: $\mathbb{R}^n \to \mathbb{R}^m$ with domain Ω_1 and **g**: $\mathbb{R}^m \to \mathbb{R}^q$ with domain Ω_2. Then the **composite function** $\mathbf{g} \circ \mathbf{f}$: $\mathbb{R}^n \to \mathbb{R}^q$ is defined by

$$\mathbf{g} \circ \mathbf{f}(\mathbf{x}) = \mathbf{g}(\mathbf{f}(\mathbf{x})). \tag{2}$$

The domain of $\mathbf{g} \circ \mathbf{f}$ is given by

$$\text{dom}(\mathbf{g} \circ \mathbf{f}) = \{\mathbf{x} \in \Omega_1: \mathbf{f}(\mathbf{x}) \in \Omega_2\}. \tag{3}$$

EXAMPLE 6 Let **f**: $\mathbb{R}^2 \to \mathbb{R}^3$ be given by $\mathbf{f}(x, y) = (x^3, xy, y)$ and **g**: $\mathbb{R}^3 \to \mathbb{R}^4$ be given by $\mathbf{g}(x, y, z) = (x^2, y^2, z^2, \sqrt{x})$. Compute $\mathbf{g} \circ \mathbf{f}$ and determine its domain.

Solution.

$$\begin{aligned}\mathbf{g} \circ \mathbf{f}(\mathbf{x}) &= \mathbf{g}(\mathbf{f}(\mathbf{x})) = \mathbf{g}(x^3, xy, y) \\ &= ([x^3]^2, (xy)^2, y^2, \sqrt{x^3}) = (x^6, x^2y^2, y^2, x^{3/2})\end{aligned}$$

and dom $\mathbf{g} \circ \mathbf{f}$ is $\{(x, y): x \geq 0\}$. ■

Definition 4 LIMIT Suppose that **f**: $\mathbb{R}^n \to \mathbb{R}^m$ is defined on a neighborhood $B_r(\mathbf{x}_0)$ of the vector $\mathbf{x}_0$ except possibly at $\mathbf{x}_0$ itself. Then

$\lim_{\mathbf{x}\to\mathbf{x}_0} \mathbf{f}(\mathbf{x}) = \mathbf{L}$ if for every $\epsilon > 0$ there is a number $\delta > 0$ such that $|\mathbf{f}(\mathbf{x}) - \mathbf{L}| < \epsilon$ whenever $0 < |\mathbf{x} - \mathbf{x}_0| < \delta$. **(4)**

The definition of a limit is a familiar one. It turns out that $\mathbf{f}(\mathbf{x})$ has a limit as $\mathbf{x} \to \mathbf{x}_0$ if and only if each of its component functions has a limit as $\mathbf{x} \to \mathbf{x}_0$.

Theorem 1 Let $\mathbf{f}: \mathbb{R}^n \to \mathbb{R}^m$ with component functions $f_1, f_2, \ldots, f_m$, and let $\mathbf{L} = (L_1, L_2, \ldots, L_m)$. Then

$$\lim_{\mathbf{x}\to\mathbf{x}_0} f(x) = \mathbf{L} \quad \text{if and only if} \quad \lim_{\mathbf{x}\to\mathbf{x}_0} f_i(\mathbf{x}) = L_i \qquad \text{for } i = 1, 2, \ldots, m. \tag{5}$$

Proof. Suppose that $\lim_{\mathbf{x}\to\mathbf{x}_0} \mathbf{f}(\mathbf{x}) = \mathbf{L}$. Let $\epsilon > 0$ be given and choose δ as in Definition 4. Then, if $0 < |\mathbf{x} - \mathbf{x}_0| < \delta$,

$$\begin{aligned} |f_i(\mathbf{x}) - L_i| &= \sqrt{(f_i(\mathbf{x}) - L_i)^2} \\ &\le \sqrt{(f_1(\mathbf{x}) - L_1)^2 + (f_2(\mathbf{x}) - L_2)^2 + \cdots + (f_m(\mathbf{x}) - L_m)^2} \\ &= |\mathbf{f}(\mathbf{x}) - \mathbf{L}| < \epsilon. \end{aligned}$$

Thus $\lim_{\mathbf{x}\to\mathbf{x}_0} f_i(\mathbf{x}) = L_i$ for $i = 1, 2, 3, \ldots, m$. Conversely, suppose that $\lim_{\mathbf{x}\to\mathbf{x}_0} f_i(\mathbf{x}) = L_i$ for $i = 1, 2, \ldots, n$. Let $\epsilon > 0$ be given. Then, by Definition 4.2.6, there is a number $\delta_i > 0$ such that

$$|f_i(\mathbf{x}) - L_i| < \frac{\epsilon}{\sqrt{m}} \qquad \text{if} \quad 0 < |\mathbf{x} - \mathbf{x}_0| < \delta_i.$$

Let $\delta =$ minimum of $\delta_1, \delta_2, \ldots, \delta_m$. Then, for $0 < |\mathbf{x} - \mathbf{x}_0| < \delta$,

$$\begin{aligned} |\mathbf{f}(\mathbf{x}) - \mathbf{L}| &= \sqrt{(f_1(\mathbf{x}) - L_1)^2 + (f_2(\mathbf{x}) - L_2)^2 + \cdots + (f_m(\mathbf{x}) - L_m)^2} \\ &\le \sqrt{\underbrace{\frac{\epsilon^2}{m} + \frac{\epsilon^2}{m} + \cdots + \frac{\epsilon^2}{m}}_{m \text{ terms}}} = \sqrt{\epsilon^2} = \epsilon. \quad \blacksquare \end{aligned}$$

EXAMPLE 7 Find $\lim_{\mathbf{x}\to(1,2)} \mathbf{f}(\mathbf{x})$ if $\mathbf{f}: \mathbb{R}^2 \to \mathbb{R}^4$ is given by

$$\mathbf{f}(x, y) = \left(x^2y - y^3, \sqrt{x + y}, \sin \pi\left(\frac{x + y}{12}\right), \frac{16x^4 - y^4}{4x^2 - y^2}\right).$$

Solution. Using Theorem 1, we need only take the limits of the component functions. We can determine the first three limits by evaluation because each component function is continuous at (1, 2). Thus

$$\lim_{(x,y)\to(1,2)} (x^2y - y^3) = -6, \qquad \lim_{(x,y)\to(1,2)} \sqrt{x + y} = \sqrt{3},$$

and

$$\lim_{(x,y)\to(1,2)} \sin \pi\left(\frac{x + y}{12}\right) = \sin\frac{\pi}{4} = \frac{\sqrt{2}}{2}.$$

The last component function is not defined at (1, 2) since the denominator is 0 there. However, we easily calculate

$$\lim_{(x,y)\to(1,2)} \frac{16x^4 - y^4}{4x^2 - y^2} = \lim_{(x,y)\to(1,2)} \frac{(4x^2 - y^2)(4x^2 + y^2)}{4x^2 - y^2}$$
$$= \lim_{(x,y)\to(1,2)} (4x^2 + y^2) = 8.$$

Hence

$$\lim_{\mathbf{x}\to(1,2)} \mathbf{f}(\mathbf{x}) = \left(-6, \sqrt{3}, \frac{\sqrt{2}}{2}, 8\right). \blacksquare$$

Definition 5 CONTINUITY Let $\mathbf{f}: \mathbb{R}^n \to \mathbb{R}^m$ be defined in a neighborhood of the vector $\mathbf{x}_0$. Then $\mathbf{f}$ is **continuous** at $\mathbf{x}_0$ if each component function f_i is continuous at $\mathbf{x}_0$.

REMARK. An alternative definition of continuity at $\mathbf{x}_0$ (for $\mathbf{f}$ defined in a neighborhood of $\mathbf{x}_0$) is

$$\lim_{\mathbf{x}\to\mathbf{x}_0} \mathbf{f}(\mathbf{x}) = \mathbf{f}(\mathbf{x}_0). \tag{6}$$

In Problems 28 and 29 you are asked to show that these two definitions are equivalent.

The following theorem enables us to compute a large number of limits by evaluation.

Theorem 2 Let $\mathbf{f}: \mathbb{R}^n \to \mathbb{R}^m$ be continuous at $\mathbf{x}_0$ and let $\mathbf{g}: \mathbb{R}^m \to \mathbb{R}^q$ be continuous at $\mathbf{f}(\mathbf{x}_0)$. Then the composite function $\mathbf{g} \circ \mathbf{f}$ is continuous at $\mathbf{x}_0$.

Proof. Let $\mathbf{y} = \mathbf{f}(\mathbf{x})$. As $\mathbf{x} \to \mathbf{x}_0$, $\mathbf{y} = \mathbf{f}(\mathbf{x}) \to \mathbf{f}(\mathbf{x}_0) = \mathbf{y}_0$, so that

$$\lim_{\mathbf{x}\to\mathbf{x}_0} (\mathbf{g}\circ\mathbf{f})(\mathbf{x}) = \lim_{\mathbf{x}\to\mathbf{x}_0} \mathbf{g}(\mathbf{f}(\mathbf{x})) = \lim_{\mathbf{y}\to\mathbf{y}_0} \mathbf{g}(\mathbf{y}) = \mathbf{g}(\mathbf{y}_0) = \mathbf{g}(\mathbf{f}(\mathbf{x}_0)) = (\mathbf{g}\circ\mathbf{f})(\mathbf{x}_0). \blacksquare$$

PROBLEMS 10.3

In Problems 1–7 find the domain of the given function.

1. $\mathbf{f}(x, y) = (x^2 + y^2, x^2 - y^2, \sqrt{x^2 + y^2})$
2. $f(x, y, z) = \left(\frac{x}{y}, \frac{y}{z}, \frac{z}{x}\right)$
3. $\mathbf{f}(x, y, z) = (e^{x+y}, \ln(x + y + z))$
4. $\mathbf{f}(x, y, z) = (x^2, y^2, z^2, \sqrt{36 - 9x^2 - 4y^2 - z^2})$
5. $\mathbf{f}(x_1, x_2, x_3, x_4, x_5) = (x_1x_2, x_2x_3, x_3x_4, x_4x_5, x_1x_5)$
6. $\mathbf{f}(x_1, x_2, x_3, x_4) = \left(\frac{1}{x_1 + x_2 + 3x_3 + x_4 - 5}, x_1 + x_2 + x_3 + x_4 - 5\right)$
7. $\mathbf{f}(x_1, x_2, x_3, x_4) = (\sqrt{1 - x_1^2 - x_2^2 - x_3^2 - x_4^2}, \sqrt[3]{1 - x_1^2 - x_2^2 - x_3^2 - x_4^2})$
8. Let $\mathbf{f}: \mathbb{R}^2 \to \mathbb{R}^2$ be defined by $\mathbf{f}(x, y) = (-x, y)$. Describe the image of the set of points lying on the unit semicircle $x^2 + y^2 = 1$, $x \geq 0$.
9. Do the same as in Problem 8 for $\mathbf{f}(x, y) = (3x, -4y)$.
10. Do the same as in Problem 8 for $\mathbf{f}(x, y) = (ax, by)$.
11. Let $\mathbf{f}: \mathbb{R}^2 \to \mathbb{R}^2$ be given by $\mathbf{f}(x, y) = (e^y \sin x, e^y \cos x)$. Describe the image of the rectangle $0 \leq x \leq 2\pi$, $0 \leq y \leq 1$ under this mapping.
12. Let **f** be as in Problem 11. Describe the image of the ellipse $(x^2/4) + (y^2/9) = 1$ under this mapping.
13. Describe the image under the polar coordinate mapping of the rectangle $0 \leq r \leq 2$, $0 \leq \theta \leq \pi/3$.
14. Do the same as in Problem 13 for the rectangle $1 \leq r \leq 3$, $\pi/6 \leq \theta \leq \pi/2$.

In Problems 15–24 find the limit of the given function at the given point (if it exists). Is **f** continuous at that point?

15. $\mathbf{f}(x, y) = (x^2 + y^2, x^2 - y^2, xy)$; $(4, 6)$

16. $\mathbf{f}(x, y, z) = \left(\frac{x^2 - y^2}{x - y}, \frac{y^2 - z^2}{y - z}\right)$; $(0, 0, 0)$

17. $\mathbf{f}(x, y, z) = \left(\sin yz, \tan^{-1}\frac{y}{x}, x^3\right)$; $\left(\frac{\pi}{6}, \frac{\pi}{6}, 2\right)$

18. $\mathbf{f}(x_1, x_2, x_3, x_4) =$

$\left(x_1^2, x_2^2, x_3^2, x_4^2, \frac{\sin(x_1^2 + x_2^2 + x_3^2 + x_4^2)}{x_1^2 + x_2^2 + x_3^2 + x_4^2}\right)$;

$(0, 0, 0, 0)$

19. $\mathbf{f}(x_1, x_2, \ldots, x_n) =$

$\left(\sum_{i=1}^{n} x_i, \left[\sum_{i=1}^{n} x_i\right]^2, \left[\sum_{i=1}^{n} x_i\right]^3, \ldots, \left[\sum_{i=1}^{n} x_i\right]^m\right)$;

$(1, 1, \ldots, 1)$

20. $\mathbf{f}(x, y, z) = (\sin xyz, \cos xyz, e^{\sqrt{x^2+y^2+z^2}})$; $(0, 0, 0)$

21. $\mathbf{f}(x, y, z) =$

$\left(\ln(1 + x + y + z), \frac{x + y + z}{\ln(x + y + z)}, e^{x+y+z}\right)$;

$(0, 0, 0)$

22. $\mathbf{f}(x, y) = \left(x, y, \frac{x^2 - 2y}{y^2 + 2x}\right)$; $(0, 0)$

23. $\mathbf{f}(x_1, x_2, x_3, x_4) = \left(x_1x_2, \sin x_3x_4, \frac{x_1 + x_2 + x_3 + x_4}{x_1 - x_2 - x_3 - x_4}\right)$;

$(0, 0, 0, 0)$

24. $\mathbf{f}(x_1, x_2, x_3, x_4) =$

$\left(\frac{5x_1^2x_2^2 - 3x_1^2x_3^2}{x_1^4 + x_4^2}, \frac{x_1x_2^2 - x_3x_4^2 + x_2x_3x_4}{x_1^2 + x_2^2 + x_3^2 + x_4^2}\right)$;

$(0, 0, 0, 0)$

In Problems 25–27 write out the composite function $\mathbf{g} \circ \mathbf{f}$. Determine its domain.

25. $\mathbf{f}(x, y) = (xy, x^2, \cos(y/x))$; $\mathbf{g}(x, y, z) = (xyz, e^{x+y-2z})$

26. $\mathbf{f}(x, y, z) = (\sqrt{1 - x^2 - y^2 - z^2}, \ln(1 - x - y + z), xy, yz)$; $\mathbf{g}(x_1, x_2, x_3, x_4) = (x_1 - x_3, x_1^3x_3)$

27. $\mathbf{f}(x, y) = (e^{x+y}, \sqrt{1 - x^2 + y^2})$; $\mathbf{g}(x, y) = (\sin(x/y), \sin(y/x), e^{\sqrt{x^2+y^2}})$

28. Suppose that $\mathbf{f}: \mathbb{R}^n \to \mathbb{R}^m$ is continuous according to Definition 5. Show that $\lim_{\mathbf{x}\to\mathbf{x}_0} \mathbf{f}(\mathbf{x}) = \mathbf{f}(\mathbf{x}_0)$.

29. Suppose that **f** is defined in a neighborhood of $\mathbf{x}_0$ and that $\lim_{\mathbf{x}\to\mathbf{x}_0} \mathbf{f}(\mathbf{x}) = \mathbf{f}(\mathbf{x}_0)$. Prove that **f** is continuous at $\mathbf{x}_0$.

30. Suppose that $\mathbf{f}: \mathbb{R}^n \to \mathbb{R}^m$ is continuous at $\mathbf{x}_0$. Show that $\alpha\mathbf{f}$ is continuous at $\mathbf{x}_0$ for every scalar α.

31. Suppose that $\mathbf{f}: \mathbb{R}^n \to \mathbb{R}^m$ and $\mathbf{g}: \mathbb{R}^n \to \mathbb{R}^m$ are continuous at $\mathbf{x}_0$. Show that $\mathbf{f} + \mathbf{g}$ is continuous at $\mathbf{x}_0$.

32. Let $\mathbf{f}: \mathbb{R}^n \to \mathbb{R}^m$ be a linear transformation. Show that **f** is continuous at every **x** in $\mathbb{R}^n$.

33. Let $\mathbf{f}: \mathbb{R}^n \to \mathbb{R}^m$ be a linear transformation with matrix representation A_1, and let $\mathbf{g}: \mathbb{R}^m \to \mathbb{R}^q$ be a linear transformation with matrix representation A_2. Show that the matrix representation of $\mathbf{g} \circ \mathbf{f}$ is A_2A_1.

10.4 DERIVATIVES AND THE JACOBIAN MATRIX

In this section we discuss the derivative of a function **f** from $\mathbb{R}^n$ to $\mathbb{R}^m$. As you might expect, the notion is closely related to the gradient of a function from $\mathbb{R}^n$ to $\mathbb{R}$. Our definition is similar to the definition of the gradient given in Section 4.5.

Definition 1 DIFFERENTIABILITY Let $f: \mathbb{R}^n \to \mathbb{R}^m$ be defined on an open set Ω in $\mathbb{R}^n$ and suppose that $\mathbf{x}_0 \in \Omega$. Then **f** is **differentiable** at $\mathbf{x}_0$ if there exists a linear transformation $\mathbf{L}: \mathbb{R}^n \to \mathbb{R}^m$ and a function $\mathbf{g}: \mathbb{R}^n \to \mathbb{R}^m$ such that for every **x** in Ω,

$$\mathbf{f}(\mathbf{x}) - \mathbf{f}(\mathbf{x}_0) = \mathbf{L}(\mathbf{x} - \mathbf{x}_0) + \mathbf{g}(\mathbf{x} - \mathbf{x}_0) \tag{1}$$

where

$$\lim_{|\mathbf{x}-\mathbf{x}_0|\to 0} \frac{g(\mathbf{x}-\mathbf{x}_0)}{|\mathbf{x}-\mathbf{x}_0|} = \mathbf{0}. \quad (2)$$

REMARK. If we write $\Delta\mathbf{x} = \mathbf{x} - \mathbf{x}_0$, then Definition 1 is virtually identical to Definitions 4.5.2 and 4.5.7 for a function from $\mathbb{R}^n \to \mathbb{R}$.

We now must answer two questions: First, under what conditions is a given function from $\mathbb{R}^n \to \mathbb{R}^m$ differentiable? Second, how do we determine the linear transformation $\mathbf{L}$? We shall answer the second question first.

If $f: \mathbb{R}^n \to \mathbb{R}$ is differentiable, then

$$\mathbf{L} = \nabla f = \left(\frac{\partial f}{\partial x_1}, \frac{\partial f}{\partial x_2}, \ldots, \frac{\partial f}{\partial x_n}\right).$$

Thus we should not be too surprised if, in this case, the linear transformation $\mathbf{L}$ involves the partial derivatives of the component functions of $\mathbf{f}$. We will soon see that this is the case.

Since $\mathbf{L}$ is a linear transformation from $\mathbb{R}^n \to \mathbb{R}^m$, there is, by Theorem 9.9.1 a unique $m \times n$ matrix, which we denote $J_{\mathbf{f}}(\mathbf{x}_0)$, such that

$$\mathbf{L}(\mathbf{x}) = J_{\mathbf{f}}(\mathbf{x}_0)\mathbf{x} \quad \text{for every } \mathbf{x} \text{ in } \mathbb{R}^n. \quad (3)$$

We assume that $\mathbf{x}$ is written as a column vector in terms of the standard basis in $\mathbb{R}^n$ and $\mathbf{L}(\mathbf{x})$ is written as a column vector in terms of the standard basis in $\mathbb{R}^m$. The notation $J_{\mathbf{f}}(\mathbf{x}_0)$ is used to emphasize that the matrix J depends on both the function $\mathbf{f}$ and the vector $\mathbf{x}_0$. With this notation, (1) can be written as

$$\mathbf{f}(\mathbf{x}) - \mathbf{f}(\mathbf{x}_0) = J_{\mathbf{f}}(\mathbf{x}_0)(\mathbf{x} - \mathbf{x}_0) + \mathbf{g}(\mathbf{x} - \mathbf{x}_0). \quad (4)$$

REMARK. In Equation (4) we assume that all vectors are written as column vectors so that the matrix multiplication makes sense.

Definition 2 JACOBIAN MATRIX Let $\mathbf{f}: \mathbb{R}^n \to \mathbb{R}^m$ be differentiable at $\mathbf{x}_0$. Then the $m \times n$ matrix $J_{\mathbf{f}}(\mathbf{x}_0)$ determined by (4) is called the **Jacobian matrix** of $\mathbf{f}$ at $\mathbf{x}_0$.[†]

REMARK. The matrix $J_{\mathbf{f}}(\mathbf{x}_0)$ is sometimes called the **derivative matrix** or, more simply, the **derivative** of the linear transformation $\mathbf{f}$.

Theorem 1 Let $\mathbf{f}: \mathbb{R}^n \to \mathbb{R}^m$ be differentiable at $\mathbf{x}_0$. Then the Jacobian matrix $J_{\mathbf{f}}(\mathbf{x}_0)$ is given by

[†]See the biographical sketch on p. 390.

$$
J_{\mathbf{f}}(\mathbf{x}_0) = \begin{pmatrix} \nabla f_1(\mathbf{x}_0) \\ \nabla f_2(\mathbf{x}_0) \\ \vdots \\ \nabla f_m(\mathbf{x}_0) \end{pmatrix} = \begin{pmatrix} \frac{\partial f_1}{\partial x_1}(\mathbf{x}_0) & \frac{\partial f_1}{\partial x_2}(\mathbf{x}_0) & \frac{\partial f_1}{\partial x_3}(\mathbf{x}_0) & \cdots & \frac{\partial f_1}{\partial x_n}(\mathbf{x}_0) \\ \frac{\partial f_2}{\partial x_1}(\mathbf{x}_0) & \frac{\partial f_2}{\partial x_2}(\mathbf{x}_0) & \frac{\partial f_2}{\partial x_3}(\mathbf{x}_0) & \cdots & \frac{\partial f_2}{\partial x_n}(\mathbf{x}_0) \\ \frac{\partial f_3}{\partial x_1}(\mathbf{x}_0) & \frac{\partial f_3}{\partial x_2}(\mathbf{x}_0) & \frac{\partial f_3}{\partial x_3}(\mathbf{x}_0) & \cdots & \frac{\partial f_3}{\partial x_n}(\mathbf{x}_0) \\ \vdots & \vdots & \vdots & & \vdots \\ \frac{\partial f_m}{\partial x_1}(\mathbf{x}_0) & \frac{\partial f_m}{\partial x_2}(\mathbf{x}_0) & \frac{\partial f_m}{\partial x_3}(\mathbf{x}_0) & \cdots & \frac{\partial f_m}{\partial x_n}(\mathbf{x}_0) \end{pmatrix}. \tag{5}
$$

NOTE. The functions $f_1, f_2, \ldots, f_m$ in (5) are the component functions of the function $\mathbf{f}$.

Proof. Let $\mathbf{J}_1, \mathbf{J}_2, \ldots, \mathbf{J}_m$ denote the rows of $J_{\mathbf{f}}(\mathbf{x}_0)$. Then, writing (4) in terms of components, we have

$$
\begin{pmatrix} f_1(\mathbf{x}) - f_1(\mathbf{x}_0) \\ f_2(\mathbf{x}) - f_2(\mathbf{x}_0) \\ \vdots \\ f_m(\mathbf{x}) - f_m(\mathbf{x}_0) \end{pmatrix} = \begin{pmatrix} \mathbf{J}_1 \cdot (\mathbf{x} - \mathbf{x}_0) \\ \mathbf{J}_2 \cdot (\mathbf{x} - \mathbf{x}_0) \\ \vdots \\ \mathbf{J}_m \cdot (\mathbf{x} - \mathbf{x}_0) \end{pmatrix} + \begin{pmatrix} g_1(\mathbf{x} - \mathbf{x}_0) \\ g_2(\mathbf{x} - \mathbf{x}_0) \\ \vdots \\ g_m(\mathbf{x} - \mathbf{x}_0) \end{pmatrix} \tag{6}
$$

where $g_1, g_2, \ldots, g_m$ are the component functions of $\mathbf{g}$. Now

$$
\lim_{|\mathbf{x}-\mathbf{x}_0| \to 0} \frac{g(\mathbf{x} - \mathbf{x}_0)}{|\mathbf{x} - \mathbf{x}_0|} = \mathbf{0}.
$$

But

$$
\frac{\mathbf{g}(\mathbf{x} - \mathbf{x}_0)}{|\mathbf{x} - \mathbf{x}_0|} = \left(\frac{g_1(\mathbf{x} - \mathbf{x}_0)}{|\mathbf{x} - \mathbf{x}_0|}, \frac{g_2(\mathbf{x} - \mathbf{x}_0)}{|\mathbf{x} - \mathbf{x}_0|}, \ldots, \frac{g_m(\mathbf{x} - \mathbf{x}_0)}{|\mathbf{x} - \mathbf{x}_0|} \right).
$$

This means that

$$
\lim_{|\mathbf{x}-\mathbf{x}_0| \to 0} \frac{g_i(\mathbf{x} - \mathbf{x}_0)}{|\mathbf{x} - \mathbf{x}_0|} = 0 \qquad \text{for} \quad i = 1, 2, \ldots, m. \tag{7}
$$

Let's look at the ith row of (6). We have

$$
f_i(\mathbf{x}) - f_i(\mathbf{x}_0) = \mathbf{J}_i \cdot (\mathbf{x} - \mathbf{x}_0) + g_i(\mathbf{x} - \mathbf{x}_0). \tag{8}
$$

Now $f_i\colon \mathbb{R}^n \to \mathbb{R}$ and, according to Definition 4.5.7, f_i is differentiable and $\mathbf{J}_i = \nabla f_i(\mathbf{x})$. This, combined with the fact that

$$\nabla f_i(\mathbf{x}_0) = \left(\frac{\partial f_i}{\partial x_1}(\mathbf{x}_0), \frac{\partial f_i}{\partial x_2}(\mathbf{x}_0), \ldots, \frac{\partial f_i}{\partial x_m}(\mathbf{x}_0)\right),$$

completes the proof. ■

EXAMPLE 1 Let $\mathbf{f}: \mathbb{R}^3 \to \mathbb{R}^2$ be defined by $\mathbf{f}(x, y, z) = (e^{x+y^2+z^3}, xy/z)$. Determine $J_\mathbf{f}(1, -1, 2)$.

Solution.

$$J_\mathbf{f}(\mathbf{x}) = \begin{pmatrix} \frac{\partial f_1}{\partial x}(\mathbf{x}) & \frac{\partial f_1}{\partial y}(\mathbf{x}) & \frac{\partial f_1}{\partial z}(\mathbf{x}) \\ \frac{\partial f_2}{\partial x}(\mathbf{x}) & \frac{\partial f_2}{\partial y}(\mathbf{x}) & \frac{\partial f_2}{\partial z}(\mathbf{x}) \end{pmatrix} = \begin{pmatrix} e^{x+y^2+z^3} & 2ye^{x+y^2+z^3} & 3z^2e^{x+y^2+z^3} \\ \frac{y}{z} & \frac{x}{z} & \frac{-xy}{z^2} \end{pmatrix}$$

Thus

$$J_\mathbf{f}(1, -1, 2) = \begin{pmatrix} e^{10} & -2e^{10} & 12e^{10} \\ -\frac{1}{2} & \frac{1}{2} & \frac{1}{4} \end{pmatrix}. \blacksquare$$

EXAMPLE 2 Determine $J_\mathbf{f}(2, \pi/6)$ for the polar coordinate transformation $\mathbf{f}(r, \theta) = (r \cos \theta, r \sin \theta)$.

Solution.

$$J_\mathbf{f}(r, \theta) = \begin{pmatrix} \frac{\partial}{\partial r} r \cos \theta & \frac{\partial}{\partial \theta} r \cos \theta \\ \frac{\partial}{\partial r} r \sin \theta & \frac{\partial}{\partial \theta} r \sin \theta \end{pmatrix} = \begin{pmatrix} \cos \theta & -r \sin \theta \\ \sin \theta & r \cos \theta \end{pmatrix},$$

so that

$$J_\mathbf{f}\left(2, \frac{\pi}{6}\right) = \begin{pmatrix} \frac{\sqrt{3}}{2} & -1 \\ \frac{1}{2} & \sqrt{3} \end{pmatrix}. \blacksquare$$

Definition 3 TOTAL DERIVATIVE Let $\mathbf{f}: \mathbb{R}^n \to \mathbb{R}^m$ be differentiable at $\mathbf{x}_0$. Then the Jacobian matrix $J_\mathbf{f}(\mathbf{x}_0)$ is called the **total derivative** of $\mathbf{f}$ at $\mathbf{x}_0$.

REMARK. Because of Definition 3, $J_\mathbf{f}(\mathbf{x})$ is often written as $\mathbf{f}'(\mathbf{x}_0)$. Thus, for example, in Example 1

$$\mathbf{f}'(1, -1, 2) = \begin{pmatrix} e^{10} & -2e^{10} & 12e^{10} \\ -\frac{1}{2} & \frac{1}{2} & \frac{1}{4} \end{pmatrix}$$

and, in Example 2,

$$\mathbf{f}'\left(2, \frac{\pi}{6}\right) = \begin{pmatrix} \frac{\sqrt{3}}{2} & -1 \\ \frac{1}{2} & \sqrt{3} \end{pmatrix}.$$

We have seen how to compute the matrix or total derivative of **f** if **f** is differentiable—but how do we know when **f** is differentiable? In Theorem 4.5.4 we pointed out that if **f**: $\mathbb{R}^n \to \mathbb{R}$ has continuous first-order partial derivatives at **x**, then f is differentiable at **x**. In light of this, the answer to the question posed above is easy to obtain.

Theorem 2 Let f: $\mathbb{R}^n \to \mathbb{R}^m$ with component functions $f_1, f_2, \ldots, f_m$. Suppose that all partial derivatives $\partial f_i/\partial x_j$ for $i = 1, 2, \ldots, m$ and $j = 1, 2, \ldots, n$ exist and are continuous at $\mathbf{x}_0$. Then **f** is differentiable at $\mathbf{x}_0$.

Proof. By Theorem 4.5.4 each component function f_i is differentiable at $\mathbf{x}_0$ and

$$f_i(\mathbf{x}) - f_i(\mathbf{x}_0) = \nabla f_i(\mathbf{x}_0) \cdot (\mathbf{x} - \mathbf{x}_0) + g_i(\mathbf{x} - \mathbf{x}_0) \tag{9}$$

where g_i: $\mathbb{R}^n \to \mathbb{R}$ and

$$\lim_{|\mathbf{x}-\mathbf{x}_0| \to 0} \frac{g_i(\mathbf{x} - \mathbf{x}_0)}{|\mathbf{x} - \mathbf{x}_0|} = 0.$$

Thus, if we define **g**: $\mathbb{R}^n \to \mathbb{R}^m$ by

$$\mathbf{g}(\mathbf{x}) = \begin{pmatrix} g_1(\mathbf{x}) \\ g_2(\mathbf{x}) \\ \vdots \\ g_m(\mathbf{x}) \end{pmatrix},$$

then

$$\lim_{|\mathbf{x}-\mathbf{x}_0| \to 0} \frac{\mathbf{g}(\mathbf{x} - \mathbf{x}_0)}{|\mathbf{x} - \mathbf{x}_0|} = \mathbf{0} \tag{10}$$

and, using (9), we have

$$\begin{aligned} \mathbf{f}(\mathbf{x}) - \mathbf{f}(\mathbf{x}_0) &= \begin{pmatrix} \nabla f_1(x_0) \cdot (\mathbf{x} - \mathbf{x}_0) \\ \nabla f_2(\mathbf{x}_0) \cdot (\mathbf{x} - \mathbf{x}_0) \\ \vdots \\ \nabla f_m(\mathbf{x}_0) \cdot (\mathbf{x} - \mathbf{x}_0) \end{pmatrix} + \mathbf{g}(\mathbf{x} - \mathbf{x}_0) \\ &= J_{\mathbf{f}}(\mathbf{x}_0)(\mathbf{x} - \mathbf{x}_0) + \mathbf{g}(\mathbf{x} - \mathbf{x}_0), \end{aligned}$$

which, in light of Definition 1 and equation (10), completes the proof. ■

We now observe that in yet another case differentiable functions are continuous.

Theorem 3 Let $\mathbf{f}: \mathbb{R}^n \to \mathbb{R}^m$ be differentiable at $\mathbf{x}_0$. Then $\mathbf{f}$ is continuous at $\mathbf{x}_0$.

Proof. This follows immediately by applying Theorem 4.5.4 to the component functions of $\mathbf{f}$. ■

Since

$$\frac{\partial}{\partial x}(f + g) = \frac{\partial f}{\partial x} + \frac{\partial g}{\partial x} \qquad \text{and} \qquad \frac{\partial}{\partial x}(\alpha f) = \alpha\frac{\partial}{\partial x}$$

for any scalar α, we can easily prove the following theorem:

Theorem 4 Let $\mathbf{f}: \mathbb{R}^n \to \mathbb{R}^m$ and $\mathbf{g}: \mathbb{R}^n \to \mathbb{R}^m$ be differentiable at $\mathbf{x}_0$ and let α be a scalar. Then $\alpha\mathbf{f}$ and $\mathbf{f} + \mathbf{g}$ are differentiable at $\mathbf{x}_0$ and

$$\text{(i)}\ J_{\alpha\mathbf{f}}(\mathbf{x}_0) = \alpha J_{\mathbf{f}}(\mathbf{x}_0) \qquad (11)$$
$$\text{(ii)}\ J_{\mathbf{f}+\mathbf{g}}(\mathbf{x}_0) = J_{\mathbf{f}}(\mathbf{x}_0) + J_{\mathbf{g}}(\mathbf{x}_0) \qquad (12)$$

or, alternatively,

$$\text{(i)}\ (\alpha\mathbf{f})'(\mathbf{x}_0) = \alpha\mathbf{f}'(\mathbf{x}_0) \qquad (13)$$
$$\text{(ii)}\ (\mathbf{f} + \mathbf{g})'(\mathbf{x}_0) = \mathbf{f}'(\mathbf{x}_0) + \mathbf{g}'(\mathbf{x}_0). \qquad (14)$$

REMARK. Theorem 4 tells us that the space of differentiable functions from $\mathbb{R}^n$ to $\mathbb{R}^m$ forms a vector space and that the **differentiable operator** $\mathbf{f} \to \mathbf{f}'(\mathbf{x}_0)$ is a linear transformation from this vector space to the mn-dimensional vector space of $m \times n$ matrices.

We close this section by describing the chain rule for functions from $\mathbb{R}^n$ to $\mathbb{R}^m$. This result generalizes Theorem 4.6.4 for functions from $\mathbb{R}^n$ to $\mathbb{R}$, which in turn generalizes the chain rule for scalar functions of one variable.

Theorem 5 CHAIN RULE Let $\mathbf{f}: \mathbb{R}^n \to \mathbb{R}^m$ be differentiable at $\mathbf{x}_0$ and let $\mathbf{g}: \mathbb{R}^m \to \mathbb{R}^q$ be differentiable at $\mathbf{f}(\mathbf{x}_0)$. Suppose further that all the component functions of $\mathbf{f}$ and $\mathbf{g}$ have continuous partial derivatives. Then the composite function $\mathbf{g} \circ \mathbf{f}$ is differentiable at $\mathbf{x}_0$ and

$$J_{\mathbf{g}\circ\mathbf{f}}(\mathbf{x}_0) = J_{\mathbf{g}}(\mathbf{f}(\mathbf{x}_0))J_{\mathbf{f}}(\mathbf{x}_0). \qquad (15)$$

Proof. Let

$$\mathbf{h}(\mathbf{x}) = \mathbf{g} \circ \mathbf{f}(\mathbf{x}) = \mathbf{g}(\mathbf{f}(\mathbf{x})) = (g_1(\mathbf{f}(\mathbf{x})), g_2(\mathbf{f}(\mathbf{x})), \ldots, g_q(\mathbf{f}(\mathbf{x}))).$$

Then $\mathbf{h}$: $\mathbb{R}^n \to \mathbb{R}^q$ and the ith component function h_i: $\mathbb{R}^n \to \mathbb{R}$ of $\mathbf{h}$ is given by

$$h_i(\mathbf{x}_0) = g_i(\mathbf{f}(\mathbf{x}_0)) = (g_i \circ \mathbf{f})(\mathbf{x}_0). \tag{16}$$

Before continuining, let's take another look at the chain rule proved in Theorem 4.6.4: If f: $\mathbb{R}^n \to \mathbb{R}$ and $\mathbf{x}$: $\mathbb{R} \to \mathbb{R}^n$, then

$$\frac{d}{dt} f(\mathbf{x}(t)) = \nabla f(\mathbf{x}) \cdot \mathbf{x}'(t), \tag{17}$$

assuming that f is differentiable at $\mathbf{x}$ and that $\mathbf{x}$ is differentiable at t. Now, in order to compute a partial derivative of a function of n variables, we treat all but one of the variables as if they were constant. Thus we can use the chain rule (17) to compute partial derivatives of the functions h_i given by (16). Letting the jth component x_j of $\mathbf{x}$ play the role of t in (17), and letting $\mathbf{f}(\mathbf{x})$ play the role of $\mathbf{x}$, we have $(\partial \mathbf{f}/\partial x_j)(\mathbf{x})$ instead of $\mathbf{x}'(t)$ and

$$\frac{\partial h_i}{\partial x_j}(\mathbf{x}) = \nabla g_i(\mathbf{f}(\mathbf{x})) \cdot \frac{\partial \mathbf{f}}{\partial x_j}(\mathbf{x}). \tag{18}$$

Note that since $\mathbf{g}$ is differentiable at $\mathbf{f}(\mathbf{x}_0)$, each of its components is differentiable at $\mathbf{f}(\mathbf{x}_0)$. Also

$$\frac{\partial \mathbf{f}}{\partial x_j}(\mathbf{x}_0) = \begin{pmatrix} \dfrac{\partial f_1}{\partial x_j}(\mathbf{x}_0) \\ \dfrac{\partial f_2}{\partial x_j}(\mathbf{x}_0) \\ \vdots \\ \dfrac{\partial f_m}{\partial x_j}(\mathbf{x}_0) \end{pmatrix}$$

exists and is continuous by assumption, so that Theorem 4.6.4 can be applied in this setting.

Now g_i: $\mathbb{R}^m \to \mathbb{R}$ and all the components of the m-vector ∇g_i are the partial derivatives of g_i, and these are continuous at $\mathbf{f}(\mathbf{x}_0)$ by assumption. But since $\mathbf{f}$ is continuous at $\mathbf{x}$, $\nabla g_i(\mathbf{f}(\mathbf{x}_0))$ is continuous at $\mathbf{x}_0$ by Theorem 4.2.2(v). This means that the scalar product in (18) consists of the sum of products of functions that are continuous at $\mathbf{x}_0$, which implies that $\partial h_i/\partial x_j$ is continuous at $\mathbf{x}_0$. Thus, since $\mathbf{h} = \mathbf{g} \circ \mathbf{f}$, we see that all the first-order partial derivatives of the component functions of $\mathbf{g} \circ \mathbf{f}$ are continuous at $\mathbf{x}_0$, so that by Theorem 2, $\mathbf{g} \circ \mathbf{f}$ is differentiable at $\mathbf{x}_0$ and the Jacobian matrix $J_{\mathbf{g}\circ\mathbf{f}}(\mathbf{x}_0)$ exists. Moreover, the ijth component of $J_{\mathbf{g}\circ\mathbf{f}}(\mathbf{x}_0)$ is

$$\nabla g_i(\mathbf{f}(\mathbf{x}_0)) \cdot \frac{\partial \mathbf{f}}{\partial x_j}(\mathbf{x}). \tag{19}$$

Now the ijth component of $J_{\mathbf{g}}(\mathbf{f}(\mathbf{x}_0))J_{\mathbf{f}}(\mathbf{x}_0)$ is the scalar product of the ith row of $J_{\mathbf{g}}(\mathbf{f}(\mathbf{x}_0))$ with the jth column of $J_{\mathbf{f}}(\mathbf{x}_0)$. But the

$$i\text{th row of } J_{\mathbf{g}}(\mathbf{f}(\mathbf{x}_0)) = \nabla g_i(\mathbf{f}(\mathbf{x}_0))$$

and the

$$j\text{th column of } J_{\mathbf{f}}(\mathbf{x}_0) = \begin{pmatrix} \dfrac{\partial f_1}{\partial x_j}(\mathbf{x}_0) \\ \dfrac{\partial f_2}{\partial x_j}(\mathbf{x}_0) \\ \vdots \\ \dfrac{\partial f_m}{\partial x_j}(\mathbf{x}_0) \end{pmatrix} = \frac{\partial \mathbf{f}}{\partial x_j}(\mathbf{x}_0).$$

Glancing at equation (19), we observe that the ijth component of $J_{\mathbf{g}\circ\mathbf{f}}(\mathbf{x}_0)$ is equal to the ijth component of $J_{\mathbf{g}}(\mathbf{f}(\mathbf{x}_0))J_{\mathbf{f}}(\mathbf{x}_0)$. Since two matrices are equal if and only if their corresponding components are equal, we see that $J_{\mathbf{g}\circ\mathbf{f}}(\mathbf{x}_0) = J_{\mathbf{g}}(\mathbf{f}(\mathbf{x}_0))J_{\mathbf{f}}(\mathbf{x}_0)$ and the proof is complete. ■

EXAMPLE 3 Verify the chain rule at the point $\mathbf{x}_0 = (4, 1)$ for the mappings $\mathbf{f}(x, y) = (xy, x + 2y, x/y)$ and $\mathbf{g}(x, y, z) = (x^2 + z^2, xyz, yz/x, 2x + 3y + 4z)$.

Solution. We have

$$J_{\mathbf{f}}(x, y) = \begin{pmatrix} y & x \\ 1 & 2 \\ \dfrac{1}{y} & -\dfrac{x}{y^2} \end{pmatrix} \qquad J_{\mathbf{f}}(4, 1) = \begin{pmatrix} 1 & 4 \\ 1 & 2 \\ 1 & -4 \end{pmatrix}$$

and $\mathbf{f}(4, 1) = (4, 6, 4)$. Similarly,

$$J_{\mathbf{g}}(x, y, z) = \begin{pmatrix} 2x & 0 & 2z \\ yz & xz & xy \\ -\dfrac{yz}{x^2} & \dfrac{z}{x} & \dfrac{y}{x} \\ 2 & 3 & 4 \end{pmatrix} \quad \text{and} \quad J_{\mathbf{g}}(4, 6, 4) = \begin{pmatrix} 8 & 0 & 8 \\ 24 & 16 & 24 \\ -\dfrac{3}{2} & 1 & \dfrac{3}{2} \\ 2 & 3 & 4 \end{pmatrix}$$

so that

$$J_{\mathbf{g}}(\mathbf{f}(\mathbf{x}_0))J_{\mathbf{f}}(\mathbf{x}_0) = \begin{pmatrix} 8 & 0 & 8 \\ 24 & 16 & 24 \\ -\dfrac{3}{2} & 1 & \dfrac{3}{2} \\ 2 & 3 & 4 \end{pmatrix} \begin{pmatrix} 1 & 4 \\ 1 & 2 \\ 1 & -4 \end{pmatrix} = \begin{pmatrix} 16 & 0 \\ 64 & 32 \\ 1 & -10 \\ 9 & -2 \end{pmatrix}.$$

Now $\mathbf{g} \circ \mathbf{f}: \mathbb{R}^2 \to \mathbb{R}^4$ is given by

$$\mathbf{g} \circ \mathbf{f}(x, y) = \left(x^2y^2 + \frac{x^2}{y^2}, (xy)(x + 2y)\left(\frac{x}{y}\right), \frac{(x + 2y)(x/y)}{xy}, 2xy + 3(x + 2y) + 4\frac{x}{y}\right)$$

$$= \left(x^2y^2 + \frac{x^2}{y^2}, x^3 + 2x^2y, \frac{x}{y^2} + \frac{2}{y}, 3x + 6y + 2xy + \frac{4x}{y}\right),$$

so that

$$J_{\mathbf{g}\circ\mathbf{f}}(x, y) = \begin{pmatrix} 2xy^2 + \dfrac{2x}{y^2} & 2x^2y - \dfrac{2x^2}{y^3} \\ 3x^2 + 4xy & 2x^2 \\ \dfrac{1}{y^2} & -\dfrac{2x}{y^3} - \dfrac{2}{y^2} \\ 3 + 2y + \dfrac{4}{y} & 6 + 2x - \dfrac{4x}{y^2} \end{pmatrix}$$

and

$$J_{\mathbf{g}\circ\mathbf{f}}(\mathbf{x}_0) = J_{\mathbf{g}\circ\mathbf{f}}(4, 1) = \begin{pmatrix} 16 & 0 \\ 64 & 32 \\ 1 & -10 \\ 9 & -2 \end{pmatrix}.$$

Thus

$$J_{\mathbf{g}\circ\mathbf{f}}(\mathbf{x}_0) = J_{\mathbf{g}}(f(\mathbf{x}_0))J_{\mathbf{f}}(\mathbf{x}_0).$$

Note that in this example, as in most others, it is easier to compute the derivative of the composite function by writing two derivatives and using the chain rule rather than by finding the composite function explicitly and then computing its derivative directly. ■

PROBLEMS 10.4

In Problems 1–10 compute the derivative $\mathbf{f}'(\mathbf{x}_0) = J_{\mathbf{f}}(\mathbf{x}_0)$ of the given function at the point $\mathbf{x}_0$.

1. $\mathbf{f}(x, y) = (x^2, y^2)$; $(1, 3)$

2. $\mathbf{f}(x, y) = (x^2 + y^2, x^2 - y^2, \sqrt{x^2 + y^2})$; $(3, 4)$

3. $\mathbf{f}(x, y, z) = \left(\dfrac{x}{y}, \dfrac{y}{z}, \dfrac{z}{x}\right)$; $(1, 4, 8)$

4. $\mathbf{f}(x, y, z) = (e^{x+y}, \ln(x + y + z))$; $(0, 1, 0)$

5. $\mathbf{f}(x, y, z) = (x^2, y^2, z^2, \sqrt{36 - 9x^2 - y^2 - z^2})$; $(-1, 5, 1)$

6. $f(x_1, x_2, x_3, x_4, x_5) = (x_1x_2, x_2x_3, x_3x_4, x_4x_5, x_1x_5)$; $(1, 2, 3, 4, 5)$

7. $f(x_1, x_2, x_3, x_4) = \left(\dfrac{1}{x_1 + x_2 + 3x_3 + x_4 - 5}, x_1 + x_2 + x_3 + x_4 - 5\right)$; $(1, -1, 0, 2)$

8. $\mathbf{f}(x_1, x_2, x_3, x_4) = \left(\sqrt{x_1^2 + x_2^2 + x_3^2 + x_4^2}, \dfrac{2x_1 + x_4}{3x_2x_3}\right)$; $(1, 2, -1, 0)$

9. $\mathbf{f}(x_1x_2, x_3, x_4) = (\sin x_1x_4, \cos x_3x_4, e^{x_1^2 - x_2^2})$; $\left(1, 2, 3, \frac{\pi}{6}\right)$

10. $\mathbf{f}(x_1, x_2, \ldots, x_n) = (x_1x_2 \cdots x_n, x_1 + x_2 + \cdots + x_n)$

In Problems 11–17 find the derivative of the composite function $\mathbf{g} \circ \mathbf{f}$ at the given point $\mathbf{x}_0$ and verify the chain rule at that point.

11. $\mathbf{f}(x, y) = (x^2, y^2)$; $\mathbf{g}(x, y) = (x + y, x - y)$; $(3, 2)$

12. $\mathbf{f}(x, y) = (x + 2y, 3 - 4x)$; $\mathbf{g}(x, y) = \left(\frac{x}{y}, \frac{y}{x}\right)$; $(-1, 1)$

13. $\mathbf{f}(x, y) = (xy, x^2y, xy^2)$; $\mathbf{g}(x, y, z) = (x^3, y^2)$; $(2, -3)$

14. $\mathbf{f}(x, y, z) = \left(xyz, \frac{yx}{z}\right)$; $\mathbf{g}(x, y) = (x, 2x^2y, -3y, xy^2)$; $(1, -1, 2)$

15. $\mathbf{f}(x, y, z) = \left(x + y, x + z, xyz, \frac{x^2y}{z^3}\right)$;

$\mathbf{g}(x_1, x_2, x_3, x_4) = \left(\frac{x_1x_2}{x_3x_4}, x_1x_2x_3x_4, 3x_1 - 5x_4\right)$;

$(2, 3, -1)$

16. $\mathbf{f}(x, y, z) = \left(xy - z^2, 3y - 2z, \frac{4z}{x}\right)$;

$\mathbf{g}(x, y, z) = \left(2x - 3y, xy^2, \frac{x - y}{y + z}\right)$; $(1, 3, -2)$

17. $\mathbf{f}(x_1, x_2, \ldots, x_n) = (x_n, x_{n-1}, x_{n-2}, \ldots, x_2, x_1)$; $\mathbf{g}(x_1, x_2, \ldots, x_n) = (x_1^2, x_2^2, \ldots, x_n^2)$; $(1, 2, 3, \ldots, n)$

18. Let $\mathbf{f}: \mathbb{R}^n \to \mathbb{R}^m$ be linear. Show that $\mathbf{f}$ is differentiable at every $\mathbf{x}_0$ in $\mathbb{R}^n$.

19. Let $f: \mathbb{R}^n \to \mathbb{R}$ be differentiable. Show that $\nabla f(\mathbf{x}_0) = [J_f(\mathbf{x}_0)]^t$. (See Section 7.6 for a discussion of the transpose of a matrix.)

20. Let $\mathbf{f}(x, y) = (g(x, y), x + y)$ where

$$g(x, y) = \begin{cases} \dfrac{xy}{y^2 - x^2}, & x \neq \pm y, \\ 0, & x = \pm y. \end{cases}$$

(a) Show that $J_\mathbf{f}(0, 0)$ exists and compute it.
(b) Show that $\mathbf{f}$ is not differentiable at $(0, 0)$.
This problem shows that the existence of the Jacobian matrix is not enough to ensure differentiability.

21. Set $\mathbf{f}(x, y) = (x, g(y))$ where

$$g(y) = \begin{cases} y^2 \sin(1/y), & y \neq 0, \\ 0, & y = 0. \end{cases}$$

(a) Show that $\mathbf{f}$ is differentiable at $(0, 0)$.
(b) Show that not all the partial derivatives of the component functions of $\mathbf{f}$ are continuous. This problem shows that differentiability of $\mathbf{f}$ does not imply that the first-order partial derivatives of $\mathbf{f}$ are continuous.

22. Suppose that $\mathbf{f}: \mathbb{R}^3 \to \mathbb{R}^3$ such that $J_\mathbf{f}(\mathbf{x})$ is the zero matrix for every $\mathbf{x} \in \mathbb{R}^3$. Show that $\mathbf{f}(\mathbf{x}) = \mathbf{v}$, a constant vector.

23. Let $\mathbf{f}: \mathbb{R}^3 \to \mathbb{R}^3$ and $\mathbf{g}: \mathbb{R}^3 \to \mathbb{R}^3$ satisfy $J_\mathbf{f}(\mathbf{x}) = J_\mathbf{g}(\mathbf{x})$ for every $\mathbf{x}$ in $\mathbb{R}^3$. Show that $\mathbf{f}$ and $\mathbf{g}$ differ by a constant vector.

24. Let $\mathbf{f}: \mathbb{R}^n \to \mathbb{R}^n$ satisfy $\mathbf{f}(\mathbf{0}) = \mathbf{0}$ and $J_\mathbf{f}(\mathbf{x}) = A$, a constant matrix. Show that $\mathbf{f}$ is a linear mapping.

25. Prove Theorem 4.

10.5 INVERSE FUNCTIONS AND THE IMPLICIT FUNCTION THEOREM: II

In this section we extend the results of Section 10.2 to mappings from $\mathbb{R}^n$ to $\mathbb{R}^m$. Our first result will be a theorem about the existence of inverse functions. To state such a theorem, we need to have a condition similar to the condition $f'(x_0) \neq 0$ in Theorem 10.2.1.

Definition 1 JACOBIAN Let $\mathbf{f}: \mathbb{R}^n \to \mathbb{R}^n$ be differentiable at a point $\mathbf{x}_0$ in $\mathbb{R}^n$. Then the **Jacobian** of the mapping is the determinant of the Jacobian matrix $J_\mathbf{f}(\mathbf{x}_0)$. That is,

$$\text{Jacobian of } \mathbf{f} = \det J_{\mathbf{f}}(\mathbf{x}_0) = \begin{vmatrix} \frac{\partial f_1}{\partial x_1}(\mathbf{x}_0) & \frac{\partial f_1}{\partial x_2}(\mathbf{x}_0) & \cdots & \frac{\partial f_1}{\partial x_n}(\mathbf{x}_0) \\ \frac{\partial f_2}{\partial x_1}(\mathbf{x}_0) & \frac{\partial f_2}{\partial x_2}(\mathbf{x}_0) & \cdots & \frac{\partial f_2}{\partial x_n}(\mathbf{x}_0) \\ \vdots & \vdots & & \vdots \\ \frac{\partial f_n}{\partial x_1}(\mathbf{x}_0) & \frac{\partial f_n}{\partial x_2}(\mathbf{x}_0) & \cdots & \frac{\partial f_n}{\partial x_n}(\mathbf{x}_0) \end{vmatrix} \quad (1)$$

NOTATION: We will denote the Jacobian of $\mathbf{f}$ at $\mathbf{x}_0$ by $|J_{\mathbf{f}}(\mathbf{x}_0)|$.

REMARK 1. The Jacobian of a mapping $\mathbf{f}: \mathbb{R}^n \to \mathbb{R}^m$ is defined only when $n = m$, since only square matrices have determinants.

REMARK 2. This definition generalizes the definitions in Section 6.9 (see pages 389 and 394).

EXAMPLE 1 Let $\mathbf{f}: \mathbb{R}^2 \to \mathbb{R}^2$ be the polar coordinate transformation $f(r, \theta) = (r \cos \theta, r \sin \theta)$. Compute $|J_{\mathbf{f}}(r, \theta)|$.

Solution. From Example 10.4.2 we have

$$J_{\mathbf{f}}(r, \theta) = \begin{pmatrix} \cos \theta & -r \sin \theta \\ \sin \theta & r \cos \theta \end{pmatrix},$$

so

$$|J_{\mathbf{f}}(r, \theta)| = \cos \theta(r \cos \theta) + (r \sin \theta) \sin \theta = r(\cos^2 \theta + \sin^2 \theta) = r. \blacksquare$$

EXAMPLE 2 Let $\mathbf{f}: \mathbb{R}^3 \to \mathbb{R}^3$ be defined by

$$\mathbf{f}(x, y, z) = \left(xy - z, \frac{x^2y}{z}, \frac{xz}{y^2}\right).$$

Compute $|J_{\mathbf{f}}(2, 1, -1)|$.

Solution.

$$J_{\mathbf{f}}(x, y, z) = \begin{pmatrix} y & x & -1 \\ \frac{2xy}{z} & \frac{x^2}{z} & -\frac{x^2y}{z^2} \\ \frac{z}{y^2} & -\frac{2xz}{y^3} & \frac{x}{y^2} \end{pmatrix}$$

So that

$$J_{\mathbf{f}}(2, -1, 1) = \begin{pmatrix} -1 & 2 & -1 \\ -4 & 4 & 4 \\ 1 & 4 & 2 \end{pmatrix}$$

and

$$|J_{\mathbf{f}}(2, -1, 1)| = 52. \blacksquare$$

Definition 2 LOCAL C^1-INVERTIBILITY Let $\mathbf{f}: \mathbb{R}^n \to \mathbb{R}^n$ be in $C^1(\Omega)$ for some open set Ω in $\mathbb{R}^n$. Then $\mathbf{f}$ is said to be **locally C^1-invertible** on Ω if there exists a function $\mathbf{g}: \mathbb{R}^n \to \mathbb{R}^n$ which is in $C^1(\mathbf{f}(\Omega))$ such that

$$(\mathbf{g} \circ \mathbf{f})(\mathbf{x}) = \mathbf{x} \quad \text{for every } \mathbf{x} \text{ in } \Omega \tag{2}$$

and

$$(\mathbf{f} \circ \mathbf{g})(\mathbf{y}) = \mathbf{y} \quad \text{for every } \mathbf{y} \text{ in } \mathbf{f}(\Omega). \tag{3}$$

In this case we say that $\mathbf{g}$ is an **inverse function** for $\mathbf{f}$ and write $\mathbf{g} = \mathbf{f}^{-1}$.

REMARK. The word "locally" in the definition refers to the region Ω. There are functions that are locally invertible in a neighborhood of every point in $\mathbb{R}^n$ but that are not invertible on all of $\mathbb{R}^n$ (see Example 3 and Problem 27).

EXAMPLE 3 Let $\mathbf{f}: \mathbb{R}^2 \to \mathbb{R}^2$ be the polar coordinate mapping $\mathbf{f}(x, y) = (r \cos \theta, r \sin \theta)$. Then $\mathbf{f}$ has continuous first-order partial derivatives of all orders. Let

$$\Omega = \{(r, \theta): 0 < r < 1, 0 < \theta < \pi\}. \tag{4}$$

Then, as we saw in Example 10.3.2 $\mathbf{f}$ maps the open rectangle Ω into the open semicircle $\mathbf{f}(\Omega) = \{(x, y): 0 < x^2 + y^2 < 1, y > 0\}$ (see Figure 10.3.1). We now define

$$\mathbf{g}(x, y) = \left(\sqrt{x^2 + y^2}, \cos^{-1} \frac{x}{\sqrt{x^2 + y^2}}\right).$$

Then, if $(r, \theta) \in \Omega$,

$$(\mathbf{g} \circ \mathbf{f})(r, \theta) = \mathbf{g}(r \cos \theta, r \sin \theta)$$
$$= \left(\sqrt{r^2 \cos^2 \theta + r^2 \sin^2 \theta}, \cos^{-1} \frac{r \cos \theta}{\sqrt{r^2 \cos^2 \theta + r^2 \sin^2 \theta}}\right) = (r, \theta)^\dagger$$

and if (x, y) is in $\mathbf{f}(\Omega)$, we have

†Note that $\sqrt{r^2} = r$ because $r > 0$ and $\cos^{-1}(\cos \theta) = \theta$ because $0 < \theta < \pi$.

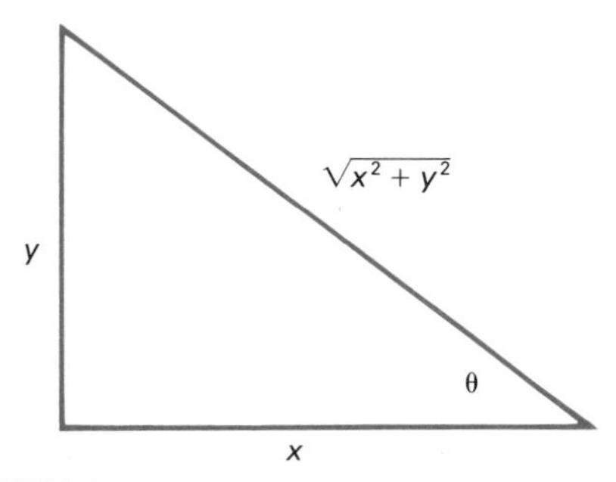

FIGURE 1

$$(\mathbf{f}\circ\mathbf{g})(x, y) = \mathbf{f}\left(\sqrt{x^2+y^2}, \cos^{-1}\frac{x}{\sqrt{x^2+y^2}}\right)$$

$$= \left(\sqrt{x^2+y^2}\cos\left(\cos^{-1}\frac{x}{\sqrt{x^2+y^2}}\right), \sqrt{x^2+y^2}\sin\left(\cos^{-1}\frac{x}{\sqrt{x^2+y^2}}\right)\right).$$

Now, from the triangle in Figure 1,

$$\sin\left(\cos^{-1}\frac{x}{\sqrt{x^2+y^2}}\right) = \frac{y}{\sqrt{x^2+y^2}}.$$

So that

$$(\mathbf{f}\circ\mathbf{g})(x, y) = (x, y)$$

and $\mathbf{g}$ is an inverse function for $\mathbf{f}$. Note that $\mathbf{g}$ is an inverse function for $\mathbf{f}$ on Ω, not on all of $\mathbb{R}^2$. For example,

$$(\mathbf{g}\circ\mathbf{f})\left(1, \frac{5\pi}{2}\right) = \mathbf{g}\left(\mathbf{f}\left(1, \frac{5\pi}{2}\right)\right) = \mathbf{g}(0, 1) = \left(1, \frac{\pi}{2}\right) \neq \left(1, \frac{5\pi}{2}\right),$$

so that $\mathbf{g}$ is *not* an inverse function for $\mathbf{f}$ everywhere in the plane. This illustrates the need for the word "local" in Definition 1. ■

EXAMPLE 4 Let $\mathbf{f}: \mathbb{R}^n \to \mathbb{R}^n$ be a linear transformation. Then there is an $n \times n$ matrix A such that

$$\mathbf{f}(\mathbf{x}) = A\mathbf{x} \tag{5}$$

where $\mathbf{x}$ is written as a column vector. Now

$$f(\mathbf{x}) - f(\mathbf{x}_0) = A(\mathbf{x} - \mathbf{x}_0) + 0,$$

so that, by Definition 10.4.1, $\mathbf{f}$ is differentiable for every $\mathbf{x}_0$ in $\mathbb{R}^n$ and $\mathbf{J}_\mathbf{f}(\mathbf{x}_0)$ is the constant matrix A. Suppose that A is invertible. Then we define a function $\mathbf{g}: \mathbb{R}^n \to \mathbb{R}^n$ by

$$\mathbf{g}(\mathbf{y}) = A^{-1}\mathbf{y}.$$

Clearly $\mathbf{g}$ is a linear transformation. Furthermore,

$$(\mathbf{g} \circ \mathbf{f})(\mathbf{x}) = \mathbf{g}(\mathbf{f}(\mathbf{x})) = \mathbf{g}(A\mathbf{x}) = A^{-1}(A\mathbf{x}) = \mathbf{x}$$

and

$$(\mathbf{f} \circ \mathbf{g})(\mathbf{y}) = \mathbf{f}(\mathbf{g}(\mathbf{y})) = \mathbf{f}(A^{-1}\mathbf{y}) = A(A^{-1}\mathbf{y}) = \mathbf{y}.$$

Thus $\mathbf{g} = \mathbf{f}^{-1}$ on all of $\mathbb{R}^n$. ■

The last result leads to the following theorem.

Theorem 1 INVERTIBILITY OF A LINEAR TRANSFORMATION Let $\mathbf{f}: \mathbb{R}^n \to \mathbb{R}^n$ be a linear transformation with matrix representation A. Then $\mathbf{f}$ has an inverse on all of $\mathbb{R}^n$ if and only if A is invertible (if and only if $\det A \neq 0$). If A is invertible, then

$$\mathbf{f}^{-1}(\mathbf{x}) = A^{-1}\mathbf{x}. \tag{6}$$

Proof. We have already proven the theorem in the case in which A is invertible. If A is not invertible, then $\det A = 0$ and the $n \times n$ system of equations $A\mathbf{x} = \mathbf{0}$ has a nontrivial solution $\mathbf{x}_0$. Pick two vectors $\mathbf{y}$ and $\mathbf{z}$ in $\mathbb{R}^n$ such that $\mathbf{y} \neq \mathbf{z}$ and $\mathbf{y} - \mathbf{z} = \mathbf{x}_0$. Then $\mathbf{0} = A\mathbf{x}_0 = A(\mathbf{y} - \mathbf{z}) = A\mathbf{y} - A\mathbf{z}$, so that $A\mathbf{y} = A\mathbf{z}$ and $\mathbf{f}\mathbf{y} = \mathbf{f}\mathbf{z} = \mathbf{w}$ for some vector $\mathbf{w}$. Thus $\mathbf{f}$ is not one-to-one and $\mathbf{f}^{-1}(\mathbf{w})$ is not defined. Thus $\mathbf{f}$ is not invertible on $\mathbb{R}^n$. ■

As we saw in Theorem 10.2.1, a function $f: \mathbb{R} \to \mathbb{R}$ is invertible in a neighborhood of x_0 if $f'(x_0) \neq 0$. We now generalize that result to nonlinear mappings from $\mathbb{R}^n$ to $\mathbb{R}^n$ (the linear case is dealt with in Theorem 1).

Theorem 2 INVERSE FUNCTION THEOREM Let $\mathbf{f}: \mathbb{R}^n \to \mathbb{R}^n$ be in $C^1(\Omega)$ on an open set Ω and let $\mathbf{x}_0$ be in Ω. If the Jacobian $|J_{\mathbf{f}}(\mathbf{x}_0)| \neq 0$, then there is an open set $\Omega_1 \subset \Omega$ such that $\mathbf{x}_0 \in \Omega_1$ and $\mathbf{f}$ is locally C^1-invertible on Ω_1; moreover, if $\mathbf{g} = \mathbf{f}^{-1}$ on $\mathbf{f}(\Omega_1)$, then

$$J_{\mathbf{f}^{-1}}(\mathbf{x}_0) = J_{\mathbf{g}}(\mathbf{f}(\mathbf{x}_0)) = [J_{\mathbf{f}}(\mathbf{x}_0)]^{-1}. \tag{7}$$

Partial proof. A proof of the first part of this theorem is best left to an advanced calculus text (see the reference cited in Section 10.2). However, we can justify formula (7); for if $\mathbf{g} = \mathbf{f}^{-1}$, then

$$(\mathbf{g} \circ \mathbf{f})(\mathbf{x}) = \mathbf{x} = I\mathbf{x}$$

where I denotes the identity transformation (see Example 9.6.5 on p. 525). Then by the chain rule (Theorem 10.4.5),

$$I = J_I(\mathbf{x}) = J_{\mathbf{g}\circ\mathbf{f}}(\mathbf{x}) = J_{\mathbf{g}}(\mathbf{f}(\mathbf{x}_0))J_{\mathbf{f}}(\mathbf{x}_0). \tag{8}$$

But since $|J_{\mathbf{f}}(\mathbf{x}_0)| \neq 0$, $J_{\mathbf{f}}(\mathbf{x}_0)$ is invertible and (8) shows that $[J_{\mathbf{f}}(\mathbf{x}_0)]^{-1} = J_{\mathbf{g}}(\mathbf{f}(\mathbf{x}_0))$. ■

EXAMPLE 5 Let $\mathbf{f}: \mathbb{R}^2 \to \mathbb{R}^2$ be the polar coordinate transformation. Then, as we saw in Example 1, $|J_{\mathbf{f}}(r, \theta)| = r \neq 0$ if $r \neq 0$. Thus, if $\mathbf{g} = \mathbf{f}^{-1}$, we can compute

$$J_{\mathbf{g}}(\mathbf{f}(r, \theta)) = [J_{\mathbf{f}}(r, \theta)]^{-1} = \begin{pmatrix} \cos\theta & -r\sin\theta \\ \sin\theta & r\cos\theta \end{pmatrix}^{-1} = \begin{pmatrix} \cos\theta & \sin\theta \\ -\dfrac{1}{r}\sin\theta & \dfrac{\cos\theta}{r} \end{pmatrix}.$$

We can verify this directly. From Example 3 we have

$$\mathbf{g}(\mathbf{f}(r, \theta)) = \mathbf{g}(r\cos\theta, r\sin\theta) = \mathbf{g}(x, y) = \left(\sqrt{x^2 + y^2}, \cos^{-1}\frac{x}{\sqrt{x^2 + y^2}}\right).$$

Since

$$\frac{d}{du}\cos^{-1} u = \frac{-1}{\sqrt{1 - u^2}},$$

$$J_{\mathbf{g}}(x, y) = \begin{pmatrix} \dfrac{x}{\sqrt{x^2 + y^2}} & \dfrac{y}{\sqrt{x^2 + y^2}} \\ \dfrac{-1}{\sqrt{1 - \dfrac{x^2}{x^2 + y^2}}}\left(\dfrac{\sqrt{x^2 + y^2} - \dfrac{x^2}{\sqrt{x^2 + y^2}}}{x^2 + y^2}\right) & \dfrac{-1}{\sqrt{1 - \dfrac{x^2}{x^2 + y^2}}}\left(\dfrac{-xy}{(x^2 + y^2)^{3/2}}\right) \end{pmatrix}$$

(After some algebra)

$$= \begin{pmatrix} \dfrac{x}{\sqrt{x^2 + y^2}} & \dfrac{y}{\sqrt{x^2 + y^2}} \\ -\dfrac{y}{x^2 + y^2} & \dfrac{x}{x^2 + y^2} \end{pmatrix} = \begin{pmatrix} \dfrac{r\cos\theta}{r} & \dfrac{r\sin\theta}{r} \\ -\dfrac{r\sin\theta}{r^2} & \dfrac{r\cos\theta}{r^2} \end{pmatrix} = \begin{pmatrix} \cos\theta & \sin\theta \\ -\dfrac{1}{r}\sin\theta & \dfrac{1}{r}\cos\theta \end{pmatrix}. \blacksquare$$

EXAMPLE 6 Let $\mathbf{f}: \mathbb{R}^3 \to \mathbb{R}^3$ be defined by

$$\mathbf{f}(x, y, z) = \left(xy - z, \frac{x^2 y}{z}, \frac{xz}{y^2}\right).$$

Show that $\mathbf{f}$ is locally C^1-invertible in a neighborhood of $(2, 1, -1)$, and compute $J_{\mathbf{f}^{-1}}(\mathbf{f}(2, 1, -1))$.

Solution. In Example 2, we found that

$$J_{\mathbf{f}}(2, -1, 1) = \begin{pmatrix} -1 & 2 & -1 \\ -4 & 4 & 4 \\ 1 & 4 & 2 \end{pmatrix}$$

and that $|J_{\mathbf{f}}(2, -1, 1)| = 52$. Thus $\mathbf{f}^{-1}$ exists and is C^1-invertible in a neighborhood of $\mathbf{f}(2, -1, 1) = (-3, -4, 2)$ and

$$J_{\mathbf{f}^{-1}}(-3, -4, 2) = \begin{pmatrix} -1 & 2 & -1 \\ -4 & 4 & 4 \\ 1 & 4 & 2 \end{pmatrix}^{-1} = \frac{1}{52}\begin{pmatrix} -8 & -8 & 12 \\ 12 & -1 & 8 \\ -20 & 6 & 4 \end{pmatrix}. \blacksquare$$

We now state a more general form of the implicit function theorem given in Section 10.2. We first need some notation. Let $\mathbf{x}_0 = (x_1^{(0)}, x_2^{(0)}, \ldots, x_n^{(0)})$ be in $\mathbb{R}^n$ and let $\mathbf{y}_0 = (y_1^{(0)}, y_2^{(0)}, \ldots, y_m^{(0)})$ be in $\mathbb{R}^m$. Then if $\mathbf{F}: \mathbb{R}^{n+m} \to \mathbb{R}^m$, the notation $\mathbf{F}(\mathbf{x}_0, \mathbf{y}_0)$ stands for

$$\mathbf{F}(x_1^{(0)}, x_2^{(0)}, \ldots, x_n^{(0)}, y_1^{(0)}, y_2^{(0)}, \ldots, y_m^{(0)}),$$

the $m \times m$ matrix $J_{\mathbf{F}}(\mathbf{y}_0)$ is given by

$$J_{\mathbf{F}}(\mathbf{y}_0) = \begin{pmatrix} \frac{\partial F_1}{\partial y_1}(\mathbf{x}_0, \mathbf{y}_0) & \frac{\partial F_1}{\partial y_2}(\mathbf{x}_0, \mathbf{y}_0) & \cdots & \frac{\partial F_1}{\partial y_m}(\mathbf{x}_0, \mathbf{y}_0) \\ \frac{\partial F_2}{\partial y_1}(\mathbf{x}_0, \mathbf{y}_0) & \frac{\partial F_2}{\partial y_2}(\mathbf{x}_0, \mathbf{y}_0) & \cdots & \frac{\partial F_2}{\partial y_m}(\mathbf{x}_0, \mathbf{y}_0) \\ \vdots & \vdots & & \vdots \\ \frac{\partial F_m}{\partial y_1}(\mathbf{x}_0, \mathbf{y}_0) & \frac{\partial F_m}{\partial y_2}(\mathbf{x}_0, \mathbf{y}_0) & \cdots & \frac{\partial F_m}{\partial y_m}(\mathbf{x}_0, \mathbf{y}_0) \end{pmatrix} \tag{9}$$

and the $m \times n$ matrix $J_{\mathbf{F}}(\mathbf{x}_0)$ is given by

$$J_{\mathbf{F}}(\mathbf{x}_0) = \begin{pmatrix} \frac{\partial F_1}{\partial x_1}(\mathbf{x}_0, \mathbf{y}_0) & \frac{\partial F_1}{\partial x_2}(\mathbf{x}_0, \mathbf{y}_0) & \cdots & \frac{\partial F_1}{\partial x_n}(\mathbf{x}_0, \mathbf{y}_0) \\ \frac{\partial F_2}{\partial x_1}(\mathbf{x}_0, \mathbf{y}_0) & \frac{\partial F_2}{\partial x_2}(\mathbf{x}_0, \mathbf{y}_0) & \cdots & \frac{\partial F_2}{\partial x_n}(\mathbf{x}_0, \mathbf{y}_0) \\ \vdots & \vdots & & \vdots \\ \frac{\partial F_m}{\partial x_1}(\mathbf{x}_0, \mathbf{y}_0) & \frac{\partial F_m}{\partial x_2}(\mathbf{x}_0, \mathbf{y}_0) & \cdots & \frac{\partial F_m}{\partial x_n}(\mathbf{x}_0, \mathbf{y}_0) \end{pmatrix} \tag{10}$$

Note that $J_{\mathbf{F}}(\mathbf{x}_0)$ consists of the first n columns of $J_{\mathbf{F}}(\mathbf{x}_0, \mathbf{y}_0)$ and $J_{\mathbf{F}}(\mathbf{y}_0)$ consists of the last m columns of $J_{\mathbf{F}}(\mathbf{x}_0, \mathbf{y}_0)$.

Theorem 3 IMPLICIT FUNCTION THEOREM† Let the function $\mathbf{F}: \mathbb{R}^{n+m} \to \mathbb{R}^m$ be C^1 in a neighborhood Ω of the vector $(\mathbf{x}_0, \mathbf{y}_0)$ in $\mathbb{R}^{n+m}$ with $\mathbf{F}(\mathbf{x}_0, \mathbf{y}_0) = \mathbf{0}$. If $|J_{\mathbf{F}}(\mathbf{y}_0)| \neq 0$, then there exists a neighborhood Ω_1 of $\mathbf{x}_0$ in $\mathbb{R}^n$, a neighborhood Ω_2 of $(\mathbf{x}_0, \mathbf{y}_0)$ in $\mathbb{R}^{n+m}$, and a function $\mathbf{f}: \mathbb{R}^n \to \mathbb{R}^m$ in $C^1(\Omega_1)$ such that $\mathbf{y} = \mathbf{f}(\mathbf{x})$ satisfies $\mathbf{F}(\mathbf{x}, \mathbf{f}(\mathbf{x})) = \mathbf{0}$ in Ω_2. Furthermore,

†A proof of this theorem can be found in C. H. Edwards, Jr., *Advanced Calculus in Several Variables* (Academic Press, New York, 1973).

$$J_{\mathbf{f}}(\mathbf{x}) = -[J_{\mathbf{F}}(\mathbf{y})]^{-1}J_{\mathbf{F}}(\mathbf{x}). \tag{11}$$

REMARK 1. If we set $m = 1$ in this theorem, then we obtain Theorem 10.2.2. Note that in this case $f: \mathbb{R}^n \to \mathbb{R}$, so $J_{\mathbf{f}}(\mathbf{x})$ is an n-component column vector, $\mathbf{y}_0 = x_{n+1}$, and $J_{\mathbf{F}}(\mathbf{y}_0)$ is a 1×1 matrix, so (11) becomes

$$\begin{pmatrix} \dfrac{\partial f}{\partial x_1}(\mathbf{x}) \\ \dfrac{\partial f}{\partial x_2}(\mathbf{x}) \\ \vdots \\ \dfrac{\partial f}{\partial x_n}(\mathbf{x}) \end{pmatrix} = \left[\frac{\partial F}{\partial x_{n+1}}(\mathbf{x}, x_{n+1})\right]^{-1} \begin{pmatrix} \dfrac{\partial F}{\partial x_1}(\mathbf{x}, x_{n+1}) \\ \dfrac{\partial F}{\partial x_2}(\mathbf{x}, x_{n+1}) \\ \vdots \\ \dfrac{\partial F}{\partial x_n}(\mathbf{x}, x_{n+1}) \end{pmatrix}$$

which is equation (10.2.5).

REMARK 2. We observe that even though it will generally be impossible to find the implicitly given function $\mathbf{f}$, we can use equation (11) to compute its derivative. This will often be all that we need.

REMARK 3. Formula (11) should not be surprising. To see why, let us treat $\mathbf{x}$ and $\mathbf{y} = f(\mathbf{x})$ as real variables. Then, if we differentiate $F(x, y) = F(x, f(x)) = 0$ and use the chain rule, we obtain

$$F'(x) = \frac{\partial F}{\partial x} + \frac{\partial F}{\partial y} f'(x) = 0,$$

so that $f'(x) = -(\partial F/\partial y)^{-1}(\partial F/\partial x)$. But going back now to the vector situation, we have

$$f'(x) = J_{\mathbf{f}}(\mathbf{x}), \qquad \frac{\partial F}{\partial y} = J_{\mathbf{F}}(\mathbf{y}), \qquad \text{and} \qquad \frac{\partial F}{\partial x} = J_{\mathbf{F}}(\mathbf{x}),$$

and substitution gives us formula (11).

EXAMPLE 7 Let

$$\mathbf{F}(x_1, x_2, x_3, x_4) = (2x_1 - 3x_2 + 4x_3 - x_4, 2x_1 - x_2 + 5x_3 - 2x_4) = (0, 0).$$

Here $\mathbf{F}: \mathbb{R}^4 \to \mathbb{R}^2$ is a linear transformation and $n = m = 2$. Then with $\mathbf{x} = (x_1, x_2)$ and $\mathbf{y} = (x_3, x_4)$,

$$J_{\mathbf{F}}(\mathbf{y}) = \begin{pmatrix} \dfrac{\partial F_1}{\partial x_3} & \dfrac{\partial F_1}{\partial x_4} \\ \dfrac{\partial F_2}{\partial x_3} & \dfrac{\partial F_2}{\partial x_4} \end{pmatrix} = \begin{pmatrix} 4 & -1 \\ 5 & -2 \end{pmatrix}$$

and $|J_{\mathbf{F}}(\mathbf{y})| = -3 \neq 0$. Since $|J_{\mathbf{F}}(\mathbf{y})|$ is nonzero for every $\mathbf{y}$, we can find a function $\mathbf{f}$: $\mathbb{R}^2 \to \mathbb{R}^2$ such that $(x_3, x_4) = \mathbf{f}(x_1, x_2)$ and

$$J_{\mathbf{F}}(\mathbf{x}) = -[J_{\mathbf{F}}(\mathbf{y})]^{-1} J_{\mathbf{F}}(\mathbf{x}).$$

But

$$J_{\mathbf{F}}(\mathbf{x}) = \begin{pmatrix} \dfrac{\partial F_1}{\partial x_1} & \dfrac{\partial F_1}{\partial x_2} \\ \dfrac{\partial F_2}{\partial x_1} & \dfrac{\partial F_2}{\partial x_2} \end{pmatrix} = \begin{pmatrix} 2 & -3 \\ 2 & -1 \end{pmatrix}.$$

Thus

$$J_{\mathbf{f}}(\mathbf{x}) = \tfrac{1}{3}\begin{pmatrix} -2 & 1 \\ -5 & 4 \end{pmatrix}\begin{pmatrix} 2 & -3 \\ 2 & -1 \end{pmatrix} = \tfrac{1}{3}\begin{pmatrix} -2 & 5 \\ -2 & 11 \end{pmatrix} = \begin{pmatrix} -\frac{2}{3} & \frac{5}{3} \\ -\frac{2}{3} & \frac{11}{3} \end{pmatrix}.$$

Since $\mathbf{F}$ is linear, we can verify this result. The equation $\mathbf{F}(x_1, x_2, x_3, x_4) = (0, 0)$ can be written as a system in augmented matrix form which we can solve by the methods of Section 7.3. Now, however, since we wish to write x_3 and x_4 in terms of x_1 and x_2, we make the coefficients of the x_3 and x_4 terms 1:

$$\left(\begin{array}{cccc|c} 2 & -3 & 4 & -1 & 0 \\ 2 & -1 & 5 & -2 & 0 \end{array}\right) \xrightarrow{M_1(\frac{1}{4})} \left(\begin{array}{cccc|c} \frac{1}{2} & -\frac{3}{4} & 1 & -\frac{1}{4} & 0 \\ 2 & -1 & 5 & -2 & 0 \end{array}\right)$$

$$\xrightarrow{A_{1,2}(-5)} \left(\begin{array}{cccc|c} \frac{1}{2} & -\frac{3}{4} & 1 & -\frac{1}{4} & 0 \\ -\frac{1}{2} & \frac{11}{4} & 0 & -\frac{3}{4} & 0 \end{array}\right) \xrightarrow{M_2(-\frac{4}{3})} \left(\begin{array}{cccc|c} \frac{1}{2} & -\frac{3}{4} & 1 & -\frac{1}{4} & 0 \\ \frac{2}{3} & -\frac{11}{3} & 0 & 1 & 0 \end{array}\right)$$

$$\xrightarrow{A_{2,1}(\frac{1}{4})} \left(\begin{array}{cccc|c} \frac{2}{3} & -\frac{5}{3} & 1 & 0 & 0 \\ \frac{2}{3} & -\frac{11}{3} & 0 & 1 & 0 \end{array}\right).$$

The system represented by the last augmented matrix is

$$\begin{aligned} \tfrac{2}{3}x_1 - \tfrac{5}{3}x_2 + x_3 \qquad &= 0 \\ \tfrac{2}{3}x_1 - \tfrac{11}{3}x_2 \qquad + x_4 &= 0 \end{aligned}$$

which we can write as

$$(x_3, x_4) = \mathbf{f}(x_1, x_2) = (-\tfrac{2}{3}x_1 + \tfrac{5}{3}x_2, -\tfrac{2}{3}x_1 + \tfrac{11}{3}x_2)$$

and

$$J_{\mathbf{f}}(x_1, x_2) = \begin{pmatrix} -\frac{2}{3} & \frac{5}{3} \\ -\frac{2}{3} & \frac{11}{3} \end{pmatrix}. \quad ■$$

REMARK. Of course, in all but the simplest problems (such as those involving linear transformations), it will be impossible to verify (except by rechecking calculations) that our computed value of $J_{\mathbf{f}}(\mathbf{x})$ is correct. However, it is nice to see that in a verifiable situation the implicit function theorem gives us the correct answer.

EXAMPLE 8 Consider the following system of equations

$$\begin{aligned} x^2 - 2xz + y^2z^3 - 7 &= 0 \\ 2xy^4 - 3y^2 + xz^2 + 5z + 2 &= 0 \end{aligned} \qquad \textbf{(12)}$$

We seek to write y and z in terms of x near the point $(2, 1, -1)$. Although a function $(y, z) = f(x)$ cannot be determined explicitly, we can think of the system (12) as a function $\mathbf{f}\colon \mathbb{R}^{1+2} \to \mathbb{R}^2$ where $n = 1$ and $m = 2$ and we can use the implicit function theorem. Here $x_0 = 2$ and $\mathbf{y}_0 = (1, -1)$. Then

$$J_{\mathbf{F}}(\mathbf{y}_0) = \begin{pmatrix} \dfrac{\partial F_1}{\partial y} & \dfrac{\partial F_1}{\partial z} \\ \dfrac{\partial F_2}{\partial y} & \dfrac{\partial F_2}{\partial z} \end{pmatrix} = \begin{pmatrix} 2yz^3 & -2x + 3y^2z^2 \\ 8xy^3 - 6y & 2xz + 5 \end{pmatrix} = \begin{pmatrix} -2 & -1 \\ 10 & 1 \end{pmatrix}$$

at $(2, 1, -1)$ and $|J_{\mathbf{F}}(\mathbf{y}_0)| = 8 \neq 0$. Thus we can write

$$(y, z) = f(x)$$

for x in a neighborhood of (an open interval containing) 2. Moreover, $J_{\mathbf{f}}(\mathbf{x}_0) = (J_{\mathbf{F}}(\mathbf{y}_0))^{-1}(J_{\mathbf{F}}(\mathbf{x}_0))$. But

$$J_{\mathbf{F}}(\mathbf{x}_0) = \begin{pmatrix} \dfrac{\partial F_1}{\partial x} \\ \dfrac{\partial F_2}{\partial x} \end{pmatrix} = \begin{pmatrix} 2x - 2z \\ 2y^4 + z^2 \end{pmatrix} = \begin{pmatrix} 6 \\ 3 \end{pmatrix}$$

at $(2, 1, -1)$. Then

$$J_{\mathbf{f}}(2) = -\tfrac{1}{8}\begin{pmatrix} 1 & 1 \\ -10 & -2 \end{pmatrix}\begin{pmatrix} 6 \\ 3 \end{pmatrix} = -\tfrac{1}{8}\begin{pmatrix} 9 \\ -66 \end{pmatrix} = \begin{pmatrix} -\frac{9}{8} \\ \frac{33}{4} \end{pmatrix}.$$

Thus, if $\mathbf{f}(x) = (f_1(x), f_2(x)) = (y, z)$, we have

$$\left.\frac{dy}{dx}\right|_{x=2} = f_1'(2) = -\tfrac{9}{8} \qquad \text{and} \qquad \left.\frac{dz}{dx}\right|_{x=2} = f_2'(2) = \tfrac{33}{4}.$$

This is a great deal of information considering that we cannot compute $\mathbf{f}$ directly. ■

EXAMPLE 9 Let

$$\mathbf{F}(x_1, x_2, x_3, x_4, x_5) = \left(x_1^4 - 2x_3x_4 + x_2x_5^3 + 9,\ x_1x_2x_3x_4x_5 + 4\frac{x_3x_5}{x_2},\ \frac{2x_2 + 5x_4}{x_3} + x_1x_3\right) = (0, 0, 0).$$

Show that (x_3, x_4, x_5) can be written as a function $\mathbf{f}$ of (x_1, x_2) in a neighborhood of $(1, -2, 3, -1, 2)$ and compute $J_{\mathbf{f}}(1, -2)$.

Solution. Here $\mathbf{F}\colon \mathbb{R}^{2+3} \to \mathbb{R}^3$, so that $n = 2$, $m = 3$, $\mathbf{x}_0 = (1, -2)$, $\mathbf{y}_0 = (3, -1, 2)$, and

$$J_{\mathbf{F}}(\mathbf{y}) = \begin{pmatrix} \frac{\partial F_1}{\partial x_3} & \frac{\partial F_1}{\partial x_4} & \frac{\partial F_1}{\partial x_5} \\ \frac{\partial F_2}{\partial x_3} & \frac{\partial F_2}{\partial x_4} & \frac{\partial F_2}{\partial x_5} \\ \frac{\partial F_3}{\partial x_3} & \frac{\partial F_3}{\partial x_4} & \frac{\partial F_3}{\partial x_5} \end{pmatrix} = \begin{pmatrix} -2x_4 & -2x_3 & 3x_2x_5^{\,2} \\ x_1x_2x_4x_5 + 4\frac{x_5}{x_2} & x_1x_2x_3x_5 & x_1x_2x_3x_4 + \frac{4x_3}{x_2} \\ \frac{-2x_2 - 5x_4}{x_3^{\,2}} + x_1 & \frac{5}{x_3} & 0 \end{pmatrix}$$

$$= \begin{pmatrix} 2 & -6 & -24 \\ 0 & -12 & 0 \\ 2 & \frac{5}{3} & 0 \end{pmatrix}$$

at $(1, -2, 3, -1, 2)$ and $|J_{\mathbf{F}}(\mathbf{y})| = -576 \neq 0$. Thus the required $\mathbf{f}$ exists, and we can write $\mathbf{y} = \mathbf{f}(\mathbf{x})$ in a neighborhood of $(1, -2)$. Finally

$$J_{\mathbf{f}}(1, -2) = -\begin{pmatrix} 2 & -6 & -24 \\ 0 & -12 & 0 \\ 2 & \frac{5}{3} & 0 \end{pmatrix}^{-1} J_{\mathbf{F}}(1, -2).$$

But

$$J_{\mathbf{F}}(\mathbf{x}) = \begin{pmatrix} \frac{\partial F_1}{\partial x_1} & \frac{\partial F_1}{\partial x_2} \\ \frac{\partial F_2}{\partial x_1} & \frac{\partial F_2}{\partial x_2} \\ \frac{\partial F_3}{\partial x_1} & \frac{\partial F_3}{\partial x_2} \end{pmatrix} = \begin{pmatrix} 4x_1^{\,3} & x_5^{\,3} \\ x_2x_3x_4x_5 & x_1x_3x_4x_5 - \frac{4x_3x_5}{x_2^{\,2}} \\ x_3 & \frac{2}{x_3} \end{pmatrix} = \begin{pmatrix} 4 & 8 \\ 12 & -12 \\ 3 & \frac{2}{3} \end{pmatrix}$$

at $(\mathbf{x}_0, \mathbf{y}_0)$, so that

$$J_{\mathbf{f}}(1, -2) = \tfrac{1}{576}\begin{pmatrix} 0 & -40 & -288 \\ 0 & 48 & 0 \\ 24 & -\frac{46}{3} & -24 \end{pmatrix}\begin{pmatrix} 4 & 8 \\ 12 & -12 \\ 3 & \frac{2}{3} \end{pmatrix}$$

$$= \tfrac{1}{576}\begin{pmatrix} -1344 & 288 \\ 576 & -576 \\ -160 & 360 \end{pmatrix} = \begin{pmatrix} -\frac{7}{3} & \frac{1}{2} \\ 1 & -1 \\ -\frac{5}{18} & \frac{5}{8} \end{pmatrix}. \quad \blacksquare$$

PROBLEMS 10.5

In Problems 1–13 compute the Jacobian of the given mapping at the given point.

1. $\mathbf{f}(x, y) = (x + y, x - y)$; $(1, 2)$
2. $\mathbf{f}(x, y) = (x^2 - y^2, x^2 + y^2)$; $(3, 2)$
3. $\mathbf{f}(x, y) = (\sin x, \cos y)$; $\left(\frac{\pi}{6}, \frac{\pi}{3}\right)$
4. $\mathbf{f}(x, y) = (e^{x+y}, e^{x-y})$; $(1, 1)$
5. $\mathbf{f}(x, y) = (\ln(x + y), \ln xy)$; $(2, 1)$
6. $\mathbf{f}(x, y) = (x \ln y, y \ln x)$; $(4, 2)$
7. $\mathbf{f}(x, y, z) = (x^2, y^2, z^2)$; $(1, 4, 2)$
8. $\mathbf{f}(x, y, z) = \left(xyz, \frac{xy}{z}, x^2 + y^2\right)$; $(2, -1, 1)$
9. $\mathbf{f}(x, y, z) = (x \cos y, y \cos z, z \sin x)$; $\left(\frac{\pi}{6}, \frac{\pi}{3}, \frac{\pi}{2}\right)$
10. $\mathbf{f}(x, y, z) = (x^2 + y^2 + z^2, x + y + z, xyz)$; $(2, -1, 4)$
11. $\mathbf{f}(x, y, z) = (x \ln(y + z), y \ln(x + z), z \ln(x + y))$; $(1, 1, 0)$
12. $\mathbf{f}(x_1, x_2, \ldots, x_n) = (x_1^2, x_2^2, \ldots, x_n^2)$; $(1, 1, \ldots, 1)$
13. $\mathbf{f}(x_1, x_2, \ldots, x_n) = (x_1, x_2^2, x_3^3, \ldots, x_n^n)$; $(1, 1, \ldots, 1)$
14. Let $\mathbf{f}: \mathbb{R}^n \to \mathbb{R}^n$ be differentiable at $\mathbf{x}_0$ and let $\mathbf{g}: \mathbb{R}^n \to \mathbb{R}^n$ be differentiable at $\mathbf{f}(\mathbf{x}_0)$. Show that

$$|J_{\mathbf{g}\circ\mathbf{f}}(\mathbf{x}_0)| = |J_{\mathbf{g}}(\mathbf{f}(\mathbf{x}_0))|\,|J_{\mathbf{f}}(\mathbf{x}_0)|.$$

In Problems 15–26 a mapping **f** and a point $\mathbf{x}_0$ are given. Show that the mapping is locally C^1-invertible at the given point and compute $J_{\mathbf{f}^{-1}}(\mathbf{x}_0)$.

15. $\mathbf{f}(x, y) = (2x - 3y, 7x + 2y)$; $(1, 3)$
16. $\mathbf{f}(x, y) = (x - 2y, -3x + y)$; (x_0, y_0)
17. $\mathbf{f}(x, y, z) = (x + y + z, x - y - z, -x + y - z)$; $(2, -1, 5)$
18. $\mathbf{f}(x, y, z) = (2x + 4y + 3z, y - z, 3x + 5y + 7z)$; (x_0, y_0, z_0)
19. $\mathbf{f}(x, y) = (x + y^2 + 1, x^2 + y + 2)$; $(1, 2)$
20. $\mathbf{f}(x, y) = (x^2 + y^2, x^2 - y^2)$; $(3, 5)$
21. $\mathbf{f}(x, y) = (e^{x+y}, e^{x-y})$; $(1, 1)$
22. $\mathbf{f}(x, y, z) = (x^2, y^2, z^2)$; $(2, -1, 3)$
23. $\mathbf{f}(x, y, z) = \left(x^2y - yz, \frac{2x + y}{z}, -3z^3\right)$; $(3, 0, 1)$
24. $\mathbf{f}(x, y, z) = \left(\sin \frac{\pi}{6}(xy), \ln(x - z), e^{y+z}\right)$; $(2, -1, 1)$
25. $\mathbf{f}(x_1, x_2, \ldots, x_n) = (x_1^2, x_2^2, \ldots, x_n^2)$; $(1, 1, 1, \ldots, 1)$
26. $\mathbf{f}(x_1, x_2, \ldots, x_n) = (x_1, x_1x_2, x_1x_2x_3, \ldots, x_1x_2\cdots x_n)$; $(1, 2, 3, \ldots, n)$
27. Show that the mapping $\mathbf{f}: \mathbb{R}^2 \to \mathbb{R}^2$ given by $\mathbf{f}(x, y) = (e^y \sin x, e^y \cos x)$ is locally C^1-invertible in a neighborhood of every point in $\mathbb{R}^2$ but is not invertible on all of $\mathbb{R}^2$.

In Problems 28–39 show that the given function $\mathbf{F}: \mathbb{R}^{n+m} \to \mathbb{R}^m$ with $\mathbf{F}(\mathbf{x}, \mathbf{y}) = \mathbf{0}$ determines $\mathbf{y} \in \mathbb{R}^m$ as a function **f** of $\mathbf{x} \in \mathbb{R}^n$ near the point $(\mathbf{x}_0, \mathbf{y}_0)$ in $\mathbb{R}^{n+m}$ and compute $J_{\mathbf{f}}(\mathbf{x}_0)$.

28. $\mathbf{F}(x_1, x_2, x_3, x_4) = (x_1 - 3x_2 + x_3 - 5x_4 - 8, 3x_1 - 2x_2 + 2x_3 - 8x_4 - 20) = (0, 0)$; $(\mathbf{x}_0, \mathbf{y}_0) = (2, 1, 4, -1)$
29. $\mathbf{F}(x_1, x_2, x_3, x_4, x_5) = (2x_1 + x_2 + 3x_3 - x_4 + 2x_5 - 2, 5x_1 - 2x_2 + x_3 - 2x_4 + 3x_5 + 1, x_1 + 6x_2 + 6x_3 + x_4 - 4x_5) = (0, 0, 0)$; $(\mathbf{x}_0, \mathbf{y}_0) = (1, 2, -1, 5, 3)$
30. $\mathbf{F}(x_1, x_2, x_3, x_4) = (x_1x_2 - x_3x_4, x_1x_3 - x_2x_4) = (0, 0)$; $(\mathbf{x}_0, \mathbf{y}_0) = (1, 1, 1, 1)$
31. $\mathbf{F}(x_1, x_2, x_3, x_4) = (x_1^2 + x_2^2 + x_3^2 + x_4^2 - 4, x_1x_2x_3x_4 - 1) = (0, 0)$; $(\mathbf{x}_0, \mathbf{y}_0) = (1, 1, 1, 1)$
32. $\mathbf{F}(x_1, x_2, x_3, x_4) = \left(2x_1^2 - 3x_1x_2 + x_3x_4^2 + 1, \frac{-4x_3}{x_2} + \frac{x_2}{x_4} + 8\right) = (0, 0)$; $(\mathbf{x}_0, \mathbf{y}_0) = (1, 2, 3, -1)$
33. $\mathbf{F}(x_1, x_2, x_3, x_4) = (x_1^2 + x_2^2 - x_3^2 - x_4^2, 2x_1x_2 + x_2^2 + 3x_3^2 - 2x_4^2 + 8) = (0, 0)$; $(\mathbf{x}_0, \mathbf{y}_0) = (2, -1, 1, 2)$
34. $\mathbf{F}(x_1, x_2, x_3, x_4) = \left(2x_1x_2^2x_3x_4^3 + 2x_2^3x_3^2 + 24, \frac{5x_3}{x_1} + \frac{2x_2}{x_4} + 1\right) = (0, 0)$; $(\mathbf{x}_0, \mathbf{y}_0) = (5, -2, 3, 1)$
35. $\mathbf{F}(x_1, x_2, x_3, x_4, x_5) = (x_1^2 + x_2^2 + x_3^2 + x_4^2 - 4, x_1x_2x_3x_4x_5 - 1) = (0, 0)$; $(\mathbf{x}_0, \mathbf{y}_0) = (1, 1, 1, 1, 1)$
36. $\mathbf{F}(x_1, x_2, x_3, x_4, x_5) = (x_1^2 + x_2^2 + x_3^2 + x_5^2 - 4, x_1x_2x_3x_4x_5 - 1, x_1 + x_2 + x_3 + x_4 - 4) = (0, 0, 0)$; $(\mathbf{x}_0, \mathbf{y}_0) = (1, 1, 1, 1, 1)$

37. $\mathbf{F}(x_1, x_2, x_3, x_4, x_5) = \left(2x_1^2 - 3x_2x_3x_4 - 44, 3x_1^2 + \frac{5x_4x_5}{x_3} + 28\right) = (0, 0)$; $(\mathbf{x}_0, \mathbf{y}_0) = (2, -3, 1, 4, -2)$

38. $\mathbf{F}(x_1, x_2, x_3, x_4, x_5) = \left(2x_1^2 - 3x_2x_3 - 17, 3x_1^2 + \frac{5x_4x_5}{x_3} + 28, 2x_3^3 - 4x_2x_5 + 22\right) = (0, 0, 0)$; $(\mathbf{x}_0, \mathbf{y}_0) = (2, -3, 1, 4, -2)$

39. $\mathbf{F}(x_1, x_2, x_3, x_4, x_5, x_6) = (e^{x1} \cos x_2 + e^{x3} \cos x_4 + e^{x5} \cos x_6 + 2x_1 - 3, e^{x1} \sin x_2 + e^{x3} \sin x_4 + e^{x5} \cos x_6 - x_2 - 1, e^{x1} \tan x_2 + e^{x3} \tan x_4 + e^{x5} \tan x_6 + x_3) = (0, 0, 0)$; $(\mathbf{x}_0, \mathbf{y}_0) = (0, 0, 0, 0, 0, 0)$

REVIEW EXERCISES FOR CHAPTER TEN

1. Find the third-degree Taylor polynomial of $\cos(x + 2y)$ at $\mathbf{x}_0 = \mathbf{0}$.

2. Find the second-degree Taylor polynomial of $\ln(1 + 5x - 4y)$ at $\mathbf{x}_0 = \mathbf{0}$.

3. Let $F(x, y, z) = x^2y + y^2z + z^5x - 3 = 0$. Show that z can be written as a function of x and y in a neighborhood of (1, 1, 1) and compute $\partial z/\partial x$ and $\partial z/\partial y$.

4. Let $F(x, y, z) = e^{2x-y+5z} + xyz - 1 = 0$. Show that z can be written as a function of x and y in a neighborhood of (0, 0, 0) and compute $\partial z/\partial x$ and $\partial z/\partial y$.

5. Let $x^2y^3 - 4xz^2 + 3z^4 - 5y + 4 = 0$. Find formulas for $\partial z/\partial x$ and $\partial z/\partial y$ and state where they are valid.

In Exercises 6–9 find the domain of the given function.

6. $\mathbf{f}(x, y) = (\sqrt{x^2 + y^2}, x^2 - y^3)$

7. $\mathbf{f}(x, y) = \left(\sec \frac{y}{x}, \frac{1}{x + y}, e^{x^2y}\right)$

8. $\mathbf{f}(x, y, z) = \left(\frac{1}{x + y + z}, \frac{1}{x + y - z}\right)$

9. $\mathbf{f}(x, y, z) = \left(x^2, y^2, \tan\left(\frac{x + y}{x - y}\right), \sin^{-1}\frac{xy}{z}\right)$

10. Compute $\lim_{\mathbf{x}\to(1,2)} \mathbf{f}(\mathbf{x})$ where $\mathbf{f}$ is as in Exercise 6.

11. Compute $\lim_{\mathbf{x}\to(1,2,4)} \mathbf{f}(\mathbf{x})$ where $\mathbf{f}$ is as in Exercise 9.

12. Let $\mathbf{f}(x, y) = (xy, x + y, \ln|y|^2)$ and $\mathbf{g}(x, y, z) = (xyz, x + 2y - 3z, e^{xy/z})$. Write out the composite function $\mathbf{g} \circ \mathbf{f}$ and determine its domain.

In Exercises 13–17 compute the derivative $\mathbf{f}'(\mathbf{x}_0) = J_\mathbf{f}(\mathbf{x}_0)$ of the given function $\mathbf{f}$ at the given point $\mathbf{x}_0$.

13. $\mathbf{f}(x, y) = (x^3, y^3)$; (2, 1)

14. $\mathbf{f}(x, y) = (xy, y/x, x^3y^2)$; (1, 4)

15. $\mathbf{f}(x, y, z) = (x^2, 2y^2, 3z^2)$; (1, 2, 3)

16. $\mathbf{f}(x, y, z) = (x + 2y - 3z, 2x - 4y + 7z, 4x - y - z, -x - y + 2z)$; (5, 10, 20)

17. $\mathbf{f}(x_1, x_2, x_3, x_4) = \left(x_1x_3, x_2x_4, x_1^2 + x_2^2 + x_3^2 + x_4^2, \frac{x_1x_4}{x_2x_3}\right)$; (3, 1, −1, 5)

18. Let $\mathbf{f}(x, y) = (x^3, y^3)$ and $\mathbf{g}(x, y) = (x + 2y, 2x - y)$. Find the derivative of the composite function $\mathbf{g} \circ \mathbf{f}$ at the point (2, −1) and verify the chain rule at that point.

19. Let $\mathbf{f}(x, y, z) = (x^2y, xz + 3y^2, 3xyz^3)$ and $\mathbf{g}(x, y, z) = (2xy, 3xz, yz, xyz)$. Find the derivative of the composite function $\mathbf{g} \circ \mathbf{f}$ at the point (2, 1, −1) and verify the chain rule at that point.

20. Compute the Jacobian of $\mathbf{f}(x, y) = (x^3, y^3)$ at the point (2, 1).

21. Compute the Jacobian of $\mathbf{f}(x, y, z) = (x^2, 2y^2, 3z^2)$ at the point (1, 2, 3).

22. Compute the Jacobian of $\mathbf{f}(x, y, z) = (x^2 - yz, 3y - 2x^3z^2, 4xyz)$ at the point (2, −1, 4).

In Exercises 23–26 a mapping f and a point $\mathbf{x}_0$ are given. Show that the mapping is locally C^1-invertible at the given point and compute $J_{\mathbf{f}^{-1}}(\mathbf{x}_0)$.

23. $\mathbf{f}(x, y) = (3x + 2y, -x + 5y)$; (4, −7)

24. $\mathbf{f}(x, y) = (x^3 + y^3, x^3 - y^3)$; (2, 1)

25. $\mathbf{f}(x, y, z) = (x^3, y^3, z^3)$; (1, −1, 2)

26. $\mathbf{f}(x, y, z) = (x^2y, xy + 3y^2, 3xyz^3)$; (−3, 2, 1)

In Exercises 27–29 show that the given function $\mathbf{F}: \mathbb{R}^{n+m} \to \mathbb{R}^m$ with $\mathbf{F}(\mathbf{x}, \mathbf{y}) = \mathbf{0}$ determines $\mathbf{y} \in \mathbb{R}^m$ as a function $\mathbf{f}$ of $\mathbf{x} \in \mathbb{R}^n$ near the point $(\mathbf{x}_0, \mathbf{y}_0)$ in $\mathbb{R}^{n+m}$ and compute $J_\mathbf{f}(\mathbf{x}_0)$.

27. $\mathbf{F}(x_1, x_2, x_3, x_4) = (2x_1 - x_2 + x_3 - 4x_4 - 8, 3x_1 + 2x_2 - 2x_3 - 4x_4 + 7) = (0, 0)$; $(\mathbf{x}_0, \mathbf{y}_0) = (3, -2, 4, 1)$

28. $\mathbf{F}(x_1, x_2, x_3, x_4) = (x_1^3 + x_2^3 + x_3^3 + x_4^3 - 4, 2x_1x_2x_3x_4 - 2) = (0, 0)$; $(\mathbf{x}_0, \mathbf{y}_0) = (1, 1, 1, 1)$

29. $\mathbf{F}(x_1, x_2, x_3, x_4, x_5) = (x_1^3 + x_2^3 + x_3^3 + x_5^3 - 4, x_1x_2x_3x_4x_5 - 1, x_1 + 2x_2 - 5x_3 + 2x_4)$; $(\mathbf{x}_0, \mathbf{y}_0) = (1, 1, 1, 1, 1)$

11 Ordinary Differential Equations

11.1 INTRODUCTION

Many of the basic laws of the physical sciences and, more recently, of the biological and social sciences are formulated in terms of mathematical relations involving certain known and unknown quantities and their derivatives. Such relations are called **differential equations.**

In your study of one-variable calculus you probably encountered the following differential equation:

$$\frac{dy}{dx} = \alpha x \tag{1}$$

where α is a given constant. Equation (1) is called the differential equation of **exponential growth** (if $\alpha > 0$) and of **exponential decay** (if $\alpha < 0$). In this chapter we will discuss this and many other types of differential equations. In Chapter 12 we will discuss systems of differential equations. We begin, in this section, by categorizing the types of differential equations that may be encountered.

The most obvious classification is based on the nature of the derivative (or derivatives) in the equation. A differential equation involving only ordinary derivatives (derivatives of functions of one variable) is called an **ordinary differential equation.** Equation (1) is an ordinary differential equation.

A differential equation containing partial derivatives is called a **partial differential equation.** Examples of partial differential equations are given in Problems 4.3.41, 4.3.42, 4.4.31, 4.4.32, 4.4.35, and 4.4.39.

In this chapter we will consider only the solution of certain ordinary differential equations, which we will refer to simply as differential equations, dropping the word "ordinary." Procedures for solving all but the most trivial partial differential equations are beyond the scope of this discussion.

ORDER

The **order** of a differential equation is the order of the highest-order derivative appearing in the equation.

EXAMPLE 1 The following are examples of differential equations with indicated orders.

(a) $dy/dx = 3x$ (first order).
(b) $x''(t) + 4x'(t) - x(t) = \sin t$ (second order).
(c) $(dy/dx)^5 - 2e^y = 6 \cos x$ (first order).
(d) $(y^{(4)})^2 - 2y''' + y' - 6y = \sqrt{x}$ (fourth order). ■

Consider the following equation having the form of equation (1)

$$\frac{dy}{dx} = 3y \tag{2}$$

together with the **initial condition**

$$y(0) = 2. \tag{3}$$

SOLUTION

A **solution** to (2), (3) is defined to be a function that

(i) is differentiable,
(ii) satisfies the differential equation (2) on some open interval containing 0, and
(iii) satisfies the initial condition (3).

Let $y = 2e^{3x}$. Then

$$\frac{dy}{dx} = 2\frac{d}{dx}e^{3x} = 2 \cdot 3e^{3x} = 3(2e^{3x}) \overset{\substack{y = 2e^{3x}\\ \downarrow}}{=} 3y$$

That is, $y = 2e^{3x}$ is a differentiable function that satisfies equation (2). Moreover,

$$y(0) = 2e^{3 \cdot 0} = 2e^0 = 2$$

so that y satisfies the initial condition (3). In fact, the function $y = 2e^{3x}$ is the *only* function that satisfies equations (2) and (3).

INITIAL VALUE PROBLEM

The system (2), (3) is called an **initial value problem.**

In Appendix 5 we will show that for a certain class of initial value problems, including the problem (2), (3), there is always a unique solution. This makes sense intuitively. For example, suppose that a bacteria population is growing at a rate proportional to itself. That is, if $P(t)$ denotes the population size at time t, then

$$P'(t) = \alpha P(t) \tag{4}$$

for some constant of proportionality α. Furthermore, we assume that the population at some given time, which we denote by $t = 0$, is 10,000. Then we have

$$P(0) = 10{,}000. \tag{5}$$

It is reasonable to expect that starting with the initial population given by (5) and assuming that population growth is governed by the equation (4), the population in the future will be completely determined. That is, there is one and only one way to write the population $P(t)$; this is a way of saying that the initial value problem (4), (5) has a unique solution.

There are other types of problems in which other than initial conditions are given. For example, let a string be held taut between two points one meter apart, as in Figure 1.

0 1 x

FIGURE 1

Suppose that the string is then plucked so that it begins to vibrate. Let $\nu(x, t)$ denote the height of the string at a distance x units from the left-hand endpoint and at a time t, where $0 \le x \le 1$ and $t \ge 0$. Then, since the two endpoints are held fixed, we have the conditions

$$\nu(0, t) = \nu(1, t) = 0 \qquad \text{for all} \quad t \ge 0. \tag{6}$$

The conditions (6) are called **boundary conditions** and if we write a differential equation governing the motion of the string (see Problem 4.4.31 on p. 198), then the equation, together with the boundary conditions (6), is called a **boundary value problem.** In general, a boundary value problem is a differential equation together with values given at two or more points, while an initial value problem is a differential equation with values given at only one point.

EXAMPLE 2 The problem

$$y'' + 2y' + 3y = 0, \qquad y(0) = 1, \quad y(1) = 2$$

is a boundary value problem. The problem

$$y'' + 2y' + 3y = 0, \qquad y(0) = 1, \quad y'(0) = 2$$

is an initial value problem. ■

In this chapter we will consider certain kinds of first- and second-order ordinary differential equations and some simple initial value problems. For a more complete discussion see an introductory book on differential equations. In Appendix 5 we shall prove that many first-order initial value problems have unique solutions in some neighborhood of the initial point $(x_0, y(x_0))$.

There is one further classification of differential equations that will be important to us in this book.

LINEAR EQUATION

An nth-order differential equation is **linear** if it can be written in the form

$$\frac{d^n y}{dx^n} + a_{n-1}(x)\frac{d^{n-1}y}{dx^{n-1}} + \cdots + a_1(x)\frac{dy}{dx} + a_0(x)y = f(x). \tag{7}$$

Hence a first-order linear equation has the form

$$\frac{dy}{dx} + a(x)y = f(x), \tag{8}$$

and a second-order linear equation can be written as

$$\frac{d^2y}{dx^2} + a(x)\frac{dy}{dx} + b(x)y = f(x). \tag{9}$$

The notation indicates that $a(x)$, $b(x)$, and $f(x)$ are functions of x alone.

A differential equation that is not linear is called **nonlinear.**

EXAMPLE 3 The following are linear differential equations.

(a) $\dfrac{dy}{dx} - 3y = 0$

(b) $y'' + 4y = 0$

(c) $y'' + 2xy' + e^x y = \sin^2 x$

(d) $y^{(4)} + 5y''' - 6y'' + 7y' + 8y = x^5 + x^3 + 10$ ■

EXAMPLE 4 The following are nonlinear differential equations.

(a) $\dfrac{dy}{dx} - y^2 = 0$

(b) $yy'' = y' + x$

(c) $\frac{dy}{dx} + \sqrt{y} = 0$

(d) $\sin xy + \cos\left(1 + \frac{dy}{dx}\right) = x^3y^4$ ■

In Sections 11.4, 11.6, and 11.8–12, we will show how solutions to a great number of linear differential equations can be found. Solutions to nonlinear equations can be found only in some special situations. Two of these are given in Sections 11.2 and 11.3.

PROBLEMS 11.1

In Problems 1–6 state the order of the differential equation.

1. $x'(t) + 2x(t) = t$
2. $(d^2y/dx^2) + 4(dy/dx)^3 - y = x^3$
3. $y''' + y = 0$
4. $(dx/dt)^5 = x^4$
5. $x'' - x^2 = 2x^{(4)}$
6. $(x''(t))^4 - (x'(t))^{3/2} = 2x(t) + e^t$

7. Verify that $y_1 = 2\sin x$ and $y_2 = -4\cos x$ are solutions to $y'' + y = 0$.
8. Verify that $y = (x/2)e^x$ is a solution to $y'' - y = e^x$.
9. Verify that $y = 3x/(4x - 3)$ is the solution to the initial value problem

$$x^2\,dy/dx + y^2 = 0, \qquad y(1) = 3.$$

10. Verify that $y = -\frac{1}{2}x^2e^{-x}$ is a solution to the equation $y''' - 3y' - 2y = 3e^{-x}$.
11. Verify that $y = e^{3x}$ satisfies the initial value problem

$$y'' - 6y' + 9y = 0, \qquad y(0) = 1, \quad y'(0) = 3.$$

[*Hint:* Check that the equation is satisfied, and then verify that the two initial conditions are also satisfied.]

*12. Verify that $(y = e^{-x/2}[\cos(\sqrt{3}/2)x + (7/\sqrt{3})\sin(\sqrt{3}/2)x])$ is the solution to the initial value problem.

$$y'' + y' + y = 0, \qquad y(0) = 1, \quad y'(0) = 3.$$

In Problems 13–23 determine whether the differential equation is linear or nonlinear.

13. $y'' + 2y' + 3y = 0$
14. $y'' + 2y' + 3y = x^3 + \ln(1 + x^2)$
15. $y'' + x^5y' + 3(\sin x)y = x$
16. $y'' + 2y' + y^2 = 0$
17. $y'' + \sqrt{y} = 0$
18. $\left(\frac{dy}{dx}\right)^2 + y = 0$
19. $y\frac{dy}{dx} = 2x$
20. $\frac{dy}{dx} + 10x^{10}y = \cos x - \frac{d^2y}{dx^2}$
21. $\frac{dy}{dx} = \frac{3}{y}$
22. $\frac{dy}{dx} = \frac{y}{x}$
23. $x^2\frac{d^2y}{dx^2} + y\frac{dy}{dx} = 2$

11.2 FIRST-ORDER EQUATIONS—SEPARATION OF VARIABLES

Consider the first-order differential equation

$$\frac{dy}{dx} = f(x, y). \tag{1}$$

Suppose that $f(x, y)$ can be written as

$$f(x, y) = \frac{g(x)}{h(y)} \tag{2}$$

where g and h are each functions of only one variable. Then (1) can be written

$$h(y)\frac{dy}{dx} = g(x)$$

or, after integration of both sides with respect to x,

$$\int h(y)\,dy = \int h(y)\frac{dy}{dx}\,dx = \int g(x)\,dx + C. \tag{3}$$

The method of solution suggested in (3) is called the method of **separation of variables.** In general, if a differential equation can be written in the form $h(y)\,dy = g(x)\,dx$, then direct integration of both sides (if possible) will produce a family of solutions.

EXAMPLE 1 Solve the differential equation $dy/dx = 4y$.

Solution. This is a special case of the linear equation (11.1.1). We have $f(x, y) = 4y$, and we can write

$$\frac{dy}{y} = 4\,dx,$$

so that

$$\int \frac{dy}{y} = \int 4\,dx$$

or

$$\ln|y| = 4x + C$$

and

$$|y| = e^{4x+C} = e^{4x}e^{C}.$$

This can be written

$$y = ke^{4x},$$

where k is some real number other than 0. In addition, the constant function $y \equiv 0$ is a solution, so 0 can be allowed as a value for k. ■

EXAMPLE 2 Solve the initial value problem $dy/dx = \alpha y$, $y(x_0) = y_0$.

Solution. Separating variables as in Example 1, we have

$$\frac{dy}{y} = \alpha\, dx \qquad \text{or} \qquad \int \frac{dy}{y} = \int \alpha\, dx$$

and

$$\ln|y| + \alpha x + C.$$

Then, as before, we can write

$$y(x) = ke^{\alpha x}.$$

But $y(x_0) = ke^{\alpha x_0} = y_0$, so that $k = y_0 e^{-\alpha x_0}$ and we obtain the unique solution

$$y(x) = y_0 e^{-\alpha x_0} e^{\alpha x} = y_0 e^{\alpha(x - x_0)}. \tag{4}$$

REMARK. As mentioned earlier, we say that y is **growing exponentially** if $\alpha > 0$ and **decaying exponentially** if $\alpha < 0$.

When $x_0 = 0$ in equation (4), we obtain the unique solution

$$y(x) = y_0 e^{\alpha x} \tag{5}$$

to the initial value problem

$$\frac{dy}{dx} = \alpha y, \qquad y(0) = y_0. \tag{6}$$

The initial value problem (6) arises in many practical situations. We illustrate this with a number of examples and then show how the technique of separation of variables can be used to solve nonlinear problems as well. ■

EXAMPLE 3 **(Population Growth)** A bacterial population is growing continuously at a rate equal to 10% of its population each day. Its initial size is 10,000 organisms. How many bacteria are present after 10 days? After 30 days?

Solution. Since the percentage growth of the population is 10% = 0.1, we have

$$\frac{dP/dt}{P} = 0.1, \qquad \text{or} \qquad \frac{dP}{dt} = 0.1P. \tag{7}$$

Here $\alpha = 0.1$, and all solutions have the form

$$P(t) = ce^{0.1t} \tag{8}$$

where t is measured in days. Since $P(0) = 10{,}000$, we have

$$ce^{(0.1)(0)} = c = 10{,}000, \quad \text{and} \quad P(t) = 10{,}000e^{0.1t}.$$

After 10 days $P(10) = 10{,}000e^{(0.1)(10)} = 10{,}000e \approx 27{,}183$, and after 30 days $P(30) = 10{,}000e^{0.1(30)} = 10{,}000e^3 \approx 200{,}855$ bacteria. ■

EXAMPLE 4 **(Newton's Law of Cooling)** **Newton's law of cooling** states that the rate of change of the temperature difference between an object and its surrounding medium is proportional to the temperature difference. If $D(t)$ denotes this temperature difference at time t and if α denotes the constant of proportionality, then we obtain

$$\frac{dD}{dt} = -\alpha D. \tag{9}$$

The minus sign indicates that this difference decreases. (If the object is cooler than the surrounding medium—usually air—it will warm up; if it is hotter, it will cool.) The solution to this differential equation is

$$D(t) = ce^{-\alpha t}.$$

If we denote the initial ($t = 0$) temperature difference by D_0, then

$$D(t) = D_0 e^{-\alpha t} \tag{10}$$

is the formula for the temperature difference for any $t > 0$. Notice that for t large $e^{-\alpha t}$ is very small, so that, as we have all observed, temperature differences tend to die out rather quickly.

We now may ask: In terms of the constant α, how long does it take for the temperature difference to decrease to half its original value?

Solution. The original value is D_0. We are therefore looking for a value of t for which $D(t) = \frac{1}{2}D_0$. That is, $\frac{1}{2}D_0 = D_0 e^{-\alpha t}$, or $e^{-\alpha t} = \frac{1}{2}$. Taking natural logarithms, we obtain

From a calculator ↙

$$-\alpha t = \ln\frac{1}{2} = -\ln 2 = -0.6931, \quad \text{and} \quad t \approx \frac{0.6931}{\alpha}.$$

Notice that this value of t does *not* depend on the initial temperature difference D_0. ■

EXAMPLE 5 With the air temperature equal to 30°C, an object with an initial temperature of 10°C warmed to 14°C in 1 hr.

(a) What was its temperature after 2 hr?
(b) After how many hours was its temperature 25°C?

Solution. Let $T(t)$ denote the temperature of the object. Then $D(t) = 30 - T(t) = D_0 e^{-\alpha t}$ [from (10)]. But $D_0 = 30 - T(0) = 30 - 10 = 20$, so that

$$D(t) = 20e^{-\alpha t}.$$

We are given that $T(1) = 14$, so $D(1) = 30 - T(1) = 16$ and

$$16 = D(1) = 20e^{-\alpha \cdot 1} = 20e^{-\alpha}, \qquad \text{or} \qquad e^{-\alpha} = 0.8.$$

Thus

$$D(t) = 20e^{-\alpha t} = 20(e^{-\alpha})^t = 20(0.8)^t,$$

and

$$T(t) = 30 - D(t) = 30 - 20(0.8)^t.$$

We can now answer the two questions.

(a) $T(2) = 30 - 20(0.8)^2 = 30 - 20(0.64) = 17.2°\text{C}$.
(b) We need to find t such that $T(t) = 25$. That is,

$$25 = 30 - 20(0.8)^t, \qquad \text{or} \qquad (0.8)^t = \tfrac{1}{4} \qquad \text{or} \qquad t \ln(0.8) = -\ln 4,$$

and

$$t = \frac{-\ln 4}{\ln(0.8)} \approx \frac{1.3863}{0.2231} \approx 6.2 \text{ hr} = 6 \text{ hr } 12 \text{ min.} \quad \blacksquare$$

EXAMPLE 6 **(Carbon Dating)** **Carbon dating** is a technique used by archaeologists, geologists, and others who want to estimate the ages of certain artifacts and fossils they uncover. The technique is based on certain properties of the carbon atom. In its natural state the nucleus of the carbon atom ^{12}C has 6 protons and 6 neutrons. An *isotope* of carbon ^{12}C is ^{14}C, which has 2 additional neutrons in its nucleus. ^{14}C is *radioactive*. That is, it emits neutrons until it reaches the stable state ^{12}C. We make the assumption that the ratio of ^{14}C to ^{12}C in the atmosphere is constant. This assumption has been shown experimentally to be approximately valid, for although ^{14}C is being constantly lost through **radioactive decay** (as this process is often termed), new ^{14}C is constantly being produced by the cosmic bombardment of nitrogen in the upper atmosphere. Living plants and animals do not distinguish between ^{12}C and ^{14}C, so at the time of death the ratio of ^{12}C to ^{14}C in an organism is the same as the ratio in the atmosphere. However, this ratio changes after death since ^{14}C is converted to ^{12}C but no further ^{14}C is taken in.

It has been observed that ^{14}C decays at a rate proportional to its mass and that its **half-life** is approximately 5580 years.[†] That is, if a substance starts with 1 g of ^{14}C,

[†]This number was first determined in 1941 by the American chemist W. S. Libby, who based his calculations on the wood from sequoia trees, whose ages were determined by rings marking years of growth. Libby's method has come to be regarded as the archaeologist's absolute measuring scale. But in truth, this scale is flawed. Libby used the assumption that the atmosphere had at all times a constant amount of ^{14}C. Recently, however, the American chemist C. W. Ferguson of the University of Arizona deduced from his study of tree rings in 4000-year-old American giant trees that before 1500 B.C. the radiocarbon content of the atmosphere was considerably higher than it was later. This result implied that objects from the pre-1500 B.C. era were much older than previously believed, because Libby's "clock" allowed for a smaller amount of ^{14}C than actually was present. For example, a find dated at 1800 B.C. was in fact from 2500 B.C. This fact has had a considerable impact on the study of prehistoric times. For a fascinating discussion of this subject, see Gerhard Herm, *The Celts* (St. Martin's Press, New York, 1975), pages 90–92.

then 5580 years later it would have $\frac{1}{2}$ g of ^{14}C, the other $\frac{1}{2}$ g having been converted to ^{12}C.

We may now pose a question typically asked by an archaeologist. A fossil is unearthed and it is determined that the amount of ^{14}C present is 40% of what it would be for a similarly sized living organism. What is the approximate age of the fossil?

Solution. Let $M(t)$ denote the mass of ^{14}C present in the fossil. Then since ^{14}C decays at a rate proportional to its mass, we have

$$\frac{dM}{dt} = -\alpha M,$$

where α is the constant of proportionality. Then $M(t) = ce^{-\alpha t}$, where $c = M_0$, the initial amount of ^{14}C present. When $t = 0$, $M(0) = M_0$; when $t = 5580$ years, $M(5580) = \frac{1}{2}M_0$, since half the original amount of ^{14}C has been converted to ^{12}C. We can use this fact to solve for α since we have

$$\tfrac{1}{2}M_0 = M_0 e^{-\alpha \cdot 5580}, \qquad \text{or} \qquad e^{-5580\alpha} = \tfrac{1}{2}.$$

Thus

$$(e^{-\alpha})^{5580} = \tfrac{1}{2}, \qquad \text{or} \qquad e^{-\alpha} = (\tfrac{1}{2})^{1/5580}, \qquad \text{and} \qquad e^{-\alpha t} = (\tfrac{1}{2})^{t/5580},$$

so

$$M(t) = M_0(\tfrac{1}{2})^{t/5580}.$$

Now we are told that after t years (from the death of the fossilized organism to the present) $M(t) = 0.4M_0$, and we are asked to determine t. Then

$$0.4M_0 = M_0(\tfrac{1}{2})^{t/5580},$$

and taking natural logarithms (after dividing by M_0), we obtain

$$\ln 0.4 = \frac{t}{5580}\ln\left(\frac{1}{2}\right), \qquad \text{or} \qquad t = \frac{5580\ln(0.4)}{\ln(\frac{1}{2})} \approx 7376 \text{ years}.$$

The carbon-dating method has been used successfully on numerous occasions. It was this technique that established that the Dead Sea scrolls were prepared and buried about two thousand years ago. ■

We now turn to some nonlinear examples.

EXAMPLE 7 Solve the initial value problem $dy/dx = y^2(1 + x^2)$, $y(0) = 1$.

Solution. We divide both sides of the differential equation by y^2 and multiply by dx. Then, successively,

$$\frac{dy}{y^2} = (1 + x^2)\,dx, \qquad \int \frac{dy}{y^2} = \int (1 + x^2)\,dx, \qquad -\frac{1}{y} = x + \frac{x^3}{3} + C,$$

or

$$y = -\frac{1}{x + (x^3/3) + C}.$$

For every number C, this is a solution to the differential equation. Moreover, the constant function $y \equiv 0$ is also a solution. When $x = 0$, $y = 1$, so that

$$1 = y(0) = -\frac{1}{0 + 0 + C} = -\frac{1}{C}.$$

This result implies that $C = -1$, and we obtain the unique solution to the initial value problem:

$$y = -\frac{1}{x + (x^3/3) - 1}.$$

REMARK. This solution (like any solution to a differential equation) can be checked by differentiation. You should *always* carry out this check. To check, we have

$$\begin{aligned}\frac{dy}{dx} &= \frac{1}{[x + (x^3/3) - 1]^2}(1 + x^2) = \left\{\frac{-1}{[x + (x^3/3) - 1]}\right\}^2 (1 + x^2) \\ &= y^2(1 + x^2).\end{aligned}$$

Also, $y(0) = -1/(0 + 0 - 1) = 1$. ■

EXAMPLE 8 Solve the differential equation $dx/dt = t\sqrt{1 - x^2}$.

Solution. We have

$$\frac{dx}{\sqrt{1 - x^2}} = t\,dt \qquad \text{or} \qquad \int \frac{dx}{\sqrt{1 - x^2}} = \int t\,dt.$$

Integration yields

$$\sin^{-1} x = \frac{t^2}{2} + C \qquad \text{or} \qquad x = \sin\left(\frac{t^2}{2} + C\right).$$

This should be checked by differentiation, using the fact that $\cos[(t^2/2) + C] = \sqrt{1 - \sin^2[(t^2/2) + C]}$. Note that, as in Example 1, there are an infinite number of solutions to the equation. We can obtain a unique solution by specifying an initial conditions. For example, if we have $x(0) = \frac{1}{2}$, then $\frac{1}{2} = \sin(0 + C) = \sin C$ and $C = \sin^{-1}\frac{1}{2} = \pi/6$. Then the unique solution would be

$$x(t) = \sin\left(\frac{t^2}{2} + \frac{\pi}{6}\right). \text{ ■}$$

EXAMPLE 9 **(Logistic Growth)** The growth rate per individual in a population is the difference between the average birth rate and the average death rate. Suppose that in a given population the average birth rate is a positive constant β, but the average death rate, because of the effects of crowding and increased competition for the available food, is proportional to the size of the population. We call the constant of proportionality δ (which is >0). If $P(t)$ denotes the population at time t, then dP/dt is the growth rate of the population. The growth rate per individual is given by

$$\frac{1}{P}\frac{dP}{dt}.$$

Then, using the conditions described above, we have

$$\frac{1}{P}\frac{dP}{dt} = \beta - \delta P$$

or

$$\frac{dP}{dt} = P(\beta - \delta P). \tag{11}$$

This differential equation, together with the condition

$$P(0) = P_0 \qquad \text{(the initial population)}, \tag{12}$$

is an initial value problem. To solve, we have

$$\frac{dP}{P(\beta - \delta P)} = dt$$

or

$$\int \frac{dP}{P(\beta - \delta P)} = \int dt = t + C. \tag{13}$$

To calculate the integral on the left, we use partial fractions. We have (verify this)

$$\frac{1}{P(\beta - \delta P)} = \frac{1}{\beta P} + \frac{\delta}{\beta(\beta - \delta P)},$$

so that

$$\int \frac{dP}{P(\beta - \delta P)} = \int \frac{1}{\beta}\frac{dP}{P} + \frac{\delta}{\beta}\int \frac{dP}{\beta - \delta P}$$

$$= \frac{1}{\beta}\ln|P| - \frac{1}{\beta}\ln|\beta - \delta P| = t + C$$

or

$$\ln\left|\frac{P}{\beta - \delta P}\right| = \ln|P| - \ln|\beta - \delta P| = \beta t - \beta C = \beta t - C_1$$

and

$$\frac{P}{\beta - \delta P} = e^{\beta t - C_1} = C_2 e^{\beta t}. \tag{14}$$

Here $C_1 = -\beta C$ and $C_2 = \pm e^{-C_1}$.

Using the initial condition (12), we have

$$\frac{P_0}{\beta - \delta P_0} = C_2 e^0 = C_2. \tag{15}$$

Finally, we insert (15) into equation (14) to obtain

$$\frac{P}{\beta - \delta P} = \frac{P_0}{\beta - \delta P_0} e^{\beta t}$$

$$P = \frac{P_0}{\beta - \delta P_0} e^{\beta t}(\beta - \delta P) = \frac{\beta P_0}{\beta - \delta P_0} e^{\beta t} - \frac{\delta P_0 e^{\beta t}}{\beta - \delta P_0} P$$

$$P\left(1 + \frac{\delta P_0}{\beta - \delta P_0} e^{\beta t}\right) = \frac{\beta P_0}{\beta - \delta P_0} e^{\beta t} = P\left[\frac{\beta - \delta P_0 + \delta P_0 e^{\beta t}}{\beta - \delta P_0}\right]$$

$$P(t) = \frac{\beta - \delta P_0}{\beta - \delta P_0 + \delta P_0 e^{\beta t}}\left[\frac{\beta P_0}{\beta - \delta P_0} e^{\beta t}\right] = \frac{\beta P_0 e^{\beta t}}{\beta - \delta P_0 + \delta P_0 e^{\beta t}}$$

Multiply and divide by $e^{\beta t}$ ↓ Divide top and bottom by P_0 ↓

$$= \frac{\beta P_0}{\delta P_0 + (\beta - \delta P_0)e^{-\beta t}} = \frac{\beta}{\delta + \left(\frac{\beta}{P_0} - \delta\right)e^{-\beta t}}.$$

That is,

$$P(t) = \frac{\beta}{\delta + [(\beta/P_0) - \delta]e^{-\beta t}}. \tag{16}$$

Equation (11) is called the **logistic equation** and we have shown that the solution to the logistic equation with initial population P_0 is given by (16). Sketches of the growth governed by the logistic equation are given in Figure 1. ■

We close this section by noting that most differential equations, even those of the first order, cannot be solved by elementary methods (although there are other techniques of solution besides the technique of separation of variables). However, even

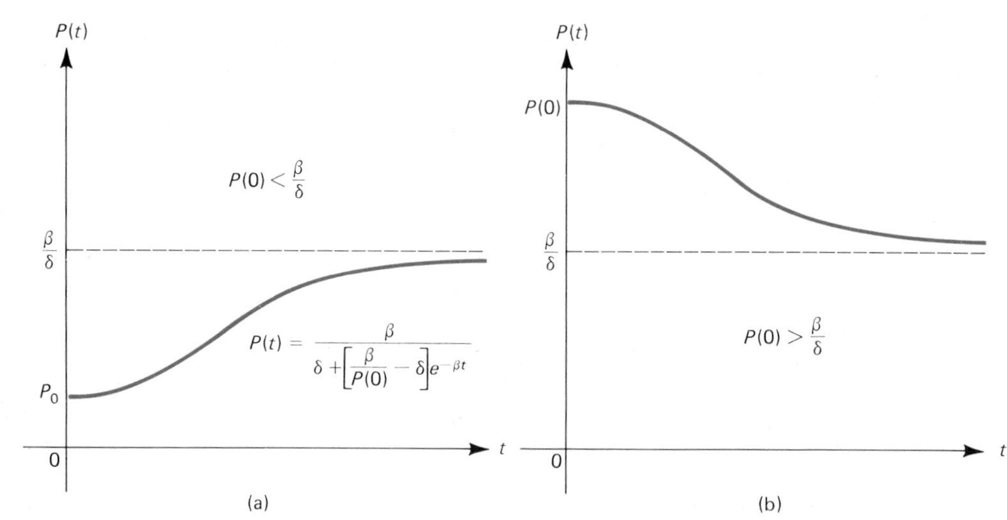

FIGURE 1

when an equation cannot be solved in a "closed form" (i.e., by writing one variable in terms of another, numerical techniques usually exist for calculating a solution to as many decimal places of accuracy as needed.

In Section 11.4 we discuss a wide class of first-order equations for which closed form solutions can be found.

PROBLEMS 11.2

In Problems 1–12 solve the given differential equation by the method of separation of variables. If an initial condition is given, find the unique solution to the initial value problem.

1. $\dfrac{dy}{dx} = -7x$

2. $\dfrac{dy}{dx} = e^{x+y},\ y(0) = 2$

3. $\dfrac{dx}{dt} = \sin x \cos t,\ x\left(\dfrac{\pi}{2}\right) = 3$

4. $\dfrac{dy}{dx} = \dfrac{1}{y^3}$

5. $x^3(y^2 - 1)\dfrac{dy}{dx} = (x + 3)y^5$

6. $\dfrac{dy}{dx} = 2x^2y^2,\ y(1) = 2$

7. $\dfrac{dx}{dt} = e^x \sin t,\ x(0) = 1$

8. $\dfrac{dy}{dx} = 1 + y^2,\ y(0) = 1$

9. $\dfrac{dx}{dt} = x(1 - \cos 2t),\ x(0) = 1$

***10.** $\dfrac{dx}{dt} = x^n t^m$, n, m integers

11. $\dfrac{dy}{dx} = x^2(1 + y^2)$

12. $\dfrac{dx}{dt} = \dfrac{e^x t}{e^x + t^2 e^x}$

13. The growth rate of a bacteria population is proportional to its size. Initially the population is 10,000; after 10 days its size is 25,000. What is the population size after 20 days? After 30 days?

14. In Problem 13 suppose instead that the population after 10 days is 6000. What is the population after 20 days? After 30 days?

15. The population of a certain city grows 6% a year. If the population in 1970 was 250,000, what would be the population in 1980? In 2000?

16. When the air temperature is 70°F, an object cools from 170°F to 140°F in $\frac{1}{2}$ hr.
(a) What will be the temperature after 1 hr?
(b) When will the temperature be 90°F? [*Hint:* Use Newton's law of cooling.]

17. A hot coal (temperature 150°C) is immersed in ice water (temperature -10°C). After 30 sec the temperature of the coal is 60°C. Assume that the ice water is kept at -10°C.
(a) What is the temperature of the coal after 2 min?
(b) When will the temperature of the coal be 0°C?

****18.** The president and vice-president sit down for coffee. They are both served a cup of hot black coffee (at the same temperature). The president takes a container of cream and immediately adds it to his coffee, stirs it, and waits. The vice-president waits ten minutes and then adds the same amount of cream (which has been kept cool) to her coffee and stirs it in. Then they both drink. Assuming that the temperature of the cream is lower than that of the air, who drinks the hotter coffee? [*Hint:* Use Newton's law of cooling. It is necessary to treat each case separately and to keep track of the volumes of coffee, cream, and the coffee-cream mixture.]†

19. A fossilized leaf contains 70% of a "normal" amount of ^{14}C. How old is the fossil?

20. Forty percent of a radioactive substance disappears in 100 years.
(a) What is its half-life?
(b) After how many years will 90% be gone?

21. Salt decomposes in water into sodium $[Na^+]$ and chloride $[Cl^-]$ ions at a rate proportional to its mass. Suppose there were initially 25 kg of salt and 15 kg after 10 hr.
(a) How much salt would be left after one day?
(b) After how many hours would there be less than $\frac{1}{2}$ kg of salt left?

22. X rays are absorbed into a uniform, partially opaque body as a function not of time but of penetration distance. The rate of change of the intensity $I(x)$ of the X ray is proportional to the intensity. Here x measures the distance of penetration. The more the X ray penetrates, the lower the intensity is. The constant of proportionality is the density D of the medium being penetrated.
(a) Formulate a differential equation describing this phenomenon.
(b) Solve for $I(x)$ in terms of x, D, and the initial (surface) intensity $I(0)$.

23. Radioactive beryllium is sometimes used to date fossils found in deep-sea sediment. The decay of beryllium satisfies the equation

$$\frac{dA}{dt} = -\alpha A, \qquad \text{where} \qquad \alpha = 1.5 \times 10^{-7}.$$

What is the half-life of beryllium?

24. In a certain medical treatment a tracer dye is injected into the pancreas to measure its function rate. A normally active pancreas will secrete 4% of the dye each minute. A physician injects 0.3 g of the dye and 30 min later 0.1 g remains. How much dye would remain if the pancreas were functioning normally?

25. Atmospheric pressure is a function of altitude above sea level and is given by $dP/da = \beta P$, where β is a constant. The pressure is measured in millibars (mbar). At sea level ($a = 0$), $P(0)$ is 1013.25 mbar which means that the atmosphere at sea level will support a column of mercury 1013.25 mm high at a standard temperature of 15°C. At an altitude of $a = 1500$ m, the pressure is 845.6 mbar.
(a) What is the pressure at $a = 4000$ m?
(b) What is the pressure at 10 km?
(c) In California the highest and lowest points are Mount Whitney (4418 m) and Death Valley (86 m below sea level). What is the difference in their atmospheric pressures?
(d) What is the atmospheric pressure at Mount Everest (elevation 8848 m)?
(e) At what elevation is the atmospheric pressure equal to 1 mbar?

---

†This is a famous old problem that keeps on popping up (with an ever-changing pair of characters) in books on games and puzzles in mathematics. The problem is hard and has stymied many a mathematician. Do not be frustrated if you cannot solve it. The trick is to write everything down and to keep track of all the variables. The fact that the air is warmer than the cream is critical. It should also be noted that guessing the correct answer is fairly easy. Proving that your guess is correct is what makes the problem difficult.

26. A bacteria population is known to grow exponentially. The following data were collected:

Number of Days	Number of Bacteria
5	936
10	2190
20	11,986

(a) What was the initial population?
(b) If the present growth rate were to continue, what would be the population after 60 days?

27. A bacteria population is declining exponentially. The following data were collected:

Number of Hours	Number of Bacteria
12	5969
24	3563
48	1269

(a) What was the initial population?
(b) How many bacteria are left after one week?
(c) When will there be no bacteria left? (i.e., when is $P(t) < 1$?)

28. The population of a certain species of bacteria is growing according to the logistic equation. Determine the **equilibrium population** P_e. It is given by

$$P_e = \lim_{t\to\infty} P(t).$$

29. Show that the growth rate of the population in Problem 28 is a maximum when the population is equal to half its equilibrium size.

30. Bacteria are supplied as food to a protozoan population at a constant rate μ. It is observed that the bacteria are consumed at a rate that is proportional to the square of their numbers. The concentration $c(t)$ of the bacteria therefore satisfies the differential equation $dc/dt = \mu - \lambda c^2$, where λ is a positive constant.
(a) Determine $c(t)$ in terms of $c(0)$.
(b) What is the equilibrium concentration of the bacteria?

31. In some chemical reactions certain products catalyze their own formation. If $x(t)$ is the amount of such a product at time t, a possible model for the reaction is given by the differential equation $dx/dt = \alpha(\beta - x)$, where α and β are positive constants. According to this model the reaction is completed when $x = \beta$, since this condition indicates that one of the chemicals has been depleted.
(a) Solve the equation in terms of the constants α, β, and $x(0)$.
(b) For $\alpha = 1$, $\beta = 200$, and $x(0) = 20$, draw a graph of $x(t)$ for $t > 0$.

***32.** On a certain day it began to snow early in the morning and the snow continued to fall at a constant rate. The velocity at which a snowplow is able to clear a road is inversely proportional to the height of the accumulated snow. The snowplow started at 11 A.M. and cleared four miles by 2 P.M. By 5 P.M. it had cleared another two miles. When did it start snowing?

****33.** A large open cistern filled with water has the shape of a hemisphere with radius 25 ft. The bowl has a circular hole of radius 1 ft in the bottom. By Torricelli's law,† water will flow out of the hole with the same speed it would attain in falling freely from the level of the water to the hole. How long will it take for all the water to flow from the cistern?

34. An object of mass m that falls from rest, starting at a point near the earth's surface, is subjected to two forces: a downward force mg and a resisting force proportional to the square of the velocity of the body. Thus

$$F = ma = m\frac{dv}{dt} = mg - \alpha v^2$$

where α is a constant of proportionality.
(a) Find $v(t)$ as a function of t. [*Hint:* Use the fact that $v(0) = 0$.]
(b) Show that the velocity does not increase indefinitely but approaches the equilibrium value $\sqrt{mg/\alpha}$. This value is called **terminal velocity.**

SKIP

11.3 EXACT EQUATIONS (OPTIONAL)

We shall now use partial derivatives to solve ordinary differential equations. The material in this section depends on material in Chapter 4. In particular, you should review Sections 4.9 and 4.11.

†Evangelista Torricelli (1608–1647) was an Italian physicist.

Suppose that we take the total differential of the equation $g(x, y) = C$:

$$dg = \frac{\partial g}{\partial x}\,dx + \frac{\partial g}{\partial y}\,dy = 0. \tag{1}$$

For example, the equation $xy = C$ has the total differential $y\,dx + x\,dy = 0$, which may be rewritten as the differential equation $y' = -y/x$.

Reversing the situation, suppose that we start with the differential equation

$$M(x, y)\,dx + N(x, y)\,dy = 0. \tag{2}$$

If we can find a function $g(x, y)$ such that

$$\frac{\partial g}{\partial x} = M \qquad \text{and} \qquad \frac{\partial g}{\partial y} = N,$$

then (2) becomes $dg = 0$, so that $g(x, y) = C$ is the general solution of (2). In this case $M\,dx + N\,dy$ is said to be an **exact differential,** and (2) is called an **exact differential equation.**

Recall Theorem 4.11.1:

Theorem 1 Let $\mathbf{F}(x, y) = P(x, y)\mathbf{i} + Q(x, y)\mathbf{j}$ and suppose that P, Q, $\frac{\partial P}{\partial y}$, and $\frac{\partial Q}{\partial x}$ are continuous in an open disk D centered at (x, y). Then, in D, $\mathbf{F}$ is the gradient of a function f if and only if

$$\frac{\partial P}{\partial y} = \frac{\partial Q}{\partial x}.$$

Now, suppose that

$$\frac{\partial M}{\partial y} = \frac{\partial N}{\partial x} \tag{3}$$

in an open disk centered at (x, y). Then, from Theorem 1, there is a differentiable function g such that

$$\nabla g = M(x, y)\mathbf{i} + N(x, y)\mathbf{j}.$$

That is,

$$\frac{\partial g}{\partial x} = M \qquad \text{and} \qquad \frac{\partial g}{\partial y} = N.$$

From the discussion above we see that, if (3) holds, then the differential equation (2) has the solution $g(x, y) = C$.

EXAMPLE 1 Solve $(1 - \sin x \tan y)\,dx + (\cos x \sec^2 y)\,dy = 0$.

Solution. Here

$$M(x, y) = 1 - \sin x \tan y \qquad \text{and} \qquad N(x, y) = \cos x \sec^2 y,$$

so that

$$\frac{\partial M}{\partial y} = -\sin x \sec^2 y = \frac{\partial N}{\partial x}$$

and the equation is exact. To find the solution, we use the technique of Section 4.11:

$$g(x, y) = \int M\,dx + h(y) = x + \cos x \tan y + h(y).$$

Taking the partial derivative with respect to y of both sides yields

$$\cos x \sec^2 y = N = \frac{\partial G}{\partial y} = \cos x \sec^2 y + h'(y).$$

Thus $h'(y) = 0$, so that $h(y)$ is constant and the general solution is

$$x + \cos x \tan y + C = 0. \quad \blacksquare$$

It should be apparent that exact equations are comparatively rare, since condition (3) requires a precise balance of the functions M and N. For example,

$$(3x + 2y)\,dx + x\,dy = 0$$

is not exact. However, if we multiply the equation by x, then the new equation

$$(3x^2 + 2xy)\,dx + x^2\,dy = 0$$

is exact. The question we now ask is: If

$$M(x, y)\,dx + N(x, y)\,dy = 0 \tag{4}$$

is not exact, under what conditions does an **integrating factor** $\mu(x, y)$ exist such that

$$\mu M\,dx + \mu N\,dy = 0$$

is exact? Surprisingly, the answer is whenever (4) has a general solution $g(x, y) = C$. To see this, we assume that such a solution exists and solve equation (4) for dy/dx:

$$\frac{dy}{dx} = -\frac{M}{N} = -\frac{\partial g/\partial x}{\partial g/\partial y},$$

from which it follows that

$$\frac{\partial g/\partial x}{M} = \frac{\partial g/\partial y}{N}.$$

Denote either side of the equation by $\mu(x, y)$. Then

$$\frac{\partial g}{\partial x} = \mu M, \qquad \frac{\partial g}{\partial y} = \mu N, \tag{5}$$

and equation (4) has at least one integrating factor μ. However, finding integrating factors is usually very difficult. There is one procedure that is sometimes successful. Since equation (5) indicates that $\mu M\,dx + \mu N\,dy = 0$ is exact, by (3) we have

$$\mu\frac{\partial M}{\partial y} + M\frac{\partial \mu}{\partial y} = \frac{\partial}{\partial y}(\mu M) = \frac{\partial}{\partial x}(\mu N) = \mu\frac{\partial N}{\partial x} + N\frac{\partial \mu}{\partial x},$$

so that

$$\frac{1}{\mu}\left\{N\frac{\partial \mu}{\partial x} - M\frac{\partial \mu}{\partial y}\right\} = \frac{\partial M}{\partial y} - \frac{\partial N}{\partial x}. \tag{6}$$

In case the integrating factor μ depends only on x, $d\mu/dy = 0$, and equation (6) becomes

$$\frac{1}{\mu}\frac{d\mu}{dx} = \frac{\partial M/\partial y - \partial N/\partial x}{N} = r(x, y).$$

Since the left-hand side of this equation consists only of functions of x, r must also be a function of x. If this is indeed true, then μ can be found by separating the variables: $\mu(x) = e^{[\int r(x)dx]}$. A similar result holds if μ is a function of y alone, in which case

$$\frac{\partial M/\partial y - \partial N/\partial x}{-M} = R(x, y)$$

is also a function of y only. In this case $\mu(y) = e^{[\int R(y)dy]}$ is an integrating factor.

EXAMPLE 2 Solve $(3x^2 - y^2)\,dy - 2xy\,dx = 0$.

Solution. In this problem $M = -2xy$ and $N = 3x^2 - y^2$, so that

$$\frac{\partial M}{\partial y} = -2x, \qquad \frac{\partial N}{\partial x} = 6x.$$

Then

$$R = \frac{\partial M/\partial y - \partial N/\partial x}{-M} = -\frac{4}{y}$$

and

$$\mu = e^{[\int -(4/y)dy]} = e^{(-4\ln y)} = y^{-4}.$$

Now we obtain

$$\left(\frac{3x^2 - y^2}{y^4}\right) dy - \frac{2x}{y^3}\,dx = 0,$$

which is exact. Thus the general solution is

$$g = \int M\,dx + h(y) = -\frac{x^2}{y^3} + h(y);$$

differentiating this equation with respect to y, we have

$$\frac{3x^2 - y^2}{y^4} = N = \frac{3x^2}{y^4} + h'(y).$$

Hence $h'(y) = -y^{-2}$, so $h(y) = y^{-1} + C$, yielding

$$g(x, y) = -\frac{x^2}{y^3} + \frac{1}{y} + C = 0 \quad \text{or} \quad Cy^3 + y^2 - x^2 = 0. \blacksquare$$

PROBLEMS 11.3

In Problems 1–9 verify that each differential equation is exact and find the general solution. Find a particular solution when an initial condition is given.

1. $2xy\,dx + (x^2 + 1)\,dy = 0$
2. $\left(4x^3y^3 + \frac{1}{x}\right) dx + \left(3x^4y^2 - \frac{1}{y}\right) dy = 0,\ x(e) = 1$
3. $\left[\frac{\ln(\ln y)}{x} + \frac{2}{3}xy^3\right] dx + \left(\frac{\ln x}{y \ln y} + x^2y^2\right) = 0$
4. $(x - y\cos x)\,dx - \sin x\,dy = 0,\ y(\pi/2) = 1$
5. $\cosh 2x \cosh 2y\,dx + \sinh 2x \sinh 2y\,dy = 0$
6. $(ye^{xy} + 4y^3)\,dx + (xe^{xy} + 12xy^2 - 2y)\,dy = 0,$ $y(0) = 2$
7. $(3x^2 \ln x + x^2 - y)\,dx - x\,dy = 0,\ y(1) = 5$
8. $(x^2 + y^2)\,dx + 2xy\,dy = 0,\ y(1) = 1$
9. $\left(\frac{1}{x} - \frac{y}{x^2 + y^2}\right) dx + \left(\frac{x}{x^2 + y^2} - \frac{1}{y}\right) dy = 0$

In Problems 10–13 find an integrating factor for each differential equation and obtain the general solution.

10. $y\,dx + (y - x)\,dy = 0$
11. $2y^2\,dx + (2x + 3xy)\,dy = 0$
12. $(x^2 + 2y)\,dx - x\,dy = 0$
13. $(x^2 + y^2)\,dx + (3xy)\,dy = 0$

14. Solve $xy\,dx + (x^2 + 2y^2 + 2)\,dy = 0$
15. Let $M = yf(xy)$ and $N = xh(xy)$, where f and h are differentiable functions of one variable. Show that $1/(xM - yN)$ is an integrating factor for $M\,dx + N\,dy = 0$.
16. Use the result of Problem 15 to solve the equation

$$2x^2y^3\,dx + x^3y^2\,dy = 0.$$

*17. Solve $(x^2 + y^2 + 1)\,dx - (xy + y)\,dy = 0$. [*Hint:* Try an integrating factor of the form $\mu(x) = (x + 1)^r$.]

skip

11.4 FIRST-ORDER LINEAR EQUATIONS

In Section 11.1 we stated that an nth-order **linear differential equation** is an equation of the form

$$\frac{d^n y}{dx^n} + a_{n-1}(x)\frac{d^{n-1}y}{dx^{n-1}} + \cdots + a_1(x)\frac{dy}{dx} + a_0(x)y = f(x). \tag{1}$$

One nice fact about linear equations is given by the following theorem:†

Theorem 1 Let $x_0, y_0, y_1, \ldots, y_{n-1}$ be real numbers and let $a_0, a_1, \ldots, a_{n-1}$ be continuous. Then there is a unique solution $y = g(x)$ to the nth-order linear differential equation (1) which satisfies

$$g(x_0) = y_0, \qquad g'(x_0) = y_1, \ldots, g^{(n-1)}(x_0) = y_{n-1}. \tag{2}$$

That is, if we specify n initial conditions, there exists a unique solution. Of course, knowing that an equation has a solution is different from finding that solution. In this section we will study the linear first-order equation

$$\frac{dy}{dx} + a(x)y = f(x). \tag{3}$$

We say that the linear equation (1) is **homogeneous** if $f(x) = 0$ for every number x in the domain of f. Otherwise we say that the equation is **nonhomogeneous.** If the functions $a_0(x), a_1(x), \ldots, a_{n-1}(x)$ are constant, then the equation is said to have **constant coefficients.** Otherwise it is said to have **variable coefficients.** It turns out that we can solve by integration *all* first-order linear equations and all nth-order linear homogeneous and nonhomogeneous equations with constant coefficients if the functions a_i and f are integrable. In this section we show how solutions to linear equations in the form (3) can always be explicitly calculated. We do this in three steps.

Case 1. **(Constant coefficients, homogeneous)** Then (3) can be written

$$\frac{dy}{dx} + ay = 0 \tag{4}$$

where a is a constant, or

$$\frac{dy}{dx} = -ay. \tag{5}$$

†For a proof, see Derrick and Grossman, *Elementary Differential Equations with Applications*, Second Edition (Addison-Wesley, Reading, MA, 1981), Chapter 10.

The variables separate, and (see Example 11.2.2) solutions are given by

$$y = Ce^{-ax} \tag{6}$$

for any constant C. We saw several examples of this type of equation in Section 11.2.

Case 2. **(Constant coefficients, nonhomogeneous)** Then (3) can be written

$$\frac{dy}{dx} + ay = f(x) \tag{7}$$

where a is a constant and f is an integrable function. It is now impossible to separate the variables. However, equation (7) can be solved by multiplying both sides of (7) by an **integrating factor.** We first note that

$$\frac{d}{dx}(e^{ax}y) = e^{ax}\frac{dy}{dx} + ae^{ax}y = e^{ax}\left(\frac{dy}{dx} + ay\right). \tag{8}$$

Thus, if we multiply both sides of (7) by e^{ax}, we obtain

$$e^{ax}\left(\frac{dy}{dx} + ay\right) = e^{ax}f(x)$$

or, using (8),

$$\frac{d}{dx}(e^{ax}y) = e^{ax}f(x)$$

and, upon integration,

$$e^{ax}y = \int e^{ax}f(x)\,dx + C$$

where $\int e^{ax}f(x)\,dx$ denotes one particular antiderivative. This leads to the general solution

$$y = e^{-ax}\int e^{ax}f(x)\,dx + Ce^{-ax}. \tag{9}$$

The term e^{ax} is called an integrating factor for (7) because it allows us, after multiplication, to solve the equation by integration.

EXAMPLE 1 Find all solutions to

$$\frac{dy}{dx} + 3y = x. \tag{10}$$

Solution. We multiply both sides of the equation by e^{3x}. Then

$$e^{3x}\left(\frac{dy}{dx} + 3y\right) = xe^{3x} \qquad \text{or} \qquad \frac{d}{dx}(e^{3x}y) = xe^{3x}$$

and

$$e^{3x}y = \int xe^{3x}\,dx.$$

But setting $u = x$ and $dv = e^{3x}\,dx$, we find that $du = dx$, $v = \frac{1}{3}e^{3x}$, and

$$\int xe^{3x}\,dx = \frac{x}{3}e^{3x} - \frac{1}{3}\int e^{3x}\,dx = \frac{x}{3}e^{3x} - \frac{1}{9}e^{3x} + C,$$

so that

$$e^{3x}y = \frac{x}{3}e^{3x} - \frac{1}{9}e^{3x} + C$$

and

$$y = \frac{x}{3} - \frac{1}{9} + Ce^{-3x}. \tag{11}$$

This answer should be checked by differentiation. ■

Let us take a closer look at the answer we just obtained. The answer (11) is given in two parts. The function Ce^{-3x} (for each real number C) is a solution to the homogeneous equation

$$\frac{dy}{dx} + 3y = 0$$

while $y = (x/3) - \frac{1}{9}$ is one particular solution to the nonhomogeneous problem (10). You should verify this. That this is no accident is suggested by the following theorem.

Theorem 2

(i) Let y_1 and y_2 be two (nonzero) solutions to the homogeneous equation

$$\frac{dy}{dx} + a(x)y = 0. \tag{12}$$

Then for some constant α,

$$y_1(x) = \alpha y_2(x) \qquad \text{for every real number } x. \tag{13}$$

(ii) Let y_p and y_q be two solutions to the nonhomogeneous equation (3). Then

$$y = y_p - y_q$$

is a solution to the homogeneous equation (12).

REMARK 1. Part (i) of the theorem states that if you know one nonzero solution to the homogeneous equation (12), then you know them all.

REMARK 2. Part (ii) of the theorem is important because it states, in effect, that if you know *one* solution to the nonhomogeneous equation (3) and all the solutions to the homogeneous equation (12), then you also know all solutions to the nonhomogeneous equation. This follows from the fact that if y_p is the known solution to (3) and if y_q is any other solution to (3), then since $y_q - y_p$ is a solution y_h to (12), we can write

$$y_q - y_p = y_h \qquad \text{or} \qquad y_q = y_p + y_h.$$

That is, any other solution to (3) can be written as the sum of the one known solution to (3) and some solution to (12). Thus, for example, in (11), we have $y_p = (x/3) - \frac{1}{9}$, and all solutions to $(dy/dx) + 3y = 0$ are given by $y_h = Ce^{-3x}$. Since, as we will see, nonhomogeneous equations are harder to solve than homogeneous ones, it is nice to know that we need find only *one* solution to a given linear nonhomogeneous equation.

Proof of Theorem 2.

(i) Assume that $y_2(x) \neq 0$ for every x. Then

$$\left(\frac{y_1}{y_2}\right)' = \frac{y_2 y_1' - y_1 y_2'}{y_2^2} = \frac{y_2(-ay_1) - y_1(-ay_2)}{y_2^2} = 0,$$

which implies that y_1/y_2 is a constant, so that $y_1/y_2 = \alpha$ or $y_1(x) = \alpha y_2(x)$.

Note that we may assume that one of the solutions y_1 or y_2 is never zero. For if both are identically zero, then $y_1(x) = \alpha y_2(x)$ for every number α. On the other hand, if y_2, say, is not identically zero, then it can never be equal to zero. For if $y_2(x_0) = 0$ for some number x_0, then by the uniqueness theorem (Theorem 1), y_2 is the zero function since the zero function satisfies (12) and the initial condition $y_2(x_0) = 0$.

(ii) Let $y = y_p - y_q$. Then

$$\begin{aligned}\frac{dy}{dx} + a(x)y &= \frac{d}{dx}(y_p - y_q) + a(x)(y_p - y_q)\\ &= \frac{dy_p}{dx} - \frac{dy_q}{dx} + a(x)y_p - a(x)y_q\\ &= \left[\frac{dy_p}{dx} + a(x)y_p\right] - \left[\frac{dy_q}{dx} + a(x)y_q\right]\\ &= f(x) - f(x) = 0\end{aligned}$$

since y_p and y_q both are solutions of equation (3). ■

EXAMPLE 2 Find the solution to $(dy/dx) + 2y = e^{-5x}$ that satisfies $y(0) = 4$.

Solution. We multiply both sides of the equation by e^{2x} to obtain

$$e^{2x}\left(\frac{dy}{dx} + 2y\right) = e^{-3x} \qquad \text{or} \qquad \frac{d}{dx}(e^{2x}y) = e^{-3x}$$

and

$$e^{2x}y = \int e^{-3x}\,dx = -\frac{1}{3}e^{-3x} + C.$$

Then

$$y = -\tfrac{1}{3}e^{-5x} + Ce^{-2x}.$$

Note that $-\frac{1}{3}e^{-5x}$ is one solution to $(dy/dx) + 2y = e^{-5x}$ and that Ce^{-2x} represents all solutions to $(dy/dx) + 2y = 0$. Finally, we have

$$4 = y(0) = -\tfrac{1}{3} + C \qquad \text{or} \qquad C = \tfrac{13}{3}$$

and the solution to the initial value problem is

$$y = -\tfrac{1}{3}e^{-5x} + \tfrac{13}{3}e^{-2x}. \quad ■$$

We now turn to the final case.

Case 3. **(Variable coefficients, nonhomogeneous)** We note the following facts, the first of which follows from the fundamental theorem of calculus (assuming that a is continuous):

(i) $\dfrac{d}{dx}\displaystyle\int a(x)\,dx = a(x)$. [This holds for *any* antiderivative $\int a(x)\,dx$.] **(14)**

(ii) $\dfrac{d}{dx}e^{\int a(x)dx} = e^{\int a(x)dx}\dfrac{d}{dx}\displaystyle\int a(x)\,dx = a(x)e^{\int a(x)dx}$. **(15)**

Now consider the equation (3):

$$\frac{dy}{dx} + a(x)y = f(x).$$

We multiply both sides by the integrating factor $e^{\int a(x)dx}$. Then we have

$$e^{\int a(x)dx}\frac{dy}{dx} + a(x)e^{\int a(x)dx}y = e^{\int a(x)dx}f(x). \tag{16}$$

Using (15), we obtain

$$\begin{aligned}\frac{d}{dx} ye^{\int a(x)\,dx} &= y\frac{d}{dx}e^{\int a(x)\,dx} + \frac{dy}{dx}e^{\int a(x)\,dx}\\ &= a(x)e^{\int a(x)\,dx}y + \frac{dy}{dx}e^{\int a(x)\,dx}\\ &= \text{the left-hand side of (16).}\end{aligned}$$

Thus, from (16),

$$\frac{d}{dx}[e^{\int a(x)\,dx}y] = e^{\int a(x)\,dx}f(x)$$

or, integrating both sides, we have

$$e^{\int a(x)\,dx}y = \int e^{\int a(x)\,dx}f(x)\,dx + C$$

and

$$y = e^{-\int a(x)\,dx}\int e^{\int a(x)\,dx}f(x)\,dx + Ce^{-\int a(x)\,dx}. \quad \textbf{(17)}$$

It is probably a waste of time to try to memorize the complicated-looking formula (17). Rather it is important to remember that multiplication by the integrating factor $e^{\int a(x)\,dx}$ will always enable us to reduce the problem of solving a differential equation to the problem of calculating an integral.

EXAMPLE 3 Find all solutions to the equation

$$\frac{dy}{dx} + \frac{4}{x}y = 3x^2.$$

Solution. Here $a(x) = 4/x$, $\int a(x)\,dx = 4\ln|x| = \ln x^4$ and $e^{\int a(x)\,dx} = e^{\ln x^4} = x^4$ (since $e^{\ln u} = u$ for all $u > 0$). Thus we can multiply both sides of the equation by the integrating factor x^4 to obtain

$$x^4\left(\frac{dy}{dx} + \frac{4}{x}y\right) = x^4\frac{dy}{dx} + 4x^3y = 3x^6 \quad \text{or} \quad \frac{d}{dx}(yx^4) = 3x^6.$$

Then we integrate to find that

$$yx^4 = \frac{3x^7}{7} + C \quad \text{or} \quad y = \frac{3x^3}{7} + \frac{C}{x^4}.$$

Note that $3x^3/7$ is one solution to the nonhomogeneous equation $(dy/dx) + (4/x)y = 3x^2$ while C/x^4 represents all solutions to the homogeneous equation $(dy/dx) + (4/x)y = 0$. This should be checked by differentiation. ■

EXAMPLE 4 Find the solution to the initial value problem $dy/dx = x^2 - 3x^2y$ that satisfies $y(1) = 2$.

Solution. We write the equation in the form

$$\frac{dy}{dx} + 3x^2y = x^2,$$

so that $a(x) = 3x^2$, $\int a(x)\,dx = x^3$, and $e^{\int a(x)\,dx} = e^{x^3}$. Then we have, after multiplication by e^{x^3},

$$\frac{d}{dx}(ye^{x^3}) = x^2e^{x^3}$$

and

$$ye^{x^3} = \int x^2e^{x^3}\,dx = \frac{1}{3}e^{x^3} + C,$$

so that

$$y = \tfrac{1}{3} + Ce^{-x^3}.$$

Then

$$2 = y(1) = \frac{1}{3} + Ce^{-1} = \frac{1}{3} + \frac{C}{e}$$

or

$$\frac{C}{e} = \frac{5}{3} \quad \text{and} \quad C = \frac{5e}{3}.$$

Thus the solution to the initial value problem is given by

$$y = \frac{1}{3} + \left(\frac{5}{3}e\right)(e^{-x^3}) = \frac{1}{3} + \frac{5}{3}e^{1-x^3}. \ ■$$

EXAMPLE 5 **(A Mixing Problem)** Consider a tank holding 100 gallons of water in which is dissolved 50 pounds of salt. Suppose that 2 gallons of brine, each containing 1 pound of dissolved salt, run into the tank per minute, and the mixture, kept uniform by high-speed stirring, runs out of the tank at the rate of 2 gallons per minute. Find the amount of salt in the tank at any time t.

Solution. Let $x(t)$ be the number of pounds of salt at the end of t minutes. Since each gallon of brine that enters the tank contains 1 pound of salt, we know that 2 pounds of salt are entering the tank each minute. However, as 2 gallons of solution out of 100 gallons are leaving the tank, $(2/100)x = 0.02x$ pounds of salt leave the tank each minute. These data lead to the differential equation

$$\frac{dx}{dt} = 2 - 0.02x$$

or

$$\frac{dx}{dt} + 0.02x = 2.$$

Multiplying both sides of this equation by the integrating factor $e^{0.02t}$ yields

$$\frac{d}{dt}(xe^{0.02t}) = 2e^{0.02t}.$$

We then integrate:

$$xe^{0.02t} = \frac{2}{0.02}e^{0.02t} + C = 100e^{0.02t} + C$$

and

$$x(t) = 100 + Ce^{-0.02t}.$$

But we are told that $x(0) = 50$. Then

$$x(0) = 100 + C = 50,$$

so that $C = -50$, and the unique solution to our equation is

$$x(t) = 100 - 50e^{-0.02t}.$$

Observe that x increases and approaches the ratio of salt to water in the input stream as time increases. That is, there is one pound of salt in each gallon of brine and this leads, eventually, to 100 pounds of salt in 100 gallons of brine. ■

EXAMPLE 6 Suppose that, in Example 5, 3 gallons of brine, each containing 1 pound of salt, run into the tank each minute, and all other facts are the same.

(a) Find the amount of salt in the tank at any time t.
(b) How much salt is in the tank after 1 hour?

Solution.

(a) Now 3 pounds of salt enter the tank each minute, but because 3 gallons of brine enter while only two leave, the amount of brine in the tank increases at a rate of 1 gal/min. Thus, after t minutes there are $100 + t$ gallons in the tank and two of these are leaving. This means that the amount of salt leaving the tank each minute is $\frac{2}{100 + t}x$ gallons, and we obtain the differential equation

$$\frac{dx}{dt} = 3 - \frac{2x}{100 + t}$$

or

$$\frac{dx}{dt} + \frac{2}{100 + t}x = 3 \tag{18}$$

Here

$$a(t) = \frac{2}{100 + t},$$

$$\int a(t)\,dt = 2\ln(100 + t) = \ln(100 + t)^2$$

and

$$e^{\int a(t)dt} = e^{\ln(100 + t)^2} = (100 + t)^2.$$

Multiplying both sides of (18) by $(100 + t)^2$ and integrating gives, successively,

$$\frac{d}{dt}[x(100 + t)^2] = 3(100 + t)^2$$

$$x(100 + t)^2 = (100 + t)^3 + C$$

$$x(t) = (100 + t) + \frac{C}{(100 + t)^2}.$$

As in Example 5, $x(0) = 50$, so

$$x(0) = 50 = 100 + \frac{C}{(100)^2}$$

$$\frac{C}{100^2} = -50$$

$$C = -50(100)^2.$$

Therefore,

Divide top and bottom by 100.

$$x(t) = 100 + t - 50\left(\frac{100}{100 + t}\right)^2 = 100 + t - 50\left(\frac{1}{1 + \frac{t}{100}}\right)^2.$$

Note that as $t \to \infty$,

$$\frac{1}{1 + (t/100)} \to 0.$$

The amount of salt in the tank is, thus for t large, approximately equal to $100 + t$ pounds.

(b) After 1 hour = (60 minutes),

$$x(60) = 100 + 60 - 50\left(\frac{100}{160}\right)^2 = 160 - 50\left(\frac{5}{8}\right)^2$$

$$= 160 - 50\left(\frac{25}{64}\right) \approx 140.5 \text{ pounds of salt.} \blacksquare$$

BERNOULLI'S EQUATION†

Certain nonlinear first-order equations can be reduced to linear equations by a suitable change of variables. The equation

$$\frac{dy}{dx} + a(x)y = f(x)y^n, \tag{19}$$

which is known as **Bernoulli's equation,** is of this type. Set $z = y^{1-n}$. Then $z' = (1 - n)y^{-n}y'$, so if we multiply both sides by $(1 - n)y^{-n}$, we obtain

$$(1 - n)y^{-n}y' + (1 - n)a(x)y^{1-n} = (1 - n)f(x)$$

or

$$\frac{dz}{dx} + (1 - n)a(x)z = (1 - n)f(x).$$

The equation is now linear and can be solved as before.

EXAMPLE 7 Solve

$$\frac{dy}{dx} - \frac{y}{x} = -\frac{5}{2}x^2y^3. \tag{20}$$

Solution. Here $n = 3$, so we let $z = y^{-2}$ and $z' = -2y^{-3}y'$. Multiplying both sides of equation (20) by $-2y^{-3}$ gives

$$-2y^{-3}y' + \frac{2}{x}y^{-2} = 5x^2$$

†See the accompanying biographical sketch.

1654–
1705

Jakob Bernoulli

Jakob Bernoulli
The Granger Collection

One of the most distinguished families in the history of mathematics and science is the Bernoulli family of Switzerland, which, from the late seventeenth century on, produced an unusual number of capable mathematicians and scientists. The family record begins with the two brothers Jakob and Johann Bernoulli. These two men both gave up earlier vocational interests to become mathematicians when Leibniz's papers began to appear in the *Acta eruditorum.* They were among the first mathematicians to recognize the surprising power of the calculus and to apply the tool to a great diversity of problems. From 1687 until his death, Jakob occupied the mathematics chair at Basel University. Often bitter rivals, the brothers maintained an almost constant exchange of ideas with Leibniz and with each other.

Among Jakob Bernoulli's contributions to mathematics are the early use of polar coordinates; the derivation in both rectangular and polar coordinates of the formula for the radius of curvature of a plane curve; the study of the catenary curve with extensions to strings of variable density and strings under the action of a central force; the study of a number of other higher plane curves; the discovery of the so-called **isochrone,** or curve along which a body falls with uniform vertical velocity; the determination of the form taken by an elastic rod fixed at one end and carrying a weight at the other; the form assumed by a flexible rectangular sheet having two opposite edges held horizontally fixed at the same height and loaded with a heavy liquid; and the shape of a rectangular sail filled with wind. He also proposed and discussed the problem of isoperimetric figures (planar closed paths of given types and fixed perimeter bounding a maximum area); he was thus one of the first mathematicians to work in the calculus of variations. He was also one of the early students of mathematical probability; his book in this field, the *Ars conjectandi,* was posthumously published in 1713. Several mathematics ideas now bear Jakob Bernoulli's name. Among these are the *Bernoulli distribution* and *Bernoulli theorem* of statistics and probability theory, the *Bernoulli equation* met by every student of a first course in differential equations, the *Bernoulli numbers* and *Bernoulli polynomials* of number-theory interest, and the *lemniscate of Bernoulli* encountered in a first course in the calculus. In Jakob Bernoulli's solution to the problem of the isochrone curve, which was published in the *Acta eruditorum* in 1690, we meet for the first time the word "integral" in a calculus sense. Leibniz had called the integral calculus *calculus summatorius;* in 1696 Leibniz and Johann Bernoulli agreed to call it *calculus integralis.*

Jakob Bernoulli was struck by the way the equiangular spiral reproduces itself under a variety of transformations and asked, in imitation of Archimedes, that such a spiral be engraved on his tombstone, along with the inscription *Eadem mutata resurgo* ("I shall arise the same, though changed").

or

$$z' + \frac{2z}{x} = 5x^2. \tag{21}$$

The integrating factor for this linear equation is

$$e^{2\int dx/x} = 3^{2\ln x} = e^{\ln x^2} = x^2.$$

Multiplying both sides of equation (21) by x^2 we have

$$x^2z' + 2xz = 5x^4$$
$$(x^2z)' = 5x^4$$

so that

$$x^2z = 5\int x^4\,dx + C = x^5 + C.$$

Hence,

$$y^{-2} = z = x^3 + Cx^{-2}$$

or

$$y = (x^3 + Cx^{-2})^{-1/2}. \blacksquare$$

A similar procedure can be used to solve

$$\frac{dy}{dx} + a(x)y = f(x)y \ln y. \tag{22}$$

We let $z = \ln y$. Then $z' = y'/y$, so that dividing (22) by y, we obtain the linear equation

$$\frac{dz}{dx} + a(x) = f(x)z.$$

PROBLEMS 11.4

In Problems 1–14 find all solutions to the given equation. If an initial condition is given, find the particular solution that satisfies that condition.

1. $\dfrac{dy}{dx} = 4x$

2. $\dfrac{dy}{dx} + xy = 0,\ y(0) = 2$

3. $\dfrac{dx}{dt} + x = \sin t,\ x(0) = 1$

4. $\dfrac{dx}{dt} + x = \dfrac{1}{1 + e^{2t}}$

*5. $\dfrac{dy}{dx} - y \ln x = x^x$

6. $\dfrac{dy}{dx} + (\tan x)y = 2x \sec x,\ y\left(\dfrac{\pi}{4}\right) = 1$

7. $\dfrac{dx}{dt} - ax = be^{at}$, a, b constant

8. $\dfrac{dx}{dt} - ax = be^{ct},\ x(0) = d;\ a, b, c, d$ constants

9. $\dfrac{dy}{dx} = x + 2y \tan 2x$

10. $\dfrac{dy}{dx} = 2y + x^2e^{2x}$

11. $(x^2 + 1)\dfrac{dy}{dx} + 2xy = x,\ y(0) = 1$

12. $\dfrac{dy}{dx} + \dfrac{y}{x^2} = \dfrac{3}{x^2},\ y(1) = 2$

13. $\dfrac{dx}{dt} + \dfrac{4t}{t^2+1}x = 3t,\ x(0) = 4$

14. $\dfrac{ds}{dt} + s \tan t = \cos t,\ s\left(\dfrac{\pi}{3}\right) = \dfrac{1}{2}$

***15.** Solve the equation $y - x\,dy/dx = (dy/dx)\,y^2e^y$ by reversing the roles of x and y (i.e., by writing x as a function of y).

16. A tank initially contains 100 liters of fresh water. Brine containing 20 grams per liter of salt flows into the tank at the rate of 4 liters per minute, and the mixture, kept uniform by stirring, runs out at the same rate. How long will it take for the quantity of salt in the tank to become 1 kilogram?

17. Given the same data as in Problem 16 determine how long it will take for the quantity of salt in the tank to increase from 1 kilogram to 1.5 kilograms.

18. A tank contains 100 gallons of fresh water. Brine containing 2 pounds per gallon of salt runs into the tank at the rate of 4 gallons per minute, and the mixture, kept uniform by stirring, runs out at the rate of 2 gallons per minute. Find:

(a) the amount of salt present when the tank has 120 gallons of brine;

(b) the concentration of salt in the tank at the end of 20 minutes.

19. A tank contains 50 liters of water. Brine containing x grams per liter of salt enters the tank at the rate of 1.5 liters per minute. The mixture, thoroughly stirred, leaves the tank at the rate of 1 liter per minute. If the concentration is to be 20 grams per liter at the end of 20 minutes, what is the value of x?

20. A tank holds 500 gallons of brine. Brine containing 2 pounds of salt per gallon flows into the tank at the rate of 5 gallons per minute, and the mixture, kept uniform, flows out at the rate of 10 gallons per minute. If the maximum amount of salt is found in the tank at the end of 20 minutes, what was the initial salt content of the tank?

In Problems 21–25 find the general solution to each equation and the unique solution when an initial condition is given.

21. $y' = -y^3xe^{-2x} + y$

22. $x\dfrac{dy}{dx} + y = x^4y^3;\ y(1) = 1$

23. $tx^2\dfrac{dx}{dt} + x^3 = t \cos t$

24. $\dfrac{dy}{dx} + \dfrac{3}{x}y = x^2y^2;\ y(1) = 2$

25. $xyy' - y^2 + x^2 = 0$

11.5 SIMPLE ELECTRIC CIRCUITS (OPTIONAL)

In this section we shall consider simple electric circuits containing a resistor and an inductor or capacitor in series with a source of electromotive force (emf). Such circuits are shown in Figure 1, and their action can be understood easily without any special knowledge of electricity.

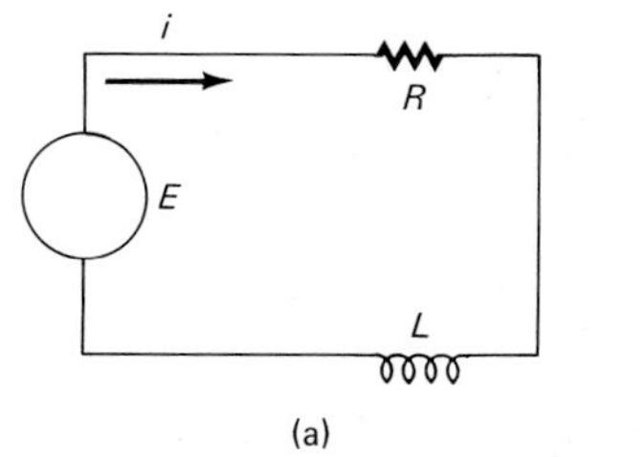

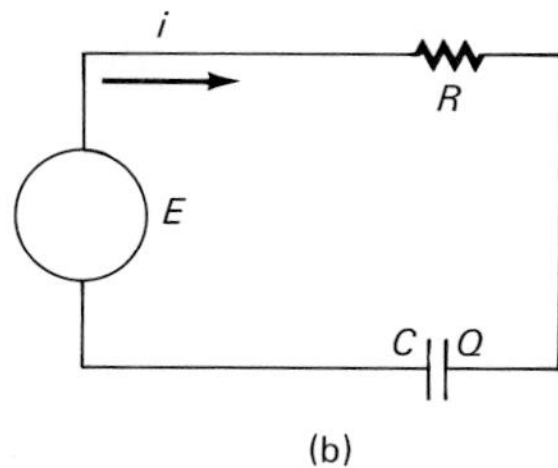

FIGURE 1

Definition 1 An **electromotive force (emf)** E (volts), usually a battery or generator, drives an electric charge Q (coulombs) and produces a current I (amperes). **Current** is defined as the rate of flow of the charge, and we can write

$$I = \frac{dQ}{dt}. \tag{1}$$

Definition 2 A **resistor** of resistance R (ohms) is a component of the circuit that opposes the current, dissipating the energy in the form of heat. It produces a drop in voltage given by **Ohm's law:**

$$E_R = IR. \tag{2}$$

Definition 3 An **inductor** of inductance L (henries) opposes any change in current by producing a voltage drop of

$$E_L = L\frac{dI}{dt}. \tag{3}$$

Definition 4 A **capacitor** of capacitance C (farads) stores charge. In so doing, it resists the flow of further charge, causing a drop in the voltage of

$$E_C = \frac{Q}{C}. \tag{4}$$

The quantities R, L, and C are usually constants associated with the particular component in the circuit; E may be a constant or a function of time. The fundamental principle guiding such circuits is given by **Kirchhoff's voltage law:**

The algebraic sum of all voltage drops around a closed circuit is zero.

In the circuit of Figure 1a the resistor and the inductor cause voltage drops of E_R and E_L, respectively. The emf, however, *provides* a voltage of E (i.e., a voltage drop of $-E$). Thus Kirchhoff's voltage law yields

$$E_R + E_L - E = 0.$$

Transposing E to the other side of the equation and using equations (2) and (3) to replace E_R and E_L, we have

$$L\frac{dI}{dt} + IR = E. \tag{5}$$

The following two examples illustrate the use of equation (5) in analyzing the circuit shown in Figure 1a.

EXAMPLE 1 An inductance of 2 henries (h) and a resistance of 10 ohms (Ω) are connected in series with an emf of 100 volts (V). If the current is zero when $t = 0$, what is the current at the end of 0.1 second (sec)?

Solution. Since $L = 2, R = 10$, and $E = 100$, equation (5) and the initial current yield the initial value problem

$$2\frac{dI}{dt} + 10I = 100, \qquad I(0) = 0. \tag{6}$$

Dividing both sides of equation (6) by 2, we note that the resulting linear first-order equation has e^{5t} as an integrating factor; that is,

$$\frac{d}{dt}(e^{5t}I) = e^{5t}\left(\frac{dI}{dt} + 5I\right) = 50^{5t}. \tag{7}$$

Integrating both ends of equation (7), we get

$$e^{5t}I(t) = 10e^{5t} + c,$$

or

$$I(t) = 10 + ce^{-5t}. \tag{8}$$

Setting $t = 0$ in equation (8) and using the initial condition $I(0) = 0$, we have

$$0 = I(0) = 10 + c,$$

which implies that $c = -10$. Substituting this value into equation (8), we obtain an equation for the current at any time t:

$$I(t) = 10(1 - e^{-5t}).$$

Thus, when $t = 0.1$, we have

$$I(0.1) = 10(1 - e^{-0.5}) \approx 3.93 \text{ amp.} \blacksquare$$

EXAMPLE 2 Suppose that the emf $E = 100 \sin 60t$ volts but all other values remain the same as those given in Example 1. Then equation (5) yields

$$2\frac{dI}{dt} + 10I = 100 \sin 60t, \qquad I(0) = 0. \tag{9}$$

Again dividing by 2 and multiplying both sides by the integrating factor e^{5t}, we have

$$\frac{d}{dt}(e^{5t}I) = e^{5t}\left(\frac{dI}{dt} + 5I\right) = 50e^{5t} \sin 60t. \tag{10}$$

Integrating both ends of equation (10) and using formula 168 of the integral table, we obtain

$$I(t) = e^{-5t}\left[50 \int (\sin 60t)e^{5t}\, dt + c\right]$$
$$= e^{-5t}\left[50e^{5t}\left(\frac{5 \sin 60t - 60 \cos 60t}{3625}\right) + c\right]$$
$$= \frac{2 \sin 60t - 24 \cos 60t}{29} + ce^{-5t}.$$

Thus setting $t = 0$, we find that $c = 24/29$ and

$$I(0.1) = \frac{2 \sin 6 - 24 \cos 6}{29} + \frac{24}{29}e^{-0.5} \approx -0.31 \text{ amp.} \blacksquare$$

For the circuit in Figure 1b we have $E_R + E_C - E = 0$, or

$$IR + \frac{Q}{C} = E.$$

Using the fact that $I = dQ/dt$, we obtain the linear first-order equation

$$R\frac{dQ}{dt} + \frac{Q}{C} = E. \tag{11}$$

The next example illustrates how to use equation (11).

EXAMPLE 3 If a resistance of 2000 ohms and a capacitance of 5×10^{-6} farad (f) are connected in series with an emf of 100 volts, what is the current at $t = 0.1$ second if $I(0) = 0.01$ ampere?

Solution. Setting $R = 2000$, $C = 5 \times 10^{-6}$, and $E = 100$ in equation (11), we have

$$2000\left(\frac{dQ}{dt} + 100Q\right) = 100 \qquad \frac{1}{5 \times 10^{-6}} = \frac{10^6}{5} = 200{,}000$$

or

$$\frac{dQ}{dt} + 100Q = \frac{1}{20}, \tag{12}$$

from which we can determine $Q(0)$, since

$$\tfrac{1}{20} = Q'(0) + 100Q(0) = I(0) + 100Q(0).$$

Thus,

$$Q(0) = \tfrac{1}{100}\left[\tfrac{1}{20} - I(0)\right] = 4 \times 10^{-4} \text{ coulombs.} \tag{13}$$

Multiplying both sides of equation (12) by the integrating factor e^{100t}, we get

$$\frac{d}{dt}(e^{100t}Q) = \frac{e^{100t}}{20}.$$

Integrating this equation yields

$$e^{100t}Q = \frac{e^{100t}}{2000} + c.$$

Dividing both sides by e^{100t} gives us

$$Q(t) = \tfrac{1}{2000} + ce^{-100t}.$$

Setting $t = 0$, we find that $c = -10^{-4}$. Thus, the charge at any time t is

$$Q(t) = (5 - e^{-100t})/10^4,$$

and the current is

$$I(t) = Q'(t) = \tfrac{1}{100}e^{-100t}.$$

Thus $I(0.1) = 10^{-2}e^{-10} \approx 4.54 \times 10^{-7}$ amp. ■

PROBLEMS 11.5

In Problems 1–5, assume that the *RL* circuit shown in Figure 1a has the given resistance, inductance, emf, and initial current. Find an expression for the current at all times t and calculate the current after one second.

1. $R = 10\ \Omega$, $L = 1$ h, $E = 12$ V, $I(0) = 0$ amp
2. $R = 8\ \Omega$, $L = 1$ h, $E = 6$ V, $I(0) = 1$ amp
3. $R = 50\ \Omega$, $L = 2$ h, $E = 100$ V, $I(0) = 0$ amp
4. $R = 10\ \Omega$, $L = 5$ h, $E = 10 \sin t$ V, $I(0) = 1$ amp
5. $R = 10\ \Omega$, $L = 10$ h, $E = e^t$ V, $I(0) = 0$ amp

In Problems 6–10, use the given resistance, capacitance, emf, and initial charge in the *RC* circuit shown in Figure 1b. Find an expression for the charge at all times t.

6. $R = 1\ \Omega$, $C = 1$ f, $E = 12$ V, $Q(0) = 0$ coulomb
7. $R = 10\ \Omega$, $C = 0.001$ f, $E = 10 \cos 60t$ V, $Q(0) = 0$ coulomb
8. $R = 1\ \Omega$, $C = 0.01$ f, $E = \sin 60t$ V, $Q(0) = 0$ coulomb
9. $R = 100\ \Omega$, $C = 10^{-4}$ f, $E = 100$ V, $Q(0) = 1$ coulomb
10. $R = 200\ \Omega$, $C = 5 \times 10^{-5}$ f, $E = 1000$ V, $Q(0) = 1$ coulomb

11. Solve the problem in Example 3 with an emf of $100 \sin 120\pi t$ V.
12. An inductance of 1 h and a resistance of 2 Ω are connected in series with a battery of $6e^{-0.0001t}$ V. No current is flowing initially. When will the current measure 0.5 amp?
13. A variable resistance $R = 1/(5 + t)\ \Omega$ and a capacitance of 5×10^{-6} f are connected in series with an emf of 100 V. If $Q(0) = 0$, what is the charge on the capacitor after one minute?
14. In the *RC* circuit (Figure 1b) with constant voltage E, how long will it take the current to decrease to one-half its original value?
15. Suppose that the voltage in an *RC* circuit is $E(t) = E_0 \cos \omega t$, where $2\pi/\omega$ is the period of the cycle. Assuming that the initial charge is zero, what are the charge and the current as functions of R, C, ω, and t?

16. Show that the current in Problem 15 consists of two parts: a steady-state term that has a period of $2\pi/\omega$ and a transient term that tends to zero as t increases.

17. In Problem 16 show that if R is small, then the transient term can be quite large for small values of t. (This is why fuses can blow when a switch is flipped.)

18. Find the steady-state current, given that a resistance of 2000 Ω and a capacitance of 3×10^{-6} f are connected in series with an alternating emf of 120 cos $2t$ V.

19. Find an expression for the current of a series RL circuit, where $R = 100\ \Omega$, $L = 2$ h, $I(0) = 0$, and the emf voltage satisfies

$$E = \begin{cases} 6 & \text{for} \quad 0 \le t \le 10, \\ 7 - e^{10-t} & \text{for} \quad t \ge 10. \end{cases}$$

20. Repeat Problem 19 with $R = 100/(1 + t)$, all other values remaining the same.

11.6 SECOND-ORDER LINEAR DIFFERENTIAL EQUATIONS: THEORY

Although there is no procedure for explicitly solving arbitrary differential equations, systematic methods do exist for certain classes of differential equations. In this section, we shall study a class of differential equations for which there are always unique solutions, and in the next four sections we shall present some methods for calculating them.

In Section 11.1 we defined an nth-order *linear* equation as an equation having the form

$$y^{(n)}(x) + a_{n-1}(x)y^{(n-1)}(x) + \cdots + a_1(x)y'(x) + a_0(x)y(x) = f(x).$$

We observe that a differential equation is linear if it does not involve nonlinear functions (squares, exponentials, etc.) or products of the dependent variable and its derivatives. Thus $y'' + (x^3 \sin x)^5 y' + y = \cos x^3$ is linear, and $y'' + (y)^2 + y = 0$ is nonlinear. The most general second-order linear equation is

$$y''(x) + a(x)y'(x) + b(x)y(x) = f(x), \tag{1}$$

and the most general third-order linear equation is

$$y'''(x) + a(x)y''(x) + b(x)y'(x) + c(x)y(x) = f(x). \tag{2}$$

In Section 11.4 we discussed the general first-order linear equation

$$y'(x) + a(x)y(x) = f(x).$$

If $a(x)$ and $f(x)$ are continuous, we saw that this equation has infinitely many solutions since the general solution involves an arbitrary constant [see equation (11.3.17).] As we have seen, this arbitrary constant can be determined if one condition $y(x_0) = y_0$ is given. In this case, the equation has a unique solution. We can restate this basic fact as follows:

> If $a(x)$ and $f(x)$ are continuous, then the equation $y'(x) + a(x)y(x) = f(x)$ has one and only one solution that satisfies the initial condition $y(x_0) = y_0$.

This is a very nice result, for it tells us that every linear first-order equation with a given initial condition has a unique solution. We need only set about finding it. It turns out that this special property holds for linear differential equations of any order. The only difference is that in order to have a unique solution to a second-order equation, we must specify two initial conditions, for a third-order equation three conditions, and so on. The result we need is Theorem 11.4.1, which we restate below.

Theorem 1 EXISTENCE–UNIQUENESS THEOREM FOR LINEAR INITIAL VALUE PROBLEMS Let $a_1(x)$, $a_2(x), \ldots, a_n(x)$, and $f(x)$ be continuous functions on the interval $[x_0, x_1]$, and let $c_0, c_1, c_2, \ldots, c_{n-1}$ be n given constants. Then there exists a unique function $y(x)$ that satisfies the linear differential equation

$$y^{(n)} + a_1(x)y^{(n-1)} + a_2(x)y^{(n-2)} + \cdots + a_n(x)y = f(x)$$

on $[x_0, x_1]$ *and* the n initial conditions

$$y(x_0) = c_0, \qquad y'(x_0) = c_1,\ y''(x_0) = c_2, \ldots, y^{(n-1)}(x_0) = c_{n-1}. \tag{3}$$

If we apply Theorem 1 to the second-order equation (1), we have the following result:

If $a(x)$, $b(x)$, and $f(x)$ are continuous functions, then the equation

$$y''(x) + a(x)y'(x) + b(x)y(x) = f(x)$$

has a unique solution that satisfies the conditions

$$y(x_0) = y_0, \quad y'(x_0) = y_1,$$

for any real numbers x_0, y_0, and y_1.

For simplicity we shall limit most of our discussion in this chapter to second-order linear equations (and associated systems). We emphasize, however, that *every* result we will prove can be extended to higher-order linear equations. This is done in Section 11.15. In the remainder of this chapter we will assume, unless otherwise stated, that all functions in each equation are continuous for all real values of x.

Definition 1 HOMOGENEOUS AND NONHOMOGENEOUS EQUATIONS If the function $f(x)$ is identically zero, we say that the general second-order linear equation, equation (1), is **homogeneous.** Otherwise, it is **nonhomogeneous.**

EXAMPLE 1

(a) The equation $y'' + 2xy' + 3y = 0$ is homogeneous.
(b) The equation $y'' + 2xy' + 3y = e^x$ is nonhomogeneous. ■

Definition 2 CONSTANT AND VARIABLE COEFFICIENTS If the coefficient functions $a(x)$ and $b(x)$ are constants, $a(x) = a$ and $b(x) = b$, then the equation is said to have **constant coefficients.** As we shall see, linear differential equations with constant coefficients are the easiest to solve. If either $a(x)$ or $b(x)$ is not constant, the equation is said to have **variable coefficients.**

EXAMPLE 2 **(a)** The equation $y'' + 3y' - 10y = 0$ has constant coefficients.
(b) The equation $y'' + 3xy' - 10x^2y = 0$ has variable coefficients. ■

Before solving a second-order differential equation, we need to know what we are seeking. A clue is provided by examining a first-order equation. Consider the equation

$$y' + 2y = 0. \tag{4}$$

In Section 11.4 we saw that one solution to (4) is $y = e^{-2x}$. In fact, $y = Ce^{-2x}$ is a solution for any constant C, and every solution to (4) has the form Ce^{-2x}. We can summarize this result by noting that once we found one nonzero solution to (4), we found all of the solutions, since every other solution to the equation is a constant multiple of this one solution.

It turns out that similar results hold for homogeneous second-order equations. The major difference is that now we have to find *two* solutions to (1) where neither solution is a multiple of the other. We now make these ideas more precise. The following definitions are similar to definitions given in Section 9.3.

LINEAR COMBINATION AND LINEAR INDEPENDENCE

Let y_1 and y_2 be any two functions. By a **linear combination** of y_1 and y_2, we mean a function $y(x)$ that can be written in the form

$$y(x) = c_1y_1(x) + c_2y_2(x)$$

for some constants c_1 and c_2. Two functions are **linearly independent** on an interval $[x_0, x_1]$ whenever the relation $c_1y_1(x) + c_2y_2(x) = 0$ for all x in $[x_0, x_1]$ implies that $c_1 = c_2 = 0$. Otherwise, they are **linearly dependent.** There is, however, an easier way to see that two functions y_1 and y_2 are linearly dependent. If $c_1y_1(x) + c_2y_2(x) = 0$ (where not both c_1 and c_2 are zero), we may suppose that $c_1 \neq 0$. Then, dividing the above expression by c_1, we obtain

$$y_1(x) + \frac{c_2}{c_1}y_2(x) = 0,$$

or

$$y_1(x) = -\frac{c_2}{c_1}y_2(x).$$

Therefore,

> Two functions are linearly dependent on the interval $[x_0, x_1]$ if and only if one of the functions is a constant multiple of the other.

The notions of linear combination and linear independence are central to the theory of linear homogeneous equations, as illustrated by the results that follow. Our first result holds for both homogeneous and nonhomogeneous equations. (Proofs of Theorem 2 and the others in this section are given at the end of the section.)

Theorem 2 Two linearly independent solutions of equation (1) can always be found.

EXAMPLE 3 Verify that e^{-5x} and e^{2x} are linearly independent solutions to the equation $y'' + 3y' - 10y = 0$.

Solution. That e^{-5x} and e^{2x} are solutions is easy to show:

$$(e^{-5x})'' + 3(e^{-5x})' - 10e^{-5x} = 25e^{-5x} + 3(-5)e^{-5x} - 10e^{-5x} = 0,$$
$$(e^{2x})'' + 3(e^{2x})' - 10e^{2x} = 4e^{2x} + 3(2)e^{2x} - 10e^{2x} = 0.$$

Since $e^{-5x} = e^{-7x} \cdot e^{2x}$ and e^{-7x} is not a constant, we see that e^{-5x} and e^{2x} are linearly independent. ■

EXAMPLE 4 Verify that $\sin x$ and $\cos x$ are linearly independent solutions to $y'' + y = 0$.

Solution. To check that $\sin x$ and $\cos x$ are solutions we write

$$(\sin x)'' + \sin x = -\sin x + \sin x = 0;$$
$$(\cos x)'' + \cos x = -\cos x + \cos x = 0.$$

Since $\sin x/\cos x = \tan x \neq$ constant, we see that $\sin x$ and $\cos x$ are, indeed, linearly independent solutions to $y'' + y = 0$. ■

Every linear homogeneous second-order equation has two linearly independent solutions. The next two theorems show us that this is all we need. That is, once we have two linearly independent solutions, we have them all.

Consider the homogeneous system

$$y'' + a(x)y' + b(x)y = 0. \tag{5}$$

Theorem 3 Let $y_1(x)$ and $y_2(x)$ be any two solutions of the homogeneous equation (5). Then any linear combination of them is also a solution of (5).

Theorem 4 Let $y_1(x)$ and $y_2(x)$ be linearly independent solutions to (5) and let $y_3(x)$ be another solution to (5). Then there exist unique constants c_1 and c_2 such that

$$y_3(x) = c_1y_1(x) + c_2y_2(x). \tag{6}$$

In other words, any solutions of (5) can be written as a linear combination of two given linearly independent solutions of (5).

We stress the importance of this theorem. It indicates that once we have found two linearly independent solutions y_1 and y_2 of equation (5), we have, in effect, found all the solutions of equation (5).

GENERAL SOLUTION TO A LINEAR SECOND-ORDER EQUATION

The general solution of equation (5) is given by the linear combination

$$y(x) = c_1 y_1(x) + c_2 y_2(x),$$

where c_1 and c_2 are arbitrary constants.

EXAMPLE 5 The general solution to $y'' + 3y' - 10y = 0$ is

$$y = c_1 e^{-5x} + c_2 e^{2x}. \blacksquare$$

See Example 3.

EXAMPLE 6 The general solution to $y'' + y = 0$ is

$$y = c_1 \cos x + c_2 \sin x. \blacksquare$$

See Example 4.

Definition 3 WRONSKIAN Let $y_1(x)$ and $y_2(x)$ be any two solutions to equation (5). The **Wronskian** of y_1 and y_2, $W(y_1, y_2)$, is defined as

$$W(y_1, y_2)(x) = y_1(x)y_2'(x) - y_1'(x)y_2(x). \tag{7}$$

There is another way to write the Wronskian. This notation involves a 2×2 determinant. We have

$$W(y_1, y_2)(x) = \begin{vmatrix} y_1(x) & y_2(x) \\ y_1'(x) & y_2'(x) \end{vmatrix} \tag{8}$$

Since

$$\begin{vmatrix} a & b \\ c & d \end{vmatrix} = ad - bc,$$

by the definition of a 2×2 determinant, we see that

$$\begin{vmatrix} y_1 & y_2 \\ y_1' & y_2' \end{vmatrix} = y_1 y_2' - y_2 y_1',$$

so that (7) and (8) are equivalent expressions.

Using the product rule of differentiation on equation (7), we see that

$$\begin{aligned} W'(y_1, y_2) &= y_1 y_2'' + y_1' y_2' - y_1' y_2' - y_1'' y_2 \\ &= y_1 y_2'' - y_1'' y_2. \end{aligned}$$

Since y_1 and y_2 are solutions of equation (1),

$$y_1'' + ay_1' + by_1 = 0 \quad \text{and} \quad y_2'' + ay_2' + by_2 = 0.$$

Multiplying the first of these equations by y_2 and the second by y_1 and subtracting, we obtain

$$y_1y_2'' - y_2y_1'' + a(y_1y_2' - y_2y_1') = 0,$$

which is just

$$W' + aW = 0. \tag{9}$$

From Example 11.2.2 we see that the solution of equation (9) can be written as

$$W(y_1, y_2) = ce^{-\int a(x)\,dx} \tag{10}$$

for some arbitrary constant c. Equation (10) is known as **Abel's formula.**

Since an exponential is never zero, we see that $W(y_1, y_2)$ is either always zero (when $c = 0$) or never zero (when $c \neq 0$). The importance of this fact is given by the following theorem.

Theorem 5 The solutions $y_1(x)$ and $y_2(x)$ of equation (5) are linearly independent on $[x_0, x_1]$ if and only if $W(y_1, y_2) \neq 0$.

Theorem 5 is useful in at least three ways. First, it provides us with an easy way to determine whether or not two solutions are linearly independent. Second, it greatly simplifies the proof of Theorem 2—as we shall see later in this section. Third, the Wronskian can be easily extended to third- and higher-order equations with similar results. It is not easy to verify directly that three functions are linearly independent, but the task is made routine by use of the Wronskian.

EXAMPLE 7 The functions $y_1 = e^{-5x}$ and $y_2 = e^{2x}$ are solutions to $y'' + 3y' - 10y = 0$. Then

$$W(y_1, y_2) = \begin{vmatrix} e^{-5x} & e^{2x} \\ -5e^{-5x} & 2e^{2x} \end{vmatrix}$$
$$= e^{-5x}(2e^{2x}) - (-5e^{-5x})(e^{2x}) = 7e^{-3x}.$$

This is nonzero for every x so y_1 and y_2 are linearly independent. ■

EXAMPLE 8 The functions $y_1 = \cos x$ and $y_2 = \sin x$ are linearly independent solutions to $y'' + y = 0$ since

$$W(y_1, y_2) = \begin{vmatrix} \cos x & \sin x \\ -\sin x & \cos x \end{vmatrix} = \cos x \cos x - (-\sin x)(\sin x)$$
$$= \cos^2 x + \sin^2 x = 1 \neq 0. \ ■$$

We now turn briefly to the nonhomogeneous equation

$$y'' + a(x)y' + b(x)y = f(x). \tag{11}$$

Let y_p be any solution to equation (11). If we know the general solution to the homogeneous equation

$$y'' + a(x)y' + b(x)y = 0, \tag{12}$$

we can find all solutions to equation (11).

Theorem 6 Let $y_p(x)$ be a solution of equation (11) and let $y^*(x)$ be any other solution. Then $y^*(x) - y_p(x)$ is a solution of equation (12); that is,

$$y^*(t) = c_1y_1(t) + c_2y_2(t) + y_p(t)$$

for some constants c_1 and c_2, where y_1, y_2 are two linearly independent solutions of equations (12).

Thus **in order to find all solutions to the nonhomogeneous problem, we need only find one solution to the nonhomogeneous problem and the general solution of the homogeneous problem.**

GENERAL SOLUTION TO A LINEAR, NONHOMOGENEOUS, SECOND-ORDER EQUATION

Let $y_p(x)$ be one solution to the nonhomogeneous equation (11) and let $y_1(x)$ and $y_2(x)$ be two linearly independent solutions to the homogeneous equation (5). Then the general solution to (11) is given by

$$y(x) = c_1y_1(x) + c_2y_2(x) + y_p(x)$$

where c_1 and c_2 are arbitrary constants.

EXAMPLE 9 It is easy to verify that $\frac{1}{2}te^t$ is a particular solution of $x'' - x = e^t$. Two linearly independent solutions of $x'' - x = 0$ are given by $x_1 = e^t$ and $x_2 = e^{-t}$. The general solution is therefore $x(t) = \frac{1}{2}te^t + c_1e^t + c_2e^{-t}$. Note that x_1 and x_2 are independent since $W(x_1, x_2) = e^t(-e^{-t}) - e^{-t}(e^t) = -2$. ■

In the next three sections we will present methods for finding the general solution of the homogeneous problem. In Sections 11.10 and 11.11 techniques for obtaining the solution of the nonhomogeneous problem will be developed.

PROOFS OF THEOREMS 2, 3, 4, 5 and 6 (OPTIONAL)

We prove the theorems in the order 3, 5, 2, 4, 6.

Theorem 3 Let $y_1(x)$ and $y_2(x)$ be any two solutions of the homogeneous equation

$$y'' + a(x)y' + b(x)y = 0. \tag{5}$$

Then any linear combination of them is also a solution of (5).

Proof. Let $y(x) = c_1y_1(x) + c_2y_2(x)$. Then

$$\begin{aligned} y'' + ay' + by &= c_1y_1'' + c_2y_2'' + c_1ay_1' + c_2ay_2' + c_1by_1 + c_2by_2 \\ &= c_1(y_1'' + ay_1' + by_1) + c_2(y_2'' + ay_2' + by_2) = 0, \end{aligned}$$

since y_1 and y_2 are solutions of the homogeneous equation (12). ■

Theorem 5 The solutions $y_1(x)$ and $y_2(x)$ of the equation (5), are linearly independent if and only if $W(y_1, y_2) \neq 0$.

Proof. We first show that if $W(y_1, y_2) = 0$, then y_1 and y_2 are linearly dependent. Let x_2 be a point in the interval $x_0 \leq x \leq x_1$. Consider the system of equations

$$\begin{aligned} c_1y_1(x_2) + c_2y_2(x_2) &= 0, \\ c_1y_1'(x_2) + c_2y_2'(x_2) &= 0. \end{aligned} \tag{13}$$

The determinant of the system (13) is

$$y_1(x_2)y_2'(x_2) - y_2(x_2)y_1'(x_2) = W(y_1, y_2)(x_2) = 0.$$

Thus according to the theory of determinants (Corollary 1 on page 472) there exists a solution (c_1, c_2) for equations (13) where c_1 and c_2 are not both equal to zero. Define $y(x) = c_1y_1(x) + c_2y_2(x)$. By Theorem 3, $y(x)$ is a solution of equation (5). Then since c_1 and c_2 are solutions of equations (13),

$$y(x_2) = c_1y_1(x_2) + c_2y_2(x_2) = 0$$

and

$$y'(x_2) = c_1y_1'(x_2) + c_2y_2'(x_2) = 0.$$

Thus $y(x)$ solves the initial value problem

$$y'' + a(x)y' + b(x)y = 0, \quad y(x_2) = y'(x_2) = 0.$$

But this initial value problem also has the solution $y_3(x) \equiv 0$ for all values of x in $x_0 \leq x \leq x_1$. By Theorem 1 the solution of this initial value problem is unique so that necessarily $y(x) = y_3(x) = 0$. Thus

$$y(x) = c_1y_1(x) + c_2y_2(x) = 0,$$

for all values of x in $x_0 \leq x \leq x_1$, which proves that y_1 and y_2 are linearly dependent.

We now assume that $W(y_1, y_2) \neq 0$ in $[x_0, x_1]$ and shall prove that y_1 and y_2 are linearly independent. If y_1 and y_2 are not linearly independent, then there is a constant c such that $y_2 = cy_1$ or $y_1 = cy_2$. Assume that $y_2 = cy_1$. Then $y_2' = cy_1'$ and

$$W(y_1, y_2) = y_1 y_2' - y_1' y_2 = y_1(cy_1') - y_1'(cy_1) = 0.$$

But this contradicts the assumption that $W \neq 0$. Hence the solutions y_1 and y_2 must be independent. ■

Theorem 2 Two linearly independent solutions of equation (1) can always be found.

Proof. The existence part of Theorem 1 guarantees that we can find a solution $y_1(x)$ to equation (1) satisfying

$$y_1(x_0) = 1 \quad \text{and} \quad y_1'(x_0) = 0.$$

Similarly, we can also find a solution $y_2(x)$ to equation (1) satisfying

$$y_2(x_0) = 0 \quad \text{and} \quad y_2'(x_0) = 1.$$

Now

$$W(y_1, y_2)(x_0) = \begin{vmatrix} 1 & 0 \\ 0 & 1 \end{vmatrix} = 1,$$

so y_1 and y_2 are linearly independent by Theorem 5. ■

Theorem 4 Let $y_1(x)$ and $y_2(x)$ be linearly independent solutions to the homogeneous system

$$y'' + a(x)y' + b(x)y = 0, \tag{5}$$

and let $y_3(x)$ be another solution to (5). Then there exist unique constants c_1 and c_2 such that

$$y_3(x) = c_1 y_1(x) + c_2 y_2(x).$$

Proof. Let $y(x_0) = a$ and $y'(x_0) = b$. Consider the following linear system of equations in two unknowns c_1 and c_2:

$$\begin{aligned} y_1(x_0)c_1 + y_2(x_0)c_2 &= a, \\ y_1'(x_0)c_1 + y_2'(x_0)c_2 &= b. \end{aligned} \tag{14}$$

As we saw earlier, the determinant of this system is $W(y_1, y_2)(x_0)$, which is nonzero since the solutions are linearly independent. Thus there is a unique solution (c_1, c_2) to equations (14) and a solution $y^*(x) = c_1 y_1(x) + c_2 y_2(x)$ that satisfies the conditions $y^*(x_0) = a$ and $y^{*\prime}(x_0) = b$. Since every initial value problem has a unique solution (by Theorem 1), it must follow that $y(x) = y^*(x)$ on the interval $x_0 \leq x \leq x_1$, and so the proof is complete. ■

Theorem 6 Let $y_p(x)$ be a solution of the nonhomogeneous equation

$$y'' + a(x)y' + b(x)y = f(x) \tag{11}$$

and let $y^*(x)$ be any other solution. Then $y^*(x) - y_p(x)$ is a solution of the corresponding homogeneous equation (5). That is,

$$y^*(t) = c_1 y_1(t) + c_2 y_2(t) + y_p(t)$$

for some constants c_1 and c_2, where y_1 and y_2 are two linearly independent solutions of equation (5).

Proof. We have

$$\begin{aligned}(y^* - y_p)'' + a(y^* - y_p)' + b(y^* - y_p) \\ &= (y^{*\prime\prime} + ay^{*\prime} + by^*) - (y_p'' + ay_p' + by_p) \\ &= f - f = 0. \quad \blacksquare\end{aligned}$$

PROBLEMS 11.6

In Problems 1–10, determine whether the given equation is linear or nonlinear. If it is linear, state whether it is homogeneous or nonhomogeneous with constant or variable coefficients.

1. $y'' + 2x^3y' + y = 0$
2. $y'' + 2y' + y^2 = x$
3. $y'' + 3y' + yy' = 0$
4. $y'' + 3y' + 4y = 0$
5. $y'' + 3y' + 4y = \sin x$
6. $y'' + y(2 + 3y) = e^x$
7. $y'' + 4xy' + 2x^3y = e^{2x}$
8. $y'' + \sin(xe^x)y' + 4xy = 0$
9. $3y'' + 16y' + 2y = 0$
10. $yy'y'' = 1$

In Problems 11–14, determine how many of the given functions are linearly independent.

11. $y_0 = 1,\ \ y_1 = 1 + x,\ \ y_2 = x^2,$
$y_3 = x(1 - x),\ \ y_4 = x$
12. $y_0 = \sin^2 x,\ \ y_1 = 1,\ \ y_2 = \sin x \cos x,$
$y_3 = \cos^2 x,\ \ y_4 = \sin 2x$
13. $y_0 = 1 + x,\ \ y_1 = 1 - x,\ \ y_2 = 1,$
$y_3 = x^2,\ \ y_4 = 1 + x^2$
14. $y_0 = 2\cos^2 x - 1,\ \ y_1 = \cos^2 x - \sin^2 x,$
$y_2 = 1 - 2\sin^2 x,\ \ y_3 = \cos 2x$

In Problems 15–18, test each of the functions 1, x, x^2, and x^3 to see which functions satisfy the given differential equation. Then construct the *general solution* to the equation by writing a linear combination of the linearly independent solutions you have found.

15. $y'' = 0$
16. $y''' = 0$
17. $xy'' - y' = 0$
18. $x^2y'' - 2xy' + 2y = 0$

19. Let $y_1(x)$ be a solution of the homogeneous equation

$$y'' + a(x)y' + b(x)y = 0$$

on the interval $\alpha \le x \le \beta$. Suppose that the curve y_1 is tangent to the x-axis at some point of this interval. Prove that y_1 must be identically zero.

20. Let $y_1(x)$ and $y_2(x)$ be two solutions of the homogeneous equation

$$y'' + a(x)y' + b(x)y = 0$$

on the interval $\alpha \le x \le \beta$. Suppose $y_1(x_0) = y_2(x_0) = 0$ for some point $\alpha \le x_0 \le \beta$. Show that y_2 is a constant multiple of y_1.

21. **(a)** Show that $y_1(x) = \sin x^2$ and $y_2(x) = \cos x^2$ are linearly independent solutions of

$$xy'' - y' + 4x^3y = 0.$$

(b) Calculate $W(y_1, y_2)$ and show that it is zero when $x = 0$. Does this result contradict Theorem 5? [*Hint:* In Theorem 5, as elsewhere in this section, $a(x)$ and $b(x)$ are assumed to be continuous.]

22. Show that two solution $y_1(x)$ and $y_2(x)$ to equation (5) are linearly dependent if and only if one is a constant multiple of the other.

23. Show that

$$y_1(x) = \sin x \quad \text{and} \quad y_2(x) = 4\sin x - 2\cos x$$

are linearly independent solutions of $y'' + y = 0$. Write the solution $y_3(x) = \cos x$ as a linear combination of y_1 and y_2.

24. Prove that $e^x \sin x$ and $e^x \cos x$ are linearly independent solutions of the equation

$$y'' - 2y' + 2y = 0.$$

(a) Find a solution that satisfies the conditions $y(0) = 1$, $y'(0) = 4$.
(b) Find another pair of linearly independent solutions.

25. Assume that some nonzero solution of

$$y'' + a(x)y' + b(x)y = 0, \quad y(0) = 0,$$

vanishes at some point x_1, where $x_1 > 0$. Prove that any other solution vanishes at $x = x_1$.

26. Define the function $s(x)$ to be the unique solution of the initial value problem

$$y'' + y = 0; \quad y(0) = 0, \quad y'(0) = 1,$$

and the function $c(x)$ as the solution of

$$y'' + y = 0; \quad y(0) = 1, \quad y'(0) = 0.$$

Without using trigonometry, prove that:

(a) $\dfrac{ds}{dx} = c(x)$;

(b) $\dfrac{dc}{dx} = -s(x)$;

(c) $s^2 + c^2 = 1$.

27. **(a)** Show that $y_1 = \sin \ln x^2$ and $y_2 = \cos \ln x^2$ are linearly independent solutions to

$$y'' + \frac{1}{x}y' + \frac{4}{x^2}y = 0 \quad (x > 0).$$

(b) Calculate $W(y_1, y_2)$.

11.7 USING ONE SOLUTION TO FIND ANOTHER

As we saw in Theorem 11.6.4, it is easy to write down the general solution of the homogeneous equation

$$y'' + a(x)y' + b(x)y = 0, \tag{1}$$

provided we know two linearly independent solutions y_1 and y_2 of equation (1). The general solution is then given by

$$y = c_1y_1 + c_2y_2,$$

where c_1 and c_2 are arbitrary constants. Unfortunately, there is no general procedure for determining y_1 and y_2. However, a standard procedure does exist for finding y_2 when y_1 is known. This method is of considerable importance, since it is often possible to find one solution by inspecting the equation or by trial and error.

We assume that y_1 is a nonzero solution of equation (1) and seek another solution y_2 such that y_1 and y_2 are linearly independent. If it can be found, then

$$\frac{y_2}{y_1} = v(x)$$

must be a nonconstant function of x, and $y_2 = vy_1$ must satisfy equation (1). Thus

$$(vy_1)'' + a(vy_1)' + b(vy_1) = 0 \tag{2}$$

But

$$(vy_1)' = vy_1' + v'y_1 \tag{3}$$

and

$$(vy_1)'' = (vy_1' + v'y_1)' = vy_1'' + v'y_1' + v'y_1' + v''y_1$$
$$= vy_1'' + 2v'y_1' + v''y_1. \tag{4}$$

Then, using (3) and (4) in (2), we have

$$(vy_1'' + 2v'y_1' + v''y_1) + a(vy_1' + v'y_1) + bvy_1 = 0$$

or

$$v(y_1'' + ay_1' + by_1) + v'(2y_1' + ay_1) + v''y_1 = 0. \tag{5}$$

The first term in parentheses in (5) vanishes since y_1 is a solution of equation (1), so we obtain the equation

$$v''y_1 + v'(2y_1' + ay_1) = 0.$$

Dividing by $v'y_1$, we can rewrite this equation in the form

$$\frac{v''}{v'} = -2\frac{y_1'}{y_1} - a. \tag{6}$$

We set $z = v'$ in (6). Then $z' = v''$, and (6) becomes

$$\frac{z'}{z} = -2\frac{y_1'}{y_1} - a. \tag{7}$$

The functions in both sides of (7) are functions of x, so we integrate with respect to x to obtain

$$\ln z = -2 \ln y_1 - \int a(x)\,dx,$$

so that, exponentiating,

$$z = e^{\ln z} = e^{-2\ln y_1 - \int a(x)\,dx} = \frac{1}{y_1^2}e^{-\int a(x)\,dx}.$$

But, as $z = v'$, we have

$$v' = \frac{1}{y_1^2}e^{-\int a(x)\,dx}.$$

Since the exponential is never zero, v is nonconstant. To find v, we perform another integration and obtain

$$y_2 = vy_1 = y_1(x)\int \frac{e^{-\int a(x)\,dx}}{y_1^2(x)}\,dx. \tag{8}$$

We shall make frequent use of formula (8) in later sections; the following examples illustrate how it can be used.

EXAMPLE 1 Note that $y_1 = x$ is a solution of

$$x^2y'' - xy' + y = 0, \quad x > 0. \tag{9}$$

To find y_2, we rewrite equation (8) as

$$y'' - \left(\frac{1}{x}\right)y' + \left(\frac{1}{x^2}\right)y = 0.$$

Then $\int a(x)\,dx = -\ln x$, so that using formula (9), we have

$$y_2 = x\int \frac{x}{x^2}\,dx = x \ln x.$$

Thus the general solution of equation (9) is $y = c_1x + c_2x \ln x$, $x > 0$. ■

EXAMPLE 2 Consider the **Legendre equation of order one:**

$$(1 - x^2)y'' - 2xy' + 2y = 0, \qquad -1 < x < 1, \tag{10}$$

or

$$y'' - \frac{2x}{1 - x^2}y' + \frac{2y}{1 - x^2} = 0.$$

You can verify that $y_1 = x$ is a solution. To find y_2, we note that $\int a(x)\,dx = \ln(1 - x^2)$, so that by equation (8) and partial fractions

$$\begin{aligned} y_2 &= x\int \frac{e^{-\ln(1-x^2)}}{x^2}\,dx = x\int \frac{dx}{x^2(1 - x^2)} \\ &= x\int \left[\frac{1}{x^2} + \frac{1}{2}\left(\frac{1}{1 + x} - \frac{1}{1 - x}\right)\right] dx \\ &= x\left[-\frac{1}{x} + \frac{1}{2}\ln\left(\frac{1 + x}{1 - x}\right)\right] = \frac{x}{2}\ln\left(\frac{1 + x}{1 - x}\right) - 1. \end{aligned}$$

Note that in this example y_2 is defined in $-1 < x < 1$ even though $v(0)$ is undefined. ■

PROBLEMS 11.7

In each of Problems 1–10 a second-order differential equation and one solution $y_1(x)$ are given. Verify that $y_1(x)$ is indeed a solution and find a second linearly independent solution.

1. $y'' - 2y' + y = 0, \quad y_1(x) = e^x$
2. $y'' + \left(\frac{3}{x}\right)y' = 0, \quad y_1(x) = 1$

3. $y'' - \left(\frac{2x}{1-x^2}\right)y' + \left(\frac{6}{1-x^2}\right)y = 0, \quad (|x| < 1)$

   $y_1(x) = \frac{3x^2 - 1}{2}$

   (This equation is called the **Legendre differential equation of order two.)**
4. $y'' - 2xy' + 2y = 0, \quad y_1(x) = x$
   [*Hint:* Write y_2 as an integral.]
5. $x^2y'' + xy' - 4y = 0, \quad y_1(x) = x^2$
   [*Hint:* Divide by x^2 first.]
6. $x^2y'' - 2xy' + (x^2 + 2)y = 0, \quad (x > 0),$
   $y_1 = x \sin x$
7. $xy'' + (2x - 1)y' - 2y = 0, \quad (x > 0), \quad y_1 = e^{-2x}$
8. $xy'' + (x - 1)y' + (3 - 12x)y = 0, \quad (x > 0),$
   $y_1 = e^{3x}$
9. $xy'' - y' + 4x^3y = 0, \quad (x > 0), \quad y_1 = \sin(x^2)$
10. $x^{1/3}y'' + y' + \left(\frac{1}{4}x^{-1/3} - \frac{1}{6x} - 6x^{-5/3}\right)y = 0,$

    $y_1 = x^3e^{-3x^{2/3}/4} \quad (x > 0)$
11. The **Bessel differential equation** is given by

    $x^2y'' + xy' + (x^2 - p^2)y = 0.$

    For $p = \frac{1}{2}$, verify that $y_1(x) = (\sin x)/\sqrt{x}$ is a solution for $x > 0$. Find a second, linearly independent solution.
12. Letting $p = 0$ in the equation of Problem 11, we obtain the **Bessel differential equation of index zero.** One solution is the **Bessel function of order zero** denoted by $J_0(x)$. In terms of $J_0(x)$, find a second, linearly independent solution.

11.8 HOMOGENEOUS EQUATIONS WITH CONSTANT COEFFICIENTS: REAL ROOTS

In this section we shall present a simple procedure for finding the general solution to the linear homogeneous equation with constant coefficients

$$y'' + ay' + by = 0. \tag{1}$$

We recall that for the comparable first-order equation $y' + ay = 0$ the general solution is $y(x) = ce^{-ax}$. It is not implausible to "guess" that there may be a solution to equation (1) of the form $y(x) = e^{\lambda x}$ for some number λ (real or complex). Setting $y(x) = e^{\lambda x}$, we obtain $y' = \lambda e^{\lambda x}$ and $y'' = \lambda^2 e^{\lambda x}$ so that equation (1) yields

$$\lambda^2 e^{\lambda x} + a\lambda e^{\lambda x} + be^{\lambda x} = 0.$$

Since $e^{\lambda x} \neq 0$, we can divide this equation by $e^{\lambda x}$ to obtain

$$\lambda^2 + a\lambda + b = 0, \tag{2}$$

where a and b are real numbers. Equation (2) is called the **characteristic equation** of the differential equation (1). It is clear that if λ satisfies equation (2), then $y(x) = e^{\lambda x}$ is a solution to equation (1). As we saw in Section 11.6, we need only obtain two linearly independent solutions. Equation (2) has the roots

$$\lambda_1 = \frac{-a + \sqrt{a^2 - 4b}}{2} \quad \text{and} \quad \lambda_2 = \frac{-a - \sqrt{a^2 - 4b}}{2} \tag{3}$$

There are three possibilities: $a^2 - 4b > 0$, $a^2 - 4b = 0$, $a^2 - 4b < 0$.

Case 1. (Roots Real and Unequal) If $a^2 - 4b > 0$, then λ_1 and λ_2 are distinct real numbers (given by equation (3)) and $y_1(x) = e^{\lambda_1 x}$ and $y_2 = e^{\lambda_2 x}$ are distinct solutions. These two solutions are linearly independent because

$$\frac{y_1}{y_2} = e^{(\lambda_1 - \lambda_2)x},$$

which is not a constant when $\lambda_1 \neq \lambda_2$. Thus we have proved the following theorem.

Theorem 1 If $a^2 - 4b > 0$, then the roots λ_1 and λ_2 to the characteristic equation are real and unequal and the general solution to equation (1) is given by

$$y(x) = c_1 e^{\lambda_1 x} + c_2 e^{\lambda_2 x}, \tag{4}$$

where c_1 and c_2 are arbitrary constants and λ_1 and λ_2 are the real roots of equation (2).

EXAMPLE 1 Consider the equation

$$y'' + 3y' - 10y = 0.$$

The characteristic equation is $\lambda^2 + 3\lambda - 10 = 0$, $a^2 - 4b = 49$, and the roots are $\lambda_1 = 2$ and $\lambda_2 = -5$ (the order in which the roots are taken is irrelevant). The general solution is

$$y(x) = c_1 e^{2x} + c_2 e^{-5x}.$$

If we specify the initial conditions $y(0) = 1$ and $y'(0) = 3$, for example, then differentiating and substituting $x = 0$ we obtain the simultaneous equations

$$\begin{aligned} c_1 + c_2 &= 1, \\ 2c_1 - 5c_2 &= 3, \end{aligned}$$

which have the unique solution $c_1 = \frac{8}{7}$ and $c_2 = -\frac{1}{7}$. The unique solution to the initial value problem is therefore

$$y(x) = \tfrac{1}{7}(8e^{2x} - e^{-5x}). \quad \blacksquare$$

Case 2. (Roots Equal) Suppose $a^2 - 4b = 0$. In this case, equation (2) has the double root $\lambda_1 = \lambda_2 = -a/2$. Thus $y_1(x) = e^{-ax/2}$ is a solution of (1). To find the second solution y_2, we make use of equation (11.7.8) since one solution is known:

$$\begin{aligned} y_2 &= y_1 \int \frac{e^{-ax}}{y_1^2}\,dx = e^{-ax/2} \int \frac{e^{-ax}}{(e^{-ax/2})^2}\,dx \\ &= e^{-ax/2} \int \frac{e^{-ax}}{e^{-ax}}\,dx = e^{-ax/2} \int dx = xe^{-ax/2}. \end{aligned}$$

Since $y_2/y_1 = x$, y_1 and y_2 are linearly independent. Hence we have the following result:

Theorem 2 If $a^2 - 4b = 0$, then the roots to the characteristic equation are equal and the general solution to (1) is given by

$$y(x) = c_1 e^{-(a/2)x} + c_2 x e^{-(a/2)x},$$

where c_1 and c_2 are arbitrary constants.

EXAMPLE 2 Consider the equation

$$y'' - 6y' + 9 = 0.$$

The characteristic equation is $\lambda^2 - 6\lambda + 9 = 0$, and $a^2 - 4b = 0$, yielding the unique double root $\lambda_1 = -a/2 = 3$. The general solution is

$$y(x) = c_1 e^{3x} + c_2 x e^{3x}.$$

If we use the initial conditions $y(0) = 1$, $y'(0) = 3$, we obtain the simultaneous equations

$$c_1 = 1,$$
$$3c_1 + c_2 = 3,$$

which yield the unique solution

$$y(x) = e^{3x}. \blacksquare$$

We will deal with the more complicated situation of complex roots ($a^2 - 4b < 0$) in the next section.

1 5 6 10 18 20 22 23 25

PROBLEMS 11.8

In Problems 1–20 find the general solution of each equation. When initial conditions are specified, give the particular solution that satisfies them.

1. $y'' - 4y = 0$
2. $x'' + x' - 6x = 0, \quad x(0) = 0, \quad x'(0) = 5$
3. $y'' - 3y' + 2y = 0$
4. $y'' + 5y' + 6y = 0, \quad y(0) = 1, \quad y'(0) = 2$
5. $4x'' + 20x' + 25x = 0, \quad x(0) = 1, \quad x'(0) = 2$
6. $y'' + 6y' + 9y = 0$
7. $x'' - x' - 6x = 0, \quad x(0) = -1, \quad x'(0) = 1$
8. $y'' - 8y' + 16y = 0, \quad y(0) = 2, \quad y'(0) = -1$
9. $y'' - 5y' = 0$
10. $y'' + 17y' = 0, \quad y(0) = 1, \quad y'(0) = 0$
11. $y'' + 2\pi y' + \pi^2 y = 0$
12. $y'' - 13y' + 42y = 0$
13. $z'' + 2z' - 15z = 0$
14. $w'' + 8w' + 12w = 0$
15. $y'' - 8y' + 16y = 0, \quad y(0) = 1, \quad y'(0) = 6$
16. $y'' + 2y' + y = 0, \quad y(1) = 2/e, \quad y'(1) = -3/e$
17. $y'' - 2y = 0$
18. $y'' + 6y' + 5y = 0$
19. $y'' - 5y = 0, \quad y(0) = 3, \quad y'(0) = -\sqrt{5}$
20. $y'' - 2y' - 2y = 0, \quad y(0) = 1, \quad y'(0) = 1 + 3\sqrt{3}$

Linear second-order differential equations may also be used in finding the solution to the **Riccati equation:**[†]

$$y' + y^2 + a(x)y + b(x) = 0. \tag{5}$$

This nonlinear first-order equation frequently occurs in physical applications. To change equation (5) into a linear second-order equation, let $y = z'/z$. Then $y' = (z''/z) - (z'/z)^2$, so (5) becomes

$$\frac{z''}{z} - \left(\frac{z'}{z}\right)^2 + \left(\frac{z'}{z}\right)^2 + a(x)\left(\frac{z'}{z}\right) + b(x) = 0.$$

Multiplying by z, we obtain the linear second-order equation

$$z'' + a(x)z' + b(x)z = 0. \tag{6}$$

If the general solution to (6) can be found, the quotient $y = z'/z$ is the general solution to equation (5).

21. Suppose $z = c_1z_1 + c_2z_2$ is the general solution to (6). Explain why the quotient z'/z involves only one arbitrary constant. [*Hint:* Divide numerator and denominator by c_1.]

22. For arbitrary constants a, b, and c, find the substitution that changes the nonlinear equation

$$y' + ay^2 + by + c = 0$$

into a linear second-order equation with constant coefficients. What second-order equation is obtained?

Use the method above to find the general solution to the Riccati equations in Problems 23 through 28. If an initial condition is specified, give the particular solution that satisfies that solution.

23. $y' + y^2 - 1 = 0, \quad y(0) = -\frac{1}{3}$

24. $\dfrac{dx}{dt} + x^2 + 1 = 0$

25. $y' + y^2 - 2y + 1 = 0$

26. $y' + y^2 + 3y + 2 = 0, \quad y(0) = 1$

27. $y' + y^2 - y - 2 = 0$

28. $y' + y^2 + 2y + 1 = 0, \quad y(1) = 0$

11.9 HOMOGENEOUS EQUATIONS WITH CONSTANT COEFFICIENTS: COMPLEX ROOTS

This section requires the material in Appendix 3 on complex numbers.

We return to our examination of the homogeneous equation with constant coefficients

$$y'' + ay' + by = 0 \tag{1}$$

where $a^2 - 4b < 0$.

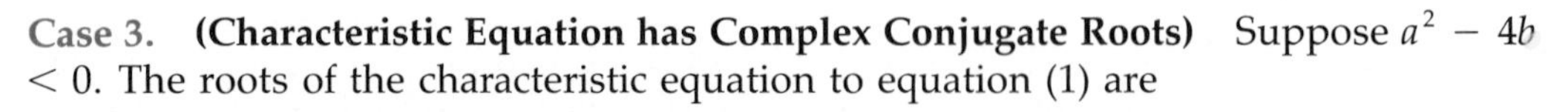

Case 3. **(Characteristic Equation has Complex Conjugate Roots)** Suppose $a^2 - 4b < 0$. The roots of the characteristic equation to equation (1) are

$$\lambda_1 = \alpha + i\beta, \quad \lambda_2 = \alpha - i\beta, \tag{2}$$

where $\alpha = -a/2$ and $\beta = \sqrt{4b - a^2}/2$. Thus $y_1 = e^{\lambda_1 x}$ and $y_2 = e^{\lambda_2 x}$ are solutions to (1). However, in this case it is useful to recall that any linear combination of solutions is also a solution (Theorem 11.6.3) and instead consider the solutions

$$y_1^* = \frac{e^{\lambda_1 x} + e^{\lambda_2 x}}{2} \quad \text{and} \quad y_2^* = \frac{e^{\lambda_1 x} - e^{\lambda_2 x}}{2i}.$$

[†]Giacomo Riccati (1676–1754) was an Italian mathematician, physicist, and philosopher. Riccati was responsible for bringing much of Newton's work on calculus to the attention of Italian mathematicians.

Since $\cos(-\theta) = \cos\theta$ and $\sin(-\theta) = -\sin\theta$, we can rewrite y_1^* as

$$\begin{aligned} y_1^* &= \frac{e^{(\alpha+i\beta)x} + e^{(\alpha-i\beta)x}}{2} = \frac{e^{\alpha x}}{2}(e^{i\beta x} + e^{-i\beta x}) \\ &= \frac{e^{\alpha x}}{2}[\cos\beta x + i\sin\beta x + \cos(-\beta x) + i\sin(-\beta x)] \\ &= e^{\alpha x}\cos\beta x. \end{aligned}$$

Similarly, $y_2^* = e^{\alpha x}\sin\beta x$, and the linear independence of y_1^* and y_2^* follows easily since

$$\frac{y_1^*}{y_2^*} = \cot\beta x, \quad \beta \neq 0,$$

which is not a constant. Alternatively, it is easy to compute $W(y_1^*, y_2^*) = \beta e^{\alpha x} \neq 0$. Thus we have proved the following theorem:

Theorem 1 If $a^2 - 4b < 0$, then the characteristic equation has complex conjugate roots and the general solution to equation (1) is given by

$$y(x) = e^{\alpha x}(c_1\cos\beta x + c_2\sin\beta x), \tag{3}$$

where c_1 and c_2 are arbitrary constants and

$$\alpha = -\frac{a}{2}, \quad \beta = \frac{\sqrt{4b - a^2}}{2}.$$

EXAMPLE 1 Let $y'' + y = 0$. Then the characteristic equation is $\lambda^2 + 1 = 0$ with roots $\pm i$. We have $\alpha = 0$ and $\beta = 1$ so that the general solution is

$$y(x) = c_1\cos x + c_2\sin x. \tag{4}$$

This is the **equation of harmonic motion.** ■

EXAMPLE 2 Consider the equation $y'' + y' + y = 0$, $y(0) = 1$, $y'(0) = 3$. We have $\lambda^2 + \lambda + 1 = 0$ with roots $\lambda_1 = (-1 + i\sqrt{3})/2$ and $\lambda_2 = (-1 - i\sqrt{3})/2$. Then $\alpha = -\frac{1}{2}$ and $\beta = \sqrt{3}/2$, so that the general solution is

$$y(x) = e^{-x/2}\left(c_1\cos\frac{\sqrt{3}}{2}x + c_2\sin\frac{\sqrt{3}}{2}x\right).$$

To solve the initial value problem, we solve the simultaneous equations

$$\begin{aligned} c_1 &= 1, \\ \frac{\sqrt{3}}{2}c_2 - \frac{1}{2}c_1 &= 3. \end{aligned}$$

Thus $c_1 = 1$, $c_2 = 7/\sqrt{3}$, and

$$y(x) = e^{-x/2}\left(\cos\frac{\sqrt{3}}{2}x + \frac{7}{\sqrt{3}}\sin\frac{\sqrt{3}}{2}x\right). \blacksquare$$

PROBLEMS 11.9

In Problems 1–12 find the general solution of each equation. When initial conditions are specified, give the particular solution that satisfies them.

1. $y'' + 2y' + 2y = 0$
2. $8y'' + 4y' + y = 0, \quad y(0) = 0, \quad y'(0) = 1$
3. $x'' + x' + 7x = 0$
4. $y'' + y' + 2y = 0$
5. $\dfrac{d^2x}{d\theta^2} + 4x = 0, \quad x\left(\dfrac{\pi}{4}\right) = 1, \quad x'\left(\dfrac{\pi}{4}\right) = 3$
6. $y'' + y = 0, \quad y(\pi) = 2, \quad y'(\pi) = -1$
7. $y'' + \frac{1}{4}y = 0, \quad y(\pi) = 1, \quad y'(\pi) = -1$
8. $y'' + 6y' + 12y = 0$
9. $y'' + 2y' + 5y = 0$
10. $y'' + 2y' + 5y = 0, \quad y(0) = 1, \quad y'(0) = -3$
11. $y'' + 2y' + 2y = 0, \quad y(\pi) = e^{-\pi}, \quad y'(\pi) = -2e^{-\pi}$
12. $y'' + 2y' + 5y = 0, \quad y(\pi) = e^{-\pi}, \quad y'(\pi) = 3e^{-\pi}$

11.10 NONHOMOGENEOUS EQUATIONS: THE METHOD OF UNDETERMINED COEFFICIENTS

In this and the following section, we shall present methods for finding a particular solution to the nonhomogeneous linear equation with constant coefficients

$$y'' + ay' + by = f(x). \tag{1}$$

First, however, we shall prove a very useful result concerning nonhomogeneous equations called the principle of superposition:

Theorem 1 PRINCIPLE OF SUPERPOSITION Suppose the function $f(x)$ in (1) is a sum of two functions $f_1(x)$ and $f_2(x)$:

$$f(x) = f_1(x) + f_2(x).$$

If $y_1(x)$ is a solution of the equation

$$y'' + ay' + by = f_1(x) \tag{2}$$

and $y_2(x)$ is a solution of the equation

$$y'' + ay' + by = f_2(x), \tag{3}$$

then $y = y_1 + y_2$ is a solution of equation (1); that is, the solution of (1) is obtained by superimposing the solution of equation (3) on that of equation (2).

Proof. Substituting $y = y_1 + y_2$ in the left-hand side of equation (1), we have

$$\begin{aligned} y'' + ay' + by &= (y_1'' + y_2'') + a(y_1' + y_2') + b(y_1 + y_2) \\ &= (y_1'' + ay_1' + by_1) + (y_2'' + ay_2' + by_2) \\ &= f_1 + f_2 = f, \end{aligned}$$

since y_1 and y_2 are solutions of equations (2) and (3), respectively. ■

Briefly, the principle of superposition tells us that if we can split the function $f(x)$ into a sum of two (or more) simpler expressions $f_k(x)$, then we can restrict our attention to solving the nonhomogeneous equations

$$y'' + ay' + by = f_k(x), \quad k = 1, 2, \ldots, m, \tag{4}$$

because the solution to equation (1) is simply the sum of the solutions of these equations.

The method we shall present in this section *requires* that the function $f(x)$ in equation (1) be of *one* of the following three forms:

(i) $P_n(x)$,
(ii) $P_n(x)e^{ax}$, or
(iii) $e^{ax}[P_n(x) \cos bx + Q_n(x) \sin bx]$,

where $P_n(x)$ and $Q_n(x)$ are polynomials in x of degree n ($n \geq 0$). The method we present below can also be used if $f(x)$ is a sum of functions $f_k(x)$ of these three forms, since by the principle of superposition we can solve each of the equations in (4) and add the solutions together. **However, if any term of $f(x)$ is not of one of these three forms, we cannot use the method of this section.**[†]

Note that the three forms involve a multitude of situations. The following three functions are all of one of these forms:

(a) $2e^{3x}$ (the polynomial is the constant 2);
(b) $e^{4x} \cos x$ (here $P_n(x) = 1$ and $Q_n(x) = 0$ are both polynomials of degree zero);
(c) $x \cos x + \sin x$ (Here $a = 0$, $b = 1$, $P_n(x) = x + 0$, and $Q_n(x) = 0 \cdot x + 1$).

The **method of undetermined coefficients** assumes that the solution to equation (1) is exactly of the same "form" as $f(x)$. The technique requires that we replace each dependent variable y in (1) with an expression of the same form as $f(x)$ having polynomial terms with **undetermined coefficients.** If we compare both sides of the resulting equation, it is then possible to "determine" the unknown coefficients. The method will be illustrated by a number of examples.

[†]Instead we must use the variation of constants method, which will be described in Section 11.11.

EXAMPLE 1 Solve the following equation:

$$y'' - y = x^2 \tag{5}$$

Solution. Since $f(x) = x^2$ is a polynomial of degree two, we "guess" that (5) has a solution $y_p(x)$ that is a polynomial of degree two, that is,

$$y_p(x) = a + bx + cx^2.$$

After calculating that $y_p'' = 2c$, we substitute $y_p(x)$ into (5) to obtain

$$2c - (a + bx + cx^2) = x^2.$$

Equating coefficients, we have

Coefficient of constant term	Coefficient of x	Coefficent of x^2
$2c - a = 0,$	$-b = 0,$	$-c = 1,$

which immediately yields $a = -2$, $b = 0$, $c = -1$, and the particular solution

$$y_p(x) = -2 - x^2.$$

This particular solution is easily verified by substitution into (5). Finally, since the general solution of the homogeneous equation $y'' - y = 0$ is given by

$$y = c_1e^x + c_2e^{-x},$$

the general solution of (5) is

$$y = c_1e^x + c_2e^{-x} - 2 - x^2. \blacksquare$$

EXAMPLE 2 Solve

$$y'' - 3y' + 2y = e^x \sin x. \tag{6}$$

Solution. Since $f(x)$ is of form (iii) with $P_n(x) = 0$ and $Q_n(x) = 1$, we "guess" that there is a solution to (6) of the form

$$y_p(x) = ae^x \sin x + be^x \cos x.$$

Then

$$y_p'(x) = (a - b)e^x \sin x + (a + b)e^x \cos x$$

and

$$y_p''(x) = 2ae^x \cos x - 2be^x \sin x.$$

Substituting these expressions into (6) we have

$$e^x(2a\cos x - 2b\sin x) - 3e^x[(a-b)\sin x + (a+b)\cos x] + 2e^x(a\sin x + b\cos x) = e^x \sin x.$$

Dividing both sides by e^x and equating the coefficients of $\sin x$ and $\cos x$, we have

$$\begin{aligned} 2a - 3(a+b) + 2b &= 0, \\ -2b - 3(a-b) + 2a &= 1, \end{aligned}$$

which yield $a = -\frac{1}{2}$ and $b = \frac{1}{2}$ so that

$$y_p = \frac{e^x}{2}(\cos x - \sin x).$$

Again this result is easily verified by substitution. Finally, the general solution of (6) is

$$y = c_1 e^{2x} + c_2 e^x + \frac{e^x}{2}(\cos x - \sin x). \blacksquare$$

EXAMPLE 3 Solve $y'' + y = xe^{2x}$.

Solution. Here $f(x)$ is of form (ii), where $P_n(x)$ is a polynomial of degree one, so we try a solution of the form

$$y_p(x) = e^{2x}(a + bx).$$

Then

$$y_p'(x) = e^{2x}(2a + b + 2bx), \quad y_p''(x) = e^{2x}(4a + 4b + 4bx),$$

and substitution yields

$$e^{2x}(4a + 4b + 4bx) + e^{2x}(a + bx) = xe^{2x}.$$

Dividing both sides by e^{2x} and equating like powers of x, we obtain the equations

$$5a + 4b = 0, \quad 5b = 1.$$

Thus $a = -\frac{4}{25}$, $b = \frac{1}{5}$, and a particular solution is

$$y_p(x) = \frac{e^{2x}}{25}(5x - 4).$$

Therefore, the general solution of this example is

$$y(x) = c_1 \sin x + c_2 \cos x + \frac{e^{2x}}{25}(5x - 4). \blacksquare$$

Difficulties arise in connection with problems of this type whenever any term of the guessed solution is a solution of the homogeneous equation

$$y'' + ay' + by = 0. \quad \textbf{(7)}$$

For example, in the equation

$$y'' + y = (1 + x + x^2)\sin x, \quad \textbf{(8)}$$

the function $f(x)$ is the sum of three functions, one of which ($\sin x$) is a solution to the homogeneous equation $y'' + y = 0$. As another example, in

$$y'' + y = (x + x^2)\sin x \quad \textbf{(9)}$$

the guessed solution is $y_p = (a_0 + a_1x + a_2x^2)\sin x + (b_0 + b_1x + b_2x^2)\cos x$ and $a_0 \sin x + b_0 \cos x$ is a solution to the homogeneous equation $y'' + y = 0$. When this situation occurs, the method of undetermined coefficients must be modified. To see why, consider the following example.

EXAMPLE 4 Find the solution to the equation

$$y'' - y = 2e^x. \quad \textbf{(10)}$$

Solution. The general solution of $y'' - y = 0$ is

$$y(x) = c_1e^x + c_2e^{-x}.$$

Here $f(x) = 2e^x$ is a solution to the homogeneous equation. If we try to find a solution of the form Ae^x, we will get nowhere since Ae^x is a solution to the homogeneous equation for every constant A and, therefore, it cannot possibly be a solution to the nonhomogeneous equation.

What do we do? Recall that if λ was a double root of the characteristic equation for a homogeneous differential equation, then two solutions are $e^{\lambda x}$ and $xe^{\lambda x}$. This suggests that we try xe^x instead of Ae^x as a possible solution to (10). Thus, we consider a particular solution of the form

$$y_p = axe^x.$$

Then $y_p' = ae^x(x + 1)$, $y_p''(x) = ae^x(x + 2)$, and

$$y_p'' - y_p = ae^x(x + 2) - axe^x = 2ae^x = 2e^x.$$

Hence $a = 1$ and $y_p = xe^x$. Thus the general solution is

$$y(x) = c_1e^x + c_2e^{-x} + xe^x. \blacksquare$$

The preceding example suggests the following rule.

MODIFICATIONS OF THE METHOD
If any term of the guessed solution $y_p(x)$ is a solution of the homogeneous equation (7), multiply $y_p(x)$ by x repeatedly until no term of the product $x^k y_p(x)$ is a solution of (7). Then, use the product $x^k y_p(x)$ to solve equation (1).

EXAMPLE 5 Find the solution to

$$y'' + y = \cos x \tag{11}$$

that satisfies $y(0) = 2$ and $y'(0) = -3$.

Solution. The general solution to $y'' + y = 0$ is $c_1 \cos x + c_2 \sin x$, and $\cos x$ is a solution, so we must use the modification of the method to find a particular solution to (11). Ordinarily we would guess a solution of the form $y_g = A \cos x + B \sin x$. Instead we multiply by x and try a solution of the form

$$y_p = Ax \cos x + Bx \sin x.$$

Note that no term of y_p is a solution to $y'' + y = 0$. Then

$$y_p' = A \cos x - Ax \sin x + B \sin x + Bx \cos x$$

and

$$\cos x \overset{\text{From (11)}}{=} y_p'' + y_p = (-2A \sin x - Ax \cos x + 2B \cos x - Bx \sin x) + (Ax \cos x + Bx \sin x)$$

$$= -2A \sin x + 2B \cos x.$$

Therefore,

$$-2A = 0,\ 2B = 1,\ B = \tfrac{1}{2}$$

and

$$y_p = \frac{1}{2}x \sin x.$$

Thus the general solution to (11) is

$$y = c_1 \cos x + c_2 \sin x + \frac{1}{2}x \sin x.$$

We are not finished yet as initial conditions were given. We have

$$y' = -c_1 \sin x + c_2 \cos x + \frac{1}{2}x \cos x + \frac{1}{2} \sin x.$$

Then

$$y(0) = c_1 = 2 \quad \text{and} \quad y'(0) = c_2 = -3,$$

which yields the unique solution

$$y(x) = 2\cos x - 3\sin x + \frac{1}{2}x\sin x. \quad \blacksquare$$

EXAMPLE 6 Find the general solution of

$$y'' - 4y' + 4y = e^{2x}.$$

Solution. The homogeneous equation $y'' - 4y' + 4y = 0$ has the independent solutions e^{2x} and xe^{2x}. Thus, multiplying $f(x) = e^{2x}$ by x twice, we look for a particular solution of the form $y_p = ax^2e^{2x}$. Then

$$y_p' = ae^{2x}(2x^2 + 2x)$$

and

$$y_p'' = ae^{2x}(4x^2 + 8x + 2),$$

so that

$$\begin{aligned} y_p'' - 4y_p' + 4y_p &= ae^{2x}(4x^2 + 8x + 2 - 8x^2 - 8x + 4x^2) \\ &= 2ae^{2x} = e^{2x}, \end{aligned}$$

or $2a = 1$ and $a = \frac{1}{2}$. Thus $y_p = \frac{1}{2}x^2e^{2x}$, and the general solution is

$$y(x) = c_1e^{2x} + c_2xe^{2x} + \tfrac{1}{2}x^2e^{2x} = e^{2x}(c_1 + c_2x + \tfrac{1}{2}x^2). \quad \blacksquare$$

EXAMPLE 7 Consider the equation

$$y'' - y = x^2 + 2e^x.$$

Using the results of Examples 1 and 4 and the principle of superposition, we find immediately that a particular solution is given by

$$y_p(x) = -2 - x^2 + xe^x. \quad \blacksquare$$

EXAMPLE 8 Find the general solution to

$$y'' + y = x\sin x.$$

Solution. The guessed solution is $y_p = (A + Bx)\cos x + (C + Dx)\sin x$. Since $A\cos x + C\sin x$ solves $y'' + y = 0$, the modification is required. We therefore multiply by x and try a solution of the form

$$y_p = (Ax^2 + Bx)\cos x + (Cx^2 + Dx)\sin x$$

Then

$$y_p' = [Cx^2 + (2A + D)x + B]\cos x + [-Ax^2 + (2C - B)x + D]\sin x$$
$$y_p'' = [-Ax^2 + (4C - B)x + 2A + 2D]\cos x + [-Cx^2 - (4A + D)x + 2C - 2B]\sin x$$

and

$$y_p'' + y_p = [4cx + 2A + 2D]\cos x + [-4Ax + 2C - 2B]\sin x \overset{\text{given}}{\downarrow =} x\sin x.$$

This yields $A = -\frac{1}{4}$, $B = 0$, $C = 0$, $D = \frac{1}{4}$ and the particular solution

$$y_p = -\tfrac{1}{4}x^2\cos x + \tfrac{1}{4}x\sin x.$$

Thus the general solution

$$y = (C_1 - \tfrac{1}{4}x^2)\cos x + (C_2 + \tfrac{1}{4}x)\sin x. \quad ■$$

Let us now summarize the results of this section as follows: Consider the nonhomogeneous equation

$$y'' + ay' + by = f(x) \tag{12}$$

and the homogeneous equation

$$y'' + ay' + by = 0. \tag{13}$$

Case 1. *No term in the guessed solution $y_p(x)$ is a solution of equation (13).* A particular solution of equation (12) will have the form $y_p(x)$ given by the table below:

$f(x)$	$y_p(x)$
$P_n(x)$	$a_0 + a_1x + a_2x^2 + \cdots + a_nx^n$
$P_n(x)e^{ax}$	$(a_0 + a_1x + a_2x^2 + \cdots + a_nx^n)e^{ax}$
$P_n(x)e^{ax}\sin bx$ + $Q_n(x)e^{ax}\cos bx$	$(a_0 + a_1x + a_2x^2 + \cdots + a_nx^n)e^{ax}\sin bx$ + $(c_0 + c_1x + c_2x + \cdots + c_nx^n)e^{ax}\cos bx$

Case 2. If any term of $y_p(x)$ is a solution of equation (12), then multiply the appropriate function $y_p(x)$ of Case 1 by x^k, where k is the smallest integer such that no term in $x^ky_p(x)$ is a solution of equation (13).

PROBLEMS 11.10

In Problems 1–13, find the general solution of each given differential equation. If initial conditions are given, then find the particular solution that satisfies them.

1. $y'' + 4y = 3 \sin x$
2. $y'' - y' - 6y = 20e^{-2x}, \quad y(0) = 0, \quad y'(0) = 6$
3. $y'' - 3y' + 2y = 6e^{3x}$
4. $y'' + y' = 3x^2, \quad y(0) = 4, \quad y'(0) = 0$
5. $y'' - 2y' + y = -4e^x$
6. $y'' - 4y' + 4y = 6xe^{2x}, \quad y(0) = 0, \quad y'(0) = 3$
7. $y'' - 7y' + 10y = 100x, \quad y(0) = 0, \quad y'(0) = 5$
8. $y'' + y = 1 + x + x^2$
9. $y'' + y' = x^3 - x^2$
10. $y'' + 4y = 16x \sin 2x$
11. $y'' - 4y' + 5y = 2e^{2x} \cos x$
12. $y'' - y' - 2y = x^2 + \cos x$
13. $y'' + 6y' + 9y = 10e^{-3x}$

Use the principle of superposition to find the general solution of each of the equations in Problems 14–17.

14. $y'' + y = 1 + 2 \sin x$
15. $y'' - 2y' - 3y = x - x^2 + e^x$
16. $y'' + 4y = 3 \cos 2x - 7x^2$
17. $y'' + 4y' + 4y = xe^x + \sin x$

18. Show by the methods of this section that a particular solution of

$$y'' + 2ay' + b^2y = A \sin \omega x \quad (a, \omega > 0)$$

is given by

$$y = \frac{A \sin (\omega x - \alpha)}{\sqrt{(b^2 - \omega^2)^2 + 4\omega^2 a^2}},$$

where

$$\alpha = \tan^{-1} \frac{2a\omega}{(b^2 - \omega^2)}, \quad (0 < \alpha < \pi).$$

19. Let $f(x)$ be a polynomial of degree n. Show that, if $b \neq 0$, then there is always a solution that is a polynomial of degree n for the equation $y'' + ay' + by = f(x)$.
20. Use the method indicated in Problem 19 to find a particular solution of

$$y'' + 3y' + 2y = 9 + 2x - 2x^2.$$

In Problems 21–24 find particular solutions to the given differential equation.

21. $y'' + y = (x + x^2) \sin x$
22. $y'' - y' = x^2$
23. $y'' - 2y' + y = x^2e^x$
24. $y'' - 4y' + 3y = x^3e^{3x}$

11.11 NONHOMOGENEOUS EQUATIONS: VARIATION OF CONSTANTS†

In this section we shall consider a procedure developed by J. L. Lagrange (see page 255) for finding a particular solution of any nonhomogeneous linear equation

$$y'' + a(x)y' + b(x)y = f(x), \tag{1}$$

where the functions $a(x)$, $b(x)$, and $f(x)$ are continuous. To use this method it is necessary to know the general solution $c_1y_1(x) + c_2y_2(x)$ of the homogeneous equation

$$y'' + a(x)y' + b(x)y = 0. \tag{2}$$

If $a(x)$ and $b(x)$ are constants, the general solution to (2) can always be obtained by the methods of Sections 11.8 and 11.9. If $a(x)$ and $b(x)$ are not both constants, it may be

†This procedure is also called the **variation of parameters method,** or **Lagrange's method.**

difficult to find this general solution; however, if one solution y_1 of (2) can be found, then the method of Section 11.7 will yield the general solution to (2).

Lagrange noticed that any particular solution y_p of (1) must have the property that y_p/y_1 and y_p/y_2 are not constants, suggesting that we look for a particular solution of (1) of the form

$$y(x) = c_1(x)y_1(x) + c_2(x)y_2(x). \tag{3}$$

This replacement of constants by variables gives the method its name. Differentiating equation (3), we obtain

$$y'(x) = c_1(x)y_1'(x) + c_2(x)y_2'(x) + c_1'(x)y_1(x) + c_2'(x)y_2(x).$$

To simplify this expression, it is convenient (but not necessary—see Problem 14) to set

$$c_1'(x)y_1(x) + c_2'(x)y_2(x) = 0. \tag{4}$$

Then

$$y'(x) = c_1(x)y_1'(x) + c_2(x)y_2'(x).$$

Differentiating once again, we obtain

$$y''(x) = c_1(x)y_1''(x) + c_2(x)y_2''(x) + c_1'(x)y_1'(x) + c_2'(x)y_2'(x).$$

Substitution of the expressions for $y(x)$, $y'(x)$ and $y''(x)$ into (1) yields

$$\begin{aligned} y'' + a(x)y' + b(x)y &= c_1(x)(y_1'' + ay_1' + by_1) + c_2(x)(y_2'' + ay_2' + by_2) \\ &\quad + c_1'y_1' + c_2'y_2' \\ &= f(x). \end{aligned}$$

But y_1 and y_2 are solutions to the homogeneous equation so that the equation above reduces to

$$c_1'y_1' + c_2'y_2' = f(x). \tag{5}$$

This gives a second equation relating $c_1'(x)$ and $c_2'(x)$, and we have the simultaneous equations

$$\begin{aligned} y_1c_1' + y_2c_2' &= 0, \\ y_1'c_1' + y_2'c_2' &= f \end{aligned} \tag{6}$$

The determinant of system (6) is

$$\begin{vmatrix} y_1 & y_2 \\ y_1' & y_2' \end{vmatrix} = W(y_1, y_2) \neq 0$$

Since y_1 and y_2 are linearly independent.

Thus, for each value of x, $c_1'(x)$ and $c_2'(x)$ are uniquely determined, and the problem has essentially been solved. We obtain,

$$\begin{aligned} y_1y_2'c_1' + y_2y_2'c_2' &= 0 && \text{First equation multiplied by } y_2'. \\ y_1'y_2c_1' + y_2y_2'c_2' &= y_2f(x) && \text{Second equation multiplied by } y_2. \\ \hline (y_1y_2' - y_1'y_2)c_1' &= -y_2f(x) \end{aligned}$$

or

$$c_1' = \frac{-y_2f(x)}{W(y_1,y_2)(x)}.$$

A similar calculation yields an expression for c_2'. Thus, we obtain

$$c_1'(x) = \frac{-f(x)y_2(x)}{y_1(x)y_2'(x) - y_1'(x)y_2(x)} = \frac{-f(x)y_2(x)}{W(y_1, y_2)(x)} \tag{7}$$

$$c_2'(x) = \frac{f(x)y_1(x)}{y_1(x)y_2'(x) - y_1'(x)y_2(x)} = \frac{f(x)y_1(x)}{W(y_1, y_2)(x)}. \tag{8}$$

Finally, if we can integrate c_1' and c_2', we can substitute c_1 and c_2 into equation (3) to obtain a particular solution to the nonhomogeneous equation.

EXAMPLE 1 Solve $y'' - y = e^{2x}$ by the variation of constants method.

Solution. Solutions to the homogeneous equation are $y_1 = e^{-x}$ and $y_2 = e^x$. We obtain $W(y_1, y_2) = 2$, so that equations (7) and (8) become

$$c_1'(x) = \frac{-e^{2x}e^x}{2} = \frac{-e^{3x}}{2}, \; c_2'(x) = \frac{e^{2x}e^{-x}}{2} = \frac{e^x}{2}.$$

Integrating these functions, we obtain $c_1(x) = -\dfrac{e^{3x}}{6}$ and $c_2(x) = \dfrac{e^x}{2}$. A particular solution is therefore

$$c_1(x)y_1(x) + c_2(x)y_2(x) = \frac{-e^{3x}}{6}e^{-x} + \frac{e^x}{2}e^x = \frac{-e^{2x}}{6} + \frac{e^{2x}}{2} = \frac{e^{2x}}{3}$$

and the general solution is

$$y(x) = c_1e^x + c_2e^{-x} + \frac{e^{2x}}{3}. \blacksquare$$

EXAMPLE 2 Determine the solution of

$$y'' + y = 4 \sin x$$

that satisfies $y(0) = 3$ and $y'(0) = -1$.

Solution. Here $y_1(x) = \cos x$, $y_2(x) = \sin x$ and $W(y_1, y_2) = 1$, so that $c_1'(x) = -4\sin^2 x$ and $c_2'(x) = 4\sin x\cos x$. Since

$$\int \sin^2 x\, dx = \frac{1}{2}(x - \sin x\cos x),$$

we see that

$$c_1(x) = 2(\sin x\cos x - x),\ c_2(x) = 2\sin^2 x$$

and a particular solution is

$$\begin{aligned} y_p = c_1(x)y_1(x) + c_2(x)y_2(x) &= 2\sin x\cos x(\cos x) - 2x\cos x + 2\sin^3 x \\ &= 2\sin x(\cos^2 x + \sin^2 x) - 2x\cos x \\ &= 2\sin x - 2x\cos x. \end{aligned}$$

Thus the general solution is

$$\begin{aligned} y &= c_1\cos x + c_2\sin x + 2\sin x - 2x\cos x \\ &= c_1\cos x + c_2^*\sin x - 2x\cos x, \end{aligned}$$

and

$$y' = -c_1\sin x + c_2^*\cos x + 2x\sin x - 2\cos x.$$

But

$$3 = y(0) = c_1 \text{ and } -1 = y'(0) = c_2^* - 2, \text{ so that } c_2^* = 1.$$

The unique solution is therefore

$$y = 3\cos x + \sin x - 2x\cos x = (3 - 2x)\cos x + \sin x. \quad \blacksquare$$

EXAMPLE 3 Solve $y'' + y = \tan x$.

Solution. The solutions to the homogeneous equation are $y_1 = \cos x$ and $y_2 = \sin x$. Also $W(y_1, y_2) = 1$, so that (7) become

$$c_1'(x) = -\tan x\sin x = -\frac{\sin^2 x}{\cos x} = \frac{\cos^2 x - 1}{\cos x} = \cos x - \sec x,$$

$$c_2'(x) = \tan x\cos x = \sin x.$$

Hence

$$c_1(x) = \sin x - \ln|\sec x + \tan x|$$

and

$$c_2(x) = -\cos x.$$

Thus the particular solution is

$$\begin{aligned} y_p(x) &= c_1(x)y_1(x) + c_2(x)y_2(x) \\ &= \cos x \sin x - \cos x \ln|\sec x + \tan x| - \sin x \cos x \\ &= -\cos x \ln|\sec x + \tan x|, \end{aligned}$$

and the general solution is

$$y(x) = c_1 \cos x + c_2 \sin x - \cos x \ln|\sec x + \tan x|. \blacksquare$$

Example 3 illustrates that there are instances in which we cannot apply the method of undetermined coefficients. (Try to "guess" a solution in this case.) As a rule, the method of undetermined coefficients is easier to use if the function $f(x)$ is in the right form. However, the method of variation of parameters is far more general, since it will yield a solution whenever the functions c_1' and c_2' have known antiderivatives.

PROBLEMS 11.11

In Problems 1–10 find the general solution of each equation by the variation of constants method.

1. $y'' - y' = \sec^2 x - \tan x$

2. $y'' + y = \cot x$

3. $y'' + 4y = \sec 2x$

4. $y'' + 4y = \sec x \tan x$

5. $y'' - 2y' + y = \dfrac{e^x}{(1-x)^2}$

6. $y'' - y = \sin^2 x$

7. $y'' - y = \dfrac{(2x-1)e^x}{x^2}$

8. $y'' - 3y' - 4y = \dfrac{e^{4x}(5x-2)}{x^3}$

9. $y'' - 4y' + 4y = \dfrac{e^{2x}}{(1+x)}$

10. $y'' + 2y' + y = e^{-x} \ln|x|$

11. Find a particular solution of

$$y'' + \frac{1}{x}y' - \frac{y}{x^2} = \frac{1}{x^2 + x^3} \quad (x > 0),$$

given that two solutions of the associated homogeneous equations are $y_1 = x$ and $y_2 = 1/x$.

12. Find a particular solution of

$$y'' - \frac{2}{x}y' + \frac{2}{x^2}y = \frac{\ln|x|}{x} \quad (x > 0),$$

given the two homogeneous solutions $y_1 = x$ and $y_2 = x^2$.

***13.** Verify that

$$y = \frac{1}{\omega}\int_0^x f(t) \sin \omega(x - t)\, dt$$

is a particular solution of $y'' + \omega^2 y = f(x)$.

***14.** This problem will show why there is no loss in generality in equation (4) by setting

$$c_1'y_1 + c_2'y_2 = 0.$$

Suppose instead that we let $c_1'y_1 + c_2'y_2 = z(x)$, with $z(x)$ an undetermined function of x.

(a) Show that we then obtain the system

$$\begin{aligned} c_1'y_1 + c_2'y_2 &= z, \\ c_1'y_1' + c_2'y_2' &= f - z' - az. \end{aligned}$$

(b) Show that the system in part (a) has the solution

$$c_1' = \frac{-y_2 f}{W(y_1, y_2)} + \frac{(e^{\int a(x)\,dx} z y_2)'}{e^{\int a(x)\,dx} W(y_1, y_2)}$$

$$c_2' = \frac{y_1 f}{W(y_1, y_2)} - \frac{(e^{\int a(x)\,dx} z y_1)'}{e^{\int a(x)\,dx} W(y_1, y_2)}.$$

(c) Integrate by parts to show that

$$\int \frac{(e^{\int a(x)\,dx} z y_i)'}{e^{\int a(x)\,dx} W(y_1, y_2)}\,dx = \frac{z y_i}{W(y_1, y_2)}, \quad i = 1, 2.$$

(d) Conclude that the particular solution obtained by letting $c_1' y_1 + c_2' y_2 = z$ is identical to that obtained by assuming equation (4).

(e) Letting t be a dummy variable of integration, show the particular solution can always be represented by the integral

$$y_p(x) = \int^x \frac{y_2(x) y_1(t) - y_1(x) y_2(t)}{W(y_1, y_2)(t)} f(t)\,dt.$$

11.12 EULER EQUATIONS

For most linear second-order equations with variable coefficients it is impossible to write solutions in terms of elementary functions (i.e., the functions we know). In most cases it is necessary to use techniques such as the power series method (Section 13.14) to obtain information about solutions. However, there is one class of such equations for which closed form solutions can be obtained.

Definition 1 EULER EQUATION† An equation of the form

$$x^2 y'' + axy' + by = f(x), \quad x \neq 0 \tag{1}$$

is called a **Euler equation.**

NOTE. Equation (1) can be written as

$$y'' + \frac{1}{x} y' + \frac{1}{x^2} y = \frac{f(x)}{x^2}$$

which is not defined for $x = 0$. This is why we make the restriction that $x \neq 0$.

We begin by solving the homogeneous Euler equation

$$x^2 y'' + axy' + by = 0, \quad x \neq 0 \tag{2}$$

If we can find two linearly independent solutions to (2), we can solve (1) by the method of variation of constants. There are two ways to solve equation (2). Each involves a trick. We give one method here and leave the other method for the problem set.

Our method involves guessing an appropriate solution to (2). We note that if $y = x^\lambda$ for some number λ, then $y' = \lambda x^{\lambda-1}$ and $y'' = \lambda(\lambda - 1)x^{\lambda-2}$. This is interesting because then $x^2 y''$, xy' and y all can be written as constant multiples of x^λ. Therefore,

†See the biographical sketch of Euler on page 265.

we guess that there is a solution having the form $y = x^\lambda$. Then, substituting this into equation (2), we obtain

$$\lambda(\lambda - 1)x^\lambda + a\lambda x^\lambda + bx^\lambda = x^\lambda[\lambda(\lambda - 1) + a\lambda + b] = 0.$$

If $x \neq 0$, we can divide by x^λ to obtain the **characteristic equation for Euler's equation:**

$$\lambda(\lambda - 1) + a\lambda + b = 0 \tag{3}$$

or

$$\lambda^2 + (a - 1)\lambda + b = 0. \tag{4}$$

As with constant-coefficient equations, there are three cases to consider.

Case 1. The characteristic equation (4) has two real, distinct roots.

EXAMPLE 1 Find the general solution to

$$x^2y'' + 2xy' - 12y = 0.$$

Solution. The characteristic equation is

$$\lambda(\lambda - 1) + 2\lambda - 12 = \lambda^2 + \lambda - 12 = 0 = (\lambda + 4)(\lambda - 3)$$

with roots $\lambda_1 = -4$ and $\lambda_2 = 3$. Thus, two solutions (that are linearly independent) are

$$y_1 = x^{-4} = \frac{1}{x^4} \quad \text{and} \quad y_2 = x^3$$

The general solution is

$$y(x) = \frac{c_1}{x^4} + c_2x^3. \quad \blacksquare$$

In general, we have the following result:

Theorem 1 If λ_1 and λ_2 are real and distinct, then the general solution to equation (2) is

$$y(x) = c_1x^{\lambda_1} + c_2x^{\lambda_2}, \quad x \neq 0 \tag{5}$$

Case 2. The roots are real and equal ($\lambda_1 = \lambda_2$).

EXAMPLE 2 Find the general solution to

$$x^2y'' - 3xy' + 4y = 0. \tag{6}$$

Solution. The characteristic equation is

$$\lambda^2 - 4\lambda + 4 = (\lambda - 2)^2 = 0$$

with the single root $\lambda = 2$. Thus one solution is $y_1(x) = x^2$. To find a second solution, we use formula (8) on page 667. First, we write (6) in the form

$$y'' - \frac{3}{x}y' + \frac{4}{x^2}y = 0.$$

As in Section 11.7, $a(x) = -\dfrac{3}{x}$ so that

$$e^{-\int a(x)\,dx} = e^{\int 3/x\,dx} = e^{3\ln x} = x^3$$

and

$$y_2(x) = y_1(x)\int \frac{e^{-\int a(x)\,dx}}{y_1^2(x)}\,dx = x^2\int \frac{x^3}{x^4}\,dx = x^2 \ln x.$$

Thus the general solution to equation (6) is

$$y(x) = c_1x^2 + c_2x^2 \ln x = x^2(c_1 + c_2 \ln x). \quad \blacksquare$$

Theorem 2 If λ is the only root of the characteristic equation (4), then the general solution to (2) is

$$y(x) = x^{\lambda}(c_1 + c_2 \ln x).$$

Case 3. The roots are complex conjugates $(\lambda_1 = \alpha + i\beta,\ \lambda_2 = \alpha - i\beta)$.

EXAMPLE 3 Find the general solution of

$$x^2y'' + 5xy' + 13y = 0. \qquad \textbf{(7)}$$

Solution. The characteristic equation is

$$\lambda^2 + 4\lambda + 13 = 0$$

and

$$\lambda = \frac{-4 \pm \sqrt{16 - 4(13)}}{2} = \frac{-4 \pm \sqrt{-36}}{2} = -2 \pm 3i.$$

Thus two linearly independent solutions are

$$y_1(x) = x^{-2+3i} \quad \text{and} \quad y_2(x) = x^{-2-3i}.$$

Using the material in Appendix 3 we can eliminate the imaginary exponents. First we note that

$$x^a = e^{\ln x^a} = e^{a \ln x}.$$

Then, by Euler's formula,

$$y_1(x) = (x^{-2})(x^{3i}) = x^{-2}e^{3i \ln x} = x^{-2}[\cos (3 \ln x) + i \sin (3 \ln x)]$$

and

$$y_2(x) = (x^{-2})(x^{-3i}) = x^{-2}e^{-3i \ln x} = x^{-2}[\cos (3 \ln x) - i \sin (3 \ln x)].$$

We now form two new solutions:

$$y_3(x) = \frac{1}{2}[y_1(x) + y_2(x)] = x^{-2} \cos (3 \ln x)$$

and

$$y_4(x) = \frac{1}{2i}[y_1(x) - y_2(x)] = x^{-2} \sin (3 \ln x)$$

These new solutions contain no complex numbers and are easier to work with. The general solution to (7) is

$$y(x) = x^{-2}[c_1 \cos (3 \ln x) + c_2 \sin (3 \ln x)]. \blacksquare$$

Theorem 3 If $\lambda_1 = \alpha + i\beta$ and $\lambda_2 = \alpha - i\beta$ are complex conjugate roots of the characteristic equation (4), then the general solution to (2) is

$$y(x) = x^{\alpha}[c_1 \cos (\beta \ln x) + c_2 \sin (\beta \ln x)] \tag{8}$$

EXAMPLE 4 Find the general solution to

$$x^2y'' + 2xy' - 12y = \sqrt{x}. \tag{9}$$

Solution. In Example 1 we found the homogeneous solutions

$$y_1 = x^{-4} \quad \text{and} \quad y_2 = x^3.$$

Then

$$W(y_1, y_2)(x) = \begin{vmatrix} x^{-4} & x^3 \\ -4x^{-5} & 3x^2 \end{vmatrix} = 3x^{-2} + 4x^{-2} = \frac{7}{x^2} \text{ and } \frac{1}{W} = \frac{x^2}{7}.$$

We rewrite (9) in the standard form:

$$y'' + \frac{1}{x}y' - \frac{12}{x^2}y = x^{-3/2}$$

This is the form for which the variation of constants formulas, formulas (7) and (8), on page 684 apply. Then $f(x) = x^{-3/2}$, and we obtain

$$c_1'(x) = \frac{-f(x)y_2(x)}{W(y_1, y_2)(x)} = \frac{x^2}{7}(-x^{-3/2})(x^3) = \frac{-x^{7/2}}{7},$$

and

$$c_2'(x) = \frac{f(x)y_1(x)}{W(y_1, y_2)(x)} = \frac{x^2}{7}(x^{-3/2})(x^{-4}) = \frac{1}{7}x^{-7/2}.$$

Hence

$$c_1(x) = -\frac{1}{7}\cdot\frac{2}{9}x^{9/2}, \quad c_2(x) = -\frac{1}{7}\cdot\frac{2}{5}x^{-5/2},$$

so that

$$y_p(x) = c_1(x)y_1(x) + c_2(x)y_2(x) = -\frac{1}{7}\left[\frac{2}{9}x^{9/2}\cdot x^{-4} + \frac{2}{5}x^{-5/2}\cdot x^3\right]$$

$$= \frac{-x^{1/2}}{7}\left(\frac{2}{9} + \frac{2}{5}\right) = -\frac{4}{45}x^{1/2}.$$

Thus the general solution is given by

$$y(x) = c_1x^{-4} + c_2x^3 - \frac{4}{45}x^{1/2}. \quad \blacksquare$$

PROBLEMS 11.12

In Problems 1–17 find the general solution to the given Euler equation. Find the unique solution when initial conditions are given.

1. $x^2y'' + xy' - y = 0$
2. $x^2y'' - 5xy' + 9y = 0$
3. $x^2y'' - xy' + 2y = 0$
4. $x^2y'' - 2y = 0, \quad y(1) = 3, \quad y'(1) = 1$
5. $4x^2y'' - 4xy' + 3y = 0, \quad y(1) = 0, \quad y'(1) = 1$
6. $x^2y'' + 3xy' + 2y = 0$
7. $x^2y'' - 3xy' + 3y = 0$
8. $x^2y'' + 5xy' + 4y = 0, \quad y(1) = 1, \quad y'(1) = 3$
9. $x^2y'' + 5xy' + 5y = 0$
10. $4x^2y'' - 8xy' + 8y = 0$
11. $x^2y'' + 2xy' - 12y = 0$
12. $x^2y'' + xy' + y = 0$
13. $x^2y'' + 7xy' + 5y = x$
14. $x^2y'' + 3xy' - 3y = 5x^2$
15. $x^2y'' - 2y = \ln x, \quad x > 0$
16. $4x^2y'' - 4xy' + 3y = \sin \ln(-x), \quad x < 0$

*17. Show that the homogeneous Euler equation (2) can be transformed into the constant coefficient equation $y'' + (a - 1)y' + by = 0$ by making the substitution $x = e^t$.

11.13 VIBRATORY MOTION (Optional)

Differential equations were first studied in attempts to describe the motion of particles. As a simple example, consider a mass m attached to a coiled spring of length l_0, the upper end of which is securely fastened (see Figure 1).

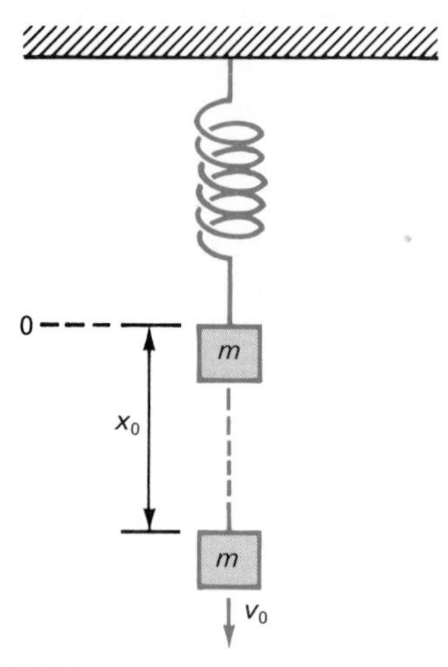

FIGURE 1

We have denoted by zero the equilibrium position of the mass on the spring, that is, the point where the mass remains at rest. Suppose that the mass is given an initial displacement x_0, and an initial velocity v_0. Can we describe the future movement of the mass? To do so, we make the following assumptions about the force[†] exerted by the spring on the mass:

(i) The force acts along a vertical line through the center of gravity of the mass (which is then treated as if it were a point mass), and its direction is always from the mass toward the point of equilibrium.

(ii) At any time t the magnitude of the force exerted on the mass is proportional to the difference between the length l of the spring and its equilibrium length l_0. The positive constant of proportionality k is called the **spring constant.** This relationship between the force and length of the spring is known as **Hooke's law.**

[†]The most common systems of units are given in the table below.

System of units	Force	Length	Mass	Time
International (SI)	Newton (N)	Meter (m)	Kilogram (kg)	Second (sec)
English	Pound (lb)	Foot (ft)	Slug	Second (sec)

1 N = 1 kg-m/sec^2 = 0.22481 lb   1 kg = 0.06852 slug
1 m = 3.28084 ft   1 lb = 1 slug-ft/sec^2 = 4.4482 N

Newton's second law of motion states that the force F acting on a particle moving with varying velocity v is equal to the time rate of change of the momentum mv. Since the mass is constant,

$$F = \frac{d(mv)}{dt} = ma.$$

Equating the two forces and applying Hooke's law, we have

$$m\frac{d^2x}{dt^2} = -kx, \tag{1}$$

where $x(t)$ denotes the displacement from equilibrium of the spring and is positive when the spring is stretched. The negative sign in equation (1) is present because the force always acts toward the equilibrium position and therefore is in the negative direction when x is positive.

Note that we have assumed that all other forces acting on the spring (such as friction, air resistance, etc.) can be ignored. Equation (1) yields the initial value problem

$$\frac{d^2x}{dt^2} + \frac{k}{m}x = 0, \quad x(0) = x_0, \quad x'(0) = v_0. \tag{2}$$

To find the solution of equation (2), the characteristic equation has the complex roots $\pm i\omega_0$, where $\omega_0 = \sqrt{k/m}$. This leads to the general solution

$$x(t) = c_1 \cos \omega_0 t + c_2 \sin \omega_0 t.$$

Using the initial conditions, we find that $c_1 = x_0$ and $c_2 = v_0/\omega_0$, so that the solution of equation (2) is given by

$$x(t) = x_0 \cos \omega_0 t + (v_0/\omega_0) \sin \omega_0 t. \tag{3}$$

It is useful to write $x(t)$ in the form

$$x(t) = A \sin(\omega_0 t + \phi).$$

To do so we use the trigonometric formula for $\sin(x + y)$:

$$x(t) = A \sin(\omega_0 t + \phi) = A \sin \omega_0 t \cos \phi + A \cos \omega_0 t \sin \phi$$

$$\overset{\text{From 3}}{\downarrow}$$
$$= x_0 \cos \omega_0 t + \frac{v_0}{\omega_0} \sin \omega_0 t.$$

Thus, equating coefficients of $\sin \omega_0 t$ and $\cos \omega_0 t$, we have

$$A \sin \phi = x_0 \qquad \text{and} \qquad A \cos \phi = \frac{v_0}{\omega_0}$$

and

$$x_0^2 + \left(\frac{v_0}{\omega_0}\right)^2 = A^2 \sin^2 \phi + A^2 \cos^2 \phi = A^2(\sin^2 \phi + \cos^2 \phi) = A^2,$$

so that

$$A = \sqrt{x_0^2 + (v_0/\omega_0)^2}.$$

Also,

$$\cos \phi = \frac{1}{A}\frac{v_0}{\omega_0} \quad \text{and} \quad \sin \phi = \frac{x_0}{A}$$

so that

$$\tan \phi = \frac{\sin \phi}{\cos \phi} = \frac{x_0/A}{v_0/\omega_0 A} = \frac{x_0\omega_0}{v_0}.$$

Thus we may write equation (3) as

$$x(t) = A \sin (\omega_0 t + \phi), \tag{4}$$

with $A = \sqrt{x_0^2 + (v_0/\omega_0)^2}$ and $\phi = \tan^{-1} \dfrac{x_0\omega_0}{v_0}$.

REMARK. The function $A \sin (\omega_0 t + \phi)$ is easy to graph. We do so in three steps:

(i) $\sin \omega_0 t$ is periodic of *period* $2\pi/\omega_0$ since

$$\sin [\omega_0(t + 2\pi/\omega_0)] = \sin (\omega_0 t + 2\pi) = \sin \omega_0 t$$

Thus the graph of $\sin \omega_0 t$ has the same shape as the graph of $\sin t$ except that it repeats every $2\pi/\omega_0$ units (instead of every 2π units).

(ii) The graph of $A \sin \omega_0 t$ is the graph of $\sin \omega_0 t$ multiplied by the *amplitude A*. That is, just as $\sin \omega_0 t$ ranges from -1 to 1, $A \sin \omega_0 t$ ranges from $-A$ to A.

(iii) The graph of $f(x + c)$, for $c > 0$, is the graph of $f(x)$ shifted c units to the left. For example, the graph of $(x + 2)^2$ is the graph of x^2 shifted two units to the left (see Figure 2).

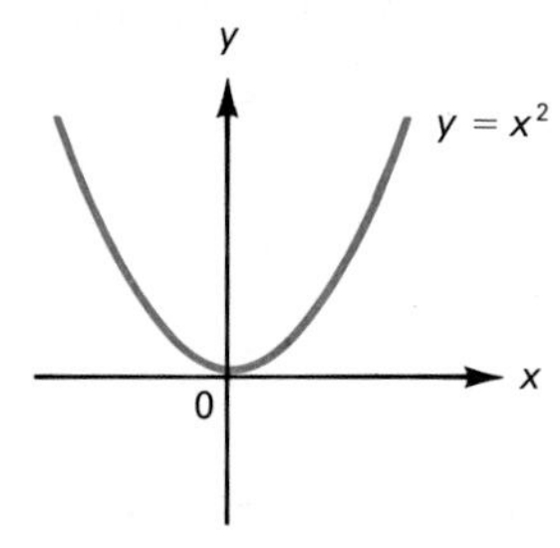

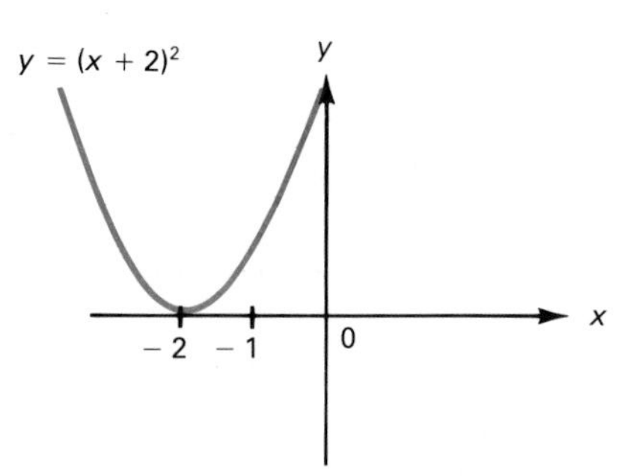

FIGURE 2

Thus the graph of $A \sin(\omega_0 t + \phi) = A \sin \omega_0\left(t + \dfrac{\phi}{\omega_0}\right)$ is the graph of $A \sin \omega_0 t$ shifted ϕ/ω_0 units to the left.

EXAMPLE 1 Sketch the graph of $x = 3(\sin 4t + \pi/6)$

Solution. We do this in three steps, starting with the graph of $\sin x$. Here $\omega_0 = 4$, so the graph is periodic of period $2\pi/4 = \pi/2$. Also, we note that

$$\frac{\phi}{\omega_0} = \frac{\pi/6}{4} = \frac{\pi}{24}.$$

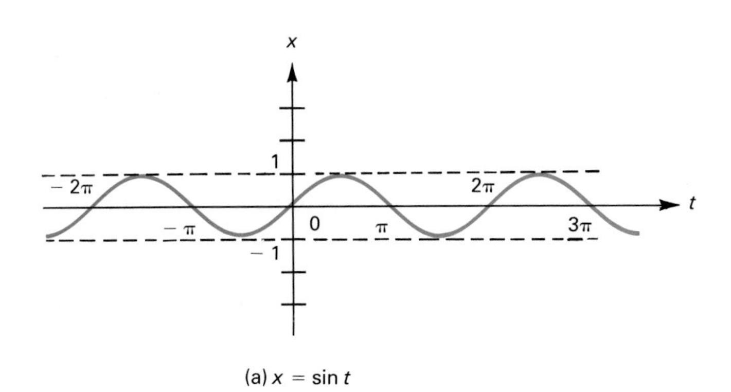

(a) $x = \sin t$

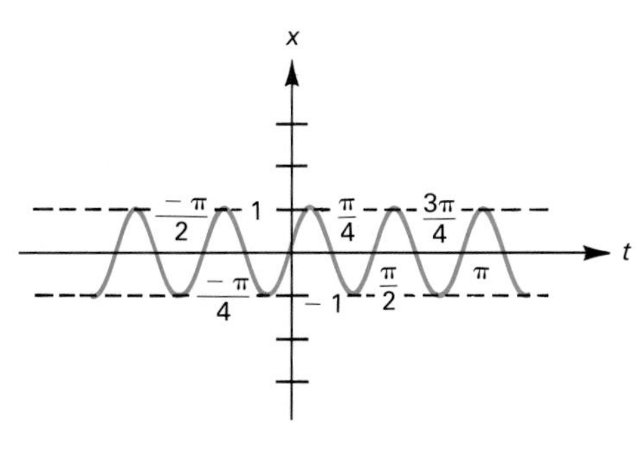

(b) $x = \sin 4t$

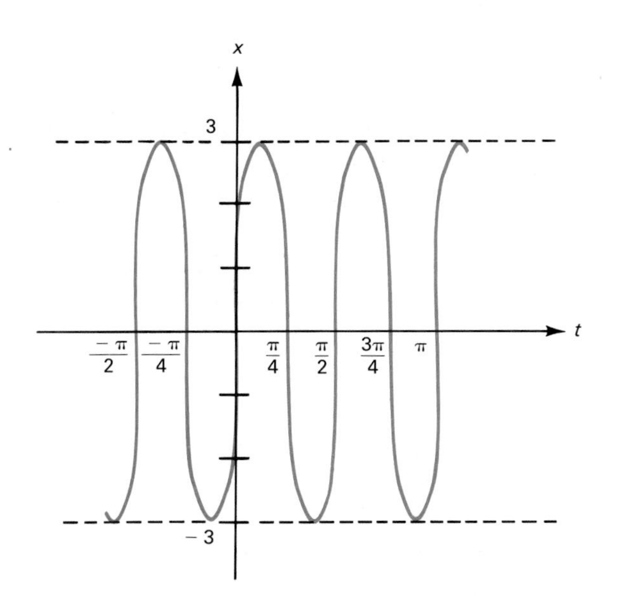

(c) $x = 3 \sin 4t$

FIGURE 3

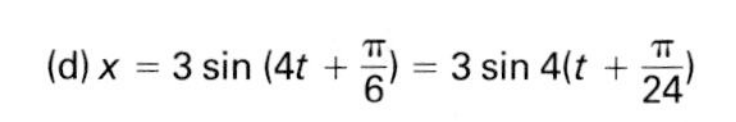
(d) $x = 3 \sin\left(4t + \frac{\pi}{6}\right) = 3 \sin 4\left(t + \frac{\pi}{24}\right)$

■

EXAMPLE 2 Suppose that $x_0 = 0.5$ m, $k = 0.4$, $m = 10$ kg and $v_0 = 0.25$ m/sec. Then $\omega_0 = \sqrt{k/m} = \sqrt{0.4/10} = \sqrt{0.04} = 0.2$, and equation (3) becomes

$$x(t) = 0.5 \cos 0.2t + \frac{0.25}{0.2} \sin 0.2t = 0.5 \cos 0.2t + 1.25 \sin 0.2t.$$

Now

$$A = \sqrt{0.5^2 + 1.25^2} = \sqrt{1.8125} \approx 1.3463$$

and

$$\phi = \tan^{-1} \frac{(0.5)(0.2)}{0.25} = \tan^{-1} 0.4 \approx 0.3805 \text{ radians } (\approx 21.8°).$$

We can therefore write

$$x(t) \approx 1.3463 \sin (0.2t + 0.3805). \blacksquare$$

Because of equation (4), the motion of the mass is called **simple harmonic motion.**

That equation indicates that the mass oscillates between the extreme positions $\pm A$; A is called the **amplitude** of the motion. Since the sine term has period $2\pi/\omega_0$, this is the time required for each complete oscillation. The **natural frequency** f of the motion is the number of complete oscillations per unit time:†

$$f = \frac{\omega_0}{2\pi}. \tag{5}$$

Note that although the amplitude depends on the initial conditions, the frequency does not.

EXAMPLE 3 Consider a spring fixed at its upper end and supporting a weight of 10 pounds at its lower end. Suppose the 10-pound weight stretches the spring by 6 inches. Find the equation of motion of the weight if it is drawn to a position 4 inches below its equilibrium position and released.

Solution. By Hooke's law, since a force of 10 lb stretches the spring by $\frac{1}{2}$ ft, $10 = k(\frac{1}{2})$ or $k = 20$ lb/ft. We are given the initial values $x_0 = \frac{1}{3}$ ft and $v_0 = 0$, so by equation (3) and the identity‡ $k/m = gk/w = 64 \text{ sec}^{-2}$, we obtain

$$x(t) = \tfrac{1}{3} \cos 8t \text{ ft}.$$

Thus the amplitude is $\frac{1}{3}$ ft ($= 4$ in.), and the frequency is $f = 4/\pi$ hertz. $\blacksquare$

---

†Cycles/sec = hertz (Hz)

‡The identity $w = mg$ may be used to convert weight to mass. Keep in mind that pounds or Newtons are a unit of weight (force) whereas slugs or kilograms are units of mass. The gravitational constant $g = 9.80 \text{ m/sec}^2 = 32 \text{ ft/sec}^2$ (approximately).

DAMPED VIBRATIONS

Throughout the discussion above we made the assumption that there were no external forces acting on the spring. This assumption, however, is not very realistic. To take care of such things as friction in the spring and air resistance, we now assume that there is a *damping* force (that tends to slow things down), which can be thought of as the resultant of all external forces (except gravity) acting on the spring. It is reasonable to assume that the magnitude of the damping force is proportional to the velocity of the particle (for example, the slower the movement, the smaller the air resistance). Therefore, to equation (1) we add the term $c(dx/dt)$, where c is the damping constant that depends on all external factors. This constant could be determined experimentally. The equation of motion then becomes

$$\frac{d^2x}{dt^2} = -\frac{k}{m}x - \frac{c}{m}\frac{dx}{dt}, \quad x(0) = x_0, \quad x'(0) = v_0. \tag{6}$$

[Of course, since c depends on external factors, it may very well not be a constant at all but may vary with time and position. In that case, c is really $c(t, x)$, and the equation becomes much harder to analyze than the constant coefficient case.]

To study equation (6), we first find the roots of the characteristic equation:

$$\frac{-c \pm \sqrt{c^2 - 4mk}}{2m}. \tag{7}$$

The nature of the general solution depends on the discriminant $\sqrt{c^2 - 4mk}$. If $c^2 > 4mk$, both roots are negative since $\sqrt{c^2 - 4mk} < c$. In this case

$$x(t) = c_1 \exp\left(\frac{-c + \sqrt{c^2 - 4mk}}{2m}t\right) + c_2 \exp\left(\frac{-c - \sqrt{c^2 - 4mk}}{2m}t\right) \tag{8}$$

becomes small as t becomes large whatever the initial conditions may be. Similarly, in the event the discriminant vanishes, then

$$x(t) = e^{(-c/2m)t}(c_1 + c_2t), \tag{9}$$

and the solution has a similar behavior. For example, if $c = 5$ (lb-sec/ft) in Example 3, then the discriminant vanishes since $4mk = 4 \cdot \frac{10}{32} \cdot 20 = 25$ and

$$x(t) = e^{-8t}(c_1 + c_2t).$$

Applying the initial conditions yields

$$x(t) = \tfrac{1}{3}e^{-8t}(1 + 8t) \text{ ft},$$

which has the graph shown in Figure 4. We observe that the solution does not oscillate. This type of motion can take place in a highly viscous medium (such as oil or water).

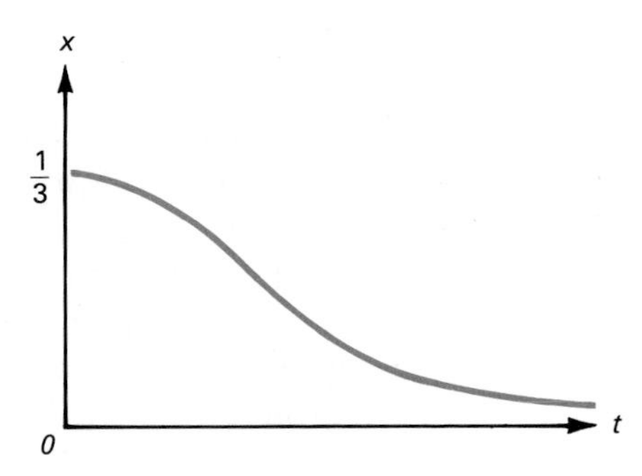

FIGURE 4

If $c^2 < 4mk$, the general solution is

$$x(t) = e^{(-c/2m)t}\left(c_1 \cos \frac{\sqrt{4mk - c^2}}{2m} t + c_2 \sin \frac{\sqrt{4mk - c^2}}{2m} t\right), \tag{10}$$

which shows an oscillation with frequency

$$f = \frac{\sqrt{4mk - c^2}}{4\pi m}.$$

The factor $e^{(-c/2m)t}$ is called the **damping factor.** In Example 3, $F = 10$ so $m = F/a = \frac{10}{32}$. If, in this example, we set $c = 4$ (lb − sec/ft), we are led to the general solution

$$x(t) = e^{-32t/5}(c_1 \cos \tfrac{24}{5}t + c_2 \sin \tfrac{24}{5}t) \text{ ft}.$$

Using the initial values, we find that $c_1 = \frac{1}{3}$, $c_2 = \frac{4}{9}$, and the motion is illustrated in Figure 5.

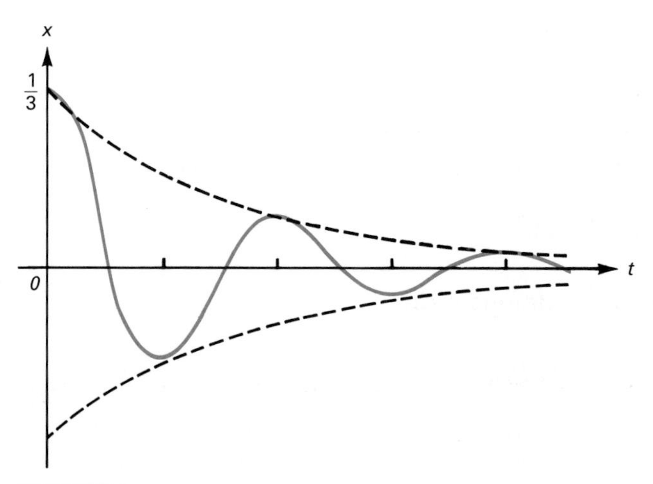

FIGURE 5

FORCED VIBRATIONS

The motion of the mass considered in the two cases above is determined by the inherent forces of the spring-weight system and the natural forces acting on the system. Accordingly, the vibrations are called **free** or **natural vibrations.** We will now assume that the mass is also subject to an external periodic force $F_0 \sin \omega t$, due to the motion of the object to which the upper end of the spring is attached (see Figure 6). In this case the mass will undergo **forced vibrations.**

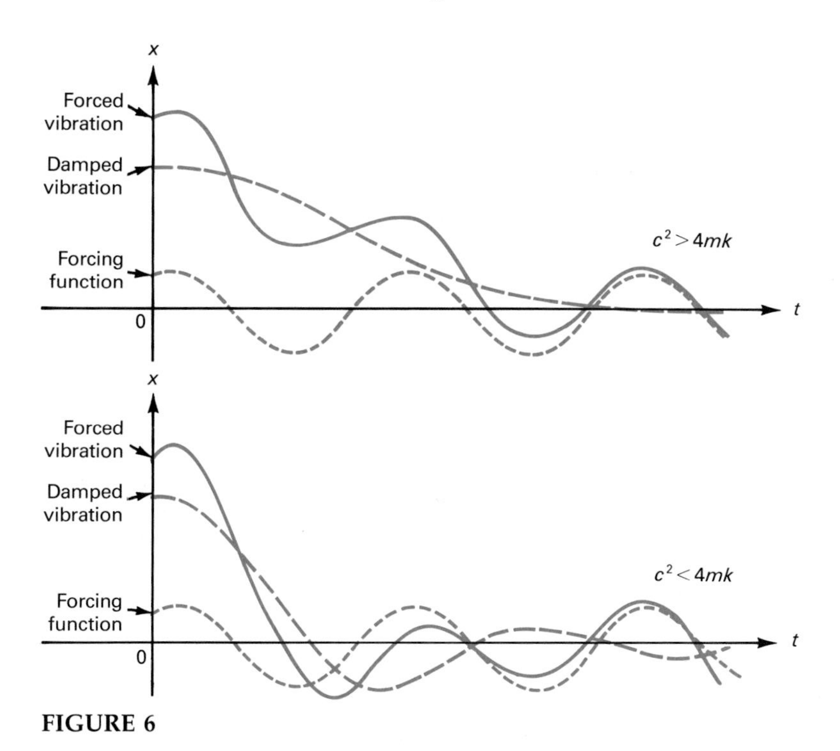

FIGURE 6

Equation (6) may be replaced by the nonhomogeneous second-order differential equation

$$m \frac{d^2x}{dt^2} = -kx - c\frac{dx}{dt} + F_0 \sin \omega t,$$

which we write in the form

$$\frac{d^2x}{dt^2} + \frac{c}{m}\frac{dx}{dt} + \frac{k}{m}x = \frac{F_0}{m} \sin \omega t. \tag{11}$$

By the method of undetermined coefficients, we know that $x(t)$ has a particular solution of the form

$$x_p(t) = b_1 \cos \omega t + b_2 \sin \omega t. \tag{12}$$

Substituting this function into equation (11) yields the simultaneous equations

$$
\begin{aligned}
(\omega_0^2 - \omega^2)b_1 + \frac{c\omega}{m}b_2 &= 0, \\
-\frac{c\omega}{m}b_1 + (\omega_0^2 - \omega^2)b_2 &= \frac{F_0}{m},
\end{aligned}
\tag{13}
$$

where $\omega_0 = \sqrt{k/m}$, from which we obtain

$$
\begin{aligned}
b_1 &= \frac{-F_0 c\omega}{m^2(\omega_0^2 - \omega^2)^2 + (c\omega)^2}, \\
b_2 &= \frac{F_0 m(\omega_0^2 - \omega^2)}{m^2(\omega_0^2 - \omega^2)^2 + (c\omega)^2}.
\end{aligned}
$$

Using the same method we used to obtain equation (4), we have

$$
x_p = A \sin(\omega t + \phi),
\tag{14}
$$

where

$$
A = \frac{F_0/k}{\sqrt{\left[1 - \left(\frac{\omega}{\omega_0}\right)^2\right]^2 + \left(2\frac{c}{c_0}\frac{\omega}{\omega_0}\right)^2}}
$$

and

$$
\tan\phi = \frac{2\frac{c}{c_0}\frac{\omega}{\omega_0}}{\left(\frac{\omega}{\omega_0}\right)^2 - 1}
$$

with $c_0 = 2m\omega_0$. Here A is the amplitude of the motion, ϕ is the **phase angle,** c/c_0 is the **damping ratio,** and ω/ω_0 is the **frequency ratio** of the motion.

The general solution is found by superimposing the periodic function (equation (14)), on the general solution, equation (8), (9), or (10), of the homogeneous equation. Since the solution of the homogeneous equation damps out as t increases, the general solution will be very close to equation (14) for large values of t. Figure 6 illustrates two typical situations.

It is interesting to see what occurs if the damping constant c vanishes. There are two cases.

Case 1. If $\omega^2 \neq \omega_0^2$, we superimpose the periodic function of equation (14), on the general solution of the homogeneous equation $x'' + \omega_0^2 x = 0$, obtaining

$$
x(t) = c_1 \cos\omega_0 t + c_2 \sin\omega_0 t + \frac{F_0/k}{1 - (\omega/\omega_0)^2}\sin\omega t.
\tag{15}
$$

Using the initial conditions, we find that

$$c_1 = x_0 \quad \text{and} \quad c_2 = \frac{v_0}{\omega_0} - \frac{(F_0/k)(\omega/\omega_0)}{1 - (\omega/\omega_0)^2}$$

so that

$$x(t) = A \sin(\omega_0 t + \phi) + \frac{F_0/k}{1 - (\omega/\omega_0)^2} \sin \omega t,$$

where

$$A = \sqrt{c_1^2 + c_2^2} \quad \text{and} \quad \tan \phi = c_1/c_2.$$

Hence the motion in this case is simply the sum of two sinusoidal curves as illustrated in Figure 7.

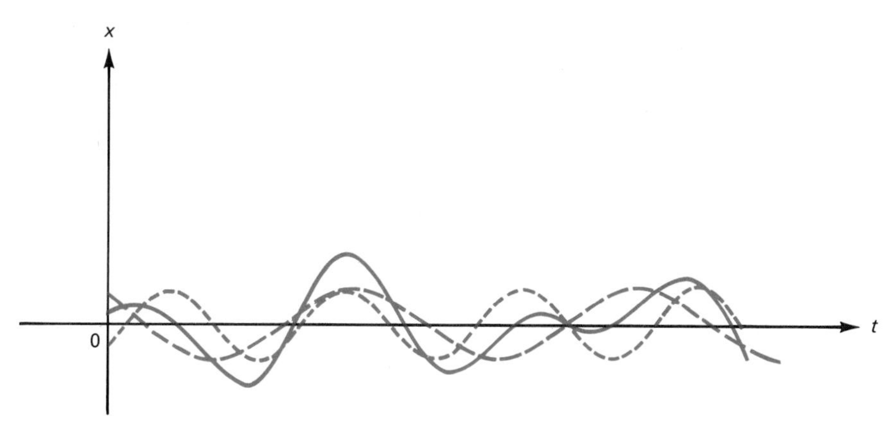

FIGURE 7

Case 2. If $\omega^2 = \omega_0^2$, we must seek a particular solution of the form

$$x_p(t) = b_1 t \cos \omega t + b_2 t \sin \omega t, \tag{16}$$

since equation (12) is a solution of the homogeneous equation (2). (See the modification of the method of undetermined coefficients on page 679.) Substituting equation (16) into equation (2), we get

$$b_1 = \frac{-F_0}{2m\omega} \quad \text{and} \quad b_2 = 0,$$

so the general solution has the form

$$x(t) = c_1 \cos \omega t + c_2 \sin \omega t - \frac{F_0}{2m\omega} t \cos \omega t. \tag{17}$$

Note that as t increases the vibrations caused by the last term in equation (17) increase *without bound*. The external force is said to be in **resonance** with the vibrating mass. It is evident that the displacement will become so large that the elastic limit of the spring will be exceeded, leading to fracture or to a permanent distortion in the spring.

Suppose that c is positive but very close to zero while $\omega^2 = \omega_0^2$. Note that equation (13) will yield $b_1 = -F_0/c\omega$ and $b_2 = 0$ when substituted in equation (12). Superimposing $x_p(t) = -(F_0/c\omega)\cos\omega t$ on equation (10) and letting $c_0 = 2m\omega_0$, we obtain

$$x(t) = e^{(-c/c_0)\omega_0 t}\left(c_1 \cos \omega_0\sqrt{1-\left(\frac{c}{c_0}\right)^2}\,t + c_2 \sin \omega_0\sqrt{1-\left(\frac{c}{c_0}\right)^2}\,t\right) - \frac{F_0}{c\omega}\cos \omega t. \tag{18}$$

Since c/c_0 is very small, for small values of t we see that equation (18) can be approximated as

$$x(t) \approx c_1 \cos \omega t + c_2 \sin \omega t - \frac{F_0}{2m\omega}\left(\frac{2m}{c}\right)\cos \omega t,$$

which bears a marked resemblance to equation (17) *when equation (17) is evaluated at large values of t* (since $2m/c$ is large). Thus, the *damped* spring problem approaches resonance. This phenomenon is extremely important in engineering since resonance may produce undesirable effects such as metal fatigue and structural fracture, as well as desirable objectives such as sound and light amplification.

PROBLEMS 11.13

In Problems 1–6 determine the equation of motion of a mass m attached to a coiled spring with spring constant k initially displaced a distance x_0 from equilibrium and released with velocity v_0 subject to

(a) no damping or external forces,
(b) a damping constant c, but no external force,
(c) an external force $F_0 \sin \omega t$, but no damping,
(d) both a damping constant and external force $F_0 \sin \omega t$.

√ **1.** $m = 10$ kg, $k = 1000$ N/m, $x_0 = 1$ m, $v_0 = 0$, $c = 200$ N/(m/s), $F_0 = 1$ N, $\omega = 10$ rad/sec

2. $m = 10$ kg, $k = 10$ N/m, $x_0 = 0$, $v_0 = 1$ m/s, $c = 20$ N/(m/sec), $F_0 = 1$ N, $\omega = 1$ rad/sec

√ **3.** $m = 10$ kg, $k = 10$ N/m, $x_0 = 3$ m, $v_0 = 4$ m/s, $c = 10\sqrt{5}$ N/(m/sec), $F_0 = 1$ N, $\omega = 1$ rad/sec

4. $m = 1$ kg, $k = 16$ N/m, $x_0 = 4$ m, $v_0 = 0$, $c = 10$ N/(m/sec), $F_0 = 4$ N, $\omega = 4$ rad/sec

√ **5.** $m = 1$ kg, $k = 25$ N/m, $x_0 = 0$ m, $v_0 = 3$ m/sec, $c = 8$ N/(m/sec), $F_0 = 1$ N, $\omega = 3$ rad/sec

6. $m = 9$ kg, $k = 1$ N/m, $x_0 = 4$ m, $v_0 = 1$ m/sec, $c = 10$ N/(m/sec), $F_0 = 2$ N, $\omega = \frac{1}{3}$ rad/sec

7. One end of a rubber band is fixed at a point A. A 1-kg mass, attached to the other end, stretches the rubber band vertically to the point B in such a way that the length AB is 16 cm greater than the natural length of the band. If the mass is further drawn to a position 8 cm below B and released, what will be its velocity (if we neglect resistance) as it passes the position B?

8. If in Problem 7 the mass is released at a position 8 cm above B, what will be its velocity as it passes 1 cm above B?

***9.** A cylindrical block of wood of radius and height 1 ft and weighing 12.48 lb floats with its axis vertical in water (62.4 lb/ft^3). If it is depressed so that the surface of the water is tangent to the block, and is then released, what will be its period of vibration and equation of motion? Neglect resistance. [*Hint:* The upward force on the block is equal to the weight of the water displaced by the block.]

***10.** A cubical block of wood, 1 ft on a side, is depressed so that its upper face lies along the surface of the water, and is then released. The period of vibration is found to be 1 sec. Neglecting resistance, what is the weight of the block of wood?

11. A 10-kg mass suspended from a spring vibrates freely, the resistance being numerically equal to half the velocity (in m/sec) at any instant. If the period of the motion is 8 sec, what is the spring constant (in kg/sec^2)?

12. A weight w (lb) is suspended from a spring whose constant is 10 lb/ft. The motion of the weight is subject to a resistance (lb) numerically equal to half the velocity (ft/sec). If the motion is to have a 1-sec period, what are the possible values of w?

13. A 1-g mass is hanging at rest on a spring that is stretched 25 cm by the weight. The upper end of the spring is given the periodic force 0.01 sin $2t$ N and air resistance has a magnitude 0.0216 (k/sec) times the velocity in meters per second. Find the equation of motion of the mass.

14. An ideal pendulum consists of a weightless rod of length l attached at one end to a frictionless hinge and supporting a body of mass m at the other end. Suppose the pendulum is displaced an angle θ_0 and released (see Figure 8). The tangential acceleration of the ideal pendulum is $l\theta''$, and must be proportional, by Newton's second law of motion, to the tangential component of gravitational force.

(a) Neglecting air resistance, show that the ideal pendulum satisfies the nonlinear initial value problem

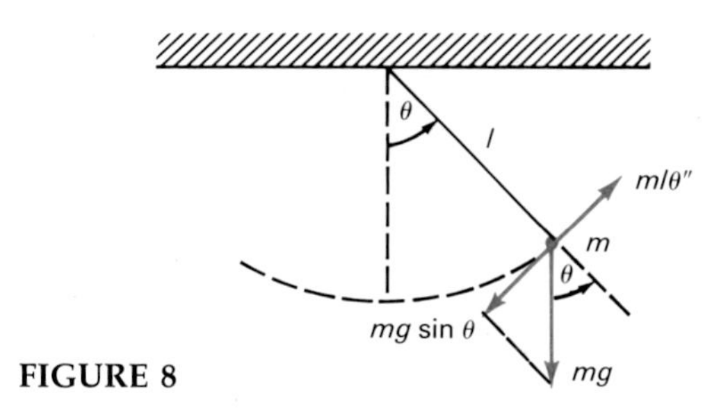

FIGURE 8

$$l\frac{d^2\theta}{dt^2} = -g \sin \theta, \quad \theta(0) = \theta_0, \quad \theta'(0) = 0. \tag{19}$$

(b) Assuming θ_0 is small, explain why equation (19) may be approximated by the linear initial value problem

$$\frac{d^2\theta}{dt^2} + \frac{g}{l}\theta = 0, \quad \theta(0) = \theta_0, \quad \theta'(0) = 0. \tag{20}$$

(c) Solve equation (20) assuming that the rod is 6 inches long and that the initial displacement $\theta_0 = 0.5$ radian. What is the frequency of the pendulum?

15. A grandfather clock has a pendulum that is one meter long. The clock ticks each time the pendulum reaches the rightmost extent of its swing. Neglecting friction and air resistance, and assuming that the motion is small, determine how many times the clock ticks in one minute.

11.14 MORE ON ELECTRIC CIRCUITS (Optional)

We shall make use of the concepts developed in Section 11.5 and the methods of this chapter to study a simple electric circuit containing a resistor, an inductor, and a capacitor in series with an electromotive force (Figure 1). Suppose that R, L, C, and E are constants. Applying Kirchhoff's law, we obtain

$$L\frac{dI}{dt} + RI + \frac{Q}{C} = E. \tag{1}$$

Since $dQ/dt = I$, we may differentiate equation (1) to get the second-order homogeneous differential equation

$$L\frac{d^2I}{dt^2} + R\frac{dI}{dt} + \frac{I}{C} = 0. \tag{2}$$

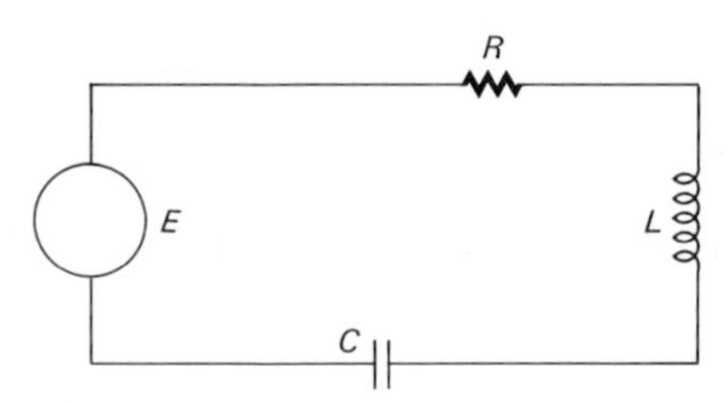

FIGURE 1

To solve this equation, we note that the characteristic equation

$$\lambda^2 + \frac{R}{L}\lambda + \frac{1}{CL} = 0$$

has the following roots:

$$\lambda_1 = \frac{-R + \sqrt{R^2 - 4L/C}}{2L}, \quad \lambda_2 = \frac{-R - \sqrt{R^2 - 4L/C}}{2L},$$

or, rewriting the radical, we have

$$\lambda_1 = \frac{R}{2L}\left(-1 + \sqrt{1 - \frac{4L}{CR^2}}\right), \quad \lambda_2 = \frac{R}{2L}\left(-1 - \sqrt{1 - \frac{4L}{CR^2}}\right). \tag{3}$$

Equation (2) can now be solved using the methods of Sections 11.8 and 11.9.

EXAMPLE 1 Let $L = 1$ henry (h), $R = 100$ ohms (Ω), $C = 10^{-4}$ farads (f), and $E = 1000$ volts (V) in the circuit shown in Figure 1. Suppose that no charge is present and no current is flowing at time $t = 0$ when E is applied. Here $R^2 - 4L/C = 10{,}000 - 4 \cdot 10^4 = -30{,}000$, $\sqrt{30{,}000} = \sqrt{(10{,}000)(3)} = 100\sqrt{3}$ and $R/2L = 50$ so $\lambda_1 = -50 + 50\sqrt{3}i$ and $\lambda_2 = -50 - 50\sqrt{3}i$. Thus

$$I(t) = e^{-50t}(c_1 \cos 50\sqrt{3}t + c_2 \sin 50\sqrt{3}t).$$

Applying the initial condition $I(0) = 0$, we have $c_1 = 0$. Hence

$$I(t) = c_2 e^{-50t} \sin 50\sqrt{3}t \quad \text{and} \quad I'(t) = c_2 e^{-50t}(50\sqrt{3} \cos 50\sqrt{3}t - 50 \sin 50\sqrt{3}t).$$

To establish the value of c_2, we must make use of equation (1) and the initial condition $Q(0) = 0$. Then

$$Q(t) = C\left(E - L\frac{dI}{dt} - RI\right) = 10^{-4}[1000 - c_2 e^{-50t}(50\sqrt{3} \cos 50\sqrt{3}t - 50 \sin 50\sqrt{3}t + 100 \sin 50\sqrt{3}t)]$$

$$50 \cdot 10^{-4} = \frac{1}{200}$$
$$\downarrow$$
$$= \frac{1}{10} - \frac{c_2}{200} e^{-50t}(\sin 50\sqrt{3}t + \sqrt{3} \cos 50\sqrt{3}t).$$

Thus $Q(0) = \dfrac{1}{10} - \dfrac{c_2}{200}\sqrt{3} = 0$, so $c_2 = 20/\sqrt{3}$, and we have

$$Q(t) = \frac{1}{10} - \frac{1}{10\sqrt{3}}e^{-50t}(\sin 50\sqrt{3}t + \sqrt{3}\cos 50\sqrt{3}t)$$

and

$$I(t) = \frac{20}{\sqrt{3}}e^{-50t}\sin 50\sqrt{3}t.$$

From these equations we observe that the current will rapidly damp out and that the charge will rapidly approach its **steady-state value** of $\frac{1}{10}$ coulomb (coul). Here $I(t)$ is called the **transient current** because it dies out. ■

EXAMPLE 2 Let the inductance, resistance, and capacitance remain the same as in Example 1, but suppose $E = 962 \sin 60t$. By equation (1) we have

$$\frac{dI}{dt} + 100I + 10^4 Q = 962 \sin 60t. \tag{4}$$

Converting equation (4) so that all expressions are in terms of $Q(t)$, we obtain

$$\frac{d^2Q}{dt^2} + 100\frac{dQ}{dt} + 10^4 Q = 962 \sin 60t. \tag{5}$$

It is evident that equation (5) has a particular solution of the form

$$Q_p(t) = A_1 \sin 60t + A_2 \cos 60t. \tag{6}$$

To determine the values A_1 and A_2, we substitute equation (6) into equation (4), obtaining the following simultaneous equations:

$$6400A_1 - 6000A_2 = 962$$
$$6000A_1 + 6400A_2 = 0$$

Thus $A_1 = \frac{2}{25}$, $A_2 = -\frac{3}{40}$, and since the general solution of the homogeneous equation is the same as that of equation (2), the general solution of equation (1) is

$$Q(t) = e^{-50t}(c_1 \cos 50\sqrt{3}t + c_2 \sin 50\sqrt{3}t) + \tfrac{2}{25}\sin 60t - \tfrac{3}{40}\cos 60t. \tag{7}$$

Differentiating equation (7), we obtain

$$I(t) = 50e^{-50t}[(\sqrt{3}c_2 - c_1)\cos 50\sqrt{3}t - (c_2 + \sqrt{3}c_1)\sin 50\sqrt{3}t] + \tfrac{24}{5}\cos 60t + \tfrac{9}{2}\sin 60t.$$

Setting $t = 0$ and using the initial conditions, we obtain the simultaneous equations

$$c_1 = \tfrac{3}{40},$$
$$50(\sqrt{3}c_2 - c_1) = -\tfrac{24}{5}.$$

Therefore, $c_1 = \frac{3}{40}$ and $c_2 = -21/1000\sqrt{3}$. ■

PROBLEMS 11.14

1. In Example 1, let $L = 10$ h, $R = 250\ \Omega$, $C = 10^{-3}$ f, and $E = 900$ V. With the same assumptions, calculate the current and charge for all values of $t \geq 0$.
2. In Problem 1, suppose instead that $E = 50 \cos 30t$. Find $Q(t)$ for $t \geq 0$.

In Problems 3–6, find the steady-state current in the *RLC* circuit of Figure 1 where:

3. $L = 5$ h, $R = 10\ \Omega$, $C = 0.1$ f, $E = 25 \sin t$ V.
4. $L = 10$ h, $R = 40\ \Omega$, $C = 0.025$ f, $E = 100 \cos 5t$ V.
5. $L = 1$ h, $R = 7\ \Omega$, $C = 0.1$ f, $E = 100 \sin 10t$ V.
6. $L = 2.5$ h, $R = 10\ \Omega$, $C = 0.08$ f, $E = 100 \cos 5t$ V.

Find the transient current in the *RLC* circuit of Fig. 1 for Problems 7–12.

7. Problem 3.
8. Problem 4.
9. Problem 5.
10. Problem 6.
11. $L = 20$ h, $R = 40\ \Omega$, $C = 10^{-3}$ f, $E = 500 \sin t$ V.
12. $L = 24$ h, $R = 48\ \Omega$, $C = 0.375$ f, $E = 900 \cos 2t$ V.

13. Given that $L = 1$ h, $R = 1200\ \Omega$, $C = 10^{-6}$ f, $I(0) = Q(0) = 0$, and $E = 100 \sin 600t$ V, determine the transient current and the steady-state current.
14. Find the ratio of the current in the circuit of Problem 13 to that which would be flowing if there were no resistance, at $t = 0.001$ sec.
15. Consider the system governed by equation (1) for the case where the resistance is zero and $E = E_0 \sin \omega t$. Show that the solution consists of two parts, a general solution with frequency $1/\sqrt{LC}$ and a particular solution with frequency ω. The frequency $1/\sqrt{LC}$ is called the **natural frequency** of the circuit. Note that if $\omega = 1/\sqrt{LC}$, then the particular solution changes form.
16. To allow for different variations of the voltage, let us assume in equation (1) that $E = E_0 e^{it}$ $(= E_0 \cos t + iE_0 \sin t)$. Assume also, as in Problem 15, that $R = 0$. Finally, for simplicity assume that $E_0 = L = C = 1$. Then $1 = \omega = 1/\sqrt{LC}$.
    (a) Show that equation (2) becomes
    $$\frac{d^2I}{dt^2} + I = e^{it}.$$
    (b) Determine λ such that $I(t) = \lambda t e^{it}$ is a solution.
    (c) Calculate the general solution and show that the magnitude of the current increases without bound as t increases. This phenomenon will produce resonance.
17. Let an inductance of L henries, a resistance of R ohms, and a capacitance of C farads be connected in series with an emf of $E_0 \sin \omega t$ volts. Suppose $Q(0) = I(0) = 0$, and $4L > R^2C$.
    (a) Find the expressions for $Q(t)$ and $I(t)$.
    (b) What value of ω will produce resonance?
18. Solve Problem 17 for $4L = R^2C$.
19. Solve Problem 17 for $4L < R^2C$.

11.15 HIGHER-ORDER LINEAR DIFFERENTIAL EQUATIONS

In this section we extend the results of the chapter to linear differential equations of order greater than two. There is little theoretical difference between second and higher order systems so we shall be relatively brief. We shall state all theorems without proof.

The **general nonhomogeneous linear nth-order equation** is

$$y^{(n)}(x) + a_{n-1}(x)y^{(n-1)}(x) + \cdots + a_1(x)y'(x) + a_0(x)y(x) = f(x) \tag{1}$$

The associated homogeneous equation is

$$y^{(n)}(x) + a_{n-1}(x)y^{(n-1)}(x) + \cdots + a_1(x)y'(x) + a_0(x)y(x) = 0. \tag{2}$$

In Theorem 11.4.1 we stated that the equation (1) has a unique solution provided that all the functions in the equation are continuous and n initial conditions are specified. Now we shall concern ourselves with finding the general solutions to equations (1) and (2). To do so we shall follow the procedures we developed for solving second-order equations.

Definition 1 LINEAR INDEPENDENCE We say that the functions $y_1, y_2, \ldots, y_n$ are **linearly independent** in $[x_0, x_1]$ if the following condition holds:

$$c_1y_1(x) + c_2y_2(x) + \cdots + c_ny_n(x) = 0 \text{ for all}$$
$$x \in [x_0, x_1] \text{ implies that } c_1 = c_2 = \cdots = c_n = 0$$

Otherwise the functions are **linearly dependent.**

Definition 2 LINEAR COMBINATION The expression $c_1y_1 + c_2y_2 + \cdots + c_ny_n$ is called a **linear combination** of the functions $y_1, y_2, \ldots, y_n$.

Definition 3 WRONSKIAN The **Wronskian** of $y_1, y_2, \ldots, y_n$ is determined by

$$W(y_1, y_2, \ldots, y_n)(x) = \begin{vmatrix} y_1 & y_2 & \cdots & y_n \\ y_1' & y_2' & \cdots & y_n' \\ y_1'' & y_2'' & \cdots & y_n'' \\ \vdots & \vdots & & \vdots \\ y_1^{(n-1)} & y_2^{(n-1)} & \cdots & y_n^{(n-1)} \end{vmatrix} \tag{3}$$

Theorem 1 Let $a_0, a_1, \ldots, a_{n-1}$ be continuous on $[x_0, x_1]$ and let $y_1, y_2, \ldots, y_n$ be n solutions of equation (2). Then

(i) $W(y_1, y_2, \ldots, y_n)(x)$ is either zero for all $x \in [x_0, x_1]$ or for no value of $x \in [x_0, x_1]$.

(ii) $y_1, y_2, \ldots, y_n$ are linearly independent if and only if $W(y_1, y_2, \ldots, y_n)(x) \neq 0$.

EXAMPLE 1 The functions 1, x, and x^2 are solutions to the equation $y'''(x) = 0$. Determine whether they are linearly independent or dependent.

Solution.

$$W(y_1, y_2, y_3) = \begin{vmatrix} 1 & x & x^2 \\ 0 & 1 & 2x \\ 0 & 0 & 2 \end{vmatrix} = 2 \neq 0$$

so the functions are linearly independent. ■

GENERAL SOLUTION

Our procedure for solving equation (1) is as follows:

step (i) Find n linearly independent solutions, $y_1, y_2, \ldots, y_n$ to the homogeneous equation (2).
step (ii) Find one solution, y_p, to the nonhomogeneous equation (1).

Then the **general solution** to (1) is given by

$$y(x) = c_1y_1(x) + c_2y_2(x) + \cdots + c_ny_n(x) + y_p \tag{4}$$

and

$$y(x) = c_1y_1(x) + c_2y_2(x) + \cdots + c_ny_n(x) \tag{5}$$

is the general solution to (2).

As in the case of second-order equations, we can generally find solutions to (2) only when a_i are constants. In this case, equations (1) and (2) are said to have **constant coefficients.**

The general nth order linear, homogeneous, constant coefficient equation is

$$y^{(n)}(x) + a_{n-1}y^{(n-1)}(x) + \cdots + a_1y'(x) + a_0y(x) = 0 \tag{6}$$

We note that

$$\frac{d^n}{dx^n}e^{\lambda x} = \lambda^n e^{\lambda x}.$$

Thus, if we substitute $y = e^{\lambda x}$ into (6) and then divide by $e^{\lambda x}$, we obtain the **characteristic equation**

$$\lambda^n + a_{n-1}\lambda^{n-1} + \cdots + a_1\lambda + a_0 = 0 \tag{7}$$

Equation (7) has n roots $\lambda_1, \lambda_2, \ldots, \lambda_n$. Some of these roots may be real and distinct, real and equal, distinct complex conjugate pairs, or equal complex conjugate pairs. If a root λ_k (real or complex) occurs m times, we say that it has multiplicity m. The following rules tell us how to find the general solution to equation (6).

PROCEDURE FOR SOLVING LINEAR HOMOGENEOUS EQUATIONS WITH CONSTANT COEFFICIENTS

(i) Obtain the characteristic equation (7).
(ii) Find the roots $\lambda_1, \lambda_2, \ldots, \lambda_n$ of (7). (This is usually the most difficult step.)

(iii) For each real root λ_k of multiplicity 1 (*simple root*), one solution to (6) is $y_k = e^{\lambda_k x}$.

(iv) For each real root λ_k of multiplicity $m > 1$, m solutions to (6) are

$$y_1 = e^{\lambda_k x},\ y_2 = xe^{\lambda_k x}, \ldots, y_m = x^{m-1}e^{\lambda_k x}.$$

(v) If $\alpha + i\beta$ and $\alpha - i\beta$ are simple roots, then two solutions to (6) are

$$y_1 = e^{\alpha x}\cos\beta x \quad \text{and} \quad y_2 = e^{\alpha x}\sin\beta x.$$

(vi) If $\alpha + i\beta$ and $\alpha - i\beta$ are roots of multiplicity $m > 1$, then $2m$ solutions to (6) are

$$y_1 = e^{\alpha x}\cos\beta x,\ y_2 = xe^{\alpha x}\cos\beta x, \ldots, y_m = x^{m-1}e^{\alpha x}\cos\beta x,$$

$$y_{m+1} = e^{\alpha x}\sin\beta x,\ y_{m+2} = xe^{\alpha x}\sin\beta x, \ldots, y_{2m} = x^{m-1}e^{\alpha x}\sin\beta x.$$

(vii) If $y_1, y_2, \ldots, y_n$ are the n solutions obtained in steps (iii), (iv), (v), and (vi), then $y_1, y_2, \ldots, y_n$ are linearly independent and the general solution to (6) is given by

$$y(x) = c_1y_1(x) + c_2y_2(x) + \cdots + c_ny_n(x).$$

EXAMPLE 2 Find the general solution of

$$y''' - 3y'' - 10y' + 24y = 0$$

Solution. The characteristic equation is

$$\lambda^3 - 3\lambda^2 - 10\lambda + 24 = (\lambda - 2)(\lambda + 3)(\lambda - 4) = 0$$

with roots

$\lambda_1 = 2$, $\lambda_2 = -3$, $\lambda_3 = 4$. Since these roots are real and distinct, three linearly independent solutions are

$$y_1 = e^{2x},\ y_2 = e^{-3x} \text{ and } y_3 = e^{4x}.$$

The general solution is

$$y(x) = c_1e^{2x} + c_2e^{-3x} + c_3e^{4x}. \blacksquare$$

EXAMPLE 3 Find the general solution of

$$y^{(4)} - 4y''' + 6y'' - 4y' + y = 0$$

Solution. The characteristic equation is

$$\lambda^4 - 4\lambda^3 + 6\lambda^2 - 4\lambda + 1 = (\lambda - 1)^4 = 0.$$

with the single root $\lambda = 1$ of multiplicity 4. Thus four linearly independent solutions are

$$y_1 = e^x,\ y_2 = xe^x,\ y_3 = x^2e^x \text{ and } y_4 = x^3e^x.$$

The general solution is

$$y(x) = e^x(c_1 + c_2x + c_3x^2 + c_4x^3). \blacksquare$$

EXAMPLE 4 Find the general solution of

$$y^{(5)} - 2y^{(4)} + 8y'' - 12y' + 8y = 0$$

Solution. The characteristic equation is

$$\lambda^5 - 2\lambda^4 + 8\lambda^2 - 12\lambda + 8 = 0.$$

This can be factored as

$$(\lambda + 2)(\lambda^2 - 2\lambda + 2)^2 = 0.$$

The solutions to $\lambda^2 - 2\lambda + 2 = 0$ are $\lambda = 1 \pm i$.

Thus the roots are

$$\lambda_1 = -2 \text{ (simple)},\ \lambda_2 = 1 + i, \text{ and } \lambda_3 = 1 - i$$

with the complex roots λ_2 and λ_3 having multiplicity 2. Thus, five linearly independent solutions are

$$y_1 = e^{-2x},\ y_2 = e^x \cos x,\ y_3 = xe^x \cos x,\ y_4 = e^x \sin x,\ y_5 = xe^x \sin x,$$

and the general solution is

$$y(x) = c_1e^{-2x} + e^x \cos x(c_2 + c_3x) + e^x \sin x(c_4 + c_5x). \blacksquare$$

REMARK. In solving the last three characteristic equations we made the factoring look easy. Finding roots of a polynomial of degree greater than two is usually very difficult.

How do we find a particular solution to the nonhomogeneous equation (1)? As with second-order equations, there are two approaches: the method of undetermined coefficients and the method of variation of constants. The method of undetermined coefficients is identical to the technique we used for second-order equations. The method of variation of constants is discussed in Problems 28 and 29.

Finally, certain equations with variable coefficients can be solved. The higher-order Euler equation is discussed in Problems 30–33.

PROBLEMS 11.15

In Problems 1–18 find the general solution to the given equation. If initial conditions are given, find the particular solutions that satisfy them.

1. $y^{(4)} + 2y'' + y = 0.$
2. $y''' - y'' - y' + y = 0$
3. $y''' - 3y'' + 3y' - y = 0, \quad y(0) = 1, \quad y'(0) = 2, \quad y''(0) = 3$
4. $x''' + 5x'' - x' - 5x = 0$
5. $y''' - 9y' = 0, \quad y(0) = 3, \quad y'(0) = 0, \quad y''(0) = 18$
6. $y''' - 6y'' + 3y' + 10y = 0$
7. $y^{(4)} = 0$
8. $y^{(4)} - 9y'' = 0$
9. $y^{(4)} - 5y'' + 4y = 0$
10. $y^{(5)} - 2y''' + y' = 0$
11. $y^{(4)} - 4y'' = 0, \quad y(0) = 1, \quad y'(0) = 3, \quad y''(0) = 0, \quad y'''(0) = 16$
12. $y^{(4)} - 4y''' - 7y'' + 22y' + 24y = 0$
13. $y''' - y'' + y' - y = 0$
14. $y''' - 3y'' + 4y' - 2y = 0, \quad y(0) = 1, \quad y'(0) = 2, \quad y''(0) = 3$
15. $y''' - 27y = 0$
16. $y^{(5)} + 2y''' + y' = 0, \quad y(\pi/2) = 0, \quad y'(\pi/2) = 1, \quad y''(\pi/2) = 0, \quad y'''(\pi/2) = -3, \quad y^{(4)}(\pi/2) = 0$
17. Show that the solutions y_1, y_2, and y_3 of the linear third-order differential equation

$$y''' + a_1(x)y'' + a_2(x)y' + a_3(x)y = 0$$

that satisfy the conditions

$$y_1(x_0) = 1, \quad y_1'(x_0) = 0, \quad y_1''(x_0) = 0,$$
$$y_2(x_0) = 0, \quad y_2'(x_0) = 1, \quad y_2''(x_0) = 0,$$

and

$$y_3(x_0) = 0, \quad y_3'(x_0) = 0, \quad y_3''(x_0) = 1,$$

respectively, are linearly independent.

18. Show that *any* solution of

$$y''' + a_1(x)y'' + a_2(x)y' + a_3(x)y = 0$$

can be expressed as a linear combination of the solutions y_1, y_2, y_3 given in Problem 17. [*Hint:* If $y(x_0) = c_1$, $y'(x_0) = c_2$, and $y''(x_0) = c_3$, consider the linear combination $c_1y_1 + c_2y_2 + c_3y_3$.]

*19. Consider the third-order equation

$$y''' + a(x)y'' + b(x)y' + c(x)y = 0,$$

and let $y_1(x)$ and $y_2(x)$ be two linearly independent solutions. Define $y_3(x) = v(x)y_1(x)$ and assume that $y_3(x)$ is a solution to the equation.

(a) Find a second-order differential equation that is satisfied by v'.

(b) Show that $(y_2/y_1)'$ is a solution of this equation.

(c) Use the result of part (b) to find a second, linearly independent solution of the equation derived in part (a).

20. Consider the equation

$$y''' - \left(\frac{3}{x^2}\right)y' + \left(\frac{3}{x^3}\right)y = 0 \quad (x > 0).$$

(a) Show that $y_1(x) = x$ and $y_2(x) = x^3$ are two linearly independent solutions.

(b) Use the results of Problem 19 to find a third linearly independent solution.

21. Consider the third-order equation

$$y''' + a(x)y'' + b(x)y' + c(x)y = 0,$$

where a, b, and c are continuous functions of x in some interval I. Prove that if $y_1(x)$, $y_2(x)$, and $y_3(x)$ are solutions to the equation, then so is any linear combination of them.

22. In Problem 21 let

$$W(y_1, y_2, y_3) = \begin{vmatrix} y_1 & y_2 & y_3 \\ y_1' & y_2' & y_3' \\ y_1'' & y_2'' & y_3'' \end{vmatrix}.$$

(a) Show that W satisfies the differential equation

$$W'(x) = -a(x)W.$$

(b) Prove that $W(y_1, y_2, y_3)(x)$ is either always zero or never zero.

23. **(a)** Prove that the solutions $y_1(x)$, $y_2(x)$, $y_3(x)$ of the equation in Problem 21 are linearly independent on $[x_0, x_1]$ if and only if $W(y_1, y_2, y_3) \neq 0$.

(b) Show that $\sin t$, $\cos t$, and e^t are linearly independent solutions of

$$y''' - y'' + y' - y = 0$$

on any interval (a, b) where $-\infty < a < b < \infty$.

24. Assume that $y_1(x)$ and $y_2(x)$ are two solutions to

$$y''' + a(x)y'' + b(x)y' + c(x)y = f(x).$$

Prove that $y_3(x) = y_1(x) - y_2(x)$ is a solution of the associated homogeneous equation.

In Problems 25–27 use the method of undetermined coefficients to find the general solution of the given equation.

25. $y''' - y'' - y' + y = e^x$
26. $y''' - y'' - y' + y = e^{-x}$
27. $y''' - 3y'' - 10y' + 24y = x + 3$

***28.** Consider the third-order equation

$$y''' + ay'' + by' + cy = f(x) \tag{8}$$

Let $y_1(x)$, $y_2(x)$, and $y_3(x)$ be three linearly independent solutions to the homogeneous equation.

Assume that there is a solution of equation (8) of the form $y(x) = c_1(x)y_1(x) + c_2(x)y_2(x) + c_3(x)y_3(x)$.

(a) Following the steps used in deriving the variation of constants procedure for second-order equations, derive a method for solving third-order equations.

(b) Find a particular solution of the equation

$$y''' - 2y' - 4y = e^{-x}\tan x.$$

29. Use the method derived in Problem 28 to find a particular solution of

$$y''' + 5y'' + 9y' + 5y = 2e^{-2x}\sec x.$$

In Problems 30–32 guess that there is a solution of the form $y = x^{\lambda}$ to solve the given Euler equation.

30. $x^3y''' + 2x^2y'' - xy' + y = 0$

31. $x^3y''' - 12xy' + 24y = 0$

32. $x^3y''' + 4x^2y'' + 3xy' + y = 0$

33. Show that the substitution $x = e^t$ can be used to solve the third-order Euler equation

$$x^3y''' + x^2y'' - 2xy' + 2y = 0.$$

11.16 NUMERICAL SOLUTION OF DIFFERENTIAL EQUATIONS: EULER'S METHODS

In this chapter we have provided a number of methods for solving differential equations. However, as we pointed out earlier, most differential equations cannot be solved by elementary methods. For that reason, a number of numerical techniques have been developed for finding solutions or, more precisely, for finding solutions at particular points. We discuss two of the most elementary numerical techniques for solving first-order equations in this section.

Before presenting these numerical techniques, we should consider *when* numerical methods could or should be employed. Such methods are used primarily when other methods are not applicable. Additionally, even when other methods do apply, there may be an advantage in having a numerical solution, as solutions in terms of more exotic special functions are sometimes difficult to interpret. There may also be computational advantages: the exact solution may be extremely tedious to obtain.

On the other hand, care must always be exercised in using any numerical scheme, as the accuracy of the solution depends not only on the "correctness" of the numerical method being used but also on the precision of the device (hand calculator or computer) used for the computations.

From the general theory (see Appendix 5) we know that, in many cases, the initial value problem

$$\frac{dy}{dx} = f(x, y), \quad y(x_0) = y_0 \tag{1}$$

has a unique solution $y(x)$. The two techniques we will describe below approximate this solution $y(x)$ only at a finite number of points

$$x_0, \quad x_1 = x_0 + h, \quad x_2 = x_0 + 2h, \ldots, \quad x_n = x_0 + nh,$$

where h is some (nonzero) real number. The methods provide a value y_k that is an approximation to the exact value $y(x_k)$ for $k = 0, 1, \ldots, n$.

EULER'S METHOD†

This procedure is crude but very simple. The idea is to obtain y_1 by assuming that $f(x, y)$ varies so little on the interval $x_0 \le x \le x_1$ that only a very small error is made by replacing it by the constant value $f(x_0, y_0)$. Integrating

$$\frac{dy}{dx} = f(x, y)$$

from x_0 to x_1, we obtain

$$y(x_1) - y_0 = y(x_1) - y(x_0) = \int_{x_0}^{x_1} f(x, y)\, dx \approx f(x_0, y_0)(x_1 - x_0), \tag{2}$$

or, since $h = x_1 - x_0$,

$$y_1 = y_0 + hf(x_0, y_0).$$

Repeating the process with (x_1, y_1) to obtain y_2, etc., we obtain the **difference equation**

$$y_{n+1} = y_n + hf(x_n, y_n). \tag{3}$$

We shall solve equation (3) iteratively—that is, by first finding y_1, then using it to find y_2, and so on.

The geometric meaning of equation (3) is easily seen by considering the solution curve of the differential equation (1): we are simply following the tangent to the solution curve passing through (x_n, y_n) for a small horizontal distance. Looking at Figure 1, where the smooth curve is the unknown exact solution to the initial value

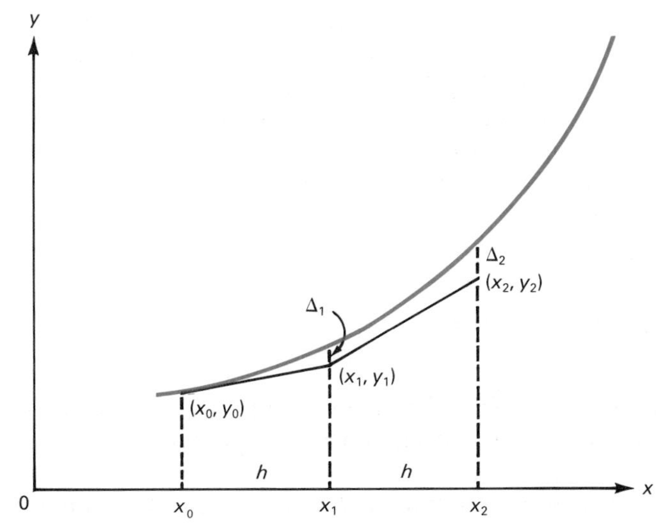

FIGURE 1

†See the biographical sketch on page 265.

problem (1), we see how equation (3) approximates the exact solution. Since $f(x_0, y_0)$ is the slope of the exact solution at (x_0, y_0), we follow this line to the point (x_1, y_1). Some solution to the differential equation passes through this point. We follow its tangent line at this point to reach (x_2, y_2), and so on. The differences Δ_k are errors at the kth stage in the process.

EXAMPLE 1 Solve

$$\frac{dy}{dx} = y + x^2, \quad y(0) = 1. \tag{4}$$

Solution. We wish to find $y(1)$ by approximating the solution at $x = 0.0, 0.2, 0.4, 0.6, 0.8$, and 1.0. Here $h = 0.2$, $f(x_n, y_n) = y_n + x_n^2$, and Euler's method [equation (3)] yields

$$y_{n+1} = y_n + h \cdot f(x_n, y_n) = y_n + h(y_n + x_n^2).$$

Since $y_0 = y(0) = 1$, we obtain

$$y_1 = y_0 + h \cdot (y_0 + x_0^2) = 1 + 0.2(1 + 0^2) = 1.2,$$

$$y_2 = y_1 + h \cdot (y_1 + x_1^2) = 1.2 + 0.2[1.2 + (0.2)^2] = 1.448 \approx 1.45,$$

$$y_3 = y_2 + h(y_2 + x_2^2) = 1.45 + 0.2[1.45 + (0.4)^2] \approx 1.77,$$

$$y_4 = y_3 + h(y_3 + x_3^2) = 1.77 + 0.2[1.77 + (0.6)^2] \approx 2.20,$$

$$y_5 = y_4 + h(y_4 + x_4^2) = 2.20 + 0.2[2.20 + (0.8)^2] \approx 2.77.$$

We arrange our work as shown in Table 1. The value $y_5 = 2.77$, corresponding to $x_5 = 1.0$, is our approximate value for $y(1)$.

TABLE 1

x_n	y_n	$f(x_n, y_n) = y_n + x_n^2$	$y_{n+1} = y_n + h \cdot f(x_n, y_n)$
0.0	1.00	1.00	1.20
0.2	1.20	1.24	1.45
0.4	1.45	1.61	1.77
0.6	1.77	2.13	2.20
0.8	2.20	2.84	2.77
1.0	2.77		

Equation (4) is a linear equation. We can solve it to obtain the exact solution $y = 3e^x - x^2 - 2x - 2$ (check this), so that $y(1) = 3e - 5 \approx 3.154$. Thus the Euler's method estimate was off by about twelve percent. This is not surprising because we treated the derivative as a constant over intervals of length of 0.2 unit. The error that arises in this way is called **discretization error,** because the "discrete" function $f(x_n, y_n)$ was substituted for the "continuously valued" function $f(x, y)$. It is true that

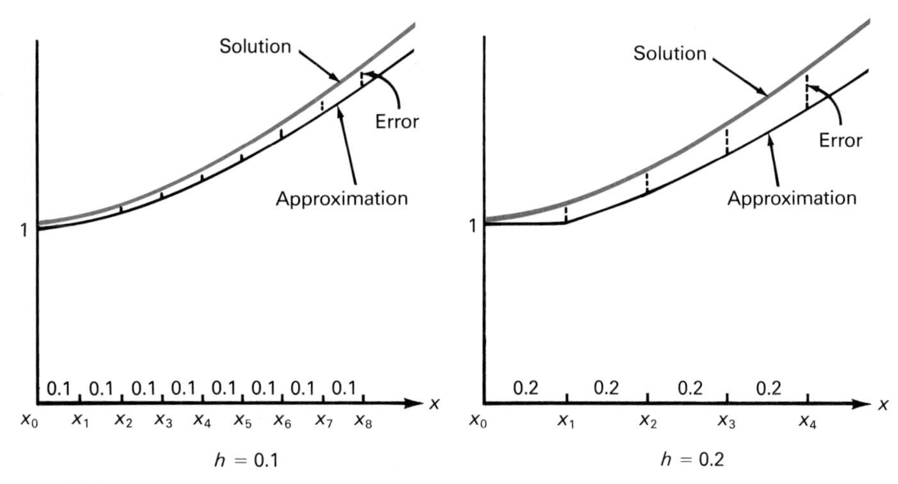

FIGURE 2

if we reduce the step size h, then we can improve the accuracy of our answer, since, then, the "discretized" function $f(x_n, y_n)$ will be closer to the true value of $f(x, y)$ over the interval $[0, 1]$. This is illustrated in Figure 2 with $h = 0.2$ and $h = 0.1$. Indeed, carrying out similar calculations with $h = 0.1$ yields an approximation of $y(1)$ of 3.07, which is a good deal more accurate (an error of about three percent). ■

Usually, reducing step size improves accuracy. However, a warning must be attached to this. Reducing the step size obviously also increases the amount of work that must be done. Moreover, at every stage of the computation **round-off errors** are introduced. For example, in our calculations with $h = 0.2$, we rounded off the exact value 1.448 to the value 1.45 (correct to two decimal places). The rounded-off value was then used to calculate further values of y_n. It is not unusual for a computer solution of a more complicated differential equation to take several thousand individual computations, thus having several thousand round-off errors. In some problems the accumulated round-off error can be so large that the resulting computed solution will be sufficiently inaccurate to invalidate the result. Fortunately, this usually does not occur since round-off errors can be positive or negative and tend to cancel one another out. This statement is made under the assumption (usually true) that the average of the round-off errors is zero. In any event, it should be clear that reducing the step size, thereby increasing the number of computations, is a procedure that should be carried out carefully. In general, each problem has an optimal step size, and a smaller than optimal step size will yield a greater error due to accumulated round-off errors.

IMPROVED EULER METHOD

This method has better accuracy than Euler's method and so is more valuable for hand computation. It is based on the fact that an improvement will result if we average the slopes at the left and right endpoints of each interval, thereby reducing the difference between $f(x, y)$ and $f(x_n, y_n)$ in each interval of the form $x_n \leq x < x_{n+1}$ (see Figure 3). This amounts to approximating the integral in equation (2) by the **trapezoidal rule:**

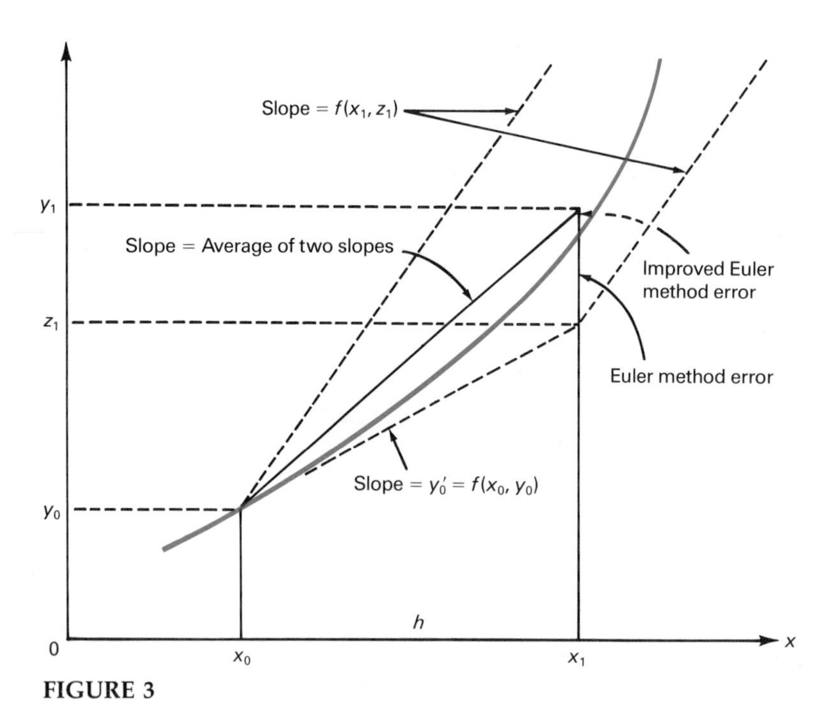

FIGURE 3

$$\int_{x_0}^{x_1} f(x, y)\, dx \approx \frac{h}{2}\{f(x_0, y_0) + f(x_1, y(x_1))\}.$$

Since $y(x_1)$ is not known, we replace it by the value found by Euler's method, which we call z_1; then equation (2) can be replaced by the system of equations

$$z_1 = y_0 + hf(x_0, y_0),$$

$$y_1 = y_0 + \frac{h}{2}[f(x_0, y_0) + f(x_1, z_1)].$$

This gives us the general procedure

$$z_{n+1} = y_n + hf(x_n, y_n),$$
$$y_{n+1} = y_n + \frac{h}{2}[f(x_n, y_n) + f(x_{n+1}, z_{n+1})]. \tag{5}$$

Using $x_0 = 0$ and $y_0 = 1$ in Example 1, we obtain, with $h = 0.2$,

$$z_1 = y_0 + hf(x_0, y_0) = 1 + 0.2(1 + 0^2) = 1.2,$$

$$y_1 = y_0 + \frac{h}{2}[f(x_0, y_0) + f(x_1, z_1)] = 1 + 0.1[(1 + 0^2) + 1.2 + 0.2^2]$$

$$= 1 + 0.1[2.24] = 1.224 \approx 1.22,$$

$$z_2 = y_1 + hf(x_1, y_1) = 1.22 + 0.2[1.22 + (0.2)^2] = 1.472 \approx 1.47,$$

$$y_2 = y_1 + \frac{h}{2}[f(x_1, y_1) + f(x_2, z_2)]$$

$$= 1.22 + 0.1[1.22 + (0.2)^2 + 1.47 + (0.4)^2] = 1.509 \approx 1.51,$$

and so on. Table 2 shows the approximating values of the solution of equation (4) used in Example 1. The error this time is less than one percent.

TABLE 2

x_n	y_n	$f(x_n, y_n) = y_n + x_n^2$	z_{n+1}	$f(x_{n+1}, z_{n+1}) = z_{n+1} + x_{n+1}^2$	y_{n+1}
0.0	1.0	1.0	1.20	1.24	1.22
0.2	1.22	1.26	1.47	1.63	1.51
0.4	1.51	1.67	1.84	2.20	1.90
0.6	1.90	2.26	2.35	2.99	2.43
0.8	2.43	3.07	3.04	4.04	3.14
1.0	3.14				

Short programs may be written in BASIC, FORTRAN or PASCAL which will carry out the calculations necessary for either Euler method. Hand calculators may also be used, the programmable type being most useful.

PROBLEMS 11.16

Solve Problems 1–10 by using

(a) The Euler method with the indicated value of h;

(b) The improved Euler method with the given value of h.

1. $\dfrac{dy}{dx} = x + y$, $y(0) = 1$. Find $y(1)$ with $h = 0.2$.

2. $\dfrac{dy}{dx} = x - y$, $y(1) = 2$. Find $y(3)$ with $h = 0.4$.

3. $\dfrac{dy}{dx} = \dfrac{x - y}{x + y}$, $y(2) = 1$. Find $y(1)$ with $h = -0.2$.

4. $\dfrac{dy}{dx} = \dfrac{y}{x} + \left(\dfrac{y}{x}\right)^2$, $y(1) = 1$. Find $y(2)$ with $h = 0.2$.

5. $\dfrac{dy}{dx} = x\sqrt{1 + y^2}$, $y(1) = 0$. Find $y(3)$ with $h = 0.4$.

***6.** $\dfrac{dy}{dx} = x\sqrt{1 - y^2}$, $y(1) = 0$. Find $y(2)$ with $h = 0.125$.

7. $\dfrac{dy}{dx} = \dfrac{y}{x} - \dfrac{5}{2}x^2y^3$, $y(1) = \dfrac{1}{\sqrt{2}}$. Find $y(2)$ with $h = 0.125$.

8. $\dfrac{dy}{dx} = \dfrac{-y}{x} + x^2y^2$, $y(1) = \dfrac{2}{9}$. Find $y(3)$ with $h = \frac{1}{3}$.

9. $\dfrac{dy}{dx} = ye^x$, $y(0) = 2$. Find $y(2)$ with $h = 0.2$.

10. $\dfrac{dy}{dx} = xe^y$, $y(0) = 0$. Find $y(1)$ with $h = 0.1$.

In Problems 11–20 use the improved Euler method to graph approximately the solution of the given initial value problem by plotting the points (x_k, y_k) over the indicated range, where $x_k = x_0 + kh$.

11. $y' = xy^2 + y^3$, $y(0) = 1$, $h = 0.02$, $0 \le x \le 0.1$

12. $y' = x + \sin(\pi y)$, $y(1) = 0$, $h = 0.2$, $1 \le x \le 2$

13. $y' = x + \cos(\pi y)$, $y(0) = 0$, $h = 0.4$, $0 \le x \le 2$

14. $y' = \cos(xy)$, $y(0) = 0$, $h = \pi/4$, $0 \le x \le \pi$

15. $y' = \sin(xy)$, $y(0) = 1$, $h = \pi/4$, $0 \le x \le 2\pi$

16. $y' = \sqrt{x^2 + y^2}$, $y(0) = 1$, $h = 0.5$, $0 \le x \le 5$

17. $y' = \sqrt{y^2 - x^2}$, $y(0) = 1$, $h = 0.1$, $0 \le x \le 1$

18. $y' = \sqrt{x + y^2}$, $y(0) = 1$, $h = 0.2$, $0 \le x \le 1$

19. $y' = \sqrt{x + y^2}$, $y(1) = 2$, $h = -0.2$, $0 \le x \le 1$

20. $y' = \sqrt{x^2 + y^2}$, $y(1) = 5$, $h = -0.2$, $0 \le x \le 1$

REVIEW EXERCISES FOR CHAPTER ELEVEN

In Exercises 1–37 find the general solution to the given differential equation. If initial conditions are given, find the unique solution to the initial value problem.

1. $\dfrac{dy}{dx} = 3x$

2. $\dfrac{dy}{dx} = e^{x-y},\ y(0) = 4$

3. $\dfrac{dx}{dt} = e^{x} \cos t,\ x(0) = 3$

4. $\dfrac{dx}{dt} = x^{13}t^{11}$

5. $\dfrac{dy}{dx} + 3y = \cos x,\ y(0) = 1$

6. $\dfrac{dx}{dt} + 3x = \dfrac{1}{1 + e^{3t}}$

7. $\dfrac{dx}{dt} = 3x + t^3e^{3t},\ x(1) = 2$

8. $\dfrac{dy}{dx} + y \cot x = \sin x,\ y\left(\dfrac{\pi}{6}\right) = \dfrac{1}{2}$

9. $(y - e^y \sec^2 x)\, dx + (x - e^y \tan x)\, dy = 0$

10. $(2x^2y^3 - y^2)\, dx + (x^3y^2 - x)\, dy = 0$

11. $y'' - 5y' + 4y = 0$

12. $y'' - 9y' + 14y = 0,\ y(0) = 2,\ y'(0) = 1$

13. $y'' - 9y = 0$

14. $y'' + 9y = 0$

15. $y'' + 6y' + 9y = 0$

16. $y'' + 8y' + 16y = 0,\ y(0) = -1,\ y'(0) = 3$

17. $y'' - 2y' + 2y = 0,\ y(0) = 0,\ y'(0) = 1$

18. $y'' + 8y' = 0,\ y(0) = 2,\ y'(0) = -3$

19. $y'' + 4y = 2 \sin x$

20. $y'' + y' - 12y = 4e^{2x},\ y(0) = 1,\ y'(0) = -1$

21. $y'' + y' + y = e^{-x/2} \sin(\sqrt{3}/2)x$

22. $y'' + 4y = 6x \cos 2x,\ y(0) = 1,\ y'(0) = 0$

23. $y'' + y = x^3 - x$

24. $y'' - 2y' + y = e^{-x}$

25. $y'' - 8y' + 16y = e^{4x},\ y(0) = 3,\ y'(0) = 1$

26. $y'' + y = x + e^x + \sin x$

27. $y'' - 2y' + 3y = e^x \cos \sqrt{2}x$

28. $y'' + 16y = \cos 4x + x^2 - 3$

29. $y'' + y = 0,\ y(0) = y'(0) = 0$

30. $y'' + y = \sec x,\ 0 < x < \pi/2$

31. $y'' - 2y' + y = 2e^x/x^3$

32. $y'' + 4y' + 4y = e^{-2x}/x^2$

33. $x^2y'' - 2xy' + 3y = 0,\ x > 0$

34. $y''' - y'' + 4y' - 4y = 0$

35. $y''' + 3y'' + 3y' + y = 0$

36. $y^{(4)} - 10y'' + 9y = 0$

37. $y^{(4)} + 18y'' + 81y = 0$

38. Suppose a constant capacitor is connected in series to an emf whose voltage is a sine wave. Show that the current is 90° out of phase with the voltage.

39. Repeat Exercise 38 with the capacitor replaced by an inductor. What can you say in this case?

40. A spring, fixed at its upper end, supports a 10-kg mass that stretches the spring 60 cm. Find the equation of motion of the mass if it is drawn to a position 10 cm below its equilibrium position and released with an initial velocity of 5 cm/sec upward.

41. What are the period, frequency, and amplitude of the motion of the mass in Exercise 40?

42. Find the equation of motion of the mass of Exercise 40 if it is subjected to damping forces having the damping constant $\mu = 3$.

43. In Exercise 40, for what minimum damping constant would the mass fail to oscillate about its equilibrium?

44. For what value of ω will the external force of $10 \sin \omega t\ nt$ produce resonance in the spring of Exercise 43?

In Exercises 45–48 solve the given initial value problem and then obtain an approximate solution at the indicated value of x using the improved Euler method.

45. $\dfrac{dy}{dx} = \dfrac{e^x}{y},\ y(0) = 2.$ Find $y(3)$ with $h = \frac{1}{2}$.

46. $\dfrac{dy}{dx} = \dfrac{e^y}{x},\ y(1) = 0.$ Find $y(\frac{1}{2})$ with $h = -0.1$.

47. $\dfrac{dy}{dx} = \dfrac{y}{\sqrt{1 + x^2}},\ y(0) = 1.$ Find $y(3)$ with $h = \frac{1}{2}$.

48. $xy\dfrac{dy}{dx} = y^2 - x^2,\ y(1) = 2.$ Find $y(3)$ with $h = \frac{1}{2}$.

12 Matrices and Systems of Differential Equations

In this chapter we discuss systems of first-order differential equations. We shall show how higher-order differential equations can be written as first-order systems and how all linear first-order systems can be written in matrix notation. We shall see that a great deal of information can be obtained by determining the eigenvalues and eigenvectors of this matrix.

The prerequisites for this chapter are:

For Sections 12.1–12.3: Sections 11.1, 11.2, 11.4, 11.6, 11.8–11.11, 11.15, and 8.1.

For Sections 12.4–12.8: all of the above plus Chapter 7 and Sections 9.1–9.4, 9.10, and 9.12.

12.1 THE METHOD OF ELIMINATION FOR LINEAR SYSTEMS WITH CONSTANT COEFFICIENTS

In Chapter 11 we discussed the problem of finding the solution to a single linear differential equation. In this section we will discuss an elementary method for solving a system of simultaneous first-order linear differential equations by converting the system into a single higher-order linear differential equation that may then be solved by the methods we have already seen. Systems of simultaneous differential equations arise in problems involving more than one unknown function, each of which is a function of a single independent variable (often time). For consistency throughout the remaining sections of this chapter, we denote the independent variable by t and the dependent variables by $x(t)$ and $y(t)$ or by the subscripted letters $x_1(t)$, $x_2(t)$, $\ldots, x_n(t)$.

Although some familiarity with the elementary properties of determinants will be helpful, we will not use matrix methods in solving systems of simultaneous linear differential equations in this section. Here we shall give some examples of how simple systems arise and shall describe an elementary procedure for finding their solution.

EXAMPLE 1 Suppose that a chemical solution flows from one container into a second container at a rate proportional to the volume of solution in the first vessel. It flows out from the second container at a constant rate. Let $x(t)$ and $y(t)$ denote the volumes of solution in the first and second containers, respectively, at time t. (The containers may be, for example, cells, in which case we are describing a diffusion process across a cell wall.) To establish the necessary equations, we note that the change in volume equals the difference between input and output in each container. The change in volume is the derivative of volume with respect to time. Since no chemical is flowing into the first container, the change in its volume equals the output:

$$\frac{dx}{dt} = -c_1x$$

where c_1 is a positive constant of proportionality. The amount of solution c_1x flowing out of the first container is the input of the second container. Let c_2 be the constant output of the second container. Then the change in volume in the second container equals the difference between its input and output:

$$\frac{dy}{dt} = c_1x - c_2.$$

Thus we can describe the flow of a solution by means of two differential equations. Since more than one differential equation is involved, we say that we have obtained a **system of differential equations:**

$$\begin{aligned} \frac{dx}{dt} &= -c_1x, \\ \frac{dy}{dt} &= c_1x - c_2 \end{aligned} \tag{1}$$

where c_1 and c_2 are positive constants. By a **solution** of the system (1) we shall mean a pair of functions $x(t)$, $y(t)$ that simultaneously satisfy both equations in (1). Since the first equation contains only x and t, it may be solved for x as in Chapter 11. The result is then substituted into the second equation, permitting its solution also. (When more complicated systems arise, this successive solution will usually not be possible.) If we denote the initial volumes in the two containers by $x(0)$ and $y(0)$, respectively, we see that the first equation has the solution

$$x(t) = x(0)e^{-c_1 t}. \tag{2}$$

Substituting equation (2) into the second equation of (1), we obtain

$$\frac{dy}{dt} = c_1 x(0)e^{-c_1 t} - c_2,$$

which, upon integration, yields the solution

$$y(t) = y(0) + x(0)(1 - e^{-c_1 t}) - c_2 t. \tag{3}$$

Equations (2) and (3) together constitute the unique solution of system (1) that satisfies the given initial conditions. ■

EXAMPLE 2 Let tank X contain 100 gallons of brine in which 100 pounds of salt is dissolved and tank Y contain 100 gallons of water. Suppose water flows into tank X at the rate of 2 gallons per minute, and the mixture flows from tank X into tank Y at 3 gallons per minute. One gallon is pumped from Y back to X (establishing **feedback**) while 2 gallons are flushed away. Find the amount of salt in each tank at any time t (see Figure 1).

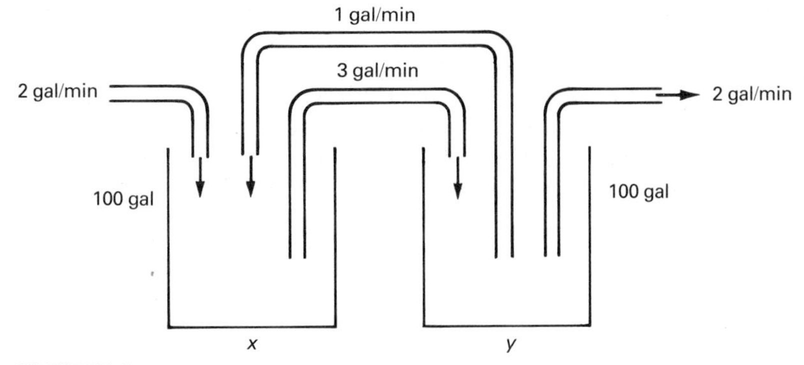

FIGURE 1

Solution. If we let $x(t)$ and $y(t)$ represent the number of pounds of salt in tanks X and Y at time t and note that the change in weight equals the difference between input and output, we can again derive a system of linear first-order equations. Tanks X and Y initially contain $x(0) = 100$ and $y(0) = 0$ pounds of salt, respectively, at time $t = 0$. The quantities $x/100$ and $y/100$ are, respectively, the amounts of salt contained in each gallon of water taken from tanks X and Y at time t. Three gallons are being removed from tank X and added to tank Y, while only one of the three gallons removed from tank Y is put in tank X. Thus we have the system

$$\begin{aligned} \frac{dx}{dt} &= -3\frac{x}{100} + \frac{y}{100}, \qquad x(0) = 100, \\ \frac{dy}{dt} &= 3\frac{x}{100} - 3\frac{y}{100}, \qquad y(0) = 0. \end{aligned} \tag{4}$$

Since both equations in the system (4) involve *both* dependent variables, we cannot immediately solve for one of the variables, as we did in Example 1. Instead, we shall use differentiation to eliminate one of the dependent variables. Suppose we begin by solving the second equation for x in terms of the dependent variable y and its derivative:

$$x = y + \frac{100}{3}\frac{dy}{dt}. \tag{5}$$

Differentiating equation (5) and equating the right-hand side to the right-hand side of the first equation in system (4), we have

$$\frac{-3x}{100} + \frac{y}{100} = \frac{dx}{dt} = \frac{dy}{dt} + \frac{100}{3}\frac{d^2y}{dt^2}. \tag{6}$$

Replacing the x-term on the left-hand side of equation (6) with equation (5) produces the second-order linear equation

$$\frac{100}{3}\frac{d^2y}{dt^2} + 2\frac{dy}{dt} + \frac{2y}{100} = 0. \tag{7}$$

The initial conditions for equation (7) are obtained directly from system (4), since $y(0) = 0$ and

$$y'(0) = 3\frac{x(0)}{100} - 3\frac{y(0)}{100} = 3. \tag{8}$$

Multiplying both sides of equation (7) by $\frac{3}{100}$, we have the initial value problem

$$y'' + \frac{6}{100}y' + \frac{6}{(100)^2}y = 0, \qquad y(0) = 0, \qquad y'(0) = 3. \tag{9}$$

The characteristic equation for (9) has the roots

$$\lambda_1 = \frac{-3 + \sqrt{3}}{100}, \qquad \lambda_2 = \frac{-3 - \sqrt{3}}{100},$$

so that the general solution is

$$y(t) = c_1 e^{[(-3+\sqrt{3})t]/100} + c_2 e^{[(-3-\sqrt{3})t]/100}.$$

Using the initial conditions, we obtain the simultaneous equations

$$\begin{aligned} c_1 + \qquad\qquad c_2 &= 0, \\ \frac{-3 + \sqrt{3}}{100}c_1 - \frac{3 + \sqrt{3}}{100}c_2 &= 3. \end{aligned}$$

These have the unique solution $c_1 = -c_2 = 50\sqrt{3}$. Hence

$$y(t) = 50\sqrt{3}\,\{e^{[(-3+\sqrt{3})t]/100} - e^{[(-3-\sqrt{3})t]/100}\}$$

and substituting this function into the right-hand side of equation (5), we obtain

$$x(t) = 50\{e^{[(-3+\sqrt{3})t]/100} + e^{[(-3-\sqrt{3})t]/100}\}.$$

As is evident from the problem, the amounts of salt in the two tanks approach zero as time tends to infinity. ■

The technique we have used in solving Example 2 is called the **method of elimination,** since all but one of the dependent variables are eliminated by repeated differentiation. The method is quite elementary but requires many calculations. In Section 12.3 we will introduce a direct way of obtaining the solution without doing the elimination procedure. Nevertheless, because of its simplicity, elimination can be a very useful tool.

EXAMPLE 3 Solve the system

$$\begin{aligned} x' &= x + y, \qquad & x(0) &= 1, \\ y' &= -3x - y, \qquad & y(0) &= 0. \end{aligned} \tag{10}$$

Solution. Differentiating the first equation and substituting from the second equation for y', we have

$$x'' = x' + (-3x - y). \tag{11}$$

Solving the first equation of system (10) for y and substituting that expression in equation (11), we obtain

$$x'' + 2x = 0.$$

Thus

$$x(t) = c_1 \cos \sqrt{2}t + c_2 \sin \sqrt{2}t.$$

Then, according to the first equation of system (10),

$$\begin{aligned} y(t) = x'(t) - x(t) &= -\sqrt{2}c_1 \sin \sqrt{2}t + \sqrt{2}c_2 \cos \sqrt{2}t - c_1 \cos \sqrt{2}t \\ &\quad - c_2 \sin \sqrt{2}t \\ &= (c_2\sqrt{2} - c_1) \cos \sqrt{2}t - (\sqrt{2}c_1 + c_2) \sin \sqrt{2}t. \end{aligned}$$

Using the initial conditions, we find that

$$x(0) = c_1 = 1, \qquad y(0) = c_2\sqrt{2} - c_1 = 0, \qquad \text{or} \qquad c_2 = 1/\sqrt{2}.$$

Therefore, the unique solution of system (10) is given by the pair of functions

$$x(t) = \cos\sqrt{2}t + \frac{1}{\sqrt{2}}\sin\sqrt{2}t,$$

$$y(t) = -\left(\sqrt{2} + \frac{1}{\sqrt{2}}\right)\sin\sqrt{2}t = -\frac{3}{\sqrt{2}}\sin\sqrt{2}t. \blacksquare$$

EXAMPLE 4 Solve the following system:

$$\begin{aligned} x' &= 4x - y + t^2 \\ y' &= x + 2y + 3t \end{aligned} \tag{12}$$

Solution. Proceeding as before, we obtain

$$\begin{aligned} x'' &= 4x' - y' + 2t = 4x' - \overbrace{(x + 2y + 3t)}^{y'} + 2t \\ &= 4x' - x - 2y - \text{t} \end{aligned}$$

From the first equation of (12), $-y = x' - 4x - t^2$, so $-2y = 2x' - 8x - 2t^2$ and

$$x'' = 4x' - x + (2x' - 8x - 2t^2) - t,$$

or, simplifying,

$$x'' - 6x' + 9x = -2t^2 - t. \tag{13}$$

The associated homogeneous equation is

$$x'' - 6x' + 9x = 0$$

with characteristic equation

$$0 = \lambda^2 - 6\lambda + 9 = (\lambda - 3)^2.$$

Thus the solution to the homogeneous equation is $x(t) = c_1e^{3t} + c_2te^{3t}$. Using the method of undetermined coefficients (Section 11.10), we obtain the particular solution

$$-\frac{2}{9}t^2 - \frac{11}{27}t - \frac{2}{9}.$$

Hence the general solution to equation (13) is

$$x(t) = c_1e^{3t} + c_2te^{3t} - \frac{2}{9}t^2 - \frac{11}{27}t - \frac{2}{9}.$$

Also,

$$x'(t) = 3c_1e^{3t} + 3c_2te^{3t} + c_2e^{3t} - \frac{4}{9}t - \frac{11}{27}.$$

From the first equation in (12),

$$y(t) = 4x - x' + t^2,$$

and we obtain

$$y(t) = c_1e^{3t} + c_2te^{3t} - c_2e^{3t} + \frac{1}{9}t^2 - \frac{32}{27}t - \frac{13}{27}. \quad ■$$

The method illustrated in the last three examples can easily be generalized to apply to linear systems with three or more equations. *A linear system of n first-order equations usually reduces to an nth-order linear differential equation,* because it generally requires one differentiation to eliminate each variable $x_2, \ldots, x_n$ from the system.

EXAMPLE 5 As a fifth example, we consider the mass–spring system of Figure 2, which is a direct generalization of the system described in Section 11.13. In this example we have two masses suspended by springs in series with spring constants k_1 and k_2 (see page 692). If the vertical displacements from equilibrium of the two masses are denoted by $x_1(t)$ and $x_2(t)$, respectively, then using assumptions (i) and (ii) (Hooke's law) of Section 11.13, we find that the net forces acting on the two masses are given by

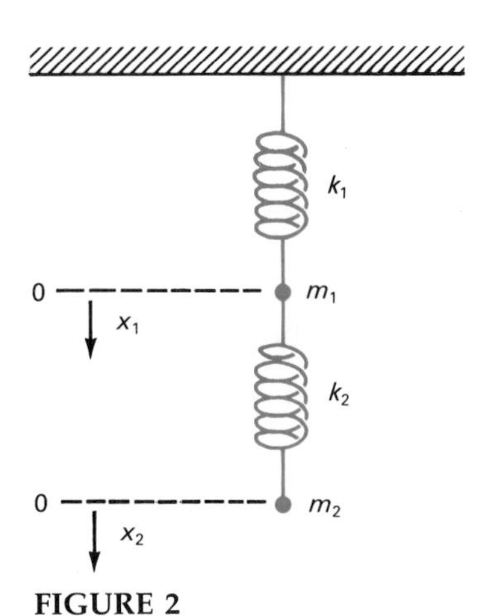

FIGURE 2

$$F_1 = -k_1x_1 + k_2(x_2 - x_1),$$
$$F_2 = -k_2(x_2 - x_1).$$

Here the positive direction is downward. The first spring is compressed when $x_1 < 0$ and the second spring is compressed when $x_1 > x_2$. The equations of motion are

$$
\begin{aligned}
m_1\frac{d^2x_1}{dt^2} &= -k_1x_1 + k_2(x_2 - x_1) = -(k_1 + k_2)x_1 + k_2x_2, \\
m_2\frac{d^2x_2}{dt^2} &= -k_2(x_2 - x_1) = k_2x_1 - k_2x_2,
\end{aligned} \tag{14}
$$

which constitute a system of two second-order linear differential equations with constant coefficients. ■

We will now show how linear differential equations (and systems) of *any* order can be converted, by the introduction of new variables, into a system of first-order differential equations. This concept is very important, since it means that the study of first-order linear systems provides a unified theory for all linear differential equations and systems. From a practical point of view it means that once we know how to solve first-order linear systems with constant coefficients, we will be able to solve any constant-coefficient linear differential equation or system.

To rewrite system (14) as a first-order system, we define the new variables $x_3 = x_1'$ and $x_4 = x_2'$. Then $x_3' = x_1''$, $x_4' = x_2''$ and (14) can be expressed as the system of four first-order equations

$$
\begin{aligned}
x_1' &= x_3, \\
x_2' &= x_4, \\
x_3' &= -\left(\frac{k_1 + k_2}{m_1}\right)x_1 + \left(\frac{k_2}{m_1}\right)x_2, \\
x_4' &= \left(\frac{k_2}{m_2}\right)x_1 - \left(\frac{k_2}{m_2}\right)x_2.
\end{aligned} \tag{15}
$$

If we wish, we can now use the method of elimination to reduce system (15) to a single fourth-order linear differential equation that can be solved by techniques of Section 11.15.

Theorem 1 The linear nth-order differential equation

$$
x^{(n)} + a_1(t)x^{(n-1)} + a_2(t)x^{(n-2)} + \cdots + a_{n-1}(t)x' + a_n(t)x = f(t) \tag{16}
$$

can be rewritten as a system of n first-order linear equations.

Proof. Define $x_1 = x$, $x_2 = x'$, $x_3 = x''$, . . . , $x_n = x^{(n-1)}$. Then we have

$$
\begin{aligned}
x_1' &= x_2, \\
x_2' &= x_3, \\
&\ \vdots \\
x_{n-1}' &= x_n, \\
x_n' &= -a_nx_1 - a_{n-1}x_2 - \cdots - a_1x_n + f. \quad ■
\end{aligned} \tag{17}
$$

In some cases, Theorem 1 can be extended to nonlinear differential equations (see Problem 28).

Suppose that n initial conditions are specified for the nth-order equation (16):

$$x(t_0) = c_1, \quad x'(t_0) = c_2, \quad \ldots, \quad x^{(n-1)}(t_0) = c_n$$

These initial conditions can be immediately transformed into an initial condition for system (17):

$$x_1(t_0) = c_1, \quad x_2(t_0) = c_2, \quad \ldots, \quad x_n(t_0) = c_n$$

EXAMPLE 6 Write the following initial value problem as a first-order system:

$$t^3x''' + 4t^2x'' - 8tx' + 8x = 0, \quad x(2) = 3, \quad x'(2) = -6, \quad x''(2) = 14.$$

Solution. Defining $x_1 = x$, $x_2 = x'$, $x_3 = x''$, we obtain the system

$$x_1' = x_2,$$

$$x_2' = x_3,$$

$$x_3' = \frac{-8}{t^3}x_1 + \frac{8}{t^2}x_2 - \frac{4}{t}x_3,$$

with the initial condition $x_1(2) = 3$, $x_2(2) = -6$, $x_3(2) = 14$. ■

PROBLEMS 12.1

In Problems 1–9 find the general solution of each system of equations. When initial conditions are given, find the unique solution.

1. $x' = x + 2y,$
   $y' = 3x + 2y$
2. $x' = x + 2y + t - 1,\ x(0) = 0$
   $y' = 3x + 2y - 5t - 2,\ y(0) = 4$
3. $x' = -4x - y,$
   $y' = x - 2y$
4. $x' = x + y,\ x(0) = 1,$
   $y' = y,\ y(0) = 0$
5. $x' = 8x - y,$
   $y' = 4x + 12y$
6. $x' = 2x + y + 3e^{2t},$
   $y' = -4x + 2y + te^{2t}$
7. $x' = 3x + 3y + t,$
   $y' = -x - y + 1$
8. $x' = 4x + y,\ x(\pi/4) = 0,$
   $y' = -8x + 8y,\ y(\pi/4) = 1$
9. $x' = 12x - 17y,$
   $y' = 4x - 4y$
10. By elimination, find a solution to the following nonlinear system:

$$x' = x + \sin x \cos x + 2y,$$

$$y' = (x + \sin x \cos x + 2y) \sin^2 x + x$$

In Problems 11–17 transform each equation into a system of first-order equations.

11. $x'' + 2x' + 3x = 0$
12. $x'' - 6tx' + 3t^3x = \cos t$
13. $x''' - x'' + (x')^2 - x^3 = t$
14. $x^{(4)} - \cos x(t) = t$
15. $x''' + xx'' - x'x^4 = \sin t$
16. $xx'x''x''' = t^5$
17. $x''' - 3x'' + 4x' - x = 0$

18. A mass m moves in xyz-space according to the following equations of motion:

$$mx'' = f(t, x, y, z),$$
$$my'' = g(t, x, y, z),$$
$$mz'' = h(t, x, y, z).$$

Transform these equations into a system of six first-order equations.

19. Consider the uncoupled system

$$x_1' = x_1, \qquad x_2' = x_2.$$

(a) What is the general solution of this system?

(b) Show that there is no second-order equation equivalent to this system. [*Hint:* Show that any second-order equation has solutions that are not solutions of this system.] This shows that first-order systems are more general than higher-order equations in the sense that any of the latter can be written as a first-order system, but not vice versa.

Use the method of elimination to solve the systems in Problems 20 and 21.

20. $x_1' = x_1,$
$x_2' = 2x_1 + x_2 - 2x_3,$
$x_3' = 3x_1 + 2x_2 + x_3$

21. $x_1' = x_1 + x_2 + x_3,$
$x_2' = 2x_1 + x_2 - x_3,$
$x_3' = -8x_1 - 5x_2 - 3x_3$

22. In Example 2, when does tank Y contain a maximum amount of salt? How much salt is in tank Y at that time?

23. Suppose in Example 2 that the rate of flow from tank Y to tank X is 2 gallons (gal) per minute (instead of one) and all other facts are unchanged. Find the equations for the amount of salt in each tank at any time t.

24. Tank X contains 500 gal of brine in which 500 lb of salt are dissolved. Tank Y contains 500 gal of water. Water flows into tank X at the rate of 30 gal/min, and the mixture flows into Y at the rate of 40 gal/min. From Y the solution is pumped back into X at the rate of 10 gal/min and into a third tank at the rate of 30 gal/min. Find the maximum amount of salt in Y. When does this concentration occur?

25. Suppose in Problem 24 that tank X contains 1000 gal of brine. Solve the problem, given that all other conditions are unchanged.

26. Consider the mass–spring system illustrated in Figure 3. Here three masses are suspended in series by three springs with spring constants k_1, k_2, and k_3, respectively. Formulate a system of second-order differential equations that describes this system.

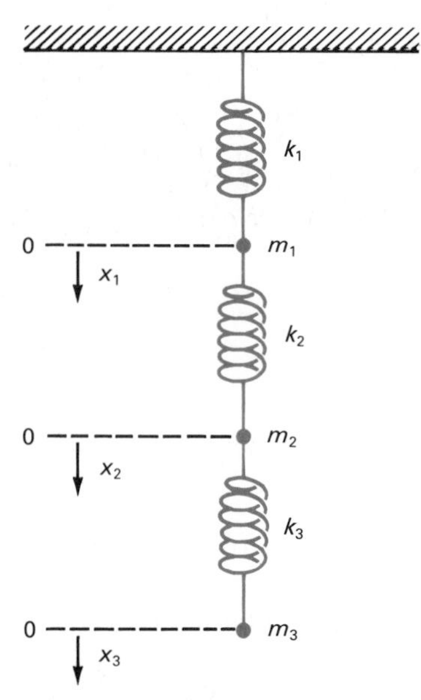

FIGURE 3

27. Find a single fourth-order linear differential equation in terms of the dependent variable x_1 for system (14). Find a solution to the system if $m_1 = 1$ kg, $m_2 = 2$ kg, $k_1 = 5$ N/m and $k_2 = 4$ N/m.

28. Show that the differential equation

$$x^{(n)} = g(t, x, x', \ldots, x^{(n-1)})$$

can be transformed into a system of n first-order equations.

***29.** In a study concerning the distribution of radioactive potassium ^{42}K between red blood cells and the plasma of human blood, C. W. Sheppard and W. R. Martin [*J. Gen. Physiol.* **33**, 703–722 (1950)] added ^{42}K to freshly drawn blood. They discovered that although the total amount of potassium (stable and radioactive) in the red cells and in the plasma remained practically constant during the experiment, the radioactivity was gradually transferred from the plasma to the red cells. Thus the behavior of the radioactivity is that of a linear closed two-compartment system. If 30.1% of the radioactive material is transferred from the plasma to the cells each hour, while 1.7% is transferred back, and the initial radioactivity was 800 counts per minute in the plasma and 25 counts per minute in the red cells, what is the number of counts per minute in the red cells after 300 minutes?

30. Temperature inversions and low wind speeds can trap air pollutants in a mountain valley for an extended period of time. Gaseous sulfur compounds are often a significant air pollution problem, but their study is complicated by their rapid oxidation. Hydrogen sulfide, H_2S, oxidizes into sulfur dioxide, SO_2, which in turn oxidizes into a sulfate. The following model has been proposed for determining the concentrations $x(t)$ and $y(t)$ of H_2S and SO_2, respectively, in a fixed airshed.[†] Let

$$\frac{dx}{dt} = -\alpha x + \gamma, \qquad \frac{dy}{dt} = \alpha x - \beta y + \delta$$

where the constants α and β are the conversion rates of H_2S into SO_2 and SO_2 into sulfate, respectively, and γ and δ are the production rates of H_2S and SO_2, respectively. Solve the equations sequentially and estimate the concentration levels that could be reached under prolonged air pollution.

12.2 LINEAR SYSTEMS: THEORY

In this section we shall consider the linear system of two first-order equations

$$\begin{aligned} x' &= a_{11}(t)x + a_{12}(t)y + f_1(t), \\ y' &= a_{21}(t)x + a_{22}(t)y + f_2(t), \end{aligned} \tag{1}$$

and the associated homogeneous system (i.e., $f_1 = f_2 = 0$)

$$\begin{aligned} x' &= a_{11}(t)x + a_{12}(t)y, \\ y' &= a_{21}(t)x + a_{22}(t)y. \end{aligned} \tag{2}$$

The point of view here will emphasize the similarities between such systems and the linear second-order equations discussed in Section 11.6. That there is a parallel between the two theories should not be surprising, since we have already shown in Section 12.1 that any linear second-order equation can be transformed into a system of the form of equation (1).

By a **solution** of system (1) (or (2)) we will mean a vector function $(x(t), y(t))$ that is differentiable and that satisfies the given equations. This is a formal statement of what we have been assuming all along. The following theorem is a consequence of the method of elimination and the existence–uniqueness theorem (Theorem 11.6.1).

Theorem 1 If the functions $a_{11}(t)$, $a_{12}(t)$, $a_{21}(t)$, $a_{22}(t)$, $f_1(t)$, and $f_2(t)$ are continuous, then given any numbers t_0, x_0, and y_0, there exists exactly one solution $(x(t), y(t))$ of system (1) that satisfies $x(t_0) = x_0$ and $y(t_0) = y_0$.

[†] R. L. Bohac, "A Mathematical Model for the Conversion of Sulphur Compounds in the Missoula Valley Airshed," *Proceedings of the Montana Academy of Science* (1974).

The vector function $(x_3(t), y_3(t))$ is a **linear combination** of the vectors $(x_1(t), y_1(t))$ and $(x_2(t), y_2(t))$ if there exist constants c_1 and c_2 such that the following two equations hold:

$$\begin{aligned} x_3(t) &= c_1x_1(t) + c_2x_2(t), \\ y_3(t) &= c_1y_1(t) + c_2y_2(t). \end{aligned} \tag{3}$$

The proof of the next theorem is left as an exercise.

Theorem 2 If $(x_1(t), y_1(t))$ and $(x_2(t), y_2(t))$ are solutions of the homogeneous system (2), then any linear combination of them is also a solution of system (2).

EXAMPLE 1 Consider the system

$$\begin{aligned} x' &= -x + 6y, \\ y' &= x - 2y. \end{aligned} \tag{4}$$

It is easy to verify that $(-2e^{-4t}, e^{-4t})$ and $(3e^t, e^t)$ are solutions of equations (4). Hence, by Theorem 2, the pair $(-2c_1e^{-4t} + 3c_2e^t, c_1e^{-4t} + c_2e^t)$ is a solution of equations (4) for any constants c_1 and c_2. ■

We define two vector functions $(x_1(t), y_1(t))$ and $(x_2(t), y_2(t))$ to be **linearly independent** if whenever the equations

$$\begin{aligned} c_1x_1(t) + c_2x_2(t) &= 0, \\ c_1y_1(t) + c_2y_2(t) &= 0 \end{aligned} \tag{5}$$

hold for all values of t, then $c_1 = c_2 = 0$. In Example 1, the two given vector solutions are linearly independent, since $c_1e^{-4t} + c_2e^t$ vanishes for all t only when $c_1 = c_2 = 0$.

Given two solutions $(x_1(t), y_1(t))$ and $(x_2(t), y_2(t))$, we define the **Wronskian**[†] of the two solutions to be the determinant:

$$W(t) = \begin{vmatrix} x_1(t) & x_2(t) \\ y_1(t) & y_2(t) \end{vmatrix} = x_1(t)y_2(t) - x_2(t)y_1(t). \tag{6}$$

We can then prove the next theorem.

Theorem 3 If $W(t) \neq 0$ for every t, then equations (3) are the **general solution** of the homogeneous system (2) in the sense that given any solution (x^*, y^*) of system (2), there exist constants c_1 and c_2 such that

$$\begin{aligned} x^* &= c_1x_1 + c_2x_2, \\ y^* &= c_1y_1 + c_2y_2. \end{aligned} \tag{7}$$

[†]Compare this with the definition of the Wronskian for a second-order equation in Chapter 11. See equation (11.6.7).

Proof. Let t_0 be given and consider the linear system of two equations in the unknown quantities c_1 and c_2:

$$\begin{aligned} c_1x_1(t_0) + c_2x_2(t_0) &= x^*(t_0), \\ c_1y_1(t_0) + c_2y_2(t_0) &= y^*(t_0). \end{aligned} \tag{8}$$

The determinant of this system is $W(t_0)$, which is nonzero by assumption. Thus there is a unique pair of constants (c_1, c_2) satisfying (8). By Theorem 2,

$$(c_1x_1(t) + c_2x_2(t), c_1y_1(t) + c_2y_2(t))$$

is a solution of equations (2). But by equations (8), this solution satisfies the same initial conditions at t_0 as the solution $(x^*(t), y^*(t))$. By the uniqueness part of Theorem 1, these solutions must be identical for all t. ■

EXAMPLE 2 In Example 1 the Wronskian $W(t)$ is

$$W(t) = \begin{vmatrix} -2e^{-4t} & e^{-4t} \\ 3e^{t} & e^{t} \end{vmatrix} = -2e^{-3t} - 3e^{-3t} = -5e^{-3t} \neq 0.$$

Hence we need look no further for the general solution of system (4). ■

In view of the condition required in Theorem 3 that the Wronskian $W(t)$ never vanish, we shall consider the properties of the Wronskian more carefully. Let (x_1, y_1) and (x_2, y_2) be two solutions of the homogeneous system (2). Since $W(t) = x_1y_2 - x_2y_1$, we have

$$\begin{aligned} W'(t) &= x_1y_2' + x_1'y_2 - x_2y_1' - x_2'y_1 \\ &= x_1(a_{21}x_2 + a_{22}y_2) + y_2(a_{11}x_1 + a_{12}y_1) - x_2(a_{21}x_1 + a_{22}y_1) - y_1(a_{11}x_2 + a_{12}y_1). \end{aligned}$$

Multiplying these expressions through and canceling like terms, we obtain

$$\begin{aligned} W' &= a_{11}x_1y_2 + a_{22}x_1y_2 - a_{11}x_2y_1 - a_{22}x_2y_1 \\ &= (a_{11} + a_{22})(x_1y_2 - x_2y_1) = (a_{11} + a_{22})W. \end{aligned}$$

Thus

$$W(t) = W(t_0) \exp\left\{\int_{t_0}^{t} [a_{11}(u) + a_{22}(u)]\, du\right\}. \tag{9}$$

We have shown the following theorem to be true:

Theorem 4 Let (x_1, y_1) and (x_2, y_2) be two solutions of the homogeneous system (2). Then the Wronskian $W(t)$ is either always zero or never zero in any interval (since $\exp u \neq 0$ for any u).

We are now ready to state the theorem that links linear independence with a nonvanishing Wronskian.

Theorem 5 Two solutions $(x_1(t), y_1(t))$ and $(x_2(t), y_2(t))$ are linearly independent if and only if $W(t) \neq 0$.

Proof. Let the solutions be linearly independent and suppose $W(t) = 0$. Then $x_1y_2 = x_2y_1$ or $x_1/x_2 = y_1/y_2 = c$ for some constant c. Then $x_1 = cx_2$ and $y_1 = cy_2$, so that the solutions are dependent, in contradiction of the original assumption. Hence $W(t) \neq 0$. Conversely, let $W(t) \neq 0$. If the solutions were dependent, then there would exist constants c_1 and c_2, not both zero, such that

$$c_1x_1 + c_2x_2 = 0,$$
$$c_1y_1 + c_2y_2 = 0.$$

Assuming that $c_1 \neq 0$, we then have $x_1 = cx_2$, $y_1 = cy_2$, where $c = -c_2/c_1$. But then

$$W(t) = x_1y_2 - x_2y_1 = cx_2y_2 - cx_2y_2 = 0.$$

Again this is a contradiction. Therefore the solutions are linearly independent. ■

We may summarize the contents of the previous four theorems in the following statement:

Let (x_1, y_1) and (x_2, y_2) be solutions of the homogeneous linear system

$$\begin{aligned} x' &= a_{11}x + a_{12}y, \\ y' &= a_{21}x + a_{22}y. \end{aligned} \tag{10}$$

Then $(c_1x_1 + c_2x_2, c_1y_1 + c_2y_2)$ will be the general solution of the system (10) provided that $W(t) \neq 0$; that is, provided the solutions (x_1, y_1) and (x_2, y_2) are linearly independent.

Finally, let us consider the nonhomogeneous system (1). The proof of the following theorem is left as an exercise:

Theorem 6 Let (x^*, y^*) be the general solution of the system (1), and let (x_p, y_p) be any solution of (1). Then $(x^* - x_p, y^* - y_p)$ is the general solution of the homogeneous equation (10). In other words, the general solution of the system (1) can be written as the sum of the general solution of the homogeneous system (10) and any particular solution of the nonhomogeneous system (1).

EXAMPLE 3 Consider the system

$$\begin{aligned} x' &= 3x + 3y + t, \\ y' &= -x - y + 1. \end{aligned} \tag{11}$$

We could solve this system by the methods given in the previous section. Here we note first that $(1, -1)$ and $(-3e^{2t}, e^{2t})$ are solutions to the homogeneous system

$$x' = 3x + 3y,$$
$$y' = -x - y.$$

A particular solution to the system (11) is $(-\frac{1}{4}(t^2 + 9t + 3), \frac{1}{4}(t^2 + 7t))$. The general solution to (11) is therefore

$$(x(t), y(t)) = (c_1 - 3c_2e^{2t} - \tfrac{1}{4}(t^2 + 9t + 3), -c_1 + c_2e^{2t} + \tfrac{1}{4}(t^2 + 7t)). \blacksquare$$

We close this section by noting that the theorems in this section can readily be generalized to systems of three or more equations.

PROBLEMS 12.2

1. (a) Show that

$$(e^{-3t}, -e^{-3t}) \quad \text{and} \quad ((1 - t)e^{-3t}, te^{-3t})$$

are solutions to

$$x' = -4x - y,$$
$$y' = x - 2y.$$

(b) Calculate the Wronskian and verify that the solutions are linearly independent.

(c) Write the general solution to the system.

2. (a) Show that $(e^{2t} \cos 2t, -2e^{2t} \sin 2t)$ and $(e^{2t} \sin 2t, 2e^{2t} \cos 2t)$ are solutions of the system

$$x' = 2x + y,$$
$$y' = -4x + 2y.$$

(b) Calculate the Wronskian of these solutions and show that they are linearly independent.

(c) Show that $(\frac{1}{4}te^{2t}, -\frac{11}{4}e^{2t})$ is a solution of the nonhomogeneous system

$$x' = 2x + y + 3e^{2t},$$
$$y' = -4x + 2y + te^{2t}.$$

(d) Combining (a) and (c), write the general solution of the nonhomogeneous equation in (c).

3. (a) Show that

$$(\sin t^2, 2t \cos t^2) \quad \text{and} \quad (\cos t^2, -2t \sin t^2)$$

are solutions of the system

$$x' = y,$$
$$y' = -4t^2x + \frac{1}{t}y.$$

(b) Show that the solutions are linearly independent.

(c) Show that $W(0) = 0$.

(d) Explain the apparent contradiction of Theorem 5.

4. (a) Show that $(\sin \ln t^2, (2/t) \cos \ln t^2)$ and $(\cos \ln t^2, -(2/t) \sin \ln t^2)$ are linearly independent solutions of the system

$$x' = y,$$
$$y' = -\frac{4}{t^2}x - \frac{1}{t}y.$$

(b) Calculate the Wronskian $W(t)$.

5. Prove Theorem 2.

6. Prove Theorem 6.

12.3 THE SOLUTION OF HOMOGENEOUS LINEAR SYSTEMS WITH CONSTANT COEFFICIENTS: THE METHOD OF DETERMINANTS

As we saw in Section 12.1, the method of elimination can be used to solve systems of linear equations with constant coefficients. Since the algebraic manipulations required can be cumbersome, we shall develop in this section a more efficient method of solving homogeneous systems. Nonhomogeneous systems are discussed in Problems 9–14. Consider the homogeneous system

$$\begin{aligned} x' &= a_{11}x + a_{12}y, \\ y' &= a_{21}x + a_{22}y \end{aligned} \tag{1}$$

where the a_{ij} are constants. Our main tool for solving second-order linear homogeneous equations with constant coefficients involved obtaining a characteristic equation by guessing that the solution had the form $y = e^{\lambda x}$.

Parallel to the method of Section 11.8, we guess that there is a solution to system (1) of the form $(\alpha e^{\lambda t}, \beta e^{\lambda t})$, where α, β, and λ are constants yet to be determined. Substituting $x(t) = \alpha e^{\lambda t}$, $y(t) = \beta e^{\lambda t}$, $x'(t) = \alpha\lambda e^{\lambda t}$ and $y'(t) = \beta\lambda e^{\lambda t}$ into equation (1), we obtain

$$\begin{aligned} \alpha\lambda e^{\lambda t} &= a_{11}\alpha e^{\lambda t} + a_{12}\beta e^{\lambda t}, \\ \beta\lambda e^{\lambda t} &= a_{21}\alpha e^{\lambda t} + a_{22}\beta e^{\lambda t}. \end{aligned}$$

After dividing by $e^{\lambda t}$, we obtain the linear system

$$\begin{aligned} (a_{11} - \lambda)\alpha + a_{12}\beta &= 0, \\ a_{21}\alpha + (a_{22} - \lambda)\beta &= 0. \end{aligned} \tag{2}$$

We would like to find values for λ such that the system of equations (2) has a solution (α, β) where α and β are not both zero. According to the theory of determinants, such a solution occurs whenever the determinant of the system

$$\begin{aligned} D &= \begin{vmatrix} a_{11} - \lambda & a_{12} \\ a_{21} & a_{22} - \lambda \end{vmatrix} \\ &= (a_{11} - \lambda)(a_{22} - \lambda) - a_{21}a_{12} \end{aligned} \tag{3}$$

is zero. Setting $D = 0$, we obtain the quadratic equation

$$\lambda^2 - (a_{11} + a_{22})\lambda + (a_{11}a_{22} - a_{21}a_{12}) = 0. \tag{4}$$

We define this equation to be the **characteristic equation** of system (1). That we are using the same term again is no accident, as we shall now demonstrate. Suppose we differentiate the first equation in system (1) and eliminate the function $y(t)$:

$$x'' = a_{11}x' + a_{12}(a_{21}x + a_{22}y).$$

Then

$$x'' - a_{11}x' - a_{12}a_{21}x = a_{22}a_{12}y = a_{22}(x' - a_{11}x),$$

and gathering like terms, we obtain the homogeneous equation

$$x'' - (a_{11} + a_{22})x' + (a_{11}a_{22} - a_{12}a_{21})x = 0. \tag{5}$$

The characteristic equation for (5), as the term was used in Section 11.8, is the same as equation (4). Hence the algebraic steps needed to obtain (5) can be avoided by setting the determinant $D = 0$.

As in Sections 11.8 and 11.9, there are three cases to consider, depending on whether the two roots λ_1 and λ_2 of the characteristic equation are real and distinct, real and equal, or complex conjugates. We will deal with each case separately.

Case 1. **Distinct real roots.** If λ_1 and λ_2 are distinct real numbers, then corresponding to λ_1 and λ_2 we have the vector solutions to system (1) $(\alpha_1 e^{\lambda_1 t}, \beta_1 e^{\lambda_1 t})$ and $(\alpha_2 e^{\lambda_2 t}, \beta_2 e^{\lambda_2 t})$, respectively. To find the constants α_1 and β_1 (not both zero), replace λ in the system of equations (2) by the value λ_1. The procedure is repeated for α_2, β_2, and λ_2. We note that the constants α_1, β_1, α_2 and β_2 are not unique. In fact, for each number λ_1 or λ_2, there are an infinite number of vectors (α, β) that satisfy system (2). To see this, we observe that if (α, β) is a solution pair, then so is $(c\alpha, c\beta)$ for any real number c. Finally, the vectors given above are linearly independent, since if not, there exists a constant c such that $\alpha_2 e^{\lambda_2 t} = c\alpha_1 e^{\lambda_1 t}$ and $\beta_2 e^{\lambda_2 t} = c\beta_1 e^{\lambda_1 t}$, which is clearly impossible because $\lambda_1 \neq \lambda_2$. We therefore have proved the following theorem:

Theorem 1 If λ_1 and λ_2 are distinct real roots of equation (4), then two linearly independent solutions of the system (1) are given by

$$(\alpha_1 e^{\lambda_1 t}, \beta_1 e^{\lambda_1 t}), \qquad (\alpha_2 e^{\lambda_2 t}, \beta_2 e^{\lambda_2 t}),$$

where the vectors (α_1, β_1) and (α_1, β_2) are solutions of system (2), with $\lambda = \lambda_1$ and $\lambda = \lambda_2$, respectively.

EXAMPLE 1 Find the general solution of the system

$$x' = -x + 6y,$$
$$y' = x - 2y.$$

Solution. Here $a_{11} = -1$, $a_{12} = 6$, $a_{21} = 1$, $a_{22} = -2$, and equation (3) becomes

$$D = \begin{vmatrix} -1 - \lambda & 6 \\ 1 & -2 - \lambda \end{vmatrix} = (\lambda + 2)(\lambda + 1) - 6 = \lambda^2 + 3\lambda - 4 = 0,$$

which has the roots $\lambda_1 = -4$, $\lambda_2 = 1$. For $\lambda_1 = -4$ the system of equations (2) becomes

$$3\alpha_1 + 6\beta_1 = 0,$$
$$\alpha_1 + 2\beta_1 = 0.$$

Ignoring the first equation because it is just a multiple of the second, select β_1 to be 1 to obtain $(-2, 1)$ as a solution of the second equation. Hence a first solution is $(-2e^{-4t}, e^{-4t})$. Similarly, with $\lambda_2 = 1$, we obtain the equations

$$-2\alpha_2 + 6\beta_2 = 0,$$
$$\alpha_2 - 3\beta_2 = 0,$$

which have a solution $\alpha_2 = 3$, $\beta_2 = 1$. Thus a second solution, linearly independent of the first, is given by the pair $(3e^t, e^t)$. By Theorem 12.2.3, the general solution is given by the pair

$$(x(t), y(t)) = (-2c_1e^{-4t} + 3c_2e^t, c_1e^{-4t} + c_2e^t).$$

NOTE. As defined in Section 9.10, the numbers -4 and 1 are eigenvalues of the matrix $\begin{pmatrix} -1 & 6 \\ 1 & -2 \end{pmatrix}$ with corresponding eigenvectors $\begin{pmatrix} -2 \\ 1 \end{pmatrix}$ and $\begin{pmatrix} 3 \\ 1 \end{pmatrix}$. ■

Case 2. Two equal roots. When $\lambda_1 = \lambda_2$, one solution is $(\alpha_1 e^{\lambda_1 t}, \beta_1 e^{\lambda_1 t})$. The other solution $(\alpha_2 e^{\lambda_2 t}, \beta_2 e^{\lambda_2 t})$ given by Theorem 1 is not independent of the first unless $a_{11} = a_{22}$ and $a_{12} = a_{21} = 0$. In this case we have the *uncoupled* system of equations

$$x' = a_{11}x, \qquad y' = a_{22}y,$$

with the linearly independent solutions $(\alpha_1 e^{\lambda_1 t}, 0)$ and $(0, \beta_2 e^{\lambda_1 t})$. (The equations are said to be uncoupled because each involves only one dependent variable.) On the basis of the results of Section 11.8, we would expect that, if the system is not uncoupled, a second linearly independent solution would have the form $(\alpha_2 te^{\lambda_1 t}, \beta_2 te^{\lambda_1 t})$. This, however, *does not turn out to be the case.* Rather, the second linearly independent solution has the form

$$(x(t), y(t)) = ((\alpha_2 + \alpha_3 t)e^{\lambda_1 t}, (\beta_2 + \beta_3 t)e^{\lambda_1 t}). \tag{6}$$

To calculate the constants α_2, α_3, β_2, and β_3, it is necessary to substitute back into the original system (1). This is best shown by an example.

EXAMPLE 2 Find the general solution of the system

$$\begin{aligned} x' &= -4x - y, \\ y' &= x - 2y. \end{aligned} \tag{7}$$

Solution. Equation (3) is

$$D = \begin{vmatrix} -4-\lambda & -1 \\ 1 & -2-\lambda \end{vmatrix} = (\lambda+4)(\lambda+2)+1 = \lambda^2+6\lambda+9=0,$$

which has the double root $\lambda_1 = \lambda_2 = -3$. From system (2), with $\lambda = -3$, we find that

$$\begin{aligned} -\alpha_1 - \beta_1 &= 0, \\ \alpha_1 + \beta_1 &= 0. \end{aligned}$$

A nontrivial solution is $\alpha_1 = 1$, $\beta_1 = -1$, yielding the solution $(e^{-3t}, -e^{-3t})$.

If we try to find a solution of the form $(\alpha_2 te^{-3t}, \beta_2 te^{-3t})$ we immediately run into trouble, since the derivatives on the left-hand side of system (7) produce terms of the form ce^{-3t} not present on the right-hand side of (7). This explains why we must seek a solution of the form

$$((\alpha_2 + \alpha_3 t)e^{-3t}, (\beta_2 + \beta_3 t)e^{-3t}). \tag{8}$$

Substituting the pair (8) into system (7), we obtain

$$\begin{aligned} e^{-3t}(\alpha_3 - 3\alpha_2 - 3\alpha_3 t) &= -4(\alpha_2 + \alpha_3 t)e^{-3t} - (\beta_2 + \beta_3 t)e^{-3t}, \\ e^{-3t}(\beta_3 - 3\beta_2 - 3\beta_3 t) &= (\alpha_2 + \alpha_3 t)e^{-3t} - 2(\beta_2 + \beta_3 t)e^{-3t}. \end{aligned}$$

Dividing by e^{-3t} and equating constant terms and coefficients of t, we obtain the system of equations

$$\begin{aligned} \alpha_3 - 3\alpha_2 &= -4\alpha_2 - \beta_2, \\ -3\alpha_3 &= -4\alpha_3 - \beta_3, \\ \beta_3 - 3\beta_2 &= \alpha_2 - 2\beta_2, \\ -3\beta_3 &= \alpha_3 - 2\beta_3. \end{aligned}$$

One solution is $\alpha_2 = 1$, $\beta_2 = -2$, $\alpha_3 = 1$, $\beta_3 = -1$. Thus a second solution of system (7) is $((1+t)e^{-3t}, (-2-t)e^{-3t})$. It is easy to verify that the two solutions are linearly independent, since $W(t) = -e^{-6t}$. ■

We summarize these results by stating the following theorem:

Theorem 2 Let equation (4) have two equal real roots $\lambda_1 = \lambda_2$. Then there exist constants α_1, α_2, α_3, β_1, β_2, and β_3 such that two linearly independent solutions of system (1) are given by

$$\begin{aligned} (x_1(t), y_1(t)) &= (\alpha_1 e^{\lambda_1 t}, \beta_1 e^{\lambda_1 t}), \\ (x_2(t), y_2(t)) &= ((\alpha_2 + \alpha_3 t)e^{\lambda_1 t}, (\beta_2 + \beta_3 t)e^{\lambda_1 t}). \end{aligned} \tag{9}$$

The constants α_1 and β_1 are found as nontrivial solutions of the homogeneous system of equations (2); the other constants are found by substituting the second equation of (9) back into system (1).

REMARK. In the substitution process, we always obtain a homogeneous system of four equations in the four unknowns α_2, β_2, α_3, and β_3. That this system has nontrivial solutions follows because the determinant of the system is zero. The proof is left as an exercise (see Problem 8).

Case 3. **Complex conjugate roots.**[†] Let $\lambda_1 = a + ib$ and $\lambda_2 = a - ib$, where a and b are real and $b \neq 0$. Then the solutions $(\alpha_1 e^{(a+ib)t}, \beta_1 e^{(a-ib)t})$ and $(\alpha_2 e^{(a-ib)t}, \beta_2 e^{(a-ib)t})$ are linearly independent. However, the constants α_1, β_1, α_2, and β_2 obtained from system (2) are complex numbers. To obtain real solution pairs we proceed as follows. Let $\alpha_1 = A_1 + iA_2$, $\beta_1 = B_1 + iB_2$ and apply Euler's formula, $e^{i\theta} = \cos\theta + i\sin\theta$, to the first complex solution, obtaining

$$\begin{aligned} x(t) &= (A_1 + iA_2)e^{at}(\cos bt + i\ sin\ bt), \\ y(t) &= (B_1 + iB_2)e^{at}(\cos bt + i\sin bt). \end{aligned} \tag{10}$$

Multiplying and remembering that $i^2 = -1$, we obtain the equations

$$\begin{aligned} x(t) &= e^{at}[(A_1\cos bt - A_2\sin bt) + i(A_1\sin bt + A_2\cos bt)], \\ y(t) &= e^{at}[(B_1\cos bt - B_2\sin bt) + i(B_1\sin bt + B_2\cos bt)]. \end{aligned}$$

Now, since the coefficients of system (1) are *real*, the only way (x, y) can be a solution is that all the real terms and, similarly, the imaginary terms, cancel out. Thus the real parts of x and y must form a solution of system (1), as must the imaginary parts:

$$\begin{aligned} (x_1(t), y_1(t)) &= (e^{at}(A_1\cos bt - A_2\sin bt), e^{at}(B_1\cos bt - B_2\sin bt)), \\ (x_2(t), y_2(t)) &= (e^{at}(A_1\sin bt + A_2\cos bt), e^{at}(B_1\sin bt + B_2\cos bt)). \end{aligned} \tag{11}$$

The Wronskian of the solutions (11) is

$$W(t) = e^{2at}(A_1B_2 - A_2B_1).$$

We want to show that the solutions (11) are linearly independent. Suppose otherwise. Then $W(t) = 0$, which means that $A_1B_2 = A_2B_1$. This implies that $B_2\alpha_1 = A_2\beta_1$ (according to the definition of α_1 and β_1). Now, neither α_1 nor β_1 vanishes. If either were zero, so would be the other, and the solution (10) would be trivial. Also A_2 cannot vanish, since if it did, so would B_2, and the first equation of system (2) would prevent λ_1 from being complex. Multiplying the first equation of system (2) by A_2, using the identity $B_2\alpha_1 = A_2\beta_1$, and dividing by α_1, we have

$$(a_{11} - \lambda_1)A_2 + a_{12}B_2 = 0.$$

[†]For this discussion we assume you are familiar with the material in Appendix 3.

But then λ_1 is not complex. Therefore, it is impossible that $W(t)$ could vanish, and we have proved the following theorem:

Theorem 3 If equation (4) has the complex roots $\lambda_1 = a + ib$ and $\lambda_2 = a - ib$, then two linearly independent solutions to system (1) are given by (11).

EXAMPLE 3 Find the general solution of the system

$$\begin{aligned} x' &= 4x + y, \\ y' &= -8x + 8y. \end{aligned} \tag{12}$$

Solution. Here

$$D = \begin{vmatrix} 4-\lambda & 1 \\ -8 & 8-\lambda \end{vmatrix} = \lambda^2 - 12\lambda + 40 = 0.$$

The roots of the characteristic equation are $\lambda_1 = 6 + 2i$ and $\lambda_2 = 6 - 2i$, so that Theorem 3 yields the linearly independent solutions

$$\begin{aligned} (x_1(t), y_1(t)) &= (e^{6t}(A_1 \cos 2t - A_2 \sin 2t), e^{6t}(B_1 \cos 2t - B_2 \sin 2t)), \\ (x_2(t), y_2(t)) &= (e^{6t}(A_1 \sin 2t + A_2 \cos 2t), e^{6t}(B_1 \sin 2t - B_2 \cos 2t)). \end{aligned}$$

Substituting the first equation into system (12) yields, after a great deal of algebra, the system of equations

$$\begin{aligned} (2A_1 - 2A_2 - B_1) \cos 2t - (2A_1 + 2A_2 - B_2) \sin 2t &= 0, \\ (8A_1 - 2B_1 - 2B_2) \cos 2t - (8A_2 + 2B_1 - 2B_2) \sin 2t &= 0. \end{aligned}$$

Since t is arbitrary and the functions $\sin 2t$ and $\cos 2t$ are linearly independent, the terms in parentheses must all vanish. A choice of values that will accomplish this is $A_1 = 1$, $A_2 = \frac{1}{2}$, $B_1 = 1$, and $B_2 = 3$. Thus two linearly independent solutions to system (12) are $(e^{6t}(\cos 2t - \frac{1}{2} \sin 2t), e^{6t}(\cos 2t - 3 \sin 2t))$ and $(e^{6t}(\sin 2t + \frac{1}{2} \cos 2t), e^{6t}(\sin 2t + 3 \cos 2t))$. The general solution of system (12) is a linear combination of these two solutions. ■

EXAMPLE 4 Most biological systems are controlled by the production of enzymes or hormones that stimulate or inhibit the secretion of some compound. For example, the pancreatic hormone *glucagon* stimulates the release of glucose from the liver to the plasma. A rise in blood glucose inhibits the secretion of glucagon but causes an increase in the production of the hormone insulin. Insulin, in turn, aids in the removal of glucose from the blood and in its conversion to glycogen in the muscle tissue. Let G and I be the deviations of plasma glucose and plasma insulin from the normal (fasting) level, respectively. We then have the system

$$\begin{aligned} \frac{dG}{dt} &= -k_{11}G - k_{12}I, \\ \frac{dI}{dt} &= k_{21}G - k_{22}I \end{aligned} \tag{13}$$

where the positive constants k_{ij} are model parameters, some of which may be determined experimentally. It is known that the system (13) exhibits a strongly damped oscillatory behavior, since direct injection of glucose into the blood will produce a fall of blood glucose to a level below fasting in about one and a half hours, followed by a rise slightly above the fasting level in about three hours. Hence, the characteristic equation of system (13),

$$D = \begin{vmatrix} -k_{11} - \lambda & -k_{12} \\ k_{21} & -k_{22} - \lambda \end{vmatrix} = (k_{11} + \lambda)(k_{22} + \lambda) + k_{12}k_{21}$$
$$= \lambda^2 + (k_{11} + k_{22})\lambda + (k_{11}k_{22} + k_{12}k_{21}) = 0,$$

must have complex conjugate roots $-a \pm ib$, with $a = (k_{11} + k_{22})/2$ and $b = \sqrt{k_{12}k_{21} - (k_{11} - k_{22})^2/4}$, since only complex roots can lead to oscillatory behavior. By Theorem 3, we have the solutions

$$\begin{aligned} (G_1, I_1) &= (e^{-at}(A_1 \cos bt - A_2 \sin bt), e^{-at}(B_1 \cos bt - B_2 \sin bt)), \\ (G_2, I_2) &= (e^{-at}(A_1 \sin bt + A_2 \cos bt), e^{-at}(B_1 \sin bt + B_2 \cos bt)). \end{aligned} \quad \textbf{(14)}$$

Since the period of the oscillation is approximately three hours, we may set $b = 2\pi/3$ and measure time in hours. Substituting the first equation of system (14) into equations (13) we obtain the following equations:

$$(-aA_1 - bA_2 + k_{11}A_1 + k_{12}B_1) \cos bt + (aA_2 - bA_1 - k_{11}A_2 - k_{12}B_2) \sin bt = 0$$
$$(-aB_1 - bB_2 - k_{21}A_1 + k_{22}B_1) \cos bt + (aB_2 - bB_1 + k_{21}A_2 - k_{22}B_2) \sin bt = 0$$

These equations must hold for all t. Thus all the terms in parentheses must vanish. A choice of values for which this occurs is $A_1 = 1$, $A_2 = 0$, $B_1 = (k_{22} - k_{11})/2k_{12}$, and $B_2 = -b/k_{12}$. Then the general solution of system (13) is given by the vector $(G(t), I(t))$ with

$$G(t) = e^{-at}(c_1 \cos bt + c_2 \sin bt),$$
$$I(t) = e^{-at}\left[\frac{k_{22} - k_{11}}{2k_{12}}(c_1 \cos bt + c_2 \sin bt) + \frac{b}{k_{12}}(c_1 \sin bt - c_2 \cos bt)\right].$$

Assume now that the glucose injection was administered at a time when plasma insulin and glucose were at fasting levels and that the glucose was diffused completely in the blood before the insulin level began to increase ($t = 0$). Also, $G(0) = G_0$ equals the ratio of the volume of glucose administered to blood volume, and $I(0) = 0$. Since $G(t)$ is at a maximum when $t = 0$, it follows that $c_1 = G_0$ and $c_2 = 0$. Hence

$$G(t) = G_0 e^{-at} \cos bt. \quad \textbf{(15)}$$

But

$$0 = I(0) = \frac{k_{22} - k_{11}}{2k_{12}} G_0,$$

so that $k_{11} = k_{22}$, $b = \sqrt{k_{12}k_{21}} = 2\pi/3$, and

$$I(t) = G_0 \frac{b}{k_{12}} e^{-at} \sin bt.$$

If the minimum level $G(\frac{3}{2})$ (< 0) is known, then by equation (15),

$$e^{3a/2} = |G(\tfrac{3}{2})|/G_0,$$

so that

$$k_{11} = a = \frac{2}{3} \ln \frac{|G(\frac{3}{2})|}{G_0}.$$

If we determine the plasma insulin at any given time $t_0 > 0$, we can then evaluate the parameters k_{12} and k_{21}. ■

PROBLEMS 12.3

In Problems 1–7 use the method of determinants to find two linearly independent solutions for each given system.

1. $x' = 4x - 3y$, $y' = 5x - 4y$

2. $x' = 7x + 6y$, $y' = 2x + 6y$

3. $x' = -x + y$, $y' = -5x + 3y$

4. $x' = x + y$, $y' = -x + 3y$

5. $x' = -4x - y$, $y' = x - 2y$

6. $x' = 4x - 2y$, $y' = 5x + 2y$

7. $x' = 4x - 3y$, $y' = 8x - 6y$

8. Substituting the second solution vector (9) into system (1), we obtain the homogeneous system of linear equations

$$\begin{aligned} (\lambda - a_{11})\alpha_2 + \alpha_3 - \alpha_{12}\beta_2 &= 0, \\ (\lambda - a_{11})\alpha_3 - a_{12}\beta_3 &= 0, \\ -a_{21}\alpha_2 + (\lambda - a_{22})\beta_2 + \beta_3 &= 0, \\ -a_{21}\alpha_3 + (\lambda - a_{22})\beta_3 &= 0. \end{aligned} \qquad \textbf{(16)}$$

(a) Show that since $\lambda_1 = \lambda_2 = (a_{11} + a_{22})/2$, the second and fourth equations of system (16) are identical.

(b) Conclude from part (a) that the determinant of system (16) is zero, and from this that (16) has nontrivial solutions.

9. Consider the nonhomogeneous equations

$$\begin{aligned} x' &= a_{11}x + a_{12}y + f_1, \\ y' &= a_{21}x + a_{22}y + f_2. \end{aligned} \qquad \textbf{(17)}$$

Let (x_1, y_1) and (x_2, y_2) be two linearly independent solutions of the homogeneous system (1). Show that

$$\begin{aligned} x_p(t) &= \nu_1(t)x_1(t) + \nu_2(t)x_2(t), \\ y_p(t) &= \nu_1(t)y_1(t) + \nu_2(t)y_2(t), \end{aligned}$$

is a particular solution of system (17) if ν_1 and ν_2 satisfy the equations

$$\begin{aligned} \nu_1'x_1 + \nu_2'x_2 &= f_1, \\ \nu_1'y_1 + \nu_2'y_2 &= f_2. \end{aligned}$$

This process for finding a particular solution of the nonhomogeneous system (16) is called the **variation of constants method for systems**. Note the close parallel between this method and the method given in Section 11.11.

In Problems 10–14 use the variation of constants method to find a particular solution for each given nonhomogeneous system.

10. $x' = 2x + y + 3e^{2t}$, $y' = -4x + 2y + te^{2t}$

11. $x' = 3x + 3y + t$, $y' = -x - y + 1$

12. $x' = -2x + y$, $y' = -3x + 2y + 2\sin t$

13. $x' = -x + y + \cos t$, $y' = -5x + 3y$

14. $x' = 3x - 2y + t$, $y' = 2x - 2y + 3e^{t}$

***15.** In an experiment on cholesterol turnover in humans, radioactive cholesterol-4–^{14}C was injected intravenously, and the total plasma cholesterol and radioactivity were measured. It was discovered that the turnover of cholesterol behaves like a two-compartment system.[†] The compartment consisting of the organs and blood has a rapid turnover, while the turnover in the other compartment is much slower. Assume that the body takes in and excretes all cholesterol through the first compartment. Let $x(t)$ and $y(t)$ denote the deviations from normal cholesterol levels in each compartment. Suppose that the daily fractional transfer from compartment x is 0.134, of which 0.036 is the input to compartment y, and that the transfer from compartment y is 0.02.

(a) Describe the problem discussed above as a system of homogeneous linear differential equations.

(b) Obtain the general solution of the system.

12.4 MATRICES AND SYSTEMS OF LINEAR FIRST-ORDER EQUATIONS

In this section we begin to use the powerful tools of matrix theory to describe the behavior of solutions to systems of differential equations.

We shall assume, from here on, that you are familiar with the elementary properties of vectors and matrices, including vector and matrix addition and scalar multiplication, matrix multiplication, the notion of linear dependence and independence of vectors, and the calculation of the inverse of an invertible matrix.

Before discussing the relationship between matrices and systems of equations, let us consider the notion of a vector and matrix function.

An n-component **vector function**

$$\mathbf{v}(t) = \begin{pmatrix} v_1(t) \\ v_2(t) \\ \vdots \\ v_n(t) \end{pmatrix} \tag{1}$$

is an n-vector, each of whose components is a function (usually assumed to be continuous). An $n \times n$ **matrix function** $A(t)$ is an $n \times n$ matrix

$$A(t) = \begin{pmatrix} a_{11}(t) & a_{12}(t) & \cdots & a_{1n}(t) \\ a_{21}(t) & a_{22}(t) & \cdots & a_{2n}(t) \\ \vdots & \vdots & & \vdots \\ a_{n1}(t) & a_{n2}(t) & \cdots & a_{nn}(t) \end{pmatrix}, \tag{2}$$

[†]D. S. Goodman and R. P. Noble, "Turnover of Plasma Cholesterol in Man," *J. Clin. Invest.* **47**, 231–241 (1968).

each of whose n^2 components is a function. We may add and multiply vector and matrix functions in the same way that we add and multiply constant vectors and matrices. Thus if, for example,

$$A(t) = \begin{pmatrix} a_{11}(t) & a_{12}(t) \\ a_{21}(t) & a_{22}(t) \end{pmatrix} \quad \text{and} \quad B(t) = \begin{pmatrix} b_{11}(t) & b_{12}(t) \\ b_{21}(t) & b_{22}(t) \end{pmatrix},$$

then

$$A(t)B(t) = \begin{pmatrix} a_{11}(t)b_{11}(t) + a_{12}(t)b_{21}(t) & a_{11}(t)b_{12}(t) + a_{12}(t)b_{22}(t) \\ a_{21}(t)b_{11}(t) + a_{22}(t)b_{21}(t) & a_{21}(t)b_{12}(t) + a_{22}(t)b_{22}(t) \end{pmatrix}.$$

We may also differentiate and integrate vector and matrix functions, component by componeent. Thus if $\mathbf{v}(t)$ is given by Equation (1), then we define

$$\mathbf{v}'(t) = \begin{pmatrix} v_1'(t) \\ v_2'(t) \\ \vdots \\ v_n'(t) \end{pmatrix} \quad \text{and} \quad \int_{t_0}^{t} \mathbf{v}(s)\, ds = \begin{pmatrix} \int_{t_0}^{t} v_1(s)\, ds \\ \int_{t_0}^{t} v_2(s)\, ds \\ \vdots \\ \int_{t_0}^{t} v_n(s)\, ds \end{pmatrix}.$$

Similarly, if $A(t)$ is given by equation (2), then we define

$$A'(t) = \begin{pmatrix} a_{11}'(t) & a_{12}'(t) & \cdots & a_{1n}'(t) \\ a_{21}'(t) & a_{22}'(t) & \cdots & a_{2n}'(t) \\ \vdots & \vdots & & \vdots \\ a_{n1}'(t) & a_{n2}'(t) & \cdots & a_{nn}'(t) \end{pmatrix}$$

and

$$\int_{t_0}^{t} A(s)\, ds = \begin{pmatrix} \int_{t_0}^{t} a_{11}(s)\, ds & \int_{t_0}^{t} a_{12}(s)\, ds & \cdots & \int_{t_0}^{t} a_{1n}(s)\, ds \\ \int_{t_0}^{t} a_{21}(s)\, ds & \int_{t_0}^{t} a_{22}(s)\, ds & \cdots & \int_{t_0}^{t} a_{2n}(s)\, ds \\ \vdots & \vdots & & \vdots \\ \int_{t_0}^{t} a_{n1}(s)\, ds & \int_{t_0}^{t} a_{n2}(s)\, ds & \cdots & \int_{t_0}^{t} a_{nn}(s)\, ds \end{pmatrix}.$$

EXAMPLE 1 Let

$$\mathbf{v}(t) = \begin{pmatrix} t \\ t^2 \\ \sin t \\ e^t \end{pmatrix}.$$

Then

$$\mathbf{v}'(t) = \begin{pmatrix} 1 \\ 2t \\ \cos t \\ e^t \end{pmatrix} \quad \text{and} \quad \int_0^t \mathbf{v}(s)\, ds = \begin{pmatrix} t^2/2 \\ t^3/3 \\ 1 - \cos t \\ e^t - 1 \end{pmatrix}. \quad \blacksquare$$

Consider the system of n first-order equations

$$\begin{aligned} x_1' &= a_{11}(t)x_1 + a_{12}(t)x_2 + \cdots + a_{1n}(t)x_n + f_1(t), \\ x_2' &= a_{21}(t)x_1 + a_{22}(t)x_2 + \cdots + a_{2n}(t)x_n + f_2(t), \\ &\vdots \\ x_n' &= a_{n1}(t)x_1 + a_{n2}(t)x_2 + \cdots + a_{nn}(t)x_n + f_n(t), \end{aligned} \tag{3}$$

which is nonhomogeneous if at least one of the functions $f_i(t)$, $i = 1, 2, \ldots, n$, is not the zero function. Consider also the associated homogeneous system

$$\begin{aligned} x_1' &= a_{11}(t)x_1 + a_{12}(t)x_2 + \cdots + a_{1n}(t)x_n, \\ x_2' &= a_{21}(t)x_1 + a_{22}(t)x_2 + \cdots + a_{2n}(t)x_n, \\ &\vdots \\ x_n' &= a_{n1}(t)x_1 + a_{n2}(t)x_2 + \cdots + a_{nn}(t)x_n. \end{aligned} \tag{4}$$

As was shown in Section 12.1, systems containing higher-order equations can always be reduced to systems of first-order equations. Hence we shall restrict our discussion to the systems of first-order equations (3) and (4).

We now define the vector function $\mathbf{x}(t)$, the matrix function $A(t)$, and the vector function $\mathbf{f}(t)$ as follows:

$$\mathbf{x}(t) = \begin{pmatrix} x_1(t) \\ x_2(t) \\ \vdots \\ x_n(t) \end{pmatrix}, \quad A(t) = \begin{pmatrix} a_{11}(t) & a_{12}(t) & \cdots & a_{1n}(t) \\ a_{21}(t) & a_{22}(t) & \cdots & a_{2n}(t) \\ \vdots & \vdots & & \vdots \\ a_{n1}(t) & a_{n2}(t) & \cdots & a_{nn}(t) \end{pmatrix}, \quad \mathbf{f}(t) = \begin{pmatrix} f_1(t) \\ f_2(t) \\ \vdots \\ f_n(t) \end{pmatrix}. \tag{5}$$

Then, using equations (5), we can rewrite system (3) as the **vector differential equation**

$$\mathbf{x}'(t) = A(t)\mathbf{x}(t) + \mathbf{f}(t). \tag{6}$$

System (4) becomes

$$\mathbf{x}'(t) = A(t)\mathbf{x}(t). \tag{7}$$

From what has already been said, any linear differential equation or system can be written in the form (7) if it is homogeneous and in the form (6) if it is nonhomogeneous. The reason for writing a system in these forms is that, besides the obvious advantage of compactness of notation, equations (6) and (7) behave very much like first-order linear differential equations, as we shall see. It will be very easy to work with systems of equations in this way once we get used to the notation.

EXAMPLE 2 Consider the system

$$\begin{aligned} x_1' &= (2t)x_1 + (\sin t)x_2 - e^t x_3 - e^t, \\ x_2' &= -t^3 x_1 + e^{\sin t} x_2 - (\ln t)x_3 + \cos t, \\ x_3' &= 2x_1 - 5tx_2 + 2tx_3 + \tan t. \end{aligned}$$

It can be rewritten as

$$\begin{pmatrix} x_1 \\ x_2 \\ x_3 \end{pmatrix}' = \begin{pmatrix} 2t & \sin t & -e^t \\ -t^3 & e^{\sin t} & -\ln t \\ 2 & -5t & 2t \end{pmatrix} \begin{pmatrix} x_1 \\ x_2 \\ x_3 \end{pmatrix} + \begin{pmatrix} -e^t \\ \cos t \\ \tan t \end{pmatrix}. \quad \blacksquare$$

EXAMPLE 3 Consider the general second-order linear differential equation

$$x'' + a(t)x' + b(t)x = f(t).$$

Defining $x_1 = x$ and $x_2 = x'$, we obtain the equivalent system

$$\begin{aligned} x_1' &= x_2, \\ x_2' &= -b(t)x_1 - a(t)x_2 + f(t), \end{aligned}$$

which can be rewritten as

$$\mathbf{x}' = A(t)\mathbf{x} + \mathbf{f}(t)$$

where

$$\mathbf{x} = \begin{pmatrix} x_1 \\ x_2 \end{pmatrix}, \quad A(t) = \begin{pmatrix} 0 & 1 \\ -b(t) & -a(t) \end{pmatrix}, \quad \mathbf{f}(t) = \begin{pmatrix} 0 \\ f(t) \end{pmatrix}. \quad \blacksquare$$

EXAMPLE 4 Consider the third-order equation with constant coefficients

$$x''' - 6x'' + 11x' - 6x = 0. \tag{8}$$

Defining $x_1 = x$, $x_2 = x'$, and $x_3 = x''$, we obtain the system

$$\begin{aligned} x_1' &= x_2, \\ x_2' &= x_3, \\ x_3' &= 6x_1 - 11x_2 + 6x_3, \end{aligned}$$

or

$$\mathbf{x}' = A\mathbf{x} \tag{9}$$

where

$$\mathbf{x} = \begin{pmatrix} x_1 \\ x_2 \\ x_3 \end{pmatrix} \quad \text{and} \quad A = \begin{pmatrix} 0 & 1 & 0 \\ 0 & 0 & 1 \\ 6 & -11 & 6 \end{pmatrix}. \quad \blacksquare$$

Let us now consider the initial value problem

$$\mathbf{x}'(t) = A(t)\mathbf{x}(t) + \mathbf{f}(t), \qquad \mathbf{x}(t_0) = \mathbf{x}_0, \tag{10}$$

where

$$\mathbf{x}(t) = \begin{pmatrix} x_1(t) \\ x_2(t) \\ \vdots \\ x_n(t) \end{pmatrix} \quad \text{and} \quad \mathbf{x}_0 = \begin{pmatrix} x_{10} \\ x_{20} \\ \vdots \\ x_{n0} \end{pmatrix}.$$

We say that a vector function

$$\varphi(t) = \begin{pmatrix} \varphi_1(t) \\ \vdots \\ \varphi_n(t) \end{pmatrix}$$

is a **solution** to the system (10) if φ is differentiable and satisfies the differential equation and the given initial condition. The following theorem is proved in most differential equation texts:†

---

†See, for example, W. Derrick and S. Grossman, *Elementary Differential Equations with Applications*, second edition (Addison-Wesley, Reading, MA, 1981), Section 10.2.

Theorem 1 Let $A(t)$ and $\mathbf{f}(t)$ be continuous matrix and vector functions, respectively, on some interval $[a, b]$ (i.e., the component functions of both $A(t)$ and $\mathbf{f}(t)$ are continuous). Then there exists a unique vector function $\varphi(t)$ that is a solution to the initial value problem (10) on the entire interval $[a, b]$.

EXAMPLE 5 Referring to Example 4, we find by the methods of Section 11.8 that e^t, e^{2t}, and e^{3t} are solutions of equation (8). Since a solution vector for this problem is

$$\varphi(t) = \begin{pmatrix} x(t) \\ x'(t) \\ x''(t) \end{pmatrix},$$

three vector solutions of equation (9) are

$$\varphi_1(t) = \begin{pmatrix} e^t \\ e^t \\ e^t \end{pmatrix} = e^t \begin{pmatrix} 1 \\ 1 \\ 1 \end{pmatrix}, \qquad \varphi_2(t) = \begin{pmatrix} e^{2t} \\ 2e^{2t} \\ 4e^{2t} \end{pmatrix} = e^{2t} \begin{pmatrix} 1 \\ 2 \\ 4 \end{pmatrix},$$

and

$$\varphi_3(t) = \begin{pmatrix} e^{3t} \\ 3e^{3t} \\ 9e^{3t} \end{pmatrix} = e^{3t} \begin{pmatrix} 1 \\ 3 \\ 9 \end{pmatrix}.$$

If we specify the initial condition

$$\mathbf{x}(0) = \begin{pmatrix} 2 \\ -3 \\ 5 \end{pmatrix},$$

then the unique solution vector is (check it)

$$\varphi = 16\varphi_1 - 23\varphi_2 + 9\varphi_3,$$

or

$$\begin{pmatrix} 16e^t - 23e^{2t} + 9e^{3t} \\ 16e^t - 46e^{2t} + 27e^{3t} \\ 16e^t - 92e^{2t} + 81e^{3t} \end{pmatrix}. \quad ■$$

EXAMPLE 6 The system

$$\begin{aligned} x_1' &= -4x_1 - x_2, & x_1(0) &= 1, \\ x_2' &= x_1 - 2x_2, & x_2(0) &= 2, \end{aligned}$$

can be written

$$\begin{pmatrix} x_1 \\ x_2 \end{pmatrix}' = \begin{pmatrix} -4 & -1 \\ 1 & -2 \end{pmatrix}\begin{pmatrix} x_1 \\ x_2 \end{pmatrix}, \qquad \begin{pmatrix} x_1(0) \\ x_2(0) \end{pmatrix} = \begin{pmatrix} 1 \\ 2 \end{pmatrix}.$$

It can be verified that

$$\varphi_1(t) = \begin{pmatrix} e^{-3t} \\ -e^{-3t} \end{pmatrix} \quad \text{and} \quad \varphi_2(t) = \begin{pmatrix} (1-t)e^{-3t} \\ te^{-3t} \end{pmatrix}$$

are solution vectors. We can also show that the unique solution vector that satisfies the given initial conditions is

$$\varphi(t) = \begin{pmatrix} (1-3t)e^{-3t} \\ (2+3t)e^{-3t} \end{pmatrix}. \quad ■$$

The central problem of the remainder of this chapter is to derive properties of vector solutions and, where possible, to calculate them. In the next section and in Section 12.6 we will show how all solutions to the homogeneous system $\mathbf{x}' = A\mathbf{x}$ can be represented in a convenient form, and in Section 12.7 we will show how information about the solutions to this homogeneous system can be used to find a particular solution to the nonhomogeneous system $\mathbf{x}' = A\mathbf{x} + \mathbf{f}$.

PROBLEMS 12.4

In Problems 1–6 write each equation or system in the matrix-vector form (6), (7), or (10).

1. $x_1' = 2x_1 + 3x_2$,
   $x_2' = 4x_1 - 6x_2$
2. $x_1' = (\cos t)x_1 - (\sin t)x_2 + e^{t^2}$,
   $x_2' = e^t x_1 + 2tx_2 - \ln t$,
   $x_1(2) = 3,\ x_2(2) = 7$
3. $x''' - 2x'' + 4tx' - x = \sin t$
4. $x^{(4)} + 2x''' - 3x'' + 4x' - 7x = 0$,
   $x(0) = 1,\ x'(0) = 2,\ x''(0) = 3,\ x'''(0) = 4$
5. $x_1' = 2tx_1 - 3t^2x_2 + (\sin t)x_3$,
   $x_2' = 2x_1 - 4x_3 - \sin t$,
   $x_3' = 17x_2 + 4tx_3 + e^t$
6. $x''' + a(t)x'' + b(t)x' + c(t)x = f(t)$,
   $x(t_0) = d_1,\ x'(t_0) = d_2,\ x''(t_0) = d_3$

In Problems 7–14, verify that each vector function is a solution to the given system.

7. $\mathbf{x}' = \begin{pmatrix} 1 & 1 \\ -3 & -1 \end{pmatrix}\mathbf{x}$,
   $\varphi(t) = \begin{pmatrix} \cos\sqrt{2}t \\ -\sqrt{2}\sin\sqrt{2}t - \cos\sqrt{2}t \end{pmatrix}$
8. $\mathbf{x}' = \begin{pmatrix} 2 & 1 \\ 1 & 2 \end{pmatrix}\mathbf{x} + \begin{pmatrix} t \\ t^2 \end{pmatrix}$,
   $\varphi(t) = \begin{pmatrix} e^t + \frac{1}{3}t^2 + \frac{2}{9}t + \frac{11}{27} \\ -e^t - \frac{2}{3}t^2 - \frac{7}{9}t - \frac{16}{27} \end{pmatrix}$
9. $\mathbf{x}' = \begin{pmatrix} -1 & 6 \\ 1 & -2 \end{pmatrix}\mathbf{x},\ \varphi(t) = \begin{pmatrix} 3e^t \\ e^t \end{pmatrix}$
10. $\mathbf{x}' = \begin{pmatrix} -4 & -1 \\ 1 & -2 \end{pmatrix}\mathbf{x},\ \varphi(t) = \begin{pmatrix} (1+t)e^{-3t} \\ (-2-t)e^{-3t} \end{pmatrix}$
11. $\mathbf{x}' = \begin{pmatrix} 4 & 1 \\ -8 & 8 \end{pmatrix}\mathbf{x},\ \varphi(t) = \begin{pmatrix} e^{6t}(\cos 2t - \frac{1}{2}\sin 2t) \\ e^{6t}(\cos 2t - 3\sin 2t) \end{pmatrix}$
12. $\mathbf{x}' = \begin{pmatrix} 1 & 1 & 1 \\ -1 & -1 & 0 \\ -1 & 0 & -1 \end{pmatrix}\mathbf{x},\ \varphi(t) = \begin{pmatrix} \sin t - \cos t \\ \cos t \\ \cos t \end{pmatrix}$
13. $\mathbf{x}' = \begin{pmatrix} 1 & -1 & 1 & -1 \\ 0 & -1 & 2 & -2 \\ 0 & 0 & 2 & -3 \\ 0 & 0 & 0 & -2 \end{pmatrix}\mathbf{x},\ \varphi(t) = \begin{pmatrix} e^{-2t} \\ 2e^{-2t} \\ 3e^{-2t} \\ 4e^{-2t} \end{pmatrix}$

14. $\mathbf{x}' = \begin{pmatrix} 3 & 2 & 1 \\ -1 & 0 & -1 \\ 1 & 1 & 2 \end{pmatrix}\mathbf{x}, \ \varphi(t) = \begin{pmatrix} e^{2t} + te^{2t} \\ -te^{2t} \\ te^{2t} \end{pmatrix}$

15. Let $\varphi_1(t)$ and $\varphi_2(t)$ be any two vector solutions of the homogeneous system $\mathbf{x}' = A(t)\mathbf{x}$. Show that $\varphi(t) = c_1\varphi_1(t) + c_2\varphi_2(t)$ is also a solution.

16. Let $\varphi_1(t)$ and $\varphi_2(t)$ be vector solutions of the nonhomogeneous system (6). Show that their difference,

$$\varphi(t) = \varphi_1(t) - \varphi_2(t),$$

is a solution to the homogeneous system (7).

17. Find the derivative and an antiderivative of each of the following vector and matrix functions:

(a) $\mathbf{x}(t) = (t, \sin t)$; **(b)** $\mathbf{y}(t) = \begin{pmatrix} e^t \\ \cos t \\ \tan t \end{pmatrix}$;

(c) $A(t) = \begin{pmatrix} \sqrt{t} & t^2 \\ e^{2t} & \sin 2t \end{pmatrix}$;

(d) $B(t) = \begin{pmatrix} \ln t & e^t \sin t & e^t \cos t \\ t^{5/2} & -\cos t & -\sin t \\ 1/t & te^{t^2} & t^2e^{t^3} \end{pmatrix}$.

12.5 FUNDAMENTAL SETS AND FUNDAMENTAL MATRIX SOLUTIONS OF A HOMOGENEOUS SYSTEM OF DIFFERENTIAL EQUATIONS

In this section we shall discuss properties of the homogeneous system

$$\mathbf{x}' = A(t)\mathbf{x} \tag{1}$$

where $\mathbf{x}(t)$ is an n-vector and $A(t)$ is an $n \times n$ matrix.

Let $\varphi_1(t), \varphi_2(t), \ldots, \varphi_m(t)$ be m vector solutions of the system (1). Recall from Section 9.3 that they are **linearly independent** if the equation

$$c_1\varphi_1(t) + c_2\varphi_2(t) + \cdots + c_m\varphi_m(t) = \mathbf{0}$$

holds only for $c_1 = c_2 = \cdots = c_m = 0$. Since system (1) is equivalent to an nth-order equation, it is natural for us to seek n linearly independent solutions to the system.

Definition 1 FUNDAMENTAL SET OF SOLUTIONS Any set of n linearly independent solutions to system (1) is called a **fundamental set of solutions**.

In Sections 12.2 and 12.3 we saw how a fundamental set of solutions (i.e., two linearly independent solutions) could be found in the case in which A was a 2×2 constant matrix. By Theorem 12.2.5 the vectors

$$\varphi_1(t) = \begin{pmatrix} x_1(t) \\ y_1(t) \end{pmatrix} \quad \text{and} \quad \varphi_2(t) = \begin{pmatrix} x_2(t) \\ y_2(t) \end{pmatrix}$$

are a fundamental set of solutions if and only if the Wronskian $W(t)$, defined by equation (12.2.6), is nonzero.

EXAMPLE 1 Let $\mathbf{x}' = A\mathbf{x}$ where

$$A = \begin{pmatrix} -1 & 6 \\ 1 & -2 \end{pmatrix}.$$

In Example 12.3.1 we verified that

$$\varphi_1(t) = \begin{pmatrix} -2e^{-4t} \\ e^{-4t} \end{pmatrix} \quad \text{and} \quad \varphi_2(t) = \begin{pmatrix} 3e^{t} \\ e^{t} \end{pmatrix}$$

are solution vectors. To show that they are fundamental (as we did in Example 12.3.1), we compute the Wronskian:

$$W(\varphi_1, \varphi_2)(t) = \begin{vmatrix} -2e^{-4t} & 3e^{t} \\ e^{-4t} & e^{t} \end{vmatrix} = -5e^{-3t} \neq 0. \blacksquare$$

EXAMPLE 2 Let $\mathbf{x}' = A\mathbf{x}$ where

$$A = \begin{pmatrix} -4 & -1 \\ 1 & -2 \end{pmatrix}.$$

In Example 12.3.2 we obtained the two solution vectors

$$\varphi_1(t) = \begin{pmatrix} e^{-3t} \\ -e^{-3t} \end{pmatrix} \quad \text{and} \quad \varphi_2(t) = \begin{pmatrix} (1+t)e^{-3t} \\ (-2-t)e^{-3t} \end{pmatrix}.$$

Then

$$W(\varphi_1, \varphi_2)(t) = \begin{vmatrix} e^{-3t} & (1+t)e^{-3t} \\ -e^{-3t} & (-2-t)e^{-3t} \end{vmatrix} = e^{-6t}\begin{vmatrix} 1 & 1+t \\ -1 & -2-t \end{vmatrix} = -e^{-6t} \neq 0.$$

Thus $\varphi_1(t)$ and $\varphi_2(t)$ form a fundamental set of solutions. $\blacksquare$

EXAMPLE 3 Consider the system $\mathbf{x}' = A\mathbf{x}$ where

$$A = \begin{pmatrix} 4 & 1 \\ -8 & 8 \end{pmatrix}$$

(see Example 12.3.3). Two solution vectors are

$$\varphi_1(t) = \begin{pmatrix} e^{6t}(\cos 2t - \frac{1}{2}\sin 2t) \\ e^{6t}(\cos 2t - 3\sin 2t) \end{pmatrix}, \qquad \varphi_2(t) = \begin{pmatrix} e^{6t}(\sin 2t + \frac{1}{2}\cos 2t) \\ e^{6t}(\sin 2t + 3\cos 2t) \end{pmatrix}.$$

To show that they are linearly independent, let $c_1\varphi_1(t) + c_2\varphi_2(t) = \mathbf{0}$ for every t. Then for $t = 0$, we have

$$\begin{pmatrix} c_1 \\ c_2 \end{pmatrix} + \begin{pmatrix} \frac{1}{2}c_2 \\ 3c_2 \end{pmatrix} = \mathbf{0}$$

or

$$c_1 + \tfrac{1}{2}c_2 = 0,$$
$$c_1 + 3c_2 = 0.$$

The only solution of this system is $c_1 = c_2 = 0$, and so φ_1 and φ_2 are, indeed, linearly independent. We can also show this by computing the Wronskian. It is (check it)

$$W(\varphi_1, \varphi_2)(t) = e^{12t}\left(\frac{5}{2}\cos^2 2t + \frac{5}{2}\sin^2 2t\right) = \frac{5}{2}e^{12t} \neq 0. \blacksquare$$

EXAMPLE 4 Let $\mathbf{x}' = A\mathbf{x}$ where

$$A = \begin{pmatrix} 1 & 1 & -2 \\ -1 & 2 & 1 \\ 0 & 1 & -1 \end{pmatrix}.$$

Then

$$\varphi_1(t) = \begin{pmatrix} e^{-t} \\ 0 \\ e^{-t} \end{pmatrix} = e^{-t}\begin{pmatrix} 1 \\ 0 \\ 1 \end{pmatrix}, \qquad \varphi_2(t) = \begin{pmatrix} 3e^{t} \\ 2e^{t} \\ e^{t} \end{pmatrix} = e^{t}\begin{pmatrix} 3 \\ 2 \\ 1 \end{pmatrix},$$

$$\varphi_3(t) = \begin{pmatrix} e^{2t} \\ 3e^{2t} \\ e^{2t} \end{pmatrix} = e^{2t}\begin{pmatrix} 1 \\ 3 \\ 1 \end{pmatrix}$$

are a fundamental set of solutions. Verify this. $\blacksquare$

Definition 2 FUNDAMENTAL MATRIX SOLUTION Let $\varphi_1, \varphi_2, \ldots, \varphi_n$ be n-vector solutions of $\mathbf{x}' = A(t)\mathbf{x}$. Let $\Phi(t)$ be the matrix whose columns are the vectors $\varphi_1, \varphi_2, \ldots, \varphi_n$. That is,

$$\Phi(t) = (\varphi_1(t), \ldots, \varphi_n(t)) = \begin{pmatrix} \varphi_{11}(t) & \varphi_{12}(t) & \cdots & \varphi_{1n}(t) \\ \varphi_{21}(t) & \varphi_{22}(t) & \cdots & \varphi_{2n}(t) \\ \vdots & \vdots & & \vdots \\ \varphi_{n1}(t) & \varphi_{n2}(t) & \cdots & \varphi_{nn}(t) \end{pmatrix}. \quad (2)$$

Such a matrix is called a **matrix solution** of the system $\mathbf{x}' = A\mathbf{x}$. Equivalently, *an $n \times n$ matrix function $\Phi(t)$ is a matrix solution of $\mathbf{x}' = A\mathbf{x}$ if and only if each of its columns is a solution vector of $\mathbf{x}' = A\mathbf{x}$.* If the vectors $\varphi_1, \varphi_2, \ldots, \varphi_n$ form a fundamental set of solutions (i.e., if they are linearly independent), then $\Phi(t)$ is called a **fundamental matrix solution**.

In what follows, we shall show that fundamental matrix solutions play a central role in the theory of linear systems of differential equations.

EXAMPLE 5 In the four previous examples of this section, fundamental matrix solutions were found to be, respectively:

Example 1. $\begin{pmatrix} -2e^{-4t} & 3e^{t} \\ e^{-4t} & e^{t} \end{pmatrix}$;

Example 2. $\begin{pmatrix} e^{-3t} & (1+t)e^{-3t} \\ -e^{-3t} & (-2-t)e^{-3t} \end{pmatrix}$;

Example 3. $\begin{pmatrix} e^{6t}(\cos 2t - \frac{1}{2}\sin 2t) & e^{6t}(\sin 2t + \frac{1}{2}\cos 2t) \\ e^{6t}(\cos 2t - 3\sin 2t) & e^{6t}(\sin 2t + 3\cos 2t) \end{pmatrix}$;

Example 4. $\begin{pmatrix} e^{-t} & 3e^{t} & e^{2t} \\ 0 & 2e^{t} & 3e^{2t} \\ e^{-t} & e^{t} & e^{2t} \end{pmatrix}$. ■

PRINCIPAL MATRIX SOLUTION

Fundamental matrix solutions are not unique, because a solution vector may be multiplied by any constant and still remain a solution. In addition, any linear combination of solutions is again a solution (see Problem 6). However, we have uniqueness for the **principal matrix solution** $\Psi(t)$, which is defined as that fundamental matrix solution that satisfies the condition

$$\Psi(t_0) = I \tag{3}$$

where I is the $n \times n$ identity matrix and t_0 is the initial value of the independent variable t.

We will show later in this section that if $A(t)$ is continuous, then $\mathbf{x}' = A\mathbf{x}$ always has a unique principal matrix solution. But first we will demonstrate an easy way to determine whether or not a given matrix solution is a fundamental matrix solution.

Let $\Phi(t)$ be a matrix solution of $\mathbf{x}' = A(t)\mathbf{x}$. We define the **Wronskian** of $\Phi(t)$, written $W(t)$, by

$$W(t) = \det \Phi(t). \tag{4}$$

We remind you of two results from matrix theory: First, a matrix A is invertible if and only if $\det A \neq 0$. Second, $\det A \neq 0$ if and only if the columns of A are linearly independent. From these facts it follows that $\Phi(t)$ will be a fundamental matrix solution if and only if $W(t)$ is nonzero for some t. We will see in the next theorem that $W(t)$ is either always zero or never zero, so that we can calculate $W(t)$ for some especially simple value of t, say $t = 0$, to determine whether Φ is a fundamental matrix solution. Note that many of these theorems are similar to those proven in Section 12.2. In particular, you should compare the present definition (4) of the Wronskian with the definition of the Wronskian in Section 12.2.

Theorem 1 ABEL'S FORMULA Let $W(t)$ be the Wronskian of the matrix solution $\Phi(t)$ of the system $\mathbf{x}' = A(t)\mathbf{x}$. Then

$$W(t) = W(t_0) \exp\left(\int_{t_0}^{t} \operatorname{tr} A(s)\, ds\right), \tag{5}$$

where the **trace** of A, written $\operatorname{tr} A(t)$, is the sum of the diagonal elements of the matrix $A(t)$:

$$\operatorname{tr} A(t) = a_{11}(t) + a_{22}(t) + \cdots + a_{nn}(t). \tag{6}$$

Proof. We prove this theorem for the case of $A(t)$ a 2×2 matrix (see Theorem 12.2.4). The proof for the $n \times n$ case is similar (but more complicated) and is left as an exercise. In the 2×2 case, the system $\mathbf{x}' = A\mathbf{x}$ and the matrix solution may be written as

$$\begin{pmatrix} x_1 \\ x_2 \end{pmatrix}' = \begin{pmatrix} a_{11}(t) & a_{12}(t) \\ a_{21}(t) & a_{22}(t) \end{pmatrix}\begin{pmatrix} x_1 \\ x_2 \end{pmatrix}, \qquad \Phi(t) = (\varphi_1, \varphi_2) = \begin{pmatrix} \varphi_{11} & \varphi_{12} \\ \varphi_{21} & \varphi_{22} \end{pmatrix}.$$

Since $W(t) = \varphi_{11}\varphi_{22} - \varphi_{12}\varphi_{21}$, the derivative

$$W' = \varphi_{11}\varphi_{22}' + \varphi_{11}'\varphi_{22} - \varphi_{12}\varphi_{21}' - \varphi_{12}'\varphi_{21}$$

may be written in determinant form as

$$W' = \begin{vmatrix} \varphi_{11} & \varphi_{12} \\ \varphi_{21}' & \varphi_{22}' \end{vmatrix} + \begin{vmatrix} \varphi_{11}' & \varphi_{12}' \\ \varphi_{21} & \varphi_{22} \end{vmatrix}. \tag{7}$$

But $\varphi_{11}' = a_{11}\varphi_{11} + a_{12}\varphi_{21}$, since φ_1 is a vector solution, and similarly for φ_{12}', φ_{21}', and φ_{22}'. Replacing these derivatives in (7), we obtain

$$\begin{aligned} W' &= \begin{vmatrix} \varphi_{11} & \varphi_{12} \\ a_{21}\varphi_{11} + a_{22}\varphi_{21} & a_{21}\varphi_{12} + a_{22}\varphi_{22} \end{vmatrix} \\ &\quad + \begin{vmatrix} a_{11}\varphi_{11} + a_{12}\varphi_{21} & a_{11}\varphi_{12} + a_{12}\varphi_{22} \\ \varphi_{21} & \varphi_{22} \end{vmatrix} \\ &= D_1 + D_2. \end{aligned} \tag{8}$$

But according to the theory of determinants (see page 460), a determinant is unchanged when a multiple of one row is added to another row. Also multiplication of every element in one row by a given constant is equivalent to multiplying the determinant by that constant. Hence we may multiply the first row of D_1 by $-a_{21}$ and add it to the second row. Then

$$D_1 = \begin{vmatrix} \varphi_{11} & \varphi_{12} \\ a_{22}\varphi_{21} & a_{22}\varphi_{22} \end{vmatrix} = a_{22}\begin{vmatrix} \varphi_{11} & \varphi_{12} \\ \varphi_{21} & \varphi_{22} \end{vmatrix} = a_{22}W.$$

Similarly, $D_2 = a_{11}W$. Thus

$$W'(t) = [a_{11}(t) + a_{22}(t)]W(t) = [\text{tr } A(t)]W(t). \tag{9}$$

Equation (9) is a first-order (scalar) differential equation that has the solution

$$W(t) = W(t_0) \exp\left(\int_{t_0}^{t} \text{tr } A(s)\, ds\right).$$

This completes the proof for the 2×2 case. This proof has in it the same ideas as the proof of Theorem 12.2.4, but it has been written in a format that can be extended to the $n \times n$ case (see Problems 13 and 14). ■

EXAMPLE 6 We consider the four matrix solutions of Example 5. Evaluating each at $t = 0$, we obtain

(i) $W(0) = \begin{vmatrix} -2 & 3 \\ 1 & 1 \end{vmatrix} = -5;$

(ii) $W(0) = \begin{vmatrix} 1 & 1 \\ -1 & -2 \end{vmatrix} = -1;$

(iii) $W(0) = \begin{vmatrix} 1 & \frac{1}{2} \\ 1 & 3 \end{vmatrix} = \frac{5}{2};$

(iv) $W(0) = \begin{vmatrix} 1 & 3 & 1 \\ 0 & 2 & 3 \\ 1 & 1 & 1 \end{vmatrix} = 6.$

Therefore, without direct verification of linear independence, we can see that all four matrix solutions are fundamental matrix solutions. ■

We are now ready to prove the theorem mentioned earlier, namely, that principal matrix solutions exist and are unique. Since principal matrix solutions are fundamental matrix solutions, this theorem also proves the existence of fundamental matrix solutions.

Theorem 2 Let $A(t)$ be continuous on some interval $[a, b]$. Then for any t_0, $a \leq t_0 \leq b$, there exists a unique fundamental matrix solution $\Psi(t)$ of the system $\mathbf{x}' = A(t)\mathbf{x}$ satisfying the condition $\Psi(t_0) = I$.

Proof. Let δ_i, $i = 1, 2, \ldots, n$, denote the n-column vector that has a one in the ith position (row) and a zero everywhere else:

$$\delta_1 = \begin{pmatrix} 1 \\ 0 \\ 0 \\ \vdots \\ 0 \end{pmatrix}, \qquad \delta_2 = \begin{pmatrix} 0 \\ 1 \\ 0 \\ \vdots \\ 0 \end{pmatrix}, \qquad \dots, \qquad \delta_n = \begin{pmatrix} 0 \\ 0 \\ \vdots \\ 0 \\ 1 \end{pmatrix}.$$

By the basic existence–uniqueness Theorem 12.4.1 with $\mathbf{f}(t) \equiv \mathbf{0}$, there exists a unique vector solution $\varphi_i(t)$ of $\mathbf{x}' = A\mathbf{x}$ that satisfies $\varphi_i(t_0) = \delta_i$, $i = 1, 2, \dots, n$. Define the matrix function

$$\Psi(t) = [\varphi_1(t), \varphi_2(t), \dots, \varphi_n(t)]. \tag{10}$$

Then $\Psi(t)$ is the matrix whose columns are the vector solutions φ_i, $i = 1, 2, \dots, n$. Clearly $\Psi(t)$ is a matrix solution and $\Psi(t_0) = (\varphi_1(t_0), \dots, \varphi_n(t_0)) = (\delta_1, \delta_2, \dots, \delta_n) = I$. It remains to be shown that $\Psi(t)$ is a fundamental matrix solution. This is easy to do. We simply note that $\det \Psi(t_0) = \det I = 1 \neq 0$. ■

ASSOCIATED MATRIX EQUATION

The calculation of a fundamental or principal matrix solution is generally impossible if $A(t)$ is nonconstant. In Section 12.6 we shall show how to compute principal matrix solutions for most constant matrices A. In the remainder of this section we will show how all solutions of $\mathbf{x}' = A(t)\mathbf{x}$ can be expressed in terms of a single fundamental matrix solution.

Definition 3 Let $X(t)$ denote an $n \times n$ matrix function. Then, we define the **associated matrix equation** to the system $\mathbf{x}' = A(t)\mathbf{x}$:

$$X'(t) = A(t)X(t). \tag{11}$$

We now seek a matrix (instead of a vector) solution of equation (11). The following result is left as an exercise: *$X(t)$ is a solution of the associated matrix equation* (11) *if and only if every column of $X(t)$ is a solution of the system $\mathbf{x}' = A(t)\mathbf{x}$.*

Theorem 3 Let Φ be a matrix solution of $\mathbf{x}' = A(t)\mathbf{x}$ and let C be a constant matrix. Then $\Phi_1 = \Phi C$ is also a matrix solution of $\mathbf{x}' = A(t)\mathbf{x}$.

Proof. Since a matrix solution of $\mathbf{x}' = A(t)\mathbf{x}$ is also a solution of equation (11), and since Φ is a solution of (11), we must show that Φ_1 is also a solution of (11). If X and Y are differentiable $n \times n$ matrix functions, then it follows from the product rule of differentiation (see Problem 25) that

$$(XY)' = X'Y + XY' \tag{12}$$

Thus, $\Phi_1' = (\Phi C)' = \Phi' C + \Phi C' = \Phi' C$, since $C' = 0$, C being constant. Finally, since Φ is a solution,

$$\Phi_1' = \Phi' C = A\Phi C = A\Phi_1. \blacksquare$$

EXAMPLE 7 Consider Example 1, with the fundamental matrix solution

$$\Phi(t) = \begin{pmatrix} -2e^{-4t} & 3e^t \\ e^{-4t} & e^t \end{pmatrix}.$$

Let

$$C_1 = \begin{pmatrix} 1 & 2 \\ 3 & 4 \end{pmatrix} \quad \text{and} \quad C_2 = \begin{pmatrix} 1 & 2 \\ 2 & 4 \end{pmatrix}.$$

Then

$$\Phi_1 = \Phi C_1 = \begin{pmatrix} -2e^{-4t} + 9e^t & -4e^{-4t} + 12e^t \\ e^{-4t} + 3e^t & 2e^{-4t} + 4e^t \end{pmatrix}$$

and

$$\Phi_2 = \Phi C_2 = \begin{pmatrix} -2e^{-4t} + 6e^t & -4e^{-4t} + 12e^t \\ e^{-4t} + 2e^t & 2e^{-4t} + 4e^t \end{pmatrix}.$$

It is not difficult to verify that both Φ_1 and Φ_2 are matrix solutions. Note that although Φ_1 is another fundamental matrix solution, Φ_2 is not, since $\det \Phi_2(0) = 0$. Can you explain this (see Problem 10)? ■

Theorem 3 gives us a way of finding a principal matrix solution when a fundamental matrix solution is known. To see this, let $\Phi(t)$ be a fundamental matrix solution. Since $\det \Phi(t_0) \neq 0$, $\Phi(t_0)$ is invertible, and we define $C = \Phi^{-1}(t_0)$ and $\Psi(t) = \Phi(t)C$. By Theorem 3, $\Psi(t)$ is a matrix solution and $\Psi(t_0) = \Phi(t_0)C = \Phi(t_0)\Phi^{-1}(t_0) = I$, so that $\Psi(t)$ is a principal matrix solution. Thus, if $\Phi(t)$ is a fundamental matrix solution, we can always obtain a principal matrix solution by multiplying $\Phi(t)$ on the right by $\Phi^{-1}(t_0)$.

EXAMPLE 8 A fundamental matrix solution of the system of Example 1 is (see part 1 of Example 5):

$$\Phi(t) = \begin{pmatrix} -2e^{-4t} & 3e^t \\ 3e^{-4t} & e^t \end{pmatrix} \quad \text{with} \quad \Phi(0) = \begin{pmatrix} -2 & 3 \\ 1 & 1 \end{pmatrix}.$$

Then

$$C = \Phi^{-1}(0) = \begin{pmatrix} -\frac{1}{5} & \frac{3}{5} \\ \frac{1}{5} & \frac{2}{5} \end{pmatrix} = \frac{1}{5}\begin{pmatrix} -1 & 3 \\ 1 & 2 \end{pmatrix},$$

so that

$$\Psi(t) = \Phi(t)C = \frac{1}{5}\begin{pmatrix} 2e^{-4t} + 3e^{t} & -6e^{-4t} + 6e^{t} \\ -e^{-4t} + e^{t} & 3e^{-4t} + 2e^{t} \end{pmatrix}$$

is a principal matrix solution. ■

Theorem 4 Let $\Phi(t)$ be a fundamental matrix solution and let $X(t)$ be any other matrix solution of the system $\mathbf{x}' = A(t)\mathbf{x}$. Then there exists a constant matrix C such that $X(t) = \Phi(t)C$. That is, *any solution vector of* $\mathbf{x}' = A(t)\mathbf{x}$ *can be written as a linear combination of vectors in a fundamental set.*

Before giving the proof, we remind you that it is important to state on which side of Φ the constant matrix C appears, since matrix multiplication may not commute.

Proof. Since $\Phi(t)$ is a fundamental matrix solution, $\det \Phi(t) \neq 0$ and $\Phi^{-1}(t)$ exists for every t. We will show that

$$\frac{d}{dt}[\Phi^{-1}(t)X(t)] = 0.$$

This will imply that $\Phi^{-1}(t)X(t)$ is a constant matrix C, and the theorem will be proved. First, we calculate

$$\frac{d}{dt}[\Phi^{-1}(t)].$$

Using the product rule of differentiation, equation (12), we have

$$0 = \frac{dI}{dt} = \frac{d}{dt}(\Phi\Phi^{-1}) = \frac{d\Phi}{dt}\Phi^{-1} + \Phi\frac{d\Phi^{-1}}{dt}. \tag{13}$$

After multiplying both sides of equation (13) on the left by Φ^{-1} and solving for $d\Phi^{-1}/dt$, we obtain

$$\frac{d\Phi^{-1}}{dt} = -\Phi^{-1}\frac{d\Phi}{dt}\Phi^{-1}. \tag{14}$$

Note the analogy between equation (14) and the identity

$$\frac{d}{dt}\left(\frac{1}{f(t)}\right) = -\frac{f'(t)}{(f(t))^2}.$$

Now, by the product formula of derivatives,

$$\frac{d}{dt}(\Phi^{-1}X) = \left(\frac{d}{dt}\Phi^{-1}\right)X + \Phi^{-1}\frac{dX}{dt}, \tag{15}$$

and since both Φ and X are solutions of equation (11), equation (15) becomes

$$\begin{aligned}\frac{d}{dt}(\Phi^{-1}X) &= \left(-\Phi^{-1}\frac{d\Phi}{dt}\Phi^{-1}\right)X + \Phi^{-1}(AX)\\ &= -\Phi^{-1}A\Phi\Phi^{-1}X + \Phi^{-1}AX\\ &= -\Phi^{-1}AX + \Phi^{-1}AX = 0. \blacksquare\end{aligned}$$

EXAMPLE 9 Consider the system $\mathbf{x}' = A\mathbf{x}$ where

$$A = \begin{pmatrix} 1 & -2 \\ 2 & -3 \end{pmatrix}.$$

It is not difficult to verify that

$$\Phi_1(t) = \begin{pmatrix} e^{-t} & (2t+2)e^{-t} \\ e^{-t} & (2t+1)e^{-t} \end{pmatrix} = e^{-t}\begin{pmatrix} 1 & 2t+2 \\ 1 & 2t+1 \end{pmatrix}$$

is a fundamental matrix solution. Another matrix solution is

$$\Phi_2(t) = e^{-t}\begin{pmatrix} 4t+7 & 8t+1 \\ 4t+5 & 8t-3 \end{pmatrix}.$$

There is a matrix C such that $\Phi_2 = \Phi_1 C$. But $\Phi_1^{-1}(t)\Phi_2(t) = C$ for every t, in particular for $t = 0$. Thus

$$C = \Phi_1^{-1}(0)\Phi_2(0) = \begin{pmatrix} -1 & 2 \\ 1 & -1 \end{pmatrix}\begin{pmatrix} 7 & 1 \\ 5 & -3 \end{pmatrix} = \begin{pmatrix} 3 & -7 \\ 2 & 4 \end{pmatrix}. \blacksquare$$

EXAMPLE 10 Consider the system $\mathbf{x}' = A\mathbf{x}$ where

$$A = \begin{pmatrix} 3 & -1 & 1 \\ -1 & 5 & -1 \\ 1 & -1 & 3 \end{pmatrix}.$$

A fundamental matrix solution is

$$\Phi(t) = \begin{pmatrix} e^{2t} & e^{3t} & e^{6t} \\ 0 & e^{3t} & -2e^{6t} \\ -e^{2t} & e^{3t} & e^{6t} \end{pmatrix}.$$

Another matrix solution is

$$X(t) = \begin{pmatrix} e^{2t} + 2e^{3t} + 3e^{6t} & e^{2t} - 3e^{3t} - 2e^{6t} & 2e^{2t} + 5e^{3t} + 7e^{6t} \\ 2e^{3t} - 6e^{6t} & -3e^{3t} + 4e^{6t} & 5e^{3t} - 14e^{6t} \\ -e^{2t} + 2e^{2t} + 3e^{6t} & -e^{2t} - 3e^{3t} - 2e^{6t} & -2e^{2t} + 5e^{3t} + 7e^{6t} \end{pmatrix}.$$

As in the previous example, a matrix C such that $X(t) = \Phi(t)C$ is given by

$$C = \Phi^{-1}(0)X(0) = \frac{1}{6}\begin{pmatrix} 3 & 0 & -3 \\ 2 & 2 & 2 \\ 1 & -2 & 1 \end{pmatrix}\begin{pmatrix} 6 & -4 & 14 \\ -4 & 1 & -9 \\ 4 & -6 & 10 \end{pmatrix}$$

$$= \frac{1}{6}\begin{pmatrix} 6 & 6 & 12 \\ 12 & -18 & 30 \\ 18 & -12 & 42 \end{pmatrix} = \begin{pmatrix} 1 & 1 & 2 \\ 2 & -3 & 5 \\ 3 & -2 & 7 \end{pmatrix}. \quad ■$$

Theorem 5 Let $\Phi(t)$ be a fundamental matrix solution and let $\mathbf{x}(t)$ be any solution of $\mathbf{x}' = A(t)\mathbf{x}$. Then there exists a constant vector $\mathbf{c}$ such that

$$\mathbf{x}(t) = \Phi(t)\mathbf{c}. \tag{16}$$

Proof. This theorem is an immediate consequence of Theorem 4 if we form the matrix solution $X(t) = (\mathbf{x}(t), \mathbf{x}(t), \ldots, \mathbf{x}(t))$ whose n columns are each the vector solution $\mathbf{x}(t)$. Then a matrix C exists such that $X(t) = \Phi(t)C$. Every column of C is a vector $\mathbf{c}$. ■

EXAMPLE 11 In Example 3 we found the fundamental matrix solution

$$\Phi(t) = e^{6t}\begin{pmatrix} \cos 2t - \frac{1}{2}\sin t & \sin 2t + \frac{1}{2}\cos 2t \\ \cos 2t - 3\sin 2t & \sin 2t + 3\cos 2t \end{pmatrix}.$$

The vector

$$\mathbf{x}(t) = e^{6t}\begin{pmatrix} -5\sin 2t \\ -10\cos 2t - 10\sin 2t \end{pmatrix}$$

is also a solution. From equation (16) we obtain

$$\mathbf{c} = \Phi^{-1}(t)\mathbf{x}(t) = \Phi^{-1}(0)\mathbf{x}(0) = \frac{2}{5}\begin{pmatrix} 3 & -\frac{1}{2} \\ -1 & 1 \end{pmatrix}\begin{pmatrix} 0 \\ -10 \end{pmatrix}$$

$$= \frac{2}{5}\begin{pmatrix} 5 \\ -10 \end{pmatrix} = \begin{pmatrix} 2 \\ -4 \end{pmatrix}. \quad ■$$

EXAMPLE 12 Consider the system

$$x_1' = \frac{-t}{1-t^2}x_1 + \frac{1}{1-t^2}x_2, \quad t \neq \pm 1$$

$$x_2' = \frac{1}{1-t^2}x_1 - \frac{t}{1-t^2}x_2, \quad t \neq \pm 1$$

or

$$\mathbf{x}' = \begin{pmatrix} \dfrac{-t}{1-t^2} & \dfrac{1}{1-t^2} \\ \dfrac{1}{1-t^2} & \dfrac{-t}{1-t^2} \end{pmatrix}\mathbf{x}, \ t \neq \pm 1 \tag{17}$$

Here

$$\Phi_1(t) = \begin{pmatrix} t \\ 1 \end{pmatrix} \quad \text{and} \quad \Phi_2(t) = \begin{pmatrix} 1 \\ t \end{pmatrix}$$

are linearly independent solutions, so

$$\Phi(t) = \begin{pmatrix} t & 1 \\ 1 & t \end{pmatrix}$$

is a fundamental matrix solution. First, note that although $W(0) = -1 \neq 0$, we also have $W(1) = 0$. A cursory inspection will help to explain this apparent contradiction of Theorem 1. The matrix $A(t)$ is undefined at $t = 1$; thus there cannot be a solution to equation (17) at $t = 1$. The theorem about Wronskians is, of course, only valid in an interval over which the solution is defined. In this example a suitable interval is $(-1, 1)$ or $(1, \infty)$, or any other interval that does not contain 1 or -1.

Continuing with the example, let us find a solution $\mathbf{x}(t)$ that satisfies the initial conditions

$$\mathbf{x}(0) = \begin{pmatrix} x_1(0) \\ x_2(0) \end{pmatrix} = \begin{pmatrix} 2 \\ -3 \end{pmatrix}.$$

Then

$$\mathbf{c} = \Phi^{-1}(0)\mathbf{x}(0) = \begin{pmatrix} 0 & 1 \\ 1 & 0 \end{pmatrix}\begin{pmatrix} 2 \\ -3 \end{pmatrix} = \begin{pmatrix} -3 \\ 2 \end{pmatrix}.$$

(Note that $\Phi(0) = \begin{pmatrix} 0 & 1 \\ 1 & 0 \end{pmatrix}$ is a matrix that is its own inverse.) Thus

$$\mathbf{x}(t) = \Phi(t)\mathbf{c} = \begin{pmatrix} t & 1 \\ 1 & t \end{pmatrix}\begin{pmatrix} -3 \\ 2 \end{pmatrix} = \begin{pmatrix} 2 - 3t \\ -3 + 2t \end{pmatrix}$$

is a solution vector of equation (17) that satisfies the given initial conditions. ■

PROBLEMS 12.5

In Problems 1–5 decide whether each set of solution vectors constitutes a fundamental set of the given system by **(a)** determining whether the vectors are linearly independent, and **(b)** using the Wronskian to determine whether or not $W(t)$ is zero.

1. $\mathbf{x}' = \begin{pmatrix} 2 & 5 \\ 0 & 2 \end{pmatrix}\mathbf{x}$, $\varphi_1(t) = \begin{pmatrix} e^{2t}(1 + 10t) \\ 2e^{2t} \end{pmatrix}$,

$$\varphi_2(t) = \begin{pmatrix} e^{2t}(-3 + 20t) \\ 4e^{2t} \end{pmatrix}$$

2. $\mathbf{x}' = \begin{pmatrix} 4 & -13 \\ 2 & -6 \end{pmatrix}\mathbf{x}$,

$$\varphi_1(t) = \begin{pmatrix} e^{-t}(13\cos t - 26\sin t) \\ e^{-t}(7\cos t - 9\sin t) \end{pmatrix},$$

$$\varphi_2(t) = \begin{pmatrix} e^{-t}(26\cos t - 52\sin t) \\ e^{-t}(14\cos t - 18\sin t) \end{pmatrix}$$

3. $\mathbf{x}' = \begin{pmatrix} 1 & 1 \\ 4 & 1 \end{pmatrix}\mathbf{x}$, $\varphi_1(t) = \begin{pmatrix} e^{3t} - e^{-t} \\ 2e^{3t} + 2e^{-t} \end{pmatrix}$,

$$\varphi_2(t) = \begin{pmatrix} 2e^{3t} \\ 4e^{3t} \end{pmatrix}$$

4. $\mathbf{x}' = \begin{pmatrix} 1 & -1 & 4 \\ 3 & 2 & -1 \\ 2 & 1 & -1 \end{pmatrix}\mathbf{x}$,

$$\varphi_1(t) = \begin{pmatrix} e^{t} + 2e^{-2t} + 3e^{3t} \\ -4e^{t} - 2e^{-2t} + 6e^{3t} \\ -e^{t} - 2e^{-2t} + 3e^{3t} \end{pmatrix},$$

$$\varphi_2(t) = \begin{pmatrix} -2e^{t} + 2e^{-2t} \\ 8e^{t} - 2e^{-2t} \\ 2e^{t} - 2e^{-2t} \end{pmatrix},$$

$$\varphi_3(t) = \begin{pmatrix} 3e^{t} - 6e^{-2t} + 3e^{3t} \\ -12e^{t} + 6e^{-2t} + 6e^{3t} \\ -3e^{t} + 6e^{-2t} + 3e^{3t} \end{pmatrix}$$

5. $\mathbf{x}' = \begin{pmatrix} 3 & 2 & 1 \\ -1 & 0 & -1 \\ 1 & 1 & 2 \end{pmatrix}\mathbf{x}$, $\varphi_1(t) = \begin{pmatrix} -e^{t} + te^{2t} + e^{2t} \\ e^{t} - te^{2t} \\ te^{2t} \end{pmatrix}$,

$$\varphi_2(t) = \begin{pmatrix} 2e^{2t} + te^{2t} \\ -e^{2t} - te^{2t} \\ e^{2t} + te^{2t} \end{pmatrix},$$

$$\varphi_3(t) = \begin{pmatrix} -e^{t} + 3e^{2t} + 2te^{2t} \\ e^{t} - e^{2t} - 2te^{2t} \\ e^{2t} + 2te^{2t} \end{pmatrix}$$

6. Let $\varphi_1(t), \varphi_2(t), \ldots, \varphi_m(t)$ be m solutions of the homogeneous system (1). Show that

$$\varphi(t) = c_1\varphi_1(t) + c_2\varphi_2(t) + \cdots + c_m\varphi_m(t)$$

is also a solution.

In each of Problems 7–9 two matrix functions Φ_1 and Φ_2 are given. Find a matrix C such that $\Phi_2(t) = \Phi_1(t)C$.

7. $\Phi_1(t) = e^{6t}\begin{pmatrix} \cos 2t - \frac{1}{2}\sin 2t & \sin 2t + \frac{1}{2}\cos 2t \\ \cos 2t - 3\sin 2t & \sin 2t + 3\cos 2t \end{pmatrix}$,

$$\Phi_2(t) = e^{6t}\begin{pmatrix} \frac{1}{2}\cos 2t - \frac{3}{2}\sin 2t & -\frac{3}{2}\cos 2t + 2\sin 2t \\ -2\cos 2t - 4\sin 2t & \cos 2t + 7\sin 2t \end{pmatrix}$$

8. $\Phi_1(t) = \begin{pmatrix} \sin e^{t} & \cos e^{t} \\ e^{t}\cos e^{t} & -e^{t}\sin e^{t} \end{pmatrix}$,

$$\Phi_2(t) = \begin{pmatrix} \cos e^{t} & 3\sin e^{t} + 2\cos e^{t} \\ -e^{t}\sin e^{t} & 3e^{t}\cos e^{t} - 2e^{t}\sin e^{t} \end{pmatrix}$$

9. $\Phi_1(t) = \begin{pmatrix} e^{-t} & 3e^{t} & e^{2t} \\ 0 & 2e^{t} & 3e^{2t} \\ e^{-t} & e^{t} & e^{2t} \end{pmatrix}$,

$$\Phi_2(t) = \begin{pmatrix} e^{-t} + e^{2t} & -e^{-t} + 3e^{t} & e^{-t} + 3e^{t} \\ 3e^{2t} & 2e^{t} & 2e^{t} \\ e^{-t} + e^{2t} & -e^{-t} + e^{t} & e^{-t} + e^{t} \end{pmatrix}$$

10. Let $\Phi_1(t)$ be a fundamental matrix solution of $\mathbf{x}' = A\mathbf{x}$. Then $\Phi_2 = \Phi_1 C$ is a matrix solution for any constant matrix C. Show that Φ_2 is a fundamental matrix solution if and only if C is nonsingular.

11. Let

$$Y(t) = \begin{vmatrix} a_{11} & a_{12} & a_{13} \\ a_{21} & a_{22} & a_{23} \\ a_{31} & a_{32} & a_{33} \end{vmatrix}$$

where a_{ij} is a differentiable function for $i, j = 1, 2, 3$. Show that

$$Y'(t) = \begin{vmatrix} a'_{11} & a'_{12} & a'_{13} \\ a_{21} & a_{22} & a_{23} \\ a_{31} & a_{32} & a_{33} \end{vmatrix} + \begin{vmatrix} a_{11} & a_{12} & a_{13} \\ a'_{21} & a'_{22} & a'_{23} \\ a_{31} & a_{32} & a_{33} \end{vmatrix} + \begin{vmatrix} a_{11} & a_{12} & a_{13} \\ a_{21} & a_{22} & a_{23} \\ a'_{31} & a'_{32} & a'_{33} \end{vmatrix}$$

***12.** Let

$$Y(t) = \begin{vmatrix} a_{11} & a_{12} & \cdots & a_{1n} \\ a_{21} & a_{22} & \cdots & a_{2n} \\ \vdots & \vdots & & \vdots \\ a_{n1} & a_{n2} & \cdots & a_{nn} \end{vmatrix}$$

where each a_{ij} is a differentiable function for $i, j = 1, 2, \ldots, n$. Use mathematical induction (see Appendix 1) to prove that

$$Y'(t) = \begin{vmatrix} a'_{11} & a'_{12} & \cdots & a'_{1n} \\ a_{21} & a_{22} & \cdots & a_{2n} \\ \vdots & \vdots & & \vdots \\ a_{n1} & a_{n2} & \cdots & a_{nn} \end{vmatrix} + \begin{vmatrix} a_{11} & a_{12} & \cdots & a_{1n} \\ a'_{21} & a'_{22} & \cdots & a'_{2n} \\ \vdots & \vdots & & \vdots \\ a_{n1} & a_{n2} & \cdots & a_{nn} \end{vmatrix} + \cdots + \begin{vmatrix} a_{11} & a_{12} & \cdots & a_{1n} \\ a_{21} & a_{22} & \cdots & a_{2n} \\ \vdots & \vdots & & \vdots \\ a'_{n1} & a'_{n2} & \cdots & a'_{nn} \end{vmatrix}$$

13. Let $\Phi(t)$ be a matrix solution of the system $\mathbf{x}' = A(t)\mathbf{x}$ where $A(t)$ is a 3×3 matrix. Prove that

$$W(t) = W(t_0) \exp\left[\int_{t_0}^{t} [a_{11}(s) + a_{22}(s) + a_{33}(s)]\, ds\right].$$

[*Hint:* Use the result of Problem 11.]

***14.** Using the result of Problem 12 prove Theorem 1 for the case in which $A(t)$ is an $n \times n$ matrix.

In Problems 15–18 find the principal matrix solution $\Psi(t)$ corresponding to each fundamental matrix solution $\Phi(t)$. Assume that $t_0 = 0$.

15. $\Phi(t) = e^{6t}\begin{pmatrix} \cos 2t - \frac{1}{2}\sin 2t & \sin 2t + \frac{1}{2}\cos 2t \\ \cos 2t - 3\sin 2t & \sin 2t + 3\cos 2t \end{pmatrix}$

16. $\Phi(t) = \begin{pmatrix} \sin e^t & \cos e^t \\ e^t \cos e^t & -e^t \sin e^t \end{pmatrix}$

17. $\Phi(t) = \begin{pmatrix} e^{-t} & 3e^t & e^{2t} \\ 0 & 2e^t & 3e^{2t} \\ e^{-t} & e^t & e^{2t} \end{pmatrix}$

18. $\Phi(t) = \begin{pmatrix} 2e^t + te^t & te^t + e^t & e^t \\ 0 & e^t & e^t \\ -3e^t - te^t & -te^t - 2e^t & -e^t \end{pmatrix}$

19. Consider the system

$$\mathbf{x}' = \begin{pmatrix} 3 & -2 \\ 2 & -1 \end{pmatrix}\mathbf{x}.$$

Verify that

$$\Phi(t) = \begin{pmatrix} 2te^t + e^t & 2te^t \\ 2te^t & -e^t + 2te^t \end{pmatrix}$$

is a fundamental matrix solution. Then find a solution that satisfies each of the following initial conditions:

(a) $\mathbf{x}(0) = \begin{pmatrix} 1 \\ 2 \end{pmatrix}$; **(b)** $\mathbf{x}(0) = \begin{pmatrix} -2 \\ 3 \end{pmatrix}$;

(c) $\mathbf{x}(1) = \begin{pmatrix} 0 \\ 1 \end{pmatrix}$; **(d)** $\mathbf{x}(-1) = \begin{pmatrix} 2 \\ 1 \end{pmatrix}$;

(e) $\mathbf{x}(3) = \begin{pmatrix} 3 \\ 3 \end{pmatrix}$; **(f)** $\mathbf{x}(a) = \begin{pmatrix} b \\ c \end{pmatrix}$.

20. In Example 5 we saw that a fundamental matrix solution to the system

$$\mathbf{x}' = \begin{pmatrix} 1 & 1 & -2 \\ -1 & 2 & 1 \\ 0 & 1 & -1 \end{pmatrix}\mathbf{x}$$

of Example 4 was

$$\Phi(t) = \begin{pmatrix} e^{-t} & 3e^t & e^{2t} \\ 0 & 2e^t & 3e^{2t} \\ e^{-t} & e^t & e^{2t} \end{pmatrix}.$$

Find a particular solution that satisfies each of the following conditions:

(a) $\mathbf{x}(0) = \begin{pmatrix} 1 \\ -1 \\ 2 \end{pmatrix}$; **(b)** $\mathbf{x}(0) = \begin{pmatrix} 3 \\ 1 \\ 2 \end{pmatrix}$;

(c) $\mathbf{x}(1) = \begin{pmatrix} 1 \\ 0 \\ 1 \end{pmatrix}$; **(d)** $\mathbf{x}(-1) = \begin{pmatrix} 2 \\ -3 \\ 5 \end{pmatrix}$.

21. Consider the second-order equation

$$x'' + a(t)x' + b(t)x = 0. \tag{18}$$

(a) Write equation (18) in the form $\mathbf{x}' = A(t)\mathbf{x}$.
(b) Given that

$$\Phi(t) = \begin{pmatrix} \varphi_1 & \varphi_2 \\ \varphi_1' & \varphi_2' \end{pmatrix}$$

is a fundamental matrix solution, show that

$$\det \Phi(t) = \det \Phi(t_0) \exp\left[-\int_{t_0}^{t} a(s)\, ds\right].$$

(c) Show that the formula in part (b) can be rearranged as

$$\varphi_2' - \frac{\varphi_1'}{\varphi_1}\varphi_2 = \frac{\det \Phi(t_0)}{\varphi_1}\exp\left[-\int_{t_0}^{t} a(s)\, ds\right]. \tag{19}$$

Therefore, if one solution $\varphi_1(t)$ of (18) is known, another solution can be calculated by solving this equation.

22. Given that $\varphi_1(t) = \sin(\ln t)$, find a second linearly independent solution of

$$x'' + \frac{1}{t}x' + \frac{1}{t^2}x = 0.$$

23. Given that $\varphi_1(t) = e^{t^2}$ is a solution of

$$x'' - 2tx' - 2x = 0,$$

find a second linearly independent solution.

24. Given that $\varphi_1(t) = \sin t^2$ is a solution of

$$tx'' - x' + 4t^3x = 0,$$

find a second linearly independent solution.

***25.** Let $X(t)$ and $Y(t)$ be differentiable $n \times n$ matrix functions. Show that XY is differentiable and that

$$(XY)' = X'Y + XY'$$

12.6 THE COMPUTATION OF THE PRINCIPAL MATRIX SOLUTION TO A HOMOGENEOUS SYSTEM OF EQUATIONS

Consider the homogeneous system of differential equations

$$\mathbf{x}' = A\mathbf{x}(t) \tag{1}$$

where A is a constant $n \times n$ matrix. In this section we show how to use matrix theory directly to compute the principal matrix solution of a system whenever A is diagonalizable.[†]

First, let us consider the scalar differential equation

$$x'(t) = ax(t). \tag{2}$$

Note that equation (1) is almost identical to equation (2). The only difference is that

[†]The principal matrix solution can also be computed in the case in which A is not diagonalizable. This technique is given in S. I. Grossman, *Elementary Linear Algebra*, second edition (Wadsworth, Belmont, CA, 1984), Section 7.7.

now we have a vector function and a matrix whereas before we had a "scalar" function and a number (1×1 matrix).

To solve equation (1), we might guess that a solution would have the form e^{At}. But what does e^{At} mean? We shall answer this question in a moment. First, let us recall the series expansion of the function e^t:

$$e^t = 1 + t + \frac{t^2}{2!} + \frac{t^3}{3!} + \frac{t^4}{4!} + \cdots .^{\dagger} \tag{3}$$

This series converges for every real number t. Then, for any real number a,

$$e^{at} = 1 + at + \frac{(at)^2}{2!} + \frac{(at)^3}{3!} + \frac{(at)^4}{4!} + \cdots . \tag{4}$$

THE MATRIX e^A

Definition 1 Let A be an $n \times n$ matrix with real (or complex) entries. Then e^A is an $n \times n$ matrix defined by

$$e^A = I + A + \frac{A^2}{2!} + \frac{A^3}{3!} + \frac{A^4}{4!} + \cdots . \tag{5}$$

REMARK. It is possible to define convergence of series of matrices and to prove that the series of matrices in equation (5) converges for every matrix A, but to do so would take us too far afield. We can, however, give an indication of why it is so. We first define $|A|_i$ to be the sum of the absolute values of the components in the ith row of A. We then define the norm[‡] of A, denoted $|A|$, by

$$|A| = \max_{1 \le i \le n} |A|_i. \tag{6}$$

It can be shown that

$$|AB| \le |A|\,|B| \tag{7}$$

and

$$|A + B| \le |A| + |B|. \tag{8}$$

Then, using (7) and (8) in (5), we obtain

$$|e^A| \le 1 + |A| + \frac{|A|^2}{2!} + \frac{|A|^3}{3!} + \frac{|A|^4}{4!} + \cdots = e^{|A|}.$$

[†]This series is derived in Section 13.13 of this book.

[‡]This is called the **max-row sum norm** of A.

Since $|A|$ is a real number, $e^{|A|}$ is finite. This shows that the series in (5) converges for any matrix A.

We shall now see the usefulness of the series in equation (5).

Theorem 1 For any constant vector $\mathbf{c}$, $\mathbf{x}(t) = e^{At}\mathbf{c}$ is a solution of (1). Moreover, the solution of (1) given by $\mathbf{x}(t) = e^{At}\mathbf{x}_0$ satisfies $\mathbf{x}(0) = \mathbf{x}_0$.

Proof. We compute, using (5),

$$\mathbf{x}(t) = e^{At}\mathbf{c} = \left(I + At + A^2\frac{t^2}{2!} + A^3\frac{t^2}{3!} + \cdots\right)\mathbf{c}. \tag{9}$$

Since A is a constant matrix, we have

$$\begin{aligned}\frac{d}{dt}A^k\frac{t^k}{k!} &= \frac{d}{dt}\frac{t^k}{k!}A^k = \frac{kt^{k-1}}{k!}A^k \\ &= \frac{A^k t^{k-1}}{(k-1)!} = A\left[A^{k-1}\frac{t^{k-1}}{(k-1)!}\right].\end{aligned} \tag{10}$$

Then, combining (9) and (10), and using the fact that $\mathbf{c}$ is a constant vector, we obtain

$$\mathbf{x}'(t) = \frac{d}{dt}e^{At}\mathbf{c} = A\left(I + At + A^2\frac{t^2}{2!} + A^3\frac{t^2}{3!} + \cdots\right)\mathbf{c} = Ae^{At}\mathbf{c} = A\mathbf{x}(t).$$

Finally, since $e^{A\cdot 0} = e^0 = I$, we have

$$\mathbf{x}(0) = e^{A\cdot 0}\mathbf{x}_0 = I\mathbf{x}_0 = \mathbf{x}_0. \blacksquare$$

REMARK. From Theorem 1 we can show that e^{At} is a solution matrix to (1). Furthermore, since $e^{A0} = I$, e^{At} is the principal matrix solution.

A major difficulty remains: How do we compute e^{At} without using a series? We begin with an example.

EXAMPLE 1 Let $A = \begin{pmatrix} 1 & 0 & 0 \\ 0 & 2 & 0 \\ 0 & 0 & 3 \end{pmatrix}$. Then

$$A^2 = \begin{pmatrix} 1 & 0 & 0 \\ 0 & 2^2 & 0 \\ 0 & 0 & 3^2 \end{pmatrix}, \quad A^3 = \begin{pmatrix} 1 & 0 & 0 \\ 0 & 2^3 & 0 \\ 0 & 0 & 3^3 \end{pmatrix}, \quad \ldots, \quad A^m = \begin{pmatrix} 1 & 0 & 0 \\ 0 & 2^m & 0 \\ 0 & 0 & 3^m \end{pmatrix}$$

and

$$e^{At} = I + At + \frac{A^2t^2}{2!} + \frac{A^3t^3}{3!} + \cdots$$

$$= \begin{pmatrix} 1 & 0 & 0 \\ 0 & 1 & 0 \\ 0 & 0 & 1 \end{pmatrix} + \begin{pmatrix} t & 0 & 0 \\ 0 & 2t & 0 \\ 0 & 0 & 3t \end{pmatrix} + \begin{pmatrix} \frac{t^2}{2!} & 0 & 0 \\ 0 & \frac{2^2t^2}{2!} & 0 \\ 0 & 0 & \frac{3^2t^2}{2!} \end{pmatrix} + \begin{pmatrix} \frac{t^3}{3!} & 0 & 0 \\ 0 & \frac{2^3t^3}{3!} & 0 \\ 0 & 0 & \frac{3^3t^3}{3!} \end{pmatrix} + \cdots$$

$$= \begin{pmatrix} 1 + t + \frac{t^2}{2!} + \frac{t^3}{3!} + \cdots & 0 & 0 \\ 0 & 1 + (2t) + \frac{(2t)^2}{2!} + \frac{(2t)^3}{3!} + \cdots & 0 \\ 0 & 0 & 1 + (3t) + \frac{(3t)^2}{2!} + \frac{(3t)^3}{3!} + \cdots \end{pmatrix}$$

$$= \begin{pmatrix} e^t & 0 & 0 \\ 0 & e^{2t} & 0 \\ 0 & 0 & e^{3t} \end{pmatrix}. \blacksquare$$

As Example 1 illustrates, if $D = \text{diag}(\lambda_1, \lambda_2, \ldots, \lambda_n)$, then $e^{Dt} = \text{diag}(e^{\lambda_1 t}, e^{\lambda_2 t}, \ldots, e^{\lambda_n t})$. It turns out that if A is diagonalizable, this is really all we need to do, as the next theorem suggests.

Theorem 2 Let A be diagonalizable and suppose that $D = C^{-1}AC$. Then $A = CDC^{-1}$ and

$$e^{At} = Ce^{Dt}C^{-1}. \qquad \textbf{(11)}$$

Proof. We first note that

$$\begin{aligned} A^n = (CDC^{-1})^n &= \overbrace{(CDC^{-1})(CDC^{-1}) \cdots (CDC^{-1})}^{n \text{ times}} \\ &= CD(C^{-1}C)D(C^{-1}C)D(C^{-1}C) \cdots (C^{-1}C)DC^{-1} \\ &= CD^nC^{-1}. \end{aligned}$$

It follows that

$$(At)^n = C(Dt)^nC^{-1}. \qquad \textbf{(12)}$$

Thus

$$e^{At} = I + (At) + \frac{(At)^2}{2!} + \cdots = CIC^{-1} + C(Dt)C^{-1} + C\frac{(Dt)^2}{2!}C^{-1} + \cdots$$

$$= C\left[I + (Dt) + \frac{(Dt)^2}{2!} + \cdots\right]C^{-1} = Ce^{Dt}C^{-1}. \blacksquare$$

We now apply our theory to a simple biological model of population growth. Suppose that in an ecosystem there are two interacting species S_1 and S_2. We denote the populations of the species at time t by $x_1(t)$ and $x_2(t)$. One system governing the relative growth of the two species is

$$\begin{aligned} x_1'(t) &= ax_1(t) + bx_2(t), \\ x_2'(t) &= cx_1(t) + dx_2(t). \end{aligned} \tag{13}$$

We can interpret the constants a, b, c, and d as follows. If the species are competing, it is reasonable to have $b < 0$ and $c < 0$. This is true because increases in the population of one species will slow the growth of the other. A second model is a **predator–prey** relationship. If S_1 is the prey and S_2 is the predator (S_2 eats S_1), then it is reasonable to have $b < 0$ and $c > 0$, since an increase in the predator species will cause a decrease in the prey species, while an increase in the prey species will cause an increase in the predator species (since it will have more food). Finally, in a **symbiotic** relationship (each species lives off the other), we would likely have $b > 0$ and $c > 0$. Of course, the constants a, b, c, and d depend on a wide variety of factors including available food, time of year, climate, limits due to overcrowding, other competing species, and so on. We shall analyze three different models by using the material in this section. We assume that t is measured in years.

EXAMPLE 2 **(A Competitive Model)** Consider the system

$$\begin{aligned} x_1'(t) &= 3x_1(t) - x_2(t), \\ x_2'(t) &= -2x_1(t) + 2x_2(t). \end{aligned}$$

Here, an increase in the population of one species causes a decline in the growth rate of another. Suppose that the initial populations are $x_1(0) = 90$ and $x_2(0) = 150$. Find the populations of both species for $t > 0$.

Solution. We have $A = \begin{pmatrix} 3 & -1 \\ -2 & 2 \end{pmatrix}$. The eigenvalues of A are $\lambda_1 = 1$ and $\lambda_2 = 4$ with corresponding eigenvectors $\mathbf{v}_1 = \begin{pmatrix} 1 \\ 2 \end{pmatrix}$ and $\mathbf{v}_2 = \begin{pmatrix} 1 \\ -1 \end{pmatrix}$. Then

$$C = \begin{pmatrix} 1 & 1 \\ 2 & -1 \end{pmatrix} \quad C^{-1} = -\frac{1}{3}\begin{pmatrix} -1 & -1 \\ -2 & 1 \end{pmatrix} \quad D = \begin{pmatrix} 1 & 0 \\ 0 & 4 \end{pmatrix} \quad e^{Dt} = \begin{pmatrix} e^t & 0 \\ 0 & e^{4t} \end{pmatrix}$$

$$e^{At} = Ce^{Dt}C^{-1} = -\frac{1}{3}\begin{pmatrix} 1 & 1 \\ 2 & -1 \end{pmatrix}\begin{pmatrix} e^t & 0 \\ 0 & e^{4t} \end{pmatrix}\begin{pmatrix} -1 & -1 \\ -2 & 1 \end{pmatrix}$$

$$= -\frac{1}{3}\begin{pmatrix} 1 & 1 \\ 2 & -1 \end{pmatrix}\begin{pmatrix} -e^t & -e^t \\ -2e^{4t} & e^{4t} \end{pmatrix}$$

$$= -\frac{1}{3}\begin{pmatrix} -e^t - 2e^{4t} & -e^t + e^{4t} \\ -2e^t + 2e^{4t} & -2e^t - e^{4t} \end{pmatrix}.$$

Finally, the solution to the system is given by

$$\mathbf{x}(t) = \begin{pmatrix} x_1(t) \\ x_2(t) \end{pmatrix} = e^{At}\mathbf{x}_0 = -\frac{1}{3}\begin{pmatrix} -e^t - 2e^{4t} & -e^t + e^{4t} \\ -2e^t + 2e^{4t} & -2e^t - e^{4t} \end{pmatrix}\begin{pmatrix} 90 \\ 150 \end{pmatrix}$$

$$= -\frac{1}{3}\begin{pmatrix} -240e^t - 30e^{4t} \\ -480e^t + 30e^{4t} \end{pmatrix} = \begin{pmatrix} 80e^t + 10e^{4t} \\ 160e^t - 10e^{4t} \end{pmatrix}.$$

For example, after six months ($t = \frac{1}{2}$ year), $x_1(t) = 80e^{1/2} + 10e^2 \approx 206$ individuals, while $x_2(t) = 160e^{1/2} - 10e^2 \approx 190$ individuals. More significantly, $160e^t - 10e^{4t} = 0$ when $16e^t = e^{4t}$ or $16 = e^{3t}$ or $3t = \ln 16$ and $t = (\ln 16)/3 \approx 2.77/3 \approx 0.92$ years $\approx$ 11 months. Thus the second species will be eliminated after only 11 months even though it started with a larger population. In Problems 10 and 11 you are asked to show that neither population will be eliminated if $x_2(0) = 2x_1(0)$ and that the first population will be eliminated if $x_2(0) > 2x_1(0)$. Thus, as was well known to Darwin, survival in this very simple model depends on the relative sizes of the competing species when competition begins. ■

EXAMPLE 3 **(A Predator–Prey Model)** Consider the predator–prey model governed by the system

$$\begin{aligned} x_1'(t) &= x_1(t) + x_2(t), \\ x_2'(t) &= -x_1(t) + x_2(t). \end{aligned}$$

If the initial populations are $x_1(0) = x_2(0) = 1000$, determine the populations of the two species for $t > 0$.

Solution. Here $A = \begin{pmatrix} 1 & 1 \\ -1 & 1 \end{pmatrix}$ with characteristic equation $\lambda^2 - 2\lambda + 2 = 0$, complex roots $\lambda_1 = 1 + i$ and $\lambda_2 = 1 - i$, and eigenvectors $\mathbf{v}_1 = \begin{pmatrix} 1 \\ i \end{pmatrix}$ and $\mathbf{v}_2 = \begin{pmatrix} 1 \\ -i \end{pmatrix}$.† Then

$$C = \begin{pmatrix} 1 & 1 \\ i & -i \end{pmatrix} \qquad C^{-1} = -\frac{1}{2i}\begin{pmatrix} -i & -1 \\ -i & 1 \end{pmatrix} = \frac{1}{2}\begin{pmatrix} 1 & -i \\ 1 & i \end{pmatrix} \qquad D = \begin{pmatrix} 1+i & 0 \\ 0 & 1-i \end{pmatrix}$$

†Note that $\lambda_2 = \overline{\lambda_1}$ and $\mathbf{v}_2 = \overline{\mathbf{v}_1}$. This should be no surprise, because, according to the result of Problem 9.10.33, eigenvalues of real matrices occur in complex conjugate pairs and their corresponding eigenvectors are complex conjugates.

and

$$e^{Dt} = \begin{pmatrix} e^{(1+i)t} & 0 \\ 0 & e^{(1-i)t} \end{pmatrix}.$$

Now, by Euler's formula (see Appendix 3), $e^{it} = \cos t + i \sin t$. Thus $e^{(1+i)t} = e^t e^{it} = e^t(\cos t + i \sin t)$. Similarly, $e^{(1-i)t} = e^t e^{-it} = e^t(\cos t - i \sin t)$. Thus

$$e^{Dt} = e^t \begin{pmatrix} \cos t + i \sin t & 0 \\ 0 & \cos t - i \sin t \end{pmatrix}$$

and

$$\begin{aligned} e^{At} = Ce^{Dt}C^{-1} &= \frac{e^t}{2}\begin{pmatrix} 1 & 1 \\ i & -i \end{pmatrix}\begin{pmatrix} \cos t + i \sin t & 0 \\ 0 & \cos t - i \sin t \end{pmatrix}\begin{pmatrix} 1 & -i \\ 1 & i \end{pmatrix} \\ &= \frac{e^t}{2}\begin{pmatrix} 1 & 1 \\ i & -i \end{pmatrix}\begin{pmatrix} \cos t + i \sin t & -i \cos t + \sin t \\ \cos t - i \sin t & i \cos t + \sin t \end{pmatrix} \\ &= \frac{e^t}{2}\begin{pmatrix} 2\cos t & 2\sin t \\ -2\sin t & 2\cos t \end{pmatrix} = e^t\begin{pmatrix} \cos t & \sin t \\ -\sin t & \cos t \end{pmatrix}. \end{aligned}$$

Finally,

$$\mathbf{x}(t) = e^{At}\mathbf{x}(0) = e^t\begin{pmatrix} \cos t & \sin t \\ -\sin t & \cos t \end{pmatrix}\begin{pmatrix} 1000 \\ 1000 \end{pmatrix} = \begin{pmatrix} 1000e^t(\cos t + \sin t) \\ 1000e^t(\cos t - \sin t) \end{pmatrix}.$$

The prey species is eliminated when $1000e^t(\cos t - \sin t) = 0$ or when $\sin t = \cos t$. The first positive solution of this last equation is $t = \pi/4 \approx 0.7854$ year ≈ 9.4 months. ■

EXAMPLE 4 **(A Model of Species Cooperation—Symbiosis)** Consider the symbiotic model governed by the system

$$\begin{aligned} x_1'(t) &= -\tfrac{1}{2}x_1(t) + x_2(t), \\ x_2'(t) &= \tfrac{1}{4}x_1(t) - \tfrac{1}{2}x_2(t). \end{aligned}$$

In this model the population of each species increases proportionally to the population of the other and decreases proportionally to its own population. Suppose that $x_1(0) = 200$ and $x_2(0) = 500$. Determine the population of each species for $t > 0$.

Solution. Here $A = \begin{pmatrix} -\frac{1}{2} & 1 \\ \frac{1}{4} & -\frac{1}{2} \end{pmatrix}$ with eigenvalues $\lambda_1 = 0$ and $\lambda_2 = -1$ and corresponding eigenvectors $\mathbf{v}_1 = \begin{pmatrix} 2 \\ 1 \end{pmatrix}$ and $\mathbf{v}_2 = \begin{pmatrix} 2 \\ -1 \end{pmatrix}$. Then

$$C = \begin{pmatrix} 2 & 2 \\ 1 & -1 \end{pmatrix}, \quad C^{-1} = -\frac{1}{4}\begin{pmatrix} -1 & -2 \\ -1 & 2 \end{pmatrix}, \quad D = \begin{pmatrix} 0 & 0 \\ 0 & -1 \end{pmatrix},$$

and $e^{Dt} = \begin{pmatrix} e^{0t} & 0 \\ 0 & e^{-t} \end{pmatrix} = \begin{pmatrix} 1 & 0 \\ 0 & e^{-t} \end{pmatrix}$. Thus

$$\begin{aligned} e^{At} &= -\frac{1}{4}\begin{pmatrix} 2 & 2 \\ 1 & -1 \end{pmatrix}\begin{pmatrix} 1 & 0 \\ 0 & e^{-t} \end{pmatrix}\begin{pmatrix} -1 & -2 \\ -1 & 2 \end{pmatrix} \\ &= -\frac{1}{4}\begin{pmatrix} 2 & 2 \\ 1 & -1 \end{pmatrix}\begin{pmatrix} -1 & -2 \\ -e^{-t} & 2e^{-t} \end{pmatrix} \\ &= -\frac{1}{4}\begin{pmatrix} -2-2e^{-t} & -4+4e^{-t} \\ -1+e^{-t} & -2-2e^{-t} \end{pmatrix} \end{aligned}$$

and

$$\begin{aligned} \mathbf{x}(t) = e^{At}\mathbf{x}(0) &= -\frac{1}{4}\begin{pmatrix} -2-2e^{-t} & -4+4e^{-t} \\ -1+e^{-t} & -2-2e^{-t} \end{pmatrix}\begin{pmatrix} 200 \\ 500 \end{pmatrix} \\ &= -\frac{1}{4}\begin{pmatrix} -2400+1600e^{-t} \\ -1200-800e^{-t} \end{pmatrix} \\ &= \begin{pmatrix} 600-400e^{-t} \\ 300+200e^{-t} \end{pmatrix}. \end{aligned}$$

We have $e^{-t} \to 0$ as $t \to \infty$. This means that as time goes on, the two cooperating species approach the **equilibrium** populations 600 and 300, respectively. Neither population is eliminated. ■

PROBLEMS 12.6

In Problems 1–9 find the principal matrix solution e^{At} of the system $\mathbf{x}'(t) = A\mathbf{x}(t)$.

1. $A = \begin{pmatrix} -2 & -2 \\ -5 & 1 \end{pmatrix}$

2. $A = \begin{pmatrix} 3 & -1 \\ -2 & 4 \end{pmatrix}$

3. $A = \begin{pmatrix} 2 & -1 \\ 5 & -2 \end{pmatrix}$

4. $A = \begin{pmatrix} 3 & -5 \\ 1 & -1 \end{pmatrix}$

5. $A = \begin{pmatrix} 3 & 2 \\ -5 & 1 \end{pmatrix}$

6. $A = \begin{pmatrix} -2 & 1 \\ 5 & 2 \end{pmatrix}$

7. $A = \begin{pmatrix} 7 & -2 & -4 \\ 3 & 0 & -2 \\ 6 & -2 & -3 \end{pmatrix}$

8. $A = \begin{pmatrix} 1 & 1 & -2 \\ -1 & 2 & 1 \\ 0 & 1 & -1 \end{pmatrix}$

9. $A = \begin{pmatrix} 3 & 2 & 4 \\ 2 & 0 & 2 \\ 4 & 2 & 3 \end{pmatrix}$

10. In Example 2, show that if the initial vector $\mathbf{x}(0) = \begin{pmatrix} a \\ 2a \end{pmatrix}$ where a is a constant, then both populations grow at a rate proportional to e^t.

11. In Example 2, show that if $x_2(0) > 2x_1(0)$, then the first population will be eliminated.

***12.** In a water desalinization plant there are two tanks of water. Suppose that tank 1 contains 1000 liters of brine in which 1000 kg of salt is dissolved and tank 2 contains 1000 liters of pure water. Suppose that water flows into tank 1 at the rate of 20 liters per minute and the mixture flows from tank 1 into tank 2 at a rate of 30 liters per minute. From tank 2, 10 liters is pumped back to tank 1 (establishing *feedback*) while 20 liters is flushed away. Find the amount of salt in both tanks at any time t. [*Hint:*

Write the information as a 2×2 system and let $x_1(t)$ and $x_2(t)$ denote the amount of salt in each tank.]

13. Consider the second-order differential equation $x''(t) + ax'(t) + bx(t) = 0$.

(a) Letting $x_1(t) = x(t)$ and $x_2(t) = x'(t)$, write the preceding equation as a first-order system in the form of equation (1), where A is a 2×2 matrix.

(b) Show that the characteristic equation of A is $\lambda^2 + a\lambda + b = 0$.

In Problems 14–17 use the result of Problem 13 to solve the given equation.

14. $x'' + 3x' + 2x = 0$; $x(0) = 1$, $x'(0) = 0$

15. $x'' + 4x' + 4x = 0$; $x(0) = 1$, $x'(0) = 2$

16. $x'' + x = 0$; $x(0) = 0$, $x'(0) = 1$

17. $x'' - 3x' - 10x = 0$; $x(0) = 3$, $x'(0) = 2$

12.7 NONHOMOGENEOUS SYSTEMS

We shall now present a method for solving the nonhomogeneous system

$$\mathbf{x}' = A(t)\mathbf{x} + \mathbf{f}(t), \tag{1}$$

given that a fundamental matrix solution $\Phi(t)$ for the homogeneous system

$$\mathbf{x}' = A(t)\mathbf{x} \tag{2}$$

is known. A solution to (2) can always be found if $A(t)$ is a constant matrix (by the methods of Section 12.3 or Section 12.6).

Theorem 1 Let $\varphi_p(t)$ and $\varphi_q(t)$ be two solutions of the system (1). Then their difference,

$$\varphi(t) = \varphi_p(t) - \varphi_q(t),$$

is a solution of equation (2).

Proof. $\varphi' = (\varphi_p - \varphi_q)' = (A\varphi_p + \mathbf{f}) - (A\varphi_q + \mathbf{f}) = A(\varphi_p - \varphi_q) = A\varphi$. Thus, as in the case of linear scalar equations (Section 11.4), it is necessary only to find one particular solution of equation (1). ■

If $\varphi_p(t)$ is such a solution, then *the general solution to the nonhomogeneous system* (1) *is of the form*

$$\varphi(t) = \Phi(t)\mathbf{c} + \varphi_p(t) \tag{3}$$

where $\mathbf{c}$ *is a vector of arbitrary constants and* $\Phi(t)$ *is a fundamental matrix solution of the homogeneous equation* (2). That equation (3) is a solution can be verified as follows:

$$\begin{aligned}\varphi'(t) &= \Phi'(t)\mathbf{c} + \varphi_p'(t)\\ &= [A(t)\Phi(t)\mathbf{c}] + [A(t)\varphi_p(t) + \mathbf{f}(t)],\end{aligned}$$

since Φ is a solution of the associated homogeneous equation and φ_p is a particular solution of equation (2). Combining terms, we have

$$\varphi'(t) = A(t)[\Phi(t)\mathbf{c} + \varphi_p(t)] + \mathbf{f}(t) = A(t)\varphi(t) + \mathbf{f}(t),$$

and φ is a solution of equation (2).

We now derive a **variation-of-constants** formula for the nonhomogeneous system

$$\mathbf{x}' = A(t)\mathbf{x} + \mathbf{f}(t). \tag{4}$$

All variation-of-constants formulas begin by assuming that a solution to the homogeneous equation $\mathbf{x}' = A(t)\mathbf{x}$ is known. Assuming that Φ is a fundamental matrix solution of the homogeneous equation, we seek a particular solution to equation (4) of the form

$$\varphi_p(t) = \Phi(t)\mathbf{k}(t) \tag{5}$$

where $\mathbf{k}(t)$ is a vector function in t. Differentiating both sides of equation (5) with respect to t, we have

$$\varphi_p{}' = \Phi'\mathbf{k} + \Phi\mathbf{k}' = A\Phi\mathbf{k} + \Phi\mathbf{k}' = A\varphi_p + \Phi\mathbf{k}'.$$

Since φ_p is a particular solution of equation (4), $\Phi\mathbf{k}' = \mathbf{f}$. But every fundamental matrix solution has an inverse, so we can integrate $\mathbf{k}' = \Phi^{-1}\mathbf{f}$, obtaining

$$\varphi_p(t) = \Phi(t)\mathbf{k}(t) = \Phi(t)\int \Phi^{-1}(t)\mathbf{f}(t)\,dt. \tag{6}$$

This is the **variation-of-constants** formula for a particular solution to the nonhomogeneous system (4). Thus, the general solution to system (4) has the form

$$\varphi(t) = \Phi(t)\mathbf{c} + \Phi(t)\int \Phi^{-1}(t)\mathbf{f}(t)\,dt \tag{7}$$

where $\mathbf{c}$ is an arbitrary *constant* vector.

For the initial value problem

$$\mathbf{x}' = A(t)\mathbf{x} + \mathbf{f}(t), \qquad \mathbf{x}(t_0) = \mathbf{x}_0, \tag{8}$$

it is convenient to choose a particular solution $\varphi_p(t)$ that vanishes at t_0. This can be done by selecting the limits of integration in equation (6) to be from t_0 to t, so that

$$\varphi(t) = \Phi(t)\mathbf{c} + \Phi(t)\int_{t_0}^{t} \Phi^{-1}(s)\mathbf{f}(s)\,ds.$$

Substituting $t = t_0$ in this equation, we obtain

$$\mathbf{x}_0 = \varphi(t_0) = \Phi(t_0)\mathbf{c},$$

which implies that $\mathbf{c} = \Phi^{-1}(t_0)\mathbf{x}_0$. Hence, the solution of the initial value problem (8) is

$$\begin{aligned} \varphi(t) &= \Phi(t)\Phi^{-1}(t_0)\mathbf{x}_0 + \Phi(t)\int_{t_0}^{t} \Phi^{-1}(s)\mathbf{f}(s)\,ds \\ &= \varphi_h(t) + \varphi_p(t) \end{aligned} \tag{9}$$

where φ_h and φ_p are the homogeneous and particular solutions, respectively. Note that if $\Psi(t)$ is the principal matrix solution of $\mathbf{x}' = A\mathbf{x}$, then $\Psi(t_0) = \Psi^{-1}(t_0) = I$. Thus, equation (9) takes the simpler form

$$\varphi(t) = \Psi(t)\mathbf{x}_0 + \Psi(t)\int_{t_0}^{t} \Psi^{-1}(s)\mathbf{f}(s)\,ds \tag{10}$$

when the principal matrix solution $\Psi(t)$ of $\mathbf{x}' = A(t)\mathbf{x}$ is used. We summarize these results in the following theorem:

Theorem 2 Let $\Phi(t)$ be a fundamental matrix solution of the homogeneous system

$$\mathbf{x}' = A(t)\mathbf{x}. \tag{11}$$

Then the solution to the initial value problem

$$\mathbf{x}' = A(t)\mathbf{x} + \mathbf{f}(t), \qquad \mathbf{x}(t_0) = \mathbf{x}_0, \tag{12}$$

is given by

$$\varphi(t) = \Phi(t)\Phi^{-1}(t_0)\mathbf{x}_0 + \varphi_p(t)$$

where

$$\varphi_p(t) = \Phi(t)\int_{t_0}^{t} \Phi^{-1}(s)\mathbf{f}(s)\,ds \tag{13}$$

is a particular solution of the nonhomogeneous system

$$\mathbf{x}' = A(t)\mathbf{x} + \mathbf{f}(t).$$

If $\Psi(t)$ is the principal matrix solution of the homogeneous system (11), then the solution of the initial value problem (12) is

$$\varphi(t) = \Psi(t)\mathbf{x}_0 + \Psi(t)\int_{t_0}^{t} \Psi^{-1}(s)\mathbf{f}(s)\,ds. \tag{14}$$

As we have already seen, the situation is simplest when $A(t)$ is a constant matrix. We now give some examples.

EXAMPLE 1 Find the unique solution to the system

$$\mathbf{x}' = \begin{pmatrix} x_1 \\ x_2 \end{pmatrix}' = \begin{pmatrix} 4 & 2 \\ 3 & 3 \end{pmatrix}\begin{pmatrix} x_1 \\ x_2 \end{pmatrix} + \begin{pmatrix} e^t \\ e^{2t} \end{pmatrix} = A\mathbf{x} + \mathbf{f}(t), \qquad \mathbf{x}(0) = \begin{pmatrix} 1 \\ 2 \end{pmatrix}.$$

Solution. A fundamental matrix solution for the homogeneous system is

$$\Phi(t) = \begin{pmatrix} 2e^t & e^{6t} \\ -3e^t & e^{6t} \end{pmatrix}.$$

(You should verify this.) Then

$$\Phi^{-1}(t) = \tfrac{1}{5}\begin{pmatrix} e^{-t} & -e^{-t} \\ 3e^{-6t} & 2e^{-6t} \end{pmatrix},$$

and, by equation (12), we have the particular solution

$$\begin{aligned} \varphi_p &= \tfrac{1}{5}\begin{pmatrix} 2e^t & e^{6t} \\ -3e^t & e^{6t} \end{pmatrix} \int_0^t \begin{pmatrix} e^{-s} & -e^{-s} \\ 3e^{-6s} & 2e^{-6s} \end{pmatrix}\begin{pmatrix} e^s \\ e^{2s} \end{pmatrix} ds \\ &= \tfrac{1}{5}\begin{pmatrix} 2e^t & e^{6t} \\ -3e^t & e^{6t} \end{pmatrix} \int_0^t \begin{pmatrix} 1 - e^s \\ 3e^{-5s} + 2e^{-4s} \end{pmatrix} ds. \end{aligned}$$

Since the integral of a vector function is the vector of integrals,

$$\begin{aligned} \varphi_p(t) &= \tfrac{1}{5}\begin{pmatrix} 2e^t & e^{6t} \\ -3e^t & e^{6t} \end{pmatrix}\begin{pmatrix} t - e^t + 1 \\ \dfrac{11}{10} - \dfrac{3}{5}e^{-5t} - \dfrac{e^{-4t}}{2} \end{pmatrix} \\ &= \begin{pmatrix} \dfrac{2}{5}te^t - \dfrac{1}{2}e^{2t} + \dfrac{7}{25}e^t + \dfrac{11}{50}e^{6t} \\ \dfrac{-3}{5}te^t + \dfrac{1}{2}e^{2t} - \dfrac{18}{25}e^t + \dfrac{11}{50}e^{6t} \end{pmatrix}. \end{aligned}$$

Note that $\varphi_p(0) = \mathbf{0}$, which must be the case from the way in which we found φ_p. Next, from equation (9), we see that the unique solution is

$$\begin{aligned} \varphi(t) &= \Phi(t)\Phi^{-1}(0)\mathbf{x}_0 + \varphi_p(t) \\ &= \tfrac{1}{5}\begin{pmatrix} 2e^t & e^{6t} \\ -3e^t & e^{6t} \end{pmatrix}\begin{pmatrix} 1 & -1 \\ 3 & 2 \end{pmatrix}\begin{pmatrix} 1 \\ 2 \end{pmatrix} + \varphi_p \\ &= \tfrac{1}{5}\begin{pmatrix} 2e^t & e^{6t} \\ -3e^t & e^{6t} \end{pmatrix}\begin{pmatrix} -1 \\ 7 \end{pmatrix} + \varphi_p \end{aligned}$$

$$= \begin{pmatrix} -\frac{2}{5}e^t + \frac{7}{5}e^{6t} \\ \frac{3}{5}e^t + \frac{7}{5}e^{6t} \end{pmatrix} + \begin{pmatrix} \frac{2}{5}te^t - \frac{1}{2}e^{2t} + \frac{7}{25}e^t + \frac{11}{50}e^{6t} \\ \frac{-3}{5}te^t + \frac{1}{2}e^{2t} - \frac{18}{25}e^t + \frac{11}{50}e^{6t} \end{pmatrix}$$

$$= \begin{pmatrix} \frac{2}{5}te^t - \frac{1}{2}e^{2t} - \frac{3}{25}e^t + \frac{81}{50}e^{6t} \\ \frac{-3}{5}te^t + \frac{1}{2}e^{2t} - \frac{3}{25}e^t + \frac{81}{50}e^{6t} \end{pmatrix}.$$

Note, as a check, that $\varphi(0) = \begin{pmatrix} 1 \\ 2 \end{pmatrix}$. ■

EXAMPLE 2 Solve the initial value problem

$$x'' + x = 8 \sin t, \qquad x(0) = 3, \qquad x'(0) = 1.$$

Solution. Using the substitution $x_1 = x$, $x_2 = x'$, we can write this in matrix form:

$$\begin{pmatrix} x_1 \\ x_2 \end{pmatrix}' = \begin{pmatrix} 0 & 1 \\ -1 & 0 \end{pmatrix}\begin{pmatrix} x_1 \\ x_2 \end{pmatrix} + \begin{pmatrix} 0 \\ 8 \sin t \end{pmatrix}.$$

The associated homogeneous system has the principal matrix solution

$$\Psi(t) = \begin{pmatrix} \cos t & \sin t \\ -\sin t & \cos t \end{pmatrix}.$$

Then

$$\Psi^{-1}(t) = \begin{pmatrix} \cos t & -\sin t \\ \sin t & \cos t \end{pmatrix}$$

and

$$\int_0^t \Psi^{-1}(s)\mathbf{f}(s)\, ds = \int_0^t \begin{pmatrix} \cos s & -\sin s \\ \sin s & \cos s \end{pmatrix}\begin{pmatrix} 0 \\ 8 \sin s \end{pmatrix} ds$$

$$= \int_0^t \begin{pmatrix} -8 \sin^2 s \\ 8 \sin s \cos s \end{pmatrix} ds = \begin{pmatrix} -4t + 4 \sin t \cos t \\ 4 \sin^2 t \end{pmatrix}.$$

Since the solution $\varphi(t)$ satisfies the initial conditions

$$\varphi(0) = \begin{pmatrix} 3 \\ 1 \end{pmatrix},$$

we obtain from equation (14)

$$\varphi(t) = \begin{pmatrix} \cos t & \sin t \\ -\sin t & \cos t \end{pmatrix}\begin{pmatrix} 3 \\ 1 \end{pmatrix} + \begin{pmatrix} \cos t & \sin t \\ -\sin t & \cos t \end{pmatrix}\begin{pmatrix} -4t + 4\sin t \cos t \\ 4\sin^2 t \end{pmatrix}$$

$$= \begin{pmatrix} -4t\cos t + 5\sin t + 3\cos t \\ 4t \sin t - 3\sin t + \cos t \end{pmatrix}.$$

Here we have used the identity

$$\sin^3 t + \cos^2 t \sin t = \sin t\,(\sin^2 t + \cos^2 t) = \sin t.$$

Thus, $x(t) = -4t\cos t + 5\sin t + 3\cos t$ is the solution to the problem. ■

EXAMPLE 3 Consider the system

$$\begin{pmatrix} x_1 \\ x_2 \\ x_3 \end{pmatrix}' = \begin{pmatrix} 0 & 1 & 0 \\ 0 & 0 & 1 \\ -2/t^3 & 2/t^2 & -1/t \end{pmatrix}\begin{pmatrix} x_1 \\ x_2 \\ x_3 \end{pmatrix} + \begin{pmatrix} 2t^2 \\ -t^3 \\ t^5 \end{pmatrix} \qquad \textbf{(15)}$$

with initial conditions $x_1(1) = 2$, $x_2(1) = 0$, and $x_3(1) = -1$. A fundamental matrix solution (check!) is

$$\Phi(t) = \begin{pmatrix} t & 1/t & t^2 \\ 1 & -1/t^2 & 2t \\ 0 & 2/t^3 & 2 \end{pmatrix}.$$

Note that this solution is valid only for $t > 0$ since $\Phi(t)$ is not defined at $t = 0$. Then

$$\Phi^{-1}(t) = \frac{1}{6}\begin{pmatrix} 6/t & 0 & -3t \\ 2t & -2t^2 & t^3 \\ -2/t^2 & 2/t & 2 \end{pmatrix}.$$

Hence, by equation (9), the solution to the initial value problem (15) is

$$\varphi(t) = \Phi(t)\Phi^{-1}(1)\begin{pmatrix} 2 \\ 0 \\ -1 \end{pmatrix} + \Phi(t)\int_1^t \Phi^{-1}(s)\mathbf{f}(s)\,ds = \Phi(t)\mathbf{c} + \varphi_p(t).$$

Setting $t = 1$, we have

$$\Phi^{-1}(1) = \frac{1}{6}\begin{pmatrix} 6 & 0 & -3 \\ 2 & -2 & 1 \\ -2 & 2 & 2 \end{pmatrix},$$

so that

$$\Phi(t)\Phi^{-1}(1)\begin{pmatrix}2\\0\\-1\end{pmatrix} = \tfrac{1}{6}\begin{pmatrix}15t + 3/t - 6t^2\\ 15 - 3/t^2 - 12t\\ 6/t^3 - 12\end{pmatrix}.$$

After a great deal of arithmetic, we arrive at

$$\varphi(t) = \tfrac{1}{6}\begin{pmatrix}15t + 3/t - 6t^2\\ 15 - 3/t^2 - 12t\\ 6/t^3 - 12\end{pmatrix} + \tfrac{1}{6}\begin{pmatrix}t & 1/t & t^2\\ 1 & -1/t^2 & 2t\\ 0 & 2/t^3 & 2\end{pmatrix}\int_1^t\begin{pmatrix}12s - 3s^6\\ 4s^3 + 2s^5 + s^8\\ -4 - 2s^2 + 2s^5\end{pmatrix}ds$$

$$= \begin{pmatrix}\frac{2}{27}t^8 - \frac{1}{14}t^7 - \frac{1}{18}t^5 - \frac{1}{2}t^3 + \frac{13}{18}t^2 + \frac{5}{2}t - \frac{13}{14} + \frac{7}{27t}\\ -\frac{17}{189}t^7 + \frac{1}{9}t^6 - \frac{5}{18}t^4 - \frac{1}{2}t^2 - \frac{5}{9}t + \frac{11}{7} - \frac{7}{27t^2}\\ \frac{4}{27}t^6 - \frac{1}{9}t^3 - t - \frac{5}{9} + \frac{14}{27t^3}\end{pmatrix}. \quad \blacksquare$$

PROBLEMS 12.7

In Problems 1–9 calculate a fundamental matrix solution for the associated homogeneous system and then use equation (13) and the variation-of-constants formula (9) to obtain a particular solution to the given nonhomogeneous system. Where initial conditions are given, find the unique solution that satisfies them.

1. $\mathbf{x}' = \begin{pmatrix}-2 & -2\\ -5 & 1\end{pmatrix}\mathbf{x} + \begin{pmatrix}e^t\\ e^{2t}\end{pmatrix}$

2. $\mathbf{x}' = \begin{pmatrix}1 & -2\\ 2 & -3\end{pmatrix}\mathbf{x} + \begin{pmatrix}t\\ 2\end{pmatrix}$, $x_1(0) = 1$, $x_2(0) = 0$

3. $\mathbf{x}' = \begin{pmatrix}2 & -1\\ 5 & -2\end{pmatrix}\mathbf{x} + \begin{pmatrix}\sin t\\ \cos t\end{pmatrix}$, $x_1(0) = 0$, $x_2(0) = 1$

4. $\mathbf{x}' = \begin{pmatrix}3 & 2\\ -5 & 1\end{pmatrix}\mathbf{x} + \begin{pmatrix}2\sin 3t\\ \cos 3t\end{pmatrix}$

5. $\mathbf{x}' = \begin{pmatrix}1 & 1 & -2\\ -1 & 2 & 1\\ 0 & 1 & -1\end{pmatrix}\mathbf{x} + \begin{pmatrix}e^t\\ e^{2t}\\ e^{2t}\end{pmatrix}$, $x_1(0) = 0$, $x_2(0) = 1$, $x_3(0) = -1$

6. $\mathbf{x}' = \begin{pmatrix}3 & -1 & 1\\ -1 & 5 & -1\\ 1 & -1 & 3\end{pmatrix}\mathbf{x} + \begin{pmatrix}t^2\\ 0\\ 1\end{pmatrix}$

7. $\mathbf{x}' = \begin{pmatrix}7 & -2 & -4\\ 3 & 0 & -2\\ 6 & -2 & -3\end{pmatrix}\mathbf{x} + \begin{pmatrix}0\\ 1\\ t\end{pmatrix}$

8. $\mathbf{x}' = \begin{pmatrix}1 & -1 & -1\\ 1 & -1 & 0\\ 1 & 0 & -1\end{pmatrix}\mathbf{x} + \begin{pmatrix}2\\ e^{-t}\\ e^{-t}\end{pmatrix}$, $x_1(0) = 1$, $x_2(0) = -1$, $x_3(0) = 0$

9. $\mathbf{x}' = \begin{pmatrix}2 & -5\\ 1 & -2\end{pmatrix}\mathbf{x} + \begin{pmatrix}0\\ \cot t\end{pmatrix}$, $0 < t < \pi$

In Problems 10–12, one homogeneous solution to a given system is given. Use the method of Problem 12.5.21 to obtain a fundamental matrix solution. Then use this solution to find the general solution of the nonhomogeneous system.

10. $\mathbf{x}' = \begin{pmatrix}0 & 1\\ -1/4t^2 & 0\end{pmatrix}\mathbf{x} + \begin{pmatrix}\sqrt{t}\\ 2\end{pmatrix}$, $\varphi_1(t) = \begin{pmatrix}\sqrt{t}\\ 1/2\sqrt{t}\end{pmatrix}$, $t > 0$

11. $\mathbf{x}' = \begin{pmatrix}0 & 1\\ -1/t^2 & -3/t\end{pmatrix}\mathbf{x} + \begin{pmatrix}t\\ e^t\end{pmatrix}$, $\varphi_1(t) = \begin{pmatrix}1/t\\ -1/t^2\end{pmatrix}$, $t > 0$

12. $\mathbf{x}' = \begin{pmatrix} \dfrac{-t}{1-t^2} & \dfrac{1}{1-t^2} \\ \dfrac{1}{1-t^2} & \dfrac{-t}{1-t^2} \end{pmatrix}\mathbf{x} + \begin{pmatrix} t^2 \\ t^3 \end{pmatrix},$

$\varphi_1(t) = \begin{pmatrix} 1 \\ t \end{pmatrix}, \quad -1 < t < 1$

13. Let $\varphi_1(t)$ be a solution to $\mathbf{x}'(t) = A\mathbf{x}(t) + \mathbf{b}_1(t)$, $\varphi_2(t)$ a solution to $\mathbf{x}'(t) = A\mathbf{x}(t) + \mathbf{b}_2(t), \ldots,$ and $\varphi_n(t)$ a solution to $\mathbf{x}'(t) = A\mathbf{x}(t) + \mathbf{b}_n(t)$. Prove that $\varphi_1(t) + \varphi_2(t) + \cdots + \varphi_n(t)$ is a solution to

$$\mathbf{x}'(t) = A\mathbf{x}(t) + \mathbf{b}_1(t) + \mathbf{b}_2(t) + \cdots + \mathbf{b}_n(t).$$

This again is called the **principle of superposition.**

12.8 AN APPLICATION OF NONHOMOGENEOUS SYSTEMS: FORCED OSCILLATIONS (OPTIONAL)

Consider the nonhomogeneous system with constant coefficients

$$\mathbf{x}'(t) = A\mathbf{x}(t) + \mathbf{b}(t). \tag{1}$$

The system (1) can be considered in the following way. If A is fixed, then given an **input vector** $\mathbf{b}(t)$, we can obtain an **output vector** $\mathbf{x}(t)$. That is, the output of the system (the solution) is determined by the input to the system. In this context, vector $\mathbf{b}(t)$ is called the **forcing vector** of the system. With this terminology, system (1) is called an **input–output system.**

It often occurs in practice that the forcing term is periodic (for example, electrical circuits forced by an alternating current). An important question may now be asked: If the forcing term is periodic, does there exist a periodic solution to equation (1) with the same period? An affirmative answer to this question would tell us that an oscillatory input function (an alternating current, for example) will lead to oscillatory behavior of the system governed by the differential equation.

Accordingly, we suppose that

$$\mathbf{b}(t) = e^{i\beta t}\mathbf{c} = (\cos \beta t + i \sin \beta t)\mathbf{c} \tag{2}$$

where $\mathbf{c}$ is a constant vector and $e^{i\beta t}$ is periodic with period $2\pi/\beta$. Suppose that there is a solution $\mathbf{x}(t)$ to equation (1) that is periodic with period $2\pi/\beta$. Then we can write

$$\mathbf{x}(t) = e^{i\beta t}\mathbf{d} \tag{3}$$

where $\mathbf{d}$ is a constant vector. Substituting equations (2) and (3) into (1), we obtain

$$i\beta e^{i\beta t}\mathbf{d} = (e^{i\beta t}\mathbf{d})' = e^{i\beta t}A\mathbf{d} + e^{i\beta t}\mathbf{c}. \tag{4}$$

Dividing equations (4) by $e^{i\beta t}$ and rearranging terms, we obtain

$$(A - i\beta I)\mathbf{d} = -\mathbf{c}. \tag{5}$$

Equation (5) is a nonhomogeneous system of n equations in n unknowns that has a unique solution if and only if

$$\det(A - i\beta I) \neq 0.$$

In other words, there is a unique solution if and only if $i\beta$ is *not* an eigenvalue of the matrix A. We therefore have the following theorem.

Theorem 1 In the system $\mathbf{x}' = A\mathbf{x} + \mathbf{b}$, let the forcing term $\mathbf{b}(t)$ be periodic with the form $\mathbf{b}(t) = e^{i\beta t}\mathbf{c}$. Then, if $i\beta$ is not an eigenvalue of the matrix A, there exists a unique periodic solution $\mathbf{x}(t) = e^{i\beta t}\mathbf{d}$ of $\mathbf{x}' = A\mathbf{x} + \mathbf{b}$ such that $\mathbf{d} = -(A - i\beta I)^{-1}\mathbf{c}$. If $i\beta$ is an eigenvalue of A, then there are either no periodic solutions or an infinite number of them.

NOTE. The differential equation will always have a solution. This theorem tells us something about the nature of the solutions.

REMARK. If $i\beta$ is not an eigenvalue of A, then a periodic (or oscillatory) input will give rise to a periodic output. This phenomenon is called **forced oscillations.**

EXAMPLE 1 Consider the system

$$\mathbf{x}'(t) = \begin{pmatrix} x_1 \\ x_2 \end{pmatrix}' = \begin{pmatrix} 2 & 1 \\ 1 & 0 \end{pmatrix}\begin{pmatrix} x_1 \\ x_2 \end{pmatrix} + e^{2it}\begin{pmatrix} 2 \\ 1 \end{pmatrix} = A\mathbf{x}(t) + \mathbf{b}(t).$$

Here the forcing term is periodic with period π. Since $2i$ is not an eigenvalue of A, we can use equation (5) to obtain a period solution of period π:

$$\mathbf{x}(t) = e^{2it}\mathbf{d}$$

where

$$\mathbf{d} = -(A - 2iI)^{-1}\begin{pmatrix} 2 \\ 1 \end{pmatrix} = -\begin{pmatrix} 2 - 2i & 1 \\ 1 & -2i \end{pmatrix}\begin{pmatrix} 2 \\ 1 \end{pmatrix}$$

$$= \frac{-1}{5 + 4i}\begin{pmatrix} 2i & 1 \\ 1 & -2 + 2i \end{pmatrix}\begin{pmatrix} 2 \\ 1 \end{pmatrix}.$$

Multiplying the numerator and denominator by $5 - 4i$, we have

$$\mathbf{d} = \frac{-5 + 4i}{41}\begin{pmatrix} 1 + 4i \\ 2i \end{pmatrix} = \frac{-1}{41}\begin{pmatrix} 21 + 16i \\ 8 + 10i \end{pmatrix}.$$

We would like to write this solution in a more illuminating form. We recall (see page 694) that

$$A + Bi = \sqrt{A^2 + B^2}\left(\frac{A}{\sqrt{A^2 + B^2}} + i\frac{B}{\sqrt{A^2 + B^2}}\right) = \sqrt{A^2 + B^2}e^{i\theta}$$

where $\tan\theta = B/A$ (and $\cos\theta = A/\sqrt{A^2 + B^2}$, $\sin\theta = B/\sqrt{A^2 + B^2}$). Therefore, we have

$$\mathbf{d} = \frac{-1}{41}\begin{pmatrix} \sqrt{697}e^{i\theta_1} \\ \sqrt{164}e^{i\theta_2} \end{pmatrix}$$

where $\tan \theta_1 = 16/21$ and $\tan \theta_2 = 5/4$. Hence

$$\mathbf{x}(t) = \frac{-1}{41}\begin{pmatrix} \sqrt{697}e^{i(\theta_1+2t)} \\ \sqrt{164}e^{i(\theta_2+2t)} \end{pmatrix}.$$

Although $\mathbf{x}(t)$ has the same period, π, as the forcing term, the coordinate functions have been shifted by θ_1 and θ_2. Such a phenomenon is called a **phase shift.** ■

Returning to equation (5), we answer the question, "What happens if $i\beta$ is an eigenvalue of A?" According to the discussion in Section 12.6, the principal matrix solution e^{At} of the homogeneous system $\mathbf{x}' = A\mathbf{x}$ contains terms of the form $e^{i\beta t}$, so that $\Phi^{-1}(t)$ contains terms of the form $e^{-i\beta t}$ (since $\Phi\Phi^{-1} = I$). Using Equation 12.7.13, we find a particular solution of equation (1):

$$\varphi_p(t) = \Phi(t)\int_{t_0}^{t} \Phi^{-1}(s)\mathbf{c}e^{i\beta s}\,ds.$$

But the product of $\Phi^{-1}(s)$ and $e^{i\beta s}$ contains constant terms (since $e^{-i\beta s}e^{i\beta s} = 1$), and the integral of these constant terms will be of the form ct, which becomes unbounded as t tends to $\pm\infty$. Such a phenomenon is called **resonance** (see page 702). If $i\beta$ is an eigenvalue of A, then β is called a **natural frequency** of the system. We can summarize the above discussion by stating that **in general, resonance will occur when the frequency of the input vector is a natural frequency of the system.**

EXAMPLE 2 Consider the circuit shown in Figure 1. Applying Kirchhoff's voltage law (see page 652), we obtain the second-order nonhomogeneous equation

$$L\frac{d^2I}{dt^2} + \frac{I}{C} = E'(t). \qquad \textbf{(6)}$$

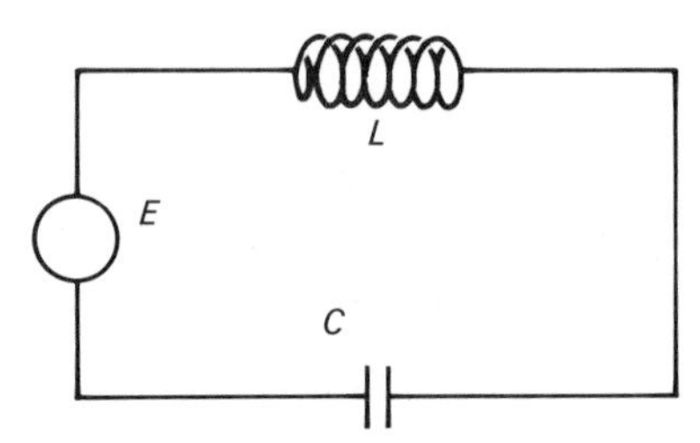

FIGURE 1

Letting $x_1 = I$ and $x_2 = I'$, we have

$$x_2' = I'' = \frac{E'}{L} - \frac{I}{LC} = \frac{E'}{L} - \frac{x_1}{LC},$$

so we can rewrite equation (6) in the form

$$\begin{pmatrix} x_1 \\ x_2 \end{pmatrix}' = \begin{pmatrix} 0 & 1 \\ -1/LC & 0 \end{pmatrix}\begin{pmatrix} x_1 \\ x_2 \end{pmatrix} + \begin{pmatrix} 0 \\ E'/L \end{pmatrix}. \tag{7}$$

Suppose that $E = E_0 e^{i\omega t}$. Then equation (7) may be treated by the method of this section, and the solution $\mathbf{x}(t)$ is of the form $e^{i\omega t}\mathbf{x}$ where

$$\mathbf{x} = -(A - i\omega I)^{-1}\mathbf{b} \qquad \text{and} \qquad \mathbf{b} = \begin{pmatrix} 0 \\ i\omega E_0/L \end{pmatrix}.$$

But

$$\det(A - i\omega I) = \det\begin{pmatrix} -i\omega & 1 \\ -1/LC & -i\omega \end{pmatrix} = \frac{1}{LC} - \omega^2,$$

which is nonzero if $\omega \neq 1/\sqrt{LC}$. Therefore,

$$\mathbf{x} = \frac{-1}{(1/LC) - \omega^2}\begin{pmatrix} -i\omega & -1 \\ 1/LC & -i\omega \end{pmatrix}\begin{pmatrix} 0 \\ i\omega E_0/L \end{pmatrix} = \frac{\omega C E_0}{LC\omega^2 - 1}\begin{pmatrix} -i \\ \omega \end{pmatrix}$$

and

$$\mathbf{x}(t) = \frac{\omega C E_0}{LC\omega^2 - 1}\begin{pmatrix} -i \\ \omega \end{pmatrix} e^{i\omega t}.$$

Thus the current is

$$x_1(t) = I(t) = -i\omega C E_0 e^{i\omega t}$$

or

$$I(t) = \omega C E_0(\sin \omega t - i \cos \omega t).$$

The complex term here simply tells us that the real and imaginary parts of the solution are the responses to the real and imaginary parts of the forcing function.

If $\omega = 1/\sqrt{LC}$ (the natural frequency), then the circuit is in resonance. Since $A^2 = (-1/LC)I$, it follows after a short computation that the principal matrix solution e^{At} is

$$\Psi(t) = \begin{pmatrix} \cos(t/\sqrt{LC}) & \sqrt{LC}\sin(t/\sqrt{LC}) \\ (-1/\sqrt{LC})\sin(t/\sqrt{LC}) & \cos(t/\sqrt{LC}) \end{pmatrix},$$

and by equation (12.7.14),

$$\mathbf{x}(t) = \Psi(t)\mathbf{x}(0) + \Psi(t)\int_0^t \Psi^{-1}(s)\mathbf{b}(s)\,ds. \tag{8}$$

To calculate the integral, we observe that since $\cos(-s) = \cos s$ and $\sin(-s) = -\sin s$, we have

$$\Psi^{-1}(s)\mathbf{b}(s) = \begin{pmatrix} \cos\frac{s}{\sqrt{LC}} & -\sqrt{LC}\sin\frac{s}{\sqrt{LC}} \\ \frac{1}{\sqrt{LC}}\sin\frac{s}{\sqrt{LC}} & \cos\frac{s}{\sqrt{LC}} \end{pmatrix} \begin{pmatrix} 0 \\ \frac{iE_0e^{is/\sqrt{LC}}}{L\sqrt{LC}} \end{pmatrix}$$

$$= \begin{pmatrix} \frac{E_0}{L}\sin\frac{s}{\sqrt{LC}}\left(\sin\frac{s}{\sqrt{LC}} - i\cos\frac{s}{\sqrt{LC}}\right) \\ \frac{E_0}{L\sqrt{LC}}\cos\frac{s}{\sqrt{LC}}\left(-\sin\frac{s}{\sqrt{LC}} + i\cos\frac{s}{\sqrt{LC}}\right) \end{pmatrix}.$$

After we perform the integration in (8), we obtain

$$\int_0^t \Psi^{-1}(s)\mathbf{b}(s)\,ds = \frac{E_0}{2L}\begin{pmatrix} t - \sqrt{LC}\sin\frac{t}{\sqrt{LC}}e^{it/\sqrt{LC}} \\ i\left(\frac{t}{\sqrt{LC}} + \sin\frac{t}{\sqrt{LC}}e^{it/\sqrt{LC}}\right) \end{pmatrix}. \tag{9}$$

Adding this vector to $\mathbf{x}(0)$ and multiplying $\Psi(t)$ by the result will give the solution to system (7). Note that the vector (9) has components that become arbitrarily large as t approaches $+\infty$. This is an effect of resonance on the system. ■

PROBLEMS 12.8

In Problems 1–4 for each system find (if possible) a solution $\mathbf{x}(t)$ that has the same period of oscillation as the forcing term. If the natural and forcing frequencies coincide, show that a particular solution is unbounded as t approaches $\pm\infty$ (i.e., resonance occurs).

1. $\mathbf{x}' = \begin{pmatrix} 2 & -5 \\ 0 & 3 \end{pmatrix}\mathbf{x} + \begin{pmatrix} 1 \\ 4 \end{pmatrix}e^{it}$

2. $\mathbf{x}' = \begin{pmatrix} 1 & -2 \\ \frac{5}{2} & -1 \end{pmatrix}\mathbf{x} + \begin{pmatrix} 2 \\ 3 \end{pmatrix}e^{2it}$

3. $\mathbf{x}' = \begin{pmatrix} 3 & 4 \\ -1 & -2 \end{pmatrix}\mathbf{x} + \begin{pmatrix} -1 \\ 4 \end{pmatrix}e^{-5it}$

4. $\mathbf{x}' = \begin{pmatrix} -1 & -7 \\ 5 & 1 \end{pmatrix}\mathbf{x} + \begin{pmatrix} -2 \\ 7 \end{pmatrix}e^{4it}$

5. Suppose that $\mathbf{b}(t) = \sum_{k=1}^{m} \mathbf{b}_k e^{i\beta_k t}$ (a finite sum of periodic inputs). Assume that $i\beta_k$ is not a root of the characteristic equation of A for $k = 1, 2, \ldots, m$. Use the principle of superposition to find a solution $\mathbf{x}(t)$ to equation (1) that can be written in the form

$$\mathbf{x}(t) = \sum_{k=1}^{m} \mathbf{x}_k e^{i\beta_k t}.$$

6. Use Euler's formula to show that

$$2\cos\beta t = e^{i\beta t} + e^{-i\beta t}$$

and

$$2i\sin\beta t = e^{i\beta t} - e^{-i\beta t}.$$

Then use the results of Problem 5 to obtain (if possible) periodic solutions to the following equations:
(a) $x'' - x' - 2x = \sin t$;
(b) $x'' + 4x = \cos t$;
(c) $x'' + 4x = \cos 2t$;
(d) $x'' + 2x' - 15x = \sin 4t$.

7. Verify the form of the principal matrix solution $\Psi(t)$ if ω equals the natural frequency in Example 2.

8. Verify equation (9).

9. **(a)** Express the forced vibration equation (11.13.11) for a spring–mass system with $c = 0$ as a 2×2 system of differential equations.

(b) Obtain a solution by the method of this section, given that $\omega \neq \sqrt{k/m}$.

(c) What happens if $\omega = \sqrt{k/m}$? Justify your answer.

REVIEW EXERCISES FOR CHAPTER TWELVE

In Exercises 1–3 transform the equation into a first-order system.

1. $x''' - 6x'' + 2x' - 5x = 0$
2. $x'' - 3x' + 4t^2x = \sin t$
3. $xx'' + x'x''' = \ln t$

In Exercises 4–8 find the general solution for each system.

4. $x' = x + y,\quad y' = 9x + y$

5. $x' = x + 2y,\quad y' = 4x + 3y$

6. $x' = 4x - y,\quad y' = x + 2y$

7. $x' = 3x + 2y,\quad y' = -5x + y$

8. $x' = x - 4y,\quad y' = x + y$

9. Find the general solution to

$$x' = -x - 3e^{-2t},$$
$$y' = -2x - y - 6e^{-2t}.$$

10. Find the unique solution to

$$x' = -4x - 6y + 9e^{-3t}, \qquad x(0) = -9,$$
$$y' = x + y - 5e^{-3t}, \qquad y(0) = 4.$$

In Exercises 11–14 write the given system of equations in vector–matrix form.

11. $x_1' = 3x_1 - 4x_2,\quad x_2' = -2x_1 + 7x_2$

12. $x_1' = (\sin t)x_1 + e^tx_2,\quad x_2' = -x_1 + (\tan t)x_2$

13. $x_1' = x_1 + x_2 + e^t,\quad x_2' = -3x_1 + 2x_2 + e^{2t}$

14. $x_1' = -tx_1 + t^2x_2 + t^3,\quad x_2' = -\sqrt{t}x_1 + \sqrt[3]{t}x_2 + t^{3/5}$

15. Consider the system

$$\mathbf{x}' = \begin{pmatrix} 4 & 2 \\ 3 & 3 \end{pmatrix}\mathbf{x}.$$

A fundamental matrix solution is

$$\Phi(t) = \begin{pmatrix} 2e^t & e^{6t} \\ -3e^t & e^{6t} \end{pmatrix}.$$

Find a solution that satisfies each of the following initial conditions.

(a) $\mathbf{x}(0) = \begin{pmatrix} 2 \\ 3 \end{pmatrix}$; **(b)** $\mathbf{x}(0) = \begin{pmatrix} -1 \\ 0 \end{pmatrix}$;

(c) $\mathbf{x}(0) = \begin{pmatrix} 0 \\ 0 \end{pmatrix}$; **(d)** $\mathbf{x}(0) = \begin{pmatrix} 7 \\ -2 \end{pmatrix}$;

(e) $\mathbf{x}(0) = \begin{pmatrix} a \\ b \end{pmatrix}$.

In Exercises 16–19 find the principal matrix solution of the given system.

16. $\mathbf{x}' = \begin{pmatrix} -18 & -15 \\ 20 & 17 \end{pmatrix}\mathbf{x}$ **17.** $\mathbf{x}' = \begin{pmatrix} -3 & 4 \\ -2 & 3 \end{pmatrix}\mathbf{x}$

18. $\mathbf{x}' = \begin{pmatrix} 1 & 1 & 1 \\ -1 & -1 & 0 \\ -1 & 0 & -1 \end{pmatrix}\mathbf{x}$ **19.** $\mathbf{x}' = \begin{pmatrix} 4 & 2 & 0 \\ 2 & 4 & 0 \\ 0 & 0 & -3 \end{pmatrix}\mathbf{x}$

20. Solve the system

$$\mathbf{x}' = \begin{pmatrix} 2 & 1 & 0 \\ -2 & -1 & 2 \\ 1 & 1 & 1 \end{pmatrix}\mathbf{x} + \begin{pmatrix} 0 \\ 1 \\ e^t \end{pmatrix}, \qquad \mathbf{x}(0) = \begin{pmatrix} 1 \\ 2 \\ 3 \end{pmatrix}.$$

21. Solve the system

$$\mathbf{x}' = \begin{pmatrix} 2 & 1 \\ -4 & 2 \end{pmatrix}\mathbf{x} + \begin{pmatrix} 3 \\ t \end{pmatrix}e^{2t}, \qquad \mathbf{x}(0) = \begin{pmatrix} 3 \\ 2 \end{pmatrix}.$$

13 Taylor Polynomials, Sequences, and Series

Many functions arising in applications are difficult to deal with. A continuous function, for example, may take a complicated form, or it may take a simple form that, nevertheless, cannot be integrated.

For this reason mathematicians and physicists have developed methods for approximating certain functions by other functions that are much easier to handle. Some of the easiest functions to deal with are the polynomials, since in addition to having other useful properties, they can be differentiated and integrated any number of times and still remain polynomials. In the first three sections of this chapter we will show how certain continuous functions can be approximated by polynomials.

13.1 TAYLOR'S THEOREM AND TAYLOR POLYNOMIALS

In this section we show how a function can be approximated as closely as desired by a polynomial, provided that the function possesses a sufficient number of derivatives.

We begin by reminding you of the factorial notation defined for each positive integer n:

$$n! = n(n-1)(n-2)\cdots 3\cdot 2\cdot 1.$$

That is, $n!$ is the product of the first n positive integers. For example, $3! = 3 \cdot 2 \cdot 1 = 6$ and $5! = 5 \cdot 4 \cdot 3 \cdot 2 \cdot 1 = 120$. By convention, we define $0!$ to be equal to 1.

Definition 1 TAYLOR[†] POLYNOMIAL Let the function f and its first n derivatives exist on the closed interval $[a, b]$. Then, for $x \in [a, b]$, the nth degree **Taylor polynomial** of f at a is the nth degree polynomial $P_n(x)$, given by

$$P_n(x) = f(a) + \frac{f'(a)}{1!}(x-a) + \frac{f''(a)}{2!}(x-a)^2 + \frac{f'''(a)}{3!}(x-a)^3 + \cdots + \frac{f^{(n)}(a)}{n!}(x-a)^n = \sum_{k=0}^{n} \frac{f^{(k)}(a)(x-a)^k}{k!}.^{\ddagger} \quad (1)$$

EXAMPLE 1 Calculate the fifth-degree Taylor polynomial of $f(x) = \sin x$ at 0.

Solution. We have $f(x) = \sin x$, $f'(x) = \cos x$, $f''(x) = -\sin x$, $f'''(x) = -\cos x$, $f^{(4)}(x) = \sin x$, and $f^{(5)}(x) = \cos x$. Then $f(0) = 0$, $f'(0) = 1$, $f''(0) = 0$, $f'''(0) = -1$, $f^{(4)}(0) = 0$, $f^{(5)}(0) = 1$, and we obtain

$$P_5(x) = f(0) + \frac{f'(0)}{1!}x + \frac{f''(0)}{2!}x^2 + \frac{f'''(0)}{3!}x^3 + \frac{f^{(4)}(0)}{4!}x^4 + \frac{f^{(5)}(0)}{5!}x^5$$

$$= x - \frac{x^3}{3!} + \frac{x^5}{5!} = x - \frac{x^3}{6} + \frac{x^5}{120}. \quad \blacksquare$$

Definition 2 REMAINDER TERM Let $P_n(x)$ be the nth degree Taylor polynomial of the function f. Then the **remainder term**, denoted $R_n(x)$, is given by

$$R_n(x) = f(x) - P_n(x). \quad (2)$$

Why do we study Taylor polynomials? Because of the following remarkable result that tells us that a Taylor polynomial provides a good approximation to a function f.

Theorem 1 TAYLOR'S THEOREM (TAYLOR'S FORMULA WITH REMAINDER) Suppose that $f^{n+1}(x)$ exists on the closed interval $[a, b]$. Let x be any number in $[a, b]$. Then there is a number c[§] in (a, x) such that

$$R_n(x) = \frac{f^{(n+1)}(c)}{(n+1)!}(x-a)^{n+1}. \quad (3)$$

[†]Named after the English mathematician Brook Taylor (1685–1731), who published what we now call *Taylor's formula* in *Methodus Incrementorum* in 1715. There was a considerable controversy over whether Taylor's discovery was, in fact, a plagiarism of an earlier result of the Swiss mathematician Jean Bernoulli (1667–1748).

[‡]In this notation we have $f^{(0)}(a) = f(a)$.

[§]c depends on x.

The expression in (3) is called **Lagrange's form of the remainder.**[†] Using (3), we can write Taylor's formula as

$$f(x) = f(a) + \frac{f'(a)}{1!}(x-a) + \frac{f''(a)}{2!}(x-a)^2 + \cdots + \frac{f^{(n)}(a)}{n!}(x-a)^n + \frac{f^{(n+1)}(c)}{(n+1)!}(x-a)^{n+1} \quad \textbf{(4)}$$

REMARK. In Section 13.2 we will show that the Taylor polynomial $P_n(x)$ is *unique* in a sense to be made precise later.

In Section 13.2 we will prove Taylor's formula. Also we will show that under certain reasonable assumptions, the remainder term $R_n(x)$ given by (3) actually approaches 0 as $n \to \infty$. Thus as n increases, $P_n(x)$ is an increasingly useful approximation to the function f over the interval $[a, b]$.

In the remainder of this section we will calculate some Taylor polynomials.

EXAMPLE 2 Calculate the fifth-degree Taylor polynomial of $f(x) = \sin x$ at $\pi/6$.

Solution. Using the derivatives found in Example 1, we have $f(\pi/6) = 1/2$, $f'(\pi/6) = \sqrt{3}/2$, $f''(\pi/6) = -1/2$, $f'''(\pi/6) = -\sqrt{3}/2$, $f^{(4)}(\pi/6) = 1/2$, and $f^{(5)}(\pi/6) = \sqrt{3}/2$, so that in this case

$$\begin{aligned} P_5(x) &= \frac{1}{2} + \frac{\sqrt{3}}{2}\left(x - \frac{\pi}{6}\right) - \frac{1}{2}\frac{[x-(\pi/6)]^2}{2!} - \frac{\sqrt{3}}{2}\frac{[x-(\pi/6)]^3}{3!} \\ &\quad + \frac{1}{2}\frac{[x-(\pi/6)]^4}{4!} + \frac{\sqrt{3}}{2}\frac{[x-(\pi/6)]^5}{5!} \\ &= \frac{1}{2} + \frac{\sqrt{3}}{2}\left(x - \frac{\pi}{6}\right) - \frac{1}{4}\left(x - \frac{\pi}{6}\right)^2 - \frac{\sqrt{3}}{12}\left(x - \frac{\pi}{6}\right)^3 \\ &\quad + \frac{1}{48}\left(x - \frac{\pi}{6}\right)^4 + \frac{\sqrt{3}}{240}\left(x - \frac{\pi}{6}\right)^5. \ \blacksquare \end{aligned}$$

Examples 1 and 2 illustrate that in many cases it is easiest to calculate the Taylor polynomial at 0. In this situation we have

$$P_n(x) = f(0) + f'(0)x + \frac{f''(0)}{2!}x^2 + \cdots + \frac{f^{(n)}(0)}{n!}x^n. \quad \textbf{(5)}$$

EXAMPLE 3 Find the eighth-degree Taylor polynomial of $f(x) = e^x$ at 0.

Solution. Here $f(x) = f'(x) = f''(x) = \cdots = f^{(8)}(x) = e^x$, and $e^0 = 1$, so that

$$P_8(x) = 1 + x + \frac{x^2}{2!} + \frac{x^3}{3!} + \frac{x^4}{4!} + \frac{x^5}{5!} + \frac{x^6}{6!} + \frac{x^7}{7!} + \frac{x^8}{8!} = \sum_{k=0}^{8} \frac{x^k}{k!}. \ \blacksquare$$

[†]See the biographical sketch of Lagrange on page 255.

EXAMPLE 4 We can extend Example 1 to see that for $f(x) = \sin x$ at 0, the Taylor polynomials for different values of n are given by

$$P_0(x) = 0$$

$$P_1(x) = x = P_2(x)$$

$$P_3(x) = x - \frac{x^3}{3!} = P_4(x)$$

$$P_5(x) = x - \frac{x^3}{3!} + \frac{x^5}{5!} = P_6(x)$$

$$P_7(x) = x - \frac{x^3}{3!} + \frac{x^5}{5!} - \frac{x^7}{7!} = P_8(x)$$

$$\vdots$$

$$P_{2n+1}(x) = x - \frac{x^3}{3!} + \frac{x^5}{5!} - \cdots + \frac{(-1)^n x^{2n+1}}{(2n+1)!} = \sum_{k=0}^{n} \frac{(-1)^k x^{2k+1}}{(2k+1)!} = P_{2n+2}(x).$$

In Figure 1 we reproduce a computer-drawn graph showing how, with smaller values of x in the interval $[0, 2\pi]$, the Taylor polynomials get closer and closer to the function $\sin x$ as n increases. We will say more about this phenomenon in Section 13.3. ■

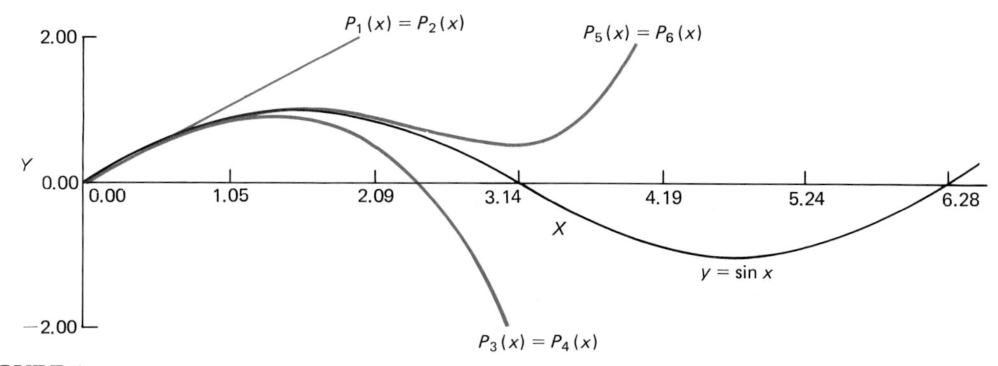

FIGURE 1

EXAMPLE 5 Find the fifth-degree Taylor polynomial of $f(x) = 1/(1 - x)$ at 0.

Solution. Here

$$f(x) = \frac{1}{1-x} \qquad f'(x) = \frac{1}{(1-x)^2} \qquad f''(x) = \frac{2}{(1-x)^3}$$

$$f'''(x) = \frac{6}{(1-x)^4} \qquad f^{(4)}(x) = \frac{24}{(1-x)^5} \qquad f^{(5)}(x) = \frac{120}{(1-x)^6}.$$

Thus $f(0) = 1$, $f'(0) = 1$, $f''(0) = 2$, $f'''(0) = 6$, $f^{(4)}(0) = 24$, $f^{(5)}(x) = 120$, and

$$P_5(x) = 1 + x + \frac{2x^2}{2!} + \frac{6x^3}{3!} + \frac{24x^4}{4!} + \frac{120x^5}{5!}$$

$$= 1 + x + x^2 + x^3 + x^4 + x^5 = \sum_{k=0}^{5} x^k.$$

In Figure 2 we reproduce a computer-drawn sketch of $1/(1 - x)$ and its first four Taylor polynomials. ■

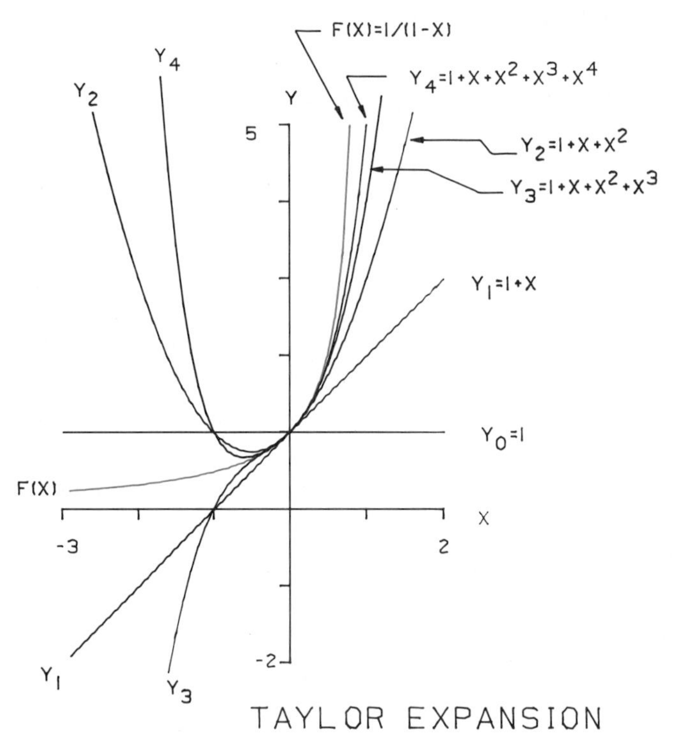

FIGURE 2

Note that in Examples 1, 2, and 3 the given function had continuous derivatives of all orders defined for all real numbers. In Example 5, $f(x)$ is defined over intervals of the form $[-b, b]$, where $b < 1$. Thus Taylor's theorem does *not* apply in any interval containing 1. *It is always necessary to check whether the hypotheses of Taylor's theorem hold over a given interval.*

Before leaving this section, we observe that *the nth-degree Taylor polynomial at a of a function is the polynomial that agrees with the function and each of its first n derivatives at a.* This follows immediately from (1):

$$P_n(a) = f(a)$$

$$P_n'(a) = \left[f'(a) + f''(a)(x - a) + f'''(a)\frac{(x - a)^2}{2!} + \cdots + \frac{f^{(n)}(a)}{(n - 1)!}(x - a)^{n-1} \right]\Bigg|_{x=a}$$

$$= f'(a),$$

$$P_n''(a) = \left[f''(a) + f'''(a)(x-a) + \cdots + \frac{f^{(n)}(a)}{(n-2)!}(x-a)^{n-2}\right]\Bigg|_{x=a} = f''(a),$$

and so on. In particular, since the $(n+1)$st derivative of an nth-degree polynomial is zero, we find that *if $Q(x)$ is a polynomial of degree n, then $P_n(x) = Q(x)$*. This follows immediately from the fact that the remainder term given by (3) will be zero since $Q^{(n+1)}(c) = 0$.

EXAMPLE 6 Let $Q(x) = 3x^4 + 2x^3 - 4x^2 + 5x - 8$. Compute $P_4(x)$ at 0.

Solution. We have

$$\begin{aligned} Q(0) &= -8 \\ Q'(0) &= (12x^3 + 6x^2 - 8x + 5)|_{x=0} = 5, \\ Q''(0) &= (36x^2 + 12x - 8)|_{x=0} = -8, \\ Q'''(0) &= (72x + 12)|_{x=0} = 12, \\ Q^{(4)}(0) &= 72. \end{aligned}$$

Therefore,

$$P_4(x) = -8 + 5x - \frac{8x^2}{2!} + \frac{12x^3}{3!} + \frac{72x^4}{4!} = -8 + 5x - 4x^2 + 2x^3 + 3x^4 = Q(x),$$

as expected. ■

PROBLEMS 13.1

In Problems 1–26, find the Taylor polynomial of the given degree n for the function f at the number a.

1. $f(x) = \cos x$; $a = \pi/4$; $n = 6$
2. $f(x) = \sqrt{x}$; $a = 1$; $n = 4$
3. $f(x) = \ln x$; $a = e$; $n = 5$
4. $f(x) = \ln(1 + x)$; $a = 0$; $n = 5$
5. $f(x) = 1/x$; $a = 1$; $n = 4$
6. $f(x) = \dfrac{1}{(1+x)}$; $a = 0$; $n = 5$
7. $f(x) = \tan x$; $a = 0$; $n = 4$
8. $f(x) = \tan^{-1} x$; $a = 0$; $n = 6$
9. $f(x) = \tan x$; $a = \pi$; $n = 4$
10. $f(x) = \tan^{-1} x$; $a = 1$; $n = 6$
11. $f(x) = \dfrac{1}{(1+x^2)}$; $a = 0$; $n = 4$
12. $f(x) = \dfrac{1}{\sqrt{x}}$; $a = 4$; $n = 3$
13. $f(x) = \sinh x$; $a = 0$; $n = 4$
14. $f(x) = \cosh x$; $a = 0$; $n = 4$
15. $f(x) = \ln \sin x$; $a = \pi/2$; $n = 3$
16. $f(x) = \ln \cos x$; $a = 0$; $n = 3$
17. $f(x) = \dfrac{1}{\sqrt{4-x}}$; $a = 0$; $n = 4$
18. $f(x) = \dfrac{1}{\sqrt{4-x}}$; $a = 3$; $n = 4$
19. $f(x) = e^{\alpha x}$; $a = 0$; $n = 6$; α real
20. $f(x) = \sin \alpha x$; $a = 0$; $n = 6$; α real
21. $f(x) = \sin^{-1} x$; $a = 0$; $n = 3$
22. $f(x) = 1 + x + x^2$; $a = 0$; $n = 10$
23. $f(x) = a_0 + a_1x + a_2x^2 + a_3x^3$, $a = 1$, $n = 10$
24. $f(x) = e^{x^2}$; $a = 0$; $n = 4$
25. $f(x) = \sin x^2$; $a = 0$; $n = 4$
26. $f(x) = \cos x^2$; $a = 0$; $n = 4$
27. Show that the nth-degree Taylor polynomial of $f(x) = 1/(1-x)$ at 0 is given by

$$P_n(x) = 1 + x + x^2 + \cdots + x^n = \sum_{k=0}^{n} x^k.$$

13.2 A PROOF OF TAYLOR'S THEOREM, ESTIMATES ON THE REMAINDER TERM, AND A UNIQUENESS THEOREM (OPTIONAL)

In this section we begin by proving Taylor's formula with remainder, as stated in Theorem 13.1.1.

Proof of Taylor's Theorem

We show that if $f^{(n+1)}(x)$ exists in $[a, b]$, then for any number $x \in [a, b]$, there is a number c in $[a, x]$ such that

$$f(x) = f(a) + \frac{f'(a)}{1!}(x - a) + \frac{f''(a)}{2!}(x - a)^2 + \cdots + \frac{f^{(n)}(a)}{n!}(x - a)^n + R_n(x), \tag{1}$$

where

$$R_n(x) = \frac{f^{(n+1)}(c)}{(n + 1)!}(x - a)^{n+1}. \tag{2}$$

Let $x \in [a, b]$ be fixed. We define a new function $h(t)$ by

$$h(t) = f(x) - f(t) - f'(t)(x - t) - \frac{f''(t)}{2!}(x - t)^2 - \cdots - \frac{f^{(n)}(t)}{n!}(x - t)^n - \frac{R_n(x)(x - t)^{n+1}}{(x - a)^{n+1}}. \tag{3}$$

Then

$$h(x) = f(x) - f(x) - f'(x)(x - x) - \frac{f''(x)}{2!}(x - x)^2 - \cdots - \frac{f^{(n)}(x)}{n!}(x - x)^n - \frac{R_n(x)(x - x)^{n+1}}{(x - a)^{n+1}} = 0$$

and

$$\begin{aligned} h(a) &= f(x) - f(a) - f'(a)(x - a) - \frac{f''(a)}{2!}(x - a)^2 - \cdots \\ &\quad - \frac{f^{(n)}(a)}{n!}(x - a)^n - \frac{R_n(x)(x - a)^{n+1}}{(x - a)^{n+1}} \\ &= f(x) - P_n(x) - R_n(x) = R_n(x) - R_n(x) = 0. \end{aligned}$$

Since $f^{(n+1)}$ exists, $f^{(n)}$ is differentiable so that h, being a sum of products of differentiable functions, is also differentiable for t in (a, x). Remember that x is fixed so h is a function of t only.

Recall Rolle's theorem which states that if h is continuous on $[a, b]$, differentiable on (a, b), and $h(a) = h(b) = 0$, then there is at least one number c in (a, b) such that $h'(c) = 0$. We see that the conditions of Rolle's theorem hold in the interval $[a, x]$ so that there is a number c in (a, x) with $h'(c) = 0$. In Problems 1–3 you are asked to show that

$$h'(t) = \frac{-f^{(n+1)}(t)(x - t)^n}{n!} + \frac{(n + 1)R_n(x)(x - t)^n}{(x - a)^{n+1}}.$$

Then, setting $t = c$, we obtain

$$0 = h'(c) = \frac{-f^{(n+1)}(c)(x - c)^n}{n!} + \frac{(n + 1)R_n(x)(x - c)^n}{(x - a)^{n+1}}.$$

Finally, dividing the equations above through by $(x - c)^n$ and solving for $R_n(x)$, we obtain

$$R_n(x) = \frac{f^{(n+1)}(c)(x - a)^n}{(n + 1)n!} = \frac{f^{(n+1)}(c)(x - a)^{n+1}}{(n + 1)!}.$$

This is what we wanted to prove. ■

REMARK. If you go over the proof, you may observe that we didn't need to assume that $x > a$; if we replace the interval $[a, x]$ with the interval $[x, a]$, then all results are still valid.

In a remark on page 786 we said that the Taylor polynomial was unique. We now make that statement more precise.

First, we note that if $f^{(n+1)}$ is continuous in $[a, b]$ and $a < c < x$ [where c is the c that appears in the formula for $R_n(x)$], we have

$$\lim_{x\to a} f^{(n+1)}(c) = f^{(n+1)}(a). \tag{4}$$

Then

$$\lim_{x\to a} \frac{R_n(x)}{(x - a)^n} \overset{\text{From (2)}}{=} \lim_{x\to a} \frac{f^{(n+1)}(c)}{(n + 1)!} \frac{(x - a)^{n+1}}{(x - a)^n} = \frac{1}{(n + 1)!} \lim_{x\to a} f^{(n+1)}(c) \lim_{x\to a} (x - a)$$

$$\overset{\text{From (4)}}{=} \frac{1}{(n + 1)!} f^{(n+1)}(a) \cdot 0 = 0.$$

That is,

$$\lim_{x\to a} \frac{R_n(x)}{(x - a)^n} = 0. \tag{5}$$

We can now state our uniqueness result.

Theorem 1 UNIQUENESS OF THE TAYLOR POLYNOMIAL If f has $n + 1$ continuous derivatives, then the Taylor polynomial $P_n(x)$ is the *only* nth-degree polynomial whose remainder term satisfies (5).

Proof. Suppose that there is another nth-degree polynomial $Q_n(x)$ such that

$$f(x) = P_n(x) + R_n(x) = Q_n(x) + S_n(x),$$

where

$$\lim_{x \to a} \frac{R_n(x)}{(x - a)^n} = 0, \quad \text{and} \quad \lim_{x \to a} \frac{S_n(x)}{(x - a)^n} = 0.$$

Let

$$D(x) = P_n(x) - Q_n(x) = S_n(x) - R_n(x).$$

We will show that $D(x) = 0$, which will imply that $P_n(x) = Q_n(x)$. This will show that $P_n(x)$ is unique. Since $D(x)$ is the difference of nth-degree polynomials, $D(x)$ is a polynomial of degree $\le n$, and it can be written in the form

$$D(x) = b_0 + b_1(x - a) + b_2(x - a)^2 + \cdots + b_n(x - a)^n.$$

Now

$$\lim_{x \to a} \frac{D(x)}{(x - a)^n} = \lim_{x \to a} \frac{S_n(x)}{(x - a)^n} - \lim_{x \to a} \frac{R_n(x)}{(x - a)^n} = 0 - 0 = 0.$$

Similarly, if $0 \le m < n$,

Multiply and divide by $(x - a)^{n-m}$

$$\lim_{x \to a} \frac{D(x)}{(x - a)^m} = \lim_{x \to a} \frac{(x - a)^{n-m} D(x)}{(x - a)^n}$$

$$= \lim_{x \to a} (x - a)^{n-m} \lim_{x \to a} \frac{D(x)}{(x - a)^n} = 0 \cdot 0 = 0,$$

and we have

$$\lim_{x \to a} \frac{D(x)}{(x - a)^m} = 0 \qquad \text{for} \qquad m = 0, 1, 2, \ldots, n. \tag{6}$$

To complete the proof, we note that

By (6) with $m = 0$

$$b_0 = D(a) = \lim_{x \to a} D(x) = 0.$$

Then

$$D(x) = b_1(x - a) + b_2(x - a)^2 + \cdots + b_n(x - a)^n,$$

so that

$$\frac{D(x)}{x - a} = b_1 + b_2(x - a) + \cdots + b_n(x - a)^{n-1}$$

and

By (6) with $m = 1$

$$b_1 = \lim_{x\to a} \frac{D(x)}{x - a} = 0.$$

Suppose we have shown that $b_0 = b_1 = \cdots = b_k = 0$, where $k < n$. Then

$$D(x) = b_{k+1}(x - a)^{k+1} + b_{k+2}(x - a)^{k+2} + \cdots + b_n(x - a)^n,$$

so that

$$\frac{D(x)}{(x - a)^{k+1}} = b_{k+1} + b_{k+2}(x - a) + \cdots + b_n(x - a)^{n-k-1}$$

and

By (6) with $m = k + 1 \leq n$

$$b_{k+1} = \lim_{x\to a} \frac{D(x)}{(x - a)^{k+1}} = 0.$$

This shows that $b_0 = b_1 = b_2 = \cdots = b_n = 0$, which means that $D(x) = 0$ for every x in $[a, b]$, so that $P_n(x) = Q_n(x)$ for every x in $[a, b]$. Thus the Taylor polynomial is unique. ■

We now prove three results that are very useful for applications.

Theorem 2 If f has $n + 1$ continuous derivatives, there exists a positive number M_n such that

$$|R_n(x)| \leq M_n \frac{|x - a|^{n+1}}{(n + 1)!}$$

for all x in $[a, b]$. Here M_n is an upper bound for the $(n + 1)$st derivative of f in the interval $[a, b]$.

Proof. Since $f^{(n+1)}$ is continuous on $[a, b]$, it is bounded above and below on that interval. That is, there is a number M_n such that $|f^{(n+1)}(x)| \leq M_n$ for every x in $[a, b]$. Since

$$R_n(x) = f^{(n+1)}(c) \frac{(x - a)^{n+1}}{(n + 1)!}$$

with c in (a, x), we see that

$$|R_n(x)| = |f^{(n+1)}(c)| \frac{(x-a)^{n+1}}{(n+1)!} \leq M_n \frac{(x-a)^{n+1}}{(n+1)!}$$

and the theorem is proved. ■

In the next section we will show how this bound on the magnitude of the remainder term can be used to find some very interesting approximations. Since many of these examples use the Taylor polynomial at 0, we restate the results of this section for that special case: If $f, f', \ldots, f^{(n+1)}$ are continuous in $[0, b]$, then for any x in $[0, b]$,

$$f(x) = f(0) + f'(0) + f''(0)\frac{x^2}{2!} + \cdots + f^{(n)}(0)\frac{x^n}{n!} + R_n(x) = \sum_{k=0}^{n} \frac{f^{(k)}(0)x^k}{k!} + R_n(x),$$

and there is a number M_n such that

$$|R_n(x)| \leq M_n \frac{|x|^{n+1}}{(n+1)!}.$$

We close this section by showing that if there is a number K such that $|f^{(n)}(x)| \leq K$ for every x in $[a, b]$ and for every positive integer n, then the remainder terms actually approach zero as n increases. First, we need the following result.

Theorem 3 Let x be any real number. Then $x^n/n! \to 0$ as $n \to \infty$.

Proof. Let N be an integer such that $N > 2|x|$. Then $|x| < N/2$, and for $n > N$ and $k > 0$,

$$\frac{|x|}{N+k} < \frac{N}{2(N+k)} = \frac{1}{2}\left(\frac{N}{N+k}\right) < \frac{1}{2}.$$

Then

$$\frac{x^n}{n!} \leq \frac{|x|^n}{n!} = \overbrace{\frac{|x| \cdot |x| \cdot \cdots \cdot |x|}{1 \cdot 2 \cdot 3 \cdot \cdots \cdot N}}^{N \text{ times}} \cdot \underbrace{\frac{|x|}{N+1} \cdot \frac{|x|}{N+2} \cdot \cdots \cdot \frac{|x|}{n!}}_{n-N \text{ terms, each of which is } <\frac{1}{2}} < \frac{|x|^N}{N!}\left(\frac{1}{2}\right)^{n-N}$$

$$= \frac{|x|^N}{N!} \cdot 2^N \cdot \left(\frac{1}{2}\right)^n.$$

Since x and N are fixed, we define $M = (|x|^N/N!) \cdot 2^N$, so that for each $n > N$, $0 < |x|^n/n! < M(\frac{1}{2})^n$, which approaches zero as $n \to \infty$. This shows that $x^n/n! \to 0$. ■

Theorem 4 Suppose that $f^{(n)}(x)$ is defined for all $n \geq 1$ and $|f^{(n)}(x)| \leq K$ for all $x \in [a, b]$. Then

$$|R_n(x)| \to 0 \quad \text{as} \quad n \to \infty.$$

Proof. We have

$$|R_n(x)| = \left| f^{(n+1)}(c)\frac{(x-a)^{n+1}}{(n+1)!} \right| = \left| f^{(n+1)}(c) \right| \frac{|x-a|^{n+1}}{(n+1)!}$$
$$\leq K\frac{|x-a|^{n+1}}{(n+1)!}.$$

But by Theorem 3 this last term approaches zero as $n \to \infty$. Thus the theorem is proved. ■

PROBLEMS 13.2

1. Show that $\frac{d}{dt}[f'(t)(x-t)] = -f'(t) + f''(t)(x-t)$.
2. Show that for $1 \leq k \leq n$,
$$\frac{d}{dt}\left[\frac{f^{(k)}(t)}{k!}(x-t)^k\right] = \frac{-f^{(k)}(t)}{(k-1)!}(x-t)^{k-1} + \frac{f^{(k+1)}(t)}{k!}(x-t)^k$$
3. Use the results of Problems 1 and 2 to show that
$$h'(t) = \frac{-f^{(n+1)}(t)}{n!}(x-t)^n + \frac{(n+1)R_n(x)(x-t)^n}{(x-a)^{n+1}}$$
where $h(t)$ is given by equation (3).

*4. Prove the following: Let f and its first $n-1$ derivatives exist in an open interval containing a, and suppose that $f^{(n)}(a)$ exists. Then $f(x) = P_n(x) + R(x)$, where $P_n(x)$ is the nth-degree Taylor polynomial centered at a and
$$\lim_{x \to a} \frac{R(x)}{(x-a)^n} = 0.$$

13.3 APPROXIMATION USING TAYLOR POLYNOMIALS

In this section we show how Taylor's formula can be used as a tool for making approximations. In many of the examples that follow, results have been obtained by making use of a hand calculator. If a hand calculator is available, we suggest that you use it to check the computations.

For convenience, we summarize the results of the preceding two sections.

Let $f, f', f'', \ldots, f^{(n+1)}$ be continuous on $[a, b]$. Then for any x in $[a, b]$,

$$f(x) = f(a) + f'(a)(x-a) + f''(a)\frac{(x-a)^2}{2!} + \cdots + f^{(n)}(a)\frac{(x-a)^n}{n!} + R_n(x)$$
$$= \sum_{k=0}^{n} \frac{f^{(k)}(a)(x-a)^k}{k!} + R_n(x), \tag{1}$$

where

$$R_n(x) = f^{(n+1)}(c)\frac{(x - a)^{n+1}}{(n + 1)!} \tag{2}$$

for some number c (which depends on x) in the interval (a, x). Moreover, there exists a positive number M_n such that for every x in $[a, x]$,

$$|R_n(x)| \leq M_n\frac{|x - a|^{n+1}}{(n + 1)!} \tag{3}$$

where M_n is the maximum value of $|f^{(n+1)}(x)|$ on the interval $[a, b]$.

We stress that (3) provides an upper bound for the error. In many cases the actual error [the difference $|f(x) - P_n(x)|$] will be considerably less than $M_n(x - a)^{n+1}/(n + 1)!$.

EXAMPLE 1 In Example 13.1.1, we found that the fifth-degree Taylor polynomial of $f(x) = \sin x$ at 0 is $P_5(x) = x - (x^3/3!) + (x^5/5!)$. We then have

$$\sin x = x - \frac{x^3}{3!} + \frac{x^5}{5!} + R_5(x), \tag{4}$$

where

$$R_5(x) = \frac{f^{(6)}(c)(x - 0)^6}{6!}.$$

But $f^{(6)}(c) = \sin^{(6)}(c) = -\sin c$ and $|-\sin c| \leq 1$. Thus for x in $[0, 1]$,

$$|R_5(x)| \leq \frac{1(x - 0)^6}{6!} = \frac{x^6}{720}.$$

For example, suppose we wish to calculate $\sin (\pi/10)$. From (4)

$$\sin\frac{\pi}{10} = \frac{\pi}{10} - \frac{\pi^3}{3!10^3} + \frac{\pi^5}{5!10^5} + R_n\left(\frac{\pi}{10}\right)$$

with

$$\left|R_5\left(\frac{\pi}{10}\right)\right| \leq \frac{(\pi/10)^6}{720} \approx 0.00000134.$$

We find that

$$\sin\frac{\pi}{10} \approx \frac{\pi}{10} - \frac{1}{3!}\frac{\pi^3}{10^3} + \frac{1}{5!}\frac{\pi^5}{10^5} \approx 0.3141593 - 0.0051677 + 0.0000255$$
$$= 0.3090171.$$

The actual value of $\sin \pi/10 = \sin 18° = 0.3090170$, correct to seven decimal places, so our actual error is 0.0000001, which is quite a bit less than 0.0000013. This illustrates the fact that the actual error (the value of the remainder term) is often quite a bit smaller than the theoretical upper bound on the error given by formula (3). ■

REMARK. In Example 1 the fifth-degree Taylor polynomial is also the sixth-degree Taylor polynomial [since $\sin^{(6)}(0) = -\sin 0 = 0$]. Thus we can use the error estimate for P_6. Since $|\sin^{(7)}(c)| = |-\cos c| \le 1$, we have

$$R_6(x) \le \frac{x^7}{7!} = \frac{x^7}{5040}.$$

If $x = \pi/10$, we obtain

$$\left|R_6\left(\frac{\pi}{10}\right)\right| \le \frac{(\pi/10)^7}{5040} \approx 0.0000000599.$$

Note that, to ten decimal places,

$$\frac{\pi}{10} - \frac{(\pi/10)^3}{3!} + \frac{(\pi/10)^5}{5!} = 0.3090170542,$$

and to ten decimal places,

$$\sin\frac{\pi}{10} = 0.3090169944,$$

with an actual error of 0.0000000598. Now we see that our estimate on the remainder term is really quite accurate.

EXAMPLE 2 Use a Taylor polynomial to estimate $e^{0.3}$ with an error of less than 0.0001.

Solution. For convenience, choose the interval [0, 1]. On [0, 1], e^x and all its derivatives have a maximum value of $e^1 = e$. Then for any n, if we use a Taylor polynomial at 0 of degree n, we have

$$|R_n(0.3)| \le e \cdot \frac{(0.3)^{n+1}}{(n+1)!}.$$

Since $e \approx 2.71828\ldots$, we use the bound $e < 2.72$. We must choose n so that $(e)(0.3)^{n+1}/(n+1)! < (2.72)(0.3)^{n+1}/(n+1)! < 0.0001$. For $n = 3$, $(2.72)(0.3)^4/4! \approx 0.00092$, while for $n = 4$, $(2.72)(0.3)^5/5! \approx 0.0000551 < 0.00006$. Thus we choose a fourth-degree Taylor polynomial for our approximation, and we know in advance that $|e^{0.3} - P_4(0.3)| < 0.00006$. We have

$$P_4(x) = 1 + x + \frac{x^2}{2!} + \frac{x^3}{3!} + \frac{x^4}{4!}$$

(see Example 13.1.3). Then

$$P_4(0.3) = 1 + 0.3 + \frac{(0.3)^2}{2!} + \frac{(0.3)^3}{3!} + \frac{(0.3)^4}{4!}$$

$$\approx 1 + 0.3 + 0.045 + 0.0045 + 0.00034 = 1.34984.$$

The actual value of $e^{0.3}$ is 1.34986, correct to five decimal places. Thus the error in our calculation is, approximately, 0.00002, one-third the calculated upper bound on the error. ■

EXAMPLE 3 Compute $\cos[(\pi/4) + 0.1]$ with an error of less than 0.00001.

Solution. In this problem it is clearly convenient to use the Taylor expansion at $\pi/4$. Since all derivatives of $\cos x$ are bounded by 1, we have, for any n,

$$|R_n(x)| \le \frac{[x - (\pi/4)]^{n+1}}{(n + 1)!},$$

so that $R_n[(\pi/4) + 0.1] \le (0.1)^{n+1}/(n + 1)!$. If $n = 2$, then $(0.1)^3/3! = 0.00017$, and for $n = 3$, $(0.1)^4/4! = 0.000004 < 0.00001$. Thus we need to calculate $P_3(x)$ at $\pi/4$. But

$$P_3(x) = \cos\frac{\pi}{4} - \left(\sin\frac{\pi}{4}\right)\left(x - \frac{\pi}{4}\right) - \left(\cos\frac{\pi}{4}\right)\frac{[x - (\pi/4)]^2}{2} + \left(\sin\frac{\pi}{4}\right)\frac{[x - (\pi/4)]^3}{6}$$

$$= \frac{1}{\sqrt{2}}\left\{1 - \left(x - \frac{\pi}{4}\right) - \frac{[x - (\pi/4)]^2}{2} + \frac{[x - (\pi/4)]^3}{6}\right\},$$

and for $x = (\pi/4) + 0.1$,

$$\cos\left(\frac{\pi}{4} + 0.1\right) \approx P_3\left(\frac{\pi}{4} + 0.1\right) = \frac{1}{\sqrt{2}}\left[1 - 0.1 - \frac{(0.1)^2}{2} + \frac{(0.1)^3}{6}\right] \approx 0.63298.$$

This is correct to five decimal places. ■

We now consider a more general example.

THE LOGARITHM [ln(1 + x)]

In Section 13.6 we will show that for any number $u \neq 1$, the formula for the sum of a **geometric progression** is given by

$$1 + u + u^2 + \cdots + u^n = \frac{1 - u^{n+1}}{1 - u} = \frac{1}{1 - u} - \frac{u^{n+1}}{1 - u}. \quad \textbf{(5)}$$

This leads to the expression

$$\frac{1}{1 - u} = 1 + u + u^2 + \cdots + u^n + \frac{u^{n+1}}{1 - u}. \quad \textbf{(6)}$$

Setting $u = -t$ in (6) gives us

$$\frac{1}{1+t} = 1 - t + t^2 - t^3 + \cdots + (-1)^n t^n + (-1)^{n+1}\frac{t^{n+1}}{1+t}. \tag{7}$$

Integration of both sides of (7) from 0 to x yields

$$\ln(1+x) = x - \frac{x^2}{2} + \frac{x^3}{3} - \cdots + (-1)^n \frac{x^{n+1}}{n+1} + (-1)^{n+1}\int_0^x \frac{t^{n+1}}{t+1}\,dt, \tag{8}$$

which is valid if $x > -1$ [since $\ln(1+x)$ is not defined for $x \le -1$].

Now let

$$R_{n+1}(x) = (-1)^{n+1}\int_0^x \frac{t^{n+1}}{t+1}\,dt.$$

We will show two things: First, we will prove that, for $-1 < x \le 1$,

$$\lim_{n\to\infty} R_{n+1}(x) = 0.$$

This will ensure that the polynomials given in (8) provide an increasingly good approximation to $\ln(1+x)$ as n increases. Second, we will show that for every $n \ge 1$,

$$\lim_{x\to 0} \frac{R_{n+1}(x)}{x^{n+1}} = 0.$$

Then, according to the uniqueness theorem (Theorem 13.2.1), we can conclude that the polynomial in (8) is *the* $(n+1)$st degree Taylor polynomial for $\ln(1+x)$ in the interval $-1 < x \le 1$.

There are two cases to consider.

Case 1. $0 \le x \le 1$. Then for t in $[0, 1]$, $1/(1+t) \le 1$, so that

$$|R_{n+1}(x)| = \int_0^x \frac{t^{n+1}}{t+1}\,dt \le \int_0^x t^{n+1}\,dt = \frac{x^{n+2}}{n+2} \le \frac{1}{n+2} \tag{9}$$

which approaches 0 as $n \to \infty$.

Moreover, from (9),

$$\lim_{x\to 0} \frac{|R_{n+1}(x)|}{x^{n+1}} \underset{x>0}{\le} \lim_{x\to 0+} \frac{x^{n+2}}{x^{n+1}(n+2)} \underset{\text{Remember, } n \text{ is fixed}}{=} \lim_{x\to 0+} \frac{x}{n+2} = 0.$$

Case 2. $-1 < x < 0$. First, we note that

$$(-1)^{n+1}\int_0^x \frac{t^{n+1}}{t+1}\,dt = (-1)^n \int_x^0 \frac{t^{n+1}}{t+1}\,dt.$$

In this latter integral t is in the interval $[x, 0]$, where $-1 < x < 0$ (see Figure 1). Then we have $1 \geq 1 + t \geq 1 + x > 0$, so that

$$1 \leq \frac{1}{1+t} \leq \frac{1}{1+x}$$

FIGURE 1

and

$$|R_{n+1}(x)| \leq \int_x^0 \frac{|t|^{n+1}}{|t+1|}\,dt \leq \int_x^0 \frac{(-t)^{n+1}}{(x+1)}\,dt \qquad \text{Since } |t| = -t \text{ in } [x, 0]$$

$$= \frac{1}{x+1}\int_x^0 (-t)^{n+1}\,dt = -\frac{1}{x+1}\frac{(-t)^{n+2}}{(n+2)}\bigg|_x^0 = \frac{(-x)^{n+2}}{(1+x)(n+2)}.$$

Thus

$$|R_{n+1}(x)| \leq \frac{(-x)^{n+2}}{(1+x)(n+2)} \leq \frac{1}{(1+x)(n+2)} \tag{10}$$

since $-1 < x < 0$ implies that $0 < -x < 1$, and we have $|R_n(x)| \to 0$ as $n \to \infty$. Moreover, from (10),

$$\lim_{x\to 0} \frac{|R_{n+1}(x)|}{x^{n+1}} \leq \lim_{x\to 0+} \frac{(-x)^{n+2}}{(-x)^{n+1}(n+2)} = \lim_{x\to 0+} \frac{-x}{n+2} = 0. \qquad (-x > 0;\ n \text{ is fixed})$$

We conclude that the Taylor polynomial

$$x - \frac{x^2}{2} + \frac{x^3}{3} - \cdots + (-1)^n \frac{x^{n+1}}{n+1}$$

is a good approximation to $\ln(1 + x)$ for sufficiently large n, provided that $-1 < x \leq 1$.

EXAMPLE 4 Calculate ln 1.4 with an error of less than 0.001.

Solution. Here $x = 0.4$, and from (9) we need to find an n such that $(0.4)^{n+2}/(n+2) < 0.001$. We have $(0.4)^5/5 \approx 0.00205$ and $(0.4)^6/6 \approx 0.00068$, so choosing $n = 4$ $(n + 2 = 6)$, we obtain the Taylor polynomial [from (8)]

$$P_5(x) = x - \frac{x^2}{2} + \frac{x^3}{3} - \frac{x^4}{4} + \frac{x^5}{5}$$

Remembering, from (8) the last term in $P_{n+1}(x)$ is $(-1)^n x^{n+1}/(n+1)$, we have

$$\ln 1.4 \approx P_5(0.4) = 0.4 - \frac{(0.4)^2}{2} + \frac{(0.4)^3}{3} - \frac{(0.4)^4}{4} + \frac{(0.4)^5}{5}$$

$$\approx 0.4 - 0.08 + 0.02133 - 0.0064 + 0.00205 = 0.33698.$$

The actual value of ln 1.4 is 0.33647, correct to five decimal places, so that the error is $0.33698 - 0.33647 = 0.00051$. This error is slightly less than the maximum possible error of 0.00068, and so in this case our error bound is fairly sharp. Note that the error bound (9) used in this problem is *not* the same as the error bound given by (3). To obtain the bound (3), we would have to compute the sixth derivative of $\ln(1 + x)$. ■

EXAMPLE 5 Calculate ln 0.9 with an error of less than 0.001.

Solution. Here we need to find $\ln(1 + x)$ for $x = -0.1$. Using (10), we have

$$|R_{n+1}(x)| \le \frac{(-x)^{n+2}}{(1 + x)(n + 2)} = \frac{(0.1)^{n+2}}{0.9(n + 2)}.$$

For $n = 1$, $(0.1)^3/(0.9)(3) = 0.00037$, so that we need only evaluate $P_{n+1}(-0.1) = P_2(-0.1) = (-0.1) - [(-0.1)^2/2] = -0.10500$. The value of ln 0.9, correct to five decimal places, is -0.10536, so our actual error of 0.00036 and our maximum possible error of 0.00037 almost coincide. ■

THE ARC TANGENT ($\tan^{-1} x$)

If in equation (6) we substitute $u = -t^2$, then for any number t we obtain

$$\frac{1}{1 + t^2} = 1 - t^2 + t^4 - t^6 + \cdots + (-1)^n t^{2n} + (-1)^{n+1} \frac{t^{2(n+1)}}{1 + t^2}. \tag{11}$$

Integration of both sides of (11) from 0 to x yields

$$\tan^{-1} x = x - \frac{x^3}{3} + \frac{x^5}{5} - \frac{x^7}{7} + \cdots + \frac{(-1)^n x^{2n+1}}{2n + 1} + (-1)^{n+1} \int_0^x \frac{t^{2(n+1)}}{1 + t^2}\, dt. \tag{12}$$

This equation is valid for any real number $x \ge 0$.

The polynomial in (12) is a polynomial of degree $2n + 1$. Since $1/(1 + t^2) \le 1$, we have

$$|R_{2n+1}(x)| = \int_0^x \frac{t^{2(n+1)}}{1 + t^2}\, dt \le \int_0^x t^{2(n+1)}\, dt = \frac{x^{2n+3}}{2n + 3}. \tag{13}$$

Thus

$$\lim_{x \to 0} \frac{|R_{2n+1}(x)|}{x^{2n+1}} = \lim_{x \to 0} \frac{x^2}{2n + 3} = 0,$$

so that again by the uniqueness theorem, the polynomial in (12) is the Taylor polynomial for $\tan^{-1} x$ at 0. We also see, from (13), that

$$R_{2n+1}(x) \to 0 \quad \text{as} \quad n \to \infty \quad \text{if} \quad |x| \le 1.$$

We conclude that the Taylor polynomial

$$x - \frac{x^3}{3} + \frac{x^5}{5} + \cdots + \frac{(-1)^n x^{2n+1}}{2n+1}$$

provides a good approximation to $\tan^{-1} x$ for sufficiently large n, provided that $|x| \le 1$.

Equation (12) gives us a formula for calculating π.[†] Setting $x = 1$, we have

$$\frac{\pi}{4} = \tan^{-1} 1 \approx 1 - \frac{1}{3} + \frac{1}{5} - \frac{1}{7} + \frac{1}{9} - \frac{1}{11} + \cdots + \frac{(-1)^n}{2n+1}.$$

The error here [given by (13)], is bounded by $1/(2n + 3)$, which approaches zero very slowly. To get an error less than 0.001, we would need $1/(2n + 3) < 1/1000$, or $2n + 3 > 1000$, and $n \ge 499$. For example, $1 - \frac{1}{3} + \frac{1}{5} - \frac{1}{7} + \frac{1}{9} - \frac{1}{11} + \frac{1}{13} - \frac{1}{15} = 0.75427$, while $\pi/4$ is 0.78540. A better way to approximate π is suggested in Problems 29 and 30.

PROBLEMS 13.3

In Problems 1–10 find a bound for $|R_n(x)|$ for x in the given interval, where $P_n(x)$ is a Taylor polynomial of degree n having terms of the form $(x - a)^k$.

1. $f(x) = \sin x;\ a = \pi/4;\ n = 6;\ x \in [0, \pi/2]$
2. $f(x) = \sqrt{x};\ a = 1;\ n = 4;\ x \in [\frac{1}{4}, 4]$
3. $f(x) = \dfrac{1}{x};\ a = 1;\ n = 4;\ x \in \left[\dfrac{1}{2}, 2\right]$
4. $f(x) = \tan x;\ a = 0;\ n = 4;\ x \in [0, \pi/4]$
5. $f(x) = \dfrac{1}{\sqrt{x}};\ a = 5;\ n = 5;\ x \in \left[\dfrac{19}{4}, \dfrac{21}{4}\right]$
6. $f(x) = \sinh x;\ a = 0;\ n = 4;\ x \in [0, 1]$
7. $f(x) = \ln \cos x;\ a = 0;\ n = 3;\ x \in [0, \pi/6]$
8. $f(x) = e^{\alpha x};\ a = 0;\ n = 4;\ x \in [0, 1]$
9. $f(x) = e^{x^2};\ a = 0;\ n = 4;\ x \in [0, \frac{1}{3}]$
10. $f(x) = \sin x^2;\ a = 0;\ n = 4;\ x \in [0, \pi/4]$

In Problems 11–26 use a Taylor polynomial to estimate the given number with the given degree of accuracy.

11. $\sin\left(\dfrac{\pi}{6} + 0.2\right)$; error < 0.001
12. $\sin 33°$; error < 0.001 [*Hint:* Convert to radians.]
13. $\tan\left(\dfrac{\pi}{4} + 0.1\right)$; error < 0.01
14. e; error < 0.0001 [*Hint:* You may assume that $2 < e < 3$.]
15. e^{-1}; error < 0.001
16. $\ln 2$; error < 0.1
17. $\ln 1.5$; error < 0.001
18. $\ln 0.5$; error < 0.0001
19. e^3; error < 0.01 [*Hint:* See Problem 14.]
20. $\tan^{-1} 0.5$; error < 0.001

*21. $\sinh \frac{1}{2}$; error < 0.01

*22. $\cosh \frac{1}{2}$; error < 0.01

[†]This formula was discovered by the Scottish mathematician James Gregory (1638–1675) and was first published in 1712.

23. sin 100°; error < 0.001

24. cos 195°; error < 0.001

25. $\dfrac{1}{\sqrt{1.1}}$; error < 0.001

***26.** ln cos 0.3; error < 0.01

27. Use the result of Problem 9 to estimate $\int_0^{1/3} e^{x^2}\,dx$. What is the maximum error of your estimate?

28. Use the result of Problem 7 to estimate $\int_0^{\pi/6} \ln \cos x\,dx$. What is the maximum error of your estimate?

***29.** Use the formula $\tan(A \pm B) = (\tan A \pm \tan B)/(1 \mp \tan A \tan B)$ to prove that $\tan(4\tan^{-1} 1/5 - \tan^{-1} 1/239) = 1$. [*Hint:* First calculate $\tan 2(\tan^{-1} 1/5)$ and $\tan(4\tan^{-1} 1/5)$.] This implies that $4\tan^{-1} 1/5 - \tan^{-1} 1/239 = \pi/4$.

***30.** Use the result of Problem 29 to calculate π to five decimal places.

31. Let $f(x) = (1 + x)^n$, where n is a positive integer.
(a) Show that $f^{(n+1)}(x) = 0$.
(b) Show that

$$f(x) = 1 + nx + \frac{n(n-1)}{2!}x^2 + \frac{n(n-1)(n-2)}{3!}x^3 + \cdots + x^n.$$

This result is called the **binomial theorem.** (Also see Appendix 2.)

32. Use the binomial theorem to calculate **(a)** $(1.2)^4$, **(b)** $(0.8)^5$, **(c)** $(1.03)^4$.

33. Let $f(x) = (1 + x)^\alpha$, where α is any real number.
(a) Show that

$$f(x) = 1 + \alpha x + \frac{\alpha(\alpha-1)}{2!}x^2 + \frac{\alpha(\alpha-1)(\alpha-2)}{3!}x^3 + \cdots + \frac{\alpha(\alpha-1)(\alpha-2)\cdots(\alpha-n+1)}{n!}x^n + R_n(x).$$

****(b)** Show that if $|x| < 1$, then $R_n(x) \to 0$ as $n \to \infty$. This result is called the **general binomial expansion**. If $x = 1$, the result is true for $\alpha > -1$. If $x = -1$, it is true for $\alpha > 0$.

34. Use the result of Problem 33 to calculate the following numbers to four decimal places of accuracy.
(a) $\sqrt{1.2}$ **(b)** $(0.9)^{3/4}$
(c) $(1.8)^{1/4}$ **(d)** $\dfrac{1}{\sqrt[3]{1.01}}$
(e) $2^{5/3}$ **(f)** $(0.4)^{1.6}$

13.4 SEQUENCES OF REAL NUMBERS

According to a popular dictionary,[†] a *sequence* is "the following of one thing after another." In mathematics we could define a sequence intuitively as a succession of numbers that never terminates. The numbers in the sequence are called the *terms* of the sequence. In a sequence there is one term for each positive integer.

EXAMPLE 1 Consider the following sequence:

1st term, 2nd term, 3rd term, 4th term, 5th term, nth term

$$\frac{1}{2}, \frac{1}{4}, \frac{1}{8}, \frac{1}{16}, \frac{1}{32}, \ldots, \frac{1}{2^n}, \ldots$$

We see that there is one term for each positive integer. The terms in this sequence form an infinite set of real numbers, which we write as

$$A = \left\{\frac{1}{2}, \frac{1}{4}, \frac{1}{8}, \ldots, \frac{1}{2^n}, \ldots\right\}. \quad \textbf{(1)}$$

That is, the set A consists of all numbers of the form $1/2^n$, where n is a positive integer.

[†]*The Random House Dictionary* (Ballantine Books, New York, 1978).

There is another way to describe this set. We define the function f by the rule $f(n) = 1/2^n$, where the domain of f is the set of positive integers. Then the set A is precisely the set of values taken by the function f. ■

In general, we have the following formal definition:

Definition 1 SEQUENCE A **sequence** of real numbers is a function whose domain is the set of positive integers. The values taken by the function are called **terms** of the sequence.

NOTATION: We will often denote the terms of a sequence by a_n. Thus if the function given in Definition 1 is f, then $a_n = f(n)$. With this notation, *we can denote the set of values taken by the sequence by* $\{a_n\}$. Also, we will use n, m, and so on as integer variables and x, y, and so on as real variables.

EXAMPLE 2 The following are sequences of real numbers:

(a) $\{a_n\} = \left\{\frac{1}{n}\right\}$ **(b)** $\{a_n\} = \{\sqrt{n}\}$ **(c)** $\{a_n\} = \left\{\frac{1}{n!}\right\}$

(d) $\{a_n\} = \{\sin n\}$ **(e)** $\{a_n\} = \left\{\frac{e^n}{n!}\right\}$ **(f)** $\{a_n\} = \left\{\frac{n-1}{n}\right\}$ ■

We sometimes denote a sequence by writing out the values $\{a_1, a_2, a_3, \ldots\}$.

EXAMPLE 3 We write out the values of the sequences in Example 2:

(a) $\left\{1, \frac{1}{2}, \frac{1}{3}, \frac{1}{4}, \ldots, \frac{1}{n}, \ldots\right\}$

(b) $\{1, \sqrt{2}, \sqrt{3}, \sqrt{4}, \ldots, \sqrt{n}, \ldots\}$

(c) $\left\{1, \frac{1}{2}, \frac{1}{6}, \frac{1}{24}, \ldots, \frac{1}{n!}, \ldots\right\}$

(d) $\{\sin 1, \sin 2, \sin 3, \sin 4, \ldots, \sin n, \ldots\}$

(e) $\left\{e, \frac{e^2}{2}, \frac{e^3}{6}, \frac{e^4}{24}, \ldots, \frac{e^n}{n!}, \ldots\right\}$

(f) $\left\{0, \frac{1}{2}, \frac{2}{3}, \frac{3}{4}, \ldots, \frac{n-1}{n}, \ldots\right\}$ ■

Because a sequence is a function, it has a graph. In Figure 1 we draw part of the graphs of four of the sequences in Example 3.

EXAMPLE 4 Find the general term a_n of the sequence

$$\{-1, 1, -1, 1, -1, 1, -1, \ldots\}.$$

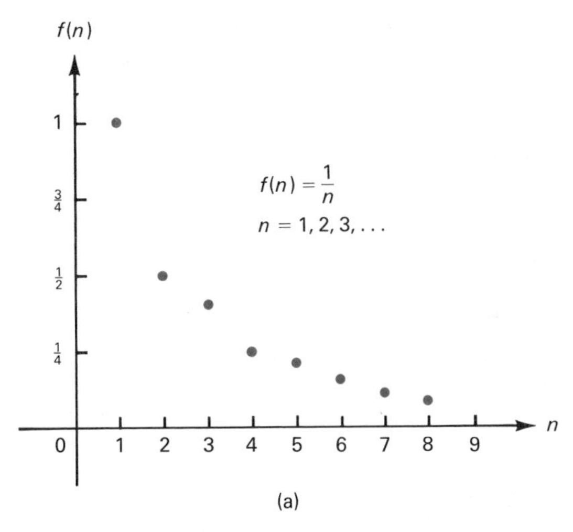

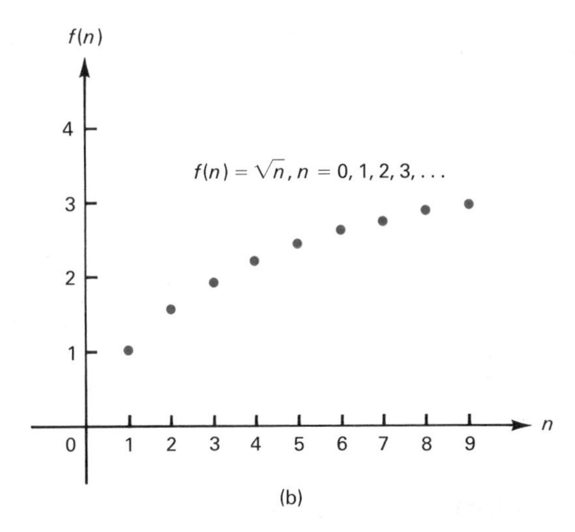

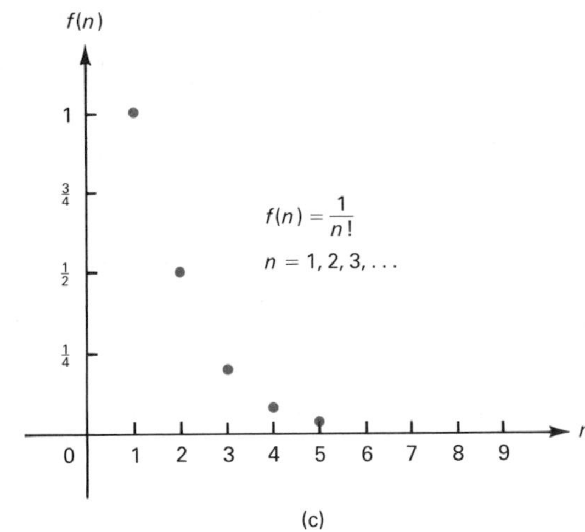

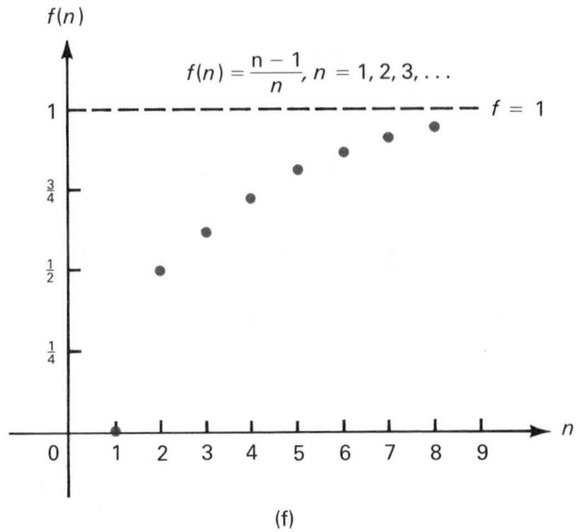

FIGURE 1

Solution. We see that $a_1 = -1$, $a_2 = 1$, $a_3 = -1$, $a_4 = 1, \ldots$. Hence

$$a_n = \begin{cases} -1 & \text{if } n \text{ is odd} \\ 1, & \text{if } n \text{ is even.} \end{cases}$$

A more concise way to write this term is

$$a_n = (-1)^n.$$

We draw the graph of this sequence in Figure 2. ■

It is evident that as n gets large, the numbers $1/n$ get small. We can write

$$\lim_{n\to\infty} \frac{1}{n} = 0.$$

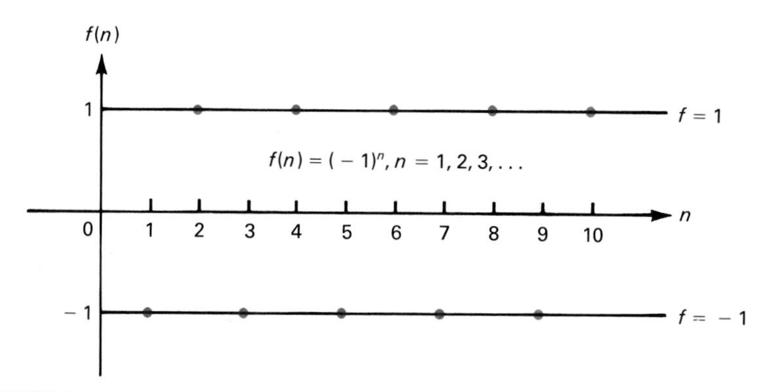

FIGURE 2

This is also suggested by the graph in Figure 1a. Similarly, it is not hard to show that as n gets large, $(n - 1)/n$ gets close to 1. We write

$$\lim_{n\to\infty} \frac{n-1}{n} = 1.$$

This is illustrated in Figure 1f.

On the other hand, it is clear that $a_n = (-1)^n$ does not get close to any one number as n increases. It simply oscillates back and forth between the numbers $+1$ and -1. This is illustrated in Figure 2.

For the remainder of this section we will be concerned with calculating the limit of a sequence as $n\to\infty$. Since a sequence is a special type of function, our formal definition of the limit of a sequence is going to be very similar to the definition of $\lim_{x\to\infty} f(x)$.

Definition 2 FINITE LIMIT OF A SEQUENCE A sequence $\{a_n\}$[†] has the limit L if for every $\epsilon > 0$ there exists an integer $N > 0$ such that if $n \geq N$, then $|a_n - L| < \epsilon$. We write

$$\lim_{n\to\infty} a_n = L. \tag{2}$$

Intuitively, this definition states that $a_n \to L$ if as n increases without bound, a_n gets arbitrarily close to L. We illustrate this definition in Figure 3.

Definition 3 INFINITE LIMIT OF A SEQUENCE The sequence $\{a_n\}$ has the limit ∞ if for every positive number M there is an integer $N > 0$ such that if $n > N$, then $a_n > M$. In this case we write

$$\lim_{n\to\infty} a_n = \infty.$$

[†]To be precise, $\{a_n\}$ denotes the set of values taken by the sequence. There is a difference between the sequence, which is a function f, and the values $a_n = f(n)$ taken by this function. However, because it is more convenient to write down the values the sequence takes, we will, from now on, use the notation $\{a_n\}$ to denote a sequence.

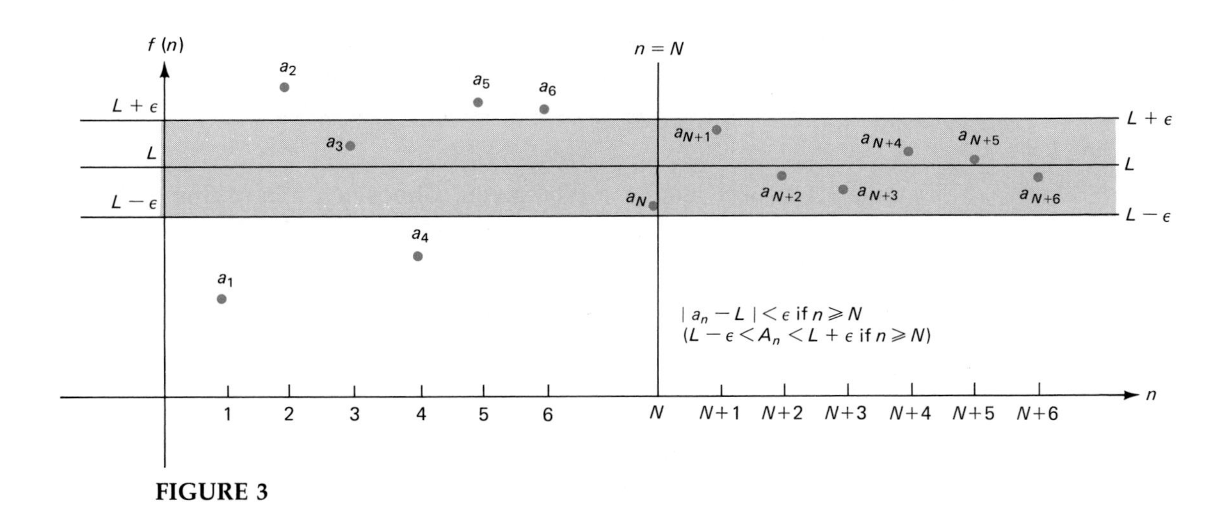

FIGURE 3

Intuitively, $\lim_{n\to\infty} a_n = \infty$ means that as n increases without bound, a_n also increases without bound.

The theorem below gives us a very useful result.

Theorem 1 Let r be a real number. Then

$$\lim_{n\to\infty} r^n = 0 \quad \text{if} \quad |r| < 1$$

and

$$\lim_{n\to\infty} |r^n| = \infty \quad \text{if} \quad |r| > 1.$$

Proof.

Case 1. $r = 0$. Then $r^n = 0$, and the sequence has the limit 0.

Case 2. $0 < |r| < 1$. For a given $\epsilon > 0$, choose N such that

$$N > \frac{\ln \epsilon}{\ln|r|}.$$

Note that since $|r| < 1$, $\ln|r| < 0$. Now if $n > N$,

$$n > \frac{\ln \epsilon}{\ln|r|} \quad \text{and} \quad n \ln|r| < \ln \epsilon.$$

The second inequality follows from the fact that $\ln|r|$ is negative, and multiplying both sides of an inequality by a negative number reverses the inequality. Thus

$$\ln|r^n - 0| = \ln|r^n| = \ln|r|^n = n \ln|r| .< \ln \epsilon.$$

Since $\ln|r^n - 0| < \ln \epsilon$ and $\ln x$ is an increasing function, we conclude that $|r^n - 0| < \epsilon$. Thus according to the definition of a finite limit of a sequence,

$$\lim_{n\to\infty} r^n = 0.$$

Case 3. $|r| > 1$. Let $M > 0$ be given. Choose $N > \ln M/\ln|r|$. Then if $n > N$,

$$\ln|r^n| = n\ln|r| \overset{n>N}{>} N\ln|r| > \left(\frac{\ln M}{\ln|r|}\right)(\ln|r|) = \ln M,$$

so that

$$|r^n| > M \qquad \text{if} \qquad n > N.$$

From the definition of an infinite limit of a sequence, we see that

$$\lim_{n\to\infty} |r^n| = \infty. \blacksquare$$

In your study of one variable calculus you saw a number of limit theorems. All those results can be applied when n, rather than x, approaches infinity. The only difference is that as n grows, it takes on integer values. For convenience, we state without proof the major limit theorems we need for sequences.

Theorem 2 Suppose that $\lim_{n\to\infty} a_n$ and $\lim_{n\to\infty} b_n$ both exist and are finite.

(i) $\lim_{n\to\infty} \alpha a_n = \alpha \lim_{n\to\infty} a_n$ for any number α. **(3)**
(ii) $\lim_{n\to\infty}(a_n + b_n) = \lim_{n\to\infty} a_n + \lim_{n\to\infty} b_n$. **(4)**
(iii) $\lim_{n\to\infty} a_n b_n = (\lim_{n\to\infty} a_n)(\lim_{n\to\infty} b_n)$. **(5)**
(iv) If $\lim_{n\to\infty} b_n \neq 0$, then

$$\lim_{n\to\infty} \frac{a_n}{b_n} = \frac{\lim_{n\to\infty} a_n}{\lim_{n\to\infty} b_n}. \tag{6}$$

Theorem 3 CONTINUITY THEOREM Suppose that L is finite and $\lim_{n\to\infty} a_n = L$. If f is continuous in an open interval containing L, then

$$\lim_{n\to\infty} f(a_n) = f(\lim_{n\to\infty} a_n) = f(L). \tag{7}$$

Theorem 4 SQUEEZING THEOREM Suppose that $\lim_{n\to\infty} a_n = \lim_{n\to\infty} b_n = L$ and that $\{c_n\}$ is a sequence having the property that for $n > N$ (a positive integer), $a_n \le c_n \le b_n$. Then

$$\lim_{n\to\infty} c_n = L. \tag{8}$$

We now give a central definition in the theory of sequences.

Definition 4 CONVERGENCE AND DIVERGENCE OF A SEQUENCE If the limit in (2) exists and if L is finite, we say that the sequence **converges** or is **convergent**. Otherwise, we say that the sequence **diverges** or is **divergent**.

EXAMPLE 5 The sequence $\{1/2^n\}$ is convergent since, by Theorem 1, $\lim_{n\to\infty} 1/2^n = \lim_{n\to\infty}(1/2)^n = 0$. ■

EXAMPLE 6 The sequence $\{r^n\}$ is divergent for $r > 1$ since $\lim_{n\to\infty} r^n = \infty$ if $r > 1$. ■

EXAMPLE 7 The sequence $\{(-1)^n\}$ is divergent since the values a_n alternate between -1 and $+1$ but do not stay close to any fixed number as n becomes large. ■

Since we have a large body of theory and experience behind us in the calculation of ordinary limits, we would like to make use of that experience to calculate limits of sequences. The following theorem, whose proof is left as a problem (see Problem 34), is extremely useful.

Theorem 5 Suppose that $\lim_{x\to\infty} f(x) = L$, a finite number, ∞, or $-\infty$. If f is defined for every positive integer, then the limit of the sequence $\{a_n\} = \{f(n)\}$ is also equal to L. That is, $\lim_{x\to\infty} f(x) = \lim_{n\to\infty} a_n = L$.

EXAMPLE 8 Calculate $\lim_{n\to\infty} 1/n^2$.

Solution. Since $\lim_{x\to\infty} 1/x^2 = 0$, we have $\lim_{n\to\infty} 1/n^2 = 0$ (by Theorem 5). ■

EXAMPLE 9 Does the sequence $\{e^n/n\}$ converge or diverge?

Solution. Since $\lim_{x\to\infty} e^x/x = \lim_{x\to\infty} e^x/1$ (by L'Hôpital's rule) $= \infty$, we find that the sequence diverges. ■

REMARK. It should be emphasized that Theorem 5 does *not* say that if $\lim_{x\to\infty} f(x)$ does not exist, then $\{a_n\} = \{f(n)\}$ diverges. For example, let

$$f(x) = \sin \pi x.$$

Then $\lim_{x\to\infty} f(x)$ does not exist, but $\lim_{n\to\infty} f(n) = \lim_{n\to\infty} \sin \pi n = 0$ since $\sin \pi n = 0$ for every integer n.

EXAMPLE 10 Let $\{a_n\} = \{[1 + (1/n)]^n\}$. Does this sequence converge or diverge?

Solution. Since $\lim_{x\to\infty}[1 + (1/x)]^x = e$, we see that a_n converges to the limit e. ■

EXAMPLE 11 Determine the convergence or divergence of the sequence $\{(\ln n)/n\}$.

Solution. $\lim_{x\to\infty}[(\ln x)/x] = \lim_{x\to\infty}[(1/x)/1] = 0$ by L'Hôpital's rule, so that the sequence converges to 0. ■

EXAMPLE 12 Let $p(x) = c_0 + c_1x + \cdots + c_mx^m$ and $q(x) = d_0 + d_1x + \cdots + d_rx^r$. It is not difficult to show that if the rational function $r(x) = p(x)/q(x)$, then, if $c_md_r \neq 0$,

$$\lim_{x\to\infty} \frac{p(x)}{q(x)} = \begin{cases} 0 & \text{if } m < r \\ \pm\infty & \text{if } m > r \\ \dfrac{c_m}{d_r} & \text{if } m = r. \end{cases}$$

Thus the sequence $\{p(n)/q(n)\}$ converges to 0 if $m < r$, converges to c_m/d_r if $m = r$, and diverges if $m > r$. ■

EXAMPLE 13 Does the sequence $\{(5n^3 + 2n^2 + 1)/(2n^3 + 3n + 4)$ converge or diverge?

Solution. Here $m = r = 3$, so that by the result of Example 12 the sequence converges to $c_3/d_3 = \frac{5}{2}$. ■

EXAMPLE 14 Does the sequence $\{n^{1/n}\}$ converge or diverge?

Solution. Since $\lim_{x\to\infty} x^{1/x} = 1$,[†] the sequence converges to 1. ■

EXAMPLE 15 Determine the convergence or divergence of the sequence $\{\sin \alpha n/n^\beta\}$, where α is a real number and $\beta > 0$.

Solution. Since $-1 \le \sin \alpha x \le 1$, we see that

$$-\frac{1}{x^\beta} \le \frac{\sin \alpha x}{x^\beta} \le \frac{1}{x^\beta} \qquad \text{for any } x > 0.$$

But $\pm\lim_{x\to\infty} 1/x^\beta = 0$, and so by the squeezing theorem $\lim_{x\to\infty}[(\sin \alpha x)/x^\beta] = 0$. Therefore the sequence $\{(\sin \alpha n)/n^\beta\}$ converges to 0. ■

As in Example 15, the squeezing theorem can often be used to calculate the limit of a sequence.

PROBLEMS 13.4

In Problems 1–9 find the first five terms of the given sequence.

1. $\left\{\dfrac{1}{3^n}\right\}$
2. $\left\{\dfrac{n+1}{n}\right\}$
3. $\left\{1 - \dfrac{1}{4^n}\right\}$
4. $\{\sqrt[3]{n}\}$
5. $\{e^{1/n}\}$
6. $\{n \cos n\}$
7. $\{\sin n\pi\}$
8. $\{\cos n\pi\}$
9. $\left\{\sin \dfrac{n\pi}{2}\right\}$

[†]If $y = x^{1/x}$, then $\ln y = \dfrac{1}{x} \ln x = \dfrac{\ln x}{x}$ and, by L'Hôpital's rule,

$$\lim_{x\to\infty} \frac{\ln x}{x} = \lim_{x\to\infty} \frac{1/x}{1} = 0$$

so $\ln y \to 0$ and $y \to e^0 = 1$ as $x \to \infty$.

In Problems 10–27 determine whether the given sequence is convergent or divergent. If it is convergent, find its limit.

10. $\left\{\frac{3}{n}\right\}$

11. $\left\{\frac{1}{\sqrt{n}}\right\}$

12. $\left\{\frac{n+1}{n^{5/2}}\right\}$

13. $\{\sin n\}$

14. $\{\sin n\pi\}$

15. $\left\{\cos\left(n+\frac{\pi}{2}\right)\right\}$

16. $\left\{\frac{n^5+3n^2+1}{n^6+4n}\right\}$

17. $\left\{\frac{4n^5-3}{7n^5+n^2+2}\right\}$

18. $\left\{\left(1+\frac{4}{n}\right)^n\right\}$

19. $\left\{\left(1+\frac{1}{4n}\right)^n\right\}$

20. $\left\{\frac{\sqrt{n}}{\ln n}\right\}$

21. $\{\sqrt{n+3}-\sqrt{n}\}$ [*Hint:* Multiply and divide by $\sqrt{n+3}+\sqrt{n}$.]

22. $\left\{\frac{2^n}{n!}\right\}$

23. $\left\{\frac{\alpha^n}{n!}\right\}$ (α real)

24. $\left\{\frac{4}{\sqrt{n^2+3}-n}\right\}$

25. $\left\{\frac{(-1)^n n^3}{n^3+1}\right\}$

26. $\{(-1)^n \cos n\pi\}$

27. $\left\{\frac{(-1)^n}{\sqrt{n}}\right\}$

In Problems 28–33 find the general term a_n of the given sequence.

28. $\{1, -2, 3, -4, 5, -6, \ldots\}$

29. $\{1, 2\cdot 5, 3\cdot 5^2, 4\cdot 5^3, 5\cdot 5^4, \ldots\}$

30. $\{\frac{1}{2}, \frac{2}{3}, \frac{3}{4}, \frac{4}{5}, \frac{5}{6}, \ldots\}$

31. $\{\frac{1}{2}, \frac{3}{4}, \frac{7}{8}, \frac{15}{16}, \frac{31}{32}, \ldots\}$

***32.** $\{\frac{1}{3}, \frac{2}{5}, \frac{3}{7}, \frac{4}{9}, \frac{5}{11}, \ldots\}$

33. $\{1, -\frac{1}{3}, \frac{1}{9}, -\frac{1}{27}, \ldots\}$

***34.** Prove Theorem 5.

35. Show that if $\{a_n\}$ and $\{b_n\}$ are two sequences such that $|a_n| \le |b_n|$ for each n, then if $|b_n|$ converges to 0, $|a_n|$ also converges to 0. [*Hint:* Use the squeezing theorem.]

36. Use the result of Problem 35 to show that the sequence $\{(a \sin bn + c \cos dn)/n^{p^2}\}$ converges to 0 for any real numbers a, b, c, d, and $p \ne 0$.

37. Prove that if $|r| < 1$, then the sequence $\{nr^n\}$ converges to 0.

***38.** Show that if $\{a_n\}$ converges, then the $\lim_{n\to\infty} a_n$ is unique. [*Hint:* Assume that $\lim_{n\to\infty} a_n = L$, $\lim_{n\to\infty} a_n = M$, and $L \ne M$. Then choose $\epsilon = \frac{1}{2}|L-M|$ to show that Definition 2 is violated.]

***39.** Suppose that $\{a_n\}$ is a sequence such that $a_{n+1} = a_n - a_n^2$, where a_0 is given. Find all values of a_0 for which $\lim_{n\to\infty} a_n = 0$.

****40.** Suppose that $a_{n+1} = 1/(2 + a_n)$. For what choices of a_0 does the sequence $\{a_k\}$ diverge?

***41.** Prove or disprove: If $\lim_{n\to\infty} n^p a_n = 0$, then there exists an $\epsilon > 0$ such that $\lim_{n\to\infty} n^{p+\epsilon} a_n$ also exists (i.e., is finite).

42. Show that $a_n = [1 + (\alpha/n)]^n$ converges to e^α as $n \to \infty$.

13.5 BOUNDED AND MONOTONIC SEQUENCES

There are certain kinds of sequences that have special properties worthy of mention.

Definition 1 BOUNDEDNESS

(i) The sequence $\{a_n\}$ is **bounded above** if there is a number M_1 such that

$$a_n \le M_1 \tag{1}$$

for every positive integer n.

(ii) It is **bounded below** if there is a number M_2 such that

$$M_2 \le a_n \tag{2}$$

for every positive integer n.

(iii) It is **bounded** if there is a number $M > 0$ such that

$$|a_n| \le M$$

for every positive integer n.

The numbers M_1, M_2, and M are called, respectively, an **upper bound**, a **lower bound**, and a **bound** for $\{a_n\}$.

(iv) If the sequence is not bounded, it is called **unbounded**.

REMARK. If $\{a_n\}$ is bounded above and below, then it is bounded. Simply set $M = \max\{|M_1|, |M_2|\}$.

EXAMPLE 1 The sequence $\{\sin n\}$ has the upper bound of 1, the lower bound of -1, and the bound of 1 since $-1 \le \sin n \le 1$ for every n. Of course, any number greater than 1 is also a bound. ■

EXAMPLE 2 The sequence $\{(-1)^n\}$ has the upper bound 1, the lower bound -1, and the bound 1. ■

EXAMPLE 3 The sequence $\{2^n\}$ is bounded below by 2 but has no upper bound and so is unbounded. ■

EXAMPLE 4 The sequence $\{(-1)^n 2^n\}$ is bounded neither below nor above. ■

It turns out that the following statement is true:

Every convergent sequence is bounded.

Theorem 1 If the sequence $\{a_n\}$ is convergent, then it is bounded.

Proof. Before giving the technical details, we remark that the idea behind the proof is easy. For if $\lim_{n\to\infty} a_n = L$, then a_n is close to the finite number L if n is large. Thus, for example, $|a_n| \le |L| + 1$ if n is large enough. Since a_n is a real number for every n, the first few terms of the sequence are bounded, and these two facts give us a bound for the entire sequence.

Now to the details: Let $\epsilon = 1$. Then there is an $N > 0$ such that (according to Definition 13.4.2)

$$|a_n - L| < 1 \quad \text{if} \quad n \ge N. \tag{3}$$

Let

$$K = \max\{|a_1|, |a_2|, \ldots, |a_N|\}. \tag{4}$$

Since each a_n is finite, K, being the maximum of a finite number of terms, is also finite. Now let

$$M = \max\{|L| + 1, K\}. \tag{5}$$

It follows from (4) that if $n \le N$, then $|a_n| \le K$. If $n \ge N$, then from (3), $|a_n| < |L| + 1$; so in either case $|a_n| \le M$, and the theorem is proved. ■

Sometimes it is difficult to find a bound for a convergent sequence.

EXAMPLE 5 Find an M such that $5^n/n! \le M$.

Solution. We know from Theorem 13.2.3 that $\lim_{n\to\infty} x^n/n! = 0$ for every real number x. In particular, $\{5^n/n!\}$ is convergent and therefore must be bounded. Perhaps the easiest way to find the bound is to tabulate a few values, as in Table 1. It is clear from the table that the maximum value of a_n occurs at $n = 4$ or $n = 5$ and is equal to 26.04. Of course, any number larger than 26.04 is also a bound for the sequence. ■

TABLE 1

n	$\frac{5^n}{n!}$	n	$\frac{5^n}{n!}$
1	5	7	15.5
2	12.5	8	9.69
3	20.83	9	5.38
4	26.04	10	2.69
5	26.04	20	0.000039
6	21.7		

Theorem 1 is useful in another way. Since every convergent sequence is bounded, it follows that:

Every unbounded sequence is divergent.

EXAMPLE 6 The following sequences are divergent since they are unbounded:

(a) $\{\ln \ln n\}$ (starting at $n = 2$) **(b)** $\{n \sin n\}$ **(c)** $\{(-\sqrt{2})^n\}$ ■

The converse of Theorem 1 is *not* true. That is, it is not true that every bounded sequence is convergent. For example, the sequences $\{(-1)^n\}$ and $\{\sin n\}$ are both bounded *and* divergent. Since boundedness alone does not ensure convergence, we need some other property. We investigate this idea now.

Definition 2 MONOTONICITY

(i) The sequence $\{a_n\}$ is **monotone increasing** if $a_n \le a_{n+1}$ for every $n \ge 1$.
(ii) The sequence $\{a_n\}$ is **monotone decreasing** if $a_n \ge a_{n+1}$ for every $n \ge 1$.
(iii) The sequence $\{a_n\}$ is **monotonic** if it is either monotone increasing or monotone decreasing.

Definition 3 STRICT MONOTONICITY

(i) The sequence $\{a_n\}$ is **strictly increasing** if $a_n < a_{n+1}$ for every $n \ge 1$.
(ii) The sequence $\{a_n\}$ is **strictly decreasing** if $a_n > a_{n+1}$ for every $n \ge 1$.
(iii) The sequence $\{a_n\}$ is **strictly monotonic** if it is either strictly increasing or strictly decreasing.

EXAMPLE 7 The sequence $\{1/2^n\}$ is strictly decreasing since $1/2^n > 1/2^{n+1}$ for every n. ■

EXAMPLE 8 Determine whether the sequence $\{2n/(3n + 2)\}$ is increasing, decreasing, or not monotonic.

Solution. If we write out the first few terms of the sequence, we find that $\{2n/(3n + 2)\} = \{\frac{2}{5}, \frac{4}{8}, \frac{6}{11}, \frac{8}{14}, \frac{10}{17}, \frac{12}{20}, \ldots\}$. Since these terms are strictly increasing, we suspect that $\{2n/(3n + 2)\}$ is an increasing sequence. To check this, we try to verify that $a_n < a_{n+1}$. We have

$$a_{n+1} = \frac{2(n + 1)}{3(n + 1) + 2} = \frac{2n + 2}{3n + 5}.$$

Then if $a_n < a_{n+1}$, we would have

$$\frac{2n}{3n + 2} < \frac{2n + 2}{3n + 5}.$$

Multiplying both sides of this inequality by $(3n + 2)(3n + 5)$, we obtain

$$(2n)(3n + 5) < (2n + 2)(3n + 2), \qquad \text{or} \qquad 6n^2 + 10n < 6n^2 + 10n + 4.$$

Since this last inequality is obviously true for all $n \geq 1$, we can reverse our steps to conclude that $a_n < a_{n+1}$, and the sequence is strictly increasing. ■

EXAMPLE 9 Determine whether the sequence $\{(\ln n)/n\}$, $n > 1$, is increasing, decreasing, or not monotonic.

Solution. Let $f(x) = (\ln x)/x$. Then $f'(x) = [x(1/x) - (\ln x)1]/x^2 = (1 - \ln x)/x^2$. If $x > e$, then $\ln x > 1$ and $f'(x) < 0$. Thus the sequence $\{(\ln n)/n\}$ is decreasing for $n \geq 3$. However, $(\ln 1)/1 = 0 < (\ln 2)/2 \approx 0.35$, so initially, the sequence is increasing. Thus the sequence is not monotonic. It is decreasing if we start with $n = 3$. ■

EXAMPLE 10 The sequence $\{[n/4]\}$ is increasing but not strictly increasing. Here $[x]$ is the "greatest integer" function. The first twelve terms are 0, 0, 0, 1, 1, 1, 1, 2, 2, 2, 2, 3. For example, $a_9 = [\frac{9}{4}] = 2$. ■

EXAMPLE 11 The sequence $\{(-1)^n\}$ is not monotonic since successive terms oscillate between $+1$ and -1. ■

In all the examples we have given, a divergent sequence diverges for one of two reasons: It goes to infinity (it is unbounded) or it oscillates [like $(-1)^n$, which oscillates between -1 and 1]. But if a sequence is bounded, it does not go to infinity. And if it is monotone, it does not oscillate. Thus the following theorem should not be surprising:

Theorem 2

A bounded monotonic sequence is convergent.

Proof. We will prove this theorem for the case in which the sequence $\{a_n\}$ is increasing. The proof of the other case is similar. Since $\{a_n\}$ is bounded, there is a number M such that $a_n \le M$ for every n. Let L be the smallest such upper bound. Now let $\epsilon > 0$ be given. Then there is a number $N > 0$ such that $a_N > L - \epsilon$. If this were not true, then we would have $a_n \le L - \epsilon$ for all $n \ge 1$. Then $L - \epsilon$ would be an upper bound for $\{a_n\}$, and since $L - \epsilon < L$, this would contradict the choice of L as the smallest such upper bound. Since $\{a_n\}$ is increasing, we have, for $n \ge N$,

$$L - \epsilon < a_N \le a_n \le L < L + \epsilon. \tag{6}$$

But the inequalities in (6) imply that $|a_n - L| \le \epsilon$ for $n \ge N$, which proves, according to the definition of convergence, that $\lim_{n\to\infty} a_n = L$. ■

The number L is called the **least upper bound** for the sequence $\{a_n\}$. It is an axiom of the real number system that every set of real numbers that is bounded above has a least upper bound and that every set of real numbers that is bounded below has a **greatest lower bound**. This axiom is called the **completeness axiom** and is of paramount importance in theoretical mathematical analysis.

We have actually proved a stronger result. Namely, that *if the sequence $\{a_n\}$ is bounded above and increasing, then it converges to its least upper bound. Similarly, if $\{a_n\}$ is bounded below and decreasing, then it converges to its greatest lower bound.*

EXAMPLE 12 In Example 8 we saw that the sequence $\{2n/(3n + 2)\}$ is strictly increasing. Also, since $2n/(3n + 2) < 3n/(3n + 2) < 3n/3n = 1$, we see that $\{a_n\}$ is bounded, so that by Theorem 2, $\{a_n\}$ is convergent. We find that $\lim_{n\to\infty} 2n/(3n + 2) = \frac{2}{3}$. ■

PROBLEMS 13.5

In Problems 1–12 determine whether the given sequence is bounded or unbounded. If it is bounded, find the smallest bound for $|a_n|$.

1. $\left\{\frac{1}{n+1}\right\}$
2. $\{\sin n\pi\}$
3. $\{\cos n\pi\}$
4. $\{\sqrt{n}\sin n\}$
5. $\left\{\frac{2^n}{1+2^n}\right\}$
6. $\left\{\frac{2^n+1}{2^n}\right\}$
7. $\left\{\frac{1}{n!}\right\}$
8. $\left\{\frac{3^n}{n!}\right\}$
9. $\left\{\frac{n^2}{n!}\right\}$
10. $\left\{\frac{2n}{2^n}\right\}$
11. $\left\{\frac{\ln n}{n}\right\}$
*12. $\{ne^{-n}\}$

13. Show that for $n > 5$, $n^{10}/n! > (n + 1)^{10}/(n + 1)!$, and use this result to conclude that $\{n^{10}/n!\}$ is bounded.

In Problems 14–28 determine whether the given sequence is monotone increasing, strictly increasing, monotone decreasing, strictly decreasing, or not monotonic.

14. $\{\sin n\pi\}$
15. $\left\{\frac{3^n}{2+3^n}\right\}$
16. $\left\{\left(\frac{n}{25}\right)^{1/3}\right\}$
17. $\{n + (-1)^n\sqrt{n}\}$
18. $\left\{\frac{\sqrt{n+1}}{n}\right\}$
19. $\left\{\frac{n!}{n^n}\right\}$
20. $\left\{\frac{n^n}{n!}\right\}$

21. $\left\{\dfrac{2n!}{1\cdot 3\cdot 5\cdot 7\cdot\cdots\cdot(2n-1)}\right\}$

***22.** $\{n + \cos n\}$

23. $\left\{\dfrac{2^{2n}}{n!}\right\}$

24. $\left\{\dfrac{\sqrt{n}-1}{n}\right\}$

25. $\left\{\dfrac{n-1}{n+1}\right\}$

26. $\left\{\ln\left(\dfrac{3n}{n+1}\right)\right\}$

27. $\{\ln n - \ln(n+2)\}$

28. $\left\{\left(1+\dfrac{3}{n}\right)^{1/n}\right\}$

***29.** Show that the sequence $\{(2^n + 3^n)^{1/n}\}$ is convergent.

***30.** Show that $\{(a^n + b^n)^{1/n}\}$ is convergent for any positive real numbers a and b. [*Hint:* First do Problem 29. Then treat the cases $a = b$ and $a \neq b$ separately.]

31. Prove that the sequence $\{n!/n^n\}$ converges.

32. Use Theorem 2 to show that $\{\ln n - \ln(n+4)\}$ converges.

33. Show that the sequence of Problem 21 is convergent.

34. Show that the sequence $\{2\cdot 5\cdot 8\cdot 11\cdot\cdots\cdot(3n-1)/3^n n!\}$ is convergent.

13.6 GEOMETRIC SERIES

Consider the sum

$$S_7 = 1 + 2 + 4 + 8 + 16 + 32 + 64 + 128.$$

This can be written as

$$S_7 = 1 + 2 + 2^2 + 2^3 + 2^4 + 2^5 + 2^6 + 2^7 = \sum_{k=0}^{7} 2^k.$$

GEOMETRIC PROGRESSION
The sum of a **geometric progression** is a sum of the form

$$S_n = 1 + r + r^2 + r^3 + \cdots + r^{n-1} + r^n = \sum_{k=0}^{n} r^k, \quad (1)$$

where r is a real number and n is a fixed positive integer.

We now obtain a formula for the sum in (1).

Theorem 1 If $r \neq 1$, the sum of a geometric progression (1) is given by

$$S_n = \frac{1 - r^{n+1}}{1 - r}. \quad (2)$$

Proof. We write

$$S_n = 1 + r + r^2 + r^3 + \cdots + r^{n-1} + r^n \quad (3)$$

and multiply both sides of (3) by r:

$$rS_n = r + r^2 + r^3 + r^4 + \cdots + r^n + r^{n+1}. \quad \textbf{(4)}$$

Subtracting (4) from (3), all terms except the first and the last cancel:

$$S_n - rS_n = 1 - r^{n+1},$$

or

$$(1 - r)S_n = 1 - r^{n+1}. \quad \textbf{(5)}$$

Finally, we divide both sides of (5) by $1 - r$ (which is nonzero) to obtain equation (2). ■

NOTE. If $r = 1$, we obtain

$$S_n = \overbrace{1 + 1 + \cdots + 1}^{n+1 \text{ terms}} = n + 1.$$

EXAMPLE 1 Calculate $S_7 = 1 + 2 + 4 + 8 + 16 + 32 + 64 + 128$, using formula (2).

Solution. Here $r = 2$ and $n = 7$, so that

$$S_7 = \frac{1 - 2^8}{1 - 2} = 2^8 - 1 = 256 - 1 = 255. \ ■$$

EXAMPLE 2 Calculate $\Sigma_{k=0}^{10} (\frac{1}{2})^k$.

Solution. Here $r = \frac{1}{2}$ and $n = 10$, so that

$$S_{10} = \frac{1 - (\frac{1}{2})^{11}}{1 - \frac{1}{2}} = \frac{1 - \frac{1}{2048}}{\frac{1}{2}} = 2\left(\frac{2047}{2048}\right) = \frac{2047}{1024}. \ ■$$

EXAMPLE 3 Calculate

$$S_6 = 1 - \frac{2}{3} + \left(\frac{2}{3}\right)^2 - \left(\frac{2}{3}\right)^3 + \left(\frac{2}{3}\right)^4 - \left(\frac{2}{3}\right)^5 + \left(\frac{2}{3}\right)^6 = \sum_{k=0}^{6} \left(-\frac{2}{3}\right)^k.$$

Solution. Here $r = -\frac{2}{3}$ and $n = 6$, so that

$$S_6 = \frac{1 - (-\frac{2}{3})^7}{1 - (-\frac{2}{3})} = \frac{1 + \frac{128}{2187}}{\frac{5}{3}} = \frac{3}{5}\left(\frac{2315}{2187}\right) = \frac{463}{729}. \ ■$$

EXAMPLE 4 Calculate the sum $1 + b^2 + b^4 + b^6 + \cdots + b^{20} = \Sigma_{k=0}^{10} b^{2k}$ for $b \neq \pm 1$.

Solution. The sum can be written $1 + b^2 + (b^2)^2 + (b^2)^3 + \cdots + (b^2)^{10}$. Here $r = b^2 \neq 1$ and $n = 10$, so that

$$S_{10} = \frac{1 - (b^2)^{11}}{1 - b^2} = \frac{b^{22} - 1}{b^2 - 1}. \ ■$$

The sum of a geometric progression is the sum of a finite number of terms. We now see what happens if the number of terms is infinite. Consider the sum

$$S = 1 + \frac{1}{2} + \frac{1}{4} + \frac{1}{8} + \frac{1}{16} + \cdots = \sum_{k=0}^{\infty} \left(\frac{1}{2}\right)^k. \tag{6}$$

What can such a sum mean? We will give a formal definition in a moment. For now, let us show why it is reasonable to say that $S = 2$. Let $S_n = \sum_{k=0}^{n} (\frac{1}{2})^k = 1 + \frac{1}{2} + \frac{1}{4} + \cdots + (\frac{1}{2})^n$. Then

$$S_n = \frac{1 - (\frac{1}{2})^{n+1}}{1 - \frac{1}{2}} = 2\left[1 - \left(\frac{1}{2}\right)^{n+1}\right].$$

Thus for any n (no matter how large), $1 \le S_n < 2$. Hence the numbers S_n are bounded. Also, since $S_{n+1} = S_n + (\frac{1}{2})^{n+1} > S_n$, the numbers S_n are monotone increasing. Thus the sequence $\{S_n\}$ converges. But

$$S = \lim_{n\to\infty} S_n.$$

Thus S has a finite sum. To compute it, we note that

$$S = \lim_{n\to\infty} S_n = \lim_{n\to\infty} 2[1 - (\tfrac{1}{2})^{n+1}] = 2 \lim_{n\to\infty} [1 - (\tfrac{1}{2})^{n+1}] = 2$$

since $\lim_{n\to\infty} (\frac{1}{2})^{n+1} = 0$.

GEOMETRIC SERIES

The infinite sum $\sum_{k=0}^{\infty} (\frac{1}{2})^k$ is an example of a *geometric series*. A **geometric series** is an infinite sum of the form

$$S = \sum_{k=0}^{\infty} r^k = 1 + r + r^2 + r^3 + \cdots. \tag{7}$$

CONVERGENCE AND DIVERGENCE OF A GEOMETRIC SERIES

Let $S_n = \sum_{k=0}^{n} r^k$. Then we say that the geometric series **converges** if $\lim_{n\to\infty} S_n$ exists and is finite. Otherwise, the series is said to **diverge**.

EXAMPLE 5 Let $r = 1$. Then

$$S_n = \sum_{k=0}^{n} 1^k = \sum_{k=0}^{n} 1 = \underbrace{1 + 1 + \cdots + 1}_{n+1 \text{ times}} = n + 1.$$

Since $\lim_{n\to\infty}(n + 1) = \infty$, the series $\sum_{k=0}^{\infty} 1^k$ diverges. ■

EXAMPLE 6 Let $r = -2$. Then

$$S_n = \sum_{k=0}^{n} (-2)^k = \frac{1 - (-2)^{n+1}}{1 - (-2)} = \frac{1}{3}[1 - (-2)^{n+1}].$$

But $(-2)^{n+1} = (-1)^{n+1}(2^{n+1}) = \pm 2^{n+1}$. As $n \to \infty$, $2^{n+1} \to \infty$. Thus the series $\Sigma_{k=0}^{\infty}(-2)^{k+1}$ diverges. ■

Theorem 2

Let $S = \Sigma_{k=0}^{\infty} r^k$ be a geometric series.

(i) The series converges to

$$\frac{1}{1-r} \quad \text{if} \quad |r| < 1.$$

(ii) The series diverges if $|r| \geq 1$.

Proof. **(i)** If $|r| < 1$, then $\lim_{n\to\infty} r^{n+1} = 0$. Thus

$$S = \lim_{n\to\infty} S_n = \lim_{n\to\infty} \frac{1 - r^{n+1}}{1-r} = \frac{1}{1-r} \lim_{n\to\infty}(1 - r^{n+1})$$
$$= \frac{1}{1-r}(1-0) = \frac{1}{1-r}.$$

(ii) If $|r| > 1$, then $\lim_{n\to\infty} |r|^{n+1} = \infty$. Thus $1 - r^{n+1}$ does not have a finite limit and the series diverges. Finally, if $r = 1$, then the series diverges, by Example 5, and if $r = -1$, then S_n alternates between the numbers 0 and 1, so that the series diverges. ■

EXAMPLE 7 $1 - \frac{2}{3} + (\frac{2}{3})^2 - \cdots = \Sigma_{k=0}^{\infty}(-\frac{2}{3})^k = 1/[1 - (\frac{2}{3})] = 1/(\frac{5}{3}) = \frac{3}{5}$. ■

EXAMPLE 8

$$1 + \frac{\pi}{4} + \left(\frac{\pi}{4}\right)^2 + \left(\frac{\pi}{4}\right)^3 + \cdots = \sum_{k=0}^{\infty}\left(\frac{\pi}{4}\right)^k = \frac{1}{1-(\pi/4)}$$
$$= \frac{4}{4-\pi} \approx 4.66. \ ■$$

PROBLEMS 13.6

In Problems 1–11 calculate the sum of the given geometric progression.

1. $1 + 3 + 9 + 27 + 81 + 243$
2. $1 + \frac{1}{4} + \frac{1}{16} + \cdots + \frac{1}{4^8}$
3. $1 - 5 + 25 - 125 + 625 - 3125$
4. $0.2 + 0.2^2 + 0.2^3 + \cdots + 0.2^9$
5. $0.3^2 - 0.3^3 + 0.3^4 - 0.3^5 + 0.3^6 - 0.3^7 + 0.3^8$
6. $1 + b^3 + b^6 + b^9 + b^{12} + b^{15} + b^{18} + b^{21}$
7. $1 - \frac{1}{b^2} + \frac{1}{b^4} - \frac{1}{b^6} + \frac{1}{b^8} - \frac{1}{b^{10}} + \frac{1}{b^{12}} - \frac{1}{b^{14}}$
8. $\pi - \pi^3 + \pi^5 - \pi^7 + \pi^9 - \pi^{11} + \pi^{13}$
9. $1 + \sqrt{2} + 2 + 2^{3/2} + 4 + 2^{5/2} + 8 + 2^{7/2} + 16$

10. $1 - \frac{1}{\sqrt{3}} + \frac{1}{3} - \frac{1}{3\sqrt{3}} + \frac{1}{9} - \frac{1}{9\sqrt{3}} + \frac{1}{27} - \frac{1}{27\sqrt{3}} + \frac{1}{81}$

11. $-16 + 64 - 256 + 1024 - 4096$

12. A bacteria population initially contains 1000 organisms, and each bacterium produces two live bacteria every 2 hr. How many organisms will be alive after 12 hr if none of the bacteria dies during the growth period?

In Problems 13–22 calculate the sum of the given geometric series.

13. $1 + \frac{1}{4} + \frac{1}{4^2} + \frac{1}{4^3} + \cdots$

14. $1 - \frac{1}{2} + \frac{1}{4} - \frac{1}{8} + \frac{1}{16} - \cdots$

15. $1 + \frac{1}{10} + \frac{1}{100} + \frac{1}{1000} + \cdots$

16. $1 - \frac{1}{10} + \frac{1}{100} - \frac{1}{1000} + \cdots$

17. $1 + \frac{1}{\pi} + \frac{1}{\pi^2} + \frac{1}{\pi^3} + \cdots$

18. $1 + 0.7 + 0.7^2 + 0.7^3 + \cdots$

19. $1 - 0.62 + 0.62^2 - 0.62^3 + 0.62^4 - \cdots$

20. $\frac{1}{4} + \frac{1}{16} + \frac{1}{64} + \cdots$ [*Hint:* Factor out the term $\frac{1}{4}$.]

21. $\frac{3}{5} - \frac{3}{25} + \frac{3}{125} - \cdots$

22. $\frac{1}{9} + \frac{1}{27} + \frac{1}{81} + \cdots$

23. How large must n be in order that $(\frac{1}{2})^n < 0.01$?

24. How large must n be in order that $(0.8)^n < 0.01$?

25. How large must n be in order that $(0.99)^n < 0.01$?

26. If $x > 1$, show that

$$1 + \frac{1}{x} + \frac{1}{x^2} + \frac{1}{x^3} + \cdots = \frac{x}{x - 1}.$$

13.7 INFINITE SERIES

In Section 13.6 we defined the geometric series $\Sigma_{k=0}^{\infty} r^k$ and showed that if $|r| < 1$, the series converges to $1/(1 - r)$. Let us again look at what we did. If S_n denotes the sum of the first $n + 1$ terms of the geometric series, then

$$S_n = 1 + r + r^2 + \cdots + r^n = \frac{1 - r^{n+1}}{1 - r} \quad \text{for} \quad r \neq 1. \tag{1}$$

For each n we obtain the number S_n, and therefore we can define a new sequence $\{S_n\}$ to be the sequence of **partial sums** of the geometric series. If $|r| < 1$, then

$$\lim_{n\to\infty} S_n = \lim_{n\to\infty} \frac{1 - r^{n+1}}{1 - r} = \frac{1}{1 - r}.$$

That is, the convergence of the geometric series is implied by the convergence of the sequence of partial sums $\{S_n\}$.

We now give a more general definition of these concepts.

Definition 1 INFINITE SERIES

Let $\{a_n\}$ be a sequence. The infinite sum

$$\sum_{k=1}^{\infty} a_k = a_1 + a_2 + a_3 + \cdots + a_n + \cdots \tag{2}$$

is called an **infinite series** (or, simply, **series**). Each a_k in (2) is called a **term** of the series. The nth **partial sums** of the series are given by

$$S_n = \sum_{k=1}^{n} a_k.$$

The term S_n is called the **nth partial sum** of the series. If the sequence of partial sums $\{S_n\}$ converges to L, then we say that the infinite series $\Sigma_{k=1}^{\infty} a_k$ **converges** to L, and we write

$$\sum_{k=1}^{\infty} a_k = L. \quad \textbf{(3)}$$

Otherwise, we say that the series $\Sigma_{k=1}^{\infty} a_k$ **diverges**.

REMARK. Occasionally a series will be written with the first term other than a_1. For example, $\Sigma_{k=0}^{\infty} (\frac{1}{2})^k$ and $\Sigma_{k=2}^{\infty} 1/(\ln k)$ are both examples of infinite series. In the second case we must start with $k = 2$ since $1/(\ln 1)$ is not defined.

EXAMPLE 1 We can write the number $\frac{1}{3}$ as

$$\frac{1}{3} = 0.33333\ldots = \frac{3}{10} + \frac{3}{100} + \frac{3}{1000} + \cdots + \frac{3}{10^n} + \cdots. \quad \textbf{(4)}$$

This expression is an infinite series. Here $a_n = 3/10^n$ and

$$S_n = \frac{3}{10} + \frac{3}{100} + \cdots + \frac{3}{10^n} = \overbrace{0.333\ldots3}^{n \text{ places}}.$$

We can formally prove that this sum converges by noting that

$$S = \frac{3}{10}\left(1 + \frac{1}{10} + \frac{1}{100} + \cdots\right) = \frac{3}{10}\sum_{k=0}^{\infty}\left(\frac{1}{10}\right)^k$$

By Theorem 13.6.2

$$= \frac{3}{10}\left[\frac{1}{1-(\frac{1}{10})}\right] = \frac{3}{10}\left(\frac{1}{\frac{9}{10}}\right) = \frac{3}{10}\cdot\frac{10}{9} = \frac{3}{9} = \frac{1}{3}. \blacksquare$$

As a matter of fact, any decimal number x can be thought of as a convergent infinite series, for if $x = 0.\,a_1a_2a_3\ldots a_n\ldots$, then

$$x = \frac{a_1}{10} + \frac{a_2}{100} + \frac{a_3}{1000} + \cdots + \frac{a_n}{10^n} + \cdots = \sum_{k=1}^{\infty}\frac{a_k}{10^k}.^{\dagger}$$

---

†Since $0 \le a_k < 10$,

$$\sum_{k=1}^{\infty}\frac{a_k}{10^k} < \sum_{k=1}^{\infty}\frac{10}{10^k} = \sum_{k=1}^{\infty}\frac{1}{10^{k-1}} = 1 + \frac{1}{10} + \left(\frac{1}{10}\right)^2 + \cdots = \frac{1}{1-\frac{1}{10}} = \frac{10}{9}.$$

In Section 13.8 we will prove the comparison test. Once we have this test, the inequality given above implies that $\Sigma_{k=1}^{\infty} (a_k/10^k)$ converges.

EXAMPLE 2 Express the **repeating decimal** 0.123123123 . . . as a rational number (the quotient of two integers).

Solution.

$$\begin{aligned}0.123123123\ldots &= 0.123 + 0.000123 + 0.000000123 + \cdots \\ &= \frac{123}{10^3} + \frac{123}{10^6} + \frac{123}{10^9} + \cdots = \frac{123}{10^3}\left[1 + \frac{1}{10^3} + \frac{1}{(10^3)^2} + \cdots\right] \\ &= \frac{123}{1000}\sum_{k=0}^{\infty}\left(\frac{1}{1000}\right)^k = \frac{123}{1000}\left[\frac{1}{1-(1/1000)}\right] = \frac{123}{1000}\cdot\frac{1}{999/1000} \\ &= \frac{123}{1000}\cdot\frac{1000}{999} = \frac{123}{999} = \frac{41}{333}. \quad ■\end{aligned}$$

In general, we can use the geometric series to write any repeating decimal in the form of a fraction by using the technique of Example 1 or 2. In fact, *the rational numbers are exactly those real numbers that can be written as repeating decimals.* Repeating decimals include numbers like $3 = 3.00000\ldots$ and $\frac{1}{4} = 0.25 = 0.25000000\ldots$.

EXAMPLE 3 **(Telescoping Series)** Consider the infinite series $\Sigma_{k=1}^{\infty} 1/k(k+1)$. We write the first three partial sums:

$$S_1 = \sum_{k=1}^{1}\frac{1}{k(k+1)} = \frac{1}{1\cdot 2} = \frac{1}{2} = 1 - \frac{1}{2},$$

$$S_2 = \sum_{k=1}^{2}\frac{1}{k(k+1)} = \frac{1}{1\cdot 2} + \frac{1}{2\cdot 3} = \frac{1}{2} + \frac{1}{6} = \frac{2}{3} = 1 - \frac{1}{3},$$

$$S_3 = \sum_{k=1}^{3}\frac{1}{k(k+1)} = \frac{1}{1\cdot 2} + \frac{1}{2\cdot 3} + \frac{1}{3\cdot 4} = \frac{1}{2} + \frac{1}{6} + \frac{1}{12} = \frac{3}{4} = 1 - \frac{1}{4}.$$

We can use partial fractions to rewrite the general term as

$$a_k = \frac{1}{k(k+1)} = \frac{1}{k} - \frac{1}{k+1},$$

from which we can get a better view of the nth partial sum:

$$\begin{aligned}S_n &= \left(\frac{1}{1} - \frac{1}{2}\right) + \left(\frac{1}{2} - \frac{1}{3}\right) + \left(\frac{1}{3} - \frac{1}{4}\right) + \cdots + \left(\frac{1}{n-1} - \frac{1}{n}\right) + \left(\frac{1}{n} - \frac{1}{n+1}\right) \\ &= 1 - \frac{1}{n+1}\end{aligned}$$

because all other terms cancel. Since $\lim_{n\to\infty} S_n = \lim_{n\to\infty}\{1 - [1/(n-1)]\} = 1$, we see that

$$\sum_{k=1}^{\infty} \frac{1}{k(k+1)} = 1.$$

When, as here, alternate terms cancel, we say that the series is a **telescoping series**. ■

REMARK. *Often, it is* not *possible to calculate the exact sum of an infinite series, even if it can be shown that the series converges.*

EXAMPLE 4 Consider the series

$$\sum_{k=1}^{\infty} \frac{1}{k} = 1 + \frac{1}{2} + \frac{1}{3} + \frac{1}{4} + \cdots + \frac{1}{n} + \cdots. \tag{5}$$

This series is called the **harmonic series**. Although $a_n = 1/n \to 0$ as $n \to \infty$, it is not difficult to show that the harmonic series diverges. To see this, we write

$$\sum_{k=1}^{\infty} \frac{1}{k} = 1 + \frac{1}{2} + \underbrace{\overbrace{\left(\frac{1}{3} + \frac{1}{4}\right)}^{2 \text{ terms}}}_{> \frac{1}{2}} + \underbrace{\overbrace{\left(\frac{1}{5} + \frac{1}{6} + \frac{1}{7} + \frac{1}{8}\right)}^{4 \text{ terms}}}_{> \frac{1}{2}} + \underbrace{\overbrace{\left(\frac{1}{9} + \cdots + \frac{1}{16}\right)}^{8 \text{ terms}}}_{> \frac{1}{2}} + \cdots.$$

Here we have written the terms in groups containing 2^n numbers. Note that $\frac{1}{3} + \frac{1}{4} > \frac{2}{4} = \frac{1}{2}$, $\frac{1}{5} + \frac{1}{6} + \frac{1}{7} + \frac{1}{8} > \frac{1}{8} + \frac{1}{8} + \frac{1}{8} + \frac{1}{8} = \frac{1}{2}$, and so on. Thus $\Sigma_{k=1}^{\infty} 1/k > 1 + \frac{1}{2} + \frac{1}{2} + \cdots$, and the series diverges. ■

■ WARNING: Example 4 clearly shows that even though the sequence $\{a_n\}$ converges to 0, the series Σa_n may, in fact, diverge. That is, if $a_n \to 0$, then $\Sigma_{k=1}^{\infty} a_k$ may or may not converge. Some additional test is needed to determine convergence or divergence.

It is often difficult to determine whether a series converges or diverges. For that reason a number of techniques have been developed to make it easier to do so. We present some easy facts here, and then we will develop additional techniques in the three sections that follow.

Theorem 1 Let c be a constant. Suppose that $\Sigma_{k=1}^{\infty} a_k$ and $\Sigma_{k=1}^{\infty} b_k$ both converge. Then $\Sigma_{k=1}^{\infty} (a_k + b_k)$ and $\Sigma_{k=1}^{\infty} ca_k$ converge, and

(i) $$\sum_{k=1}^{\infty} (a_k + b_k) = \sum_{k=1}^{\infty} a_k + \sum_{k=1}^{\infty} b_k, \tag{6}$$

(ii) $$\sum_{k=1}^{\infty} ca_k = c \sum_{k=1}^{\infty} a_k. \tag{7}$$

This theorem should not be surprising. Since the sum in a series is the limit of a sequence (the sequence of partial sums), the first part, for example, simply restates the fact that the limit of the sum is the sum of the limits. That is Theorem 13.4.2(ii).

Proof.

(i) Let $S = \sum_{k=1}^{\infty} a_k$ and $T = \sum_{k=1}^{\infty} b_k$. The partial sums are given by $S_n = \sum_{k=1}^{n} a_k$ and $T_n = \sum_{k=1}^{n} b_k$. Then

$$\sum_{k=1}^{\infty} (a_k + b_k) = \lim_{n\to\infty} \sum_{k=1}^{n} (a_k + b_k) = \lim_{n\to\infty} \left(\sum_{k=1}^{n} a_k + \sum_{k=1}^{n} b_k\right) = \lim_{n\to\infty} (S_n + T_n)$$

$$= \lim_{n\to\infty} S_n + \lim_{n\to\infty} T_n = S + T = \sum_{k=1}^{\infty} a_k + \sum_{k=1}^{\infty} b_k.$$

(ii) $$\sum_{k=1}^{\infty} ca_k = \lim_{n\to\infty} \sum_{k=1}^{n} ca_k = \lim_{n\to\infty} c \sum_{k=1}^{n} a_k = \lim_{n\to\infty} cS_n$$

$$= c \lim_{n\to\infty} S_n = cS = c \sum_{k=1}^{\infty} a_k. \blacksquare$$

EXAMPLE 5 Show that $\sum_{k=1}^{\infty} \{[1/k(k+1)] + (\frac{5}{6})^k\}$ converges.

Solution. This follows since $\sum_{k=1}^{\infty} 1/k(k+1)$ converges (Example 3) and $\sum_{k=1}^{\infty} (\frac{5}{6})^k$ converges because $\sum_{k=1}^{\infty} (\frac{5}{6})^k = \sum_{k=0}^{\infty} (\frac{5}{6})^k - (\frac{5}{6})^0$ [we added and subtracted the term $(\frac{5}{6})^0 = 1$] $= 1/(1 - \frac{5}{6}) - 1 = 5$. $\blacksquare$

EXAMPLE 6 Does $\sum_{k=1}^{\infty} 1/50k$ converge or diverge?

Solution. We show that the series diverges by assuming that it converges to obtain a contradiction. If $\sum_{k=1}^{\infty} 1/50k$ did converge, then $50 \sum_{k=1}^{\infty} 1/50k$ would also converge by Theorem 1. But then $50 \sum_{k=1}^{\infty} 1/50k = \sum_{k=1}^{\infty} 50 \cdot 1/50k = \sum_{k=1}^{\infty} 1/k$, and this series is the harmonic series, which we know diverges. Hence $\sum_{k=1}^{\infty} 1/50k$ diverges. $\blacksquare$

Another useful test is given by the following theorem and corollary.

Theorem 2 If $\sum_{k=1}^{\infty} a_k$ converges, then $\lim_{n\to\infty} a_n = 0$.

Proof. Let $S = \sum_{k=1}^{\infty} a_k$. Then the partial sums S_n and S_{n-1} are given by

$$S_n = \sum_{k=1}^{n} a_k = a_1 + a_2 + \cdots + a_{n-1} + a_n$$

and

$$S_{n-1} = \sum_{k=1}^{n-1} a_k = a_1 + a_2 + \cdots + a_{n-1},$$

so that

$$S_n - S_{n-1} = a_n.$$

Then

$$\lim_{n\to\infty} a_n = \lim_{n\to\infty} (S_n - S_{n-1}) = \lim_{n\to\infty} S_n - \lim_{n\to\infty} S_{n-1} = S - S = 0. \blacksquare$$

We have already seen that the converse of this theorem is false. The convergence of $\{a_n\}$ to 0 does *not* imply that $\Sigma_{k=1}^{\infty} a_k$ converges. For example, the harmonic series does not converge, but the sequence $\{1/n\}$ does converge to zero.

Corollary

> If $\{a_n\}$ does not converge to 0, then $\Sigma_{k=1}^{\infty} a_k$ diverges.

EXAMPLE 7 $\Sigma_{k=1}^{\infty} (-1)^k$ diverges since the sequence $\{(-1)^k\}$ does not converge to zero. ■

EXAMPLE 8 $\Sigma_{k=1}^{\infty} k/(k+100)$ diverges since $\lim_{n\to\infty} a_n = \lim_{n\to\infty} n/(n+100) = 1 \neq 0$. ■

PROBLEMS 13.7

In Problems 1–15 a convergent infinite series is given. Find its sum.

1. $\displaystyle\sum_{k=0}^{\infty} \frac{1}{4^k}$

2. $\displaystyle\sum_{k=0}^{\infty} \left(-\frac{2}{3}\right)^k$

3. $\displaystyle\sum_{k=2}^{\infty} \frac{1}{2^k}$

4. $\displaystyle\sum_{k=1}^{\infty} \frac{1}{2^{k-1}}$

5. $\displaystyle\sum_{k=-3}^{\infty} \frac{1}{2^{k+3}}$

6. $\displaystyle\sum_{k=3}^{\infty} \left(\frac{2}{3}\right)^k$

7. $\displaystyle\sum_{k=0}^{\infty} \frac{100}{5^k}$

8. $\displaystyle\sum_{k=0}^{\infty} \frac{5}{100^k}$

9. $\displaystyle\sum_{k=2}^{\infty} \frac{1}{k(k+1)}$

10. $\displaystyle\sum_{k=3}^{\infty} \frac{1}{k(k-1)}$

11. $\displaystyle\sum_{k=0}^{\infty} \frac{1}{(k+1)(k+2)}$

12. $\displaystyle\sum_{k=-1}^{\infty} \frac{1}{(k+3)(k+4)}$

13. $\displaystyle\sum_{k=2}^{\infty} \frac{2^{k+3}}{3^k}$

14. $\displaystyle\sum_{k=2}^{\infty} \frac{2^{k+4}}{3^{k-1}}$

15. $\displaystyle\sum_{k=4}^{\infty} \frac{5^{k-2}}{6^{k+1}}$

In Problems 16–24 write the repeating decimals as rational numbers.

16. 0.666 . . .

17. 0.353535 . . .

18. 0.282828 . . .

19. 0.717171 . . .

20. 0.214214214 . . .

21. 0.501501501 . . .

22. 0.124242424 . . .

23. 0.11362362362 . . .

24. 0.513651365136 . . .

25. Give a new proof, using the corollary to Theorem 2, that the geometric series diverges if $|r| \geq 1$.

In Problems 26–30 use Theorem 1 to calculate the sum of the convergent series.

26. $\displaystyle\sum_{k=0}^{\infty} \left[\frac{1}{2^k} + \frac{1}{5^k}\right]$

27. $\displaystyle\sum_{k=1}^{\infty} \left[\frac{1}{k(k+1)} + \frac{1}{(k+1)(k+2)}\right]$

28. $\displaystyle\sum_{k=0}^{\infty} \left[\frac{3}{5^k} - \frac{7}{4^k}\right]$

29. $\displaystyle\sum_{k=1}^{\infty} \left[\frac{8}{5^k} - \frac{7}{(k+3)(k+4)}\right]$

30. $\displaystyle\sum_{k=3}^{\infty} \left[\frac{12 \cdot 2^{k+1}}{3^{k-2}} - \frac{15 \cdot 3^{k+1}}{4^{k+2}}\right]$

31. Show that for any nonzero real numbers a and b, $\Sigma_{k=1}^{\infty} a/bk$ diverges.

32. Show that if the sequences $\{a_k\}$ and $\{b_k\}$ differ only for a finite number of terms, then $\Sigma_{k=1}^{\infty} a_k$ and $\Sigma_{k=1}^{\infty} b_k$ either both converge or both diverge.

33. Use the result of Problem 32 to show that $\Sigma_{k=1}^{\infty} 1/(k+6)$ diverges.

***34.** Show that if $\Sigma_{k=1}^{\infty} a_k$ converges and $\Sigma_{k=1}^{\infty} b_k$ diverges, then $\Sigma_{k=1}^{\infty} (a_k + b_k)$ diverges. [*Hint:* Assume that $\Sigma_{k=1}^{\infty} (a_k + b_k)$ converges and then show that this leads to a contradiction of Theorem 1.]

35. Use the result of Problem 34 to show that $\Sigma_{k=1}^{\infty} (3/2^k + 2 \cdot 5^k)$ diverges.

36. Give an example in which $\sum_{k=1}^{\infty} a_k$ and $\sum_{k=1}^{\infty} b_k$ both diverge but $\sum_{k=1}^{\infty} (a_k + b_k)$ converges.

37. Use the geometric series to show that

$$\frac{1}{1+x} = \sum_{k=0}^{\infty} (-1)^k x^k$$

for any real number x with $|x| < 1$.

38. Show that $1/(1 + x^2) = \sum_{k=0}^{\infty} (-1)^k x^{2k}$ if $|x| < 1$.

***39.** At what time between 1 P.M. and 2 P.M. is the minute hand of a clock exactly over the hour hand? [*Hint:* The minute hand moves 12 times as fast as the hour hand. Start at 1:00 P.M. When the minute hand has reached 1, the hour hand points to $1 + \frac{1}{12}$; when the minute hand has reached $1 + \frac{1}{12}$, the hour hand has reached $1 + \frac{1}{12} + \frac{1}{12} \cdot \frac{1}{12}$; etc. Now add up the geometric series.]

***40.** At what time between 7 A.M. and 8 A.M. is the minute hand exactly over the hour hand?

41. A ball is dropped from a height of 8 m. Each time it hits the ground, it rebounds to a height of two-thirds the height from which it fell. Find the total distance traveled by the ball until it comes to rest (i.e., until it stops bouncing).

***42.** Pick a_0 and a_1. For $n \geq 2$, compute a_n recursively by $n(n-1)a_n = (n-1)(n-2)a_{n-1} - (n-3)a_{n-2}$. Evaluate $\sum_{n=0}^{\infty} a_n$.

13.8 SERIES WITH NONNEGATIVE TERMS I: TWO COMPARISON TESTS AND THE INTEGRAL TEST

In this section and the next one we consider series of the form $\sum_{k=1}^{\infty} a_k$, where each a_k is nonnegative. Such series are often easier to handle than others. One fact is easy to prove. The sequence $\{S_n\}$ of partial sums is a monotone increasing sequence since $S_{n+1} = S_n + a_{n+1}$ and $a_{n+1} \geq 0$ for every n. Then if $\{S_n\}$ is bounded, it is convergent by Theorem 14.2.2, and we have the following theorem:

Theorem 1 An infinite series of nonnegative terms is convergent if and only if its sequence of partial sums is bounded.

EXAMPLE 1 Show that $\sum_{k=1}^{\infty} 1/k^2$ is convergent.

Solution. We group the terms as follows:

$$\sum_{k=1}^{\infty} \frac{1}{k^2} = \frac{1}{1^2} + \underbrace{\frac{1}{2^2} + \frac{1}{3^2}}_{\text{2 terms}} + \underbrace{\frac{1}{4^2} + \frac{1}{5^2} + \frac{1}{6^2} + \frac{1}{7^2}}_{\text{4 terms}} + \underbrace{\frac{1}{8^2} + \cdots + \frac{1}{15^2}}_{\text{8 terms}} + \cdots$$

$$\leq 1 + \underbrace{\frac{1}{2^2} + \frac{1}{2^2}}_{\text{2 terms}} + \underbrace{\frac{1}{4^2} + \frac{1}{4^2} + \frac{1}{4^2} + \frac{1}{4^2}}_{\text{4 terms}} + \underbrace{\frac{1}{8^2} + \cdots + \frac{1}{8^2}}_{\text{8 terms}} + \cdots$$

$$= 1 + \frac{2}{2^2} + \frac{4}{4^2} + \frac{8}{8^2} + \cdots = 1 + \frac{1}{2} + \frac{1}{4} + \frac{1}{8} + \cdots$$

$$= \sum_{k=0}^{\infty} \frac{1}{2^k} = 2.$$

The sequence of partial sums is bounded by 2 and is therefore convergent. ■

With Theorem 1 the convergence or divergence of a series of nonnegative terms depends on whether or not its partial sums are bounded. There are several tests that can be used to determine whether or not the sequence of partial sums of a series is bounded. We will deal with these one at a time.

Theorem 2 COMPARISON TEST Let $\Sigma_{k=1}^{\infty} a_k$ be a series with $a_k \geq 0$ for every k.

(i) If there exists a convergent series $\Sigma_{k=1}^{\infty} b_k$ and a number N such that $a_k \leq b_k$ for every $k \geq N$, then $\Sigma_{k=1}^{\infty} a_k$ converges.
(ii) If there exists a divergent series $\Sigma_{k=1}^{\infty} c_k$ and a number N such that $a_k \geq c_k \geq 0$ for every $k \geq N$, then $\Sigma_{k=1}^{\infty} a_k$ diverges.

Proof. In either case the sum of the first N terms is finite, so we need only consider the series $\Sigma_{k=N+1}^{\infty} a_k$ since if this is convergent or divergent, then the addition of a finite number of terms does not affect the convergence or divergence.

(i) $\Sigma_{k=N+1}^{\infty} b_k$ is a nonnegative series (since $b_k \geq a_k \geq 0$ for $k > N$) and is convergent. Thus the partial sums $T_n = \Sigma_{k=N+1}^{n} b_k$ are bounded. If $S_n = \Sigma_{k=N+1}^{n} a_k$, then $S_n \leq T_n$, and so the partial sums of $\Sigma_{k=N+1}^{\infty} a_k$ are also bounded, implying that $\Sigma_{k=N+1}^{\infty} a_k$ is convergent.
(ii) Let $U_n = \Sigma_{k=N+1}^{n} c_k$. By Theorem 1 these partial sums are unbounded since $\Sigma_{k=N+1}^{\infty} c_k$ diverges. Since in this case $S_n \geq U_n$, the partial sums of $\Sigma_{k=N+1}^{\infty} a_k$ are also unbounded, and the series $\Sigma_{k=N+1}^{\infty} a_k$ diverges. ■

REMARK. One idea appearing in the proof of (i) is important enough to state again: *If for some positive integer N, $\Sigma_{k=N+1}^{\infty} a_k$ converges, then $\Sigma_{k=1}^{\infty} a_k$ also converges. If $\Sigma_{k=N+1}^{\infty} a_k$ diverges, then $\Sigma_{k=1}^{\infty} a_k$ diverges. That is, the addition of a finite number of terms does not affect convergence or divergence.*

EXAMPLE 2 Determine whether $\Sigma_{k=1}^{\infty} 1/\sqrt{k}$ converges or diverges.

Solution. Since $1/\sqrt{k} \geq 1/k$ for $k \geq 1$, and since $\Sigma_{k=1}^{\infty} 1/k$ diverges, we see that by the comparison test, $\Sigma_{k=1}^{\infty} 1/\sqrt{k}$ diverges. ■

EXAMPLE 3 Determine whether $\Sigma_{k=1}^{\infty} 1/k!$ converges or diverges.

Solution. If $k \geq 4$, $k! \geq 2^k$. To see this, note that $4! = 24$ and $2^4 = 16$. Then $5! = 5 \cdot 24$ and $2^5 = 2 \cdot 16$ and since $5 > 2$, $5! > 2^5$, and so on. Then since $\Sigma_{k=1}^{\infty} 1/2^k$ converges, we see that $\Sigma_{k=1}^{\infty} 1/k!$ converges. In fact, as we will show in Section 13.12, it converges to $e - 1$. That is,

$$e = 1 + 1 + \frac{1}{2!} + \frac{1}{3!} + \frac{1}{4!} + \cdots. \quad \blacksquare \tag{1}$$

Theorem 3 THE INTEGRAL TEST Let f be a function that is continuous, positive, and decreasing for all $x \geq 1$. Then the series

$$\sum_{k=1}^{\infty} f(k) = f(1) + f(2) + f(3) + \cdots + f(n) + \cdots \tag{2}$$

converges if $\int_1^{\infty} f(x)\,dx$ converges, and diverges if $\int_1^n f(x)\,dx \to \infty$ as $n \to \infty$.

Proof. The idea behind this proof is described below. Take a look at Figure 1. Comparing areas, we immediately see that

$$f(2) + f(3) + \cdots + f(n) \le \int_1^n f(x)\,dx \le f(1) + f(2) + \cdots + f(n-1).$$

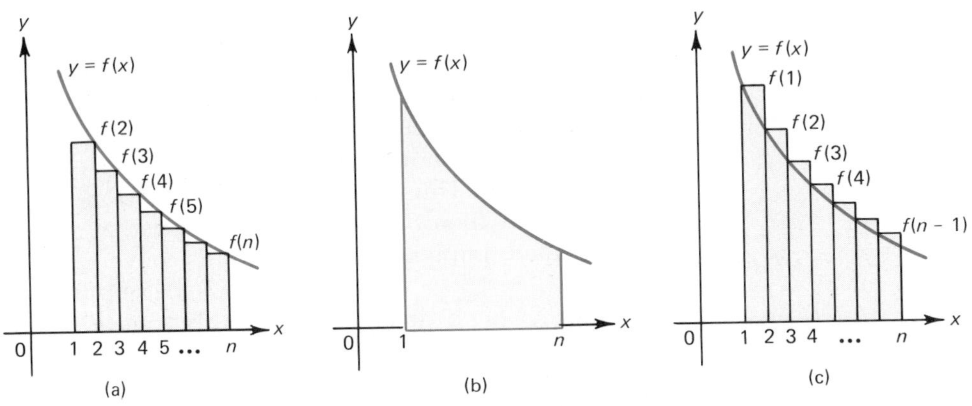

FIGURE 1

If $\lim_{n\to\infty} \int_1^n f(x)\,dx$ is finite, then the partial sums $[f(2) + f(3) + \cdots + f(n)]$ are bounded and the series converges. On the other hand, if $\lim_{n\to\infty} \int_1^n f(x)\,dx = \infty$, then the partial sums $[f(1) + f(2) + \cdots + f(n-1)]$ are unbounded and the series diverges. ■

EXAMPLE 4 Consider the series $\sum_{k=1}^{\infty} 1/k^{\alpha}$ with $\alpha > 0$. We have already seen that this series diverges for $\alpha = 1$ (the harmonic series) and converges for $\alpha = 2$ (Example 1). Now let $f(x) = 1/x^{\alpha}$. Then for $\alpha \neq 1$

$$\int_1^n f(x)\,dx = \int_1^n \frac{1}{x^{\alpha}}\,dx = \left.\frac{x^{1-\alpha}}{1-\alpha}\right|_1^n = \frac{1}{1-\alpha}(n^{1-\alpha} - 1).$$

This last expression converges to $1/(\alpha - 1)$ if $\alpha > 1$ and diverges if $\alpha < 1$. For $\alpha = 1$

$$\int_1^n f(x)\,dx = \int_1^n \frac{1}{x}\,dx = \ln x\bigg|_1^n = \ln n,$$

which diverges. (This is another proof that the harmonic series diverges.) Hence

$$\sum_{k=1}^{\infty} \frac{1}{k^\alpha} \quad \begin{cases} \text{diverges if } \alpha \leq 1, \\ \text{converges if } \alpha > 1. \end{cases}$$ ■

EXAMPLE 5 Determine whether $\Sigma_{k=1}^{\infty} (\ln k)/k^2$ converges or diverges.

Solution. We see, using L'Hôpital's rule, that

$$\lim_{x\to\infty} \frac{\ln x}{\sqrt{x}} = \lim_{x\to\infty} \frac{1/x}{1/2\sqrt{x}} = 2 \lim_{x\to\infty} \frac{\sqrt{x}}{x} = 2 \lim_{x\to\infty} \frac{1}{\sqrt{x}} = 0,$$

so that for k sufficiently large, $\ln k \leq \sqrt{k}$. Thus

$$\frac{\ln k}{k^2} \leq \frac{\sqrt{k}}{k^2} = \frac{1}{k^{3/2}}.$$

But $\Sigma_{k=1}^{\infty} 1/k^{3/2}$ converges by the result of Example 4, and therefore by the comparison test, $\Sigma_{k=1}^{\infty} (\ln k)/k^2$ also converges.

NOTE. The integral test can also be used directly here since $\int_1^\infty (\ln x/x^2)\,dx$ can be integrated by parts with $u = \ln x$. ■

EXAMPLE 6 Determine whether $\Sigma_{k=1}^{\infty} 1/[k \ln(k + 5)]$ converges or diverges.

Solution. First, we note that $1/[k \ln(k + 5)] > 1/[(k + 5) \ln(k + 5)]$. Also,

$$\int_1^n \frac{dx}{(x + 5)\ln(x + 5)} = \ln \ln(x + 5)\Big|_1^n = \ln \ln(n + 5) - \ln \ln 6,$$

which diverges, so that $\Sigma_{k=1}^{\infty} 1/[k \ln(k + 5)]$ also diverges. ■

We now give another test that is an extension of the comparison test.

Theorem 4 THE LIMIT COMPARISON TEST Let $\Sigma_{k=1}^{\infty} a_k$ and $\Sigma_{k=1}^{\infty} b_k$ be series with positive terms.

If there is a number $c > 0$ such that

$$\lim_{k\to\infty} \frac{a_k}{b_k} = c, \tag{3}$$

then either both series converge or both series diverge.

Proof. We have $\lim_{k\to\infty}(a_k/b_k) = c > 0$. In the definition of a limit on page 806, let $\epsilon = c/2$. Then there is a number $N > 0$ such that

$$\left|\frac{a_k}{b_k} - c\right| < \frac{c}{2} \quad \text{if} \quad k \geq N. \tag{4}$$

Equation (4) is equivalent to

$$-\frac{c}{2} < \frac{a_k}{b_k} - c < \frac{c}{2}$$

or

$$\frac{c}{2} < \frac{a_k}{b_k} < \frac{3c}{2}. \tag{5}$$

From the right inequality in (5), we obtain

$$a_k < \frac{3c}{2}\, b_k. \tag{6}$$

If $\Sigma\, b_k$ is convergent, then so is $(3c/2)\, \Sigma\, b_k = \Sigma\, (3c/2)b_k$. Thus from (6) and the comparison test, $\Sigma\, a_k$ is convergent. From the left inequality in (5), we have

$$a_k > \frac{c}{2}\, b_k. \tag{7}$$

If $\Sigma\, b_k$ is divergent, then so is $(c/2)\, \Sigma\, b_k = \Sigma\, (c/2)\, b_k$. Then using (7) and the comparison test, we find that $\Sigma\, a_k$ is also divergent. Thus if $\Sigma\, b_k$ is convergent, then $\Sigma\, a_k$ is convergent; and if $\Sigma\, b_k$ is divergent, then $\Sigma\, a_k$ is divergent. This proves the theorem. ■

EXAMPLE 7 Show that $\Sigma_{k=1}^{\infty}\, 1/(ak^2 + bk + c)$ is convergent, where a, b, and c are positive real numbers.

Solution. We know from Example 1 that $\Sigma_{k=1}^{\infty}\, 1/k^2$ is convergent. If

$$a_k = \frac{1}{k^2} \qquad \text{and} \qquad b_k = \frac{1}{ak^2 + bk + c}$$

then

$$\frac{a_k}{b_k} = \frac{1/k^2}{1/(ak^2 + bk + c)} = \frac{ak^2 + bk + c}{k^2} = a + \frac{b}{k} + \frac{c}{k^2}$$

so that $\lim_{k\to\infty} (a_k/b_k) = a > 0$. Thus by the limit comparison test,

$$\sum_{k=1}^{\infty} \frac{1}{ak^2 + bk + c}$$

is convergent. ■

EXAMPLE 8 Determine whether $\Sigma_{k=1}^{\infty}\, (k + 1)/(k + 3)(k + 5)$ converges or diverges.

Solution. For k large, $k + 1 \approx k$ (since $(k + 1)/k \to 1$ as $k \to \infty$), $k + 3 \approx k$, and $k + 5 \approx k$. Thus

$$\frac{k+1}{(k+3)(k+5)} \approx \frac{k}{k^2} = \frac{1}{k},$$

and $\Sigma_{k=1}^{\infty} 1/k$ diverges. Therefore we suspect that $\Sigma_{k=1}^{\infty} (k + 1)/(k + 3)(k + 5)$ also diverges. This, of course, is not a proof, but it helps us to guess what answer we might expect. Let $a_k = 1/k$ (this is a term of a natural series with which to compare). If $b_k = (k + 1)/(k + 3)(k + 5)$, then

$$\frac{a_k}{b_k} = \frac{1/k}{(k+1)/(k+3)(k+5)} = \frac{(k+3)(k+5)}{(k)(k+1)}$$

$$= \frac{k^2 + 8k + 15}{k^2 + k} = \frac{1 + (8/k) + (15/k^2)}{1 + (1/k)} \to 1 \quad \text{as} \quad k \to \infty.$$

Thus by the limit comparison test, $\Sigma_{k=1}^{\infty} (k + 1)/(k + 3)(k + 5)$ diverges because $\Sigma_{k=1}^{\infty} (1/k)$ diverges. ■

PROBLEMS 13.8

In Problems 1–33 determine the convergence or divergence of the given series.

1. $\sum_{k=1}^{\infty} \frac{1}{k^2+1}$

2. $\sum_{k=10}^{\infty} \frac{1}{k(k-3)}$

3. $\sum_{k=4}^{\infty} \frac{1}{5k+50}$

4. $\sum_{k=1}^{\infty} \frac{1}{\sqrt{k^2+2k}}$

5. $\sum_{k=1}^{\infty} \frac{1}{\sqrt{k^3+1}}$

6. $\sum_{k=1}^{\infty} \frac{\ln k}{k^3}$

7. $\sum_{k=2}^{\infty} \frac{1}{k^2+1}$

8. $\sum_{k=2}^{\infty} \frac{4}{k \ln k}$

9. $\sum_{k=0}^{\infty} ke^{-k}$

10. $\sum_{k=3}^{\infty} k^3 e^{-k^4}$

11. $\sum_{k=5}^{\infty} \frac{1}{k(\ln k)^3}$

12. $\sum_{k=4}^{\infty} \frac{1}{k^2\sqrt{\ln k}}$

13. $\sum_{k=1}^{\infty} \frac{1}{(3k-1)^{3/2}}$

14. $\sum_{k=1}^{\infty} \frac{1}{\sqrt{k^2+3}}$

15. $\sum_{k=2}^{\infty} \frac{1}{k\sqrt{\ln k}}$

16. $\sum_{k=1}^{\infty} \frac{1}{50+\sqrt{k}}$

17. $\sum_{k=3}^{\infty} \left(\frac{k}{k+1}\right)^k$

18. $\sum_{k=1}^{\infty} \left(\frac{k}{k+1}\right)^{1/k}$

19. $\sum_{k=4}^{\infty} \frac{1}{k \ln \ln k}$

20. $\sum_{k=1}^{\infty} \sin \frac{1}{k}$

21. $\sum_{k=1}^{\infty} \frac{1}{(k+2)\sqrt{\ln(k+1)}}$

22. $\sum_{k=1}^{\infty} \frac{e^{1/k}}{k^2}$

23. $\sum_{k=1}^{\infty} \frac{\tan^{-1} k}{1+k^2}$

24. $\sum_{k=1}^{\infty} \operatorname{sech} k$

25. $\sum_{k=10}^{\infty} \frac{1}{k(\ln k)(\ln \ln k)}$

26. $\sum_{k=1}^{\infty} \frac{k^2}{50k}$

27. $\sum_{k=1}^{\infty} \frac{1}{\cosh^2 k}$

28. $\sum_{k=1}^{\infty} \tan^{-1} k$

*29. $\sum_{k=2}^{\infty} \frac{1}{\sqrt{k} \ln^{10} k}$

30. $\sum_{k=1}^{\infty} \frac{\sqrt{k}}{3k^2+2k+20}$

31. $\sum_{k=1}^{\infty} \frac{(k+1)^{7/8}}{k^3+k^2+3}$

32. $\sum_{k=1}^{\infty} \frac{1}{(k+1)\ln(k+1)}$

33. $\sum_{k=1}^{\infty} \frac{k^5+2k^4+3k+7}{k^6+3k^4+2k^2+1}$

*34. Let $p(x)$ be a polynomial of degree n with positive coefficients and let $q(x)$ be a polynomial of degree $\leq n + 1$ with positive coefficients. Show that $\Sigma_{k=1}^{\infty} p(k)/q(k)$ diverges.

***35.** With $p(x)$ as in Problem 34 and $r(x)$ a polynomial of degree $\geq n + 2$ with positive coefficients, show that $\Sigma_{k=1}^{\infty} p(k)/r(k)$ converges.

36. Determine whether $\Sigma_{n=1}^{\infty} (1/n)^{1+(1/n)}$ converges or diverges.

37. Let $S_n = \ln n! = \ln 2 + \ln 3 + \cdots + \ln n$. By calculating $\int_1^n \ln x\, dx$ and comparing areas, as in the proof of the integral test, show that for $n \geq 2$

$$\ln(n - 1)! < n \ln n - n + 1 < \ln n!$$

and that

$$(n - 1)! < n^n e^{-(n-1)} < n!.$$

***38.** Let $S_n = \Sigma_{k=1}^{n} 1/k$. Show that

$$\ln(n + 1) < S_n < \ln n + 1.$$

[*Hint:* Use the inequality $1/(k + 1) \leq 1/x \leq 1/k$ if $0 < k \leq x \leq k + 1$, integrate, and add, as in the proof of the integral test.]

39. Let $S_n = \Sigma_{k=1}^{n} 1/k$.

(a) Show that the sequence $\{S_n - \ln(n + 1)\}$ is increasing.

***(b)** Show that this sequence is bounded by 1. [*Hint:* $S_n - \ln(n + 1) =$ the sum of the areas of "triangular" shaped regions in the region. Move those "triangles" over so that they all fit into the first rectangle.]

(c) Show that $\lim_{n\to\infty} [S_n - \ln(n + 1)]$ exists. This limit is denoted by γ, which is called the **Euler constant**:

$$\gamma = \lim_{n\to\infty} \left[1 + \frac{1}{2} + \frac{1}{3} + \cdots + \frac{1}{n} - \ln(n + 1)\right]$$

$$= \lim_{n\to\infty} \left(1 + \frac{1}{2} + \frac{1}{3} + \cdots + \frac{1}{n} - \ln n\right).$$

This number arises in physical applications. To seven decimal places, $\gamma = 0.5772157$.

13.9 SERIES WITH NONNEGATIVE TERMS II: THE RATIO AND ROOT TESTS

In this section we discuss two more tests that can be used to determine whether an infinite series converges or diverges. The first of these, the ratio test, is useful in a wide variety of applications.

Theorem 1 THE RATIO TEST Let $\Sigma_{k=1}^{\infty} a_k$ be a series with $a_k > 0$ for every k, and suppose that

$$\lim_{n\to\infty} \frac{a_{n+1}}{a_n} = L. \tag{1}$$

(i) If $L < 1$, $\Sigma_{k=1}^{\infty} a_k$ converges.
(ii) If $L > 1$, $\Sigma_{k=1}^{\infty} a_k$ diverges.
(iii) If $L = 1$, $\Sigma_{k=1}^{\infty} a_k$ may converge or diverge and the ratio test is inconclusive; some other test must be used.

Proof.

(i) Pick $\epsilon > 0$ such that $L + \epsilon < 1$. By the definition of the limit in (1), there is a number $N > 0$ such that if $n \geq N$, we have

$$\frac{a_{n+1}}{a_n} < L + \epsilon.$$

Then

$$a_{n+1} < a_n(L + \epsilon), \qquad a_{n+2} < a_{n+1}(L + \epsilon) < a_n(L + \epsilon)^2,$$

and

$$a_{n+k} < a_n(L + \epsilon)^k \tag{2}$$

for each $k \geq 1$ and each $n \geq N$. In particular, for $k \geq N$ we use (2) to obtain

$$a_k = a_{(k-N)+N} \leq a_N(L + \epsilon)^{k-N}.$$

Then

$$S_n = \sum_{k=N}^{n} a_k \leq \sum_{k=N}^{n} a_N(L + \epsilon)^{k-N} = \frac{a_N}{(L + \epsilon)^N} \sum_{k=N}^{n} (L + \epsilon)^k.$$

But since $L + \epsilon < 1$, $\Sigma_{k=0}^{\infty} (L + \epsilon)^k = 1/[1 - (L + \epsilon)]$ (since this last sum is the sum of a geometric series). Thus

$$S_n \leq \frac{a_N}{(L + \epsilon)^N} \cdot \frac{1}{1 - (L + \epsilon)}$$

and so the partial sums of $\Sigma_{k=N}^{\infty} a_k$ are bounded, implying that $\Sigma_{k=N}^{\infty} a_k$ converges. Thus $\Sigma_{k=1}^{\infty} a_k = \Sigma_{k=1}^{N-1} a_k + \Sigma_{k=N}^{\infty} a_k$ also converges.

(ii) If $1 < L < \infty$, pick ϵ such that $L - \epsilon > 1$. Then for $n \geq N$, the same proof as before (with the inequalities reversed) shows that

$$a_k \geq a_N(L - \epsilon)^{k-N}$$

and that

$$S_n = \sum_{k=N}^{n} a_k > \frac{a_N}{(L - \epsilon)^N} \sum_{k=N}^{n} (L - \epsilon)^k.$$

But since $L - \epsilon > 1$, $\Sigma_{k=N}^{\infty} (L - \epsilon)^k$ diverges, so that the partial sums S_n are unbounded and $\Sigma_{k=N}^{\infty} a_k$ diverges. The proof in the case $L = \infty$ is suggested in Problem 33.

(iii) To illustrate (iii), we show that $L = 1$ can occur for a converging or diverging series.

(a) The harmonic series $\Sigma_{k=1}^{\infty} 1/k$ diverges. But

$$\lim_{n\to\infty} \frac{a_{n+1}}{a_n} = \lim_{n\to\infty} \frac{1/(n + 1)}{1/n} = \lim_{n\to\infty} \frac{n}{n + 1} = 1.$$

(b) The series $\Sigma_{k=1}^{\infty} 1/k^2$ converges. Here

$$\lim_{n\to\infty} \frac{a_{n+1}}{a_n} = \lim_{n\to\infty} m\frac{1/(n+1)^2}{1/n^2} = \lim_{n\to\infty} \left(\frac{n}{n+1}\right)^2 = 1. \blacksquare$$

REMARK. The ratio test is very useful. But in those cases where $L = 1$, we must try another test to determine whether the series converges or diverges.

EXAMPLE 1 We have used the comparison test to show that $\Sigma_{k=1}^{\infty} 1/k!$ converges. Using the ratio test, we find that

$$\lim_{n\to\infty} \frac{a_{n+1}}{a_n} = \lim_{n\to\infty} \frac{1/(n+1)!}{1/n!} = \lim_{n\to\infty} \frac{n!}{(n+1)!} = \lim_{n\to\infty} \frac{1}{n+1} = 0 < 1,$$

so that the series converges. ■

EXAMPLE 2 Determine whether the series $\Sigma_{k=0}^{\infty}(100)^k/k!$ converges or diverges.

Solution. Here

$$\lim_{n\to\infty} \frac{a_{n+1}}{a_n} = \lim_{n\to\infty} \frac{(100)^{n+1}/(n+1)!}{(100)^n/n!} = \lim_{n\to\infty} \frac{100}{n+1} = 0,$$

so that the series converges. ■

EXAMPLE 3 Determine whether the series $\Sigma_{k=1}^{\infty} k^k/k!$ converges or diverges.

Solution.

$$\lim_{n\to\infty} \frac{a_{n+1}}{a_n} = \lim_{n\to\infty} \frac{[(n+1)^{n+1}/(n+1)!]}{n^n/n!} = \lim_{n\to\infty} \frac{(n+1)^{n+1}}{(n+1)n^n}$$

$$= \lim_{n\to\infty} \left(\frac{n+1}{n}\right)^n = \lim_{n\to\infty} \left(1 + \frac{1}{n}\right)^n = e > 1,$$

so that the series diverges. ■

EXAMPLE 4 Determine whether the series $\Sigma_{k=1}^{\infty} (k+1)/[k(k+2)]$ converges or diverges.

Solution. Here

$$\lim_{n\to\infty} \frac{a_{n+1}}{a_n} = \lim_{n\to\infty} \frac{(n+2)/(n+1)(n+3)}{(n+1)/n(n+2)} = \lim_{n\to\infty} \frac{n(n+2)^2}{(n+1)^2(n+3)}$$

$$= \lim_{n\to\infty} \frac{n^3 + 4n^2 + 4n}{n^3 + 5n^2 + 7n + 3} = 1. \qquad \text{By Example 13.4.12}$$

Thus the ratio test fails. However, $\lim_{k\to\infty}[(k+1)/k(k+2)]/(1/k) = 1$, so that $\Sigma_{k=1}^{\infty} (k+1)/[k(k+2)]$ diverges by the limit comparison test. ■

Theorem 2 THE ROOT TEST Let $\Sigma_{k=1}^{\infty} a_k$ be a series with $a_k > 0$ and suppose that $\lim_{n\to\infty}(a_n)^{1/n} = R$.

(i) If $R < 1$, $\Sigma_{k=1}^{\infty} a_k$ converges.
(ii) If $R > 1$, $\Sigma_{k=1}^{\infty} a_k$ diverges.
(iii) If $R = 1$, the series either converges or diverges, and no conclusions can be drawn from this test.

The proof of this theorem is similar to the proof of the ratio test and is left as an exercise (see Problems 27–29).

EXAMPLE 5 Determine whether $\Sigma_{k=2}^{\infty} 1/(\ln k)^k$ converges or diverges.

Solution. Note first that we start at $k = 2$ since $1/(\ln 1)^1$ is not defined.

$$\lim_{n\to\infty}\left[\frac{1}{(\ln n)^n}\right]^{1/n} = \lim_{n\to\infty}\frac{1}{\ln n} = 0,$$

so that the series converges. ■

EXAMPLE 6 Determine whether the series $\Sigma_{k=1}^{\infty}(k^k/3^{4k+5})$ converges or diverges.

Solution. $\lim_{n\to\infty}(n^n/3^{4n+5})^{1/n} = \lim_{n\to\infty}(n/3^{4+5/n}) = \infty$, since $\lim_{n\to\infty}3^{4+5/n} = 3^4 = 81$. Thus the series diverges. ■

EXAMPLE 7 Determine whether the series $\Sigma_{k=1}^{\infty}(1/2 + 1/k)^k$ converges or diverges.

Solution.

$$\lim_{n\to\infty}\left[\left(\frac{1}{2}+\frac{1}{n}\right)^n\right]^{1/n} = \lim_{n\to\infty}\left(\frac{1}{2}+\frac{1}{n}\right) = \frac{1}{2} < 1,$$

so the series converges. ■

PROBLEMS 13.9

In Problems 1–25, determine whether the given series converges or diverges.

1. $\sum_{k=1}^{\infty}\frac{2^k}{k^2}$
2. $\sum_{k=1}^{\infty}\frac{5^k}{k^5}$
3. $\sum_{k=1}^{\infty}\frac{r^k}{k^r}$, $0 < r < 1$
4. $\sum_{k=1}^{\infty}\frac{r^k}{k^r}$, $r > 1$
5. $\sum_{k=2}^{\infty}\frac{k!}{k^k}$
6. $\sum_{k=1}^{\infty}\frac{k^k}{(2k)!}$
7. $\sum_{k=1}^{\infty}\frac{e^k}{k^5}$
8. $\sum_{k=1}^{\infty}\frac{e^k}{k!}$
9. $\sum_{k=1}^{\infty}\frac{k^{2/3}}{10^k}$
10. $\sum_{k=1}^{\infty}\frac{3^k + k}{k! + 2}$
11. $\sum_{k=2}^{\infty}\frac{k}{(\ln k)^k}$
12. $\sum_{k=1}^{\infty}\frac{4^k}{k^3}$
13. $\sum_{k=2}^{\infty}\left(1+\frac{1}{k}\right)^k$
14. $\sum_{k=1}^{\infty}\frac{\sqrt{k}\ln k}{k^3 + 1}$

15. $\sum_{k=1}^{\infty} \frac{3^{4k+5}}{k^k}$

16. $\sum_{k=1}^{\infty} \frac{a^{mk+b}}{k^k}$, $a > 1$, b real

17. $\sum_{k=1}^{\infty} \frac{k^k}{a^{mk+b}}$, $a > 1$, b real

18. $\sum_{k=1}^{\infty} \frac{k^6 5^k}{(k+1)!}$

19. $\sum_{k=1}^{\infty} \frac{k^2 k!}{(2k)!}$

20. $\sum_{k=1}^{\infty} \frac{(2k)!}{k^2 k!}$

***21.** $\sum_{k=1}^{\infty} \left(\frac{k!}{k^k}\right)^k$

22. $\sum_{k=1}^{\infty} \left(\frac{k^k}{k!}\right)^k$

23. $\sum_{k=2}^{\infty} \frac{e^k}{(\ln k)^k}$

24. $\sum_{k=1}^{\infty} \frac{(\ln k)^k}{k^2}$

25. $\sum_{k=1}^{\infty} \left(\frac{k}{3k+2}\right)^k$

26. Show that $\Sigma_{k=0}^{\infty} x^k/k!$ converges for every real number x.

***27.** Prove part (i) of the root test (Theorem 2). [*Hint:* If $R < 1$, choose $\epsilon > 0$ so that $R + \epsilon < 1$. Show that there is an N such that if $n \geq N$, then $a_n < (R + \epsilon)^n$. Then complete the proof by comparing $\Sigma\, a_k$ with the sum of a geometric series.]

28. Prove part (ii) of the root test. [*Hint:* Follow the steps of Problem 27.]

29. Show that if $a_n^{1/n} \to 1$, then $\Sigma_{k=1}^{\infty} a_k$ may converge or diverge. [*Hint:* Consider $\Sigma\, 1/k$ and $\Sigma\, 1/k^2$.]

30. Prove that $k!/k^k \to 0$ as $k \to \infty$.

31. Let $a_k = 3/k^2$ if k is even and $a_k = 1/k^2$ if k is odd. Show that $\lim_{n\to\infty} (a_{n+1}/a_n)$ does not exist, but $\Sigma_{k=1}^{\infty} a_k$ converges.

32. Construct a series of positive terms for which $\lim_{n\to\infty} (a_{n+1}/a_n)$ does not exist but for which $\Sigma_{k=1}^{\infty} a_k$ diverges.

33. Prove that if $a_n > 0$ and $\lim_{n\to\infty} (a_{n+1}/a_n) = \infty$, then $\Sigma_{k=1}^{\infty} a_k$ diverges.

13.10 ABSOLUTE AND CONDITIONAL CONVERGENCE: ALTERNATING SERIES

In Sections 13.8 and 13.9 all the series we dealt with had positive terms. In this section we consider special types of series that have positive and negative terms.

Definition 1 ABSOLUTE CONVERGENCE The series $\Sigma_{k=1}^{\infty} a_k$ is said to **converge absolutely** if the series $\Sigma_{k=1}^{\infty} |a_k|$ converges.

EXAMPLE 1 The series

$$\sum_{k=1}^{\infty} \frac{(-1)^{k+1}}{k^2} = \frac{1}{1^2} - \frac{1}{2^2} + \frac{1}{3^2} - \frac{1}{4^2} + \cdots$$

converges absolutely because $\Sigma_{k=1}^{\infty} |(-1)^{k+1}/k^2| = \Sigma_{k=1}^{\infty} 1/k^2$ converges. ■

EXAMPLE 2 The series

$$\sum_{k=1}^{\infty} \frac{(-1)^{k+1}}{k} = \frac{1}{1} - \frac{1}{2} + \frac{1}{3} - \frac{1}{4} + \frac{1}{5} + \cdots$$

does not converge absolutely because $\Sigma_{k=1}^{\infty} 1/k$ diverges. ■

The importance of absolute convergence is given in the theorem below.

Theorem 1 If $\Sigma_{k=1}^{\infty} |a_k|$ converges, then $\Sigma_{k=1}^{\infty} a_k$ also converges. That is:

Absolute convergence implies convergence.

REMARK. The converse of this theorem is false. That is, there are series that are convergent but not absolutely convergent. We will see examples of this phenomenon shortly.

Proof. Since $a_k \le |a_k|$, we have

$$0 \le a_k + |a_k| \le 2|a_k|.$$

Since $\Sigma_{k=1}^{\infty} |a_k|$ converges, we see that $\Sigma_{k=1}^{\infty} (a_k + |a_k|)$ converges by the comparison test. Then since $a_k = (a_k + |a_k|) - |a_k|$, $\Sigma_{k=1}^{\infty} a_k$ converges because it is the sum of two convergent series. ■

EXAMPLE 3 The series $\Sigma_{k=1}^{\infty} (-1)^{k+1}/k^2$ considered in Example 1 converges since it converges absolutely. ■

Definition 2 ALTERNATING SERIES A series in which successive terms have opposite signs is called an **alternating series**.

EXAMPLE 4 The series

$$\sum_{k=1}^{\infty} \frac{(-1)^{k+1}}{k} = 1 - \frac{1}{2} + \frac{1}{3} - \frac{1}{4} + \frac{1}{5} - \frac{1}{6} + \cdots$$

is an alternating series. ■

EXAMPLE 5 The series $1 + \frac{1}{2} - \frac{1}{3} - \frac{1}{4} + \frac{1}{5} + \frac{1}{6} - \cdots$ is not an alternating series because two successive terms have the same sign. ■

Let us consider the series of Example 4:

$$S = 1 - \tfrac{1}{2} + \tfrac{1}{3} - \tfrac{1}{4} + \tfrac{1}{5} - \tfrac{1}{6} + \cdots.$$

Calculating successive partial sums, we find that

$$S_1 = 1, \qquad S_2 = \tfrac{1}{2}, \qquad S_3 = \tfrac{5}{6}, \qquad S_4 = \tfrac{7}{12}, \qquad S_5 = \tfrac{47}{60}, \qquad \ldots.$$

It is clear that this series is not diverging to infinity (indeed, $\frac{1}{2} \le S_n \le 1$) and that the partial sums are getting "narrowed down." At this point it is reasonable to suspect that the series converges. But it does *not* converge absolutely (since the series of absolute values is the harmonic series), and we cannot use any of the tests of the previous section since the terms are not nonnegative. The result we need is given in the theorem below.

Theorem 2 ALTERNATING SERIES TEST Let $\{a_k\}$ be a decreasing sequence of positive numbers such that $\lim_{k\to\infty} a_k = 0$. Then the alternating series $\Sigma_{k=1}^{\infty} (-1)^{k+1} a_k = a_1 - a_2 + a_3 - a_4 + \cdots$ converges.

Proof. Looking at the odd-numbered partial sums of this series, we find that

$$S_{2n+1} = (a_1 - a_2) + (a_3 - a_4) + (a_5 - a_6) + \cdots + (a_{2n-1} - a_{2n}) + a_{2n+1}.$$

Since $\{a_k\}$ is decreasing, all the terms in parentheses are nonnegative, so that $S_{2n+1} \geq 0$ for every n. Moreover,

$$S_{2n+3} = S_{2n+1} - a_{2n+2} + a_{2n+3} = S_{2n+1} - (a_{2n+2} - a_{2n+3}),$$

and since $a_{2n+2} - a_{2n+3} \geq 0$, we have

$$S_{2n+3} \leq S_{2n+1}.$$

Hence the sequence of odd-numbered partial sums is bounded below by 0 and is decreasing and is therefore convergent by Theorem 13.5.2. Thus S_{2n+1} converges to some limit L. Now let us consider the sequence of even-numbered partial sums. We find that $S_{2n+2} = S_{2n+1} - a_{2n+2}$ and since $a_{2n+2} \to 0$,

$$\lim_{n\to\infty} S_{2n+2} = \lim_{n\to\infty} S_{2n+1} - \lim_{n\to\infty} a_{2n+2} = L - 0 = L,$$

so that the even partial sums also converge to L. Since both the odd and even sums converge to L, we see that the partial sums converge to L, and the proof is complete. ■

EXAMPLE 6 The following alternating series are convergent by the alternating series test:

(a) $1 - \frac{1}{2} + \frac{1}{3} - \frac{1}{4} + \frac{1}{5} - \frac{1}{6} + \cdots$

(b) $1 - \frac{1}{\sqrt{2}} + \frac{1}{\sqrt{3}} - \frac{1}{\sqrt{4}} + \frac{1}{\sqrt{5}} - \frac{1}{\sqrt{6}} + \frac{1}{\sqrt{7}} - \cdots$

(c) $\frac{1}{\ln 2} - \frac{1}{\ln 3} + \frac{1}{\ln 4} - \frac{1}{\ln 5} + \frac{1}{\ln 6} - \cdots$

(d) $1 - \frac{1}{2} + \frac{1}{2^2} - \frac{1}{2^3} + \frac{1}{2^4} - \frac{1}{2^5} + \frac{1}{2^6} - \frac{1}{2^7} + \cdots$ ■

Definition 3 CONDITIONAL CONVERGENCE An alternating series is said to be **conditionally convergent** if it is convergent but not absolutely convergent.

In Example 6 all the series are conditionally convergent except the last one, which is absolutely convergent.

It is not difficult to estimate the sum of a convergent alternating series. We again consider the series

$$S = 1 - \tfrac{1}{2} + \tfrac{1}{3} - \tfrac{1}{4} + \tfrac{1}{5} - \cdots.$$

Suppose we wish to approximate S by its nth partial sum S_n. Then

$$S - S_n = \pm\left(\frac{1}{n+1} - \frac{1}{n+2} + \frac{1}{n+3} - \frac{1}{n+4} + \cdots\right) = R_n.$$

But we can estimate the remainder term R_n:

$$|R_n| = \left|\left[\frac{1}{n+1} - \left(\frac{1}{n+2} - \frac{1}{n+3}\right) - \left(\frac{1}{n+4} - \frac{1}{n+5}\right) - \cdots\right]\right| \le \frac{1}{n+1}.$$

That is, the error is less than the first term that we left out! For example, $|S - S_{20}| \le \frac{1}{21} \approx 0.0476$.

In general, we have the following result, whose proof is left as an exercise (see Problem 31).

Theorem 3 If $S = \sum_{k=1}^{\infty}(-1)^{k+1}a_k$ is a convergent alternating series with terms monotone decreasing in absolute value, then for any n,

$$|S - S_n| \le |a_{n+1}|. \quad \textbf{(1)}$$

EXAMPLE 7 The series

$$\sum_{k=1}^{\infty} \frac{(-1)^{k+1}}{\ln(k+1)} = \frac{1}{\ln 2} - \frac{1}{\ln 3} + \frac{1}{\ln 4} - \frac{1}{\ln 5} + \cdots$$

can be approximated by S_n with an error of less than $1/\ln(n + 2)$. For example, with $n = 10$, $1/\ln(n + 2) = 1/\ln 12 \approx 0.4$. Hence the sum

$$\sum_{k=1}^{\infty} \frac{(-1)^{k+1}}{\ln(k+1)} = \frac{1}{\ln 2} - \frac{1}{\ln 3} + \cdots$$

can be approximated by

$$S_{10} = \frac{1}{\ln 2} - \frac{1}{\ln 3} + \frac{1}{\ln 4} - \frac{1}{\ln 5} + \frac{1}{\ln 6} - \frac{1}{\ln 7} + \frac{1}{\ln 8} - \frac{1}{\ln 9} + \frac{1}{\ln 10} - \frac{1}{\ln 11}$$
$$\approx 0.7197,$$

with an error of less than 0.4. ■

By modifying Theorem 3 we can significantly improve on the last result.

Theorem 4 Suppose that the hypotheses of Theorem 3 hold and that, in addition, the sequence $\{|a_n - a_{n+1}|\}$ is monotone decreasing. Let $T_n = S_{n-1} - (-1)^n \frac{1}{2}a_n$, Then

$$|S - T_n| \le \tfrac{1}{2}|a_n - a_{n+1}|. \quad \textbf{(2)}$$

This result follows from Theorem 3 and is also left as an exercise (see Problem 45).

EXAMPLE 8 We can improve the estimate in Example 7. We may approximate $\Sigma_{k=1}^{\infty}(-1)^{k+1}/\ln(k+1)$ by

$$\frac{1}{\ln 2}-\frac{1}{\ln 3}+\frac{1}{\ln 4}-\frac{1}{\ln 5}+\frac{1}{\ln 6}-\frac{1}{\ln 7}+\frac{1}{\ln 8}-\frac{1}{\ln 9}+\frac{1}{\ln 10}-\frac{1}{2\ln 11}\approx 0.9282.$$

With $n = 10$ (so that $n + 1 = 11$),

$$T_{10} = S_9 - \frac{1}{2}\left(\frac{1}{\ln 11}\right),$$

which is precisely the sum given above. Thus

$$|S - T_{10}| < \frac{1}{2}\left|a_{10} - a_{11}\right| = \frac{1}{2}\left(\frac{1}{\ln 11} - \frac{1}{\ln 12}\right) \approx 0.0073.$$

This result is a considerable improvement.

However, to justify this result, we must verify that $|a_n - a_{n+1}|$ is monotone decreasing. But here

$$|a_n - a_{n+1}| = \frac{1}{\ln(n+1)} - \frac{1}{\ln(n+2)}.$$

Let

$$f(x) = \frac{1}{\ln(x+1)} - \frac{1}{\ln(x+2)}.$$

Then

$$f'(x) = -\frac{1}{(x+1)\ln^2(x+1)} + \frac{1}{(x+2)\ln^2(x+2)} < 0.$$

Thus f is a decreasing function, which shows that $f(n+1) < f(n)$. ■

There is one fascinating fact about an alternating series that is conditionally but not absolutely convergent:

> By reordering the terms of a conditionally convergent alternating series, the new series of rearranged terms can be made to converge to any real number.

Let us illustrate this fact with the series

$$S = 1 - \tfrac{1}{2} + \tfrac{1}{3} - \tfrac{1}{4} + \tfrac{1}{5} - \tfrac{1}{6} + \cdots.$$

The odd-numbered terms diverge:

$$1 + \tfrac{1}{3} + \tfrac{1}{5} + \tfrac{1}{7} + \cdots. \tag{3}$$

The even-numbered terms are likewise a divergent series:

$$-\tfrac{1}{2} - \tfrac{1}{4} - \tfrac{1}{6} - \cdots. \tag{4}$$

If either of these series converged, then the other one would too (by Theorem 13.7.1(i), and then the entire series would be absolutely convergent (which we know to be false). Now choose any real number, say 1.5. Then:

(i) Choose enough terms from the series (3) so that the sum exceeds 1.5. We can do so since the series diverges.

$$1 + \tfrac{1}{3} + \tfrac{1}{5} = 1.53333\ldots.$$

(ii) Add enough negative terms from (4) so that the sum is now just under 1.5.

$$1 + \tfrac{1}{3} + \tfrac{1}{5} - \tfrac{1}{2} = 1.0333\ldots.$$

(iii) Add more terms from (3) until 1.5 is exceeded.

$$1 + \tfrac{1}{3} + \tfrac{1}{5} - \tfrac{1}{2} + \tfrac{1}{7} + \tfrac{1}{9} + \tfrac{1}{11} + \tfrac{1}{13} + \tfrac{1}{15} = 1.5218.$$

(iv) Again subtract terms from (4) until the sum is under 1.5.

$$1 + \tfrac{1}{3} + \tfrac{1}{5} - \tfrac{1}{2} + \tfrac{1}{7} + \tfrac{1}{9} + \tfrac{1}{11} + \tfrac{1}{13} + \tfrac{1}{15} - \tfrac{1}{4} = 1.2718.$$

We continue the process to "converge" to 1.5. Since the terms in each series are decreasing to 0, the amount above or below 1.5 will approach 0 and the partial sums converge.

We will indicate in Section 13.13 that without rearranging, we have

$$\sum_{k=1}^{\infty} \frac{(-1)^{k+1}}{k} = 1 - \frac{1}{2} + \frac{1}{3} - \frac{1}{4} + \frac{1}{5} - \frac{1}{6} + \cdots = \ln 2 \approx 0.693147. \tag{5}$$

REMARK. It should be noted that *any* rearrangement of the terms of an *absolutely converging* series converges to the same number.

We close this section by providing in Table 1 a summary of the convergence tests we have discussed.

TABLE 1 TESTS OF CONVERGENCE

Test	First discussed on page	Description	Examples and Comments
Convergence test for a geometric series	819	$\Sigma_{k=0}^{\infty} r^k$ converges to $1/(1-r)$ if $\lvert r\rvert < 1$ and diverges if $\lvert r\rvert > 1$.	$\Sigma_{k=0}^{\infty} (\frac{1}{2})^k$ converges to 2; $\Sigma_{k=0}^{\infty} 2^k$ diverges.
Look at the terms of the series—the limit test	825	If $\lvert a_k\rvert$ does not converge to 0, then $\Sigma\, a_n$ diverges.	If $a_k \to 0$, then $\Sigma_0^{\infty} a_k$ may converge ($\Sigma_{k=0}^{\infty} 1/k^2$) or it may not (the harmonic series $\Sigma_{k=0}^{\infty} 1/k$).
Comparison test	827	If $0 \le a_k \le b_k$ and $\Sigma\, b_k$ converges, then $\Sigma\, a_k$ converges. If $a_k \ge b_k \ge 0$ and $\Sigma\, b_k$ diverges, then $\Sigma\, a_k$ diverges.	It is not necessary that $a_k \le b_k$ or $a_k \ge b_k$ for *all* k, only for $k \ge N$ for some integer N; convergence or divergence of a series is not affected by the values of the first few terms.
Integral test	827	If $a_k = f(k) \ge 0$, then $\Sigma_{k=1}^{\infty} a_k$ converges if $\int_1^{\infty} f(x)\,dx$ converges and $\Sigma_{k=1}^{\infty} a_k$ diverges if $\int_1^{\infty} f(x)\,dx$ diverges.	Use this test whenever $f(x)$ can easily be integrated.
$\Sigma_{k=1}^{\infty} 1/k^{\alpha}$	828	$\Sigma_{k=1}^{\infty} 1/k^{\alpha}$ diverges if $0 \le \alpha \le 1$ and converges if $\alpha > 1$.	
Limit comparison test	829	If $a_k \ge 0$, $b_k \ge 0$ and there is a number c such that $\lim_{k\to\infty} a_k/b_k = c$, then either both series converge or both series diverge.	Use the limit comparison test when a series $\Sigma\, b_k$ can be found such that (a) it is known whether $\Sigma\, b_k$ converges or diverges and (b) it appears that a_k/b_k has an easily computed limit; (b) will be true, for instance, when $a_k = 1/p(k)$ and $b_k = 1/q(k)$ where $p(k)$ and $q(k)$ are polynomials.
Ratio test	832	If $a_k > 0$ and $\lim_{n\to\infty} a_{n+1}/a_n = L$, then $\Sigma_{k=1}^{\infty} a_k$ converges if $L < 1$ and diverges when $L > 1$.	This is often the easiest test to apply; note that if $L = 1$, then the series may either converge ($\Sigma\, 1/k^2$) or diverge ($\Sigma\, 1/k$).

TABLE 1 TESTS OF CONVERGENCE (Continued)

Test	First discussed on page	Description	Examples and Comments
Root test	835	If $a_k > 0$ and $\lim_{n\to\infty}(a_n)^{1/n} = R$, then $\sum_{k=1}^{\infty} a_k$ converges if $R < 1$ and diverges if $R > 1$.	If $R = 1$, the series may either converge ($\Sigma\ 1/k^2$) or diverge ($\Sigma\ 1/k$); the root test is the hardest test to apply; it is most useful when a_k is something raised to the kth power [$\Sigma\ 1/(\ln k)^k$, for example].
Alternating series test	837	$\Sigma\ (-1)^k\ a_k$ with $a_k \geq 0$ converges if (a) $a_k \to 0$ as $k \to \infty$ and (b) $\{a_k\}$ is a decreasing sequence; also, $\Sigma\ (-1)^k\ a_k$ diverges if $\lim_{k\to\infty} a_k \neq 0$.	This test can only be applied when the terms are alternately positive and negative; if there are two or more positive (or negative) terms in a row, then try another test.
Absolute convergence test for a series with both positive and negative terms	836	$\Sigma\ a_k$ converges absolutely if $\Sigma\ \lvert a_k \rvert$ converges.	To determine whether $\Sigma\ \lvert a_k \rvert$ converges, try any of the tests that apply to series with nonnegative terms.

PROBLEMS 13.10

In Problems 1–30, determine whether the given series is absolutely convergent, conditionally convergent, or divergent.

1. $\sum_{k=1}^{\infty} (-1)^k$

2. $\sum_{k=1}^{\infty} \frac{(-1)^{k+1}}{2k}$

3. $\sum_{k=2}^{\infty} \frac{(-1)^k}{k \ln k}$

4. $\sum_{k=1}^{\infty} \frac{(-1)^k}{k^{3/2}}$

5. $\sum_{k=2}^{\infty} \frac{(-1)^k k}{\ln k}$

6. $\sum_{k=1}^{\infty} \frac{(-1)^k \ln k}{k}$

7. $\sum_{k=1}^{\infty} \frac{(-1)^{k+1}}{5k - 4}$

8. $\sum_{k=1}^{\infty} \sin\frac{k\pi}{2}$

9. $\sum_{k=0}^{\infty} \cos\frac{k\pi}{2}$

10. $\sum_{k=1}^{\infty} \frac{(-3)^k}{k!}$

11. $\sum_{k=1}^{\infty} \frac{k!}{(-3)^k}$

12. $\sum_{k=1}^{\infty} \frac{(-2)^k}{k^2}$

13. $\sum_{k=1}^{\infty} \frac{k^2}{(-2)^k}$

14. $\sum_{k=2}^{\infty} \frac{(-1)^{k+1}}{\sqrt{k(k-1)}}$

15. $\sum_{k=2}^{\infty} \frac{(-1)^k k^2}{k^3 + 1}$

16. $\sum_{k=1}^{\infty} \frac{\cos(k\pi/6)}{k^2}$

17. $\sum_{k=3}^{\infty} \frac{\sin(k\pi/7)}{k^3}$

18. $\sum_{k=1}^{\infty} \frac{(-1)^k(k+2)}{k(k+1)}$

19. $\sum_{k=2}^{\infty} \frac{(-1)^k k(k+1)}{(k+2)^3}$

20. $\sum_{k=2}^{\infty} \frac{(-1)^k k(k+1)}{(k+2)^4}$

21. $\sum_{k=1}^{\infty} \frac{(-1)^k 2^k}{k}$

22. $\sum_{k=1}^{\infty} \frac{(-1)^{k+1}}{k!}$

23. $\sum_{k=1}^{\infty} \frac{(-1)^k k^k}{k!}$

24. $\sum_{k=1}^{\infty} \frac{(-1)^k \sqrt{k}}{k + 3}$

25. $\sum_{k=2}^{\infty} \frac{(-1)^k(k^2 + 3)}{k^3 + 4}$

26. $\sum_{k=2}^{\infty} \frac{(-1)^k}{\sqrt[3]{\ln k}}$

27. $\sum_{k=1}^{\infty} \frac{(-1)^k k^2}{4+k^2}$

28. $\sum_{k=1}^{\infty} (-1)^k \left(1+\frac{1}{k}\right)^k$

29. $\sum_{k=2}^{\infty} \frac{(-1)^k}{k\sqrt{\ln k}}$

30. $\sum_{k=2}^{\infty} \frac{(-1)^k k^3}{k^3+2k^2+k-1}$

***31.** Prove Theorem 3. [*Hint:* Assume that the odd-numbered terms are positive. Show that the sequence $\{S_{2n}\}$ is increasing and that $S_{2n} < S_{2n+2} < S$ for all $n \geq 1$. Then show that the sequence of odd-numbered partial sums is decreasing and that $S < S_{2n+1} < S_{2n-1}$ for all $n \geq 1$. Conclude that (a) $0 < S - S_{2n} < a_{2n+1}$ for all $n \geq 1$ and that (b) $0 < S_{2n-1} - S < -a_{2n}$. Use inequalities (a) and (b) to prove the theorem.]

In Problems 32–37, use the result of Theorem 3 or Theorem 4 to estimate the given sum to within the indicated accuracy.

32. $\sum_{k=1}^{\infty} \frac{(-1)^{k+1}}{k!}$; error < 0.001

33. $\sum_{k=1}^{\infty} \frac{(-1)^{k+1}}{k^2}$; error < 0.01

34. $\sum_{k=1}^{\infty} \frac{(-1)^{k+1}}{k^4}$; error < 0.0001

35. $\sum_{k=2}^{\infty} \frac{(-1)^{k+1}}{k \ln k}$; error < 0.05

36. $\sum_{k=1}^{\infty} \frac{(-1)^{k+1}}{k^k}$; error < 0.0001

37. $\sum_{k=1}^{\infty} \frac{(-1)^{k+1}}{\sqrt{k}}$; error < 0.1

38. Find the first ten terms of a rearrangement of the series $\Sigma_{k=1}^{\infty}(-1)^{k+1}/k$ that converges to 0.

39. Find the first ten terms of a rearrangement of the series $\Sigma_{k=1}^{\infty}(-1)^{k+1}/k$ that converges to 0.3.

40. Explain why there is no rearrangement of the series $\Sigma_{k=1}^{\infty}(-1)^{k}/k^2$ that converges to -1.

41. Prove that if $\Sigma_{k=1}^{\infty} a_k$ is a convergent series of nonzero terms, then $\Sigma_{k=1}^{\infty} 1/a_k$ diverges.

42. Show that if $\Sigma_{k=1}^{\infty} a_k$ is absolutely convergent, then $\Sigma_{k=1}^{\infty} a_k{}^p$ is convergent for any integer $p \geq 1$.

43. Give an example of a sequence $\{a_k\}$ such that $\Sigma_{k=1}^{\infty} a_k{}^2$ converges but $\Sigma_{k=1}^{\infty} a_k$ diverges.

***44.** Give an example of a sequence $\{a_k\}$ such that $\Sigma_{k=1}^{\infty} a_k$ converges but $\Sigma_{k=1}^{\infty} a_k{}^3$ diverges.

***45.** Prove Theorem 4. [*Hint:* Write the series as $S = \frac{1}{2}a_1 + \frac{1}{2}(a_1 - a_2) - \frac{1}{2}(a_2 - a_3) + \frac{1}{2}(a_3 - a_4) - \cdots = \frac{1}{2}a_1 + \Sigma_{k=1}^{\infty}(-1)^{k+1}(a_k - a_{k+1})/2$. Then apply Theorem 3.]

13.11 POWER SERIES

In previous sections we discussed infinite series of real numbers. Here we discuss series of functions.

Definition 1 POWER SERIES

(i) A **power series** in x is a series of the form

$$\sum_{k=0}^{\infty} a_k x^k = a_0 + a_1 x + a_2 x^2 + \cdots + a_n x^n + \cdots. \tag{1}$$

(ii) A power series in $(x - x_0)$ is a series of the form

$$\sum_{k=0}^{\infty} a_k (x-x_0)^k = a_0 + a_1(x-x_0) + a_2(x-x_0)^2 + \cdots + a_n(x-x_0)^n + \cdots, \tag{2}$$

where x_0 is a real number.

A power series in $(x - x_0)$ can be converted to a power series in u by the change of variables $u = x - x_0$. Then $\Sigma_{k=0}^{\infty} a_k(x - x_0)^k = \Sigma_{k=0}^{\infty} a_k u^k$. For example, consider

$$\sum_{k=0}^{\infty} \frac{(x-3)^k}{k!} \tag{3}$$

If $u = x - 3$, then the power series in $(x - 3)$ given by (3) can be written as

$$\sum_{k=0}^{\infty} \frac{u^k}{k!}$$

which is a power series in u.

Definition 2 CONVERGENCE AND DIVERGENCE OF A POWER SERIES

(i) A power series is said to **converge** at x if the series of real numbers $\Sigma_{k=0}^{\infty} a_k x^k$ converges. Otherwise, it is said to **diverge** at x.
(ii) A power series is said to converge in a set D of real numbers if it converges for every real number x in D.

EXAMPLE 1 For what real numbers does the power series

$$\sum_{k=0}^{\infty} \frac{x^k}{3^k} = 1 + \frac{x}{3} + \frac{x^2}{3^2} + \frac{x^3}{3^3} + \cdots$$

converge?

Solution. The nth term in this series is $x^n/3^n$. Using the ratio test, we find that

$$\lim_{n\to\infty} \frac{|a_{n+1}|}{|a_n|} = \lim_{n\to\infty} \frac{|x^{n+1}|/3^{n+1}}{|x^n/3^n|} = \lim_{n\to\infty} \left|\frac{x}{3}\right| = \left|\frac{x}{3}\right|.$$

We put in the absolute value bars since the ratio test only applies to a series of *positive* terms. However, as this example shows, we can use the ratio test to test for the absolute convergence of any series of nonzero terms by inserting absolute value bars, thereby making all the terms positive.

Thus the power series converges absolutely if $|x/3| < 1$ or $|x| < 3$ and diverges if $|x| > 3$. The case $|x| = 3$ has to be treated separately. For $x = 3$

$$\sum_{k=0}^{\infty} \frac{x^k}{3^k} = \sum_{k=0}^{\infty} \frac{3^k}{3^k} = \sum_{k=0}^{\infty} 1^k,$$

which diverges. For $x = -3$

$$\sum_{k=0}^{\infty} \frac{x^k}{3^k} = \sum_{k=0}^{\infty} (-1)^k,$$

which also diverges. Thus the series converges in the open interval $(-3, 3)$. We will show in Theorem 1 that since the series diverges for $x = 3$, it diverges for $|x| > 3$, so that conditional convergence at any x for which $|x| > 3$ is ruled out. ■

EXAMPLE 2 For what values of x does the series $\Sigma_{k=0}^{\infty} x^k/(k+1)$ converge?

Solution. Here $a_n = x^n/(n+1)$ so that

$$\lim_{n\to\infty} \frac{|a_{n+1}|}{|a_n|} = \lim_{n\to\infty} \left| \frac{x^{n+1}/(n+2)}{x^n/(n+1)} \right| = |x| \lim_{n\to\infty} \frac{n+1}{n+2} \overset{\lim_{n\to\infty} \frac{n+1}{n+2} = 1}{\downarrow}{=} |x|.$$

Thus the series converges absolutely for $|x| < 1$ and diverges for $|x| > 1$. If $x = 1$, then

$$\sum_{k=0}^{\infty} \frac{x^k}{k+1} = \sum_{k=0}^{\infty} \frac{1}{k+1},$$

which diverges since this series is the harmonic series. If $x = -1$, then

$$\sum_{k=0}^{\infty} \frac{x^k}{k+1} = \sum_{k=0}^{\infty} \frac{(-1)^k}{k+1} = 1 - \frac{1}{2} + \frac{1}{3} - \frac{1}{4} + \cdots,$$

which converges conditionally by the alternating series test ([see Example 13.10.6(a)]). In sum, the power series $\Sigma_{k=0}^{\infty} x^k/(k+1)$ converges in the half-open interval $[-1, 1)$. ■

The following theorem is of great importance in determining the range of values over which a power series converges.

Theorem 1

(i) If $\Sigma_{k=0}^{\infty} a_k x^k$ converges at x_0, $x_0 \neq 0$, then it converges absolutely at all x such that $|x| < |x_0|$.
(ii) If $\Sigma_{k=0}^{\infty} a_k x^k$ diverges at x_0, then it diverges at all x such that $|x| > |x_0|$.

Proof.

(i) Since $\Sigma_{k=0}^{\infty} a_k x_0^k$ converges, $a_k x_0^k \to 0$ as $k \to \infty$ by Theorem 13.7.2. This implies that for all k sufficiently large, $|a_k x_0^k| < 1$. Then if $|x| < |x_0|$ and if k is sufficiently large,

$$|a_k x^k| = \left| a_k \frac{x_0^k x^k}{x_0^k} \right| = |a_k x_0^k| \left| \frac{x}{x_0} \right|^k < \left| \frac{x}{x_0} \right|^k.$$

Since $|x| < |x_0|$, $|x/x_0| < 1$, and the geometric series $\Sigma_{k=0}^{\infty} |x/x_0|^k$ converges. Thus $\Sigma_{k=0}^{\infty} |a_k x^k|$ converges by the comparison test.

(ii) Suppose $|x| > |x_0|$ and $\Sigma_{k=0}^{\infty} a_k x_0^k$ diverges. If $\Sigma_{k=0}^{\infty} a_k x^k$ did converge, then by part (i), $\Sigma_{k=0}^{\infty} a_k x_0^k$ would also converge. This contradiction completes the proof of the theorem. ■

Theorem 1 is very useful for it enables us to place all power series in one of three categories:

Definition 3 RADIUS OF CONVERGENCE

Category 1: $\Sigma_{k=0}^{\infty} a_k x^k$ converges only at 0.
Category 2: $\Sigma_{k=0}^{\infty} a_k x^k$ converges for all real numbers.
Category 3: There exists a positive real number R, called the **radius of convergence** of the power series, such that $\Sigma_{k=0}^{\infty} a_k x^k$ converges if $|x| < R$ and diverges if $|x| > R$. At $x = R$ and at $x = -R$, the series may converge or diverge.

We can extend the notion of radius of convergence to Categories 1 and 2:

1. In Category 1 we say that the radius of convergence is 0.
2. In Category 2 we say that the radius of convergence is ∞.

NOTE. The series in Examples 1 and 2 both fall into Category 3. In Example 1, $R = 3$; and in Example 2, $R = 1$.

EXAMPLE 3 For what values of x does the series $\Sigma_{k=0}^{\infty} k!x^k$ converge?

Solution. Here

$$\lim_{n\to\infty}\left|\frac{a_{n+1}}{a_n}\right| = \lim_{n\to\infty}\left|\frac{(n+1)!x^{n+1}}{n!x^n}\right| = |x| \lim_{n\to\infty}(n+1) = \infty,$$

so that if $x \neq 0$, the series diverges. Thus $R = 0$ and the series falls into Category 1. ■

EXAMPLE 4 For what values of x does the series $\Sigma_{k=0}^{\infty} x^k/k!$ converge?

Solution. Here

$$\lim_{n\to\infty}\left|\frac{a_{n+1}}{a_n}\right| = \lim_{n\to\infty}\left|\frac{x^{n+1}/(n+1)!}{x^n/n!}\right| = |x| \lim_{n\to\infty}\frac{1}{n+1} = 0,$$

so that the series converges for every real number x. Here $R = \infty$ and the series falls into Category 2. ■

In going through these examples, we find that there is an easy way to calculate the radius of convergence. The proof of the following theorem is left as an exercise (see Problems 36 and 37).

Theorem 2 Consider the power series $\Sigma_{k=0}^{\infty} a_k x^k$ and suppose that $\lim_{n\to\infty}|a_{n+1}/a_n|$ exists and is equal to L or that $\lim_{n\to\infty}|a_n|^{1/n}$ exists and is equal to L.

(i) If $L = \infty$, then $R = 0$ and the series falls into Category 1.
(ii) If $L = 0$, then $R = \infty$ and the series falls into Category 2.
(iii) If $0 < L < \infty$, then $R = 1/L$ and the series falls into Category 3.

Definition 4 INTERVAL OF CONVERGENCE The **interval of convergence** of a power series is the interval over which the power series converges.

Using Theorem 2, we can calculate the interval of convergence of a power series in one or two steps:

(i) Calculate R. If $R = 0$, the series converges only at 0; and if $R = \infty$, the interval of convergence is $(-\infty, \infty)$.
(ii) If $0 < R < \infty$, check the values $x = -R$ and $x = R$. Then the interval of convergence is $(-R, R)$, $[-R, R)$, $(-R, R]$, or $[-R, R]$, depending on the convergence or divergence of the series at $x = R$ and $x = -R$.

NOTE. In Example 1, the interval of convergence is $(-3, 3)$ and in Example 2 the interval of convergence is $[-1, 1)$.

EXAMPLE 5 Find the radius of convergence and interval of convergence of the power series $\sum_{k=0}^{\infty} 2^k x^k/\ln(k+2)$.

Solution. Here $a_n = 2^n/\ln(n+2)$ and

$$L = \lim_{n\to\infty}\left|\frac{a_{n+1}}{a_n}\right| = \lim_{n\to\infty}\left|\frac{2^{n+1}/\ln(n+3)}{2^n/\ln(n+2)}\right| = 2\lim_{n\to\infty}\frac{\ln(n+2)}{\ln(n+3)} = 2.$$

Thus $R = 1/L = \frac{1}{2}$. For $x = \frac{1}{2}$,

$$\sum_{k=0}^{\infty}\frac{2^k x^k}{\ln(k+2)} = \sum_{k=0}^{\infty}\frac{1}{\ln(k+2)},$$

which diverges by comparison with the harmonic series since $1/\ln(k+2) > 1/(k+2)$ if $k \geq 1$. If $x = -\frac{1}{2}$, then

$$\sum_{k=0}^{\infty}\frac{2^k x^k}{\ln(k+2)} = \sum_{k=0}^{\infty}\frac{(-1)^k}{\ln(k+2)}$$

which converges by the alternating series test. Thus the interval of convergence is $[-\frac{1}{2}, \frac{1}{2})$. ■

EXAMPLE 6 Find the radius of convergence and interval of convergence of the power series $\sum_{k=0}^{\infty}(-1)^k(x-3)^k/(k+1)^2$.

Solution. We make the substitution $u = x - 3$. The series then becomes $\sum_{k=0}^{\infty}(-1)^k u^k/(k+1)^2$, and

$$L = \lim_{n\to\infty}\left|\frac{a_{n+1}}{a_n}\right| = \lim_{n\to\infty}\frac{(n+1)^2}{(n+2)^2} = 1,$$

so that $R = 1$. If $u = -1$,

$$\sum_{k=0}^{\infty} \frac{(-1)^k u^k}{(k+1)^2} = \sum_{k=0}^{\infty} \frac{1}{(k+1)^2},$$

which converges. If $u = 1$,

$$\sum_{k=0}^{\infty} \frac{(-1)^k u^k}{(k+1)^2} = \sum_{k=0}^{\infty} \frac{(-1)^k}{(k+1)^2},$$

which also converges. Thus the interval of convergence of the series $\Sigma_{k=0}^{\infty}(-1)^k u^k/(k+1)^2$ is $[-1, 1]$. Since $u = x - 3$, the original series converges for $-1 \le x - 3 \le 1$, or $2 \le x \le 4$. Hence the interval of convergence is $[2, 4]$. ■

EXAMPLE 7 Find the radius of convergence and interval of convergence of the power series $\Sigma_{k=0}^{\infty} x^{2k} = 1 + x^2 + x^4 + \cdots$.

Solution.

$$\sum_{k=0}^{\infty} x^{2k} = 1 + 0 \cdot x + 1 \cdot x^2 + 0 \cdot x^3 + 1 \cdot x^4 + 0 \cdot x^5 + 1 \cdot x^6 + \cdots.$$

This example illustrates the pitfalls of blindly applying formulas. We have $a_0 = 1$, $a_1 = 0$, $a_2 = 1$, $a_3 = 0, \ldots$. Thus the ratio a_{n+1}/a_n is 0 if n is even and is undefined if n is odd. The simplest thing to do here is to apply the ratio test directly. The ratio of consecutive terms is $x^{2k+2}/x^{2k} = x^2$. Thus the series converges if $|x| < 1$, diverges if $|x| > 1$, and the radius of convergence is 1. If $x = \pm 1$, then $x^2 = 1$ and the series diverges. Finally, the interval of convergence is $(-1, 1)$. ■

PROBLEMS 13.11

In Problems 1–33, find the radius of convergence and interval of convergence of the given power series.

1. $\sum_{k=0}^{\infty} \frac{x^k}{6^k}$
2. $\sum_{k=0}^{\infty} \frac{(-1)^k x^k}{8^k}$
3. $\sum_{k=0}^{\infty} \frac{(x+1)^k}{3^k}$
4. $\sum_{k=0}^{\infty} \frac{(-1)^k (x-3)^k}{4^k}$
5. $\sum_{k=0}^{\infty} (3x)^k$
6. $\sum_{k=0}^{\infty} \frac{x^k}{k^2+1}$
7. $\sum_{k=0}^{\infty} \frac{(x-1)^k}{k^3+3}$
8. $\sum_{k=2}^{\infty} \frac{x^k}{(\ln k)^2}$
9. $\sum_{k=0}^{\infty} \frac{(x+17)^k}{k!}$
10. $\sum_{k=2}^{\infty} \frac{x^k}{k \ln k}$
11. $\sum_{k=0}^{\infty} x^{2k}$
12. $\sum_{k=1}^{\infty} \frac{x^{2k}}{k}$
13. $\sum_{k=1}^{\infty} \frac{(-1)^k x^{2k}}{k^k}$
14. $\sum_{k=1}^{\infty} \frac{kx^k}{\ln(k+1)}$
15. $\sum_{k=0}^{\infty} \frac{(-1)^k k x^k}{\sqrt{k+1}}$
16. $\sum_{k=1}^{\infty} \frac{x^k}{k^k}$

*17. $\sum_{k=2}^{\infty} \frac{x^k}{(\ln k)^k}$ [*Hint:* Use the root test.]

*18. $\sum_{k=1}^{\infty} \frac{3^k x^k}{k^5}$

19. $\sum_{k=1}^{\infty} \frac{(-2x)^k}{k^4}$
20. $\sum_{k=0}^{\infty} \frac{(2x+3)^k}{k!}$
21. $\sum_{k=0}^{\infty} \frac{(2x+3)^k}{5^k}$
22. $\sum_{k=0}^{\infty} \frac{(3x-5)^k}{3^{2k}}$
23. $\sum_{k=0}^{\infty} \left(\frac{k}{15}\right)^k x^k$
24. $\sum_{k=0}^{\infty} (-1)^k x^{2k}$
25. $\sum_{k=0}^{\infty} (-1)^k x^{2k+1}$

26. $\sum_{k=1}^{\infty} \frac{(\ln k)(x+3)^k}{k+1}$

***27.** $\sum_{k=1}^{\infty} k^k(x+1)^k$

***28.** $\sum_{k=1}^{\infty} \frac{k^k}{k!}x^k$ [*Hint:* See Example 13.9.3, and use Stirling's formula][†]

29. $\sum_{k=0}^{\infty} \frac{(x+10)^k}{(k+1)3^k}$

***30.** $\sum_{k=1}^{\infty} \frac{k!}{k^k}x^k$

31. $\sum_{k=0}^{\infty} [1+(-1)^k]x^k$

32. $\sum_{k=0}^{\infty} \frac{[1+(-1)^k]}{k!}x^k$

33. $\sum_{k=1}^{\infty} \frac{[1+(-1)^k]}{k}x^k$

34. Show that the interval of convergence of the power series $\Sigma_{k=0}^{\infty} (ax+b)^k/c^k$ with $a > 0$ and $c > 0$ is $((-c-b)/a, (c-b)/a)$.

35. Prove that if the interval of convergence of a power series is $[a, b]$, then the power series is conditionally convergent at b.

36. Prove the ratio limit part of Theorem 2. [*Hint:* Show that if $|x| < 1/L$, then the series converges absolutely by applying the ratio test. Then show that if $|x| > 1/L$, the series diverges.]

37. Show that if $\lim_{n\to\infty} |a_n|^{1/n} = L$, then the radius of convergence of $\Sigma_{k=0}^{\infty} a_k x^k$ is $1/L$.

38. Show that if the radius of convergence of $\Sigma_{k=0}^{\infty} a_k x^k$ is R, and if $m > 0$ is an integer, then the radius of convergence of the power series $\Sigma_{k=0}^{\infty} a_k x^{mk}$ is $R^{1/m}$.

13.12 DIFFERENTIATION AND INTEGRATION OF POWER SERIES

Consider the power series

$$\sum_{k=0}^{\infty} a_k(x-x_0)^k = a_0 + a_1(x-x_0) + a_2(x-x_0)^2 + \cdots \tag{1}$$

with interval of convergence I. For each x in I we may define a new function f by

$$f(x) = \sum_{k=0}^{\infty} a_k(x-x_0)^k. \tag{2}$$

As we will see in this section and in Section 13.13, many familiar functions can be written as power series. In this section we will discuss some properties of a function given in the form of equation (2).

EXAMPLE 1 We know that

$$\sum_{k=0}^{\infty} x^k = 1 + x + x^2 + \cdots = \frac{1}{1-x} \quad \text{if} \quad |x| < 1. \tag{3}$$

Thus the function $1/(1-x)$ for $|x| < 1$ can be defined by

$$f(x) = \frac{1}{1-x} = \sum_{k=0}^{\infty} x^k, \qquad |x| < 1. \blacksquare$$

[†]**Stirling's formula** states that when n is large,

$$n! \approx \sqrt{2\pi n}\left(\frac{n}{e}\right)^n.$$

EXAMPLE 2 Substituting x^4 for x in (3) leads to the equality

$$f(x) = \frac{1}{1 - x^4} = \sum_{k=0}^{\infty} x^{4k} = 1 + x^4 + x^8 + x^{12} + \cdots \quad \text{if} \quad |x| < 1. \blacksquare$$

EXAMPLE 3 Substituting $-x$ for x in (3) leads to the equality

$$f(x) = \frac{1}{1 - (-x)} = \frac{1}{1 + x} = \sum_{k=0}^{\infty} (-1)^k x^k$$
$$= 1 - x + x^2 - x^3 + x^4 - \cdots \quad \text{if} \quad |x| < 1. \blacksquare$$

Once we see that certain functions can be written as power series, a question which might arise is whether such functions can be differentiated and integrated. The remarkable theorem given next ensures that every function represented as a power series can be differentiated and integrated at any x such that $|x| < R$, the radius of convergence. Moreover, we see how the derivative and integral can be calculated. The proof of this theorem is long (but not conceptually difficult) and so is omitted.[†]

Theorem 1 Let the power series $\Sigma_{k=0}^{\infty} a_k x^k$ have the radius of convergence $R > 0$. Let

$$f(x) = \sum_{k=0}^{\infty} a_k x^k = a_0 + a_1 x + a_2 x^2 + \cdots \quad \text{for} \quad |x| < R.$$

Then for $|x| < R$ we have the following:

(i) $f(x)$ is continuous.
(ii) The derivative $f'(x)$ exists, and

$$f'(x) = \frac{d}{dx} a_0 + \frac{d}{dx} a_1 x + \frac{d}{dx} a_2 x^2 + \cdots$$
$$= a_1 + 2a_2 x + 3a_3 x^2 + \cdots = \sum_{k=1}^{\infty} k a_k x^{k-1}.$$

(iii) The antiderivative $\int f(x)\,dx$ exists and

$$\int f(x)\,dx = \int a_0\,dx + \int a_1 x\,dx + \int a_2 x^2\,dx + \cdots$$
$$= a_0 x + a_1 \frac{x^2}{2} + a_2 \frac{x^3}{3} + \cdots + C = \sum_{k=0}^{\infty} a_k \frac{x^{k+1}}{k+1} + C.$$

Moreover, the two series $\Sigma_{k=1}^{\infty} k a_k x^{k-1}$ and $\Sigma_{k=0}^{\infty} a_k x^{k+1}/(k+1)$ both have radius of convergence R.

[†]See R. C. Buck *Advanced Calculus*, McGraw-Hill, New York, 1965.

Simply put, this theorem says that the derivative of a converging power series is the series of derivatives of its terms and that the integral of a converging power series is the series of integrals of its terms.

A more concise statement of Theorem 1 is: *A power series may be differentiated and integrated term by term within its radius of convergence.*

EXAMPLE 4 From Example 3 we have

$$\frac{1}{1+x} = 1 - x + x^2 - x^3 + \cdots = \sum_{k=0}^{\infty} (-1)^k x^k \tag{4}$$

for $|x| < 1$. Substituting $u = x + 1$, we have $x = u - 1$ and $1/(1 + x) = 1/u$. If $-1 < x < 1$, then $0 < u < 2$, and we obtain

$$\frac{1}{u} = 1 - (u - 1) + (u - 1)^2 - (u - 1)^3 + \cdots = \sum_{k=0}^{\infty} (-1)^k (u - 1)^k$$

for $0 < u < 2$. Integration then yields

$$\ln u = \int \frac{du}{u} = u - \frac{(u-1)^2}{2} + \frac{(u-1)^3}{3} - \cdots + C.$$

Since $\ln 1 = 0$, we immediately find that $C = -1$, so that

$$\ln u = \sum_{k=0}^{\infty} (-1)^k \frac{(u-1)^{k+1}}{k+1} \tag{5}$$

for $0 < u < 2$. Here we have expressed the logarithmic function defined on the interval $(0, 2)$ as a power series. ■

EXAMPLE 5 The series

$$f(x) = 1 + x + \frac{x^2}{2!} + \frac{x^3}{3!} + \cdots = \sum_{k=0}^{\infty} \frac{x^k}{k!} \tag{6}$$

converges for every real number x (i.e., $R = \infty$; see Example 13.11.4). But

$$f'(x) = \frac{d}{dx}1 + \frac{d}{dx}x + \frac{d}{dx}\frac{x^2}{2!} + \cdots = 1 + x + \frac{x^2}{2!} + \cdots = f(x).$$

Thus f satisfies the differential equation

$$f' = f,$$

and so from the discussion in Section 11.2, we find that

$$f(x) = ce^x \tag{7}$$

for some constant c. Substituting $x = 0$ into equations (6) and (7) yields

$$f(0) = 1 = ce^0 = c,$$

so that $f(x) = e^x$. We have obtained an important expansion which is valid for any real number x:

$$e^x = 1 + x + \frac{x^2}{2!} + \frac{x^3}{3!} + \cdots = \sum_{k=0}^{\infty} \frac{x^k}{k!}. \tag{8}$$

For example, if we substitute the value $x = 1$ into (8), we obtain partial sum approximations for $e = 1 + 1 + 1/2! + 1/3! + \cdots$ (see Table 1). The last value $(\Sigma_{k=0}^{8} 1/k!)$ is correct to five decimal places.

TABLE 1

n	0	1	2	3	4	5	6	7	8
$S_n = \sum_{k=0}^{n} \frac{1}{k!}$	1	2	2.5	2.66667	2.70833	2.71667	2.71806	2.71825	2.71828

■

EXAMPLE 6 Substituting $-x$ for x in (8), we obtain

$$e^{-x} = 1 - x + \frac{x^2}{2!} - \frac{x^3}{3!} + \cdots = \sum_{k=0}^{\infty} (-1)^k \frac{x^k}{k!}. \tag{9}$$

Since this is an alternating series if $x > 0$, Theorem 13.10.3 tells us that the error $|S - S_n|$ in approximating e^{-x} for $x > 0$ is bounded by $|a_{n+1}| = x^{n+1}/(n + 1)!$.[†] For example, to calculate e^{-1} with an error of less than 0.0001, we must have $|S - S_n| \le 1/(n + 1)! < 0.0001 = 1/10{,}000$, or $(n + 1)! > 10{,}000$. If $n = 7$, $(n + 1)! = 8! = 40{,}320$, so that $\Sigma_{k=0}^{7} (-1)^k/k!$ will approximate e^{-1} correct to four decimal places. We obtain

$$\begin{aligned} e^{-1} &\approx 1 - 1 + \frac{1}{2!} - \frac{1}{3!} + \frac{1}{4!} - \frac{1}{5!} + \frac{1}{6!} - \frac{1}{7!} \\ &= \frac{1}{2} - \frac{1}{6} + \frac{1}{24} - \frac{1}{120} + \frac{1}{720} - \frac{1}{5040} \\ &= 0.5 - 0.16667 + 0.04167 - 0.00833 + 0.00139 - 0.0002 \\ &\approx 0.36786. \end{aligned}$$

It is true that $e^{-1} \approx 0.36788$ correct to five decimal places. ■

[†]We can apply Theorem 13.10.3 here because the terms in the sequence $\{x^n/n!\}$ are monotone decreasing as long as $x < n + 1$.

EXAMPLE 7 Consider the series

$$f(x) = 1 - \frac{x^2}{2!} + \frac{x^4}{4!} - \frac{x^6}{6!} + \cdots = \sum_{k=0}^{\infty} (-1)^k \frac{x^{2k}}{(2k)!}. \tag{10}$$

It is easy to see that $R = \infty$ since the series is absolutely convergent for every x by comparison with the series (8) for e^x. [The series (8) is larger than the series (10) since it contains the terms $x^n/n!$ for n both even and odd, not just for n even, as in (10).] Differentiating, we obtain

$$f'(x) = -x + \frac{x^3}{3!} - \frac{x^5}{5!} + \frac{x^7}{7!} - \cdots = \sum_{k=0}^{\infty} (-1)^{k+1} \frac{x^{2k+1}}{(2k+1)!}. \tag{11}$$

Since this series has a radius of convergences $R = \infty$, we can differentiate once more to obtain

$$f''(x) = -1 + \frac{x^2}{2!} - \frac{x^4}{4!} + \frac{x^6}{6!} - \cdots = \sum_{k=0}^{\infty} \frac{(-1)^{k+1} x^{2k}}{(2k)!} = -f(x).$$

Thus we see that f satisfies the differential equation

$$f'' + f = 0. \tag{12}$$

Moreover, from equations (10) and (11) we see that

$$f(0) = 1 \quad \text{and} \quad f'(0) = 0. \tag{13}$$

We discussed this differential equation in Section 11.9. There it was seen that the function $f(x) = \cos x$ is the unique solution to the initial value problem (12), (13). Thus we have

$$\cos x = 1 - \frac{x^2}{2!} + \frac{x^4}{4!} - \frac{x^6}{6!} + \cdots = \sum_{k=0}^{\infty} (-1)^k \frac{x^{2k}}{(2k)!}. \tag{14}$$

Since

$$\frac{d}{dx} \cos x = -\sin x,$$

we obtain, from (14), the series

$$\sin x = x - \frac{x^3}{3!} + \frac{x^5}{5!} - \frac{x^7}{7!} + \cdots = \sum_{k=0}^{\infty} (-1)^k \frac{x^{2k+1}}{(2k+1)!}. \quad ■ \tag{15}$$

Power series expansions can be very useful for approximate integration.

EXAMPLE 8 Calculate $\int_0^1 e^{-t^2}\,dt$ with an error < 0.0001.

Solution. Substituting t^2 for x in (9), we find that

$$e^{-t^2} = 1 - t^2 + \frac{t^4}{2!} - \frac{t^6}{3!} + \cdots = \sum_{k=0}^{\infty} (-1)^k \frac{t^{2k}}{k!}. \tag{16}$$

Then we integrate to find that

$$\int_0^x e^{-t^2}\,dt = x - \frac{x^3}{3} + \frac{x^5}{5\cdot 2!} - \frac{x^7}{7\cdot 3!} + \cdots = \sum_{k=0}^{\infty} \frac{(-1)^k x^{2k+1}}{(2k+1)k!}. \tag{17}$$

The error $|S - S_n|$ is bounded by

$$|a_{n+1}| = \left|\frac{x^{2(n+1)+1}}{[2(n+1)+1](n+1)!}\right|.^{\dagger}$$

In our example $x = 1$, so we need to choose n so that

$$\frac{1}{(2n+3)(n+1)!} < 0.0001, \qquad \text{or} \qquad (2n+3)(n+1)! > 10{,}000.$$

If $n = 6$, then $(2n + 3)(n + 1)! = (15)(7!) = 75{,}600$. With this choice of n ($n = 5$ is too small), we obtain

$$\begin{aligned}\int_0^1 e^{-t^2}\,dt &\approx 1 - \frac{1}{3} + \frac{1}{5\cdot 2!} - \frac{1}{7\cdot 3!} + \frac{1}{9\cdot 4!} - \frac{1}{11\cdot 5!} + \frac{1}{13\cdot 6!} \\ &\approx 1 - 0.33333 + 0.1 - 0.02381 + 0.00463 - 0.00076 + 0.00011 \\ &= 0.74684,\end{aligned}$$

and to four decimal places,

$$\int_0^1 e^{-t^2}\,dt = 0.7468. \blacksquare$$

The trapezoidal rule and Simpson's rule are examples of some more general ways to calculate definite integrals that do not require the existence of a series expansion of the function being integrated. However, as shown in Example 8, a power series provides an easy method of numerical integration when the power series representation of a function is available.

---

† As in Example 6, we can apply Theorem 13.10.3 as long as the terms are decreasing. Here $a_{n-1} < a_n$ if $x^2 < (2n + 3)(n + 1)/(2n + 1)$. This holds for every n when $x = 1$.

PROBLEMS 13.12

1. By substituting x^2 for x in (4), find a series expansion for $1/(1 + x^2)$ that is valid for $|x| < 1$.
2. Integrate the series obtained in Problem 1 to obtain a series expansion for $\tan^{-1} x$.
3. Use the result of Problem 2 to obtain an estimate of π that is correct to two decimal places.
4. Use the series expansion for $\ln x$ to calculate the following to two decimal places of accuracy:
   (a) $\ln 0.5$ **(b)** $\ln 1.6$

In Problems 5–13, estimate the given integral to within the given accuracy.

5. $\int_0^1 e^{-t^2}\,dt$; error < 0.01
6. $\int_0^1 e^{-t^3}\,dt$; error < 0.001
7. $\int_0^{1/2} \cos t^2\,dt$; error < 0.001
8. $\int_0^{1/2} \sin t^2\,dt$; error < 0.0001
9. $\int_0^1 t^2 e^{-t^2}\,dt$; error < 0.01

   [*Hint:* The series expansion of $t^2e^{-t^2}$ is obtained by multiplying each term of the series expansion of e^{-t^2} by t^2.]
10. $\int_0^{1/4} t^5 e^{-t^5}\,dt$; error < 0.0001
11. $\int_0^1 \cos\sqrt{t}\,dt$; error < 0.01
12. $\int_0^1 t\sin\sqrt{t}\,dt$; error < 0.001
13. $\int_0^{1/2} \frac{dt}{1 + t^8}$; error < 0.0001
14. Find a series expansion for xe^x that is valid for all real values of x.
15. Use the result of Problem 14 to find a series expansion for $\int_0^x te^t\,dt$.
16. Use the result of Problem 15 to show that $\sum_{k=0}^{\infty} 1/(k + 2)k! = 1$.
17. Find a power series expansion for $\int_0^x [\ln(1 + t)/t]\,dt$.
18. Expand $1/x$ as a power series of the form $\sum_{k=0}^{\infty} a_k(x - 1)^k$. What is the interval of convergence of this series?

*19. Define the function $J(x)$ by

$$J(x) = \sum_{k=0}^{\infty} [(-1)^k/(k!)^2](x/2)^{2k}.$$

(a) What is the interval of convergence of this series?

(b) Show that $J(x)$ satisfies the differential equation

$$x^2J''(x) + xJ'(x) + x^2J(x) = 0.$$

The function $J(x)$ is called a **Bessel function of order zero.**†

13.13 TAYLOR AND MACLAURIN SERIES

In the last two sections we used the fact that within its interval of convergence, the function

$$f(x) = \sum_{k=0}^{\infty} a_k(x - x_0)^k$$

†Named after the German physicist and mathematician Wilhelm Bessel (1784–1846), who used the function in his study of planetary motion. The Bessel functions of various orders arise in many applications in modern physics and engineering.

is differentiable and integrable. In this section we look more closely at the coefficients a_k and show that they can be represented in terms of derivatives of the function f.

We begin with the case $x_0 = 0$ and assume that $R > 0$, so that the theorem on power series differentiation applies. We have

$$f(x) = \sum_{k=0}^{\infty} a_k x^k = a_0 + a_1 x + a_2 x^2 + \cdots + a_n x^n + \cdots, \tag{1}$$

and clearly,

$$f(0) = a_0 + 0 + 0 + \cdots + 0 + \cdots = a_0. \tag{2}$$

If we differentiate (1), we obtain

$$f'(x) = \sum_{k=1}^{\infty} k a_k x^{k-1} = a_1 + 2a_2 x + 3a_3 x^2 + \cdots + n a_n x^{n-1} + \cdots \tag{3}$$

and

$$f'(0) = a_1. \tag{4}$$

Continuing to differentiate, we obtain

$$\begin{aligned} f''(x) &= \sum_{k=2}^{\infty} k(k-1) a_k x^{k-2} \\ &= 2a_2 + 3 \cdot 2a_3 x + 4 \cdot 3a_4 x^2 + \cdots + n(n-1)a_n x^{n-2} + \cdots \end{aligned}$$

and

$$f''(0) = 2a_2,$$

or

$$a_2 = \frac{f''(0)}{2} = \frac{f''(0)}{2!}. \tag{5}$$

Similarly,

$$\begin{aligned} f'''(x) &= \sum_{k=3}^{\infty} k(k-1)(k-2) a_k x^{k-3} \\ &= 3 \cdot 2a_3 + 4 \cdot 3 \cdot 2a_4 x + 5 \cdot 4 \cdot 3a_5 x^2 + \cdots + n(n-1)(n-2)a_n x^{n-3} + \cdots \end{aligned}$$

and

$$f'''(0) = 3 \cdot 2a_3,$$

or

$$a_3 = \frac{f'''(0)}{3 \cdot 2} = \frac{f'''(0)}{3!}. \tag{6}$$

It is not difficult to see that this pattern continues and that for every positive integer n

$$a_n = \frac{f^{(n)}(0)}{n!}. \tag{7}$$

For $n = 0$ we use the convention $0! = 1$ and $f^{(0)}(x) = f(x)$. Then formula (7) holds for every nonnegative n, and we have the following:

If

$$f(x) = \sum_{k=0}^{\infty} a_k x^k,$$

then

$$f(x) = \sum_{k=0}^{\infty} \frac{f^{(k)}(0)}{k!} x^k$$

$$= f(0) + f'(0)x + f''(0)\frac{x^2}{2!} + \cdots + f^{(n)}(0)\frac{x^n}{n!} + \cdots \tag{8}$$

for every x in the interval of convergence.

In case $x_0 \neq 0$, similar behavior occurs. If

$$f(x) = \sum_{k=0}^{\infty} a_k (x - x_0)^k$$

$$= a_0 + a_1(x - x_0) + a_2(x - x_0)^2 + \cdots + a_n(x - x_0)^n + \cdots, \tag{9}$$

then

$$f(x_0) = a_0,$$

and differentiating as before, we find that

$$a_n = \frac{f^{(n)}(x_0)}{n!}. \tag{10}$$

Thus we have the following: If

$$f(x) = \sum_{k=0}^{\infty} a_k (x - x_0)^k,$$

then

$$f(x) = \sum_{k=0}^{\infty} \frac{f^{(k)}(x_0)}{k!}(x - x_0)^k$$

$$= f(x_0) + f'(x_0)(x - x_0) + f''(x_0)\frac{(x - x_0)^2}{2!} + \cdots$$

$$+ f^{(n)}(x_0)\frac{(x - x_0)^n}{n!} + \cdots \qquad \textbf{(11)}$$

for every x in the interval of convergence.

Definition 1 TAYLOR AND MACLAURIN SERIES The series in (11) is called the **Taylor series**† of the function f at x_0. The special case $x_0 = 0$ in (8) is called a **Maclaurin series.**‡ We see that the first n terms of the Taylor series of a function are simply the Taylor polynomial described in Section 13.1.

■ WARNING: We have shown here that *if* $f(x) = \sum_{k=0}^{\infty} a_k(x - x_0)^k$, then f is infinitely differentiable (i.e., f has derivatives of all orders) and that the series for f is the Taylor series (or Maclaurin series if $x_0 = 0$) of f. What we have *not* shown is that if f is infinitely differentiable at x_0, then f has a Taylor series expansion at x_0. In general, this last statement is false, as we will see in Example 3.

EXAMPLE 1 Find the Maclaurin series for e^x.

Solution. If $f(x) = e^x$, then $f(0) = f'(0) = \cdots = f^{(k)}(0) = 1$, and

$$e^x = \sum_{k=0}^{\infty} \frac{x^k}{k!} = 1 + x + \frac{x^2}{2!} + \frac{x^3}{3!} + \cdots + \frac{x^n}{n!} + \cdots. \qquad \textbf{(12)}$$

This series is the series we obtained in Example 13.12.5. It is important to comment here that what this example shows is that *if* e^x has a Maclaurin series expansion, then the series must be the series (12). It does not show that e^x actually does have such a series expansion. To prove that the series in (12) is really equal to e^x, we differentiate, as in Example 13.12.5, and use the fact that the only continuous function that satisfies

$$f'(x) = f(x), \qquad f(0) = 1,$$

is the function e^x. ■

EXAMPLE 2 Assuming that the function $f(x) = \cos x$ can be written as a Maclaurin series, find that series.

---

†The history of the Taylor series is somewhat muddied. It has been claimed that the basis for its development was found in India before 1550! (Taylor published the result in 1715.) For an interesting discussion of this controversy, see the paper by C. T. Rajagopal and T. V. Vedamurthi, "On the Hindu proof of Gregory's series," *Scripta Mathematica* **17**, 65–74 (1951).

‡See the accompanying biographical sketch.

1698–
1746

Colin Maclaurin

Considered the finest British mathematician of the generation after Newton, Colin Maclaurin was certainly one of the best mathematicians of the eighteenth century.

Born in Scotland, Maclaurin was a mathematical prodigy and entered Glasgow University at the age of eleven. By the age of nineteen he was a professor of mathematics in Aberdeen and later obtained a post at the University of Edinburgh.

Maclaurin is best known for the term *Maclaurin series,* which is the Taylor series in the case $x_0 = 0$. He used this series in his 1742 work, *Treatise of Fluxions.* (Maclaurin acknowledged that the series had first been used by Taylor in 1715.) The *Treatise of Fluxions* was most significant in that it presented the first logical description of Newton's method of fluxions. This work was written to defend Newton from the attacks of the powerful Bishop George Berkeley (1685–1753). Berkeley was troubled (as are many of today's calculus students) by the idea of a quotient that takes the form 0/0. This, of course, is what we obtain when we take a derivative. Berkeley wrote:

And what are these fluxions? The velocities of evanescent increments. And what are these same evanescent increments? They are neither finite quantities nor quantities infinitely small nor yet nothing. May we not call them ghosts of departed quantities?

Maclaurin answered Berkeley using geometric arguments. Later, Newton's calculus was put on an even firmer footing by the work of Lagrange in 1797 (see page 255).

Maclaurin made many other contributions to mathematics—especially in the areas of geometry and algebra. He published his *Geometria organica* when only 21 years old. His posthumous work *Treatise of Algebra,* published in 1748, contained many important results, including the well-known *Cramer's rule* for solving a system of equations (Cramer published the result in 1750).

In 1745, when "Bonnie Prince Charlie" marched against Edinburgh, Maclaurin helped defend the city. When the city fell, Maclaurin escaped, fleeing to York, where he died in 1746 at the age of 48.

Solution. If $f(x) = \cos x$, then $f(0) = 1$, $f'(0) = 0$, $f''(0) = -1$, $f'''(0) = 0$, $f^{(4)}(0) = 1$, and so on, so that if

$$\cos x = \sum_{k=0}^{\infty} a_k x^k,$$

then

$$\cos x = f(0) + f'(0) + \frac{f''(0)x^2}{2!} + \frac{f'''(0)x^3}{3!} + \frac{f^{(4)}(0)x^4}{4!} + \cdots,$$

and

$$\cos x = 1 - \frac{x^2}{2!} + \frac{x^4}{4!} - \frac{x^6}{6!} + \cdots = \sum_{k=0}^{\infty} \frac{(-1)^k x^{2k}}{(2k)!}. \quad \textbf{(13)}$$

This series is the series found in Example 13.12.7.

NOTE. Again, this does not prove that the equality in (13) is correct. It only shows that *if* cos x has a Maclaurin expansion, then the expansion must be given by (13). We will show that cos x has a Maclaurin series in Example 5. ■

EXAMPLE 3 Let

$$f(x) = \begin{cases} e^{-1/x^2}, & \text{if } x \neq 0 \\ 0, & \text{if } x = 0. \end{cases}$$

Find a Maclaurin expansion for f if one exists.

Solution. First, we note that since $\lim_{x \to 0} e^{-1/x^2} = 0$, f is continuous. Now recall that $\lim_{x \to \infty} x^a e^{-bx} = 0$ if $b > 0$. Let $y = 1/x^2$. Then as $x \to 0$, $y \to \infty$. Also, $1/x^n = (1/x^2)^{n/2}$, so that $\lim_{x \to 0} (e^{-1/x^2}/x^n) = \lim_{y \to \infty} y^{n/2}e^{-y} = 0$.

Now for $x \neq 0$, $f'(x) = (2/x^3)e^{-1/x^2} \to 0$ as $x \to 0$, so that f' is continuous at 0. Similarly, $f''(x) = [(4/x^6) - (6/x^4)]e^{-1/x^2}$, which also approaches 0 as $x \to 0$ by the limit result above. In fact, *every* derivative of f is continuous and $f^{(n)}(0) = 0$ for every n. Thus f is infinitely differentiable, and *if* it had a Maclaurin series that represented the function, then we would have

$$f(x) = f(0) + f'(0)x + f''(0)\frac{x^2}{2!} + \cdots.$$

But $f(0) = f'(0) = f''(0) = \cdots = 0$, so that the Maclaurin series would be the zero series. But since f is obviously not the zero function, we can only conclude that there is *no* Maclaurin series that represents f at any point other than 0. ■

Example 3 illustrates that infinite differentiability is not sufficient to guarantee that a given function can be represented by its Taylor series. Something more is needed.

Definition 2 ANALYTIC FUNCTION We say that a function f is **analytic** at x_0 if f can be represented by a Taylor series in some neighborhood of x_0.

We see that the functions e^x and $\cos x$ are analytic at 0, while the function

$$f(x) = \begin{cases} e^{-1/x^2}, & x \neq 0 \\ 0, & x = 0 \end{cases}$$

is not. A condition that guarantees analyticity of an infinitely differentiable function is given below.

Theorem 1 Suppose that the function f has continuous derivatives of all orders in a neighborhood $N(x_0)$ of the number x_0, and let $R_n(x)$ denote the remainder term in Taylor's theorem.

Then f is analytic at x_0 if and only if

$$\lim_{n \to \infty} R_n(x) = \lim_{n \to \infty} \frac{f^{(n+1)}(c_n)}{(n+1)!}(x - x_0)^{n+1} = 0 \qquad \textbf{(14)}$$

for every x in $N(x_0)$ where c_n is between x_0 and x.

REMARK. The expression between the equal signs in (14) is simply the remainder term given by Taylor's theorem, Theorem 13.1.1.

Proof. The hypotheses of Taylor's theorem apply, so that we can write, for any n,

$$f(x) = P_n(x) + R_n(x), \tag{15}$$

where $P_n(x)$ is the nth degree Taylor polynomial for f. To show that f is analytic, we must show that

$$\lim_{n\to\infty} P_n(x) = f(x) \tag{16}$$

for every x in $N(x_0)$. But if x is in $N(x_0)$, we obtain, from (14) and (15),

$$\lim_{n\to\infty} P_n(x) = \lim_{n\to\infty} [f(x) - R_n(x)] = f(x) - \lim_{n\to\infty} R_n(x) = f(x) - 0 = f(x).$$

Conversely, if f is analytic, then $f(x) = \lim_{n\to\infty} P_n(x)$ so $R_n(x) \to 0$ as $n \to \infty$. ■

EXAMPLE 4 If $f(x) = e^x$, then $f^{(n)}(x) = e^x$, and

$$\lim_{n\to\infty} \left| \frac{f^{(n+1)}(c_n)}{(n+1)!}(x - 0)^{n+1} \right| = \lim_{n\to\infty} \frac{e^{c_n}|x|^{n+1}}{(n+1)!} \le e^{|x|} \lim_{n\to\infty} \frac{|x|^{n+1}}{(n+1)!} \to 0, \quad (0 < c_n < |x|)$$

since $|x|^{n+1}/(n+1)!$ is the $(n+2)$nd term in the converging power series $\Sigma_{k=0}^{\infty} |x|^k/k!$ and the terms in a converging power series $\to 0$ by the Theorem 13.7.2. Since this result is true for any $x \in \mathbb{R}$, we may take $N = (-\infty, \infty)$ to conclude that the series (12) is valid for every real number x. ■

EXAMPLE 5 Let $f(x) = \cos x$. Since all derivatives of $\cos x$ are equal to $\pm \sin x$ or $\pm \cos x$, we see that $|f^{(n+1)}(c_n)| \le 1$. Then for $x_0 = 0$, $|R_n(x)| \le |x|^{n+1}/(n+1)!$, which $\to 0$ as $n \to \infty$, so that the series (13) is also valid for every real number x. ■

EXAMPLE 6 It is evident for the function in Example 3 that, $R_n(x) \not\to 0$ if $x \ne 0$. This follows from the fact that $R_n(x) = f(x) - P_n(x) = e^{-1/x^2} - 0 = e^{-1/x^2} \ne 0$ if $x \ne 0$. ■

EXAMPLE 7 Find the Taylor expansion for $f(x) = \ln x$ at $x = 1$.

Solution. Since $f'(x) = 1/x$, $f''(x) = -1/x^2$, $f'''(x) = 2/x^3$, $f^{(4)}(x) = -6/x^4, \ldots,$ $f^{(n)}(x) = (-1)^{n+1}(n-1)!/x^n$, we find that $f(1) = 0$, $f'(1) = 1$, $f''(1) = -1$, $f'''(1) = 2$, $f^{(4)}(1) = -6, \ldots, f^{(n)}(1) = (-1)^{n+1}(n-1)!$. Then wherever valid,

$$\begin{aligned}\ln x &= \sum_{k=0}^{\infty} f^{(k)}(1)\frac{(x-1)^k}{k!} \\ &= 0 + (x-1) - \frac{(x-1)^2}{2} + \frac{2(x-1)^3}{3!} - \frac{3!(x-1)^4}{4!} + \frac{4!(x-1)^5}{5!} + \cdots,\end{aligned}$$

or

$$\ln x = (x-1) - \frac{(x-1)^2}{2} + \frac{(x-1)^3}{3} - \frac{(x-1)^4}{4} + \cdots$$
$$= \sum_{k=1}^{\infty} \frac{(-1)^{k+1}(x-1)^k}{k}. \quad \blacksquare \tag{17}$$

In Section 13.3 [see equation (13.3.8)] we showed that

$$\ln(1+x) = \sum_{k=1}^{n+1} \frac{(-1)^{k+1}x^k}{k} + R_{n+1}(x), \tag{18}$$

where $R_{n+1}(x) \to 0$ as $n \to \infty$ whenever $-1 < x \le 1$ [see equations (13.3.9) and (13.3.10)]. From (18) we see that

$$\ln u = \ln[1 + (u-1)] = \sum_{k=1}^{n+1} \frac{(-1)^{k+1}(u-1)^k}{k} + R_{n+1}(u-1), \tag{19}$$

where $R_{n+1}(u-1) \to 0$ as $n \to \infty$ whenever $-1 < u - 1 \le 1$, or $0 < u \le 2$. But this implies that the series (17) converges to $\ln(x)$ for $0 < x \le 2$. When $x = 2$, we obtain, from (17),

$$\ln 2 = 1 - \frac{1}{2} + \frac{1}{3} - \frac{1}{4} + \frac{1}{5} - \cdots = \sum_{k=1}^{\infty} \frac{(-1)^{k+1}}{k}. \tag{20}$$

EXAMPLE 8 Find a Taylor series for $f(x) = \sin x$ at $x = \pi/3$.

Solution. Here $f(\pi/3) = \sqrt{3}/2$, $f'(\pi/3) = 1/2$, $f''(\pi/3) = -\sqrt{3}/2$, $f'''(\pi/3) = -1/2$, and so on, so that

$$\sin x = \frac{\sqrt{3}}{2} + \frac{1}{2}\left(x - \frac{\pi}{3}\right) - \frac{\sqrt{3}}{2}\frac{[x-(\pi/3)]^2}{2!} - \frac{1}{2}\frac{[x-(\pi/3)]^3}{3!} + \frac{\sqrt{3}}{2}\frac{[x-(\pi/3)]^4}{4!} + \cdots.$$

The proof that this series is valid for every real number x is similar to the proof in Example 5 and is therefore omitted. ■

We provide here a list of useful Maclaurin series:

$$e^x = \sum_{k=0}^{\infty} \frac{x^k}{k!} = 1 + x + \frac{x^2}{2!} + \frac{x^3}{3!} + \cdots \tag{21}$$

$$\cos x = \sum_{k=0}^{\infty} \frac{(-1)^k x^{2k}}{(2k)!} = 1 - \frac{x^2}{2!} + \frac{x^4}{4!} - \frac{x^6}{6!} + \cdots \tag{22}$$

$$\sin x = \sum_{k=0}^{\infty} \frac{(-1)^k x^{2k+1}}{(2k+1)!} = x - \frac{x^3}{3!} + \frac{x^5}{5!} - \frac{x^7}{7!} + \cdots \tag{23}$$

$$\cosh x = \sum_{k=0}^{\infty} \frac{x^{2k}}{(2k)!} = 1 + \frac{x^2}{2!} + \frac{x^4}{4!} + \frac{x^6}{6!} + \cdots \tag{24}$$

$$\sinh x = \sum_{k=0}^{\infty} \frac{x^{2k+1}}{(2k+1)!} = x + \frac{x^3}{3!} + \frac{x^5}{5!} + \frac{x^7}{7!} + \cdots \tag{25}$$

$$\frac{1}{1-x} = \sum_{k=0}^{\infty} x^k = 1 + x + x^2 + x^3 + \cdots, \text{ for } |x| < 1 \tag{26}$$

$$\ln(1+x) = \sum_{k=0}^{\infty} \frac{(-1)^k x^{k+1}}{k+1} = x - \frac{x^2}{2} + \frac{x^3}{3} - \frac{x^4}{4} + \cdots, \text{ for } |x| < 1 \tag{27}$$

$$\tan^{-1} x = \sum_{k=0}^{\infty} \frac{(-1)^k x^{2k+1}}{2k+1} = x - \frac{x^3}{3} + \frac{x^5}{5} - \frac{x^7}{7} + \cdots, \text{ for } |x| < 1 \tag{28}$$

You are asked to prove, in Problems 1 and 2, that the series (23), (24), and (25) are valid for every real number x.

BINOMIAL SERIES

We close this section by deriving another series that is quite useful. Let $f(x) = (1+x)^r$, where r is a real number not equal to an integer. We have

$$\begin{aligned}
f'(x) &= r(1+x)^{r-1}, \\
f''(x) &= r(r-1)(1+x)^{r-2}, \\
f'''(x) &= r(r-1)(r-2)(1+x)^{r-3}, \\
&\vdots \\
f^{(n)}(x) &= r(r-1)(r-2)\cdots(r-n+1)(1+x)^{r-n}.
\end{aligned}$$

Since r is not an integer, $r - n$ is never equal to 0, and all derivatives exist and are nonzero as long as $x \neq -1$. Then

$$\begin{aligned}
f(0) &= 1, \\
f'(0) &= r, \\
f''(0) &= r(r-1), \\
&\vdots \\
f^{(n)}(0) &= r(r-1)\cdots(r-n+1),
\end{aligned}$$

and we can write

$$(1+x)^r = 1 + rx + \frac{r(r-1)}{2!}x^2 + \frac{r(r-1)(r-2)}{3!}x^3 + \cdots$$
$$+ \frac{r(r-1)\cdots(r-n+1)}{n!}x^n + \cdots \quad \textbf{(29)}$$
$$= 1 + \sum_{k=1}^{\infty} \frac{r(r-1)\cdots(r-k+1)}{k!}x^k$$

The series (29) is called the **binomial series.**

Some applications of the binomial series are given in Problems 25–29.

PROBLEMS 13.13

1. Prove that the series (23) represents $\sin x$ for all real x.
2. **(a)** Prove that the series (24) represents $\cosh x$ for all real x.
   (b) Use the fact that $\sinh x = (d/dx) \cosh x$ to derive the series in (25).
3. Find the Taylor series for e^x at 1.
4. Find the Maclaurin series for e^{-x}.
5. Find the Taylor series for $\cos x$ at $\pi/4$.
6. Find the Taylor series for $\sinh x$ at $\ln 2$.
7. Find the Maclaurin series for $e^{\alpha x}$, α real.
8. Find the Maclaurin series for xe^x.
9. Find the Maclaurin series for $x^2e^{-x^2}$.
10. Find the Maclaurin series for $(\sin x)/x$.
11. Find the Taylor series for e^x at $x = -1$.
12. Find the Maclaurin series for $\sin^2 x$. [*Hint:* $\sin^2 x = (1 - \cos 2x)/2$.]
13. Find the Taylor series for $(x - 1)\ln x$ at 1. Over what interval is this representation valid?
14. Find the first three nonzero terms of the Maclaurin series for $\tan x$. What is its interval of convergence?
15. Find the first four terms of the Taylor series for $\csc x$ at $\pi/2$. What is its interval of convergence?
16. Find the first three nonzero terms of the Maclaurin series for $\ln |\cos x|$. What is its interval of convergence? [*Hint:* $\int \tan x\, dx = -\ln |\cos x|$.]
17. Find the Taylor series of $\sqrt{x}$ at $x = 4$. What is its radius of convergence?
18. Find the Maclaurin series of $\sin^{-1} x$. What is its radius of convergence? [*Hint:* Expand $1/\sqrt{1 - x^2}$ and integrate.]
19. Use the Maclaurin series for $\sin x$ to obtain the Maclaurin series for $\sin x^2$.
20. Find the Maclaurin series for $\cos x^2$.
21. Differentiate the Maclaurin series for $\sin x$ and show that it is equal to the Maclaurin series for $\cos x$.
22. Differentiate the Maclaurin series for $\sinh x$ and show that it is equal to the Maclaurin series for $\cosh x$.
23. Using the fact that if f has a Taylor series at x_0, then the Taylor series is given by (11), show that $1 + x + x^2 + \cdots$ is the Taylor series for $1/(1 - x)$ when $|x| < 1$.
24. Find the Maclaurin series for $\tan^{-1} x$. What is its interval of convergence? [*Hint:* Integrate the series for $1/(1 + x^2)$.]
25. Show that for any real number r
$$1 + \frac{r}{2} + \frac{r(r-1)}{2^2 2!} + \cdots + \frac{r(r-1)\cdots(r-n+1)}{2^n n!} + \cdots = \left(\frac{3}{2}\right)^r.$$
26. Use equation (29) to find a power series representation for $\sqrt[4]{1 + x}$.
27. Use the result of Problem 26 to find a power series representation for $\sqrt[4]{1 + x^3}$.
28. Use the result of Problem 27 to estimate $\int_0^{0.5} \sqrt[4]{1 + x^3}\, dx$ to four significant figures.
29. Using the technique suggested in Problems 26–28, estimate $\int_0^{1/4} (1 + \sqrt{x})^{3/5}\, dx$ to four significant figures.
30. The **error function** (which arises in mathematical statistics) is defined by
$$\text{erf}(x) = \frac{2}{\sqrt{\pi}} \int_0^x e^{-t^2}\, dt.$$
   (a) Find a Maclaurin series for $\text{erf}(x)$ by integrating the Maclaurin series for e^{-x^2}.
   (b) Use the series obtained in (a) to estimate, with an error < 0.0001, $\text{erf}(1)$ and $\text{erf}(\frac{1}{2})$.

31. The **complementary error function** is defined by

$$\operatorname{erfc}(x) = 1 - \operatorname{erf}(x) = 1 - \frac{2}{\sqrt{\pi}}\int_0^x e^{-t^2}\,dt$$
$$= \frac{2}{\sqrt{\pi}}\int_x^\infty e^{-t^2}\,dt.$$

Find a Maclaurin series for $\operatorname{erfc}(x)$ and use it to estimate $\operatorname{erfc}(1)$ and $\operatorname{erfc}(\frac{1}{2})$ with a maximum error of 0.0001. Note that for large values of x, $\operatorname{erfc}(x)$ can be estimated by integrating the last integral by parts.

***32.** The **sine integral** is defined by

$$\operatorname{Si}(x) = \int_0^x \frac{\sin t}{t}\,dt.$$

(a) Show that $\operatorname{Si}(x)$ is defined and continuous for all real x.
(b) Find a Maclaurin series expansion for $\operatorname{Si}(x)$.
(c) Estimate $\operatorname{Si}(1)$ and $\operatorname{Si}(\frac{1}{2})$ with a maximum error of 0.0001.

13.14 USING POWER SERIES TO SOLVE ORDINARY DIFFERENTIAL EQUATIONS (OPTIONAL)

Consider the second-order linear equation

$$y'' + a(x)y' + b(x)y = f(x).$$

If the functions a and b are not constant functions, then there is, in general, no way to obtain a closed form solution to the equation even in the homogeneous case ($f = 0$). In this section we show how power series can be used to obtain series solutions to the equation above in some cases. The examples we present are merely illustrative of a technique that *sometimes* works. For a more complete discussion you should consult a book on differential equations.

The fundamental assumption used in solving a differential equation by the power series method is that the solution of the differential equation can be expressed in the form of a power series, say,

$$y = \sum_{n=0}^{\infty} c_n x^n. \tag{1}$$

Once this assumption has been made, power series expansions for y', y'', . . . can be obtained by differentiating equation (1) term by term:

$$y' = \sum_{n=1}^{\infty} n c_n x^{n-1}, \tag{2}$$

$$y'' = \sum_{n=2}^{\infty} n(n-1)c_n x^{n-2}, \text{ etc.} \tag{3}$$

These can then be substituted into the given differential equation. After all the indicated operations have been carried out, and like powers of x have been collected, we obtain an expression of the form

$$k_0 + k_1x + k_2x^2 + \cdots = \sum_{n=0}^{\infty} k_n x^n = 0 \tag{4}$$

where the coefficients $k_0, k_1, k_2, \ldots$ are expressions involving the unknown coefficients $c_0, c_1, c_2, \ldots$. Since equation (4) must hold for all values of x in some interval, all the coefficients $k_0, k_1, k_2, \ldots$ must vanish. From the equations

$$k_0 = 0, \qquad k_1 = 0, \qquad k_2 = 0, \quad \ldots$$

it is then possible to determine successively the coefficients $c_0, c_1, c_2, \ldots$. In this section we illustrate this procedure by means of several examples, without concerning ourselves with questions of the convergence of the power series under consideration or the inherent limitations of the method. We shall see that power series provide a powerful method for solving certain linear differential equations with variable coefficients. First, however, in order to check that the power series method does provide the required solution, we shall solve three problems that could be solved more easily by other methods.

EXAMPLE 1 Consider the initial value problem

$$y' = y + x^2, \qquad y(0) = 1. \tag{5}$$

Inserting equations (1) and (2) into the equation, we have

$$c_1 + 2c_2x + 3c_3x^2 + 4c_4x^3 + \cdots = (c_0 + c_1x + c_2x^2 + c_3x^3 + \cdots) + x^2.$$

Collecting like powers of x, we obtain

$$(c_1 - c_0) + (2c_2 - c_1)x + (3c_3 - c_2 - 1)x^2 + (4c_4 - c_3)x^3 + \cdots = 0.$$

Equating each of the coefficients to zero, we obtain the identities

$$c_1 - c_0 = 0, \qquad 2c_2 - c_1 = 0, \qquad 3c_3 - c_2 - 1 = 0, \qquad 4c_4 - c_3 = 0, \quad \ldots,$$

from which we find that

$$c_1 = c_0, \qquad c_2 = \frac{c_1}{2} = \frac{c_0}{2!}, \qquad c_3 = \frac{c_2 + 1}{3} = \frac{c_0 + 2}{3!}, \qquad c_4 = \frac{c_3}{4} = \frac{c_0 + 2}{4!}, \quad \ldots.$$

With these values, equation (1) becomes

$$\begin{aligned} y &= c_0 + c_0x + \frac{c_0}{2!}x^2 + \frac{c_0 + 2}{3!}x^3 + \frac{c_0 + 2}{4!}x^4 + \frac{c_0 + 2}{5!}x^5 + \cdots \\ &= (c_0 + 2)\left(1 + x + \frac{x^2}{2!} + \frac{x^3}{3!} + \frac{x^4}{4!} + \cdots\right) - 2\left(1 + x + \frac{x^2}{2!}\right). \end{aligned}$$

Looking carefully at the series in parentheses, we recognize the expansion for e^x, so we have the general solution

$$y = (c_0 + 2)e^x - x^2 - 2x - 2.$$

To solve the initial value problem, we set $x = 0$ to obtain

$$1 = y(0) = c_0 + 2 - 2 = c_0.$$

Thus the solution of the initial value problem (5) is given by the equation

$$y = 3e^x - x^2 - 2x - 2. \blacksquare$$

EXAMPLE 2 Solve.

$$y'' + y = 0. \tag{6}$$

Using equations (1) and (3), we have

$$(2c_2 + 3 \cdot 2c_3x + 4 \cdot 3c_4x^2 + \cdots) + (c_0 + c_1x + c_2x^2 + \cdots) = 0.$$

Gathering like powers of x yields

$$(2c_2 + c_0) + (3 \cdot 2c_3 + c_1)x + (4 \cdot 3c_4 + c_2)x^2 + \cdots = 0.$$

Setting each of the coefficients to zero, we obtain

$$2c_2 + c_0 = 0, \qquad 3 \cdot 2c_3 + c_1 = 0, \qquad 4 \cdot 3c_4 + c_2 = 0, \qquad 5 \cdot 4c_5 + c_3 = 0, \quad \ldots,$$

and

$$c_2 = -\frac{c_0}{2!}, \qquad c_3 = -\frac{c_1}{3!}, \qquad c_4 = -\frac{c_2}{4 \cdot 3} = \frac{c_0}{4!}, \qquad c_5 = -\frac{c_3}{5 \cdot 4} = \frac{c_1}{5!}, \quad \ldots.$$

Substituting these values into the power series (1) for y yields

$$y = c_0 + c_1x - \frac{c_0}{2!}x^2 - \frac{c_1}{3!}x^3 + \frac{c_0}{4!}x^4 + \frac{c_1}{5!}x^5 + \cdots.$$

Splitting this series into two parts, we have

$$y = c_0\left(1 - \frac{x^2}{2!} + \frac{x^4}{4!} - \cdots\right) + c_1\left(x - \frac{x^3}{3!} + \frac{x^5}{5!} - \cdots\right).$$

Using equations (13.13.22) and (13.13.23) reveals the familiar general solution

$$y = c_0 \cos x + c_1 \sin x.$$

We observe that in this case the power series method produces two arbitrary constants c_0, c_1, and yields the general solution for equation (6). ■

So far we have considered only linear equations with constant coefficients. We turn now to linear equations with variable coefficients.

EXAMPLE 3 Consider the initial value problem

$$(1 + x^2)y' = 2pxy, \qquad y(0) = 1, \tag{7}$$

where p is a constant. Applying equations (1) and (2), we have

$$(1 + x^2) \sum_{n=1}^{\infty} nc_n x^{n-1} = 2px \sum_{n=0}^{\infty} c_n x^n.$$

Equation (7) can be rewritten in the form

$$\sum_{n=1}^{\infty} nc_n x^{n-1} + \sum_{n=1}^{\infty} nc_n x^{n+1} = \sum_{n=0}^{\infty} 2pc_n x^{n+1}. \tag{8}$$

We would like to rewrite each of the sums in equation (8) so that each general term contains the same power of x. This can be done by assuming that each general term contains the term x^k. For the first sum, this amounts to substituting $k = n - 1$. Since n ranges from 1 to ∞, $k = n - 1$ will range from 0 to ∞. Substituting $k = n + 1$ with k ranging from 2 to ∞ into the second sum, and $k = n + 1$ with k ranging from 1 to ∞ into the third sum, allows us to rewrite these sums so that the general term will involve the power x^k. We then obtain

$$\sum_{k=0}^{\infty} (k + 1)c_{k+1}x^k + \sum_{k=2}^{\infty} (k - 1)c_{k-1}x^k = \sum_{k=1}^{\infty} 2pc_{k-1}x^k.$$

Now we can gather like terms in x. We take out the $k = 0$ and $k = 1$ terms first:

$$c_1 + (2c_2 - 2pc_0)x + \sum_{k=2}^{\infty} \{(k + 1)c_{k+1} + [(k - 1) - 2p]c_{k-1}\}x^k = 0.$$

Equating each coefficient to zero yields

$$c_1 = 0, \qquad 2c_2 - 2pc_0 = 0, \qquad 3c_3 + (1 - 2p)c_1 = 0,$$

and in general

$$(k + 1)c_{k+1} + [(k - 1) - 2p]c_{k-1} = 0, \qquad k \geq 1. \tag{9}$$

We note that equation (9) is a difference equation with variable coefficients. This equation is called a **recursion formula** and can be used to evaluate the constants $c_0, c_1, c_2, \ldots$ successively. We see that

$$c_1 = 0, \qquad c_2 = pc_0, \qquad c_3 = 0,$$

and by equation (9), in general

$$c_{k+1} = \frac{(2p - k + 1)}{k + 1}c_{k-1}.$$

Thus

$$c_4 = \frac{2p-2}{4}c_2 = \frac{p(p-1)}{1\cdot 2}c_0, \qquad c_5 = 0,$$

$$c_6 = \frac{2p-4}{6}c_4 = \frac{p(p-1)(p-2)}{1\cdot 2\cdot 3}c_0, \qquad c_7 = 0, \ldots,$$

since $c_3 = 0$. Thus the coefficients with odd-numbered subscripts vanish and the power series for y is given by

$$\begin{aligned} y &= c_0 + \frac{p}{1}c_0x^2 + \frac{p(p-1)}{1\cdot 2}c_0x^4 + \frac{p(p-1)(p-2)}{1\cdot 2\cdot 3}c_0x^6 + \cdots \\ &= c_0\left(1 + \frac{p}{1}x^2 + \frac{p(p-1)}{1\cdot 2}x^4 + \frac{p(p-1)(p-2)}{1\cdot 2\cdot 3}x^6 + \cdots\right). \end{aligned}$$

The binomial series (see equation (13.13.26)), states that

$$(1+x)^p = 1 + \frac{p}{1}x + \frac{p(p-1)}{1\cdot 2}x^2 + \frac{p(p-1)(p-2)}{1\cdot 2\cdot 3}x^3 + \cdots.$$

Replacing x by x^2 in this equation yields the general solution of the differential equation:

$$y = c_0(1+x^2)^p.$$

Since $y(0) = 1$, it follows that $c_0 = 1$ and $y = (1+x^2)^p$. ■

EXAMPLE 4 Consider the differential equation

$$y'' + xy' + y = 0. \qquad \textbf{(10)}$$

Using equations (1), (2), and (3), we obtain the equation

$$\sum_{n=2}^{\infty} n(n-1)c_nx^{n-2} + x\sum_{n=1}^{\infty} nc_nx^{n-1} + \sum_{n=0}^{\infty} c_nx^n = 0.$$

Reindexing to obtain equal powers of x, we have

$$\sum_{k=0}^{\infty} (k+2)(k+1)c_{k+2}x^k + \sum_{k=1}^{\infty} kc_kx^k + \sum_{k=0}^{\infty} c_kx^k = 0.$$

Note that the second sum can also be allowed to range from 0 to ∞. Gathering like terms in x produces the equation

$$\sum_{k=0}^{\infty} [(k+2)(k+1)c_{k+2} + (k+1)c_k]x^k = 0.$$

Setting the coefficients equal to zero, we obtain the general recursion formula

$$(k+2)(k+1)c_{k+2} + (k+1)c_k = 0.$$

Therefore $(k+2)c_{k+2} = -c_k$, and

$$c_2 = -\frac{c_0}{2}, \qquad c_3 = -\frac{c_1}{3}, \qquad c_4 = -\frac{c_2}{4} = \frac{c_0}{2 \cdot 4},$$

$$c_5 = -\frac{c_3}{5} = \frac{c_1}{3 \cdot 5}, \qquad c_6 = -\frac{c_4}{6} = -\frac{c_0}{2 \cdot 4 \cdot 6}, \qquad \text{etc.}$$

Hence the power series for y can be written in the form

$$\begin{aligned} y &= c_0 + c_1 x - \frac{c_0}{2}x^2 - \frac{c_1}{3}x^3 + \frac{c_0}{2 \cdot 4}x^4 + \frac{c_1}{3 \cdot 5}x^5 - \cdots \\ &= c_0\left(1 - \frac{x^2}{2} + \frac{x^4}{2 \cdot 4} - \frac{x^6}{2 \cdot 4 \cdot 6} + \cdots\right) + c_1\left(x - \frac{x^3}{3} + \frac{x^5}{3 \cdot 5} - \frac{x^7}{3 \cdot 5 \cdot 7} + \cdots\right) \end{aligned} \tag{11}$$

by separating the terms that involve c_0 and c_1. At this point we try to see whether we recognize the two series that have been obtained by the power series method. Very frequently this is an unproductive task, but in this instance we are fortunate:

$$\begin{aligned} 1 - \frac{x^2}{2} + \frac{x^4}{2 \cdot 4} - \frac{x^6}{2 \cdot 4 \cdot 6} + \cdots &= 1 + \left(-\frac{x^2}{2}\right) + \frac{1}{2!}\left(-\frac{x^2}{2}\right)^2 + \frac{1}{3!}\left(-\frac{x^2}{2}\right)^3 + \cdots \\ &= e^{-x^2/2}. \end{aligned}$$

The second series is not a familiar one, so we use the method given in Section 11.7 of finding one solution when another is known. By equation (11.7.8), we have

$$\begin{aligned} y_2 &= y_1 \int \frac{e^{-\int x\,dx}}{y_1^2}\,dx = e^{-x^2/2} \int \frac{e^{-x^2/2}}{(e^{-x^2/2})^2}\,dx \\ &= e^{-x^2/2} \int e^{x^2/2}\,dx. \end{aligned} \tag{12}$$

The integral in equation (12) does not have a closed form solution. That this is indeed the second series in equation (11) can be verified by integrating the series for $e^{x^2/2}$ term by term and multiplying the result by the series for $e^{-x^2/2}$. Hence the general solution of equation (10) is given by

$$y = c_0 e^{-x^2/2} + c_1 e^{-x^2/2} \int e^{x^2/2}\,dx. \quad \blacksquare$$

EXAMPLE 5 Solve the equation

$$xy'' + y' + xy = 0. \tag{13}$$

Solution. Using the power series (1), (2), and (3) for equation (13) yields the equation

$$\sum_{n=2}^{\infty} n(n-1)c_n x^{n-1} + \sum_{n=1}^{\infty} nc_n x^{n-1} + \sum_{n=0}^{\infty} c_n x^{n+1} = 0.$$

Reindexing the series to obtain like powers of x, we have

$$\sum_{k=1}^{\infty} (k+1)kc_{k+1}x^k + \sum_{k=0}^{\infty} (k+1)c_{k+1}x^k + \sum_{k=1}^{\infty} c_{k-1}x^k = 0.$$

Condensing the three series in one yields, after some algebra,

$$c_1 + \sum_{k=1}^{\infty} [(k+1)^2 c_{k+1} + c_{k-1}]x^k = 0.$$

Setting the coefficients equal to zero, we have $c_1 = 0$ and

$$(k+1)^2 c_{k+1} = -c_{k-1}, \qquad k = 1, 2, 3, \ldots. \tag{14}$$

The recursion formula (14) together with $c_1 = 0$ implies that all coefficients with odd-numbered subscripts vanish, and

$$c_2 = -\frac{c_0}{2^2}, \qquad c_4 = -\frac{c_2}{4^2} = \frac{c_0}{2^2 4^2}, \qquad c_6 = -\frac{c^4}{6^2} = -\frac{c_0}{2^2 4^2 6^2}, \qquad \ldots.$$

Hence

$$\begin{aligned} y &= c_0 - \frac{c_0}{2^2}x^2 + \frac{c_0}{2^2 4^2}x^4 - \frac{c_0}{2^2 4^2 6^2}x^6 + \cdots \\ &= c_0 \sum_{n=0}^{\infty} \frac{1}{(n!)^2}\left(-\frac{x^2}{4}\right)^n. \end{aligned} \tag{15}$$

It is unlikely that you are familiar with the series in equation (15). This series is often used in applied mathematics and is known as the **Bessel function of index zero, $J_0(x)$.** Note also that the power series method has produced only *one* of the solutions of equation (13). To find the other solution, we can again proceed as in Example 4. Thus

$$y_2(x) = J_0(x) \int \frac{dx}{xJ_0^2(x)}.$$

Finally, the general solution of equation (13) is given by

$$y(x) = AJ_0(x) + BJ_0(x) \int \frac{dx}{xJ_0^2(x)}. \quad ■$$

In our next example we meet a situation in which the power series method fails to yield any solution.

EXAMPLE 6 Solve the Euler equation

$$x^2y'' + xy' + y = 0. \tag{16}$$

Solution. Making use of series (1), (2), and (3), and multiplying by the appropriate powers of x, we have

$$\sum_{n=2}^{\infty} n(n-1)c_nx^n + \sum_{n=1}^{\infty} nc_nx^n + \sum_{n=0}^{\infty} c_nx^n = 0$$

or

$$\sum_{n=0}^{\infty} (n^2+1)c_nx^n = 0. \tag{17}$$

Clearly, if we equate each of the coefficients of equation (17) to zero, all the coefficients c_n vanish and $y \equiv 0$. Thus in this case the power series method fails completely in helping us find the general solution

$$y = A\cos(\ln|x|) + B\sin(\ln|x|)$$

of equation (16) (check!). ■

When initial conditions are given, there is another method based on the Taylor series that can also be used.

EXAMPLE 1 **(Revisited)** Consider again the initial value problem

$$y' = y + x^2, \qquad y(0) = 1. \tag{18}$$

Differentiating both sides of the differential equation repeatedly and evaluating each derivative at the initial value of $x = 0$, we have

$$\begin{aligned}
y' &= y + x^2|_{x=0} = y(0) + (0)^2 = 1,\\
y'' &= y' + 2x|_{x=0} = y'(0) + 2(0) = 1,\\
y''' &= y'' + 2|_{x=0} = y''(0) + 2 = 3,\\
y^{(4)} &= y'''|_{x=0} = y'''(0) = 3, \ldots.
\end{aligned}$$

Substituting these derivatives in the Taylor series

$$y(x) = \sum_{n=0}^{\infty} \frac{y^{(n)}(x_0)}{n!}(x - x_0)^n \tag{19}$$

with $x_0 = 0$, we have

$$y(x) = 1 + x + \frac{x^2}{2!} + \frac{3x^3}{3!} + \frac{3x^4}{4!} + \cdots = 3e^x - 2 - 2x - x^2,$$

which is the result that we obtained before. ■

Taylor's method is easily adapted to higher-order initial value problems by rewriting the differential equation so that the highest-order derivative is expressed in terms of the other derivatives and the independent variable. Successive differentiations again yields the values $y^{(n)}(x_0)$ for substitution into the Taylor series.

PROBLEMS 13.14

In Problems 1–16 find the general solution of each equation by the power series method. When initial conditions are specified, give the solution that satisfies them.

1. $y' = y - x$, $y(0) = 2$
2. $y' = x^3 - 2xy$, $y(0) = 1$
3. $y'' + y = x$
4. $y'' + 4y = 0$, $y(0) = 1$, $y'(0) = 0$
5. $(1 + x^2)y'' + 2xy' - 2y = 0$
6. $xy'' - xy' + y = e^x$, $y(0) = 1$, $y'(0) = 2$
7. $xy'' - x^2y' + (x^2 - 2)y = 0$, $y(0) = 0$, $y'(0) = 1$
8. $(1 - x)y'' - y' + xy = 0$, $y(0) = y'(0) = 1$
9. $y'' - 2xy' + 4y = 0$, $y(0) = 1$, $y'(0) = 0$
10. $(1 - x^2)y'' - xy' + y = 0$, $y(0) = 0$, $y'(0) = 1$
11. $y'' - xy' + y = -x \cos x$, $y(0) = 0$, $y'(0) = 2$
12. $y'' - xy' + xy = 0$, $y(0) = 2$, $y'(0) = 1$
13. $(1 - x)^2y'' - (1 - x)y' - y = 0$, $y(0) = y'(0) = 1$
14. $y'' - 2xy' + 2y = 0$
15. $y'' - 2xy' - 2y = x$, $y(0) = 1$, $y'(0) = -\frac{1}{4}$
16. $y'' - x^2y = 0$
17. **Airy's equation**

$$y'' - xy = 0$$

has applications in the theory of diffraction. Find the general solution of this equation.

18. **Hermite's equation**

$$y'' - 2xy' + 2py = 0,$$

where p is constant, arises in quantum mechanics in connection with the Schrödinger equation for a harmonic oscillator. Show that if p is a positive integer, one of the two linearly independent solutions of Hermite's equation is a polynomial, called the **Hermite polynomial** $H_p(x)$.

19. Use the Taylor series method to solve Airy's equation

$$y'' - xy = 0, \qquad y(1) = 1, \qquad y'(1) = 0.$$

20. Use the Taylor series method to solve

$$y'' - xy' - y = 0, \qquad y(0) = 1, \qquad y'(0) = 0.$$

21. Does the power series method yield a solution to the equation
(a) $x^2y' = y$? (b) $x^3y' = y$?

*22. Solve $y' = y\sqrt{y^2 - 1}$ by squaring the power series for y.

23. Show that the power series method fails for

$$x^2y'' + x^2y' + y = 0.$$

REVIEW EXERCISES FOR CHAPTER THIRTEEN

In Exercises 1–8 find the Taylor polynomial of the given degree at the number a.

1. $f(x) = e^x$; $a = 0$; $n = 3$
2. $f(x) = \ln x$; $a = 1$; $n = 4$
3. $f(x) = \sin x$; $a = \pi/6$; $n = 3$
4. $f(x) = \cos x$; $a = \pi/2$; $n = 5$
5. $f(x) = \cot x$; $a = \pi/2$; $n = 4$
6. $f(x) = \sinh x$; $a = 0$; $n = 3$
7. $f(x) = x^3 - x^2 + 2x + 3$; $a = 0$; $n = 8$
8. $f(x) = e^{-x^2}$; $a = 0$; $n = 5$

In Exercises 9–13 find a bound for $|R_n(x)|$ for x in the given interval.

9. $f(x) = \cos x;\ a = \pi/6;\ n = 5;\ x \in [0, \pi/2]$
10. $f(x) = \sqrt[3]{x};\ a = 1;\ n = 4;\ x \in [\frac{7}{8}, \frac{9}{8}]$
11. $f(x) = e^x;\ a = 0;\ n = 6;\ x \in [-\ln e, \ln e]$
12. $f(x) = \cot x;\ a = \pi/2;\ n = 2;\ x \in [\pi/4, 3\pi/4]$
13. $f(x) = e^{-x^2};\ a = 0;\ n = 4;\ x \in [-1, 1]$

In Exercises 14–20 use a Taylor polynomial to estimate the given number with the given degree of accuracy.

14. $\cos\left(\frac{\pi}{3} + 0.1\right)$; error < 0.001

15. $\cos 43°$; error < 0.001

16. $\cot\left(\frac{\pi}{4} + 0.1\right)$; error < 0.001

***17.** $\ln 2$; error < 0.0001 [*Hint:* Look at $\ln[(1 + x)/(1 - x)]$.]
18. e^2; error < 0.0001
19. $\tan^{-1} 0.3$; error < 0.0001

20. $\ln \sin\left(\frac{\pi}{2} + 0.2\right)$; error < 0.01

21. Use a Taylor polynomial of degree 4 to estimate $\int_0^{1/2} \cos x^2\,dx$. What is the maximum error of your estimate?
22. Use a Taylor polynomial of degree 4 to estimate $\int_0^1 e^{-x^3}\,dx$. What is the maximum error of your estimate?
23. Find the first five terms of the sequence $\{(n - 2)/n\}$.
24. Find the first seven terms of the sequence $\{n^2 \sin n\}$.

In Exercises 25–30 determine whether the given sequence is convergent or divergent. If it is convergent, find its limit.

25. $\left\{\frac{-7}{n}\right\}$ **26.** $\{\cos \pi n\}$

27. $\left\{\frac{\ln n}{\sqrt{n}}\right\}$ **28.** $\left\{\frac{7^n}{n!}\right\}$

29. $\left\{\left(1 - \frac{2}{n}\right)^n\right\}$ **30.** $\left\{\frac{3}{\sqrt{n^2 + 8} - n}\right\}$

31. Find the general term of the sequence $\frac{1}{8}, \frac{3}{16}, \frac{5}{32}, \frac{7}{64}, \ldots$.

32. Find the general term of the sequence $1, -\frac{1}{5}, \frac{1}{25}, -\frac{1}{125}, \frac{1}{625}, \ldots$.

In Exercises 33–41 determine whether the given sequence is bounded or unbounded and increasing, decreasing, or not monotonic.

33. $\{\sqrt{n} \cos n\}$ **34.** $\left\{\frac{3}{n + 2}\right\}$

35. $\left\{\frac{2^n}{1 + 2^n}\right\}$ **36.** $\left\{\frac{n^n}{n!}\right\}$

37. $\left\{\frac{n!}{n^n}\right\}$ **38.** $\left\{\frac{\sqrt{n} + 1}{n}\right\}$

39. $\left\{\left(1 - \frac{1}{n}\right)^{1/n}\right\}$ **40.** $\left\{\frac{n - 7}{n + 4}\right\}$

41. $\{(3^n + 5^n)^{1/n}\}$

42. Write the following using the Σ notation:
$$1 - 2^{1/2} + 3^{1/3} - 4^{1/4} + 5^{1/5} - \cdots + (-1)^{n+1} n^{1/n}.$$
43. Evaluate $\sum_{k=2}^{10} 4^k$.
44. Evaluate $\sum_{k=1}^{\infty} 1/3^k$.
45. Evaluate $\sum_{k=3}^{\infty} [(\frac{3}{4})^k - (\frac{2}{5})^k]$.
46. Evaluate $\sum_{k=2}^{\infty} 1/k(k - 1)$.
47. Write 0.79797979 . . . as a rational number.
48. Write 0.142314231423 . . . as a rational number.
49. At what time between 9 P.M. and 10 P.M. is the minute hand of a clock exactly over the hour hand?

In Exercises 50–62 determine the convergence or divergence of the given series.

50. $\sum_{k=1}^{\infty} \frac{1}{k^3 - 5}$ **51.** $\sum_{k=5}^{\infty} \frac{1}{k(k + 6)}$

52. $\sum_{k=1}^{\infty} \frac{1}{\sqrt{k^3 + 4}}$ **53.** $\sum_{k=2}^{\infty} \frac{3}{\ln k}$

54. $\sum_{k=4}^{\infty} \frac{1}{\sqrt[3]{k^3 + 50}}$ **55.** $\sum_{k=1}^{\infty} \frac{r^k}{k^r},\ 0 < r < 1$

56. $\sum_{k=1}^{\infty} \frac{k!}{k^k}$ **57.** $\sum_{k=2}^{\infty} \frac{10^k}{k^5}$

58. $\sum_{k=1}^{\infty} \frac{k^{6/5}}{8^k}$ **59.** $\sum_{k=1}^{\infty} \frac{\sqrt{k} \ln(k + 3)}{k^2 + 2}$

60. $\sum_{k=1}^{\infty} \operatorname{csch} k$ **61.** $\sum_{k=2}^{\infty} \frac{e^{1/k}}{k^{3/2}}$

62. $\sum_{k=1}^{\infty} \frac{k(k + 6)}{(k + 1)(k + 3)(k + 5)}$

In Exercises 63–74 determine whether the given alternating series is absolutely convergent, conditionally convergent, or divergent.

63. $\sum_{k=1}^{\infty} \frac{(-1)^{k+1}}{50k}$

64. $\sum_{k=2}^{\infty} \frac{(-1)^k \sqrt{k}}{\ln k}$

65. $\sum_{k=2}^{\infty} \frac{(-1)^{k+1}}{\sqrt{k(k-1)}}$

66. $\sum_{k=2}^{\infty} \frac{(-1)^k k^2}{k^3+1}$

67. $\sum_{k=2}^{\infty} \frac{(-1)^k k^2}{k^4+1}$

68. $\sum_{k=2}^{\infty} \frac{(-1)^k k^3}{k^3+1}$

69. $\sum_{k=3}^{\infty} \frac{(-1)^k (k+2)(k+3)}{(k+1)^3}$

70. $\sum_{k=2}^{\infty} \frac{(-1)^k 3^k}{3^k}$

71. $\sum_{k=1}^{\infty} \frac{(-1)^k k^k}{k!}$

72. $\sum_{k=1}^{\infty} \frac{(-1)^k k^4}{k^4 + 20k^3 + 17k + 2}$

73. $\sum_{k=1}^{\infty} (-1)^k \left(1 + \frac{1}{k}\right)^k$

74. $\sum_{k=1}^{\infty} \frac{(-1)^k k!}{k^k}$

75. Calculate $\sum_{k=1}^{\infty} (-1)^{k+1}/k^3$ with an error of less than 0.001.

76. Calculate $\sum_{k=0}^{\infty} (-1)^k/k!$ with an error of less than 0.0001.

77. Find the first ten terms of a rearrangement of the series $\sum_{k=1}^{\infty} (-1)^{k+1}/k$ that converges to $\frac{1}{2}$.

In Exercises 78–89 find the radius of convergence and interval of convergence of the given power series.

78. $\sum_{k=0}^{\infty} \frac{x^k}{3^k}$

79. $\sum_{k=0}^{\infty} \frac{(-1)^k x^k}{3^k}$

80. $\sum_{k=0}^{\infty} \frac{x^k}{k^2+2}$

81. $\sum_{k=1}^{\infty} \frac{x^k}{k^k}$

82. $\sum_{k=2}^{\infty} \frac{x^k}{(2 \ln k)^k}$

83. $\sum_{k=0}^{\infty} \frac{(3x+5)^k}{k!}$

84. $\sum_{k=0}^{\infty} \frac{(3x-5)^k}{3^k}$

85. $\sum_{k=0}^{\infty} \left(\frac{k}{6}\right)^k x^k$

86. $\sum_{k=0}^{\infty} (-1)^k x^{3k}$

87. $\sum_{k=1}^{\infty} \frac{\ln k (x-1)^k}{k+2}$

88. $\sum_{k=1}^{\infty} \frac{[1-(-1)^{k+1}]}{k} x^k$

89. $\sum_{k=0}^{\infty} \frac{(x+8)^k}{(k+1)3^k}$

90. Estimate $\int_0^{1/2} e^{-t^4}\, dt$ with an error of less than 0.00001.

91. Estimate $\int_0^{1/2} \sin t^2\, dt$ with an error of less than 0.001.

92. Estimate $\int_0^{1/2} t^3 e^{-t^3}\, dt$ with an error of less than 0.001.

93. Estimate $\int_0^{1/2} [1/(1+t^4)]\, dt$ with an error of less than 0.00001.

94. Find the Maclaurin series for $x^2 e^x$.

95. Find the Taylor series for e^x at ln 2.

96. Find the Maclaurin series for $\cos^2 x$. [*Hint:* $\cos^2 x = (1 + \cos 2x)/2$.]

97. Find the Maclaurin series for $\sin \alpha x$, α real.

In Exercises 98–100 use the power series method to obtain the solution to the initial value problem.

98. $\frac{dy}{dx} = 3y;\ y(0) = 2$

99. $y'' + 9y = 0;\ y(0) = 1,\ y'(0) = 0$

100. $y'' - 2y' + y = \cos x;\ y(0) = 0,\ y'(0) = 3$

101. Use the power series method to obtain the general solution to $y'' - xy' - y = 0$.

Appendix I
Mathematical Induction

Mathematical induction[†] is the name given to a logical principle that can be used to prove a certain type of mathematical statement. Typically, we use mathematical induction to prove that a certain statement or equation holds for every positive integer. For example, we may need to prove that $2^n > n$ for all integers $n \geq 1$.

To do so, we proceed in two steps:

(i) We prove that the statement is true for some integer N (usually $N = 1$).
(ii) We *assume* that the statement is true for an integer k and then *prove* that it is true for the integer $k + 1$.

If we can complete these two steps, we will have demonstrated the validity of the statement for *all* positive integers greater than or equal to N. To convince you of this fact, we reason as follows: Since the statement is true for N [by step (i)] it is true for the integer $N + 1$ [by step (ii)]. Then it is also true for the integer $(N + 1) + 1 = N + 2$ [again by step (ii)], and so on. We now demonstrate the procedure with some examples.

[†]This technique was first used in a mathematical proof by the great French mathematician Pierre de Fermat (1601–1665).

EXAMPLE 1 Show that $2^n > n$ for all integers $n \geq 1$.

Solution.
(i) If $n = 1$, then $2^n = 2^1 = 2 > 1 = n$, so $2^n > n$ for $n = 1$.
(ii) Assume that $2^k > k$, where $k \geq 1$ is an integer. Then

$$2^{k+1} = 2 \cdot 2^k = 2^k + 2^k \overset{\text{Since } 2^k > k}{\downarrow\!>} k + k \geq k + 1.$$

This completes the proof since we have shown that $2^1 > 1$, which implies, by step (ii), that $2^2 > 2$, so that, again by step (ii), $2^3 > 3$, so that $2^4 > 4$, and so on. ■

EXAMPLE 2 Use mathematical induction to prove the formula for the sum of the first n positive integers:

$$1 + 2 + 3 + \cdots + n = \frac{n(n+1)}{2}. \qquad \textbf{(1)}$$

Solution.
(i) If $n = 1$, then the sum of the first one integer is 1. But $(1)(1 + 1)/2 = 1$, so that equation (1) holds in the case in which $n = 1$.
(ii) Assume that (1) holds for $n = k$; that is,

$$1 + 2 + 3 + \cdots + k = \frac{k(k+1)}{2}.$$

We must now show that it holds for $n = k + 1$. That is, we must show that

$$1 + 2 + 3 + \cdots + k + (k+1) = \frac{(k+1)(k+2)}{2}.$$

But

$$\begin{aligned} 1 + 2 + 3 + \cdots + k + (k+1) &= (1 + 2 + 3 + \cdots + k) + (k+1) \\ &= \frac{k(k+1)}{2} + (k+1) \\ &= \frac{k(k+1) + 2(k+1)}{2} = \frac{(k+1)(k+2)}{2}, \end{aligned}$$

and the proof is complete. ■

You may wish to try a few examples to illustrate that formula (1) really works. For example,

$$1 + 2 + 3 + 4 + 5 + 6 + 7 + 8 + 9 + 10 = \frac{10(11)}{2} = 55.$$

EXAMPLE 3 Use mathematical induction to prove this formula for the sum of the squares of the first n positive integers:

$$1^2 + 2^2 + 3^2 + \cdots + n^2 = \frac{n(n+1)(2n+1)}{6}. \quad \textbf{(2)}$$

Solution.

(i) Since $1(1 + 1)(2 \cdot 1 + 1)/6 = 1 = 1^2$, equation (2) is true for $n = 1$.

(ii) Suppose that equation (2) holds for $n = k$; that is,

$$1^2 + 2^2 + 3^2 + \cdots + k^2 = \frac{k(k+1)(2k+1)}{6}.$$

Then to prove that (2) is true for $n = k + 1$, we have

$$\begin{aligned}
1^2 + 2^2 + 3^2 + \cdots + k^2 + (k+1)^2 &= \frac{k(k+1)(2k+1)}{6} + (k+1)^2 \\
&= \frac{k(k+1)(2k+1) + 6(k+1)^2}{6} \\
&= \frac{k+1}{6}[k(2k+1) + 6(k+1)] \\
&= \frac{k+1}{6}(2k^2 + 7k + 6) = \frac{k+1}{6}[(k+2)(2k+3)] \\
&= \frac{(k+1)(k+2)[2(k+1)+1]}{6},
\end{aligned}$$

which is equation (2) for $n = k + 1$, and the proof is complete. ■

Again you may wish to experiment with this formula. For example,

$$1^2 + 2^2 + 3^2 + 4^2 + 5^2 + 6^2 + 7^2 = \frac{7(7+1)(2 \cdot 7 + 1)}{6} = \frac{7 \cdot 8 \cdot 15}{6} = 140.$$

EXAMPLE 4 For $a \neq 1$, use mathematical induction to prove the formula for the sum of a geometric progression:

$$1 + a + a^2 + \cdots + a^n = \frac{1 - a^{n+1}}{1 - a}. \quad \textbf{(3)}$$

Solution.

(i) If $n = 0$, then

$$\frac{1 - a^{0+1}}{1 - a} = \frac{1 - a}{1 - a} = 1.$$

Thus equation (3) holds for $n = 0$. (We use $n = 0$ instead of $n = 1$ since $a^0 = 1$ is the first term.)

(ii) Assume that (3) holds for $n = k$; that is,

$$1 + a + a^2 + \cdots + a^k = \frac{1 - a^{k+1}}{1 - a}.$$

Then

$$\begin{aligned} 1 + a + a^2 + \cdots + a^k + a^{k+1} &= \frac{1 - a^{k+1}}{1 - a} + a^{k+1} \\ &= \frac{1 - a^{k+1} + (1 - a)a^{k+1}}{1 - a} = \frac{1 - a^{k+2}}{1 - a}, \end{aligned}$$

so that equation (3) also holds for $n = k + 1$, and the proof is complete. ■

EXAMPLE 5 Let $f_1, f_2, \ldots, f_n$ be differentiable functions. Use mathematical induction to prove that

$$\frac{d}{dx}(f_1 + f_2 + \cdots + f_n) = \frac{df_1}{dx} + \frac{df_2}{dx} + \cdots + \frac{df_n}{dx}. \tag{4}$$

Solution.
(i) For $n = 2$, equation (4) is a basic theorem in one-variable calculus.
(ii) Assume that equation (4) is valid for $n = k$; that is,

$$\frac{d}{dx}(f_1 + f_2 + \cdots + f_k) = \frac{df_1}{dx} + \frac{df_2}{dx} + \cdots + \frac{df_k}{dx}.$$

Let $g(x) = f_1(x) + f_2(x) + \cdots + f_k(x)$. Then

$$\begin{aligned} \frac{d}{dx}(f_1 + f_2 + \cdots + f_k + f_{k+1}) &= \frac{d}{dx}(g + f_{k+1}) \\ &= \frac{dg}{dx} + \frac{df_{k+1}}{dx} \qquad \text{(By the case } n = 2\text{)} \\ &= \frac{d}{dx}(f_1 + f_2 + \cdots + f_k) + \frac{df_{k+1}}{dx} \\ &= \frac{df_1}{dx} + \frac{df_2}{dx} + \cdots + \frac{df_k}{dx} + \frac{df_{k+1}}{dx}, \end{aligned}$$

which is equation (4) in the case $n = k + 1$, and the theorem is proved. ■

PROBLEMS

1. Use mathematical induction to prove that the sum of the cubes of the first n positive integers is given by

$$1^3 + 2^3 + 3^3 + \cdots + n^3 = \frac{n^2(n + 1)^2}{4}. \tag{5}$$

2. Let the functions $f_1, f_2, \ldots, f_n$ be integrable on [0, 1]. Show that $f_1 + f_2 + \cdots + f_n$ is integrable on [0, 1] and that

$$\int_0^1 [f_1(x) + f_2(x) + \cdots + f_n(x)]\,dx = \int_0^1 f_1(x)\,dx + \int_0^1 f_2(x)\,dx + \cdots + \int_0^1 f_n(x)\,dx.$$

3. Use mathematical induction to prove that the nth derivative of the nth-order polynomial

$$P_n(x) = x^n + a_{n-1}x^{n-1} + a_{n-2}x^{n-2} + \cdots + a_1x^1 + a_0$$

is equal to $n!$

4. Show that if $a \neq 1$,

$$1 + 2a + 3a^2 + \cdots + na^{n-1} = \frac{1 - (n+1)a^n + na^{n+1}}{(1-a)^2}.$$

*5. Prove, using mathematical induction, that there are exactly 2^n subsets of a set containing n elements.

6. Use mathematical induction to prove that

$$\ln(a_1a_2a_3 \ldots a_n) = \ln a_1 + \ln a_2 + \cdots + \ln a_n,$$

if $a_k > 0$ for $k = 1, 2, \ldots, n$.

7. Let $\mathbf{u}, \mathbf{v}_1, \mathbf{v}_2, \ldots, \mathbf{v}_n$ be $n + 1$ vectors in $\mathbb{R}^2$. Prove that

$$\mathbf{u} \cdot (\mathbf{v}_1 + \mathbf{v}_2 + \cdots + \mathbf{v}_n) = \mathbf{u} \cdot \mathbf{v}_1 + \mathbf{u} \cdot \mathbf{v}_2 + \cdots + \mathbf{u} \cdot \mathbf{v}_n.$$

Appendix 2
The Binomial Theorem

The binomial theorem provides a useful device for multiplying expressions of the form $(x + y)^n$, where n is a positive integer. You are all familiar with the identity

$$(x + y)^2 = x^2 + 2xy + y^2.$$

Multiplying the previous equation on both sides by $(x + y)$ results in

$$(x + y)^3 = x^3 + 3x^2y + 3xy^2 + y^3.$$

To calculate larger powers of $x + y$, we first define the **binomial coefficient** $\binom{n}{k}$ for n and k positive integers by

$$\binom{n}{k} = \frac{n(n-1)\cdots(n-k+1)}{k(k-1)\cdots 3\cdot 2\cdot 1} = \frac{n!}{k!(n-k)!} \qquad \textbf{(1)}$$

where $n! = n(n-1)(n-2)\cdots 3\cdot 2\cdot 1$, and, by convention, $0! = 1$.

EXAMPLE 1 Evaluate (a) $\binom{4}{2}$, (b) $\binom{8}{5}$, (c) $\binom{7}{0}$.

Solution.

(a) $\binom{4}{2} = 4!/2!2! = (4\cdot 3\cdot 2)/(2\cdot 2) = 6.$
(b) $\binom{8}{5} = 8!/5!3! = (8\cdot 7\cdot 6\cdot 5!)/5!3! = (8\cdot 7\cdot 6)/6 = 56.$
(c) $\binom{7}{0} = 7!/7!0! = 1/0! = 1/1 = 1.$ ■

Theorem 1 THE BINOMIAL THEOREM Let n be a positive integer. Then

$$(x + y)^n = x^n + \binom{n}{1}x^{n-1}y + \binom{n}{2}x^{n-2}y^2 + \cdots + \binom{n}{n-1}xy^{n-1} + y^n, \tag{2}$$

or more concisely,

$$(x + y)^n = \sum_{j=0}^{n} \binom{n}{j} x^{n-j}y^j. \tag{3}$$

Proof. We prove this theorem by mathematical induction (see Appendix 1).

(i) If $n = 1$, then

$$\sum_{j=0}^{n} \binom{n}{j} x^{n-j}y^j = \sum_{j=0}^{1} \binom{1}{j} x^{1-j}y^j = \binom{1}{0}x^1y^0 + \binom{1}{1}x^0y^1 = x + y,$$

implying the validity of equation (3) for $n = 1$.

(ii) We assume that equation (3) holds for $n = k$ and prove it for $n = k + 1$. By assumption, we have

$$(x + y)^k = \sum_{j=0}^{k} \binom{k}{j} x^{k-j}y^j. \tag{4}$$

Then

$$(x + y)^{k+1} = (x + y)^k(x + y) = \left[\sum_{j=0}^{k} \binom{k}{j} x^{k-j}y^j\right](x + y),$$

or

$$(x + y)^{k+1} = \sum_{j=0}^{k} \binom{k}{j} x^{k+1-j}y^j + \sum_{j=0}^{k} \binom{k}{j} x^{k-j}y^{j+1}. \tag{5}$$

We now write out the sums in (5):

$$\begin{aligned}(x + y)^{k+1} = {} & \binom{k}{0}x^{k+1} + \binom{k}{1}x^k y + \binom{k}{2}x^{k-1}y^2 + \cdots + \binom{k}{j}x^{k+1-j}y^j \\ & + \binom{k}{j+1}x^{k-j}y^{j+1} + \cdots + \binom{k}{k}xy^k \\ & + \binom{k}{0}x^k y + \binom{k}{1}x^{k-1}y^2 + \binom{k}{2}x^{k-2}y^3 + \cdots + \binom{k}{j-1}x^{k+1-j}y^j \\ & + \binom{k}{j}x^{k-j}y^{j+1} + \cdots + \binom{k}{k}y^{k+1}.\end{aligned} \tag{6}$$

In (6) we see that there are two terms containing the expressions $x^{k+1-j}y^j$ (for $j = 1, 2, \ldots, k$), namely,

$$\binom{k}{j}x^{k+1-j}y^j + \binom{k}{j-1}x^{k+1-j}y^j. \tag{7}$$

But

$$\binom{k}{j} + \binom{k}{j-1} = \frac{k(k-1)\cdots(k-j+2)(k-j+1)}{j!} + \frac{k(k-1)\cdots(k-j+2)}{(j-1)!}. \tag{8}$$

We multiply the second term in (8) by j/j to obtain

$$\binom{k}{j} + \binom{k}{j-1} = \frac{k(k-1)\cdots(k-j+2)}{j!}[(k-j+1)+j] \tag{9}$$

Since $j(j-1)! = j!$

$$= \frac{(k+1)k(k-1)\cdots(k-j+2)}{j!}$$

$$= \frac{(k+1)k\cdots(k-j+2)(k+1-j)!}{j!(k+1-j)!} = \binom{k+j}{j}.$$

Hence

$$\binom{k}{j}x^{k+1-j}y^j + \binom{k}{j-1}x^{k+1-j}y^j = \binom{k+1}{j}x^{k+1-j}y^j.$$

Finally from (6), (7), and (9), we have

$$(x+y)^{k+1} = \sum_{j=0}^{k+1}\binom{k+1}{j}x^{k+1-j}y^j,$$

which is equation (3) for $n = k + 1$, and the theorem is proved. ■

EXAMPLE 2 Calculate $(x + y)^5$.

Solution.

$$(x+y)^5 = \sum_{k=0}^{5}\binom{5}{k}x^{5-k}y^k$$

$$= \binom{5}{0}x^5 + \binom{5}{1}x^4y + \binom{5}{2}x^3y^2 + \binom{5}{3}x^2y^3 + \binom{5}{4}xy^4 + \binom{5}{5}y^5$$

$$= x^5 + 5x^4y + 10x^3y^2 + 10x^2y^3 + 5xy^4 + y^5$$ ■

EXAMPLE 3 Find the coefficient of the term containing x^3y^6 in the expansion of $(x + y)^9$.

Solution. In (3) we obtain the term x^3y^6 by setting $j = 6$ (so that $9 - j = 3$). The coefficient is

$$\binom{9}{6} = \frac{9!}{6!3!} = \frac{9 \cdot 8 \cdot 7}{3!} = \frac{9 \cdot 8 \cdot 7}{3 \cdot 2} = 84. \blacksquare$$

EXAMPLE 4 Calculate $(2x - 3y)^4$.

Solution.

$$\begin{aligned}(2x - 3y)^4 &= \binom{4}{0}(2x)^4 + \binom{4}{1}(2x)^3(-3y)^1 + \binom{4}{2}(2x)^2(-3y)^2 \\ &\quad + \binom{4}{3}(2x)^1(-3y)^3 + \binom{4}{4}(-3y)^4 \\ &= 16x^4 + 4(8x^3)(-3y) + 6(4x^2)(9y^2) + 4(2x)(-27y^3) + 81y^4 \\ &= 16x^4 - 96x^3y + 216x^2y^2 - 216xy^3 + 81y^4 \blacksquare\end{aligned}$$

PROBLEMS

In Problems 1–8, calculate the binomial coefficients.

1. $\binom{5}{3}$
2. $\binom{7}{4}$
3. $\binom{9}{2}$
4. $\binom{10}{5}$
5. $\binom{11}{3}$
6. $\binom{20}{0}$
7. $\binom{41}{41}$
8. $\binom{12}{7}$
9. Prove that $\binom{n}{k} = \binom{n}{n-k}$ for any integers $0 \le k \le n$.
10. Calculate $(x + y)^6$.
11. Calculate $(a + b)^7$.
12. Calculate $(u - w)^6$.
13. Calculate $(x - 2y)^5$.
14. Calculate $(x^2 - y^3)^4$.
15. Calculate $(ax - by)^5$.
16. Calculate $(x^n + y^n)^5$.
17. Calculate $[(x/2) + (y/3)]^5$.
18. Find the coefficient of x^5y^7 in the expansion of $(x + y)^{12}$.
19. Find the coefficient of x^8y^3 in the expansion of $(x + y)^{11}$.
20. Show that in the expansion of $(x + y)^n$ the coefficient of x^ky^{n-k} is equal to the coefficient of $x^{n-k}y^k$.
21. Show that for any nonnegative integer n

$$\binom{n}{0} + \binom{n}{1} + \binom{n}{2} + \cdots + \binom{n}{n} = 2^n.$$

[*Hint:* Expand $(1 + 1)^n$.]

22. Show that for any nonnegative integer n,

$$\binom{n}{0} - \binom{n}{1} + \binom{n}{2} - \binom{n}{3} + \cdots + (-1)^n\binom{n}{n} = 0.$$

[*Hint:* Expand $(1 - 1)^n$.]

Appendix 3
Complex Numbers

In algebra we encounter the problem of finding the roots of the polynomial

$$\lambda^2 + a\lambda + b = 0. \tag{1}$$

To find the roots, we use the quadratic formula to obtain

$$\lambda = \frac{-a \pm \sqrt{a^2 - 4b}}{2}. \tag{2}$$

If $a^2 - 4b > 0$, there are two real roots. If $a^2 - 4b = 0$, we obtain the single root (of multiplicity 2) $\lambda = -a/2$. To deal with the case $a^2 - 4b < 0$, we introduce the **imaginary number**[†]

$$i = \sqrt{-1}. \tag{3}$$

[†]You should not be troubled by the term "imaginary." It's just a name. The British mathematician Alfred North Whitehead, in the chapter on imaginary numbers in his *Introduction to Mathematics,* wrote:

> At this point it may be useful to observe that a certain type of intellect is always worrying itself and others by discussion as to the applicability of technical terms. Are the incommensurable numbers properly called numbers? Are the positive and negative numbers really numbers? Are the imaginary numbers imaginary, and are they numbers?—are types of such futile questions. Now, it cannot be too clearly understood that, in science, technical terms are names arbitrarily assigned, like Christian names to children. There can be no question of the names being right or wrong. They may be judicious or injudicious; for they can sometimes be so arranged as to be easy to remember, or so as to suggest relevant and important ideas. But the essential principle involved was quite clearly enunciated in Wonderland to Alice by Humpty Dumpty, when he told her, apropos of his use of words, 'I pay them extra and make them mean what I like'. So we will not bother as to whether imaginary numbers are imaginary, or as to whether they are numbers, but will take the phrase as the arbitrary name of a certain mathematical idea, which we will now endeavour to make plain.

Then for $a^2 - 4b < 0$

$$\sqrt{a^2 - 4b} = \sqrt{(4b - a^2)(-1)} = \sqrt{4b - a^2}\sqrt{-1} = \sqrt{4b - a^2}\, i,$$

and the two roots of (1) are given by

$$\lambda_1 = -\frac{a}{2} + \frac{\sqrt{4b - a^2}}{2} i \qquad \text{and} \qquad \lambda_2 = -\frac{a}{2} - \frac{\sqrt{4b - a^2}}{2} i.$$

EXAMPLE 1 Find the roots of the quadratic equation $\lambda^2 + 2\lambda + 5 = 0$.

Solution. We have $a = 2$, $b = 5$, and $a^2 - 4b = -16$. Thus $\sqrt{a^2 - 4b} = \sqrt{-16} = \sqrt{16}\sqrt{-1} = 4i$, and the roots are

$$\lambda_1 = \frac{-2 + 4i}{2} = -1 + 2i \qquad \text{and} \qquad \lambda_2 = -1 - 2i. \blacksquare$$

Definition 1 COMPLEX NUMBER A **complex number** is a number of the form

$$z = \alpha + i\beta \tag{4}$$

where α and β are real numbers; α is called the **real part** of z and is denoted Re z. β is called the **imaginary part** of z and is denoted Im z. Representation (4) is sometimes called the **Cartesian form** of the complex number z.

REMARK. If $\beta = 0$ in equation (4), then $z = \alpha$ is a real number. In this context we can regard the set of real numbers as a subset of the set of complex numbers.

EXAMPLE 2 In Example 1, Re $\lambda_1 = -1$ and Im $\lambda_1 = 2$. ■

We can add and multiply complex numbers by using the standard rules of algebra.

EXAMPLE 3 Let $z = 2 + 3i$ and $w = 5 - 4i$. Calculate (a) $z + w$, (b) $3w - 5z$, and (c) zw.

Solution.

(a) $z + w = (2 + 3i) + (5 - 4i) = (2 + 5) + (3 - 4)i = 7 - i$.
(b) $3w = 3(5 - 4i) = 15 - 12i$, $5z = 10 + 15i$, and $3w - 5z = (15 - 12i) - (10 + 15i) = (15 - 10) + i(-12 - 15) = 5 - 27i$.
(c) $zw = (2 + 3i)(5 - 4i) = (2)(5) + 2(-4i) + (3i)(5) + (3i)(-4i) = 10 - 8i + 15i - 12i^2 = 10 + 7i + 12 = 22 + 7i$. Here we used the fact that $i^2 = -1$. ■

We can plot a complex number z in the xy-plane by plotting Re z along the x-axis and Im z along the y-axis. Thus each complex number can be thought of as a point in the xy-plane. With this representation the xy-plane is called the **complex plane.** Some representative points are plotted in Figure 1.

If $z = \alpha + i\beta$, then we define the **conjugate** of z, denoted $\bar{z}$, by

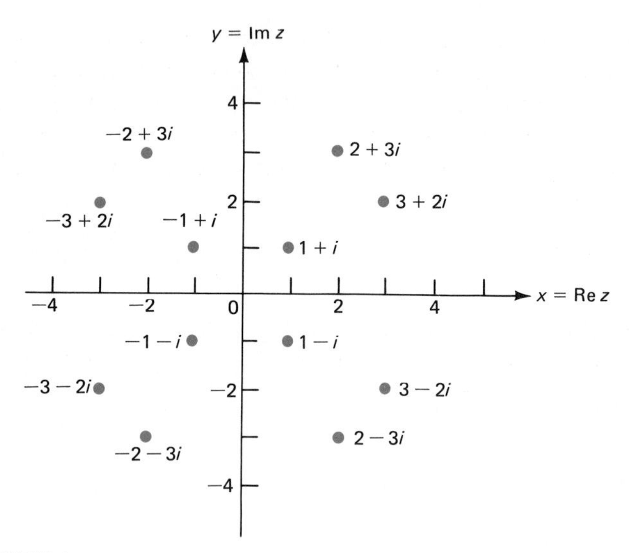

FIGURE 1

$$\bar{z} = \alpha - i\beta. \tag{5}$$

Figure 2 depicts a representative value of z and $\bar{z}$.

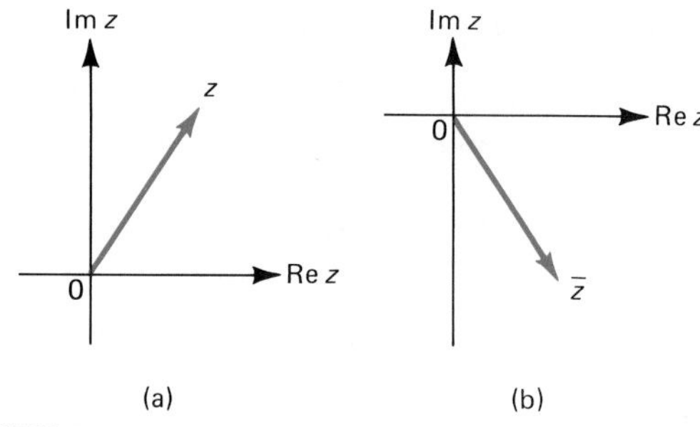

FIGURE 2

EXAMPLE 4 Compute the conjugate of **(a)** $1 + i$, **(b)** $3 - 4i$, **(c)** $-7 + 5i$, and **(d)** -3.

Solution. **(a)** $\overline{1 + i} = 1 - i$. **(b)** $\overline{3 - 4i} = 3 + 4i$. **(c)** $\overline{-7 + 5i} = -7 - 5i$. **(d)** $\overline{-3} = -3$. ■

It is not difficult to show (see Problem 35) that

$$\bar{z} = z \qquad \text{if and only if } z \text{ is real.} \tag{6}$$

If $z = \beta i$ with β real, then z is said to be **pure imaginary.** We can then show (see Problem 36) that

$$\bar{z} = -z \qquad \text{if and only if } z \text{ is pure imaginary.} \tag{7}$$

Let $p_n(x) = a_0 + a_1x + a_2x^2 + \cdots + a_nx^n$ be a polynomial with real coefficients. Then it can be shown (see Problem 41) that the complex roots of the equation $p_n(x) = 0$ occur in complex conjugate pairs. That is, if z is a root of $p_n(x) = 0$, then so is $\bar{z}$. We saw this fact illustrated in Example 1 in the case in which $n = 2$.

For $z = \alpha + i\beta$ we define the **magnitude** of z, denoted $|z|$, by

$$|z| = \sqrt{\alpha^2 + \beta^2}, \tag{8}$$

and we define the **argument** of z, denoted arg z, as the angle θ between the line 0z and the positive x-axis. From Figure 3 we see that $r = |z|$ is the distance from z to the origin, and

$$\theta = \arg z = \tan^{-1}\frac{\beta}{\alpha}. \tag{9}$$

By convention, we always choose a value of $\tan^{-1} \beta/\alpha$ that lies in the interval

$$-\pi < \theta \leq \pi. \tag{10}$$

From Figure 4 we see that

$$|\bar{z}| = |z| \tag{11}$$

and

$$\arg \bar{z} = -\arg z. \tag{12}$$

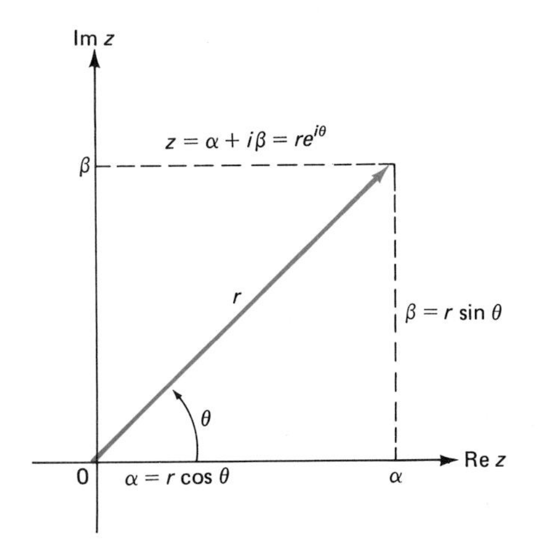

FIGURE 3

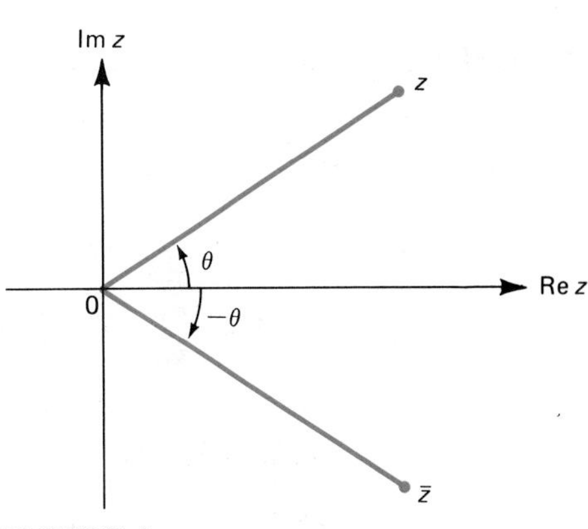

FIGURE 4

We can use $|z|$ and arg z to describe what is often a more convenient way to represent complex numbers.[†] From Figure 3 it is evident that if $z = \alpha + i\beta$, $r = |z|$, and $\theta = \arg z$, then

$$\alpha = r\cos\theta \quad \text{and} \quad \beta = r\sin\theta. \tag{13}$$

We shall see at the end of this appendix that

$$e^{i\theta} = \cos\theta + i\sin\theta. \tag{14}$$

Since $\cos(-\theta) = \cos\theta$ and $\sin(-\theta) = -\sin\theta$, we also have

$$e^{-i\theta} = \cos(-\theta) + i\sin(-\theta) = \cos\theta - i\sin\theta. \tag{14'}$$

Formula (14) is called **Euler's formula.**[‡] Using Euler's formula and equation (13), we have

$$z = \alpha + i\beta = r\cos\theta + ir\sin\theta = r(\cos\theta + i\sin\theta),$$

or

$$z = re^{i\theta}. \tag{15}$$

Representation (15) is called the **polar form** of the complex number z.

EXAMPLE 5 Determine the polar forms of the following complex numbers: **(a)** 1 **(b)** -1 **(c)** i **(d)** $1 + i$ **(e)** $-1 - \sqrt{3}$ **(f)** $-2 + 7i$

Solution. The six points are plotted in Figure 5.

(a) From Figure 5a it is clear that $\arg 1 = 0$. Since $\text{Re } 1 = 1$, we see that, in polar form,

$$1 = 1e^{i0} = 1e^{0} = 1.$$

(b) Since $\arg(-1) = \pi$ (Figure 5b) and $|-1| = 1$, we have

$$-1 = 1e^{\pi i} = e^{i\pi}.$$

(c) From Figure 5c we see that $\arg i = \pi/2$. Since $|i| = \sqrt{0^2 + 1^2} = 1$, it follows that

[†]From your study of polar coordinates, you will find this new representation very familiar.

[‡]Named for the Swiss mathematician Leonhard Euler (1707–1783).

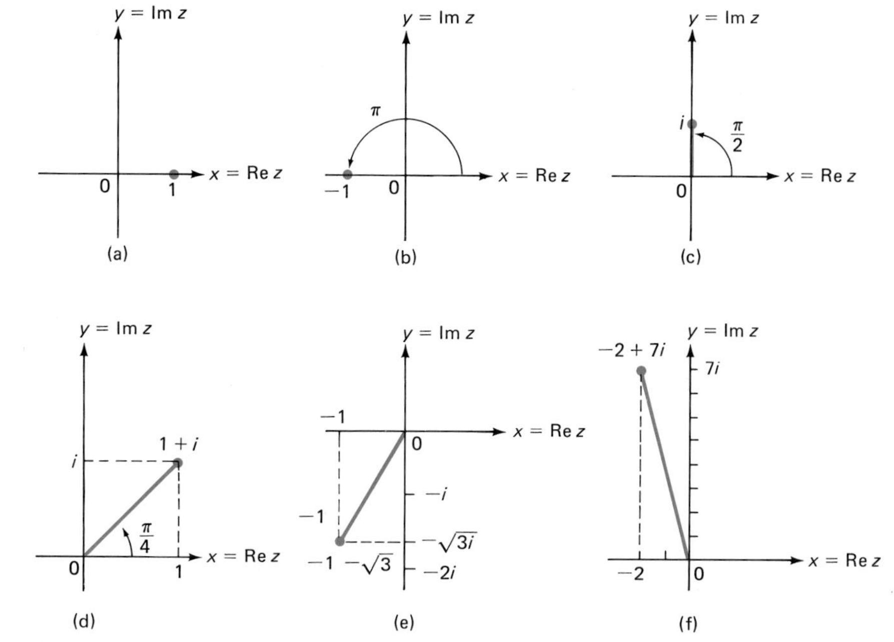

FIGURE 5

$$i = e^{i\pi/2}.$$

(d) $\arg(1 + i) = \tan^{-1} 1/1 = \pi/4$, and $|1 + i| = \sqrt{1^2 + 1^2} = \sqrt{2}$, so that

$$1 + i = \sqrt{2}\, e^{i\pi/4}.$$

(e) Here $\tan^{-1} \beta/\alpha = \tan^{-1}\sqrt{3} = \pi/3$. However, arg z is in the third quadrant, so $\theta = (\pi/3) - \pi = -2\pi/3$. Also, $|-1 - \sqrt{3}i| = \sqrt{1^2 + (\sqrt{3})^2} = \sqrt{1 + 3} = 2$, so that $-1 - \sqrt{3} = 2e^{-2\pi i/3}$.

(f) To compute this complex number, we need a calculator. A calculator indicates that

$$\arg z = \tan^{-1}(-\tfrac{7}{2}) = \tan^{-1}(-3.5) \approx -1.2925.$$

But $\tan^{-1} x$ is defined as a number in the interval $(-\pi, \pi]$. Since from Figure 5f θ is in the second quadrant, we see that $\arg z = \tan^{-1}(-3.5) + \pi \approx 1.8491$. Next, we see that

$$|-2 + 7i| = \sqrt{(-2)^2 + 7^2} = \sqrt{53}.$$

Hence

$$-2 + 7i \approx \sqrt{53}\, e^{1.8491i}. \ ■$$

EXAMPLE 6 Convert the following complex numbers from polar to Cartesian form: **(a)** $2e^{i\pi/3}$, **(b)** $4e^{3\pi i/2}$.

Solution.
(a) $e^{i\pi/3} = \cos \pi/3 + i \sin \pi/3 = \frac{1}{2} + (\sqrt{3}/2)i$. Thus $2e^{i\pi/3} = 1 + \sqrt{3}i$.
(b) $e^{3\pi i/2} = \cos 3\pi/2 + i \sin 3\pi/2 = 0 + i(-1) = -i$. Thus $4e^{3\pi/2} = -4i$. ■

If $\theta = \arg z$, then by equation (12), $\arg \bar{z} = -\theta$. Thus since $|\bar{z}| = |z|$, we have the following:

$$\text{If } z = re^{i\theta}, \text{ then } \bar{z} = re^{-i\theta}. \tag{16}$$

Suppose we write a complex number in its polar form $z = re^{i\theta}$. Then

$$z^n = (re^{i\theta})^n = r^n(e^{i\theta})^n = r^n e^{in\theta} = r^n(\cos n\theta + i \sin n\theta). \tag{17}$$

Formula (17) is useful for a variety of computations. In particular, when $r = |z| = 1$, we obtain the **De Moivre formula:**†

$$(\cos \theta + i \sin \theta)^n = \cos n\theta + i \sin n\theta. \tag{18}$$

EXAMPLE 7 Compute $(1 + i)^5$.

Solution. In Example 5(d) we showed that $1 + i = \sqrt{2}\, e^{\pi i/4}$. Then

$$(1 + i)^5 = (\sqrt{2}\, e^{\pi i/4})^5 = (\sqrt{2})^5 e^{5\pi i/4} = 4\sqrt{2}\left(\cos \frac{5\pi}{4} + i \sin \frac{5\pi}{4}\right)$$

$$= 4\sqrt{2}\left(-\frac{1}{\sqrt{2}} - \frac{1}{\sqrt{2}} i\right) = -4 - 4i.$$

This can be checked by direct calculation. If the direct calculation seems no more difficult, then try to compute $(1 + i)^{20}$ directly. Proceeding as above, we obtain

$$(1 + i)^{20} = (\sqrt{2})^{20} e^{20\pi i/4} = 2^{10}(\cos 5\pi + i \sin 5\pi)$$
$$= 2^{10}(-1 + 0) = -1024. \quad ■$$

Proof of Euler's Formula. We will show that

$$e^{i\theta} = \cos \theta + i \sin \theta \tag{19}$$

by using power series. We have

---

†Abraham De Moivre (1667–1754) was a French mathematician well known for his work in probability theory, infinite series, and trigonometry. He was so highly regarded that Newton often told those who came to him with questions on mathematics. "Go to M. De Moivre; he knows these things better than I do."

$$e^x = 1 + x + \frac{x^2}{2!} + \frac{x^3}{3!} + \cdots,^{\dagger} \tag{20}$$

$$\sin x = x - \frac{x^3}{3!} + \frac{x^5}{5!} - \cdots, \tag{21}$$

$$\cos x = 1 - \frac{x^2}{2!} + \frac{x^4}{4!} - \cdots. \tag{22}$$

Then

$$e^{i\theta} = 1 + (i\theta) + \frac{(i\theta)^2}{2!} + \frac{(i\theta)^3}{3!} + \frac{(i\theta)^4}{4!} + \frac{(i\theta)^5}{5!} + \cdots. \tag{23}$$

Now $i^2 = -1$, $i^3 = -i$, $i^4 = 1$, $i^5 = i$, and so on. Thus (23) can be written

$$\begin{aligned} e^{i\theta} &= 1 + i\theta - \frac{\theta^2}{2!} - \frac{i\theta^3}{3!} + \frac{\theta^4}{4!} + \frac{i\theta^5}{5!} - \cdots \\ &= \left(1 - \frac{\theta^2}{2!} + \frac{\theta^4}{4!} - \cdots\right) + i\left(\theta - \frac{\theta^3}{3!} + \frac{\theta^5}{5!} - \cdots\right) \\ &= \cos\theta + i\sin\theta. \end{aligned}$$

This completes the proof. ■

PROBLEMS

In Problems 1–5 perform the indicated operation.

1. $(2 - 3i) + (7 - 4i)$
2. $3(4 + i) - 5(-3 + 6i)$
3. $(1 + i)(1 - i)$
4. $(2 - 3i)(4 + 7i)$
5. $(-3 + 2i)(7 + 3i)$

In Problems 6–15 convert the complex number to its polar form.

6. $5i$
7. $5 + 5i$
8. $-2 - 2i$
9. $3 - 3i$
10. $2 + 2\sqrt{3}i$
11. $3\sqrt{3} + 3i$
12. $1 - \sqrt{3}i$
13. $4\sqrt{3} - 4i$
14. $-6\sqrt{3} - 6i$
15. $-1 - \sqrt{3}i$

In Problems 16–25 convert from polar to Cartesian form.

16. $e^{3\pi i}$
17. $2e^{-7\pi i}$
18. $\frac{1}{2}e^{3\pi i/4}$
19. $\frac{1}{2}e^{-3\pi i/4}$
20. $6e^{\pi i/6}$
21. $4e^{5\pi i/6}$
22. $4e^{-5\pi i/6}$
23. $3e^{-2\pi i/3}$
24. $\sqrt{3}e^{23\pi i/4}$
25. e^{i}

In Problems 26–34 compute the conjugate of the given number.

26. $3 - 4i$
27. $4 + 6i$
28. $-3 + 8i$
29. $-7i$
30. 16
31. $2e^{\pi i/7}$
32. $4e^{3\pi i/5}$
33. $3e^{-4\pi i/11}$
34. $e^{0.012i}$

---

†Although we will not prove it here, these series expansions are also valid when x is a complex number.

35. Show that $z = \alpha + i\beta$ is real if and only if $z = \bar{z}$. [*Hint:* if $z = \bar{z}$, show that $\beta = 0$.]

36. Show that $z = \alpha + i\beta$ is pure imaginary if and only if $z = -\bar{z}$. [*Hint:* if $z = -\bar{z}$, show that $\alpha = 0$.]

37. For any complex number z, show that $z\bar{z} = |z|^2$.

38. Show that the circle of radius 1 centered at the origin (the *unit circle*) is the set of points in the complex plane that satisfy $|z| = 1$.

39. For any complex number z_0 and real number a, describe $\{z: |z - z_0| = a\}$.

40. Describe $\{z: |z - z_0| \leq a\}$, where z_0 and a are as in Problem 39.

***41.** Let $p(\lambda) = \lambda^n + a_{n-1}\lambda^{n-1} + a_{n-2}\lambda^{n-2} + \cdots + a_1\lambda + a_0$ with $a_0, a_1, \ldots, a_{n-1}$ real numbers. Show that if $p(z) = 0$, then $p(\bar{z}) = 0$. That is, *The roots of polynomials with real coefficients occur in complex conjugate pairs.*

42. Derive expressions for $\cos 4\theta$ and $\sin 4\theta$ by comparing the De Moivre formula and the expansion of $(\cos\theta + i\sin\theta)^4$.

***43.** Prove De Moivre's formula by mathematical induction. [*Hint:* Recall the trigonometric identities $\cos(x + y) = \cos x \cos y - \sin x \sin y$ and $\sin(x + y) = \sin x \cos y + \cos x \sin y$.]

Appendix 4
Proof of the Basic Theorem About Determinants

Theorem 1 BASIC THEOREM Let $A = (a_{ij})$ be an $n \times n$ matrix. Then

$$\det A = a_{11}A_{11} + a_{12}A_{12} + \cdots + a_{1n}A_{1n}$$

$$= a_{i1}A_{i1} + a_{i2}A_{i2} + \cdots + a_{in}A_{in} \quad \textbf{(1)}$$

$$= a_{1j}A_{1j} + a_{2j}A_{2j} + \cdots + a_{nj}A_{nj} \quad \textbf{(2)}$$

for $i = 1, 2, \ldots, n$ and $j = 1, 2, \ldots, n$.

NOTE. The first equality is Definition 8.1.3 of the determinant by cofactor expansion in the first row; the second equality says that the expansion by cofactors in any other row yields the determinant; the third equality says that expansion by cofactors in any column gives the determinant.

Proof. We prove equality (1) by mathematical induction. For the 2×2 matrix $A = \begin{pmatrix} a_{11} & a_{12} \\ a_{21} & a_{22} \end{pmatrix}$, we first expand the first row by cofactors:

$$\det A = a_{11}A_{11} + a_{12}A_{12} = a_{11}(a_{22}) + a_{12}(-a_{21}) = a_{11}a_{22} - a_{12}a_{21}.$$

Similarly, expanding in the second row, we obtain

$$a_{21}A_{21} + a_{22}A_{22} = a_{21}(-a_{12}) + a_{22}(a_{11}) = a_{11}a_{22} - a_{12}a_{21}.$$

Thus we get the same result by expanding in any row of a 2×2 matrix, and this proves equality (1) in the 2×2 case.

We now assume that equality (1) holds for all $(n-1) \times (n-1)$ matrices. We must show that it holds for $n \times n$ matrices. Our procedure will be to expand by cofactors in the first and ith rows and show that the expansions are identical. If we expand in the first row, then a typical term in the cofactor expansion is

$$a_{1k}A_{1k} = (-1)^{1+k}a_{1k}|M_{1k}|. \tag{3}$$

Note that this is the only place in the expansion of $|A|$ that the term a_{1k} occurs, since another typical term is $a_{1m}A_{1m} = (-1)^{1+m}|M_{1m}|$, $k \neq m$, and M_{1m} is obtained by deleting the first row and mth column of A (and a_{1k} is in the first row of A). Since M_{1k} is an $(n-1) \times (n-1)$ matrix, we can, by the induction hypothesis, calculate $|M_{1k}|$ by expanding in the ith row of A (which is the $(i-1)$st row of M_{1k}). A typical term in this expansion is

$$a_{il} \text{ (cofactor of } a_{il} \text{ in } M_{1k}) \qquad (k \neq l). \tag{4}$$

For the reasons outlined above, this is the only term in the expansion of $|M_{1k}|$ in the ith row of A that contains the term a_{il}. Substituting (4) into (3), we find that

$$(-1)^{1+k}a_{1k}a_{il} \text{ (cofactor of } a_{il} \text{ in } M_{1k}) \qquad (k \neq l) \tag{5}$$

is the only occurrence of the term $a_{1k}a_{il}$ in the cofactor expansion of det A in the first row.

Now if we expand by cofactors in the ith row of A (where $i \neq 1$), a typical term is

$$(-1)^{i+l}a_{il}|M_{il}| \tag{6}$$

and a typical term in the expansion of $|M_{il}|$ in the first row of M_{il} is

$$a_{1k} \text{ (cofactor of } a_{1k} \text{ in } M_{il}) \qquad (k \neq l) \tag{7}$$

and inserting (7) in (6), we find that the only occurrence of the term $a_{il}a_{1k}$ in the expansion of det A along its ith row is

$$(-1)^{i+l}a_{1k}a_{il} \text{ (cofactor of } a_{1k} \text{ in } M_{il}) \qquad (k \neq l). \tag{8}$$

If we can show that the expressions in (5) and (8) are the same, then (1) will be proved, for the term in (5) is the only occurrence of $a_{1k}a_{il}$ in the first row expansion, the term in (8) is the only occurrence of $a_{1k}a_{il}$ in the ith row expansion, and k, i, and l are arbitrary. This will show that the sums of the terms in the first and ith row expansions are the same.

Now let $M_{1i,kl}$ denote the $(n-2) \times (n-2)$ matrix obtained by deleting the first and ith rows and kth and lth columns of A. (This is called a **second-order minor** of A.) We first suppose that $k < l$. Then

$$M_{1k} = \begin{pmatrix} a_{21} & \cdots & a_{2,k-1} & a_{2,k+1} & \cdots & a_{2l} & \cdots & a_{2n} \\ \vdots & & \vdots & \vdots & & \vdots & & \vdots \\ a_{i1} & \cdots & a_{i,k-1} & a_{i,k+1} & \cdots & a_{il} & \cdots & a_{in} \\ \vdots & & \vdots & \vdots & & \vdots & & \vdots \\ a_{n1} & \cdots & a_{n,k-1} & a_{n,k+1} & \cdots & a_{nl} & \cdots & a_{nn} \end{pmatrix}, \tag{9}$$

$$M_{il} = \begin{pmatrix} a_{11} & \cdots & a_{1k} & \cdots & a_{1,l-1} & a_{1,l+1} & \cdots & a_{1n} \\ \vdots & & \vdots & & \vdots & \vdots & & \vdots \\ a_{i-1,1} & \cdots & a_{i-1,k} & \cdots & a_{i-1,l-1} & a_{i-1,l+1} & \cdots & a_{i-1,n} \\ a_{i+1,1} & \cdots & a_{i+1,k} & \cdots & a_{i+1,l-1} & a_{i+1,l+1} & \cdots & a_{i+1,n} \\ \vdots & & \vdots & & \vdots & \vdots & & \vdots \\ a_{n1} & & a_{nk} & & a_{n,l-1} & a_{n,l+1} & & a_{nn} \end{pmatrix}. \tag{10}$$

From (9) and (10), we see that the

$$\text{cofactor of } a_{il} \text{ in } M_{1k} = (-1)^{(i-1)+(l-1)}|M_{1i,kl}|, \tag{11}$$

$$\text{cofactor of } a_{1k} \text{ in } M_{il} = (-1)^{1+k}|M_{1i,kl}|. \tag{12}$$

Thus (5) becomes

$$(-1)^{1+k}a_{1k}a_{il}(-1)^{(i-1)+(l-1)}|M_{1i,kl}| = (-1)^{i+k+l-1}a_{1k}a_{il}|M_{1i,kl}| \tag{13}$$

and (8) becomes

$$(-1)^{i+l}a_{1k}a_{il}(-1)^{1+k}|M_{1i,kl}| = (-1)^{i+k+l+1}a_{1k}a_{il}|M_{1i,kl}|. \tag{14}$$

But $(-1)^{i+k+l-1} = (-1)^{i+k+l+1}$, so the right-hand sides of equations (13) and (14) are equal. Hence expressions (5) and (8) are equal and (1) is proved in the case in which $k < l$. If $k > l$, then by similar reasoning we find that the

$$\text{cofactor of } a_{il} \text{ in } M_{1k} = (-1)^{(i-1)+l}|M_{1i,kl}|,$$

$$\text{cofactor of } a_{1k} \text{ in } M_{il} = (-1)^{1+(k-1)}|M_{1i,kl}|,$$

so that (5) becomes

$$(-1)^{1+k}a_{1k}a_{il}(-1)^{(i-1)+l}|M_{1i,kl}| = (-1)^{i+k+l}a_{1k}a_{il}|M_{1i,kl}|$$

and (8) becomes

$$(-1)^{i+l}a_{1k}a_{il}(-1)^{k}|M_{1i,kl}| = (-1)^{i+k+l}a_{1k}a_{il}|M_{1i,kl}|.$$

This completes the proof of equation (1).

To prove equation (2) we go through a similar process. If we expand in the kth and lth columns, we find that the only occurrences of the term $a_{1k}a_{il}$ will be given by (5) and (8). (See Problems 1 and 2.) This shows that the expansion by cofactors in any two columns is the same and that each is equal to the expansion along any row. This completes the proof. ■

PROBLEMS

1. Show that if A is expanded along its kth column, then the only occurrence of the term $a_{1k}a_{il}$ is given by equation (5).
2. Show that if A is expanded along its lth column, then the only occurrence of the term $a_{1k}a_{il}$ is given by equation (8).
3. Show that if A is expanded along its kth column, then the only occurrence of the term $a_{ik}a_{jl}$ is $(-1)^{i+k}a_{ik}a_{jl}$ (cofactor of a_{jl} in M_{ik}) for $l \neq k$.
4. Let $A = \begin{pmatrix} 1 & 5 & 7 \\ 2 & -1 & 3 \\ 4 & 5 & -2 \end{pmatrix}$. Compute det A by expanding in each of the rows and columns.
5. Do the same for the matrix $A = \begin{pmatrix} 1 & -1 & 4 \\ 0 & 1 & 5 \\ -3 & 7 & 2 \end{pmatrix}$.

Appendix 5
Existence and Uniqueness for First-Order Initial Value Problems

In Chapters 11 and 12 we sought techniques for finding solutions to differential equations, assuming the *existence* of *unique* solutions for ordinary differential equations with specified initial conditions. The aim of this appendix is to prove some basic existence and uniqueness theorems for solutions of initial value problems and to show that the solution depends continuously on the initial conditions. At this point three questions suggest themselves:

(i) Why do we need to prove the existence of a solution, particularly if we know that the differential equation arises from a physical problem that has a solution?
(ii) Why must we worry about uniqueness?
(iii) Why is continuous dependence on initial conditions important?

To answer the first question, it is important to remember that a differential equation is only a model of a physical problem. It is possible that the differential equation is a very bad model, so bad in fact that it has no solution. Countless hours could be spent using the techniques we have developed looking for a solution that may not even exist. Thus existence theorems are not only of theoretical value in telling us which equations have solutions, but are also of value in developing mathematical models of physical problems.

Similarly, uniqueness theorems have both theoretical and practical implications. If we know that a problem has a unique solution, then once we have found one solution, we are done. If the solution is not unique, then we cannot talk about "the" solution, but must instead, worry about *which* solution is being discussed. In practice, if the physical problem has a unique solution, so should any mathematical model of the problem.

Finally, continuous dependence on the initial conditions is very important, since some inaccuracy is always present in practical situations. We need to know that if the initial conditions are slightly changed, the solution of the differential equation will change only slightly. Otherwise, slight inaccuracies could yield very different solutions. This property will be discussed in Problem 18.

Before continuing with our discussion, we point out that in this chapter we shall use theoretical tools from calculus that have not been widely used earlier in the text. In particular, we shall need the following facts about continuity and convergence of functions. They are discussed in most intermediate and advanced calculus texts:[†]

(i) Let $f(t, x)$ be a continuous function of the two variables t and x and let the closed, bounded region D be defined by

$$D = \{(t, x)\colon a \le t \le b,\ c \le x \le d\}$$

where a, b, c, and d are finite real numbers. Then $f(t, x)$ is bounded for (t, x) in D. That is, there is a number $M > 0$ such that $|f(t, x)| \le M$ for every pair (t, x) in D.

(ii) Let $f(x)$ be continuous on the closed interval $a \le x \le b$ and differentiable on the open interval $a < x < b$. Then the *mean value theorem* of differential calculus states that there is a number ξ between a and b ($a < \xi < b$) such that

$$f(b) - f(a) = f'(\xi)(b - a).$$

The equation can be written as

$$\frac{f(b) - f(a)}{b - a} = f'(\xi),$$

which says, geometrically, that the slope of the tangent to the curve $y = f(x)$ at the point ξ between a and b is equal to the slope of the secant line passing through the points $(a, f(a))$ and $(b, f(b))$ (see Figure 1).

(iii) Let $\{x_n(t)\}$ be a sequence of functions. Then $x_n(t)$ is said to **converge uniformly** to a (limit) function $x(t)$ on the interval $a \le t \le b$ if for every real number $\epsilon > 0$ there exists an integer $N > 0$ such that whenever $n \ge N$, we have

$$|x_n(t) - x(t)| < \epsilon$$

for every t, $a \le t \le b$.

[†]See, for example, R. C. Buck, *Advanced Calculus* (McGraw-Hill, New York, 1965).

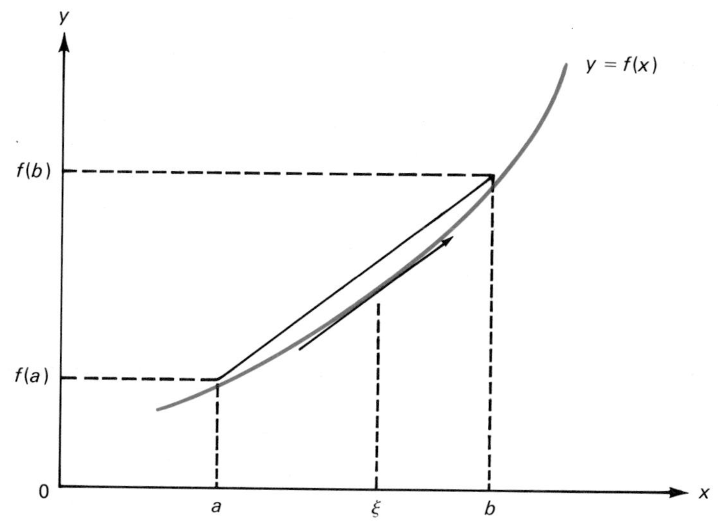

FIGURE 1

(iv) If the functions $\{x_n(t)\}$ of statement (iii) are continuous on the interval $a \le t \le b$, then the limit function $x(t)$ is also continuous there. This fact is often stated as "the uniform limit of continuous functions is continuous."

(v) Let $f(t, x)$ be a continuous function in the variable x and suppose that $x_n(t)$ converges to $x(t)$ uniformly as $n \to \infty$. Then

$$\lim_{n\to\infty} f(t, x_n(t)) = f(t, x(t)).$$

(vi) Let $f(t)$ be an integrable function on the interval $a \le t \le b$. Then

$$\left|\int_a^b f(t)\,dt\right| \le \int_a^b |f(t)|\,dt,$$

and if $|f(t)| \le M$, then

$$\int_a^b |f(t)|\,dt \le M \int_a^b dt = M(b - a).$$

(vii) Let $\{x_n(t)\}$ be a sequence of functions with $|x_n(t)| \le M_n$ for $a \le t \le b$. Then, if $\Sigma_{n=0}^{\infty} |M_n| < \infty$ (i.e., if $\Sigma_{n=0}^{\infty} M_n$ converges absolutely), then $\Sigma_{n=0}^{\infty} x_n(t)$ converges uniformly on the interval $a \le t \le b$ to a unique limit function $x(t)$. This is often called the **Weierstrass *M*-test** for uniform convergence.

(viii) Let $\{x_n(t)\}$ converge uniformly to $x(t)$ on the interval $a \le t \le b$ and let $f(t, x)$ be a continuous function of t and x in the region D defined in statement (i). Then

$$\lim_{n\to\infty} \int_a^b f(s, x_n(s))\,ds = \int_a^b \lim_{n\to\infty} f(s, x_n(s))\,ds = \int_a^b f(s, x(s))\,ds.$$

Using the facts stated above, we shall prove a general theorem about the existence and uniqueness of solutions of the first-order initial value problem

$$x'(t) = f(t, x(t)), \qquad x(t_0) = x_0, \tag{1}$$

where t_0 and x_0 are real numbers. Equation (1) includes all the first-order equations we have discussed in this book. For example, for the linear nonhomogeneous equation $x' + a(t)x = b(t)$,

$$f(t, x) = -a(t)x + b(t).$$

We shall show that if $f(t, x)$ and $(\partial f/\partial x)(t, x)$ are continuous in some region containing the point (t_0, x_0), then there is an interval (containing t_0) on which a unique solution of equation (1) exists. First, we need some preliminary results.

Theorem 1 Let $f(t, x)$ be continuous for all values t and x. Then the initial value problem (1) is equivalent to the integral equation

$$x(t) = x_0 + \int_{t_0}^{t} f(s, x(s))\, ds \tag{2}$$

in the sense that $x(t)$ is a solution of equation (1) if and only if $x(t)$ is a solution of equation (2).

Proof. If $x(t)$ satisfies equation (1), then

$$\int_{t_0}^{t} f(s, x(s))\, ds = \int_{t_0}^{t} x'(s)\, ds = x(s)\Big|_{t_0}^{t} = x(t) - x_0,$$

which shows that $x(t)$ satisfies equation (2). Conversely, if $x(t)$ satisfies equation (2), then differentiating equation (2), we have

$$x'(t) = \frac{d}{dt}\int_{t_0}^{t} f(s, x(s))\, ds = f(t, x(t))$$

and

$$x(t_0) = x_0 + \int_{t_0}^{t_0} f(s, x(s))\, ds = x_0.$$

Hence $x(t)$ also satisfies equation (1). ■

Let D denote the rectangular region in the tx-plane defined by

$$D: a \le t \le b,\ c \le x \le d, \tag{3}$$

where $-\infty < a < b < +\infty$ and $-\infty < c < d < +\infty$. (See Figure 2.) We say that the function $f(t, x)$ is **Lipschitz continuous** *in x over D* if there exists a constant k, $0 < k < \infty$, such that

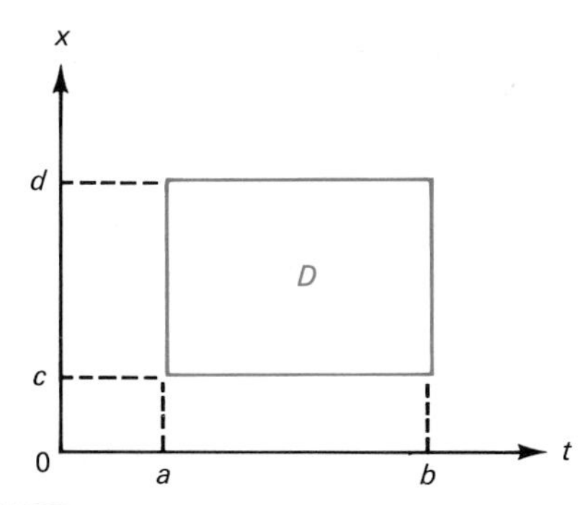

FIGURE 2

$$|f(t, x_1) - f(t, x_2)| \leq k|x_1 - x_2| \tag{4}$$

whenever (t, x_1) and (t, x_2) belong to D. The constant k is called a **Lipschitz constant.** Clearly, according to equation (4), every Lipschitz continuous function is continuous in x for each fixed t. However, *not every continuous function is Lipschitz continuous.*

EXAMPLE 1 Let $f(t, x) = \sqrt{x}$ on the set $0 \leq t \leq 1$, $0 \leq x \leq 1$. Then $f(t, x)$ is certainly continuous on this region. But

$$|f(t, x) - f(t, 0)| = |\sqrt{x} - 0| = \frac{1}{\sqrt{x}}|x - 0|$$

for all $0 < x < 1$, and $x^{-1/2}$ tends to ∞ as x approaches zero. Thus no finite Lipschitz constant can be found to satisfy equation (4). ■

However, Lipschitz continuity is not a rare occurrence, as shown by the following theorem.

Theorem 2 Let $f(t, x)$ and $(\partial f/\partial x)(t, x)$ be continuous on D. Then $f(t, x)$ is Lipschitz continuous in x over D.

Proof. Let (t, x_1) and (t, x_2) be points in D. For fixed t, $(\partial f/\partial x)(t, x)$ is a function of x, and so we may apply the mean value theorem of differential calculus [statement (ii)] to obtain

$$|f(t, x_1) - f(t, x_2)| = \left|\frac{\partial f}{\partial x}(t, \xi)\right| |x_1 - x_2|$$

where $x_1 < \xi < x_2$. But since $\partial f/\partial x$ is continuous in D, it is bounded there [according to statement (i)]. Hence there is a constant k, $0 < k < \infty$, such that

$$\left|\frac{\partial f}{\partial x}(t, x)\right| \leq k$$

for all (t, x) in D. ■

EXAMPLE 2 If $f(t, x) = tx^2$ on $0 \leq t \leq 1$, $0 \leq x \leq 1$, then

$$\left|\frac{\partial f}{\partial x}\right| = |2tx| \leq 2,$$

so that

$$|f(t, x_1) - f(t, x_2)| \leq 2|x_1 - x_2|. \blacksquare$$

We now define a sequence of functions $\{x_n(t)\}$, called **Picard**[†] **iterations,** by the successive formulas

$$\begin{aligned} x_0(t) &= x_0, \\ x_1(t) &= x_0 + \int_{t_0}^{t} f(s, x_0(s))\, ds, \\ x_2(t) &= x_0 + \int_{t_0}^{t} f(s, x_1(s))\, ds, \\ &\vdots \\ x_n(t) &= x_0 + \int_{t_0}^{t} f(s, x_{n-1}(s))\, ds. \end{aligned} \tag{5}$$

We shall show that under certain conditions the Picard iterations defined by (5) converge uniformly to a solution of equation (2). First we illustrate the process of this iteration by a simple example.

EXAMPLE 3 Consider the initial value problem

$$x'(t) = x(t), \qquad x(0) = 1. \tag{6}$$

As we know, equation (6) has the unique solution $x(t) = e^t$. In this case, the function $f(t, x)$ in equation (1) is given by $f(t, x(t)) = x(t)$, so that the Picard iterations defined by (5) yield successively

$$\begin{aligned} x_0(t) &= x_0 = 1, \\ x_1(t) &= 1 + \int_0^t (1)\, ds = 1 + t, \\ x_2(t) &= 1 + \int_0^t (1 + s)\, ds = 1 + t + \frac{t^2}{2}, \\ x_3(t) &= 1 + \int_0^t \left(1 + s + \frac{s^2}{2}\right) ds = 1 + t + \frac{t^2}{2!} + \frac{t^3}{3!}, \end{aligned}$$

[†]Emile Picard (1856–1941), one of the most eminent French mathematicians of the past century, made several outstanding contributions to mathematical analysis.

and clearly,

$$x_n(t) = 1 + t + \frac{t^2}{2!} + \cdots + \frac{t^n}{n!} = \sum_{k=0}^{n} \frac{t^k}{k!}.$$

Hence

$$\lim_{n\to\infty} x_n(t) = \sum_{k=0}^{\infty} \frac{t^k}{k!} = e^t$$

by equation (13.13.21). ■

We now state and prove the main result of this appendix.

Theorem 3 EXISTENCE THEOREM Let $f(t, x)$ be Lipschitz continuous in x with the Lipschitz constant k on the region D of all points (t, x) satisfying the inequalities

$$|t - t_0| \leq a, \qquad |x - x_0| \leq b.$$

(See Figure 3.) Then there exists a number $\delta > 0$ with the property that the initial value problem

$$x' = f(t, x) \qquad x(t_0) = x_0,$$

has a solution $x = x(t)$ on the interval $|t - t_0| \leq \delta$.

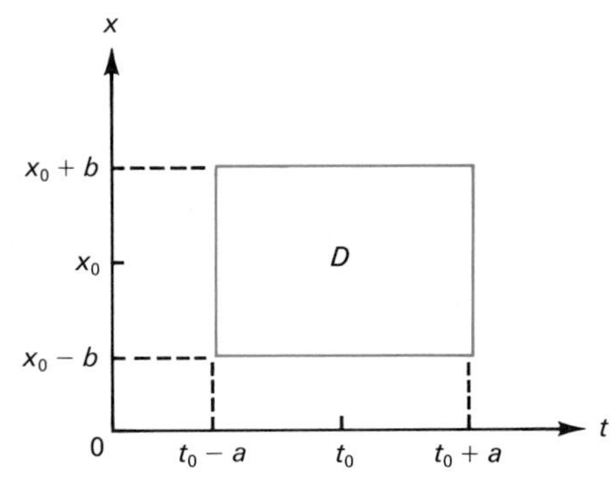

FIGURE 3

Proof. The proof of this theorem is complicated and will be done in several stages. However, the basic idea is simple: We need only justify that the Picard iterations converge uniformly and yield, in the limit, the solution of the integral equation (2).

Since f is continuous on D, it is bounded there [statement (i)] and we may begin by letting M be a finite upper bound for $|f(t, x)|$ on D. We then define

$$\delta = \min\{a, b/M\}. \tag{7}$$

1. We first show that the iterations $\{x_n(t)\}$ are continuous and satisfy the inequality

$$|x_n(t) - x_0| \le b. \tag{8}$$

Inequality (8) is necessary in order that $f(t, x_n(t))$ be defined for $n = 0, 1, 2, \ldots$. To show the continuity of $x_n(t)$, we first note that $x_0(t) = x_0$ is continuous (a constant function is always continuous). Then

$$x_1(t) = x_0 + \int_{t_0}^{t} f(t, x_0(s))\, ds.$$

But $f(t, x_0)$ is continuous (since $f(t, x)$ is continuous in t and x), and the integral of a continuous function is continuous. Thus $x_1(t)$ is continuous. In a similar fashion, we can show that

$$x_2(t) = x_1(t) + \int_{t_0}^{t} f(t, x_1(s))\, ds$$

is continuous and so on for $n = 3, 4, \ldots$.

Obviously the inequality (8) holds when $n = 0$, because $x_0(t) = x_0$. For $n \neq 0$, we use the definition (5) and equation (7) to obtain

$$|x_n(t) - x_0| = \left|\int_{t_0}^{t} f(s, x_{n-1}(s))\, ds\right| \le \left|\int_{t_0}^{t} |f(s, x_{n-1}(s))|\, ds\right|$$

$$\le M\left|\int_{t_0}^{t} ds\right| = M|t - t_0| \le M\delta \le b.$$

These inequalities follow from statement (vi). Note that the last inequality helps explain the choice of δ in equation (7).

2. Next, we show by mathematical induction that

$$|x_n(t) - x_{n-1}(t)| \le Mk^{n-1}\frac{|t - t_0|^n}{n!} \le \frac{Mk^{n-1}\delta^n}{n!}. \tag{9}$$

If $n = 1$, we obtain

$$|x_1(t) - x_0(t)| \le \left|\int_{t_0}^{t} f(s, x_0(s))\, ds\right| \le M\left|\int_{t_0}^{t} ds\right|$$

$$= M|t - t_0| \le M\delta.$$

Thus the result is true for $n = 1$.

We assume that the result is true for $n = m$ and prove that it holds for $n = m + 1$. That is, we assume that

$$|x_m(t) - x_{m-1}(t) \le \frac{Mk^{m-1}|t - t_0|^m}{m!} \le \frac{Mk^{m-1}\delta^m}{m!}.$$

Then, since $f(t, x)$ is Lipschitz continuous in x over D,

$$\begin{aligned}|x_{m+1}(t) - x_m(t)| &= \left|\int_{t_0}^{t} f(s, x_m(s))\, ds - \int_{t_0}^{t} f(s, x_{m-1}(s))\, ds\right| \\ &\le \left|\int_{t_0}^{t} |f(s, x_m(s)) - f(s, x_{m-1}(s))|\, ds\right| \\ &\le k\left|\int_{t_0}^{t} |x_m(s) - x_{m-1}(s)|\, ds\right| \\ &\le \frac{Mk^m}{m!}\left|\int_{t_0}^{t} (s - t_0)^m\, ds\right|^{\dagger} = \frac{Mk^m|t - t_0|^{m+1}}{(m+1)!} \le \frac{Mk^m\delta^{m+1}}{(m+1)!},\end{aligned}$$

which is what we wanted to show.

3. We will now show that $x_n(t)$ converges uniformly to a limit function $x(t)$ on the interval $|t - t_0| \le \delta$. By statement (iv), this will show that $x(t)$ is continuous.

We first note that

$$\begin{aligned}x_n(t) - x_0(t) &= x_n(t) - x_{n-1}(t) + x_{n-1}(t) - x_{n-2}(t) + \cdots + x_1(t) - x_0(t) \\ &= \sum_{k=0}^{n} [x_m(t) - x_{m-1}(t)].\end{aligned} \tag{10}$$

But by inequality (9),

$$|x_m(t) - x_{m-1}(t)| \le \frac{Mk^{m-1}\delta^m}{m!} = \frac{M}{k}\frac{k^m\delta^m}{m!},$$

so that

$$\sum_{m=1}^{\infty} |x_m(t) - x_{m-1}(t)| \le \frac{M}{k}\sum_{m=1}^{\infty}\frac{(k\delta)^m}{m!} = \frac{M}{k}(e^{k\delta} - 1),$$

since

$$e^{k\delta} = \sum_{m=0}^{\infty}\frac{(k\delta)^m}{m!} = 1 + \sum_{m=1}^{\infty}\frac{(k\delta)^m}{m!}.$$

By the Weierstrass M-test [(statement (vii))], we conclude that the series

$$\sum_{m=1}^{\infty} [x_m(t) - x_{m-1}(t)]$$

converges absolutely and uniformly on $|t - t_0| \le \delta$ to a unique limit function $y(t)$.

---

†This inequality follows from the induction assumption that the inequality (9) holds for $n = m$.

But

$$y(t) = \lim_{n\to\infty} \sum_{m=1}^{\infty} [x_m(t) - x_{m-1}(t)]$$
$$= \lim_{n\to\infty} [x_n(t) - x_0(t)] = \lim_{n\to\infty} x_n(t) - x_0(t)$$

or

$$\lim_{n\to\infty} x_n(t) = y(t) + x_0(t).$$

We denote the right-hand side of this equation by $x(t)$. Thus the limit of the Picard iterations $x_n(t)$ exists and the convergence $x_n(t) \to x(t)$ is uniform for all t in the interval $|t - t_0| \leq \delta$.

4. It remains to be shown that $x(t)$ is a solution to equation (2) for $|t - t_0| < \delta$. Since $f(t, x)$ is a continuous function of x and $x_n(t) \to x(t)$ as $n \to \infty$, we have, by statement (v),

$$\lim_{n\to\infty} f(t, x_n(t)) = f(t, x(t)).$$

Hence, by equation (5),

$$x(t) = \lim_{n\to\infty} x_{n+1}(t) = x_0 + \lim_{n\to\infty} \int_{t_0}^{t} f(s, x_n(s))\, ds$$
$$= x_0 + \int_{t_0}^{t} \lim_{n\to\infty} f(s, x_n(s))\, ds = x_0 + \int_{t_0}^{t} f(s, x(s))\, ds.$$

The step in which we interchange the limit and integral is justified by statement (viii). Thus $x(t)$ solves equation (2), and therefore it solves the initial value problem (1). ■

It turns out that the solution obtained in Theorem 3 is unique. Before proving this, however, we shall derive a simple version of a very useful result known as **Gronwall's inequality.**

Theorem 4 GRONWALL'S INEQUALITY Let $x(t)$ be a continuous nonnegative function and suppose that

$$x(t) \leq A + B\left|\int_{t_0}^{t} x(s)\, ds\right|, \tag{11}$$

where A and B are positive constants, for all values of t such that $|t - t_0| \leq \delta$. Then

$$x(t) \leq Ae^{B|t-t_0|} \tag{12}$$

for all t in the interval $|t - t_0| \leq \delta$.

Proof. We shall prove this result for $t_0 \le t \le t_0 + \delta$. The proof for $t_0 - \delta \le t \le t_0$ is similar (see Problem 14). We define

$$y(t) = B \int_{t_0}^{t} x(s)\, ds.$$

Then

$$y'(t) = Bx(t) \le B\left[A + B \int_{t_0}^{t} x(s)\, ds\right] = AB + By$$

or

$$y'(t) - By(t) \le AB. \tag{13}$$

We note that

$$\frac{d}{dt}[y(t)e^{-B(t-t_0)}] = e^{-B(t-t_0)}[y'(t) - By(t)].$$

Therefore, multiplying both sides of equation (13) by the integrating factor $e^{-B(t-t_0)}$ (which is greater than zero), we have

$$\frac{d}{dt}[y(t)e^{-B(t-t_0)}] \le ABe^{-B(t-t_0)}.$$

An integration of both sides of the inequality from t_0 to t yields

$$y(s)e^{-B(s-t_0)}\Big|_{t_0}^{t} \le AB \int_{t_0}^{t} e^{-B(s-t_0)}\, ds = -Ae^{-B(s-t_0)}\Big|_{t_0}^{t}.$$

But $y(t_0) = 0$, so that

$$y(t)e^{-B(t-t_0)} \le A(1 - e^{-B(t-t_0)}),$$

from which, after multiplying both sides by $e^{B(t-t_0)}$, we obtain

$$y(t) \le A[e^{B(t-t_0)} - 1].$$

Then by equation (11),

$$x(t) \le A + y(t) \le Ae^{B(t-t_0)}. \quad \blacksquare$$

Theorem 5 UNIQUENESS THEOREM Let the conditions of Theorem 3 (existence theorem) hold. Then $x(t) = \lim_{n\to\infty} x_n(t)$ is the only continuous solution of the initial value problem (1) in $|t - t_0| \le \delta$.

Proof. Let $x(t)$ and $y(t)$ be two continuous solutions of equation (2) in the interval $|t - t_0| \leq \delta$ and suppose that $(t, y(t))$ belongs to the region D for all t in that interval.[†] Define $v(t) = |x(t) - y(t)|$. Then $v(t) \geq 0$ and $v(t)$ is continuous. Since $f(t, x)$ is Lipschitz continuous in x over D,

$$\begin{aligned} v(t) &= \left| \left[x_0 + \int_{t_0}^{t} f(s, x(s))\, ds \right] - \left[x_0 + \int_{t_0}^{t} f(s, y(s))\, ds \right] \right| \\ &\leq k \left| \int_{t_0}^{t} |x(s) - y(s)|\, ds \right| = k \left| \int_{t_0}^{t} v(s)\, ds \right| \\ &\leq \epsilon + k \left| \int_{t_0}^{t} v(s)\, ds \right| \end{aligned}$$

for every $\epsilon > 0$. By Gronwall's inequality, we have

$$v(t) \leq \epsilon\, e^{k|t - t_0|}.$$

But $\epsilon > 0$ can be chosen arbitrarily close to zero, so that $v(t) \leq 0$. Since $v(t) \geq 0$, it follows that $v(t) \equiv 0$, implying that $x(t)$ and $y(t)$ are identical. Hence the limit of the Picard iterations is the only continuous solution. ■

Theorem 6 Let $f(t, x)$ and $\partial f/\partial x$ be continuous on D. Then there exists a constant $\delta > 0$ such that the Picard iteration $\{x_n(t)\}$ converges to a unique continuous solution of the initial value problem (1) on $|t - t_0| \leq \delta$.

Proof. This theorem follows directly from Theorem 2 and the existence and uniqueness theorems. ■

We note that Theorems 3, 5, and 6 are *local* results. By this we mean that unique solutions are guaranteed to exist only "near" the initial point (t_0, x_0).

EXAMPLE 4 Let

$$x'(t) = x^2(t), \qquad x(1) = 2.$$

Without solving this equation, we can show that there is a unique solution in some interval $|t - t_0| = |t - 1| \leq \delta$. Let $a = b = 1$. Then $|f(t, x)| = x^2 \leq 9\ (=M)$ for all $|x - x_0| \leq 1$, $x_0 = x(1) = 2$. Therefore, $\delta = \min\{a, b/M\} = \frac{1}{9}$, and Theorem 6 guarantees the existence of a unique solution on the interval $|t - 1| \leq \frac{1}{9}$. The solution of this initial value problem is easily found by a separation of variables to be $x(t) = 2/(3 - 2t)$. This solution exists so long as $t \neq \frac{3}{2}$. Starting at $t_0 = 1$, we see that the maximum interval of existences is $|t - t_0| < \frac{1}{2}$. Hence the value $\delta = \frac{1}{9}$ is not the best possible. ■

[†]Note that without this assumption, the function $f(t, y(t))$ may not even be defined at points where $(t, y(t))$ is not in D.

EXAMPLE 5 Consider the initial value problem

$$x' = \sqrt{x}, \qquad x(0) = 0.$$

As we saw in Example 1, $f(t, x) = \sqrt{x}$ does *not* satisfy a Lipschitz condition in any region containing the point (0, 0). By a separation of variables, it is easy to calculate the solution

$$x(t) = \left(\frac{t}{2}\right)^2.$$

However, $y(t) = 0$ is also a solution. Hence without a Lipschitz condition, the solution to an initial value problem (if one exists) may fail to be unique. ■

The last two examples illustrate the local nature of our existence–uniqueness theorem. We ask you to show (in Problem 17) that it is possible to derive *global* existence–uniqueness results for certain linear differential equations. That is, we can show that a unique solution exists for every real number t.

PROBLEMS

For each initial value problem of Problems 1–10, determine whether a unique solution can be guaranteed. If so, let $a = b = 1$ if possible, and find the number δ as given by equation (7). When possible, solve the equation and find a better value for δ, as in Example 4.

1. $x' = x^3,\ x(2) = 5$

2. $x' = x^3,\ x(1) = 2$

3. $x' = \dfrac{x}{t - x},\ x(0) = 1$

4. $x' = x^{1/3},\ x(1) = 0$

5. $x' = \sin x,\ x(1) = \pi/2$

6. $x' = \sqrt{x(x - 1)},\ x(1) = 2$

7. $x' = \ln|\sin x|,\ x(\pi/2) = 1$

8. $x' = \sqrt{x(x - 1)},\ x(2) = 3$

9. $x' = |x|,\ x(0) = 1$

10. $x' = tx,\ x(5) = 10$

11. Compute a Lipschitz constant for each of the following functions on the indicated region D:

(a) $f(t, x) = te^{-2x/t}, \quad t > 0,\ x > 0$

(b) $f(t, x) = \sin tx, \quad |t| \le 1,\ |x| \le 2$

(c) $f(t, x) = e^{-t^2} x^2 \sin(1/x)$, for all t, $-1 \le x < 1,\ x \ne 0$

(Part (c) shows that a Lipschitz constant may exist even when $\partial f/\partial x$ is not bounded in D.)

(d) $f(t, x) = (t^2x^3)^{3/2}, \quad |t| \le 2,\ |x| \le 3$

12. Consider the initial value problem

$$x' = x^2, \qquad x(1) = 3.$$

Show that the Picard iterations converge to the unique solution of this problem.

13. Construct the sequence $\{x_n(t)\}$ of Picard iterations for the initial value problem

$$x' = -x, \qquad x(0) = 3,$$

and show that it converges to the unique solution $x(t) = 3e^{-t}$.

14. Prove Gronwall's inequality (Theorem 4) for $t_0 - \delta \le t \le t_0$. [*Hint:* For $t < t_0$, $y(t) \le 0$.]

15. Let $v(t)$ be a positive function that satisfies the inequality

$$v(t) \le A + \int_{t_0}^{t} r(s)v(s)\,ds \tag{14}$$

where $A \ge 0$, $r(t)$ is a continuous positive function, and $t \ge t_0$. Prove that

$$v(t) \le A \exp\left[\int_{t_0}^{t} r(s)\,ds\right] \tag{15}$$

for all $t \ge t_0$. What kind of result holds for $t \le t_0$? This is a general form of Gronwall's inequality. [*Hint:* Define $y(t) = \int_{t_0}^{t} r(s)v(s)\,ds$ and show that $y'(t) = r(t)v(t) \le r(t)[A + y(t)]$. Finish the proof by following the steps of the proof of Theorem 4, using the integrating factor $\exp[-\int_{t_0}^{t} r(s)\,ds]$.]

16. Consider the initial value problem

$$x'' = f(t, x), \qquad x(0) = x_0, \ x'(0) = x_1, \tag{16}$$

where f is defined on the rectangle D: $|t| \le a$, $|x - x_0| \le b$. Prove under appropriate hypotheses that if a solution exists, then it must be unique. [*Hint:* Let $x(t)$ and $y(t)$ be continuous solutions of (16). Verify by differentiation that

$$x(t) = x_0 + x_1 t - \int_0^t (t - s) f(s, x(s))\, ds,$$

$$y(t) = x_0 + x_1 t - \int_0^t (t - s) f(s, y(s))\, ds$$

in some interval $|t| \le \delta$, $\delta > 0$. Then subtract these two expressions, use an appropriate Lipschitz condition, and apply the Gronwall inequality of Problem 15.]

17. Consider the first-order linear problem

$$x'(t) = a(t)x + b(t), \qquad x(t_0) = x_0, \tag{17}$$

where $a(t)$ and $b(t)$ are continuous in the interval and $a \le t_0 \le b$.

(a) Using $f(t, x) = a(t)x + b(t)$, show that the Picard iterations $\{x_n(t)\}$ exist and are continuous for all t in the interval $a \le t \le b$.

(b) Show that $f(t, x)$ defined in part (a) is Lipschitz continuous in x over the region D of points (t, x) satisfying the conditions $a \le t \le b$, $|x| < \infty$.

***(c)** Modify the proofs of Theorems 3 and 5 to show that (17) has a unique solution in the *entire* interval $[a, b]$.

(d) Show that if $a(t)$ and $b(t)$ are continuous for all values t, then (17) has a unique solution that is defined for *all* values $-\infty < t < \infty$. (This is a *global* existence–uniqueness theorem.)

18. Consider the initial value problem

$$x' = f(t, x), \qquad x(t_0) = x_0. \tag{18}$$

Show that if $f(t, x)$ satisfies the conditions of Theorem 3, then the solution $x(t) \equiv x(t, x_0)$ depends continuously on the value of x_0. Thus a small change in x_0 produces a small change in the solution over the interval $|t - t_0| \le \delta$. This result is often termed **continuous dependence on initial conditions.** [*Hint:* Define $x_0(t)$ to be the unique solution of (18) and $x_1(t)$ to be the unique solution of

$$x' = f(t, x), \qquad x(t_0) = x_1. \tag{19}$$

Rewrite the initial value problems (18) and (19) as integrals (2) and subtract one from the other to obtain

$$|x_1(t) - x_0(t)| \le |x_1 - x_0| + \left| \int_{t_0}^t |f(s, x_1(s)) - f(s, x_0(s))|\, ds \right|.$$

Then apply the Lipschitz condition to show that

$$|x_1(t) - x_0(t)| \le |x_1 - x_0| + k\left| \int_{t_0}^t |x_1(s) - x_0(s)|\, ds \right|.$$

Finally, apply Gronwall's inequality (Theorem 4) to show that

$$|x_1(t) - x_0(t)| \le |x_1 - x_0|\, e^{k|t - t_0|} \le |x_1 - x_0|\, e^{k\delta}$$

for $|t - t_0| \le \delta$, from which the desired result is immediately obtained.]

Table of Integrals†

STANDARD FORMS

1. $\int a\,dx = ax + C$

2. $\int af(x)\,dx = a\int f(x)\,dx + C$

3. $\int u\,dv = uv - \int v\,du$ (integration by parts)

4. $\int u^n\,du = \dfrac{u^{n+1}}{n+1} + C, \quad n \neq -1$

5. $\int \dfrac{du}{u} = \ln u$ if $u > 0$ or $\ln(-u)$ if $u < 0$

   $= \ln|u| + C$

6. $\int e^u\,du = e^u + C$

7. $\int a^u\,du = \int e^{u \ln a}\,du$

   $= \dfrac{e^{u \ln a}}{\ln a} = \dfrac{a^u}{\ln a} + C, \quad a > 0, a \neq 1$

8. $\int \sin u\,du = -\cos u + C$

9. $\int \cos u\,du = \sin u + C$

10. $\int \tan u\,du = \ln|\sec u| = -\ln|\cos u| + C$

11. $\int \cos u\,du = \ln|\sin u| + C$

12. $\int \sec u\,du = \ln|\sec u + \tan u|$

    $= \ln \tan\left|\dfrac{u}{2} + \dfrac{\pi}{4}\right| + C$

†All angles are measured in radians.

13. $\int \csc u\, du = \ln|\csc u - \cot u| = \ln\left|\tan \frac{u}{2}\right| + C$

14. $\int \sec^2 u\, du = \tan u + C$

15. $\int \csc^2 u\, du = -\cot u + C$

16. $\int \sec u \tan u\, du = \sec u + C$

17. $\int \csc u \cot u\, du = -\csc u + C$

18. $\int \frac{du}{u^2 + a^2} = \frac{1}{a} \tan^{-1} \frac{u}{a} + C$

19. $\int \frac{du}{u^2 - a^2} = \frac{1}{2a} \ln\left|\frac{u - a}{u + a}\right| + C$

$= -\frac{1}{a} \coth^{-1} \frac{u}{a} + C, \quad u^2 > a^2$

20. $\int \frac{du}{a^2 - u^2} = \frac{1}{2a} \ln\left|\frac{a + u}{a - u}\right| + C$

$= \frac{1}{a} \tanh^{-1} \frac{u}{a} + C, \quad u^2 < a^2$

21. $\int \frac{du}{\sqrt{a^2 - u^2}} = \sin^{-1} \frac{u}{|a|} + C$

22. $\int \frac{du}{\sqrt{u^2 + a^2}} = \ln(u + \sqrt{u^2 + a^2}) + C$

23. $\int \frac{du}{\sqrt{u^2 - a^2}} = \ln|u + \sqrt{u^2 - a^2}| + C$

24. $\int \frac{du}{u\sqrt{u^2 - a^2}} = \frac{1}{|a|} \sec^{-1}\left|\frac{u}{a}\right| + C$

25. $\int \frac{du}{u\sqrt{u^2 + a^2}} = -\frac{1}{a} \ln\left|\frac{a + \sqrt{u^2 + a^2}}{u}\right| + C$

26. $\int \frac{du}{u\sqrt{a^2 - u^2}} = -\frac{1}{a} \ln\left|\frac{a + \sqrt{a^2 - u^2}}{u}\right| + C$

INTEGRALS INVOLVING $au + b$

27. $\int \frac{du}{au + b} = \frac{1}{a} \ln|au + b| + C$

28. $\int \frac{u\, du}{au + b} = \frac{u}{a} - \frac{b}{a^2} \ln|au + b| + C$

29. $\int \frac{u^2\, du}{au + b} = \frac{(au + b)^2}{2a^3} - \frac{2b(au + b)}{a^3} + \frac{b^2}{a^3} \ln|au + b| + C$

30. $\int \frac{du}{u(au + b)} = \frac{1}{b} \ln\left|\frac{u}{au + b}\right| + C$

31. $\int \frac{du}{u^2(au + b)} = -\frac{1}{bu} + \frac{a}{b^2} \ln\left|\frac{au + b}{u}\right| + C$

32. $\int \frac{du}{(au + b)^2} = \frac{-1}{a(au + b)} + C$

33. $\int \frac{u\, du}{(au + b)^2} = \frac{b}{a^2(au + b)} + \frac{1}{a^2} \ln|au + b| + C$

34. $\int \frac{du}{u(au + b)^2} = \frac{1}{b(au + b)} + \frac{1}{b^2} \ln\left|\frac{u}{au + b}\right| + C$

35. $\int (au + b)^n\, du = \frac{(au + b)^{n+1}}{(n + 1)a} + C, \ n \neq -1$

36. $\int u(au + b)^n\, du = \frac{(au + b)^{n+2}}{(n + 2)a^2} - \frac{b(au + b)^{n+1}}{(n + 1)a^2} + C, \quad n \neq -1, -2$

37. $$\int u^m(au + b)^n\, du = \begin{cases} \dfrac{u^{m+1}(au + b)^n}{m + n + 1} + \dfrac{nb}{m + n + 1} \displaystyle\int u^m(au + b)^{n-1}\, du \\ \dfrac{u^m(au + b)^{n+1}}{(m + n + 1)a} - \dfrac{mb}{(m + n + 1)a} \displaystyle\int u^{m-1}(au + b)^n\, du \\ \dfrac{-u^{m+1}(au + b)^{n+1}}{(n + 1)b} + \dfrac{m + n + 2}{(n + 1)b} \displaystyle\int u^m(au + b)^{n+1}\, du \end{cases}$$

INTEGRALS INVOLVING $\sqrt{au + b}$

38. $\int \frac{du}{\sqrt{au + b}} = \frac{2\sqrt{au + b}}{a} + C$

39. $\int \frac{u\, du}{\sqrt{au + b}} = \frac{2(au - 2b)}{3a^2} \sqrt{au + b} + C$

40. $$\int \frac{du}{u\sqrt{au+b}} = \begin{cases} \dfrac{1}{\sqrt{b}} \ln\left|\dfrac{\sqrt{au+b}-\sqrt{b}}{\sqrt{au+b}+\sqrt{b}}\right| + C, & b > 0 \\[2ex] \dfrac{2}{\sqrt{-b}} \tan^{-1}\sqrt{\dfrac{au+b}{-b}} + C, & b < 0 \end{cases}$$

41. $$\int \sqrt{au+b}\, du = \frac{2\sqrt{(au+b)^3}}{3a} + C$$

42. $$\int u\sqrt{au+b}\, du = \frac{2(3au-2b)}{15a^2}\sqrt{(au+b)^3} + C$$

43. $$\int \frac{\sqrt{au+b}}{u}\, du = 2\sqrt{au+b} + b\int \frac{du}{u\sqrt{au+b}} \quad \text{(See 40.)}$$

INTEGRALS INVOLVING $u^2 + a^2$

44. $$\int \frac{du}{u^2+a^2} = \frac{1}{a}\tan^{-1}\frac{u}{a} + C$$

45. $$\int \frac{u\,du}{u^2+a^2} = \frac{1}{2}\ln(u^2+a^2) + C$$

46. $$\int \frac{u^2\,du}{u^2+a^2} = u - a\tan^{-1}\frac{u}{a} + C$$

47. $$\int \frac{du}{u(u^2+a^2)} = \frac{1}{2a^2}\ln\left(\frac{u^2}{u^2+a^2}\right) + C$$

48. $$\int \frac{du}{u^2(u^2+a^2)} = -\frac{1}{a^2u} - \frac{1}{a^3}\tan^{-1}\frac{u}{a} + C$$

49. $$\int \frac{du}{(u^2+a^2)^n} = \frac{u}{2(n-1)a^2(u^2+a^2)^{n-1}} + \frac{2n-3}{(2n-2)a^2}\int \frac{du}{(u^2+a^2)^{n-1}}$$

50. $$\int \frac{u\,du}{(u^2+a^2)^n} = \frac{-1}{2(n-1)(u^2+a^2)^{n-1}} + C,\ n \neq 1$$

51. $$\int \frac{du}{u(u^2+a^2)^n} = \frac{1}{2(n-1)a^2(u^2+a^2)^{n-1}} + \frac{1}{a^2}\int \frac{du}{u(u^2+a^2)^{n-1}},\ n \neq 1$$

INTEGRALS INVOLVING $u^2 - a^2$, $u^2 > a^2$

52. $$\int \frac{du}{u^2-a^2} = \frac{1}{2a}\ln\left|\frac{u-a}{u+a}\right| + C$$

53. $$\int \frac{u\,du}{u^2-a^2} = \frac{1}{2}\ln(u^2-a^2) + C$$

54. $$\int \frac{u^2\,du}{u^2-a^2} = u + \frac{a}{2}\ln\left|\frac{u-a}{u+a}\right| + C$$

55. $$\int \frac{du}{u(u^2-a^2)} = \frac{1}{2a^2}\ln\left|\frac{u^2-a^2}{u^2}\right| + C$$

56. $$\int \frac{du}{u^2(u^2-a^2)} = \frac{1}{a^2u} + \frac{1}{2a^3}\ln\left|\frac{u-a}{u+a}\right| + C$$

57. $$\int \frac{du}{(u^2-a^2)^2} = \frac{-u}{2a^2(u^2-a^2)} - \frac{1}{4a^3}\ln\left|\frac{u-a}{u+a}\right| + C$$

58. $$\int \frac{du}{(u^2-a^2)^n} = \frac{-u}{2(n-1)a^2(u^2-a^2)^{n-1}} - \frac{2n-3}{(2n-2)a^2}\int \frac{du}{(u^2-a^2)^{n-1}}$$

59. $$\int \frac{u\,du}{(u^2-a^2)^n} = \frac{-1}{2(n-1)(u^2-a^2)^{n-1}} + C$$

60. $$\int \frac{du}{u(u^2-a^2)^n} = \frac{-1}{2(n-1)a^2(u^2-a^2)^{n-1}} - \frac{1}{a^2}\int \frac{du}{u(u^2-a^2)^{n-1}}$$

INTEGRALS INVOLVING $a^2 - u^2$, $u^2 < a^2$

61. $\int \frac{du}{a^2 - u^2} = \frac{1}{2a} \ln\left|\frac{a + u}{a - u}\right| + C$ or $\frac{1}{a} \tanh^{-1}\frac{u}{a} + C$

62. $\int \frac{u\,du}{a^2 - u^2} = -\frac{1}{2}\ln|a^2 - u^2| + C$

63. $\int \frac{u^2\,du}{a^2 - u^2} = -u + \frac{a}{2}\ln\left|\frac{a + u}{a - u}\right| + C$

64. $\int \frac{du}{u(a^2 - u^2)} = \frac{1}{2a^2}\ln\left|\frac{u^2}{a^2 - u^2}\right| + C$

65. $\int \frac{du}{(a^2 - u^2)^2} = \frac{u}{2a^2(a^2 - u^2)} + \frac{1}{4a^3}\ln\left|\frac{a + u}{a - u}\right| + C$

66. $\int \frac{u\,du}{(a^2 - u^2)^2} = \frac{1}{2(a^2 - u^2)} + C$

INTEGRALS INVOLVING $\sqrt{u^2 + a^2}$

67. $\int \frac{du}{\sqrt{u^2 + a^2}} = \ln(u + \sqrt{u^2 + a^2}) + C$ or $\sinh^{-1}\frac{u}{|a|} + C$

68. $\int \frac{u\,du}{\sqrt{u^2 + a^2}} = \sqrt{u^2 + a^2} + C$

69. $\int \frac{u^2\,du}{\sqrt{u^2 + a^2}} = \frac{u\sqrt{u^2 + a^2}}{2} - \frac{a^2}{2}\ln(u + \sqrt{u^2 + a^2}) + C$

70. $\int \frac{du}{u\sqrt{u^2 + a^2}} = -\frac{1}{a}\ln\left|\frac{a + \sqrt{u^2 + a^2}}{u}\right| + C$

71. $\int \sqrt{u^2 + a^2}\,du = \frac{u\sqrt{u^2 + a^2}}{2} + \frac{a^2}{2}\ln(u + \sqrt{u^2 + a^2}) + C$

72. $\int u\sqrt{u^2 + a^2}\,du = \frac{(u^2 + a^2)^{3/2}}{3} + C$

73. $\int u^2\sqrt{u^2 + a^2}\,du = \frac{u(u^2 + a^2)^{3/2}}{4} - \frac{a^2u\sqrt{u^2 + a^2}}{8} - \frac{a^4}{8}\ln(u + \sqrt{u^2 + a^2}) + C$

74. $\int \frac{\sqrt{u^2 + a^2}}{u}\,du = \sqrt{u^2 + a^2} - a\ln\left|\frac{a + \sqrt{u^2 + a^2}}{u}\right| + C$

75. $\int \frac{\sqrt{u^2 + a^2}}{u^2}\,du = -\frac{\sqrt{u^2 + a^2}}{u} + \ln(u + \sqrt{u^2 + a^2}) + C$

INTEGRALS INVOLVING $\sqrt{u^2 - a^2}$

76. $\int \frac{du}{\sqrt{u^2 - a^2}} = \ln|u + \sqrt{u^2 - a^2}| + C$

77. $\int \frac{u\,du}{\sqrt{u^2 - a^2}} = \sqrt{u^2 - a^2} + C$

78. $\int \frac{u^2\,du}{\sqrt{u^2 - a^2}} = \frac{u\sqrt{u^2 - a^2}}{2} + \frac{a^2}{2}\ln|u + \sqrt{u^2 - a^2}| + C$

79. $\int \frac{du}{u\sqrt{u^2 - a^2}} = \frac{1}{|a|}\sec^{-1}\left|\frac{u}{a}\right| + C$

80. $\displaystyle\int \sqrt{u^2 - a^2}\, du = \frac{u\sqrt{u^2 - a^2}}{2} - \frac{a^2}{2}\ln|u + \sqrt{u^2 - a^2}| + C$

81. $\displaystyle\int u\sqrt{u^2 - a^2}\, du = \frac{(u^2 - a^2)^{3/2}}{3} + C$

82. $\displaystyle\int u^2\sqrt{u^2 - a^2}\, du = \frac{u(u^2 - a^2)^{3/2}}{4} + \frac{a^2u\sqrt{u^2 - a^2}}{8} - \frac{a^4}{8}\ln|u + \sqrt{u^2 - a^2}| + C$

83. $\displaystyle\int \frac{\sqrt{u^2 - a^2}}{u}\, du = \sqrt{u^2 - a^2} - |a| \sec^{-1}\left|\frac{u}{a}\right| + C$

84. $\displaystyle\int \frac{\sqrt{u^2 - a^2}}{u^2}\, du = -\frac{\sqrt{u^2 - a^2}}{u} + \ln|u + \sqrt{u^2 - a^2}| + C$

85. $\displaystyle\int \frac{du}{(u^2 - a^2)^{3/2}} = -\frac{u}{a^2\sqrt{u^2 - a^2}} + C$

INTEGRALS INVOLVING $\sqrt{a^2 - u^2}$

86. $\displaystyle\int \frac{du}{\sqrt{a^2 - u^2}} = \sin^{-1}\frac{u}{|a|} + C$

87. $\displaystyle\int \frac{u\, du}{\sqrt{a^2 - u^2}} = -\sqrt{a^2 - u^2} + C$

88. $\displaystyle\int \frac{u^2\, du}{\sqrt{a^2 - u^2}} = -\frac{u\sqrt{a^2 - u^2}}{2} + \frac{a^2}{2}\sin^{-1}\frac{u}{|a|} + C$

89. $\displaystyle\int \frac{du}{u\sqrt{a^2 - u^2}} = -\frac{1}{a}\ln\left|\frac{a + \sqrt{a^2 - u^2}}{u}\right| + C$

90. $\displaystyle\int \frac{du}{u^2\sqrt{a^2 - u^2}} = -\frac{\sqrt{a^2 - u^2}}{a^2u} + C$

91. $\displaystyle\int \sqrt{a^2 - u^2}\, du = \frac{u\sqrt{a^2 - u^2}}{2} + \frac{a^2}{2}\sin^{-1}\frac{u}{|a|} + C$

92. $\displaystyle\int u\sqrt{a^2 - u^2}\, du = -\frac{(a^2 - u^2)^{3/2}}{3} + C$

93. $\displaystyle\int u^2\sqrt{a^2 - u^2}\, du = -\frac{u(a^2 - u^2)^{3/2}}{4} + \frac{a^2u\sqrt{a^2 - u^2}}{8} + \frac{a^4}{8}\sin^{-1}\frac{u}{|a|} + C$

94. $\displaystyle\int \frac{\sqrt{a^2 - u^2}}{u}\, du = \sqrt{a^2 - u^2} - a\ln\left|\frac{a + \sqrt{a^2 - u^2}}{u}\right| + C$

95. $\displaystyle\int \frac{\sqrt{a^2 - u^2}}{u^2}\, du = -\frac{\sqrt{a^2 - u^2}}{u} - \sin^{-1}\frac{u}{|a|} + C$

INTEGRALS INVOLVING THE TRIGONOMETRIC FUNCTIONS

96. $\displaystyle\int \sin au\, du = -\frac{\cos au}{a} + C$

97. $\displaystyle\int u\sin au\, du = \frac{\sin au}{a^2} - \frac{u\cos au}{a} + C$

98. $\displaystyle\int u^2\sin au\, du = \frac{2u}{a^2}\sin au + \left(\frac{2}{a^3} - \frac{u^2}{a}\right)\cos au + C$

99. $\displaystyle\int \frac{du}{\sin au} = \frac{1}{a}\ln(\csc au - \cot au)$

$\displaystyle= \frac{1}{a}\ln\left|\tan\frac{au}{2}\right| + C$

100. $\displaystyle\int \sin^2 au\, du = \frac{u}{2} - \frac{\sin 2au}{4a} + C$

101. $\int u \sin^2 au \, du = \frac{u^2}{4} - \frac{u \sin 2au}{4a} - \frac{\cos 2au}{8a^2} + C$

102. $\int \frac{du}{\sin^2 au} = -\frac{1}{a} \cot au + C$

103. $\int \sin pu \sin qu \, du = \frac{\sin(p - q)u}{2(p - q)} - \frac{\sin(p + q)u}{2(p + q)} + C, \quad p \neq \pm q$

104. $\int \frac{du}{1 - \sin au} = \frac{1}{a} \tan\left(\frac{\pi}{4} + \frac{au}{2}\right) + C$

105. $\int \frac{u \, du}{1 - \sin au} = \frac{u}{a} \tan\left(\frac{\pi}{4} + \frac{au}{2}\right) + \frac{2}{a^2} \ln\left|\sin\left(\frac{\pi}{4} - \frac{au}{2}\right)\right| + C$

106. $\int \frac{du}{1 + \sin au} = -\frac{1}{a} \tan\left(\frac{\pi}{4} - \frac{au}{2}\right) + C$

107. $$\int \frac{du}{p + q \sin au} = \begin{cases} \dfrac{2}{a\sqrt{p^2 - q^2}} \tan^{-1} \dfrac{p \tan \frac{1}{2}au + q}{\sqrt{p^2 - q^2}} + C, & |p| > |q| \\[2ex] \dfrac{1}{a\sqrt{q^2 - p^2}} \ln\left|\dfrac{p \tan \frac{1}{2}au + q - \sqrt{q^2 - p^2}}{p \tan \frac{1}{2}au + q + \sqrt{q^2 - p^2}}\right| + C, & |p| < |q| \end{cases}$$

108. $\int u^m \sin au \, du = -\frac{u^m \cos au}{a} + \frac{mu^{m-1} \sin au}{a^2} - \frac{m(m - 1)}{a^2} \int u^{m-2} \sin au \, du$

109. $\int \sin^n au \, du = -\frac{\sin^{n-1} au \cos au}{an} + \frac{n - 1}{n} \int \sin^{n-2} au \, du$

110. $\int \frac{du}{\sin^n au} = \frac{-\cos au}{a(n - 1)\sin^{n-1} au} + \frac{n - 2}{n - 1} \int \frac{du}{\sin^{n-2} au}, \quad n \neq 1$

111. $\int \cos au \, du = \frac{\sin au}{a} + C$

112. $\int u \cos au \, du = \frac{\cos au}{a^2} + \frac{u \sin au}{a} + C$

113. $\int u^2 \cos au \, du = \frac{2u}{a^2} \cos au + \left(\frac{u^2}{a} - \frac{2}{a^3}\right) \sin au + C$

114. $\int \frac{du}{\cos au} = \frac{1}{a} \ln(\sec au + \tan au) = \frac{1}{a} \ln\left|\tan\left(\frac{\pi}{4} + \frac{au}{2}\right)\right| + C$

115. $\int \cos^2 au \, du = \frac{u}{2} + \frac{\sin 2au}{4a} + C$

116. $\int u \cos^2 au \, du = \frac{u^2}{4} + \frac{u \sin 2au}{4a} + \frac{\cos 2au}{8a^2} + C$

117. $\int \frac{du}{\cos^2 au} = \frac{\tan au}{a} + C$

118. $\int \cos qu \cos pu \, du = \frac{\sin(q - p)u}{2(q - p)} + \frac{\sin(q + p)u}{2(q + p)} + C, \quad q \neq \pm p$

119. $$\int \frac{du}{p + q \cos au} = \begin{cases} \dfrac{2}{a\sqrt{p^2 - q^2}} \tan^{-1}\sqrt{(p - q)/(p + q)} \tan \frac{1}{2}au + C, & |p| > |q| \\[2ex] \dfrac{1}{a\sqrt{q^2 - p^2}} \ln\left[\dfrac{\tan \frac{1}{2}au + \sqrt{(q + p)/(q - p)}}{\tan \frac{1}{2}au - \sqrt{(q + p)/(q - p)}}\right] + C, & |p| < |q| \end{cases}$$

120. $\int u^m \cos au \, du = \frac{u^m \sin au}{a} + \frac{mu^{m-1}}{a^2} \cos au - \frac{m(m - 1)}{a^2} \int u^{m-2} \cos au \, du$

121. $\int \cos^n au\, du = \dfrac{\sin au \cos^{n-1} au}{an} + \dfrac{n-1}{n}\int \cos^{n-2} au\, du$

122. $\int \dfrac{du}{\cos^n au} = \dfrac{\sin au}{a(n-1)\cos^{n-1} au} + \dfrac{n-2}{n-1}\int \dfrac{du}{\cos^{n-2} au}$

123. $\int \sin au \cos au\, du = \dfrac{\sin^2 au}{2a} + C$

124. $\int \sin pu \cos qu\, du = -\dfrac{\cos(p-q)u}{2(p-q)} - \dfrac{\cos(p+q)u}{2(p+q)} + C, \quad p \neq \pm q$

125. $\int \sin^n au \cos au\, du = \dfrac{\sin^{n+1} au}{(n+1)a} + C, \quad n \neq -1$

126. $\int \cos^n au \sin au\, du = -\dfrac{\cos^{n+1} au}{(n+1)a} + C, \quad n \neq -1$

127. $\int \sin^2 au \cos^2 au\, du = \dfrac{u}{8} - \dfrac{\sin 4au}{32a} + C$

128. $\int \dfrac{du}{\sin au \cos au} = \dfrac{1}{a}\ln|\tan au| + C$

129. $\int \dfrac{du}{\cos au(1 \pm \sin au)} = \mp\dfrac{1}{2a(1 \pm \sin au)} + \dfrac{1}{2a}\ln\left|\tan\left(\dfrac{au}{2} + \dfrac{\pi}{4}\right)\right| + C$

130. $\int \dfrac{du}{\sin au(1 \pm \cos au)} = \pm\dfrac{1}{2a(1 \pm \cos au)} + \dfrac{1}{2a}\ln\left|\tan\dfrac{au}{2}\right| + C$

131. $\int \dfrac{du}{\sin au \pm \cos au} = \dfrac{1}{a\sqrt{2}}\ln\left|\tan\left(\dfrac{au}{2} \pm \dfrac{\pi}{8}\right)\right| + C$

132. $\int \dfrac{\sin au\, du}{\sin au \pm \cos au} = \dfrac{u}{2} \mp \dfrac{1}{2a}\ln|\sin au \pm \cos au| + C$

133. $\int \dfrac{\cos au\, du}{\sin au \pm \cos au} \pm \dfrac{u}{2} + \dfrac{1}{2a}\ln|\sin au \pm \cos au| + C$

134. $\int \dfrac{\sin au\, du}{p + q\cos au} = -\dfrac{1}{aq}\ln|p + q\cos au| + C$

135. $\int \dfrac{\cos au\, du}{p + q\sin au} = \dfrac{1}{aq}\ln|p + q\sin au| + C$

136. $\int \sin^m au \cos^n au\, du = \begin{cases} -\dfrac{\sin^{m-1} au \cos^{n+1} au}{a(m+n)} + \dfrac{m-1}{m+n}\displaystyle\int \sin^{m-2} au \cos^n au\, du, \; m \neq -n \\ \dfrac{\sin^{m+1} au \cos^{n-1} au}{a(m+n)} + \dfrac{n-1}{m+n}\displaystyle\int \sin^m au \cos^{n-2} au\, du, \; m \neq -n \end{cases}$

137. $\int \tan au\, du = -\dfrac{1}{a}\ln|\cos au| = \dfrac{1}{a}\ln|\sec au| + C$

138. $\int \tan^2 au\, du = \dfrac{\tan au}{a} - u + C$

139. $\int \tan^n au \sec^2 au\, du = \dfrac{\tan^{n+1} au}{(n+1)a} + C, n \neq -1$

140. $\int \tan^n au\, du = \dfrac{\tan^{n-1} au}{(n-1)a} - \int \tan^{n-2} au\, du + C, n \neq 1$

141. $\int \cot au\, du = \dfrac{1}{a}\ln|\sin au| + C$

142. $\int \cot^2 au\, du = -\dfrac{\cot au}{a} - u + C$

143. $\int \cot^n au \csc^2 au\, du = -\dfrac{\cot^{n+1} au}{(n+1)a} + C, n \neq -1$

144. $\int \cot^n au\, du = -\dfrac{\cot^{n-1} au}{(n-1)a} - \int \cot^{n-2} au\, du, n \neq 1$

145. $\displaystyle\int \sec au\, du = \frac{1}{a}\ln|\sec au + \tan au| = \frac{1}{a}\ln\left|\tan\left(\frac{au}{2} + \frac{\pi}{4}\right)\right| + C$

146. $\displaystyle\int \sec^2 au\, du = \frac{\tan au}{a} + C$

147. $\displaystyle\int \sec^3 au\, du = \frac{\sec au \tan au}{2a} + \frac{1}{2a}\ln|\sec au + \tan au| + C$

148. $\displaystyle\int \sec^n au \tan au\, du = \frac{\sec^n au}{na} + C$

149. $\displaystyle\int \sec^n au\, du = \frac{\sec^{n-2} au \tan au}{a(n-1)} + \frac{n-2}{n-1}\int \sec^{n-2} au\, du,\ n \neq 1$

150. $\displaystyle\int \csc au\, du = \frac{1}{a}\ln|\csc au - \cot au| = \frac{1}{a}\ln\left|\tan\frac{au}{2}\right| + C$

151. $\displaystyle\int \csc^2 au\, du = -\frac{\cot au}{a} + C$

152. $\displaystyle\int \csc^n au \cot au\, du = -\frac{\csc^n au}{na} + C$

153. $\displaystyle\int \csc^n au\, du = -\frac{\csc^{n-2} au \cot au}{a(n-1)} + \frac{n-2}{n-1}\int \csc^{n-2} au\, du,\ n \neq 1$

INTEGRALS INVOLVING INVERSE TRIGONOMETRIC FUNCTIONS

154. $\displaystyle\int \sin^{-1}\frac{u}{a}\, du = u \sin^{-1}\frac{u}{a} + \sqrt{a^2 - u^2} + C$

155. $\displaystyle\int u \sin^{-1}\frac{u}{a}\, du = \left(\frac{u^2}{2} - \frac{a^2}{4}\right)\sin^{-1}\frac{u}{a} + \frac{u\sqrt{a^2 - u^2}}{4} + C$

156. $\displaystyle\int \cos^{-1}\frac{u}{a}\, du = u \cos^{-1}\frac{u}{a} - \sqrt{a^2 - u^2} + C$

157. $\displaystyle\int u \cos^{-1}\frac{u}{a}\, du = \left(\frac{u^2}{2} - \frac{a^2}{4}\right)\cos^{-1}\frac{u}{a} - \frac{u\sqrt{a^2 - u^2}}{4} + C$

158. $\displaystyle\int \tan^{-1}\frac{u}{a}\, du = u \tan^{-1}\frac{u}{a} - \frac{a}{2}\ln(u^2 + a^2) + C$

159. $\displaystyle\int u \tan^{-1}\frac{u}{a}\, du = \frac{1}{2}(u^2 + a^2)\tan^{-1}\frac{u}{a} - \frac{au}{2} + C$

160. $\displaystyle\int u^m \sin^{-1}\frac{u}{a}\, du = \frac{u^{m+1}}{m+1}\sin^{-1}\frac{u}{a} - \frac{1}{m+1}\int \frac{u^{m+1}}{\sqrt{a^2 - u^2}}\, du$

161. $\displaystyle\int u^m \cos^{-1}\frac{u}{a}\, du = \frac{u^{m+1}}{m+1}\cos^{-1}\frac{u}{a} + \frac{1}{m+1}\int \frac{u^{m+1}}{\sqrt{a^2 - u^2}}\, du$

162. $\displaystyle\int u^m \tan^{-1}\frac{u}{a}\, du = \frac{u^{m+1}}{m+1}\tan^{-1}\frac{u}{a} - \frac{a}{m+1}\int \frac{u^{m+1}}{u^2 + a^2}\, du$

INTEGRALS INVOLVING e^{au}

163. $\displaystyle\int e^{au}\, du = \frac{e^{au}}{a} + C$

164. $\displaystyle\int ue^{au}\, du = \frac{e^{au}}{a}\left(u - \frac{1}{a}\right) + C$

165. $\displaystyle\int u^2 e^{au}\, du = \frac{e^{au}}{a}\left(u^2 - \frac{2u}{a} + \frac{2}{a^2}\right) + C$

166. $\displaystyle\int u^n e^{au}\,du = \frac{u^n e^{au}}{a} - \frac{n}{a}\int u^{n-1}e^{au}\,du$

$\displaystyle = \frac{e^{au}}{a}\left[u^n - \frac{nu^{n-1}}{a} + \frac{n(n-1)u^{n-2}}{a^2} - \cdots + \frac{(-1)^n n!}{a^n}\right] + C \quad \text{if } n \text{ is a positive integer}$

167. $\displaystyle\int \frac{du}{p + qe^{au}} = \frac{u}{p} - \frac{1}{ap}\ln|p + qe^{au}| + C$

168. $\displaystyle\int e^{au}\sin bu\,du = \frac{e^{au}(a\sin bu - b\cos bu)}{a^2 + b^2} + C$

169. $\displaystyle\int e^{au}\cos bu\,du = \frac{e^{au}(a\cos bu + b\sin bu)}{a^2 + b^2} + C$

170. $\displaystyle\int ue^{au}\sin bu\,du = \frac{ue^{au}(a\sin bu - b\cos bu)}{a^2 + b^2} - \frac{e^{au}[(a^2 - b^2)\sin bu - 2ab\cos bu]}{(a^2 + b^2)^2} + C$

171. $\displaystyle\int ue^{au}\cos bu\,du = \frac{ue^{au}(a\cos bu + b\sin bu)}{a^2 + b^2} - \frac{e^{au}[(a^2 - b^2)\cos bu + 2ab\sin bu]}{(a^2 + b^2)^2} + C$

172. $\displaystyle\int e^{au}\sin^n bu\,du = \frac{e^{au}\sin^{n-1} bu}{a^2 + n^2b^2}(a\sin bu - nb\cos bu) + \frac{n(n-1)b^2}{a^2 + n^2b^2}\int e^{au}\sin^{n-2} bu\,du$

173. $\displaystyle\int e^{au}\cos^n bu\,du = \frac{e^{au}\cos^{n-1} bu}{a^2 + n^2b^2}(a\cos bu + nb\sin bu) + \frac{n(n-1)b^2}{a^2 + n^2b^2}\int e^{au}\cos^{n-2} bu\,du$

INTEGRALS INVOLVING ln *u*

174. $\displaystyle\int \ln u\,du = u\ln u - u + C$

175. $\displaystyle\int u\ln u\,du = \frac{u^2}{2}\left(\ln u - \tfrac{1}{2}\right) + C$

176. $\displaystyle\int u^m\ln u\,du = \frac{u^{m+1}}{m+1}\left(\ln u - \frac{1}{m+1}\right) \quad \text{if } m \neq -1$

177. $\displaystyle\int \frac{\ln u}{u}\,du = \frac{1}{2}\ln^2 u + C$

178. $\displaystyle\int \frac{\ln^n u\,du}{u} = \frac{\ln^{n+1} u}{n+1} + C \quad \text{if } n \neq -1$

179. $\displaystyle\int \frac{du}{u\ln u} = \ln|\ln u| + C$

180. $\displaystyle\int \ln^n u\,du = u\ln^n u - n\int \ln^{n-1} u\,du + C$

181. $\displaystyle\int u^m\ln^n u\,du = \frac{u^{m+1}\ln^n u}{m+1} - \frac{n}{m+1}\int u^m\ln^{n-1} u\,du + C \quad \text{if } m \neq -1$

182. $\displaystyle\int \ln(u^2 + a^2)\,du = u\ln(u^2 + a^2) - 2u + 2a\tan^{-1}\frac{u}{a} + C$

183. $\displaystyle\int \ln|u^2 - a^2|\,du = u\ln|u^2 - a^2| - 2u + a\ln\left|\frac{u+a}{u-a}\right| + C$

INTEGRALS INVOLVING HYPERBOLIC FUNCTIONS

184. $\displaystyle\int \sinh au\,du = \frac{\cosh au}{a} + C$

185. $\displaystyle\int u\sinh au\,du = \frac{u\cosh au}{a} - \frac{\sinh au}{a^2} + C$

186. $\displaystyle\int \cosh au\,du = \frac{\sinh au}{a} + C$

187. $\displaystyle\int u\cosh au\,du = \frac{u\sinh au}{a} - \frac{\cosh au}{a^2} + C$

188. $\displaystyle\int \cosh^2 au\, du = \frac{u}{2} + \frac{\sinh au \cosh au}{2a} + C$

189. $\displaystyle\int \sinh^2 au\, du = \frac{\sinh au \cosh au}{2a} - \frac{u}{2} + C$

190. $\displaystyle\int \sinh^n au\, du = \frac{\sinh^{n-1} au \cosh au}{an} - \frac{n-1}{n}\int \sinh^{n-2} au\, du$

191. $\displaystyle\int \cosh^n au\, du = \frac{\cosh^{n-1} au \sinh au}{an} + \frac{n-1}{n}\int \cosh^{n-2} au\, du$

192. $\displaystyle\int \sinh au \cosh au\, du = \frac{\sinh^2 au}{2a} + C$

193. $\displaystyle\int \sinh pu \cosh qu\, du = \frac{\cosh(p+q)u}{2(p+q)} + \frac{\cosh(p-q)u}{2(p-q)} + C$

194. $\displaystyle\int \tanh au\, du = \frac{1}{a}\ln \cosh au + C$

195. $\displaystyle\int \tanh^2 au\, du = u - \frac{\tanh au}{a} + C$

196. $\displaystyle\int \tanh^n au\, du = \frac{-\tanh^{n-1} au}{a(n-1)} + \int \tanh^{n-2} au\, du$

197. $\displaystyle\int \coth au\, du = \frac{1}{a}\ln|\sinh au| + C$

198. $\displaystyle\int \coth^2 au\, du = u - \frac{\coth au}{a} + C$

199. $\displaystyle\int \operatorname{sech} au\, du = \frac{2}{a}\tan^{-1} e^{au} + C$

200. $\displaystyle\int \operatorname{sech}^2 au\, du = \frac{\tanh au}{a} + C$

201. $\displaystyle\int \operatorname{sech}^n au\, du = \frac{\operatorname{sech}^{n-2} au \tanh au}{a(n-1)} + \frac{n-2}{n-1}\int \operatorname{sech}^{n-2} au\, du$

202. $\displaystyle\int \operatorname{csch} au\, du = \frac{1}{a}\ln\left|\tanh\frac{au}{2}\right| + C$

203. $\displaystyle\int \operatorname{csch}^2 au\, du = -\frac{\coth au}{a} + C$

204. $\displaystyle\int \operatorname{sech} u \tanh u\, du = -\operatorname{sech} u + C$

205. $\displaystyle\int \operatorname{csch} u \coth u\, du = -\operatorname{csch} u + C$

SOME DEFINITE INTEGRALS

Unless otherwise stated, all letters stand for positive numbers.

206. $\displaystyle\int_0^\infty \frac{dx}{x^2 + a^2} = \frac{\pi}{2a}$

207. $\displaystyle\int_0^\infty \frac{x^{p-1}}{1+x}\, dx = \frac{\pi}{\sin p\pi}$

208. $\displaystyle\int_0^a \frac{dx}{\sqrt{a^2 - x^2}} = \frac{\pi}{2}$

209. $\displaystyle\int_0^a \sqrt{a^2 - x^2}\, dx = \frac{\pi a^2}{4}$

210. $\displaystyle\int_0^\pi \sin mx \sin nx\, dx = \begin{cases} 0, & \text{if } m, n \text{ integers and } m \neq n \\ \dfrac{\pi}{2}, & \text{if } m, n \text{ integers and } m = n \end{cases}$

211. $\displaystyle\int_0^\pi \cos mx \cos nx\, dx = \begin{cases} 0, & \text{if } m, n \text{ integers and } m \neq n \\ \dfrac{\pi}{2}, & \text{if } m, n \text{ integers and } m = n \end{cases}$

212. $\displaystyle\int_0^\pi \sin mx \cos nx\, dx = \begin{cases} 0, & \text{if } m, n \text{ integers and } m + n \text{ is odd, or } m = \pm n \\ \dfrac{2m}{(m^2 - n^2)}, & \text{if } m, n \text{ integers and } m + n \text{ is even} \end{cases}$

213. $\int_0^{\pi/2} \sin^2 x \, dx = \int_0^{\pi/2} \cos^2 x \, dx = \frac{\pi}{4}$

214. $\int_0^{\infty} e^{-ax} \cos bx \, dx = \frac{a}{a^2 + b^2}$

215. $\int_0^{\infty} e^{-ax} \sin bx \, dx = \frac{b}{a^2 + b^2}$

216. $\int_0^{\infty} e^{-a^2x^2} \, dx = \frac{\sqrt{\pi}}{2a}$

217. $\int_0^{\pi/2} \sin^{2m} x \, dx = \int_0^{\pi/2} \cos^{2m} x \, dx = \frac{1 \cdot 3 \cdot 5 \cdot \cdots \cdot (2m - 1)}{2 \cdot 4 \cdot 6 \cdot \cdots \cdot 2m} \frac{\pi}{2}, \quad m = 1, 2, 3, \ldots$

218. $\int_0^{\pi/2} \sin^{2m+1} x \, dx = \int_0^{\pi/2} \cos^{2m+1} x \, dx = \frac{2 \cdot 4 \cdot 6 \cdot \cdots \cdot 2m}{1 \cdot 3 \cdot 5 \cdot \cdots \cdot (2m + 1)}, \quad m = 1, 2, 3, \ldots$

219. $\int_0^{\infty} \frac{e^{-x}}{\sqrt{x}} \, dx = \sqrt{\pi}$

220. $\int_0^{1} x^m (\ln x)^n \, dx = \frac{(-1)^n n!}{(m + 1)^{n+1}}$

Answers to Odd-Numbered Problems and Review Exercises

CHAPTER ONE

PROBLEMS 1.1

1.

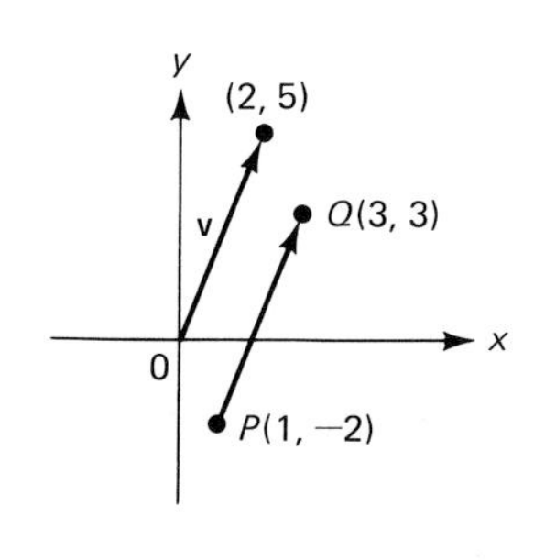

3.

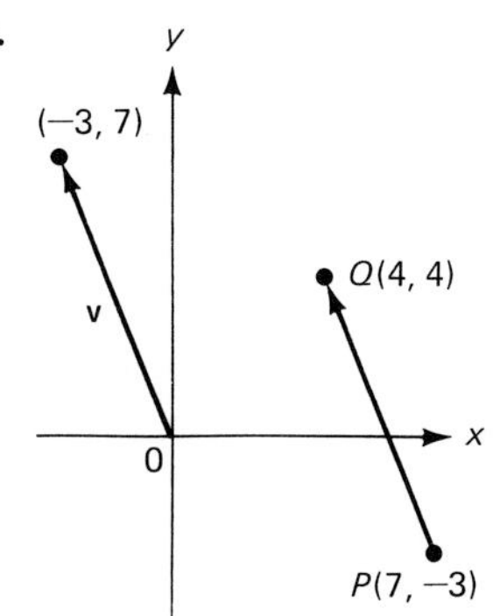

5.

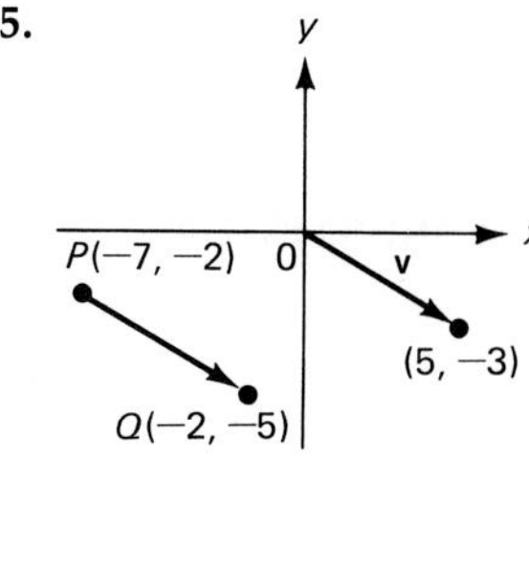

7. $|\mathbf{v}| = 4\sqrt{2},\ \theta = \pi/4$
9. $|\mathbf{v}| = 4\sqrt{2},\ \theta = 7\pi/4$ **11.** $|\mathbf{v}| = 2,\ \theta = \pi/6$ **13.** $|\mathbf{v}| = 2,\ \theta = 2\pi/3$
15. $|\mathbf{v}| = 2,\ \theta = 4\pi/3$ **17.** $|\mathbf{v}| = \sqrt{89},\ \theta = \pi + \tan^{-1}(-\frac{8}{5}) \approx 2.13$ (second quadrant)

19.

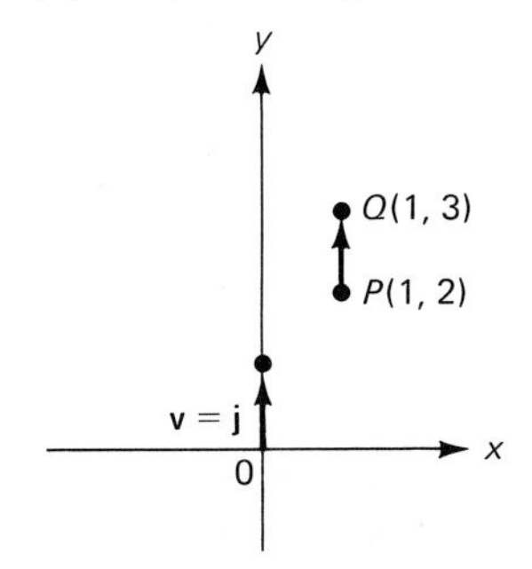

21.

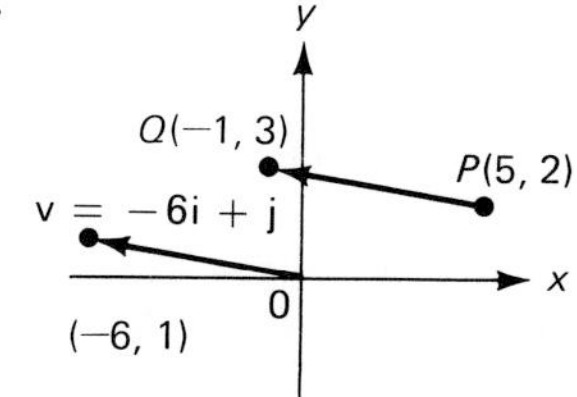

23.

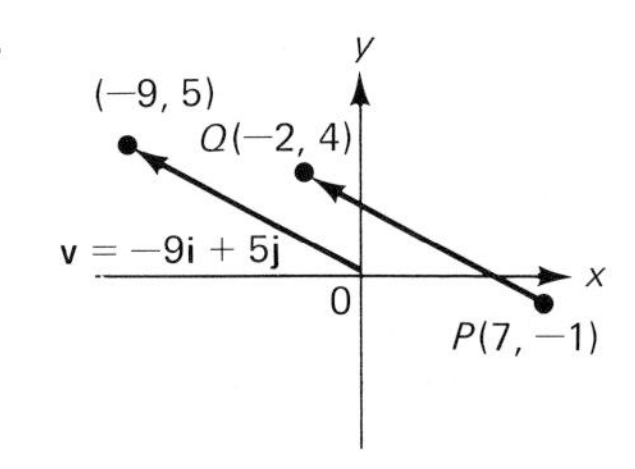

25.

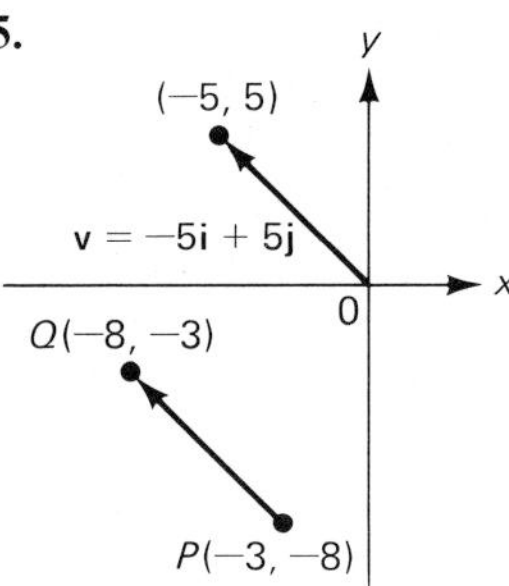

27. (a) $(6, 9)$ (b) $(-3, 7)$ (c) $(-7, 1)$ (d) $(39, -22)$ **33.** $(1/\sqrt{2})\mathbf{i} - (1/\sqrt{2})\mathbf{j}$

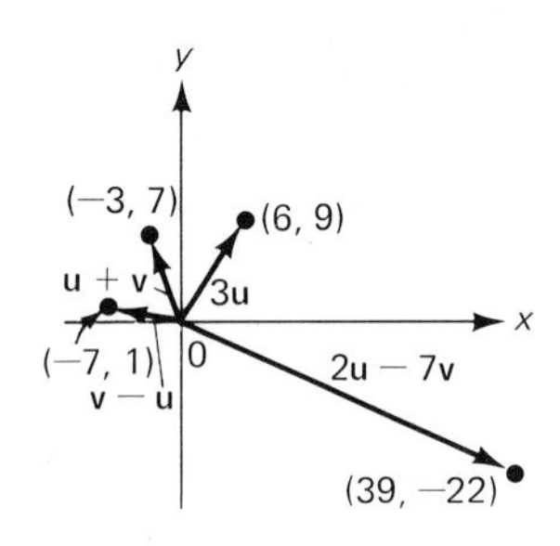

35. $(\frac{3}{5}, -\frac{4}{5})$ **37.** $(1/\sqrt{2}, 1/\sqrt{2})$ if $a > 0$, $(-1/\sqrt{2}, -1/\sqrt{2})$ if $a < 0$
39. $\sin\theta = -3/\sqrt{13}$, $\cos\theta = 2/\sqrt{13}$ **41.** $-(1/\sqrt{2})\mathbf{i} - (1/\sqrt{2})\mathbf{j}$ **43.** $(\frac{3}{5}, -\frac{4}{5})$
45. $\frac{3}{5}\mathbf{i} + \frac{4}{5}\mathbf{j}$
47. (a) $(1/\sqrt{2})\mathbf{i} - (1/\sqrt{2})\mathbf{j}$, (b) $(7/\sqrt{193})\mathbf{i} - (12/\sqrt{193})\mathbf{j}$, (c) $-(2/\sqrt{53})\mathbf{i} + (7/\sqrt{53})\mathbf{j}$
51. $4\mathbf{i} + 4\sqrt{3}\mathbf{j}$ **53.** $4\mathbf{j}$ **55.** $-3\mathbf{i} + 3\sqrt{3}\mathbf{j}$ **57.** $3\sqrt{3}\mathbf{i} - 3\mathbf{j}$

PROBLEMS 1.2

1. 0; 0 **3.** 0; 0 **5.** 20; $\frac{20}{29}$ **7.** -22; $-22/(5\sqrt{53})$ **9.** 100; $20/\sqrt{481}$
11. $\mathbf{u} \cdot \mathbf{v} = \alpha\beta - \beta\alpha = 0$
13. parallel **15.** neither **17.** orthogonal **19.** parallel

13. **15.** **17.**

21. (a) $-\frac{3}{4}$ (b) $\frac{4}{3}$ (c) $\frac{1}{7}$ (d) $(-96 + \sqrt{7500})/78 \approx -0.12$ **25.** $\frac{3}{2}\mathbf{i} + \frac{3}{2}\mathbf{j}$ **27.** $\mathbf{0}$
29. $-\frac{2}{13}\mathbf{i} + \frac{3}{13}\mathbf{j}$ **31.** $\frac{14}{5}\mathbf{i} + \frac{28}{5}\mathbf{j}$ **33.** $-\frac{14}{5}\mathbf{i} + \frac{28}{5}\mathbf{j}$ **35.** $[(\alpha + \beta)/2]\mathbf{i} + [(\alpha + \beta)/2]\mathbf{j}$
37. $[(\alpha - \beta)/2]\mathbf{i} + [(\alpha - \beta)/2]\mathbf{j}$ **39.** $a_1a_2 + b_1b_2 > 0$
41. $\text{Proj}_{\overrightarrow{PQ}}\ \overrightarrow{RS} = \frac{51}{25}\mathbf{i} + \frac{68}{25}\mathbf{j}$; $\text{Proj}_{\overrightarrow{RS}}\ \overrightarrow{PQ} = -\frac{17}{26}\mathbf{i} + \frac{85}{26}\mathbf{j}$
47. $-27/(5\sqrt{34}) \approx -0.9261$; $61/(\sqrt{34}\sqrt{113}) \approx 0.9841$; $52/(5\sqrt{113}) \approx 0.9783$
49. when the vectors (a_1, a_2) and (b_1, b_2) have the same or opposite direction **51.** $\sqrt{5}$

PROBLEMS 1.3

7. $-2\mathbf{i} - 5\mathbf{j}$ N; $2\mathbf{i} + 5\mathbf{j}$ N **9.** $-2\mathbf{i} - 3\mathbf{j}$ N; $2\mathbf{i} + 3\mathbf{j}$ N
11. $-2\sqrt{3}\mathbf{i} - 7\mathbf{j}$ lb; $2\sqrt{3}\mathbf{i} + 7\mathbf{j}$ lb
13. $(3/\sqrt{2})\mathbf{i} - [(3/\sqrt{2}) + 2]\mathbf{j}$ N; $-(3/\sqrt{2})\mathbf{i} + [(3/\sqrt{2}) + 2]\mathbf{j}$ N
15. $(7\sqrt{2} - \frac{7}{2}\sqrt{3} - \frac{7}{2})(\mathbf{i} + \mathbf{j})$ N; $(-7\sqrt{2} + \frac{7}{2}\sqrt{3} + \frac{7}{2})(\mathbf{i} + \mathbf{j})$ N **17.** -12 J
19. $(8\sqrt{3} + 4)$ J **21.** $3\sqrt{2}$ J **23.** $12/\sqrt{13}$ J **25.** $500(\sin 20°/\sin 30°) \approx 342$ N
27. tugboat 1: $(500)(\cos 20°)(750) \approx 352{,}385$ J; tugboat 2: $(342)(\cos 30°)(750) \approx 222{,}136$ J

REVIEW EXERCISES FOR CHAPTER ONE

1. $|\mathbf{v}| = 3\sqrt{2}$, $\theta = \pi/4$ **3.** $|\mathbf{v}| = 4$, $\theta = 5\pi/3$ **5.** $|\mathbf{v}| = 12\sqrt{2}$, $\theta = 5\pi/4$
7. $2\mathbf{i} + 2\mathbf{j}$ **9.** $4\mathbf{i} + 2\mathbf{j}$ **11.** (a) $(10, 5)$, (b) $(5, -3)$, (c) $(-31, 12)$
13. $(1/\sqrt{2})\mathbf{i} + (1/\sqrt{2})\mathbf{j}$ **15.** $(2/\sqrt{29})\mathbf{i} + (5/\sqrt{29})\mathbf{j}$ **17.** $\frac{3}{5}\mathbf{i} + \frac{4}{5}\mathbf{j}$
19. $(1/\sqrt{2})\mathbf{i} - (1/\sqrt{2})\mathbf{j}$ if $a > 0$ and $-(1/\sqrt{2})\mathbf{i} + (1/\sqrt{2})\mathbf{j}$ if $a < 0$
21. $-(5/\sqrt{29})\mathbf{i} - (2/\sqrt{29})\mathbf{j}$ **23.** $-(10/\sqrt{149})\mathbf{i} + (7/\sqrt{149})\mathbf{j}$ **25.** $\mathbf{j}$
27. $-\frac{7}{2}\sqrt{3}\mathbf{i} + \frac{7}{2}\mathbf{j}$ **29.** 0; 0 **31.** -14, $-14/(\sqrt{5}\sqrt{41})$ **33.** neither **35.** parallel
37. parallel **39.** $7\mathbf{i} + 7\mathbf{j}$ **41.** $\frac{15}{13}\mathbf{i} + \frac{10}{13}\mathbf{j}$ **43.** $-\frac{3}{2}\mathbf{i} - \frac{7}{2}\mathbf{j}$
45. $\text{Proj}_{\overrightarrow{RS}}\, \overrightarrow{PQ} = -\frac{99}{25}\mathbf{i} + \frac{132}{25}\mathbf{j}$; $\text{Proj}_{\overrightarrow{PQ}}\, \overrightarrow{RS} = -\frac{33}{82}\mathbf{i} - \frac{297}{82}\mathbf{j}$ **47.** $-3\mathbf{i} - 3\mathbf{j}$
49. $-\sqrt{2}$ J **51.** $-11(1 + 3\sqrt{3})$ J

CHAPTER TWO

PROBLEMS 2.1

1. $\mathbb{R} - \{0, 1\}$ **3.** $\mathbb{R} - \{-1, 1\}$ **5.** $(0, 1)$ **7.** $\mathbb{R} - \{n\pi/2: n = 0, \pm 1, \pm 2, \ldots\}$
9. $y^2 = 4x$ **11.** $x = y^{2/3}$ **13.** $y = 2x + 5$ **15.** $x = y^2 + y + 1$, $y \geq 0$
17. $y = x^3 - 1$ **19.** $y = (\ln x)^2$, $x > 0$
21. $x/y = \tan[\frac{1}{2}\ln(x^2 + y^2)]$. The graph of this equation is a spiral.
23. $y = x^2$, $x > 0$ **25.** $(2 - t)\mathbf{i} + (4 + 2t)\mathbf{j}$ / $(1 + t)\mathbf{i} + (6 - 2t)\mathbf{j}$ **27.** $(3 - 4t)\mathbf{i} + (5 - 12t)\mathbf{j}$ / $(-1 + 4t)\mathbf{i} + (-7 + 12t)\mathbf{j}$
29. $(-2 + 6t)\mathbf{i} + (3 + 4t)\mathbf{j}$
$(4 - 6t)\mathbf{i} + (7 - 4t)\mathbf{j}$
33. $(650\sqrt{2})^2/16 = 52{,}812.5$ ft (using the value $g = 32$ ft/sec²)
35. $x(t) = 50\sqrt{3}t$; $y(t) = -(9.81t^2/2) - 50t + 150$
37. $x = \frac{1}{2}\alpha - \frac{1}{4}\sin\alpha$ m, $y = \frac{1}{2}\alpha - \frac{1}{4}\cos\alpha$ m

47.

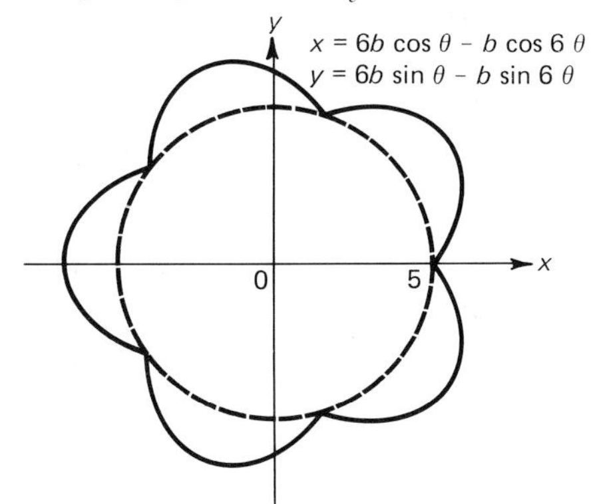

PROBLEMS 2.2

1. $\frac{2}{3}$ **3.** $-1/\sqrt{2}$ **5.** $-e^{-4}$ **7.** 0 **9.** $\sqrt{2}$ **11.** 0 **13.** $-1/\sqrt{3}$
15. $-16/\pi^2$ **17.** $y = \frac{2}{3}x + \frac{22}{3}$ **19.** $y = -e^{-4}x + 2e^{-2}$ **21.** $y = \sqrt{2}$
23. none (note that condition (1) does not hold at $t = 0$ and that the Cartesian equation of the curve is the ray $y = x - 3$, $x \geq -1$, which has neither horizontal nor vertical tangents)

25. V: $((-1)^k, \cos\frac{5}{6}(2k+1)\pi$ for $k = 0, 1, 2, 3, 4, 5$;
H: $(\sin\frac{3}{5}k\pi, (-1)^k)$ for $k = 0, 1, 2, 3, 4, 5, 6, 7, 8, 9$
27. none **29.** V: (1, 0); H: none **31.** none (condition (1) does not hold at $t = 0$)
35. -1 **37.** $(4 + 3\sqrt{3})/(4\sqrt{3} + 3)$ **39.** $\frac{1}{2}$ **41.** 0
43. $y = -x$; $y = x - 2$; $y = -\frac{1}{2}x - \frac{1}{2}$ **45.** when $\cos\alpha = r/s$

PROBLEMS 2.3

1. $\mathbf{f}' = \mathbf{i} - 5t^4\mathbf{j}$; $\mathbf{f}'' = -20t^3\mathbf{j}$
3. $\mathbf{f}' = (2\cos 2t)\mathbf{i} - 3(\sin 3t)\mathbf{j}$; $\mathbf{f}'' = (-4\sin 2t)\mathbf{i} - (9\cos 3t)\mathbf{j}$
5. $\mathbf{f}' = (1/t)\mathbf{i} + 3e^{3t}\mathbf{j}$; $\mathbf{f}'' = -(1/t^2)\mathbf{i} + 9e^{3t}\mathbf{j}$, $t > 0$
7. $\mathbf{f}' = (\sec^2 t)\mathbf{i} + (\sec t)(\tan t)\mathbf{j}$; $\mathbf{f}'' = 2(\sec^2 t)(\tan t)\mathbf{i} + [\sec^3 t + (\sec t)(\tan^2 t)]\mathbf{j}$
9. $\mathbf{f}' = -(\tan t)\mathbf{i} + (\cot t)\mathbf{j}$; $\mathbf{f}'' = -(\sec^2 t)\mathbf{i} - (\csc^2 t)\mathbf{j}$ **11.** $(2/\sqrt{13})\mathbf{i} + (3/\sqrt{13})\mathbf{j}$
13. $\mathbf{j}$ **15.** $-(1/\sqrt{2})\mathbf{i} + (1/\sqrt{2})\mathbf{j}$ **17.** $\mathbf{i}$ **19.** $(4/\sqrt{97})\mathbf{i} - (9/\sqrt{97})\mathbf{j}$
21. $\frac{8}{3}\mathbf{i} + \frac{32}{5}\mathbf{j}$ **23.** $(2\sqrt{t} + C_1)\mathbf{i} + (\frac{2}{3}t^{3/2} + C_2)\mathbf{j}$ **25.** $(\cosh 1 - 1)\mathbf{i} - (\sinh 1)\mathbf{j}$
27. $(t\ln t - t + C_1)\mathbf{i} + (te^t - e^t + C_2)\mathbf{j}$ **29.** $[(t^4/4) + 2]\mathbf{i} + [5 - (t^6/6)]\mathbf{j}$
31. $(\sin t)\mathbf{i} - (\cos t)\mathbf{j}$ **33.** $-(a/\sqrt{a^2+b^2})\mathbf{i} + (b/\sqrt{a^2+b^2})\mathbf{j}$ **35.** $(1/\sqrt{2})\mathbf{i} + (1/\sqrt{2})\mathbf{j}$
37. $$\frac{1}{\sqrt{18 + (9/\sqrt{2})(1 - \sqrt{3})}}\left[\left(-\frac{3}{2} - \frac{3}{\sqrt{2}}\right)\mathbf{i} + \left(\frac{3\sqrt{3}}{2} - \frac{3}{\sqrt{2}}\right)\mathbf{j}\right]$$

PROBLEMS 2.4

1. $(2 + \sec^2 t)\mathbf{i} - [\sin t + (\sec t)(\tan t)]\mathbf{j}$ **3.** $3/(2\sqrt{t}) - 2t^{-3/2}$ **5.** 0
7. $(3\sin t\cosh t + 3\cos t\sinh t)\mathbf{i} + (3\sin t\sinh t + 3\cos t\cosh t)\mathbf{j}$
9. $10t^9 + 9t^8 + \{(10t + 9)/[(t+1)^2 t^{10}]\}$
11. $\ln t\,\mathrm{sech}^2\, t + [(\tanh t)/t] - 3\ln t\,\mathrm{sech}\, t\tanh t + [(3\,\mathrm{sech}\, t)/t]$
13. $[1/(\sqrt{1-t^2}) - \sin t]\mathbf{i} + [\cos t - (1/\sqrt{1-t^2})]\mathbf{j}$
15. $[(\sin^{-1} t)/(1+t^2)] + [(\tan^{-1} t)/\sqrt{1-t^2}] - (\cos^{-1} t)\sin t - [(\cos t)/\sqrt{1-t^2}]$
17. $\mathbf{T}(t) = -(\sin 5t)\mathbf{i} + (\cos 5t)\mathbf{j}$; $\mathbf{T}(\pi/2) = -\mathbf{i}$; $\mathbf{n}(t) = -(\cos 5t) - (\sin 5t)\mathbf{j}$; $\mathbf{n}(\pi/2) = -\mathbf{j}$
19. $\mathbf{T}(t) = (\sin 10t)\mathbf{i} - (\cos 10t)\mathbf{j}$; $\mathbf{T}(\pi) = -\mathbf{j}$; $\mathbf{n}(t) = (\cos 10t)\mathbf{i} + (\sin 10t)\mathbf{j}$; $\mathbf{n}(\pi) = \mathbf{i}$
21. $\mathbf{T}(t) = (1/\sqrt{1+t^2})\mathbf{i} + (t/\sqrt{1+t^2})\mathbf{j}$; $\mathbf{T}(1) = (1/\sqrt{2})\mathbf{i} + (1/\sqrt{2})\mathbf{j}$;
$\mathbf{n}(t) = -(t/\sqrt{1+t^2})\mathbf{i} + (1/\sqrt{1+t^2})\mathbf{j}$; $\mathbf{n}(1) = -(1/\sqrt{2})\mathbf{i} + (1/\sqrt{2})\mathbf{j}$
23. $\mathbf{T}(t) = -(7/\sqrt{74})\mathbf{i} + (5/\sqrt{74})\mathbf{j}$ (constant); $\mathbf{n}(t) = (5/\sqrt{74})\mathbf{i} + (7/\sqrt{74})\mathbf{j}$ or
$-(5/\sqrt{74})\mathbf{i} - (7/\sqrt{74})\mathbf{j}$
25. $\mathbf{T}(t) = (\mathrm{sech}\, t)\mathbf{i} + (\tanh t)\mathbf{j}$; $\mathbf{T}(0) = \mathbf{i}$; $\mathbf{n}(t) = -(\tanh t)\mathbf{i} + (\mathrm{sech}\, t)\mathbf{j}$; $\mathbf{n}(0) = \mathbf{j}$
27. $\mathbf{T}(t) = \sqrt{\dfrac{1+\sin t}{2}}\,\mathbf{i} - \dfrac{\cos t}{\sqrt{2 + 2\sin t}}\,\mathbf{j}$; $\mathbf{T}(\pi) = \dfrac{1}{\sqrt{2}}\mathbf{i} + \dfrac{1}{\sqrt{2}}\mathbf{j}$;
$\mathbf{n}(t) = \dfrac{\cos t}{\sqrt{2+2\sin t}}\,\mathbf{i} + \sqrt{\dfrac{1+\sin t}{2}}\,\mathbf{j}$; $\mathbf{n}(\pi) = -(1/\sqrt{2})\mathbf{i} + (1/\sqrt{2})\mathbf{j}$
29. $\mathbf{T}(\pi/6) = (3/\sqrt{10})\mathbf{i} - (1/\sqrt{10})\mathbf{j}$; $\mathbf{n}(\pi/6) = -(1/\sqrt{10})\mathbf{i} - (3/\sqrt{10})\mathbf{j}$

PROBLEMS 2.5

1. $\frac{1}{27}(148^{3/2} - 13^{3/2})$ **3.** $16\sqrt{2} - 8$ **5.** $y = \frac{1}{2}(1 - x^2)$;
$$\int_{1/\sqrt{5}}^{1}\sqrt{1+x^2}\,dx = \int_{\tan^{-1}(1/\sqrt{5})}^{\pi/4}\sec^3\theta\,d\theta = \frac{1}{2}\left[\sqrt{2} - \left(\frac{\sqrt{6}}{5}\right) + \ln\left(\frac{\sqrt{5}(1+\sqrt{2})}{1+\sqrt{6}}\right)\right] \approx 0.6861$$
7. $\sqrt{2}$ **9.** $8a$ (see Example 3) **11.** $\frac{2}{27}[13^{3/2} - 8)$ **13.** $|a|\pi$ **15.** $\sqrt{2}(e^3 - 1)$
17. $6\sqrt{2}$ **19.** $\sqrt{2}(e^{\pi/2} - 1)$ **21.** $4\int_0^{\pi/2}\sqrt{a^2\sin^2\theta + b^2\cos^2\theta}\,d\theta$

PROBLEMS 2.6

1. $\mathbf{f} = 3\{[(s+2)/2]^{2/3} - 1\}\mathbf{i} + 2\{[(s+2)/2]^{2/3} - 1\}^{3/2}\mathbf{j}$
3. $\mathbf{f} = \frac{1}{27}\{[(27s+8)^{2/3} - 4]^{3/2} + 27\}\mathbf{i} + \frac{1}{9}\{(27s+8)^{2/3} - 13\}\mathbf{j}$
5. $\mathbf{f} = 3\cos(s/3)\mathbf{i} + 3\sin(s/3)\mathbf{j}$ **7.** $\mathbf{f} = a\cos(s/|a|)\mathbf{i} + a\sin(s/|a|)\mathbf{j}$
9. $\mathbf{f} = (a + b\cos(s/|b|))\mathbf{i} + (c + b\sin(s/|b|))\mathbf{j}$
11. $\mathbf{f} = a[1 - (2s/3a)]^{3/2}\mathbf{i} + a(2s/3a)^{3/2}\mathbf{j}$; here $\theta = \sin^{-1}\sqrt{2s/3a}$

PROBLEMS 2.7

1. $\mathbf{v} = 3\mathbf{j}$, $|\mathbf{v}| = 3$, $\mathbf{a} = -9\mathbf{i}$, $|\mathbf{a}| = 9$ **3.** $\mathbf{v} = -4\sqrt{3}\mathbf{i} - 4\mathbf{j}$; $|\mathbf{v}| = 8$;

$\mathbf{a} = 16\mathbf{i} - 16\sqrt{3}\mathbf{j}$; $|\mathbf{a}| = 32$ **5.** $\mathbf{v} = 4\mathbf{i} + 4\mathbf{j}$; $|\mathbf{v}| = 4\sqrt{2}$; $\mathbf{a} = 4\mathbf{j}$; $|\mathbf{a}| = 4$
7. $\mathbf{v} = -7\mathbf{i} + 5\mathbf{j}$; $|\mathbf{v}| = \sqrt{74}$; $\mathbf{a} = 0$; $|\mathbf{a}| = 0$
9. $\mathbf{v} = 2\mathbf{i}$; $|\mathbf{v}| = 2$; $\mathbf{a} = 2\mathbf{j}$; $|\mathbf{a}| = 2$ **11.** $\mathbf{v} = 2\mathbf{i}$; $|\mathbf{v}| = 2$; $\mathbf{a} = \mathbf{j}$; $|\mathbf{a}| = 1$
13. $\mathbf{v} = [1 + (1/\sqrt{2})]\mathbf{i} - (1/\sqrt{2})\mathbf{j}$; $|\mathbf{v}| = \sqrt{2 + \sqrt{2}}$; $\mathbf{a} = (1/\sqrt{2})\mathbf{i} + (1/\sqrt{2})\mathbf{j}$; $|\mathbf{a}| = 1$
15. (a) $s = \int_0^{t_f} \sqrt{(600\sqrt{2})^2 + (600\sqrt{2} - gt)^2}\, dt = \frac{1200^2}{9.81}\left[\frac{1}{\sqrt{2}} + \frac{1}{4}\ln\left(\frac{\sqrt{2}+1}{\sqrt{2}-1}\right)\right]$
$\approx 168{,}483$ m ≈ 168.5 km where $t_f = 1200\sqrt{2}/g$ sec. This can be integrated via the substitution $600\sqrt{2} - gt = 600\sqrt{2}\tan\theta$. (b) $(1200)^2/g \approx 146{,}789\ m$
(c) $(600)^2/g \approx 36{,}697\ m$ (d) 1200 m/sec
17. (a) $s = \int_0^{20/\sqrt{g}} \sqrt{400 + g^2t^2}\, dt = (200/g)[\sqrt{g}\sqrt{1+g} + \ln(\sqrt{g} + \sqrt{1+g})] \approx 247.9$ m
(b) $400/\sqrt{g} \approx 127.7$ m (c) $\tan^{-1}(-gt_f/20) \approx 107.7°$ (d) $20\sqrt{1+g} \approx 65.75\ m/\text{sec}$
19. $20\sqrt{3} - (g/2)(\frac{2}{3})^2 + 4 = 20\sqrt{3} + \frac{4}{9} \approx 32$ ft **21.** $\ln 2.5 \approx 0.9163$ km/sec
23. $5000g \approx 49{,}050$ N

PROBLEMS 2.8

1. $\kappa = \frac{1}{2}, \rho = 2$ **3.** $\kappa = 2/5^{3/2}, \rho = 5^{3/2}/2$ **5.** $\kappa = \frac{3}{16}, \rho = \frac{16}{3}$
7. $\kappa = 6/\pi, \rho = \pi/6$ **9.** $\kappa = 2/5^{3/2}, \rho = 5^{3/2}/2$ **11.** $\kappa = 1/2\sqrt{2}, \rho = 2\sqrt{2}$
13. $\kappa = 1/2\sqrt{2}, \rho = 2\sqrt{2}$ **15.** $\kappa = 2|a|/(1 + b^2)^{3/2}, \rho = (1 + b^2)^{3/2}/2|a|$
17. $\kappa = 1, \rho = 1$ **19.** $\kappa = 0$, ρ is undefined **21.** at the origin
23. minimum of $2/3|a|$ for $t = \pi/4$, no maximum (approaches ∞ as $t \to 0^+$ or $t \to \pi/2^-$).
25. $13\sqrt{2}/(5\sqrt{5a})$ **27.** $1/a$ **29.** $3/4a$ **31.** $a_T = 0, a_n = 4$
33. $a_T = 4|t|/\sqrt{1 + 4t^2}, a_n = 2/\sqrt{1 + 4t^2}$
35. $a_T = |18t^3 - 14t|/\sqrt{9t^4 - 14t^2 + 9}, a_n = 6(1 + t^2)/\sqrt{9t^4 - 14t^2 + 9}$
37. $a_T = (4 + 18t^2)/\sqrt{4 + 9t^2}, a_n = 6|t|/\sqrt{4 + 9t^2}$
41. $[(10{,}000)(80{,}000)^2/(3600)^2] \cdot (1/\sqrt{2}) \approx 3{,}491{,}885.3$ N
43. $v = 150/\sqrt{2} \approx 106$ km/hr (reduce speed by a factor of $\sqrt{2}$)

REVIEW EXERCISES FOR CHAPTER TWO

1. $y = 2x$ **3.** $x = [(y/2) + 3]^2$ **5.** $x^2 + y^2 = 1$ **7.** $x = y^3, x \ge 0$
9. 2; V: (0, 0); no H **11.** $\sqrt{3}$; V: (1, 0), (–1, 0); H: (0, 1), (0, –1)
13. undefined; V: (1, 0); no H **15.** $-4/(3\sqrt{3})$; V: (3, 0), (–3, 0); H: (0, 4), (0, –4)
17. $1/\sqrt{3}$ **19.** π **21.** $\mathbf{f}' = 2\mathbf{i} - 3t^2\mathbf{j}$; $\mathbf{f}'' = -6t\mathbf{j}$
23. $\mathbf{f}' = (-5\sin 5t)\mathbf{i} + (8\cos 4t)\mathbf{j}$: $\mathbf{f}'' = (-25\cos 5t)\mathbf{i} - (32\sin 4t)\mathbf{j}$
25. $\mathbf{T} = (4/\sqrt{41})\mathbf{i} + (5/\sqrt{41})\mathbf{j}$; $\mathbf{n} = (-5/\sqrt{41})\mathbf{i} + (4/\sqrt{41})\mathbf{j}$
27. $\mathbf{T} = (\sqrt{3}/2)\mathbf{i} - \frac{1}{2}\mathbf{j}$; $\mathbf{n} = -\frac{1}{2}\mathbf{i} - (\sqrt{3}/2)\mathbf{j}$
29. $\mathbf{T} = (2/\sqrt{5})\mathbf{i} + (1/\sqrt{5})\mathbf{j}$; $\mathbf{n} = -(1/\sqrt{5})\mathbf{i} + (2/\sqrt{5})\mathbf{j}$ **31.** $\frac{81}{4}\mathbf{i} + \frac{243}{2}\mathbf{j}$
33. $(\frac{2}{3}t^{3/2} + C_1)\mathbf{i} + (\frac{3}{4}t^{4/3} + C_2)\mathbf{j}$ **35.** $(\frac{1}{8}t^8 - 1)\mathbf{i} + (3 - \frac{1}{7}t^7)\mathbf{j}$ **37.** $-2\sin t\cosh t$
39. $(t^2 + t + 1)[e^t/(1 + t)^2 + e^{-t}/t^2]$ **41.** $\pi/3$ **43.** 16
45. $\frac{1}{3}(29^{3/2} - 5^{3/2})$ **47.** $\mathbf{f} = \frac{3}{4}[(2s + 1)^{2/3} - 1]\mathbf{i} + \frac{1}{2}[(2s + 1)^{2/3} - 1]^{3/2}\mathbf{j}$
49. $\mathbf{f} = 2\cos(s/2)\mathbf{i} + 2\sin(s/2)\mathbf{j}$
51. $\mathbf{v} = -\sqrt{3}\mathbf{i} + \mathbf{j}$; $|\mathbf{v}| = 2$; $\mathbf{a} = -2\mathbf{i} - 2\sqrt{3}\mathbf{j}$; $|\mathbf{a}| = 4$
53. $\mathbf{v} = (\ln 2 - 1)\mathbf{i} + 2\mathbf{j}$; $|\mathbf{v}| = \sqrt{(\ln 2 - 1)^2 + 4}$; $\mathbf{a} = (\ln^2 2 + 1)\mathbf{i}$; $|\mathbf{a}| = \ln^2 2 + 1$
55. $\mathbf{v} = 2(\cosh 1)\mathbf{i} + 4\mathbf{j}$; $|\mathbf{v}| = \sqrt{4\cosh^2 1 + 16}$; $\mathbf{a} = 2(\sinh 1)\mathbf{i}$; $|\mathbf{a}| = 2\sinh 1$
57. (a) $s = \int_0^{t_f} \sqrt{675 + (15 - gt)^2}\, dt$ m ≈ 86.6 m, where
$t_f = (30 + \sqrt{900 + 12g})/2g \approx 3.155$ sec (b) $15\sqrt{3}t_f \approx 81.97$ m
(c) $1.5 + 15(15/g) - (g/2)(15/g)^2 \approx 12.97$ m (d) $\sqrt{675 + (15 - gt_f)^2} \approx 30.49$ m/sec
(e) $\tan^{-1}[(15 - gt_f)/15\sqrt{3}] \approx 148.45°$
59. (a) (1500)(40) – (9.81)(2500) = 35,475 N (b) (1500)(40) – (9.81)(1500) = 45,285 N
61. $\kappa = \rho = 1$ **63.** $\kappa = 36/(97/2)^{3/2}, \rho = (97/2)^{3/2}/36$
65. $\kappa = 16/17^{3/2}, \rho = 17^{3/2}/16$ **67.** $\kappa = 2/17^{3/2}, \rho = 17^{3/2}/2$ **69.** $\kappa = \frac{3}{4}, \rho = \frac{4}{3}$
71. $a_T = 0, a_n = 2$ **73.** $a_T = 6(1 + 2t^2)/\sqrt{1 + t^2}$; $a_n = 6|t|/\sqrt{1 + t^2}$
75. $[1300(175{,}000)^2/(3600)^2] \cdot (1/65) \approx 47{,}261$ N

CHAPTER THREE

PROBLEMS 3.1

1.

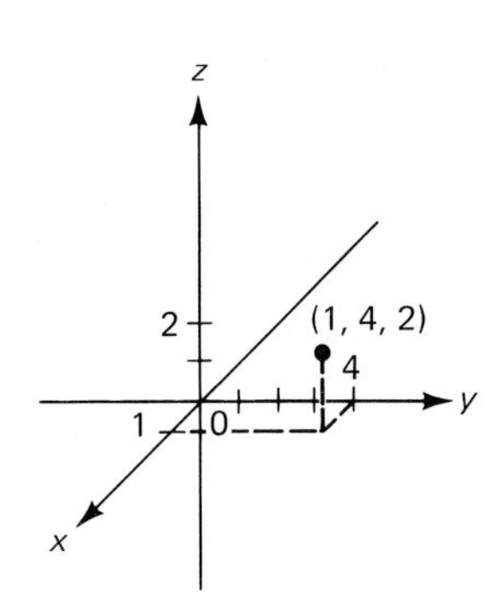

3.

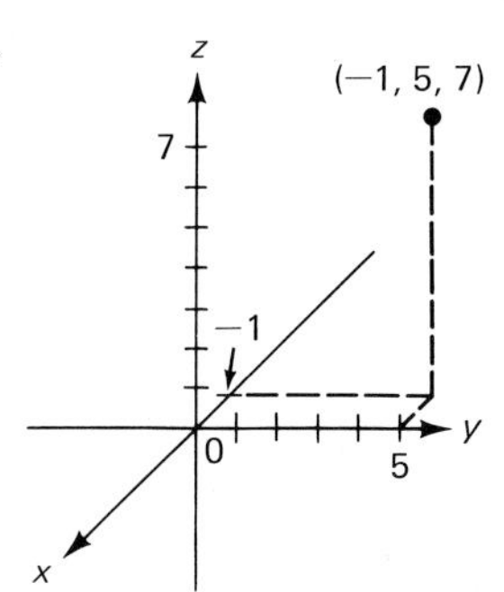

5.

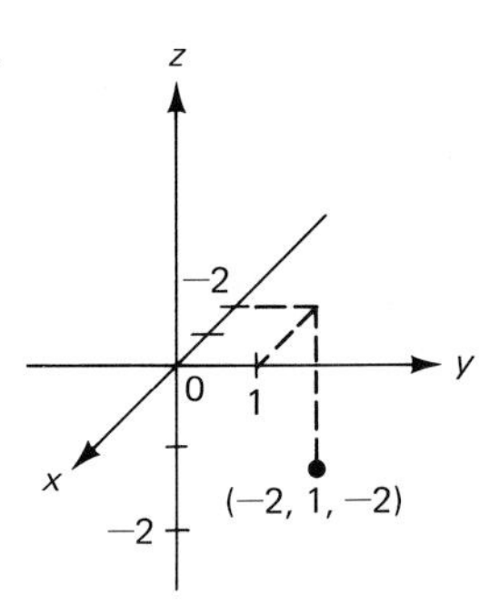

7.

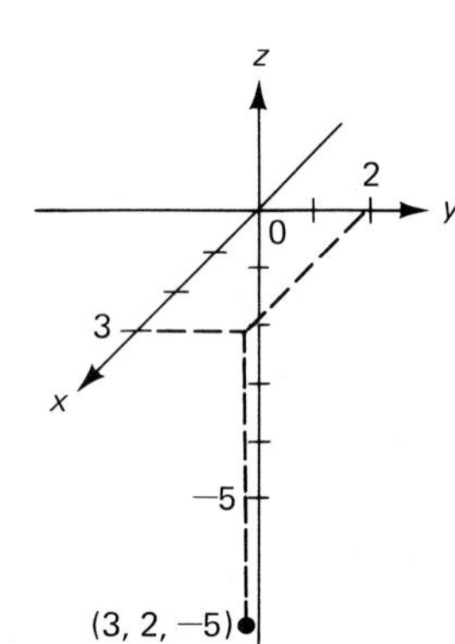

9.

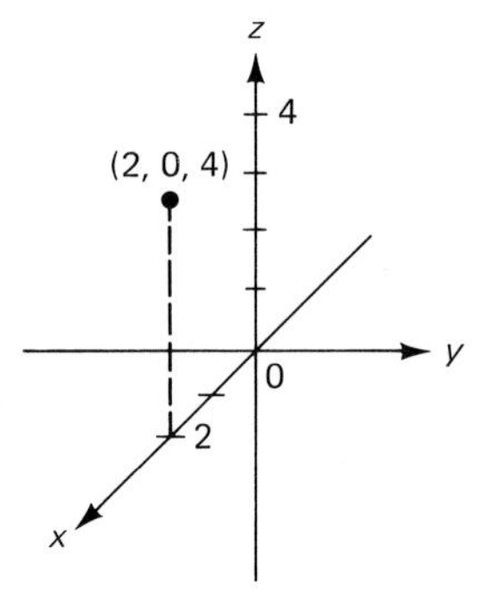

11.

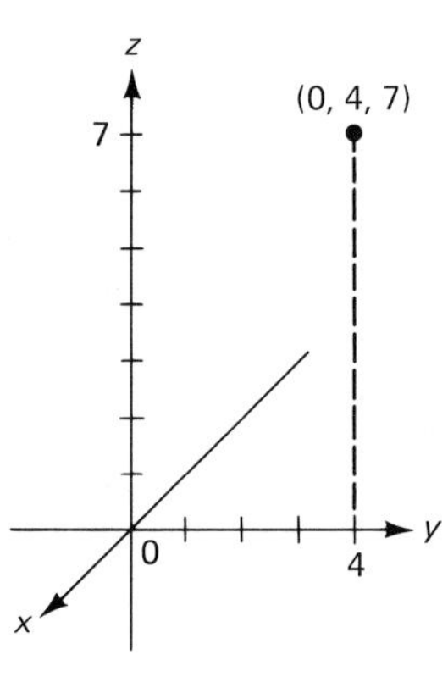

13.

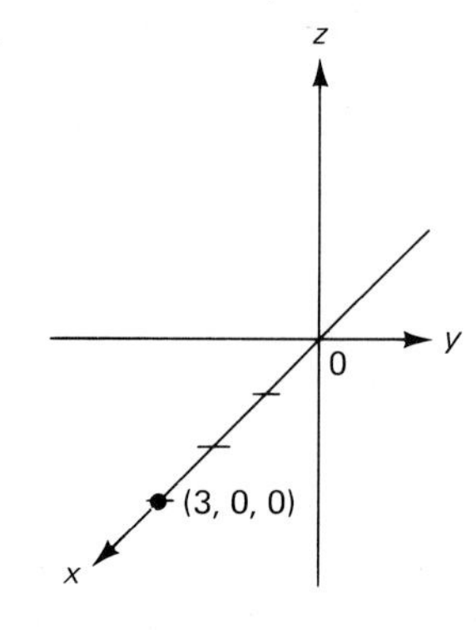

15.

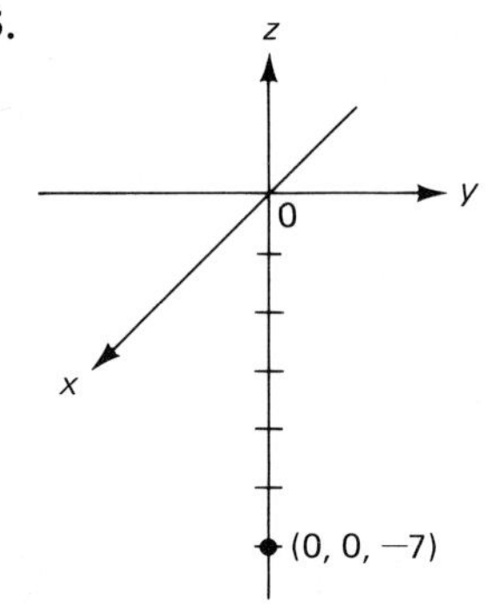

17. $\sqrt{40}$ **19.** 6 **21.** 5 **23.** $\sqrt{329}$ **25.** $\sqrt{250}$
27. $(x + 1)^2 + (y - 8)^2 + (z + 3)^2 = 5$ **29.** center $(-\frac{3}{2}, \frac{1}{2}, -1)$, $r = 3/\sqrt{2}$
31. $\alpha > 23\frac{1}{4}$ **37.** $(x - \frac{7}{2})^2 + (y - 1)^2 + (z - 2)^2 = \frac{65}{4}$
33. $\overline{RP} + \overline{PQ} = \sqrt{14} + 3\sqrt{14} = 4\sqrt{14} = \overline{RQ}$ **39.** $\frac{1}{6}\pi(65)^{3/2}$; $\frac{4}{3}\pi$

PROBLEMS 3.2

1. $|\mathbf{v}| = 3$; 0, 1, 0 **3.** $|\mathbf{v}| = 14$; 0, 0, 1 **5.** $|\mathbf{v}| = \sqrt{17}$; $4/\sqrt{17}$, $-1/\sqrt{17}$, 0
7. $|\mathbf{v}| = \sqrt{13}$; $-2/\sqrt{13}$, $3/\sqrt{13}$, 0 **9.** $|\mathbf{v}| = \sqrt{3}$; $1/\sqrt{3}$, $-1/\sqrt{3}$, $1/\sqrt{3}$
11. $|\mathbf{v}| = \sqrt{3}$; $-1/\sqrt{3}$, $1/\sqrt{3}$, $1/\sqrt{3}$ **13.** $|\mathbf{v}| = \sqrt{3}$; $-1/\sqrt{3}$, $1/\sqrt{3}$, $-1/\sqrt{3}$
15. $|\mathbf{v}| = \sqrt{3}$; $-1/\sqrt{3}$, $-1\sqrt{3}$, $-1/\sqrt{3}$
17. $|\mathbf{v}| = \sqrt{222}$; $-7/\sqrt{222}$; $2/\sqrt{222}$, $-13/\sqrt{222}$
19. $|\mathbf{v}| = \sqrt{82}$; $-3/\sqrt{82}$, $-3/\sqrt{82}$, $8/\sqrt{82}$ **21.** $(1/\sqrt{3})\mathbf{i} + (1/\sqrt{3})\mathbf{j} + (1/\sqrt{3})\mathbf{k}$
23. $\cos^2(\pi/6) + \cos^2(\pi/3) + \cos^2(\pi/4) = \frac{3}{4} + \frac{1}{4} + \frac{1}{2} = 1.5 > 1$ **25.** $-\mathbf{i}$

29. $10\mathbf{i} + 3\mathbf{j} - 7\mathbf{k}$ **31.** $\mathbf{i} - \mathbf{j} - 6\mathbf{k}$ **33.** $-13\mathbf{i} + 28\mathbf{j} + 12\mathbf{k}$ **35.** 25 **37.** 45
39. $\cos^{-1}(-10/\sqrt{59}\sqrt{50}) \approx 1.76$ rad $\approx 100.6°$ **41.** $\frac{25}{38}\mathbf{v} = -\frac{25}{19}\mathbf{i} - \frac{75}{38}\mathbf{j} + \frac{125}{38}\mathbf{k}$
43. $-\frac{1}{5}\mathbf{t} = -\frac{3}{5}\mathbf{i} - \frac{4}{5}\mathbf{j} - \mathbf{k}$ **45.** $\frac{35}{59}\mathbf{w} = \frac{35}{59}\mathbf{i} - \frac{245}{59}\mathbf{j} + \frac{105}{59}\mathbf{k}$
47. $\sqrt{5610}/51 = \sqrt{110/51}$ (Note that the distance is given by $|\overrightarrow{QP} - \text{Proj}_{\overrightarrow{QR}}\, \overrightarrow{QP}|$. Draw a picture.)
49. $\overrightarrow{PQ} \cdot \overrightarrow{PR} = -6 + 12 - 6 = 0$ **53.** $14\sqrt{3}$ J **63.** $(\pm 1/\sqrt{180})(8, 10, 4)$

PROBLEMS 3.3

1. $x\mathbf{i} + y\mathbf{j} + z\mathbf{k} = (2\mathbf{i} + \mathbf{j} + 3\mathbf{k}) + t(-\mathbf{i} + \mathbf{j} - 4\mathbf{k})$; $x = 2 - t, y = 1 + t, z = 3 - 4t$; $(x - 2)/(-1) = y - 1 = (z - 3)/(-4)$
3. $x\mathbf{i} + y\mathbf{j} + z\mathbf{k} = (\mathbf{i} + 3\mathbf{j} + 2\mathbf{k}) + t(\mathbf{i} + \mathbf{j} - 4\mathbf{k})$; $x = 1 + t, y = 3 + t, z = 2 - 4t$; $x - 1 = y - 3 = (z - 2)/(-4)$
5. $x\mathbf{i} + y\mathbf{j} + z\mathbf{k} = (-4\mathbf{i} + \mathbf{j} + 3\mathbf{k}) + t(-\mathbf{j} - 2\mathbf{k})$; $x = -4, y = 1 - t, z = 3 - 2t$; $x = -4$ and $(y - 1)/(-1) = (z - 3)/(-2)$
7. $x\mathbf{i} + y\mathbf{j} + z\mathbf{k} = (\mathbf{i} + 2\mathbf{j} + 3\mathbf{k}) + t(2\mathbf{i} - 2\mathbf{k})$; $x = 1 + 2t, y = 2, z = 3 - 2t$; $y = 2$ and $(x - 1)/2 = (z - 3)/(-2)$
9. $x\mathbf{i} + y\mathbf{j} + z\mathbf{k} = (\mathbf{i} + 2\mathbf{j} + 4\mathbf{k}) + t(3\mathbf{k})$; $x = 1, y = 2, z = 4 + 3t$; $x = 1$ and $y = 2$
11. $x\mathbf{i} + y\mathbf{j} + z\mathbf{k} = (2\mathbf{i} + 2\mathbf{j} + \mathbf{k}) + t(2\mathbf{i} - \mathbf{j} - \mathbf{k})$; $x = 2 + 2t, y = 2 - t, z = 1 - t$; $(x - 2)/2 = (y - 2)/(-1) = (z - 1)/(-1)$
13. $x\mathbf{i} + y\mathbf{j} + z\mathbf{k} = (\mathbf{i} + 3\mathbf{k}) + t(\mathbf{i} - \mathbf{j})$; $x = 1 + t, y = -t, z = 3$; $z = 3$ and $(x - 1)/1 = y/(-1)$
15. $x\mathbf{i} + y\mathbf{j} + z\mathbf{k} = (-\mathbf{i} - 2\mathbf{j} + 5\mathbf{k}) + t(-3\mathbf{j} + 7\mathbf{k})$; $x = -1, y = -2 - 3t, z = 5 + 7t$; $x = -1$ and $(y + 2)/(-3) = (z - 5)/7$
17. $x\mathbf{i} + y\mathbf{j} + z\mathbf{k} = (-\mathbf{i} - 3\mathbf{j} + \mathbf{k}) + t(-7\mathbf{j})$; $x = -1, y = -3 - 7t, z = 1$; $x = -1$ and $z = 1$
19. $x\mathbf{i} + y\mathbf{j} + z\mathbf{k} = (a\mathbf{i} + b\mathbf{j} + c\mathbf{k}) + t(d\mathbf{i} + e\mathbf{j})$; $x = a + dt, y = b + et, z = c$; $z = c$ and $(x - a)/d = (y - b)/e$
21. $x\mathbf{i} + y\mathbf{j} + z\mathbf{k} = (4\mathbf{i} + \mathbf{j} - 6\mathbf{k}) + t(3\mathbf{i} + 6\mathbf{j} + 2\mathbf{k})$; $x = 4 + 3t, y = 1 + 6t, z = -6 + 2t$; $(x - 4)/3 = (y - 1)/6 = (z + 6)/2$
27. no point of intersection **29.** none **31.** $(-3, 2, 7)$ **33.** none
35. (a) $\sqrt{186}/3$ $(t = \frac{1}{3})$ (b) $\sqrt{1518}/11 = \sqrt{138/11}$ $(t = -\frac{4}{11})$; (c) $\sqrt{750}/6 = 5\sqrt{30}/6$ $(t = -\frac{1}{6})$
37. the lines do intersect because

$$\begin{vmatrix} 1 & 1 & 2-9 \\ 2 & -1 & -1-(-2) \\ 4 & -2 & 3-1 \end{vmatrix} = 0$$

PROBLEMS 3.4

1. $-6\mathbf{i} - 3\mathbf{j}$ **3.** $-\mathbf{i} - \mathbf{j} + \mathbf{k}$ **5.** $12\mathbf{i} + 8\mathbf{j} - 21\mathbf{k}$ **7.** $(bc - ad)\mathbf{j}$
9. $-5\mathbf{i} - \mathbf{j} + 7\mathbf{k}$ **11.** 0 **13.** $42\mathbf{i} + 6\mathbf{j}$ **15.** $-9\mathbf{i} + 39\mathbf{j} + 61\mathbf{k}$
17. $-4\mathbf{i} + 8\mathbf{k}$ **19.** 0
21. $-(9/\sqrt{181})\mathbf{i} - (6/\sqrt{181})\mathbf{j} + (8/\sqrt{181})\mathbf{k}$; $(9/\sqrt{181})\mathbf{i} + (6/\sqrt{181})\mathbf{j} - (8/\sqrt{181})\mathbf{k}$
23. $\sqrt{30}/(\sqrt{6}\sqrt{29}) \approx 0.415$ **25.** $(x - 1)/2 = (y + 3)/(-26) = (z - 2)/(-22)$
27. $x = -2 + 13t, y = 3 + 22t, z = 4 - 8t$ **29.** $5\sqrt{5}$ **31.** $\sqrt{523}$
33. $\sqrt{a^2b^2 + a^2c^2 + b^2c^2}$ **39.** $\frac{1}{2}\sqrt{595}$ **43.** 23 **45.** $48/\sqrt{437}$ **47.** $23/\sqrt{27}$

PROBLEMS 3.5

1. $x = 0$ (yz-plane)) **3.** $z = 0$ (xy-plane) **5.** $x + z = 4$ **7.** $3x - y + 2z = 19$
9. $4x + y - 7z = -15$ **11.** $x + y + z = -6$ **13.** $20x + 13y - 3z = 58$
15. $x + y + z = 1$ **17.** coincident **19.** orthogonal **21.** orthogonal
23. coincident **25.** $x = t, y = -11 - 37t, z = -2 - 10t$ **27.** $33/\sqrt{59}$
29. $11/\sqrt{68}$ **33.** $\cos^{-1}(18/\sqrt{26}\sqrt{69}) \approx 1.132 \approx 64.9°$ **37.** $x - 22y - 17z = 0$
39. not coplanar **41.** $x - 8y + 10z = 0$

PROBLEMS 3.6

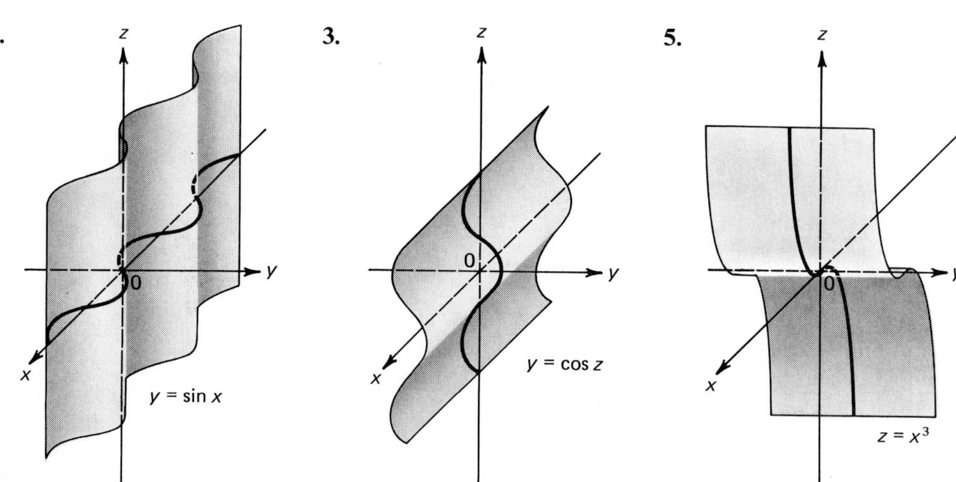

7. right circular cylinder, radius 1, centered on line $x = 0$, $y = -1$
9. right circular cylinder, radius 2, centered on x-axis **11.** hyperbolic cylinder
13. elliptic cylinder **15.** hyperbolic cylinder **17.** hyperboloid of one sheet
19. ellipsoid **21.** hyperbolic paraboloid (cross sections parallel to the xz-plane are hyperbolas)
23. hyperboloid of two sheets centered at (2, 0, 0) ($[(x - 2)^2/4] - (y^2/4) - (z^2/4) = 1$); cross sections parallel to the yz-plane are circles; surface only defined for $|x - 2| \geq 2$
25. hyperboloid of two sheets (like the surface in Problem 23)
27. hyperboloid of two sheets; cross sections parallel to the xy-plane are circles; surface defined for $|z| \geq \sqrt{2}$ **29.** sphere, radius $\sqrt{3}$, center (1, 1, 1)
31. hyperboloid of one sheet, cross sections parallel to the yz-plane are ellipses
33. hyperbolic paraboloid; like the surface in Figure except that (i) the point O is (1, 3, −8) instead of the origin and (ii) the line $x = 1$, $z = -8$ instead of the x-axis passes through O
35. ellipsoid with equation $(x^2 + z^2)/(a^2) + (y^2/b^2) = 1$

PROBLEMS 3.7

1. $\begin{pmatrix} 2 \\ -3 \\ 11 \end{pmatrix}$ **3.** $\begin{pmatrix} -4 \\ 0 \\ 4 \end{pmatrix}$ **5.** $\begin{pmatrix} -31 \\ 22 \\ -27 \end{pmatrix}$ **7.** $\begin{pmatrix} 0 \\ 0 \\ 0 \end{pmatrix}$ **9.** $\begin{pmatrix} -11 \\ 11 \\ -10 \end{pmatrix}$ **11.** (1, 2, 5, 7)
13. (−8, 12, 4, 20) **15.** (8, −5, 7, −1) **17.** (7, 2, 4, 11) **19.** (−11, 9, 18, 18)
25. $\mathbf{d}_1 + \mathbf{d}_2$ represents the combined demand of the two factories for each of the four raw materials needed to produce one unit of each factory's product; $2\mathbf{d}_1$ represents the demand of factory 1 for each of the four raw materials needed to produce two units of its product.
27. −14 **29.** 1 **31.** $ac + bd$ **33.** 51 **35.** $\mathbf{a} = \mathbf{0}$ **37.** 4 **39.** 28
41. orthogonal **43.** orthogonal **45.** orthogonal
47. all α and β that satisfy $5\alpha + 4\beta = 25$ ($\beta = (25 - 5\alpha)/4$, α arbitrary) **49.** $\sqrt{30}$
51. $\sqrt{a^2 + b^2 + c^2 + d^2 + e^2}$ **53.** $\sqrt{10}$

PROBLEMS 3.8

1. $(1/\sqrt{14})\mathbf{i} + (2/\sqrt{14})\mathbf{j} + (3/\sqrt{14})\mathbf{k}$ **3.** $(1/\sqrt{3})\mathbf{i} + (1/\sqrt{3})\mathbf{j} - (1/\sqrt{3})\mathbf{k}$
5. $(-8/\sqrt{65})\mathbf{i} + (1/\sqrt{65})\mathbf{k}$
7. $\int_0^{10} \sqrt{36 + 4t^2}\, dt = 18 \int_0^{\tan^{-1}(10/3)} \sec^3 \theta\, d\theta = 10\sqrt{109} + 9 \ln[(\sqrt{109} + 10)/3]$
9. $\frac{4}{729}(328^{3/2} - 8) \approx 32.55$
11. $\mathbf{v} = -(\sin 1)\mathbf{i} + (\cos 1)\mathbf{j} + 4\mathbf{k}$; $|\mathbf{v}| = \sqrt{17}$; $\mathbf{a} = -(\cos 1)\mathbf{i} - (\sin 1)\mathbf{j} + 12\mathbf{k}$; $|\mathbf{a}| = \sqrt{145}$
13. $\mathbf{v} = \mathbf{j} + \mathbf{k}$; $|\mathbf{v}| = \sqrt{2}$, $\mathbf{a} = \mathbf{i}$; $|\mathbf{a}| = 1$
15. $\mathbf{v} = \frac{1}{2}\mathbf{i} + \frac{1}{3}\mathbf{j} + \frac{1}{4}\mathbf{k}$; $|\mathbf{v}| = \sqrt{61}/12$; $\mathbf{a} = -\frac{1}{4}\mathbf{i} - \frac{2}{9}\mathbf{j} - \frac{3}{16}\mathbf{k}$; $|\mathbf{a}| = \sqrt{3049}/144$

17. $\mathbf{T} = \frac{\sqrt{3}a}{2\sqrt{a^2+1}}\mathbf{i} - \frac{a}{2\sqrt{a^2+1}}\mathbf{j} + \frac{1}{\sqrt{a^2+1}}\mathbf{k}$; $\kappa = \frac{a}{a^2+1}$; $\mathbf{n} = -\frac{1}{2}\mathbf{i} - \frac{\sqrt{3}}{2}\mathbf{j}$;
$\mathbf{B} = \frac{\sqrt{3}}{2\sqrt{a^2+1}}\mathbf{i} - \frac{1}{2\sqrt{a^2+1}}\mathbf{j} - \frac{a}{\sqrt{a^2+1}}\mathbf{k}$

19. $\mathbf{T} = \frac{-a}{\sqrt{1+a^2}}\mathbf{i} + \frac{1}{\sqrt{1+a^2}}\mathbf{k}$; $\kappa = \frac{b}{1+a^2}$ $\left(\text{since } \frac{d\mathbf{T}}{dt} = -\frac{b}{\sqrt{1+a^2}}\mathbf{j},\ \frac{ds}{dt} = \sqrt{1+a^2} \text{ and } \frac{d\mathbf{T}}{ds} = \frac{d\mathbf{T}/dt}{ds/dt} = \frac{-b}{1+a^2}\mathbf{j}\right)$; $\mathbf{n} = -\mathbf{j}$; $\mathbf{B} = \frac{1}{\sqrt{1+a^2}}\mathbf{i} + \frac{a}{\sqrt{1+a^2}}\mathbf{k}$

21. $\mathbf{T} = (1/\sqrt{6})\mathbf{i} + (2/\sqrt{6})\mathbf{j} + (1/\sqrt{6})\mathbf{k}$; $\kappa = \sqrt{5}/3$, $\mathbf{n} = -(2/\sqrt{5})\mathbf{i} + (1/\sqrt{5})\mathbf{j}$;
$\mathbf{B} = -1/(\sqrt{30})\mathbf{i} - (2/\sqrt{30})\mathbf{j} + (5/\sqrt{30})\mathbf{k}$

23. $\mathbf{T} = (1/\sqrt{5})\mathbf{j} + (2/\sqrt{5})\mathbf{k}$; $\kappa = \sqrt{20}/25 = 2/5^{3/2}$; $\mathbf{n} = -(2/\sqrt{5})\mathbf{j} + (1/\sqrt{5})\mathbf{k}$; $\mathbf{B} = \mathbf{i}$

31. $\frac{1}{2}$ **33.** $\mathbf{f}'(t) = (1, \cos t, -\sin t, 3t^2)$ **35.** $\mathbf{f}'(t) = (e^t, 2te^{t^2}, \ldots, nt^{n-1}e^{t^n})$

37. $\mathbf{f}'(t) = (1/t, 1/t, \ldots, 1/t)$, $(t > 0)$ **39.** $\mathbf{f}'(t) = 2t\mathbf{x}$

41. $\mathbf{h}'(t) = (3e^t, 6te^{t^2}, \ldots, 3nt^{n-1}e^{t^n})$

43. $h'(t) = \sum_{k=1}^{n} \frac{ke^{t^n}(t^k - 1)}{t^{k+1}}$ **45.** $\mathbf{T}(t) = \left(\sum_{k=1}^{n} k^2t^{2k-2}\right)^{-1/2}(1, 2t, \ldots, nt^{n-1})$

47. $\mathbf{T}(t) = \left(\sum_{k=1}^{n} k^2e^{2kt}\right)^{-1/2}(e^t, 2e^{2t}, \ldots, ne^{nt})$ **49.** $4\sqrt{\frac{n(n+1)(2n+1)}{6}}$

PROBLEMS 3.9

1. $(1, \sqrt{3}, 5)$ **3.** $(-4, 4\sqrt{3}, 1)$ **5.** $(-3/\sqrt{2}, 3/\sqrt{2}, 2)$ **7.** $(-10, 0, -3)$
9. $(-7/2, -7/\sqrt{2}, 2)$ **11.** $(1, \pi/2, 0)$ **13.** $(\sqrt{2}, \pi/4, 2)$ **15.** $(4, \pi/3, -5)$
17. $(4, 5\pi/3, 4)$ **19.** $(\sqrt{3}, 0, 1)$ **21.** $(0, 3\sqrt{3}, 3)$ **23.** $(\frac{7}{2}, -\frac{7}{2}, -7/\sqrt{2})$
25. $(\sqrt{3}, 3, -2)$ **27.** $(5\sqrt{3}/4, -\frac{5}{4}, -5\sqrt{3}/2)$ **29.** $(2, \pi/4, \pi/4)$
31. $(2, 5\pi/4, \pi/4)$ **33.** $(2\sqrt{2}, 7\pi/6, \pi/4)$
35. $(\sqrt{23}, \cos^{-1}(-2/\sqrt{7}), \cos^{-1}(-4/\sqrt{23}))$
37. cylindrical: $r^2 + z^2 = 25$; spherical: $\rho = 5$ **39.** $x^2 + y^2 - 9y = 0$
41. cylindrical: $r^2 - z^2 = 1$; spherical: $\rho^2(\sin^2\phi - \cos^2\phi) = \rho^2(1 - 2\cos^2\phi) = 1$
43. $z = x^2 + y^2$ **45.** $z\sqrt{x^2+y^2} = 1$ **47.** $z = 2xy$ **49.** $\rho = 6 \sin\phi \sin\theta$

REVIEW EXERCISES FOR CHAPTER THREE

1. $\sqrt{68}$ **3.** $\sqrt{216}$ **5.** $(x+1)^2 + (y-4)^2 + (z-2)^2 = 9$
9. $\sqrt{130}$; $0, 3/\sqrt{130}, 11/\sqrt{130}$ **11.** $\sqrt{53}$; $-4/\sqrt{53}, 1/\sqrt{53}, 6/\sqrt{53}$
13. $(2/\sqrt{6})\mathbf{i} - (1/\sqrt{6})\mathbf{j} + (1/\sqrt{6})\mathbf{k}$ **15.** $\mathbf{i} - 14\mathbf{j} + 20\mathbf{k}$ **17.** $\frac{26}{21}\mathbf{i} - \frac{52}{21}\mathbf{j} + \frac{13}{21}\mathbf{k}$
19. 22 **21.** $\alpha = -21/4$ **21.** $\cos^{-1}(-9/\sqrt{798}) \approx 108.6°$ **23.** $68/\sqrt{3}$ J
25. $x\mathbf{i} + y\mathbf{j} + z\mathbf{k} = (-4\mathbf{i} + \mathbf{j}) + t(7\mathbf{i} - \mathbf{j} + 7\mathbf{k})$; $x = -4 + 7t$, $y = 1 - t$, $z = 7t$; $(x+4)/7 = (y-1)/(-1) = z/7$
27. $x\mathbf{i} + y\mathbf{j} + z\mathbf{k} = (\mathbf{i} - 2\mathbf{j} - 3\mathbf{k}) + t(5\mathbf{i} - 3\mathbf{j} + 2\mathbf{k})$; $x = 1 + 5t$, $y = -2 - 3t$, $z = -3 + 2t$; $(x-1)/5 = (y+2)/(-3) = (z+3)/2$ **29.** $\sqrt{165}/3$ **31.** $-7\mathbf{i} - 7\mathbf{k}$
33. $-26\mathbf{i} - 8\mathbf{j} + 7\mathbf{k}$ **35.** $(x+1)/14 = (y-2)/(-26) = (z-4)/(-11)$ **37.** $2\sqrt{22}$
39. $x + z = -1$ **41.** $2x - 3y + 5z = 19$ **43.** $x = \frac{1}{2} - \frac{9}{2}t$, $y = \frac{7}{2} - \frac{11}{2}t$, $z = t$
45. $x = \frac{4}{3} - \frac{5}{3}t$, $y = -4 - \frac{7}{3}t$, $z = t$ **47.** $\cos^{-1}|-1/\sqrt{207}| \approx 1.501 \approx 86.01°$

49.

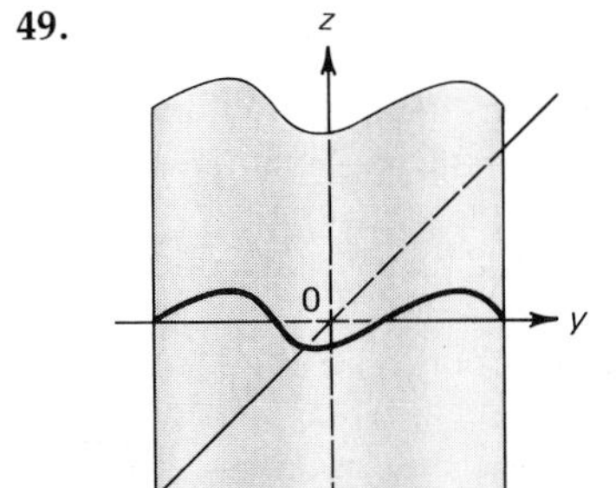

51.

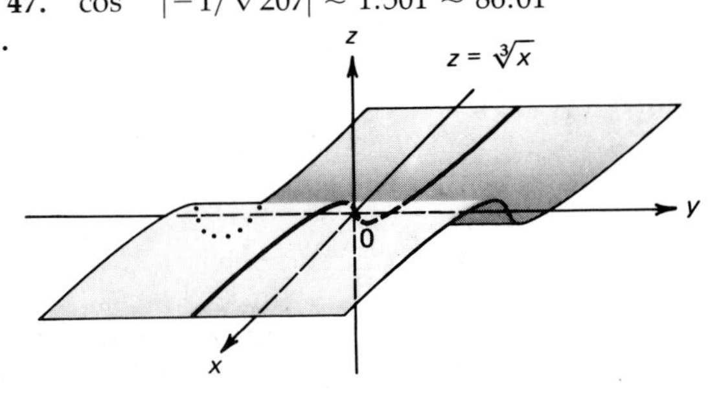

53. hyperboloid of one sheet; cross sections parallel to xz-plane are ellipses
55. hyperboloid of two sheets; cross sections parallel to xz-plane are circles; only defined for $|y| \geq \frac{5}{4}$
57. hyperbolic paraboloid; cross sections parallel to yz-plane are parabolas.
59. $(6/\sqrt{53})\mathbf{i} - (1/\sqrt{53}\mathbf{j} + (4\sqrt{53})\mathbf{k}$ **61.** $-(3\sqrt{3}/2\sqrt{10})\mathbf{i} - (3/2\sqrt{10})\mathbf{j} + (1/\sqrt{10})\mathbf{k}$
63. $\mathbf{v} = \frac{3}{16}\pi^2\mathbf{i} - (1/\sqrt{2})\mathbf{j} - (1/\sqrt{2})\mathbf{k}$; $|\mathbf{v}| = \sqrt{1 + (9\pi^4/256)}$;
$\mathbf{a} = (3\pi/2)\mathbf{i} - (1/\sqrt{2})\mathbf{j} + (1/\sqrt{2})\mathbf{k}$; $|\mathbf{a}| = \sqrt{1 + (9\pi^2/4)}$
65. $\mathbf{T} = -(1/\sqrt{5})\mathbf{i} + (\sqrt{3/5})\mathbf{j} + (1/\sqrt{5})\mathbf{k}$; $\kappa = \frac{2}{5}$; $\mathbf{n} = -(\sqrt{3}/2)\mathbf{i} - \frac{1}{2}\mathbf{j}$;
$\mathbf{B} = (1/2\sqrt{5})\mathbf{i} - (\sqrt{3}/2\sqrt{5})\mathbf{j} + (2/\sqrt{5})\mathbf{k}$
67. $\mathbf{T} = (1/\sqrt{3})\mathbf{i} + (1/\sqrt{3})\mathbf{j} + (1/\sqrt{3})\mathbf{k}$; $\kappa = \sqrt{2}/3$; $\mathbf{n} = -(1/\sqrt{2})\mathbf{j} + (1/\sqrt{2})\mathbf{k}$;
$\mathbf{B} = (2/\sqrt{6})\mathbf{i} - (1/\sqrt{6})\mathbf{j} - (1/\sqrt{6})\mathbf{k}$ **69.** $\mathbf{f}'(t) = (3t^2, -\sin t, 1/t, 2e^{2t})$
71. $\mathbf{T}(1) = \dfrac{1}{\sqrt{120}}(2, 4, 6, 8) = \dfrac{1}{\sqrt{30}}(1, 2, 3, 4)$; $\mathbf{n}(1) = \dfrac{1}{\sqrt{186}}(-7, -8, -3, 8)$
73. $(3\sqrt{3}/2, \frac{3}{2}, -1)$ **75.** $(-1, \sqrt{3}, 4)$ **77.** $(3\sqrt{2}/4, 3\sqrt{6}/4, 3\sqrt{2}/2)$
79. $(2, 3\pi/4, 3\pi 4)$ **81.** cylindrical: $r^2 + z^2 = 25$; spherical: $\rho = 5$
83. cylindrical: $r^2(\cos^2\theta - \sin^2\theta) + z^2 = r^2(1 - 2\sin^2\theta) + z^2 = 1$;
spherical: $\rho^2(\sin^2\phi\cos^2\theta - \sin^2\phi\sin^2\theta + \cos^2\phi) = \rho^2(1 - 2\sin^2\phi\sin^2\theta) = 1$
85. $x^2 + y^2 = 1$

CHAPTER FOUR

PROBLEMS 4.1

1. $\mathbb{R}^2$; $[0, \infty]$ **3.** $\{(x, y): y \neq 0\}$; $\mathbb{R}$ **5.** $\{(x, y): x^2 - 4y^2 \leq 1\}$; $[0, \infty)$ **7.** $\mathbb{R}^2$; $(0, \infty)$
9. $\{(x, y): x - y \neq (n + \frac{1}{2})\pi, n = 0, \pm 1, \pm 2, \ldots\}$; $\mathbb{R}$ **11.** $\{(x, y): |x| \geq |y|$ and $x \neq -y\}$; $[0, \infty)$ **13.** $\{(x, y): |x - y| \leq 1\}$: $[0, \pi]$ **15.** $\{(x, y): x \neq -y\}$; $\mathbb{R}$
17. $\{(x, y): x \neq 0$ and $y \neq 0\}$; $(-\infty, -2] \cup [2, \infty)$ (The range is obtained from $f(x, y) = \pm\sqrt{2 + x^2/4y^2 + 4y^2/x^2}$
19. $\{(x, y, z): x + y + z \geq 0\}$–this is the half-space "in front of" the plane $x + y + z = 0$; $[0, \infty)$
21. $\{(x, y, z): y^2 < x^2 + z^2\}$–this is the region "outside" the cone $y^2 = x^2 + z^2$; $(0, \infty)$
23. $\{(0, 0, 0)\}$; $\{0\}$ **25.** $\{(x, y, z): z \neq 0\}$; $\mathbb{R}$ **27.** $\{(x, y, z): |x + y - z| \leq 1\}$–this is the part of $\mathbb{R}^3$ between the planes $x + y - z = -1$ and $x + y - z = 1$; $[-\pi/2, \pi/2]$
29. $\{(x, y, z): y \neq 0\}$; $(-\pi/2, \pi/2)$ **31.** $\mathbb{R}^3$; $(0, \infty)$
33. $\{(x, y, z): x \neq 0, y \neq 0$ and $z \neq 0\}$; $\mathbb{R} - \{0\}$ **35.** $\mathbb{R}^3$; $[-3, 3]$
37. $\{(x_1, x_2, x_3, x_4): x_1 + 2x_2 + x_3 - x_4 - 1 \geq 0\}$; $[0, \infty)$
39. $\{(x_1, x_2, x_3, x_4): x_2x_4 \neq 0\}$; $\mathbb{R}$ **41.** $\{(x_1, x_2, x_3, x_4): x_2^2 + x_4^2 \neq 0\}$; $\mathbb{R}$
43. $\{(x_1, x_2, x_3, x_4, x_5); x_3^2 + x_5^2 \neq 0\}$; $\mathbb{R}$ **45.** $\mathbb{R}^n$; $[0, \infty)$
47. elliptic paraboloid opening around the y-axis; cross sections parallel to the xz-plane are ellipses
49. hyperbolic paraboloid; cross sections parallel to the xy-plane are hyperbolas
51. right half of hyperboloid of one sheet; cross sections parallel to xy-plane are hyperbolas
53. these are the parallel straight lines (with slopes of -1) $y = -x + (z^2 - 1)$
55. these are concentric ellipses (centered at the origin) with equations $x^2 + 4y^2 = 1 - z^2$, $|z| \leq 1$; for $z = 1$ we obtain the single point $(0, 0, 0)$
57. parallel straight lines (with slopes of 1) $y = x - \cos z$
59. for each value of z we get a family of straight lines all of which have a slope of -1: for $z = 0$ we obtain $x + y = n\pi$, n an integer; for $z = 1$ we obtain $x + y = (\pi/4) + n\pi$; for $z = -1$ we obtain $x + y = -(\pi/4) + n\pi$; for $z = \sqrt{3}$ we obtain $x + y = (\pi/3) + n\pi$
61. concentric ellipses (centered at the origin); $a = \sqrt{T - 20}$, $b = \frac{1}{2}\sqrt{T - 20}$, for $T > 20$
63. concentric ellipses (centered at the origin); $a = \sqrt{(P - 100)/2}$, $b = \sqrt{(P - 100)/3}$, for $P > 100$; for $P = 100$ we obtain the single point $(0, 0)$
65. (h) **67.** (a) **69.** (e) **71.** (g) **73.** (d)

PROBLEMS 4.2

13. along the line $x = at$, $y = bt$, $xy/(x^2 - y^2) = ab/(a^2 - b^2)$; the limit depends on the choice of a and b[†] **15.** along the line $x = at$, $y = bt$, $xy^3/(x^4 + y^4) = ab^3/(a^4 + b^4)$ **17.** along the line $ax + by = 0$, $(x^2 + y^2)^2/(x^4 + y^4) = (a^2 + b^2)^2/(a^4 + b^4)$ **19.** along the line $y = kx$, $(ax^2 + by)/(cy^2 + dx) = (ax + bk)/(ck^2x + d) \to bk/d$ as $(x, y) \to (0, 0)$ **21.** along the line $y = kx = z$, $xyz/(x^3 + y^3 + z^3) = k^2/(1 + 2k^3)$ **23.** 0; since $0 \le 5x^2y^2/(x^4 + y^2) = 5x^2y^2/[(x^2 - |y|)^2 + 2x^2|y|] \le 5x^2y^2/2x^2|y| = \frac{5}{2}|y|$ **25.** 0; since $0 < |(yx^2 + z^3)/(x^2 + y^2 + z^2)| \le (|y|x^2 + |z|z^2)/(x^2 + y^2 + z^2) \le |y| + |z| \to 0$. **27.** $-\frac{61}{25}$ **29.** $\ln(4\pi/3)$ **31.** sinh 1 **33.** $\frac{1}{6}$ **35.** ln 769 **37.** $\mathbb{R}^2 - \{(0, 0)\}$ **39.** $\{(x, y): y \ne 4 \text{ and } x \ne -3\}$ **41.** $\mathbb{R}^2$ **43.** $\{(x, y): xy \ne 1\}$ **45.** $\{(x, y, z): x > 0)$ **47.** $\{(x, y, z): xz > 0\}$ **49.** $\{(x, y, z): x^2 + y^2 + z^2 < 1\}$ = the "interior" of the unit sphere **51.** $c = 0$ **53.** Along the line $y = kx = z$, $(yz - x^2)/(x^2 + y^2 + z^2) = (k^2 - 1)/(1 + 2k^2)$ which shows that $\lim_{(x, y, z)\to(0, 0, 0)} (yz - x^2)/(x^2 + y^2 + z^2)$ does not exist. The function is, therefore, not continuous at the origin. **61.** 0 **63.** 1 **65.** 0 **67.** $\ln\left(\frac{8}{3}\right)$ **69.** $\sqrt{6139/76}$ **71.** $\{(x_1, x_2, x_3, x_4): x_3x_4 \ne 0\}$ **73.** $\{(x_1, x_2, x_3, x_4, x_5): x_1^2 + x_2^2 + x_3^2 + x_4^2 + x_5^2 < 1\}$. This is the interior (inside) of the unit sphere in $\mathbb{R}^5$ (also called a **hypersphere**).

PROBLEMS 4.3

1. $2xy;\ x^2$ **3.** $3y^3e^{xy^3};\ 9xy^2e^{xy^3}$ **5.** $4/y^5;\ -20x/y^6$ **7.** $3x^2y^5/(x^3y^5 - 2)$; $5x^3y^4/(x^3y^5 - 2)$ **9.** $\frac{4}{3}(1 + 5y\cos x)(x + 5y\sin x)^{1/3};\ \frac{20}{3}\sin x(x + 5y\sin x)^{1/3}$ **11.** $[2y/(x^2 + y^2)](xy/|xy|);\ [-2x/(x^2 + y^2)](xy/|xy|)$ **13.** 3 **15.** 0 **17.** 1 **19.** $\frac{48}{169}$ **21.** $yz;\ xz;\ xy$ **23.** $1/z;\ 1/z;\ -(x + y)/z^2$ **25.** $3x^2/(x^3 + y^2 + z);\ 2y/(x^3 + y^2 + z);\ 1/(x^3 + y^2 + z)$ **27.** $yz\cos xyz;\ xz\cos xyz;\ xy\cos xyz$ **29.** $\dfrac{\sinh\sqrt{x + 2y + 5z}}{2\sqrt{x + 2y + 5z}};\ \dfrac{\sinh\sqrt{x + 2y + 5z}}{\sqrt{x + 2y + 5z}};\ \dfrac{5\sinh\sqrt{x + 2y + 5z}}{2\sqrt{x + 2y + 5z}}$ **31.** $\frac{1}{3}$ **33.** $-2\cos 1$ **35.** $-\frac{1}{2}$ **37.** $-1/(3\sqrt{3})$ **39.** $a/\sqrt{a^2 + b^2 + c^2}$ **43.** (a) $x = 1$, $(z - 5)/(-12) = (y + 1)/1$; (b) $y = -1$, $(z - 5)/3 = (x - 1)/1$ (The first line is obtained as in $\mathbb{R}^2$ by setting $(z - 5)/[y - (-1)] = -12$ = the slope.) **45.** $y = 1$, $(x - 1)/(-4) = z - 1$ **49.** $\partial R/\partial x = [50/(1 + 50x + 75y)] + (20/\sqrt{1 + 40x + 125y})$, $\partial R/\partial y = [75/(1 + 50x + 75y)] + (125/2\sqrt{1 + 40x + 125y})$ **59.** $f_1 = x_2x_3x_4$, $f_2 = x_1x_3x_4$, $f_3 = x_1x_2x_4$, $f_4 = x_1x_2x_3$ **61.** $f_1 = \dfrac{x_3}{x_2x_4} + \dfrac{x_3}{x_2}e^{x_1x_4/x_2}$, $f_2 = -\dfrac{x_1x_3}{x_2^2x_4} - \dfrac{x_1x_3}{x_2^2}e^{x_1x_4/x_2}$, $f_3 = \dfrac{x_1}{x_2x_4} + \dfrac{x_1}{x_2}e^{x_1x_4/x_2}$, $f_4 = -\dfrac{x_1x_3}{x_2x_4^2}$ **63.** $f_j = x_j\left(\sum_{i=1}^{n} x_i^2\right)^{-1/2}$ for $j = 1, 2, \ldots, n$ **65.** $f_j = -x_j\left(\sum_{i=1}^{n} x_i^2\right)^{-3/2}$ for $j = 1, 2, \ldots, n$

PROBLEMS 4.4

In Problems 1–11 the answers are given in the order f_{xx}, $f_{xy}(= f_{yx})$, f_{yy}.
1. $2y;\ 2x;\ 0$ **3.** $3y^6e^{xy^3};\ 9y^2(1 + xy^3)e^{xy^3};\ 9xy(2 + 3xy^3)e^{xy^3}$ **5.** $0;\ -20/y^6;\ 120x/y^7$ **7.** $\dfrac{-3xy^5(4 + x^3y^5)}{(x^3y^5 - 2)^2};\ \dfrac{-30x^2y^4}{(x^3y^5 - 2)^2};\ \dfrac{-5x^3y^3(8 + x^3y^5)}{(x^3y^5 - 2)^2}$ **9.** $\frac{4}{9}(1 + 5y\cos x)^2(x + 5y\sin x)^{-2/3} - \frac{20}{3}y\sin x(x + 5y\sin x)^{1/3}$; $\frac{20}{9}\sin x(1 + 5y\cos x)(x + 5y\sin x)^{-2/3} + \frac{20}{3}\cos x(x + 5y\sin x)^{1/3}$; $\frac{100}{9}\sin^2 x(x + 5y\sin x)^{-2/3}$ **11.** $[-4xy/(x^2 + y^2)^2](xy/|xy|);\ 2[(x^2 - y^2)/(x^2 + y^2)^2][(xy/|xy|)];\ [4xy/(x^2 + y^2)^2][(xy/|xy|)]$

[†]In Problems 4.2.13, 19, 21, it is immediately evident that the limit does not exist because the function is not defined in any neighborhood of the origin.

13. $f_{xx} = 0;\ f_{xy} = z = f_{yx};\ f_{xz} = y = f_{zx};\ f_{yy} = 0;\ f_{yz} = x = f_{zy};\ f_{zz} = 0$
15. $f_{xx} = 0;\ f_{xy} = 0 = f_{yz};\ f_{xz} = -1/z^2 = f_{zx};\ f_{yy} = 0;\ f_{yz} = -1/z^2 = f_{zy};\ f_{zz} = 2(x+y)/z^3$
17. $f_{xx} = \dfrac{-2xyz^3}{(y^2+x^2z^2)^2};\ f_{xy} = \dfrac{x^2z^3 - y^2z}{(y^2+x^2z^2)^2} = f_{yx};\ f_{xz} = \dfrac{y^3 - x^2yz^2}{(y^2+x^2z^2)^2} = f_{zx};$
$f_{yy} = \dfrac{2xyz}{(y^2+x^2z^2)^2};\ f_{yz} = \dfrac{x^3z^2 - xy^2}{(y^2+x^2z^2)^2} = f_{zy};\ f_{zz} = \dfrac{-2x^3yz}{(y^2+x^2z^2)^2}$
19. $f_{xx} = 9y^2e^{3xy}\cos z;\ f_{xy} = 3(1+3xy)e^{3xy}\cos z = f_{yz};\ f_{xz} = -3ye^{3xy}\sin z = f_{zx};$
$f_{yy} = 9x^2e^{3xy}\cos z;\ f_{yz} = -3xe^{3xy}\sin z = f_{zy};\ f_{zz} = -e^{3xy}\cos z$
21. $f_{xx} = \dfrac{\cosh\sqrt{x+yz}}{4(x+yz)} - \dfrac{\sinh\sqrt{x+yz}}{4(x+yz)^{3/2}};\ f_{xy} = zf_{xx} = f_{yz};\ f_{xz} = yf_{xx} = f_{zx};$
$f_{yy} = z^2f_{xx};\ f_{yz} = \dfrac{\sinh\sqrt{x+yz}}{2\sqrt{x+yz}} + yzf_{xx} = f_{zy};\ f_{zz} = y^2f_{xx}$ **23.** (a) 16 (b) 81
25. $6y^2$ **27.** $24/(3x-2y)^3$ **29.** $9\sin(x+2y+3z)$
41. $f_{13} = x_2x_4,\ f_{11} = 0,\ f_{134} = x_2$
43. $f_{jk} = 3x_jx_k\left(\sum_{i=1}^n x_i^2\right)^{-5/2}$ if $j \neq k$: $f_{ii} = -\left(\sum_{i=1}^n x_i^2\right)^{-3/2} + 3x_i^2\left(\sum_{i=1}^n x_i^2\right)^{-5/2}$
45. $f_{2211} = e^{x_1x_2x_3}(2x_3^2 + 4x_1x_2x_3^3 + x_1^2x_2^2x_3^4)$ **47.** (a) $3^3 = 27$ (b) $3^4 = 81$ (c) 3^m

PROBLEMS 4.5

5. $(e/2)\mathbf{i} + (e/2)\mathbf{j}$ **7.** $[2/(2x-y+1)]\mathbf{i} + [-1/(2x-y+1)]\mathbf{j}$ **9.** $-\frac{1}{6}\mathbf{i} + \frac{1}{6}\mathbf{j}$
11. $2x(\sinh y)\mathbf{i} + x^2(\cosh y)\mathbf{j}$ **13.** $\frac{1}{8}\mathbf{i} - \frac{3}{8}\mathbf{j}$
15. $(2xe^{x^2}/3y)\mathbf{i} + [-e^{x^2} + (1+2y^2)e^{-y^2}]/3y^2\mathbf{j}$ **17.** $(3\sqrt{2}/4)\mathbf{i} - (\sqrt{6}/4)\mathbf{j} + \sqrt{2}\mathbf{k}$
19. $[\ln y - (z/x)]\mathbf{i} + (x/y)\mathbf{j} - (\ln x)\mathbf{k}$ **21.** $-4e^3\mathbf{i} - 7e^3\mathbf{j} - 13e^3\mathbf{k}$ **23.** $\mathbf{i} - \mathbf{k}$
25. $\nabla f = (x_2x_3x_4, x_1x_3x_4, x_1x_2x_4, x_1x_2x_3)$ **27.** $\nabla f = (2x_1, 2x_2, 2x_3, 2x_4, 2x_5) = 2\mathbf{x}$
33. $f(x, y) = \frac{1}{2}x^2 + \frac{1}{2}y^2 + C$

PROBLEMS 4.6

1. $3e^{3t}$ **3.** 1 **5.** $[5(\cos 5t)\cos 3t + 3(\sin 5t)\sin 3t]/(\sin^2 5t + \cos^2 3t)$ **7.** $2t$
9. $(-t-2)/t^3$ **11.** $[2e^t - (3/t) + 4\sinh t]/(2e^t - 3\ln t + 4\cosh t)$
13. $\partial z/\partial r = -2\sin(r+s)\cos(r+s) + 2\sin(r-s)\cos(r-s) = -2\cos 2r\sin 2s;$
$\partial z/\partial s = -2\sin(r+s)\cos(r+s) - 2\sin(r-s)\cos(r-s) = -2\sin 2r\cos 2s$
15. $\partial z/\partial r = -(2s^2/r^3)\cos(s^2/r^2);\ \partial z/\partial s = (2s/r^2)\cos(s^2/r^2)$
17. $\partial z/\partial r = 2(r-s^2)(2s+r)^3 + 3(r-s^2)^2(2s+r)^2;$
$\partial z/\partial s = -4s(r-s^2)(2s+r)^3 + 6(r-s^2)^2(2s+r)^2$
19. $\partial w/\partial r = s/t;\ \partial w/\partial s = r/t;\ \partial w/\partial t = -rs/t^2$
21. $\partial w/\partial r = (4s^3r^3 - 3s^4r^2)\cos[s^3r^3(r-s)];\ \partial w/\partial s = (3s^2r^4 - 4s^3r^3)\cos[s^3r^3(r-s)]$
23. $\partial w/\partial r = 3r^2s^3t - 1/(r^2st^3);\ \partial w/\partial s = 3r^3s^2t - 1/(rs^2t^3);\ \partial w/\partial t = r^3s^3 - 3/(rst^4)$
25. $\partial w/\partial r = 2r[(s^2-t^2)/(r^2+s^2)]^2w;\ \partial w/\partial s = 2s[(r^2+t^2)/(r^2+s^2)]^2w;$
$\partial w/\partial t = 2t[(s^2 - r^2 - 2t^2)/(r^2+s^2)]w$ **27.** $V_t = 7500\pi$ in^3/min; increasing
29. $P_t = (10R + 75)/1000$ N/cm^2/min; increasing
31. $(2\pi - 38 - 2\sqrt{3})/\sqrt{676 - 240\sqrt{3}} \approx -2.18$ cm/sec; decreasing
33. $(\partial z/\partial x)^2 + (\partial z/\partial y)^2 = [(1/r)(\partial z/\partial\theta)]^2 + (\partial z/\partial r)^2$
35. $\partial w/\partial\rho = (\partial w/\partial x)\sin\phi\cos\theta + (\partial w/\partial y)\sin\phi\sin\theta + (\partial w/\partial z)\cos\phi$
$\partial w/\partial\phi = (\partial w/\partial x)\rho\cos\phi\cos\theta + (\partial w/\partial y)\rho\cos\phi\sin\theta - (\partial w/\partial z)\rho\sin\phi$
$\partial w/\partial\theta = -(\partial w/\partial x)\rho\sin\phi\sin\theta + (\partial w/\partial y)\rho\sin\phi\cos\theta$
39. $f(\mathbf{x}) = \dfrac{1}{2}\sum_{i=1}^n x_i^2 + C$

PROBLEMS 4.7

1. $x = 1;\ y = 0,\ z = 0$ **3.** $z = 1;\ x = 0,\ y = 0$
5. $(1/a)(x-a) + (1/b)(y-b) + (1/c)(z-c) = 0;\ a(x-a) = b(y-b) = c(z-c)$

7. $\frac{1}{4}(x-4)+\frac{1}{2}(y-1)+\frac{1}{6}(z-9)=0$; $4(x-4)=2(y-1)=6(z-9)$
9. $2(x-1)+(y-2)+(z-2)=0$; $(x-1)/2=y-2=z-2$
11. $24(x-3)-2(y-1)+20(z+2)=0$; $(x-3)/24=(y-1)/(-2)=(z+2)/20$
13. $(\pi/\sqrt{3})(y-1)+\sqrt{3}[z-(\pi/3)]=0$; $x=\pi/2$, $(\sqrt{3}/\pi)(y-1)=(1/\sqrt{3})[z-(\pi/3)]$
15. $(\ln 5)(x-1)+(\ln 5)(y-1)+(z-\ln 5)=0$; $(x-1)/(\ln 5)=(y-1)(\ln 5)=z-\ln 5$
17. $z=1+(x-1)+2(y-1)$; $x-1=(y-1)/2=(z-1)/(-1)$
19. $z=1$; $x=\pi/8$, $y=\pi/20$ **21.** $z=-(\pi/4)-\frac{1}{4}(x+2)-\frac{1}{4}(y-2)$; $4(x+2)=4(y-2)=z+\pi/4$ **23.** $z=2+2\sqrt{3}(x-\pi/2)-2\sqrt{3}(y-\pi/6)$; $(x-\pi/2)/(2\sqrt{3})=(y-\pi/6)/(-2\sqrt{3})=(z-2)/(-1)$ **25.** (0, 2, 4), (1, 1, 2)
29. All tangent planes pass through the origin. (Note that the surface cannot pass through the origin since $f(y/x)$ is not defined when $x=0$.)
33. $[(x-x_0)\mathbf{i}+(y-y_0)\mathbf{j}]\cdot\nabla F(x_0,y_0)=0$ **35.** $3\mathbf{i}-8\mathbf{j}$; $3(x-1)-8(y+2)=0$
37. $\mathbf{i}+\mathbf{j}$; $(x-1)+y=0$ **39.** $2\mathbf{i}+2\mathbf{j}$; $(x-\pi/4)+y=0$

PROBLEMS 4.8

1. $9/\sqrt{10}$ **3.** $\sqrt{2}/7$ **5.** $-5/(4\sqrt{13})$ **7.** $(2e^2+3e)/\sqrt{2}$ **9.** $2\sqrt{3}$
11. $1/(5\sqrt{6})$ **13.** $61/\sqrt{26}$ **15.** 0 **17.** (a) at the origin
(b) $\nabla T=200e^{-(x^2+y^2+z^2)}(-x\mathbf{i}-y\mathbf{j}-z\mathbf{k})$
19. $(x/a)^{a^2}=(y/b)^{b^2}$
21. (a) ascend, because $\nabla h\cdot\mathbf{u}>0$ when $\mathbf{u}=(-1/\sqrt{2})\mathbf{i}-(1/\sqrt{2})\mathbf{j}$
(b) $(-120\mathbf{i}+40\mathbf{j})/\sqrt{120^2+40^2}=(-3\mathbf{i}+\mathbf{j})/\sqrt{10}$; i.e., $\approx 161.1°$ (toward the southeast)
23. $\sqrt{5}$ **25.** $1/\sqrt{2}$ **29.** $12/\sqrt{6}=2\sqrt{6}$ **31.** $\dfrac{2}{\sqrt{n}}\cdot\dfrac{n(n+1)}{2}=\sqrt{n}(n+1)$

PROBLEMS 4.9

1. $F(x,y,z)=(\alpha/2)\ln(x^2+y^2+z^2)=\alpha\ln|\mathbf{x}|$
3. If $k=2$, look at Problem 1. If $k\neq 2$, a potential function is given by $F(x,y,z)=[-a/(k-2)](1/|\mathbf{x}|^{k-2})$.

PROBLEMS 4.10

1. $y^3\,\Delta x+3xy^2\,\Delta y$ **3.** $(x-y)^{-1/2}(x+y)^{-3/2}(y\,\Delta x-x\,\Delta y)$
5. $[1/(2x+3y)](2\,\Delta x+3\,\Delta y)$ **7.** $y^2z^5\,\Delta x+2xyz^5\,\Delta y+5xy^2z^4\,\Delta z$
9. $[1/(x+2y+3z)](\Delta x+2\,\Delta y+3\,\Delta z)$ **11.** $\sinh(xy-z)(y\,\Delta x+x\,\Delta y-\Delta z)$
13. (a) $x(\Delta y)^2+2y(\Delta x)(\Delta y)+(\Delta x)(\Delta y)^2$
(b) $\Delta f-df=[(0.99)(2.03)^2-(1)(2)^2]-[2^2(-0.01)+(2)(1)(2)(0.03)]=[0.079691]-[0.08]=-0.000309$ **15.** $\frac{3}{6}+\frac{1}{6}(0.01)-[3(-0.01)/6^2]=0.5025$ (actual value ≈ 0.50250417)
17. $\sqrt{36}\sqrt[3]{64}+(\frac{1}{2})(\frac{1}{6})(4)(-0.4)+(\frac{1}{3})(6)(\frac{1}{16})(0.08)=23.87666\ldots$ (actual value ≈ 23.87623437)
19. $\sqrt{\frac{1}{9}}+\frac{1}{27}[4(0.02)-5(-0.04)]=0.3437037037\ldots$ (actual value ≈ 0.3435696277)
21. $(3\cdot 2)/\sqrt{9})+(2/\sqrt{9})(0.02)+(3/\sqrt{9})(-0.03)-(3\cdot 2/2(9)^{3/2})(-0.05)=1.9888\ldots$ (actual value ≈ 1.988665097)
23. $8\sqrt{5-1^2}+(\sqrt{5-1^2})(-0.08)+(8/2\sqrt{5-1^2})(0.01)-[(8)(1)/\sqrt{5-1^2}](-0.02)=15.94$ (actual value ≈ 15.93790543) **25.** (a) $\sqrt{9100}\approx 95.3939$ m
(b) $(1/\sqrt{9100})\{|50-110(\frac{1}{2})|[0.3]+[110-50(\frac{1}{2})][0.3]+[(50)(110)(\sqrt{3}/2](\pi/180)\}\approx 0.899768$ m
27. (a) $\frac{8}{3}=2.6666\ldots$ ohms (b) $(\frac{8}{3})^2[(0.1/6^2)+(0.03/8^2)+(0.15/12^2)]\approx 0.03049$

PROBLEMS 4.11

1. x^2y+y+C **3.** not exact **5.** not exact **7.** $(x^2/2)-y\sin x+C$
9. $(3x^2/2)\ln x-(3x^2/4)+(x^6/6)-xy+C$ **11.** $\tan x-x+\tan^{-1}(y/x)-e^y+C$
13. not exact **15.** $x+y+z+C$ **17.** $(xy/z)+(x^3/3)+\cos y+\sin z+C$
19. not exact **21.** $\mathbf{f}(t)=-\alpha\mathbf{x}/|\mathbf{x}|^4$ and a potential function for $\mathbf{F}$ is $f(\mathbf{x})=-\alpha/(2|\mathbf{x}|^2)$ $(-\nabla f=\mathbf{F})$ **23.** $\mathbf{F}=-\alpha\mathbf{x}/|\mathbf{x}|$ and $f(\mathbf{x})=\alpha|\mathbf{x}|$ $(-\nabla f=\mathbf{F})$

PROBLEMS 4.12

1. (0, 0) is a global minimum **3.** $(-2, 1)$ is a global minimum **5.** $(-2, 1)$ is a global minimum **7.** $(\sqrt{5}, 0)$ is a local minimum; $(-\sqrt{5}, 0)$ is a local maximum; $(-1, 2)$ and $(-1, -2)$ are saddle points† **9.** $(\frac{1}{2}, 1)$ and $(-\frac{1}{2}, -1)$ are saddle points; $(-\frac{1}{2}, 1)$ is a local minimum; $(\frac{1}{2}, -1)$ is a local maximum† **11.** $(-2, -2)$ is a global minimum **13.** (0, 0), (0, 4) and (4, 0) are saddle points; $(\frac{4}{3}, \frac{4}{3})$ is a local maximum† **15.** (1, 1) and $(-1, 1)$ are local minima† **17.** (0, 0) is a saddle point. (Note that $D = 0$, but it is evident that f can take positive and negative values near the origin.) **19.** $x = y = z = \frac{50}{3}$ **21.** $x = y = z = \frac{50}{3}$ **23.** $3\sqrt{3}$ **25.** Corners are at $(\pm a/\sqrt{3}, \pm b/\sqrt{3}, \pm c/\sqrt{3})$. (The problem is to maximize $(2x)(2y)(2z)$ with $(x^2/a^2) + (y^2/b^2) + (z^2/c^2) = 1$.) **27.** The profit is given by $P(x, y) = 40(8xy + 32x + 40y - 4x^2 - 6y^2) - 10x - 4y$, which is maximized when $x \approx 22$ and $y \approx 18$ ($x = 21.88125$, $y = 17.9125$). **29.** $y = -\frac{54}{52}x + \frac{120}{52} \approx -1.04x + 2.3$

PROBLEMS 4.13

1. $3/\sqrt{13}$ at the point $(\frac{7}{13}, \frac{17}{13})$ **3.** $5/\sqrt{3}$ at the point $(\frac{8}{3}, \frac{2}{3}, \frac{1}{3})$ **5.** $|d|/\sqrt{a^2 + b^2 + c^2}$ at the point $(ad/(a^2 + b^2 + c^2), bd/(a^2 + b^2 + c^2), cd/(a^2 + b^2 + c^2))$ **7.** minimum is $\frac{15}{32}$ at $(\frac{9}{16}, \frac{1}{8})$; there is no maximum **9.** maximum is 8 at (2, 0, 0), (0, 2, 0) and (0, 0, 2); minimum is -8 at $(-2, 0, 0)$, $(0, -2, 0)$ and $(0, 0, -2)$ **11.** maximum is $2/\sqrt{3}$ at $(1/\sqrt{3}, 2/\sqrt{3}, 3/\sqrt{3})$ and at 3 other points with signs $+--$, $-+-$, $--+$; minimum is $-2/\sqrt{3}$ at $(-1/\sqrt{3}, 2/\sqrt{3}, 3/\sqrt{3})$ and at 3 other points with signs $+-+$, $++-$, $---$. **13.** $\frac{148}{45}$ at $(\frac{74}{45}, \frac{-20}{45}, \frac{28}{45})$ **15.** maximum is the larger of $|a|$ and $|b|$; minimum is the smaller of $|a|$ and $|b|$ **17.** 2 at $(\frac{1}{2}, \frac{1}{2}, \frac{1}{2}, \frac{1}{2})$ **21.** maximum is $2/(3\sqrt{3})$ at $(\sqrt{\frac{2}{3}}, \sqrt{\frac{2}{3}}, \sqrt{\frac{1}{3}})$ and $(-\sqrt{\frac{2}{3}}, -\sqrt{\frac{2}{3}}, \sqrt{\frac{1}{3}})$ minimum is $-2/(3\sqrt{3})$ at $(\sqrt{\frac{2}{3}}, \sqrt{\frac{2}{3}}, -\sqrt{\frac{1}{3}})$ and $(-\sqrt{\frac{2}{3}}, -\sqrt{\frac{2}{3}}, -\sqrt{\frac{1}{3}})$ **23.** height of cone is $12/\sqrt{5}$ m; height of cylinder is $(50/3\pi) - (9/\sqrt{5})$ m **25.** base is a square: 2×2 m; height is 3 m **27.** $x/y = \frac{1}{2}$

REVIEW EXERCISES FOR CHAPTER FOUR

1. $\{(x, y): |x| \geq |y|\}$; $[0, \infty)$ **3.** $\mathbb{R}^2$; $[-1, 1]$ **5.** $\{(x, y, z): x^2 + y^2 + z^2 > 1\}$ = the "exterior" of the unit sphere; $(0, \infty)$ **7.** parallel straight lines with slopes of -1; $x + y = 1 - z^2$ **9.** parallel straight lines with slopes of $\frac{1}{3}$; $x - 3y = e^z$ **13.** along $y = kx$ limit is $k/(k^2 - 1)$ **15.** Since $|4xy^3/(x^2 + y^4)| = 4|xy^3|/[(|x| - y^2)^2 + 2|x|y^2] \leq 4|xy^3|/(2|x|y^2) = 2|y|$ **17.** $-\frac{1}{4}$ **19.** 5 **21.** $\{(x, y, z): z < x - y + 4\}$, i.e., the half-space "below" but not including the plane $x - y - z = -4$ **23.** 0 **25.** $f_x = -\sin(x - 3y)$; $f_y = 3\sin(x - 3y)$ **27.** $f_x = -y/[(1 + x)^2 + y^2]$; $f_y = (1 + x)/[(1 + x)^2 + y^2]$ **29.** $f_x = -x(x^2 + y^2 + z^2)^{-3/2}$; $f_y = -y(x^2 + y^2 + z^2)^{-3/2}$; $f_z = -z(x^2 + y^2 + z^2)^{-3/2}$

31. $f_x = \dfrac{1}{z}\sec\left(\dfrac{x - y}{z}\right)\tan\left(\dfrac{x - y}{z}\right)$; $f_y = \dfrac{-1}{z}\sec\left(\dfrac{x - y}{z}\right)\tan\left(\dfrac{x - y}{z}\right)$;

$f_z = \dfrac{y - x}{z^2}\sec\left(\dfrac{x - y}{z}\right)\tan\left(\dfrac{x - y}{z}\right)$

33. $f_x = \dfrac{2x}{y^3 + z^4}$; $f_y = \dfrac{-3y^2(x^2 + z^4)}{(y^3 + z^4)^2}$; $f_z = \dfrac{4z^3(y^3 - x^2)}{(y^3 + z^4)^2}$

---

† f is unbounded from above and below, so there is no global maximum or minimum.

35. $f_x = \dfrac{1}{y+z}f;\ f_y = \dfrac{w-x}{(y+z)^2}f;\ f_z = \dfrac{w-x}{(y+z)^2}f;\ f_w = \dfrac{-1}{y+z}f$
37. $f_{xx} = \dfrac{2xy}{(x^2+y^2)^2};\ f_{xy} = \dfrac{y^2-x^2}{(x^2+y^2)^2} = f_{yx};\ f_{yy} = \dfrac{-2xy}{(x^2+y^2)^2}$
39. $f_{xx} = \dfrac{4y}{(x-y)^3};\ f_{xy} = \dfrac{-2(x+y)}{(x-y)^3} = f_{yx};\ f_{yy} = \dfrac{4x}{(x-y)^3}$
41. Let $R = \sqrt{1-x^2-y^2-z^2}$, then $f_{xx} = \dfrac{1+2x^2-y^2-z^2}{R^5};\ f_{xy} = \dfrac{3xy}{R^5} = f_{yx};$
$f_{xz} = \dfrac{3xz}{R^5} = f_{zx};\ f_{yy} = \dfrac{1-x^2+2y^2-z^2}{R^5};\ f_{yz} = \dfrac{3yz}{R^5} = f_{zy};\ f_{zz} = \dfrac{1-x^2-y^2+2z^2}{R^5}$
43. $f_{zxx} = 0 = f_{zxxy} = f_{zxxyz}$ **45.** $\frac{1}{2}\mathbf{i} - \frac{1}{2}\mathbf{j}$ **47.** $-\frac{1}{2}\mathbf{i} + \mathbf{j}$ **49.** $\frac{1}{12}\mathbf{i} - \frac{1}{12}\mathbf{j} - \frac{1}{48}\mathbf{k}$
51. $-3\mathbf{j} - 4\mathbf{k}$ **53.** $(t^2+2t)/(1+t)^2\sqrt{1-t^4/(1+t)^2}$ **55.** $2r/(r-s)^2$
57. $[\cos(e^{r+s}-e^{r-s})][e^{r+s}+e^{r-s}]$ **59.** $\partial w/\partial s = 4r^4s^3t^2 + 2r^4st^{-2}$;
$\partial w/\partial t = 2r^4s^4t - 2r^4s^2t^{-3}$ **61.** $x+y+z=3;\ x=y=z$ **63.** $3x-y+5z=15$;
$(x+1)/3 = (y-2)/(-1) = (z-4)/5$ **65.** $-3x+6y-2z=18$;
$(x+2)/(-3) = (y-1)/6 = (z+3)/(-2)$ **67.** $-3/\sqrt{2}$ **69.** $-1/(2\sqrt{13})$
71. $9/(6^{3/2}\sqrt{14})$ **73.** $-1/|\mathbf{x}|^3$ is a potential function **75.** $3x^2y^2\,\Delta x + 2x^3y\,\Delta y$
77. $\Delta x/2\sqrt{(x+1)(y-1)} - (\Delta y)\sqrt{x+1}/2(y-1)^{3/2}$
79. $[1/(x-y+4z)](\Delta x - \Delta y + 4\,\Delta z)$
81. $\frac{4}{7} + (0.03/7) - \frac{4}{49}(-0.03) \approx 0.578163$ (actual value ≈ 0.578192)
83. $54 + \frac{27}{4}(-0.03) + 9(0.05-0.03) = 53.9775$ (actual value ≈ 53.97654)
85. $xy^2 + x + C$ **87.** not exact **89.** $xy^3z^5 + \frac{3}{2}x^2 - \frac{8}{3}y^{3/2} - \cos z + C$
91. $(0, 0)$ is a local minimum
93. $(\sqrt{2}, 1/\sqrt{2})$ is a local minimum; $(\sqrt{2}, -1/\sqrt{2})$ and $(-\sqrt{2}, 1/\sqrt{2})$ are saddle points; $(-\sqrt{2}, -1/\sqrt{2})$ is a local maximum. **95.** $(2^{-1/5}, 2^{3/10})$ and $(2^{-1/5}, -2^{3/10})$ are local minima.
97. $8/\sqrt{11}$ at $(\frac{14}{11}, \frac{-3}{11}, \frac{20}{11})$ **99.** max $V = \frac{1}{2}(\frac{10}{3})^{3/2}$ if base is a square $\sqrt{\frac{10}{3}} \times \sqrt{\frac{10}{3}}$ and height $= \frac{1}{2}\sqrt{\frac{10}{3}}$ **101.** vertices are $(\pm 6/\sqrt{3}, \pm 2/\sqrt{3}, \pm 3/\sqrt{3})$; dimensions are $12/\sqrt{3} \times 4/\sqrt{3} \times 6/\sqrt{3}$
103. dom $f = \mathbb{R}^4$, range $f = \mathbb{R}^+$ **105.** dom $f = \{(x_1, x_2, x_3, x_4): x_3^2 + x_4^2 \neq 0\}$, range $f = \mathbb{R}^+$
109. -21 **111.** $\dfrac{\partial f}{\partial x_1} = 3x_1^2x_2^2x_3 - \dfrac{1}{x_1+2x_2-x_4} - \dfrac{x_4^5}{x_1^2};\ \dfrac{\partial f}{\partial x_2} = 2x_1^3x_2x_3 - \dfrac{2}{x_1+2x_2-x_4};$
$\dfrac{\partial f}{\partial x_3} = x_1^3x_2^2;\ \dfrac{\partial f}{\partial x_4} = \dfrac{1}{x_1+2x_2-x_4} + \dfrac{5x_4^4}{x_1}$
113. $f_{12} = 6x_1^2x_2x_3 + 2(x_1+2x_2-x_4)^{-2};\ f_{31} = 3x_1^2x_2^2;\ f_{124} = 4(x_1+2x_2-x_4)^{-3}$
115. $\mathbf{f}'(t) = \left(\dfrac{1}{2\sqrt{t}}, \dfrac{5}{3}t^{2/3}, -\dfrac{1}{t^2}, 5t^4 - 2\right)$ **117.** $\displaystyle\int_0^1 \sqrt{2+8t^2}\,dt = \frac{\sqrt{10}}{2} + \frac{1}{\sqrt{8}}\ln(2+\sqrt{5})$
119. $\nabla f = \left(\dfrac{x_2}{x_3-x_4}, \dfrac{x_1}{x_3-x_4}, -\dfrac{x_1x_2}{(x_3-x_4)^2}, \dfrac{x_1x_2}{(x_3-x_4)^2}\right)$

CHAPTER FIVE

PROBLEMS 5.1

1. $\frac{45}{2}$ **3.** 0 **5.** 16 **7.** 39 **9.** 0 **11.** $\frac{15}{2}$ **13.** 10 **15.** $0 \le I \le 6 (I = \frac{41}{36})$
17. $-\sqrt{2}\pi/3 \le I \le \sqrt{2}\pi/3$ (These are crude bounds obtained from $|x-y| \le \sqrt{2}$ and $1/(4-x^2-y^2) \le \frac{1}{3}$ on the unit disk.); by symmetry, $I = 0$.
19. $0 \le I \le \frac{1}{4}\ln 2 \approx 0.173 (I = \frac{1}{8} = .125)$

PROBLEMS 5.2

1. $\frac{2}{3}$ **3.** $e^{-5} - e^{-1} + e^2 + e^{-2}$ **5.** -31 **7.** $\frac{162}{5}$ **9.** $\frac{16}{3}$ **11.** $\frac{20}{3}$
13. $\frac{1}{2}(e^{19} - e^{17} - e^3 + e)$ **15.** $\frac{1}{3} + \pi/16$ **17.** $\int_0^{1/2}\int_x^{1-x} (x+2y)\,dy\,dx = \frac{7}{24}$

19. $\int_0^{1/\sqrt{2}} \int_{x^2}^{1-x^2} (x^2 + y)\, dy\, dx = \sqrt{2}/5$
21. $\int_1^2 \int_1^y (y/\sqrt{x^2+y^2})\, dx\, dy = \int_1^2 \int_x^2 (y/\sqrt{x^2+y^2})\, dy\, dx = \int_1^2 (\sqrt{x^2+4} - x\sqrt{2})\, dx$
$= 1/\sqrt{2} - \frac{1}{2}\sqrt{5} + 2\ln[(2+2\sqrt{2})/(1+\sqrt{5})]$
23. $\int_0^\infty \int_0^\infty (x+y)e^{-(x+y)}\, dy\, dx = \int_0^\infty (1+y)e^{-y}\, dy = 2$ **25.** (a) $\int_{-5}^8 \int_0^4 (x+y)\, dx\, dy$
(b) 182 (c) region is a rectangle **27.** $\int_0^1 \int_y^1 dx\, dy$ (b) $\frac{1}{2}$
(c) triangle with vertices at (0, 0), (1, 0) and (1, 1) **29.** $\int_0^1 \int_0^{\cos^{-1} x} y\, dy\, dx$ (b) $(\pi/2) - 1$
(c) region in first quadrant bounded by x-axis, y-axis, and curve $y = \cos^{-1} x$
(vertices are (0, 0), $(\pi/2, 0)$, (1, 0)) **31.** (a) $\int_0^1 \int_{y^3}^{y^2} (1 + y^6)\, dx\, dy$ (b) $\frac{17}{180}$
(c) region bounded by $y = x^{1/2}$ and $y = x^{1/3}$ (the curves meet at (0, 0) and (1, 1))
33. (a) $\int_0^\infty \int_0^y (1+y^2)^{-7/5}\, dx\, dy$ (b) $\frac{5}{4}$
(c) region is the "triangular" part of the first quadrant above the line $y = x$
35. $\frac{9}{2}$ **37.** $\frac{128}{3}$ **39.** 48π **41.** $2\int_0^2 \int_0^x \sqrt{2-x}\, dy\, dx = 32\sqrt{2}/15$
43. $\int_{(-1-\sqrt{13})/2}^{(-1+\sqrt{13})/2} \int_{x^2}^{3-x} 2\sqrt{y - x^2}\, dy\, dx = \frac{4}{3}\int_{(-1-\sqrt{13})/2}^{(-1+\sqrt{13})/2} (3 - x - x^2)^{3/2}\, dx = 169\pi/32$
(Note: Limits on x were obtained by first setting $z = 0$ to obtain $y = x^2$ and then equating this with the line $y = -x + 3$.)
45. $\int_{-\sqrt{3}}^{-1} \int_0^{3-x^2} dy\, dx + \int_{-1}^1 \int_{x^3+1}^{3-x^2} dy\, dx = 2\sqrt{3} + \frac{2}{3}$;
also $\int_{-1}^1 \int_0^{x^3+1} dy\, dx + \int_1^{\sqrt{3}} \int_0^{3-x^2} dy\, dx = 2\sqrt{3} - \frac{2}{3}$
49. The surface $z = \sqrt{x^2+y^2}$ is a cone; the region in the first quadrant bounded by the curves $y = \sqrt{x}$ and $y = x^2$ is the directrix of a cylinder whose generatrix is the z-axis; the solid in question is bounded below by the xy-plane, above by the cone, and whose "walls" are the sides of the cylinder.

PROBLEMS 5.3

1. $\frac{16}{3}$; $(\frac{51}{32}, 0)$ (See Problem 19.2.9)
3. $(\sqrt{3} - 1)/12$; $(\frac{1}{2} + (\pi/6) - [\pi/6(\sqrt{3} - 1)], \frac{1}{3} - [\pi/18(\sqrt{3} - 1)]) =$
$(\frac{1}{2} - \pi(\sqrt{3} - 1)/12, \frac{1}{3} - \pi(\sqrt{3} + 1)/36)$
5. $(3e^4 + 1)(e^{-1} - e^{-3})$; $\left(\dfrac{10e^4 - 2}{3e^4 + 1}, \dfrac{2e^{-2} - 4e^{-3}}{e^{-1} - e^{-3}}\right)$
7. $(\pi/16) + \frac{1}{3}$; $(62/(80 + 15\pi), (16 + 15\pi)/(80 + 15\pi))$ **9.** $\frac{5}{24}$; $(\frac{11}{20}, \frac{1}{5})$ **11.** $\frac{1}{2}$; (1, 1)
13. $4\sqrt{2}\pi^2$ **15.** $6\pi^2 a^2 b$

PROBLEMS 5.4

1. $\frac{2}{3}\pi a^3$ **3.** $\int_{-\pi/2}^{\pi/2} \int_0^{2\cos\theta} (3 - r)r\, dr\, d\theta = 3\pi - \frac{32}{9}$ (Note that $\int_0^\pi \int_0^{2\cos\theta} (3 - r)r\, dr\, d\theta =$ 3π is *wrong* since the cone $z = 3 - r$ is defined only for $r > 0$.)
5. $\frac{32}{15}\pi^6 a^5$ **7.** $(4^2)(3\pi/2) = 24\pi$ **9.** 11π **11.** $2\int_0^{\pi/2} \int_0^{2\sqrt{\sin 2\theta}} r\, dr\, d\theta = 6$
17. 6 **19.** $4\int_0^{\pi/2} \int_0^2 [r - (r^2/2)]r\, dr\, d\theta = 4\pi/3$ **21.** $\pi(1 - e^{-1})$
23. area $= 19\pi/2$; centroid $= (0, \frac{37}{38})_{\text{rectangular}} = (\frac{37}{38}, \pi/2)_{\text{polar}}$

PROBLEMS 5.5

1. $2\sqrt{6}$ **3.** $(\pi/2)\sqrt{1 + a^2 + b^2}$ **5.** $2\int_0^1 \int_1^2 \sqrt{1 + \frac{4}{9}x^{-2/3}}\, dy\, dx = \frac{2}{27}(13^{3/2} - 8)$
(Note that $\int_{-1}^1 \int_1^2 \sqrt{1 + \frac{4}{9}x^{-2/3}}\, dy\, dx$ is *wrong* since the integrand is not defined at $x = 0$
7. 132 **9.** $\int_0^1 \int_0^2 \sqrt{1 + 9x}\, dy\, dx = \frac{4}{27}(10^{3/2} - 1)$ (This is the surface area for $z \geq -1$.)
11. $2\pi a^2$ **13.** 16
15. $\int_{-4}^4 \int_{-\sqrt{16-y^2}}^{\sqrt{16-y^2}} \sqrt{1 + (x^2/4) + (y^2/4)}\, dx\, dy = \frac{1}{2}\int_0^{2\pi} \int_0^4 r\sqrt{4 + r^2}\, dr\, d\theta = (\pi/3)(20^{3/2} - 8)$
17. $\int_{-1}^1 \int_{-\sqrt{1-y^2}}^{\sqrt{1-y^2}} \sqrt{1 + 9x^4 + 9y^4}\, dx\, dy$
19. $\int_0^2 \int_x^{4-x} \sqrt{1 + 1/2(1 + x + y)}\, dy\, dx = (\int_0^2 \int_0^y + \int_2^4 \int_0^{4-y}) \sqrt{1 + 1/2(1 + x + y)}\, dx\, dy$
21. $2\displaystyle\int_{-a}^{a} \int_{-b\sqrt{1-x^2/a^2}}^{b\sqrt{1-x^2/a^2}} \sqrt{\frac{1 + (c^2/a^2 - 1)(x^2/a^2) + (c^2/b^2 - 1)(y^2/b^2)}{1 - (x^2/a^2) - (y^2/b^2)}}\, dy\, dx$

PROBLEMS 5.6

1. $\frac{1}{8}$ **3.** $(a_2 - a_1)(b_2 - b_1)(c_2 - c_1)$ **5.** $6(\frac{31}{5} + \frac{127}{7})$

7. $\int_{-\pi/2}^{\pi/2}\int_0^1\int_{-\sqrt{1-r^2}}^{\sqrt{1-r^2}} z^2\, r\, dz\, dr\, d\theta = 2\pi/15$ **9.** (a) $\int_0^1\int_0^{\sqrt{1-z^2}}\int_z^{\sqrt{1-y^2}} yz\, dx\, dy\, dz$

(b) $\int_0^1\int_0^x\int_0^{\sqrt{1-x^2}} yz\, dy\, dz\, dx$ **11.** $\frac{1}{6}$ **13.** $\int_0^3\int_0^{\sqrt{9-x^2}}\int_0^{4-x} dy\, dz\, dx = 9(\pi - 1)$

15. $\int_{-\sqrt{12}}^{\sqrt{12}}\int_{-\sqrt{12-x^2}}^{\sqrt{12-x^2}}\int_2^{\sqrt{16-x^2-y^2}} dz\, dy\, dx = 40\pi/3$ **17.** $\frac{4}{3}\pi abc$

19. $\mu = \frac{1}{24}$; $(\frac{2}{5}, \frac{1}{5}, \frac{1}{5})$ **21.** $\mu = \frac{207}{8}$; $(\frac{108}{115}, \frac{176}{115}, \frac{18}{23}\pi - \frac{72}{115})$

23. $\mu = (4\pi/15)abc(\alpha a^2 + \beta b^2 + \gamma c^2)$; $(0, 0, 0)$ **25.** $(a/4, b/4, c/4)$ **27.** $(0, \frac{15}{7}, \frac{20}{7})$

29. $\mu = \int_0^1\int_0^y\int_0^{xy} (1 + 2z)\, dz\, dx\, dy = \frac{13}{72}$ (a) $(\bar{x}, \bar{y}, \bar{z}) = (\frac{258}{455}, \frac{372}{455}, \frac{7}{26})$
(b) $\bar{x} \approx 0.57$, $\bar{y} \approx 0.82$, $\bar{z} \approx 0.27$, and $(\bar{x})(\bar{y}) \approx 0.46$ so that $0 < \bar{z} < \overline{xy}$, $0 < \bar{x} < \bar{y}$, and $0 < \bar{y} < 1$

PROBLEMS 5.7

1. $\int_0^{\pi}\int_0^{2\cos\theta}\int_{-\sqrt{4-r^2}}^{\sqrt{4-r^2}} r\, dz\, dr\, d\theta = \frac{16}{3}[\pi - \frac{4}{3}]$

3. $\mu = \int_0^{\pi}\int_0^{2\cos\theta} (2\alpha/3)r(4 - r^2)^{3/2}\, dr\, d\theta = \frac{64}{15}\alpha(\pi - \frac{16}{15})$; $(32/[21(\pi - \frac{16}{15})], 0, 0)$

5. $\mu = \int_0^{2\pi}\int_0^2\int_0^{4-r^2} \alpha rz\, dz\, dr\, d\theta = \frac{32}{3}\alpha\pi$; $(0, 0, 2)$

7. $V = \int_0^{\pi}\int_0^{\sin\theta}\int_{r^2}^{r\sin\theta} r\, dz\, dr\, d\theta = \pi/32$; $(0, \frac{1}{2}, \frac{5}{12})$

9. $\mu = \int_0^{2\pi}\int_0^2\int_{z/4}^{z} \alpha r^2\, dr\, dz\, d\theta = \frac{21}{8}\alpha\pi$; $(0, 0, \frac{8}{5})$ **11.** $\int_0^{\pi/2}\int_0^1\int_r^{\sqrt{2-r^2}} rz^3\, dz\, dr\, d\theta = \frac{\pi}{8}$

13. $V = \int_0^2\int_0^{2\pi}\int_{\pi/4}^{3\pi/4} \rho^2 \sin\varphi\, d\varphi\, d\theta\, d\rho = (16\sqrt{2}/3)\pi$

15. $\int_0^{2\pi}\int_0^{\pi}\int_a^b (\alpha/\rho)\,\rho^2 \sin\varphi\, d\rho\, d\varphi\, d\theta = 2\pi\alpha(b^2 - a^2)$. Here the density is given by α/ρ, where α is constant of proportionality.

17. $V = \int_0^1\int_0^{\pi}\int_0^{\pi/6} \rho^2 \sin\varphi\, d\theta\, d\varphi\, d\rho = \pi/9$

19. $I = \int_0^3\int_0^{2\pi}\int_0^{\pi} \rho^3 \cdot \rho^2 \sin\varphi\, d\varphi\, d\theta\, d\rho = 486\pi$

21. $I = \int_0^a\int_0^{\pi/2}\int_0^{\pi/2} \frac{(\rho\cos\varphi)^3}{\rho\sin\varphi} \cdot \rho^2 \sin\varphi\, d\varphi\, d\theta\, d\rho = \pi a^5/15$

REVIEW EXERCISES FOR CHAPTER FIVE

1. $\frac{4}{3}$ **3.** $\frac{67}{3}$ **5.** -24 **7.** $\frac{1}{3} + \pi/16$ **9.** $-\pi \le I \le \pi$

11. $\int_0^3\int_0^{\sqrt{9-y^2}} (9 - y^2)^{3/2}\, dx\, dy = \frac{648}{5}$ **13.** 6 (See Problem 5.6.12)

15. $400\pi/3$ (See Problem 5.6.17) **17.** $(0, 0)$ **19.** $\mu = \frac{5}{2}$; $(\frac{28}{75}, \frac{19}{15})$

21. 6π **23.** $\int_0^{2\pi}\int_0^4\int_{r^2/4}^{r} r\, dz\, dr\, d\theta = 32\pi/3$ **25.** $\sqrt{35}/12$ **27.** 32π

29. $[(96)(31)/5] + [(127)(24)/7] = 36072/35$ **31.** 1 (See Problem 5.6.12)

33. $\frac{11}{10}$ **35.** $V = \frac{128}{15}$; $(\frac{12}{7}, 0, \frac{16}{7})$ **37.** $V = \frac{2}{3}a^3(\pi - \frac{4}{3})$; $(0, 4a/5(\pi - 4/3), 0)$
(Note that the equation of the cylinder is $r = a\sin\theta$.)

39. $\int_0^1 \int_{x^2}^x \int_{-\sqrt{z-x^2}}^{\sqrt{z-x^2}} dy\, dz\, dx = \int_0^1 \frac{4}{3}(x - x^2)^{3/2}\, dx = \pi/32$

41. $\int_{\pi/4}^{3\pi/4} \int_0^{2\pi} \int_0^3 \rho^2 \sin\phi\, d\rho\, d\theta\, d\phi = 18\sqrt{2}\pi$ **43.** $\int_0^{\pi} \int_0^{\pi/3} \int_0^2 \rho^2 \sin\phi\, d\rho\, d\theta\, d\phi = 16\pi/9$

CHAPTER SIX

PROBLEMS 6.1

1. $2(x+y)(\mathbf{i}+\mathbf{j})$ **3.** $\sin(x-y)(-\mathbf{i}+\mathbf{j})$ **5.** $(x/\sqrt{x^2+y^3})\mathbf{i} + (3y^2/2\sqrt{x^2+y^3})\mathbf{j}$
7. $-y\sec^2(y-x)\mathbf{i} + [\tan(y-x) + y\sec^2(y-x)]\mathbf{j}$
9. $\sec(x+3y)\tan(x+3y)(\mathbf{i}+3\mathbf{j})$ **11.** $[4xy/(x^2+y^2)^2](y\mathbf{i} - x\mathbf{j})$
13. $yz\mathbf{i} + xz\mathbf{j} + xy\mathbf{k}$ **15.** $\left(\frac{1}{3xy}\right)\left[\left(\frac{x^2+y^2-z^2}{x}\right)\mathbf{i} - \left(\frac{x^2+y^2+z^2}{y}\right)\mathbf{j} + 2z\mathbf{k}\right]$
17. $y^2\mathbf{i} + 2y(x+z^3)\mathbf{j} + 3y^2z^2\mathbf{k}$ **19.** $\sin y \ln z\,\mathbf{i} + x\cos y \ln z\,\mathbf{j} + (x/z)\sin y\,\mathbf{k}$
21. $(\cosh z - y\cos x)\mathbf{i} - \sin x\mathbf{j} + x\sinh z\mathbf{k}$ **23.** $-\omega y\mathbf{i} + \omega x\mathbf{j}$

PROBLEMS 6.2

1. $\mathbf{x} = (t, 2t);\ I = \int_0^2 9t^2\, dt = 24$ **3.** $\mathbf{x} = (t, 2t-4);\ I = \int_1^2 (2t^2 - 2t - 8)\, dt = -\frac{19}{3}$
5. $\mathbf{x} = (\cos\theta, \sin\theta);\ I = \int_0^{2\pi} [(\cos\theta)(\sin\theta)(-\sin\theta) + (\sin\theta - \cos\theta)\cos\theta]\, d\theta = -\pi$
7. $\int_0^1 0\, dt + \int_0^1 (t-1)\, dt - \int_0^1 t^2\, dt = -\frac{5}{6}$ **9.** $\mathbf{x} = (3\cos\theta, -\sin\theta);$
$I = \int_{\pi/2}^{3\pi/2} [(9\cos^2\theta - 2\sin\theta)(-3\sin\theta) + (-\sin^2\theta)(-\cos\theta)]\, d\theta = 3\pi - \frac{2}{3}$
11. $\int_0^1 e^t\, dt + \int_0^1 (-e + e^{1-2t})\, dt + \int_1^0 e^{-t}\, dt = \frac{1}{2}(e + e^{-1}) - 2 = \cosh 1 - 2$
13. $\mathbf{x} = (2+2t, 1+5t);\ I = \int_0^1 \left\{\frac{1+5t}{[2(1+t)]^2}\cdot 2 + \frac{2(1+t)}{[1+5t]^2}\cdot 5\right\} dt = \frac{5}{2}\ln 2 + \frac{2}{5}\ln 6 + \frac{1}{3}$
15. $W = \int_0^{\pi/2} [\sin^3 t\cos t + e^{2t}\sin t]\, dt = \frac{9}{20} + (2e^{\pi}/5) \approx 9.71$ J
17. $W = -\int_0^{2\pi} [(\cos\theta\sin\theta)(-\sin\theta) + (2\cos^3\theta - \sin\theta)(\cos\theta)]\, d\theta = -3\pi/2$ J
19. $W = \int_0^1 0\, dt + \int_0^1 t^2\, dt + \int_1^0 (2t^2 + t^2)\, dt = -\frac{2}{3}$ J
21. $\mathbf{F} = \nabla(x^2y + y);\ I = 14$ **23.** $\mathbf{F} = \nabla x\sin(x+y);\ I = \pi/6$
25. $\mathbf{F} = \nabla x^2\cos y;\ I = \pi^2/4$ **27.** $\mathbf{F} = \nabla xe^y;\ I = 5e^7$

PROBLEMS 6.3

1. 2 **3.** $\int_0^1 \int_0^{1-y} 2e^x \sin y\, dx\, dy = \cos 1 - \sin 1 + e - 2$ **5.** $2(e^2 - 1)(1 - \cos 1)$
7. $\iint_{\text{disk}} (-4y) = 0$ **9.** $(\pi/8) - (\pi/3) + (\pi/3\sqrt{2}) = \pi[(4\sqrt{2} - 5)/24]$
11. $\frac{242}{5} - \frac{26}{3} = \frac{596}{15}$ **13.** $\iint_\Omega (b-a)\, dA = (b-a)(\text{area of } \Omega)$
15. 0 (exact differential with $f = -2x^2/\sqrt{1+y^2}$) **17.** Use (14). Note that the line from (a_1, b_1) to (a_2, b_2) is parametrized as $\mathbf{x} = (a_1 + t(a_2 - a_1), b_1 + t(b_2 - b_1))$ for $0 \le t \le 1$. This yields area as $\frac{1}{2}|(a_1 + a_2)(b_2 - b_1) + (a_2 + a_3)(b_3 - b_2) + (a_3 + a_1)(b_1 - b_3)|$
19. $\frac{1}{2}|(a_1 + a_2)(b_2 - b_1) + (a_2 + a_3)(b_3 - b_2) + (a_3 + a_4)(b_4 - b_3) + (a_4 + a_1)(b_1 - b_4)|$ (See answer to Problem 17.) **21.** (a) $2x - 2y$ (b) 0 (c) 0 (d) 0
23. (a) $3(x^2 - y^2)$ (b) 0 (c) 0 (d) 0 **25.** (a) -2 (b) -2π (c) 0 (d) 0
27. curl $\mathbf{F} = 0$ **29.** curl $\mathbf{F}$ is an odd function **31.** curl $\mathbf{F} = 0$ **33.** div $\mathbf{F} = 0$

PROBLEMS 6.4

1. $\int_0^1 [(t)(1) + (t^2)(2t) + (t^3)(3t^2)]\, dt = \frac{3}{2}$
3. $\int_0^{\pi/2} [(\cos^2 t)(-\sin t) + (\sin^2 t)(\cos t) + (t^2)(1)]\, dt = \pi^3/24$ **5.** $\mathbf{F} = \nabla xy^2z^4;\ I = 12$
7. $\mathbf{F} = \nabla(\ln|x + 2y + 3z| - \frac{3}{2}x^2 + \frac{1}{3}y^3);\ I = \ln 20 + \frac{9}{2}$

PROBLEMS 6.5

1. $(5^{3/2} - 1)/6$ **3.** $(9\sqrt{5}/16) - (\frac{1}{32})\ln(2 + \sqrt{5})$ **5.** $\frac{14}{3}$ **7.** 0 **9.** $-7\sqrt{14}/108$
11. $76\sqrt{2}\pi$ **13.** $64 + 32\sqrt{21}/3$; note that a tetrahedron has four faces. The integral over the sloping face is $32\sqrt{21}/3$. The other three integrals are 0, $\frac{64}{3}$, and $\frac{128}{3}$.
15. 0 **17.** 54π **19.** $\alpha\sqrt{3}/12$ (α is proportionally constant) **23.** $-\frac{32}{3}$
25. 2π **27.** 0 **29.** $\pi/3$; note that $\mathbf{n}\cdot\mathbf{j} < 0$ on the lateral surface of the cone.
33. $\frac{419}{3360}$

PROBLEMS 6.6

1. div $\mathbf{F} = 2(x + y + z)$; curl $\mathbf{F} = \mathbf{0}$ **3.** div $\mathbf{F} = 0$; curl $\mathbf{F} = \mathbf{0}$
5. div $\mathbf{F} = x + y + z$; curl $\mathbf{F} = -y\mathbf{i} - z\mathbf{j} - x\mathbf{k}$
7. div $\mathbf{F} = 0$; curl $\mathbf{F} = x(e^{xy} - e^{xz})\mathbf{i} + y(e^{yz} - e^{xy})\mathbf{j} + z(e^{xz} - e^{yz})\mathbf{k}$
9. div $\mathbf{F} = (1/y) + (1/z) + (1/x)$; curl $\mathbf{F} = (y/z^2)\mathbf{i} + (z/x^2)\mathbf{j} + (x/y^2)\mathbf{k}$ **19.** 0 **21.** 0
23. Problems 19 and 21 have harmonic functions; 20 and 22 do not **29.** (b) $\mathbf{F} = \nabla(2x^2yz)$

PROBLEMS 6.7

1. -3π **3.** -6 **5.** 108π **7.** $21 - \sin 3$
11. the integral of $\mathbf{F}$ along the straight line segment from (6, 0, 0) to (0, 3, 0) is -18, from (0, 3, 0) to (0, 0, 2) is -4, from (0, 0, 2) to (6, 0, 0) is -24; the integral of $\mathbf{F}$ along the triangle is -46 **13.** the integral equals -18π

PROBLEMS 6.8

1. 4π **3.** 36π **5.** 0 **7.** 2 **9.** 108π **11.** $\frac{1}{6}$ **13.** $\frac{1}{2}$ **15.** $\frac{184}{35}$

PROBLEMS 6.9

1. $[\partial(x, y)/\partial(u, v)] = -2$ **3.** $[\partial(x, y)/\partial(u, v)] = 4(u^2 + v^2)$
5. $[\partial(x, y)/\partial(u, v)] = -2$ **7.** $[\partial(x, y)/\partial(u, v)] = -(a^2 + b^2)$
9. $[\partial(x, y)/\partial(u, v)] = (1 - uv)e^{u+v}$ **11.** $[\partial(x, y)/\partial(u, v)] = [1/(u + v)][(1/v) - (1/u)]$
13. $[\partial(x, y)/\partial(u, v)] = \csc u \sec v(1 + uv \cot u \tan v)$ **15.** $[\partial(x, y, z)/\partial(u, v, w)] = 0$
17. $[\partial(x, y, z)/\partial(u, v, w)] = 2(u^2v - uv^2 + v^2w - vw^2 + w^2u - wu^2)$
19. $[\partial(x, y, z)/\partial(u, v, w)] = e^{u+v+w}$
21. $\int_0^1 \int_y^1 xy\, dx\, dy = \int_{-1/2}^0 \int_{-v}^{1+v} (u^2 - v^2)(2)\, du\, dv = \frac{1}{8}$ **23.** 2π **25.** $\frac{128}{15}$ **27.** $4\pi abc/3$
29. $\frac{594}{5}$

REVIEW EXERCISES FOR CHAPTER SIX

1. $3(x + y)^2(\mathbf{i} + \mathbf{j})$ **3.** $\frac{1}{2}\sqrt{y/x}\,\mathbf{i} + \frac{1}{2}\sqrt{x/y}\,\mathbf{j}$ **5.** $2x\mathbf{i} + 2y\mathbf{j} + 2z\mathbf{k}$
9. $\mathbf{F} = \nabla((x^3/3) + (y^3/3))$; $I = \frac{2}{3}$ **11.** $-\frac{1}{2}$ **13.** $\frac{5}{3}$ **15.** (a) $\mathbf{F} = \nabla(xe^{xy})$ (b) 1
17. $\int_0^1 \int_x^1 (y^2 - x^2)\, dy\, dx = \frac{1}{6}$ (using Green's theorem)
19. $\int_{-1}^1 \int_{x^2}^1 (x/\sqrt{1 + x^2})\, dy\, dx = 0$ **21.** (a) curl $\mathbf{F} = 0$ (b) 0 (c) div $\mathbf{F} = y^2 + x^2$ (d) 8π
23. $\text{curl}[(\cos x^2)\mathbf{i} + e^y\mathbf{j}] = 0$ **25.** $\frac{3}{2}$ **27.** $\mathbf{F} = \nabla\left(\frac{-xy}{z}\right)$; $I = \frac{5}{3}$
29. $(\frac{91}{48})\sqrt{5} + (\frac{61}{96}) \ln(2 + \sqrt{5})$ **31.** 0 **33.** $567\sqrt{2}\pi/4$
35. $\alpha\pi^2/2$ (α is proportionality constant) **37.** 4π **39.** 0
41. div $\mathbf{F} = 3$; curl $\mathbf{F} = \mathbf{i} + \mathbf{j} + \mathbf{k}$ **43.** div $\mathbf{F} = (1/x) + (1/y) + (1/z)$; curl $\mathbf{F} = \mathbf{0}$
45. div $\mathbf{F} = -\sin z$; curl $\mathbf{F} = (-\sin x + \sin y)\mathbf{k}$ **47.** 0 **49.** 0
51. $4(a + b + c)\pi/3$ **53.** 0 **55.** $\frac{5}{4}$ **57.** $[\partial(x, y)/\partial(u, v)] = 18u^2v^2$
59. $[\partial(x, y)/(u, v)] = (uv - 1)e^{u+v}$ **61.** $[\partial(x, y)/\partial(u, v)] = v[\sec^2(uv)][u \sec^2 u - \tan u]$
63. $[\partial(x, y, z)/\partial(u, v, w)] = 2uvw$ **65.** $\int_2^3 \int_{-v}^{v} e^{(u/v)} \frac{1}{2}\, du\, dv = \frac{5}{4}(e - e^{-1}) = \frac{5}{2} \sinh 1$

CHAPTER SEVEN

PROBLEMS 7.1

1. $\begin{pmatrix} 3 & 9 \\ 6 & 15 \\ -3 & 6 \end{pmatrix}$ **3.** $\begin{pmatrix} 2 & 2 \\ -2 & -1 \\ 6 & -1 \end{pmatrix}$ **5.** $\begin{pmatrix} 0 & 0 \\ 0 & 0 \\ 0 & 0 \end{pmatrix}$ **7.** $\begin{pmatrix} -2 & 4 \\ 7 & 15 \\ -15 & 10 \end{pmatrix}$ **9.** $\begin{pmatrix} 4 & 10 \\ 17 & 22 \\ -9 & 1 \end{pmatrix}$

11. $\begin{pmatrix} 0 & 6 \\ 5 & 14 \\ -9 & 9 \end{pmatrix}$ **13.** $\begin{pmatrix} 1 & -5 & 0 \\ -3 & 4 & -5 \\ -14 & 13 & -1 \end{pmatrix}$ **15.** $\begin{pmatrix} 1 & 1 & 5 \\ 9 & 5 & 10 \\ 7 & -7 & 3 \end{pmatrix}$ **17.** $\begin{pmatrix} -1 & -1 & -1 \\ -3 & -3 & -10 \\ -7 & 3 & 5 \end{pmatrix}$

19. $\begin{pmatrix} -1 & -1 & -5 \\ -9 & -5 & -10 \\ -7 & 7 & -3 \end{pmatrix}$ **25.** $\begin{pmatrix} 1 & 1 & 1 & 0 \\ 1 & 1 & 1 & 0 \\ 1 & 1 & 1 & 1 \\ 0 & 0 & 1 & 1 \end{pmatrix}$ Note: a point is connected to itself

PROBLEMS 7.2

1. $\begin{pmatrix} 8 & 20 \\ -4 & 11 \end{pmatrix}$ **3.** $\begin{pmatrix} -3 & -3 \\ 1 & 3 \end{pmatrix}$ **5.** $\begin{pmatrix} 13 & 35 & 18 \\ 20 & 26 & 20 \end{pmatrix}$ **7.** $\begin{pmatrix} 19 & -17 & 34 \\ 8 & -12 & 20 \\ -8 & -11 & 7 \end{pmatrix}$

9. $\begin{pmatrix} 18 & 15 & 35 \\ 9 & 21 & 13 \\ 10 & 9 & 9 \end{pmatrix}$ **11.** $(7 \quad 16)$ **13.** $\begin{pmatrix} 3 & 15 & -9 & 24 \\ -1 & -5 & 3 & -8 \\ 10 & 50 & -30 & 80 \\ 2 & 10 & -6 & 16 \end{pmatrix}$ **15.** $\begin{pmatrix} a & b & c \\ d & e & f \\ g & h & j \end{pmatrix}$

17. If $D = a_{11}a_{22} - a_{12}a_{21}$, then

$$\begin{pmatrix} b_{11} & b_{21} \\ b_{21} & b_{22} \end{pmatrix} = \begin{pmatrix} a_{22}/D & -a_{12}/D \\ -a_{21}/D & a_{11}/D \end{pmatrix}.$$

19. (a) 3 in Group 1, 4 in Group 2, 5 in Group 3 (b) $\begin{pmatrix} 2 & 1 & 1 & 0 & 0 \\ 1 & 1 & 0 & 1 & 0 \\ 1 & 0 & 2 & 0 & 1 \end{pmatrix}$

21. (a) $\begin{pmatrix} 80{,}000 & 45{,}000 & 40{,}000 \\ 50 & 20 & 10 \end{pmatrix}$ (b) $\begin{pmatrix} 1 \\ 3 \\ 1 \end{pmatrix}$ (c) Money: \$255,000; Shares: 120

23. $\begin{pmatrix} 0 & -8 \\ 32 & 32 \end{pmatrix}$ **25.** $\begin{pmatrix} 11 & 38 \\ 57 & 106 \end{pmatrix}$

27. $A^2 = \begin{pmatrix} 0 & 0 & 1 & 0 & 0 \\ 0 & 0 & 0 & 1 & 0 \\ 0 & 0 & 0 & 0 & 1 \\ 0 & 0 & 0 & 0 & 0 \\ 0 & 0 & 0 & 0 & 0 \end{pmatrix}$ $A^3 = \begin{pmatrix} 0 & 0 & 0 & 1 & 0 \\ 0 & 0 & 0 & 0 & 1 \\ 0 & 0 & 0 & 0 & 0 \\ 0 & 0 & 0 & 0 & 0 \\ 0 & 0 & 0 & 0 & 0 \end{pmatrix}$

$A^4 = \begin{pmatrix} 0 & 0 & 0 & 0 & 1 \\ 0 & 0 & 0 & 0 & 0 \\ 0 & 0 & 0 & 0 & 0 \\ 0 & 0 & 0 & 0 & 0 \\ 0 & 0 & 0 & 0 & 0 \end{pmatrix}$ $A^5 = \begin{pmatrix} 0 & 0 & 0 & 0 & 0 \\ 0 & 0 & 0 & 0 & 0 \\ 0 & 0 & 0 & 0 & 0 \\ 0 & 0 & 0 & 0 & 0 \\ 0 & 0 & 0 & 0 & 0 \end{pmatrix}$

29. $\begin{pmatrix} \frac{11}{90} & \frac{41}{90} & \frac{19}{45} \\ \frac{11}{120} & \frac{71}{120} & \frac{19}{60} \\ \frac{1}{5} & \frac{1}{5} & \frac{3}{5} \end{pmatrix}$; all entries are nonnegative and $\frac{11}{90} + \frac{41}{90} + \frac{19}{45} = \frac{11}{120} + \frac{71}{120} + \frac{19}{60} = \frac{1}{5} + \frac{1}{5} + \frac{3}{5} = 1$

33. (a) $R + \frac{1}{2}R^2 = \begin{pmatrix} 0 & 1 & \frac{1}{2} & \frac{1}{2} \\ 1 & 0 & \frac{3}{2} & 1 \\ 1 & \frac{1}{2} & 0 & 0 \\ \frac{3}{2} & \frac{1}{2} & 1 & 0 \end{pmatrix}$; player 2 > player 4 > player 1 > player 3 (b) score = number of games won plus one-half the number of games that were won by each player that this given player beat

35. $A(B + C) = \begin{pmatrix} 1 & 2 & 4 \\ 3 & -1 & 0 \end{pmatrix}\begin{pmatrix} 1 & 9 \\ 2 & 11 \\ 10 & 1 \end{pmatrix} = \begin{pmatrix} 45 & 35 \\ 1 & 16 \end{pmatrix}$; $AB + AC = \begin{pmatrix} 24 & 14 \\ 7 & 17 \end{pmatrix} + \begin{pmatrix} 21 & 20 \\ -6 & -1 \end{pmatrix} = \begin{pmatrix} 45 & 35 \\ 1 & 16 \end{pmatrix}$

PROBLEMS 7.3

1. $x_1 = -\frac{13}{5}, x_2 = -\frac{11}{5}$ **3.** no solutions
5. $x_1 = \frac{11}{2}, x_2 = -30$ *Note:* Where there were an infinite number of solutions, we wrote the solutions with the last variable chosen arbitrarily. The solutions can be written in other ways as well. **7.** $(2, -3, 1)$ **9.** $(3 + \frac{2}{9}x_3, \frac{8}{9}x_3, x_3)$, x_3 arbitrary
11. $(-9, 30, 14)$ **13.** no solution **15.** $(-\frac{4}{5}x_3, \frac{9}{5}x_3, x_3)$, x_3 arbitrary
17. $(-1, \frac{5}{2} + \frac{1}{2}x_3, x_3)$, x_3 arbitrary **19.** no solution
21. $(\frac{20}{13} - \frac{4}{13}x_4, -\frac{28}{13} + \frac{3}{13}x_4, -\frac{45}{13} + \frac{9}{13}x_4, x_4)$ arbitrary
23. $(18 - 4x_4, -\frac{15}{2} + 2x_4, -31 + 7x_4, x_4)$, x_4 arbitrary **25.** no solution
27. $(0, 0, 0)$ **29.** $(\frac{1}{6}x_3, \frac{5}{6}x_3, x_3)$, x_3 arbitrary **31.** $(0, 0)$
33. $(-4x_4, 2x_4, 7x_4, x_4)$, x_4 arbitrary **35.** $(0, 0)$
37. row echelon form **39.** reduced row echelon form **41.** neither
43. reduced row echelon form **45.** neither

47. row echelon form: $\begin{pmatrix} 1 & -6 \\ 0 & 1 \end{pmatrix}$: reduced row echelon form: $\begin{pmatrix} 1 & 0 \\ 0 & 1 \end{pmatrix}$

49. row echelon form: $\begin{pmatrix} 1 & -2 & 4 \\ 0 & 1 & -\frac{4}{11} \\ 0 & 0 & 1 \end{pmatrix}$: reduced row echelon form: $\begin{pmatrix} 1 & 0 & 0 \\ 0 & 1 & 0 \\ 0 & 0 & 1 \end{pmatrix}$

51. row echelon form: $\begin{pmatrix} 1 & -\frac{7}{2} \\ 0 & 1 \\ 0 & 0 \end{pmatrix}$: reduced row echelon form: $\begin{pmatrix} 1 & 0 \\ 0 & 1 \\ 0 & 0 \end{pmatrix}$

53. $x_1 = 30{,}000 - 5x_3$
$x_2 = x_3 - 5000$
$5000 \le x_3 \le 6000$; no

55. Infinite number of solutions; x = no. of cups; y = no. of saucers; solutions are $(160 - \frac{2}{3}y, y)$, $0 \le y \le 240$ **57.** 32 sodas, 128 milk shakes
59. No unique solution (2 equations in 3 unknowns); if 200 shares of McDonalds', then 300 shares of Eastern and 100 shares of Hilton
61. The row echelon form of the augmented matrix representing this system is

$$\left(\begin{array}{ccc|c} 1 & -\frac{1}{2} & \frac{3}{2} & a/2 \\ 0 & 1 & -\frac{19}{5} & \frac{2}{5}(b - \frac{3}{2}a) \\ 0 & 0 & 0 & -2a + 3b + c \end{array}\right)$$

which is inconsistent if $-2a + 3b + c \ne 0$ or $c \ne 2a - 3b$.
63. $a_{11}a_{22}a_{33} + a_{12}a_{23}a_{31} + a_{13}a_{32}a_{21} - a_{13}a_{22}a_{31} - a_{12}a_{21}a_{33} - a_{11}a_{32}a_{23} \ne 0$
65. $(1.900812947, 4.194110816, -11.348518338)$

PROBLEMS 7.4

1. $\begin{pmatrix} 2 & -1 \\ 4 & 5 \end{pmatrix}\begin{pmatrix} x_1 \\ x_2 \end{pmatrix} = \begin{pmatrix} 3 \\ 7 \end{pmatrix}$ **3.** $\begin{pmatrix} 3 & 6 & -7 \\ 2 & -1 & 3 \end{pmatrix}\begin{pmatrix} x_1 \\ x_2 \\ x_3 \end{pmatrix} = \begin{pmatrix} 0 \\ 1 \end{pmatrix}$

5. $\begin{pmatrix} 0 & 1 & -1 \\ 1 & 0 & 1 \\ 3 & 2 & 0 \end{pmatrix}\begin{pmatrix} x_1 \\ x_2 \\ x_3 \end{pmatrix} = \begin{pmatrix} 7 \\ 2 \\ -5 \end{pmatrix}$

7. $\begin{aligned} x_1 + x_2 - x_3 &= 7 \\ 4x_1 - x_2 + 5x_3 &= 4 \\ 6x_1 + x_2 + 3x_3 &= 20 \end{aligned}$ **9.** $\begin{aligned} 2x_1 \qquad + x_3 &= 2 \\ -3x_1 + 4x_2 \qquad &= 3 \\ 5x_2 + 6x_3 &= 5 \end{aligned}$ **11.** $\begin{aligned} x_1 &= 2 \\ x_2 &= 3 \\ x_3 &= -5 \\ x_4 &= 6 \end{aligned}$

13. $\begin{aligned} 6x_1 + 2x_2 + x_3 &= 2 \\ -2x_1 + 3x_2 + x_3 &= 4 \\ 0x_1 + 0x_2 + 0x_3 &= 2 \end{aligned}$ **15.** $\begin{aligned} 7x_1 + 2x_2 &= 1 \\ 3x_1 + x_2 &= 2 \\ 6x_1 + 9x_2 &= 3 \end{aligned}$

17. $x_1 = 4 - 2x_2 + 4x_3$; x_2, x_3 arbitrary **19.** $x_1 = x_2 = x_3 = 0$ **21.** $\begin{pmatrix} 1 \\ -\frac{1}{3} \\ \frac{1}{2} \\ 4 \end{pmatrix}$

23. $A = \begin{pmatrix} 2 & 0 & 0 \\ 0 & 4 & 0 \\ 0 & 0 & -5 \end{pmatrix}$, $\mathbf{x} = \begin{pmatrix} x_1 \\ x_2 \\ x_3 \end{pmatrix}$, $\mathbf{b} = \begin{pmatrix} 3 \\ 5 \\ 2 \end{pmatrix}$, $\mathbf{x} = \begin{pmatrix} \frac{3}{2} \\ \frac{5}{2} \\ -\frac{2}{5} \end{pmatrix}$

PROBLEMS 7.5

1. $\begin{pmatrix} 2 & -1 \\ -3 & 2 \end{pmatrix}$ **3.** $\begin{pmatrix} 0 & 1 \\ 1 & 0 \end{pmatrix}$ **5.** not invertible **7.** $\begin{pmatrix} \frac{1}{3} & -\frac{1}{3} & -\frac{1}{3} \\ 0 & \frac{1}{2} & 1 \\ 0 & 0 & -1 \end{pmatrix}$

9. not invertible **11.** not invertible **13.** $\begin{pmatrix} \frac{7}{3} & -\frac{1}{3} & -\frac{1}{3} & -\frac{2}{3} \\ \frac{4}{9} & -\frac{1}{9} & -\frac{4}{9} & \frac{1}{9} \\ -\frac{1}{9} & -\frac{2}{9} & \frac{1}{9} & \frac{2}{9} \\ -\frac{5}{3} & \frac{2}{3} & \frac{2}{3} & \frac{1}{3} \end{pmatrix}$

15. $\begin{pmatrix} 0 & 1 & 0 & 2 \\ 1 & -1 & -2 & 2 \\ 0 & 1 & 3 & -3 \\ -2 & 2 & 3 & -2 \end{pmatrix}$

23. $\begin{pmatrix} \sin\theta & \cos\theta & 0 \\ \cos\theta & -\sin\theta & 0 \\ 0 & 0 & 1 \end{pmatrix}$ is its own inverse (since $\sin^2\theta + \cos^2\theta = 1$).

27. $\begin{pmatrix} \frac{1}{2} & -\frac{1}{6} & \frac{7}{30} \\ 0 & \frac{1}{3} & -\frac{4}{15} \\ 0 & 0 & \frac{1}{5} \end{pmatrix}$ **31.** any nonzero multiple of (1, 2) **33.** 3 chairs and 2 tables

35. 4 units of A and 5 units of B

37. (a) $A = \begin{pmatrix} 0.293 & 0 & 0 \\ 0.014 & 0.207 & 0.017 \\ 0.044 & 0.010 & 0.216 \end{pmatrix}$ $I - A = \begin{pmatrix} 0.707 & 0 & 0 \\ -0.014 & 0.793 & -0.017 \\ -0.044 & -0.010 & 0.784 \end{pmatrix}$

(b) $\begin{pmatrix} 18{,}688.826 \\ 22{,}597.857 \\ 3{,}615.161 \end{pmatrix}$ **39.** $\begin{pmatrix} 1 & \frac{1}{2} \\ 0 & 1 \end{pmatrix}$; yes **41.** $\begin{pmatrix} 1 & \frac{2}{3} & \frac{1}{3} \\ 0 & 1 & 1 \\ 0 & 0 & 1 \end{pmatrix}$; yes

43. $\begin{pmatrix} 1 & -\frac{1}{2} & 2 \\ 0 & 1 & -14 \\ 0 & 0 & 0 \end{pmatrix}$; no **45.** $\begin{pmatrix} 1 & 0 & 2 & 3 \\ 0 & 1 & 2 & 7 \\ 0 & 0 & 1 & \frac{10}{7} \\ 0 & 0 & 0 & 0 \end{pmatrix}$; no

PROBLEMS 7.6

1. $\begin{pmatrix} -1 & 6 \\ 4 & 5 \end{pmatrix}$ **3.** $\begin{pmatrix} 2 & -1 & 1 \\ 3 & 2 & 4 \end{pmatrix}$ **5.** $\begin{pmatrix} 1 & -1 & 1 \\ 2 & 0 & 5 \\ 3 & 4 & 5 \end{pmatrix}$ **7.** $\begin{pmatrix} 1 & 0 \\ 0 & 1 \\ 1 & 0 \\ 0 & 1 \end{pmatrix}$ **9.** $\begin{pmatrix} a & d & g \\ b & e & h \\ c & f & j \end{pmatrix}$

REVIEW EXERCISES FOR CHAPTER SEVEN

1. $(\frac{1}{7}, \frac{10}{7})$ **3.** no solution **5.** $(0, 0, 0)$ **7.** $(-\frac{1}{2}, 0, \frac{5}{2})$ **9.** $(\frac{1}{3}x_3, \frac{7}{3}x_3, x_3)$, x_3 arbitrary
11. no solution **13.** $(0, 0, 0, 0)$ **15.** reduced row echelon form **17.** neither
19. row echelon form

21. row echelon form: $\begin{pmatrix} 1 & -1 & 2 & 4 \\ 0 & 1 & 2 & 7 \\ 0 & 0 & 1 & \frac{14}{5} \end{pmatrix}$; reduced row echelon form: $\begin{pmatrix} 1 & 0 & 0 & -\frac{1}{5} \\ 0 & 1 & 0 & \frac{7}{5} \\ 0 & 0 & 1 & \frac{14}{5} \end{pmatrix}$

23. $\begin{pmatrix} 3 & 0 & 7 \\ 0 & 4 & 14 \end{pmatrix}$ **25.** $\begin{pmatrix} 16 & 19 \\ 3 & 29 \end{pmatrix}$ **27.** $\begin{pmatrix} -26 & 16 & 35 \\ -10 & 19 & 30 \\ -42 & 17 & 32 \end{pmatrix}$ **29.** $\begin{pmatrix} 7 \\ 29 \\ 5 \end{pmatrix}$

31. $\begin{pmatrix} 1 & \frac{3}{2} \\ 0 & 1 \end{pmatrix}$; inverse is $\begin{pmatrix} \frac{4}{11} & -\frac{3}{11} \\ \frac{1}{11} & \frac{2}{11} \end{pmatrix}$ **33.** $\begin{pmatrix} 1 & 2 & 0 \\ 0 & 1 & \frac{1}{3} \\ 0 & 0 & 1 \end{pmatrix}$; inverse is $\begin{pmatrix} -\frac{1}{4} & \frac{1}{4} & \frac{1}{4} \\ \frac{5}{8} & -\frac{1}{8} & -\frac{1}{8} \\ \frac{1}{8} & -\frac{5}{8} & \frac{3}{8} \end{pmatrix}$

35. $\begin{pmatrix} 1 & 0 & 2 \\ 0 & 1 & 1 \\ 0 & 0 & 1 \end{pmatrix}$; inverse is $\begin{pmatrix} \frac{5}{6} & \frac{2}{3} & -2 \\ \frac{1}{3} & \frac{2}{3} & -1 \\ -\frac{1}{6} & -\frac{1}{3} & 1 \end{pmatrix}$

37. $\begin{pmatrix} 1 & 2 & 0 \\ 2 & 1 & -1 \\ 3 & 1 & 1 \end{pmatrix}\begin{pmatrix} x_1 \\ x_2 \\ x_3 \end{pmatrix} = \begin{pmatrix} 3 \\ -1 \\ 7 \end{pmatrix}$: A^{-1} is given in Exercise 33; $x_1 = \frac{3}{4}$, $x_2 = \frac{9}{8}$, $x_3 = \frac{29}{8}$

39. $\begin{pmatrix} 2 & -1 \\ 3 & 0 \\ 1 & 2 \end{pmatrix}$; neither **41.** $\begin{pmatrix} 2 & 3 & 1 \\ 3 & -6 & -5 \\ 1 & -5 & 9 \end{pmatrix}$; symmetric

43. $\begin{pmatrix} 1 & -1 & 4 & 6 \\ -1 & 2 & 5 & 7 \\ 4 & 5 & 3 & -8 \\ 6 & 7 & -8 & 9 \end{pmatrix}$; symmetric

CHAPTER EIGHT

PROBLEMS 8.1

1. -10 **3.** 47 **5.** 4 **7.** 56 **9.** 274

13. Almost any example will work. For instance, $\det\begin{pmatrix} 1 & 0 \\ 0 & 1 \end{pmatrix} = 1$, but $\det\begin{pmatrix} 1 & 0 \\ 0 & 0 \end{pmatrix} + \det\begin{pmatrix} 0 & 0 \\ 0 & 1 \end{pmatrix} = 0 + 0 \neq 1$. As another example, let $A = \begin{pmatrix} 1 & 2 \\ 3 & 4 \end{pmatrix}$ and $B = \begin{pmatrix} 5 & 6 \\ 7 & 8 \end{pmatrix}$; then $(A + B) = \begin{pmatrix} 6 & 8 \\ 10 & 12 \end{pmatrix}$, $\det A = -2$, $\det B = -2$, and $\det(A + B) = -8 \neq \det A + \det B$.

PROBLEMS 8.2

1. 28 **3.** 2 **5.** 32 **7.** -36 **9.** -260 **11.** -183 **13.** 24 **15.** -296
17. 138 **19.** *abcde* **21.** -8 **23.** 16 **25.** -16 **27.** -16

PROBLEMS 8.3

1. $\begin{pmatrix} \frac{1}{2} & -\frac{1}{2} \\ -\frac{1}{4} & \frac{3}{4} \end{pmatrix}$ **3.** $\begin{pmatrix} 0 & 1 \\ 1 & 0 \end{pmatrix}$ **5.** $\begin{pmatrix} \frac{1}{3} & -\frac{1}{4} & -\frac{1}{6} \\ 0 & \frac{1}{4} & \frac{1}{2} \\ 0 & \frac{1}{4} & -\frac{1}{2} \end{pmatrix}$ **7.** $\begin{pmatrix} 0 & 1 & -1 \\ 2 & -2 & -1 \\ -1 & 1 & 1 \end{pmatrix}$

9. not invertible **11.** $\begin{pmatrix} \frac{7}{3} & -\frac{1}{3} & -\frac{1}{3} & -\frac{2}{3} \\ \frac{4}{9} & -\frac{1}{9} & -\frac{4}{9} & \frac{1}{9} \\ -\frac{1}{9} & -\frac{2}{9} & \frac{1}{9} & \frac{2}{9} \\ -\frac{5}{3} & \frac{2}{3} & \frac{2}{3} & \frac{1}{3} \end{pmatrix}$

13. det = 13; trivial solution only **15.** det = 0; infinite number of solutions

19. $A^{-1} = \begin{pmatrix} \frac{1}{14} & \frac{1}{14} & \frac{9}{28} \\ -\frac{5}{7} & \frac{2}{7} & -\frac{3}{14} \\ \frac{1}{14} & \frac{1}{14} & -\frac{5}{28} \end{pmatrix}$, $\det A = -28$, $\det A^{-1} = -\frac{1}{28}$

21. no inverse if α is any real number (row 1 + 2 × row 2 = row 3)

PROBLEMS 8.4

1. $x_1 = -5, x_2 = 3$ **3.** $x_1 = 2, x_2 = 5, x_3 = -3$ **5.** $x_1 = \frac{45}{13}, x_2 = -\frac{11}{13}, x_3 = \frac{22}{13}$
7. $x_1 = \frac{3}{2}, x_2 = \frac{3}{2}, x_3 = \frac{1}{2}$ **9.** $x_1 = \frac{21}{29}, x_2 = \frac{171}{29}, x_3 = -\frac{284}{29}, x_4 = -\frac{182}{29}$

REVIEW EXERCISES FOR CHAPTER EIGHT

1. -4 **3.** 24 **5.** 60 **7.** 34 **9.** $\begin{pmatrix} -\frac{1}{11} & \frac{4}{11} \\ \frac{2}{11} & \frac{3}{11} \end{pmatrix}$ **11.** not invertible

13. $\begin{pmatrix} \frac{1}{11} & \frac{1}{11} & 0 & \frac{3}{11} \\ \frac{9}{11} & -\frac{2}{11} & 0 & -\frac{6}{11} \\ \frac{3}{11} & \frac{3}{11} & 0 & -\frac{2}{11} \\ \frac{1}{22} & \frac{1}{22} & -\frac{1}{2} & \frac{3}{22} \end{pmatrix}$ **15.** $x_1 = \frac{11}{7}, x_2 = \frac{1}{7}$ **17.** $x_1 = \frac{1}{4}, x_2 = \frac{5}{4}, x_3 = -\frac{3}{4}$

CHAPTER NINE

PROBLEMS 9.1

1. yes **3.** no; (iv); also (vi) does not hold if $\alpha < 0$ **5.** yes **7.** yes
9. no; (i), (iii), (iv), (vi), do not hold **11.** yes **13.** yes
15. no; (i), (iii), (iv), (vi) do not hold **17.** yes **19.** yes

PROBLEMS 9.2

1. no; because $\alpha(x, y) \notin H$ if $\alpha < 0$ **3.** yes **5.** yes **7.** yes **9.** yes
11. yes **13.** yes **15.** no; the zero polynomial $\notin H$
17. no; the function $f(x) \equiv 0 \notin V$
19. yes

PROBLEMS 9.3

1. independent **3.** dependent; $-2\begin{pmatrix} 2 \\ -1 \\ 4 \end{pmatrix} + \begin{pmatrix} 4 \\ -2 \\ 8 \end{pmatrix} = \begin{pmatrix} 0 \\ 0 \\ 0 \end{pmatrix}$ **5.** dependent
7. independent **9.** independent **11.** independent **13.** independent
15. independent **17.** dependent **19.** independent **21.** independent
23. $ad - bc = 0$ **25.** $\alpha = -\frac{13}{2}$ **33.** yes
35. no; for example, $x \notin \text{span}\,\{1 - x, 3 - x^2\}$ **37.** yes **39.** yes
61. $1 - x^2, 1 + x^2, x$

PROBLEMS 9.4

1. no; does not span **3.** no; dependent; does not span **5.** no; does not span
7. yes **9.** yes

11. $\left\{\begin{pmatrix}0\\1\\-1\end{pmatrix}, \begin{pmatrix}1\\0\\2\end{pmatrix}\right\}$ **13.** $\left\{\begin{pmatrix}2\\3\\4\end{pmatrix}\right\}$ **19.** $\left\{\begin{pmatrix}1\\1\end{pmatrix}\right\}$ **21.** $\left\{\begin{pmatrix}-2\\-3\\1\end{pmatrix}\right\}$

23. $\left\{\begin{pmatrix}3\\1\\0\end{pmatrix}, \begin{pmatrix}-2\\0\\1\end{pmatrix}\right\}$ **25.** n

PROBLEMS 9.5

1. $\frac{x+y}{2}\begin{pmatrix}1\\1\end{pmatrix} + \frac{x-y}{2}\begin{pmatrix}1\\-1\end{pmatrix} = \begin{pmatrix}x\\y\end{pmatrix}$ **3.** $\frac{4x+3y}{41}\begin{pmatrix}5\\7\end{pmatrix} + \frac{7x-5y}{41}\begin{pmatrix}3\\-4\end{pmatrix} = \begin{pmatrix}x\\y\end{pmatrix}$

5. $\frac{dx-by}{ad-bc}\begin{pmatrix}a\\b\end{pmatrix} + \frac{-cx+ay}{ad-bc}\begin{pmatrix}b\\d\end{pmatrix} = \begin{pmatrix}x\\y\end{pmatrix}$

7. $(x-y)\begin{pmatrix}1\\0\\0\end{pmatrix} + (y-z)\begin{pmatrix}1\\1\\0\end{pmatrix} + z\begin{pmatrix}1\\1\\1\end{pmatrix} = \begin{pmatrix}x\\y\\z\end{pmatrix}$

9. $\frac{6x-11y+10z}{31}\begin{pmatrix}2\\1\\3\end{pmatrix} + \frac{2x+17y-7z}{31}\begin{pmatrix}-1\\4\\5\end{pmatrix} + \frac{7x+13y-9z}{31}\begin{pmatrix}3\\-2\\-4\end{pmatrix} = \begin{pmatrix}x\\y\\z\end{pmatrix}$

11. $(a_0+a_1+a_2)1 + a_1(x-1) + a_2(x^2-1) = a_0 + a_1x + a_2x^2$

13. $\frac{(a_0+a_1+a_2)}{2}(x+1) + \frac{(a_1-a_0-a_2)}{2}(x-1) + a_2(x^2-1) = a_0 + a_1x + a_2x^2$

15. $2(x^3+x^2) - 5(x^2+x) + 10(x+1) - 16(1) = 2x^3 - 3x^2 + 5x - 6$

17. $\mathbf{x} = \begin{pmatrix}-\frac{1}{3}\\0\end{pmatrix}_{B_2}$ **19.** $\mathbf{x} = \begin{pmatrix}\frac{86}{33}\\-\frac{20}{11}\\\frac{7}{11}\end{pmatrix}_{B_2}$ **21.** independent since det $= -55$

23. dependent since det $= 0$ **25.** independent since det $= -260$
27. independent since det $= -183$

35. $\begin{pmatrix}\cos(\pi/4) & -\sin(\pi/4)\\ \sin(\pi/4) & \cos(\pi/4)\end{pmatrix}\begin{pmatrix}2\\-7\end{pmatrix} = \begin{pmatrix}1/\sqrt{2} & -1/\sqrt{2}\\ 1/\sqrt{2} & 1/\sqrt{2}\end{pmatrix}\begin{pmatrix}2\\-7\end{pmatrix} = \begin{pmatrix}9/\sqrt{2}\\-5/\sqrt{2}\end{pmatrix}$

PROBLEMS 9.6

1. linear **3.** linear

5. not linear, since $T\left(\alpha\begin{pmatrix}x\\y\\z\end{pmatrix}\right) = T\begin{pmatrix}\alpha x\\\alpha y\\\alpha z\end{pmatrix} = \begin{pmatrix}1\\\alpha z\end{pmatrix}$ while $\alpha T\begin{pmatrix}x\\y\\z\end{pmatrix} = \alpha\begin{pmatrix}1\\z\end{pmatrix} = \begin{pmatrix}\alpha\\\alpha z\end{pmatrix}$

7. linear

9. not linear, since $T\left(\alpha\begin{pmatrix}x\\y\end{pmatrix}\right) = T\begin{pmatrix}\alpha x\\\alpha y\end{pmatrix} = (\alpha x)(\alpha y) = \alpha^2xy$ while $\alpha T\begin{pmatrix}x\\y\end{pmatrix} = \alpha xy$

11. linear

13. not linear, since $T\left(\alpha\begin{pmatrix}x\\y\\z\\w\end{pmatrix}\right) = \alpha^2\begin{pmatrix}xz\\yw\end{pmatrix} = \alpha^2 T\begin{pmatrix}x\\y\\z\\w\end{pmatrix} \neq \alpha T\begin{pmatrix}x\\y\\z\\w\end{pmatrix}$ if $\alpha \neq 1$ or 0

15. not linear, since $T(\alpha A) = (\alpha A)^t(\alpha A) = \alpha^2 A^t A$ while $\alpha T(A) = \alpha A^t A$

17. not linear, since $T(\alpha D) = (\alpha D)^2 = \alpha^2 D^2 \neq \alpha T(D) = \alpha D^2$ unless $\alpha = 1$ or 0 **19.** linear

21. linear **23.** not linear, since $T(f+g) = (f+g)^2 \neq f^2 + g^2 = T(f) + T(g)$

25. linear **27.** linear

29. not linear, since $T(\alpha A) = \det(\alpha A) = \alpha^n \det A \neq \alpha \det A = \alpha T(A)$ unless $\alpha = 0$ or 1 ($\det \alpha A = \alpha^n \det A$ by Problem 8.2.28). Also, in general, $\det(A+B) \neq \det A + \det B$.

31. (a) $\begin{pmatrix}-14\\4\\26\end{pmatrix}$ (b) $\begin{pmatrix}-31\\-6\\26\end{pmatrix}$

33. It rotates a vector counterclockwise around the z-axis through an angle of θ in a plane parallel to the xy-plane.

PROBLEMS 9.7

1. kernel $= \{(0, y): y \in \mathbb{R}\}$, i.e., the y-axis; range $= \{(x, 0); x \in \mathbb{R}\}$, i.e., the x-axis; $\rho(T) = \nu(T) = 1$

3. kernel $= \{(x, -x); x \in \mathbb{R}\}$—this is the line $x + y = 0$; range $= \mathbb{R}$; $\rho(T) = \nu(T) = 1$

5. kernel $= \left\{\begin{pmatrix}0&0\\0&0\end{pmatrix}\right\}$; range $= M_{22}$; $\rho(T) = 4$, $\nu(T) = 0$

7. kernel $= \{A: A^t = -A\} = \{A: A$ is skew-symmetric$\}$; range $= \{A: A$ is symmetric$\}$; $\rho(T) = (n^2 + n)/2$; $\nu(T) = (n^2 - n)/2$

9. kernel $= \{f \in C[0, 1]: f(\frac{1}{2}) = 0\}$; range $= \mathbb{R}$; $\rho(T) = 1$; the kernel is an infinite-dimensional space, so that $\nu(T) = \infty$. For example, the linearly independent functions $x - \frac{1}{2}$, $(x - \frac{1}{2})^2$, $(x - \frac{1}{2})^3$, $(x - \frac{1}{2})^4, \ldots, (x - \frac{1}{2})^n, \ldots$ all satisfy $f(\frac{1}{2}) = 0$.

15. $T\mathbf{x} = A\mathbf{x}$ where $A = \begin{pmatrix}0&a\\b&c\end{pmatrix}$, a, b, c *real* **17.** $T\mathbf{x} = A\mathbf{x}$ where $A = \begin{pmatrix}2&-1&1\\2&-1&1\\2&-1&1\end{pmatrix}$

PROBLEMS 9.8

1. $\rho = 2, \nu = 0$ **3.** $\rho = 1, \nu = 2$ **5.** $\rho = 2, \nu = 1$ **7.** $\rho = 2, \nu = 2$
9. $\rho = 2, \nu = 0$ **11.** $\rho = 2, \nu = 2$ **13.** $\rho = 3, \nu = 1$ **15.** $\rho = 2, \nu = 1$

17. range basis $= \left\{\begin{pmatrix}1\\3\\5\end{pmatrix}, \begin{pmatrix}-1\\1\\-1\end{pmatrix}\right\}$; these are the first two columns of A.

null space basis $= \left\{\begin{pmatrix}-3\\1\\2\end{pmatrix}\right\}$

19. range basis $= \left\{\begin{pmatrix}1\\0\\1\end{pmatrix}, \begin{pmatrix}-1\\1\\0\end{pmatrix}, \begin{pmatrix}3\\3\\5\end{pmatrix}\right\}$; these are the first three (linearly independent) columns of A; null space basis $= \left\{\begin{pmatrix}-6\\-4\\1\\0\end{pmatrix}\right\}$

21. range basis $= \left\{\begin{pmatrix}1\\-2\\2\\3\end{pmatrix}\right\}$; null space basis $= \left\{\begin{pmatrix}1\\1\\0\\0\end{pmatrix}, \begin{pmatrix}-2\\0\\1\\0\end{pmatrix}\begin{pmatrix}-3\\0\\0\\1\end{pmatrix}\right\}$ **23.** no **25.** yes

PROBLEMS 9.9

1. $\begin{pmatrix}1 & -2\\-1 & 1\end{pmatrix}$; $\ker T = \{\mathbf{0}\}$; range $T = \mathbb{R}^2$; $\nu(T) = 0$, $\rho(T) = 2$

3. $\begin{pmatrix}1 & -1 & 1\\-2 & 2 & -2\end{pmatrix}$; range $T = \text{span}\left\{\begin{pmatrix}1\\-2\end{pmatrix}\right\}$; $\ker T = \text{span}\left\{\begin{pmatrix}1\\1\\0\end{pmatrix}, \begin{pmatrix}-1\\0\\1\end{pmatrix}\right\}$; $\rho(T) = 1$, $\nu(T) = 2$

5. $\begin{pmatrix}1 & -1 & 2\\3 & 1 & 4\\5 & -1 & 8\end{pmatrix}$; range $T = \text{span}\left\{\begin{pmatrix}1\\3\\5\end{pmatrix}, \begin{pmatrix}-1\\1\\-1\end{pmatrix}\right\}$; $\ker T = \text{span}\left\{\begin{pmatrix}-3\\1\\2\end{pmatrix}\right\}$; $\rho(T) = 2$, $\nu(T) = 1$

7. $\begin{pmatrix}1 & -1 & 2 & 3\\0 & 1 & 4 & 3\\1 & 0 & 6 & 6\end{pmatrix}$; range $T = \text{span}\left\{\begin{pmatrix}1\\0\\1\end{pmatrix}, \begin{pmatrix}-1\\1\\0\end{pmatrix}\right\}$; $\ker T = \text{span}\left\{\begin{pmatrix}-6\\-4\\1\\0\end{pmatrix}, \begin{pmatrix}-6\\-3\\0\\1\end{pmatrix}\right\}$; $\rho(T) = 2$, $\nu(T) = 2$

9. $\begin{pmatrix}\frac{5}{4} & -\frac{13}{4}\\ \frac{5}{4} & \frac{3}{4}\end{pmatrix}$; range $T = \mathbb{R}^2$; $\ker T = \{\mathbf{0}\}$; $\rho(T) = 2$, $\nu(T) = 0$

11. $\begin{pmatrix}3 & \frac{7}{5} & \frac{16}{5}\\ 0 & \frac{4}{5} & \frac{2}{5}\end{pmatrix}$; range $T = \mathbb{R}^2$; $\ker T = \text{span}\left\{\begin{pmatrix}5\\3\\-6\end{pmatrix}_{B_1}\right\}$; $\rho(T) = 2$, $\nu(T) = 1$

13. $\begin{pmatrix}0 & 1 & 0\\0 & -1 & 0\\0 & 0 & 0\\1 & 0 & 0\end{pmatrix}$; range $T = \text{span}\,\{1 - x, x^3\}$; $\ker T = \text{span}\,\{x^2\}$; $\rho(T) = 2$, $\nu(T) = 1$

15. $(0, 0, 1, 0)$; range $T = \mathbb{R}$; $\ker T = \text{span}\,\{1, x, x^3\}$; $\rho(T) = 1$, $\nu(T) = 3$

17. $\begin{pmatrix}1 & -1 & 2 & 3\\0 & 1 & 4 & 3\\1 & 0 & 6 & 5\end{pmatrix}$; $\rho(T) = 3$; range $T = P_2$; Ker $T = \text{span}\,\{x^2 - 4x - 6\}$; $\nu(T) = 1$

19. $\begin{pmatrix}1 & 1 & 1 & 1\\1 & 1 & 1 & 0\\1 & 1 & 0 & 0\\1 & 0 & 0 & 0\end{pmatrix}$; range $T = M_{22}$; $\ker T = \left\{\begin{pmatrix}0 & 0\\0 & 0\end{pmatrix}\right\}$; $\rho(T) = 4$, $\nu(T) = 0$

21. $\begin{pmatrix}0 & 1 & 0 & 0 & 0\\0 & 0 & 2 & 0 & 0\\0 & 0 & 0 & 3 & 0\\0 & 0 & 0 & 0 & 4\end{pmatrix}$; range $D = P_3$; $\ker D = \mathbb{R}$; $\rho(D) = 4$, $\nu(D) = 1$

23. $\begin{pmatrix} 0 & 1 & 0 & 0 & \cdots & 0 \\ 0 & 0 & 2 & 0 & \cdots & 0 \\ 0 & 0 & 0 & 3 & \cdots & 0 \\ \vdots & \vdots & \vdots & \vdots & & \vdots \\ 0 & 0 & 0 & 0 & \cdots & n \end{pmatrix}$; range $D = P_{n-1}$; ker $D = \mathbb{R}$; $\rho(D) = n$, $\nu(D) = 1$

25. $\begin{pmatrix} 2 & 0 & 2 & 0 & 0 \\ 0 & 3 & 0 & 6 & 0 \\ 0 & 0 & 4 & 0 & 12 \\ 0 & 0 & 0 & 5 & 0 \\ 0 & 0 & 0 & 0 & 6 \end{pmatrix}$; $\rho(T) = 5$, range $T = P_4$; ker $T = \{\mathbf{0}\}$; $\nu(T) = 0$

27. $A_T = \text{diag}\,(b_0, b_1, b_2, \ldots, b_n)$ where $b_i = 1 + j + j(j-1) + j(j-1)(j-2) + \cdots + j!$
range $T = P_n$; ker $T = \{\mathbf{0}\}$; $\rho(T) = n + 1$, $\nu(T) = 0$

29. $\begin{pmatrix} 1 & 0 & 0 \\ 0 & 1 & 0 \\ 0 & 0 & 1 \end{pmatrix}$; range $T = P_2$; ker $T = \{\mathbf{0}\}$, $\rho(T) = 3$, $\nu(T) = 0$

31. For example, in M_{34},

$$A_T = \begin{pmatrix} 1 & 0 & 0 & 0 & 0 & 0 & 0 & 0 & 0 & 0 & 0 & 0 \\ 0 & 0 & 0 & 0 & 1 & 0 & 0 & 0 & 0 & 0 & 0 & 0 \\ 0 & 0 & 0 & 0 & 0 & 0 & 0 & 0 & 1 & 0 & 0 & 0 \\ 0 & 1 & 0 & 0 & 0 & 0 & 0 & 0 & 0 & 0 & 0 & 0 \\ 0 & 0 & 0 & 0 & 0 & 1 & 0 & 0 & 0 & 0 & 0 & 0 \\ 0 & 0 & 0 & 0 & 0 & 0 & 0 & 0 & 0 & 1 & 0 & 0 \\ 0 & 0 & 1 & 0 & 0 & 0 & 0 & 0 & 0 & 0 & 0 & 0 \\ 0 & 0 & 0 & 0 & 0 & 0 & 1 & 0 & 0 & 0 & 0 & 0 \\ 0 & 0 & 0 & 0 & 0 & 0 & 0 & 0 & 0 & 0 & 1 & 0 \\ 0 & 0 & 0 & 1 & 0 & 0 & 0 & 0 & 0 & 0 & 0 & 0 \\ 0 & 0 & 0 & 0 & 0 & 0 & 0 & 1 & 0 & 0 & 0 & 0 \\ 0 & 0 & 0 & 0 & 0 & 0 & 0 & 0 & 0 & 0 & 0 & 1 \end{pmatrix}$$

In general, $A_T = (a_{ij})$ where $ij = 0, 1, 2, \ldots, mn - 1$ and

$$a_{ij} = \begin{cases} 1, \text{ if } i < mn - 1 \\ \text{and } j \text{ is the remainder} \\ \text{when } ni \text{ is divided by } mn - 1 \\ \text{or } i = j = mn - 1 \\ 0, \text{ otherwise.} \end{cases}$$

33. $\begin{pmatrix} 0 & 0 & 0 \\ 0 & 0 & -1 \\ 0 & 1 & 0 \end{pmatrix}$; range $D = \text{span}\,\{\sin x, \cos x\}$; ker $D = \mathbb{R}$;
$\rho(D) = 2$, $\nu(D) = 1$

PROBLEMS 9.10

1. $-4, 3$; $E_{-4} = \text{span}\left\{\begin{pmatrix}1\\1\end{pmatrix}\right\}$; $E_3 = \text{span}\left\{\begin{pmatrix}2\\-5\end{pmatrix}\right\}$

3. $i, -i$; $E_i = \text{span}\left\{\begin{pmatrix}2+i\\5\end{pmatrix}\right\}$; $E_{-i} = \text{span}\left\{\begin{pmatrix}2-i\\5\end{pmatrix}\right\}$

5. $-3, -3$; $E_{-3} = \text{span}\left\{\begin{pmatrix}1\\0\end{pmatrix}\right\}$; geom. mult. is 1

7. $0, 1, 3$; $E_0 = \text{span}\left\{\begin{pmatrix}1\\1\\1\end{pmatrix}\right\}$; $E_1 = \left\{\begin{pmatrix}-1\\0\\1\end{pmatrix}\right\}$; $E_3 = \text{span}\left\{\begin{pmatrix}1\\-2\\1\end{pmatrix}\right\}$

9. $1, 1, 10$; $E_1 = \text{span}\left\{\begin{pmatrix}1\\0\\-2\end{pmatrix}, \begin{pmatrix}0\\1\\-2\end{pmatrix}\right\}$; $E_{10} = \text{span}\left\{\begin{pmatrix}2\\2\\1\end{pmatrix}\right\}$; geom. mult. of 1 is 2

11. $1, 1, 1$; $E_1 = \text{span}\left\{\begin{pmatrix}1\\1\\1\end{pmatrix}\right\}$; geom. mult. is 1 (alg. mult. is 3)

13. $-1, i, -i$; $E_{-1} = \text{span}\left\{\begin{pmatrix}0\\-1\\1\end{pmatrix}\right\}$; $E_i = \text{span}\left\{\begin{pmatrix}1+i\\1\\1\end{pmatrix}\right\}$; $E_{-i} = \text{span}\left\{\begin{pmatrix}1-i\\1\\1\end{pmatrix}\right\}$

15. $1, 2, 2$; $E_i = \text{span}\left\{\begin{pmatrix}4\\1\\-3\end{pmatrix}\right\}$; $E_2 = \text{span}\left\{\begin{pmatrix}3\\1\\-2\end{pmatrix}\right\}$; geom. mult. of 2 is 1

17. a, a, a, a; $E_a = \mathbb{R}^4$; geom. mult. of a = alg. mult. of $a = 4$

19. a, a, a, a; $E_a = \text{span}\left\{\begin{pmatrix}1\\0\\0\\0\end{pmatrix}, \begin{pmatrix}0\\0\\0\\1\end{pmatrix}\right\}$; alg. mult. of $a = 4$; geom. mult. of $a = 2$.

21. Eigenvalues are $a \pm ib$. Then $[A - (a+ib)I]\begin{pmatrix}1\\i\end{pmatrix} = \begin{pmatrix}-ib & b\\-b & -ib\end{pmatrix}\begin{pmatrix}1\\i\end{pmatrix} = \begin{pmatrix}0\\0\end{pmatrix}$.

Similarly, $[A - (a-ib)I]\begin{pmatrix}1\\-i\end{pmatrix} = \begin{pmatrix}0\\0\end{pmatrix}$.

PROBLEMS 9.11

1.

n	$P_{j,n}$	$P_{a,n}$	T_n	$P_{j,n}/P_{a,n}$	T_n/T_{n-1}
0	0	0	12	0	—
1	36	7	43	5	3.583
2	21	18	40	1.15	0.930
5	103	44	147	2.32	—
10	587	283	870	2.07	—
19	15,299	7321	22,621	2.089	—
20	21,965	10,512	32,478	2.089	1.435

Note that the eigenvalues are 1.435 and -0.836. The corresponding eigenvectors are $\begin{pmatrix}2.089\\1\end{pmatrix}$ and $\begin{pmatrix}-3.589\\1\end{pmatrix}$.

3.

n	$P_{j,n}$	$P_{a,n}$	T_n	$P_{j,n}/P_{a,n}$	T_n/T_{n-1}
0	0	20	20	0	—
1	80	16	96	5	4.8
2	64	69	133	0.930	1.385
5	1090	498	1587	2.19	—
10	42,354	22,775	65,129	1.86	—
19	3.704×10^7	1.963×10^7	5.667×10^7	1.887	—
20	7.852×10^7	4.163×10^7	12.02×10^7	1.886	2.120

The eigenvalues are 2.120 and -1.320 with corresponding eigenvectors $\begin{pmatrix} 1.886 \\ 1 \end{pmatrix}$ and $\begin{pmatrix} -3.029 \\ 1 \end{pmatrix}$.

PROBLEMS 9.12

1. yes; $C = \begin{pmatrix} 1 & 2 \\ 1 & -5 \end{pmatrix}$, $C^{-1}AC = \begin{pmatrix} -4 & 0 \\ 0 & 3 \end{pmatrix}$

3. yes; $C = \begin{pmatrix} 1 & 1 \\ 2-i & 2+i \end{pmatrix}$; $C^{-1}AC = \begin{pmatrix} i & 0 \\ 0 & -i \end{pmatrix}$

5. yes; $C = \begin{pmatrix} 2 & 2 \\ -1+3i & -1-3i \end{pmatrix}$; $C^{-1}AC = \begin{pmatrix} 2+3i & 0 \\ 0 & 2-3i \end{pmatrix}$

7. yes; $C = \begin{pmatrix} 3 & 1 & 1 \\ 2 & 3 & 0 \\ 1 & 1 & 1 \end{pmatrix}$; $C^{-1}AC = \begin{pmatrix} 1 & 0 & 0 \\ 0 & 2 & 0 \\ 0 & 0 & -1 \end{pmatrix}$

9. yes; $C = \begin{pmatrix} 0 & 0 & 1 \\ 1 & 1 & 0 \\ 0 & 2 & 0 \end{pmatrix}$; $C^{-1}AC = \begin{pmatrix} 0 & 0 & 0 \\ 0 & 2 & 0 \\ 0 & 0 & 3 \end{pmatrix}$

11. $C = \begin{pmatrix} 1 & 0 & 2 \\ 3 & -2 & 1 \\ 0 & 1 & 2 \end{pmatrix}$; $C^{-1}AC = \begin{pmatrix} 1 & 0 & 0 \\ 0 & 1 & 0 \\ 0 & 0 & 2 \end{pmatrix}$

13. No, since 1 is an eigenvalue of algebraic multiplicity 3 and geometric multiplicity 1.

15. yes; $C = \begin{pmatrix} 0 & -1 & 1 & 1 \\ 0 & 1 & 1 & 1 \\ 1 & 0 & 1 & -1 \\ 0 & 1 & -1 & 1 \end{pmatrix}$; $C^{-1}AC = \begin{pmatrix} 2 & 0 & 0 & 0 \\ 0 & 2 & 0 & 0 \\ 0 & 0 & 4 & 0 \\ 0 & 0 & 0 & 6 \end{pmatrix}$ **21.** $\begin{pmatrix} 1 & 0 \\ 0 & 1 \end{pmatrix}$

REVIEW EXERCISES FOR CHAPTER NINE

1. yes; dimension 2; basis $\{(1, 0, 1), (0, 1, 2)\}$
3. yes; dimension 3; basis $\{(1, 0, 0, -1), (0, 1, 0, -1), (0, 0, 1, -1)\}$
5. yes; dimension $[n(n+1)]/2$; basis $\{(E_{ij}: j \geq i\}$ where E_{ij} is the matrix with 1 in the i,j position and 0 everywhere else
7. no; for example, $(x^5 - 2x) + (-x^5 + x^2) = x^2 - 2x$, which is not a polynomial of degree 5, so the set is not closed under addition.
9. no; for example, $\begin{pmatrix} 1 & 1 \\ 0 & 2 \\ 3 & 1 \end{pmatrix} + \begin{pmatrix} 2 & 1 \\ -1 & 2 \\ 1 & 0 \end{pmatrix} = \begin{pmatrix} 3 & 2 \\ -1 & 4 \\ 4 & 1 \end{pmatrix}$, which does not satisfy $a_{12} = 1$.
11. dependent **13.** independent **15.** dimension 2; basis $\{(2, 0, 1), (0, 4, 3)\}$
17. dimension 3; basis $\{(1, 0, 3, 0), (0, 1, -1, 0), (0, 0, 1, 1)\}$

19. dimension 4; basis $\{D_1, D_2, D_3, D_4\}$ where D_i is the matrix with a 1 in the i, i position and 0 everywhere else

21. $\frac{3}{4}\begin{pmatrix} 1 \\ 2 \end{pmatrix} - \frac{5}{4}\begin{pmatrix} -1 \\ 2 \end{pmatrix} = \begin{pmatrix} 2 \\ -1 \end{pmatrix}$ **23.** $1(1 + x^2) + 0(1 + x) + 3(1) = 4 + x^2$

25. not linear; $T[\alpha(x, y, z)] = T(\alpha x, \alpha y, \alpha z) = (1, \alpha y, \alpha z)$, but $\alpha T(x, y, z) = \alpha(1, y, z) = (\alpha, \alpha y, \alpha z)$ **27.** linear

29. linear **31.** $\ker A = \text{span}\left\{\begin{pmatrix} -2 \\ 1 \\ 1 \end{pmatrix}\right\}$, range $A = \text{span}\left\{\begin{pmatrix} 1 \\ 2 \\ 0 \end{pmatrix}, \begin{pmatrix} -1 \\ 0 \\ -2 \end{pmatrix}\right\}$, $\nu(A) = 1, \rho(A) = 2$

33. $\ker A = \text{span}\left\{\begin{pmatrix} -2 \\ 1 \\ 0 \end{pmatrix}, \begin{pmatrix} 1 \\ 0 \\ 1 \end{pmatrix}\right\}$, range $A = \text{span}\left\{\begin{pmatrix} 2 \\ -1 \end{pmatrix}\right\}$, $\nu(A) = 2, \rho(A) = 1$

35. $\ker A = \text{span}\left\{\begin{pmatrix} -1 \\ 1 \\ 1 \\ 0 \end{pmatrix}, \begin{pmatrix} -3 \\ 0 \\ 0 \\ 1 \end{pmatrix}\right\}$, range $A = \text{span}\left\{\begin{pmatrix} 1 \\ 0 \\ 1 \\ 2 \end{pmatrix}, \begin{pmatrix} -1 \\ 1 \\ -2 \\ -3 \end{pmatrix}\right\}$, $\nu(A) = \rho(A) = 2$

37. $A_T = \begin{pmatrix} 0 & 1 & 0 \\ 0 & 0 & 1 \end{pmatrix}$, $\ker T = \text{span}\left\{\begin{pmatrix} 1 \\ 0 \\ 0 \end{pmatrix}\right\}$, range $T = \mathbb{R}^2$, $\nu(T) = 1, \rho(T) = 2$

39. $A_T = \begin{pmatrix} 0 & 0 & 0 & 0 \\ 1 & 0 & 0 & 0 \\ 0 & 1 & 0 & 0 \\ 0 & 0 & 1 & 0 \\ 0 & 0 & 0 & 1 \end{pmatrix}$, $\ker T = \{\mathbf{0}\}$, range $T = \text{span}\,\{x, x^2, x^3, x^4\}$, $\nu(T) = 0, \rho(T) = 4$

41. $4, -2$; $E_4 = \text{span}\left\{\begin{pmatrix} 1 \\ 1 \end{pmatrix}\right\}$; $E_{-2} = \text{span}\left\{\begin{pmatrix} 2 \\ 1 \end{pmatrix}\right\}$

43. $1, 7, -5$; $E_1 = \text{span}\left\{\begin{pmatrix} -6 \\ 3 \\ 4 \end{pmatrix}\right\}$; $E_7 = \text{span}\left\{\begin{pmatrix} 0 \\ 3 \\ 1 \end{pmatrix}\right\}$; $E_{-5} = \text{span}\left\{\begin{pmatrix} 0 \\ 0 \\ 1 \end{pmatrix}\right\}$

45. $1, 3, 3 + \sqrt{2}i, 3 - \sqrt{2}i$; $E_1 = \text{span}\left\{\begin{pmatrix} 1 \\ 2 \\ 0 \\ 0 \end{pmatrix}\right\}$; $E_3 = \text{span}\left\{\begin{pmatrix} 1 \\ 1 \\ 0 \\ 0 \end{pmatrix}\right\}$; $E_{3-\sqrt{2}i} = \text{span}\left\{\begin{pmatrix} 0 \\ 0 \\ -1 \\ \sqrt{2}i \end{pmatrix}\right\}$;

$E_{3-\sqrt{2}i} = \text{span}\left\{\begin{pmatrix} 0 \\ 0 \\ 1 \\ \sqrt{2}i \end{pmatrix}\right\}$

47. $C = \begin{pmatrix} -3 & 1 \\ 4 & -1 \end{pmatrix}$; $C^{-1}AC = \begin{pmatrix} 2 & 0 \\ 0 & -3 \end{pmatrix}$

49. $C = \begin{pmatrix} 0 & -1 - i & -1 + i \\ 1 & 1 & 1 \\ -1 & 1 & 1 \end{pmatrix}$; $C^{-1}AC = \begin{pmatrix} -1 & 0 & 0 \\ 0 & i & 0 \\ 0 & 0 & -i \end{pmatrix}$

51. not diagonalizable

53. $C = \begin{pmatrix} 1 & 0 & 1 \\ 1 & 0 & -1 \\ 0 & 1 & 0 \end{pmatrix}$; $C^{-1}AC = \begin{pmatrix} 4 & 0 & 0 \\ 0 & -3 & 0 \\ 0 & 0 & 0 \end{pmatrix}$

55. $C = \begin{pmatrix} 1 & 1 & 1 & 0 \\ -1 & 0 & 0 & 0 \\ 0 & 1 & 0 & 0 \\ -1 & -1 & 0 & 1 \end{pmatrix}$; $C^{-1}AC = \begin{pmatrix} -1 & 0 & 0 & 0 \\ 0 & -1 & 0 & 0 \\ 0 & 0 & 3 & 0 \\ 0 & 0 & 0 & 3 \end{pmatrix}$

PROBLEMS 10.1

1. $x + y$ **3.** $1 + x_1 - 4x_2 + x_3$ **5.** $2 + \frac{1}{4}(x - 2) + \frac{1}{4}(y - 2)$
7. $\frac{5}{2} + \frac{3}{4}(x - 2) - \frac{3}{2}(y - 1)$ **9.** $x^2 + y^2$ **11.** $xyz - (x^3y^3z^3/3!)$ **13.** $y + xy$

15. $\frac{e^2\sqrt{2}}{2} + \frac{e^2\sqrt{2}}{2}(x - 2) + \frac{e^2\sqrt{2}}{2}\left(y - \frac{\pi}{4}\right) + \frac{e^2\sqrt{2}}{4}(x - 2)^2 + \frac{e^2\sqrt{2}}{2}(x - 2)\left(y - \frac{\pi}{4}\right) - \frac{e^2\sqrt{2}}{4}\left(y - \frac{\pi}{4}\right)^2$

19. (b) It is still equal to f. This may be hard to see, however. For example, the second-degree Taylor polynomial of $f(x) = x^2$ around $x = 3$ is $9 + 6(x - 3) + (x - 3)^2 = x^2$.

21. $\frac{1}{5!}[f_{xxxxx}(x_0, y_0)(x - x_0)^5 + 5f_{xxxxy}(x_0, y_0)(x - x_0)^4(y - y_0) + 10f_{xxxyy}(x_0, y_0)(x - x_0)^3(y - y_0)^2 + 10f_{xxyyy}(x_0, y_0)(x - x_0)^2(y - y_0)^3 + 5f_{xyyyy}(x_0, y_0)(x - x_0)(y - y_0)^4 + f_{yyyyy}(x_0, y_0)(y - y_0)^5]$

PROBLEMS 10.2

1. $\frac{\partial y}{\partial x} = -1$ at $(1, 1)$ **3.** $\frac{\partial y}{\partial x} = -\frac{e}{e + 1}$ at $(1, 0)$ **5.** $\frac{\partial z}{\partial x} = \frac{\partial z}{\partial y} = -1$ at $(0, 0, 0)$

7. $\frac{\partial z}{\partial x} = -\frac{1}{\frac{4}{3} - e^3} = \frac{3}{3e^3 - 4}$ and $\frac{\partial z}{\partial y} = 0$ at $(1, 3, 0)$

9. $\frac{\partial x_{n+1}}{\partial x_i} = -1$ at $(1, 1, 1, \ldots, 1)$ for $i = 1, 2, \ldots, n$

11. $\frac{\partial z}{\partial x} = -\frac{x^3}{z^3}$, $\frac{\partial z}{\partial y} = -\frac{y^3}{z^3}$; valid for $z \neq 0$

13. $\frac{\partial z}{\partial x} = -\frac{(3x^2 - y^4z)\sin xyz + x^3yz\cos xyz}{x^4y\cos xyz - xy^4\sin xyz}$, $\frac{\partial z}{\partial y} = -\frac{(x^4z + 3y^2)\cos xyz - xy^3z\sin xyz}{x^4y\cos xyz - xy^4\sin xyz}$;
valid over $\{(x, y, z): x^4y\cos xyz - xy^4\sin xyz \neq 0\}$

15. $\frac{\partial z}{\partial x} = -\frac{z^2ye^{xy}\ln(z/x) - (z^2/x)e^{xy} - 5x^4z^2\sin[x^5 - (4y/z)]}{ze^{xy} - 4y\sin(x^5 - (4y/z))}$,
$\frac{\partial z}{\partial y} = -\frac{xz^2e^{xy}\ln(z/x) + 4z\sin[x^5 - (4y/z)]}{ze^{xy} - 4y\sin(x^5 - (4y/z))}$; valid over $\{(x, y, z): ze^{xy} \neq 4y\sin[x^5 - (4y/z)]\}$
Note that the function is not defined when $z = 0$ or $x = 0$.

17. $\frac{\partial z}{\partial x} = -\frac{\frac{1}{2}\sqrt{z/x}e^{\sqrt{zx}} - 6xy^3z^4 - (5z/(x + y)^2)\cos[5z/(x + y)]}{\frac{1}{2}\sqrt{x/z}e^{\sqrt{zx}} - 12x^2y^3z^3 + [5/(x + y)]\cos[5z/(x + y)]}$,
$\frac{\partial z}{\partial y} = -\frac{-9x^2y^2z^4 - [5z/(x + y)^2]\cos[5z/(x + y)]}{\frac{1}{2}\sqrt{x/z}e^{\sqrt{zx}} - 12x^2y^3z^3 + [5/(x + y)]\cos[5z/(x + y)]}$; valid over
$\left\{(x, y, z): \frac{1}{2}\sqrt{\frac{x}{z}}e^{\sqrt{zx}} - 12x^2y^3z^3 + \frac{5}{x + y}\cos\left(\frac{5z}{x + y}\right) \neq 0 \text{ and } zx \neq 0\right\}$
(*Note:* The function is defined only when $zx \geq 0$ and $x \neq -y$.)

PROBLEMS 10.3

1. dom $\mathbf{f} = \mathbb{R}^2$ **3.** dom $\mathbf{f} = \{(x, y, z): x + y + z > 0\}$. This is a half-space. **5.** $\mathbb{R}^5$
7. dom $\mathbf{f} = \{(x_1, x_2, x_3, x_4): x_1^2 + x_2^2 + x_3^2 + x_4^2 \le 1\}$. This is the set of points on and "inside" the "unit sphere" in $\mathbb{R}^4$.
9. Let $u = 3x$, $v = -4y$. Then $(v/3)^2 + (-v/4)^2 = 1$. Since $x \ge 0$, $u \ge 0$. Thus, the image is half of an ellipse in the uv-plane.
11. Note that if $u = e^y \sin x$ and $v = e^y \cos x$, then $u^2 + v^2 = e^{2y}$. Therefore, each value of y in the interval $[0, 1]$ gives rise to a circle of radius e^y in the uv-plane. Thus the image of the rectangle in the xy-plane is an annulus. That is, it is a ring-shaped region in the uv-plane including and between the circles $u^2 + v^2 = 1$ and $u^2 + v^2 = e^2$.
13.

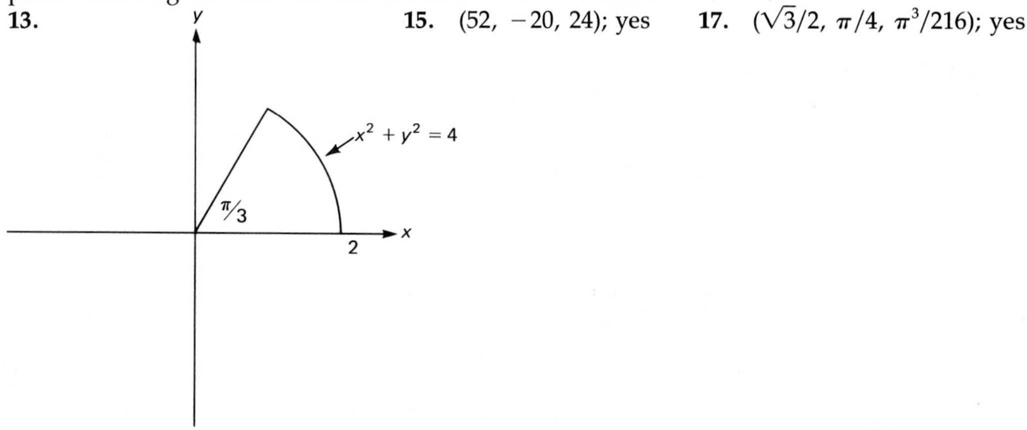

15. $(52, -20, 24)$; yes **17.** $(\sqrt{3}/2, \pi/4, \pi^3/216)$; yes

19. $(n, n^2, n^3, \ldots, n^m)$; yes **21.** limit does not exist since f is not defined in a deleted neighborhood of $(0, 0, 0)$. Note that $\lim_{t \to 0^+} \dfrac{t}{\ln t} = 0$; no **23.** limit does not exist; no
25. $\mathbf{g} \circ \mathbf{f}(x, y) = (x^3 y \cos(y/x), e^{xy + x^2 - \cos(y/x)})$; $\text{dom}(\mathbf{g} \circ \mathbf{f}) = \{(x, y): x \ne 0\}$
27. $\mathbf{g} \circ \mathbf{f}(x, y) = \left(\sin \dfrac{e^{x+y}}{\sqrt{1 - x^2 + y^2}}, \sin \dfrac{\sqrt{1 - x^2 - y^2}}{e^{x+y}}, \mathrm{e}^{\sqrt{e^{2x+y+(1-x^2+y^2)}}}\right)$;
$\text{dom}(\mathbf{g} \circ \mathbf{f}) = \{(x, y): x^2 - y^2 < 1\}$

PROBLEMS 10.4

1. $\begin{pmatrix} 2 & 0 \\ 0 & 6 \end{pmatrix}$ **3.** $\begin{pmatrix} \frac{1}{4} & -\frac{1}{16} & 0 \\ 0 & \frac{1}{8} & -\frac{1}{16} \\ -8 & 0 & 1 \end{pmatrix}$ **5.** $\begin{pmatrix} -2 & 0 & 0 \\ 0 & 10 & 0 \\ 0 & 0 & 2 \\ 9 & -5 & -1 \end{pmatrix}$ **7.** $\begin{pmatrix} -\frac{1}{9} & -\frac{1}{9} & -\frac{1}{3} & -\frac{1}{9} \\ 1 & 1 & 1 & 1 \end{pmatrix}$

9. $\begin{pmatrix} \pi\sqrt{3}/12 & 0 & 0 & \sqrt{3}/2 \\ 0 & 0 & -\pi/6 & -3 \\ 2e^{-3} & -4e^{-3} & 0 & 0 \end{pmatrix}$ **11.** $\begin{pmatrix} 6 & 4 \\ 6 & -4 \end{pmatrix} = \begin{pmatrix} 1 & 1 \\ 1 & -1 \end{pmatrix} \begin{pmatrix} 6 & 0 \\ 0 & 4 \end{pmatrix}$

13. $\begin{pmatrix} -324 & 216 \\ 288 & -96 \end{pmatrix} = \begin{pmatrix} 108 & 0 & 0 \\ 0 & -24 & 0 \end{pmatrix} \begin{pmatrix} -3 & 2 \\ -12 & 4 \\ 9 & -12 \end{pmatrix}$

15. $\begin{pmatrix} \frac{3}{16} & \frac{13}{216} & \frac{5}{24} \\ 972 & 312 & 1080 \\ -57 & -17 & -180 \end{pmatrix} = \begin{pmatrix} \frac{1}{72} & \frac{5}{72} & -\frac{5}{432} & -\frac{5}{864} \\ 72 & 360 & -60 & -30 \\ 3 & 0 & 0 & 5 \end{pmatrix} \begin{pmatrix} 1 & 1 & 0 \\ 1 & 0 & 1 \\ -3 & -2 & 6 \\ -12 & -4 & -36 \end{pmatrix}$

17. $\begin{pmatrix} 0 & 0 & \cdots & 0 & 2n \\ 0 & 0 & \cdots & 2(n-1) & 0 \\ \vdots & \vdots & & \vdots & \vdots \\ 0 & 2 & \cdots & 0 & 0 \\ 1 & 0 & \cdots & 0 & 0 \end{pmatrix} = \begin{pmatrix} 2n & 0 & \cdots & 0 & 0 \\ 0 & 2(n-1) & \cdots & 0 & 0 \\ \vdots & \vdots & & \vdots & \vdots \\ 0 & 0 & \cdots & 2 & 0 \\ 0 & 0 & \cdots & 0 & 1 \end{pmatrix} \begin{pmatrix} 0 & 0 & \cdots & 0 & 1 \\ 0 & 0 & \cdots & 1 & 0 \\ \vdots & \vdots & & \vdots & \vdots \\ 0 & 1 & \cdots & 0 & 0 \\ 1 & 0 & \cdots & 0 & 0 \end{pmatrix}$

PROBLEMS 10.5

1. -2 **3.** $-\frac{3}{4}$ **5.** $\frac{1}{6}$ **7.** 64 **9.** $\pi^3/48$ **11.** $-\ln 2$ **13.** $n!$

15. $\frac{1}{25}\begin{pmatrix} 2 & 3 \\ -7 & 2 \end{pmatrix}$ **17.** $\begin{pmatrix} \frac{1}{2} & \frac{1}{2} & 0 \\ \frac{1}{2} & 0 & \frac{1}{2} \\ 0 & -\frac{1}{2} & -\frac{1}{2} \end{pmatrix}$ **19.** $-\frac{1}{7}\begin{pmatrix} 1 & -4 \\ -2 & 1 \end{pmatrix}$ **21.** $-\frac{1}{2e^2}\begin{pmatrix} -1 & -e^2 \\ -1 & e^2 \end{pmatrix}$

23. $\frac{1}{144}\begin{pmatrix} -9 & 72 & -48 \\ 18 & 0 & 0 \\ 0 & 0 & -16 \end{pmatrix}$ **25.** $\frac{1}{2}I$

29. $\mathbf{x}_0 = (1, 2)$, $\mathbf{y}_0 = (-1, 5, 3)$, $J_\mathbf{f}(\mathbf{x}_0) = -\frac{1}{19}\begin{pmatrix} 5 & -2 & 1 \\ 22 & -24 & -7 \\ 13 & -9 & -5 \end{pmatrix}\begin{pmatrix} 2 & 1 \\ 5 & -2 \\ 1 & 6 \end{pmatrix} = -\frac{1}{19}\begin{pmatrix} 1 & 15 \\ -83 & 28 \\ -24 & 1 \end{pmatrix}$

31. $\mathbf{x}_0 = (1, 1)$, $\mathbf{y}_0 = (1, 1)$, $J_\mathbf{f}(\mathbf{x}_0) = -I$

33. $\mathbf{x}_0 = (2, -1)$, $\mathbf{y}_0 = (1, 2)$, $J_\mathbf{f}(\mathbf{x}_0) = -\frac{1}{56}\begin{pmatrix} -16 & 4 \\ -6 & -2 \end{pmatrix}\begin{pmatrix} 4 & -2 \\ 2 & 2 \end{pmatrix} = \begin{pmatrix} 1 & -\frac{5}{7} \\ \frac{1}{2} & -\frac{1}{7} \end{pmatrix}$

35. $\mathbf{x}_0 = (1, 1, 1)$, $\mathbf{y}_0 = (1, 1)$, $J_\mathbf{f}(\mathbf{x}_0) = -\frac{1}{2}\begin{pmatrix} 1 & 0 \\ -1 & 2 \end{pmatrix}\begin{pmatrix} 2 & 2 & 2 \\ 1 & 1 & 1 \end{pmatrix} = \begin{pmatrix} -1 & -1 & -1 \\ 0 & 0 & 0 \end{pmatrix}$

37. $\mathbf{x}_0 = (2, -3, 1)$, $\mathbf{y}_0 = (4, -2)$, $J_\mathbf{f}(\mathbf{x}_0) = -\frac{1}{180}\begin{pmatrix} 20 & 0 \\ 10 & 9 \end{pmatrix}\begin{pmatrix} 8 & -12 & 36 \\ 12 & 0 & 40 \end{pmatrix}$

$= -\frac{1}{180}\begin{pmatrix} 160 & -240 & 720 \\ 188 & -120 & 720 \end{pmatrix}$

39. $\mathbf{x}_0 = (0, 0, 0)$, $\mathbf{y}_0 = (0, 0, 0)$, $J_\mathbf{f}(\mathbf{x}_0) = \begin{pmatrix} 1 & -1 & 0 \\ -1 & 0 & 0 \\ -1 & 1 & -1 \end{pmatrix}\begin{pmatrix} 3 & 0 & 1 \\ 0 & 0 & 0 \\ 0 & 1 & 1 \end{pmatrix} = \begin{pmatrix} 3 & 0 & 1 \\ -3 & 0 & -1 \\ -3 & -1 & -2 \end{pmatrix}$

REVIEW EXERCISES FOR CHAPTER TEN

1. $1 - \frac{(x+2y)^2}{2!}$ **3.** $\frac{\partial z}{\partial x} = \frac{\partial z}{\partial y} = -\frac{1}{2}$

5. $\frac{\partial z}{\partial x} = -\frac{2xy^3 - 4z^2}{-8xz + 12z^3}$; $\frac{\partial z}{\partial y} = -\frac{3x^2y^2 - 5}{-8xz + 12z^3}$; valid over $\{(x, y, z): 8xz \neq 12z^3\}$

7. dom $\mathbf{f} = \{(x, y, z): x \neq 0 \text{ and } x \neq -y\}$

9. dom $\mathbf{f} = \{(x, y, z): x \neq y \text{ and } z \neq 0\}$

11. $(1, 4, -\tan 3, \pi/6)$ **13.** $\begin{pmatrix} 12 & 0 \\ 0 & 3 \end{pmatrix}$ **15.** $\begin{pmatrix} 2 & 0 & 0 \\ 0 & 8 & 0 \\ 0 & 0 & 18 \end{pmatrix}$

17. $\begin{pmatrix} -1 & 0 & 3 & 0 \\ 0 & 5 & 0 & 1 \\ 6 & 2 & -2 & 10 \\ -5 & 15 & -15 & -3 \end{pmatrix}$ **19.** $\begin{pmatrix} 0 & 56 & 16 \\ -108 & -144 & 216 \\ 3 & -42 & 6 \\ -12 & -192 & 24 \end{pmatrix} = \begin{pmatrix} 2 & 8 & 0 \\ -18 & 0 & 12 \\ 0 & -6 & 1 \\ -6 & -24 & 4 \end{pmatrix}\begin{pmatrix} 4 & 4 & 0 \\ -1 & 6 & 2 \\ -3 & -6 & 18 \end{pmatrix}$

21. 288 **23.** $\frac{1}{17}\begin{pmatrix} 5 & -2 \\ 1 & 3 \end{pmatrix}$ **25.** $\begin{pmatrix} \frac{1}{3} & 0 & 0 \\ 0 & \frac{1}{3} & 0 \\ 0 & 0 & \frac{1}{12} \end{pmatrix}$

27. $\mathbf{x}_0 = (3, -2)$, $\mathbf{y}_0 = (4, 1)$, $J_f(\mathbf{x}_0) = \frac{1}{12}\begin{pmatrix} -4 & 4 \\ 2 & 1 \end{pmatrix}\begin{pmatrix} 2 & -1 \\ 3 & 2 \end{pmatrix} = \frac{1}{12}\begin{pmatrix} 4 & 12 \\ 7 & 0 \end{pmatrix}$

29. $\mathbf{x}_0 = (1, 1)$, $\mathbf{y}_0 = (1, 1, 1)$, $J_f(\mathbf{x}_0) = -\frac{1}{15}\begin{pmatrix} -2 & 6 & -3 \\ -5 & 15 & 0 \\ 7 & -6 & 3 \end{pmatrix}\begin{pmatrix} 3 & 3 \\ 1 & 1 \\ 1 & 2 \end{pmatrix} = -\frac{1}{15}\begin{pmatrix} -3 & -6 \\ 0 & 0 \\ 18 & 21 \end{pmatrix}$

CHAPTER ELEVEN

PROBLEMS 11.1

1. first **3.** third **5.** fourth **13.** linear **15.** linear **17.** nonlinear **19.** nonlinear **21.** nonlinear **23.** nonlinear

PROBLEMS 11.2

1. $y = -(7x^2/2) + C$ **3.** $\sin t + \ln|\csc x + \cot x| = 1 + \ln(\csc 3 + \cot 3)$ **5.** $y^{-4} - 2y^{-2} = -4x^{-1} - 6x^{-2} + C$ **7.** $x = -\ln(\cos t + e^{-1} - 1)$ **9.** $x = e^{t-(\sin 2t)/2}$ **11.** $y = \tan(\frac{1}{3}x^3 + C)$ **13.** 62,500 after 20 days; 156,250 after 30 days **15.** $250{,}000e^{0.6} \approx 455{,}530$ in 1980; $250{,}000e^{1.8} \approx 1{,}512{,}412$ in 2000 **17.** (a) $-10 + 160e^{-2\alpha} \approx -4.14$ °C, where $\alpha = 2 \ln \frac{16}{7}$ (b) $(\ln 16)/\alpha \approx 1.6769$ min ≈ 100.62 sec **19.** $[(\ln 0.7)/(\ln 0.5)]5580 \approx 2871$ yr **21.** (a) $25(0.6)^{24/10} \approx 7.3367$ kg (b) $[(\ln 0.02)/(\ln 0.6)]10 \approx 76.58$ hr **23.** $(\ln 0.5)/(-\alpha) \approx 4.62 \times 10^6$ yr **25.** $\beta = \frac{1}{1500}\ln(845.6/1013.25) \approx -1.205812 \times 10^{-4}$ (a) 625.526 mbar (b) 303.416 mbar (c) 408.018 mbar (d) 348.632 mbar (e) $-(1/\beta)\ln 1013.25 \approx 57.4$ km **27.** (a) 10,000 (b) 7 (c) 214.19 hr $\approx$ 8.9 days **31.** (a) $x(t) = \beta + [x(0) - \beta]e^{-\alpha t}$ (b)

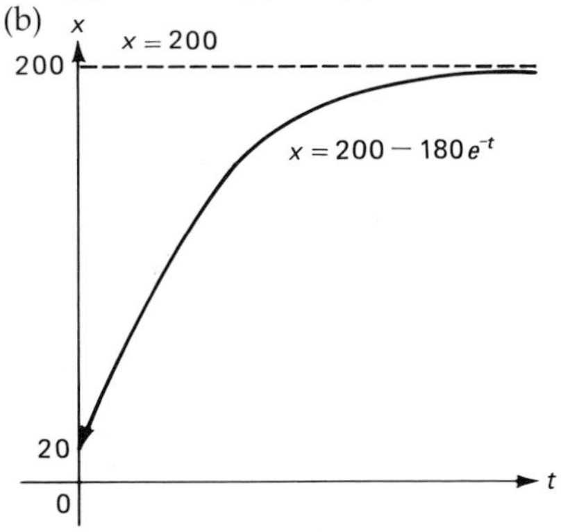

x
$x = 200$
200
$x = 200 - 180e^{-t}$
20
0
t

33. $(2(25^2) - (\frac{16}{15})a^2)\sqrt{a}/\sqrt{2g} \approx 365.1$ sec where $a = \sqrt{25^2 - 1^2}$ is the initial height of the water level above the hole

PROBLEMS 11.3

1. $y = C/(1 + x^2)$ **3.** $(\ln|x|)\ln|\ln y| + \frac{1}{3}x^2y^3 = C$ **5.** $\frac{1}{2}\sinh 2x \cosh 2y = C$ **7.** $y = x^2 \ln x + (5/x)$ **9.** $\ln|x/y| - \tan^{-1}(x/y) = C$ **11.** $2\ln|x| - (2/y) + 3\ln|y| = C$; $x \equiv 0$ and $y \equiv 0$ are also solutions (integrating factor is $1/xy^2$) **13.** $x^{8/3} + 4y^2x^{2/3} = C$; $x \equiv 0$ is also a solution (integrating factor is $x^{-1/3}$) **17.** $\ln|x + 1| + [2/(x + 1)] - [(2 + y^2)/2(x + 1)^2] = C$; $x \equiv -1$ is also a solution (integrating factor is $(x + 1)^{-3}$)

PROBLEMS 11.4

1. $y = 2x^2 + C$ **3.** $x = \frac{1}{2}(\sin t - \cos t) + \frac{3}{2}e^{-t}$ **5.** $y = x^x(1 + Ce^{-x})$ **7.** $x = (bt + C)e^{at}$ **9.** $y = \frac{1}{4} + \frac{1}{2}x \tan 2x + C \sec 2x$ **11.** $y = \frac{1}{2} + [1/2(x^2 + 1)]$ **13.** $x = (t^2 + 1)/2 + [7/2(t^2 + 1)^2]$ **15.** $x = ye^y + Cy$ **17.** $t = 25 \ln 2 \approx 17.3$ min. **19.** $47\frac{43}{91}$ g/liter **21.** $y^2 = e^{2x}/(x^2 + C)$ **23.** $x^3 = 3\sin t + \frac{9\cos t}{t} - \frac{18\sin t}{t^2} - \frac{18\cos t}{t^3} + \frac{C}{t^3}$ **25.** $y^2 = x^2(C - 2\ln|x|)$

PROBLEMS 11.5

1. $I = \frac{6}{5}(1 - e^{-10t})$ **3.** $I = 2(1 - e^{-25t})$ **5.** $I = \frac{1}{20}(e^{t} - e^{-t})$

7. $Q = \frac{1}{680}(3 \sin 60t + 5 \cos 60t - 5e^{-100t})$ **9.** $Q = \frac{1}{100}(1 + 99e^{-100t})$

11. $I(0.1) = \dfrac{e^{-10}}{100} + \dfrac{600\pi(1 - e^{-10})}{(100)^2 + (120\pi)^2}$ **13.** $Q(60) = \frac{1}{2000}(1 - e^{-10^5(600+3600)})$

15. $Q(t) = \dfrac{E_0 C}{1 + (\omega RC)^2}[(\cos \omega t - e^{-t/RC}) + \omega RC \sin \omega t]$;

$I(t) = \dfrac{E_0 C}{1 + (\omega RC)^2}\left(RC\omega^2 \cos \omega t - \omega \sin \omega t + \dfrac{1}{RC} e^{-t/RC}\right)$

17. $I_{\text{transient}}(0) = \dfrac{E_0}{R[1 + (\omega RC)^2]} \approx \dfrac{E_0}{R}$ for very small R

19. $I(t) = \begin{cases} \dfrac{3}{50}(1 - e^{-50t}), & 0 \le t \le 10 \\ \left(\dfrac{e^{500..}}{(70)^2} - \dfrac{3}{50}\right)e^{-50t} + \dfrac{7}{100} - \dfrac{e^{10-t}}{98}, & t \ge 10 \end{cases}$

PROBLEMS 11.6

1. linear, homogeneous, variable coefficients **3.** nonlinear

5. linear, nonhomogeneous, constant coefficients

7. linear, nonhomogeneous, variable coefficients

9. linear, homogeneous, constant coefficients **11.** 3 **13.** 3 **15.** $y = c_1 + c_2 x$

17. $y = c_1 + c_2 x^2$

19. Note that $y_1(x_0) = y_1'(x_0) = 0$ for some x_0 in the interval $[\alpha, \beta]$. By Theorem 1 there is a unique solution satisfying the differential equation and both conditions. Obviously $y \equiv 0$ satisfies the equation and the two conditions. Thus $y_1 \equiv 0$.

21. (b) $a(x) = -1/x$ is not defined at $x = 0$. **23.** $y_3 = 2y_1 - \frac{1}{2}y_2$

25. Let y_1 be the given solution and let y_2 be some other solution. Then $W(y_1, y_2)(0) = 0$, implying that y_1 and y_2 are linearly dependent. **27.** (b) $-2/x$

PROBLEMS 11.7

1. $y_2(x) = xe^x$ **3.** $y_2(x) = \dfrac{3x^2 - 1}{4} \ln \left|\dfrac{1 + x}{1 - x}\right| - \dfrac{3x}{2}$ **5.** $y_2(x) = \dfrac{1}{x^2}$

7. $y_2(x) = 2x - 1$ **9.** $y_2(x) = \cos(x^2)$ **11.** $y_2(x) = (\cos x)/\sqrt{x}$

PROBLEMS 11.8

1. $y = c_1 e^{2x} + c_2 e^{-2x}$ **3.** $y = c_1 e^{x} + c_2 e^{2x}$ **5.** $x = (1 + \frac{9}{2}t)e^{-5t/2}$

7. $x = -\frac{1}{5}(e^{3t} + 4e^{-2t})$ **9.** $y = c_1 + c_2 e^{5x}$ **11.** $y = (c_1 + c_2 x)e^{-\pi x}$

13. $z = c_1 e^{-5x} + c_2 e^{3x}$ **15.** $y = (1 + 2x)e^{4x}$ **17.** $y = c_1 e^{\sqrt{2}x} + c_2 e^{-\sqrt{2}x}$

19. $y = e^{\sqrt{5}x} + 2e^{-\sqrt{5}x}$ **23.** $y = \dfrac{e^x - 2e^{-x}}{e^x + 2e^{-x}}$ **25.** $y = 1 + (c + x)^{-1}$

27. $y = \dfrac{2ce^{3x} - 1}{ce^{3x} + 1}$

PROBLEMS 11.9

1. $y = e^{-x}(c_1 \cos x + c_2 \sin x)$ **3.** $x = e^{-t/2}\left(c_1 \cos \dfrac{3\sqrt{3}}{2} t + c_2 \sin \dfrac{3\sqrt{3}}{2} t\right)$

5. $x = -\frac{3}{2} \cos 2\theta + \sin 2\theta$ **7.** $y = \sin(x/2) + 2 \cos(x/2)$

9. $y = e^{-x}(c_1 \cos 2x + c_2 \sin 2x)$ **11.** $y = e^{-x}(\sin x - \cos x)$

PROBLEMS 11.10

1. $y = c_1 \sin 2x + c_2 \cos 2x + \sin x$ **3.** $y = c_1 e^{x} + c_2 e^{2x} + 3e^{3x}$

5. $y = e^{x}(c_1 + c_2 x - 2x^2)$ **7.** $y = 3e^{5x} - 10e^{2x} + 10x + 7$

9. $y = c_1 + c_2 e^{-x} + \dfrac{x^4}{4} - \dfrac{4}{3}x^3 + 4x^2 - 8x$

11. $y = e^{2x}(c_1 \cos x + c_2 \sin x) + xe^{2x} \sin x$
13. $y = e^{-3x}(c_1 + c_2x + 5x^2)$ **15.** $y = c_1e^{3x} + c_2e^{-x} + \frac{20}{27} - \frac{7}{9}x + \frac{1}{3}x^2 - \frac{1}{4}e^x$
17. $y = (c_1 + c_2x)e^{-2x} + (\frac{1}{9}x - \frac{2}{27})e^x + \frac{3}{25} \sin x - \frac{4}{25} \cos x$
21. $y_p = \left(\frac{x}{4} + \frac{x^2}{4}\right) \sin x + \left(\frac{x}{4} - \frac{x^2}{4} - \frac{x^3}{6}\right) \cos x$ **23.** $y_p = \frac{x^4e^x}{12}$

PROBLEMS 11.11

1. $y = c_1 + c_2e^x - \ln|\cos x|$
3. $y = c_1 \cos 2x + c_2 \sin 2x + \frac{1}{2}x \sin 2x + \frac{1}{4} \cos 2x \ln|\cos 2x|$
5. $y = e^x(c_1 + c_2x - \ln|1 - x|)$ **7.** $y = c_1e^x + c_2e^{-x} + e^x \ln|x|$
9. $y = e^{2x}(c_1 + c_2x + (x + 1)\ln|x + 1|)$ **11.** $y_p = -\frac{1}{2} - \frac{x}{2}\ln|x| + \left(\frac{x}{2} - \frac{1}{2x}\right)\ln|x + 1|$

PROBLEMS 11.12

1. $y = c_1x + \frac{c_2}{x}$ **3.** $y = x(c_1 \cos(\ln|x|) + c_2 \sin(\ln|x|))$
5. $y = x^{3/2} - x^{1/2}$ **7.** $y = c_1x + c_2x^3$
9. $y = \frac{1}{x^2}(c_1 \cos(\ln|x|) + c_2 \sin(\ln|x|))$ **11.** $y = c_1x^3 + c_2x^{-4}$
13. $y = c_1x^{-5} + c_2x^{-1} + \frac{1}{12}x$ **15.** $y = c_1x^2 + c_2x^{-1} - \frac{1}{2}\ln x + \frac{1}{4}, x > 0$

PROBLEMS 11.13

1. (a) $x = \cos 10t$; (b) $x = (1 + 10t)e^{-10t}$; (c) $x = \cos 10t + \frac{1}{2000}\sin 10t - \frac{t}{200}\cos 10t$;
(d) $x = \left(\frac{201}{2000} + \frac{201t}{200}\right)e^{-10t} - \frac{1}{2000}\cos 10t$
3. (a) $x = 5\sin(t + \alpha)$, $\tan\alpha = 3/4$;
(b) $x = \left(\frac{3\sqrt{5} + 11}{2}\right)e^{[(1-\sqrt{5})t/2]} - \left(\frac{3\sqrt{5} + 5}{2}\right)e^{[(-1-\sqrt{5})t/2]}$;
(c) $x = 3\cos t + \frac{81}{20}\sin t - \frac{t}{20}\cos t$;
(d) $x = \frac{1}{100}[(555 + 151\sqrt{5})e^{(1-\sqrt{5})t/2} - (255 + 149\sqrt{5})e^{(-1-\sqrt{5})t/2} - 2\sqrt{5}\cos t]$
5. (a) $x = \frac{3}{5}\sin 5t$; (b) $x = e^{-4t}\sin 3t$; (c) $x = \frac{9}{16}\sin 5t + \frac{1}{16}\sin 3t$;
(d) $x = \frac{1}{104}[e^{-4t}(106\sin 3t + 3\cos 3t) - 3\cos 3t + 2\sin 3t]$ **7.** $\pm(\sqrt{g}/5)$ m/s
9. $\frac{2\pi}{\sqrt{5\pi g}}$ s; $x = \left(\frac{1}{5\pi} - 1\right)\cos\sqrt{5\pi g}\,t$ ft **11.** $k = \frac{100\pi^2 + 1}{160}$ N/m = kg/s^2
13. $x = \frac{25}{19{,}408}\left(\frac{1213}{11}e^{-2t} - \frac{250}{11}e^{-19.6t} + 88\sin 2t - 108\cos 2t\right)$
15. $\frac{\sqrt{9.8}}{2\pi}$ Hz $\approx 0.498 \approx 29.89$ ticks/min

PROBLEMS 11.14

1. $I(t) = 6(e^{-5t} - e^{-20t})$; $Q(t) = \frac{1}{10}(9 - 12e^{-5t} + 3e^{-20t})$ **3.** $I_{\text{steady state}} = \cos t + 2\sin t$
5. $I_{\text{steady state}} = \frac{1}{13}(70\sin 10t - 90\cos 10t)$ **7.** $I_{\text{transient}} = e^{-t}(c_1\cos t + c_2\sin t)$
9. $I_{\text{transient}} = c_1e^{-2t} + c_2e^{-5t}$ **11.** $I_{\text{transient}} = e^{-t}(c_1\cos 7t + c_2\sin 7t)$
13. $I_{\text{transient}} = -\frac{e^{-600t}}{2320}(153\sin 800t + 96\cos 800t)$; $I_{\text{steady state}} = \frac{1}{580}(24\cos 600t + 27\sin 600t)$
15. $$Q = \begin{cases} c_1\cos\frac{t}{\sqrt{LC}} + c_2\sin\frac{t}{\sqrt{LC}} + \left(\frac{CE_0}{1 - LCW^2}\right)\sin\omega t, & \omega \neq \frac{1}{\sqrt{LC}}, \\ \left(c_1 - \frac{E_0t}{2\omega L}\right)\cos\omega t + c_2\sin\omega t, & \omega = \frac{1}{\sqrt{LC}} \end{cases}$$

17. (b) $\omega = \dfrac{\sqrt{(4L/C) - R^2}}{2L}$ **19.** (b) No ω produces resonance.

PROBLEMS 11.15

1. $y = (c_1 + c_2x)\cos x + (c_3 + c_4x)\sin x$ **3.** $y = (1 + x)e^x$
5. $y = 1 + e^{3x} + e^{-3x}$ **7.** $y = c_1 + c_2x + c_3x^2 + c_4x^3$
9. $y = c_1e^x + c_2e^{-x} + c_3e^{2x} + c_4e^{-2x}$ **11.** $y = 1 - x + e^{2x} - e^{-2x}$
13. $y = c_1e^x + c_2\cos x + c_3\sin x$
15. $y = c_1e^{3x} + e^{-3x/2}\left(c_2\cos\dfrac{3\sqrt{3}}{2}x + c_3\sin\dfrac{3\sqrt{3}}{2}x\right)$
19. (a) $y_1(v')'' + (3y_1' + ay_1)(v')' + (3y_1'' + 2ay_1' + by_1)v' = 0$;
(c) $v' = \dfrac{W(y_1, y_2)}{y_1^2}\displaystyle\int \dfrac{y_1e^{-\int a(x)dx}}{W^2(y_1, y_2)}\,dx$
25. $y = c_1e^x + c_2xe^x + c_3e^{-x} + \dfrac{1}{4}x^2e^x$
27. $y = c_1e^{2x} + c_2e^{-3x} + c_3e^{4x} + \dfrac{1}{24}x + \dfrac{41}{288}$
29. $y_p = e^{-x}\displaystyle\int e^{-x}\sec x\,dx + e^{-2x}[(\cos x - \sin x)\ln|\sec x| - x(\cos x + \sin x)]$
31. $y = c_1x^2 + c_2x^{-3} + c_3x^4$ **33.** $y = c_1x + c_2x^2 + c_3x^{-1}$

PROBLEMS 11.16

1. $y = 2e^x - x - 1$, $y(1) = 2(e - 1) \approx 3.44$; (a) $y_E = 2.98$; (b) $y_{IE} = 3.41$
3. $y^2 + 2xy - x^2 = 1$, $y(1) = \sqrt{3} - 1 \approx 0.73$; (a) $y_E = 0.71$; (b) $y_{IE} = 0.73$
5. $y + \sqrt{y^2 + 1} = e^{(x^2-1)/2}$, $y(3) = (e^4 - e^{-4})/2 \approx 27.29$; (a) $y_E = 8.31$; (b) $y_{IE} = 21.67$
7. $y = (x^3 + x^{-2})^{-1/2}$, $y(2) = 2/\sqrt{33} \approx 0.348$; (a) $y_E = 0.34$; (b) $y_{IE} = 0.349$
9. $y = 2e^{e^x-1}$, $y(2) = 2e^{e^2-1} \approx 1190.59$; (a) $y_E = 156.45$; (b) $y_{IE} = 781.56$
11. $y_1 = 1.02$, $y_2 = 1.04$, $y_3 = 1.07$, $y_4 = 1.09$, $y_5 = 1.12$
13. $y_1 = 0.34$, $y_2 = 0.56$, $y_3 = 0.76$, $y_4 = 0.98$, $y_5 = 1.34$
15. $y_1 = 1.28$, $y_2 = 1.65$, $y_3 = 1.46$, $y_4 = 1.09$, $y_5 = 0.87$, $y_6 = 0.79$, $y_7 = 0.93$, $y_8 = 0.95$
17. $y_1 = 1.10$, $y_2 = 1.22$, $y_3 = 1.35$, $y_4 = 1.48$, $y_5 = 1.63$, $y_6 = 1.79$, $y_7 = 1.97$, $y_8 = 2.16$, $y_9 = 2.37$, $y_{10} = 2.60$
19. $y_1 = 1.60$, $y_2 = 1.27$, $y_3 = 1.00$, $y_4 = 0.80$, $y_5 = 0.64$

REVIEW EXERCISES FOR CHAPTER ELEVEN

1. $y = \frac{3}{2}x^2 + C$ **3.** $x = -\ln(e^{-3} - \sin t)$ **5.** $y = 0.7e^{-3x} + 0.3\cos x + 0.1\sin x$
7. $x = (2e^{-3} - \frac{1}{4} + \frac{1}{4}t^4)e^{3t}$ **9.** $xy + c = e^y\tan x$ **11.** $y = c_1e^x + c_2e^{4x}$
13. $y = c_1e^{3x} + c_2e^{-3x}$ **15.** $y = (c_1 + c_2x)e^{-3x}$ **17.** $y = e^x\sin x$
19. $y = \frac{2}{3}\sin x + c_1\cos 2x + c_2\sin 2x$
21. $y = c_1e^{-x/2}\sin(\sqrt{3}/2)x + c_2e^{-x/2}\cos(\sqrt{3}/2)x - (x/\sqrt{3})[\cos(\sqrt{3}/2)x]e^{-x/2}$
23. $y = x^3 - 7x + c_1\cos x + c_2\sin x$ **25.** $y = (3 - 11x + x^2/2)e^{4x}$
27. $y = c_1e^x\cos\sqrt{2}x + c_2e^x\sin\sqrt{2}x + (1/2\sqrt{2})xe^x\sin\sqrt{2}x$
29. $y \equiv 0$ (This is a harmonic oscillator that never gets going.)
31. $y = c_1e^x + c_2xe^x + (e^x/x)$
33. $y = c_1x^{3/2}\cos\left(\dfrac{\sqrt{3}}{2}\ln x\right) + c_2x^{3/2}\sin\left(\dfrac{\sqrt{3}}{2}\ln x\right)$ **35.** $y = e^{-x}(c_1 + c_2x + c_3x^2)$
37. $y = (c_1 + c_2x)\cos 3x + (c_3 + c_4x)\sin 3x$ **39.** $I = I(0) - (E_0/\text{WL})\cos \text{wt}$
41. period $= 2\pi\sqrt{60/g}$; frequency $= (\sqrt{g/60}/2\pi$; amplitude $= \sqrt{10^2 + (-5)^2\,10/(g/6)} \approx 10.076$ cm **43.** $\mu = \sqrt{4(10)(g/6)} \approx 80.87$
45. $y^2 = 2(e^x + 1)$, $y(3) = \sqrt{2(e^3 + 1)} \approx 6.4939$; $y_{\text{approx}}(3) = 6.56$
47. $y = x + \sqrt{1 + x^2}$, $y(3) = 3 + \sqrt{10} \approx 6.1623$; $y_{\text{approx}}(3) = 5.96$

CHAPTER TWELVE

PROBLEMS 12.1

1. $x = c_1e^{-t} + c_2e^{4t},\ y = -c_1e^{-t} + \frac{3}{2}c_2e^{4t}$
3. $x = c_1e^{-3t} + c_2te^{-3t},\ y = -(c_1 + c_2)e^{-3t}\ c_2te^{-3t}$
5. $x = c_1e^{10t} + c_2te^{10t},\ y = -(2c_1 + c_2)e^{10t} - 2c_2te^{10t}$
7. $x = -\frac{1}{4}(t^2 + 9t + 3) + c_1 - 3c_2e^{2t},\ y = \frac{1}{4}(t^2 + 7t) - c_1 + c_2e^{2t}$
9. $x = e^{4t}(17c_1 \cos 2t + 17c_2 \sin 2t),\ y = e^{4t}[(8c_1 - 2c_2)\cos 2t + (8c_2 + 2c_1)\sin 2t]$
11. $x_1' = x_2,\ x_2' = -3x_1 - 2x_2$ **13.** $x_1' = x_2,\ x_2' = x_3,\ x_3' = x_1^3 - x_2^2 + x_3 + t$
15. $x_1' = x_2,\ x_2' = x_3,\ x_3' = x_1^4x_2 - x_1x_3 + \sin t$
17. $x_1' = x_2,\ x_2' = x_3,\ x_3' = x_1 - 4x_2 + 3x_3$
19. (a) $x_1 = c_1e^t,\ x_2 = c_2e^t$; c_1, c_2 constant
21. $x_1 = 4c_1e^{-2t} + 3c_2e^{-t},\ x_2 = -5c_1e^{-2t} - 4c_2e^{-t} + c_3e^{2t},\ x_3 = -7c_1e^{-2t} - 2c_2e^{-t} - c_3e^{2t}$
23. $x' = \dfrac{-3x}{100 + t} + \dfrac{2y}{100 - t},\ y' = \dfrac{3x}{100 + t} - \dfrac{4y}{100 - t}$
25. $y_{\max} = \dfrac{500}{\sqrt{3}}[(2 + \sqrt{3})^{(1-\sqrt{3})/2} - (2 + \sqrt{3})^{(-1-\sqrt{3})/2}]$ lb, $t_{\max} = \dfrac{25}{\sqrt{3}}\ln(2 + \sqrt{3})$ min
27. $m_1m_2x_1^{(4)} + [k_1m_2 + k_2(m_2 + m_1)]x_1'' + k_1k_2x_1 = 0$
$x_1 = c_1 \cos\sqrt{10}t + c_2 \sin\sqrt{10}t + c_3 \cos t + c_4 \sin t$
$x_2 = -(c_1/4)\cos\sqrt{10}t - (c_2/4)\sin\sqrt{10}t + 2c_3 \cos t + 2c_4 \sin t$
29. 780.9 counts per minute

PROBLEMS 12.2

1. (b) $W = e^{-6t}$; (c) $\{c_1e^{-3t} + c_2(1 - t)e^{-3t}, -c_1e^{-3t} + c_2te^{-3t}\}$

PROBLEMS 12.3

1. $\{e^t, e^t\}, \{3e^{-t}, 5e^{-t}\}$ **3.** $\{e^t \cos t, e^t(2\cos t - \sin t)\}, \{e^t \sin t, e^t(2\sin t + \cos t)\}$
5. $\{e^{-3t}, -e^{-3t}\}, \{(t - 1)e^{-3t}, -te^{-3t}\}$ **7.** $\{3, 4\}, \{e^{-2t}, 2e^{-2t}\}$
11. $\{-\frac{1}{4}(t^2 + 9t + 3), \frac{1}{4}(t^2 + 7t)\}$ **13.** $\{\sin t - \cos t, 2\sin t - \cos t\}$
15. (a) $x' = -0.134x + 0.02y,\ y' = 0.036x - 0.02y$
(b) $\{10c_1e^{-0.14t} + c_2e^{-0.014t}, -3c_1e^{-0.14t} + 6c_2e^{-0.014t}\}$

PROBLEMS 12.4

1. $\mathbf{x}'(t) = A\mathbf{x}(t);\ A = \begin{pmatrix} 2 & 3 \\ 4 & -6 \end{pmatrix}$

3. $\mathbf{x}'(t) = A(t)x(t) + \mathbf{f}(t);\ A(t) = \begin{pmatrix} 0 & 1 & 0 \\ 0 & 0 & 1 \\ 1 & -4t & 2 \end{pmatrix};\ \mathbf{f}(t) = \begin{pmatrix} 0 \\ 0 \\ \sin t \end{pmatrix}$

5. $\mathbf{x}'(t) = A(t)\mathbf{x}(t) + \mathbf{f}(t);\ A(t) = \begin{pmatrix} 2t & -3t^2 & \sin t \\ 2 & 0 & -4 \\ 0 & 17 & 4t \end{pmatrix};\ \mathbf{f}(t) = \begin{pmatrix} 0 \\ -\sin t \\ e^t \end{pmatrix}$

17. (a) $\mathbf{x}'(t) = (1, \cos t);\ \int \mathbf{x}(t)\,dt = \left(\frac{t^2}{2}, -\cos t\right)$ (b) $\mathbf{y}'(t) = \begin{pmatrix} e^t \\ -\sin t \\ \sec^2 t \end{pmatrix}$;

$\int \mathbf{y}(t)\,dt = \begin{pmatrix} e^t \\ \sin t \\ -\ln|\cos t| \end{pmatrix}$ (c) $A'(t) = \begin{pmatrix} 1/2\sqrt{t} & 2t \\ 2e^{2t} & 2\cos 2t \end{pmatrix};\ \int A(t)\,dt = \begin{pmatrix} \frac{2}{3}t^{3/2} & t^3/3 \\ e^{2t}/2 & -(\cos 2t)/2 \end{pmatrix}$

(d) $B'(t) = \begin{pmatrix} 1/t & e^t(\sin t + \cos t) & e^t(-\sin t + \cos t) \\ \frac{5}{2}t^{3/2} & \sin t & -\cos t \\ -(1/t^2) & e^{t^2}(1 + 2t^2) & e^{t^3}(2t + 3t^4) \end{pmatrix}$;

$$\int B(t)\,dt = \begin{pmatrix} t\ln t - t & \frac{1}{2}e^t(\sin t - \cos t) & \frac{1}{2}e^t(\sin t + \cos t) \\ \frac{2}{7}t^{7/2} & -\sin t & \cos t \\ \ln t & e^{t^2}/2 & e^{t^3}/3 \end{pmatrix}$$

PROBLEMS 12.5

1. fundamental **3.** fundamental **5.** not fundamental **7.** $C = \begin{pmatrix} 1 & -2 \\ -1 & 1 \end{pmatrix}$

9. $C = \begin{pmatrix} 1 & -1 & 1 \\ 0 & 1 & 1 \\ 1 & 0 & 0 \end{pmatrix}$ **15.** $\Psi(t) = e^{6t}\begin{pmatrix} \cos 2t - \sin t & \frac{1}{2}\sin t \\ -4\sin t & \cos 2t + \sin 2t \end{pmatrix}$

17. $\Psi(t) = \dfrac{1}{6}\begin{pmatrix} -e^{-t} + 9e^t - 2e^{2t} & -2e^{-t} + 2e^{2t} & 7e^{-t} - 9e^t + 2e^{2t} \\ 6e^t - 6e^{2t} & 6e^{2t} & -6e^t + 6e^{2t} \\ -e^{-t} + 3e^t - 2e^{2t} & -2e^{-t} + 2e^{2t} & 7e^{-t} - 3e^t + 2e^{2t} \end{pmatrix}$

19. (a) $e^t\begin{pmatrix} 1 - 2t \\ 2 - 2t \end{pmatrix}$ (b) $e^t\begin{pmatrix} -10t - 2 \\ -10t + 3 \end{pmatrix}$ (c) $e^{t-1}\begin{pmatrix} 2 - 2t \\ 3 - 2t \end{pmatrix}$ (d) $e^{t+1}\begin{pmatrix} 4 + 2t \\ 3 + 2t \end{pmatrix}$ (e) $e^{t-3}\begin{pmatrix} 3 \\ 3 \end{pmatrix}$

(f) $e^{t-a}\begin{pmatrix} 2(c-b)(a-t) + b \\ 2(c-b)(a-t) + c \end{pmatrix}$ **21.** (a) $\mathbf{x}' = A(t)\mathbf{x}$, where $A(t) = \begin{pmatrix} 0 & 1 \\ -b(t) & -a(t) \end{pmatrix}$

23. $\phi_2(t) = e^{t^2}\int e^{-t^2}\,dt$

PROBLEMS 12.6

1. $\dfrac{1}{7}\begin{pmatrix} 5e^{-4t} + 2e^{3t} & 2e^{-4t} - 2e^{3t} \\ 5e^{-4t} - 5e^{3t} & 2e^{-4t} + 5e^{3t} \end{pmatrix}$ **3.** $\begin{pmatrix} 2\sin t + \cos t & -\sin t \\ 5\sin t & -2\sin t + \cos t \end{pmatrix}$

5. $\frac{2}{3}e^{2t}\begin{pmatrix} \frac{3}{2}\cos 3t + \frac{1}{2}\sin 3t & \sin 3t \\ -\frac{5}{2}\sin 3t & -\frac{1}{2}\sin 3t + \frac{3}{2}\cos 3t \end{pmatrix}$ **7.** $\begin{pmatrix} -5e^t + 6e^{2t} & 2e^t - 2e^{2t} & 4e^t - 4e^{2t} \\ -3e^t + 3e^{2t} & 2e^t - e^{2t} & 2e^t - 2e^{2t} \\ -6e^t + 6e^{2t} & 2e^t - 2e^{2t} & 5e^t - 4e^{2t} \end{pmatrix}$

9. $-\dfrac{1}{9}\begin{pmatrix} -4e^{8t} - 5e^{-t} & -2e^{8t} + 2e^{-t} & -4e^{8t} + 4e^{-t} \\ -2e^{8t} + 2e^{-t} & -e^{8t} - 8e^{-t} & -2e^{8t} + 2e^{-t} \\ -4e^{8t} + 4e^{-t} & -2e^{8t} + 2e^{-t} & -4e^{8t} - 5e^{-t} \end{pmatrix}$

15. $(1 + 4t)e^{-2t}$ **17.** $\frac{8}{7}e^{5t} + \frac{13}{7}e^{-2t}$

PROBLEMS 12.7

1. $\phi_p(t) = \begin{pmatrix} \frac{1}{3}e^{2t} \\ \frac{1}{2}e^t - \frac{2}{3}e^{2t} \end{pmatrix}$ **3.** $\phi(t) = \begin{pmatrix} -t\cos t \\ -t\sin t - 2t\cos t + \cos t + \sin t \end{pmatrix}$

5. $\phi(t) = \begin{pmatrix} \frac{3}{2}te^t + \frac{1}{3}te^{2t} + \frac{5}{2}e^t - \frac{115}{72}e^{-t} - \frac{7}{9}e^{2t} - \frac{1}{8}e^{3t} \\ te^t + te^{2t} + \frac{5}{2}e^t - 2e^{2t} + \frac{1}{2}e^{3t} \\ \frac{1}{2}te^t + \frac{1}{3}te^{2t} + e^t - \frac{115}{72}e^{-t} - \frac{7}{9}e^{2t} + \frac{3}{8}e^{3t} \end{pmatrix}$ **7.** $\phi_p(t) = \begin{pmatrix} 6e^t - 2e^{2t} - 2t - 4 \\ 4e^t - e^{2t} - t - 3 \\ 7e^t - 2e^{2t} - 3t - 5 \end{pmatrix}$

9. $\phi_p(t) = \begin{pmatrix} 5\sin t\ln|\csc t + \cot t| \\ 1 + (2\sin t - \cos t)\ln|\csc t + \cot t| \end{pmatrix}$ **11.** $\Phi(t) = \begin{pmatrix} 1/t & \ln t/t \\ -1/t^2 & (1 - \ln t)/t^2 \end{pmatrix}$

PROBLEMS 12.8

1. $\mathbf{x}(t) = -\dfrac{1}{5}\begin{pmatrix} 12 + 11i \\ 6 + 2i \end{pmatrix}e^{it}$ **3.** $\mathbf{x}(t) = -\dfrac{1}{754}\begin{pmatrix} 353 + 205i \\ -197 - 595i \end{pmatrix}e^{-5it}$

9. (a) Let $x_1 = x$, $x_2 = x'$; then $\begin{pmatrix} x_1 \\ x_2 \end{pmatrix}' = \begin{pmatrix} 0 & 1 \\ -k/m & 0 \end{pmatrix}\begin{pmatrix} x_1 \\ x_2 \end{pmatrix} + \begin{pmatrix} 0 \\ F_0/m \end{pmatrix}e^{i\omega t}$

(b) $\mathbf{x}(t) = \dfrac{F_0}{k - \omega^2 m}\begin{pmatrix} 1 \\ i\omega \end{pmatrix}e^{i\omega t}$ (c) $\Psi(t) = \begin{pmatrix} \cos\sqrt{k/m}t & \sqrt{m/k}\sin\sqrt{k/m}t \\ -\sqrt{k/m}\sin\sqrt{k/m}t & \cos\sqrt{k/m}t \end{pmatrix}$.

Resonance.

REVIEW EXERCISES FOR CHAPTER TWELVE

1. $x_1' = x_2,\ x_2' = x_3,\ x_3' = 6x_3 - 2x_2 + 5x_1$ **3.** $x_1' = x_2,\ x_2' = x_3,\ x_3' = (\ln t - x_1x_3)/x_2$
5. $x = c_1e^{5t} + c_2e^{-t},\ y = 2c_1e^{5t} - c_2e^{-t}$
7. $x = e^{2t}(c_1 \cos 3t + c_2 \sin 3t),\ y = \frac{1}{2}e^{2t}((3c_2 - c_1)\cos 3t - (c_2 + 3c_1)\sin 3t)$
9. $x = c_1e^{-t} + 3e^{-2t},\ y = (c_2 - 2c_1t)e^{-t} + 12e^{-2t}$ **11.** $\begin{pmatrix} x_1 \\ x_2 \end{pmatrix}' = \begin{pmatrix} 3 & -4 \\ -2 & 7 \end{pmatrix}\begin{pmatrix} x_1 \\ x_2 \end{pmatrix}$

13. $\begin{pmatrix} x_1 \\ x_2 \end{pmatrix}' = \begin{pmatrix} 1 & 1 \\ -3 & 2 \end{pmatrix}\begin{pmatrix} x_1 \\ x_2 \end{pmatrix} + \begin{pmatrix} e^t \\ e^{2t} \end{pmatrix}$

15. (a) $\frac{1}{5}\begin{pmatrix} -2e^t + 12e^{6t} \\ 3e^t + 12e^{6t} \end{pmatrix}$ (b) $\frac{1}{5}\begin{pmatrix} -2e^t - 3e^{6t} \\ 3e^t - 3e^{6t} \end{pmatrix}$ (c) $\begin{pmatrix} 0 \\ 0 \end{pmatrix}$ (d) $\frac{1}{5}\begin{pmatrix} 18e^t + 17e^{6t} \\ -27e^t + 17e^{6t} \end{pmatrix}$
(e) $\frac{1}{5}\begin{pmatrix} 2(a - b)e^t + (3a + 2b)e^{6t} \\ -3(a - b)e^t + (3a + 2b)e^{6t} \end{pmatrix}$

17. $\begin{pmatrix} 2e^{-t} - e^t & -2e^{-t} + 2e^t \\ e^{-t} - e^t & -e^{-t} + 2e^t \end{pmatrix}$ **19.** $\begin{pmatrix} \frac{1}{2}e^{2t} + \frac{1}{2}e^{6t} & -\frac{1}{2}e^{2t} + \frac{1}{2}e^{6t} & 0 \\ -\frac{1}{2}e^{2t} + \frac{1}{2}e^{6t} & \frac{1}{2}e^{2t} + \frac{1}{2}e^{6t} & 0 \\ 0 & 0 & e^{-3t} \end{pmatrix}$

21. $\phi(t) = e^{2t}\begin{pmatrix} 3\cos 2t + \frac{19}{8}\sin 2t + \frac{1}{4}t \\ -6\sin 2t + \frac{19}{4}\cos 2t - \frac{11}{4} \end{pmatrix}$

CHAPTER THIRTEEN

PROBLEMS 13.1

1. $(1/\sqrt{2})[1 - (x - \pi/4) - (x - \pi/4)^2/2! + (x - \pi/4)^3/3! + (x - \pi/4)^4/4! - (x - \pi/4)^5/5! - (x - \pi/4)^6/6!]$
3. $1 + (1/e)(x - e) - (1/2e^2)(x - e)^2 + (1/3e^3)(x - e)^3 - (1/4e^4)(x - e)^4 + (1/5e^5)(x - e)^5$
5. $1 - (x - 1) + (x - 1)^2 - (x - 1)^3 + (x - 1)^4$ **7.** $x + \frac{1}{3}x^3$ **9.** $(x - \pi) + \frac{1}{3}(x - \pi)^3$
11. $1 - x^2 + x^4$ **13.** $x + (x^3/3!)$ **15.** $-\frac{1}{2}[x - (\pi/2)]^2$
17. $\frac{1}{2} + \frac{1}{2^4}x + \frac{3}{2^7 \cdot 2!}x^2 + \frac{3 \cdot 5}{2^{10} \cdot 3!}x^3 + \frac{3 \cdot 5 \cdot 7}{2^{13} \cdot 4!}x^4$
19. $1 + \alpha x + (\alpha^2/2!)x^2 + (\alpha^3/3!)x^3 + (\alpha^4/4!)x^4 + (\alpha^5/5!)x^5 + (\alpha^6/6!)x^6$ **21.** $x + x^3/6$
23. $(a_0 + a_1 + a_2 + a_3) + (a_1 + 2a_2 + 3a_3)(x - 1) + (a_2 + 3a_3)(x - 1)^2 + a_3(x - 1)^3$
25. x^2

PROBLEMS 13.3

1. $(1/7!)(\pi/4)^7 \approx 0.0000366$ **3.** $|R_n(x)| \le (120/5!)(\frac{1}{2})^{-6}(\frac{1}{2})^5 = 2$ on $[\frac{1}{2}, 1]$ and $|R_n(x)| \le (120/5!)|1^{-6}1^5 = 1$ on $[1, 2]$, so $|R_n(x)| \le 2$ on $[\frac{1}{2}, 2]$
5. $(1 \cdot 3 \cdot 5 \cdot 7 \cdot 9 \cdot 11/64)(\frac{19}{4})^{-13/2}(\frac{1}{4})^6(1/6!) \approx 2.2 \times 10^{-9}$
7. $[2\sec^4(\pi/6) + 4\sec^2(\pi/6)\tan^2(\pi/6)](1/4!)(\pi/6)^4 \approx 0.0167$
9. $(1/5!)(\frac{1}{3})^5(e^{1/9})(\frac{120}{3} + \frac{160}{27} + \frac{32}{243}) \approx 0.0018$
11. $\frac{1}{2} + (\sqrt{3}/2)(0.2) - (\frac{1}{2})(\frac{1}{2})(0.2)^2 - (\sqrt{3}/2)(\frac{1}{6})(0.2)^3 \approx 0.66205$
13. $\tan(\pi/4) + \sec^2(\pi/4)(0.1) + [2\sec^2(\pi/4)\tan(\pi/4)/2](0.1)^2 + [4\sec^2(\pi/4)\tan^2(\pi/4) + 2\sec^4(\pi/4)](0.1)^3/3! \approx 1.2227$
15. $1 + (-1) + \frac{1}{2!}(-1)^2 + \frac{1}{3!}(-1)^3 + \frac{1}{4!}(-1)^4 + \frac{1}{5!}(-1)^5 + \frac{1}{6!}(-1)^6 \approx 0.36806$
17. $0.5 - \frac{(0.5)^2}{2} + \frac{(0.5)^3}{3} - \frac{(0.5)^4}{4} + \frac{(0.5)^5}{5} - \frac{(0.5)^6}{6} + \frac{(0.5)^7}{7} \approx 0.4058$
19. $1 + 3 + 3^2/2! + \cdots + 3^{12}/12! \approx 20.0852$ **21.** $\frac{1}{2} + (\frac{1}{2})^3/6 \approx 0.5208$
23. $\sin[(\pi/2) + (\pi/18)] \approx 1 - \frac{1}{2}(\pi/18)^2 \approx 0.98477$ **25.** $1 - \frac{1}{2}(0.1) + \frac{3}{8}(0.1)^2 = 0.95375$
27. $\int_0^{1/3} e^{x^2}\,dx \approx \int_0^{1/3}[1 + x^2 + (x^4/2)]\,dx = \frac{1}{3} + \frac{1}{3}(\frac{1}{3})^3 + \frac{1}{10}(\frac{1}{3})^5 \approx 0.3461$; error $< \frac{1}{3}(0.0018) = 0.0006$

PROBLEMS 13.4

1. $\frac{1}{3}, \frac{1}{9}, \frac{1}{27}, \frac{1}{81}, \frac{1}{243}$ **3.** $\frac{3}{4}, \frac{15}{16}, \frac{63}{64}, \frac{255}{256}, \frac{1023}{1024}$ **5.** $e, e^{1/2}, e^{1/3}, e^{1/4}, e^{1/5}$ **7.** 0, 0, 0, 0, 0 **9.** 1, 0, −1, 0, 1 **11.** 0 **13.** divergent **15.** divergent **17.** $\frac{4}{7}$ **19.** $e^{1/4}$ **21.** 0 **23.** 0 **25.** divergent **27.** 0 **29.** $a_n = n \cdot 5^{n-1}$ **31.** $a_n = 1 - 1/2^n$ **33.** $a_n = (-\frac{1}{3})^{n-1}$ **39.** choose a_0 from [0, 1] **41.** disprove; $p = 1$ and $a_n = 1/(n \ln n)$ is a counterexample

PROBLEMS 13.5

1. $\frac{1}{2}$ **3.** 1 **5.** 1 **7.** 1 **9.** 2 **11.** $(\ln 3)/3 \approx 0.366$ **15.** strictly increasing **17.** not monotonic **19.** strictly decreasing **21.** strictly decreasing **23.** not monotonic, but is strictly decreasing for $n \geq 4$ **25.** strictly increasing **27.** strictly increasing

PROBLEMS 13.6

1. 364 **3.** $[1 - (-5)^6]/[1 - (-5)] = -2604$ **5.** $(0.3)^2[1 - (-0.3)^7]/1.3 \approx 0.07$ **7.** $(b^{16} - 1)/(b^{16} + b^{14})$ **9.** $(16\sqrt{2} - 1)/(\sqrt{2} - 1) = 31 + 15\sqrt{2}$ **11.** −3280 **13.** $\frac{4}{3}$ **15.** $\frac{10}{9}$ **17.** $\pi/(\pi - 1)$ **19.** $1/1.62 \approx 0.61728$ **21.** $\frac{1}{2}$ **23.** $n = 7$ **25.** $n = 459$

PROBLEMS 13.7

1. $\frac{4}{3}$ **3.** $\frac{1}{2}$ **5.** 2 **7.** 125 **9.** $\frac{1}{2}$ **11.** 1 **13.** $\frac{32}{3}$ **15.** $25/6^4 = \frac{25}{1296}$ **17.** $\frac{35}{99}$ **19.** $\frac{71}{99}$ **21.** $\frac{501}{999} = \frac{167}{333}$ **23.** $\frac{11351}{99900}$ **27.** $\frac{3}{2}$ **29.** $\frac{1}{4}$ **39.** $1\frac{1}{11}$ hr = 1:05 $\frac{5}{11}$ P.M. **41.** $8 + 8 \cdot 2 \cdot \frac{2}{3} + 8 \cdot 2 \cdot (\frac{2}{3})^2 + 8 \cdot 2 \cdot (\frac{2}{3})^3 + \cdots = 40$ m

PROBLEMS 13.8

Note: C = series converges, D = series diverges

1. C **3.** D **5.** C **7.** C **9.** C **11.** C **13.** C **15.** D **17.** D **19.** D **21.** D **23.** C **25.** D **27.** C **29.** D **31.** C **33.** D

PROBLEMS 13.9

Note: C = series converges, D = series diverges

1. D **3.** C **5.** C **7.** D **9.** C **11.** C **13.** D **15.** C **17.** D **19.** C **21.** C **23.** C **25.** C

PROBLEMS 13.10

Note: AC = absolutely convergent, CC = conditionally convergent, D = divergent

1. D **3.** CC **5.** D **7.** CC **9.** D **11.** D **13.** AC **15.** CC **17.** AC **19.** CC **21.** D **23.** D **25.** CC **27.** D **29.** CC

33. $S_{10} \approx 0.818$ with error ≤ 0.0083, $T_5 \approx 0.819$ with error ≤ 0.0061 $(S = \pi^2/12)$

35. $S_9 \approx -0.503$ with error ≤ 0.043, $T_3 \approx -0.508$ with error ≤ 0.028

37. $S_{100} \approx 0.555$ with error ≤ 0.1, $T_2 \approx 0.646$ with error ≤ 0.065 (note that $T_9 \approx 0.6003$ with error ≤ 0.0086) **39.** $1 - \frac{1}{2} - \frac{1}{4} + \frac{1}{3} - \frac{1}{6} - \frac{1}{8} + \frac{1}{5} - \frac{1}{10} - \frac{1}{12} - \frac{1}{14}$

43. $\Sigma a_k = \Sigma 1/k$ is the most obvious example.

PROBLEMS 13.11

1. 6; (−6, 6) **3.** 3; (−4, 2) **5.** $\frac{1}{3}$; $(-\frac{1}{3}, \frac{1}{3})$ **7.** 1; [0, 2] **9.** ∞; $(-\infty, \infty)$ **11.** 1; (−1, 1) **13.** ∞; $(-\infty, \infty)$ **15.** 1; (−1, 1) **17.** ∞; $(-\infty, \infty)$ **19.** $\frac{1}{2}$; $[-\frac{1}{2}, \frac{1}{2}]$ **21.** $\frac{5}{2}$; (−4, 1) **23.** 0; $x = 0$ **25.** 1; (−1, 1) **27.** 0; $x = -1$ **29.** 3; [−13, −7) **31.** 1; (−1, 1) **33.** 1; (−1, 1)

PROBLEMS 13.12

1. $\Sigma_{k=0}^{\infty} (-1)^k x^{2k}$

3. $\pi = 4 \tan^{-1} 1 = 4 \Sigma_{k=0}^{\infty}[(-1)^k/(2k + 1)]$; $T_{10} \approx 3.1371$ with error ≤ 0.00828, $S_{199} \approx 3.1366$ with error ≤ 0.00998

5. $S_3 \approx 0.743$ with error ≤ 0.0046, $T_3 \approx 0.755$ with error ≤ 0.0096

7. $S_2 \approx 0.496875$ with error ≤ 0.000009, $T_3 \approx 0.496795$ with error ≤ 0.000005
9. $S_3 \approx 0;.1862$ with error ≤ 0.0038, $T_3 \approx 0.1955$ with error ≤ 0.0074
11. $S_2 \approx 0.7639$ with error ≤ 0.0003, $T_2 \approx 0.7569$ with error ≤ 0.0068
13. $S_1 \approx 0.499783$ with error ≤ 0.0000004, $T_2 \approx 0.4997832$ with error ≤ 0.0000002
15. $\Sigma_{k=0}^{\infty} x^{k+2}/[(k+2)k!]$ **17.** $\Sigma_{k=1}^{\infty} (-1)^{k+1}x^k/k^2$ **19.** (a) $(-\infty, \infty)$

PROBLEMS 13.13

3. $\Sigma_{k=0}^{\infty} e(x-1)^k/k!$

5. $\dfrac{1}{\sqrt{2}}\left[1 - \left(x - \dfrac{\pi}{4}\right) - \dfrac{[x-(\pi/4)]^2}{2} + \dfrac{[x-(\pi/4)]^3}{3!} + \dfrac{[x-(\pi/4)]^4}{4!} - \cdots\right]$

7. $\Sigma_{k=0}^{\infty} \alpha^k x^k/k!$ **9.** $\Sigma_{k=0}^{\infty} (-1)^k x^{2k+2}/k!$ **11.** $\Sigma_{k=0}^{\infty} (x+1)^k/ek!$

13. $(x-1)\ln x = (x-1)\displaystyle\sum_{k=1}^{\infty} \frac{(-1)^{k+1}(x-1)^k}{k} = \sum_{k=1}^{\infty} \frac{(-1)^{k+1}(x-1)^{k+1}}{k}$; $(0, 2]$

15. $1 + 0 + \dfrac{[x-(\pi/2)]^2}{2!} + 0$; $(0, \pi)$

17. $2 + \displaystyle\sum_{k=1}^{\infty} \frac{(-1)^{k+1}\, 1\cdot 3\cdot \cdots \cdot (2k-3)}{(2^{3k-1})k!}(x-4)^k$; 4 **19.** $\displaystyle\sum_{k=1}^{\infty} \frac{(-1)^{k+1}x^{4k-2}}{(2k-1)!}$

27. $1 + \frac{1}{4}x^3 - \frac{3}{32}x^6 + \frac{7}{128}x^9 - \frac{77}{2048}x^{12} + \frac{231}{8192}x^{15} - \cdots$
29. $\int_0^{1/4} (1+\sqrt{x})^{3/5}\,dx = x + \frac{2}{5}x^{3/2} - \frac{3}{50}x^2 + \frac{14}{625}x^{5/2} - \frac{7}{625}x^3 + \frac{102}{15625}x^{7/2} - \cdots \big|_0^{1/4}$. Using the first five terms (through x^3), we obtain the estimate 0.296775 with error bounded by 0.000051

31. $\text{erfc}(x) = 1 - \dfrac{2}{\sqrt{\pi}}\displaystyle\sum_{k=0}^{\infty} \frac{(-1)^k x^{2k+1}}{(2k+1)k!}$

$\text{erfc}(1) \approx 1 - \dfrac{2}{\sqrt{\pi}}\left(1 - \dfrac{1}{3} + \dfrac{1}{5\cdot 2} - \dfrac{1}{7\cdot 3!} + \dfrac{1}{9\cdot 4!} - \dfrac{1}{11\cdot 5!} + \dfrac{1}{13\cdot 6!}\right) \approx 0.1572858$
with error ≤ 0.0000149,
$\text{erfc}\left(\dfrac{1}{2}\right) \approx 1 - \dfrac{2}{\sqrt{\pi}}\left(\dfrac{1}{2} - \dfrac{(\frac{1}{2})^3}{3} + \dfrac{(\frac{1}{2})^5}{5\cdot 2} - \dfrac{(\frac{1}{2})^7}{7\cdot 3!}\right) \approx 0.4795099$ with error ≤ 0.00001

PROBLEMS 13.14

1. $y = e^x + x + 1$ **3.** $y = c_0 \cos x + (c_1 - 1)\sin x + x$
5. $y = c_0(1 + x\tan^{-1} x) + c_1 x$ **7.** $y = xe^x$ **9.** $y = 1 - 2x^2$ **11.** $y = x + \sin x$
13. $y = 1/(1-x)$ **15.** $y = e^{x^2} - (x/4)$

17. $y = c_0\displaystyle\sum_{n=0}^{\infty} \frac{1\cdot 4\cdot \cdots \cdot (3n-2)}{(3n)!}x^{3n} + c_1\sum_{n=1}^{\infty} \frac{2\cdot 5\cdot \cdots \cdot (3n-1)}{(3n+1)!}x^{3n+1}$

19. $y = 1 + \dfrac{(x-1)^2}{2!} + \dfrac{(x-1)^3}{3!} + \dfrac{(x-1)^4}{4!} + \dfrac{4(x-1)^5}{5!} + \cdots$ **21.** (a) no; (b) no

REVIEW EXERCISES FOR CHAPTER THIRTEEN

1. $1 + x + (x^2/2!) + (x^3/3!)$
3. $\frac{1}{2} + (\sqrt{3}/2)[x - (\pi/6)] - \frac{1}{4}[x - \pi/6)]^2 - (\sqrt{3}/12)[x - (\pi/6)]^3$
5. $-[x - (\pi/2)] - \frac{1}{3}[x - (\pi/2)]^3$ **7.** $3 + 2x - x^2 + x^3$ **9.** $(\pi/3)^6/6! \approx 0.00183$
11. $e/7! < 3/7! \approx 0.000595$
13. $f^{(5)}(x) = -8x(15 - 20x^2 + 4x^4)e^{-x^2}$ and crude bound is $(8)(15)/5! = 1$
15. $(1/\sqrt{2}) - (1/\sqrt{2})(-\pi/90) \approx 0.73179$
17. $\ln[(1+x)/(1-x)] = 2(x + \frac{1}{3}x^3 + \frac{1}{5}x^5 + \cdots)$ and, with $x = \frac{1}{3}$,
$\ln 2 \approx 2(\frac{1}{3} + \frac{1}{81} + \frac{1}{1215} + \frac{1}{15309}) \approx 0.69313$
19. $0.3 - \frac{1}{3}(0.3)^3 + \frac{1}{5}(0.3)^5 = 0.291486$ (actual error ≈ 0.000029)
21. $\int_0^{1/2} [1 - (x^4/2)]\,dx = \frac{159}{320} = 0.496875$; error $\le \frac{1}{2}[(1/2^8)/4!] \approx 0.00008$
23. $-1, 0, \frac{1}{3}, \frac{2}{4}, \frac{3}{5}$ **25.** 0 **27.** 0 **29.** e^{-2} **31.** $a_n = (2n-1)/2^{n+2}$

33. unbounded; not monotonic **35.** bounded below by $\frac{2}{3}$ and above by 1; strictly increasing
37. bounded above by 1 and below by 0; strictly decreasing
39. bounded below by 0 and above by 1; strictly increasing
41. bounded above by 8 and below by 5; strictly decreasing **43.** 1,398,096
45. $(\frac{27}{16}) - (\frac{8}{75}) = \frac{1897}{1200} \approx 1.58$ **47.** $\frac{79}{99}$ **49.** $9\frac{9}{11}$ hr $\approx$ 9:49:05 P.M.
51. convergent **53.** divergent **55.** convergent **57.** divergent
59. convergent **61.** convergent **63.** conditionally convergent
65. conditionally convergent **67.** absolutely convergent
69. conditionally convergent **71.** divergent **73.** divergent
75. $T_6 = 1 - (1/2^3) + (1/3^3) - (1/4^3) + (1/5^3) + (1/2)(-1/6^3) \approx 0.9021$ with error ≤ 0.00086; alternatively compute
$S_{10} = 1 - (1/2^3) + (1/3^3) - (1/4^3) + (1/5^3) - (1/6^3) + (1/7^3) - (1/8^3) + (1/9^3) - (1/10^3) \approx$ 0.9011 with error ≤ 0.00075
77. $1 - \frac{1}{2} - \frac{1}{4} + \frac{1}{3} - \frac{1}{6} + \frac{1}{5} - \frac{1}{8} + \frac{1}{7} - \frac{1}{10} - \frac{1}{12}$ **79.** 3; $(-3, 3)$
81. ∞, $(-\infty, \infty)$ **83.** ∞; $(-\infty, \infty)$ **85.** 0; $x = 0$ **87.** 1; $[0, 2)$
89. 3; $[-11, -5)$ **91.** $\int_0^{1/2} t^2\, dt = \frac{1}{24}$ with error $\le \int_0^{1/2} (t^6/3!)\, dt \approx 0.00019$
93. $\int_0^{1/2} (1 - t^4 + t^8)\, dt = \frac{11381}{23040} \approx 0.493967$ with error $\le \int_0^{1/2} t^{12}\, dt \approx 0.000009$
95. $\sum_{k=0}^{\infty} 2(x - \ln 2)^k/k!$ **97.** $\sum_{k=0}^{\infty} (-1)^k \alpha^{2k+1}x^{2k+1}/(2k+1)!$ **99.** $y = \cos 3x$
101. $y = c_0 e^{x^2/2} + c_1\left(x + \frac{x^3}{3} + \frac{x^5}{3\cdot 5} + \frac{x^7}{3\cdot 5\cdot 7} + \cdots\right)$

APPENDIX 1

1. *Hint:* Verify the identity $[n^2(n+1)^2/4] + (n+1)^3 = (n+1)^2(n+2)^2/4$.
3. *Hint:* Apply the product rule to $xP_n(x) + b$.
5. *Hint:* Pick one element from the set and color it orange. Now count how many subsets contain this special orange element; count how many subsets do not contain it. Can there be any other subsets?

APPENDIX 2

1. 10 **3.** 36 **5.** 165 **7.** 1
11. $a^7 + 7a^6b + 21a^5b^2 + 35a^4b^3 + 35a^3b^4 + 21a^2b^5 + 7ab^6 + b^7$
13. $x^5 - 10x^4y + 40x^3y^2 - 80x^2y^3 + 80xy^4 - 32y^5$
15. $a^5x^5 - 5a^4bx^4y + 10a^3b^2x^3y^2 - 10a^2b^3x^2y^3 + 5ab^4xy^4 - b^5y^5$
17. $\frac{1}{32}x^5 + \frac{5}{48}x^4y + \frac{5}{36}x^3y^2 + \frac{5}{54}x^2y^3 + \frac{5}{162}xy^4 + \frac{1}{243}y^5$ **19.** 165

APPENDIX 3

1. $9 - 7i$ **3.** 2 **5.** $-27 + 5i$ **7.** $5\sqrt{2}e^{\pi i/4}$ **9.** $3\sqrt{2}e^{-\pi i/4}$ **11.** $6e^{\pi i/6}$
13. $8e^{-\pi i/6}$ **15.** $2e^{-2\pi i/3}$ **17.** -2 **19.** $-(\sqrt{2}/4) - (\sqrt{2}/4)i$
21. $-2\sqrt{3} + 2i$ **23.** $-\frac{3}{2} - (3\sqrt{3}/2)i$ **25.** $\cos 1 + i \sin 1 \approx 0.54 + 0.84i$
27. $4 - 6i$ **29.** $7i$ **31.** $2e^{-\pi i/7}$ **33.** $3e^{4\pi i/11}$
39. circle of radius a centered at z_0 if $a > 0$, single point z_0 if $a = 0$, empty set if $a < 0$

APPENDIX 4

5. -6

APPENDIX 5

1. yes; $\delta = 1/216$ (better; $\delta = 1/50$)
3. Yes, but $a = b = 1$ is not possible. You need $a + b < 1$. If $a = b = 1/3$, then $\delta = 1/4$.
5. yes; $\delta = 1$ **7.** Yes, but you need $b < 1$. If $b = 1/2$, then $\delta = 1$
9. Yes; $\delta = 1/2$ (better; $\delta = +\infty$; there is a unique solution defined for $-\infty < t < \infty$).
11. (a) $k = 1$ (b) $k = 1$ (c) $k = 3$ $\left(\text{A smaller constant is } \sup_{|x|\le 1}\left|2x \sin\frac{1}{x} - \cos\frac{1}{x}\right|.\right)$
(d) $972\sqrt{3} \approx 1683$ **13.** $x_n(t) = 3\sum_{k=0}^{n} \frac{(t-3)^k}{k!}$

Index

6
B 7
C 8
D 9
E 0

Tina Seto